Eastern Hemisphere

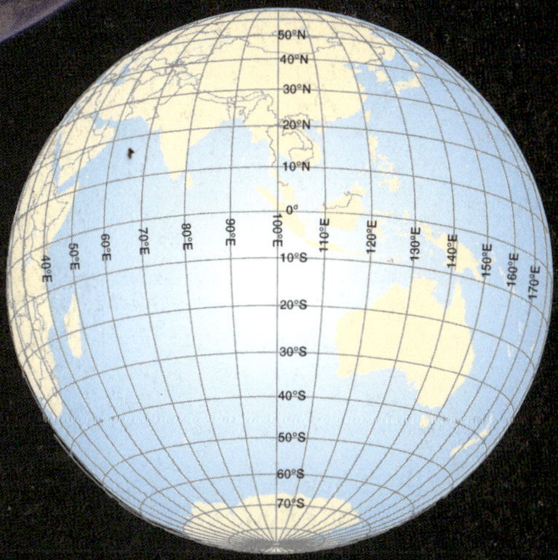

Geosystems

Super Typhoon Haiyan made landfall in the central Philippines on the morning of November 7, 2013, with sustained winds over 306 kmph (190 mph), the strongest ever recorded for a tropical cyclone at landfall using satellite measurements. In *Geosystems*, we discuss tropical cyclones and other severe weather events on Earth, including the effects of Superstorm Sandy on the U.S. East Coast in 2012 (see Focus Study 8.1 in Chapter 8). [NOAA.]

Sandstone cliffs along the Virgin River in Zion National Park, Utah. [GeoStills/Alamy.]

PEARSON ALWAYS LEARNING

Robert W. Christopherson • Ginger H. Birkeland

Geosystems

An Introduction to Physical Geography

Custom Edition for University of Nevada Las Vegas

Taken from:
Geosystems: An Introduction to Physical Geography, Ninth Edition
by Robert W. Christopherson and Ginger H. Birkeland

Geosystems: An Introduction to Physical Geography, Eighth Edition
by Robert W. Christopherson and Ginger H. Birkeland

Cover Art: Courtesy of Corbis, PhotoDisc/Getty Images.

Taken from:

Geosystems: An Introduction to Physical Geography, Ninth Edition
by Robert W. Christopherson and Ginger H. Birkeland

Upper Saddle River, New Jersey 07458

Geosystems: An Introduction to Physical Geography, Eighth Edition
by Robert W. Christopherson and Ginger H. Birkeland

Upper Saddle River, New Jersey 07458

This special edition published in cooperation with Pearson Learning Solutions.

Pearson Learning Solutions, 501 Boylston Street, Suite 900, Boston, MA 02116
A Pearson Education Company
www.pearsoned.com

Printed in the United States of America

4 5 6 7 8 9 10 V0UD 18 17 16 15 14

000200010271887328

AL

PEARSON

ISBN 10: 1-269-89178-2
ISBN 13: 978-1-269-89178-3

dedication

To the students and teachers of Earth, and to all the children and grandchildren, for it is their future and home planet.

The land still provides our genesis, however we might like to forget that our food comes from dank, muddy Earth, that the oxygen in our lungs was recently inside a leaf, and that every newspaper or book we may pick up is made from the hearts of trees that died for the sake of our imagined lives. What you hold in your hands right now, beneath these words, is consecrated air and time and sunlight.

—Barbara Kingsolver

---

brief contents

contents

PART II The Water, Weather, and Climate Systems 164

8 Weather 190

9 Water Resources 222

10 Global Climate Systems 256

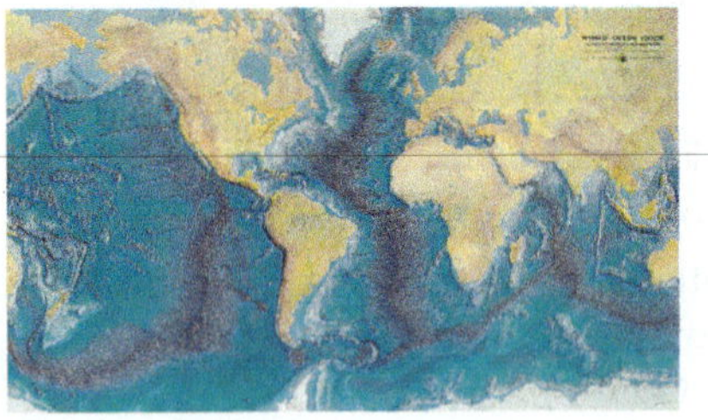

12 Tectonics, Earthquakes, and Volcanism 356

13 Weathering, Karst Landscapes, and Mass Movement 392

14 River Systems 420

15 Eolian Processes and AridLandscapes 454

16 Oceans, Coastal Systems, and Wind Processes 480

Chapters 1–14, 16–20 taken from *Geosystems: An Introduction to Physical Geography*, Ninth Edition by Robert W. Christopherson and Ginger H. Birkeland

Chapter 15 taken from *Geosystems: An Introduction to Physical Geography*, Ninth Edition by Robert W. Christopherson and Ginger H. Birkeland

preface

Welcome to the Ninth Edition of *Geosystems*. This edition marks the addition of Dr. Ginger Birkeland as a coauthor to Robert Christopherson. This Ninth Edition features significant revision, with a new chapter on climate change, new features, updated content, and many new photos and illustrations. We continue to build on the success of the first eight editions, as well as the companion texts, *Elemental Geosystems*, now in its Seventh Edition, and *Geosystems, Canadian Edition*, Third Edition. Students and teachers appreciate the systems organization, scientific accuracy, integration of figures and text, clarity of the summary and review sections, and overall relevancy to what is happening to Earth systems in real time. *Geosystems* continues to tell Earth's story in student-friendly language.

The goal of physical geography is to explain the spatial dimension of Earth's dynamic systems—its energy, air, water, weather, climate, tectonics, landforms, rocks, soils, plants, ecosystems, and biomes. Understanding human–Earth relations is part of physical geography as it seeks to understand and link the planet and its inhabitants. Welcome to physical geography!

New to the Ninth Edition

Nearly every page of *Geosystems*, Ninth Edition, presents updated material, new content in text and figures, and new features. A sampling of new features includes:

- A **new chapter on climate change**. Although climate change science affects all systems and is discussed to some extent in every chapter of *Geosystems*, we now present a stand-alone chapter covering this topic — Chapter 10B, Climate Change. This chapter covers paleoclimatology and mechanisms for past climatic change (expanding on topics covered in Chapter 17 in previous editions), climate feedbacks and the global carbon budget, the evidence and causes of present climate change, climate models and projections, and actions that we can take to moderate Earth's changing climate. This new Chapter 10B expands on the climate change discussion that was formerly part of Chapter 10, Climate Systems and Climate Change, in previous editions.
- A new ***Geosystems in Action*** feature focusing on key topics, processes, systems, or human–Earth connections. In every chapter, *Geosystems in Action* is a one- to two-page highly visual presentation of a topic central to the chapter, with active learning questions and links to media in *MasteringGeography*, as well as a GeoQuiz to aid student learning. Throughout each part of the *Geosystems in Action* figure, students are asked to analyze, explain, infer, or predict based on the information presented. Topics include Earth–Sun Relations (Chapter 2), Air Pollution (Chapter 3), Earth–Atmosphere Energy Balance (Chapter 4), The Global Carbon Budget (Chapter 10B), Glaciers as Dynamic Systems (Chapter 17), and Biological Activity in Soils (Chapter 18).
- A new feature, ***The Human Denominator***, that links chapter topics to human examples and applications. At the end of Chapters 2 through 20, this new feature includes maps, photos, graphs, and other diagrams to provide visual examples of many human–Earth interactions. This feature replaces and expands on the former Chapter 21 in previous *Geosystems* editions, called *Earth and the Human Denominator*.
- New and revised illustrations and maps to improve student learning. More than 250 new photos and images bring real-world scenes into the classroom. Our photo and remote sensing program, updated for this edition, exceeds 500 items, integrated throughout the text.
- New images and photos for the 20 chapter openers, and redesigned schematics and photos for the 4 part openers.
- **Learning Catalytics**, a "bring your own device" student engagement, assessment, and classroom intelligence system, integrated with *MasteringGeography*.

Continuing in the Ninth Edition

- Twenty ***Focus Studies***, with either updated or new content, explore relevant applied topics in greater depth and are a popular feature of the *Geosystems* texts. In the Ninth Edition, these features are grouped by topic into five categories: Pollution, Climate Change, Natural Hazards, Sustainable Resources, and Environmental Restoration.

  Nine new Focus Study topics include:

  Heat Waves (Chapter 5)
  Hurricanes Katrina and Sandy: Storm Development and Links to Climate Change (Chapter 8)
  Thawing Methane Hydrates—Another Arctic Methane Concern (Chapter 10B)
  Earthquakes in Haiti, Chile, and Japan: A Comparative Analysis (Chapter 12)
  Stream Restoration: Merging Science and Practice (Chapter 14)
  The 2011 Japan Tsunami (Chapter 16)
  Snow Avalanches (Chapter 17)
  Wildfire and Fire Ecology (Chapter 19)
  Global Conservation Strategies (Chapter 20)

- The chapter-opening *Geosystems Now* case study feature presents current issues in geography and Earth systems science. These original, unique essays, updated for the Ninth Edition, immediately engage readers into the chapter with relevant, real-world examples of physical geography. New *Geosystems Now* topics in the Ninth Edition include shale gas as an energy resource in the United States (Chapter 1), coastal redwood trees and declining summer moisture in California (Chapter 7), the effects of proposed dams on rivers in China (Chapter 14), and coastal erosion caused by Hurricane Sandy (Chapter 16). Many of these features emphasize linkages across chapters and Earth systems, exemplifying the Geosystems approach.

- *Geo Reports* continue to describe timely and relevant events or facts related to the discussion in the chapter, provide student action items, and offer new sources of information. The 75 *Geo Reports* in the Ninth Edition, placed along the bottom of pages, are updated, with many new to this edition. Example topics include:

  Did light refraction sink the *Titanic*? (Chapter 4)
  The hottest temperature on Earth (Chapter 5)
  Storm causes Hawai'i hailstorm and tornado (Chapter 8)
  Satellite GRACE enables groundwater measurements (Chapter 9)
  Tropical climate zones advance to higher latitudes (Chapter 10)
  Surprise waves flood a cruise ship (Chapter 16)
  Greenland ice sheet melting (Chapter 17)
  Overgrazing effects on Argentina's grasslands (Chapter 18)

- *Critical Thinking* exercises are integrated throughout the chapters. These carefully crafted action items bridge students to the next level of learning, placing students in charge of further inquiry. Example topics include:

  Applying Energy-Balance Principles to a Solar Cooker
  What Causes the North Australian Monsoon?
  Identify Two Kinds of Fog
  Analyzing a Weather Map
  Allocating Responsibility and Cost for Coastal Hazards
  Tropical Forests: A Global or Local Resource?

- The *Geosystems Connection* feature at the end of each chapter provides a preview "bridge" between chapters, reinforcing connections between chapter topics.
- *Key Learning Concepts* appear at the outset of each chapter, many rewritten for clarity. Each chapter concludes with *Key Learning Concepts Review*, which summarizes the chapter using the opening objectives.
- *Geosystems* continues to embed Internet URLs within the text. More than 200 appear in this edition. These allow students to pursue topics of interest to greater depth, or to obtain the latest information about weather and climate, tectonic events, floods, and the myriad other subjects covered in the book.
- The *MasteringGeography*™ online homework and tutoring system delivers self-paced tutorials that provide individualized coaching, focus on course objectives, and are responsive to each student's progress. Instructors can assign activities built around Geoscience Animations, *Encounter* Google Earth" activities, MapMaster" interactive maps, *Thinking Spatially and Data Analysis* activities, new *GeoTutors* on the most challenging topics in physical geography, end-of-chapter questions, Test Bank questions, and more. Students now have access to new *Dynamic Study Modules* that provide each student with a customized learning experience. Students also have access to a text-specific Study Area with study resources, including a Pearson eText version of *Geosystems*, Geoscience Animations, MapMaster interactive maps, new videos, Satellite Loops, Author Notebooks, additional content to support materials for the text, photo galleries, *In the News* RSS feeds, web links, career links, physical geography case studies, flashcard glossary, quizzes, and more—all at www.masteringgeography.com.

Author Acknowledgments

After all these years, the strength of a publishing team remains ever essential. Continuing thanks to President Paul Corey for his leadership since 1990 and to Frank Ruggirello, Senior Vice-president and Editorial Director for Geosciences, for his vision. Thanks to Senior Geography Editor Christian Botting for his guidance and for the attention devoted to the Geosystems texts; and to Program Manager Anton Yakovlev and Assistant Editors Bethany Sexton and Kristen Sanchez for their careful attention. Maya Melenchuk, Image Management, is a great addition to the team and a help to us. Thanks to Managing Editor Gina Cheselka, Program Manager Janice Stangel, and Director of Development Jennifer Hart, for their skills and continuing support.

Our appreciation to designers Mark Ong and Jeanne Calabrese, for such skill in a complex book design. Thanks to the late Randall Goodall for his design work over the years on the previous several editions of *Geosystems* and *Elemental Geosystems*. Thanks also to Maureen McLaughlin, Senior Marketing Manager, Nicola Houston, Senior Marketing Assistant, and the many publisher representatives who spend months in the field communicating the *Geosystems* approach. Our gratitude is extended to the entire "*Geosystems* Team" for allowing us to participate in the publishing process.

Our sincere appreciation for production coordination goes to Editorial Director for Higher Education Cindy Miller of Cenveo LLC for our friendship and sustaining care through eight books, and to Senior Project Manager Suganya Karuppasamy for her ability to respond to our feedback as she oversees manuscript, copy editing, complex compositing, and page proofs. With so many changes in this edition, her skills make it work. To photo researcher Erica Gordon, copy editor Kathy Pruno, proofreader Jeff Georgeson, and indexer Robert Swanson we give thanks for quality work. We offer special thanks to development editor Moira Lerner Nelson for advice and suggestions that improved many aspects of this edition. Our appreciation also goes to Jay McElroy and Jonathan Cheney for their creative talent in helping develop the new *Geosystems in Action* features, and to Jay for his detailed work improving the art program.

Thanks to all the colleagues who served as reviewers on one or more editions of each book, or who offered helpful suggestions in conversations at our national and regional geography meetings. Thanks to the accuracy reviewers of all ninth edition chapters: Todd Fagin, *Oklahoma University*; Giraldo Mario, *California State University, Northridge*; Stephen Cunha, *Humboldt State University*; Charlie Thomsen, *American River College*. And thanks for special reviews for the new Chapter 10B in this edition from Jason Allard, *Valdosta State University*; Marshall Shepherd, *University of Georgia*; Scott Mandia, *Suffolk County Community College, Long Island*; David Kitchen, *University of Richmond*.

We are grateful for the generosity of ideas and sacrifice of time. Thanks to all reviewers who have provided valuable feedback on *Geosystems* over the years:

Ted J. Alsop, *Utah State University*
Michael Allen, *Kent State University*
Philip P. Allen, *Frostburg State University*
Ted J. Alsop, *Utah State University*
Ward Barrett, *University of Minnesota*
Steve Bass, *Mesa Community College*
Stefan Becker, *University of Wisconsin–Oshkosh*
Daniel Bedford, *Weber State University*
David Berner, *Normandale Community College*
Trent Biggs, *San Diego State University*
Franco Biondi, *University of Nevada, Reno*
Peter D. Blanken, *University of Colorado, Boulder*
Patricia Boudinot, *George Mason University*
Anthony Brazel, *Arizona State University*
David R. Butler, *Southwest Texas State University*
Mary-Louise Byrne, *Wilfred Laurier University*
Janet Cakir, *Rappahannock Community College*
Ian A. Campbell, *University of Alberta–Edmonton*
Randall S. Cerveny, *Arizona State University*
Fred Chambers, *University of Colorado, Boulder*
Philip Chaney, *Auburn University*
Muncel Chang, *Butte College* Emeritus
Jordan Clayton, *Georgia State University*
Andrew Comrie, *University of Arizona*
C. Mark Cowell, *Indiana State University*
Richard A. Crooker, *Kutztown University*
Stephen Cunha, *Humboldt State University*
Armando M. da Silva, *Towson State University*
Dirk H. de Boer, *University of Saskatchewan*
Dennis Dahms, *University of Northern Iowa*
J. Michael Daniels, *University of Denver*
Shawna Dark, *California State University, Northridge*
Stephanie Day, *University of Kansas*
Lisa DeChano-Cook, *Western Michigan University*
Mario P. Delisio, *Boise State University*
Joseph R. Desloges, *University of Toronto*
Lee R. Dexter, *Northern Arizona University*
Don W. Duckson, Jr., *Frostburg State University*
Daniel Dugas, *New Mexico State University*
Kathryn Early, *Metropolitan State College*
Christopher H. Exline, *University of Nevada–Reno*
Todd Fagin, *Oklahoma University*
Michael M. Folsom, *Eastern Washington University*
Mark Francek, *Central Michigan University*
Glen Fredlund, *University of Wisconsin–Milwaukee*
Dorothy Friedel, *Sonoma State University*
William Garcia, *University of N. Carolina–Charlotte*
Doug Goodin, *Kansas State University*
Mark Goodman, *Grossmont College*
David E. Greenland, *University of N. Carolina–Chapel Hill*
Duane Griffin, *Bucknell University*
John W. Hall, *Louisiana State University–Shreveport*
Barry N. Haack, *George Mason University*
Roy Haggerty, *Oregon State University*
Vern Harnapp, *University of Akron*
John Harrington, *Kansas State University*
Blake Harrison, *Southern Connecticut University*
Jason "Jake" Haugland, *University of Colorado, Boulder*
Gail Hobbs, *Pierce College*
Thomas W. Holder, *University of Georgia*
David H. Holt, *University of Southern Mississippi*
Robert Hordon, *Rutgers University*
David A. Howarth, *University of Louisville*
Patricia G. Humbertson, *Youngstown State University*
David W. Icenogle, *Auburn University*
Philip L. Jackson, *Oregon State University*
J. Peter Johnson, Jr., *Carleton University*
Gabrielle Katz, *Appalachian State University*
Guy King, *California State University–Chico*
Ronald G. Knapp, *SUNY–The College at New Paltz*
Peter W. Knightes, *Central Texas College*
Jean Kowal, *University of Wisconsin, Whitewater*
Thomas Krabacher, *California State University–Sacramento*
Richard Kurzhals, *Grand Rapids Junior College*
Hsiang-te Kung, *University of Memphis*
Kara Kuvakas, *Hartnell College*
Steve Ladochy, *California State University, Los Angeles*
Charles W. Lafon, *Texas A & M University*
Paul R. Larson, *Southern Utah University*
Robert D. Larson, *Southwest Texas State University*
Derek Law, *University of Kentucky*
Elena Lioubimtseva, *Grand Valley State University*
Joyce Lundberg, *Carleton University*
W. Andrew Marcus, *Montana State University*
Giraldo Mario, *California State University, Northridge*
Brian Mark, *Ohio State University*
Nadine Martin, *University of Arizona*
Elliot G. McIntire, *California State University, Northridge*
Norman Meek, *California State University, San Bernardino*
Leigh W. Mintz, *California State University–Hayward, Emeritus*
Sherry Morea-Oaks, Boulder, CO
Debra Morimoto, *Merced College*
Patrick Moss, *University of Wisconsin, Madison*
Steven Namikas, *Louisiana State University*

Lawrence C. Nkemdirim, *University of Calgary*
Andrew Oliphant, *San Francisco State University*
John E. Oliver, *Indiana State University*
Bradley M. Opdyke, *Michigan State University*
Richard L. Orndorff, *University of Nevada, Las Vegas*
FeiFei Pan, *University of North Texas*
Patrick Pease, *East Carolina University*
James Penn, *Southeastern Louisiana University*
Rachel Pinker, *University of Maryland, College Park*
Greg Pope, *Montclair State University*
Robin J. Rapai, *University of North Dakota*
Philip Reeder, *University of South Florida*
Philip D. Renner, *American River College*
William C. Rense, *Shippensburg University*
Leslie Rigg, *Northern Illinois University*
Dar Roberts, *University of California–Santa Barbara*
Wolf Roder, *University of Cincinnati*
Robert Rohli, *Louisiana State University*
Bill Russell, *L.A. Pierce College*
Dorothy Sack, *Ohio University*
Erinanne Saffell, *Arizona State University*
Randall Schaetzl, *Michigan State University*
Glenn R. Sebastian, *University of South Alabama*
Daniel A. Selwa, *U.S.C. Coastal Carolina College*
Debra Sharkey, *Cosumnes River College*
Peter Siska, *Austin Peay State University*
Lee Slater, *Rutgers University*
Thomas W. Small, *Frostburg State University*
Daniel J. Smith, *University of Victoria*
Richard W. Smith, *Hartford Community College*
Stephen J. Stadler, *Oklahoma State University*
Michael Talbot, *Pima Community College*
Paul E. Todhunter, *University of North Dakota*
Susanna T.Y. Tong, *University of Cincinnati*
Liem Tran, *Florida Atlantic University*
Suzanne Traub-Metlay, *Front Range Community College*
Alice V. Turkington, *The University of Kentucky*
Jon Van de Grift, *Metropolitan State College*
David Weide, *University of Nevada–Las Vegas*
Forrest Wilkerson, *Minnesota State University, Mankato*
Thomas B. Williams, *Western Illinois University*
Brenton M. Yarnal, *Pennsylvania State University*
Catherine H. Yansa, *Michigan State University*
Keith Yearwood, *Georgia State University*
Stephen R. Yool, *University of Arizona*
Don Yow, *Eastern Kentucky University*
Susie Zeigler-Svatek, *University of Minnesota*

From Robert: I thank my family for believing in this work, especially considering the next generation: Chavon, Bryce, Payton, Brock, Trevor, Blake, Chase, Téyenna, and Cade. When I look into our grandchildren's faces, I see why we work toward a sustainable future.

I give special gratitude to all the students during my 30 years teaching at American River College, for it is in the classroom crucible that the *Geosystems* books were forged. Special continued thanks to Charlie Thomsen for his creative work and collaboration on *Encounter Geosystems*, the *Applied Physical Geography* lab manual, work on *MasteringGeography* media and assessments, and ancillaries. Thanks and admiration go to the many authors and scientists who published research that enriches this work. Thanks for all the dialogue received from students and teachers shared with me through e-mails from across the globe.

I offer a special thanks to Ginger Birkeland, Ph.D., my new coauthor on this edition and previous collaborator and developmental editor, for her essential work, attention to detail, and geographic sense. The challenge of such a text project is truly met by her strengths and talents. She is truly a valuable colleague and makes the future of the *Geosystems* franchise look bright as we view the path ahead. She has worked as a river guide operating boats on the Colorado River, and at times I felt her at the helm of *Geosystems*!

As you read this book, you will learn from more than 300 content-specific, beautiful photographs made by my wife, photographer, and expedition partner, Bobbé Christopherson. Her contribution to the success of *Geosystems* is obvious and begins with the spectacular cover photo and continues through the text. Please visit the photo galleries at *MasteringGeography* and learn more from her camera work. Bobbé is my expedition partner, colleague, wife, and best friend.

From Ginger: Many thanks to my husband, Karl Birkeland, for his ongoing patience, support, and inspiration throughout the many hours of work on this book. I also thank my daughters, Erika and Kelsey, who endured my absence throughout a ski season and a rafting season as I sat at my desk. My gratitude also goes to William Graf, my academic advisor from so many years ago, for always exemplifying the highest standard of research and writing, and for helping transform my love of rivers into a love of science and all things geography. Special thanks to Robert Christopherson, who took a leap of faith to bring me on this *Geosystems* journey. It is a privilege to work with him, and I hope our raft runs smoothly and stays upright on the voyage ahead!

From us both: Physical geography teaches us a holistic view of the intricate supporting web that is Earth's environment and our place in it. Dramatic global change is underway in human–Earth relations as we alter physical, chemical, and biological systems. Our attention to climate change science and applied topics is in response to the impacts we are experiencing and the future we are shaping. All things considered, this is a critical time for you to be enrolled in a physical geography course! The best to you in your studies—and ***carpe diem!***

Robert W. Christopherson
P. O. Box 128
Lincoln, California 95648-0128
E-mail: bobobbe@aol.com

Ginger H. Birkeland
Bozeman, Montana

digital and print resources

For Students and Teachers

***MasteringGeography*™ with Pearson eText.** The **Mastering** platform is the most widely used and effective online homework, tutorial, and assessment system for the sciences. It delivers self-paced tutorials that provide individualized coaching, focus on course objectives, and are responsive to each student's progress. The Mastering system helps teachers maximize class time with customizable, easy-to-assign, and automatically graded assessments that motivate students to learn outside of class and arrive prepared for lecture. *MasteringGeography* offers:

- **Assignable activities** that include MapMaster" interactive map activities, *Encounter* Google Earth Explorations, video activities, Geoscience Animation activities, Map Projections activities, GeoTutor coaching activities on the toughest topics in geography, Dynamic Study Modules that provide each student with a customized learning experience, end-of-chapter questions and exercises, reading quizzes, Test Bank questions, and more.
- **A student Study Area** with MapMaster™ interactive maps, videos, Geoscience Animations, web links, glossary flashcards, "In the News" RSS feeds, chapter quizzes, PDF downloads of outline maps, an optional Pearson eText including versions for iPad and Android devices, and more.

Pearson eText gives students access to the text whenever and wherever they can access the Internet. The eText pages look exactly like the printed text and include powerful interactive and customization functions, including links to the multimedia.

Television for the Environment Earth Report Geography Videos on DVD (0321662989). This three-DVD set helps students visualize how human decisions and behavior have affected the environment and how individuals are taking steps toward recovery. With topics ranging from the poor land management promoting the devastation of river systems in Central America to the struggles for electricity in China and Africa, these 13 videos from Television for the Environment's global *Earth Report* series recognize the efforts of individuals around the world to unite and protect the planet.

Geoscience Animation Library 5th edition DVD-ROM (0321716841). Created through a unique collaboration among Pearson's leading geoscience authors, this resource offers over 100 animations covering the most difficult-to-visualize topics in physical geology, physical geography, oceanography, meteorology, and earth science. The animations are provided as Flash files and preloaded into PowerPoint(R) slides for both Windows and Mac.

Practicing Geography: Careers for Enhancing Society and the Environment by Association of American Geographers (0321811151). This book examines career opportunities for geographers and geospatial professionals in the business, government, nonprofit, and education sectors. A diverse group of academic and industry professionals shares insights on career planning, networking, transitioning between employment sectors, and balancing work and home life. The book illustrates the value of geographic expertise and technologies through engaging profiles and case studies of geographers at work.

Teaching College Geography: A Practical Guide for Graduate Students and Early Career Faculty by Association of American Geographers (0136054471). This two-part resource provides a starting point for becoming an effective geography teacher from the very first day of class. Part One addresses "nuts-and-bolts" teaching issues. Part Two explores being an effective teacher in the field, supporting critical thinking with GIS and mapping technologies, engaging learners in large geography classes, and promoting awareness of international perspectives and geographic issues.

Aspiring Academics: A Resource Book for Graduate Students and Early Career Faculty by Association of American Geographers (0136048919). Drawing on several years of research, this set of essays is designed to help graduate students and early career faculty start their careers in geography and related social and environmental sciences. *Aspiring Academics* stresses the interdependence of teaching, research, and service—and the importance of achieving a healthy balance of professional and personal life—while doing faculty work. Each chapter provides accessible, forward-looking advice on topics that often cause the most stress in the first years of a college or university appointment.

For Students

***Applied Physical Geography—Geosystems in the Laboratory,* Ninth Edition** (0321987284) by Charlie Thomsen and Robert Christopherson. A variety of exercises provides flexibility in lab assignments. Each exercise includes key terms and learning concepts linked to *Geosystems.* The ninth edition includes new exercises on climate change, a fully updated exercise on basic GIS using ArcGIS online, and more integrated media, including Google Earth and Quick Response (QR) codes. Supported by a website with media resources needed for exercises, as well as a downloadable Solutions Manual for teachers.

Companion website for *Applied Physical Geography: Geosystems in the Laboratory.* The website for lab manual provides online worksheets as well as KMZ files for all of the Google Earth" exercises found in the lab manual. www.mygeoscienceplace.com

***Goode's World Atlas,* 22nd Edition** (0321652002). *Goode's World Atlas* has been the world's premiere educational atlas since 1923—and for good reason. It features over 250 pages of maps, from definitive physical and political maps to important thematic maps that illustrate the spatial aspects of many important topics. The 22nd Edition includes 160 pages of digitally produced reference maps, as well as thematic maps on global climate change, sea-level rise, CO_2 emissions, polar ice fluctuations, deforestation, extreme weather events, infectious diseases, water resources, and energy production.

Pearson's Encounter Series provides rich, interactive explorations of geoscience concepts through Google Earth" activities, covering a range of topics in regional, human, and physical geography. For those who do not use *MasteringGeography*, all chapter explorations are available in print workbooks, as well as in online quizzes at www.mygeoscienceplace.com, accommodating different classroom needs. Each exploration consists of a worksheet, online quizzes whose results can be emailed to teachers, and a corresponding Google Earth" KMZ file.

- *Encounter Physical Geography* by Jess C. Porter and Stephen O'Connell (0321672526)
  Encounter Geosystems by Charlie Thomsen (0321636996)
  Encounter World Regional Geography by Jess C. Porter (0321681754)
  Encounter Human Geography by Jess C. Porter (0321682203)
- *Encounter Earth* by Steve Kluge (0321581296)

Dire Predictions: Understanding Global Warming by Michael Mann, Lee R. Kump (0136044352) Appropriate for any science or social science course in need of a basic understanding of the reports from the Intergovernmental Panel on Climate Change (IPCC). These periodic reports evaluate the risk of climate change brought on by humans. But the sheer volume of scientific data remains inscrutable to the general public, particularly to those who still question the validity of climate change. In just over 200 pages, this practical text presents and expands upon the essential findings in a visually stunning and undeniably powerful way to the lay reader. Scientific findings that provide validity to the implications of climate change are presented in clear-cut graphic elements, striking images, and understandable analogies.

For Teachers

Learning Catalytics is a "bring your own device" student engagement, assessment, and classroom intelligence system. With Learning Catalytics, you can:

- Assess students in real time, using open-ended tasks to probe student understanding.
- Understand immediately where students are and adjust your lecture accordingly.
- Improve your students' critical-thinking skills.
- Access rich analytics to understand student performance.
- Add your own questions to make Learning Catalytics fit your course exactly.
- Manage student interactions with intelligent grouping and timing.

Learning Catalytics is a technology that has grown out of twenty years of cutting-edge research, innovation, and implementation of interactive teaching and peer instruction. Available integrated with *MasteringGeography*.

***Instructor Resource Manual* (Download)** (032197249X) by Charlie Thomsen includes lecture outlines and key terms, additional source materials, teaching tips, and a complete annotation of chapter review questions. Available from www.pearsonhighered.com/irc, and in the Instructor Resources area of *MasteringGeography*.

***TestGen® Test Bank* (Download)** (032197252X) by Tod Fagin. TestGen® is a computerized test generator that lets you view and edit *Test Bank* questions, transfer questions to tests, and print tests in a variety of customized formats. This *Test Bank* includes around 3,000 multiple-choice, true/false, and short answer/essay questions. All questions are correlated against the National Geography Standards, textbook key concepts, and Bloom's Taxonomy. The *Test Bank* is also available in Microsoft Word® and importable into Blackboard and WebCT. Available from www.pearsonhighered.com/irc, and in the Instructor Resources area of *MasteringGeography*.

Instructor Resource DVD (0321972538). The *Instructor Resource DVD* provides a collection of resources to help teachers make efficient and effective use of their time. All digital resources can be found in one well-organized, easy-to-access place. The IRDVD includes:

- All textbook images as JPEGs, PDFs, and PowerPoint™ Presentations
- Pre-authored Lecture Outline PowerPoint™ Presentations (by FeiFei Pan), which outline the concepts of each chapter with embedded art and can be customized to fit teachers' lecture requirements
- CRS "Clicker" Questions (by FeiFei Pan) in PowerPoint™, which correlate to the book's Learning Objectives, U.S. National Geography Standards, chapter-specific learning outcomes, and Bloom's Taxonomy
- The TestGen software, *Test Bank* questions, and answers for both MACs and PCs
- Electronic files of the *Instructor Resource Manual* and *Test Bank*

This *Instructor Resource* content is also available online via the Instructor Resources section of *MasteringGeography* and www.pearsonhighered.com/irc.

Exploring Earth's Dynamic Systems

Geosystems **is organized around the natural flow of energy, materials, and information, presenting subjects in the same sequence in which they occur in nature—an organic, holistic Earth systems approach that is unique in this discipline. Offering current examples and modern science, Geosystems combines a structured learning path, student-friendly writing, current applications, outstanding visuals, and a strong multimedia program for a truly unique physical geography experience.**

▼ NEW! Chapter 10B: **Climate Change.** Incorporating the latest climate change science and data, this new chapter covers paleoclimatology and mechanisms for past climatic change, climate feedbacks and the global carbon budget, the evidence and causes of present climate change, climate forecasts and models, and actions that we can take to moderate Earth's changing climate.

10B Climate Change

KEY LEARNING concepts

After reading the chapter, you should be able to:

- ***Describe*** scientific tools used to study paleoclimatology.
- ***Discuss*** several natural factors that influence Earth's climate, and ***describe*** climate feedbacks, using examples.
- ***List*** the key lines of evidence for present global climate change, and ***summarize*** the scientific evidence for anthropogenic forcing of climate.
- ***Discuss*** climate models, and ***summarize*** several climate projections.
- ***Describe*** several mitigation measures to slow rates of climate change.

In March 2013, scientists began the fifth year of Operation IceBridge, NASA's airborne, multi-instrument survey of Earth's rapidly changing polar ice. This view of Saunders Island and Wolstenholme Fjord in northwest Greenland in April 2013 shows the characteristic of Arctic sea ice as air and ocean temperature warm. Thinner seasonal ice appears clearer in the foreground; thicker multi-year ice appears whiter in the distance. Much of the Arctic Ocean is now dominated by seasonal ice, which melts rapidly every summer. Ice melt in the polar regions and at high altitudes is an important indicator of Earth's changing climate, the subject of this chapter [NASA/Michael Studinger.]

GEOSYSTEMS now

Greenhouse Gases Awaken in the Arctic

In the subarctic and tundra climate regions of the Northern Hemisphere, perennially frozen soils and sediment, known as permafrost, cover about 24% of the land area. With Arctic air temperatures currently rising at a rate more than two times that of the midlatitudes, ground temperatures are increasing, causing permafrost thaw. This results in changes to land surfaces, primarily sinking and slumping, that damage buildings, forests, and coastlines (Figure GN 11.1). Permafrost thaw also leads to the decay of soil material, a process that releases vast amounts of carbon, in the form of the greenhouse gases carbon dioxide (CO_2) and methane (CH_4), into the atmosphere.

Carbon in Permafrost Soils Permafrost is, by definition, soil and sediment that have remained frozen for two or more consecutive years. The "active layer" is the seasonally frozen ground on top of subsurface permafrost. This thin layer of soil and sediment thaws every summer, providing substrate for seasonal grasses and other plants that absorb CO_2 from the atmosphere. In winter, the active layer freezes, trapping plant and animal material before it can decompose completely. Over hundreds of thousands of years, this carbon-rich material has become incorporated into permafrost and now makes up roughly half of all the organic matter stored in Earth's soils—twice the amount of carbon that is stored in the atmosphere. In terms of real numbers, the latest estimate of the amount of carbon stored in Arctic permafrost soils is 1700 gigatonnes (or 1700 billion tons).

A Positive Feedback Loop As summers become warmer in the Arctic, heat radiating through the ground thaws the permafrost layers. Microbial activity in these layers increases, enhancing the breakdown of organic matter. As this occurs, bacteria and other organisms release CO_2 into the atmosphere in a process known as *microbial respiration*. In anaerobic (oxygen-free) environments, such as lakes and wetlands, the process releases methane. Studies show that thousands of methane seeps can develop under a single lake, a huge amount when multiplied by hundreds of thousands of lakes across the northern latitudes (Figure GN 11.2).

Carbon dioxide and methane are major greenhouse gases, which absorb outgoing longwave radiation and radiate it back toward Earth, enhancing the greenhouse effect and leading to atmospheric warming. Methane is especially important because, although its relative percentage is small in the atmosphere, it is over 20 times more effective than CO_2 at trapping atmospheric heat. Thus, a positive feedback loop forms: As temperatures rise, permafrost thaws, causing a release of CO_2 and CH_4 into the atmosphere, which causes more warming, leading to more permafrost thaw.

Melting Ground Ice In addi-

▲Figure GN 11.2 **Blocks of melting permafrost collapse into the Beaufort Sea, Alaska.** [USGS Alaska Science Center.]

supporting structure provided by the ice is removed, land surfaces collapse and slump. Subsurface soils are then exposed to sunlight, which speeds up microbial processes, and to water erosion, which moves organic carbon into streams and lakes, where it is mobilized into the atmosphere. Research suggests that this process may release bursts of CO_2 and CH_4 into the atmosphere, in contrast to the slower top-down melting of permafrost.

Permafrost soils are now warming at a rate faster than Arctic air temperatures, releasing vast amounts of "ancient" carbon into the atmosphere. Scientists are actively researching the locations and amounts of vulnerable permafrost, the current and projected rates of thaw, and the potential impacts to the permafrost-carbon positive feedback. The thawing Arctic is one of many immediate concerns we discuss in this chapter regarding the causes and impacts of changing climate on Earth systems.

GEOSYSTEMS NOW ONLINE Go to Chapter 10B on the *MasteringGeography* website (www.masteringgeography.com) for more on the permafrost thaw and climate change. To learn about NASA's Carbon in Arctic Reservoirs Vulnerability Experiment (CARVE), which measures CO_2 and CH_4 gas emissions in permafrost regions, go to http://science1.nasa.gov/missions/carve/

▶ NEW! ***The Human Denominator*** summarizes Human-Earth relationships, interactions, challenges for the 21st century through dynamic visuals, including maps, photos, graphs, and diagrams.

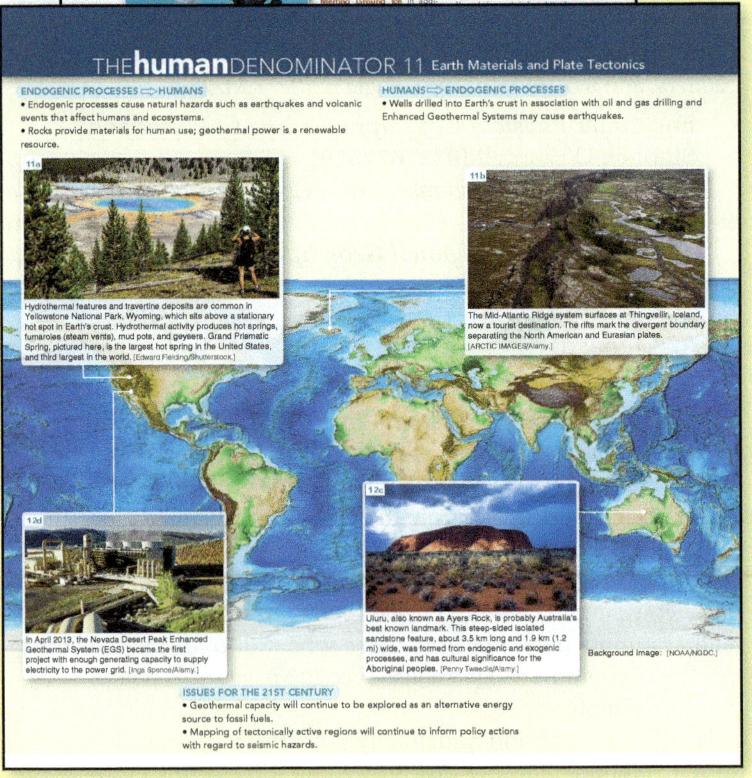

THE**human**DENOMINATOR 11 Earth Materials and Plate Tectonics

ENDOGENIC PROCESSES ⇨ HUMANS
- Endogenic processes cause natural hazards such as earthquakes and volcanic events that affect humans and ecosystems.
- Rocks provide materials for human use; geothermal power is a renewable resource.

HUMANS ⇨ ENDOGENIC PROCESSES
- Wells drilled into Earth's crust in association with oil and gas drilling and Enhanced Geothermal Systems may cause earthquakes.

11a Hydrothermal features and travertine deposits are common in Yellowstone National Park, Wyoming, which sits above a stationary hot spot in Earth's crust. Hydrothermal activity produces hot springs, fumaroles (steam vents), mud pots, and geysers. Grand Prismatic Spring, pictured here, is the largest hot spring in the United States, and third largest in the world. [Edward Fielding/Shutterstock.]

11b The Mid-Atlantic Ridge system surfaces at Thingvellir, Iceland, now a tourist destination. The rifts mark the divergent boundary separating the North American and Eurasian plates. [ARCTIC IMAGES/Alamy.]

12d In April 2013, the Nevada Desert Peak Enhanced Geothermal System (EGS) became the first project with enough generating capacity to supply electricity to the power grid. [Inga Spence/Alamy.]

12c Uluru, also known as Ayers Rock, is probably Australia's best known landmark. This steep-sided isolated sandstone feature, about 3.5 km long and 1.9 km (1.2 mi) wide, was formed from endogenic and exogenic processes, and has cultural significance for the Aboriginal peoples. [Penny Tweedie/Alamy.]

Background Image: [NOAA/NGDC.]

ISSUES FOR THE 21ST CENTURY
- Geothermal capacity will continue to be explored as an alternative energy source to fossil fuels.
- Mapping of tectonically active regions will continue to inform policy actions with regard to seismic hazards.

Visualizing Processes and Landscapes

▼ NEW! ***Geosystems in Action*** present highly-visual presentations of core physical processes and critical chapter concepts. These features include links to mobile-ready media and MasteringGeography, as well as GeoQuizzes and integrated active learning tasks that ask students to analyze, explain, infer, or predict based on the information presented.

An unparalleled visual program includes a variety of illustrations, maps, photographs, and composites, providing authoritative examples and applications of physical geography and Earth systems science.

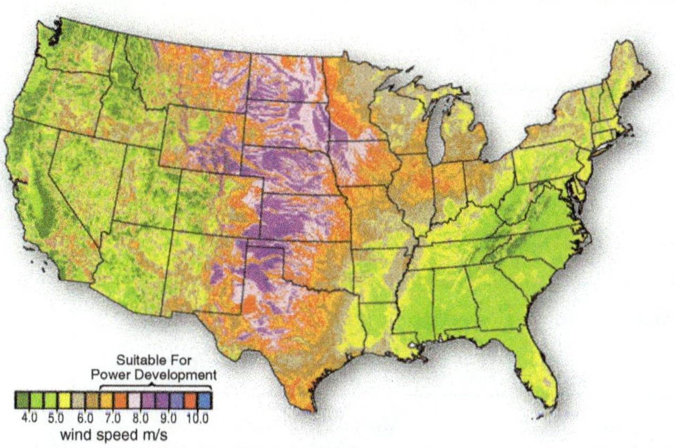

▲Figure 6.1.1 **Wind-speed map of the contiguous United States.** Map shows predicted average wind speeds at a height of 80 m (164 ft) above the ground. Areas with wind speeds greater than 6.5 m/s are considered suitable for power development. Map has a spatial resolution of 2.5 km. [NREL and AWS Truepower.]

▼Figure 17.11 **The geomorphic handiwork of alpine glaciers.** As the glaciers retreat, the new landscape is unveiled. Inset photos are surface and aerial views from Norway. [Photos by Bobbé Christopherson; waterfall by Author.]

Physical Geography in the Real World

Geosystems integrates current real events and phenomena and presents the most thorough and integrated treatment of systems trends and climate change science, giving students compelling reasons for learning physical geography.

▼ ***Geosystems*** **Now** open each chapter with interesting, current applications of physical geography and Earth systems science. New **Geosystems Now Online** features direct students online to related resources.

GEOSYSTEMS now

Sand Dunes Prevent Coastline Erosion during Hurricane Sandy

During the winter of 2013, in the ongoing aftermath of Hurricane Sandy, many residents along New Jersey's coastline added their discarded Christmas trees to carefully stacked lines of trees intended to act as "seeds" for new sand dune formation along several area beaches. The hope is that the trees will catch windblown sand to begin the dune formation process, in one of many such restoration efforts along the Atlantic coast. In the face of Sandy's winds, houses and neighborhoods with protective dunes in place experienced less damage than those that were more exposed to and closer to the ocean.

Dune Protection versus Ocean Views The effectiveness of dune systems as protection from wave erosion and storm surge during Hurricane Sandy, far from being a subtle statistical phenomenon, was easily observed by local residents. However, the fostering of large and sometimes obtrusive sand dunes near the shoreline is controversial in coastal communities with million-dollar homes. For such dunes to function as barriers to erosion, they must sit between oceanfront property and the sea, thus blocking ocean views and decreasing property values (Figure GN 16.1). For many landowners, establishing dunes for storm protection means financial loss, in the short term, even if long-term protection is the result.

Coastal Dune Geomorphology Coastal sand dunes originate from sediment supplied by the work of ocean waves and by fluvial processes that move sediment onto deltas and estuaries. Once sand is deposited on shore, it is reworked by wind processes into the shape of dunes. Dunes along seacoasts are either *foredunes*, where sand is pushed up the seaward-facing slope, or *backdunes*, which form further away from the beach and are protected from onshore winds; backdunes are more stable and may be hundreds of years old. Most areas of coastal dunes are relatively small in size (especially when compared with desert dune fields that may cover large portions of continents).

Along the Atlantic coast, foredunes are moving inland as sea level rises and storm energy increases with climate change. In developed areas, the foredunes cannot retreat inland without impinging on human development. When storms occur, dune movement is intensified, and either dune erosion or sand deposition, or both, occurs within the developed area of the coast (Figure GN 16.2).

Dune Restoration Efforts The establishment of new foredunes replenishes the sand supply and protects structures and infrastructure, making this a potentially worthwhile investment of money and effort for communities along the New Jersey shoreline. Many experts point out that dunes are *not* a guarantee of storm protection and that Sandy's winds and storm surge were strong enough to erode some large natural dune systems along the Atlantic Seaboard. However, in Bradley Beach, New Jersey, where the storm eroded several miles of restored dunes about 4.6 m (15 ft) in height, the community still escaped excessive damage since the dunes absorbed much of the storm's impact.

Thus, local communities are supporting dune restoration, as evidenced by the Christmas tree initiative. Because vegetation is important for dune stabilization, the planting of grasses is another protective strategy being embraced by New Jersey residents. In this chapter, we discuss coastal systems, wind processes, and dune formation processes.

May 21, 2009

November 5, 2012

▲Figure GN 16.2 **Coastal damage from Sandy in Mantoloking, New Jersey.** View looking west before and after Hurricane Sandy. The yellow arrow points to the same feature in each image. [USGS.]

▲Figure GN 16.1 **Constructed dunes.** Restored sand dunes shield homes in Mantoloking, New Jersey, from an incoming nor'easter a few weeks after Hurricane Sandy. [Sharon Karr/FEMA.]

GEOSYSTEMS NOW ONLINE Go to Chapter 16 on the *MasteringGeography* website (www.masteringgeography.com) for resources and activities. For information and links to research on dunes in New Jersey and along the Atlantic coast, see http://marine.rutgers.edu/geomorph/geomorph/_pages/dunes.html. MG

481

▼ ***Focus Studies*** present detailed discussions of critical physical geography topics, emphasizing the applied relevance of physical geography today.

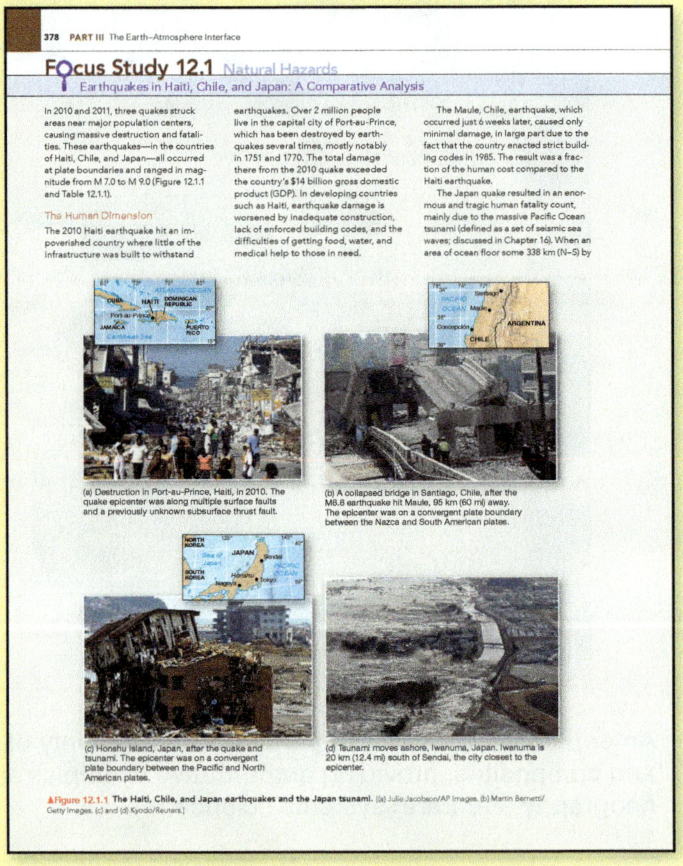
378 PART III The Earth–Atmosphere Interface

Focus Study 12.1 Natural Hazards

Earthquakes in Haiti, Chile, and Japan: A Comparative Analysis

In 2010 and 2011, three quakes struck areas near major population centers, causing massive destruction and fatalities. These earthquakes—in the countries of Haiti, Chile, and Japan—all occurred at plate boundaries and ranged in magnitude from M 7.0 to M 9.0 (Figure 12.1.1 and Table 12.1.1).

The Human Dimension

The 2010 Haiti earthquake hit an impoverished country where little of the infrastructure was built to withstand earthquakes. Over 2 million people live in the capital city of Port-au-Prince, which has been destroyed by earthquakes several times, mostly notably in 1751 and 1770. The total damage there from the 2010 quake exceeded the country's $14 billion gross domestic product (GDP). In developing countries such as Haiti, earthquake damage is worsened by inadequate construction, lack of enforced building codes, and the difficulties of getting food, water, and medical help to those in need.

The Maule, Chile, earthquake, which occurred just 6 weeks later, caused only minimal damage, in large part due to the fact that the country enacted strict building codes in 1985. The result was a fraction of the human cost compared to the Haiti earthquake.

The Japan quake resulted in an enormous and tragic human fatality count, mainly due to the massive Pacific Ocean tsunami (defined as a set of seismic sea waves; discussed in Chapter 16). When an area of ocean floor some 338 km (N–S) by

(a) Destruction in Port-au-Prince, Haiti, in 2010. The quake epicenter was along multiple surface faults and a previously unknown subsurface thrust fault.

(b) A collapsed bridge in Santiago, Chile, after the M8.8 earthquake hit Maule, 95 km (60 mi) away. The epicenter was on a convergent plate boundary between the Nazca and South American plates.

(c) Honshu Island, Japan, after the quake and tsunami. The epicenter was on a convergent plate boundary between the Pacific and North American plates.

(d) Tsunami moves ashore, Iwanuma, Japan. Iwanuma is 20 km (12.4 mi) south of Sendai, the city closest to the epicenter.

▲Figure 12.1.1 **The Haiti, Chile, and Japan earthquakes and the Japan tsunami.** [(a) Julie Jacobson/AP Images. (b) Martin Bernetti/Getty Images. (c) and (d) Kyodo/Reuters.]

► **GeoReports** offer a wide variety of brief interesting facts, examples, and applications to complement and enrich the chapter reading.

GEOreport 8.2 Mountains cause record rains

Mount Waialeale, on the island of Kaua'i, Hawai'i, rises 1569 m (5147 ft) above sea level. On its windward slope, rainfall averaged 1234 cm (486 in., or 40.5 ft) a year for the years 1941–1992. In contrast, the rain-shadow side of Kaua'i receives only 50 cm (20 in.) of rain annually. If no islands existed at this location, this portion of the Pacific Ocean would receive only an average 63.5 cm (25 in.) of precipitation a year. (These statistics are from established weather stations with a consistent record of weather data; several stations claim higher rainfall values but do not have dependable measurement records.)

Cherrapunji, India, is 1313 m (4309 ft) above sea level at 25° N latitude, in the Assam Hills south of the Himalayas. Summer monsoons pour in from the Indian Ocean and the Bay of Bengal, producing 930 cm (366 in., or 30.5 ft) of rainfall in 1 month. Not surprisingly, Cherrapunji is the all-time precipitation record holder for a single year, 2647 cm (1042 in., or 86.8 ft), and for every other time interval from 15 days to 2 years. Th

GEOreport 11.3 Large earthquakes affect Earth's axial tilt

Scientific evidence is mounting that Earth's largest earthquake events have a global influence. Both the 2004 Sumatran–Andaman quake and the 2011 Tohoku quake in Japan caused Earth's axial tilt to shift several centimeters. NASA scientists estimate that the redistribution of mass in each quake shortened daylength by 6.8 millionths of a second for the 2004 event and 1.8 millionths

GEOreport 19.2 Sea turtles navigate using Earth's magnetic field

The fact that birds and bees can detect Earth's magnetic field and use it for finding direction is well established. Small amounts of magnetically sensitive particles in the skull of the bird and the abdomen of the bee provide compass directions. Recently, scientists found that sea turtles detect magnetic fields of different strengths and inclinations (angles). This means that the turtles have a built-in navigation system that helps them find certain locations on Earth. Loggerhead turtles hatch in Florida, crawl into the water, and spend the next 70 years traveling thousands of miles between North America and Africa around the subtropical high-pressure gyre in the Atlantic Ocean. The females return to where they were hatched to lay their eggs. In turn, the hatchlings are imprinted with magnetic data unique to the location of their birth and then develop a more global sense of position as they live a life swimming across the ocean.

Tools for Structured Learning

Geosystems **provides a structured learning path that helps students achieve a deeper understanding of physical geography through active learning.**

KEY LEARNING **concepts**

After reading the chapter, you should be able to:

- ***Sketch*** a basic drainage basin model, and ***identify*** different types of drainage patterns by visual examination.
- ***Explain*** the concepts of stream gradient and base level, and ***describe*** the relationship between stream velocity, depth, width, and discharge.
- ***Explain*** the processes involved in fluvial erosion and sediment transport.
- ***Describe*** common stream channel patterns, and ***explain*** the concept of a graded stream.
- ***Describe*** the depositional landforms associated with floodplains and alluvial fan environments.
- ***List*** and ***describe*** several types of river deltas, and ***explain*** flood probability estimates.

◀ ***Key Learning Concepts*** at the beginning of every chapter help students identify the key knowledge and skills they will acquire through study of the chapter.

▼ ***Critical Thinking Activities*** integrated throughout chapter sections give students an opportunity to stop, check, and apply their understanding.

CRITICAL**thinking 14.1**
Locate Your Drainage Basin

Determine the name of the drainage basin within which your campus is located. Where are its headwaters? Where is the river's mouth? If you are in the United States or Canada, use Figure 14.3 to locate the larger drainage basins and divides for your region, and then take a look at this region on Google Earth™. Does any regulatory organization oversee planning and coordination for the drainag… you find topographic maps onlin…

CRITICAL**thinking 14.2**
Identifying Drainage Patterns

Examine the photograph in Figure CT 14.2.1, where you see two distinct drainage patterns. Of the seven types illustrated in Figure 14.5, which two patterns are most like those in the aerial photo? Looking back to Figure 14.1a, which drainage pattern is prevalent in the area around Mount Mismi in Brazil? Explain your answer. The next time you fly in an airplane, look out the window to observe the various drainage patterns across the landscape. ●

▲Figure CT 14.2.1 **Two drainage patterns dominate this scene from central Montana, in response to rock structure and local relief.** [Bobbé Christopherson.]

▼ ***Key Learning Concepts Review*** at the end of each chapter concludes the learning path and features summaries, narrative definitions, a list of key terms with page numbers, and review questions.

KEY LEARNING **concepts review**

■ ***List*** and ***describe*** several types of river deltas, and ***explain*** flood probability estimates.

A depositional plain formed at the mouth of a river is called a **delta**. Deltas may be arcuate or bird's foot in shape, or estuarine in nature. Some rivers have no deltas. When the mouth of a river enters the sea and is inundated by seawater in a mix with freshwater, it is called an **estuary**. Despite historical devastation by floods, floodplains and deltas are important sites of human activity and settlement. Efforts to reduce flooding include the construction of artificial levees, bypasses, straightened channels, diversions, dams, and reservoirs.

A **flood** occurs when high water overflows the natural bank along any portion of a stream. Human-constructed **artificial levees** are common features along many rivers of the United States, where flood protection is needed for developed floodplains. Both floods and the floodplains they occupy are rated statistically for the expected time interval between floods of given discharges. For example, a 10-year flood has the statistical probability of happening once every 10 years. Flood probabilities are useful for floodplain zoning.

delta (p. 444)
estuary (p. 444)
flood (p. 446)
artificial levee (p. 447)

20. What is a river delta? What are the various deltaic forms? Give some examples.
21. Describe the Ganges River delta. What factors upstream explain its form and pattern? Assess the consequences of settlement on this delta.
22. What is meant by the statement, "The Nile River delta is disappearing"?
23. Specifically, what is a flood? How are such flows measured and tracked, and how are they used in floodplain management?
24. What is channel avulsion, and how does it occur?

▶ ***Geosystems Connection*** at the end of chapters help students bridge concepts between chapters, reminding them where they have been and where they are going.

GEOSYSTEMS**connection**

While following the flow of water through streams, we examined fluvial processes and landforms and the river-system outputs of discharge and sediment. We saw that a scientific understanding of river dynamics, floodplain landscapes, and related flood hazards is integral to society's ability to perceive hazards in the familiar environments we inhabit. In the next chapter, we examine the erosional activities of waves, tides, currents, and wind as they sculpt Earth's coastlines and desert regions. A significant portion of the human population lives in coastal areas, making the difficulties of hazard perception and the need to plan for the future, given a rising sea level, important aspects of Chapter 16.

MasteringGeography™

MasteringGeography delivers engaging, dynamic learning opportunities—focusing on course objectives and responsive to each student's progress—that are proven to help students absorb geography course material and understand difficult physical processes and geographic concepts.

Visualize the Processes and Landscapes that form Earth's Physical Environment

▶ **Encounter Activities** provide rich, interactive explorations of geography concepts using the dynamic features of Google Earth™ to visualize and explore Earth's physical landscape. Available with multiple-choice and short answer questions. All Explorations include corresponding Google Earth KMZ media files, and questions include hints and specific wrong-answer feedback to help coach students towards mastery of the concepts.

MG
Animation
Seafloor Spreading
Subduction

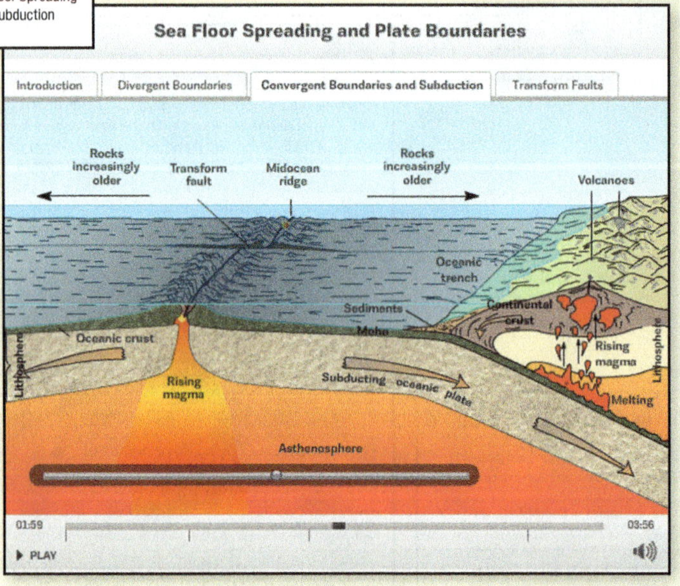

◀ **Geoscience Animations** illuminate the most difficult-to-visualize topics from across the physical geosciences, such as solar system formation, hydrologic cycle, plate tectonics, glacial advance and retreat, global warming, etc. Animations include audio narration, a text transcript, and assignable multiple-choice quizzes with specific wrong-answer feedback to help guide students towards mastery of these core physical process concepts. Icons integrated throughout the text indicate to students when they can login to the Study Area of **MasteringGeography** to access the animations.

▶ NEW! **Quick Response Codes** in the ***Geosystems in Action*** features link to select videos,t animations, and web sites, providing students with just-in-time access to visualization and data resources from their mobile devices.

Tour of the Cryosphere

Engage in Map Reading, Data Analysis, and Critical Thinking

MapMaster™ is a powerful tool that presents assignable layered thematic and place name interactive maps at world and regional scales for students to test their geographic literacy, map reading, data analysis, and spatial reasoning skills.

▶ **MapMaster** Layered Thematic Interactive Map Activities allow students to layer various thematic maps to analyze spatial patterns and data at regional and global scales. Available with assignable and customizable multiple-choice and short-answer questions organized around the textbook topics and concepts. This GIS-like tool includes zoom and annotation functionality, with hundreds of map layers leveraging recent data from sources such as NOAA, NASA, USGS, U.S. Census Bureau, United Nations, CIA, World Bank, and the Population Reference Bureau.

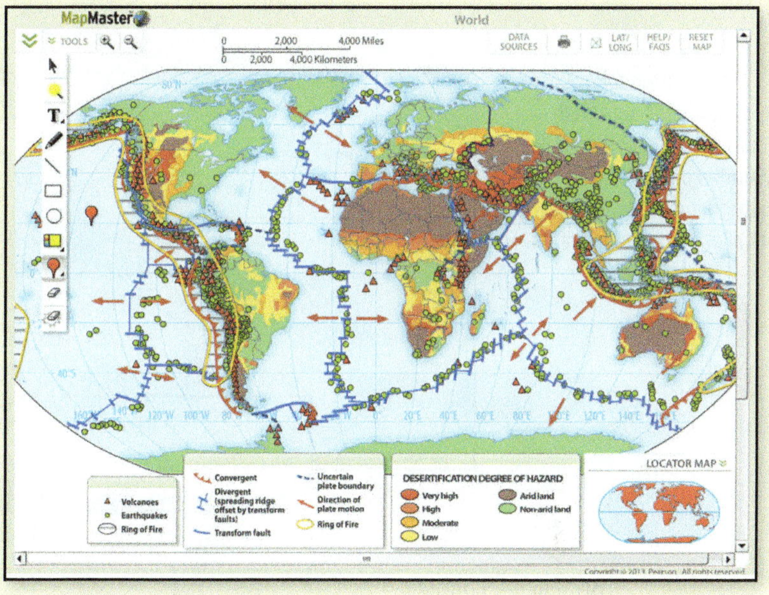

▼ **Thinking Spatially & Data Analysis and NEW GeoTutor Activities** help students master the toughest geographic concepts and develop both spatial reasoning and critical thinking skills. Students identify and label features from maps, illustrations, graphs, and charts, examine related data sets, and answer higher-order conceptual questions, which include hints and specific wrong-answer feedback.

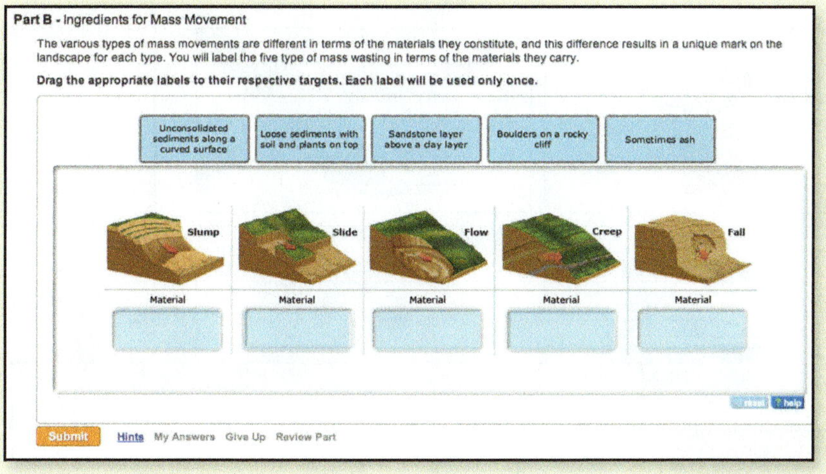

▼ **Videos** provide students with a sense of place and allow them to explore a range of locations and topics. Covering physical processes and critical issues such as climate and climate change, renewable energy resources, economy and development, culture, and globalization, these video activities include assignable questions, with many including hints and specific wrong-answer feedback.

Student Study Area Resources in MasteringGeography:

- Geoscience Animations
- MapMaster™ interactive maps
- Videos
- Practice quizzes
- "In the News" RSS feeds
- Glossary flashcards
- Optional Pearson eText and more

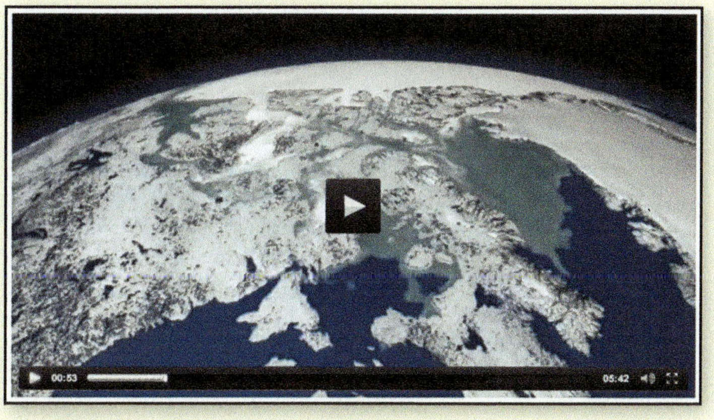

NEW! Dynamic Study Modules

Personalize each student's learning experience with Dynamic Study Modules. Created to allow students to study on their own and be better prepared to achieve higher scores on their tests. Mobile app available for iOS and Android devices for study on the go.

MasteringGeography™

With the Mastering gradebook and diagnostics, you'll be better informed about your students' progress than ever before. Mastering captures the step-by-step work of every student—including wrong answers submitted, hints requested, and time taken at every step of every problem—all providing unique insight into the most common misconceptions of your class.

▶ The **Gradebook** records all scores for automatically graded assignments. Shades of red highlight struggling students and challenging **assignments.**

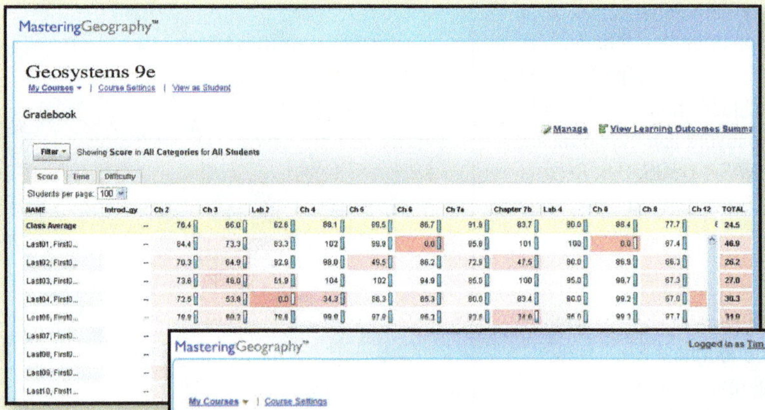

▶ **Diagnostics** provide unique insight into class and student performance. With a single click, charts summarize the most difficult questions, vulnerable students, grade distribution, and score improvement over the duration of the course.

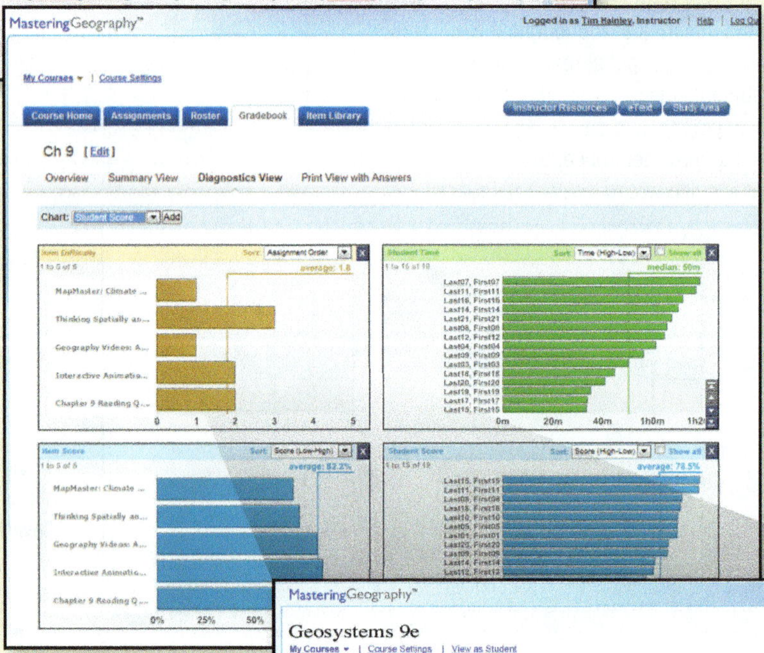

▶ With a single click, Individual Student Performance Data provide at-a-glance statistics into each individual student's performance, including time spent on the question, number of hints opened, and number of wrong and correct answers submitted.

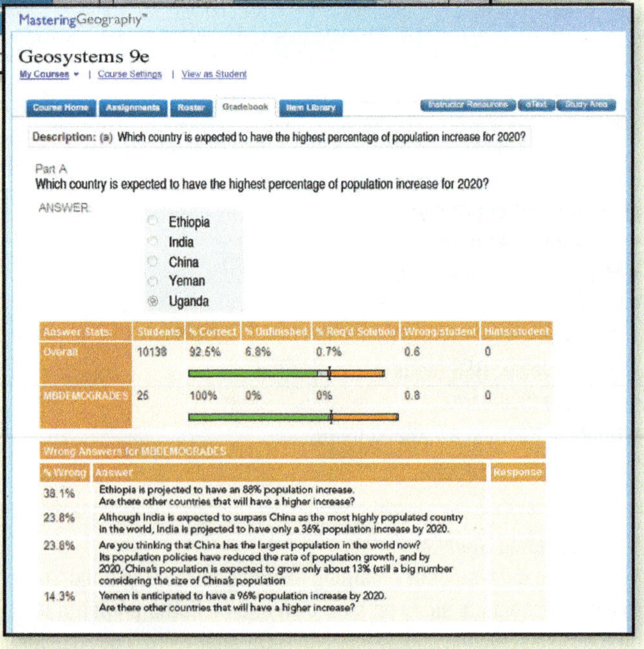

Learning Outcomes

MasteringGeography provides quick and easy access to information on student performance against your learning outcomes and makes it easy to share those results.

- Quickly add your own learning outcomes, or use publisher provided ones, to track student performance and report it to your administration.
- View class and individual student performance against specific learning outcomes.
- Effortlessly export results to a spreadsheett that you can further customize and/or share with your chair, dean, administrator, and/or accreditation board.

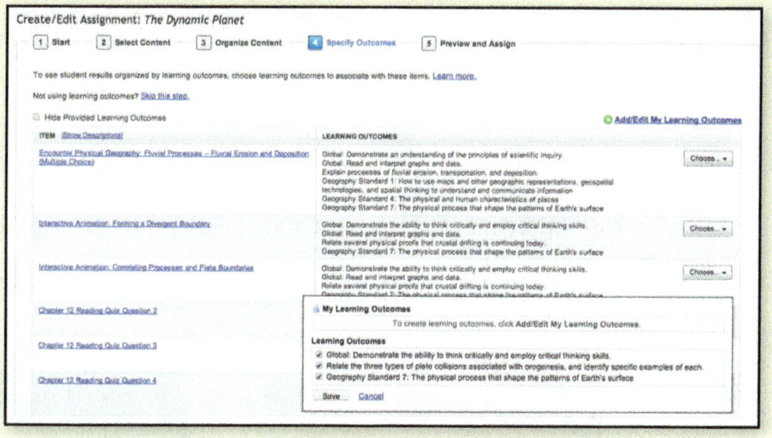

Easy to customize

Customize publisher-provided items or quickly add your own. MasteringGeography makes it easy to edit any questions or answers, import your own questions, and quickly add images, links, and files to further enhance the student experience.

Upload your own video and audio files from your hard drive to share with students, as well as record video from your computer's webcam directly into MasteringGeography—no plugins required. Students can download video and audio files to their local computer or launch them in Mastering to view the content.

learning | catalytics

Learning Catalytics is a "bring your own device" student engagement, assessment, and classroom intelligence system. With Learning Catalytics you can:

- Assess students in real time, using open-ended tasks to probe student understanding.
- Understand immediately where students are and adjust your lecture accordingly.
- Improve your students' critical-thinking skills.
- Access rich analytics to understand student performance.
- Add your own questions to make Learning Catalytics fit your course exactly.
- Manage student interactions with intelligent grouping and timing.

Learning Catalytics is a technology that has grown out of twenty years of cutting edge research, innovation, and implementation of interactive teaching and peer instruction. Available integrated with MasteringGeography or standalone.

Pearson eText gives students access to *Geosystems, 9th Edition* whenever and wherever they can access the Internet. The eText pages look exactly like the printed text, and include powerful interactive and customization functions. Users can create notes, highlight text in different colors, create bookmarks, zoom, click hyperlinked words and phrases to view definitions, and view as a single page or as two pages. Pearson eText also links students to associated media files, enabling them to view an animation as they read the text, and offers a full-text search and the ability to save and export notes. The Pearson eText also includes embedded URLs in the chapter text with active links to the Internet.

The Pearson eText app is a great companion to Pearson's eText browser-based book reader. It allows existing subscribers who view their Pearson eText titles on a Mac or PC to additionally access their titles in a bookshelf on the iPad and Android devices either online or via download.

Pearson Choices

for texts with Mastering

Pearson offers content in a variety of formats designed to meet the needs of today's cost-conscious students.

BOUND TEXTBOOK

Pearson texts are written by authors you trust, and undergo a rigorous editorial process to ensure quality and accuracy.

A LA CARTE

A binder-ready, loose-leaf, 3-hole-punched version of the text is available as a budget-friendly option for students.

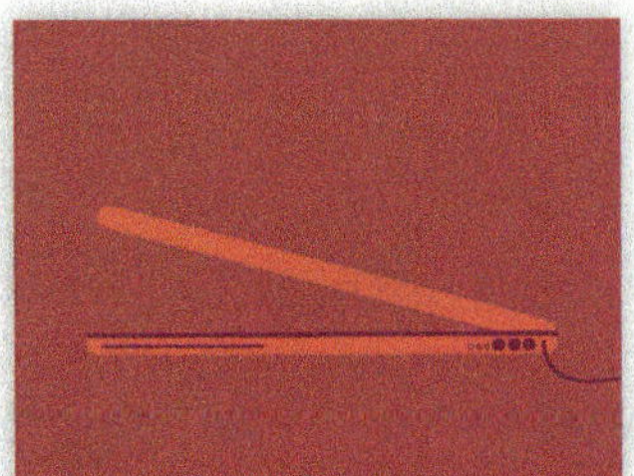

eTEXT IN MASTERING

Our new eTexts in Mastering have dynamic interactive features such as note-taking, searching, bookmarking, and more.

CUSTOM OPTIONS

CUSTOM EDITION

School-specific editions help reduce costs while aligning closely with your course goals.

PEARSON CUSTOM LIBRARY*

Pearson Custom Library is an option for many science titles, allowing you to build a custom text for as few as 25 students.
www.pearsoncustomlibrary.com

about our sustainability initiatives

Pearson recognizes the environmental challenges facing this planet, as well as acknowledges our responsibility in making a difference. This book has been carefully crafted to minimize environmental impact. The binding, cover, and paper come from facilities that minimize waste, energy consumption, and the use of harmful chemicals. Pearson closes the loop by recycling every out-of-date text returned to our warehouse.

Along with developing and exploring digital solutions to our market's needs, Pearson has a strong commitment to achieving carbon-neutrality. As of 2009, Pearson became the first carbon- and climate-neutral publishing company. Since then, Pearson remains strongly committed to measuring, reducing, and offsetting our carbon footprint.

The future holds great promise for reducing our impact on Earth's environment, and Pearson is proud to be leading the way. We strive to publish the best books with the most up-to-date and accurate content, and to do so in ways that minimize our impact on Earth. To learn more about our initiatives, please visit www.pearson.com/responsibility.

1 Essentials of Geography

KEY LEARNING **concepts**

After reading the chapter, you should be able to:

- ***Define*** geography in general and physical geography in particular.
- ***Discuss*** human activities and human population growth as they relate to geographic science, and ***summarize*** the scientific process.
- ***Describe*** systems analysis, open and closed systems, and feedback information, and ***relate*** these concepts to Earth systems.
- ***Explain*** Earth's reference grid: latitude and longitude and latitudinal geographic zones and time.
- ***Define*** cartography and mapping basics: map scale and map projections.
- ***Describe*** modern geoscience techniques—the Global Positioning System (GPS), remote sensing, and geographic information systems (GIS)—and ***explain*** how these tools are used in geographic analysis.

A snow avalanche roars down Mount Timpanogos, the second highest peak in Utah's Wasatch Mountains. Snow avalanches are a significant hazard in mountainous environments worldwide, killing hundreds of people annually. Avalanches result from the combination of steep, open slopes and unstable snow. The dramatic vertical relief of the Wasatch Range, which rises 2301 m (7,550 ft) above the Great Salt Lake, interacts with moist Pacific air masses, resulting in an average of 160 m (525 in.) of snowfall each winter. Winter storms set the stage for dangerous conditions. New snow and wind that blows snow onto lee slopes are the primary factors contributing to avalanche formation. This January 2005 avalanche stopped short of the houses in the foreground. [Bruce Tremper, Utah Avalanche Center.]

Shale Gas: An Energy Resource for the Future?

In an area stretching 965 km (600 mi) from Ohio to western New York, methane lies deeply buried in a sedimentary rock deposit, the Marcellus Shale. Methane is the primary constituent of natural gas, and scientists suggest that this ancient rock layer, underlying 60% of Pennsylvania, may be one of the most significant reservoirs of natural gas in the world. Pennsylvania alone is dotted with nearly 6000 shale gas wells extracting pressurized methane (Figure GN 1.1).

What Is Methane? Methane is a chemical compound with a formula of CH_4 and is a by-product of several natural processes: digestive activity of animals (cattle, sheep, bison) and termites; melting of arctic permafrost; burning associated with wildfires; and bacterial activity in bogs, swamps, and wetlands. Nearly 60% of the methane in our atmosphere comes from human sources, including natural gas production, beef and dairy production, rice cultivation, coal and oil extraction and burning, landfills, and wastewater treatment. In the United States, the natural gas industry makes up the largest percentage of U.S. methane emissions.

Drilling for Methane To release methane trapped within shale layers, the rock must be broken up so that gas diffuses into the cracks and flows upward. Over the past 20 years, advances in horizontal drilling techniques, combined with the process of hydraulic fracturing, or "fracking," opened access to large amounts of natural gas previously deemed too expensive or difficult to tap. A typical shale gas well descends vertically 2.4 km (1.5 mi), then turns and drills horizontally into the rock strata. Horizontal drilling exposes a greater area of the rock, allowing more of it to be broken up and more gas to be released (Figure GN 1.2).

A pressurized fluid is pumped into the well to break up the rock—90% water, 9% sand or glass beads to prop open the fissures, and 1% chemical additives as lubricants. The specific chemicals used are as yet undisclosed by the industry. This use of an injected fluid to fracture the shale is the process of fracking. Gas then flows up the well to be collected at the surface.

Fracking uses massive quantities of water: approximately 15 million liters (4 million gallons) for each well system, flowing at a rate of 16,000 L (4200 gal) per minute—far more than could be provided by a public water system. In southwestern Pennsylvania, storage ponds hold the water pumped to well sites for fracking operations.

The U.S. Energy Information Administration (EIA) projects a boom in shale gas extraction and production from fracking over the next 20 years, with U.S. production rising from 30% of all natural gas production in 2010 to 49% in 2030.

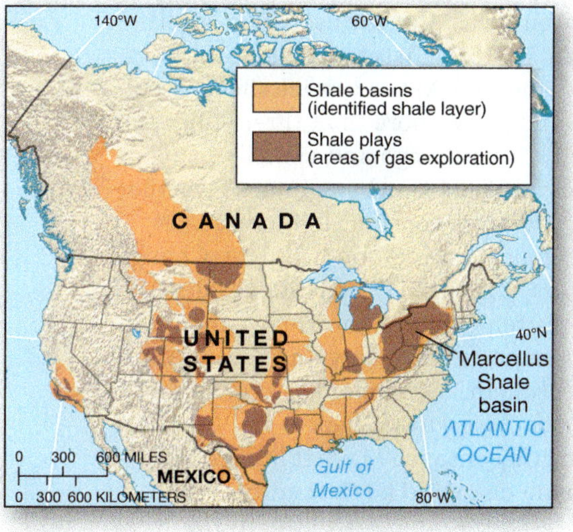

▲Figure GN 1.1 Shale deposits and areas of exploration for natural gas extraction, United States and Canada. [U.S. Energy Information Administration]

Environmental Effects As with other resource-extraction techniques, fracking leaves hazardous by-products. It produces large amounts of toxic wastewater, often held in wells or containment ponds. Any leak or failure of pond retaining walls spills pollutants into surface water supplies and groundwater. Methane gas may leak around well casings, which tend to crack during the fracking process. Leaks can cause buildup of methane in groundwater, leading to contaminated drinking water wells, flammable tap water, methane accumulation in barns and homes, and possible explosions.

Methane adds to air pollution as a constituent in smog and is a potent greenhouse gas, absorbing heat from the Sun near Earth's surface and contributing to global climate change. In addition, scientists linked the injection of fluid into wastewater wells to increased ground instability and earthquake activity in Ohio, West Virginia, Texas, Oklahoma, and parts of the Midwest.

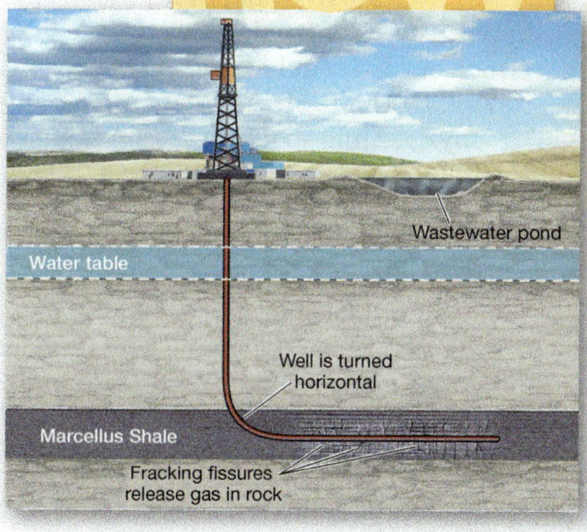

▲Figure GN 1.2 Horizontal drilling for hydraulic fracturing (fracking) and shale gas extraction.

This rapidly expanding energy resource has varied impacts on air, water, land, and living Earth systems. However, many of the environmental effects of shale gas extraction remain unknown; further scientific study is critical.

Shale Gas and *Geosystems* Resource location and distribution and human–environment interactions not only are important issues associated with shale gas extraction, but also are at the heart of geographic science. In this chapter, you work with several "Essentials of Geography": the scientific process, Earth systems thinking, spatial concepts, and mapping. Throughout *Geosystems*, we will expand the story of shale gas and its far-reaching effects on global climate, surface water and groundwater resources, and ecosystem functions.

GEOSYSTEMS NOW ONLINE Go to Chapter 1 on the *MasteringGeography* website (www.masteringgeography.com) for resources and activities regarding shale gas as an energy resource. Explore shale gas online at http://ngm.nationalgeographic.com/2012/12/methane/lavelle-text for an interactive diagram called "Breaking Fuel from the Rock" and links to articles. For another perspective, go to http://www.energyfromshale.org/shale-extraction-process, which presents shale gas extraction from the energy industry's point of view. Should the United States and other countries expand shale gas as an energy resource for the future? MG

Welcome to the Ninth Edition of *Geosystems* and the study of physical geography! In this text, we examine the powerful Earth systems that influence our lives and the many ways humans impact those systems. This is an important time to study physical geography, learning about Earth's environments, including the systems that form the landscapes, seascapes, atmosphere, and ecosystems on which humans depend. In this second decade of the 21st century, a century that will see many changes to our natural world, scientific study of the Earth and environment is more crucial than ever.

Consider the following events, among many similar ones we could mention, and the questions they raise for the study of Earth's systems and physical geography. This text provides tools for answering these questions and addressing the underlying issues.

- In October 2012, Hurricane Sandy made landfall along the U.S. East Coast, hitting New York and New Jersey at high tide with hurricane force winds and record storm surges. The storm cost 110 human lives and over $42 billion in New York State alone, approaching $100 billion in damages overall. What atmospheric processes explain the formation and movement of this storm? Why the unprecedented size and intensity? How is this storm related to record air and ocean temperatures?
- In March 2011, a magnitude 9.0 earthquake and resultant 10- to 20-m (33- to 66-ft) tsunami devastated Honshu Island, Japan—at $309 billion (U.S. dollars), Earth's most expensive natural disaster. Why do earthquakes occur in particular locations across the globe? What produces tsunami, and how far and fast do they travel? This event caused the worst multiple nuclear power plant catastrophe in history, with three core meltdowns, releasing dangerous quantities of radioactivity over land and into the atmosphere and ocean, and eventually reaching the food supply. How will prevailing winds and currents disperse the radiation across the globe?
- By the end of 2012, the removal of two dams on the Elwha River in Washington was almost complete—the largest dam removals in the world to date (Figure 1.1). The project will restore a free-flowing river for fisheries and associated ecosystems. In Brazil, construction of the controversial Belo Monte hydroelectric dam on the Xingu River continues, despite court orders and violent protests. The dam will displace nearly 20,000 people and, when completed, will be the world's third largest hydroelectric project, one of 60 planned to generate power for Brazil's rapidly expanding economy. How do dams change river environments?
- In 2011, the world released 2.4 million pounds of carbon dioxide (CO_2) into the atmosphere every second, mainly from the burning of fossil fuels; China's 1.3 billion people produce 10 billion tons of CO_2 annually. This "greenhouse gas" contributes to climate change by trapping heat near Earth's surface. Each year atmospheric CO_2 levels rise to a new record, altering Earth's climate. What are the effects and what do climate forecasts tell us?

▲**Figure 1.1 Dam removal for river restoration.** Removal of Glines Canyon Dam on the Elwha River, Washington, began in November 2012 to restore river ecosystems. [Brian Cluer/NOAA.]

Physical geography uses a *spatial* perspective to examine processes and events happening at specific locations and follow their effects across the globe. Why does the environment vary from equator to midlatitudes, and between deserts and polar regions? How does solar energy influence the distribution of trees, soils, climates, and lifestyles? What produces the patterns of wind, weather, and ocean currents? Why are global sea levels on the rise? How do natural systems affect human populations, and, in turn, what impact are humans having on natural systems? Why are record levels of plants and animals facing extinction? In this book, we explore those questions, and more, through geography's unique perspective.

Perhaps more than any other issue, climate change has become an overriding focus of the study of Earth systems. The past decade experienced the highest temperatures over land and water in the instrumental record. The year 2010 tied 2005 as the warmest for global temperatures. In response, the extent of sea ice in the Arctic Ocean continues to decline to record lows—the 2012 summer sea ice extent was the lowest since satellite measurements began in 1979. Between 1992 and 2011, melting of the Greenland and Antarctica ice sheets accelerated; together they now lose more than three times the ice they lost annually 20 years ago and contribute about 20% of current sea-level rise. Elsewhere, intense weather events, drought, and flooding continue to increase.

The Intergovernmental Panel on Climate Change (IPCC; http://www.ipcc.ch/), the lead international scientific body assessing the current state of knowledge about climate change and its impacts on society and the environment, completed its *Fourth Assessment Report* in 2007, and released the *Fifth Assessment Report* in 2014. The overwhelming scientific consensus is that human activities are forcing climate change. The first edition of *Geosystems* in 1992 featured the findings of the initial *First Assessment Report* from the IPCC, and the current edition continues to survey climate change evidence and consider its implications. In every chapter, *Geosystems* presents up-to-date science and information to help you understand our dynamic Earth systems. Welcome to an exploration of physical geography!

In this chapter: Our study of geosystems—Earth systems—begins with a look at the science of physical geography and the geographic tools it uses. Physical geography uses an integrative spatial approach, guided by the scientific process, to study entire Earth systems. The role of humans is an increasingly important focus of physical geography, as are questions of global sustainability as Earth's population grows.

Physical geographers study the environment by analyzing air, water, land, and living systems. Therefore, we discuss systems and the feedback mechanisms that influence system operations. We then consider location on Earth as determined by the coordinated grid system of latitude and longitude, and the determination of world time zones. Next, we examine maps as critical tools that geographers use to display physical and cultural information. This chapter concludes with an overview of new and widely accessible technologies that are adding exciting new dimensions to geographic science: Global Positioning System, remote sensing from space, and geographic information systems.

The Science of Geography

A common idea about geography is that it is chiefly concerned with place names. Although location and place are important geographic concepts, geography as a science encompasses much more. **Geography** (from *geo*, "Earth," and *graphein*, "to write") is the science that studies the relationships among natural systems, geographic areas, society, and cultural activities, and the interdependence of all of these, *over space*. These last two words are key, for geography is a science that is in part defined by its method—a special way of analyzing phenomena over space. In geography, the term **spatial** refers to the nature and character of physical space, its measurement, and the distribution of things within it.

Geographic concepts pertain to distributions and movement across Earth. For example, to the patterns of air and ocean currents over Earth's surface, and how these currents affect the dispersal of pollutants, such as nuclear radiation or oil spills. Geography, then, is the spatial consideration of Earth processes interacting with human actions.

Although geography is not limited to place names, maps and location are central to the discipline and are important tools for conveying geographic data. Evolving technologies such as geographic information systems (GIS) and the Global Positioning System (GPS) are widely used for scientific applications and in today's society as hundreds of millions of people access maps and locational information every day on computers and mobile devices.

For educational purposes, the concerns of geographic science have traditionally been divided into five spatial themes: **location**, **region**, **human–Earth relationships**, **movement**, and **place**, each illustrated and defined in Figure 1.2. These themes, first implemented in 1984, are still used as a framework for understanding geographic concepts at all levels, and *Geosystems* draws on each. At the same time, the National Center for Geographic Education (NCGE) has updated the geography education guidelines (most recently in 2012) in response to increasing globalization and environmental change, redefining the essential elements of geography and expanding their number to six: *the spatial world, places and regions, physical systems, human systems, environment and society,* and *uses of geography in today's society.* These categories emphasize the spatial and environmental perspectives within the discipline and reflect the growing importance of human–environment interactions.

The Geographic Continuum

Because many subjects can be examined geographically, geography is an eclectic science that integrates subject matter from a wide range of disciplines. Even so, it splits broadly into two primary fields: *physical geography,* comprising specialty areas that draw largely on the physical and life sciences; and *human geography,* comprising specialty areas that draw largely on the social and cultural sciences. Prior to this century, scientific studies tended to fall onto one end of this continuum or the other. Humans tended at times to think of themselves as exempt from physical Earth processes—like actors not paying attention to their stage, props, and lighting.

However, as global population, communication, and movement increase, so does awareness that we all depend on Earth's systems to provide oxygen, water, nutrients, energy, and materials to support life. The growing complexity of the human–Earth relationship in the twenty-first century has shifted the study of geographic processes toward the center of the continuum in Figure 1.3 to attain a more balanced perspective—such is the thrust of *Geosystems.* This more balanced synthesis is reflected in geographic subfields such as natural resource geography and environmental planning, and in technologies such as geographic information science (GISci), used by both physical and human geographers.

Within physical geography, research now emphasizes human influences on natural systems in all specialty areas, effectively moving this end of the continuum closer to the middle. For example, physical geographers monitor air pollution, examine the vulnerability of human populations to climate change, study impacts of human activities on forest health and the movement of invasive species, study changes in river systems caused by dams and dam removal, and examine the response of glacial ice to changing climate.

Geographic Analysis

As mentioned earlier, the science of geography is unified more by its method than by a specific body of knowledge.

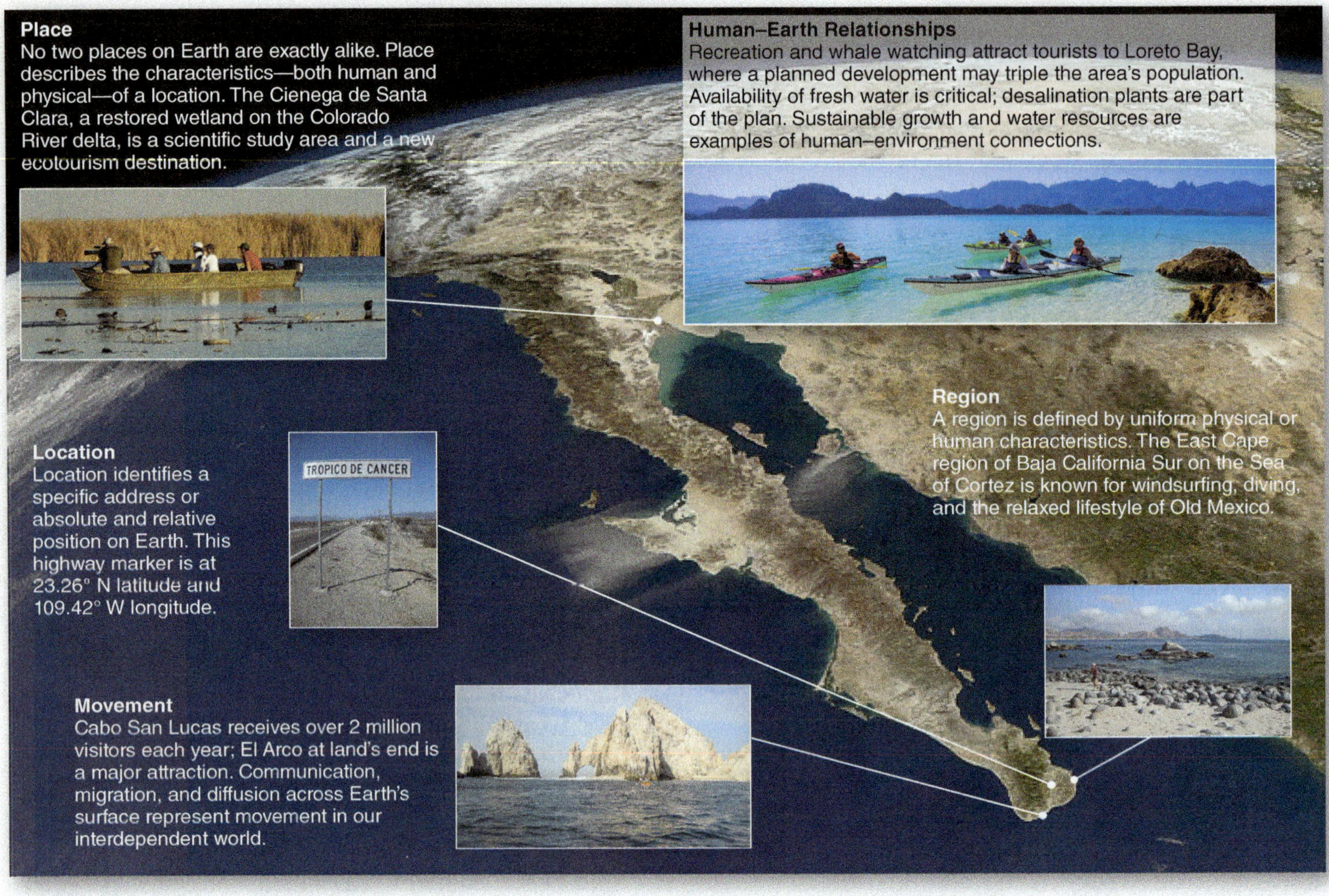

▲**Figure 1.2 Five themes of geographic science.** Drawing from your own experience, can you think of examples of each theme? This 2011 satellite image shows the entire length of Mexico's Baja peninsula, including Earth's curvature. [Photos by Karl Birkeland, except Place by Cheryl Zook/National Geographic and Human–Earth by Gary Luhm/garyluhm.net. Image from *Aqua* satellite/Norman Kuring, Ocean Color Team. NASA/GSFC.]

The method is **spatial analysis**. Using this method, geography synthesizes (brings together) topics from many fields, integrating information to form a whole-Earth concept. Geographers view phenomena as occurring across spaces, areas, and locations. The language of geography reflects this spatial view: territory, zone, pattern, distribution, place, location, region, sphere, province, and distance. Geographers analyze the differences and similarities between places.

Process, a set of actions or mechanisms that operate in some special order, is a central concept of geographic analysis. Among the examples you encounter in *Geosystems* are the numerous processes involved in Earth's vast water–atmosphere–weather system; in continental crust movements and earthquake occurrences; in ecosystem functions; or in river channel dynamics. Geographers use spatial analysis to examine how Earth's processes interact through space or over areas.

Therefore, **physical geography** is the spatial analysis of all the physical elements, processes, and systems that make up the environment: energy, air, water, weather, climate, landforms, soils, animals, plants, microorganisms, and Earth itself. Today, in addition to its place in the geographic continuum, physical geography also

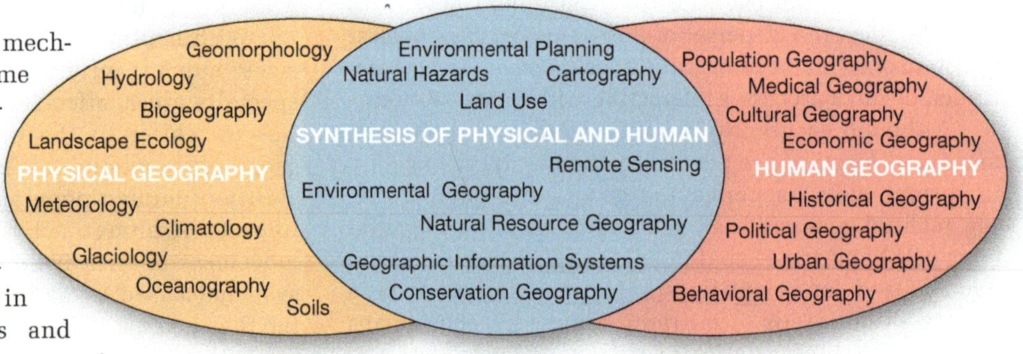

▲**Figure 1.3 The content of geography.** Geography synthesizes Earth topics and human topics, blending ideas from many different sciences. This book focuses on physical geography, but integrates pertinent human and cultural content for a whole-Earth perspective.

forms part of the broad field of **Earth systems science**, the area of study that seeks to understand Earth as a complete entity, an interacting set of physical, chemical, and biological systems. With these definitions in mind, we now discuss the general process and methods used by scientists, including geographers.

The Scientific Process

The process of science consists of observing, questioning, testing, and understanding elements of the natural world. The **scientific method** is the traditional recipe of a scientific investigation; it can be thought of as simple, organized steps leading toward concrete, objective conclusions. A scientist observes and asks questions, makes a general statement to summarize the observations, formulates a hypothesis (a logical explanation), conducts experiments or collects data to test the hypothesis, and interprets results. Repeated testing and support of a hypothesis leads to a scientific theory. Sir Isaac Newton (1642–1727) developed this method of discovering the patterns of nature, although the term *scientific method* was applied later.

While the scientific method is of fundamental importance in guiding scientific investigation, the real process of science is more dynamic and less linear, leaving room for questioning and thinking "out of the box." Flexibility and creativity are essential to the scientific process, which may not always follow the same sequence of steps or use the same methods for each experiment or research project. There is no single, definitive method for doing science; scientists in different fields and even in different subfields of physical geography may approach their scientific testing in different ways. However, the end result must be a conclusion that can be tested repeatedly and possibly shown as true, or as false. Without this characteristic, it is not science.

Using the Scientific Method

Figure 1.4 illustrates steps of the scientific method and outlines a simple application examining cottonwood tree distributions. The scientific method begins with our perception of the real world. Scientists who study the physical environment begin with the clues they see in nature. The process begins as scientists question and analyze their observations and explore the relevant published scientific literature on their topic. Brainstorming with others, continued observation, and preliminary data collection may occur at this stage.

Questions and observations identify variables, which are the conditions that change in an experiment or model. Scientists often seek to reduce the number of variables when formulating a *hypothesis*—a tentative explanation for the phenomena observed. Since natural systems are complex, controlling or eliminating variables helps simplify research questions and predictions.

Scientists test hypotheses using experimental studies in laboratories or natural settings. Correlational studies, which look for associations between variables, are common in many scientific fields, including physical geography. The methods used for these studies must be reproducible so that repeat testing can occur. Results may support or disprove the hypothesis, or predictions made according to it may prove accurate or inaccurate. If the results disprove the hypothesis, the researcher will need to adjust data-collection methods or refine the hypothesis statement. If the results support the hypothesis, repeated testing and verification may lead to its elevation to the status of a *theory*.

Reporting research results is also part of the scientific method. For scientific work to reach other scientists and eventually the public at large, it must be described in a scientific paper and published in one of many scientific journals. Critical to the process is *peer review*, in which other members of the scientific or professional community critique the methods and interpretation of results. This process also helps detect any personal or political bias by the scientist. When a paper is submitted to a scientific journal, it is sent to reviewers, who may recommend rejecting the paper or accepting and revising it for publication. Once a number of papers are published with similar results and conclusions, the building of a theory begins.

The word *theory* can be confusing as used by the media and general public. A scientific theory is constructed on the basis of several extensively tested hypotheses and can be reevaluated or expanded according to new evidence. Thus, a scientific theory is not absolute truth; the possibility always exists that the theory could be proved wrong. However, theories represent truly broad general principles—unifying concepts that tie together the laws that govern nature. Examples include the theory of relativity, theory of evolution, and plate tectonics theory. A scientific theory reinforces our perception of the real world and is the basis for predictions to be made about things not yet known. The value of a scientific theory is that it stimulates continued observation, testing, understanding, and pursuit of knowledge within scientific fields.

Applying Scientific Results

Scientific studies described as "basic" are designed largely to help advance knowledge and build scientific theories. Other research is designed to produce "applied" results tied directly to real-world problem solving. Applied scientific research may advance new technologies, affect natural resource policy, or directly impact management strategies. Scientists share the results of both basic and applied research at conferences as well as in published papers, and they may take leadership roles in policy and planning. For example, the awareness that human activity is producing global climate change places increasing pressure on scientists to participate in decision making. Numerous editorials in scientific journals have called for such practical scientific involvement.

The nature of science is objective and does not make value judgments. Instead, pure science provides people and their institutions with objective information on which to base their own value judgments. Social and political judgments about the applications of science are increasingly important as Earth's natural systems respond to the impacts of modern civilization.

(a) **Scientific Process Flow Chart**

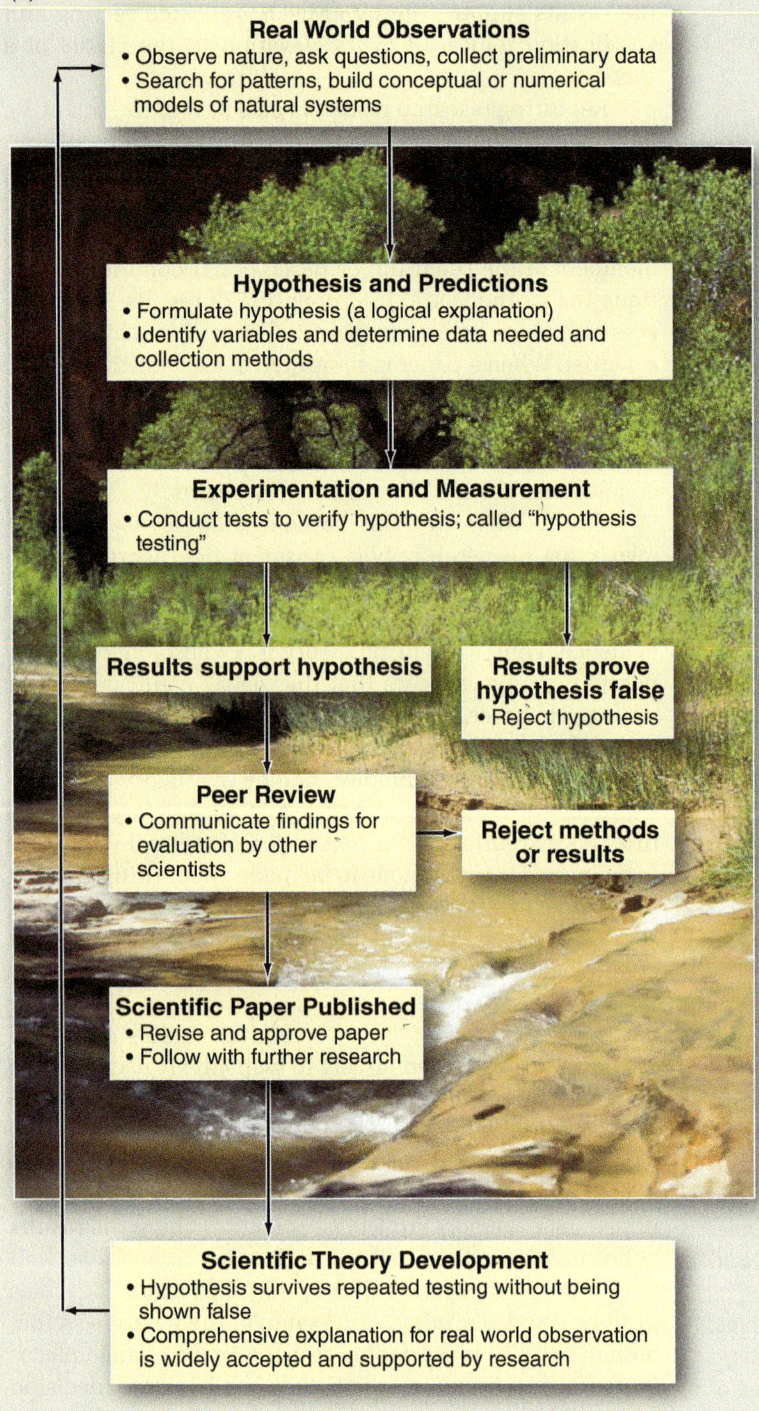

(b) **Using the Scientific Process to Study Cottonwood Tree Distribution**

1. Observations
In the dry climates of the western United States, cottonwood trees grow only along rivers. These trees are not found away from watercourses. What environmental factors influence their spatial distribution?

2. Questions and Variables
Are temperatures near rivers favorable for cottonwood growth?
Is consistent moisture needed for tree survival?
Do cottonwood roots grow only in river gravels or only in sediments with specific nutrients?
Have humans removed all the cottonwoods except along rivers?
Cottonwood trees are the *dependent variable* because their distribution is dependent on some environmental factor. Temperature, sunlight, moisture, sediment type, nutrients, and human actions are *independent variables*; any or all of these may be found to determine patterns of cottonwood distribution.

3. Hypothesis
One possible explanation for the observed pattern of tree distribution is that cottonwoods require consistent moisture in their root zone.
We can test the hypothesis that the number of cottonwoods decreases as one moves away from a river channel because there the tree roots are out of the reach of surface flows and groundwater.

4. Testing
Collect data from natural systems for a natural experiment. Establish vegetation plots (small areas of ground). Sample, or count, trees within plots and measure the distance of each tree from the main channel. Control other variables as much as possible.

5. Results
A natural experiment often reveals a *correlation*, or a statistical relationship. If a correlation shows that the number of cottonwoods decreases away from the stream channel, then the hypothesis is supported. Continued investigation might repeat the same procedure in a different environment or expand the study to a larger region, and lead to a theory. However, if results show that cottonwoods grow at a variety of distances from the main channel, then we reject the hypothesis, replacing or refining it with another possible explanation (see questions above).

6. Theory Development
If we find that the distribution of cottonwoods is correlated with the presence of surface or subsurface water, we may also conclude that cottonwoods are an easily observable indicator of surface flow and available groundwater in dry or semi-dry regions.

▲**Figure 1.4 The scientific process.** (a) Scientific method flow chart and (b) example application to cottonwood distribution. [Ginger Birkeland photograph.]

Human–Earth Interactions in the 21st Century

Issues surrounding the growing influence of humans on Earth systems are central concerns of physical geography; we discuss them in every chapter of *Geosystems*. Human influence on Earth is now pervasive. The global human population passed 6 billion in August 1999 and continued to grow at the rate of 82 million per year, adding another billion by 2011, when the 7 billionth human was born. More people are alive today than at any previous moment in the planet's long history, unevenly distributed among 193 countries and numerous colonies. Virtually all new population growth is in the less-developed countries (LDCs), which now possess 81%, or about 5.75 billion, of the total population. Over the span of human history, billion-mark milestones occurred at ever closer intervals through the sixth-billion milestone; the interval is now slightly increasing. (Figure 1.5).

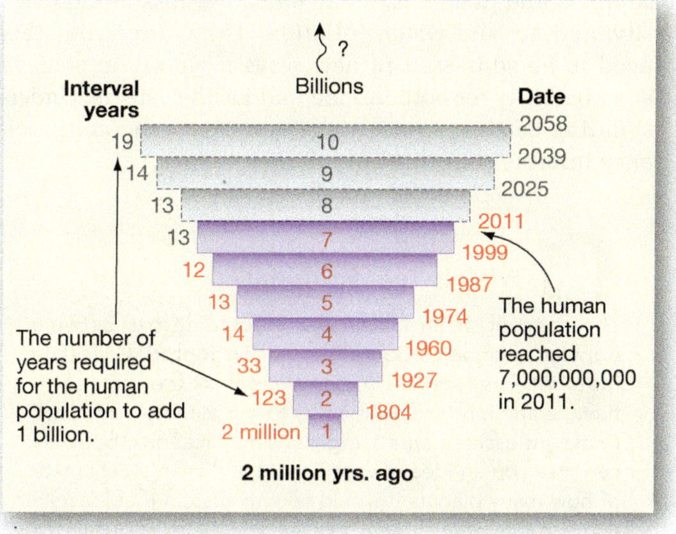

▲**Figure 1.5 Human population growth.** Note the population forecasts for the next half century.

The Human Denominator We consider the totality of human impact on Earth as the *human denominator.* Just as the denominator in a fraction tells how many parts a whole is divided into, so the growing human population and its increasing demand for resources and rising planetary impact suggest the stresses on the whole Earth system to provide support. Yet Earth's resource base remains relatively fixed.

The population in just two countries makes up 37% of Earth's human count: 19.1% live in China and 17.9% in India—2.61 billion people combined. Considered overall, the planetary population is young, with some 26% still under the age of 15 (2012 data from the Population Reference Bureau, at http://www.prb.org and the U.S. Census Bureau's *POPClock Projection,* at http://www.census.gov/popclock).

Population in most of the more-developed countries (MDCs) is no longer increasing. In fact, some European countries are actually declining in growth or are near replacement levels. However, people in these developed countries have a greater impact on the planet per person and therefore constitute a population impact crisis. The United States and Canada, with about 5% of the world's population, produce more than 25.8% ($14.7 trillion and $1.6 trillion in 2010, respectively) of the world's gross domestic product (GDP), the United States increasing to $15,094 billion GDP for 2011. These two countries use more than 2 times the energy per capita of Europeans, more than 7 times that of Latin Americans, 10 times that of Asians, and 20 times that of Africans. Therefore, the impact of this 5% on the state of Earth systems, natural resources, and sustainability of current practices in the MDCs is critical.

Global Sustainability Recently, **sustainability science** emerged as a new, integrative discipline, broadly based on concepts of sustainable development related to functioning Earth systems. Geographic concepts are fundamental to this new science, with its emphasis on human well-being, Earth systems, and human–environment interactions. Geographers are leading the effort to articulate this emerging field that seeks to directly link science and technology with sustainability.

Geographer Carol Harden, geomorphologist and past president of the Association of American Geographers, pointed out the important role of geographical concepts in sustainability science in 2009. She wrote that the idea of a human "footprint," representing the human impact on Earth systems, relates to sustainability and geography. When the human population of over 7 billion is taken into account, the human footprint on Earth is enormous, both in terms of its spatial extent and the strength of its influence. Shrinking this footprint ties to sustainability science in all of its forms—for example, sustainable development, sustainable resources, sustainable energy, and sustainable agriculture. Especially in the face of today's rapidly changing technological and environmental systems, geographers are poised to contribute to this emerging field.

If we consider some of the key issues for this century, many of them fall beneath the umbrella of sustainability science, such as feeding the world's population, energy

GEOreport 1.1 Welcome to the Anthropocene

The human population on Earth reached 7 billion in 2011. Many scientists now agree that the *Anthropocene*, a term coined by Nobel Prize–winning scientist Paul Crutzen, is an appropriate name for the most recent years of geologic history, when humans have influenced Earth's climate and ecosystems. Some scientists mark the beginning of agriculture, about 5000 years ago, as the start of the Anthropocene; others place the start at the dawn of the Industrial Revolution, in the 18th century. To see a video charting the growth of humans as a planetary force, go to http://www.anthropocene.info.

supplies and demands, climate change, loss of biodiversity, and air and water pollution. These are issues that need to be addressed in new ways if we are to achieve sustainability for both human and Earth systems. Understanding Earth's physical geography and geographic science informs your thinking on these issues.

> **CRITICALthinking 1.1**
> **What is Your Footprint?**
>
> The concept of an individual's "footprint" has become popular—ecological footprint, carbon footprint, lifestyle footprint. The term has come to represent the costs of affluence and modern technology to our planetary systems. Footprint assessments are gross simplifications, but they can give you an idea of your impact and even an estimate of how many planets it would take to sustain that lifestyle and economy if everyone lived like you. Calculate your carbon footprint online at http://www.epa.gov/climatechange/ghgemissions/ind-calculator.html, one of many such websites, for housing, transportation, or food consumption. How can you reduce your footprint at home, at school, at work, or on the road? How does your footprint compare to the U.S. and worldwide average footprints? •

Earth Systems Concepts

The word *system* is in our lives daily: "Check the car's cooling system"; "How does the grading system work?"; "A weather system is approaching." *Systems analysis* techniques in science began with studies of energy and temperature (thermodynamics) in the 19th century and were further developed in engineering studies during World War II. Systems methodology is an important analytical tool. In this book's 4 parts and 20 chapters, the content is organized along logical flow paths consistent with systems thinking.

Systems Theory

Simply stated, a **system** is any set of ordered, interrelated components and their attributes, linked by flows of energy and matter, as distinct from the surrounding environment outside the system. The elements within a system may be arranged in a series or intermingled. A system may comprise any number of subsystems. Within Earth's systems, both matter and energy are stored and retrieved, and energy is transformed from one type to another. (Remember: *Matter* is mass that assumes a physical shape and occupies space; *energy* is a capacity to change the motion of, or to do work on, matter.)

Open Systems Systems in nature are generally not self-contained: Inputs of energy and matter flow into the system, and outputs of energy and matter flow from the system. Such a system is an **open system** (Figure 1.6). Within a system, the parts function in an interrelated manner, acting together in a way that gives each system its operational character. Earth is an open system in terms of energy because solar energy enters freely and heat energy leaves, going back into space.

Within the Earth system, many subsystems are interconnected. Free-flowing rivers are open systems: inputs consist of solar energy, precipitation, and soil and rock particles; outputs are water and sediments to the ocean. Changes to a river system may affect the nearby coastal system; for example, an increase in a river's sediment load may change the shape of a river mouth or spread pollutants along a coastline. Most natural systems are open in terms of energy. Examples of open atmospheric subsystems include hurricanes and tornadoes.

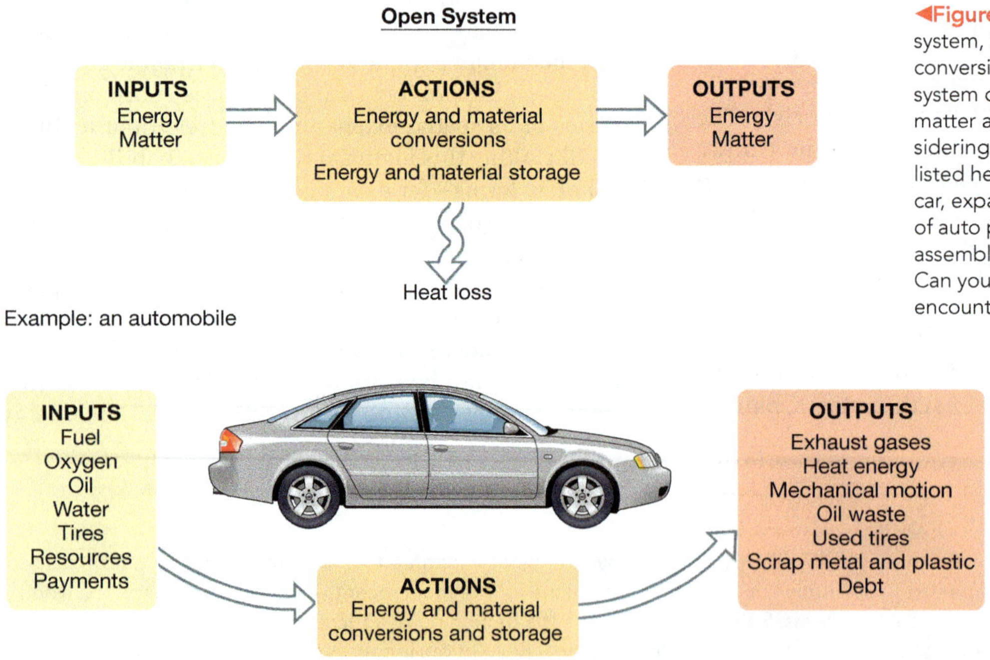

◀**Figure 1.6 An open system.** In an open system, inputs of energy and matter undergo conversions and are stored or released as the system operates. Outputs include energy and matter and heat energy (waste). After considering how the various inputs and outputs listed here are related to the operation of the car, expand your thinking to the entire system of auto production, from raw materials to assembly to sales to car accidents to junkyards. Can you identify other open systems that you encounter in your daily life?

Earth systems are dynamic (energetic, in motion) because of the tremendous infusion of radiant energy from the Sun. As this energy passes through the outermost edge of Earth's atmosphere, it is transformed into various kinds of energy that power terrestrial systems, such as kinetic energy (of motion), potential energy (of position), or chemical or mechanical energy—setting the fluid atmosphere and ocean in motion. Eventually, Earth radiates this energy back to the cold vacuum of space as heat energy.

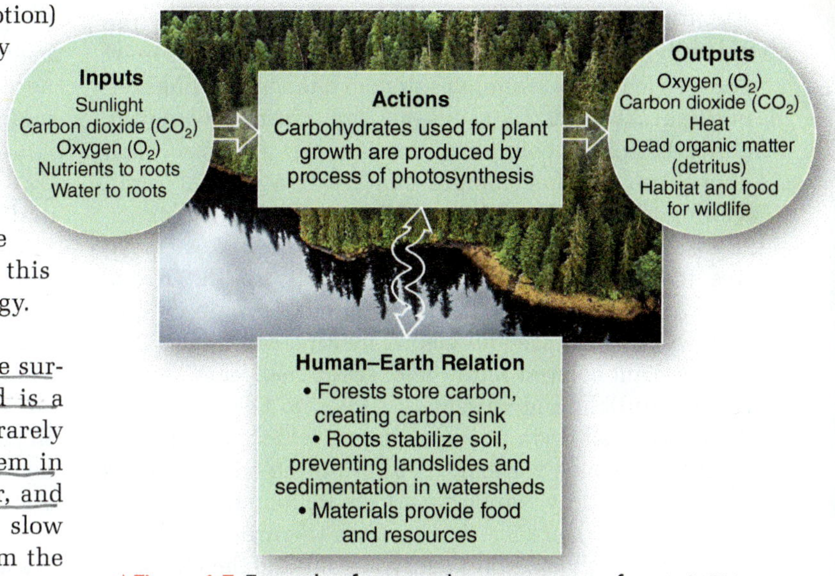

▲**Figure 1.7 Example of a natural open system: a forest.** [USDA Forest Service.]

Closed Systems A system that is shut off from the surrounding environment so that it is self-contained is a **closed system**. Although such closed systems are rarely found in nature, Earth is essentially a closed system in terms of physical matter and resources—air, water, and material resources. The only exceptions are the slow escape of lightweight gases (such as hydrogen) from the atmosphere into space and the input of frequent, but tiny, meteors and cosmic dust. The fact that Earth is a closed material system makes recycling efforts inevitable if we want a sustainable global economy.

Natural System Example A forest is an example of an open system (Figure 1.7). Through the process of photosynthesis, trees and other plants use sunlight as an energy input and water, nutrients, and carbon dioxide as material inputs. The photosynthetic process converts these inputs to stored chemical energy in the form of plant sugars (carbohydrates). The process also releases an output from the forest system: the oxygen that we breathe.

Forest outputs also include products and activities that link to other broad-scale Earth systems. For example, forests store carbon and are thus referred to as "carbon sinks." A 2011 study found that forests absorb about one-third of the carbon dioxide released through the burning of fossil fuels, making them a critical part of the climate system as global carbon dioxide levels rise. Forest roots stabilize soil on hillslopes and stream banks, connecting them to land and water systems. Finally, the food and habitat resources provided by forests link them closely to other living systems, including humans. (Chapters 10, 12, 19, and 20 discuss these processes and interactions.)

The connection of human activities to inputs, actions, and outputs of forest systems is indicated by the double-headed arrow in Figure 1.7. This interaction has two causal directions, since forest processes affect humans, and humans influence forests. Forests affect humans through the outputs of carbon storage (which mitigates climate change), soil stabilization (which prevents erosion and sedimentation into source areas for drinking water), and food and resources. Human influences on forests include direct impacts such as logging for wood resources, burning to make way for agriculture, and clearing for development, as well as indirect impacts from human-caused climate change, which may enhance the spread of disease and insects and pollution, which affects tree health.

System Feedback As a system operates, it generates outputs that influence its own operations. These outputs function as "information" that returns to various points in the system via pathways called **feedback loops**. Feedback information can guide, and sometimes control, further system operations. For the forest system in Figure 1.7, any increase or decrease in daylength (sunlight availability), carbon dioxide, or water produces feedback that causes specific responses in the individual trees and plants. For example, decreasing the water input slows the growth process; increasing daylength increases the growth process, within limits.

If the feedback information discourages change in the system, it is **negative feedback**. Further production of such feedback opposes system changes and leads to stability. Such negative feedback causes self-regulation in a natural system. Negative feedback loops are common in nature. In our forest, for example, healthy trees produce roots that stabilize hillslopes and inhibit erosion, providing a negative feedback. If the forest is damaged or removed, perhaps by fire or logging practices, the hillslope may become unstable and subject to landslides or mudslides. This instability affects nearby systems as sediment is deposited into streams, along coastlines, or into developed areas.

In many ecosystems, predator populations provide negative feedback for populations of other animals; the size of the prey population tends to achieve a balance with the number of predators. If a predator population drops abruptly, prey populations increase and cause ecosystem instability. After wolves were exterminated from Yellowstone National Park in Wyoming and Montana in the late 1800s, the unnaturally high elk population stripped many areas of natural vegetation. After the 1995 reintroduction of Canadian wolves into Yellowstone, elk numbers declined with wolf predation. Since then, aspens and willow are returning, improving habitat for birds and small mammals and providing other ecosystem benefits.

If feedback information encourages change in the system, it is **positive feedback**. Further production of positive feedback stimulates system changes. Unchecked positive feedback in a system can create a runaway ("snowballing") condition. In natural systems, such unchecked system changes can reach a critical limit, leading to instability, disruption, or death of organisms.

Global climate change creates an example of positive feedback as summer sea ice melts in the Arctic Region (discussed in Chapter 4). As arctic temperatures rise, summer sea ice and glacial melting accelerate. This causes light-colored snow and sea-ice surfaces, which reflect sunlight and so remain cooler, to be replaced by darker-colored open ocean surfaces, which absorb sunlight and become warmer. As a result, the ocean absorbs more solar energy, which raises the temperature, which, in turn, melts more ice, and so forth (Figure 1.8). This is a positive feedback loop, further enhancing the effects of higher temperatures and warming trends.

The acceleration of change in a positive feedback loop can be dramatic. Scientists have found that the *extent* of sea ice has decreased in area, and that the *volume* has dropped at an accelerating rate. Volume, a better indicator than extent for existing sea ice, has dropped by half since 1980; however, the rate of decrease was 2.5 times faster during the decade from 2000 to 2012 than it was from 1980 to 1990. As the feedback loop accelerates, the possibility of complete summer ice melt in the Arctic may become reality sooner than predicted—September is normally the month for lowest sea-ice extent; in 2012 this happened in August.

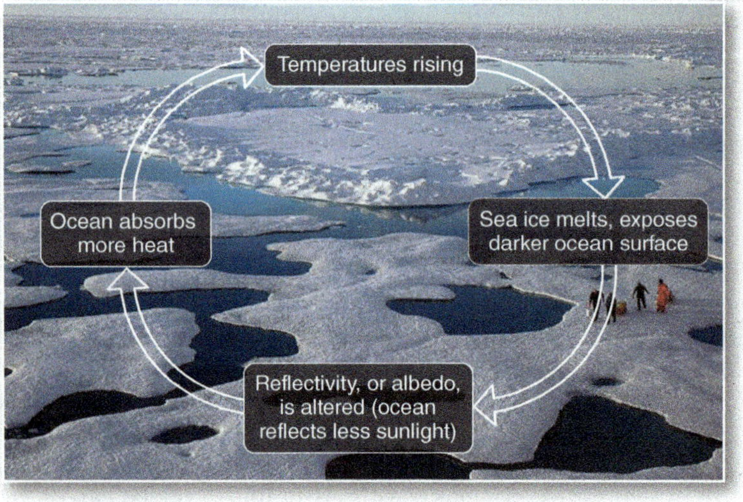

▲Figure 1.8 **The Arctic sea ice–albedo positive feedback loop.** Average ice thickness in the Arctic summer has dropped dramatically, leaving thinner ice that melts more easily. Since 2000, 70% of the September ice volume has disappeared. If this rate of ice volume loss continues, the first ice-free Arctic September might happen before 2017. [NOAA.]

System Equilibrium Most systems maintain structure and character over time. An energy and material system that remains balanced over time, in which conditions are constant or recur, is in a *steady-state condition*. When the rates of inputs and outputs in the system are equal and the amounts of energy and matter in storage within the system are constant (or more realistically, fluctuate around a stable average), the system is in **steady-state equilibrium**. For example, river channels commonly adjust their form in response to inputs of water and sediment; these inputs may change in amount from year to year, but the channel form represents a stable average—a steady-state condition.

However, a steady-state system may demonstrate a changing trend over time, a condition described as **dynamic equilibrium**. These changing trends may appear gradually and are compensated for by the system. A river may tend toward channel widening as it adjusts to greater inputs of sediment over some time scale, but the overall system will adjust to this new condition and thus maintain a dynamic equilibrium. Figure 1.9 illustrates these two equilibrium conditions, steady-state and dynamic.

Note that systems in equilibrium tend to maintain their functional operations and resist abrupt change. However, a system may reach a **threshold**, or *tipping point*, where it can no longer maintain its character, so it lurches to a new operational level. A large flood in a river system may push the river channel to a threshold where it abruptly shifts, carving a new channel. Another example of such a condition is a hillside or coastal bluff that adjusts after a sudden landslide. A new equilibrium is eventually achieved among slope, materials, and energy over time. High-latitude climate change has caused threshold events such as the relatively sudden collapse of ice shelves surrounding a portion of Antarctica and the crack-up of ice shelves on the north coast of Ellesmere Island, Canada, and Greenland.

Also, plant and animal communities can reach thresholds. After 1997, warming conditions in oceans combined with pollution to accelerate the bleaching of living coral reefs worldwide—taking coral systems to a threshold. Bleaching is the loss of colorful

GEOreport 1.2 Amphibians at thresholds

Amphibian species are a threatened group of animals, with approximately one-third of recognized species now at risk of extinction. According to the International Union for Conservation of Nature (IUCN) Amphibian Specialist Group, two new initiatives are aimed at stopping the amphibian decline: increased habitat protection for species that are found in only a single location and stepped-up efforts at testing antifungal drugs to halt the killer frog disease favored by the temperature increases. Read more about the current amphibian extinction crisis at http://www.amphibians.org/.

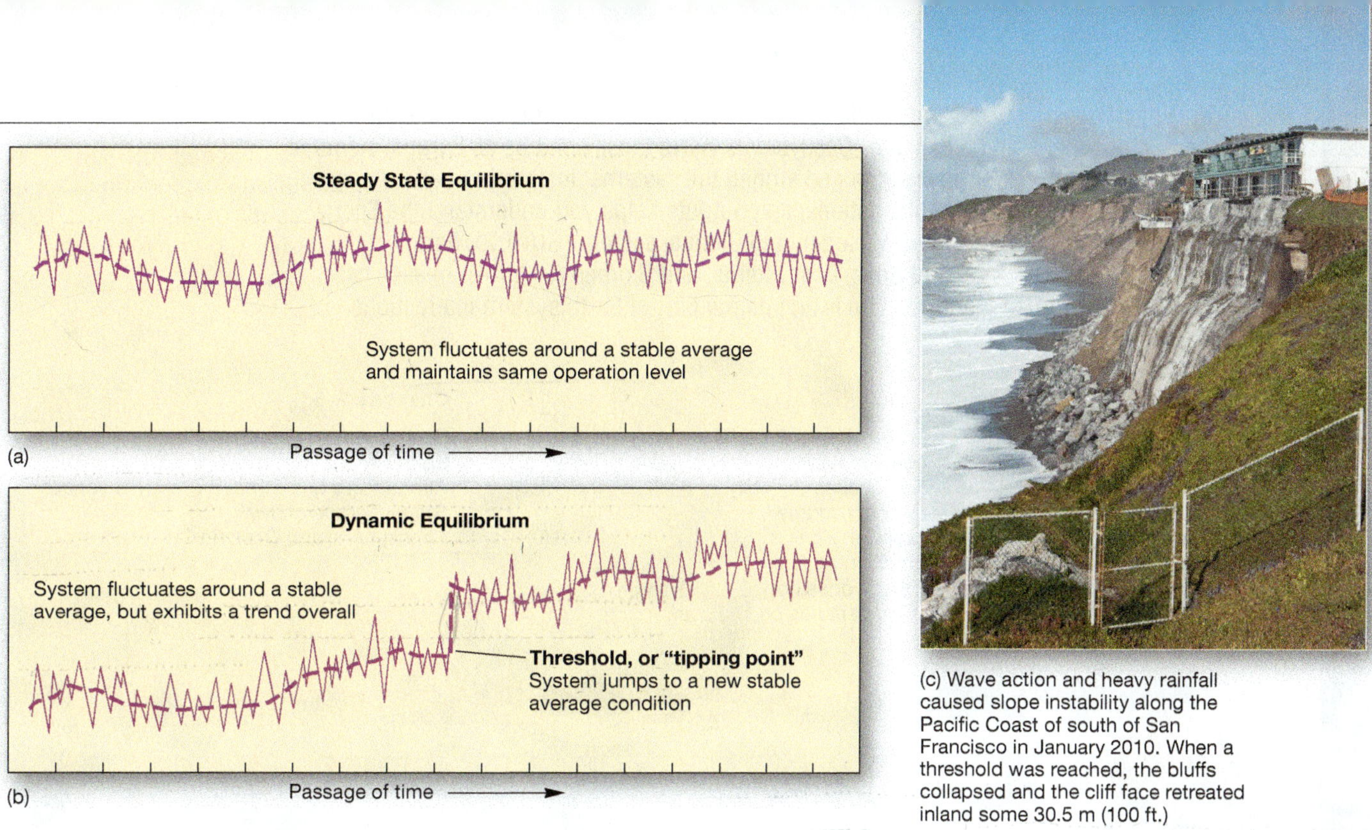

(c) Wave action and heavy rainfall caused slope instability along the Pacific Coast of south of San Francisco in January 2010. When a threshold was reached, the bluffs collapsed and the cliff face retreated inland some 30.5 m (100 ft.)

▲Figure 1.9 **System equilibria: steady-state and dynamic.** The vertical axis represents the value of a typical systems variable, such as stream channel width or hillslope angle. [Bobbé Christopherson photograph.]

algae, food source for the coral, causing the eventual death of the coral colonies making up the reef. In some areas, 50% of regional coral reefs experienced bleaching. On the Great Barrier Reef in Australia, coral die-off of up to 90% occurred during the worst years. Today, about 50% of the corals on Earth are ailing; more on this in Chapter 16. Harlequin frogs of tropical Central and South America are another example of species reaching a tipping point, with increased extinctions since 1986 related directly to climate change; discussed in Chapter 19 and GeoReport 1.2.

Models of Systems A **model** is a simplified, idealized representation of part of the real world. Scientists design models with varying degrees of specificity. A conceptual model is usually the most generalized and focuses on how processes interact within a system. A numerical model is more specific and is usually based on data collected from field or laboratory work. The simplicity of a model makes a system easier to understand and to simulate in experiments. A good example is a model of the *hydrologic system*, which represents Earth's entire water system, its related energy flows, and the atmosphere, surface, and subsurface environments through which water moves (see Figure 9.4 in Chapter 9). Predictions associated with climate change are often based on computer models of atmospheric processes, discussed in Chapter 10B. We discuss many system models in this text.

Adjusting the variables in a model simulates differing conditions and allows predictions of possible system operations. However, predictions are only as good as the assumptions and accuracy built into the model. A model is best viewed for what it is—a simplification to help us understand complex processes.

Systems Organization in *Geosystems*

From general layout to presentation of specific topics, *Geosystems* follows a systems flow. The part structure is designed around Earth systems pertaining to air, water, land, and living organisms. These are Earth's four "spheres" and represent the broadest level of organization within the book. Within each part, chapters and topics are arranged according to systems thinking, focusing on inputs, actions, and outputs, with an emphasis on human–Earth interactions and on interrelations among the parts and chapters. Specific subjects, such as the eruption of Mount Pinatubo in the Philippines discussed just ahead, recur in many chapters, illustrating systems connections. The *Geosystems in Action* illustration on the following pages outlines the part structure and chapter content within Earth's four spheres.

Earth's Four "Spheres" Earth's surface is a vast area of 500 million km² (193 million mi²) where four immense open systems interact. The Geosystems in Action feature (pp. 8–9) shows the three **abiotic**, or nonliving, systems forming the realm of the **biotic**, or living, system. The abiotic spheres are the *atmosphere*, *hydrosphere*, and *lithosphere*. The biotic sphere is the *biosphere*. Together, these spheres form a simplified model of Earth systems.

(text continued on page 14)

Earth is often described as being made up of four "spheres"—the atmosphere, hydrosphere, lithosphere, and biosphere. *Geosystems* views these spheres as Earth systems in which energy and matter flow within and among the systems' interacting parts. Analyzing Earth systems in terms of their inputs, actions, and outputs helps you understand the Energy-Atmosphere system (GIA 1.1), Water, Weather, and Climate system (GIA 1.2), Earth-Atmosphere interface (GIA 1.3), and Soils, Ecosystems, and Biomes (GIA 1.4). In each case, you will see that the human–Earth relation is an integral part of Earth system interactions.

1.1 PART I: THE ENERGY– ATMOSPHERE SYSTEM

Incoming solar energy arrives at the top of Earth's atmosphere, providing the energy input that drives Earth's physical systems and influences our daily lives. The Sun is the ultimate energy source for most life processes in our biosphere. Earth's atmosphere acts as an efficient filter, absorbing most harmful radiation so that it does not reach Earth's surface. Each of us depends on these interacting systems.

Chapters 2–6

Solar Energy to Earth and the Seasons
Earth's Modern Atmosphere
Atmosphere and Surface Energy Balances
Global Temperatures
Atmospheric and Oceanic Circulations

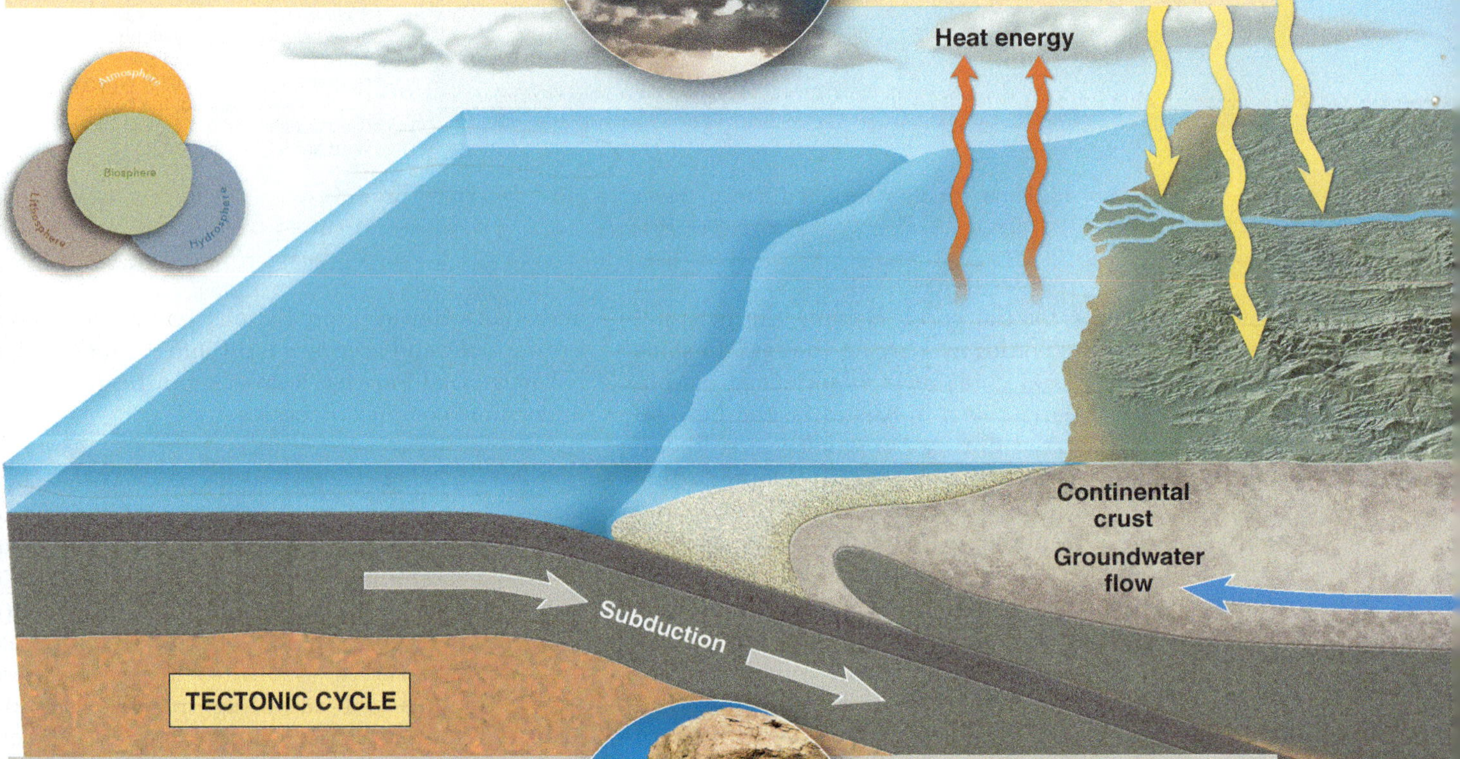

1.3 PART III: THE EARTH–ATMOSPHERE INTERFACE

Earth is a dynamic planet whose surface is changing. Two broad systems—endogenic and exogenic—organize these agents in Part III. In the endogenic system, internal processes produce flows of heat and material from deep below Earth's crust. The exogenic system involves external processes that set into motion air, water, and ice, all powered by solar energy. Thus, Earth's surface is the interface between two vast open systems: one that builds the landscape and one that tears it down.

Chapters 11–14, 16–17

The Dynamic Planet
Tectonics, Earthquakes, and Volcanism
Weathering, Karst Landscapes, and Mass Movement
River Systems
Oceans, Coastal Systems, and Wind Processes
Glacial and Periglacial Systems

Video
The Changing Face of Earth

MasteringGeography™

Visit the Study Area in MasteringGeography™ to explore Earth systems.

Visualize: Study geosciences animations of Earth's radiation balance, the hydrologic cycle, and the rock cycle.

Assess: Demonstrate understanding of Earth-system interactions (if assigned by instructor).

1.2 PART II: WATER, WEATHER, AND CLIMATE SYSTEMS

Earth is the water planet. Part II describes the distribution of water on Earth, including water circulation in the hydrologic cycle. Part II also describes the daily dynamics of the atmosphere as it interacts with the hydrosphere to produce weather. Outputs of the water–weather system range from climate patterns to our water resources. Part II concludes with an examination of global climate change, the impacts that are occurring, and forecasts of future changes.

Chapters 7–10B

Water and Atmospheric Moisture
Weather
Water Resources
Global Climate Systems
Climate Change

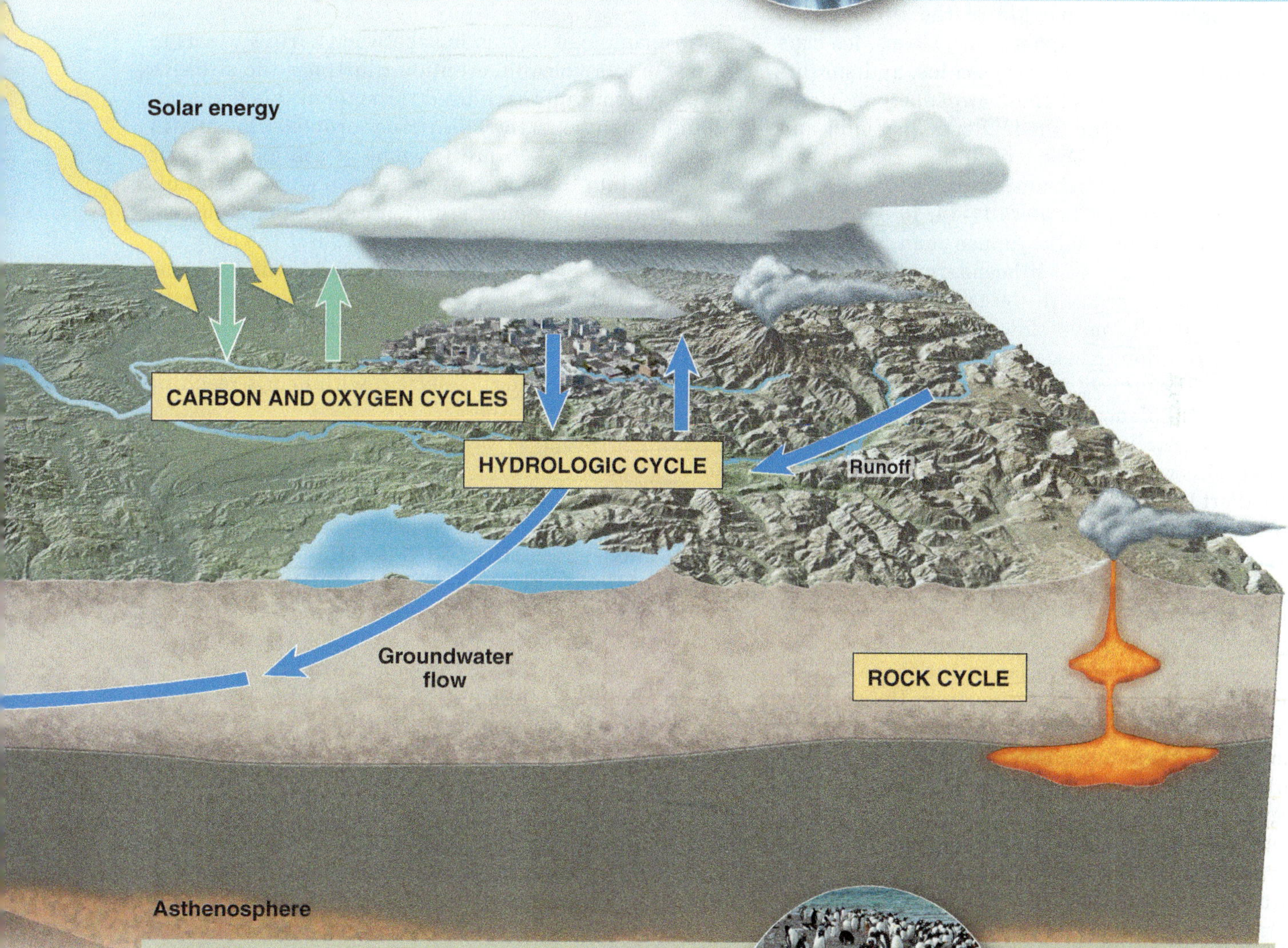

1.4 PART IV: SOILS, ECOSYSTEMS, AND BIOMES

Energy enters the biosphere through conversion of solar energy by photosynthesis in the leaves of plants. Soil is the essential link among the lithosphere, plants, and the rest of Earth's physical systems. Thus, soil helps sustain life and is a bridge between Parts III and IV of this text. Together, soils, plants, animals, and the physical environment make up ecosystems, which are grouped together in biomes.

Chapters 18–20

The Geography of Soils
Ecosystem Essentials
Terrestrial Biomes
Human–Earth Interactions: all chapters

GEOquiz

1. **Explain:** Explain how the Sun is involved in Earth's physical systems. What is the Sun's role in the biosphere?

2. **Compare:** How are endogenic and exogenic systems similar? How are they different? Based on the illustration, give an example of each type of system.

- **Atmosphere (Part I, Chapters 2–6)** The **atmosphere** is a thin, gaseous veil surrounding Earth, held to the planet by the force of gravity. Formed by gases arising from within Earth's crust and interior and the exhalations of all life over time, the lower atmosphere is unique in the Solar System. It is a combination of nitrogen, oxygen, argon, carbon dioxide, water vapor, and trace gases.
- **Hydrosphere (Part II, Chapters 7–10B)** Earth's waters exist in the atmosphere, on the surface, and in the crust near the surface. Collectively, these waters form the **hydrosphere**. That portion of the hydrosphere that is frozen is the **cryosphere**—ice sheets, ice caps and fields, glaciers, ice shelves, sea ice, and subsurface ground ice. Water of the hydrosphere exists in three states: liquid, solid (the frozen cryosphere), and gaseous (water vapor). Water occurs in two general chemical conditions, fresh and saline (salty).
- **Lithosphere (Part III, Chapters 11–14, 16–17)** Earth's crust and a portion of the upper mantle directly below the crust form the **lithosphere**. The crust is quite brittle compared with the layers deep beneath the surface, which move slowly in response to an uneven distribution of heat energy and pressure. In a broad sense, the term *lithosphere* sometimes refers to the entire solid planet. The soil layer is the *edaphosphere* and generally covers Earth's land surfaces. In this text, soils represent the bridge between the lithosphere (Part III) and biosphere (Part IV).
- **Biosphere (Part IV, Chapters 18–20)** The intricate, interconnected web that links all organisms with their physical environment is the **biosphere**, or **ecosphere**. The biosphere is the area in which physical and chemical factors form the context of life. The biosphere exists in the overlap of the three abiotic, or nonliving, spheres, extending from the seafloor, the upper layers of the crustal rock, to about 8 km (5 mi) into the atmosphere. Life is sustainable within these natural limits. The biosphere evolves, reorganizes itself at times, undergoes extinctions, and manages to flourish.

Within each part, the sequence of chapters generally follows a systems flow of energy, materials, and information. Each of the four part-opening page spreads summarizes the main system linkages; these diagrams are presented together in Figure 1.10. As an example of our systems organization, Part I, "The Energy–Atmosphere System," begins with the Sun (Chapter 2). The Sun's energy flows across space to the top of the atmosphere and through the atmosphere to Earth's surface, where it is balanced by outgoing energy from Earth (Chapters 3 and 4). Then we look at system outputs of temperature (Chapter 5) and winds and ocean currents (Chapter 6). Note the same logical systems flow in the other three parts of this text. The organization of many chapters also follows this systems flow.

Mount Pinatubo—Global System Impact A dramatic example of interactions between Earth systems in response to a volcanic eruption illustrates the strength of the systems approach used throughout this textbook. Mount Pinatubo in the Philippines erupted violently in 1991, injecting 15–20 million tons of ash and sulfuric acid mist into the upper atmosphere (Figure 1.11). This was the second greatest eruption during the 20th century; Mount Katmai in Alaska in 1912 was the only one greater. The eruption materials from Mount Pinatubo affected Earth systems in several ways, as noted on the map. For comparison, the 2010 eruption of Eyjafjallajökull in Iceland was about 100 times smaller in terms of the volume of material ejected, with debris reaching only the lower atmosphere.

As you progress through this book, you see the story of Mount Pinatubo and its implications woven through eight chapters: Chapter 1 (discussion of systems theory), Chapter 4 (effects on energy budgets in the atmosphere), Chapter 6 (satellite images of the spread of debris by atmospheric winds), Chapter 10B (temporary effect on global atmospheric temperatures), Chapter 12 (volcanic process) and Chapter 19 (effects on net photosynthesis). Instead of simply describing the eruption, we see the linkages and global impacts of such a volcanic explosion.

Earth's Dimensions

We all have heard that some people in the past believed Earth was flat. Yet Earth's *sphericity*, or roundness, is not as modern an idea as many think. For instance, more than two millennia ago, the Greek mathematician and philosopher Pythagoras (ca. 580–500 B.C.) determined through observation that Earth is spherical. We do not know what observations led Pythagoras to this conclusion. Can you guess at what he saw to deduce Earth's roundness?

GEOreport 1.3 Earth's unique hydrosphere

The hydrosphere on Earth is unique among the planets in the Solar System: only Earth possesses surface water in such quantity, some 1.36 billion km^3 (0.33 billion mi^3). Subsurface water exists on other planets, discovered on the Moon and on the planet Mercury in their polar areas, Mars, Jupiter's moon Europa, and Saturn's moons Enceladus and Titan. In the Martian polar region, remote spacecraft are studying ground ice and patterned ground phenomena caused by freezing and thawing water, as discussed for Earth in Chapter 17. The *Curiosity* rover in 2012 landed in an area of Mars that billions of years ago was flooded with waist-deep water. In the Universe, deep-space telescopes reveal traces of water in nebulae and on distant planetary objects.

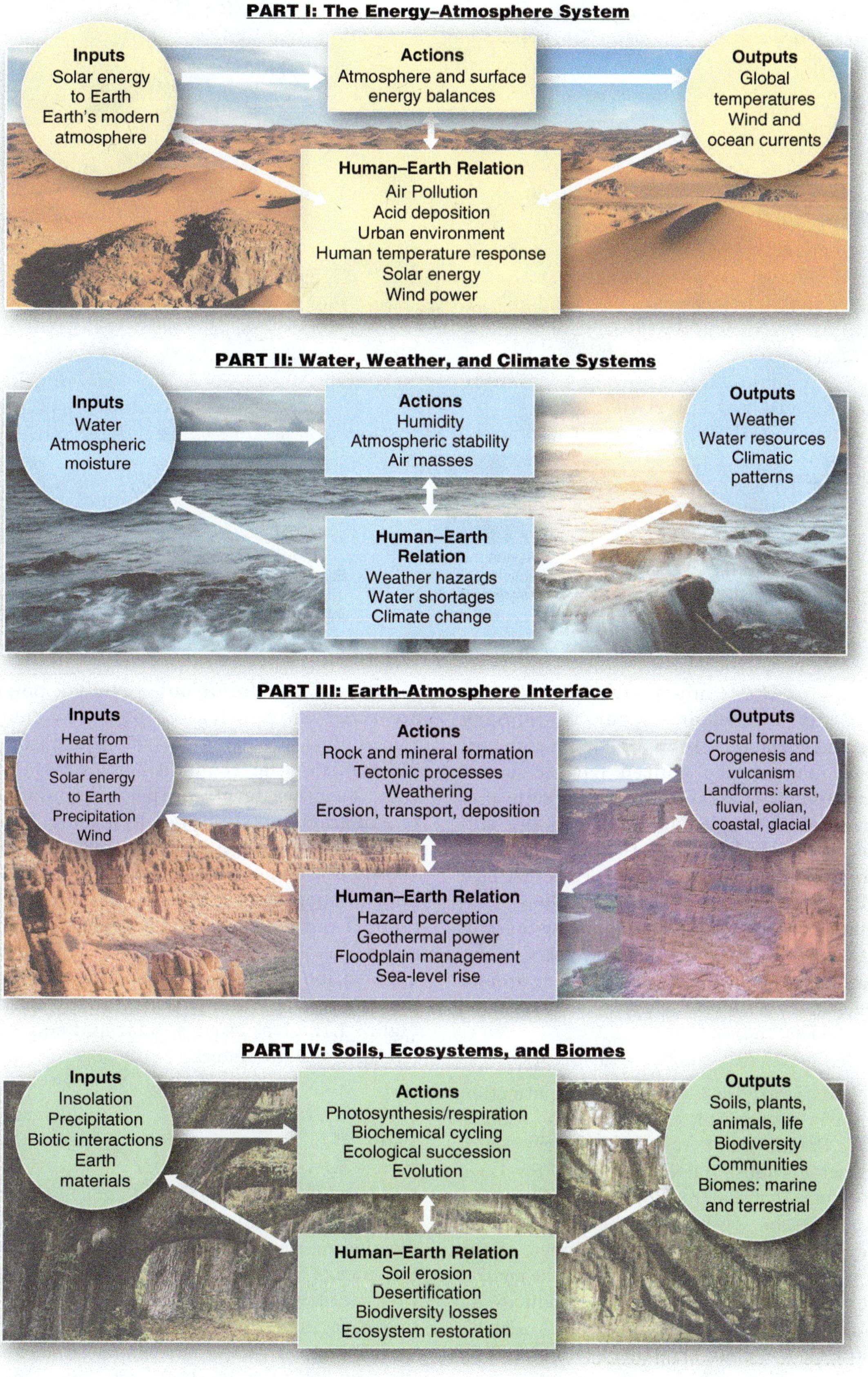

▲Figure 1.10 **The systems in *Geosystems*.** The sequence of systems flow in the organization of Parts I, II, III, and IV.

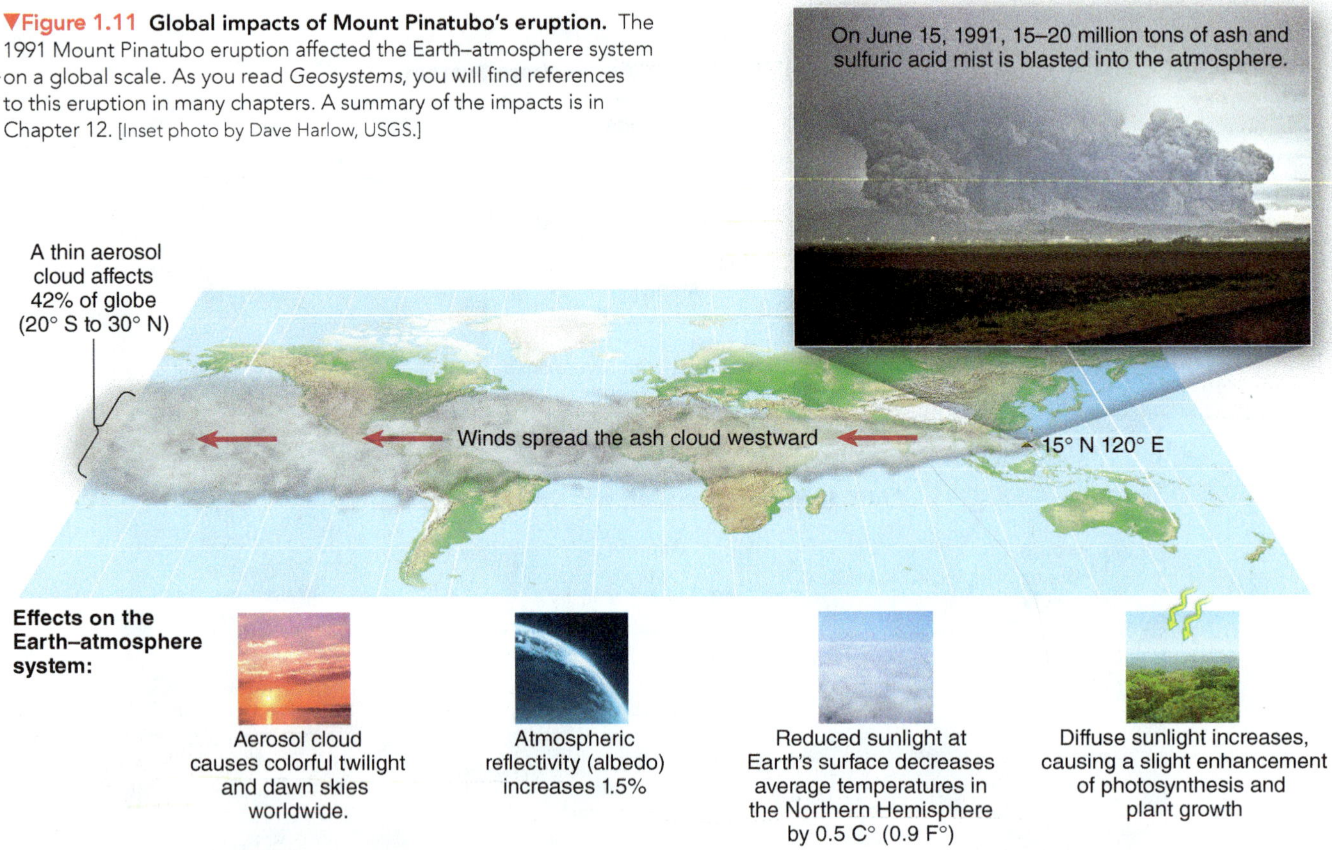

▼**Figure 1.11 Global impacts of Mount Pinatubo's eruption.** The 1991 Mount Pinatubo eruption affected the Earth–atmosphere system on a global scale. As you read *Geosystems,* you will find references to this eruption in many chapters. A summary of the impacts is in Chapter 12. [Inset photo by Dave Harlow, USGS.]

Evidence of Sphericity He might have noticed ships sailing beyond the horizon and apparently sinking below the water's surface, only to arrive back at port with dry decks. Perhaps he noticed Earth's curved shadow cast on the lunar surface during an eclipse of the Moon. He might have deduced that the Sun and Moon are not really the flat disks they appear to be in the sky, but are spherical, and that Earth must be a sphere as well.

Earth's sphericity was generally accepted by the educated populace as early as the first century A.D. Christopher Columbus, for example, knew he was sailing around a sphere in 1492; this is one reason why he thought he had arrived in the East Indies.

Earth as a Geoid Until 1687, the spherical-perfection model was a basic assumption of **geodesy**, the science that determines Earth's shape and size by surveys and mathematical calculations. But in that year, Sir Isaac Newton postulated that Earth, along with the other planets, could not be perfectly spherical. Newton reasoned that the more rapid rotational speed at the equator—the part of the planet farthest from the central axis and therefore the fastest moving—produces an equatorial bulge as centrifugal force pulls Earth's surface outward. He was convinced that Earth is slightly misshapen into an *oblate spheroid*, or, more correctly, an *oblate ellipsoid* (*oblate* means "flattened"), with the oblateness occurring at the poles.

Earth's equatorial bulge and its polar oblateness are today universally accepted and confirmed with tremendous precision by satellite observations. The unique, irregular shape of Earth's surface, coinciding with mean sea level and perpendicular to the direction of gravity, is described as a **geoid**. Imagine Earth's geoid as a sea-level surface that extends uniformly worldwide, beneath the continents. Both heights on land and depths in the oceans measure from this hypothetical surface. Think of the geoid surface as a balance among the gravitational attraction of Earth's mass, the distribution of water and ice along its surface, and the outward centrifugal pull caused by Earth's rotation. Figure 1.12 gives Earth's polar and equatorial circumferences and diameters.

Location and Time on Earth

Fundamental to geographic science is a coordinated grid system that is internationally accepted to determine location on Earth. The terms *latitude* and *longitude* for the lines of this grid were in use on maps as early as the first century A.D., with the concepts themselves dating to earlier times.

The geographer, astronomer, and mathematician Ptolemy (ca. A.D. 90–168) contributed greatly to the development of modern maps, and many of his terms are still used today. Ptolemy divided the circle into

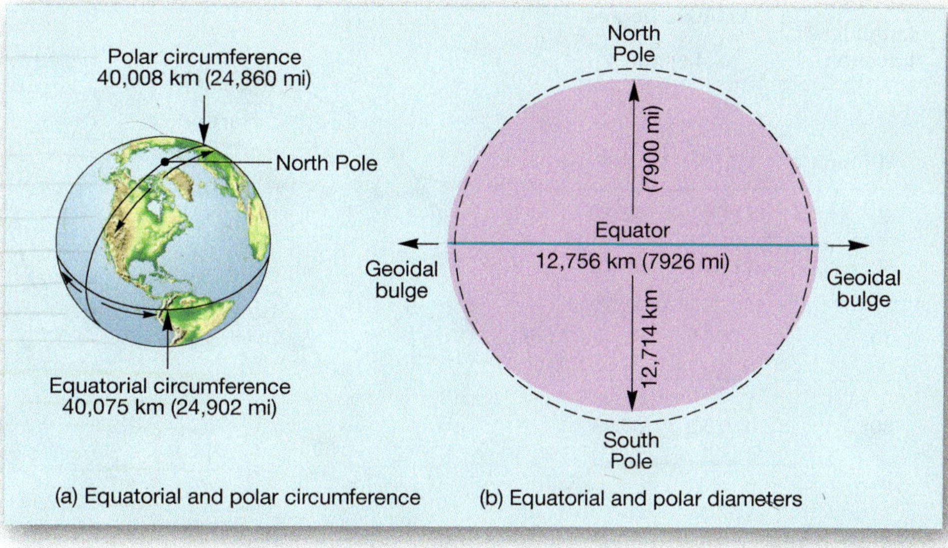

▶**Figure 1.12 Earth's dimensions.** The dashed line is a perfect circle for comparison to Earth's geoid.

360 degrees (360°), with each degree having 60 minutes (60′) and each minute having 60 seconds (60″) in a manner adapted from the ancient Babylonians. He located places using these degrees, minutes, and seconds. However, the precise length of a degree of latitude and a degree of longitude remained unresolved for the next 17 centuries.

Latitude

Latitude is an angular distance north or south of the equator, measured from the center of Earth (Figure 1.13a). On a map or globe, the lines designating these angles of latitude run east and west, parallel to the equator (Figure 1.13b). Because Earth's equator divides the distance between the North Pole and the South Pole exactly in half, it is assigned the value of 0° latitude. Thus, latitude increases from the equator northward to the North Pole, at 90° north latitude, and southward to the South Pole, at 90° south latitude.

A line connecting all points along the same latitudinal angle is a **parallel**. In the figure, an angle of 49° north latitude is measured, and, by connecting all points at this latitude, we have the 49th parallel. Thus, *latitude* is the name of the angle (49° north latitude), *parallel* names the line (49th parallel), and both indicate distance north of the equator.

From equator to poles, the distance represented by a degree of latitude is fairly consistent, about 100 km (69 mi); at the poles, a degree of latitude is only slightly larger (about 1.12 km, or 0.70 mi) than at the equator (Figure 1.14). To pinpoint location more precisely, we divide degrees into 60 minutes, and minutes into 60 seconds. For example, Cabo San Lucas, Baja California, Mexico, in Figure 1.2 is located at 22 degrees, 53 minutes, 23 seconds (22° 53′ 23″) south latitude. Alternatively, many geographic information systems (GIS) and Earth visualization programs such as Google Earth™ use decimal notation for latitude and longitude degrees (an online conversion is at http://www.csgnetwork.com/gpscoordconv.html). In decimal units, Cabo San Lucas is at +22.8897° latitude—the positive sign is for north latitude, a negative sign for south latitude.

Latitude is readily determined by observing fixed celestial objects such as the Sun or the stars, a method

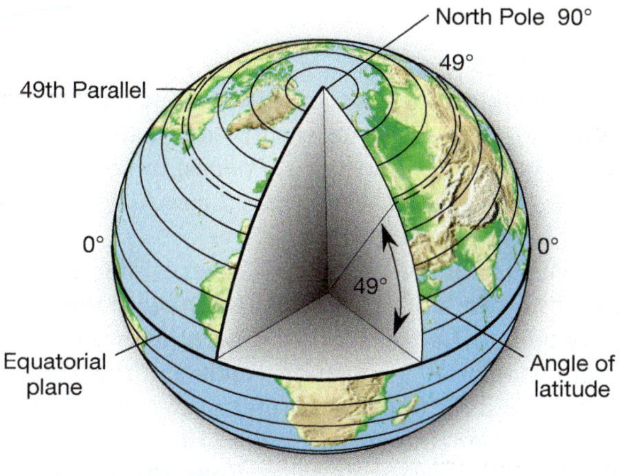

(a) Latitude is measured in degrees north or south of the Equator (0°). Earth's poles are at 90°. Note the measurement of 49° latitude.

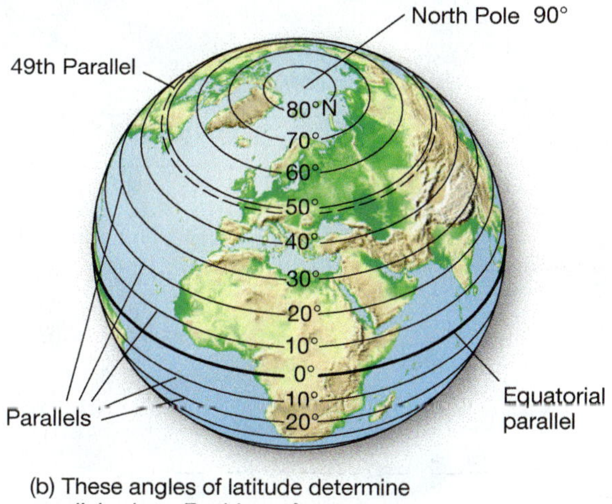

(b) These angles of latitude determine parallels along Earth's surface.

▲**Figure 1.13 Parallels of latitude.** Do you know your present latitude?

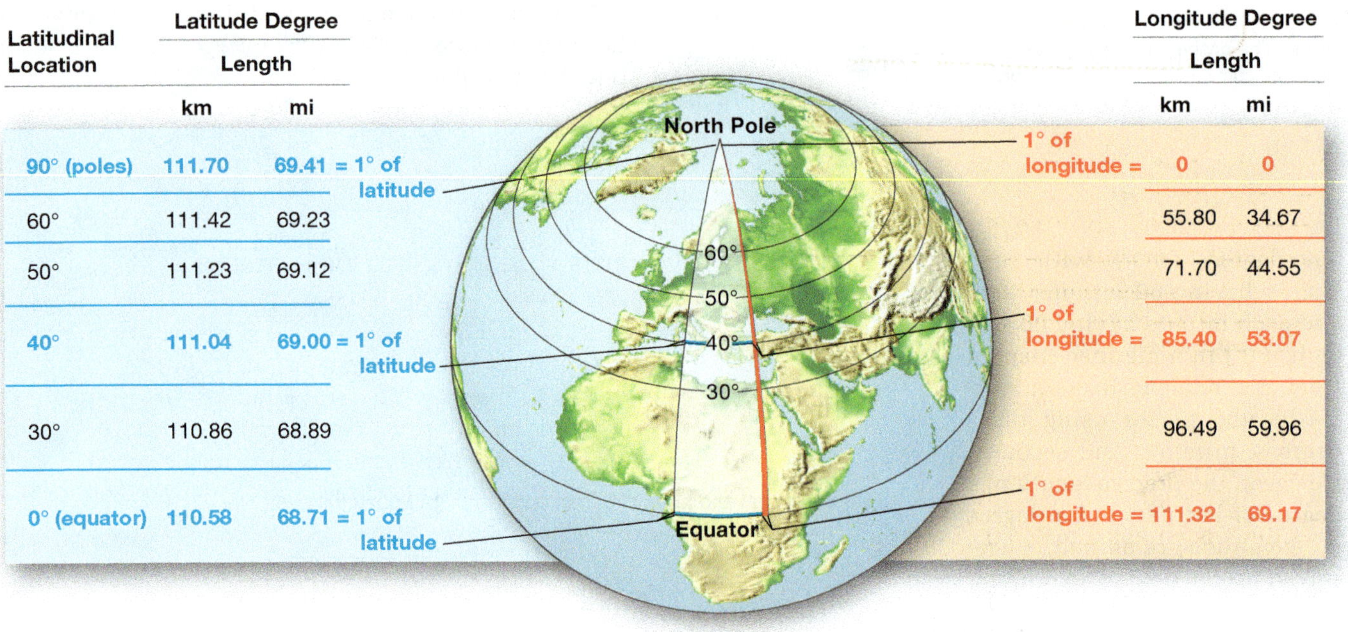

▲**Figure 1.14 Physical distances represented by degrees of latitude and longitude.**

dating to ancient times. Go to this chapter on our *MasteringGeography* website to learn more about determining latitude using the north star (*Polaris*) in the northern hemisphere and the Southern Cross (*Crux Australis*) constellation in the southern hemisphere.

"Lower latitudes" are those nearer the equator, whereas "higher latitudes" are those nearer the poles. You may be familiar with other general names describing regions related to latitude, such as "the tropics" and "the Arctic." Such terms refer to natural environments that differ dramatically from the equator to the poles. These differences result from the amount of solar energy received, which varies by latitude and season of the year.

Figure 1.15 displays the names and locations of the *latitudinal geographic zones* used by geographers: *equatorial* and *tropical*, *subtropical*, *midlatitude*, *subarctic* or *subantarctic*, and *arctic* or *antarctic*. These generalized latitudinal zones are useful for reference and comparison, but they do not have rigid boundaries; rather, think of them as transitioning one to another. We will discuss specific lines of latitude, such as the Tropic of Cancer and the Arctic Circle, in Chapter 2 as we learn about the seasons.

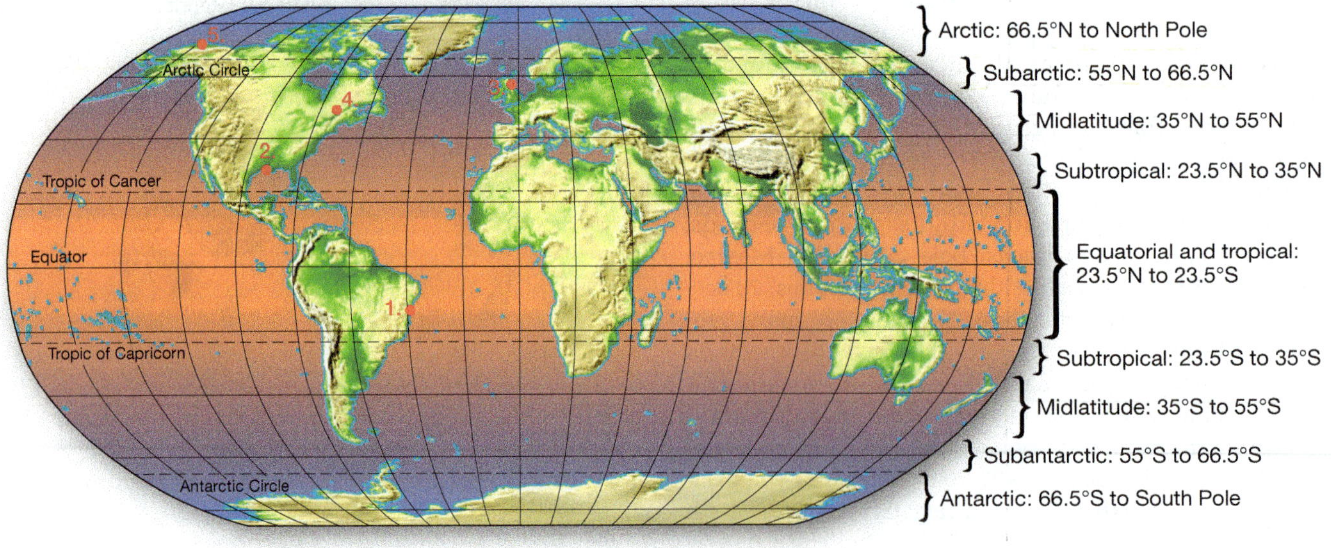

▲**Figure 1.15 Latitudinal geographic zones.** Geographic zones are generalizations that characterize various regions by latitude. Noted cities: 1. Salvador, Brazil; 2. New Orleans, Louisiana; 3. Edinburgh, Scotland; 4. Montreal, Quebec; 5. Barrow, Alaska; see Critical Thinking 1.2.

CRITICALthinking 1.2 Latitudinal Geographic Zones and Temperature

Refer to the graph in Figure 5.5 that plots annual temperature data for five cities from near the equator to beyond the Arctic Circle. Note the geographic location for each of the five cities on the latitudinal geographic zone map in Figure 1.15. In which zone is each city located? Roughly characterize changing temperature patterns through the seasons as you move away from the equator. Describe what you discover. ●

Longitude

Longitude is an angular distance east or west of a point on Earth's surface, measured from the center of Earth (Figure 1.16a). On a map or globe, the lines designating these angles of longitude run north and south (Figure 1.16b). A line connecting all points along the same longitude is a **meridian.** In the figure, a longitudinal angle of 60° E is measured. These meridians run at right angles (90°) to all parallels, including the equator.

Thus, *longitude* is the name of the angle, *meridian* names the line, and both indicate distance east or west of an arbitrary **prime meridian**—a meridian designated as 0° (Figure 1.16b). Earth's prime meridian passes through the old Royal Observatory at Greenwich, England, as set by an 1884 treaty; this is the *Greenwich prime meridian.* Because meridians of longitude converge toward the poles, the actual distance on the ground spanned by a degree of longitude is greatest at the equator (where meridians separate to their widest distance apart) and diminishes to zero at the poles (where meridians converge; Figure 1.14). As with latitude, longitude is expressed in degrees, minutes, and seconds or in decimal degrees. Cabo San Lucas in Figure 1.2 is located at 109° 54′ 56″ W longitude, or −109.9156°; east longitude has a positive value, while west longitude is negative.

We noted that latitude is determined easily by sighting the Sun or the North Star as a pointer. In contrast, a method of accurately determining longitude, especially at sea, remained a major difficulty in navigation until after 1760. The key to measuring the longitude of a place lies in accurately knowing time and required the invention of a clock without a pendulum. On the *MasteringGeography* website, read "The Timely Search for Longitude," about an important invention that solved the problem of time and longitude.

Great Circles and Small Circles

Great circles and small circles are important navigational concepts that help summarize latitude and longitude (Figure 1.17). A **great circle** is any circle of Earth's circumference whose center coincides with the center of Earth. An infinite number of great circles can be drawn on Earth. Every meridian is one-half of a great circle that passes through the poles. On flat maps, airline and shipping routes appear to arch their way across oceans and landmasses. These are *great circle routes,* tracing the shortest distances between two points on Earth (see Figure 1.24).

In contrast to meridians, only one parallel is a great circle—the *equatorial parallel.* All other parallels diminish in length toward the poles and, along with any other non–great circles that one might draw, constitute **small**

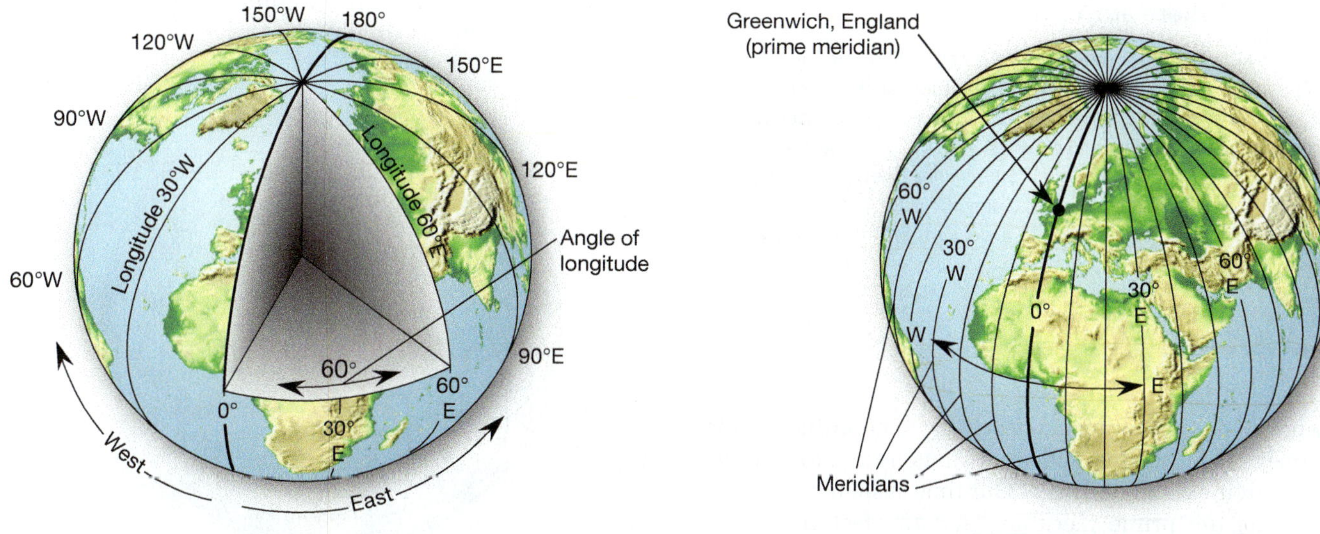

(a) Longitude is measured in degrees east or west of a 0° starting line, the prime meridian. Note the measurement of 60°E longitude

(b) Angles of longitude measured from the prime meridian determine other meridians. North America is west of Greenwich; therefore, it is in the Western Hemisphere.

▲**Figure 1.16 Meridians of longitude.** Do you know your present longitude?

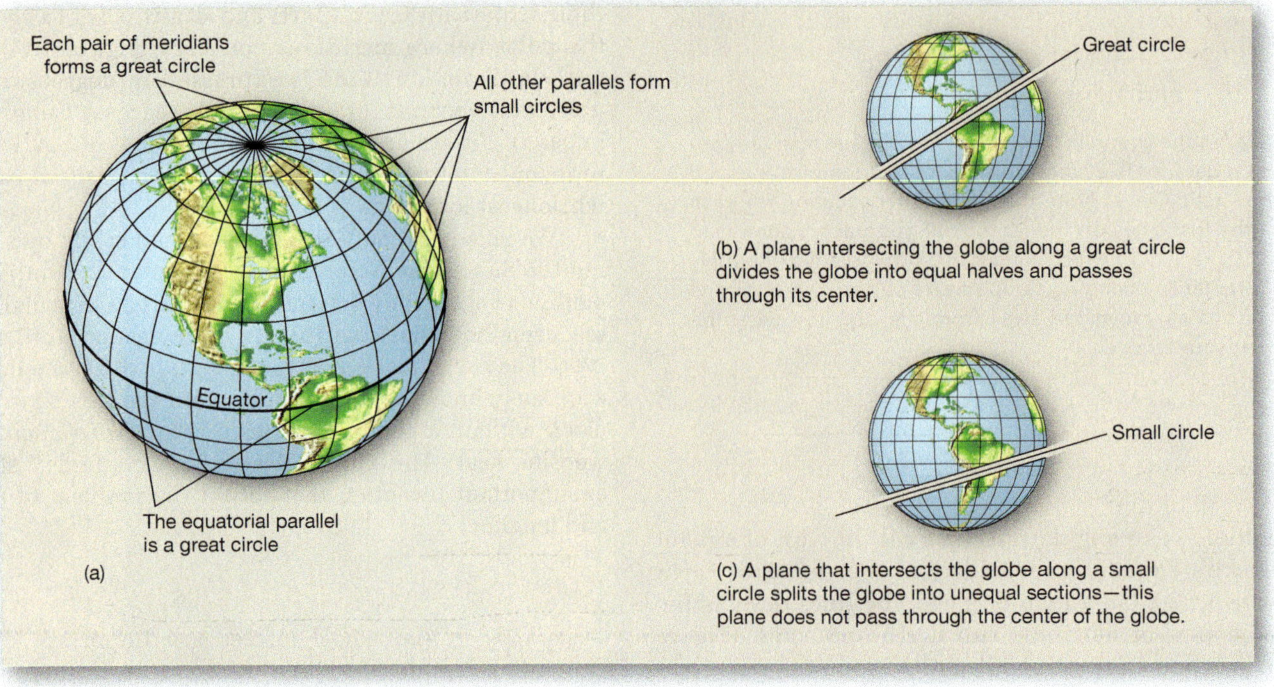

▲Figure 1.17 **Great circles and small circles.**

circles. These circles have centers that do not coincide with Earth's center.

Figure 1.18 combines latitude and parallels with longitude and meridians to illustrate Earth's complete coordinate grid system. Note the red dot that marks our measurement of 49° N and 60° E, a location in western Kazakhstan. Next time you look at a world globe, follow the parallel and meridian that converge on your location.

CRITICALthinking 1.3
Where are You?

Select a location (for example, your campus, home, or workplace or a city) and determine its latitude and longitude—both in degrees, minutes, and seconds and as decimal degrees. Describe the resources you used to gather this geographic information, such as an atlas, website, Google Earth™, or GPS measurement. Consult Figure 1.14 to find the approximate lengths of the latitude and longitude degrees at that location. ●

Meridians and Global Time

A worldwide time system is necessary to coordinate international trade, airline schedules, business and agricultural activities, and daily life. Our time system is based on longitude, the prime meridian, and the fact that Earth rotates on its own axis, revolving 360° every 24 hours, or 15° per hour (360° ÷ 24 = 15°).

In 1884 at the International Meridian Conference in Washington, DC, the prime meridian was set as the official standard for the world time zone system—**Greenwich Mean Time (GMT)** (see http://wwp.greenwichmeantime.com/). This standard time system established 24 standard meridians around the globe at equal intervals from the prime meridian, with a time zone of 1 hour spanning 7.5° on either side of these *central meridians*. Before this universal system, time zones were problematic, especially in large countries. In 1870, railroad travelers going from Maine to San Francisco made 22 adjustments to their watches to stay consistent with local time! Today, only three adjustments are needed in the continental United States—from Eastern Standard Time to Central, Mountain, and Pacific—and four changes across Canada.

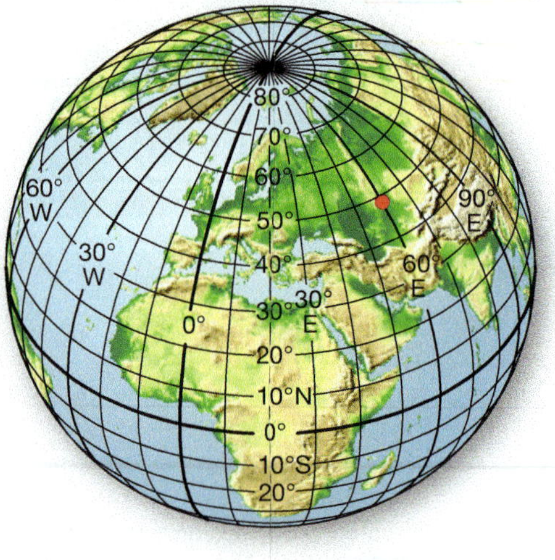

▲Figure 1.18 **Earth's coordinate grid system.** Latitude and parallels and longitude and meridians allow us to locate all places on Earth precisely. The red dot is at 49° N latitude and 60° E longitude.

As illustrated in Figure 1.19, when it is 9:00 P.M. in Greenwich, then it is 4:00 P.M. in Baltimore (–5 hr), 3:00 P.M. in Oklahoma City (–6 hr), 2:00 P.M. in Salt Lake City (–7 hr), 1:00 P.M. in Seattle and Los Angeles (–8 hr), noon in Anchorage (–9 hr), and 11:00 A.M. in Honolulu (–10 hr). To the east, it is midnight in Ar Riyāḍ, Saudi Arabia (+3 hr). The designation A.M. is for *ante meridiem*, "before noon," whereas P.M. is for *post meridiem*, "after noon." A 24-hour clock avoids the use of these designations: 3 P.M. is stated as 15:00 hours; 3 A.M. is 3:00 hours.

As you can see from the modern international time zones in Figure 1.19, national or state boundaries and political considerations distort time boundaries. For example, China spans four time zones, but its government decided to keep the entire country operating at the same time. Thus, in some parts of China clocks are several hours off from what the Sun is doing. In the United States, parts of Florida and west Texas are in the same time zone.

Coordinated Universal Time For decades, GMT was determined using the Royal Observatory's astronomical clocks and was the world's standard for accuracy. However, Earth rotation, on which those clocks were based, varies slightly over time, making it unreliable as a basis for timekeeping. Note that 150 million years ago, a "day" was 22 hours long, and 150 million years in the future, a "day" will be approaching 27 hours in length.

The invention of a quartz clock in 1939 and atomic clocks in the early 1950s improved the accuracy of measuring time. In 1972, the **Coordinated Universal Time (UTC)*** time-signal system replaced GMT and became the legal reference for official time in all countries. UTC is based on average time calculations from atomic clocks collected worldwide. You might still see official UTC referred to as GMT or Zulu time.

International Date Line An important corollary of the prime meridian is the 180° meridian on the opposite side of the planet. This meridian is the **International Date Line (IDL)**, which marks the place where each day officially begins (at 12:01 A.M.). From this "line," the new day sweeps westward. This *westward* movement of time is created by Earth's turning *eastward* on its axis. Locating the date line in the sparsely populated Pacific Ocean minimizes most local confusion (Figure 1.20).

At the IDL, the west side of the line is always one day ahead of the east side. No matter what time of day it is when the line is crossed, the calendar changes a day

*UTC is in use because agreement was not reached on whether to use English word order, CUT, or the French order, TUC. UTC was the compromise and is recommended for all timekeeping applications; use of the term *GMT* is discouraged.

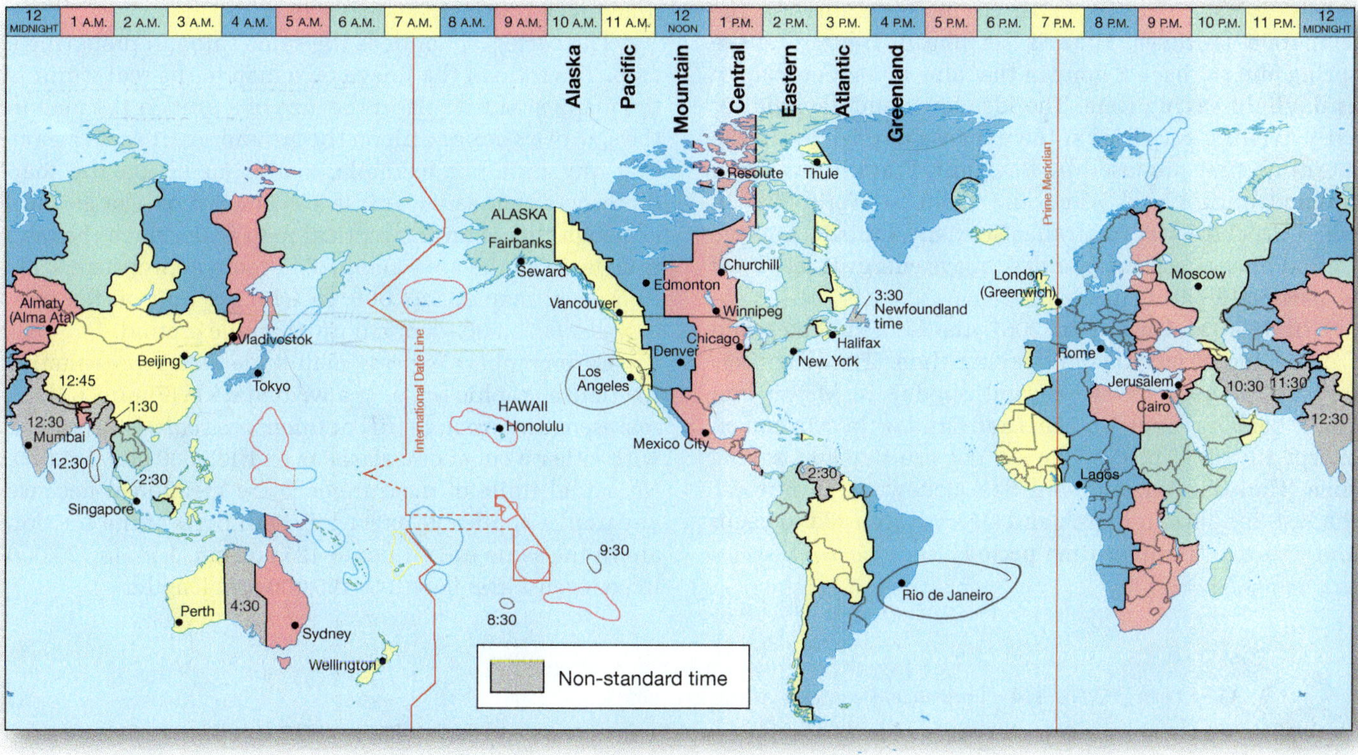

▲**Figure 1.19 Modern international standard time zones.** If it is 7 p.m. in Greenwich, determine the present time in Moscow, London, Halifax, Chicago, Winnipeg, Denver, Los Angeles, Fairbanks, Honolulu, Tokyo, and Singapore. [Adapted from Defense Mapping Agency. See http://aa.usno.navy.mil/faq/docs/world_tzones.html.]

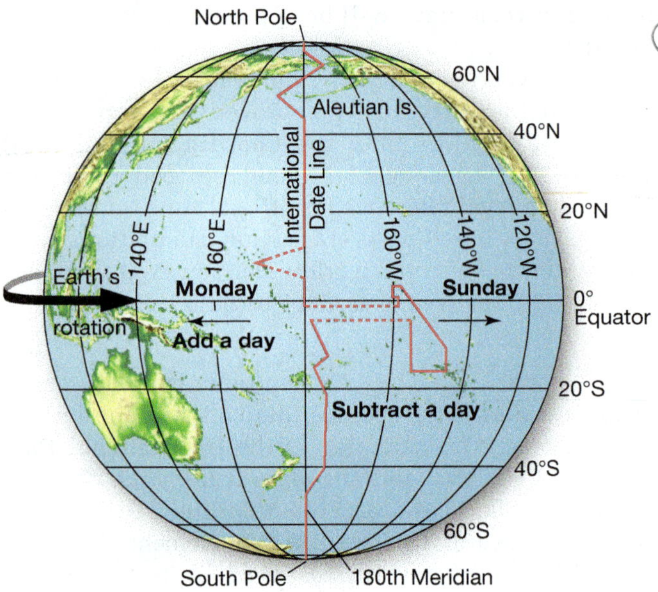

▲**Figure 1.20 International Date Line.** The IDL location is approximately along the 180th meridian (see Figure 1.19). The dotted lines on the map show where island countries have set their own time zones, but their political control extends only 3.5 nautical miles (4 mi) offshore. Officially, you gain 1 day crossing the IDL from east to west. (See GeoReport 1.5)

(Figure 1.20). Note in the illustration the departures from the IDL and the 180° meridian; this deviation is due to local administrative and political preferences.

Daylight Saving Time In 70 countries, mainly in the temperate latitudes, time is set ahead 1 hour in the spring and set back 1 hour in the fall—a practice known as **daylight saving time**. The idea to extend daylight for early evening activities at the expense of daylight in the morning, first proposed by Benjamin Franklin, was not adopted until World War I and again in World War II, when Great Britain, Australia, Germany, Canada, and the United States used the practice to save energy (1 less hour of artificial lighting needed).

In 1986 and again in 2007, the United States and Canada increased daylight saving time. Time "springs forward" 1 hour on the second Sunday in March and "falls back" 1 hour on the first Sunday in November, except in a few places that do not use daylight saving time (Hawai'i, Arizona, and Saskatchewan). In Europe, the last Sundays in March and October are used to begin and end the "summer-time period" (see http://webexhibits.org/daylightsaving/).

Maps and Cartography

For centuries, geographers have used maps as tools to display spatial information and analyze spatial relationships. A **map** is a generalized view of an area, usually some portion of Earth's surface, as seen from above and greatly reduced in size. A map usually represents a specific characteristic of a place, such as rainfall, airline routes, or political features such as state boundaries and place names. **Cartography** is the science and art of mapmaking, often blending aspects of geography, engineering, mathematics, computer science, and art. It is similar in ways to architecture, in which aesthetics and utility combine to produce a useful product.

We all use maps to visualize our location in relation to other places, or maybe to plan a trip, or to understand a news story or current event. Maps are wonderful tools! Understanding a few basics about maps is essential to our study of physical geography.

The Scale of Maps

Architects, toy designers, and mapmakers have something in common: They all represent real things and places with the convenience of a model; examples are a drawing; a pretend car, train, or plane; a diagram; or a map. In most cases, the model is smaller than the reality. For example, an architect renders a blueprint of a structure to guide the building contractors, preparing the drawing so that a centimeter (or inch) on the blueprint represents so many meters (or feet) on the proposed building. Often, the drawing is 1/50 or 1/100 real size.

The cartographer does the same thing in preparing a map. The ratio of the image on a map to the real world is the map's **scale**; it relates the size of a unit on the map to the size of a similar unit on the ground. A 1:1 scale means that any unit (for example, a centimeter) on the map represents that same unit (a centimeter) on the ground, although this is an impractical map scale, since the map is as large as the area mapped! A more appropriate scale for a local map is 1:24,000, in which 1 unit on the map represents 24,000 identical units on the ground.

Cartographers express map scale as a representative fraction, a graphic scale, or a written scale (Figure 1.21). A *representative fraction* (*RF*, or *fractional scale*) is expressed with either a colon or a slash, as in 1:125,000 or 1/125,000. No actual units of measurement are mentioned because any unit is applicable as long as both parts of the fraction are in the same unit: 1 cm to 125,000 cm, 1 in. to 125,000 in., or even 1 arm length to 125,000 arm lengths.

GEOreport 1.4 The world's most accurate clock

Time and Frequency Services of the National Institute for Standards and Technology (NIST), U.S. Department of Commerce, operates several of the most advanced clocks currently in use. The NIST-F1 cesium atomic clock is the most accurate clock in the world. Housed in Boulder, Colorado, it will not gain or lose a second in nearly 20 million years. For accurate time in the United States, see http://www.time.gov, operated by NIST and the U.S. Naval Observatory. In Canada, the Institute for Measurement Standards, National Research Council Canada, participates in determining UTC; see http://time5.nrc.ca/webclock_e.shtml.

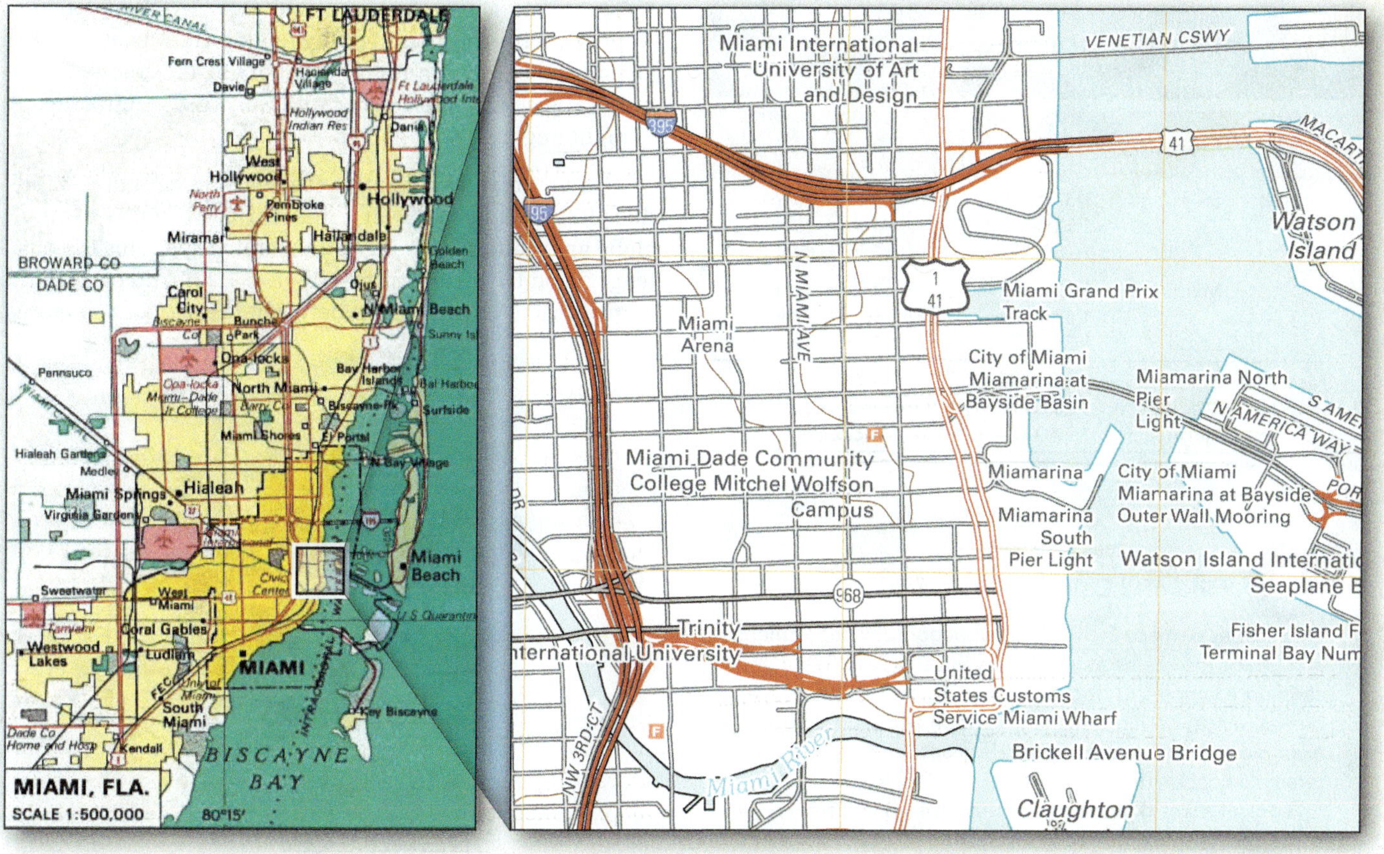

	(a)	(b)
Representative fraction:	1:500,000 or 1/500,000	1:24,000 or 1/24,000
Written scale:	1 inch = 8 miles 1 cm = 5.0 km	1 inch = 2000 feet 1 cm = 0.25 km
Graphic scale:	0 — 4 — 8 MILES 0 — 5 — 10 KILOMETERS	0 — 1000 — 2000 FEET 0 — 0.25 — 0.5 KILOMETERS

(a) Relatively small scale map of Miami area shows less detail.

(b) Relatively large scale map of the same area shows a higher level of detail.

▲**Figure 1.21 Map scale.** Examples of maps at different scales, with three common expressions of map scale—representative fraction, graphic scale, and written scale. Both maps are enlarged to show detail. [USGS. Courtesy of the University of Texas Libraries, The University of Texas at Austin.]

A *graphic scale*, or *bar scale*, is a bar graph with units to allow measurement of distances on the map. An important advantage of a graphic scale is that, if the map is enlarged or reduced, the graphic scale enlarges or reduces along with the map. In contrast, written and fractional scales become incorrect with enlargement or reduction. As an example, if you shrink a map from 1:24,000 to 1:63,360, the written scale "1 in. to 2000 ft" will no longer be correct. The new correct written scale is 1 in. to 5280 ft (1 mi).

Scales are *small*, *medium*, and *large*, depending on the ratio described. In relative terms, a scale of 1:24,000 is a large scale, whereas a scale of 1:50,000,000 is a small scale. The greater the denominator in a fractional scale (or the number on the right in a ratio expression), the smaller the scale of the map. Table 1.1 lists examples of selected representative fractions and written scales for small-, medium-, and large-scale maps.

Small-scale maps show a greater area in less detail; a small-scale map of the world is little help in finding an exact location, but works well for illustrating global wind patterns or ocean currents. Large-scale maps show a smaller area in more detail and are useful for applications needing precise location or navigation over short distances.

GEOreport 1.5 Magellan's crew loses a day

Early explorers had a problem before the date-line concept was developed. For example, Magellan's crew returned from the first circumnavigation of Earth in 1522, confident from their ship's log that it was Wednesday, September 7. They were shocked when informed by local residents that it was actually Thursday, September 8. Without an International Date Line, they had no idea that they must advance their calendars by a day when sailing around the world in a westward direction.

TABLE 1.1 Sample Representative Fractions and Written Scales for Small-, Medium-, and Large-Scale Maps

System	Scale Size	Representative Fraction	Written System Scale
English	Small	1:3,168,000	1 in. = 50 mi
		1:1,000,000	1 in. = 16 mi
		1:250,000	1 in. = 4 mi
	Medium	1:125,000	1 in. = 2 mi
		1:63,360 (or 1:62,500)	1 in. = 1 mi
	Large	1:24,000	1 in. = 2000 ft
Metric	Small	1:1,000,000	1 cm = 10.0 km
	Medium	1:25,000	1 cm = 0.25 km
	Large	1:10,000	1 cm = 0.10 km

CRITICALthinking 1.4
Find and Calculate Map Scales

Find globes or maps in the library or geography department and check the scales at which they were drawn. See if you can find examples of fractional, graphic, and written scales on wall maps, on highway maps, and in atlases. Find some examples of small- and large-scale maps, and note the different subject matter they portray.

Look at a world globe that is 61 cm (24 in.) in diameter (or adapt the following values to whatever globe you are using). We know that Earth has an equatorial diameter of 12,756 km (7926 mi), so the scale of such a globe is the ratio of 61 cm to 12,756 km. To calculate the representative fraction for the globe in centimeters, divide Earth's actual diameter by the globe's diameter (12,756 km ÷ 61 cm). (Hint: 1 km = 1000 m, 1 m = 100 cm; therefore, Earth's diameter of 12,756 km represents 1,275,600,000 cm; and the globe's diameter is 61 cm.) In general, do you think a world globe is a small- or a large-scale map of Earth's surface? ●

Map Projections

A globe is not always a helpful map representation of Earth. When you go on a trip, you need more-detailed information than a globe can provide. To provide local detail, cartographers prepare large-scale *flat maps*, which are two-dimensional representations (scale models) of our three-dimensional Earth. Unfortunately, such conversion from three dimensions to two causes distortion.

A globe is the only true representation of *distance*, *direction*, *area*, *shape*, and *proximity* on Earth. A flat map distorts these properties. Therefore, in preparing a flat map, the cartographer must decide which characteristic to preserve, which to distort, and how much distortion is acceptable. To understand this problem, consider these important properties of a globe:

- Parallels always are parallel to each other, always are evenly spaced along meridians, and always decrease in length toward the poles.
- Meridians always converge at both poles and always are evenly spaced along any individual parallel.
- The distance between meridians decreases toward poles, with the spacing between meridians at the 60th parallel equal to one-half the equatorial spacing.
- Parallels and meridians always cross each other at right angles.

The problem is that all these qualities cannot be reproduced simultaneously on a flat surface. Simply taking a globe apart and laying it flat on a table illustrates the challenge faced by cartographers (Figure 1.22). You can see the empty spaces that open up between the sections, or gores,

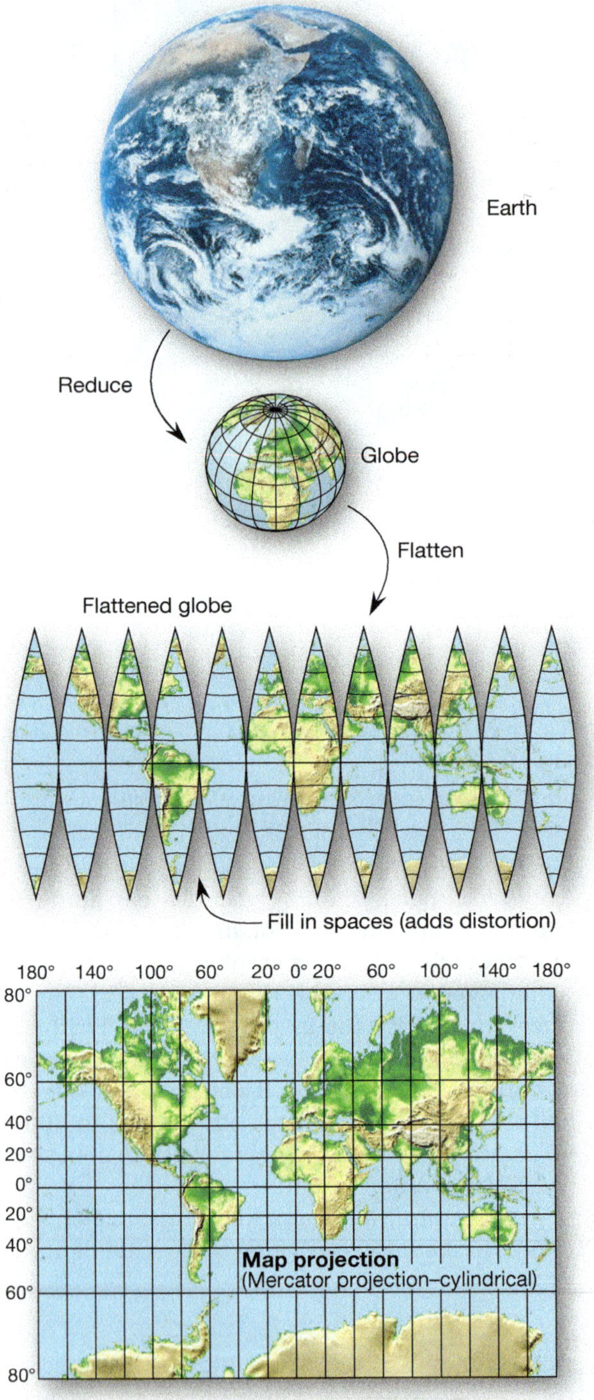

▲**Figure 1.22 From globe to flat map.** Conversion of the globe to a flat map projection requires a decision about which properties to preserve and the amount of distortion that is acceptable. [NASA astronaut photo from *Apollo 17*, 1972.]

MG Map Projections

of the globe. This reduction of the spherical Earth to a flat surface is a **map projection**, and no flat map projection of Earth can ever have all the features of a globe. Flat maps always possess some degree of distortion—much less for large-scale maps representing a few kilometers; much more for small-scale maps covering individual countries, continents, or the entire world.

Equal Area or True Shape? There are four general classes of map projections, shown in Figure 1.23. The best projection is always determined by the intended use of the map. The major decisions in selecting a map projection involve the properties of **equal area** (equivalence) and **true shape** (conformality). A decision favoring one property sacrifices the other, for they cannot be shown together on the same flat map.

If a cartographer selects equal area as the desired trait—for example, for a map showing the distribution of world climates—then true shape must be sacrificed by stretching and shearing, which allow parallels and meridians to cross at other than right angles. On an equal-area map, a coin covers the same amount of surface area no matter where you place it on the map. In contrast, if a cartographer selects the property of true shape, such as for a map used for navigational purposes, then equal area must be sacrificed, and the scale will actually change from one region of the map to another.

Classes of Projections Figure 1.23 illustrates the classes of map projections and the perspective from which each class is generated. Despite the fact that modern cartographic technology uses mathematical constructions and computer-assisted graphics, the word *projection* is still used. The term comes from times past, when geographers actually projected the shadow of a wire-skeleton globe onto a geometric surface, such as

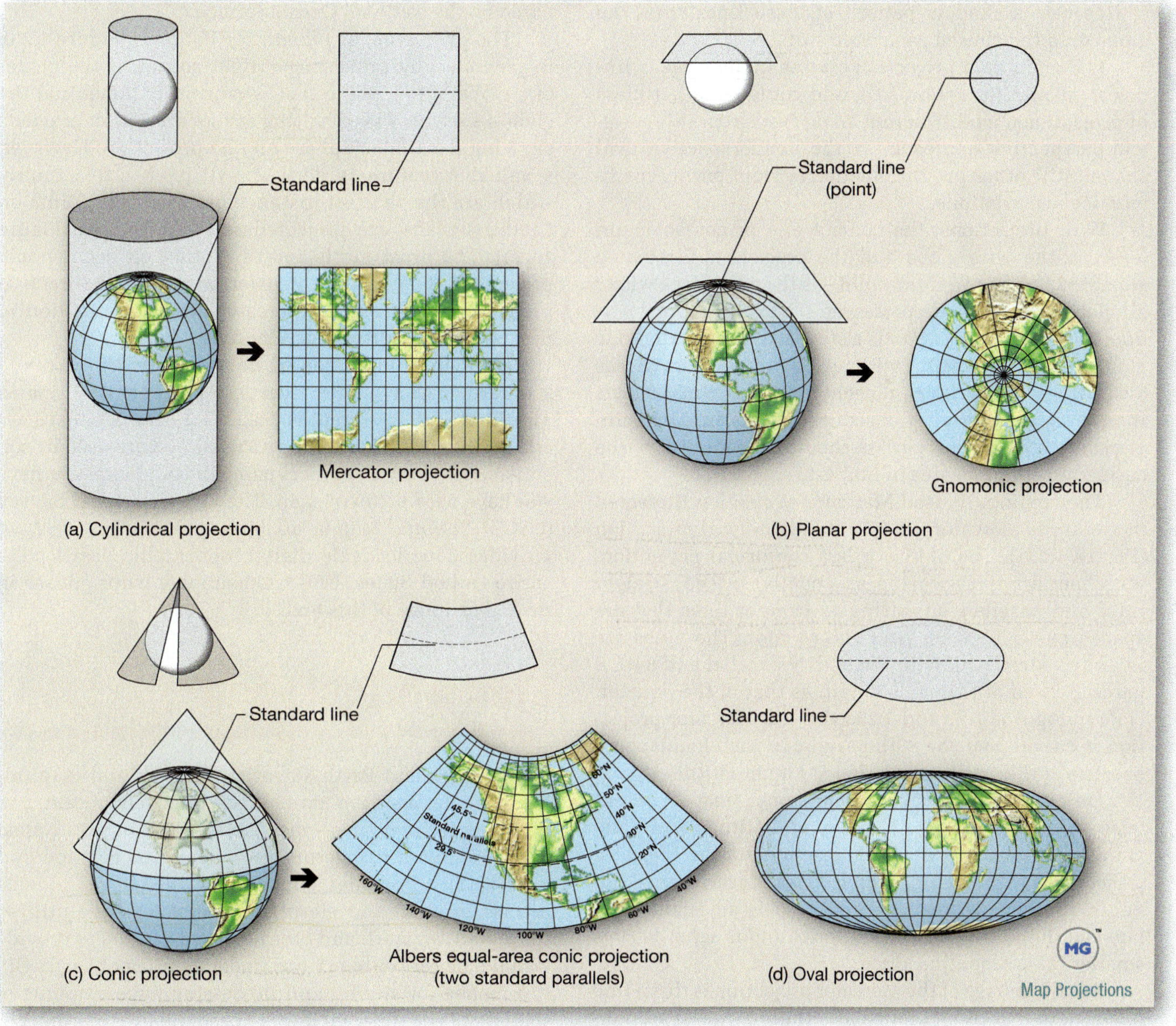

▲**Figure 1.23** **Classes of map projections.**

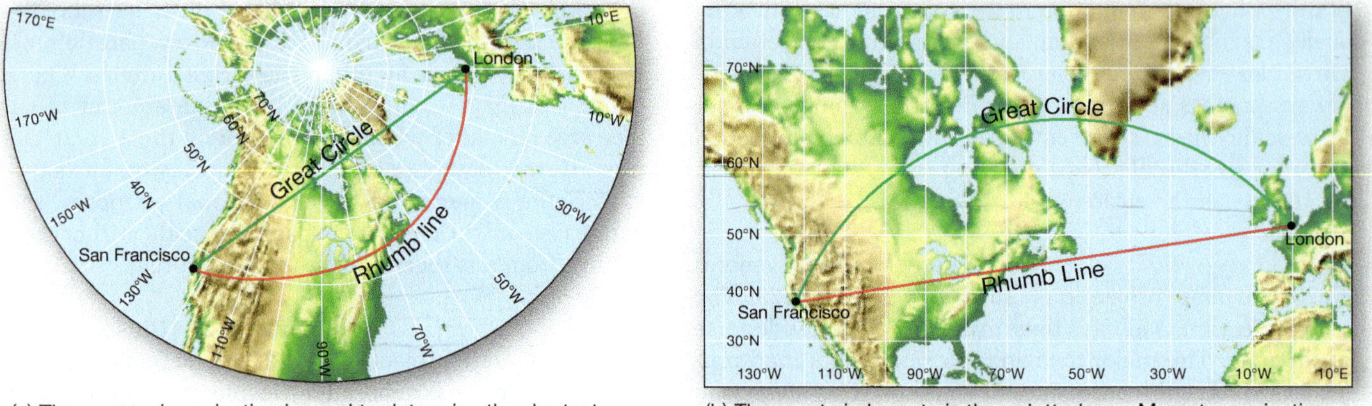

(a) The gnomonic projection is used to determine the shortest distance (great circle route) between San Francisco and London because on this projection the arc of a great circle is a straight line.

(b) The great circle route is then plotted on a Mercator projection, which has true compass direction. Note that straight lines or bearings on a Mercator projection (rhumb lines) are not the shortest route.

▲Figure 1.24 **Determining great circle routes.**

a *cylinder*, *plane*, or *cone*. The wires represented parallels, meridians, and outlines of the continents. A light source cast a shadow pattern of these lines from the globe onto the chosen geometric surface.

The main map projection classes include the cylindrical, planar (or azimuthal), and conic. Another class of projections, which cannot be derived from this physical-perspective approach, is the nonperspective oval shape. Still other projections derive from purely mathematical calculations.

With projections, the contact line or contact point between the wire globe and the projection surface—a *standard line* or *standard point*—is the only place where all globe properties are preserved. Thus, a *standard parallel* or *standard meridian* is a standard line true to scale along its entire length without any distortion. Areas away from this critical tangent line or point become increasingly distorted. Consequently, this line or point of accurate spatial properties should be centered by the cartographer on the area of interest.

The commonly used **Mercator projection** (invented by Gerardus Mercator in 1569) is a cylindrical projection (Figure 1.23a). The Mercator is a conformal projection, with meridians appearing as equally spaced straight lines and parallels appearing as straight lines that are spaced closer together near the equator. The poles are infinitely stretched, with the 84th N parallel and 84th S parallel fixed at the same length as that of the equator. Note in Figures 1.22 and 1.23a that the Mercator projection is cut off near the 80th parallel in each hemisphere because of the severe distortion at higher latitudes.

Unfortunately, Mercator classroom maps present false notions of the size (area) of midlatitude and poleward landmasses. A dramatic example on the Mercator projection is Greenland, which looks bigger than all of South America. In reality, Greenland is an island only one-eighth the size of South America and is actually 20% smaller than Argentina alone.

The advantage of the Mercator projection is that a line of constant direction, known as a **rhumb line**, is straight and therefore facilitates plotting directions between two points (Figure 1.24). Thus, the Mercator projection is useful in navigation and is standard for nautical charts prepared by the National Ocean Service.

The *gnomonic*, or *planar, projection* in Figure 1.23b is generated by projecting a light source at the center of a globe onto a plane that is tangent to (touching) the globe's surface. The resulting severe distortion prevents showing a full hemisphere on one projection. However, a valuable feature is derived: All great circle routes, which are the shortest distance between two points on Earth's surface, are projected as straight lines (Figure 1.24a). The great circle routes plotted on a gnomonic projection then can be transferred to a true-direction projection, such as the Mercator, for determination of precise compass headings (Figure 1.24b).

For more information on maps used in this text and standard map symbols, turn to Appendix A, Maps in This Text and Topographic Maps. Topographic maps are essential tools for landscape analysis and are used by scientists, travelers, and others using the outdoors—perhaps you have used a "topo" map. The U.S. Geological Survey (USGS) *National Map* (available at http://nationalmap.gov/) provides downloadable digital topographic data for the entire United States. USGS topographic maps appear in several chapters of this text.

Modern Tools and Techniques for Geoscience

Geographers and Earth scientists analyze and map our home planet using a number of relatively recent and evolving technologies—the Global Positioning System (GPS), remote sensing, and geographic information systems (GIS). GPS relies on satellites in orbit to provide precise location and elevation. Remote sensing utilizes spacecraft, aircraft, and ground-based sensors to provide visual data that enhance our understanding of Earth. GIS is a means for storing and processing large amounts of spatial data as separate layers of geographic information; GISci is the geographic subfield that uses this technique.

Global Positioning System

Using an instrument that receives radio signals from satellites, you can accurately determine latitude, longitude, and elevation anywhere on or near the surface of Earth. The **Global Positioning System (GPS)** comprises at least 27 orbiting satellites, in 6 orbital planes, that transmit navigational signals to Earth-bound receivers (backup GPS satellites are in orbital storage as replacements). Think of the satellites as a constellation of navigational beacons with which you interact to determine your unique location. As we know, every possible square meter of Earth's surface has its own address relative to the latitude–longitude grid.

A GPS receiver senses signals from at least four satellites—a minimum of three satellites for location and a fourth to determine accurate time. The distance between each satellite and the GPS receiver is calculated using clocks built into each instrument that time radio signals traveling at the speed of light between them (Figure 1.24). The receiver calculates its true position using trilateration so that it reports latitude, longitude, and elevation. GPS units also report accurate time to within 100 billionths of a second. This allows GPS base stations to have perfectly synchronized timing, essential to worldwide communication, finance, and many industries.

GPS receivers are built into many smartphones, wristwatches, and motor vehicles, and can be bought as handheld units. Standard cell phones not equipped with a GPS receiver determine location based on the position of cell phone towers—a process not as accurate as GPS measurement.

The GPS is useful for diverse applications, such as navigating on the ocean, managing the movement of fleets of trucks, mining and mapping of resources, tracking wildlife migration and behavior, carrying out police and security work, and conducting environmental planning. Commercial airlines use the GPS to improve accuracy of routes flown and thus increase fuel efficiency.

Scientific applications of GPS technology are extensive. Consider these examples:

- In geodesy, GPS helps refine knowledge of Earth's exact shape and how this shape is changing.
- Scientists used GPS technology in 1998 to accurately determine the height of Mount Everest in the Himalayan Mountains, raising its elevation by 2 m (6 ft).
- On Mount St. Helens in Washington, a network of GPS stations measure ground deformation associated with earthquake activity (Figure 1.26). In

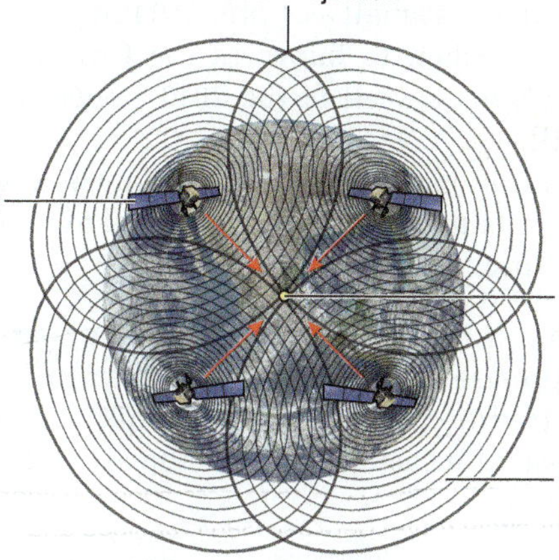

▲**Figure 1.25 Using satellites to determine location through GPS.** Imagine a ranging sphere around each of four GPS satellites. These spheres intersect at two points, one easily rejected because it is some distance above Earth and the other at the true location of the GPS receiver. In this way, signals from four satellites can reveal the receiver's location and elevation. [Based on J. Amos, "Galileo sat-nav in decisive phase," BBC News, March 2007, available at http://news.bbc.co.uk/2/hi/science/nature/6450367.stm]

▲**Figure 1.26 GPS application on Mount St. Helens.** [Mike Poland, USGS.]

GEOreport 1.6 GPS origins

Originally devised in the 1970s by the U.S. Department of Defense for military purposes, GPS is now commercially available worldwide. In 2000, the Pentagon shut down its Pentagon Selective Availability security control, making commercial resolution the same as military applications. Additional frequencies were added in 2003 and 2006, which increased accuracy significantly, to less than 10 m (33 ft). *Differential GPS (DGPS)* achieves accuracy of 1 to 3 m by comparing readings with another base station (reference receiver) for a differential correction. For a GPS overview, see http://www.gps.gov/.

southern California, a similar GPS system can record fault movement as small as 1 mm (0.04 in.).

- GPS units attached to buoys in the Gulf of Mexico helped track the spread of the 2010 *Deepwater Horizon* oil spill.
- In Virunga National Park, Rwanda, rangers use handheld GPS units to track and protect mountain gorillas from poaching.

For scientists, this important technology provides a convenient, precise way to determine location, reducing the need for traditional land surveys requiring point-to-point line-of-sight measurements on the ground. In your daily life and travels, have you ever used a GPS unit? How did GPS assist you?

Remote Sensing

The acquisition of information about distant objects without having physical contact is **remote sensing**. In this era of observations from satellites outside the atmosphere, from aircraft within it, and from remote submersibles in the oceans, scientists obtain a wide array of remotely sensed data (Figure 1.27). Remote sensing is nothing new to humans; we do it with our eyes as we scan the environment, sensing the shape, size, and color of objects

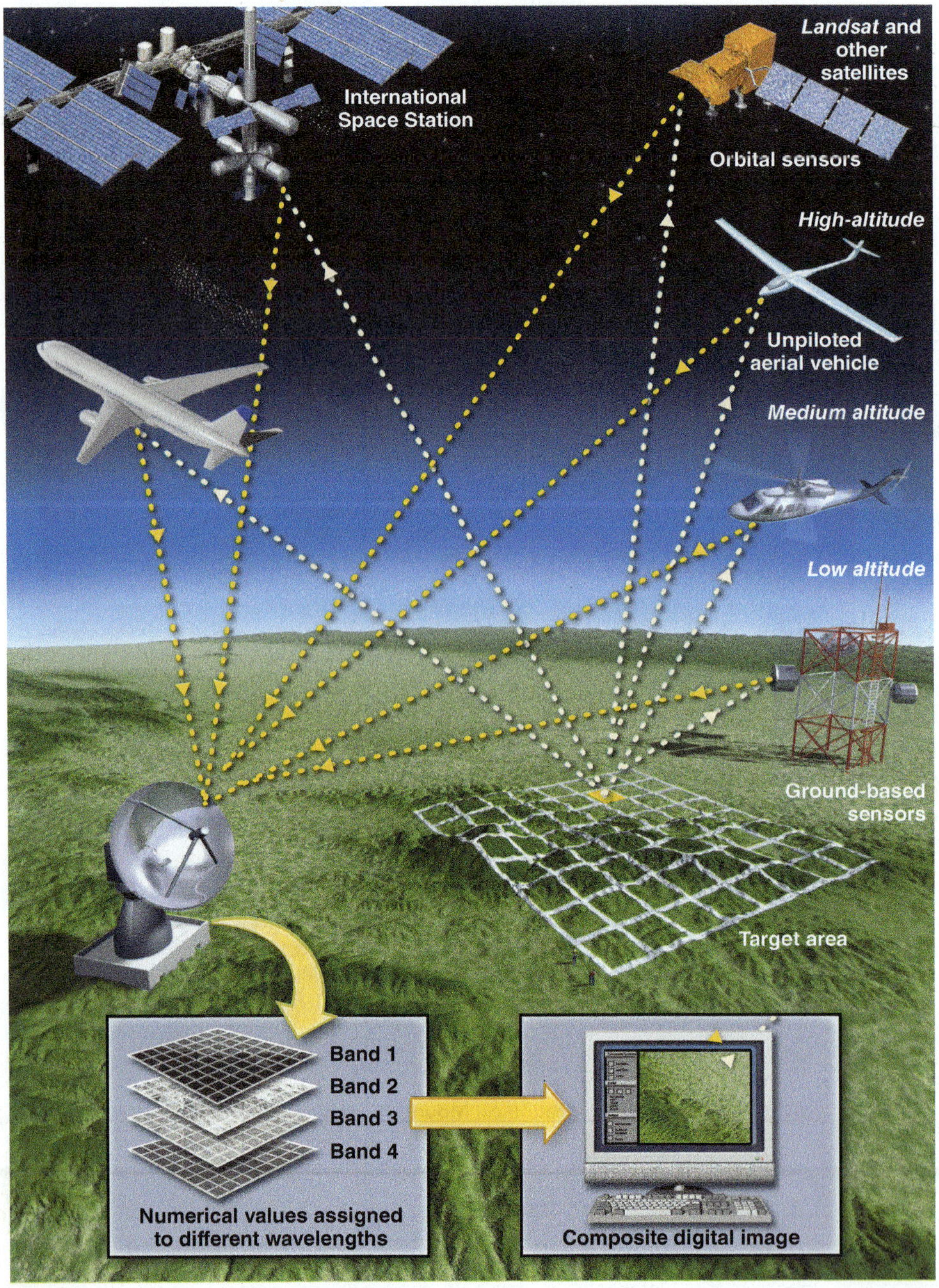

A sample of orbital platforms:
CloudSat: Studies cloud extent, distribution, radiative properties, and structure.
ENVISAT: ESA environment-monitoring satellite; 10 sensors, including next generation radar.
GOES: Weather monitoring and Forecasting; *GOES-11*, *-12*, *-13*, and *-14*.
GRACE: Accurately maps Earth's gravitational field.
JASON-1, -2: Measures sea-level heights.
Landsat: *Landsat-1* in 1972 to *Landsat-7* in 1999, and *Landsat-8* in 2013, provides millions of images for Earth systems science and global change.
NOAA: First in 1978 through *NOAA-15*, *-16*, *-17*, *-18*, and *-19* now in operation, global data gathering, short- and long-term weather forecasts.
RADARSAT-1, -2: Synthetic Aperture Radar in near-polar orbit, operated by Canadian Space Agency.
SciSat-1: Analyzes trace gases, thin clouds, atmospheric aerosols with Arctic focus.
SeaStar: Carries the *SeaWiFS* (Sea-viewing Wide Field-of-View instrument) to observe Earth's oceans and microscopic marine plants.
Terra* and *Aqua: Environmental change, error-free surface images, cloud properties, through five instrument packages.
TOMS-EP: Total Ozone Mapping Spectrometer, monitoring stratospheric ozone, similar instruments on *NIMBUS-7* and *Meteor-3*.
***TOPEX-POSEIDON*:** Measures sea-level heights.
***TRMM*:** Tropical Rainfall Measuring Mission, includes lightning detection and global energy budget measurements.

For more info see: http://www.nasa.gov/centers/goddard/missions/index.html

▲**Figure 1.27 Remote-sensing technology.** Remote-sensing technology measures and monitors Earth's systems from orbiting spacecraft, aircraft, and ground-based sensors. Various wavelengths (bands) are collected from sensors; computers process these data and produce digital images for analysis. A sample of remote-sensing platforms is listed along the side of the illustration. (Illustration is not to scale.)

from a distance by registering energy from the visible-wavelength portion of the electromagnetic spectrum (discussed in Chapter 2). Similarly, when a camera views the wavelengths for which its film or sensor is designed, it remotely senses energy that is reflected or emitted from a scene.

Aerial photographs from balloons and aircraft were the first type of remote sensing, used for many years to improve the accuracy of surface maps more efficiently than can be done by on-site surveys. Deriving accurate measurements from photographs is the realm of **photogrammetry**, an important application of remote sensing. Later, remote sensors on satellites, the International Space Station, and other craft were used to sense a broader range of wavelengths beyond the visible range of our eyes. These sensors can be designed to "see" wavelengths shorter than visible light (such as ultraviolet) and wavelengths longer than visible light (such as infrared and microwave radar). As examples, infrared sensing produces images based on the temperature of objects on the ground, microwave sensing reveals features below Earth's surface, and radar sensing shows land-surface elevations, even in areas that are obscured by clouds.

Satellite Imaging During the last 50 years, satellite remote sensing has transformed Earth observation. Physical elements of Earth's surface emit radiant energy in wavelengths that are sensed by satellites and other craft and sent to receiving stations on the ground. The receiving stations sort these wavelengths into specific bands, or ranges. A scene is scanned and broken down into pixels (*pic*ture *el*ements), each identified by coordinates named *lines* (horizontal rows) and *samples* (vertical columns). For example, a grid of 6000 lines and 7000 samples forms 42,000,000 pixels, providing an image of great detail when the pixels are matched to the wavelengths they emit.

A large amount of data is needed to produce a single remotely sensed image; these data are recorded in digital form for later processing, enhancement, and image generation. Digital data are processed in many ways to enhance their utility: with simulated natural color, "false" color to highlight a particular feature, enhanced contrast, signal filtering, and different levels of sampling and resolution.

Satellites can be set in specific orbital paths (Figure 1.28) that affect the type of data and imagery produced. Geostationary (or geosynchronous) orbits, typically at an altitude of 35,790 km (22,239 mi), are *high Earth orbits* that effectively match Earth's rotation speed so that one orbit is completed in about 24 hours. Satellites can therefore remain "parked" above a specific location, usually the equator (Figure 1.28a). This "fixed" position means that satellite antennas on Earth can be pointed permanently at one position in the sky where the satellite is located; many communications and weather satellites use these high Earth orbits.

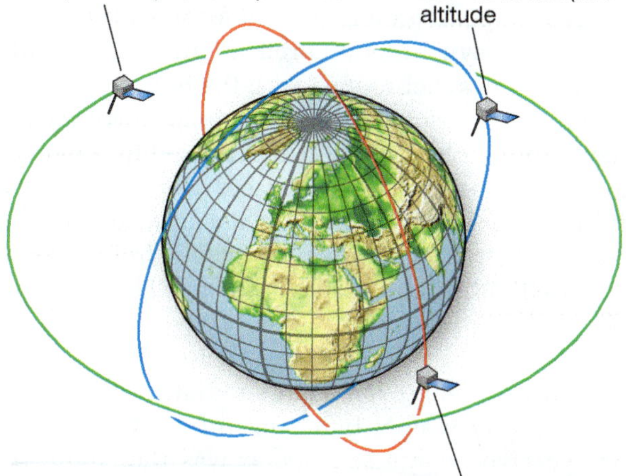

▲**Figure 1.28** **Three satellite orbital paths.**

Some satellites orbit at lower altitudes. The pull of Earth's gravity means that the closer to Earth they are, the faster their orbiting speed. For example, GPS satellites, at altitudes of about 20,200 km (12,552 mi), have *medium Earth orbits* that move more quickly than high Earth orbits. *Low Earth orbits,* at altitudes less than 1000 km (621 mi), are the most useful for scientific monitoring. Several of the National Aeronautics and Space Administration (NASA) environmental satellites in low Earth orbit are at altitudes of about 700 km (435 mi), completing one orbit every 99 minutes.

The angle of a satellite's orbit in relation to Earth's equator is its *inclination,* another factor affecting remotely sensed data. Some satellites orbit near the equator to monitor Earth's tropical regions; this low-inclination orbit acquires data only from low latitudes. An example is the *Tropical Rainfall Measuring Mission* (TRMM) satellite, which provides data for mapping water vapor and rainfall patterns in the tropics and subtropics. Monitoring the polar regions requires

GEOreport 1.7 Polar-orbiting satellites predict Hurricane Sandy's path

Scientists at the European Centre for Medium-Range Weather Forecasts report that polar-orbiting satellites, such as the National Oceanic and Atmospheric Administration (NOAA) *Suomi NPP* satellite, were critical for predicting Hurricane Sandy's track. Without data from these satellites, predictions for Hurricane Sandy would have been off by hundreds of miles, showing the storm heading out to sea rather than turning toward the New Jersey coast. *Suomi* orbits Earth about 14 times each day, collecting data from nearly the entire planet (find out more at http://npp.gsfc.nasa.gov/).

a satellite in polar orbit, with a higher inclination of about 90° (Figure 1.28b).

One type of polar orbit important for scientific observation is a Sun-synchronous orbit (Figure 1.28c). This low Earth orbit is synchronous with the Sun, so that the satellite crosses the equator at the same local solar time each day. Ground observation is maximized in Sun-synchronous orbit because Earth surfaces viewed from the satellite are illuminated by the Sun at a consistent angle. This enables better comparison of images from year to year because lighting and shadows do not change.

Passive Remote Sensing Passive remote-sensing systems record wavelengths of energy radiated from a surface, particularly visible light and infrared. Our own eyes are passive remote sensors, as was the *Apollo 17* astronaut camera that made the film photograph of Earth on the back cover of this book.

A number of satellites carry passive remote sensors for weather forecasting. The *Geostationary Operational Environmental Satellites*, known as *GOES*, became operational in 1994 and provide the images you see on television weather reports. *GOES-12*, *-13*, and *-15* are operational; *GOES-12* sits at 60° W longitude to monitor the Caribbean and South America. Think of these satellites as hovering over these meridians for continuous coverage, using visual wavelengths for daylight hours and infrared for nighttime views. *GOES-14*, parked in orbit since 2009, replaced *GOES-13* in 2012 when that satellite experienced technical problems.

Landsat satellites, which began imaging Earth in the 1970s, are used extensively for comparison of changing Earth landscapes over time, among other applications (Figure 1.29; more images of changing Earth systems at http://earthobservatory.nasa.gov). *Landsat-5* was retired in 2012 after 29 years, the longest-running Earth-observing mission in history; *Landsat-7* remains operational as part of NASA's ongoing Landsat Data Continuity Mission. Landsat-8 was launched in 2013, beginning the new Landsat program managed by the U.S. Geological Survey.

Although the *Landsat* satellites far surpassed their predicted lifespans, most satellites are removed from orbit after 3 to 5 years. Launched in 2011, *Suomi NPP* is part of the National Polar-orbiting Partnership (NPP), the next generation of satellites that will replace NASA's aging Earth Observation Satellite satellite fleet. *Suomi NPP*'s Visible Infrared Imaging Radiometer Suite (VIIRS) imaged Hurricane Sandy along the U.S. East Coast in October 2012, as seen on the half title page of this book. Many of the beautiful NASA "Blue Marble" Earth composite images are from the *Suomi* satellite (Figure 1.30).

Active Remote Sensing Active remote-sensing systems direct a beam of energy at a surface and analyze the energy reflected back. An example is *radar* (*ra*dio *d*etection *a*nd *r*anging). A radar transmitter emits short bursts of energy that have relatively long wavelengths (0.3 to 10 cm) toward the subject terrain, penetrating clouds and darkness. Energy reflected back to a radar receiver for analysis is known as *backscatter*. Radar images collected in a time series allow scientists to make pixel-by-pixel comparisons to detect Earth movement, such as elevation changes along earthquake faults (images and discussion in Chapter 12).

Another active remote-sensing technology is airborne LiDAR, or light detection and ranging. LiDAR systems collect highly detailed and accurate data for surface terrain using a laser scanner, with up to 150,000 pulses per second, 8 pulses or more per square meter, providing 15-m (49.2-ft) resolution. GPS and navigation systems onboard the aircraft determine the location of

(a) A year after the 1988 wildfires, burned land is deep red. Old pine forest is dark green; geyser fields are pale blue; lakes are dark blue.

(b) By the end of 2011, fire scars have faded to orange as grass and young trees emerge. Recovery is slow on this high-elevation plateau.

▲**Figure 1.29 Landscape recovery from fire in Yellowstone National Park, Wyoming.** *Landsat-5* images contrast the 1989 and 2011 landscapes of Yellowstone, using a combination of visible and infrared light to highlight the burned area and changes in vegetation. [NASA.]

▲Figure 1.30 ***Suomi NPP* Blue Marble image.** This composite view of Earth was imaged January 2, 2012. NASA scientist Norman Kuring combined VIIRS instrument data from 6 orbits of the *Suomi NPP* satellite. VIIRS acquires data in 22 bands covering visible, near-infrared, and thermal infrared wavelengths. [NASA.]

each pulse. LiDAR datasets are often shared between private, public, and scientific users for multiple applications. Scientists are presently studying the effects of Hurricane Sandy along the U.S. East Coast using LiDAR (see discussion in Chapter 16).

For more on remote-sensing platforms, see the "Remote-Sensing Status Report" on the *MasteringGeography* website, or go to the USGS Global Visualization Viewer (http://glovis.usgs.gov/).

Geographic Information Systems

Techniques such as remote sensing acquire large volumes of spatial data that must be stored, processed, and retrieved in useful ways. A **geographic information system (GIS)** is a computer-based data-processing tool for gathering, manipulating, and analyzing geographic information. Today's sophisticated computer systems allow the integration of geographic information from direct surveys (on-the-ground mapping) and remote sensing in complex ways never before possible. Whereas printed maps are fixed at time of publication, GIS maps can be easily modified and evolve instantly.

In a GIS, spatial data can be arranged in layers, or planes, containing different kinds of data (Figure 1.31). The beginning component for any GIS is a map, with its associated coordinate system, such as latitude–longitude provided by GPS locations or digital surveys (the top layer in Figure 1.31a). This map establishes reference points against which to accurately position other data, such as remotely sensed imagery.

CRITICALthinking 1.5
Test Your Knowledge about Satellite Imagery

Go to the NASA, USGS (http://eros.usgs.gov/), or ESA (European Space Agency; http://www.esa.int/Our_Activities/Observing_the_Earth) website and view some satellite images. Then examine the image in Figure CT 1.5.1. Was this made by an aircraft, a ground-based sensor, or a satellite? Is this a natural-color or false-color image? Can you determine what the colors represent? Can you identify the location, the land and water bodies, and other physical features? Finally, based on your research and examples in this text, can you determine the specific source (LiDAR aircraft, *GOES* or *Landsat* satellite, etc.) of the data that made this image? (Find the answers at the end of the Chapter 1 Key Learning Concept Review.) ●

▲Figure CT 1.5.1 **Can you describe this image?** [NASA.]

A GIS is capable of analyzing patterns and relationships within a single data plane, such as the floodplain or soil layer in Figure 1.31a. A GIS also can generate an overlay analysis where two or more data planes interact. When the layers are combined, the resulting synthesis—a *composite overlay*—is ready for use in analyzing complex problems. The utility of a GIS is its ability to manipulate data and bring together several variables for analysis.

Figures 1.31b and c show two GIS map layers included in the Northwest Gap Analysis Program (GAP), a five-state assessment of terrestrial species and habitat, land stewardship, and management status, which began in 2004. The Northwest project is part of a nationwide GAP environmental quality assessment by the USGS for species conservation; critical data layers are vegetation

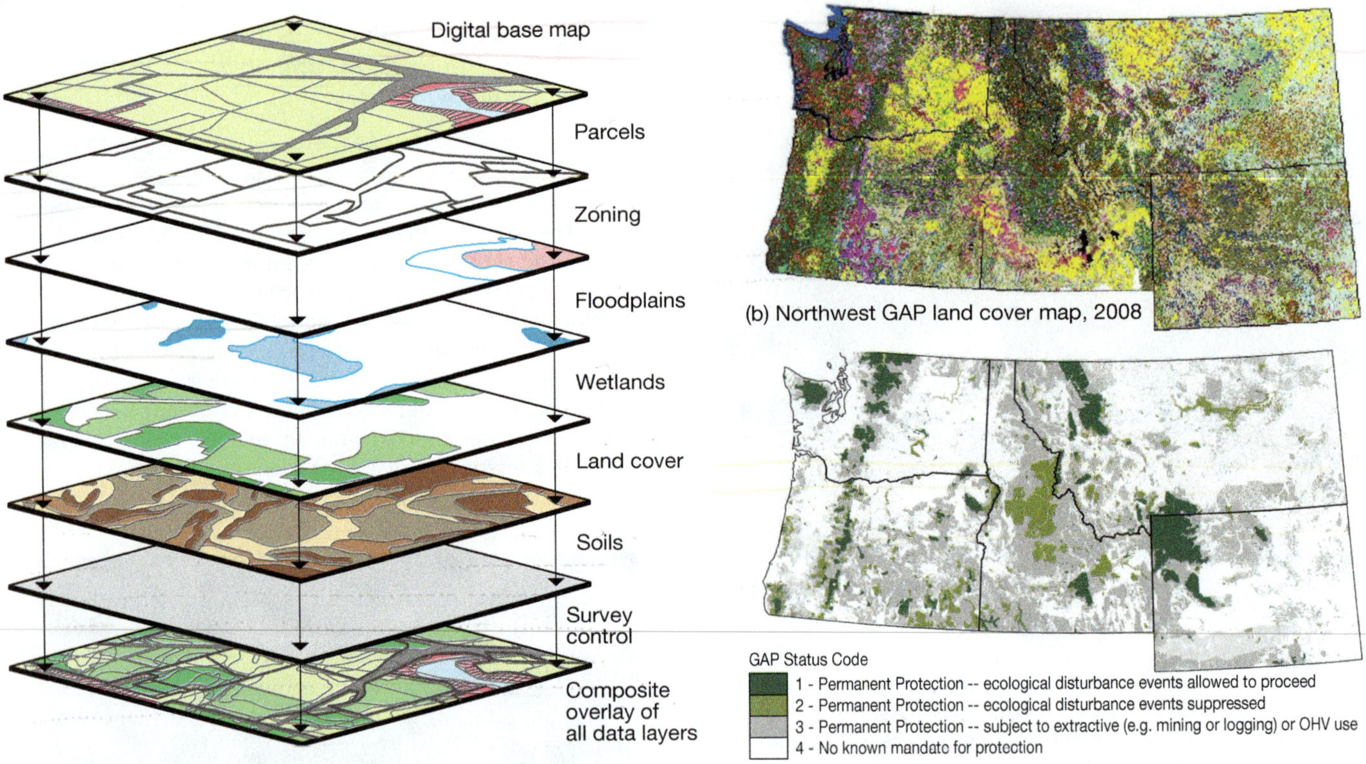

(a) Layered spatial data in a GIS format.

(c) Northwest GAP protected areas map, 2008

▲**Figure 1.31 A geographic information system (GIS) model.** [(a) After USGS. (b) and (c) University of Idaho, Northwest Gap Analysis Project.]

type, species range, and land ownership, among other environmental variables. (See http://gapanalysis.usgs.gov/ or http://gap.uidaho.edu for a project overview and to access interactive maps.)

Geographic information science (GISci) is the field that develops the capabilities of GIS for use within geography and other disciplines. GISci analyzes Earth and human phenomena over time. This can include the study and forecasting of diseases, the population displacement caused by Hurricane Sandy, the destruction from the Japan earthquake and tsunami in 2011, or the status of endangered species and ecosystems, to name a few examples.

One widely used application of GIS technology is in the creation of maps with a three-dimensional perspective. These maps are produced by combining *digital elevation models* (DEMs), which provide the base elevation data, with satellite-image overlay (see examples in Chapter 12). Through the GIS, these data are available for multiple displays, animations, and other scientific analyses.

Google Earth™ and similar programs that can be downloaded from the Internet provide three-dimensional viewing of the globe, as well as geographic information (http://earth.google.com/). Google Earth™ allows the user to "fly" anywhere on Earth and zoom in on landscapes and features of interest, using satellite imagery and aerial photography at varying resolutions. Users can select layers, as in a GIS model, depending on the task at hand and the composite overlay displayed. NASA's World Wind software is another open-source browser with access to high-resolution satellite images and multiple data layers suitable for scientific applications.

Geovisualization is the technique of adjusting geospatial data sets in real time, so that users can instantly make changes to maps and other visual models. Geovisual tools are important for translating scientific knowledge into resources that nonscientists can use for decision making and planning. At East Carolina University, scientists are developing geovisual tools to assess the effects of sea level rise along the North Carolina coast, in partnership with state, local, and nonprofit organizations (see http://www.ecu.edu/renci/Technology/GIS.html).

Access to GIS is expanding and becoming more user-friendly with the increased availability of numerous open-source GIS software packages. These are usually free, have online support systems, and are updated frequently (see http://opensourcegis.org/). In addition, public access to large remote-sensing data sets for analyses and display is now available, without the need to download large amounts of data (see examples of research applications at http://disc.sci.gsfc.nasa.gov/).

GEOSYSTEMS**connection**

With this overview of geography, the scientific process, and the *Geosystems* approach in mind, we now embark on a journey through each of Earth's four spheres—Part I, Atmosphere; Part II, Hydrosphere; Part III, Lithosphere; and Part IV, Biosphere. Chapter 2 begins at the Sun, including its place in the Universe and seasonal changes in the distribution of its energy flow to Earth. In Chapter 3, we follow solar energy through Earth's atmosphere to the surface, and in Chapters 4 through 6 examine global temperature patterns and the circulation of air and water in Earth's vast wind and ocean currents.

At the end of each chapter, you find a *Geosystems Connection* to act as a bridge from one chapter to the next, helping you to cross to the next topic.

KEY LEARNING concepts review

Here is a handy summary designed to help you review the Key Learning Concepts listed on this chapter's title page. The recap of each concept concludes with a list of the key terms from that portion of the chapter, their page numbers, and review questions pertaining to the concept. Similar summary and review sections follow each chapter in the book.

■ *Define* geography in general and physical geography in particular.

Geography combines disciplines from the physical and life sciences with the human and cultural sciences to attain a holistic view of Earth. Geography's **spatial** viewpoint examines the nature and character of physical space and the distribution of things within it. Geography integrates a wide range of subject matter, and geographic education recognizes five major themes: **location**, **region**, **human–Earth relationships**, **movement**, and **place**. A method, **spatial analysis**, ties together this diverse field, focusing on the interdependence among geographic areas, natural systems, society, and cultural activities over space or area. The analysis of **process**—a set of actions or mechanisms that operate in some special order—is also central to geographic understanding.

Physical geography applies spatial analysis to all the physical components and process systems that make up the environment: energy, air, water, weather, climate, landforms, soils, animals, plants, microorganisms, and Earth itself. Physical geography is an essential aspect of **Earth systems science.** The science of physical geography is uniquely qualified to synthesize the spatial, environmental, and human aspects of our increasingly complex relationship with our home planet—Earth.

geography (p. 3)
spatial (p. 3)
location (p. 3)
region (p. 3)
human–Earth relationships (p. 3)
movement (p. 3)
place (p. 3)
spatial analysis (p. 4)
process (p. 4)
physical geography (p. 4)
Earth systems science (p. 5)

1. On the basis of information in this chapter, define physical geography and review the approach that characterizes the geographic sciences.
2. Suggest a representative example for each of the five geographic themes; for example, atmospheric and oceanic circulation spreading radioactive contamination is an example of the movement theme.
3. Have you made decisions today that involve geographic concepts discussed within the five themes presented? Explain briefly.
4. In general terms, how might a physical geographer analyze water pollution in the Great Lakes?

■ *Discuss* human activities and human population growth as they relate to geographic science, and *summarize* the scientific process.

Understanding the complex relations between Earth's physical systems and human society is important to human survival. Hypotheses and theories about the Universe, Earth, and life are developed through the scientific process, which relies on a general series of steps that make up the **scientific method**. Results and conclusions from scientific experiments can lead to basic theories, as well as applied uses for the general public.

Awareness of the human denominator, the role of humans on Earth, has led to physical geography's increasing emphasis on human–environment interactions. Recently, **sustainability science** has become an important new discipline, integrating sustainable development and functioning Earth systems.

scientific method (p. 5)
sustainability science (p. 7)

5. Sketch a flow diagram of the scientific process and method, beginning with observations and ending with the development of theories and laws.
6. Summarize population-growth issues: population size, the impact per person, and future projections. What strategies do you see as important for global sustainability?

■ *Describe* systems analysis, open and closed systems, and feedback information, and *relate* these concepts to Earth systems.

A **system** is any ordered set of interacting components and their attributes, as distinct from their surrounding environment. Systems analysis is an important organizational and analytical tool used by geographers. Earth is an **open system** in terms of energy, receiving energy from the Sun, but it is essentially a **closed system** in terms of matter and physical resources.

As a system operates, "information" is returned to various points in the operational process via pathways of **feedback loops**. If the feedback information discourages change in the system, it is **negative feedback**. Further production of such feedback opposes system changes. Such negative feedback causes self-regulation in a natural system, stabilizing the system. If feedback information encourages change in the system, it is **positive feedback**. Further production of positive feedback stimulates system changes. Unchecked positive feedback in a system can create a runaway ("snowballing") condition. When the rates of inputs and outputs in the system are equal and the amounts of energy and matter in storage within the system are constant (or when they fluctuate around a stable average), the system is in **steady-state equilibrium**. A system showing a steady increase or decrease in some operation over time (a trend) is in **dynamic equilibrium**. A **threshold**, or tipping point, is the moment at which a system can no longer maintain its character and lurches to a new operational level. Geographers often construct a simplified **model** of natural systems to better understand them.

Four immense open systems powerfully interact at Earth's surface: three **abiotic**, or nonliving, systems—the **atmosphere**, **hydrosphere** (including the **cryosphere**), and **lithosphere**—and a **biotic**, or living, system—the **biosphere**, or **ecosphere**.

system (p. 10)
open system (p. 10)
closed system (p. 11)
feedback loop (p. 11)
negative feedback (p. 11)
positive feedback (p. 12)
steady-state equilibrium (p. 12)
dynamic equilibrium (p. 12)
threshold (p. 12)
model (p. 13)
abiotic (p. 13)
biotic (p. 13)
atmosphere (p. 14)
hydrosphere (p. 14)
cryosphere (p. 14)
lithosphere (p. 14)
biosphere (p. 14)
ecosphere (p. 14)

7. Define systems theory as an analytical strategy. What are open systems, closed systems, and negative feedback? When is a system in a steady-state equilibrium condition? What type of system (open or closed) is the human body? A lake? A wheat plant?
8. Describe Earth as a system in terms of both energy and matter; use simple diagrams to illustrate your description.
9. What are the three abiotic spheres that make up Earth's environment? Relate these to the biotic sphere, the biosphere.

■ *Explain* Earth's reference grid: latitude and longitude and latitudinal geographic zones and time.

The science that studies Earth's shape and size is **geodesy**. Earth bulges slightly through the equator and is oblate (flattened) at the poles, producing a misshapen spheroid, or **geoid**. Absolute location on Earth is described with a specific reference grid of **parallels** of **latitude** (measuring distances north and south of the equator) and **meridians** of **longitude** (measuring distances east and west of a prime meridian). A historic breakthrough in navigation occurred with the establishment of an international **prime meridian** (0° through Greenwich, England). A **great circle** is any circle of Earth's circumference whose center coincides with the center of Earth. Great circle routes are the shortest distance between two points on Earth. **Small circles** are those whose centers do not coincide with Earth's center.

The prime meridian provided the basis for **Greenwich Mean Time** (**GMT**), the world's first universal time system. Today, **Coordinated Universal Time** (**UTC**) is the worldwide standard and the basis for international time zones. A corollary of the prime meridian is the 180° meridian, the **International Date Line (IDL)**, which marks the place where each day officially begins. **Daylight saving time** is a seasonal change of clocks by 1 hour in summer months.

geodesy (p. 16)
geoid (p. 16)
latitude (p. 17)
parallel (p. 17)
longitude (p. 19)
meridian (p. 19)
prime meridian (p. 19)
great circle (p. 19)
small circle (p. 19)
Greenwich Mean Time (GMT) (p. 20)
Coordinated Universal Time (UTC) (p. 21)
International Date Line (IDL) (p. 21)
daylight saving time (p. 22)

10. Draw a simple sketch describing Earth's shape and size.
11. Define latitude and parallel and define longitude and meridian using a simple sketch with labels.
12. Define a great circle, great circle routes, and a small circle. In terms of these concepts, describe the equator, other parallels, and meridians.
13. Identify the various latitudinal geographic zones that roughly subdivide Earth's surface. In which zone do you live?
14. What does timekeeping have to do with longitude? Explain this relationship. How is Coordinated Universal Time (UTC) determined on Earth?
15. What and where is the prime meridian? How was the location originally selected? Describe the meridian that is opposite the prime meridian on Earth's surface.

■ *Define* cartography and mapping basics: map scale and map projections.

A **map** is a generalized depiction of the layout of an area, usually some portion of Earth's surface, as seen from above and greatly reduced in size. **Cartography** is the science and art of mapmaking. For the spatial portrayal of Earth's physical systems, geographers use maps. **Scale** is the ratio of the image on a map to the real world; it relates a unit on the map to a corresponding unit on the ground. When creating a **map projection**, cartographers select the class of projection that is the best compromise for the map's specific purpose. Compromise is always necessary because Earth's roughly spherical three-dimensional surface cannot be exactly duplicated on a flat, two-dimensional map. Relative abilities to portray **equal area** (equivalence), **true shape** (conformality), true direction, and true distance are all considerations in selecting a projection. The **Mercator projection** is in the cylindrical class; it has true-shape qualities and straight lines that show constant direction. A **rhumb line** denotes constant direction and appears as a straight line on the Mercator.

map (p. 22)
cartography (p. 22)
scale (p. 22)
map projection (p. 25)
equal area (p. 25)
true shape (p. 25)
Mercator projection (p. 26)
rhumb line (p. 26)

16. Define cartography. Explain why it is an integrative discipline.
17. Assess your geographic literacy by examining atlases and maps. What types of maps have you used: Political? Physical? Topographic? Do you know what map projections they employed? Do you know the names and locations of the four oceans, the seven continents, and most individual countries? Can you identify the new countries that have emerged since 1990?
18. What is map scale? In what three ways may it be expressed on a map?
19. State whether the following ratios are large scale, medium scale, or small scale: 1:3,168,000; 1:24,000; 1:125,000.
20. Describe the differences between the characteristics of a globe and those that result when a flat map is prepared.
21. What type of map projection is used in Figure 1.15? In Figure 1.19? (See Appendix A.)

- *Describe* modern geoscience techniques—the Global Positioning System (GPS), remote sensing, and the geographic information system (GIS)—and *explain* how these tools are used in geographic analysis.

Latitude, longitude, and elevation are accurately measured using a handheld **Global Positioning System (GPS)** instrument that reads radio signals from satellites. Orbital and aerial **remote sensing** obtains information about Earth systems from great distances without the need for physical contact. Satellites do not take photographs but instead record images that are transmitted to Earth-based receivers. Satellite data are recorded in digital form for later processing, enhancement, and image generation. Aerial photographs are used to improve the accuracy of surface maps, an application of remote sensing called **photogrammetry**.

Satellite and other data may be analyzed using **geographic information system (GIS)** technology. Computers process geographic information from direct ground surveys and remote sensing in complex layers of spatial data. Digital elevation models are three-dimensional products of GIS technology. Open-source GIS is increasingly available to scientists and the public for many applications, including spatial analysis in geography and the better understanding of Earth's systems.

Global Positioning System (GPS) (p. 27)
remote sensing (p. 28)
photogrammetry (p. 29)
geographic information system (GIS) (p. 31)

22. What is GPS and how does it assist you in finding location and elevation on Earth? Give several examples of GPS technology used for scientific purposes.
23. What is remote sensing? What are you viewing when you observe a weather satellite image on TV or in the newspaper? Explain.
24. If you were in charge of planning the human development of a large tract of land, how would GIS methodologies assist you? How might planning and zoning be affected if a portion of the tract in the GIS is a floodplain or prime agricultural land?

Answer for Critical Thinking 1.5, Figure CT 1.5.1: This natural, true-color image is a composite mosaic of numerous images captured between 2000 and 2002 from NASA satellite *Terra*. The location it depicts is the meeting of the European and African continents at the Strait of Gibraltar, extending from France and Spain across the Mediterranean to Morocco and Algeria.

MasteringGeography™

Looking for additional review and test prep materials? Visit the Study Area in *MasteringGeography*™ to enhance your geographic literacy, spatial reasoning skills, and understanding of this chapter's content by accessing a variety of resources, including MapMaster interactive maps, geoscience animations, satellite loops, author notebooks, videos, RSS feeds, flashcards, web links, self-study quizzes, and an eText version of *Geosystems*.

I
The Energy–Atmosphere System

For more than 4.6 billion years, solar energy has traveled across interplanetary space to Earth, where a small portion of the solar output is intercepted. Our planet and our lives are powered by this radiant energy from the Sun. Because of Earth's curvature, the arriving energy is unevenly distributed at the top of the atmosphere, creating energy imbalances over the Earth's surface—the equatorial region experiences surpluses, receiving more energy than

▼ The Sahara Desert near Tadrart, Algeria. [Pichugin Dmitry/Shutterstock.]

it emits; the polar regions experience deficits, emitting more energy than they receive. Also, the annual pulse of seasonal change varies the distribution of energy during the year.

Earth's atmosphere acts as an efficient filter, absorbing most harmful radiation, charged particles, and space debris so that they do not reach Earth's surface. In the lower atmosphere the unevenness of daily energy receipt gives rise to global patterns of temperature and the circulation of wind and ocean currents, driving weather and climate. Each of us depends on these interacting systems that are set into motion by energy from the Sun. These are the systems of Part I.

2

Solar Energy to Earth and the Seasons

KEY LEARNING concepts

After reading the chapter, you should be able to:

- ***Distinguish*** between galaxies, stars, and planets, and ***locate*** Earth.
- ***Summarize*** the origin, formation, and development of Earth, and ***reconstruct*** Earth's annual orbit about the Sun.
- ***Describe*** the Sun's operation, and ***explain*** the characteristics of the solar wind and the electromagnetic spectrum of radiant energy.
- ***Illustrate*** the interception of solar energy and its uneven distribution at the top of the atmosphere.
- ***Define*** solar altitude, solar declination, and daylength, and ***describe*** the annual variability of each—Earth's seasonality.

On Fogo Island in the Cape Verde archipelago, a small village lies within the Cha Caldera, a large depression formed when the volcanic summit of Pico de Fogo collapsed after an eruption. On the date of this photo, your author and his wife were close to the subsolar point, the latitude on Earth where the Sun's rays are perpendicular to the surface at local noon. Looking closely, you can see shadows cast directly below the trees in this courtyard. The caldera wall is visible in the background.

Geosystems Now describes the author's chase of the subsolar point from aboard a ship in the Atlantic Ocean; the chase ended here at Fogo Island on May 1 at 14.8 N latitude. Chapter 2 discusses energy from the Sun and the seasonal changes we experience on Earth. [Bobbé Christopherson.]

Chasing the Subsolar Point

April 2010 in the Atlantic Ocean: After a month at sea traveling from the Antarctic region, our ship moves northward toward the equator in the Atlantic Ocean. We have no readily accessible news or Wi-Fi; our views are of ocean horizons in every direction. Our research ship carries crew and 48 passengers. On 24 April, I swam at the Earth's equator, no land in sight for thousands of kilometers and water 3 to 4 km deep. Looking through a mask toward the seafloor the view is an infinite blue, creating both an awesome and scary feeling.

On our 5-week expedition, your author and his wife traveled from the Weddell Sea, Antarctica, at 63° S latitude, to the Cape Verde islands off the West African coast, at 14° N latitude (Figure GN 2.1). As we passed over the equator into the Northern Hemisphere, our chase of the subsolar point began.

▲**Figure GN 2.1 The 2010 author's expedition map.** See *Geosystems Now* in Chapters 6 and 20 for a description of events occurring on the island of Tristan de Cunha, another stop on the expedition.

What is the subsolar point? Every day at noon, there is some latitude on Earth at which the Sun is "directly" overhead at nearly a 90° angle. During the spring months (March–June), the latitude receiving the "direct" rays of the Sun shifts from the equator, at 0°, to the Tropic of Cancer, at 23.5° N. The exact latitude receiving these direct 90° rays is the *subsolar point*. Think of this point as the latitude where the Sun is highest in the sky and its rays are perpendicular to the Earth's surface.

Each year, around March 22, the subsolar point is on the equator; this is the *March equinox*, when daylength is equal for all latitudes on Earth. In summer, around June 21, the subsolar point is on the Tropic of Cancer; this is the *June solstice*, when daylength is longest for northern hemisphere latitudes and shortest for southern hemisphere latitudes. Around September 22, the Sun's subsolar point returns to the equator (the September equinox), and by December it is on the Tropic of Capricorn, at 23.5° S (the December solstice). Outside of the tropics, the Sun is never directly overhead. For example, at 40° N latitude the noon Sun's altitude ranges from 26° above the horizon in December to 73° in June, and is never at 90°.

Catching up to the Sun's direct rays On our expedition ship, we chased the subsolar point as it moved from the equator to the Tropic of Cancer between the March equinox and the June solstice. As we traveled, we tracked our route and that of the Sun to determine the closest we could get to this point, either on our ship or on an island in the Atlantic Ocean.

▲**Figure GN 2.2 Near the subsolar point, Fogo Island, Cape Verde.** Note that the boys' shadow is cast almost directly beneath them. [Bobbé Christopherson.]

The subsolar point occurs at 1° N latitude on March 23, moving close to 15° N on May 1. When did we come closest? On May 1, we arrived at 14.8° N, on Fogo Island, Cape Verde. We saw two boys and a donkey hauling water around local noon. Note that their shadow is cast directly beneath them, under the nearly perpendicular rays of the overhead Sun (Figure GN 2.2 and the chapter-opening photo).

In this chapter we track the march of the seasons, marked by changes in daylength and the angle of the Sun's rays. We can calculate the latitude of the subsolar point at any time during the year using a chart called the *analemma*, which appears on most globes in the area of the Southeast Pacific. An example of this figure-8 shaped chart is provided toward the end of the chapter (see Figure CT 2.3.1, p. 54).). After learning about Earth's seasonality in relation to Sun angle, you can use the analemma to determine the subsolar point for any day of the year. Check this analemma for May 1st.

GEOSYSTEMS NOW ONLINE Go to Chapter 2 on the *MasteringGeography* website (www.masteringgeography.com) for more on the subsolar point. For the current location of the subsolar point, go to http://www.timeanddate.com/worldclock/sunearth.html.

The Universe is populated with at least 125 billion galaxies. One of these is our own Milky Way Galaxy, and it contains about 300 billion stars. Among these stars is an average-size yellow star, the Sun, although the dramatic satellite image in Figure 2.2 seems anything but average! Our Sun radiates energy in all directions and upon its family of orbiting planets. Of special interest to us is the solar energy that falls on the third planet, our immediate home.

In this chapter: Incoming solar energy arrives at the top of Earth's atmosphere, establishing the pattern of energy input that drives Earth's physical systems and influences our daily lives. This solar energy input to the atmosphere, combined with Earth's tilt and rotation, produces daily, seasonal, and annual patterns of changing daylength and Sun angle. The Sun is the ultimate energy source for most life processes in our biosphere.

The Solar System, Sun, and Earth

Our Solar System is located on a remote, trailing edge of the **Milky Way Galaxy**, a flattened, disk-shaped collection of stars in the form of a barred-spiral—a spiral with a slightly barred, or elongated, core (Figure 2.1a, b). Our Solar System is embedded more than halfway out from the galactic center, in one of the Milky Way's spiral arms—the Orion Spur of the Sagittarius Arm. A supermassive black hole some 2 million solar masses in size, named *Sagittarius A** (pronounced "*Sagittarius A Star*"), sits in the galactic center. Our Solar System of eight planets, four dwarf planets, and asteroids is some 30,000 light-years from this black hole at the center of the Galaxy, and about 15 light-years above the plane of the Milky Way.

From our Earth-bound perspective in the Milky Way, the Galaxy appears to stretch across the night sky like a narrow band of hazy light. On a clear night, the unaided eye can see only a few thousand of these billions of stars gathered about us in our "neighborhood."

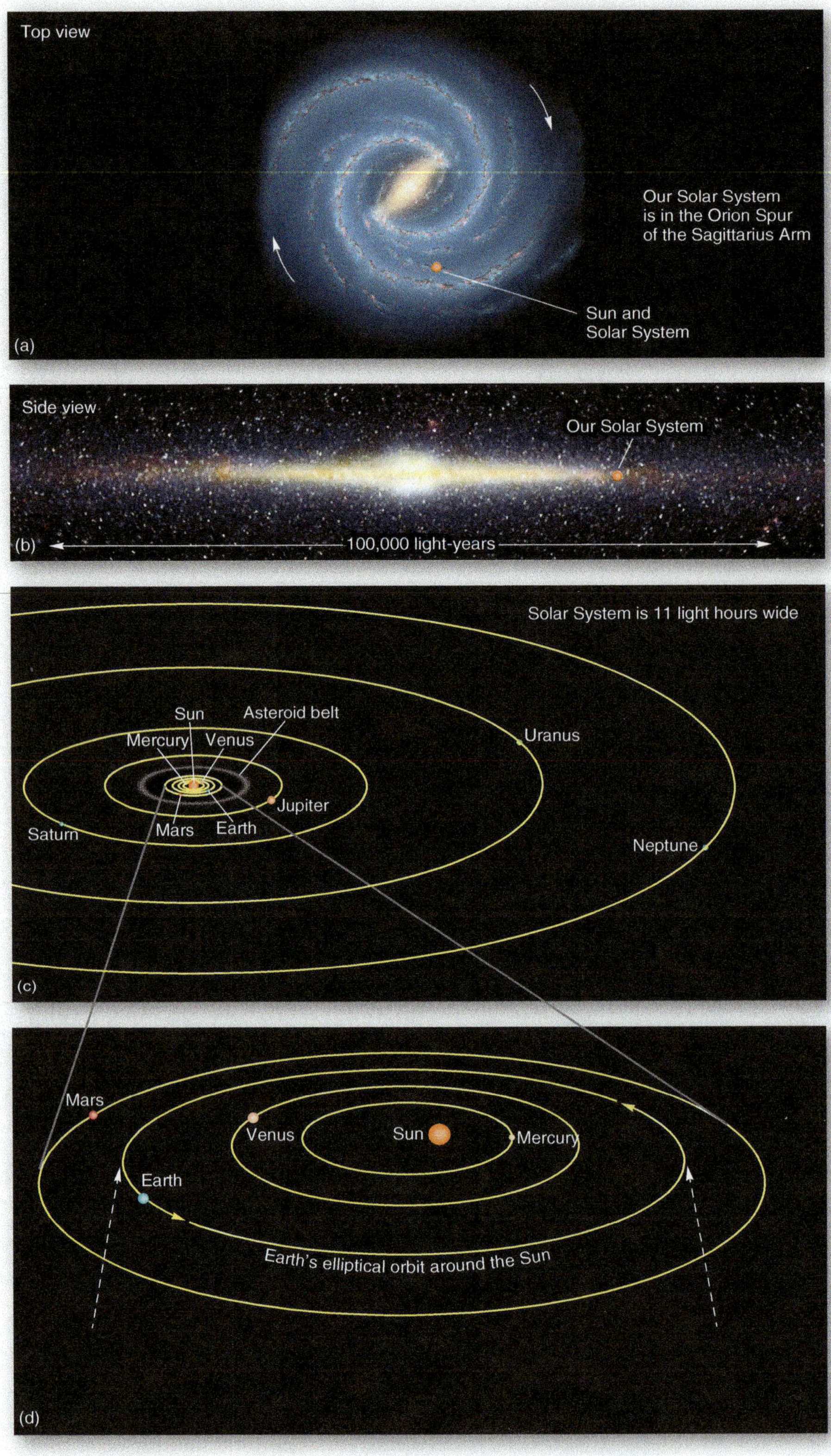

▲**Figure 2.1 Milky Way Galaxy, Solar System, and Earth's orbit.** (a) The Milky Way Galaxy viewed from above in an artist's conception, and (b) an image in cross-section side view. (c) All of the planets have orbits closely aligned to the plane of the ecliptic. Pluto, considered the ninth planet for over 70 years, was reclassified as a dwarf planet, part of the Kuiper asteroid belt, in 2006. (d) The four inner terrestrial planets and the structure of Earth's elliptical orbit, illustrating perihelion (closest) and aphelion (farthest) positions during the year. Have you ever observed the Milky Way Galaxy in the night sky? [(a) and (b) courtesy of NASA/JPL]

MG Animation **Nebular Hypothesis**

Solar System Formation

According to prevailing theory, our Solar System condensed from a large, slowly rotating and collapsing cloud of dust and gas, a *nebula*. **Gravity**, the mutual attraction exerted by every object upon all other objects in proportion to their mass, was the key force in this condensing solar nebula. As the nebular cloud organized and flattened into a disk shape, the early *protosun* grew in mass at the center, drawing more matter to it. Small eddies of accreting material swirled at varying distances from the center of the solar nebula; these were the *protoplanets*.

The **planetesimal hypothesis**, or *dust-cloud hypothesis*, explains how suns condense from nebular clouds. In this hypothesis, small grains of cosmic dust and other solids accrete to form planetesimals that may grow to become protoplanets and eventually planets; these formed in orbits about the developing Solar System's central mass.

Astronomers study this formation process in other parts of the Galaxy, where planets are observed orbiting distant stars. In fact, by 2013, astronomers had discovered more than 4400 candidate exoplanets orbiting other stars, nearly 1000 confirmed. Initial results from the orbiting Kepler telescope estimate the number of planets in the Milky Way at 50 billion, with some 500 million planets in habitable zones (with moderate temperatures and liquid water)—a staggering discovery.

Closer to home, in our Solar System, some 165 moons (planetary satellites) are in orbit about six of the eight planets. As of 2012, the new satellite count for the four outer planets, including several awaiting confirmation, was Jupiter, 67 moons; Saturn, 62 moons; Uranus, 27 moons; and Neptune, 13 moons.

Dimensions and Distances

The **speed of light** is 300,000 kmps (kilometers per second), or 186,000 mps (miles per second)*—in other words, about 9.5 trillion kilometers per year, or nearly 6 trillion miles per year. This is the tremendous distance captured by the term *light-year*, used as a unit of measurement for the vast Universe.

For spatial comparison, our Moon is an average distance of 384,400 km (238,866 mi) from Earth, or about 1.28 seconds in terms of light speed; for the *Apollo* astronauts this was a 3-day space voyage. Our entire Solar System is approximately 11 hours in diameter, measured by light speed (Figure 2.1c). In contrast, the Milky Way is about 100,000 light-years from side to side, and the known Universe that is observable from Earth stretches approximately 12 billion light-years in all directions. (See a Solar System simulator at http://space.jpl.nasa.gov/.)

Earth's average distance from the Sun is approximately 150 million km (93 million mi), which means that light reaches Earth from the Sun in an average of 8 minutes and 20 seconds. Earth's orbit around the Sun is presently elliptical—a closed, oval path (Figure 2.1d). At **perihelion**, which is Earth's closest position to the Sun, occurring on January 3 during the Northern Hemisphere winter, the Earth–Sun distance is 147,255,000 km (91,500,000 mi). At **aphelion**, which is Earth's farthest position from the Sun, occurring on July 4 during the Northern Hemisphere summer, the distance is 152,083,000 km (94,500,000 mi). This seasonal difference in distance from the Sun causes a slight variation in the solar energy incoming to Earth but is not an immediate reason for seasonal change.

The structure of Earth's orbit is not constant but instead changes over long periods. As shown in Chapter 10B, Figure 10B.16, Earth's distance from the Sun varies more than 17.7 million km (11 million mi) during a 100,000-year cycle, causing the perihelion and aphelion to be closer or farther at different times in the cycle.

*In more precise numbers, light speed is 299,792 kmps, or 186,282 mps.

Solar Energy: From Sun to Earth

Our Sun is unique to us but is a commonplace star in the Galaxy. It is average in temperature, size, and color when compared with other stars, yet it is the ultimate energy source for most life processes in our biosphere.

The Sun captured about 99.9% of the matter from the original solar nebula. The remaining 0.1% of the matter formed all the planets, their satellites, asteroids, comets, and debris. Consequently, the dominant object in our region of space is the Sun. In the entire Solar System, it is the only object having the enormous mass needed to sustain a nuclear reaction in its core and produce radiant energy.

The solar mass produces tremendous pressure and high temperatures deep in its dense interior. Under these conditions, the Sun's abundant hydrogen atoms are forced together and pairs of hydrogen nuclei are joined in the process of **fusion**. In the fusion reaction, hydrogen nuclei form helium, the second-lightest element in nature, and enormous quantities of energy are liberated—literally, disappearing solar mass becomes energy.

GEOreport 2.1 Sun and Solar System on the move

During the 4.6-billion-year existence of our Solar System, the Sun, Earth, and other planets have completed 27 orbital trips around the Milky Way Galaxy. When you combine this travel distance with Earth's orbital-revolution speed about the Sun of 107,280 kmph (66,660 mph) and Earth's equatorial rotation on its axis of 1675 kmph (1041 mph), you get an idea that "sitting still" is a relative term.

A sunny day can seem so peaceful, certainly unlike the violence taking place on the Sun. The Sun's principal outputs consist of the *solar wind* and of radiant energy spanning portions of the *electromagnetic spectrum.* Let us trace each of these emissions across space to Earth.

Solar Activity and Solar Wind

Telescopes and satellite images reveal solar activity to us in the form of sunspots and other surface disturbances. The *solar cycle* is the periodic variation in the Sun's activity and appearance over time. Since telescopes first allowed sunspot observation in the 1800s, scientists have used these solar surface features to define the solar cycle. Solar observation has recently improved significantly through data collected by satellites and spacecraft, including NASA's *SDO* (*Solar Dynamics Observatory*) and *SOHO* (*Solar and Heliospheric Observatory*) (Figure 2.2). (Find real-time *SOHO* images at http://sohowww.nascom.nasa.gov; information on all space weather is at http://spaceweather.com/.)

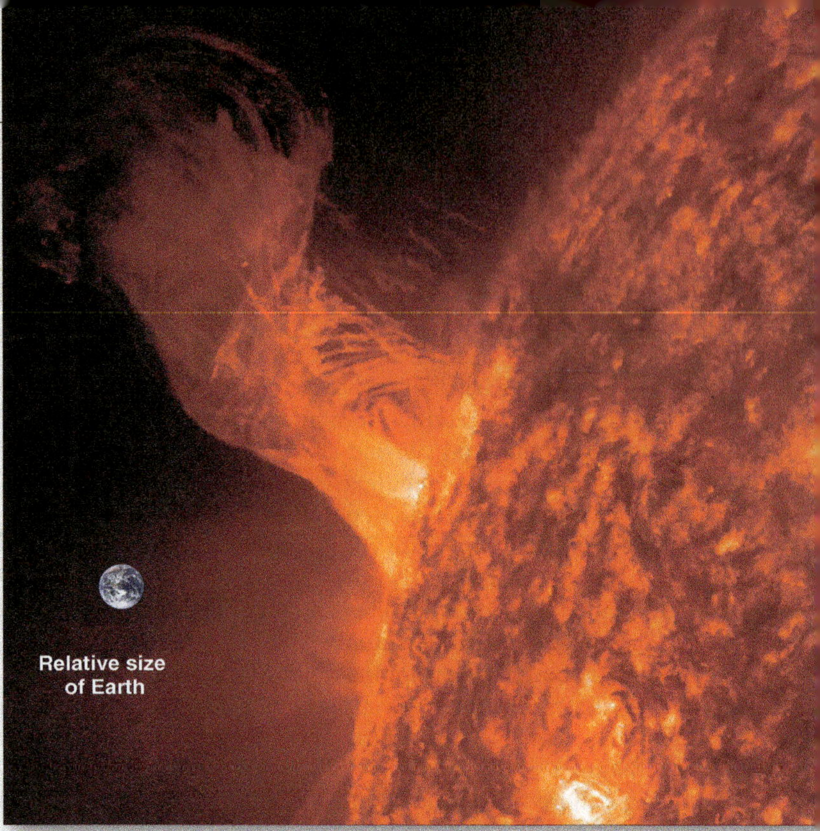

(a) Solar eruption, December 31, 2012

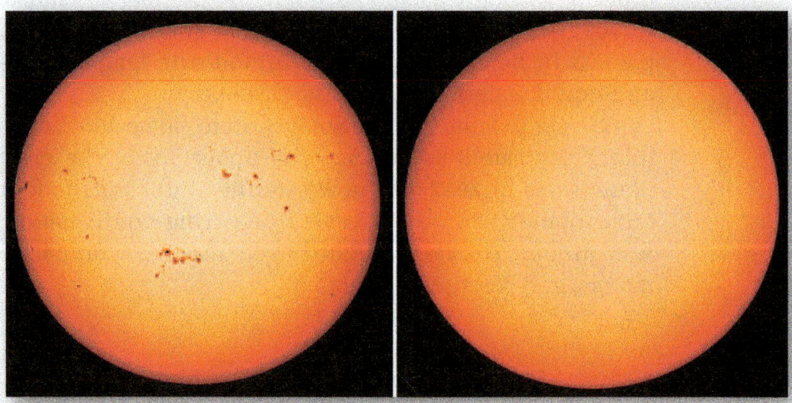
(b) Sunspot maximum in July 2000 and minimum in March 2009

▲**Figure 2.2 Image of the Sun and sunspots.** The prominence eruption rising into the Sun's corona was captured by NASA's *Solar Dynamics Observatory*. This relatively minor 2012 eruption was about 20 times the diameter of Earth, shown for scale. Earth is actually far smaller than the average sunspot. [(a) NASA/*SDO*/Steele Hill, 2012. (b) *SOHO*/EIT Consortium (NASA and ESA).]

Sunspots The Sun's most conspicuous features are large **sunspots**, surface disturbances caused by magnetic storms. Sunspots appear as dark areas on the solar surface, ranging in diameter from 10,000 to 50,000 km (6200 to 31,000 mi), with some as large as 160,000 km (100,000 mi), more than 12 times Earth's diameter.

A *solar minimum* is a period of years when few sunspots are visible; a *solar maximum* is a period during which sunspots are numerous. Over the last 300 years, sunspot occurrences have cycled fairly regularly, averaging 11 years from maximum to maximum (Figure 2.2b). A minimum in 2008 and a forecasted maximum in 2013 roughly maintain the average. (For more on the sunspot cycle, see http://solarscience.msfc.nasa.gov/SunspotCycle.shtml.) Scientists have ruled out solar cycles as a cause for increasing temperature trends on Earth over the past few decades.

Activity on the Sun is highest during solar maximum (Figure 2.2a). *Solar flares,* magnetic storms that cause surface explosions, and *prominence eruptions,* outbursts of gases arcing from the surface, often occur in active regions near sunspots (for videos and news about recent solar activity, go to http://www.nasa.gov/mission_pages/sunearth/news/). Although much of the material from these eruptions is pulled back toward the Sun by gravity, some moves into space as part of the solar wind.

Solar Wind Effects The Sun constantly emits clouds of electrically charged particles (principally, hydrogen nuclei and free electrons) that surge outward in all directions from the Sun's surface. This stream of energetic material travels more slowly than light—at about 50 million km (31 million mi) a day—taking approximately 3 days to

GEOreport 2.2 Recent solar cycles

In recent sunspot cycles, a solar minimum occurred in 1976 and a solar maximum in 1979, with more than 100 sunspots visible. Another minimum was reached in 1986, and an extremely active solar maximum followed in 1990–1991, with more than 200 sunspots visible at some time during the year. A sunspot minimum occurred in 1997, followed by an intense maximum of more than 200 in 2000–2001. The forecasted 2013 maximum is expected to have about 69 sunspots, the smallest maximum since 1906. The present solar cycle began in 2008 and carries the name *Cycle 24*.

▲**Figure 2.3 Astronaut and solar wind experiment.** Without a protective atmosphere, the lunar surface receives charged solar wind particles and all the Sun's electromagnetic radiation. Edwin "Buzz" Aldrin, one of three *Apollo XI* astronauts in 1969, deploys a sheet of foil in the solar wind experiment. Earth-bound scientists analyzed the foil upon the astronauts' return. Why wouldn't this experiment work if we set it up on Earth's surface? [NASA.]

reach Earth. This phenomenon is the **solar wind**, originating from the Sun's extremely hot solar corona, or outer atmosphere. The corona is the Sun's rim, observable with the naked eye from Earth during a solar eclipse.

As the charged particles of the solar wind approach Earth, they first interact with Earth's magnetic field. This **magnetosphere**, which surrounds the Earth and extends beyond Earth's atmosphere, is generated by dynamo-like motions within our planet. The magnetosphere deflects the solar wind toward both of Earth's poles so that only a small portion of it enters the upper atmosphere.

Because the solar wind does not reach Earth's surface, research on this phenomenon must be conducted in space. In 1969, the *Apollo XI* astronauts exposed a piece of foil on the lunar surface as a solar wind experiment (Figure 2.3). When examined back on Earth, the exposed foil exhibited particle impacts that confirmed the character of the solar wind.

In addition, massive outbursts of charged material, referred to as *coronal mass ejections* (CMEs), contribute to the flow of solar wind material from the Sun into space. CMEs that are aimed toward Earth often cause spectacular **auroras** in the upper atmosphere near the poles. These lighting effects, known as the *aurora borealis* (northern lights) and *aurora australis* (southern lights), occur 80–500 km (50–300 mi) above Earth's surface through the interaction of the solar wind with the upper layers of Earth's atmosphere. They appear as folded sheets of green, yellow, blue, and red light that undulate across the skies of high latitudes poleward of 65° (Figure 2.4; see http://www.swpc.noaa.gov/Aurora/ for tips on viewing auroras). In 2012, auroras were visible as far south as Colorado and Arkansas; these sightings may occur farther south with the 2013 solar maximum.

The solar wind disrupts certain radio broadcasts and some satellite transmissions and can cause overloads on Earth-based electrical systems. Astronauts working on the International Space Station in 2003 had to take shelter in the shielded Service Module during a particularly strong outburst. Research continues into possible links between solar activity and weather, with no conclusive results. Using satellites to harness the solar wind for power generation is another area of research; major challenges remain.

Electromagnetic Spectrum of Radiant Energy

The essential solar input to life is electromagnetic energy of various wavelengths, traveling at the speed of light to Earth. Solar radiation occupies a portion of the

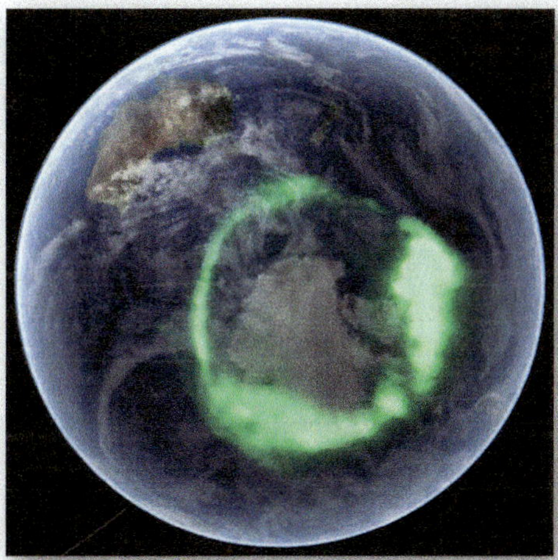

(a) *Aurora australis* as seen from orbit.

(b) *Aurora borealis* over Whitehorse, Yukon Canada, caused by solar activity. On August 31, 2012, a coronal mass ejection erupted from the Sun into space, traveling at over 900 miles per second. The CME glanced off Earth's magnetosphere, causing this aurora four days later.

▲**Figure 2.4 Auroras from orbital and ground perspectives.** [(a) Image spacecraft GSFC/NASA; (b) David Cartier, Sr., courtesy of GSFC/NASA.]

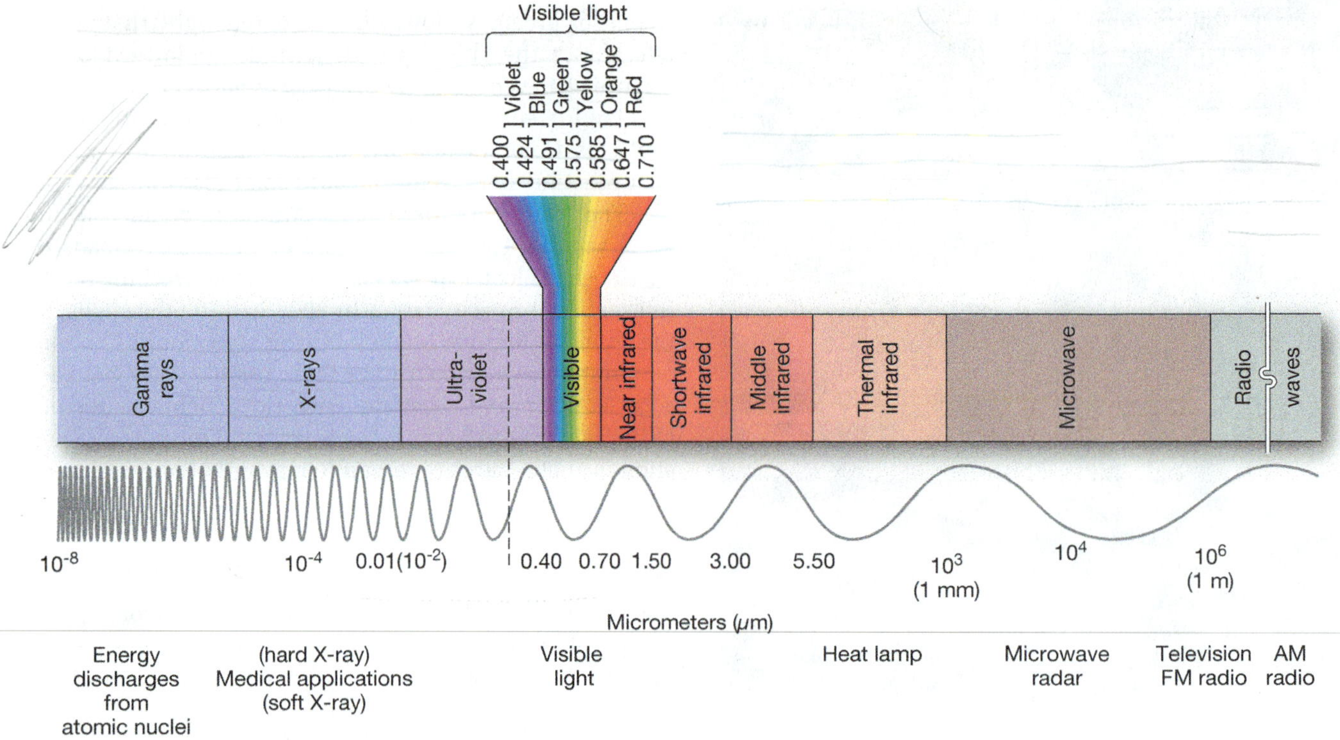

▲**Figure 2.5 A portion of the electromagnetic spectrum of radiant energy.** Wavelength and frequency are two ways of describing electromagnetic wave motion. Short wavelengths (at left) are higher in frequency; long wavelengths (at right) are lower in frequency.

electromagnetic spectrum, which is the spectrum of all possible wavelengths of electromagnetic energy. A **wavelength** is the distance between corresponding points on any two successive waves. The number of waves passing a fixed point in 1 second is the *frequency*. Note the wavelength plot below the chart in Figure 2.5.

The Sun emits radiant energy composed of 8% ultraviolet, X-ray, and gamma-ray wavelengths; 47% visible light wavelengths; and 45% infrared wavelengths. Figure 2.5 shows a portion of the electromagnetic spectrum, with wavelengths increasing from the left to right side of the illustration. Note the wavelengths at which various phenomena and human applications of energy occur.

An important physical law states that all objects radiate energy in wavelengths related to their individual surface temperatures: The hotter the object, the shorter the wavelengths emitted. This law holds true for the Sun and Earth. Figure 2.6 shows that the hot Sun radiates shorter wavelength energy, concentrated around 0.5 μm (micrometer).

The Sun's surface temperature is about 6000 K (6273°C, 11,459°F), and its emission curve is similar to that predicted for an idealized 6000-K surface, or *blackbody radiator* (shown in Figure 2.6).* A blackbody is a perfect absorber of radiant energy; it absorbs and subsequently emits all the radiant energy that it receives. A

*The Kelvin scale for measuring temperature starts at absolute zero temperature, or 0 K, so that subsequent readings are proportional to the actual kinetic energy in the material. On this scale, the melting point for ice is 273 K; the boiling point for water is 373 K.

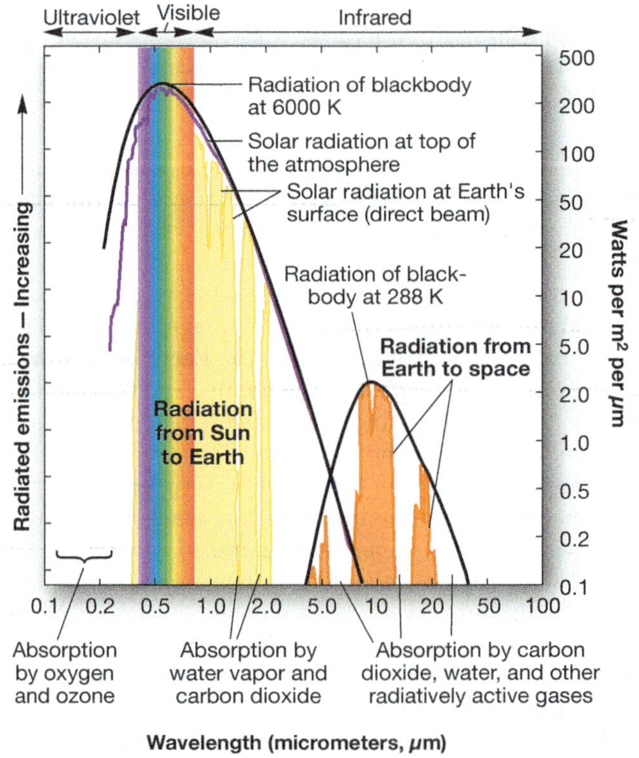

▲**Figure 2.6 Solar and terrestrial energy distribution by wavelength.** A hotter Sun radiates shorter wavelengths, whereas a cooler Earth emits longer wavelengths. Dark lines represent ideal blackbody curves for the Sun and Earth. The dropouts in the plot lines for solar and terrestrial radiation represent absorption bands of water vapor, water, carbon dioxide, oxygen, ozone (O_3), and other gases. [Adapted from W. D. Sellers, *Physical Climatology* (Chicago: University of Chicago Press), p. 20. Used by permission.]

MG Animation Electromagnetic Spectrum and Plants

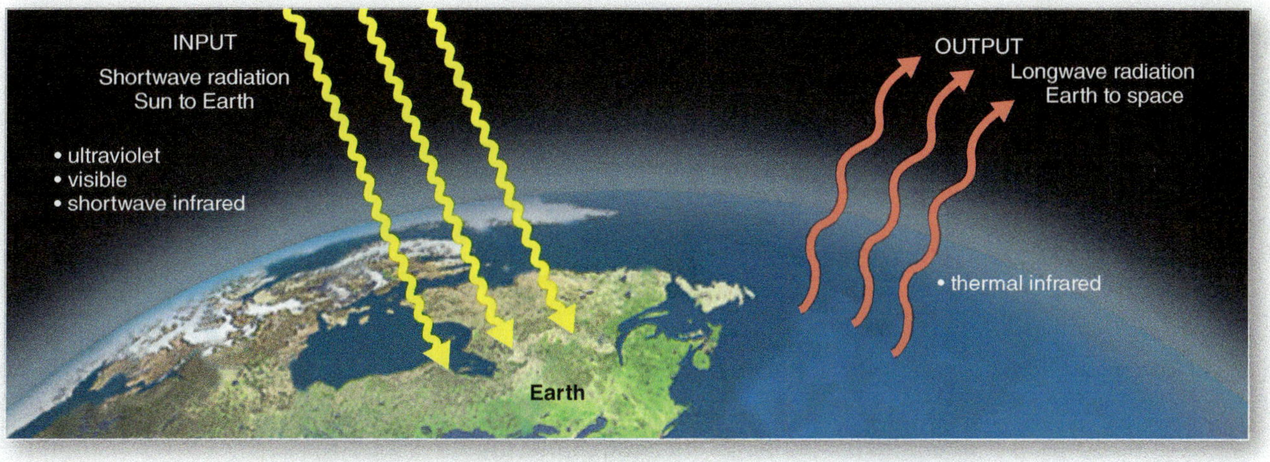

▲Figure 2.7 **Earth's energy budget simplified.**

hotter object like the Sun emits a much greater amount of energy per unit area of its surface than does a similar area of a cooler object like Earth. Shorter wavelength emissions are dominant at these higher temperatures.

Although cooler than the Sun, Earth also acts as a blackbody, radiating nearly all that it absorbs (also shown in Figure 2.6). Because Earth is a cooler radiating body, it emits longer wavelengths, mostly in the infrared portion of the spectrum, centered around 10.0 μm. Atmospheric gases, such as carbon dioxide and water vapor, vary in their response to radiation received, being transparent to some while absorbing others.

Figure 2.7 illustrates the flows of energy into and out of Earth systems. To summarize, the Sun's radiated energy is *shortwave radiation* that peaks in the short visible wavelengths, whereas Earth's radiated energy is *longwave radiation* concentrated in infrared wavelengths. In Chapter 4, we see that Earth, clouds, sky, ground, and all things that are terrestrial radiate longer wavelengths in contrast to the Sun, thus maintaining the overall energy budget of the Earth and atmosphere.

Incoming Energy at the Top of the Atmosphere

The region at the top of the atmosphere, approximately 480 km (300 mi) above Earth's surface, is the **thermopause** (see Figure 3.1). It is the outer boundary of Earth's energy system and provides a useful point at which to assess the arriving solar radiation before it is diminished by scattering and absorption in passage through the atmosphere.

Earth's distance from the Sun results in its interception of only one two-billionth of the Sun's total energy output. Nevertheless, this tiny fraction of energy from the Sun is an enormous amount of energy flowing into Earth's systems. Solar radiation that is intercepted by Earth is **insolation**, derived from the words *in*coming *sol*ar radi*ation*. Insolation specifically applies to radiation arriving at Earth's atmosphere and surface; it is measured as the rate of radiation delivery to a horizontal surface, specifically, as watts per square meter (W/m^2).

Solar Constant Knowing the amount of insolation incoming to Earth is important to climatologists and other scientists. The **solar constant** is the average insolation received at the thermopause when Earth is at its average distance from the Sun, a value of 1372 W/m^2.* As we follow insolation through the atmosphere to Earth's surface (Chapters 3 and 4), we see that its amount is reduced by half or more through reflection, scattering, and absorption of shortwave radiation.

Uneven Distribution of Insolation Earth's curved surface presents a continually varying angle to the incoming parallel rays of insolation (Figure 2.8). Differences in the angle at which solar rays meet the surface at each latitude result in an uneven distribution of insolation and heating. The only point where insolation arrives perpendicular to the surface (hitting it from directly overhead) is the **subsolar point**.

During the year, this point occurs only at lower latitudes, between the tropics (about 23.5° N and 23.5° S), and as a result, the energy received there is more concentrated. All other places, away from the subsolar point, receive insolation at an angle less than 90° and thus experience more diffuse energy; this effect becomes more pronounced at higher latitudes. Remember the photo in this chapter's *Geosystems Now*, Figure GN 2.2; the shadows are cast not at an angle but directly below the boys hauling water—this was May 1 at 14.8° N latitude.

The thermopause above the equatorial region receives 2.5 times more insolation annually than the thermopause above the poles. Of lesser importance is the fact that, because they meet the Earth from a lower

*A *watt* is equal to 1 joule (a unit of energy) per second and is the standard unit of power in the International System of Units (SI). (See the conversion tables in Appendix C of this text for more information on measurement conversions.) In nonmetric *calorie* heat units, the solar constant is expressed as approximately 2 calories per square centimeter per minute, or 2 *langleys* per minute (a langley being 1 cal/cm^2). A calorie is the amount of energy required to raise the temperature of 1 gram of water (at 15°C) 1 degree Celsius and is equal to 4.184 joules.

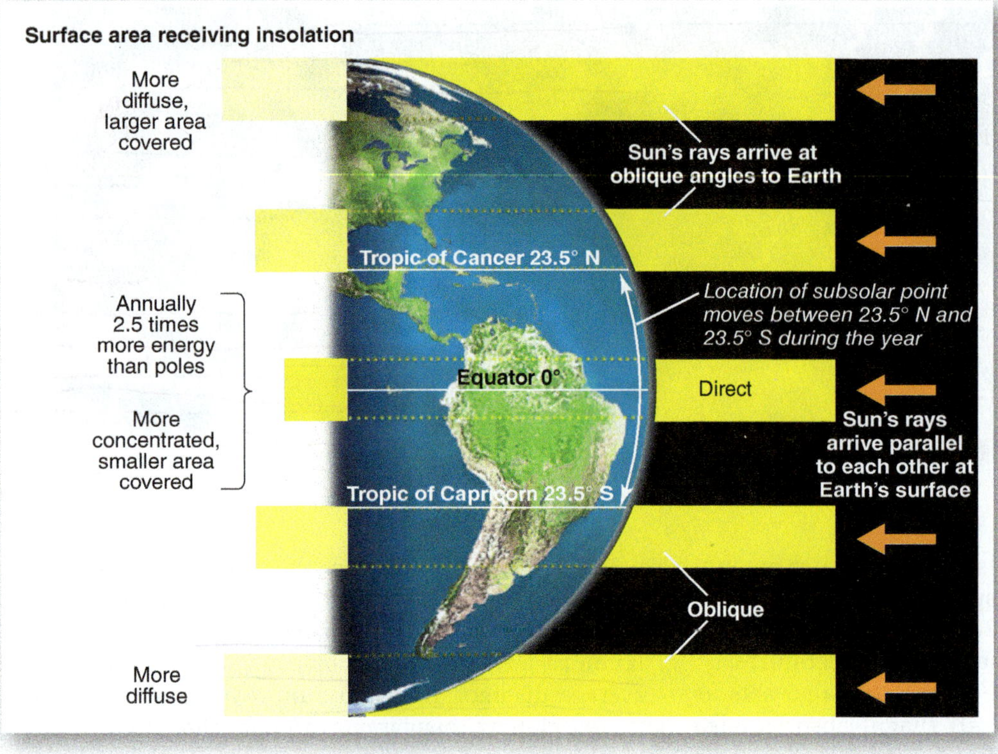

◀**Figure 2.8 Insolation receipts and Earth's curved surface.** The angle at which insolation arrives from the Sun determines the concentration of energy receipts by latitude. The subsolar point, where the Sun's rays arrive perpendicular to Earth, moves between the tropics during the year.

angle, the solar rays arriving toward the poles must pass through a greater thickness of atmosphere, resulting in greater losses of energy due to scattering, absorption, and reflection.

Figure 2.9 illustrates the daily variations throughout the year of energy at the top of the atmosphere for four locations in watts per square meter (W/m^2). The graphs show the seasonal changes in insolation from the equatorial regions northward and southward to the poles. In June, the North Pole receives slightly more than 500 W/m^2 per day, which is more than is ever received at 40° N latitude or at the equator. Such high values result from long 24-hour daylengths at the poles in summer, compared with only 15 hours of daylight at 40° N latitude and 12 hours at the equator. However, at the poles the summertime Sun at noon is low in the sky, so a daylength twice that of the equator yields only about a 100 W/m^2 difference.

In December, the pattern reverses, as shown on the graphs. Note that the top of the atmosphere at the South Pole receives even more insolation than the North Pole does in June (more than 550 W/m^2). This is a function of Earth's closer location to the Sun at perihelion (January 3 in Figure 2.1d). Along the equator, two slight maximum-insolation periods of approximately 430 W/m^2 occur at the spring and fall equinoxes, when the subsolar point is at the equator.

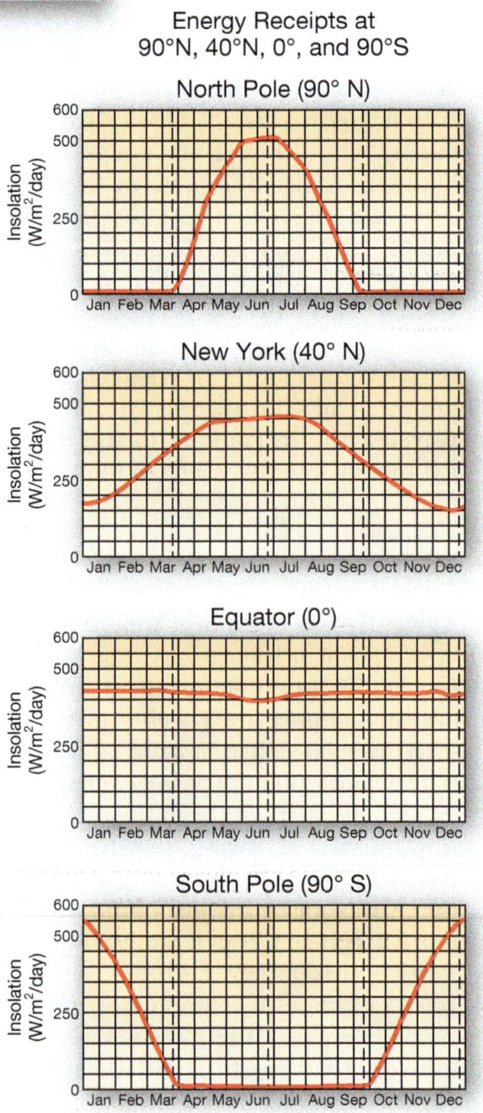

▶**Figure 2.9 Daily insolation received at the top of the atmosphere.** The total daily insolation received at the top of the atmosphere is charted in watts per square meter per day for four locations (1 W/m^2/day = 2.064 cal/cm^2/day). The vertical dashed lines mark the equinoxes and solstices, two of each during the year.

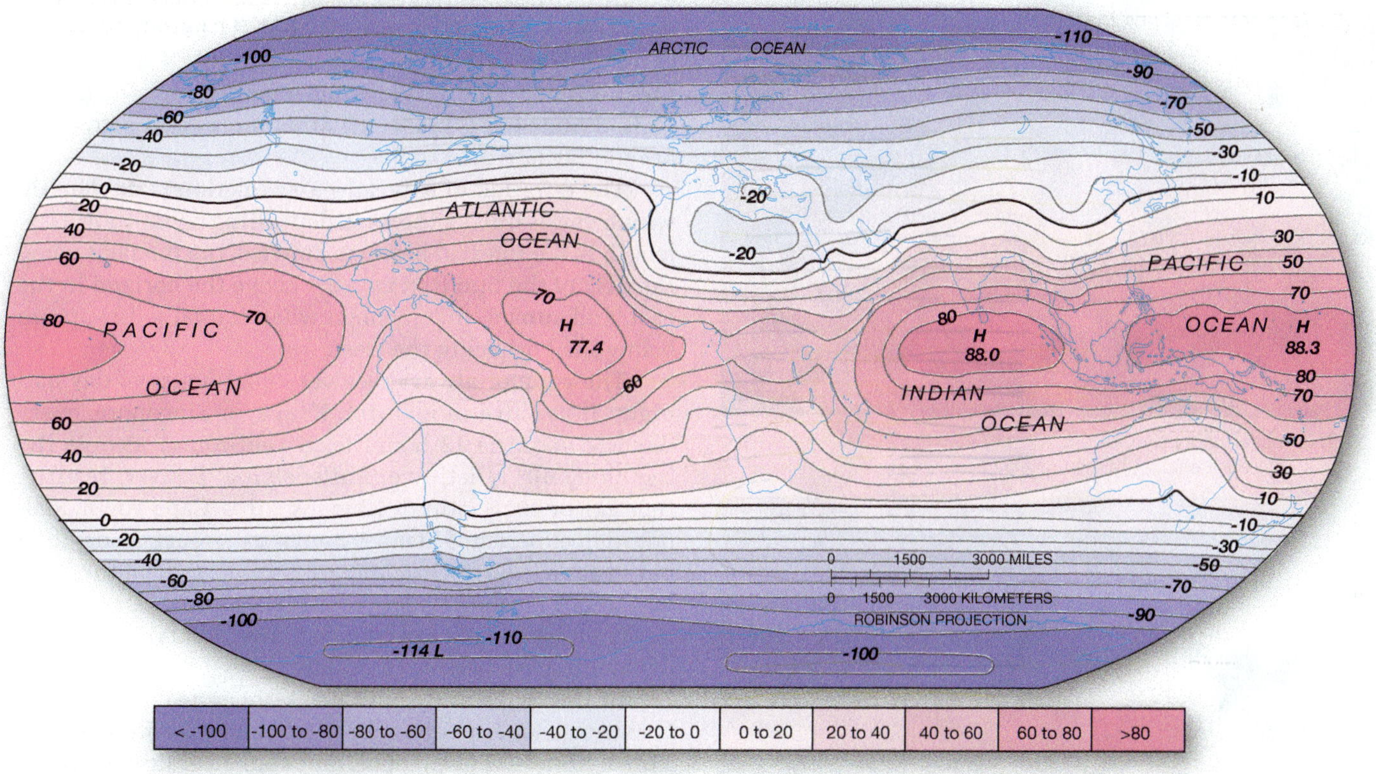

▲Figure 2.10 **Daily net radiation patterns at the top of the atmosphere.** Averaged daily net radiation flows measured at the top of the atmosphere by the Earth Radiation Budget Experiment (ERBE). Units are W/m². [Data for map courtesy of GSFC/NASA.]

Global Net Radiation Figure 2.10 shows patterns of *net radiation*, which is the balance between incoming shortwave energy from the Sun and all outgoing radiation from Earth and the atmosphere—energy inputs minus energy outputs. The map uses *isolines*, or lines connecting points of equal value, to show radiation patterns. Following the line for 70 W/m² on the map shows that the highest positive net radiation is in equatorial regions, especially over oceans.

Note the latitudinal energy imbalance in net radiation on the map—positive values in lower latitudes and negative values toward the poles. In middle and high latitudes, poleward of approximately 36° N and S latitudes, net radiation is negative. This occurs in these higher latitudes because Earth's climate system loses more energy to space than it gains from the Sun, as measured at the top of the atmosphere. In the lower atmosphere, these polar energy deficits are offset by flows of energy from tropical energy surpluses (as we see in Chapters 4 and 6). The largest net radiation values, averaging 80 W/m², are above the tropical oceans along a narrow equatorial zone. Net radiation minimums are lowest over Antarctica.

Of interest is the –20 W/m² area over the Sahara region of North Africa. Here, typically clear skies—which permit great longwave radiation losses from Earth's surface—and light-colored reflective surfaces work together to reduce net radiation values at the thermopause. In other regions, clouds and atmospheric pollution in the lower atmosphere also affect net radiation patterns at the top of the atmosphere by reflecting more shortwave energy to space.

This latitudinal imbalance in energy (discussed in Chapter 4) is critical because it drives global circulation in the atmosphere and the oceans. Think of the atmosphere and ocean forming a giant heat engine, driven by differences in energy from place to place that cause major circulations within the lower atmosphere and in the ocean. These circulations include global winds, ocean currents, and weather systems—subjects to follow in Chapters 6 and 8. As you go about your daily activities, let these dynamic natural systems remind you of the constant flow of solar energy through the environment.

Having examined the flow of solar energy to the top of Earth's atmosphere, let us now look at how seasonal changes affect the distribution of insolation as Earth orbits the Sun during the year.

The Seasons

Earth's periodic rhythms of warmth and cold, dawn and daylight, twilight and night, have fascinated humans for centuries. In fact, many ancient societies demonstrated an intense awareness of seasonal change and formally commemorated these natural energy rhythms with festivals, monuments, ground markings, and calendars (Figure 2.11). Such seasonal monuments and calendar markings are found worldwide, including thousands of

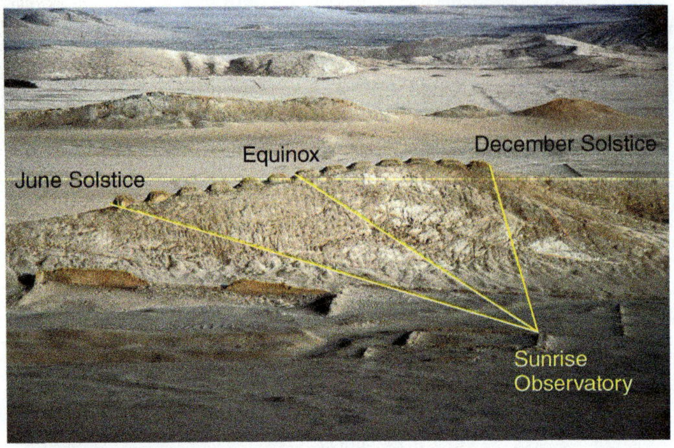

(a) Solar observatory and towers

(b) Sunrise at first tower, June solstice

▲**Figure 2.11 Solar observatory at Chankillo, Peru.** The Thirteen Towers are part of the Chankillo temple complex built in coastal Peru over 2000 years ago, the oldest known solar observatory in the Americas. Sunrise aligns with certain towers at different dates during the year. Research and preservation of the monument are ongoing; see http://www.wmf.org/project/chankillo. [(a) Ivan Ghezzi/Reuters.]

sites in North America, demonstrating an ancient awareness of seasons and astronomical relations. Many seasonal rituals and practices persist in this modern era.

Seasonality

Seasonality refers both to the seasonal variation of the Sun's position above the horizon and to changing daylengths during the year. Seasonal variations are a response to changes in the Sun's **altitude**, or the angle between the horizon and the Sun. At sunrise or sunset, the Sun is at the horizon, so its altitude is 0°. If during the day, the Sun reaches halfway between the horizon and directly overhead, it is at 45° altitude. If the Sun reaches the point directly overhead, it is at 90° altitude.

The Sun is found directly overhead (90° altitude, or *zenith*) only at the subsolar point, where insolation is at a maximum, as demonstrated in *Geosystems Now* and the chapter-opening photo. At all other surface points, the Sun is at a lower altitude angle, producing more diffuse insolation.

The Sun's **declination** is the latitude of the subsolar point. Declination annually migrates through 47° of latitude, moving between the Tropic of Cancer and Tropic of Capricorn latitudes. Although it passes through Hawai'i, which is between 19° N and 22° N, the subsolar point does not reach the continental United States or Canada; all other states and provinces are too far north.

The duration of exposure to insolation is **daylength**, which varies during the year, depending on latitude. Daylength is the interval between **sunrise**, the moment when the disk of the Sun first appears above the horizon in the east, and **sunset**, that moment when it totally disappears below the horizon in the west.

The equator always receives equal hours of day and night: If you live in Ecuador, Kenya, or Singapore, every day and night is 12 hours long, year-round. People living along 40° N latitude (Philadelphia, Denver, Madrid, Beijing), or 40° S latitude (Buenos Aires, Cape Town, Melbourne), experience about 6 hours' difference in daylight between winter (9 hours) and summer (15 hours). At 50° N or S latitude (Winnipeg, Paris, Falkland, or Malvinas Islands), people experience almost 8 hours of annual daylength variation.

At the North and South poles, the range of daylength is extreme, with a 6-month period of no insolation, beginning with weeks of twilight, then darkness, then weeks of pre-dawn. Following sunrise, daylight lasts for a 6-month period of continuous 24-hour insolation—literally, the poles experience one long day and one long night each year!

CRITICALthinking 2.1
A Way to Calculate Sunrise and Sunset

For a useful sunrise and sunset calculator for any location, go to http://www.srrb.noaa.gov/highlights/sunrise/sunrise.html, select a city near you or select "Enter lat/long" and enter your coordinates, enter the difference ("offset") between your time and UTC and whether you are on daylight saving time, and enter the date you are checking. Then click "Calculate" to see the solar declination and times for sunrise and sunset. Give this a try. Make a note of your finding, then revisit this site over the course of a full year and see the change that occurs where you live. ●

Reasons for Seasons

Seasons result from variations in the Sun's *altitude* above the horizon, the Sun's *declination* (latitude of the subsolar point), and *daylength* during the year. These in turn are created by several physical factors that operate in concert: Earth's *revolution* in orbit around the Sun, its daily *rotation* on its axis, its *tilted* axis, the unchanging *orientation of its axis*, and its *sphericity* (summarized in Table 2.2, p. 51). Of course, the essential ingredient is having a single source of radiant energy—the Sun. We now look at each of these factors individually. As we do, please note the distinction between revolution—Earth's travel around the Sun—and rotation—Earth's spinning on its axis (Figure 2.12).

TABLE 2.1 Speed of Rotation at Selected Latitudes

	Speed		
Latitude	kmph	(mph)	Representative Geographic Location near Each Latitude
90°	0	(0)	North Pole
60°	838	(521)	Seward, Alaska; Oslo, Norway; Saint Petersburg, Russia
50°	1078	(670)	Chibougamau, Québec; Kyyîv (Kiev), Ukraine
40°	1284	(798)	Columbus, Ohio; Beijing, China; Valdivia, Chile
30°	1452	(902)	New Orleans, Louisiana; Pôrto Alegre, Brazil
0°	1675	(1041)	Pontianak, Indonesia; Quito, Ecuador

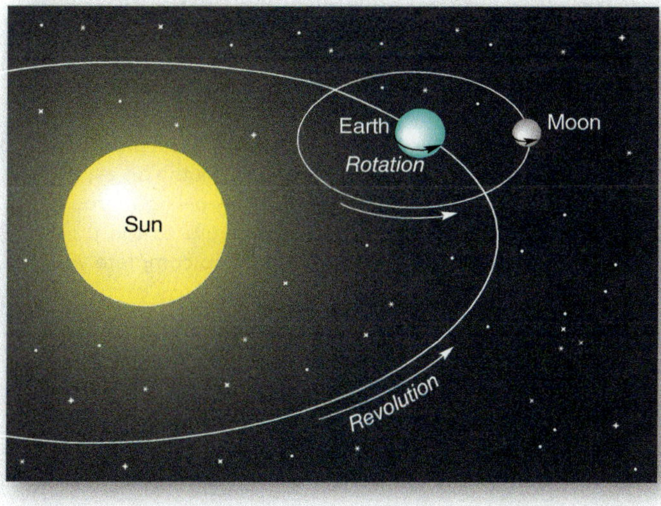

▲**Figure 2.12 Earth's revolution and rotation.** Earth's revolution about the Sun and rotation on its axis, viewed from above Earth's orbit. Note the Moon's rotation on its axis and revolution are counterclockwise, as well.

Revolution Earth's orbital **revolution** about the Sun is shown in Figure 2.1d and 2.12. Earth's speed in orbit averages 107,280 kmph (66,660 mph). This speed, together with Earth's distance from the Sun, determines the time required for one revolution around the Sun and, therefore, the length of the year and duration of the seasons. Earth completes its annual revolution in 365.2422 days. This number is based on a *tropical year*, measured from equinox to equinox, or the elapsed time between two crossings of the equator by the Sun.

The Earth-to-Sun distance from aphelion to perihelion might seem a seasonal factor, but it is not significant. It varies about 3% (4.8 million km, or 3 million mi) during the year, amounting to a 50-W/m^2 difference between summers at the different poles. Remember that the Earth–Sun distance averages 150 million km (93 million mi).

Rotation Earth's **rotation**, or turning on its axis, is a complex motion that averages slightly less than 24 hours in duration. Rotation determines daylength, creates the apparent deflection of winds and ocean currents, and produces the twice-daily rise and fall of the ocean tides in relation to the gravitational pull of the Sun and the Moon.

When viewed from above the North Pole, Earth rotates counterclockwise about its **axis**, an imaginary line extending through the planet from the geographic North Pole to the South Pole. Viewed from above the equator, Earth rotates west to east, or eastward. This eastward rotation creates the Sun's *apparent* westward daily journey from sunrise in the east to sunset in the west. Of course, the Sun actually remains in a fixed position in the center of our Solar System.

Although every point on Earth takes the same 24 hours to complete one rotation, the linear velocity of rotation at any point on Earth's surface varies dramatically with latitude. The equator is 40,075 km (24,902 mi) long; therefore, rotational velocity at the equator must be approximately 1675 kmph (1041 mph) to cover that distance in one day. At 60° latitude, a parallel is only half the length of the equator, or 20,038 km (12,451 mi) long, so the rotational velocity there is 838 kmph (521 mph). At the poles, the velocity is 0. This variation in rotational velocity establishes the effect of the Coriolis force, discussed in Chapter 6. Table 2.1 lists the speed of rotation for several selected latitudes.

Earth's rotation produces the diurnal (daily) pattern of day and night. The dividing line between day and night is the **circle of illumination** (as illustrated in *Geosystems in Action*). Because this day–night dividing circle of illumination intersects the equator (and because both are great circles, and any two great circles on a sphere bisect one another), *daylength at the equator is always evenly divided*—12 hours of day and

GEOreport 2.3 Why do we always see the same side of the Moon?

Note in Figure 2.12 that the Moon both revolves around Earth and rotates on its axis in a counterclockwise direction when viewed from above Earth's North Pole. It does both these motions in the same amount of time. The Moon's speed in orbit varies slightly during the month, whereas its rotation speed is constant, so we see about 59% of the lunar surface during the month, or exactly 50% at any one moment—always the same side facing Earth.

12 hours of night. All other latitudes experience uneven daylength through the seasons, except for 2 days a year, on the equinoxes.

The length of a true day varies slightly from 24 hours throughout the year. However, by international agreement a day is defined as exactly 24 hours, or 86,400 seconds, an average called *mean solar time*. Since Earth's rotation is gradually slowing, partially owing to the drag of lunar tidal forces, a "day" on Earth today is many hours longer than it was 4 billion years ago.

Tilt of Earth's Axis To understand Earth's **axial tilt**, imagine a plane (a flat surface) that intersects Earth's elliptical orbit about the Sun, with half of the Sun and Earth above the plane and half below. Such a plane, touching all points of Earth's orbit, is the **plane of the ecliptic**. Earth's tilted axis remains fixed relative to this plane as Earth revolves around the Sun. The plane of the ecliptic is important to our discussion of Earth's seasons. Now, imagine a perpendicular (at a 90° angle) line passing through the plane. From this perpendicular, Earth's axis is tilted about 23.5°. It forms a 66.5° angle from the plane itself (Figure 2.13). The axis through Earth's two poles points just slightly off Polaris, which is named, appropriately, the *North Star*.

The tilt angle was described above as "about" 23.5° because Earth's axial tilt changes over a complex 41,000-year cycle (see Figure 10B.16). The axial tilt ranges roughly between 22° and 24.5° from a perpendicular to the plane of the ecliptic. The present tilt is 23.45°. For convenience, this is rounded off to a 23.5° tilt (or 66.5° from the plane) in most usage. Scientific evidence shows that the angle of tilt is currently lessening in its 41,000-year cycle.

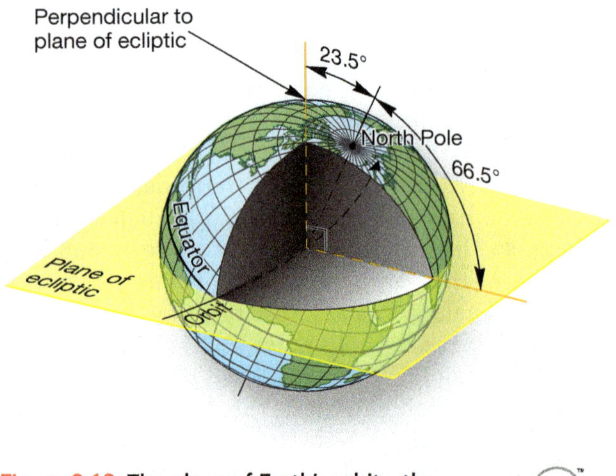

▲**Figure 2.13 The plane of Earth's orbit—the ecliptic—and Earth's axial tilt.** Note on the illustration that the plane of the equator is inclined to the plane of the ecliptic at about 23.5°.

MG™ Animation Earth–Sun Rotations, Seasons

CRITICALthinking 2.2 Astronomical Factors Vary over Long Time Frames

The variability of Earth's axial tilt and orbit about the Sun, as well as a wobble to the axis, are described in Figure 10B.16. Please refer to this figure and compare these changing conditions to the information in Table 2.2 and the related figures in this chapter.

What do you think the effect on Earth's seasons would be if the tilt of the axis was decreased? Or if the tilt was increased a little? Or if Earth was lying on its side? You can take a ball or piece of round fruit, mark the poles, and then move it around a light bulb as if revolving it around the Sun. Note where the light falls relative to the poles with no tilt and then with a 90° tilt to help complete your analysis. Now, what if Earth's orbit was more circular as opposed to its present elliptical shape? Earth's elliptical orbit actually does vary throughout a 100,000-year cycle. (Check the answer at the end of the Key Learning Concept Review.) •

Axial Parallelism Throughout our annual journey around the Sun, Earth's axis *maintains the same alignment* relative to the plane of the ecliptic and to Polaris and other stars. You can see this consistent alignment in *Geosystems in Action*, Figure GIA 2.2. If we compared the axis in different months, it would always appear parallel to itself, a condition known as **axial parallelism**.

Sphericity Even though Earth is not a perfect sphere, as discussed in Chapter 1, we can still refer to Earth's *sphericity* as contributing to seasonality. Earth's approximately spherical shape causes the parallel rays of the Sun to fall at uneven angles on Earth's surface. As we saw in Figures 2.8, 2.9, and 2.10, Earth's curvature means that insolation angles and net radiation received vary between the equator and the poles.

All five reasons for seasons are summarized in Table 2.2: revolution, rotation, tilt, axial parallelism, and sphericity. Now, considering all these factors operating together, we explore the march of the seasons.

GEOreport 2.4 Measuring Earth's rotation

A slight "wobble" along Earth's rotating axis causes it to migrate irregularly along a circular path with a radius up to about 9 m (30 ft). The accuracy of modern navigation systems such as GPS relies on measuring this "wobble." Scientists at the International Earth Rotation Service track Earth's rotation indirectly by monitoring fixed objects in space using radio telescopes (see http://www.iers.org/). In 2011, a research group made the first accurate, direct measurements of Earth's annual rotation using two counter-rotating lasers stored deep underground. The next goal is to precisely measure changes in Earth's rotation over a single day.

TABLE 2.2 Five Reasons for Seasons

Factor	Description
Revolution	Orbit around the Sun; requires 365.24 days to complete at 107,280 kmph (66,660 mph)
Rotation	Earth turning on its axis; takes approximately 24 hours to complete
Tilt	Alignment of axis at about 23.5° angle from perpendicular to the plane of the ecliptic (the plane of Earth's orbit)
Axial parallelism	Unchanging (fixed) axial alignment, with Polaris directly overhead at the North Pole throughout the year
Sphericity	Oblate spheroidal shape lit by Sun's parallel rays; the geoid

Annual March of the Seasons

During the march of the seasons on Earth, daylength is the most obvious way of sensing changes in season at latitudes away from the equator. The extremes of daylength occur in December and June. The times around December 21 and June 21 are *solstices*. Strictly speaking, the solstices are specific points in time at which the Sun's declination is at its position farthest north at the **Tropic of Cancer**, or south at the **Tropic of Capricorn**. "Tropic" is from *tropicus*, meaning a turn or change, so a tropic latitude is where the Sun's declination appears to stand still briefly (Sun stance, or *sol stice*) and then "turn" and head toward the other tropic.

During the year, places on Earth outside of the equatorial region experience a continuous but gradual shift in daylength, a few minutes each day, and the Sun's altitude increases or decreases a small amount. You may have noticed that these daily variations become more pronounced in spring and autumn, when the Sun's declination changes at a faster rate.

The *Geosystems In Action* (Figure GIA 2) illustration on the following pages summarizes the annual march of the seasons and Earth's relationship to the Sun during the year, using a side view (Figure GIA 2.1) and a top-view (Figure GIA 2.2). On December 21 or 22, at the moment of the **December solstice**, or Northern Hemisphere *winter solstice* ("winter sun stance"), the circle of illumination excludes the North Pole region from sunlight but includes the South Pole region. The subsolar point is about 23.5° S latitude, the Tropic of Capricorn parallel. The Northern Hemisphere is tilted away from these more direct rays of sunlight—our northern winter—thereby creating a lower angle for the incoming solar rays and thus a more diffuse pattern of insolation.

For locations between about 66.5° N and 90° N (the North Pole), the Sun remains below the horizon the entire day. The parallel at about 66.5° N marks the **Arctic Circle**; this is the southernmost parallel (in the Northern Hemisphere) that experiences a 24-hour period of darkness. During this period, twilight and dawn provide some lighting for more than a month at the beginning and end of the Arctic night.

During the following 3 months, daylength and solar angles gradually increase in the Northern Hemisphere as Earth completes one-fourth of its orbit. The moment of the **March equinox**, or *vernal equinox* in the northern hemisphere, occurs on March 20 or 21. At that time, the circle of illumination passes through both poles, so that all locations on Earth experience a 12-hour day and a 12-hour night. People living around 40° N latitude (New York, Denver) have gained 3 hours of daylight since the December solstice. At the North Pole, the Sun peeks above the horizon for the first time since the previous September; at the South Pole, the Sun is setting—a dramatic 3-day "moment" for the people working the Amundsen–Scott South Pole Station.

From March, the seasons move on to June 20 or 21 and the moment of the **June solstice**, or *summer solstice* in the Northern Hemisphere. The subsolar point migrates from the equator to 23.5° N latitude, the Tropic of Cancer. Because the circle of illumination now includes the North Polar region, everything north of the Arctic Circle receives 24 hours of daylight—the *Midnight Sun*. In contrast, the region from the **Antarctic Circle** to the South Pole (66.5°–90° S latitude) is in darkness. Those working in Antarctica call the June solstice *Midwinter's Day*.

September 22 or 23 is the time of the **September equinox**, or *autumnal equinox* in the Northern Hemisphere, when Earth's orientation is such that the circle of illumination again passes through both poles, so that all parts of the globe experience a 12-hour day and a 12-hour night. The subsolar point returns to the equator, with days growing shorter to the north and longer to the south. Researchers stationed at the South Pole see the disk of the Sun just rising, ending their 6 months of darkness. In the Northern Hemisphere, autumn arrives, a time of many colorful changes in the landscape, whereas in the Southern Hemisphere it is spring.

Dawn and Twilight

Dawn is the period of diffused light that occurs before sunrise. The corresponding evening time after sunset is *twilight*. During both periods, light is scattered by molecules of atmospheric gases and reflected by dust and moisture in the atmosphere. The duration of both is a function of latitude, because the angle of the Sun's path above the horizon determines the thickness of the atmosphere through which the Sun's rays must pass. The illumination may be enhanced by the presence of pollution aerosols and suspended particles from volcanic eruptions or forest and grassland fires.

(text continued on page 54)

At the equator, where the Sun's rays are almost directly above the horizon throughout the year, dawn and twilight are limited to 30–45 minutes each. These times increase to 1–2 hours each at 40° latitude, and at 60° latitude they each range upward from 2.5 hours, with little true night in summer. The poles experience about 7 weeks of dawn and 7 weeks of twilight, leaving only 2.5 months of "night" during the 6 months when the Sun is completely below the horizon.

Seasonal Observations In the midlatitudes of the Northern Hemisphere, the position of sunrise on the horizon migrates from day to day, from the southeast in December to the northeast in June. Over the same period, the point of sunset migrates from the southwest to the northwest. The Sun's altitude at local noon at 40° N latitude increases from a 26° angle above the horizon at the winter (December) solstice to a 73° angle above the horizon at the summer (June) solstice—a range of 47° (Figure GIA 2.3).

Seasonal change is quite noticeable across landscapes away from the equator. Think back over the past year. What seasonal changes have you observed in vegetation, temperatures, and weather? Recently, the timing of seasonal patterns in the biosphere is shifting with global climate change. In the middle and high latitudes, spring and leafing out are occurring as much as 3 weeks earlier than in previous human experience. Likewise, fall is happening later. Ecosystems are changing in response.

CRITICALthinking 2.3
Use the Analemma to Find the Subsolar Point

If you marked the location of the Sun in the sky at noon each day throughout the year, you would find that the Sun takes a figure-8 shaped path called an *analemma*. On the analemma chart in Figure CT 2.3.1, you can locate any date, then trace horizontally to the *y*-axis and find the Sun's declination, which is the latitude of the subsolar point. Along the Tropic of Capricorn, the subsolar point occurs on December 21–22, at the lower end of the analemma. Following the chart, you see that by March 20–21, the Sun's declination reaches the equator, and then moves on to the Tropic of Cancer in June. As an example, use the chart to calculate the subsolar point location on your birthday.

The shape of the analemma as the Sun's declination moves between the tropics is a result of Earth's axial tilt and elliptical orbit. As Earth revolves in its elliptical orbit around the Sun, it moves faster during December and January, and slower in June and July. This is reflected in the *equation of time* at the top of the chart.

An average day of 24 hours (86,400 seconds) is the basis for *mean solar time*, time measured by a clock (and introduced earlier in the chapter). However, *observed solar time* is the observed movement of the Sun crossing your meridian each day at noon. This sets the *apparent solar day*. You see on the chart that in October and November, *fast-Sun times* occur and the Sun arrives ahead of local noon (12:00), as noted on the *x-axis* along the top of the chart. In February and March the Sun arrives later than local noon, causing *slow-Sun times*. What was the equation of time on your birthday?

Search online for more information about the analemma (begin at http://www.analemma.com). Can you explain the figure-8 shape? ●

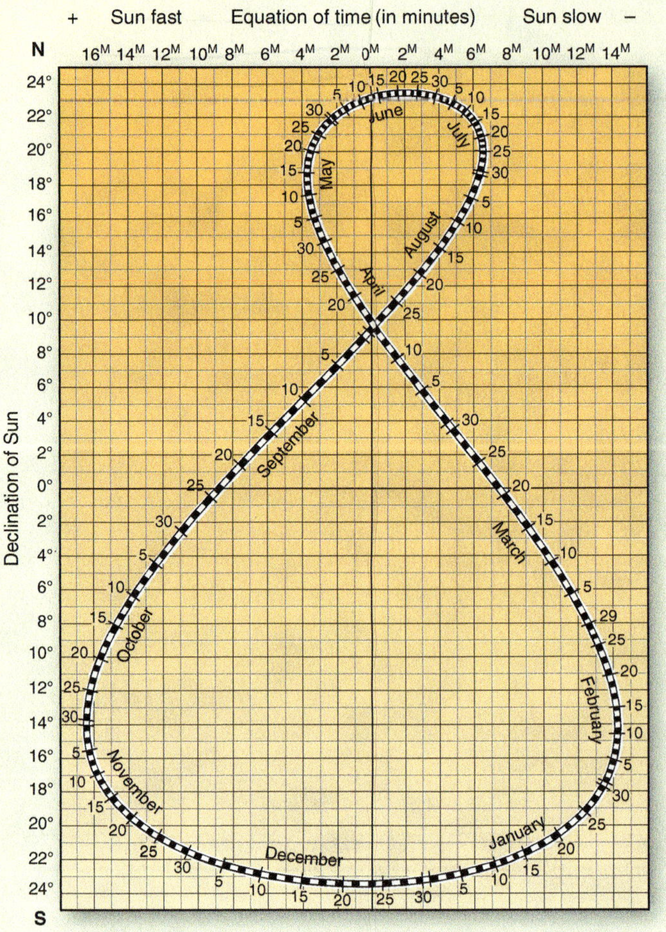

▲Figure CT 2.3.1 **The analemma chart.**

In Chapter 1, we emphasized the impacts of humans and their activities on Earth systems and processes. Humans affect, and are affected by, the systems comprising Earth's four spheres. The *Human Denominater* feature (Figure HD 2) illustrates important examples of human–Earth interactions, with summary text highlighting the direction of influence. For example, seasons affect humans by determining the rhythm of life for many societies. Humans are affecting seasonal patterns through the activities that cause climate change, and the related impacts on biological cycles. Throughout *Geosystems*, you will find similar illustrations reviewing human interactions with the systems and processes presented in each chapter.

THE**human**DENOMINATOR 2 Solar Energy and the Seasons

SOLAR ENERGY/SEASONS ⇨ HUMANS

- Solar energy drives Earth systems.
- Solar wind affects satellites, spacecraft, communication systems, and power grids on Earth.
- Seasonal change is the foundation of many human societies; it determines rhythm of life and food resources.

HUMANS ⇨ SOLAR ENERGY/SEASONS

- Climate change affects timing of the seasons. Changing temperature and rainfall patterns mean that spring is coming earlier and fall is starting later. Prolonged summer temperatures heat water bodies, promote early thawing and later refreezing of seasonal ice cover, alter animal migrations, shift vegetation patterns to higher latitudes—virtually all ecosystems on Earth are affected.

2a Solar energy drives Earth systems, including winds, ocean currents, weather, and living ecosystems. [NASA/JSC.]

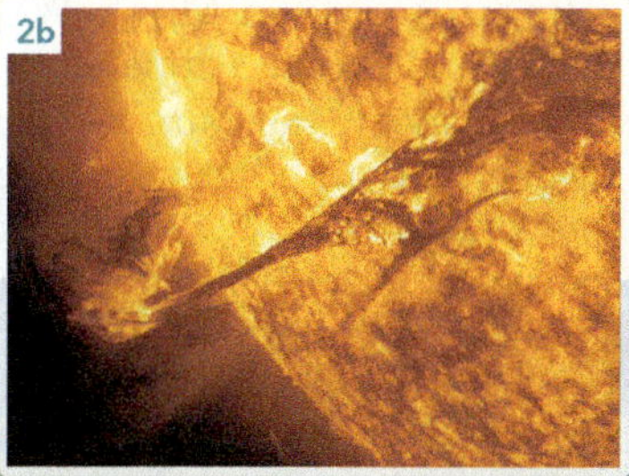

2b During coronal mass ejections (CMEs) and other solar storms, International Space Station astronauts take cover in sheltered areas of the spacecraft. [NASA/SDO.]

2c Seasonal shifts in subtropical high pressure in Africa are leading to decreased rainfall. Drought tolerant maize is one development that will help farmers adapt to climate change. [Philimon Bulawayao/Reuters]

2e As summers get longer in Alaska, moose migrations no longer coincide with the hunting seasons of native people, who depend on the meat. Shifting animal migrations and vegetation patterns will affect ecosystems across the globe. [Steve Bower/Shutterstock.]

2d Timing of last spring frost and first fall frost 1895–2011

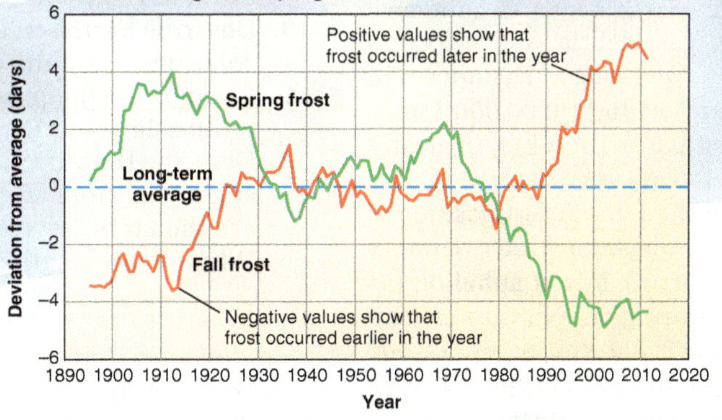

Data is for the contiguous United States. The overall trend is toward a longer growing season, with a longer fall and an earlier spring. [EPA, data courtesy K.E. Kunkel 2012; www.epa.gov/climatechange/indicators.]

ISSUES FOR THE 21ST CENTURY

- Ongoing climate change will continue to alter Earth systems. Societies will need to adapt their resource base as timing of seasonal patterns changes.

GEOSYSTEMS**connection**

In this chapter, we found our place in the Universe and in relation to the Milky Way Galaxy, Sun, other planets, and planetary satellites. We saw that, from the Sun, the solar wind and electromagnetic energy flow across space to Earth; we looked at the equator-to-pole distribution of this radiant energy at the top of the atmosphere and at how it changes in a seasonal rhythm.

Next we construct Earth's atmosphere and examine its composition, temperature, and functions. Electromagnetic energy cascades toward Earth's surface through the layers of the atmosphere, where harmful wavelengths are filtered out. Also, we examine human impacts on the atmosphere: the depletion of the ozone layer, acid deposition, and variable atmospheric components, including human air pollution.

KEY LEARNING concepts review

■ *Distinguish* between galaxies, stars, and planets, and *locate* Earth.

Our Solar System—the Sun and eight planets—is located on a remote, trailing edge of the **Milky Way Galaxy**, a flattened, disk-shaped collection estimated to contain up to 300 billion stars. **Gravity**, the mutual attracting force exerted by all objects upon all other objects in proportion to their mass, is an organizing force in the Universe. The **planetesimal hypothesis** describes the formation of solar systems as a process in which stars (like our Sun) condense from nebular dust and gas, with planetesimals and then protoplanets forming in orbits around these central masses.

Milky Way Galaxy (p. 40)
gravity (p. 41)
planetesimal hypothesis (p. 41)

1. Describe the Sun's status among stars in the Milky Way Galaxy. Describe the Sun's location, size, and relationship to its planets.
2. If you have seen the Milky Way at night, briefly describe it. Use specifics from the text in your description.
3. Compare the locations of the eight planets of the Solar System.

■ *Summarize* the origin, formation, and development of Earth, and *reconstruct* Earth's annual orbit about the Sun.

The Solar System, planets, and Earth began to condense from a nebular cloud of dust, gas, debris, and icy comets approximately 4.6 billion years ago. Distances in space are so vast that the **speed of light** (300,000 kmps, or 186,000 mps, which is about 9.5 trillion km, or nearly 6 trillion mi, per year) is used to express distance.

In its orbit, Earth is at **perihelion** (its closest position to the Sun) during our Northern Hemisphere winter (January 3 at 147,255,000 km, or 91,500,000 mi). It is at **aphelion** (its farthest position from the Sun) during our Northern Hemisphere summer (July 4 at 152,083,000 km, or 94,500,000 mi). Earth's average distance from the Sun is approximately 8 minutes and 20 seconds in terms of light speed.

speed of light (p. 41)
perihelion (p. 41)
aphelion (p. 41)

4. Briefly describe Earth's origin as part of the Solar System.
5. How far is Earth from the Sun in terms of light speed? In terms of kilometers and miles?
6. Briefly describe the relationship among these entities: Universe, Milky Way Galaxy, Solar System, Sun, Earth, and Moon.
7. Diagram in a simple sketch Earth's orbit about the Sun. How much does it vary during the course of a year?

■ *Describe* the Sun's operation, and *explain* the characteristics of the solar wind and the electromagnetic spectrum of radiant energy.

The **fusion** process—hydrogen nuclei forced together under tremendous temperature and pressure in the Sun's interior—generates incredible quantities of energy. **Sunspots** are magnetic disturbances on the solar surface; solar cycles are fairly regular, 11-year periods of sunspot activity. Solar energy in the form of charged particles of **solar wind** travels out in all directions from magnetic disturbances and solar storms. Solar wind is deflected by Earth's **magnetosphere**, producing various effects in the upper atmosphere, including spectacular **auroras**, the northern and southern lights, which surge across the skies at higher latitudes. Another effect of the solar wind in the atmosphere is its possible influence on weather.

Radiant energy travels outward from the Sun in all directions, representing a portion of the total **electromagnetic spectrum** made up of different energy wavelengths. A **wavelength** is the distance between corresponding points on any two successive waves. Eventually, some of this radiant energy reaches Earth's surface.

fusion (p. 41)
sunspots (p. 42)
solar wind (p. 42)
magnetosphere (p. 43)
auroras (p. 43)
electromagnetic spectrum (p. 43)
wavelength (p. 44)

8. How does the Sun produce such tremendous quantities of energy?
9. What is the sunspot cycle? At what stage was the cycle in the year 2013?
10. Describe Earth's magnetosphere and its effects on the solar wind and the electromagnetic spectrum.
11. Summarize the presently known effects of the solar wind relative to Earth's environment.
12. Describe the various segments of the electromagnetic spectrum, from shortest to longest wavelength. What are the main wavelengths produced by the Sun? Which wavelengths does Earth radiate to space?

■ *Illustrate* the interception of solar energy and its uneven distribution at the top of the atmosphere.

Electromagnetic radiation from the Sun passes through Earth's magnetic field to the top of the atmosphere—the **thermopause**, at approximately 500 km (300 mi) altitude. Incoming solar radiation is **insolation**, measured as energy delivered to a horizontal surface area over some unit of time. The **solar constant** is a general measure of insolation at the top of the atmosphere: The average insolation received at the thermopause when Earth is at its average distance from the Sun is approximately 1372 W/m^2 (2.0 $cal/cm^2/min$; 2 langleys/min). The place receiving maximum insolation is the **subsolar point**, where solar rays are perpendicular to the Earth's surface (radiating from directly overhead). All other locations away from the subsolar point receive slanting rays and more diffuse energy.

thermopause (p. 45)
insolation (p. 45)
solar constant (p. 45)
subsolar point (p. 45)

13. What is the solar constant? Why is it important to know?
14. Study the graph for New York at 40° N latitude on Figure 2.9. How do the trends of daily insolation throughout the year compare with those at the North and South Poles?
15. If Earth were flat and oriented at right angles to incoming solar radiation (insolation), what would be the latitudinal distribution of solar energy at the top of the atmosphere?

■ *Define* solar altitude, solar declination, and daylength, and *describe* the annual variability of each—Earth's seasonality.

The angle between the Sun and the horizon is the Sun's **altitude**. The Sun's **declination** is the latitude of the subsolar point. Declination annually migrates through 47° of latitude, moving between the Tropic of Cancer at about 23.5° N (June) and the Tropic of Capricorn at about 23.5° S latitude (December). Seasonality means an annual pattern of change in the Sun's altitude and changing **daylength**, or duration of exposure. Daylength is the interval between **sunrise**, the moment when the disk of the Sun first appears above the horizon in the east, and **sunset**, that moment when it totally disappears below the horizon in the west.

Earth's distinct seasons are produced by interactions of **revolution** (annual orbit about the Sun) and **rotation** (Earth's turning on its **axis**). As Earth rotates, the boundary that divides daylight and darkness is the **circle of illumination**. Other reasons for seasons include **axial tilt** (at about 23.5° from a perpendicular to the **plane of the ecliptic**), **axial parallelism** (the parallel alignment of the axis throughout the year), and *sphericity*.

Earth rotates about its axis, an imaginary line extending through the planet from the geographic North Pole to the South Pole. In the Solar System, an imaginary plane touching all points of Earth's orbit is the *plane of the ecliptic*. The **Tropic of Cancer** parallel marks the farthest north the subsolar point migrates during the year, about 23.5° N latitude. The **Tropic of Capricorn** parallel marks the farthest south the subsolar point migrates during the year, about 23.5° S latitude. Throughout the march of the seasons, Earth experiences the **December solstice**, **March equinox**, **June solstice**, and **September equinox** (illustrated in *Geosystems In Action*). At the moment of the December solstice, the area above the **Arctic Circle** at about 66.5° N latitude is in darkness for the entire day. At the June solstice, the area from the **Antarctic Circle** to the South Pole (66.5°–90° S latitude) experiences a 24-hour period of darkness.

altitude (p. 48)
declination (p. 48)
daylength (p. 48)
sunrise (p. 48)
sunset (p. 48)
revolution (p. 49)
rotation (p. 49)
axis (p. 49)
circle of illumination (p. 49)
axial tilt (p. 50)
plane of the ecliptic (p. 50)
axial parallelism (p. 50)
Tropic of Cancer (p. 51)
Tropic of Capricorn (p. 51)
December solstice (p. 51)
Arctic Circle (p. 51)
March equinox (p. 51)
June solstice (p. 51)
Antarctic Circle (p. 51)
September equinox (p. 51)

16. Assess the 12-month Gregorian calendar, with its months of different lengths, and leap years, and its relation to the annual seasonal rhythms—the march of the seasons. What do you find?
17. The concept of seasonality refers to what specific phenomena? How do these two aspects of seasonality change during a year at 0° latitude? At 40°? At 90°?
18. Differentiate between the Sun's altitude and its declination at Earth's surface.
19. For the latitude at which you live, how does daylength vary during the year? How does the Sun's altitude vary? Does your local newspaper publish a weather calendar containing such information?
20. List the five physical factors that operate together to produce seasons.
21. Describe Earth's revolution and rotation and differentiate between them.
22. Define Earth's present axial tilt—what is the angle? Does the axial tilt change as Earth orbits about the Sun?
23. Describe seasonal conditions at each of the four key seasonal anniversary dates during the year. What are the solstices and equinoxes, and what is the Sun's declination at these times?

Answer for Critical Thinking 2.2: Hypothetically, if Earth were tilted on its side, with its axis parallel to the plane of the ecliptic, we would experience a maximum variation in seasons worldwide. In contrast, if Earth's axis were perpendicular to the plane of its orbit—that is, with no tilt—we would experience no seasonal changes, with something like a perpetual spring or fall season, and all latitudes would experience 12-hour days and nights.

MasteringGeography™

Looking for additional review and test prep materials? Visit the Study Area in *MasteringGeography*™ to enhance your geographic literacy, spatial reasoning skills, and understanding of this chapter's content by accessing a variety of resources, including MapMaster interactive maps, geoscience animations, satellite loops, author notebooks, videos, RSS feeds, flashcards, web links, self-study quizzes, and an eText version of *Geosystems*.

3 Earth's Modern Atmosphere

As the Sun rises over Mono Lake in eastern California, rock spires are silhouetted by dawn's light. These limestone rock formations are known as "tufa towers," formed by the interaction of underwater springs with the saline lake water. The promise of each sunrise reminds us of the Sun's energy and the work of the atmosphere that sustains life—subjects of this chapter. [Bobbé Christopherson.]

KEY LEARNING concepts

After reading the chapter, you should be able to:

- ***Draw*** a diagram showing atmospheric structure based on three criteria for analysis—composition, temperature, and function.
- ***List*** and ***describe*** the components of the modern atmosphere, giving their relative percentage contributions by volume.
- ***Describe*** conditions within the stratosphere; specifically, ***review*** the function and status of the ozonosphere, or ozone layer.
- ***Distinguish*** between natural and anthropogenic pollutants in the lower atmosphere.
- ***Construct*** a simple diagram illustrating the pollution from photochemical reactions in motor vehicle exhaust, and ***describe*** the sources and effects of industrial smog.

Humans Help Define the Atmosphere

Astronaut Mark Lee, on a spacewalk from the Space Shuttle *Discovery* in 1994 (mission *STS-64*), was 241 km (150 mi) above Earth's surface, in orbit beyond the protective shield of the atmosphere (Figure GN 3.1). He was traveling at 28,165 kmph (17,500 mph), almost nine times faster than a high-speed rifle bullet, the vacuum of space all around him. Where the Sun hit his spacesuit, temperatures reached +120°C (+248°F); in the shadows, they dropped to –150°C (–238°F). Radiation and solar wind struck his pressure suit. To survive at such an altitude is an obvious challenge, one that relies on the ability of National Aeronautics and Space Administration (NASA) spacesuits to duplicate the Earth's atmosphere.

Protection in a Spacesuit For human survival, a spacesuit must block radiation and particle impacts, as does the atmosphere. It must also protect the wearer from thermal extremes.

Earth's oxygen–carbon dioxide processing systems must also be replicated in the suit, as must fluid-delivery and waste-management systems. The suit must maintain an internal air pressure against the space vacuum; for pure oxygen, this is 4.7 psi (32.4 kPa), which roughly equals the pressure that oxygen, water vapor, and CO_2 gases combined exert at sea level. All 18,000 parts of the modern spacesuit work to duplicate what the atmosphere does for us on a daily basis.

▲**Figure GN 3.1 Astronaut Mark Lee, untethered, on a working spacewalk in 1994.** [NASA.]

Kittinger's Record-Setting Jump In an earlier era, before orbital flights, scientists didn't know how a human could survive in space or how to produce an artificial atmosphere inside a spacesuit. In 1960, Air Force Captain Joseph Kittinger, Jr., stood at the opening of a small, unpressurized compartment, floating at 31.3 km (19.5 mi) altitude, dangling from a helium-filled balloon. The air pressure was barely measurable—this altitude is the beginning of space in experimental-aircraft testing.

Kittinger then leaped into the stratospheric void, at tremendous personal risk, for an experimental reentry into the atmosphere (Figure GN 3.2). He carried an instrument pack on his seat, his main chute, and pure oxygen for his breathing mask.

Initially frightened, he heard nothing, no rushing sound, for there was not enough air to produce any sound. The fabric of his pressure suit did not flutter, for there was not enough air to create friction against the cloth. His speed was remarkable, quickly accelerating to 988 kmph (614 mph)—nearly the speed of sound at sea level—owing to the lack of air resistance in the stratosphere.

When his free fall reached the stratosphere and its ozone layer, the frictional drag of denser atmospheric gases slowed his body. He then dropped into the lower atmosphere, finally falling below airplane flying altitudes.

Kittinger's free fall lasted 4 minutes and 25 seconds to the opening of his main chute at 5500 m (18,000 ft). The parachute lowered him safely to Earth's surface. This remarkable 13-minute, 35-second voyage through 99% of the atmospheric mass remained a record for 52 years.

▲**Figure GN 3.2 A remotely triggered camera captures a stratospheric leap into history.** [National Museum of the U.S. Air force.]

Baumgartner Breaks the Record On October 14, 2012, sponsored by Red Bull's Project Stratos, Felix Baumgartner ascended by helium balloon to 39.0 km (24.3 mi) altitude and then jumped (Figure GN 3.3). Guided by Colonel Kittinger's voice from mission control, Baumgartner survived an out-of-control spin early in his fall, reaching a top free-fall speed of 199.2 kmph (834 mph). Watched live online by millions around the globe, his fall lasted 4 minutes, 20 seconds; shorter than Kittinger's free fall by 5 seconds.

The experiences of these men illustrate the evolution of our understanding of upper-atmosphere survival. From events such as Kittinger's dangerous leap of discovery, the now routine spacewalks of astronauts such as Mark Lee, and Baumgartner's record-breaking descent, scientists have gained the ability to duplicate the atmosphere. This chapter describes our current knowledge—the source of design specifications for our spacesuits—as we explore Earth's atmosphere.

GEOSYSTEMS NOW ONLINE Go to Chapter 3 on the *Mastering Geography* website (www.masteringgeography.com) for resources and activities. Go to http://www.redbullstratos.com/ to watch the highlights of Baumgartner's record-breaking jump and learn about Captain Kittinger's involvement in the project. Do you think Baumgartner's recent feat makes Kittinger's accomplishment less important? MG

▼**Figure GN 3.3 Felix Baumgartner's 2012 jump sets new free-fall height and speed records** [Red Bull Stratos/AP Images.]

Earth's atmosphere is a unique reservoir of life-sustaining gases, the product of 4.6 billion years of development. Some of the gases are crucial components in biological processes; some protect us from hostile radiation and particles from the Sun and beyond. As shown in *Geosystems Now*, when humans venture away from the lower regions of the atmosphere, they must wear elaborate protective spacesuits that provide services the atmosphere performs for us all the time.

In this chapter: We examine the modern atmosphere using the criteria of composition, temperature, and function. Our look at the atmosphere also includes the spatial impacts of both natural and human-produced air pollution. We all interact with the atmosphere with each breath we take, the energy we consume, the traveling we do, and the products we buy. Human activities cause stratospheric ozone losses and the blight of acid deposition on ecosystems. These matters are essential to physical geography, for they are influencing the atmospheric composition of the future.

Atmospheric Composition, Temperature, and Function

The modern atmosphere probably is the fourth general atmosphere in Earth's history. A gaseous mixture of ancient origin, it is the sum of all the exhalations and inhalations of life interacting on Earth throughout time. The principal substance of this atmosphere is air, the medium of life as well as a major industrial and chemical raw material. Air is a simple mixture of gases that is naturally odorless, colorless, tasteless, and formless, blended so thoroughly that it behaves as if it were a single gas.

As a practical matter, we consider the top of our atmosphere to be around 480 km (300 mi) above Earth's surface, the same altitude we used in Chapter 2 for measuring the solar constant and insolation received. Beyond that altitude is the **exosphere**, which means "outer sphere," where the rarefied, less dense atmosphere is nearly a vacuum. It contains scarce lightweight hydrogen and helium atoms, weakly bound by gravity as far as 32,000 km (20,000 mi) from Earth.

Atmospheric Profile

Think of Earth's modern atmosphere as a thin envelope of imperfectly shaped concentric "shells" or "spheres" that grade into one another, all bound to the planet by gravity. To study the atmosphere, we view it in layers, each with distinctive properties and processes. Figure 3.1 charts the atmosphere in a vertical cross-section profile, or side view. Scientists use three atmospheric criteria—*composition*, *temperature*, and *function*—to define layers for distinct analytical purposes. These criteria are discussed just ahead, after a brief consideration of air pressure. (As you read the criteria discussions, note that they repeatedly follow the path of incoming solar radiation as it travels through the atmosphere to Earth's surface.)

Air pressure changes throughout the atmospheric profile. Air molecules create **air pressure** through their motion, size, and number, exerting a force on all surfaces they come in contact with. The pressure of the atmosphere (measured as force per unit area) pushes in on all of us. Fortunately, that same pressure also exists inside us, pushing outward; otherwise, we would be crushed by the mass of air around us.

Earth's atmosphere also presses downward under the pull of gravity and therefore has weight. Gravity compresses air, making it denser near Earth's surface (Figure 3.2). The atmosphere exerts an average force of approximately 1 kg/cm^2 (14.7 lb/in^2) at sea level. With increasing altitude, density and pressure decrease—this is the "thinning" of air that humans feel on high mountain summits as less oxygen is carried in each breath a person takes. This makes breathing more difficult at the top of Mount Everest, where air pressure is about 30% of that on Earth's surface. More information on air pressure and the role it plays in generating winds is in Chapter 6.

Over half the total mass of the atmosphere, compressed by gravity, lies below 5.5 km (3.14 mi, or 18,000 ft) altitude. Only 0.1% of the atmosphere remains above an altitude of 50 km (31 mi), as shown in the pressure profile in Figure 3.2b (percentage column is farthest to the right).

At sea level, the atmosphere exerts a pressure of 1013.2 mb (millibar, or mb; a measure of force per square meter of surface area) or 29.92 in. of mercury (symbol,

GEOreport 3.1 Earth's first atmospheres

The first atmosphere on Earth was probably formed from *outgassing*, or the release of gases trapped within Earth's interior. We still see outgassing today in the form of volcanic activity. This atmosphere was high in sulfuric gases, low in nitrogen, and devoid of oxygen. The second atmosphere formed when Earth cooled and water vapor condensed to form clouds and rain. Oceans formed, nitrogen increased, but oxygen was still not present in the atmosphere. As oceanic life evolved, bacteria began the process of photosynthesis, using the Sun's energy to convert atmospheric carbon dioxide to oxygen. Oxygen became significant in the atmosphere about 2.2 billion years ago, but it took another billion years before atmospheric oxygen levels were stable. Our modern atmosphere formed when oxygen molecules absorbed sunlight and formed ozone, the protective layer in the stratosphere that shields all life from ultraviolet radiation.

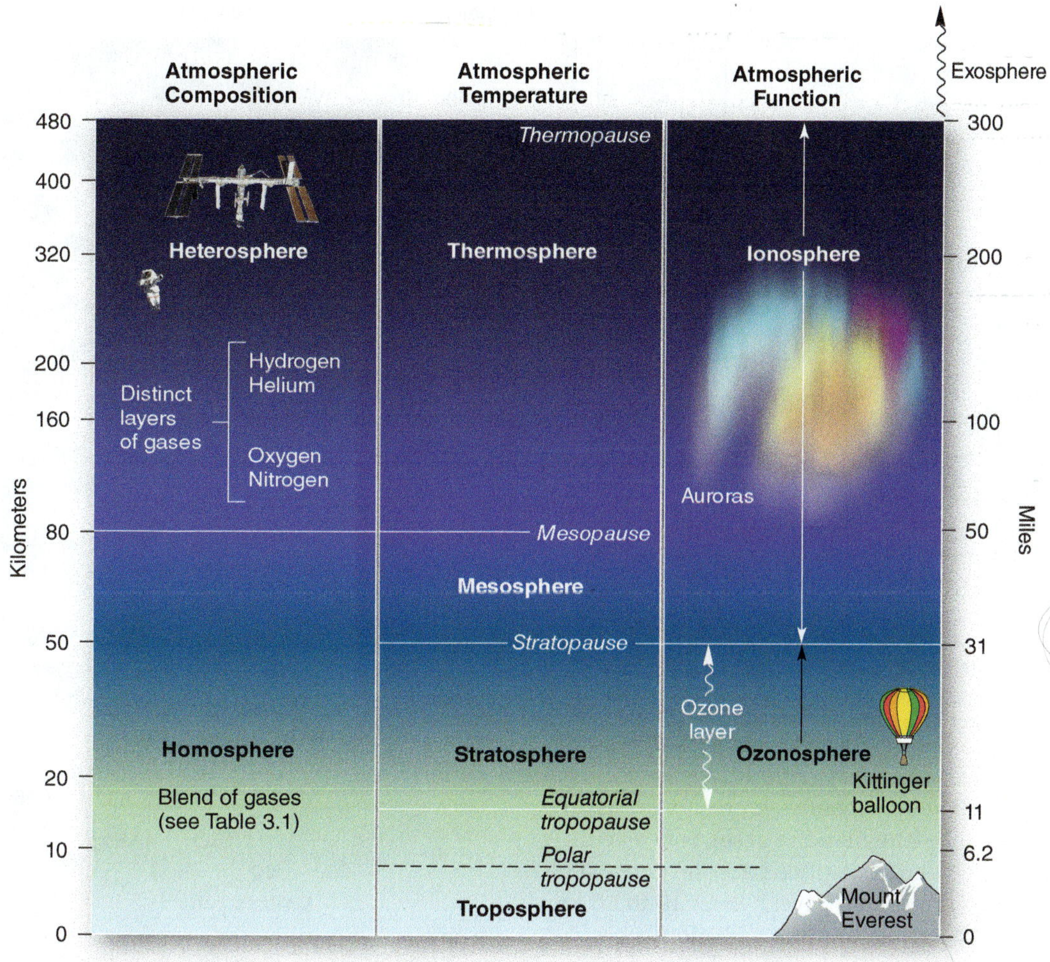

▲Figure 3.1 Profile of the modern atmosphere. Note the small astronaut and balloon showing the altitude achieved by astronaut Mark Lee and by Joseph Kittinger, discussed in *Geosystems Now.*

Hg), as measured by a barometer. In Canada and certain other countries, normal air pressure is expressed as 101.32 kPa (kilopascal; 1 kPa = 10 mb). (See Chapter 6 for further discussion.)

Atmospheric Composition Criterion

By the criterion of chemical *composition*, the atmosphere divides into two broad regions (Figure 3.1), the *heterosphere* (80 to 480 km altitude) and the *homosphere* (Earth's surface to 80 km altitude).

Heterosphere The **heterosphere** is the outer atmosphere in terms of composition. It begins at about 80 km (50 mi) altitude and extends outward to the exosphere and interplanetary space. Less than 0.001% of the atmosphere's mass is in this rarefied heterosphere. The International Space Station (ISS) orbits in the middle to upper heterosphere (note the ISS altitude in Figure 3.1).

As the prefix *hetero-* implies, this region is not uniform—its gases are not evenly mixed. Gases in the heterosphere occur in distinct layers sorted by gravity according to their atomic weight, with the lightest elements (hydrogen and helium) at the margins of outer space and the heavier elements (oxygen and nitrogen) dominant in the lower heterosphere. This distribution is quite different from the blended gases we breathe in the homosphere, near Earth's surface.

Homosphere Below the heterosphere is the **homosphere**, extending from an altitude of 80 km (50 mi) to Earth's surface. Even though the atmosphere rapidly

GEOreport 3.2 Outside the airplane

Next time you are in an airplane, think about the air pressure outside. Few people are aware that in routine air travel they are sitting above 80% of the total atmospheric volume and that the air pressure at that altitude is only about 10% of surface air pressure. Only 20% of the atmospheric mass is above you. If your plane is at about 11,000 m (36,090 ft), think of the men who jumped from stratospheric heights, described in *Geosystems Now;* Felix Baumgartner started another 28 km (17 mi) higher in altitude than your plane!

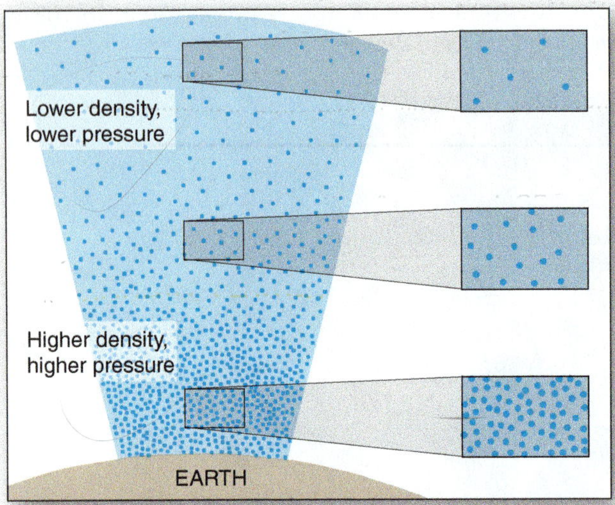

(a) Density is higher nearer Earth's surface, and decreases with altitude.

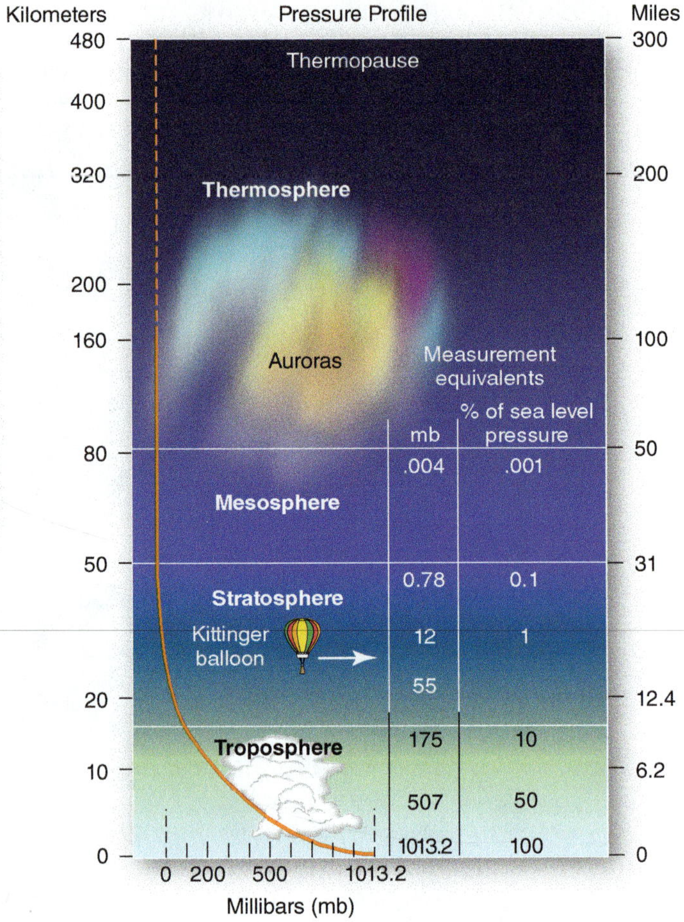

(b) Pressure profile plots the decrease in pressure with increasing altitude. Pressure is in millibars and as a percentage of sea level pressure. Note that the troposphere holds about 90% of the atmospheric mass (far-right % column).

▲**Figure 3.2 Density decreasing with altitude.** Have you experienced pressure changes that you could feel on your eardrums? How high above sea level were you at the time?

changes density in the homosphere—increasing pressure toward Earth's surface—the blend of gases is nearly uniform throughout. The only exceptions are the concentration of ozone (O_3) in the "ozone layer," from 19 to 50 km (12 to 31 mi) above sea level, and the variations in water vapor, pollutants, and some trace chemicals in the lowest portion of the atmosphere.

The present blend of gases evolved approximately 500 million years ago. Table 3.1 lists by volume the gases that constitute dry, clean air in the homosphere, divided into *constant* (showing almost no change throughout Earth's history) and *variable* (present in small but variable amounts). Air sampling occurs at the Mauna Loa Observatory, Hawai'i, operational since 1957. A marine inversion layer keeps volcanic emissions from nearby Kīlauea volcano and dust to a minimum.

The air of the homosphere is a vast reservoir of relatively inert *nitrogen*, originating principally from volcanic sources. A key element of life, nitrogen integrates into our bodies not from the air we breathe but through compounds in food. In the soil, nitrogen-fixing bacteria incorporate nitrogen from air into compounds that can be used by plants; later, the nitrogen returns to the atmosphere through the work of denitrifying bacteria that remove nitrogen from organic materials. A complete discussion of the nitrogen cycle is in Chapter 19.

Oxygen, a by-product of photosynthesis, also is essential for life processes. The percentage of atmospheric oxygen varies slightly over space with changes in photosynthetic rates of vegetation with latitude, season, and the lag time as atmospheric circulation slowly mixes the air. Although it makes up about one-fifth of the atmosphere, oxygen forms compounds that compose

TABLE 3.1 Composition of the Modern Homosphere

Gas (Symbol)	Percentage by Volume	Parts per Million (ppm)
Constant Gases		
Nitrogen (N_2)	78.084	780,840
Oxygen (O_2)	20.946	209,460
Argon (Ar)	0.934	9,340
Neon (Ne)	0.001818	18
Helium (He)	0.000525	5.2
Krypton (Kr)	0.00010	1.0
Xenon (Xe)	Trace	~0.1
Variable Gases (change over time and space)	0–4% (max at tropics, min at poles)	
Water vapor (H_2O)		
Carbon dioxide (CO_2)*	0.0399	399
Methane (CH_4)	0.00018	1.8
Hydrogen (H)	Trace	~0.6
Nitrous oxide (N_2O)	Trace	~0.3
Ozone (O_3)	Variable	

*May 2013 average CO_2 measured at Mauna Loa, Hawai'i (see ftp://ftp.cmdl.noaa.gov/ccg/co2/trends/co2_mm_mlo.txt).

about half of Earth's crust. Oxygen readily reacts with many elements to form these materials. Both nitrogen and oxygen reserves in the atmosphere are so extensive that, at present, they far exceed human capabilities to disrupt or deplete them.

The gas *argon*, constituting less than 1% of the homosphere, is completely inert (an unreactive "noble" gas) and unusable in life processes. All the argon present in the modern atmosphere comes from slow accumulation over millions of years. Because industry has found uses for inert argon (in lightbulbs, welding, and some lasers), it is extracted or "mined" from the atmosphere, along with nitrogen and oxygen, for commercial, medical, and industrial uses.

Of the variable atmospheric gases in the homosphere, we examine carbon dioxide in the next section, and we discuss ozone later in this chapter. Water vapor is in Chapter 7; methane is discussed in Chapter 10B. The homosphere also contains variable amounts of *particulates*, solids and liquid droplets that enter the air from natural and human sources. These particles, also known as aerosols, range in size from the relatively large liquid water droplets, salt, and pollen visible with the naked eye to relatively small, even microscopic, dust and soot. These particles affect the Earth's energy balance (see Chapter 4), as well as human health (discussed later in the chapter).

Carbon Dioxide *Carbon dioxide (CO_2)* is a natural by-product of life processes, a variable gas that is increasing rapidly. Although its present percentage in the atmosphere is small, CO_2 is important to global temperatures.

The study of past atmospheres trapped in samples of glacial ice reveals that the present levels of atmospheric CO_2 are higher than at any time in the past 800,000 years. Over the past 200 years, and especially since the 1950s, the CO_2 percentage increased as a result of human activities, principally the burning of fossil fuels and deforestation.

This increase in CO_2 appears to be accelerating (see graphs for atmospheric CO_2 concentrations in Chapter 10B). From 1990 to 1999, CO_2 emissions rose at an average of 1.1% per year; compare this to the average emissions increase since 2000 of 3.1% per year—a 2- to 3-ppm-per-year increase. Overall, atmospheric CO_2 increased 16% from 1992 to 2012. During May 2013, CO_2 levels reached 400 ppm. Today CO_2 far exceeds the natural range of 180 to 300 ppm over the last 800,000 years. A distinct climatic threshold is approaching at 450 ppm, forecasted for sometime in the decade of the 2020s. Beyond this tipping point, the warming associated with CO_2 increases is expected to bring irreversible ice-sheet and species losses. Chapters 4, 5, and 10B discuss the role of carbon dioxide as an important greenhouse gas and the implications of CO_2 increases for climate change.

Atmospheric Temperature Criterion

By the criterion of temperature, the atmospheric profile can be divided into four distinct zones—thermosphere, mesosphere, stratosphere, and troposphere (labeled in Figure 3.1). We begin with the zone that is highest in altitude.

Thermosphere The **thermosphere** ("heat sphere") roughly corresponds to the heterosphere (from 80 km out to 480 km, or 50–300 mi). The upper limit of the thermosphere is the **thermopause** (the suffix *-pause* means "to change"). During periods of a less active Sun, with fewer sunspots and eruptions from the solar surface, the thermopause may lower in altitude from the average 480 km (300 mi) to only 250 km (155 mi). During periods of a more active Sun, the outer atmosphere swells to an altitude of 550 km (340 mi), where it can create frictional drag on satellites in low orbit.

The temperature profile in Figure 3.3a (yellow curve) shows that temperatures rise sharply in the thermosphere, to 1200°C (2200°F) and higher. Despite such high temperatures, however, the thermosphere is not "hot" in the way you might expect. Temperature and heat are different concepts. The intense solar radiation in this portion of the atmosphere excites individual molecules (principally nitrogen and oxygen) to high levels of vibration. This **kinetic energy**, the energy of motion, is the vibrational energy that we measure as *temperature*. (Temperature is a measure of the average kinetic energy of individual molecules in matter, and is the focus of Chapter 5.)

In contrast, *heat* is created when kinetic energy is transferred between molecules, and thus between bodies or substances. (By definition, heat is the flow of kinetic energy from one body to another resulting from a temperature difference between them, and is discussed in Chapter 4.) Heat is therefore dependent on the density or mass of a substance; where little density or mass exists, the amount of heat will be small. The thermosphere is not "hot" in the

GEOreport 3.3 Human sources of atmospheric carbon dioxide

Carbon dioxide occurs naturally in the atmosphere, as part of the Earth's carbon cycle (a focus of Chapters 10B and 19). However, the recent acceleration in atmospheric CO_2 concentrations is from human sources, primarily fossil fuel combustion for energy and transportation. The main sources of U.S. CO_2 emissions are electric power plants (40%) and transportation (31%); of the sources that produce electricity, coal burning produces more CO_2 than oil or natural gas (see http://www.epa.gov/climatechange/ghgemissions/). China is now the leading emitter of CO_2, responsible for 28% of all global CO_2 emissions in 2011. However, the U.S. still leads in per capita (per person) CO_2 emissions. The use of coal to meet energy demands is increasing worldwide, especially in China and India. Some estimates predict that coal will be supplying 50% of the world's energy by 2035, an increase in 20% from 2012. Since CO_2, is linked to rising global temperatures, scientists think that such an increase could have irreversible impacts on climate.

way we are familiar with, because the density of molecules is so low there that little actual heat is produced. The thermosphere would actually feel cold to us because the number of molecules is not great enough to transfer heat to our skin. (Scientists measure temperature indirectly at those altitudes, using the low density, which is measured by the amount of drag on satellites.) Closer to Earth's surface, the atmosphere is denser. The greater number of molecules transmits their kinetic energy as *sensible heat*, meaning that we can measure and feel it. Further discussion of heat and temperature is in Chapters 4 and 5; discussion of heat as it relates to density is in Chapter 7.

Mesosphere The **mesosphere** is the area from 50 to 80 km (30 to 50 mi) above Earth and is within the homosphere. As Figure 3.3 shows, the mesosphere's outer boundary, the *mesopause*, is the coldest portion of the atmosphere, averaging −90°C (−130°F), although that temperature may vary considerably (by 25–30 C°, or 45–54 F°). Note in Figure 3.2b the extremely low pressures (low density of molecules) in the mesosphere.

The mesosphere sometimes receives cosmic or meteoric dust particles, which act as nuclei around which fine ice crystals form. At high latitudes, an observer at night may see these bands of ice crystals glow in rare and unusual **noctilucent clouds**, which are so high in altitude that they still catch sunlight after sunset. For reasons not clearly understood, these unique clouds are on the increase and may be seen in the midlatitudes. See, among several sites, http://lasp.colorado.edu/science/atmospheric/.

Figure 3.3 Temperature profile of the atmosphere, highlighting the troposphere. [NASA.]

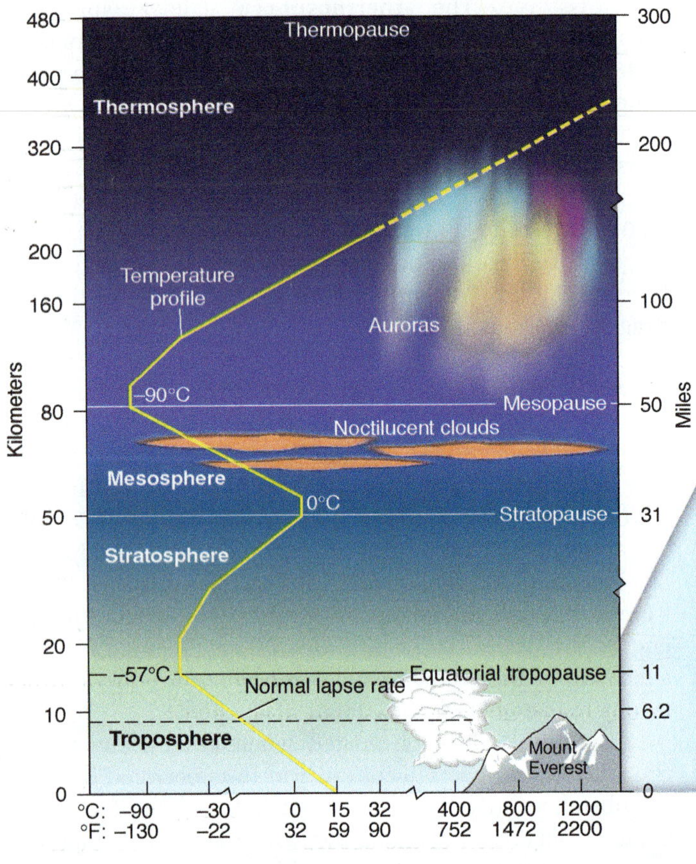

(a) Temperature profile plots temperature changes with altitude.

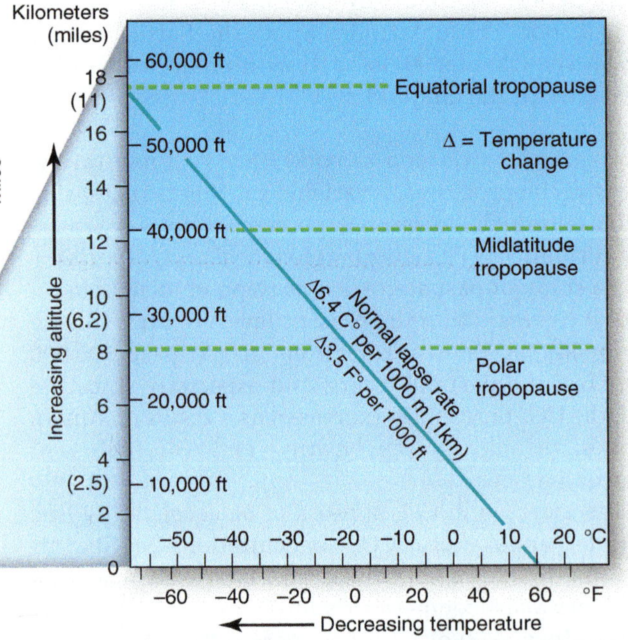

(b) Temperature decreases with increased altitude at the *normal lapse rate*.

(c) A sunset from orbit shows a silhouetted cumulonimbus thunderhead cloud topping out at the tropopause.

Stratosphere The **stratosphere** extends from about 18 to 50 km (11 to 31 mi) above Earth's surface. Temperatures increase with altitude throughout the stratosphere, from −57°C (−70°F) at 18 km to 0°C (32°F) at 50 km, the stratosphere's outer boundary, called the *stratopause*. The stratosphere is the location of the ozone layer. Stratospheric changes measured over the past 25 years show that chlorofluorocarbons are increasing and ozone concentrations decreasing, discussed in Focus Study 3.1 later in this chapter. Greenhouse gases are also on the increase; a noted stratospheric cooling is the response.

Troposphere The **troposphere** is the final layer encountered by incoming solar radiation as it surges through the atmosphere to the surface. This atmospheric layer supports life, the biosphere, and is the region of principal weather activity.

Approximately 90% of the total mass of the atmosphere and the bulk of all water vapor, clouds, and air pollution are within the troposphere. An average temperature of −57°C (−70°F) defines the **tropopause**, the troposphere's upper limit, but its exact altitude varies with the season, latitude, and surface temperatures and pressures. Near the equator, because of intense heating from the surface, the tropopause occurs at 18 km (11 mi); in the middle latitudes, it occurs at an average of 12 km (8 mi); and at the North and South Poles, it averages only 8 km (5 mi) or less above Earth's surface (Figure 3.3b). The marked warming with increasing altitude in the stratosphere above the tropopause causes the tropopause to act like a lid, generally preventing whatever is in the cooler (denser) air below from mixing into the warmer (less dense) stratosphere (Figure 3.3c).

Figure 3.3b illustrates the normal temperature profile within the troposphere during daytime. As the graph shows, temperatures decrease rapidly with increasing altitude at an average of 6.4 C° per km (3.5 F° per 1000 ft), a rate known as the **normal lapse rate**.

The normal lapse rate is an average. The actual lapse rate may vary considerably because of local weather conditions and is called the **environmental lapse rate**. This variation in temperature gradient in the lower troposphere is central to our discussion of weather processes in Chapters 7 and 8.

CRITICALthinking 3.1
Where is Your Tropopause?

On your next flight, as the plane reaches its highest altitude, see if you can learn what the temperature is outside the plane. (Some planes display it on video screens; or the flight attendant may have time to check it for you.) By definition, the tropopause is wherever the temperature −57°C (−70°F) occurs. Depending on the season of the year, altitude and temperature make an interesting comparison. Is the tropopause at a higher altitude in summer or winter?

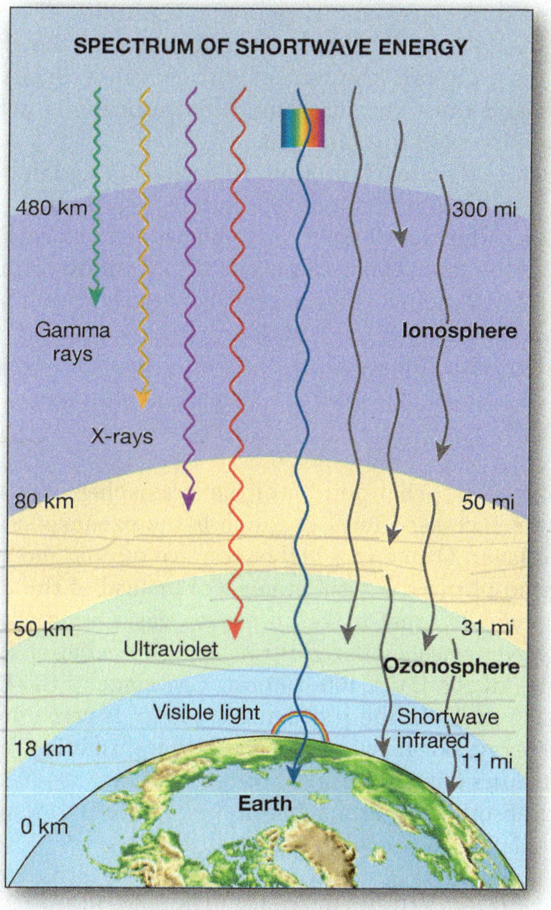

▲**Figure 3.4 Absorption of wavelengths above Earth's surface.** As shortwave solar energy passes through the atmosphere, the shortest wavelengths are absorbed. Only a fraction of the ultraviolet radiation, but most of the visible light and shortwave infrared, reaches Earth's surface.

MG Animation Ozone breakdown Ozone Hole

Atmospheric Function Criterion

According to our final atmospheric criterion of function, the atmosphere has two specific zones, the ionosphere and the ozonosphere (ozone layer), which together remove most of the harmful wavelengths of incoming solar radiation and charged particles. Figure 3.4 gives a general depiction of the absorption of radiation by these functional layers of the atmosphere.

Ionosphere The outer functional layer, the **ionosphere**, extends throughout the thermosphere and into the mesosphere below (Figure 3.1). The ionosphere absorbs cosmic rays, gamma rays, X-rays, and shorter wavelengths of ultraviolet radiation, changing atoms to positively charged ions and giving the ionosphere its name. The glowing auroral lights discussed in Chapter 2 occur principally within the ionosphere.

Distinct regions within the ionosphere, known as the D, E, F1, and F2 layers, are important for broadcast communications and GPS signals. These regions reflect certain radio wavelengths, including AM radio and other shortwave radio broadcasts, especially at night. Activity such as solar flares can trigger radio blackouts. This also affects airplanes that fly over the Arctic; over the North Pole, these aircraft lose contact with geosynchronous satellites and must rely on radio communications, which can be disrupted in a blackout.

Before they reach ground, GPS satellite signals must first pass through the ionosphere, where gases bend and weaken radio waves. Solar and geomagnetic storms that disturb the ionosphere can cause GPS position errors as large as 100 m (328 ft). (Find out how to fly through the ionosphere using NASA's *4D Ionosphere* program in conjunction with Google Earth™ at http://science.nasa.gov/science-news/science-at-nasa/2008/30apr_4dionosphere_launch/.)

Ozonosphere That portion of the stratosphere that contains an increased level of ozone is the **ozonosphere, or ozone layer**. Ozone is a highly reactive oxygen molecule made up of three oxygen atoms (O_3) instead of the usual two atoms (O_2) that make up oxygen gas. Ozone absorbs the shorter wavelengths of ultraviolet (UV) radiation (principally, all the UVC, 100–290 nm, and some of the UVB, 290–320 nm).* In the process, UV energy is converted to heat energy, safeguarding life on Earth by "filtering" some of the Sun's harmful rays. However, UVA, at 320–400 nm, is not absorbed by ozone and makes up about 95% of all UV radiation that reaches Earth's surface.

Although UVA, with its longer wavelengths, is less intense than UVB, it penetrates skin more deeply. Studies over the past 20 years have shown that UVA causes significant damage in the basal (lowest) part of the epidermis, the outer layer of skin, where most skin cancers occur. UVA levels are fairly constant throughout the year during daylight hours and can penetrate glass and clouds. In contrast, UVB intensity varies by latitude, season, and time of day. UVB damages the skin's more superficial epidermal layers and is the chief cause of skin reddening and sunburn. It, too, can lead to skin cancer.

*Nanometer (nm) = one-billionth of a meter; 1 nm = 10^{-9} m. For comparison, a micrometer, or micron (μm) = one-millionth of a meter; 1 μm = 10^{-6} m. A millimeter (mm) = one-thousandth of a meter; 1 mm = 10^{-3}.

The total amount of ozone in the stratosphere is presumed to have been relatively stable over the past several hundred million years (allowing for daily and seasonal fluctuations). Today, however, ozone is being depleted beyond the changes expected due to natural processes. Focus Study 3.1 presents an analysis of the crisis in this critical portion of our atmosphere.

CRITICALthinking 3.2
Finding Your Local Ozone

To determine the total column ozone amount at your present location, go to "What was the total ozone column at your house?" at http://ozoneaq.gsfc.nasa.gov/ozone_overhead_all_v8.md. Total column ozone is the total amount of ozone in a column from the surface to the top of the atmosphere.

Select a point on the map or enter your latitude and longitude and the date you want to check. The ozone column is currently measured by the Ozone Monitoring Instrument (OMI) sensor aboard the *Aqua* satellite and is mainly sensitive to stratospheric ozone. (Note also the limitations listed on the extent of data availability.) If you check for several different dates, when do the lowest values occur? The highest values? Briefly explain and interpret the values you found. •

UV Index Helps Save Your Skin Weather reports regularly include the *UV Index*, or *UVI*, in daily forecasts to alert the public of the need to use sun protection, especially for children. (Table 3.2). The UV Index is a simple way of describing the daily danger of solar UV-radiation intensity, using a scale from 1 to 11+. A higher number indicates a greater risk of UV exposure; an index of 0 indicates no risk, such as at night. Higher risk means that UV damage to skin and eyes can occur over a shorter time period. UV radiation varies spatially according to the amount of ozone depletion overhead, the season, and the local weather conditions.

TABLE 3.2 UV Index*

Exposure Risk Category	UVI Range	Comments
Low	less than 2	Low danger for average person. Wear sunglasses on bright days. Watch out for reflection off snow.
Moderate	3–5	Take covering precautions, such as sunglasses, sunscreen, hats, and protective clothing, and stay in shade during midday hours.
High	6–7	Use sunscreens with SPF of 15 or higher. Reduce time in the Sun between 11 A.M. and 4 P.M. Use protections mentioned above.
Very high	8–10	Minimize Sun exposure 10 A.M. to 4 P.M. Use sunscreens with SPF ratings of over 15. Use protections mentioned above.
Extreme	11+	Unprotected skin is at risk of burn. Sunscreen application every 2 hours if out-of-doors. Avoid direct Sun exposure during midday hours. Use protections listed above.

*The National Weather Service (NWS) and Environmental Protection Agency (EPA) began reporting the UVI in 1991. A revision in 2004 aligned the UVI with guidelines adopted by the World Health Organization (WHO) and World Meteorological Organization (WMO).

With stratospheric ozone levels at thinner-than-normal conditions, surface exposure to cancer-causing radiation increases. Remember, skin damage accumulates, and it may be decades before you experience the ill effects triggered by this summer's sunburn. See http://www.epa.gov/sunwise/uvindex.html for the UVI at U.S. locations; in Canada, go to http://www.ec.gc.ca/ozone/.

Pollutants in the Atmosphere

At certain times or places, the troposphere contains natural and human-caused gases, particles, and other substances in amounts that are harmful to humans or cause environmental damage. Study of the spatial aspects of these atmospheric **pollutants** is an important application of physical geography with far-reaching human-health implications.

Air pollution is not a new problem. Historically, air pollution has collected around population centers and is closely linked to human production and consumption of energy and resources. Romans complained more than 2000 years ago about the foul air of their cities. Filling Roman air was the stench of open sewers, smoke from fires, and fumes from ceramic-making kilns and from smelters (furnaces) that converted ores into metals.

Solutions to air quality issues require regional, national, and international strategies because the pollution sources often are distant from the observed impact. Pollution crosses political boundaries and even oceans. Regulations to curb human-caused air pollution have had great success, although much remains to be done. Before discussing these topics, we examine natural pollution sources.

Natural Sources of Air Pollution

Natural sources produce greater quantities of air pollutants—nitrogen oxides, carbon monoxide, hydrocarbons from plants and trees, and carbon dioxide—than do sources attributable to humans. Table 3.3 lists some of these natural sources and the substances they contribute to the air. Volcanoes, forest fires, and dust storms are the most significant sources, based on the volume of smoke and particulates produced and blown over large areas. However, pollen from crops, weeds, and other plants can

TABLE 3.3 Sources of Natural Pollutants

Source	Contribution
Volcanoes	Sulfur oxides, particulates
Forest fires	Carbon monoxide and dioxide, nitrogen oxides, particulates
Plants	Hydrocarbons, pollens
Decaying plants	Methane, hydrogen sulfides
Soil	Dust and viruses
Ocean	Salt spray, particulates

▲**Figure 3.5 Smoke from California wildfires.** Wildfires related to drought and high temperatures in southern California burned more than 117,000 ha (50,000 acres) in October and November 2007. Over half a million people were evacuated from their homes during these fires. Similar evacuations were repeated during wildfires in subsequent years. [*Terra* MODIS, NASA/GSFC.]

also cause high amounts of particle pollution, triggering asthma as well as other adverse human health effects. The particulates produced by these events are also known as **aerosols**, and include the liquid droplets and suspended solids that range in size from visible water droplets and pollen to microscopic dust. (Aerosols produced from human sources are discussed in the next section.)

A dramatic natural source of air pollution was the 1991 eruption of Mount Pinatubo in the Philippines (discussed in Chapter 1), probably the 20th century's second-largest eruption. This event injected nearly 20 million tons of sulfur dioxide (SO_2) into the stratosphere. The spread of these emissions is shown in a sequence of satellite images in Chapter 6, Figure 6.1.

Wildfires are another source of natural air pollution and occur frequently on several continents (Figure 3.5). Soot, ash, and gases darken skies and damage health in affected regions. Wind patterns spread the pollution from the fires to nearby cities, closing airports and forcing evacuations to avoid the health-related dangers. Satellite data shows smoke plumes traveling horizontally for distances up to 1600 km (1000 mi). Smoke, soot, and particulates can be propelled vertically as high as the stratosphere.

Wildfire smoke contains particulates (dust, soot, ash), nitrogen oxides, carbon monoxide, and volatile organic compounds (discussed in the next section). In southern California, recent wildfire smoke was linked to respiratory problems and increased hospital admissions, as well as lower birth weight for babies born to women living in smoke-exposed areas.

In 2006, scientists established a connection between climate change and wildfire occurrence in the western United States, where higher spring and summer temperatures and earlier snowmelt result in a longer fire season. These connections occur across the globe, as in drought-plagued Australia, where thousands of wildfires burned

(text continued on page 70)

Focus Study 3.1 Pollution

Stratospheric Ozone Losses: A Continuing Health Hazard

More UVB radiation than ever before is breaking through Earth's protective ozone layer, with detrimental effects on human health, plants, and marine ecosystems. In humans, UVB causes skin cancer, cataracts (a clouding of the eye lens), and a weakening of the immune system. UVB alters plant physiology in complex ways that lead to decreased agricultural productivity. In marine ecosystems, scientists have documented 10% declines in phytoplankton productivity in areas of ozone depletion around Antarctica—these organisms are the primary producers that form the basis of the ocean's food chain.

If all the ozone in our atmosphere were brought down to Earth's surface and compressed to surface pressure, the ozone layer would only be 3 mm thick (about one-tenth inch). At an altitude of 29 km (18 mi), where the ozone layer is densest, it contains only 1 part ozone per 4 million parts of air. Yet this relatively thin layer was in steady-state equilibrium for several hundred million years, absorbing intense ultraviolet radiation and permitting life to proceed safely on Earth.

Scientists have monitored the ozone layer from ground stations since the 1920s. Satellite measurements began in 1978. Using data from these instruments, scientists have monitored ozone with increasing accuracy over the past 35 years (Figure 3.1.1). (See http://ozonewatch.gsfc.nasa.gov/.) Table 3.1.1 provides a chronological summary of events relating to ozone depletion.

Ozone Losses Explained

In 1974, two atmospheric chemists, F. Sherwood Rowland and Mario Molina, hypothesized that some synthetic chemicals were releasing chlorine atoms that decompose ozone. These ***chlorofluorocarbons***, or *CFCs*, are synthetic molecules of chlorine, fluorine, and carbon.

CFCs are stable, or inert, under conditions at Earth's surface, and they possess remarkable heat properties. Both qualities made them valuable as propellants in aerosol sprays and as refrigerants. Also, some 45% of CFCs were used as solvents in the electronics industry and as foaming agents. Being inert, CFC molecules do not dissolve in water and do not break down in biological processes. (In contrast, chlorine compounds derived from volcanic eruptions and ocean sprays are water soluble and rarely reach the stratosphere.)

Researchers Rowland and Molina hypothesized that stable CFC molecules slowly migrate into the stratosphere, where intense ultraviolet radiation splits them, freeing chlorine (Cl) atoms. This process produces a complex set of reactions that breaks up ozone molecules (O_3) and leaves oxygen gas molecules (O_2) in their place. The effect is severe, for a single chlorine atom can decompose more than 100,000 ozone molecules.

The long residence time of chlorine atoms in the ozone layer (40 to 100 years) means the chlorine already in place is likely to have long-term consequences. Tens of millions of tons of CFCs were sold worldwide since 1950 and subsequently released into the atmosphere.

An International Response

The United States banned selling and production of CFCs in 1978. However, sales increased again when a 1981 presidential order permitted the export and sale of banned products. CFC sales hit a new peak in 1987 of 1.2 million metric tons (1.32 million tons), at which point an international agreement halted further sales growth. The *Montreal Protocol on Substances That Deplete the Ozone Layer* (1987) aims to reduce and eliminate all ozone-depleting substances. With 189 signatory countries, the protocol is regarded as the most successful international agreement in history (see http://ozone.unep.org/new_site/en/index.php).

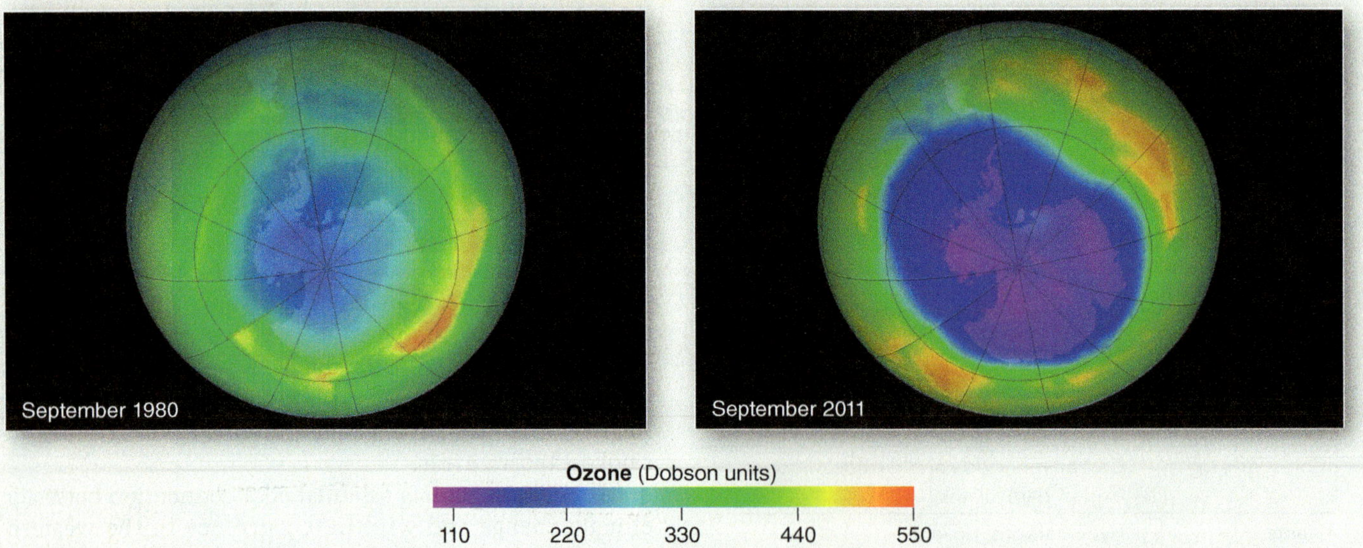

▲**Figure 3.1.1 The Antarctic ozone hole.** Images show the areal extent of the ozone "hole" in 1980 and 2011. Blues and purples show low ozone (the "hole"); greens, yellows, and reds denote more ozone. [NASA; 1979 to 2011 yearly images are at http://earthobservatory.nasa.gov/Features/WorldOfChange/ozone.php.]

CFC sales declined until all production of harmful CFCs ceased in 2010. Concern remains over some of the substitute compounds and a robust black market for banned CFCs. In 2007, the protocol instituted an aggressive phasedown of HCFCs, or *hydrochlorofluorocarbons*, one of the CFC-replacement compounds. If the protocol is fully enforced, scientists estimate that the stratosphere will return to more normal conditions in a century.

For their work, Doctors Rowland (who passed away in 2012) and Molina and another colleague, Paul Crutzen, received the 1995 Nobel Prize for Chemistry.

Ozone Losses over the Poles

How do Northern Hemisphere CFCs become concentrated over the South Pole? Chlorine freed in the Northern Hemisphere midlatitudes concentrates over Antarctica through the work of atmospheric winds. Persistent cold temperatures during the long, dark winter create a tight atmospheric circulation pattern—the polar vortex—that remains in place for several months. Chemicals and water in the stratosphere freeze out to form thin, icy *polar stratospheric clouds.*

Within these clouds, ice particle surfaces allow the chemicals to react, releasing chlorine. The chlorine cannot destroy ozone without the addition of UV light, which arrives with the spring in September (Figure 3.1.2). UV light sets off the reaction that depletes ozone and forms the ozone "hole." As the polar vortex breaks up and temperatures warm, ozone levels return to normal over the Antarctic region.

Over the North Pole, conditions are more changeable, so the hole is smaller, although growing each year. The Arctic ozone hole in 2011 was the largest on record (see http://earthobservatory.nasa.gov/IOTD/view.php?id=49874).

Make a note to check some of the listed Internet data sources for periodic updates on the stratosphere. The effects of ozone-depleting substances will be with us for the rest of this century.

TABLE 3.1.1 Significant Events in the History of Ozone Depletion

1960s	• Experts express concern that human-made chemicals in the atmosphere may affect ozone.
1970s	• Scientists hypothesize that synthetic chemicals, primarily *chlorofluorocarbons (CFCs)*, release chlorine atoms that react chemically to break down ozone. • The United States and other countries ban CFCs and aerosol propellants in 1978 (although the U.S. ban is weakened in 1981).
1980s	• Satellite measurements confirm a large ozone "hole" above Antarctica from September through November. • In the Arctic region, where stratospheric conditions and temperatures differ from the Antarctic, ozone depletion occurs on a smaller scale. • Scientific consensus confirms CFCs as the cause of ozone depletion, raising public awareness of the global effects of human activity on the atmosphere. • 189 countries sign the 1987 *Montreal Protocol* to phase out the use of ozone-depleting substances.
1990s	• International organizations standardize the reporting of UV radiation to the public—the *UV Index*.
Since 2000	• The largest areas of ozone depletion over Antarctica on record occur in September 2006. • In 2010, the U.S. Environmental Protection Agency (EPA) bans all production of CFCs. • In 2011, scientists find a significant ozone hole over the Arctic region. • In 2012, scientists report that intense summer storms over the United States are increasing atmospheric water vapor in the lower stratosphere, causing chemical reactions that deplete ozone. Climate change is driving the more frequent occurrence of these storms.

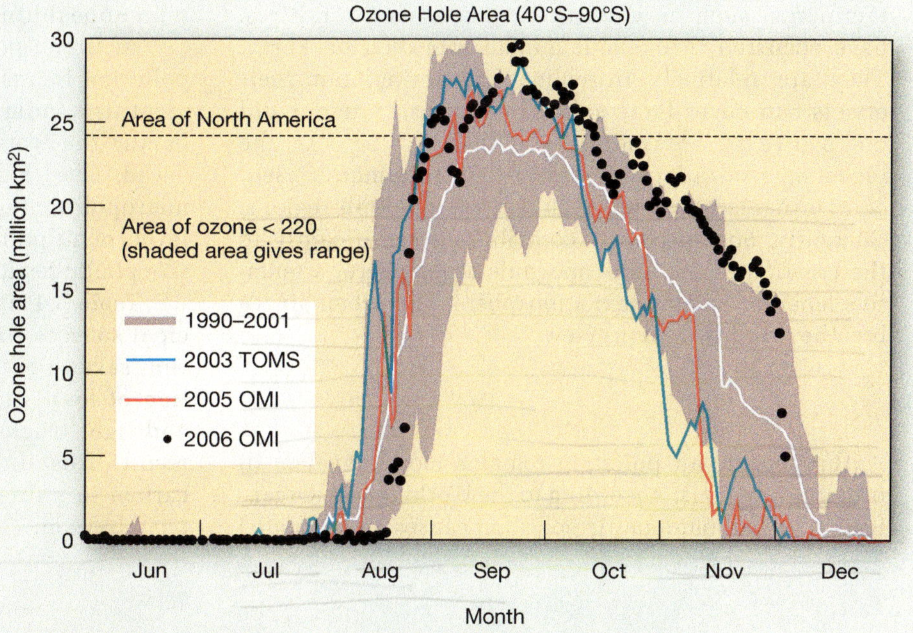

▲**Figure 3.1.2 Timing and extent of ozone depletion over Antarctica.** The area of ozone depletion generally peaks in late September; it reached record size in 2006. NASA's Total Ozone Mapping Spectrometer (TOMS) instrument began measuring ozone in 1978. Since 2004, the Ozone Monitoring Instrument (OMI) on board *Aura* has tracked ozone depletion. The *Suomi-NPP* satellite launched in 2011 carries the *Ozone Mapper Profiler Suite* (OMPS) to track ozone recovery. [Data from GSFC/NASA.]

MG™
Satellite
Southern Hemisphere
Ozone 2002–2003

TABLE 3.4 Major Pollutants over Urban Areas

Name	Symbol	Sources	Description and Effects
Carbon monoxide	CO	Incomplete combustion of fuels, mainly vehicle emissions	Odorless, colorless, tasteless gas. Toxic due to affinity for hemoglobin. Displaces O_2 in bloodstream; 50 to 100 ppm causes headaches and vision and judgment losses.
Nitrogen oxides	NO_x (NO, NO_2)	Agricultural practices, fertilizers, and high temperature/pressure combustion, mainly from vehicle emissions	Reddish-brown choking gas. Inflames respiratory system, destroys lung tissue. Leads to acid deposition.
Volatile organic compounds	VOCs	Incomplete combustion of fossil fuels such as gasoline; cleaning and paint solvents	Prime agents of surface ozone formation.
Ozone	O_3	Photochemical reactions related to motor vehicle emissions	Highly reactive, unstable gas. Ground-level ozone irritates human eyes and respiratory system. Damages plants.
Peroxyacetyl nitrates	PANs	Photochemical reactions related to motor vehicle emissions	No human health effects. Major damage to plants, forests, crops.
Sulfur oxides	SO_x (SO_2, SO_3)	Combustion of sulfur-containing fuels	Colorless, but with irritating smell. Impairs breathing and taste threshold. Causes human asthma, bronchitis, emphysema. Leads to acid deposition.
Particulate matter	PM	Industrial activities, fuel combustion, vehicle emissions, agriculture	Complex mixture of solid and liquid particles including dust, soot, salt, metals, and organics. Dust, smoke, and haze affect visibility. Black carbon may have a critical role in climate change. Various health effects: bronchitis, pulmonary function.
Carbon dioxide	CO_2	Complete combustion of fossil fuels	Principal greenhouse gas (see Chapter 10B)

millions of hectares in recent years (See the Chapter 5 opening photo and caption.)

The natural events that produce atmospheric contaminants, such as volcanic eruptions and wildfires, have occurred throughout human evolution on Earth. They are relatively infrequent in timing, but their effects can cover large areas. In contrast, humans did not evolve in environments with anything like the recent concentrations of *anthropogenic* (human-caused) contaminants now present in our metropolitan regions. Currently, our species is contributing significantly to the creation of the **anthropogenic atmosphere**, a tentative label for Earth's next atmosphere. The urban air we breathe may just be a preview.

Anthropogenic Pollution

Anthropogenic air pollution remains most prevalent in urbanized regions. According to the World Health Organization, urban outdoor air pollution causes an estimated 1.3 million deaths worldwide. In the United States, although air quality is improving in many places, 41% of the country's population (over 127 million people) live with unhealthful levels of air pollution.

As urban populations grow, human exposure to air pollution increases. A recent study using satellite data identified India, with an urban population of 31%, as having the worst air quality out of 132 countries surveyed. Over half the world's population now lives in metropolitan regions, some one-third with unhealthful levels of air pollution. This represents a potentially massive public-health issue in this century.

Table 3.4 lists the names, chemical symbols, principal sources, and impacts of urban air pollutants. The first seven pollutants in the table result from combustion of fossil fuels in transportation (specifically cars and light trucks). For example, **carbon monoxide (CO)** results from incomplete combustion, a failure of the carbon in a fuel to burn completely because of insufficient oxygen. The toxicity of carbon monoxide is due to

GEOreport 3.4 NASA's Global Hawks make scientific flights

Small unpiloted aerial vehicles (UAVs) are lightweight aircraft guided by remote control. These instrumental drones can gather all types of data and carry various remote-sensing devices. Since 2010, NASA's unpiloted Global Hawk aircraft have completed numerous scientific missions, flying for up to 32 hours or more on preprogrammed flight paths to altitudes above 18.3 km (60,000 ft) carrying instruments that sample ozone-depleting substances, aerosols, and other indicators of air quality in the upper troposphere and lower stratosphere. The first Global Hawk flew over Hurricane Karl in the Atlantic Ocean, collecting new and unprecedented data for hurricane research. Two 2012 Global Hawks are part of a multiyear study of hurricane formation and intensification in the Pacific.

its affinity for blood hemoglobin, as explained in Geo Report 3.5. The main anthropogenic source of carbon monoxide is vehicle emissions.

Motor vehicle transportation is still the largest source of air pollution in the United States and Canada, despite improvements in vehicle emissions over the past 30 years. Reducing air pollution from the transportation sector involves available technologies that result in monetary savings for consumers and lead to significant health benefits. This makes the continuing reluctance of the industry to work at achieving better fuel efficiency all the more confusing. Further improvements in fuel efficiency and reductions in fuel emissions, either through technological innovation or by promoting other forms of transportation, are critical for reducing air pollution.

Stationary pollution sources, such as electric power plants and industrial plants that use fossil fuels, contribute the most sulfur oxides and particulates. Concentrations are focused in the Northern Hemisphere, especially over eastern China and northern India.

Photochemical Smog Although not generally present in human environments until the advent of the automobile, photochemical smog is now the major component of anthropogenic air pollution; it is responsible for the hazy sky and reduced sunlight in many of our cities. **Photochemical smog** results from the interaction of sunlight and the combustion products in automobile exhaust, primarily nitrogen oxides and **volatile organic compounds (VOCs)**, such as hydrocarbons that evaporate from gasoline. Although the term *smog*—a combination of the words *sm*oke and f*og*—is generally used to describe this pollution, this is a misnomer.

The connection between automobile exhaust and photochemical smog was not determined until 1953 in Los Angeles, long after society had established its dependence upon cars and trucks. Despite this discovery, widespread mass transit declined, the railroads dwindled, and the polluting, inefficient individual automobile remains America's preferred transportation.

The high temperatures in automobile engines produce **nitrogen dioxide (NO_2)**, a chemical also emitted to a lesser extent from power plants. NO_2 is involved in several important reactions that affect air quality:

- Interactions with water vapor to form nitric acid (HNO_3), a contributor to acid deposition by precipitation, the subject of Focus Study 3.2
- Interactions with VOCs to produce **peroxyacetyl nitrates, or PANs**, pollutants that damage agricultural crops and forests, although they have no human health effects.
- Interactions with oxygen (O_2) and VOCs to form *ground-level ozone*, the principal component of photochemical smog.

Geosystems in Action, Figure GIA 3 on pages 76–77, summarizes how car exhaust is converted into photochemical smog. In the photochemical reaction, ultraviolet radiation liberates atomic oxygen (O) and a nitric oxide (NO) molecule from the NO_2. The free oxygen atom combines with an oxygen molecule, O_2, to form the oxidant ozone, O_3. The ozone in photochemical smog is the same gas that is beneficial to us in the stratosphere in absorbing ultraviolet radiation. However, ground-level ozone is a reactive gas that damages biological tissues and has a variety of detrimental human health effects, including lung irritation, asthma, and susceptibility to respiratory illnesses.

For several reasons, children are at greatest risk—one in four children in U.S. cities may develop health problems from ozone pollution. This ratio is significant; it means that more than 12 million children are vulnerable in those metropolitan regions with the highest ground-level ozone (Los Angeles, Bakersfield, Sacramento, San Diego, and other California cities; Houston; Dallas–Fort Worth; Washington–Baltimore). For more information and city rankings of the worst and the best, see http://www.lung.org (enter "city rankings" in the website search box).

VOCs, which react with nitrogen oxides to form photochemical smog, include a variety of chemicals—the outdoor pollutants from gasoline and combustion at electric utilities, as well as indoor pollutants emitted from paints and other household materials—that are important factors in ozone formation. States such as California base their standards for control of ozone pollution on VOC-emission controls—a scientifically accurate emphasis.

Industrial Smog and Sulfur Oxides Over the past 300 years, except in some developing countries, coal slowly replaced wood as society's basic fuel to provide the high-grade energy needed to run machines. The Industrial Revolution led to the conversion from *animate* energy (from animal sources, such as animal-powered farm equipment) to *inanimate* energy (from nonliving sources, such as coal, steam, and water). Pollution

GEOreport 3.5 Carbon monoxide—the colorless, odorless pollutant

As your car idles at a downtown intersection or you walk through a parking garage, you may be exposed to between 50 and 100 ppm of carbon monoxide without being aware of inhaling this colorless, odorless gas. In the burning ash of a smoking cigarette, CO levels reach 42,000 ppm. No wonder secondhand smoke affects CO levels in the blood of anyone breathing nearby, producing measurable health effects. What happens physiologically? CO combines with the oxygen-carrying hemoglobin of human blood, displacing the oxygen. The result is that the hemoglobin no longer transports adequate oxygen to vital organs such as the heart and brain; too much exposure to CO causes sudden illness and death (see http://www.cdc.gov/nceh/airpollution/.)

Focus Study 3.2 Pollution

Acid Deposition: Damaging to Ecosystems

Acid deposition is a major environmental issue in some areas of the United States, Canada, Europe, and Asia. Such deposition is most familiar as "acid rain," but it also occurs as "acid snow" and in dry form as dust or aerosols. Government estimates of damage from acid deposition in the United States, Canada, and Europe exceed $50 billion annually.

Acid deposition is causally linked to serious environmental problems: declining fish populations and fish kills, widespread forest damage, changes in soil chemistry, and damage to buildings, sculptures, and historic artifacts. Regions that have suffered most are the northeastern United States, southeastern Canada, Sweden, Norway, Germany, much of eastern Europe, and China.

Acid Formation

The problem begins when sulfur dioxide and nitrogen oxides are emitted as by-products of fertilizers and fossil fuel combustion. Winds may carry these gases many kilometers from their sources. Once in the atmosphere, the chemicals are converted to nitric acid (HNO_3) and sulfuric acid (H_2SO_4). These acids are removed from the atmosphere by wet and dry deposition processes, falling as rain or snow or attached to other particulate matter. The acid then settles on the landscape and eventually enters streams and lakes, carried by runoff and groundwater flows.

The acidity of precipitation is measured on the pH scale, which expresses the relative abundance of free hydrogen ions (H^+) in a solution—these are what make an acid corrosive, for they easily combine with other ions. The pH scale is logarithmic: Each whole number represents a tenfold change. A pH of 7.0 is neutral (neither acidic nor basic). Values less than 7.0 are increasingly acidic, and values greater than 7.0 are increasingly basic, or alkaline. (See Figure 18.7 for a graphic representation of the scale.)

Natural precipitation dissolves carbon dioxide from the atmosphere to form carbonic acid. This process releases hydrogen ions and produces an average pH reading for precipitation of 5.65. The normal range for precipitation is 5.3–6.0. Thus, normal precipitation is always slightly acidic. Scientists have measured precipitation as acidic as pH 2.0 in the eastern United States, Scandinavia, and Europe. By comparison, vinegar and lemon juice register slightly less than 3.0. In lakes, aquatic plant and animal life perishes when pH drops below 4.8.

Effects on Natural Systems

More than 50,000 lakes and some 100,000 km (62,000 mi) of streams in the United States and Canada are at a pH level below normal (that is, below pH 5.3), with several hundred lakes incapable of supporting any aquatic life. Acid deposition causes the release of aluminum and magnesium from clay minerals in the soil, and both of these are harmful to fish and plant communities.

In addition, relatively harmless mercury deposits in lake-bottom sediments convert in acidified lake waters to form highly toxic *methylmercury*, which is deadly to aquatic life and moves throughout biological systems. Local health advisories are regularly issued in two Canadian provinces and 22 U.S. states to warn those who fish of the methylmercury problem.

Acid deposition affects soils by killing microorganisms and causing a decline in soil nutrients. In New Hampshire's Hubbard Brook Experimental Forest (http://www.hubbardbrook.org/), an ongoing study from 1960 to the present found that half the nutrient calcium and magnesium base cations were leached from the soil; excess acids are the cause of the decline.

Damage to forests results from deficiencies in soil nutrients. The most advanced impact is seen in forests in eastern Europe, principally because of the area's long history of burning coal and the density of industrial activity. In Germany and

▲**Figure 3.2.1 Acid deposition blight.** Stressed forests on Mount Mitchell in the Appalachians. [Photo by Will and Deni McIntyre/Photo Researchers.]

generated by industry and coal-fired electrical generation differs from that produced by transportation.

Air pollution associated with coal-burning industries is known as **industrial smog** (see Figure GIA 3.1). The term *smog,* mentioned earlier, was coined by a London physician in 1900 to describe the combination of fog and smoke containing sulfur gases (sulfur is an impurity in fossil fuels); in the case of industrial air pollution, the use of *smog* is correct.

Industrial pollution has high concentrations of carbon dioxide, *particulates* (discussed just ahead), and sulfur oxides. Once in the atmosphere, **sulfur dioxide (SO_2)**

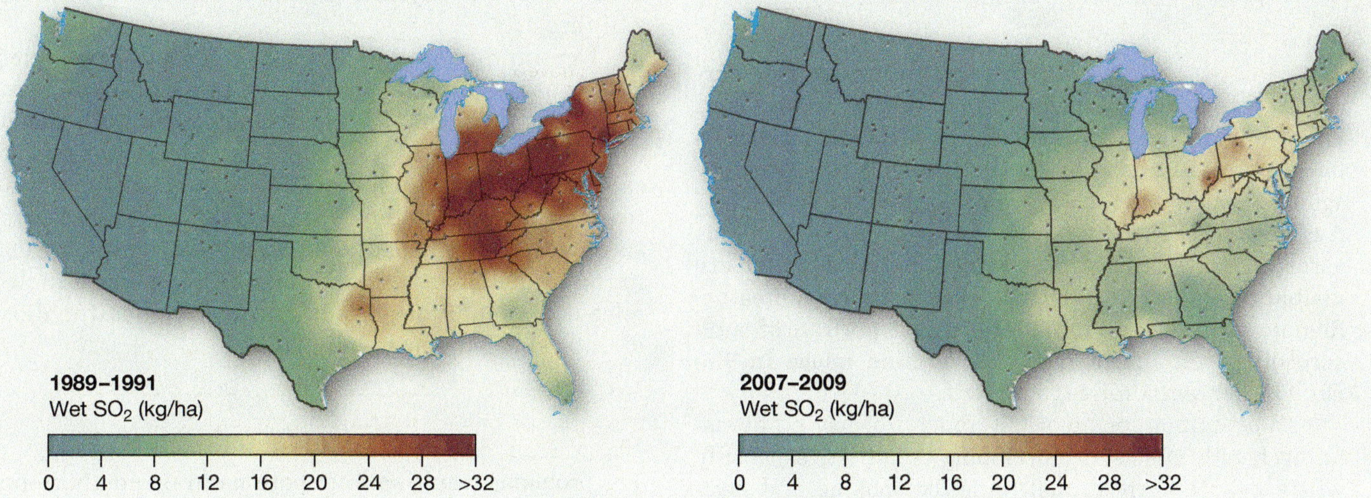

▲Figure 3.2.2 **Improvement in U.S. annual average wet sulfate deposition.** Spatial portrayal of wet sulfate deposition on the landscape, within two separate 3-year periods, in kilograms per hectare, shows a marked decrease since 1989 in the sulfate that falls as rain, snow, or fog. Improvements result from combined Canadian and U.S. regulatory action on emissions that add acids to the environment. [Source: National Atmospheric Deposition Program, 2010.]

Poland, up to 50% of the forests are dead or damaged.

In the eastern United States, some of the worst forest decline has occurred in the spruce and fir forests of North Carolina and Tennessee (Figure 3.2.1). In the Adirondacks, red spruce and sugar maples have been especially hard hit. Affected trees are susceptible to winter cold, insects, and droughts. In red spruce, decline is evidenced by poor crown condition, reduced growth shown in tree-ring analyses, and unusually high levels of tree mortality. For sugar maples, an indicator of forest damage is the reduction by almost half of the annual production of U.S. and Canadian maple sugar.

Nitrogen Oxides: A Worsening Cause

In the United States, the problem of acid deposition appeared to be solved with the passage of the 1990 Clean Air Act (CAA) amendments, which targeted industrial emissions of sulfur dioxide and nitrogen oxides. From 1990 to 2008, sulfur emissions from power plants decreased almost 70%, and wet sulfate deposition rates dropped across the eastern United States (Figure 3.2.2). However, nitrogen emissions declined little and in some areas have increased. Nitric oxide (NO) emissions come from three main sources: agricultural operations (specifically, fertilizers and animal-feeding operations, which produce nitrogen and ammonia), motor vehicles, and coal-combustion power plants.

Europe shows similar trends: Several studies have recently found areas with high levels of atmospheric nitrogen oxides in Switzerland and northern Italy, presumably associated with intensive agriculture and fossil fuel burning, as well as in Norway, where acid deposition in streams could have impacts on the economically important salmon industry. However, 49 European countries began regulating nitrogen emissions in 1999, resulting in a regional decrease in nitrogen emissions of about one-third.

In China, overuse of nitrogen fertilizers (an increase of 191% from 1991 to 2007) has led to acid deposition in soils, lessening crop production by 30% to 50% in some regions. If this trend continues, scientists fear that soil pH could drop as low as 3.0, far below the optimal level of 6.0 to 7.0 required for cereals such as rice.

Acid deposition is an issue of global spatial significance. Because wind and weather patterns are international, efforts at addressing the problem also must be international in scope. Reductions in acid-causing emissions are closely tied to energy conservation and therefore directly related to production of greenhouse gases and global climate-change concerns. Recent research pinpointing nitrogen as the leading cause of acid deposition links this issue to food production and global sustainability issues, as well.

reacts with oxygen (O) to form sulfur trioxide (SO_3), which is highly reactive and, in the presence of water or water vapor, forms tiny particles known as **sulfate aerosols**. Sulfuric acid (H_2SO_4) can also form, even in moderately polluted air at normal temperatures. Coal-burning electric utilities and steel manufacturing are the main sources of sulfur dioxide.

Sulfur dioxide–laden air is dangerous to health, corrodes metals, and deteriorates stone building materials at accelerated rates. Sulfuric acid deposition, added to nitric acid deposition, has increased in severity since it was first described in the 1970s. Focus Study 3.2 discusses this vital atmospheric issue and recent progress.

Particulates/Aerosols The diverse mixture of fine particles, both solid and liquid, that pollute the air and affect human health is referred to as **particulate matter (PM)**, a term used by meteorologists and regulatory agencies such as the U.S. Environmental Protection Agency. Other scientists refer to these particulates as aerosols. Examples are haze, smoke, and dust, which are visible reminders of particulates in the air we breathe. Remote sensing now provides a global portrait of such aerosols—look ahead to the background image in *The Human Denominator*, Figure HD 3.

Black carbon, or "soot," is an aerosol having devastating health effects in developing countries, especially where people burn animal dung for cooking and heating. This fine particulate is not necessarily prevalent over urban areas; black carbon is mainly produced in small villages, but winds can spread it over the globe. In Africa, Asia, and South America, cooking stoves produce the highest concentrations, with diesel engines and coal plants having a smaller role. Black carbon is both an indoor and outdoor pollutant, made up of pure carbon in several forms; it absorbs heat in the atmosphere and changes the reflectivity of snow and ice surfaces, giving it a critical role in climate change (discussed in Chapter 4).

The effects of PM on human health vary with the size of the particle. $PM_{2.5}$ is the designation for particulates 2.5 microns (2.5 μm) or less in diameter; these pose the greatest health risk. Sulfate aerosols are an example, with sizes about 0.1 to 1 μm in diameter. For comparison, a human hair can range from 50 to 70 μm in diameter. These fine particles, such as combustion particles, *organics* (biological materials such as pollens), and metallic aerosols, can get into the lungs and bloodstream. Coarse particles (PM_{10}) are of less concern, although they can irritate a person's eyes, nose, and throat.

New studies are implicating even smaller particles, known as *ultrafines* at a size of $PM_{0.1}$, as a cause of serious health problems. These are many times more potent than the larger $PM_{2.5}$ and PM_{10} particles because they can get into smaller channels in lung tissue and cause scarring, abnormal thickening, and damage called *fibrosis*. Asthma prevalence has nearly doubled since 1980 in the United States, a major cause being motor vehicle–related air pollution—specifically, ozone, sulfur dioxide, and fine particulate matter.

Natural Factors That Affect Pollutants

The problems resulting from both natural and anthropogenic atmospheric contaminants are made worse by several important natural factors. Among these, wind, local and regional landscape characteristics, and temperature inversions in the troposphere dominate.

Winds Winds gather and move pollutants, sometimes reducing the concentration of pollution in one location while increasing it in another. Dust, defined as particles less than 62 μm, or 0.0025 in., is often moved by wind, sometimes in dramatic episodes. Dust can come from natural sources, such as dry lakebeds, or human sources, such as overgrazed or over-irrigated lands. Dust can be tracked by chemical analysis to determine its source areas.

Traveling on prevailing winds, dust from Africa contributes to the soils of South America and Europe, and Texas dust ends up across the Atlantic (Figure 3.7).

(a) Typical wind-carried dust moves off West Africa in October 2012 over the Cape Verde islands and across the Atlantic.

(b) Dust plumes rise out of the White Sands dune field, New Mexico, in 2012. Driven by winter winds, these plumes stretch eastward more than 120 km (74 mi) over the Sacramento Mountains.

▲**Figure 3.7 Winds carrying dust in the atmosphere.** [(a) LANCE MODIS Rapid Response Team, NASA GSFC. (b) ISS astronaut photograph ISS030-E-174652, Image Science and Analysis Laboratory, NASA/JSC10.]

Imagine at any one moment a billion tons of dust aloft in the atmospheric circulation, carried by the winds!

Winds make the atmosphere's condition an international issue. For example, prevailing winds transport air pollution from the United States to Canada, causing much complaint and negotiation between the two governments. Pollution from North America adds to European air problems. In Europe, the cross-boundary drift of pollution is common because of the small size and proximity of many countries, a factor in the formation of the European Union (EU).

Arctic haze is a term from the 1950s, when pilots noticed decreased visibility, either on the horizon ahead or when looking down at an angle from their aircraft, across the Arctic region. Haze is a concentration of microscopic particles and air pollution that diminishes air clarity. Since there is no heavy industry at these high latitudes and only sparse population, this seasonal haze is the remarkable product of industrialization elsewhere in the Northern Hemisphere, especially Eurasia. Recent increases in wildfires in the Northern Hemisphere and agricultural burning in the midlatitudes contributes to Arctic haze.

There is no comparable haze over the Antarctic continent. Based on the above discussion, can you think of why Antarctica lacks such a condition?

Local and Regional Landscapes Local and regional landscapes are another important factor affecting the movement and concentration of air pollutants. Mountains and hills can form barriers to air movement or can direct the movement of pollutants from one area to another. Some of the worst air quality results when local landscapes trap and concentrate air pollution.

Places with volcanic landscapes such as Iceland and Hawai'i have their own natural pollution. During periods of sustained volcanic activity at Kīlauea, some 2000 metric tons (2200 tons) of sulfur dioxide are produced a day. Concentrations are sometimes high enough to merit broadcast warnings about health concerns, as occurred in 2011, 2012, and 2013. The resulting acid rain and volcanic smog, called *vog* by Hawaiians (for *v*olcanic sm*og*), cause losses to agriculture as well as other economic impacts.

Temperature Inversions Vertical differences in temperature and atmospheric density in the troposphere also can worsen pollution conditions. A **temperature inversion** occurs when the normal temperature, which usually decreases with altitude (normal lapse rate), reverses trend and begins to increase at some point. This can happen at any elevation from ground level to several thousand meters.

Figure 3.8 compares a normal temperature profile with that of a temperature inversion. In the normal

(text continued on page 78)

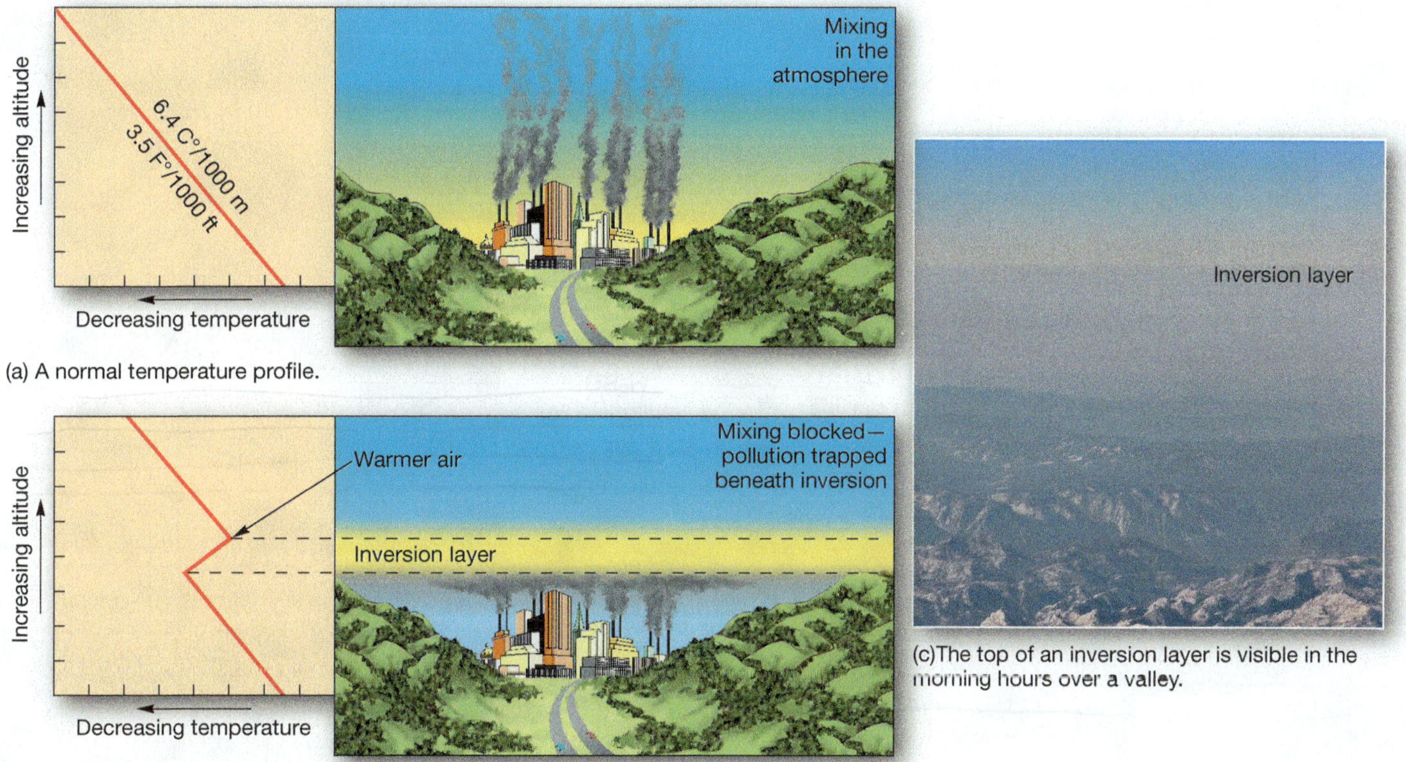

(a) A normal temperature profile.

(b) A temperature inversion in the lower atmosphere prevents the cooler air below the inversion layer from mixing with air above. Pollution is trapped near the ground.

(c)The top of an inversion layer is visible in the morning hours over a valley.

▲**Figure 3.8 Normal and inverted temperature profiles.** [(c) Bobbé Christopherson.]

Anthropogenic air pollution occurs mainly in urbanized regions, where it impacts human health. Photochemical smog produced by motor vehicle emissions is a major form of air pollution in cities. However, even in rural or isolated areas, air quality may be affected by pollution, such as from coal-fired power plants. Industrial smog affects the immediate area, and can also be transported over long distances, especially when tall smokestacks emit pollutants high into the atmosphere. Adverse impacts, such as acid deposition, can thus occur far from the pollution source.

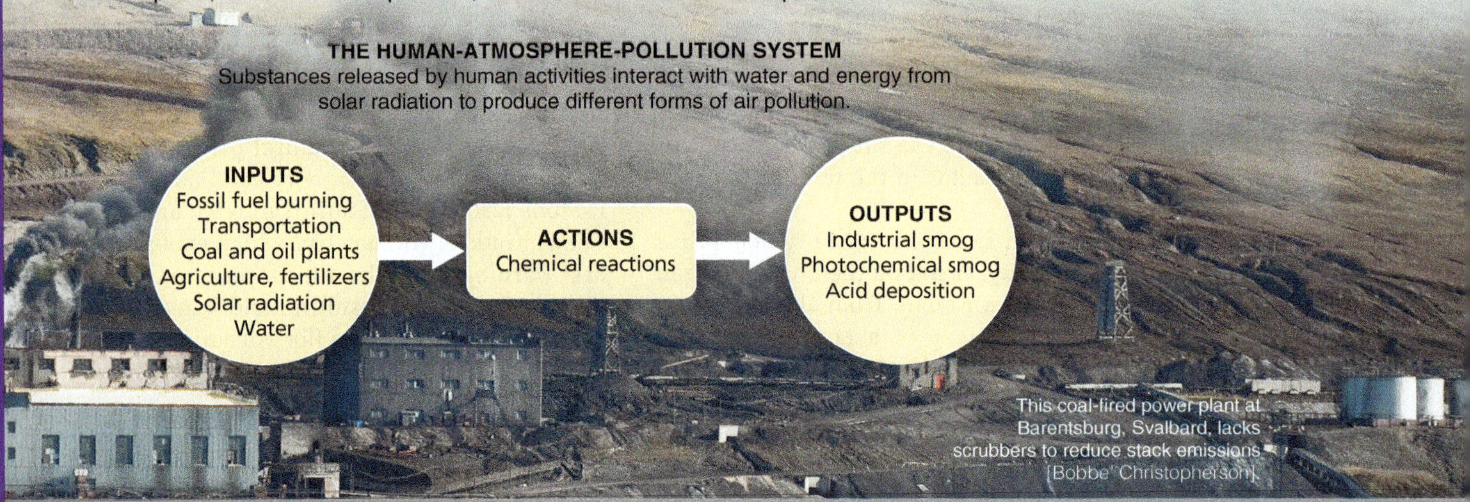

This coal-fired power plant at Barentsburg, Svalbard, lacks scrubbers to reduce stack emissions. [Bobbe' Christopherson]

3.1 INDUSTRIAL SMOG

Air pollution from coal-burning industries, including electrical power generation, is known as industrial smog. Industrial pollution has high concentrations of sulfur oxides, particulates, and carbon dioxide.

Sulfur oxides (SO_x, SO_2, SO_3)
Colorless gas with irritating smell produced by combustion of sulfur-containing fuels
In the environment: *Leads to acid deposition*
Health effects: *Impairs breathing, causes human asthma, bronchitis, emphysema*

Nitrogen oxides NO_x (NO, NO_2)
Reddish-brown, choking gas given off by agricultural activities, fertilizers, and gasoline-powered vehicles
In the environment: *Leads to acid deposition*
Health effects: *Inflames respiratory system, destroys lung tissue*

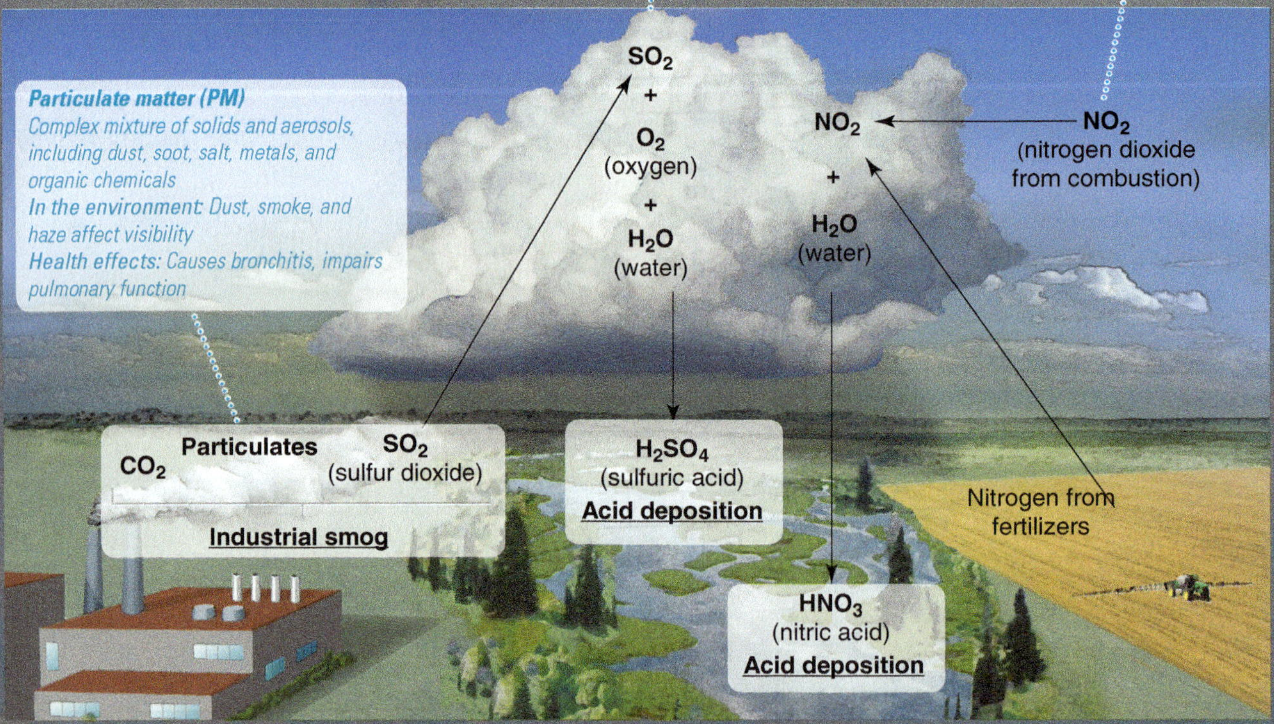

Analyze: Which pollutants could be reduced by switching to low-sulfur coal and installing scrubbers in smokestacks?

Premature Mortality Due to Air Pollution

MasteringGeography™

Visit the Study Area in MasteringGeography™ to explore Geosystems in Action.

Visualize: Study a geosciences animation of air pollution.

Assess: Demonstrate understanding of air pollution (if assigned by instructor).

3.2 PHOTOCHEMICAL SMOG

Car exhaust contains pollutants such as carbon monoxide (CO), nitrogen dioxide (NO_2), and volatile organic compounds (VOCs). When exhaust and the ultraviolet radiation in sunlight interact, photochemical reactions produce pollutants such as ozone and nitric acid, and peroxyacetyl nitrates (PANs). Ground-level ozone, the primary ingredient in photochemical smog, damages biological tissues. One in four children in U.S. cities is at risk of developing health problems from ozone pollution.

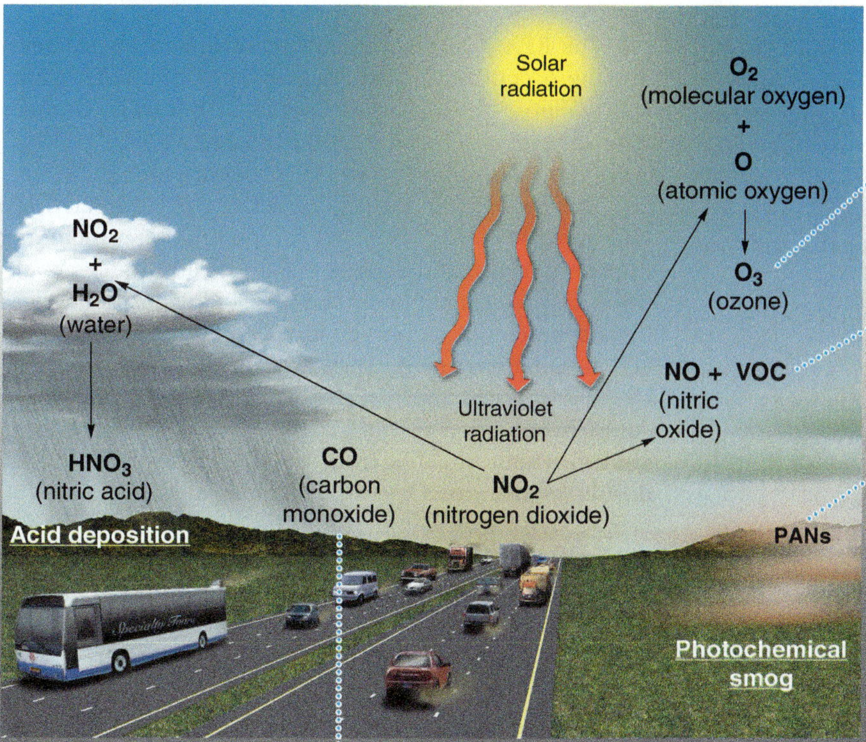

Ozone (O_3)
Highly reactive, unstable gas
In the environment: *Damages plants*
Health effects: *Irritates human eyes, nose, and throat*

Volatile organic compounds (VOCs)
In the environment: *Prime agents of surface ozone formation*

Peroxyacetyl nitrates (PANs)
In the environment: *Major damage to plants, forests, crops*
Health effects: *No human health effects*

Carbon monoxide (CO)
Odorless, colorless, tasteless gas
Health effects: *Toxic; displaces O_2 in bloodstream; 50 to 100 ppm causes headaches and vision and judgment losses*

3.3 AIR POLLUTION: A GLOBAL PROBLEM

Stationary pollution sources, such as electric power plants and industrial plants that burn fossil fuels, produce large amounts of sulfur oxides and particulates. Concentrations of these pollutants are focused in the Northern Hemisphere, especially over eastern China and northern India.

Describe: List the steps in the process by which vehicle exhaust leads to the increased ozone levels in smog.

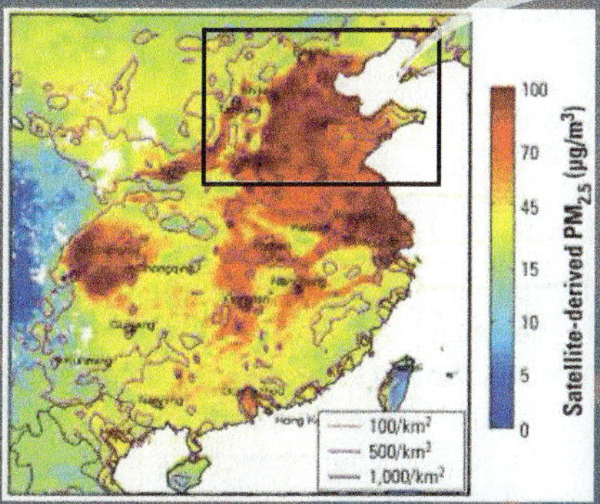

Particulate levels and population density in east Asia.

October 2012 haze over eastern China [NASA].

Photochemical smog, Mexico City [Daily Mail/Rex/Alamy.]

GEOquiz

1. Predict: In relation to a coal-fired power plant, where might industrial smog and serious air pollution problems occur? What mitigation strategies exist? Explain.

2. Compare and Contrast: How are the pollutants carbon monoxide and nitrogen dioxide similar? How are they different?

profile (Figure 3.8a), air at the surface rises because it is warmer (less dense) than the surrounding air. This ventilates the valley and moderates surface pollution by allowing air at the surface to mix with the air above. When an inversion occurs, colder (more dense) air lies below a warmer air layer (Figure 3.8b) that halts the vertical mixing of pollutants with other atmospheric gases. Thus, instead of being carried away, pollutants are trapped under the inversion layer. Inversions most often result from certain weather conditions, discussed in Chapters 7 and 8, or from topographic situations such as when cool mountain air drains into valley bottoms at night (see discussion of local winds in Chapter 6).

Benefits of the Clean Air Act

The concentration of many air pollutants declined over the past several decades because of Clean Air Act (CAA) legislation (1970, 1977, 1990), saving trillions of dollars in health, economic, and environmental losses. Despite this well-documented relationship, air pollution regulations are subject to a continuing political debate.

Since 1970 and the CAA, there have been significant reductions in atmospheric concentrations of carbon monoxide (–82%), nitrogen dioxide (–52%), volatile organic compounds (–48%), PM_{10} particulates (–75%), sulfur oxides (–76%; see Focus Study 3.2, Figure 3.2.2), and lead (–90%). Prior to the CAA, lead was added to gasoline, emitted in exhaust, and dispersed over great distances, finally settling in living tissues, especially in children. These remarkable reductions show the successful linking of science and public policy.

For abatement (mitigation and prevention) costs to be justified, they must not exceed the financial benefits derived from reducing pollution damage. Compliance with the CAA affected patterns of industrial production, employment, and capital investment. Although these expenditures were investments that generated benefits, the dislocation and job loss in some regions caused hardships—reductions in high-sulfur coal mining and cutbacks in polluting industries such as steel, for example.

In 1990, Congress therefore asked the EPA to analyze the overall health, ecological, and economic benefits of the CAA compared with the costs of implementing the law. In response, the EPA's Office of Policy, Planning, and Evaluation performed an exhaustive cost–benefit analysis and published a report in 1997: *The Benefits of the Clean Air Act, 1970 to 1990.* The analysis found a 42-to-1 benefit-over-cost ratio and provides a good lesson in cost–benefit analysis, valid to this day. The EPA report calculated the following:

- The *total direct cost* to implement all the federal, state, and local Clean Air Act rules from 1970 to 1990 was *$523 billion* (in 1990 dollars). This cost was borne by businesses, consumers, and government entities.
- The estimate of *direct economic benefits* from the Clean Air Act from 1970 to 1990 falls into the range of $5.6 to $49.4 trillion, with a central mean of **$22.2 trillion.**
- Therefore, the estimated *net financial benefit* of the Clean Air Act is **$21.7 trillion**!

The EPA has ongoing econometric studies to measure these continuing benefits. In 2010, the CAA saved U.S. society $110 billion and some 23,000 lives, whereas related costs of implementation in 2010 were $27 billion. More than 4 million lost workdays were avoided.

The benefits to society, directly and indirectly, are widespread across the entire population, including improved health and environment, less lead to harm children, lowered cancer rates, and less acid deposition. These CAA benefits were achieved during a period in which the U.S. population grew by 22% and the economy expanded by 70%. Political efforts to weaken CAA regulations seem counterproductive to the progress made so far. Given a choice, which benefits from the CAA might the public decide to give up? Or would an informed public opt to keep all the benefits in place?

In December 2009, the EPA enacted an *Endangerment Finding* that will guide planning at national, state, and local levels. This finding declares that greenhouse gases "pose a threat to human health and welfare." However, translating such a significant federal finding into local regulations is challenging.

The Human Denominator, Figure HD 3, presents examples of human–Earth interactions relating to the atmosphere. As you reflect on this chapter and our modern atmosphere, the treaties to protect stratospheric ozone, and the benefits from the CAA, you should feel encouraged. Scientists did the research and society made decisions, took action, and reaped enormous economic and health benefits. Decades ago scientists learned to sustain and protect Astronaut Mark Lee by designing a spacesuit that served as an "atmosphere." Today, we must learn to sustain and protect Earth's atmosphere to ensure our own survival.

CRITICALthinking 3.3
Evaluating Costs and Benefits

In the scientific study *The Benefits of the Clean Air Act, 1970 to 1990,* the EPA determined that the Clean Air Act provided health, social, ecological, and economic benefits 42 times greater than its costs. In 2010 alone, estimated benefits exceeded costs at a ratio of 4 to 1. In your opinion, why is the public generally unaware of these details? What are the difficulties in informing the public?

Do you think a similar benefit pattern results from laws such as the Clean Water Act, or hazard zoning and planning, or action on global climate change and the Endangerment Finding by the EPA? Take a moment and brainstorm recommendations for action, education, and public awareness on these issues. •

ATMOSPHERE ⇨ HUMANS

- Earth's atmosphere protects humans by filtering harmful wavelengths of light, such as ultraviolet radiation.
- Natural pollution from wildfires, volcanoes, and wind-blown dust are detrimental to human health.

HUMANS ⇨ ATMOSPHERE

- Depletion of the ozone layer continues as a result of human-made chemicals that break down ozone. Winds concentrate the pollutants over Antarctica, where the ozone hole is largest.
- Anthropogenic air pollution is concentrated over urban areas, which are home to half the world's population. Unhealthy pollution levels are becoming more widespread in some areas, such as northern India and eastern China; other regions have improved air quality, as in the Los Angeles metropolitan area.

3a New low emissions standards for diesel vehicles were implemented in London in 2012. Owners must comply or face a daily penalty fee. Stricter regulation is one strategy to control increasing air pollution from the transportation sector. [Daniel Berehulak/Getty Images.]

3b China burns more coal and emits more CO_2 than any country on Earth. Air pollution is severe, with $PM_{2.5}$ levels regularly reaching harmful levels above cities. American companies are testing clean energy technologies here, with hopes of using them around the world. [Jo Miyake/Alamy.]

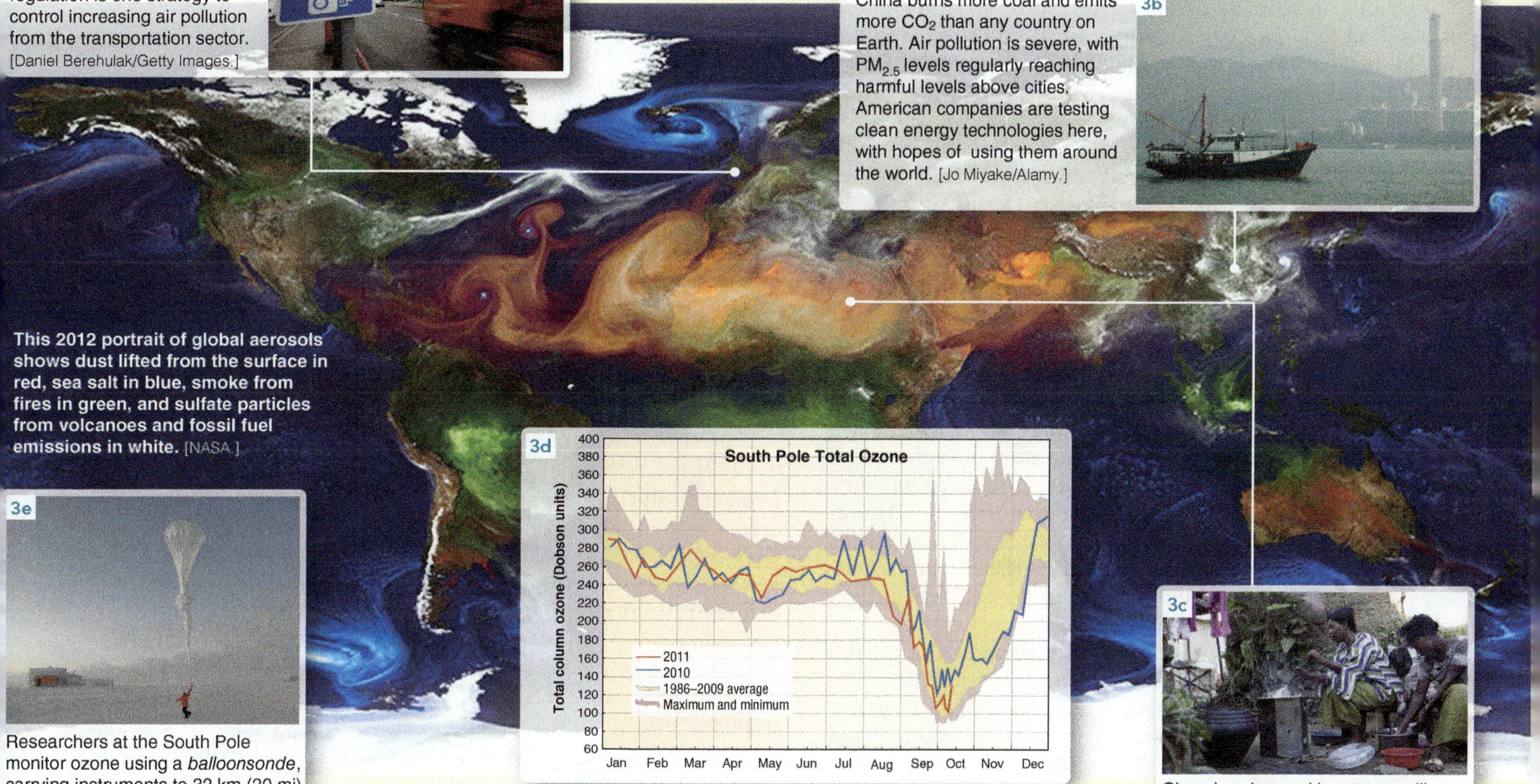

This 2012 portrait of global aerosols shows dust lifted from the surface in red, sea salt in blue, smoke from fires in green, and sulfate particles from volcanoes and fossil fuel emissions in white. [NASA.]

3e Researchers at the South Pole monitor ozone using a *balloonsonde*, carrying instruments to 32 km (20 mi) altitude. Significant seasonal ozone depletion still occurs, despite decreases in ozone-depleting chemicals. [NOAA.]

3c Clean burning cooking stoves will reduce the amount of fine particulates such as black carbon in developing countries. Several international initiatives are working toward this goal (see http://www.projectsurya.org/). [Per-Anders Pettersson/Getty Images.]

ISSUES FOR THE 21ST CENTURY

- Air quality will worsen in Asia unless human-made emissions are reduced. Air pollution will continue to improve in regions where emissions are regulated, such as in Europe, the United States, and Canada.
- Nitrogen emissions may worsen acid deposition worldwide unless regulatory action is taken.
- Alternative, clean energy sources are vital for reducing industrial pollution worldwide.
- Increases in fuel efficiency, vehicle-emissions regulations, and use of alternative and public transportation are critical to reduce urban pollution.
- Such reductions lower CO_2 emissions and slow rates of climate change.

GEOSYSTEMS**connection**

We traversed the atmosphere from the thermosphere to Earth's surface, examining its composition, temperature, and functions. Electromagnetic energy cascades toward Earth's surface through the layers of the atmosphere where harmful wavelengths are filtered out. Also we examined human impacts on the atmosphere, including the depletion of the ozone layer and air pollution, one example being acid deposition on the landscape.

In the next chapter, we focus on the flow of energy through the lower portions of the atmosphere as insolation makes its way to the surface. We establish the Earth–atmosphere energy balance and examine how surface energy budgets are powered by this arrival of energy. We also begin exploring the outputs of the energy–atmosphere system, looking at temperature concepts, temperature controls, and global temperature patterns.

KEY LEARNING concepts review

■ *Draw* a diagram showing atmospheric structure based on three criteria for analysis—composition, temperature, and function.

The principal substance of Earth's atmosphere is air—the medium of life. **Air** is naturally odorless, colorless, tasteless, and formless.

Above an altitude of 480 km (300 mi), the atmosphere is rarefied (nearly a vacuum) and is called the **exosphere**, which means "outer sphere." The weight (force over a unit area) of the atmosphere, exerted on all surfaces, is **air pressure**. It decreases rapidly with altitude.

By *composition*, we divide the atmosphere into the **heterosphere**, extending from 480 km (300 mi) to 80 km (50 mi), and the **homosphere**, extending from 80 km to Earth's surface. Using *temperature* as a criterion, we identify the **thermosphere** as the outermost layer, corresponding roughly to the heterosphere in location. Its upper limit, the **thermopause**, is at an altitude of approximately 480 km. **Kinetic energy**, the energy of motion, is the vibrational energy that we measure as temperature. *Heat* is the flow of kinetic energy between molecules and from one body to another because of a temperature difference between them. The amount of heat actually produced in the thermosphere is very small because the density of molecules is so low there. Nearer Earth's surface, the greater number of molecules in the denser atmosphere transmits their kinetic energy as *sensible heat*, meaning that we can feel it as a change in temperature.

In the homosphere, temperature criteria define the **mesosphere**, **stratosphere**, and **troposphere**. Within the mesosphere, cosmic or meteoric dust particles act as nuclei around which fine ice crystals form to produce rare and unusual night clouds, the **noctilucent clouds**.

The top of the troposphere is wherever a temperature of −57°C (−70°F) is recorded, a transition known as the **tropopause**. The normal temperature profile within the troposphere during the daytime decreases rapidly with increasing altitude at an average of 6.4 C° per km (3.5 F° per 1000 ft), a rate known as the **normal lapse rate**. The actual lapse rate at any particular time and place may deviate considerably because of local weather conditions and is called the **environmental lapse rate**.

The outermost region we distinguish by function is the **ionosphere,** extending through the heterosphere and partway into the homosphere. It absorbs cosmic rays, gamma rays, X-rays, and shorter wavelengths of ultraviolet radiation and converts them into kinetic energy. A functional region within the stratosphere is the **ozonosphere**, or **ozone layer**, which absorbs life-threatening ultraviolet radiation, subsequently raising the temperature of the stratosphere.

air (p. 60)
exosphere (p. 60)
air pressure (p. 60)
heterosphere (p. 61)
homosphere (p. 61)
thermosphere (p. 63)
thermopause (p. 63)
kinetic energy (p. 63)
mesosphere (p. 64)
noctilucent clouds (p. 64)
stratosphere (p. 65)
troposphere (p. 65)
tropopause (p. 65)
normal lapse rate (p. 65)
environmental lapse rate (p. 65)
ionosphere (p. 65)
ozonosphere, ozone layer (p. 66)

1. What is air? Where generally did the components in Earth's present atmosphere originate?
2. Characterize the various functions the atmosphere performs that protect the surface environment.
3. What three distinct criteria are employed in dividing the atmosphere for study?
4. Describe the overall temperature profile of the atmosphere and list the four layers defined by temperature.
5. Describe the two divisions of the atmosphere on the basis of composition.
6. What are the two primary functional layers of the atmosphere and what does each do?

■ *List* and *describe* the components of the modern atmosphere, giving their relative percentage contributions by volume.

Even though the atmosphere's density decreases with increasing altitude in the homosphere, the blend of gases (by proportion) is nearly uniform. This mixture of gases has evolved slowly and includes constant gases, with concentrations that have remained stable over time, and variable gases, that change over space and time.

The homosphere is a vast reservoir of relatively inert nitrogen, originating principally from volcanic sources and from bacterial action in the soil; oxygen, a by-product of photosynthesis; argon, constituting about 1% of the homosphere and completely inert; and *carbon dioxide*, a natural by-product of life processes and fuel combustion.

7. Name the four most prevalent stable gases in the homosphere. Where did each originate? Is the amount of any of these changing at this time?

■ *Describe* conditions within the stratosphere; specifically, *review* the function and status of the ozonosphere, or ozone layer.

The overall reduction of the stratospheric ozonosphere, or ozone layer, during the past several decades represents a hazard for society and many natural systems and is caused by chemicals introduced into the atmosphere by humans. Since World War II, quantities of human-made **chlorofluorocarbons** (**CFCs**) have made their way into the stratosphere. The increased ultraviolet light at those altitudes breaks down these stable chemical compounds, thus freeing chlorine atoms. These atoms act as catalysts in reactions that destroy ozone molecules.

chlorofluorocarbons (CFCs) (p. 68)

8. Why is stratospheric ozone so important? Describe the effects created by increases in ultraviolet light reaching the surface.

9. Summarize the ozone predicament, and describe treaties to protect the ozone layer.
10. Evaluate Crutzen, Rowland, and Molina's use of the scientific method in investigating stratospheric ozone depletion and the public reaction to their findings.

■ *Distinguish* between natural and anthropogenic pollutants in the lower atmosphere.

Within the troposphere, both natural and human-caused **pollutants**, defined as gases, particles, and other chemicals in amounts that are harmful to human health or cause environmental damage, are part of the atmosphere. Volcanoes, fires, and dust storms are sources of smoke and particulates, also known as **aerosols**, consisting of suspended solids and liquid droplets such as pollens, dust, and soot from natural and human sources. We coevolved with natural "pollution" and thus are adapted to it. But we are not adapted to cope with our own anthropogenic pollution. It constitutes a major health threat, particularly where people are concentrated in cities. Earth's next atmosphere most accurately may be described as the **anthropogenic atmosphere** (human-influenced atmosphere).

pollutants (p. 67)
aerosols (p. 67)
anthropogenic atmosphere (p. 70)

11. Describe two types of natural air pollution. What regions of Earth commonly experience this type of pollution?
12. What are pollutants? What is the relationship between air pollution and urban areas?

■ *Construct* a simple diagram illustrating the pollution from photochemical reactions in motor vehicle exhaust, and describe the sources and effects of industrial smog.

Transportation is the major human-caused source for carbon monoxide and nitrogen dioxide. Odorless, colorless, and tasteless, **carbon monoxide (CO)** is produced by incomplete combustion (burning with limited oxygen) of fuels or other carbon-containing substances; it is toxic because it de-oxygenates human blood.

Photochemical smog results from the interaction of sunlight and the products of automobile exhaust, the single largest contributor of air pollution over urban areas in the United States and Canada. The *nitrogen dioxide* and *volatile organic compounds* (VOCs) from car exhaust, in the presence of ultraviolet light in sunlight, convert into the principal photochemical by-products—*ozone, peroxyacetyl nitrates* (PANs), and *nitric acid*. The **volatile organic compounds (VOCs)**, including hydrocarbons from gasoline, surface coatings such as paint, and electric utility combustion, are important factors in ozone formation.

Ground-level ozone (O_3) has negative effects on human health and kills or damages plants. **Peroxyacetyl nitrates (PANs)** have no known effect on human health but are particularly harmful to plants, including both agricultural crops and forests. **Nitrogen dioxide (NO_2)** inflames human respiratory systems, destroys lung tissue, and damages plants. Nitric oxides participate in reactions that produce nitric acid (HNO_3) in the atmosphere, forming both wet and dry acidic deposition.

The distribution of human-produced **industrial smog** over North America, Europe, and Asia is related to coal-burning power plants. Such pollution contains **sulfur dioxide (SO_2)**, which reacts in the atmosphere to produce **sulfate aerosols**, which in turn produce sulfuric acid (H_2SO_4) deposition. This deposition has detrimental effects on living systems when it settles on the landscape. **Particulate matter (PM)** consists of dirt, dust, soot, and ash from industrial and natural sources.

Vertical temperature and atmospheric density distribution in the troposphere can worsen pollution conditions. A **temperature inversion** occurs when the normal temperature decrease with altitude (normal lapse rate) reverses, and temperature begins to increase at some altitude. This can cause cold air and pollutants to be trapped near Earth's surface, temporarily unable to mix with the air above the inversion layer.

carbon monoxide (CO) (p. 70)
photochemical smog (p. 71)
volatile organic compounds (VOCs) (p. 71)
nitrogen dioxide (NO_2) (p. 71)
peroxyacetyl nitrates (PANs) (p. 71)
industrial smog (p. 72)
sulfur dioxide (SO_2) (p. 72)
sulfate aerosols (p. 73)
particulate matter (PM) (p. 74)
temperature inversion (p. 75)

13. What is the difference between industrial smog and photochemical smog?
14. Describe the relationship between automobiles and the production of ozone and PANs in city air. What are the principal negative impacts of these gases?
15. How are sulfur impurities in fossil fuels related to the formation of acid in the atmosphere and acid deposition on the land?
16. In what ways does a temperature inversion worsen an air pollution episode? Why?
17. In summary, what are the cost–benefit results from the first 20 years under Clean Air Act regulations?

MasteringGeography™

Looking for additional review and test prep materials? Visit the Study Area in *MasteringGeography*™ to enhance your geographic literacy, spatial reasoning skills, and understanding of this chapter's content by accessing a variety of resources, including MapMaster interactive maps, geoscience animations, satellite loops, author notebooks, videos, RSS feeds, flashcards, web links, self-study quizzes, and an eText version of *Geosystems*.

4

Atmosphere and Surface Energy Balances

The "green roof" on top of City Hall in Chicago, Illinois, helps mitigate the urban heat-island effect, an energy balance phenomenon that causes temperatures in cities to be warmer than those of surrounding regions. This type of living roof provides insulation for the building below, reducing heating and cooling costs by as much as 20%. The roof also absorbs less insolation than one made of black asphalt, lessening the overall warming of city temperatures. Urban heat islands are one component of the Earth-atmosphere energy balance discussed in this chapter. [Diane Cook and Len Jenshel/Getty Images.]

KEY LEARNING concepts

After reading the chapter, you should be able to:

- ***Define*** energy and heat, and ***explain*** four types of heat transfer: radiation, conduction, convection, and advection.
- ***Identify*** alternative pathways for solar energy on its way through the troposphere to Earth's surface—transmission, scattering, refraction, and absorption—and ***review*** the concept of albedo (reflectivity).
- ***Analyze*** the effect of clouds and aerosols on atmospheric heating and cooling, and ***explain*** the greenhouse concept as it applies to Earth.
- ***Review*** the Earth–atmosphere energy balance and the patterns of global net radiation.
- ***Plot*** typical daily radiation and temperature curves for Earth's surface—including the daily temperature lag.
- ***List*** typical urban heat island conditions and their causes, and ***contrast*** the microclimatology of urban areas with that of surrounding rural environments.

Melting Sea Ice Opens Arctic Shipping Lanes, However...

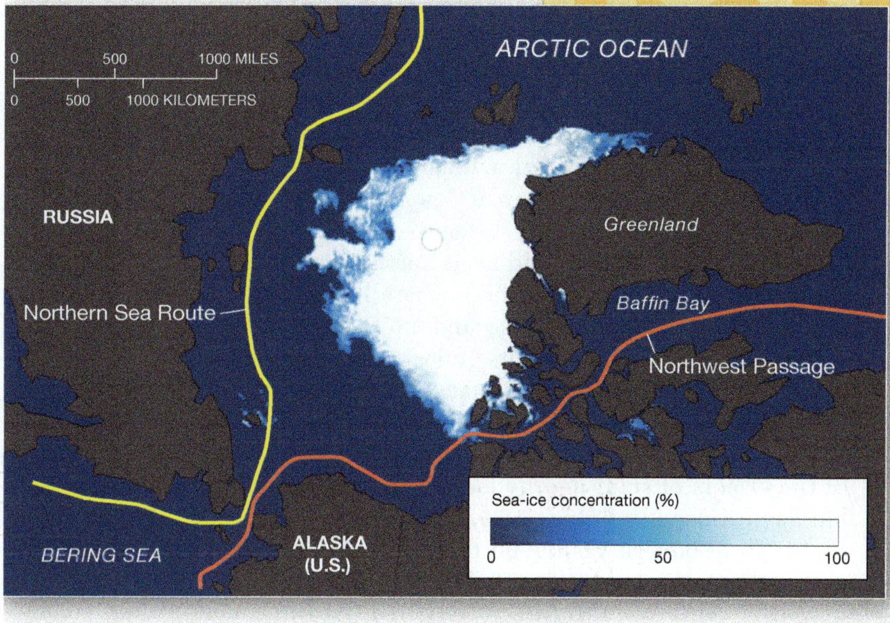

▲**Figure GN 4.1 Arctic summer sea-ice extent, 2012.** [NASA/GSFC.]

Sought for hundreds of years by explorers trying to navigate the icy waters of the Arctic, the "Northwest Passage" is a sea route connecting the Atlantic and Pacific Oceans through the inland waterways of the Canadian Archipelago. The Northern Sea Route, also known as the "Northeast Passage," is another Arctic sea route traversing the Russian coast and linking Europe and Asia. These northern passages offer ships a shorter alternative to the long route through the Panama and Suez canals.

During most years, Arctic Ocean sea ice blocked these sea routes, making them unavailable for shipping—that is, until recently. With a large ice melt in the summer of 2007, the Northwest Passage opened along its entire length for 36 days. In 2009, two German container ships completed the first commercial navigation of the Northern Sea Route. Higher ocean and air temperatures associated with climate change cause the ice losses that opened these northern passages. The 2012 summer sea-ice melt was the largest on record since 1979, allowing over 40 freighters and tankers to traverse the route (Figure GN 4.1).

Albedo and Sea-Ice Melting In Chapter 1, we described the positive feedback loop between sunlight and the reflectivity of sea-ice surfaces. Lighter surfaces reflect sunlight and remain cooler, whereas darker surfaces absorb sunlight and heat up. Snow- and ice-covered surfaces are natural reflectors; in fact, sea ice reflects about 80%–95% of the solar energy it receives. The ocean surface is darker, reflecting only an average of about 10% of insolation. This percentage is *albedo*, or the reflective value of a surface. As the ice-covered area in the Arctic retreats, darker water or land receives direct sunlight and absorbs more heat, which decreases albedo and adds to warming—a positive feedback.

Particulates from the atmosphere appear to decrease albedo even further. Scientists now have evidence that surface accumulations of black carbon (soot) and other atmospheric particulates are significant causes of glacial snow and ice losses across the Himalayas. On the Greenland Ice Sheet, black carbon accumulation from distant wildfires may be a major factor in the darkening of interior surfaces on the ice sheet.

Sea Ice and Arctic Shipping The prospect of an Arctic Ocean with long ice-free periods for freighter and oil-tanker traffic is triggering commercial excitement. Use of the Northern Sea Route is increasing: 4 vessels in 2010, 34 in 2011 (including a full-sized tanker), and 46 vessels in 2012. But while the Arctic shipping lanes save time and money in the transport of goods, their use has potentially huge environmental costs—ice loss, climate impacts, oil spills, and risk of ship groundings. The stack emissions from new freighter and tanker traffic adds soot and particulate material into the Arctic atmosphere. As this material settles on snow-covered glaciers and ice shelves, it darkens their surfaces and worsens albedo-reducing feedback loops that drive climate change.

▲**Figure GN 4.2 Ship traffic and shipments in the Arctic Ocean.** [Canadian Coast Guard]

Computer models now show that with continued significant losses of sea ice, the Arctic Ocean could be ice-free during the summer within a decade. As shipping traffic increases, the ice–albedo positive feedback will accelerate through impacts of air pollution on surface albedos and surface energy budgets, the topics of Chapter 4.

GEOSYSTEMS NOW ONLINE Go to Chapter 4 on the *MasteringGeography* website (www.masteringgeography.com) for more maps and images regarding Arctic sea ice and albedo feedback loops. For daily updates on the status of sea ice, go to the National Snow and Ice Data Center website at http://nsidc.org/arcticseaicenews/. More images of Greenland are at http://ocean.dmi.dk/arctic/modis.uk.php. MG

Earth's biosphere pulses with flows of solar energy that cascade through the atmosphere and sustain our lives. Earth's shifting seasonal rhythms, described in Chapter 2, are driven by the concentrations of solar energy on Earth's surface, which vary with latitude throughout the year. Although solar energy sustains life, it can also be harmful without the protection of Earth's atmosphere, as we saw in Chapter 3. Solar energy is the engine for functioning systems on Earth—it drives wind and ocean currents and heats Earth's surface, driving moisture into the atmosphere where it forms clouds and precipitation. These energy and moisture exchanges between Earth's surface and its atmosphere are essential elements of weather and climate, discussed in later chapters.

Photographs of Earth taken from space reveal some of the effects of solar energy (see the NASA Blue Marble Earth image in Figure 1.30, and the images on the inside front cover). Visible on these images are clouds above the Atlantic Ocean off the coast of Florida and over the Amazon rainforest, swirling weather patterns above the Pacific Ocean, and clear skies above the southwestern United States. These patterns result from regional differences in solar energy receipts and the processes driven by insolation on Earth.

In this chapter: We follow solar energy through the troposphere to Earth's surface, looking at processes that affect insolation pathways. We discuss the balance between solar radiation inputs and outputs from Earth—the energy balance in the atmosphere—and apply the "greenhouse" concept to Earth. We also examine surface energy and daily radiation patterns, analyzing the transfer of net radiation that maintains Earth's energy balance. Focus Study 4.1 discusses applications for solar energy, a renewable energy resource of great potential. The chapter concludes with a look at the energy environment in our cities, where the air shimmers as summer heat radiates skyward over traffic and pavement.

Energy-Balance Essentials

In Chapter 2, we introduced the concept of Earth's energy budget, the overall balance between shortwave solar radiation to Earth and shortwave and longwave radiation to space (please review Figure 2.7). A *budget* in terms of energy is a balance sheet of energy income and expenditure. For Earth, energy income is insolation, and energy expenditure is radiation to space, with an overall balance maintained between the two.

Transmission refers to the uninterrupted passage of shortwave and longwave energy through either the atmosphere or water. Our Earth–atmosphere energy budget comprises shortwave radiation *inputs* (ultraviolet light, visible light, and near-infrared wavelengths) and longwave radiation *outputs* (thermal infrared wavelengths) that pass through the atmosphere by transmission. Since solar energy is unevenly distributed by latitude and fluctuates seasonally, the energy budget is not the same at every location on Earth's surface, even though the overall energy system remains in steady-state equilibrium (see a simplified diagram in Figure 4.1 and discussion ahead; the more detailed energy balance is in *Geosystems In Action*, Figure GIA 4, on pages 92 and 93).

Energy and Heat

For the purpose of studying Earth's energy budget, *energy* can be defined as the capacity to do *work*, or move matter. (*Matter* is mass that assumes a physical

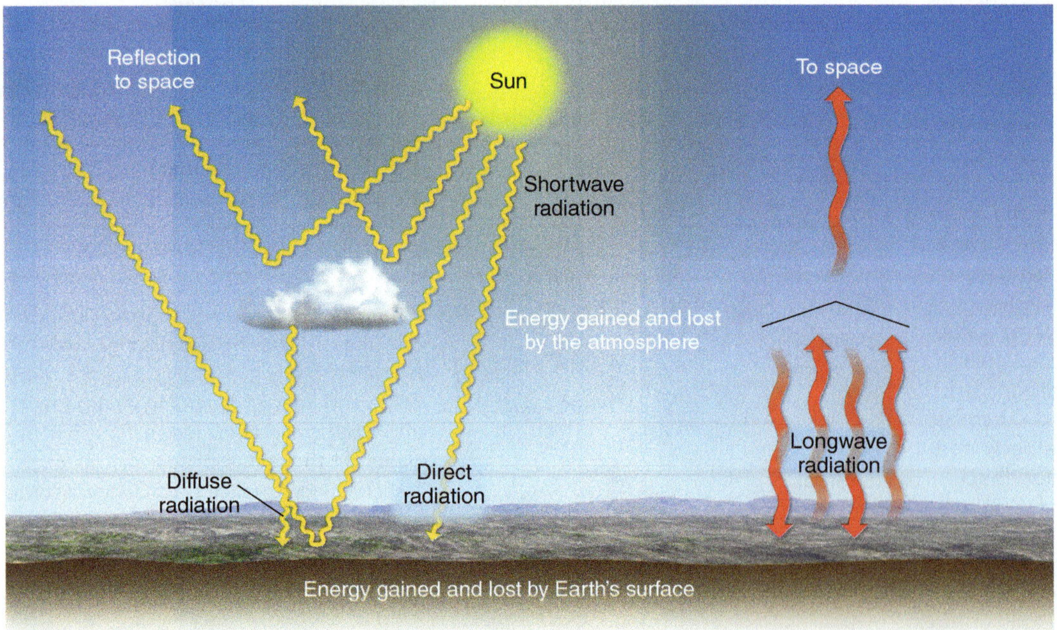

Figure 4.1 Simplified view of the Earth–atmosphere energy system. Energy gained and lost by Earth's surface and atmosphere includes incoming and reflected shortwave radiation, energy absorbed at Earth's surface, and outgoing longwave radiation. Refer back to this diagram as you read about energy pathways and principles in the atmosphere. A more complete illustration of the Earth–atmosphere energy balance is in Figure GIA 4.

MG Animation Global Warming, Climate Change

MG Animation Earth–Atmosphere Energy Balance

shape and occupies space.) Humans have learned to manipulate energy so that it does work for our benefit, such as the chemical energy used to run motor vehicles or the gravitational energy used within dams for hydropower production.

Kinetic energy is the energy of motion, produced when you run, walk, or ride a bicycle, and produced by the vibrational energy of molecules that we measure as temperature. *Potential energy* is stored energy (stored either due to composition or position) that has the capacity to do work under the right conditions. Petroleum has potential energy that is released when gasoline is burned in a car's engine. The water in the reservoir above a hydropower dam has potential energy that is released when the pull of gravity impels it through the turbines and into the river downstream. (Potential energy is converted into kinetic energy in both these examples.) Both kinetic energy and potential energy produce work, in which matter is moved into a new position or location.

Types of Heat In Chapter 3, we mentioned that **heat** is the flow of kinetic energy between molecules and from one body or substance to another resulting from a temperature difference between them. Heat always flows from an area of higher temperature into an area of lower temperature; an example is the transfer of heat when you wrap your warm hand around a snowball, or a piece of ice, and it melts. Heat flow stops when the temperatures—that is, when the amounts of kinetic energy—become equal.

Two types of heat energy are important for understanding Earth–atmosphere energy budgets. **Sensible heat** can be "sensed" by humans as temperature, because it comes from the kinetic energy of molecular motion. *Latent heat* ("hidden" heat) is the energy gained or lost when a substance changes from one state to another, such as from water vapor to liquid water (gas to liquid) or from water to ice (liquid to solid). Latent heat transfer differs from sensible heat transfer in that as long as a physical change in state is taking place, the substance itself does not change temperature (although in Chapter 7 we see that the surroundings do gain or lose heat).

Methods of Heat Transfer Heat energy can be transferred in a number of ways throughout Earth's atmosphere, land, and water bodies. *Radiation* is the transfer of heat in electromagnetic waves, such as that from the Sun to Earth, discussed in Chapter 2, or as from a fire or a burner on the stove (Figure 4.2). The temperature of the object or substance determines the wavelength of radiation it emits (we saw this when comparing Sun and Earth in Chapter 2); the hotter an object, the shorter the wavelengths that are emitted. Waves of radiation do not need to travel through a medium, such as air or water, in order to transfer heat.

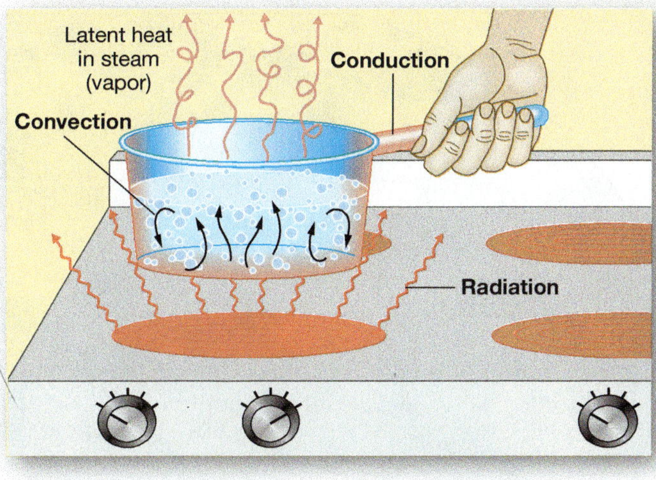

▲**Figure 4.2 Heat-transfer processes.** Infrared energy *radiates* from the burner to the saucepan and the air. Energy *conducts* through the molecules of the pan and the handle. The water physically mixes, carrying heat energy by *convection*. Latent heat is the energy absorbed when liquid water changes to steam (water vapor).

Conduction is the molecule-to-molecule transfer of heat energy as it diffuses through a substance. As molecules warm, their vibration increases, causing collisions that produce motion in neighboring molecules, thus transferring heat from warmer to cooler material. An example is energy conducted through the handle of a pan on a kitchen stove. Different materials (gases, liquids, and solids) conduct sensible heat directionally from areas of higher temperature to those of lower temperature. This heat flow transfers energy through matter at varying rates, depending on the conductivity of the material—Earth's land surface is a better conductor than air; moist air is a slightly better conductor than dry air.

Gases and liquids also transfer energy by **convection**, the transfer of heat by mixing or circulation. An example is a convection oven, in which a fan circulates heated air to uniformly cook food, or the movement of boiling water on a stove. In the atmosphere or in bodies of water, warmer (less dense) masses tend to rise and cooler (denser) masses tend to sink, establishing patterns of convection. This physical mixing usually involves a strong vertical motion. When horizontal motion dominates, the term **advection** applies.

These physical transfer mechanisms are important for many concepts and processes in physical geography: *Radiation* and *conduction* pertain to surface energy budgets, temperature differences between land and water bodies and between darker and lighter surfaces, and temperature variation in Earth materials such as soils; *convection* is important in atmospheric and oceanic circulation, air mass movements and weather systems, internal motions deep within Earth, and movements in Earth's crust; *advection* relates to the horizontal movement of winds from land to sea and sea to land, the formation and movement of fog, and air mass movements from source regions.

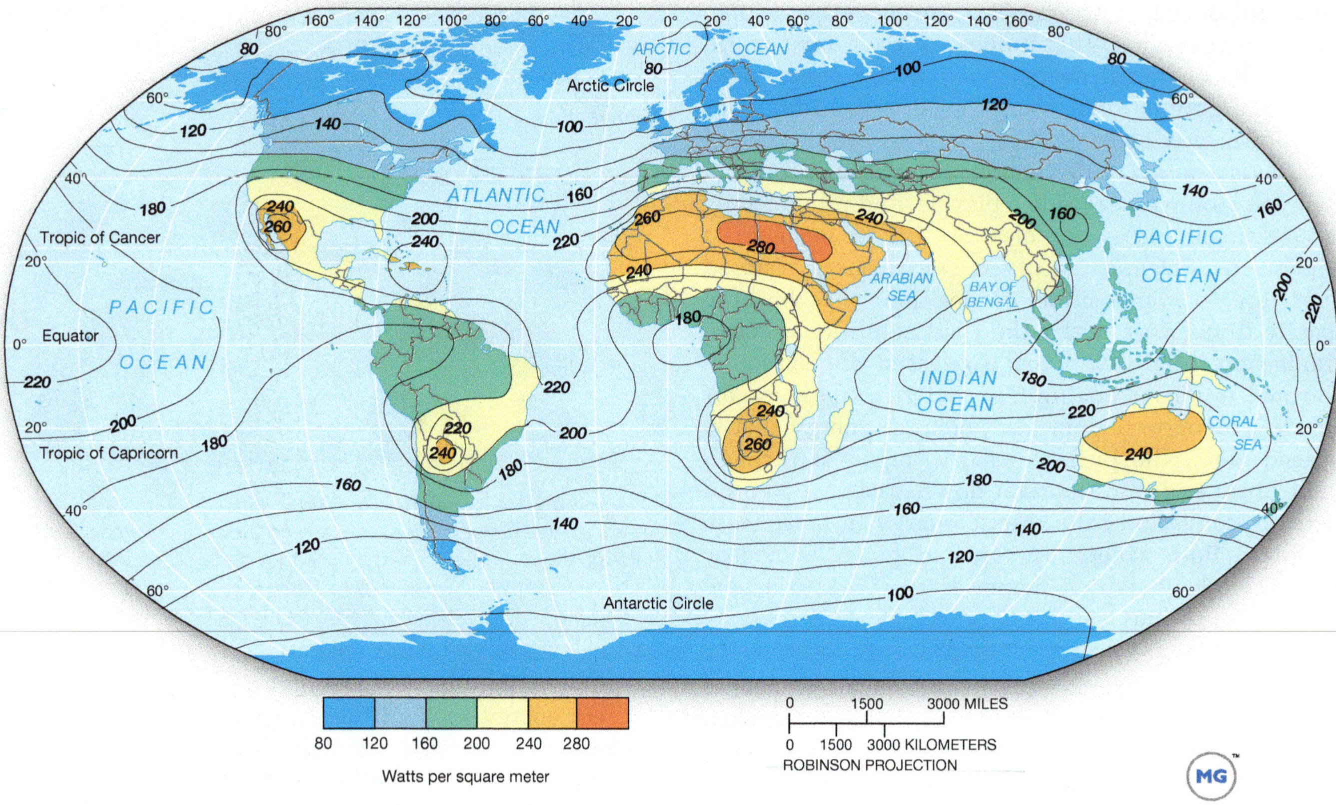

▲**Figure 4.3 Insolation at Earth's surface.** Average annual solar radiation received on a horizontal surface at ground level in watts per square meter (100 W/m^2 = 75 $kcal/cm^2$/year). [Based on M. I. Budyko, The Heat Balance of the Earth's Surface (Washington, DC: U.S. Department of Commerce, 1958).]

Satellite
Global Shortwave Radiation

Energy Pathways and Principles

Insolation, or incoming solar radiation, is the single energy input driving the Earth–atmosphere system, yet it is not equal at all surfaces across the globe (Figure 4.3). Consistent daylength and high Sun altitude produce fairly consistent insolation values (about 180–220 watts per square meter; W/m^2) throughout the equatorial and tropical latitudes. Insolation decreases toward the poles, from about 25° latitude in both the Northern and the Southern Hemispheres. In general, greater insolation at the surface (about 240–280 W/m^2) occurs in low-latitude deserts worldwide because of frequently cloudless skies. Note this energy pattern in the subtropical deserts in both hemispheres (for example, the Sonoran desert in the American southwest, the Sahara in North Africa, and the Kalahari in South Africa).

Scattering and Diffuse Radiation

Insolation encounters an increasing density of atmospheric molecules as it travels toward Earth's surface. These atmospheric gases, as well as dust, cloud droplets, water vapor, and pollutants, physically interact with insolation to redirect radiation, changing the direction of the light's movement without altering its wavelengths. **Scattering** is the name for this phenomenon, which accounts for a percentage of the insolation that does not reach Earth's surface but is instead reflected back to space.

Incoming energy that reaches Earth's surface after scattering occurs is **diffuse radiation** (labeled in Figure 4.1). This weaker, dispersed radiation is composed of waves traveling in different directions, and thus casts shadowless light on the ground. In contrast, *direct radiation* travels in a straight line to Earth's surface without being scattered or otherwise affected by materials in the atmosphere. (The values on the surface-insolation map in Figure 4.3 combine both direct and diffuse radiation.)

Have you wondered why Earth's sky is blue? And why sunsets and sunrises are often red? These common questions are answered using a principle known as Rayleigh scattering (named for English physicist Lord Rayleigh, 1881). This principle applies to radiation scattered by small gas molecules and relates the amount of scattering in the atmosphere to wavelengths of light—shorter wavelengths are scattered more, longer wavelengths are scattered less.

Looking back to Chapter 2, Figure 2.5, we see that blues and violets are the shorter wavelengths of visible light. According to the Rayleigh scattering principle, these wavelengths are scattered more than longer wavelengths such as orange or red. When we look at the sky with the sun overhead, we see the wavelengths that are scattered the most throughout the atmosphere. Although both blues and violets are scattered, our human eye perceives this color mix as blue, resulting in the common observation of a blue sky.

For atmospheric particles larger than the wavelengths of light (such as many pollutants), the Rayleigh scattering principle does not apply. Mie scattering is the process that works on these particles. In a sky filled with smog and haze, the larger particles scatter all wavelengths of visible light evenly, making the sky appear almost white.

The altitude of the Sun determines the thickness of the atmosphere through which its rays must pass to reach an observer. Direct rays (from overhead) pass through less atmosphere and experience less scattering than do low, oblique-angle rays, which must travel farther through the atmosphere. When the Sun is low on the horizon at sunrise or sunset, shorter wavelengths (blue and violet) are scattered out, leaving only the residual oranges and reds to reach our eyes.

Refraction As insolation enters the atmosphere, it passes from one medium to another, from virtually empty space into atmospheric gases. A change of medium also occurs when insolation passes from air into water. Such transitions subject the insolation to a change of speed, which also shifts its direction—this is the bending action of **refraction**. In the same way, a crystal or prism refracts light passing through it, bending different wavelengths to different angles, separating the light into its component colors to display the spectrum. A rainbow is created when visible light passes through myriad raindrops and is refracted and reflected toward the observer at a precise angle (Figure 4.4).

Another example of refraction is a **mirage**, an image that appears near the horizon when light waves are refracted by layers of air at different temperatures (and consequently of different densities). The atmospheric distortion of the setting Sun in Figure 4.5 is also a product of refraction. When the Sun is low in the sky, light must penetrate more air than when the Sun is high; thus light is refracted through air layers of different densities on its way to the observer.

Refraction adds approximately 8 minutes of daylight that we would lack if Earth had no atmosphere. We see the Sun's image about 4 minutes before the Sun actually peeks over the horizon. Similarly, as the Sun sets, its refracted image is visible from over the horizon for about 4 minutes afterward. These extra minutes vary with atmospheric temperature, moisture, and pollutants.

▲**Figure 4.4 A rainbow.** Raindrops—and in this photo, moisture droplets from the Niagara River—refract and reflect light to produce a primary rainbow. Note that in the primary rainbow the colors with the shortest wavelengths are on the inside and those with the longest wavelengths are on the outside. In the secondary bow, note that the color sequence is reversed because of an extra angle of reflection within each moisture droplet. [Bobbé Christopherson.]

Reflection and Albedo A portion of arriving energy bounces directly back into space—this is **reflection**. The reflective quality, or intrinsic brightness, of a surface is **albedo**, an important control over the amount of insolation that reaches Earth. We report albedo as the

▲**Figure 4.5 Sun refraction.** The distorted appearance of the Sun as it sets over the ocean is produced by refraction of the Sun's image in the atmosphere. Have you ever noticed this effect? [Author.]

GEOreport 4.1 Did light refraction sink the Titanic?

An unusual optical phenomenon called "super refraction" may explain why the *Titanic* struck an iceberg in 1912, and why the *California* did not come to her aid during that fateful April night. Recently, a British historian combined weather records, survivors' testimony, and ships' logs to determine that atmospheric conditions were conducive to a bending of light that causes objects to be obscured in a mirage in front of a "false" horizon. Under these conditions, the *Titanic's* lookouts could not see the iceberg until too late to turn, and the nearby *California* could not identify the sinking ship. Read the full story at http://www.smithsonianmag.com/science-nature/Did-the-Titanic-Sink-Because-of-an-Optical-Illusion.html?.

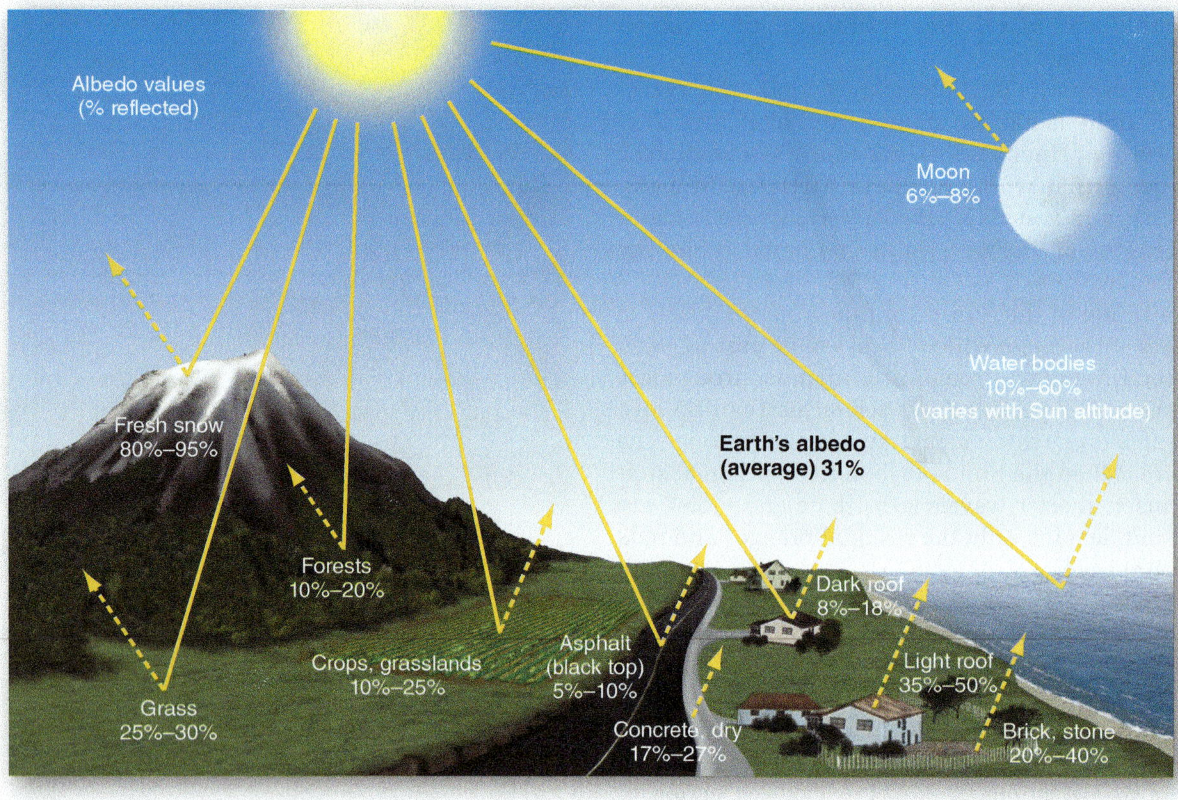

▲Figure 4.6 Various albedo values. In general, light surfaces are more reflective than dark surfaces and thus have higher albedo values.

MG Satellite Global Albedo Values, Global Shortwave Radiation

percentage of insolation that is reflected—0% is total absorption; 100% is total reflectance.

In terms of visible wavelengths, darker colored surfaces (such as asphalt) have lower albedos, and lighter-colored surfaces (such as snow) have higher albedos (Figure 4.6). On water surfaces, the angle of the solar rays also affects albedo values: Lower angles produce more reflection than do higher angles. In addition, smooth surfaces increase albedo, whereas rougher surfaces reduce it.

Individual locations can experience highly variable albedo values during the year in response to changes in cloud and ground cover. Satellite data reveal that albedos average 19%–38% for all surfaces between the tropics (23.5° N to 23.5° S), whereas albedos for the polar regions may be as high as 80% as a result of ice and snow. Tropical forests with frequent cloud cover are characteristically low in albedo (15%), whereas generally cloudless deserts have higher albedos (35%).

Earth and its atmosphere reflect 31% of all insolation when averaged over a year. The glow of Earth's albedo, or the sunshine reflected off Earth, is called *earthshine*. By comparison, a full Moon, which is bright enough to read by under clear skies, has only a 6%–8% albedo value. Thus, with earthshine being four times brighter than moonlight (four times the albedo), and with Earth four times greater in diameter than the Moon, it is no surprise that astronauts report how startling our planet looks from space.

Absorption Insolation, both direct and diffuse, that is not part of the 31% reflected from Earth's surface and atmosphere is absorbed, either in the atmosphere or by Earth's surface. **Absorption** is the assimilation of radiation by molecules of matter, converting the radiation from one form of energy to another. Solar energy is absorbed by land and water surfaces (about 45% of incoming insolation), as well as by atmospheric gases, dust, clouds, and stratospheric ozone (together about 24% of incoming insolation). It is converted into either longwave radiation or chemical energy (the latter by plants, in photosynthesis). The process of absorption raises the temperature of the absorbing surface.

The atmosphere does not absorb as much radiation as Earth's surface because gases are selective about the wavelengths they absorb. For example, oxygen and ozone effectively absorb ultraviolet radiation in the stratosphere. None of the atmospheric gases absorb the wavelengths of visible light, which pass through the atmosphere to Earth as direct radiation. Several gases—water vapor and carbon dioxide, in particular—are good absorbers of longwave radiation emitted by Earth. These gases absorb heat in the lower troposphere, a process that explains why Earth's atmosphere is warmer at the surface, acting like a natural greenhouse.

Clouds, Aerosols, and Atmospheric Albedo Clouds and aerosols are unpredictable factors in the tropospheric energy budget. The presence or absence of

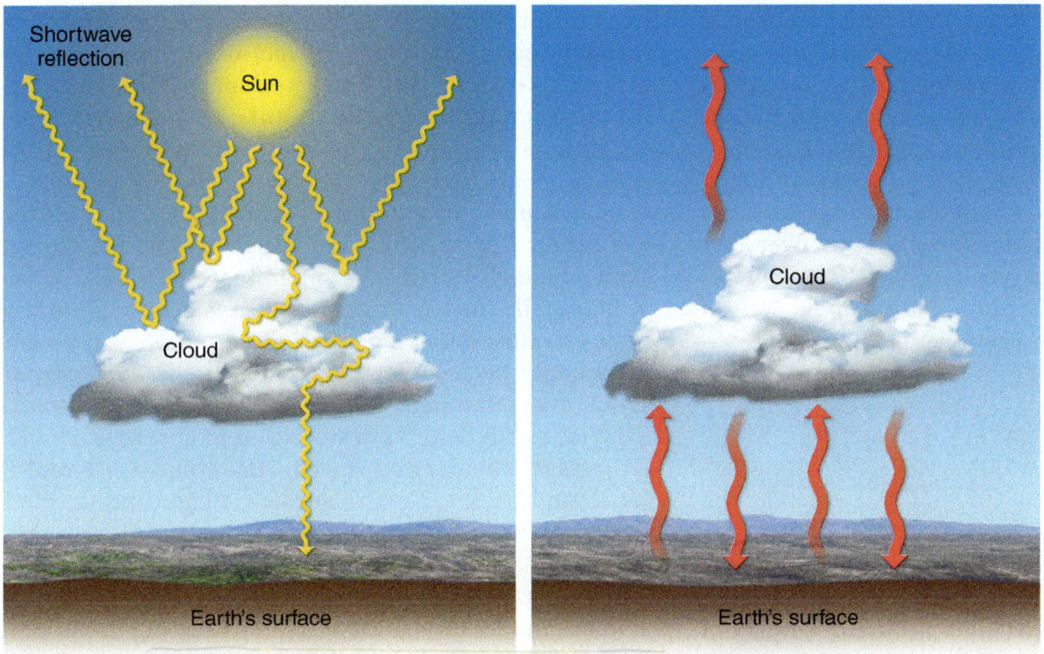

(a) Clouds reflect and scatter shortwave radiation, returning a high percentage to space.

(b) Clouds absorb and reradiate longwave radiation emitted by Earth; some longwave energy returns to space and some toward the surface.

Figure 4.7 The effects of clouds on shortwave and longwave radiation.

clouds may make a 75% difference in the amount of energy that reaches the surface. Clouds reflect shortwave insolation, so that less insolation reaches Earth's surface, and they absorb longwave radiation leaving Earth (Figure 4.7). Longwave radiation trapped by an insulating cloud layer can create a warming of Earth's atmosphere called the *greenhouse effect* (discussed in the next section).

Air pollutants from both natural and anthropogenic sources affect atmospheric albedo. The 1991 eruption of Mount Pinatubo in the Philippines, introduced in Chapter 1, injected approximately 15–20 megatons of sulfur dioxide droplets into the stratosphere. Winds rapidly spread these aerosols worldwide (see the images in Figure 6.1), resulting in an increase in global atmospheric albedo and a temporary average cooling of 0.5 C° (0.9 F°). Scientists have correlated similar cooling trends with other large volcanic eruptions throughout history.

In rapidly developing regions such as China and India, industrial pollutants such as sulfate aerosols are increasing the reflectivity of the atmosphere. These aerosols act as an insolation-reflecting haze in clear-sky conditions, cooling Earth's surface. However, some aerosols (especially black carbon) readily absorb radiation and reradiate heat back toward Earth—with warming effects.

Global dimming is the general term describing the pollution-related decline in insolation to Earth's surface. This process is difficult to incorporate into climate models, although evidence shows that it is causing an underestimation of the actual amount of warming happening in Earth's lower atmosphere. One recent study estimates that aerosols reduced surface insolation by 20% during the first decade of this century.

The connections between pollution and cooling of Earth's surface provide an example of how components of the Earth energy budget affect other Earth systems. Scientists have discovered that air pollution over the northern Indian Ocean causes a reduction in average summer monsoon precipitation. The increased presence of aerosols in this region increases cloud cover, causing surface cooling. This in turn causes less evaporation, which leads to less moisture in the atmosphere

GEOreport 4.2 Aerosols cool and warm Earth's climate

Ground and satellite measurements indicate that global stratospheric aerosols increased 7% from 2000 to 2010. The effect of aerosols on climate depends on whether they reflect or absorb insolation, which in turn depends in part on the composition and color of the particles. In general terms, brightly colored or translucent particles such as sulfates and nitrates tend to reflect radiation, with cooling effects. Darker aerosols such as black carbon absorb radiation, with warming effects (although black carbon also shades the surface, with slight cooling effects). The effects of dust are variable, depending in part on whether it is coated with black or organic carbon. One recent climate model showed that removing all aerosols over the eastern United States could lead to a slight average increase in warming and an increase in the severity of annual heat waves. Research into the complex role of these tiny atmospheric particles in energy budgets is critical for understanding climate change.

and in the monsoonal flow (read more about monsoons in Chapter 6). A weakening of the seasonal monsoon will have a negative impact on regional water resources and agriculture in this densely populated area. Recent research by an international team confirmed connections between aerosols and rainfall patterns, and suggests that increased aerosols make some regions more prone to extreme precipitation events. (An overview of the effects of aerosols is at http://earthobservatory.nasa.gov/Features/Aerosols/page1.php; information about aerosol monitoring is at http://www.esrl.noaa.gov/gmd/aero/.)

Energy Balance in the Troposphere

The Earth–atmosphere energy system budget naturally balances itself in a steady-state equilibrium. The inputs of shortwave energy to Earth's atmosphere and surface from the Sun are eventually balanced by the outputs of shortwave energy reflected and longwave energy emitted from Earth's atmosphere and surface back to space. Think of cash flows into and out of a checking account, and the balance that results when cash deposits and withdrawals are equal (see Figure GIA 4).

Certain gases in the atmosphere effectively delay longwave energy losses to space and act to warm the lower atmosphere. In this section, we examine this "greenhouse" effect and then develop an overall, detailed energy budget for the troposphere.

The Greenhouse Effect and Atmospheric Warming

In Chapter 2, we characterized Earth as a cooler blackbody radiator than the Sun, emitting energy in longer wavelengths from its surface and atmosphere toward space. However, some of this longwave radiation is absorbed by carbon dioxide, water vapor, methane, nitrous oxide, chlorofluorocarbons (CFCs), and other gases in the lower atmosphere and then emitted back, or *reradiated*, toward Earth. This process affects the heating of Earth's atmosphere. The rough similarity between this process and the way a greenhouse operates gives the process its name—the **greenhouse effect**. The gases associated with this process are collectively termed **greenhouse gases**.

The "Greenhouse" Concept

In a greenhouse, the glass is transparent to shortwave insolation, allowing light to pass through to the soil, plants, and materials inside, where absorption and conduction take place. The absorbed energy is then emitted as longwave radiation, warming the air inside the greenhouse. The glass physically traps both the longer wavelengths and the warmed air inside the greenhouse, preventing it from mixing with cooler outside air. Thus, the glass acts as a one-way filter, allowing the shortwave energy in, but not allowing the longwave energy out except through conduction or convection by opening the greenhouse-roof vent. You experience the same process in a car parked in direct sunlight. Opening the car windows allows the air inside to mix with the outside environment, thereby removing heated air physically from one place to another by convection. The interior of a car gets surprisingly hot with the windows closed, even on a day with mild temperatures outside.

Overall, the atmosphere behaves a bit differently. In the atmosphere, the greenhouse analogy does not fully apply because longwave radiation is not trapped as in a greenhouse. Rather, its passage to space is delayed as the longwave radiation is absorbed by certain gases, clouds, and dust in the atmosphere and is reradiated back to Earth's surface. According to scientific consensus, today's increasing carbon dioxide concentration in the lower atmosphere is absorbing more longwave radiation, some of which gets reradiated back toward Earth, thus producing a warming trend and related changes in the Earth–atmosphere energy system.

Clouds and Earth's "Greenhouse"

As discussed earlier, clouds sometimes cause cooling and other times cause heating of the lower atmosphere, in turn affecting Earth's climate (Figure 4.7). The effect of clouds is dependent on the percentage of cloud cover, as well as cloud type, altitude, and thickness (water content and density). Low, thick stratus clouds reflect about 90% of insolation. The term **cloud-albedo forcing** refers to an increase in albedo caused by such clouds, and the resulting cooling of Earth's climate (albedo effects exceed greenhouse effects, shown in Figure 4.8a). High-altitude, ice-crystal clouds reflect only about 50% of incoming insolation. These cirrus clouds act as insulation, trapping longwave radiation from Earth and raising minimum temperatures. This is **cloud-greenhouse forcing**, which causes warming of Earth's climate (greenhouse effects exceed albedo effects, shown in Figure 4.8b).

Jet contrails (condensation trails) produce high cirrus clouds stimulated by aircraft exhaust—sometimes called *false cirrus clouds*, or *contrail cirrus* (Figure 4.8c and d). Contrails both cool and warm the atmosphere, and these opposing effects make it difficult for scientists to determine their overall role in Earth's energy budget. Recent research indicates that contrail cirrus trap outgoing radiation from Earth at a slightly greater rate than they reflect insolation, suggesting that their overall effect is a positive radiative forcing, or warming, of climate. When numerous contrails merge and spread in size, their effect on Earth's energy budget may be significant.

The three-day grounding of commercial air traffic following the September 11, 2001, terrorist attacks on the World Trade Center provided researchers with an opportunity to assess contrail effects on temperatures. Dr. David Travis and his research team compared

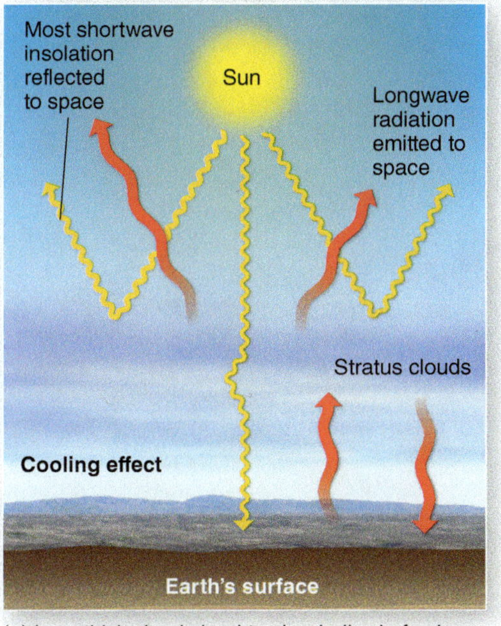

(a) Low, thick clouds lead to cloud-albedo forcing and atmospheric cooling.

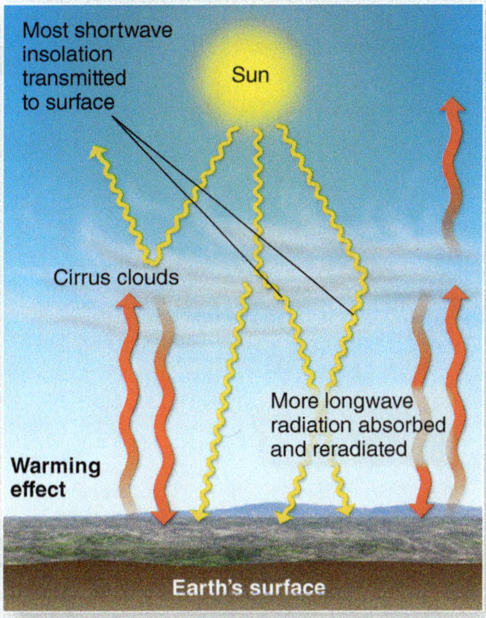

(b) High, thin clouds lead to cloud-greenhouse forcing and atmospheric warming.

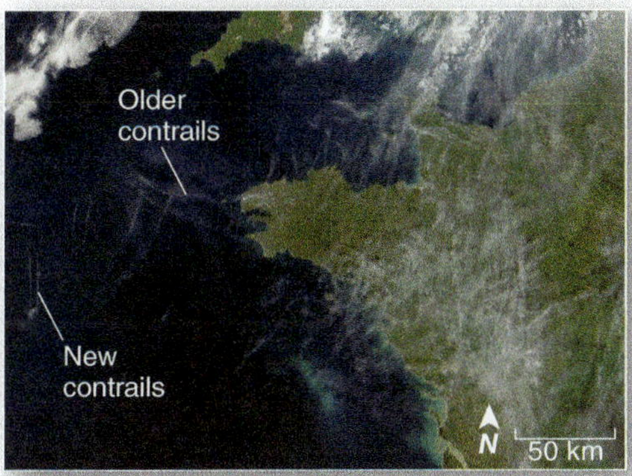

(c) Contrails over Brittany, France, in 2004. Newer contrails are thin; older contrails have widened and formed thin, high cirrus clouds, with an overall warming effect on Earth.

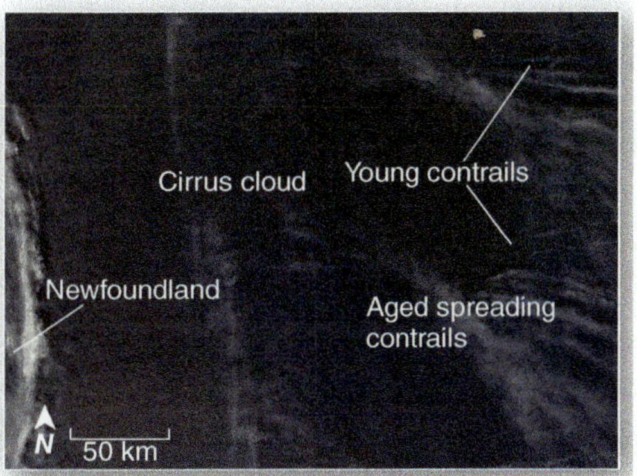

(d) Contrails off the Canadian east coast in 2012 expand with age under humid conditions, spreading outward until they are difficult to differentiate from natural cirrus clouds.

Figure 4.8 Energy effects of cloud types. [(c) NASA *Terra* MODIS. (d) NASA *Aqua* MODIS.]

MG Animation Global Warming, Climate Change

MG Animation Earth–Atmosphere Energy Balance

weather data from 4000 stations for the three-day shutdown with data from the past 30 years. Their research suggests that contrails reduce the diurnal temperature range (daytime maximum to nighttime minimum) in regions with a high density of aircraft. Active research continues on these ongoing effects of aviation.

Earth–Atmosphere Energy Balance

If Earth's surface and its atmosphere are considered separately, neither exhibits a balanced radiation budget in which inputs equal outputs. The average annual energy distribution is positive (an energy surplus, or gain) for Earth's surface and negative (an energy deficit, or loss) for the atmosphere as it radiates energy to space. However, when considered together, these two equal each other, making it possible for us to construct an overall energy balance.

The *Geosystems in Action* feature (pages 92–93) summarizes the Earth–atmosphere radiation balance, bringing together all the elements discussed in this chapter by following 100% of arriving insolation through the troposphere. Incoming energy is on the left in the illustration; outgoing energy is on the right.

Summary of Inputs and Outputs Follow along the illustration on the next two pages as you read this section. Out of 100% of the solar energy arriving at the top of the atmosphere, 31% is reflected back to space—this is Earth's average albedo. This includes scattering (7%), reflection by clouds and aerosols (21%), and reflection by Earth's surface (3%). Another 21% of arriving solar energy is absorbed by the atmosphere—3% by clouds, 18% by atmospheric gases and dust. Stratospheric ozone absorption accounts for another 3%. About 45% of the incoming insolation transmits through to Earth's surface as direct and diffuse shortwave radiation. In sum, Earth's atmosphere and surface absorb 69% of incoming shortwave radiation: 21% (atmosphere heating) + 45% (surface heating) + 3% (ozone absorption) = 69%. Earth eventually emits this 69% as longwave radiation back into space.

(text continued on page 94)

Incoming solar energy in the form of shortwave radiation interacts with both the atmosphere and Earth's surface (GIA 4.1). The surface reflects or absorbs some of the energy, reradiating the absorbed energy as longwave radiation (GIA 4.2). Averaged over a year, Earth's surface has an energy gain, or surplus, while the atmosphere has an energy deficit, or loss. These two amounts of energy equal each other, maintaining an overall balance in Earth's energy "budget."

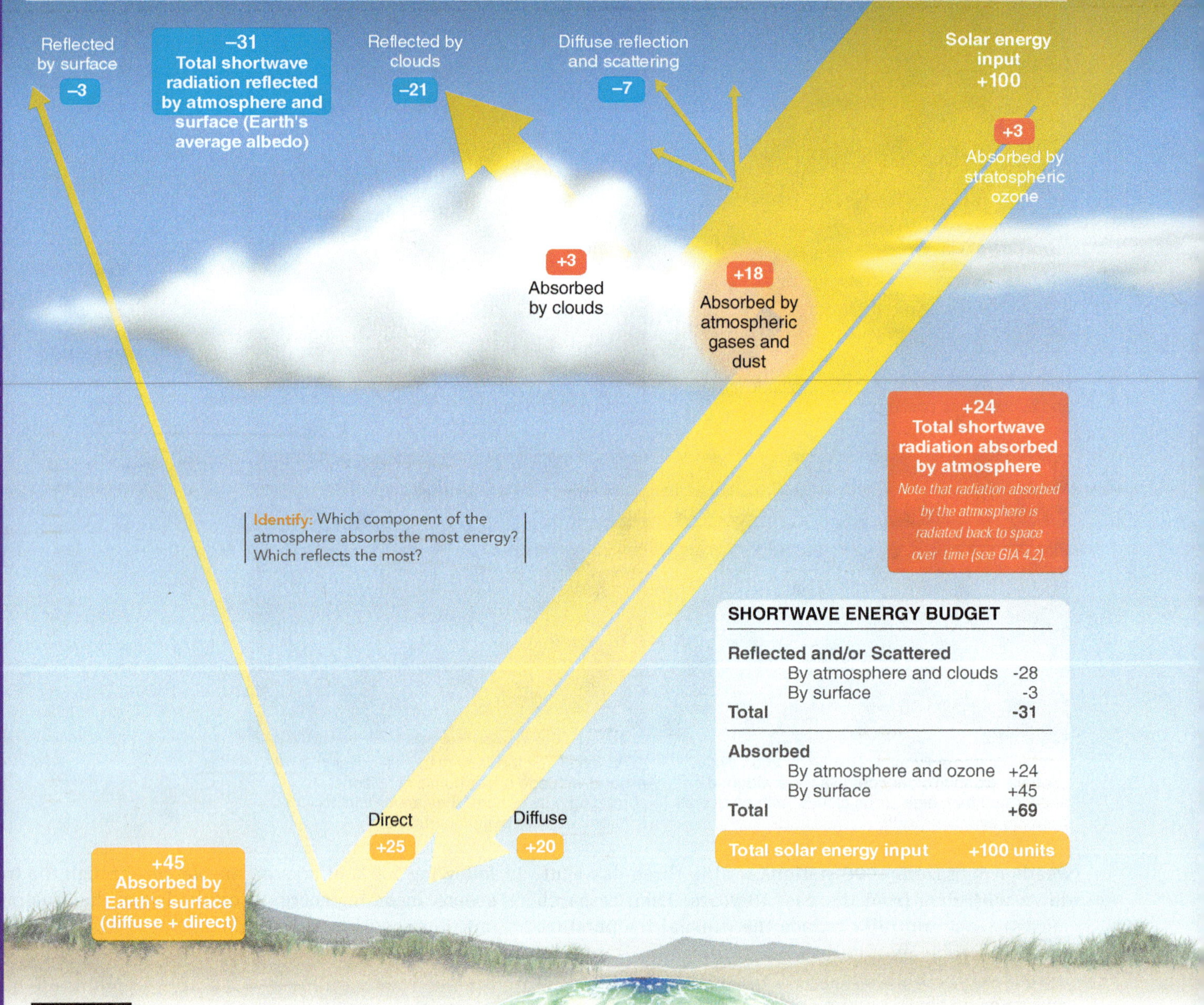

SHORTWAVE ENERGY BUDGET

Reflected and/or Scattered	
By atmosphere and clouds	-28
By surface	-3
Total	**-31**
Absorbed	
By atmosphere and ozone	+24
By surface	+45
Total	**+69**
Total solar energy input	**+100 units**

4.1

SHORTWAVE RADIATION INPUTS AND ALBEDO

Solar energy cascades through the lower atmosphere where it is absorbed, reflected, and scattered. Clouds, atmosphere, and the surface reflect 31% of this insolation back to space. Atmospheric gases and dust and Earth's surface absorb energy and radiate longwave radiation (as shown in GIA 4.2).

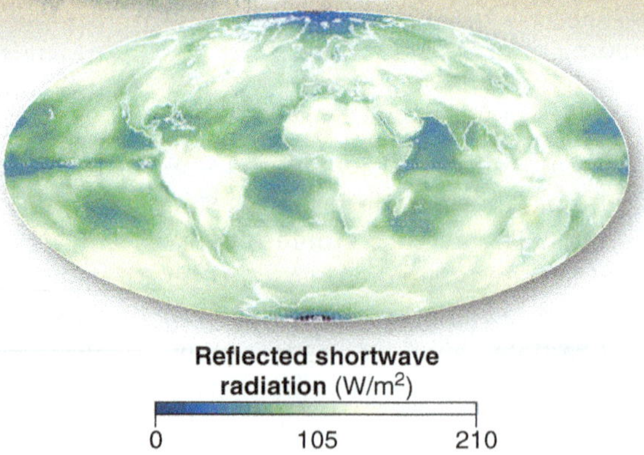

Outgoing shortwave energy reflected from atmosphere, clouds, land, and water, equivalent to Earth's albedo. [NASA, CERES.]

Animation
Atmospheric Energy Balance

MasteringGeography™

Visit the Study Area in MasteringGeography™ to explore the Earth-atmosphere energy balance.

Visualize: Study geosciences animations of atmospheric energy balance.

Assess: Demonstrate understanding of the Earth-atmosphere energy balance (if assigned by instructor).

LONGWAVE ENERGY BUDGET	
Radiative transfer	
Ozone layer	-3
Gases and dust	-18
Clouds	-3
Direct from surface	-8
Through greenhouse effect	-14
Nonradiative transfer	
Latent heat transfer	-19
Convection	-4
Total longwave	**-69**
[Total initially reflected]	**-31**
Total energy output	**-100 units**

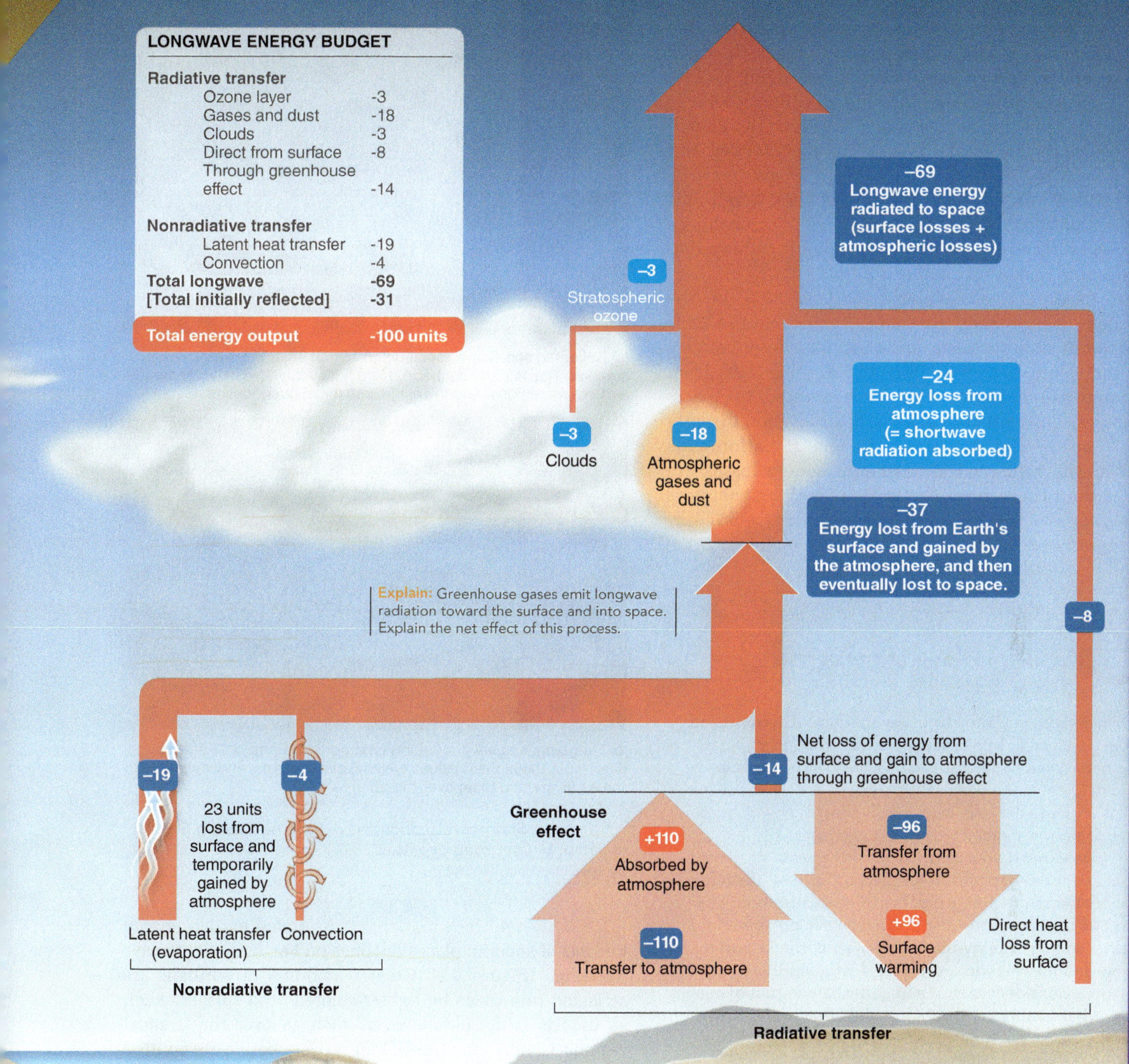

Explain: Greenhouse gases emit longwave radiation toward the surface and into space. Explain the net effect of this process.

4.2 OUTGOING LONGWAVE RADIATION

Over time, Earth emits, on average, 69% of incoming energy to space. When added to the amount of energy reflected (31%), this equals the total energy input from the Sun (100%). Outgoing energy transfers from the surface are both *radiative* (consisting of longwave radiation directly to space) and *non-radiative* (involving convection and the energy released by latent heat transfer).

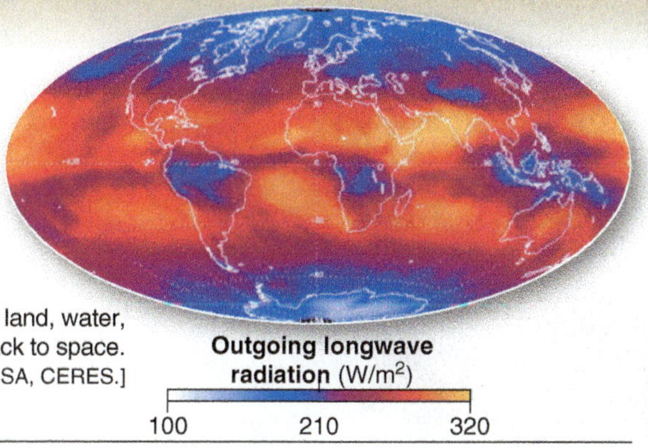

Longwave energy emitted by land, water, atmosphere, and cloud surfaces back to space. [NASA, CERES.]

GEOquiz

1. Infer: Which is more important in heating Earth's atmosphere: incoming shortwave radiation or outgoing longwave radiation? Explain.

2. Predict: Suppose that latent heat energy is released into the atmosphere as water evaporates from a lake. How is this energy involved in the atmosphere's energy balance and what eventually happens to it?

Outgoing energy transfers from the surface are both *nonradiative* (involving physical, or mechanical, motion) and *radiative* (consisting of radiation). Nonradiative transfer processes include convection (4%) and the energy released by *latent heat transfer*, the energy absorbed and dissipated by water as it evaporates and condenses (19%). Radiative transfer is by longwave radiation between the surface, the atmosphere, and space (represented on the right in Figure 4.9 as the greenhouse effect and direct loss to space). Stratospheric ozone radiation to space makes up another 3%.

In total, the atmosphere radiates 58% of the absorbed energy back to space, including the 21% absorbed by clouds, gases, and dust; 23% from convective and latent heat transfers; and another 14% from net longwave radiation that is reradiated to space. Earth's surface emits 8% of absorbed radiation directly back to space, and stratospheric ozone radiation adds another 3%. Note that atmospheric energy losses are greater than those from Earth. However, the energy is in balance overall: 61% atmospheric losses + 8% surface losses = 69%.

CRITICAL thinking 4.1
A Kelp Indicator of Surface Energy Dynamics

In Antarctica, on Petermann Island off the Graham Land Coast, a piece of kelp (a seaweed) was dropped by a passing bird. When photographed (Figure CT 4.1.1), the kelp lay about 10 cm (4 in.) deep in the snow, in a hole about the same shape as the kelp. In your opinion, what energy principles or pathways interacted to make this scene?

Now, expand your conclusion to the issue of mining coal and other deposits in Antarctica. For now, the international Antarctic Treaty blocks mining exploitation. Construct a case for continuing the ban on mining operations based on your energy-budget analysis of the kelp in the snow and information on surface energy budgets in this chapter. Consider the dust and particulate output of mining. What factors can you think of that might favor such mining? •

▲**Figure CT 4.1.1** [Bobbé Christopherson.]

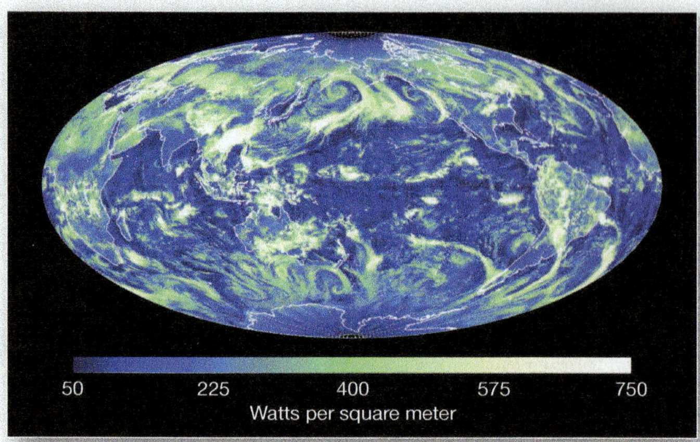

(a) Reflected shortwave radiation on March 18, 2011, near vernal equinox. This is Earth's albedo. Note high values (white) over cloudy and snowy regions, and lowest values (blue) over oceans.

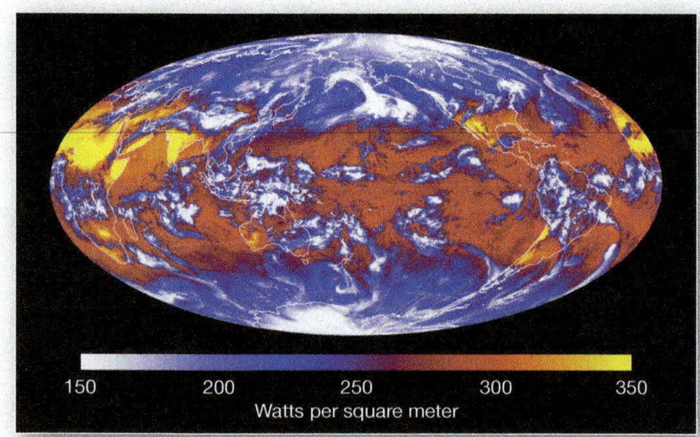

(b) Outgoing longwave radiation emitted from Earth on the same date. Note the highest values (yellow) over deserts, and lowest values (white and blue) over the polar regions.

▲**Figure 4.9 Shortwave and longwave images, showing Earth's radiation-budget components.** [CERES instrument aboard *Aqua*, Langley Research Center, NASA.]

Latitudinal Energy Imbalances As stated earlier, energy budgets at specific places or times on Earth are not always the same (Figure 4.9). Greater amounts of sunlight are reflected into space by lighter-colored land surfaces such as deserts or by cloud cover, such as over the tropical regions. Greater amounts of longwave radiation are emitted from Earth to space in subtropical desert regions where little cloud cover is present over surfaces that absorb a lot of heat. Less longwave energy is emitted over the cooler polar regions and over tropical lands covered in thick clouds (in the equatorial Amazon region, in Africa, and in Indonesia).

Figure 4.10 summarizes the Earth–atmosphere energy budget by latitude:

- Between the tropics, the angle of incoming insolation is high and daylength is consistent, with little seasonal variation, so more energy is gained than lost—*energy surpluses dominate.*
- In the polar regions, the Sun is low in the sky, surfaces are light (ice and snow) and reflective, and for up to 6 months during the year no insolation is received, so more energy is lost than gained—*energy deficits prevail.*

▼**Figure 4.10 Energy budget by latitude.** Earth's energy surpluses and deficits produce poleward transport of energy and mass in each hemisphere, through atmospheric circulation and ocean currents. Outside of the tropics, atmospheric winds are the dominant means of energy transport toward each pole.

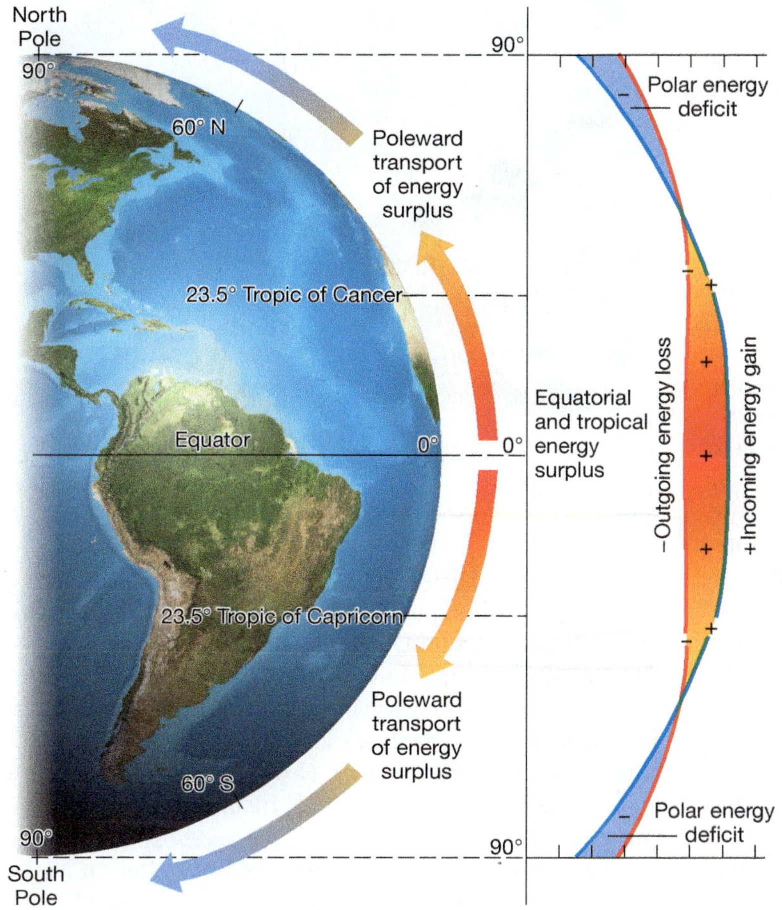

- At around 36° latitude, a balance exists between energy gains and losses for the Earth–atmosphere system.

The imbalance of energy from the *tropical surpluses* and the *polar deficits* drives a vast global circulation pattern. The meridional (north–south) transfer agents are winds, ocean currents, dynamic weather systems, and other related phenomena. Dramatic examples of such energy and mass transfers are tropical cyclones (hurricanes and typhoons) discussed in Chapter 8. After forming in the tropics, these powerful storms mature and migrate to higher latitudes, carrying with them water and energy that redistributes across the globe.

Energy Balance at Earth's Surface

Solar energy is the principal heat source at Earth's surface; the surface environment is the final stage in the Sun-to-Earth energy system. The radiation patterns at Earth's surface—inputs of diffuse and direct radiation and outputs of evaporation, convection, and radiated longwave energy—are important in forming the environments where we live.

Daily Radiation Patterns

Figure 4.11 shows the daily pattern of absorbed incoming shortwave energy and resulting air temperature. This graph represents idealized conditions for bare soil on a cloudless day in the middle latitudes. Incoming energy arrives during daylight, beginning at sunrise, peaking at noon, and ending at sunset.

The shape and height of this insolation curve vary with season and latitude. The maximum heights for such a curve occur at the time of the summer solstice (around June 21 in the Northern Hemisphere and December 21 in the Southern Hemisphere). The air temperature plot also responds to seasons and variations in insolation input. Within a 24-hour day, air temperature generally peaks between 3:00 and 4:00 P.M. and dips to its lowest point right at or slightly after sunrise.

Note that the insolation curve and the air temperature curve on the graph do not align; there is a *lag* between them. The warmest time of day occurs not at the moment of maximum insolation but at the moment when a maximum of insolation has been absorbed and emitted to the atmosphere from the ground. As long as the incoming energy exceeds the outgoing energy, air temperature continues to increase, not peaking until the incoming energy begins to diminish as the afternoon Sun's altitude decreases. If you have ever gone camping in the mountains, you no doubt experienced the coldest time of day with a wake-up chill at sunrise.

The annual pattern of insolation and air temperature exhibits a similar lag. For the

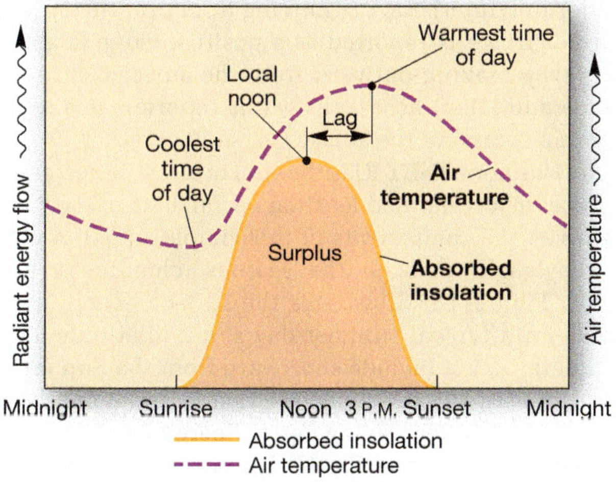

▲**Figure 4.11 Daily radiation and temperature curves.** Sample radiation plot for a typical day shows the changes in insolation (solid line) and air temperature (dashed line). Comparing the curves reveals a lag between local noon (the insolation peak for the day) and the warmest time of day.

Northern Hemisphere, January is usually the coldest month, occurring after the December solstice and the shortest days. Similarly, the warmest months of July and August occur after the June solstice and the longest days.

A Simplified Surface Energy Budget

Energy and moisture are continually exchanged with the lower atmosphere at Earth's surface—this is the *boundary layer* (also known as the atmospheric, or planetary, boundary layer). The energy balance in the boundary layer is affected by the specific characteristics of Earth's surface, such as the presence or absence of vegetation and local topography. The height of the boundary layer is not constant over time or space.

Microclimatology is the science of physical conditions, including radiation, heat, and moisture, in the boundary layer at or near Earth's surface. *Microclimates* are local climate conditions over a relatively small area, such as in a park, or on a particular slope, or in your backyard. Thus, our discussion now focuses on small-scale (the lowest few meters of the atmosphere) rather than large-scale (the troposphere) energy-budget components. (See Chapters 10 and 10B for further discussion of climates and climate change.)

The surface in any given location receives and loses shortwave and longwave energy according to the following simple scheme:

$$\underset{\text{(Insolation)}}{+SW\downarrow} - \underset{\text{(Reflection)}}{SW\uparrow} + \underset{\text{(Infrared)}}{LW\downarrow} - \underset{\text{(Infrared)}}{LW\uparrow} = \underset{\text{(Net radiation)}}{\text{NET R}}$$

We use SW for shortwave and LW for longwave for simplicity.*

Figure 4.12 shows the components of a surface energy balance over a soil surface. The soil column continues to a depth at which energy exchange with surrounding materials or with the surface becomes negligible, usually less than a meter. Heat is transferred by conduction through the soil, predominantly downward during the day (or in summer) and toward the surface at night (or in winter). Energy moving from the atmosphere into the surface is reported as a positive value (a gain), and energy moving outward from the surface, through sensible and latent heat transfers, is reported as a negative value (a loss) in the surface account.

Net radiation (NET R) is the sum of all radiation gains and losses at any defined location on Earth's surface. NET R varies as the components of this simple equation vary with daylength through the seasons, cloudiness, and latitude. Figure 4.13 illustrates the surface energy components for a typical summer day at a midlatitude location. Energy gains include shortwave from the Sun (both diffuse and direct) and longwave that is reradiated from the atmosphere after leaving Earth. Energy losses include reflected shortwave and Earth's longwave emissions that pass through to the atmosphere and space.

Different symbols are used in the microclimatology literature, such as K for shortwave, L for longwave, and Q for NET R (net radiation).

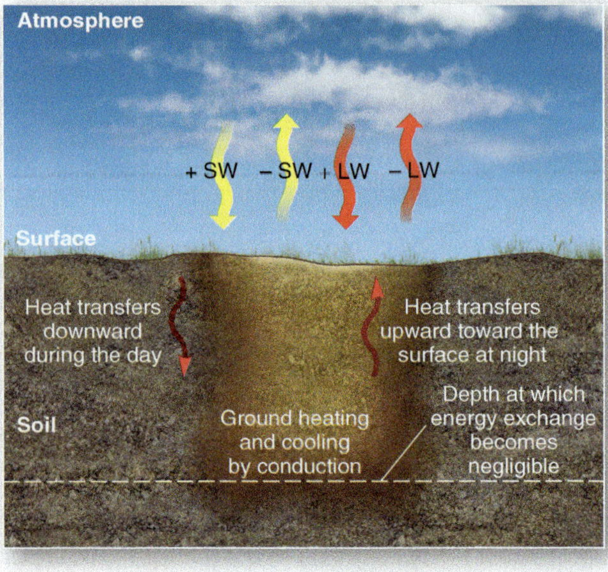

▲**Figure 4.12 Surface energy-budget components over soil column.** Idealized input and output of energy at the surface and within a column of soil (SW = shortwave, LW = longwave).

On a daily basis, NET R values are positive during the daylight hours, peaking just after noon with the peak in insolation; at night, values become negative because the shortwave component ceases at sunset and the surface continues to lose longwave radiation to the atmosphere. The surface rarely reaches a zero NET R value—a perfect balance—at any one moment. However, over time, Earth's total surface naturally balances incoming and outgoing energies.

The principles and processes of net radiation at the surface have a bearing on the design and use of solar energy technologies that concentrate shortwave energy

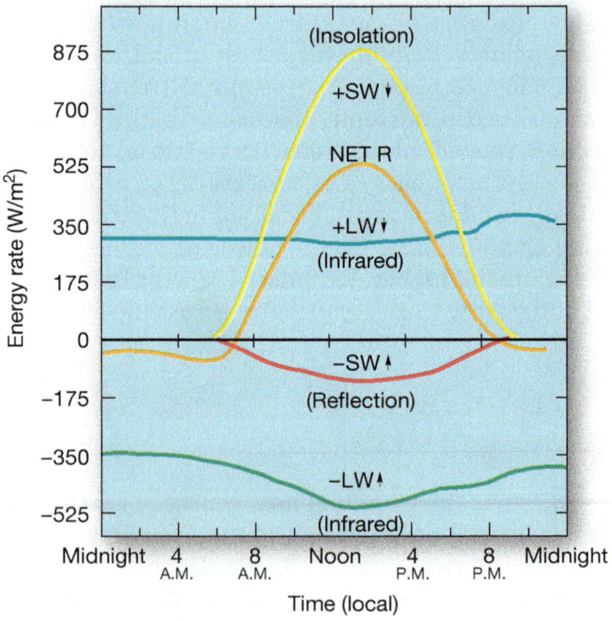

▲**Figure 4.13 Daily radiation budget.** Radiation budget on a typical July summer day at a midlatitude location (Matador in southern Saskatchewan, about 51° N). [Adapted by permission from T. R. Oke, *Boundary Layer Climates* (New York: Methuen & Co., 1978), p. 21.]

for human use. Solar energy offers great potential worldwide and is presently the fastest-growing form of energy conversion by humans. Focus Study 4.1 briefly reviews this direct application of surface energy budgets.

CRITICALthinking 4.2
Applying Energy-Balance Principles to a Solar Cooker

In Focus Study 4.1, you learn about solar cookers (see Figure 4.1.1c). Based on what you have learned about energy balance in this chapter, what principles are most important for making a solar cooker work? Can you diagram the energy flows involved (inputs, outputs, the role of albedo)? How would you position the cooker to maximize its productivity in terms of sunlight? At what time of day is the cooker most effective? ●

Global and Seasonal Net Radiation The net radiation available at Earth's surface is the final outcome of the entire energy-balance process discussed in this chapter. On a global scale, average annual net radiation is positive over most of Earth's surface (Figure 4.14). Negative net radiation values probably occur only over ice-covered surfaces poleward of 70° latitude in both hemispheres. Note the abrupt differences in net radiation between ocean and land surfaces on the map. The highest net radiation value, 185 W/m² per year, occurs north of the equator in the Arabian Sea. Aside from the obvious interruptions caused by landmasses, the pattern of values appears generally zonal, or parallel, decreasing away from the equator.

The seasonal rhythm of net radiation throughout the year influences the patterns of life on Earth's surface. Seasonal net radiation shifts between the solstice months of December and June and the equinoxes, with net radiation gains near the equator at the equinoxes shifting toward the South Pole during the December solstice and toward the North Pole during the June solstice.

Net Radiation Expenditure As we have learned, in order for the energy budget at Earth's surface to balance over time, areas that have positive net radiation must somehow dissipate, or lose, heat. This happens through nonradiative processes that move energy from the ground into the boundary layer.

- The *latent heat of evaporation* (*LE*) is the energy that is stored in water vapor as water evaporates. Water absorbs large quantities of this *latent heat* as it changes state to water vapor, thereby removing this heat energy from the surface. Conversely, this heat energy releases to the environment when water vapor changes state back to a liquid (discussed in Chapter 7). Latent heat is the dominant expenditure of Earth's entire NET R, especially over water surfaces.
- *Sensible heat* (*H*) is the heat transferred back and forth between air and surface in turbulent eddies through convection and conduction within materials. This activity depends on surface and boundary-layer temperature differences and on the intensity of convective motion in the atmosphere. About one-fifth of Earth's entire NET R is mechanically radiated as sensible heat from the surface, especially over land. The bulk of NET R is expended as sensible heat in these dry regions.

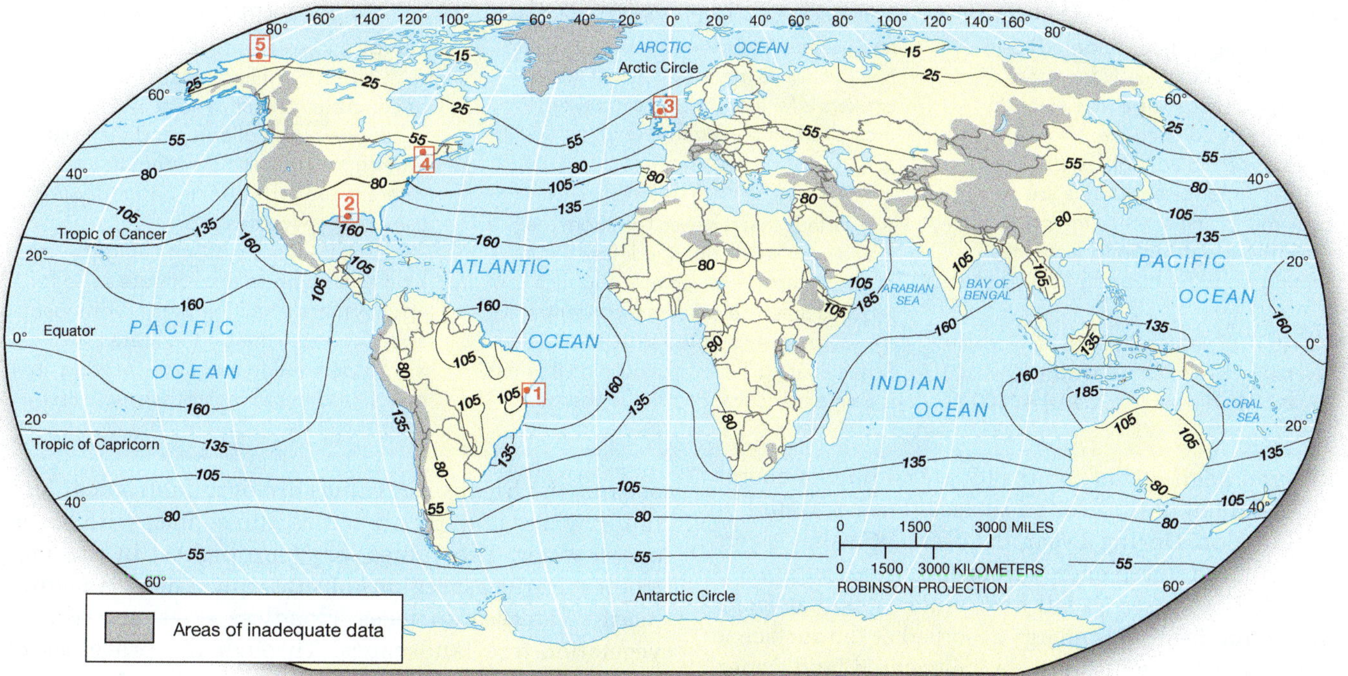

▲**Figure 4.14 Global net radiation at Earth's surface.** Distribution of mean annual net radiation (NET R) at the surface in watts per square meter (100 W/m² = 75 kcal/cm²/year). The highest net radiation, 185 W/m² per year, occurs north of the equator in the Arabian Sea. Temperatures for the five cities noted on the map are graphed in Figure 5.5. [Based on M. I. Budyko, *The Heat Balance of the Earth's Surface* (Washington, DC: U.S. Department of Commerce, 1958), p. 106.]

MG Satellite Global Net Radiation

1. Salvador, Brazil
2. New Orleans, Louisiana
3. Edinburgh, Scotland
4. Montreal, Canada
5. Barrow, Alaska

(see Figure 5.5)

Focus Study 4.1 Sustainable Resources
Solar Energy Applications

Consider the following:

- Earth receives 100,000 terawatts (TW) of solar energy per hour, enough to meet world power needs for a year.* In the United States, the energy produced by fossil fuels in a year arrives in equivalent insolation every 35 minutes.
- An average commercial building in the United States receives 6 to 10 times more energy from the Sun hitting its exterior than is required to heat the inside.
- Photographs from the early 1900s show solar (flat-plate) water heaters used on rooftops. Today, installed solar water heating is making a comeback, serving 50 million households worldwide, more than 300,000 of these in the United States.
- Photovoltaic capacity is more than doubling every 2 years. Solar-electric power passed 89,500 megawatts installed in 2012.

(a) Women in East Africa carry home solar-box cookers that they made at a workshop. Construction is easy, using cardboard components.

(b) Kenyan women in training to use solar-panel cookers.

(c) These simple cookers collect insolation through transparent glass or plastic and trap longwave radiation in an enclosing box or cooking bag.

▲**Figure 4.1.1 The solar cooking solution.** [(a) and (b) Solar Cookers International, Sacramento, California; (c) Bobbé Christopherson.]

Not only does insolation warm Earth's surface, but it also provides an inexhaustible supply of energy for humanity. Sunlight is directly and widely available, and solar power installations are decentralized, labor intensive, and renewable. Although collected for centuries using various technologies, sunlight remains underutilized.

Rural villages in developing countries could benefit greatly from the simplest, most cost-effective solar application—the solar cooker (Figure 4.1.1). With access to solar cookers, people in rural Latin America and Africa are able to cook meals and sanitize their drinking water without walking long distances collecting wood for cooking fires. These solar devices are simple yet efficient, reaching temperatures between 107°C and 127°C (225°F and 260°F). See http://solarcookers.org/ for more information.

In developing countries, the pressing need is for decentralized energy sources, appropriate in scale to everyday needs, such as cooking, heating and boiling water, and pasteurizing other liquids and food. Net per capita (per person) cost for solar cookers is far less than for centralized electrical production, regardless of fuel source.

Collecting Solar Energy

Any surface that receives light from the Sun is a *solar collector.* But the diffuse nature of solar energy received at the surface requires that it be collected, concentrated, transformed, and stored to be most useful. Space heating (heating of building interiors) is a simple application of solar energy. It can be accomplished by careful design and placement of windows so that sunlight will shine into a building and be absorbed and converted into sensible heat—an everyday application of the greenhouse effect.

A *passive solar system* captures heat energy and stores it in a "thermal mass," such as a water-filled tank, adobe, tile, or concrete. An *active solar system* heats water or air in a collector and then pumps it through a plumbing system to a tank, where it can provide hot water for direct use or for space heating.

Solar energy systems can generate heat energy on an appropriate scale for approximately half the present domestic applications in the United States, including space heating and water heating. In marginal climates, solar-assisted water

*Terawatt (10^{12} watts) = 1 trillion W; gigawatt (10^9 watts) = 1 billion W; megawatt (10^6 watts) = 1 million W; kilowatt (10^3 watts) = 1000 W.

- *Ground heating and cooling* (*G*) is the flow of energy into and out of the ground surface (land or water) by conduction. During a year, the overall *G* value is zero because the stored energy from spring and summer is equaled by losses in fall and winter. Another factor in ground heating is energy absorbed at the surface to melt snow or ice. In snow- or ice-covered landscapes, most available energy is in sensible and latent heat used in the melting and warming process.

On land, the highest annual values for *LE* occur in the tropics and decrease toward the poles. Over the oceans, the highest *LE* values are over subtropical latitudes, where hot, dry air comes into contact with warm ocean water. The values for *H* are highest in the subtropics. Here vast regions of subtropical deserts feature nearly waterless surfaces, cloudless skies, and almost vegetation-free landscapes. Through the processes of latent, sensible, and ground heat transfer, the energy from net radiation is able to do the "work" that ultimately produces the global climate system—work such as raising temperatures in the boundary layer, melting ice, or evaporating water from the oceans.

and space heating is feasible as a backup; even in New England and the Northern Plains states, solar collection systems prove effective.

Kramer Junction, California, in the Mojave Desert near Barstow, California, has the world's largest operating solar-electric generating system, with a capacity of 150 MW (megawatts), or 150 million watts. Long troughs of computer-guided curved mirrors concentrate sunlight to create temperatures of 390°C (735°F) in vacuum-sealed tubes filled with synthetic oil. The heated oil heats water; the heated water produces steam that rotates turbines to generate cost-effective electricity. The facility converts 23% of the sunlight it receives into electricity during peak hours (Figure 4.1.2a), and operation and maintenance costs continue to decrease.

Electricity Directly from Sunlight

Photovoltaic (PV) cells were first used to produce electricity in spacecraft in 1958. Today, they are the solar cells in pocket calculators and are also used in rooftop solar panels that provide electricity. When light shines upon a semiconductor material in these cells, it stimulates a flow of electrons (an electrical current) in the cell.

The efficiency of these cells, often assembled in large arrays, has improved to the level that they are cost competitive, and would be more so if government support and subsidies were balanced evenly among all energy sources. The residential installation in Figure 4.1.2b features 36 panels, generating 205 W each, or 9680 W total, at a 21.5% conversion efficiency. This solar array generates enough surplus energy to run the residential electric meters in reverse and supply electricity to the power grid.

The National Renewable Energy Laboratory (NREL; http://www.nrel.gov/solar_radiation/facilities.html) and the National Center for Photovoltaics were established in 1974 to coordinate solar energy research, development, and testing in partnership with private industry. Testing is ongoing at NREL's Outdoor Test Facility in Golden, Colorado, where solar cells have been developed that broke the 40%-conversion-efficiency barrier!

Rooftop photovoltaic electrical generation is now cheaper than power line construction to rural sites. PV roof systems provide power to hundreds of thousands of homes in Mexico, Indonesia, Philippines, South Africa, India, and Norway. (See the "Photovoltaic Home Page" at http://www.eere.energy.gov/solar/sunshot/pv.html.)

Obvious drawbacks of both solar heating and solar electric systems are periods of cloudiness and night, which inhibit operations. Research is under way to enhance energy storage and to improve battery technology.

The Promise of Solar Energy

Solar energy is a wise choice for the future. It is economically preferable to further development of our decreasing fossil-fuel reserves, or further increases in oil imports and tanker and offshore oil-drilling spills, investment in foreign military incursions, or development of nuclear power, especially in a world with security issues.

Whether to pursue the development of solar energy is more a matter of political choice than a question of technological possibilities. Much of the technology is ready for installation and is cost-effective when all the direct and indirect costs of other energy resources are considered.

On the *MasteringGeography* website for Chapter 4, you can find several listings of URLs relating to solar energy applications. Take some time to learn more about these necessary technologies (solar-thermal, solar-electric photovoltaic cells, solar-box cookers, and the like). As we near the climatic limitations of the fossil-fuel era and the depletion of fossil-fuel resources, renewable energy technologies are essential to the fabric of our lives. What kinds of solar technology are available and in use in your area?

(a) Kramer Junction solar-electric generating system in southern California.

(b) A residence with a 9680-W rooftop solar photovoltaic array of 46 panels. Excess electricity is fed to the grid for credits that offset 100% of the home's electric bill, plus producing surplus electricity for charging an electric vehicle.

▲Figure 4.1.2 **Solar thermal and photovoltaic energy production.** [Bobbé Christopherson.]

Two Sample Stations Variation in the expenditure of NET R among the processes just described produces the variety of environments we experience in nature. Let us examine the daily radiation budget at two locations, El Mirage in California and Pitt Meadows in British Columbia (Figure 4.15).

El Mirage, at 35° N, is a hot desert location characterized by bare, dry soil with sparse vegetation. The data in the graph is from a clear summer day, with a light wind in the late afternoon. The NET R value is lower than might be expected, considering the Sun's position close to zenith (June solstice) and the absence of clouds. But the income of energy at this site is countered by surfaces with relatively high albedo and by hot soil at the ground surface that radiates longwave energy back to the atmosphere throughout the afternoon.

El Mirage has little or no energy expenditure through evaporation (*LE*). With little water and sparse vegetation, most of the available radiant energy dissipates through turbulent transfer of sensible heat (*H*), warming air and soil to high temperatures. Over a

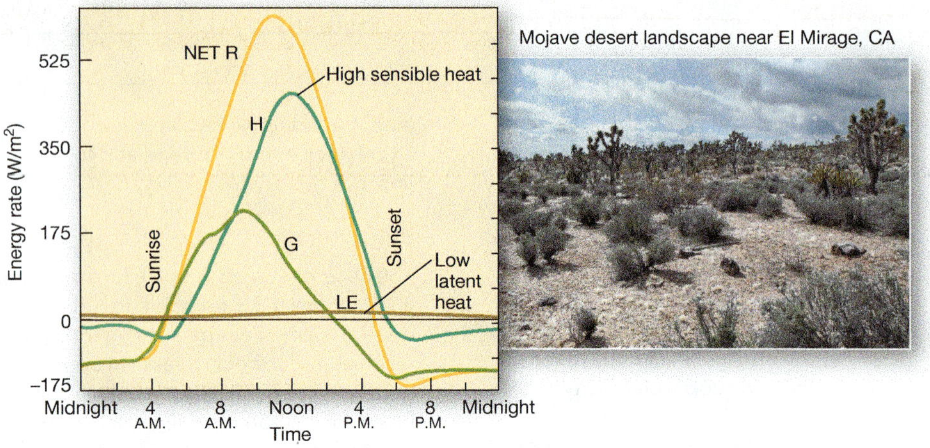

(a) Daily net radiation budget at El Mirage, California, near 35°N latitude

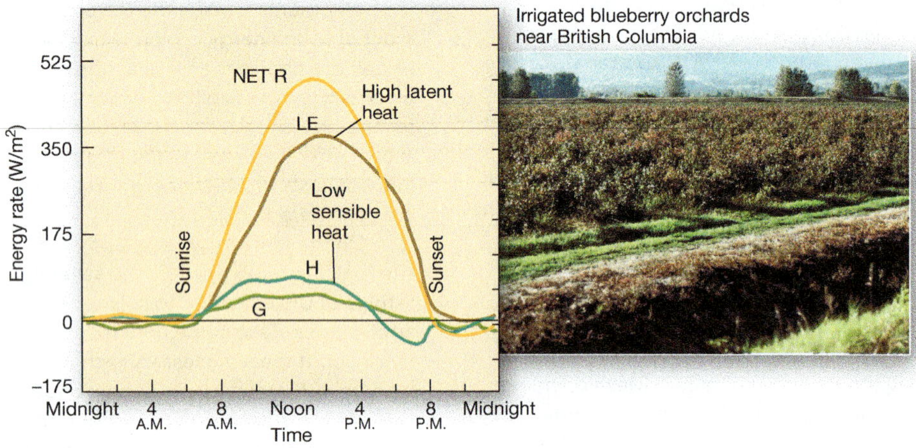

(b) Daily net radiation budget for Pitt Meadows, British Columbia, near 49°N latitude

Figure 4.15 Radiation budgets for two stations. *H* = turbulent sensible heat transfer; *LE* = latent heat of evaporation; *G* = ground heating and cooling. [(a) Based on W. D. Sellers, *Physical Climatology*, © 1965 University of Chicago. (b) Based on T. R. Oke, *Boundary Layer Climates*, © 1978 Methuen & Co. Photos by Bobbé Christopherson.]

24-hour period, *H* is 90% of NET R expenditure; the remaining 10% is for ground heating (*G*). The *G* component is greatest in the morning, when winds are light and turbulent transfers are lowest. In the afternoon, heated air rises off the hot ground, and convective heat expenditures are accelerated as winds increase.

Compare graphs for El Mirage (Figure 4.15a) and Pitt Meadows (Figure 4.15b). Pitt Meadows is midlatitude (49° N), vegetated, and moist, and its energy expenditures differ greatly from those at El Mirage. The Pitt Meadows landscape is able to retain much more of its energy because of lower albedo values (less reflection), the presence of more water and plants, and lower surface temperatures than those of El Mirage.

The energy-balance data for Pitt Meadows are from a cloudless summer day. Higher *LE* values result from the moist environment of rye grass and irrigated mixed orchards, contributing to the more moderate *H* levels during the day.

The Urban Environment

Urban microclimates generally differ from those of nearby nonurban areas, with urban areas regularly reaching temperatures as much as 6 C° (10 F°) hotter than surrounding suburban and rural areas. In fact, the surface energy characteristics of urban areas are similar to those of desert locations, mainly because vegetation is lacking in both environments.

The physical characteristics of urbanized regions produce an **urban heat island (UHI)** that has, on average, both maximum and minimum temperatures higher than nearby rural settings (Figure 4.16). A UHI has increasing temperatures toward the downtown central business district and lower temperatures over areas of trees and parks. Sensible heat is less in urban forests than in other parts of the city because of shading from tree canopies and due to plant processes such as transpiration that move moisture into the air. In New York City, daytime temperatures average 5–10 C° (9–18 F°) cooler in Central Park than in the greater metropolitan area.

Ongoing studies show that UHI effects are greater in bigger cities than smaller ones. The difference between urban and rural heating is more pronounced in cities surrounded by forest rather than by dry, sparsely vegetated environments. UHI effects also tend to be highest in cities with dense population, and slightly lower in cities with more urban sprawl. Go to http://www.nasa.gov/topics/earth/features/heat-island-sprawl.html for an interesting study on UHIs in the U.S. Northeast.

In the average city in North America, heating is increased by modified urban surfaces such as asphalt and glass, building geometry, pollution, and human activity such as industry and transportation. For example, an average car (25 mpg or 10 km/l) produces enough heat to melt a 14-lb bag of ice for every mile driven (4.5 kg ice per km driven). The removal of vegetation and the increase in human-made materials that retain heat are two of the most significant UHI causes. Urban surfaces (metal, glass, concrete, asphalt) conduct up to three times more energy than wet, sandy soil.

Most major cities also produce a **dust dome** of airborne pollution trapped by certain characteristics of air circulation in UHIs: The pollutants collect with a decrease in wind speed in urban centers; they then rise as the surface heats and remain in the air above the city, affecting urban energy budgets. Table 4.1

TABLE 4.1 Urban Heat Islands: Driving Factors and Climatic Response

Driving Factor	Climatic Element Affected	Urban Compared to Rural	Explanation
Thermal properties of urban surfaces: metal, glass, asphalt, concrete, brick	Net radiation	Higher	• Urban surfaces conduct more energy than natural surfaces such as soil.
Reflective properties of urban surfaces	Albedo	Lower	• Urban surfaces often have low albedo, so they absorb and retain heat, leading to high net radiation values.
Urban canyon effect	Wind speed • annual mean • extreme gusts	Less	• Reflected insolation in canyons is conducted into surface materials, thus increasing temperatures.
	Calm periods	More	• Buildings interrupt wind flows, diminishing heat loss through advection (horizontal movement), and block nighttime radiation to space. • Maximum UHI effects occur on calm, clear days and nights.
Anthropogenic heating	Temperature • annual mean • winter minima • summer maxima	Higher	• Heat is generated by homes, vehicles, and factories. • Heat output may surge in winter with power for heating or in summer with power for air conditioning.
Pollution	Air pollution • condensation nuclei • particulates Cloudiness, including fog	More	• Airborne pollutants (dust, particulates, aerosols) in urban dust dome raise temperatures by absorbing insolation and reradiating heat to surface.
	Precipitation Snowfall, inner city Snowfall, downwind	More More Less More	• Increased particulates are condensation nuclei for water vapor, increasing cloud formation and precipitation; heating enhances convection processes.
Urban desert effect: less plant cover and more sealed surfaces	Relative humidity • annual mean Infiltration Runoff Evaporation	Less Less More Less	• Cooling effect of evaporation and plant transpiration is reduced or absent. • Water cannot infiltrate through sealed surfaces to soil; more water flows as runoff. • Urban surfaces respond as desert landscapes—storms may cause "flash floods."

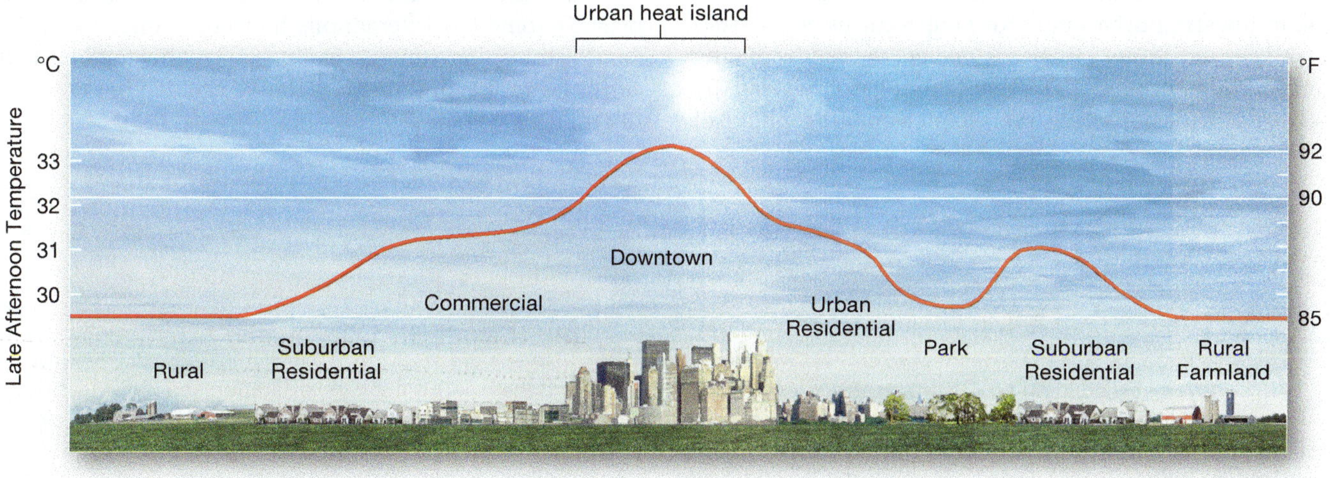

▲**Figure 4.16 Typical urban heat island profile.** On average, urban temperatures may be 1–3 C° (2–5 F°) warmer than nearby rural areas on a sunny summer day. Temperatures are highest at the urban core. Note the cooling over the park and rural areas.
[Based on "Heat Island," Urban Climatology and Air Quality, available at http://weather.msfc.nasa.gov/urban/urban_heat_island.html.]

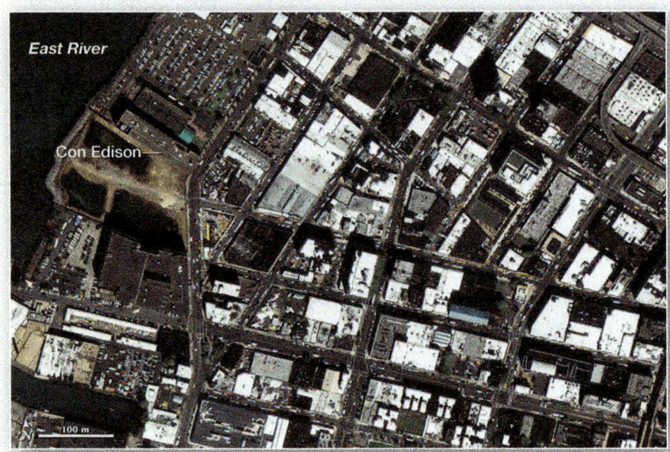

(a) Rooftops in Queens, New York, in September 2011

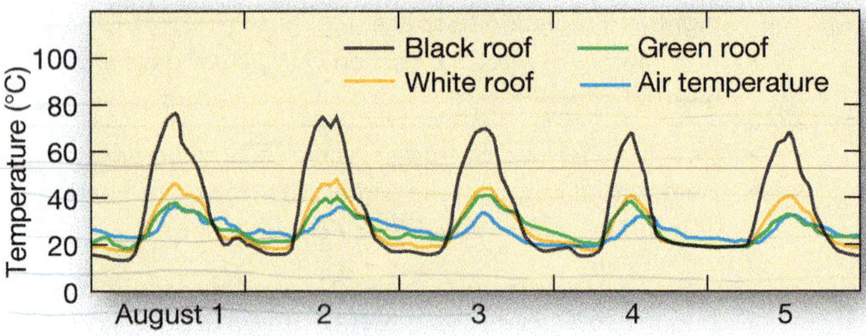

(b) Graph shows the higher temperature of black roofs as compared to bright white or green (vegetated) roofs.

▲ **Figure 4.17 Effects of rooftop materials on urban heat islands.** [(a) Natural color image from Digital Globe's *Worldview-2* satellite. (b) Columbia University and NASA Goddard Institute for Space Studies.]

lists some of the factors that cause UHIs and compares selected climatic elements of rural and urban environments.

City planners and architects use a number of strategies to mitigate UHI effects, including planting of vegetation in parks and open space (urban forests), "green" roofs (rooftop gardens), "cool" roofs (high-albedo roofs), and "cool" pavements (lighter-colored materials such as concrete, or lighter surface coatings for asphalt). In addition to lowering urban outdoor temperatures, such strategies keep buildings' interiors cooler, thereby reducing energy consumption and greenhouse gases caused by fossil-fuel emissions (see the chapter-opening photo of the living roof on Chicago's City Hall). During the hottest day of the 2011 New York City summer, temperature measurements for a white roof covering were 24 C° (42 F°) cooler than for a traditional black roof nearby (Figure 4.17). Other studies have shown that for structures with solar panel arrays, the roof temperatures under the shade of the panels dropped dramatically.

With predictions that 60% of the global population will live in cities by the year 2030, and with air and water temperatures rising because of climate change, UHI issues are emerging as a significant concern both for physical geographers and for the public at large. Studies have found a direct correlation between peaks in UHI intensity and heat-related illness and fatalities. More information is available from the Environmental Protection Agency, including mitigation strategies, publications, and hot-topic discussions, at http://www.epa.gov/hiri/.

CRITICALthinking 4.3
Looking at Your Surface Energy Budget

Given what you now know about reflection, absorption, and net radiation expenditures, assess your wardrobe (fabrics and colors); house, apartment, or dorm (colors of exterior walls, especially south- and west-facing, or in the Southern Hemisphere, north- and west-facing); roof (orientation relative to the Sun and roof color); automobile (color, use of sun shades); bicycle seat (color); and other aspects of your environment to improve your personal energy budget. Are you using colors and materials to save energy and enhance your personal comfort? Do you save money as a result of any of these strategies? What grade do you give yourself? In Chapter 1, you assessed your carbon footprint. How does that relate to these energy-budget considerations (orientation, shade, color, form of transportation)? ●

The Earth and the human denominator feature summarizes some key interactions between humans and the Earth–atmosphere energy balance. The effects of human activities on Earth's energy balance are driving changes in all Earth systems.

GEOreport 4.3 Phoenix leads in urban heat island research

In Phoenix, Arizona, a team of scientists, architects, and urban planners are studying urban climate elements—wind movement, building arrangement, street-level shading effects, and the incorporation of vegetation and water features into urban planning. In 2008, the team (sponsored by Arizona State University, in partnership with the National Weather Service and key stakeholders such as private energy firms and the city government) launched a sustainable urban development plan aimed at UHI mitigation and pedestrian comfort. One strategy: Planting a double row of broad-canopy, low-water-demand trees to increase shading along streets, in conjunction with low shrubs to reduce pedestrian exposure to longwave radiation emitted from asphalt roads.

ENERGY BALANCE ⇨ HUMANS

- The Earth–atmosphere system balances itself naturally, maintaining planetary systems that support Earth, life, and human society, although the steady-state equilibrium appears to be changing.
- Solar energy is harnessed for power production worldwide by technology ranging from small solar cookers to large-scale photovoltaic arrays.

HUMANS ⇨ ENERGY BALANCE

- Thawing of Arctic sea ice illustrates albedo and temperature effects of human-caused climate change.
- Aerosols and other particulates affect the energy budget and have both cooling and warming effects on climate.
- Fossil fuel burning produces carbon dioxide and other greenhouse gases that force warming of the lower atmosphere.
- Urban heat island effects accelerate warming in cities, which house more than half the global human population.

4a

NASA scientists collect data from U.S. Coast Guard platforms in the Atlantic Ocean. Although oceans cover 70% of Earth's surface, scientists have limited knowledge of the oceans' role in the Earth–atmosphere heat balance. [S. Smith, NASA Langley Research Center.]

4b

The International Maritime Organization, made up of 170 countries, is developing policies to reduce diesel ship emissions, especially black carbon. In 2015, improved energy-efficiency standards will be enforced for construction of all large ships. [Justin Kasezninez/Alamy.]

4e

Urban surfaces absorb more radiation than natural surfaces, adding to increased temperatures in urban heat islands. Increased use of lighter-colored pavement lowers surface albedo, reducing this effect. [Arizona State University National Center of Excellence on SMART Innovations.]

4d

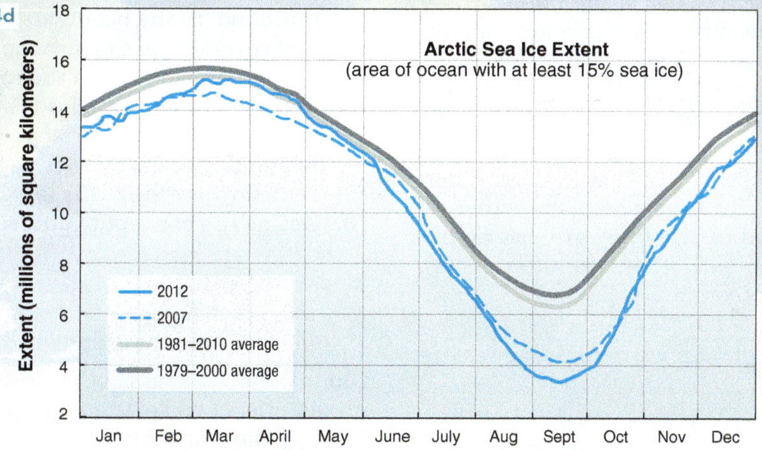

Arctic sea ice reached its lowest extent on record in 2012. Continued loss of summer ice will exacerbate the positive feedback loop between lower albedo, ice melt, and atmospheric warming. [Courtesy National Snow and Ice Data Center.]

4c

The Gujarat Solar Park in western India covers 1214 hectares (3000 acres) and is one of the world's largest photovoltaic solar facilities. The park will reach a capacity of 500 MW when completed in 2014. [Ajit Solanki/AP.]

ISSUES FOR THE 21ST CENTURY

- Continuing albedo effects associated with melting sea ice and aerosols will hasten air- and ocean-warming trends.
- Increased burning of fossil fuels will add more greenhouse gases. Solar energy can offset fossil-fuel use.
- Continuing urbanization will worsen heat-island effects and contribute to the general warming trend.

GEOSYSTEMS**connection**

We tracked the inputs of insolation, followed the cascade of energy through the atmosphere to Earth's surface, and examined the resulting movements and processes. We ended this journey from Sun to Earth by analyzing the distribution of energy at the surface and the concepts of energy balance and net radiation. In the next two chapters, we shift our focus to other outputs in the energy–atmosphere system—the results and consequences of global temperature patterns, and the wind and ocean current circulations driven by these flows of energy.

KEY LEARNING concepts review

Define energy and heat, and *explain* four types of heat transfer: radiation, conduction, convection, and advection.

Radiant energy from the Sun that cascades to the surface powers Earth's biosphere. **Transmission** is the uninterrupted passage of shortwave and longwave energy through either the atmosphere or water. Our budget of atmospheric energy comprises shortwave radiation *inputs* (ultraviolet light, visible light, and near-infrared wavelengths) and shortwave and longwave (thermal infrared) radiation *outputs*. *Energy* is the capacity to do work, or move matter. The energy of motion is *kinetic energy*, produced by molecular vibrations and measured as temperature. *Potential energy* is stored energy that has the capacity to do work under the right conditions, such as when an object moves or falls with gravity. The flow of kinetic energy from one body to another resulting from a temperature difference between them is **heat**. Two types are **sensible heat**, energy that we can feel and measure, and *latent heat*, "hidden" heat that is gained or lost in phase changes, such as from solid to liquid to gas and back, while the substance's temperature remains unchanged.

One mechanism of heat transfer is *radiation*, which flows in electromagnetic waves and does not require a medium such as air or water. **Conduction** is the molecule-to-molecule transfer of heat as it diffuses through a substance. Heat also is transferred in gases and liquids by **convection** (physical mixing that has a strong vertical motion) or **advection** (mixing in which the dominant motion is horizontal). In the atmosphere or bodies of water, warmer portions tend to rise (they are less dense) and cooler portions tend to sink (they are more dense), establishing patterns of convection.

transmission (p. 84)
heat (p. 85)
sensible heat (p. 85)
conduction (p. 85)
convection (p. 85)
advection (p. 85)

1. Give several examples of each type of heat transfer. Do you observe any of these processes on a daily basis?

Identify alternative pathways for solar energy on its way through the troposphere to Earth's surface—scattering, refraction, and absorption—and *review* the concept of albedo (reflectivity).

The molecules and particles of the atmosphere may redirect radiation, changing the direction of the light's movement *without altering its wavelengths*. This **scattering** represents 7% of Earth's reflectivity, or albedo. Dust particles, pollutants, ice, cloud droplets, and water vapor produce further scattering. Some incoming insolation is scattered by clouds and atmosphere and is transmitted to Earth as **diffuse radiation**, the downward component of scattered light.

The speed of insolation entering the atmosphere changes as it passes from one medium to another; the change of speed causes a bending action called **refraction**. **Mirage** is a refraction effect in which an image appears near the horizon when light waves are refracted by layers of air at different temperatures (and consequently of different densities). **Reflection** is the process in which a portion of arriving energy bounces directly back into space without reaching Earth's surface. **Albedo** is the reflective quality (intrinsic brightness) of a surface. Albedo can greatly reduce the amount of insolation that is available for absorption by a surface. We report albedo as the percentage of insolation that is reflected. Earth and its atmosphere reflect 31% of all insolation when averaged over a year. **Absorption** is the assimilation of radiation by molecules of a substance, converting the radiation from one form to another—for example, visible light to infrared radiation.

scattering (p. 86)
diffuse radiation (p. 86)
refraction (p. 87)
mirage (p. 87)
reflection (p. 87)
albedo (p. 87)
absorption (p. 88)

2. What would you expect the sky color to be at an altitude of 50 km (30 mi)? Why? What factors explain the lower atmosphere's blue color?
3. Define *refraction*. How is it related to daylength? To a rainbow? To the beautiful colors of a sunset?
4. List several types of surfaces and their albedo values. What determines the reflectivity of a surface?
5. Using Figure 4.6, explain the differences in albedo values for various surfaces. Based on albedo alone, which of two surfaces is cooler? Which is warmer? Why do you think this is?
6. Define the concept of absorption.

Analyze the effect of clouds and aerosols on atmospheric heating and cooling, and *explain* the greenhouse concept as it applies to Earth.

Clouds, aerosols, and other atmospheric pollutants have mixed effects on solar energy pathways, either cooling or heating the atmosphere. **Global dimming** describes the decline in sunlight reaching Earth's surface owing to pollution, aerosols, and clouds and is perhaps masking the actual degree of global warming.

Carbon dioxide, water vapor, methane, and other gases in the lower atmosphere absorb infrared radiation that is then emitted to Earth, thus delaying energy loss to space—this process is the **greenhouse effect**. In the atmosphere, longwave radiation is not actually trapped, as it would be in a greenhouse, but its passage to space is delayed (heat energy is detained in the atmosphere) through absorption and reradiation by **greenhouse gases**.

Cloud-albedo forcing is the increase in albedo, and therefore in reflection of shortwave radiation, caused by clouds, resulting in a cooling effect at the surface. Also, clouds can act as insulation, thus trapping longwave radiation and raising minimum temperatures. An increase in greenhouse warming caused by clouds is **cloud-greenhouse forcing**. Clouds' effects on the heating of the lower atmosphere depend on cloud type, height, and thickness (water content and density). High-altitude, ice-crystal clouds reflect insolation, producing a net cloud greenhouse forcing (warming); thick, lower cloud cover reflects about 90%, producing a net cloud albedo forcing (cooling). **Jet contrails**, or condensation trails, are produced by aircraft exhaust, particulates, and water vapor and can form high cirrus clouds, sometimes called *false cirrus clouds*.

global dimming (p. 89)
greenhouse effect (p. 90)
greenhouse gases (p. 90)
cloud albedo forcing (p. 90)
cloud greenhouse forcing (p. 90)
jet contrails (p. 90)

7. What is the effect of aerosols on heating and cooling of Earth's atmosphere?
8. What are the similarities and differences between an actual greenhouse and the gaseous atmospheric greenhouse? Why is Earth's greenhouse effect changing?
9. What role do clouds play in the Earth–atmosphere radiation balance? Is cloud type important? Compare high, thin cirrus clouds and lower, thick stratus clouds.
10. Jet contrails affect the Earth–atmosphere balance in what ways? Describe the recent scientific findings.

■ *Review* the Earth–atmosphere energy balance and the patterns of global net radiation.

The Earth–atmosphere energy system naturally balances itself in a steady-state equilibrium. It does so through energy transfers that are *nonradiative* (convection, conduction, and the latent heat of evaporation) and *radiative* (longwave radiation traveling between the surface, the atmosphere, and space).

In the tropical latitudes, high insolation angle and consistent daylength cause more energy to be gained than lost, producing energy surpluses. In the polar regions, an extremely low insolation angle, highly reflective surfaces, and up to 6 months of no insolation annually cause more energy to be lost, producing energy deficits. This imbalance of net radiation from tropical surpluses to polar deficits drives a vast global circulation of both energy and mass. Average monthly net radiation at the top of the atmosphere varies seasonally, highest in the northern hemisphere during the June solstice and highest in the southern atmosphere during the December solstice.

Surface energy measurements are used as an analytical tool of **microclimatology**. **Net radiation (NET R)** is the value reached by adding and subtracting the energy inputs and outputs at some location on the surface; it is the sum of all shortwave (SW) and longwave (LW) radiation gains and losses. Average annual net radiation varies over Earth's surface and is highest at lower latitudes. Net radiation is the energy available to do the "work" of running the global climate system, by melting ice, raising temperatures in the atmosphere, and evaporating water from the oceans.

microclimatology (p. 96)
net radiation (NET R) (p. 96)

11. Sketch a simple energy-balance diagram for the troposphere. Label each shortwave and longwave component and the directional aspects of related flows.
12. In terms of surface energy balance, explain the *net radiation* (*NET R*) and its general pattern on a global scale.

■ *Plot* typical daily radiation and temperature curves for Earth's surface—including the daily temperature lag.

Air temperature responds to seasons and variations in insolation input. Within a 24-hour day, air temperature peaks between 3:00 and 4:00 P.M. and dips to its lowest point right at or slightly after sunrise. Air temperature lags behind each day's peak insolation. The warmest time of day occurs not at the moment of maximum insolation but at the moment when a maximum of insolation is absorbed.

13. Why is there a temperature lag between the highest Sun altitude and the warmest time of day? Relate your answer to the insolation and temperature patterns during the day.
14. What are the nonradiative processes for the expenditure of surface net radiation on a daily basis?
15. Compare the daily surface energy balances of El Mirage, California, and Pitt Meadows, British Columbia. Explain the differences.

■ *List* typical urban heat island conditions and their causes, and *contrast* the microclimatology of urban areas with that of surrounding rural environments.

A growing percentage of Earth's people live in cities and experience a unique set of altered microclimatic effects: increased conduction by urban surfaces, lower albedos, higher NET R values, increased water runoff, complex radiation and reflection patterns, anthropogenic heating, and the gases, dusts, and aerosols of urban pollution. All of these combine to produce an **urban heat island**. Air pollution, including gases and aerosols, is greater in urban areas, producing a **dust dome** that adds to the urban heat island effects.

urban heat island (p. 100)
dust dome (p. 100)

16. What observations form the basis for the urban heat island concept? Describe the climatic effects attributable to urban as compared with nonurban environments.
17. Which of the items in Table 4.1 have you yourself experienced? Explain.
18. Assess the potential for solar energy applications in our society. What are some negatives? What are some positives?

5 Global Temperatures

Wildfires raged across Australia during the 2013 summer heat wave, fueled by drought and record-breaking high temperatures. In Australia's island state of Tasmania, bushfires destroyed over 80 homes. Pictured here, wildfire smoke consumes the horizon over the beaches near Carlton, Tasmania, about 20 km (12 mi) east of Hobart on January 4th. Over a two-day period, new records for Hobart included the hottest January night ever (23.4°C, 74.1°F) followed by the highest maximum daytime temperature in 130 years (41.8°C, 107.2°F). This heat wave fits the overall pattern of global climate change that is underway. [Joanne Giuliani/Reuters.]

KEY LEARNING concepts

After reading the chapter, you should be able to:

- ***Define*** the concept of temperature, and ***distinguish*** between Kelvin, Celsius, and Fahrenheit temperature scales and how they are measured.
- ***Explain*** the effects of latitude, altitude and elevation, and cloud cover on global temperature patterns.
- ***Review*** the differences in heating of land versus water that produce continental effects and marine effects on temperatures, and ***utilize*** a pair of cities to illustrate these differences.
- ***Interpret*** the pattern of Earth's temperatures from their portrayal on January and July temperature maps and on a map of annual temperature ranges.
- ***Discuss*** heat waves and the heat index as a measure of human heat response.

The Mystery of St. Kilda's Shrinking Sheep

On the remote, windswept island of Hirta in Scotland's Outer Hebrides, something odd is happening to the animals. The wild Soay sheep have shrunk in body size over the past quarter century, a trend that puzzled scientists until recently.

The Islands of St. Kilda Some 180 km (110 mi) west of the Scottish mainland, the St. Kilda Archipelago comprises several islands centered at 57.75° N latitude (Figure GN 5.1). These islands have a mild climate with moderate summer and winter temperatures, rain all year long (an average of 1400 mm, or 55 in., annually), and snow occurring only rarely. The mild climate at this latitude is caused in part by the Gulf Stream, a warm ocean current that flows in a clockwise direction around the North Atlantic to northern Europe (see Figure 5.8).

St. Kilda's principal island, Hirta, was continuously occupied for over 3000 years, with people living in stone shelters with sod roofs, and later in stone-walled houses. Domestic sheep were introduced to these islands about 2000 years ago and evolved in isolation as a small, primitive breed now known as Soay sheep. In 1930, the last residents left Hirta as their small society failed. The Soay sheep remained, unmanaged to this day (Figure GN 5.2).

The Problem of the Shrinking Sheep Hirta's sheep provide scientists an ideal opportunity to study an isolated population with no significant competitors or predators. According to evolutionary theory, wild sheep should gradually increase in size over many generations because larger, stronger sheep are more likely to survive winter and reproduce in the spring. This expected trend follows the principle of natural selection. Scientists began studying Hirta's Soay sheep in 1985. Surprisingly, they found that these sheep are not getting larger in size but instead are following the opposite trend: The sheep are getting smaller . . . but why?

The answer, apparently, relates to temperature. Recent temperature shifts have caused summers on Hirta to become longer, with spring arriving a few weeks earlier and fall a few weeks later. Winters have become milder and shorter in length.

Research Provides an Answer In a 2009 publication, scientists reported, first, that female sheep are giving birth at younger ages, when they can only produce offspring that are smaller than they themselves were at birth. This "young mum effect" explains why sheep size is not increasing. But why are sheep shrinking? Since 1985, female Soay sheep of all ages dropped 5% in body mass, with leg length and body weight decreasing. In addition, lambs are not growing as quickly. The explanation ties to rising temperatures: Longer summers increase the availability of grass feed, and milder winters mean that lambs do not need to put on as much weight in their first months, allowing even slower-growing individuals to survive. With smaller lambs surviving winter and reaching breeding age, smaller individuals are becoming more common in the population.

This study indicates that local factors, such as temperature, are significant in the interaction between a species' genetic makeup and its environment. Such factors, even over a short time period, can override the pressures of natural selection and evolution. In this chapter, we discuss temperature concepts, examine global temperature patterns, and explore some of the effects of temperature change on Earth systems. Animal body size is only one of many changes now occurring as a result of climate change.

▲**Figure GN 5.1 Hirta island, part of the St. Kilda Archipelago, Scotland.** [Colin Wilson/Alamy photo.]

▲**Figure GN 5.2 Wild Soay sheep on Hirta Island.** [Bobbé Christopherson.]

▲**Figure 5.1 Wind chill in the south polar region.** Antarctic scientists keep skin surfaces covered in wind-chill conditions reaching −46°C (−51°F) even in summer. [Ted Scambos & Rob Bauer, NSIDC.]

What is the temperature—both indoors and outdoors—as you read these words? How is temperature measured, and what does the measured value mean? How is air temperature influencing your plans for the day?

Our bodies subjectively sense temperature and react to temperature changes. *Apparent temperature* is the general term for the outdoor temperature as it is perceived by humans. Both humidity (the water vapor content of the air) and wind affect apparent temperature and an individual's sense of comfort. Humidity is significant in determining the effect of high temperatures, which may cause heat stress, illness, and loss of life (we discuss heat waves later in the chapter). In cold regions, wind is a more important influence. When strong winds combine with cold temperature, the effects can be deadly (Figure 5.1).

On a cold, windy day, the air feels colder because wind increases evaporative heat loss from our skin, producing a cooling effect. The *wind-chill factor* quantifies the enhanced rate at which body heat is lost to the air. As wind speeds increase, heat loss from the skin increases, and the wind-chill factor rises. To track the effects of wind on apparent temperature, the National Weather Service (NWS) uses the *wind-chill temperature index*, a chart plotting the temperature we feel as a function of actual air temperature and wind speed (Figure 5.2). The goal of the chart is to provide a simple, accurate tool for assessing the dangers to humans from winter winds and freezing temperatures.

The lower wind-chill values on the chart present a serious freezing hazard, called *frostbite*, to exposed flesh. A downhill ski racer going 130 kmph (80 mph) during a 2-minute run can easily incur this type of injury. Another danger is *hypothermia*, a condition of abnormally low body temperature that occurs when the human body is losing heat faster than it can be produced. Even without wind, frostbite and hypothermia are potential dangers in the extreme cold found on high mountain summits and in the polar regions. Moreover, hypothermia can arise in any situation where humans become chilled; it is not exclusively related to freezing temperatures.

Actual Air Temperature in °C (°F)

Wind speed, kmph (mph)									
Calm	4° (40°)	−1° (30°)	−7° (20°)	−12° (10°)	−18° (0°)	−23° (−10°)	−29° (−20°)	−34° (−30°)	−40° (−40°)
8 (5)	2° (36°)	−4° (25°)	−11° (13°)	−17° (1°)	−24° (−11°)	−30° (−22°)	−37° (−34°)	−43° (−46°)	−49° (−57°)
16 (10)	1° (34°)	−6° (21°)	−13° (9°)	−20° (−4°)	−27° (−16°)	−33° (−28°)	−41° (−41°)	−47° (−53°)	−54° (−66°)
24 (15)	0° (32°)	−7° (19°)	−14° (6°)	−22° (−7°)	−28° (−19°)	−36° (−32°)	−43° (−45°)	−50° (−58°)	−57° (−71°)
32 (20)	−1° (30°)	−8° (17°)	−16° (4°)	−23° (−9°)	−30° (−22°)	−37° (−35°)	−44° (−48°)	−52° (−61°)	−59° (−74°)
40 (25)	−2° (29°)	−9° (16°)	−16° (3°)	−24° (−11°)	−31° (−24°)	−38° (−37°)	−46° (−51°)	−53° (−64°)	−61° (−78°)
48 (30)	−2° (28°)	−9° (15°)	−17° (−1°)	−24° (−12°)	−32° (−26°)	−39° (−39°)	−47° (−53°)	−55° (−67°)	−62° (−80°)
56 (35)	−2° (28°)	−10° (14°)	−18° (0°)	−26° (−14°)	−33° (−27°)	−41° (−41°)	−48° (−55°)	−56° (−69°)	−63° (−82°)
64 (40)	−3° (27°)	−11° (13°)	−18° (−1°)	−26° (−15°)	−34° (−29°)	−42° (−43°)	−49° (−57°)	−57° (−71°)	−64° (−84°)
72 (45)	−3° (26°)	−11° (12°)	−19° (−2°)	−27° (−16°)	−34° (−30°)	−42° (−44°)	−50° (−58°)	−58° (−72°)	−66° (−86°)
80 (50)	−3° (26°)	−11° (12°)	−19° (−3°)	−27° (−17°)	−35° (−31°)	−43° (−45°)	−51° (−60°)	−59° (−74°)	−67° (−88°)

Frostbite times: 30 min. 10 min. 5 min.

◄**Figure 5.2 Wind-chill temperature index.** This index uses wind speed and actual air temperature to determine apparent temperature, and was developed by the National Weather Service (http://www.nws.noaa.gov/os/windchill/) and the Meteorological Service of Canada.

The wind-chill index does not account for sunlight intensity, a person's physical activity, or the use of protective clothing, all of which mitigate wind-chill intensity. Imagine living in some of the coldest regions of the world. To what degree would you need to adjust your personal wardrobe and make other comfort adaptations?

Air temperature plays a remarkable role in human life at all levels, affecting not only personal comfort but also environmental processes across Earth. Land temperatures interact with atmospheric moisture and precipitation to determine vegetation patterns and their associated habitats—for example, the high temperatures in deserts limit the presence of many plant and animal species. Ocean temperatures affect atmospheric moisture and weather, as well as ocean ecosystems such as coral reefs. Variations in average temperatures of air and water have far-ranging effects on Earth systems.

In addition to these considerations is the fact that temperatures are rising across the globe in response to carbon dioxide emissions, mainly from the burning of fossil fuels and removal of forests, as discussed in Chapter 3. Present CO_2 levels in the atmosphere are higher than at any time in the past 800,000 years, and they are steadily increasing. Thus, temperature concepts are at the forefront of understanding climate change and its far-reaching effects on Earth.

In this chapter: The temperature concepts presented in this chapter provide the foundation for our study of weather and climate systems. We first examine the principal temperature controls of latitude, altitude and elevation, cloud cover, and land–water heating differences as they interact to produce Earth's temperature patterns. We then look at a series of temperature maps illustrating Earth's temperature patterns and discuss current temperature trends associated with global warming. Finally, we examine the effect of high air temperatures and humidity on the human body, with a look at heat waves and their increasing occurrence across the globe.

Temperature Concepts and Measurement

In Chapter 4, we discussed types of heat and mechanisms of heat transfer, such as conduction, convection, and radiation. In this chapter, we focus on temperature, which is a different, though related, concept. We learned that *heat* is a form of energy that transfers among particles in a substance or system by means of the kinetic energy, or energy of motion, of individual molecules. In a relatively hotter substance, molecules are moving with higher energy. The addition of more heat adds more energy, with associated increases in kinetic energy and molecular motion.

Unlike heat, temperature is not a form of energy; however, temperature is related to the amount of energy in a substance. **Temperature** is a measure of the *average* kinetic energy of individual molecules in matter. (Remember that *matter* is mass that assumes a physical shape and occupies space.) Thus, temperature is a measure of heat.

Remember that heat always flows from matter at a higher temperature to matter at a lower temperature, and heat transfer usually results in a change in temperature. For example, when you jump into a cool lake, kinetic energy leaves your body and flows to the water, causing a transfer of heat and a lowering of the temperature of your skin. Heat transfer can also occur without a change in temperature when a substance changes state (as in latent heat transfer, discussed further in Chapter 7).

Temperature Scales

The temperature at which atomic and molecular motion in matter completely stops is *absolute zero*, or *0 absolute temperature*. This value on three commonly encountered temperature-measuring scales is –273° Celsius (C), –59.4° Fahrenheit (F), and 0 Kelvin (K; see Figure 5.3). (Formulas for converting between Celsius, SI (Système International), and English units are in Appendix C.)

The Fahrenheit scale is named for its developer, Daniel G. Fahrenheit, a German physicist (1686–1736). This temperature scale places the melting point of ice at 32°F, separated by 180 subdivisions from the boiling point of water at 212°F. Note that ice has only one melting point, but water has many freezing points, ranging from 32°F down to –40°F, depending on its purity, its volume, and certain conditions in the atmosphere.

About a year after the adoption of the Fahrenheit scale, Swedish astronomer Anders Celsius (1701–1744) developed the Celsius scale (formerly centigrade). He placed the melting point of ice at 0°C and the boiling temperature of water at sea level at 100°C, dividing his scale into 100 degrees using a decimal system.

British physicist Lord Kelvin (born William Thomson, 1824–1907) proposed the Kelvin scale in 1848.

GEOreport 5.1 The hottest temperature on Earth

In 1922, a record-breaking temperature was reported on a hot summer day at Al 'Azīzīya, Libya—an almost unimaginable 58°C (136°F). In 2012, an international panel of scientists assembled by the World Meteorological Organization (WMO) concluded that this record for Earth's hottest temperature, in place for 90 years, was invalid. After an in-depth investigation, the panel identified several reasons for uncertainty regarding the 1922 measurement, including instrument problems and poor matching of the temperature to nearby locations. Their final decision rejected this temperature extreme, reinstating the 57°C (134°F) temperature recorded in Death Valley, California, in 1913 as the record for hottest temperature ever measured on Earth. The station where this temperature was recorded is one of the lowest on Earth, at –54.3 m, or –178 ft (the minus indicates meters or feet below sea level). One hundred years later on June 29, 2013, the official National Park Service thermometer reached 129°F (53.9°C)—a U.S. record for June, although 2.78 C° (5 F°) lower than 1913. What do you think surface energy budgets were like on that day?

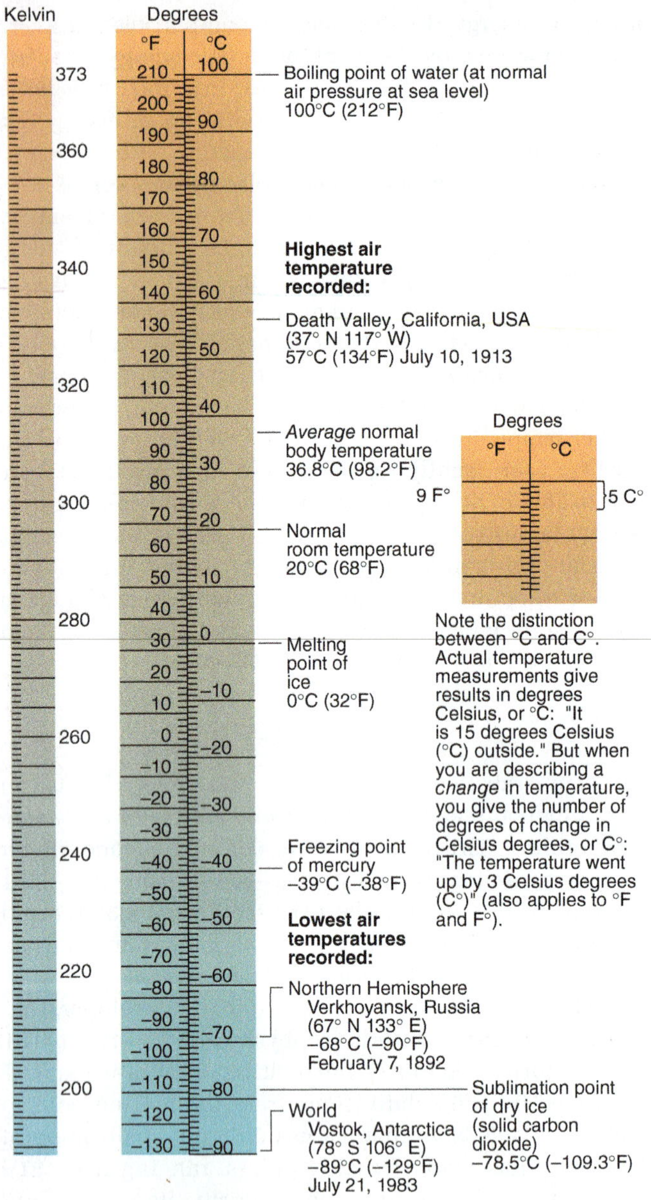

▲**Figure 5.3 Temperature scales.** Scales for expressing temperature in Kelvin (K) and degrees Celsius (°C) and Fahrenheit (°F), including significant temperatures and temperature records. Note the distinction between temperature (indicated by the color gradation on the scale) and units of temperature expression, and the placement of the degree symbol.

Science uses this scale because it starts at absolute zero, making temperature readings proportional to the actual kinetic energy in a material. The Kelvin scale's melting point for ice is 273 K, and its boiling point of water is 373 K, 100 units higher. Therefore, the size of one Kelvin unit is the same size as one Celsius degree.

Most countries use the Celsius scale to express temperature—the United States is an exception. Continuing pressure from the international scientific community and other organizations makes adoption of Celsius and SI units inevitable in the United States. This textbook presents Celsius (with Fahrenheit equivalents in parentheses) throughout to help bridge this transition.

▲**Figure 5.4 Instrument shelter.** This standard thermistor shelter is white and louvered, installed above a turf surface. [Bobbé Christopherson.]

Measuring Temperature

A familiar instrument for measuring temperature is a thermometer, a sealed glass tube containing a fluid that expands and contracts according to whether heat is added or removed—when the fluid is heated, it expands; upon cooling, it contracts. Both *mercury thermometers* and *alcohol thermometers* are used to measure outdoor temperatures—however, mercury thermometers have the limitation that mercury freezes at –39°C (–38.2°F). Therefore in Earth's colder climates, alcohol thermometers are preferred (alcohol freezes at a much lower temperature of –112°C, or –170°F). The principle of these thermometers is simple: A thermometer stores fluid in a small reservoir at one end and is marked with calibrations to measure the expansion or contraction of the fluid, which reflects the temperature of the thermometer's environment.

Devices for taking standardized official temperature readings are placed outdoors in small shelters that are white (for high albedo) and louvered (for ventilation) to avoid overheating of the instruments (Figure 5.4). They are placed at least 1.2–1.8 m (4–6 ft) above the ground surface, usually on turf; in the United States, a 1.2-m (4-ft) height is standard. Official temperature measurements are made in the shade to prevent the effect of direct insolation. Standard instrument shelters contain a *thermistor*, which measures temperature by sensing the electrical resistance of a semiconducting material. Since resistance changes at a rate of 4% per C°, the measure can be converted to temperature and reported electronically to the weather station. *Geosystems In Action*, Figure GIA 5, shows some of the hottest temperatures ever recorded on Earth.

Temperature readings occur daily, sometimes hourly, at more than 16,000 weather stations worldwide. Some stations also report the duration of a temperature, the

Earth's hottest locations are determined by air temperature, which is measured at least 1.2–1.8 m (4–6 ft) above a surface in an instrument shelter (shown in Figure 5.4). The MODIS sensor aboard several satellites provides another temperature measurement, known as land surface temperature (LST), or land skin temperature, which is often significantly hotter than air temperature (GIA 5.1). One main factor influencing LST is surface albedo (GIA 5.2)

5.1 EARTH'S HOTTEST "HOT SPOTS"

In terms of air temperature, Death Valley, California, is considered the hottest place on Earth. Based on land surface temperature, locations in Iran, China, and Australia are hotter.

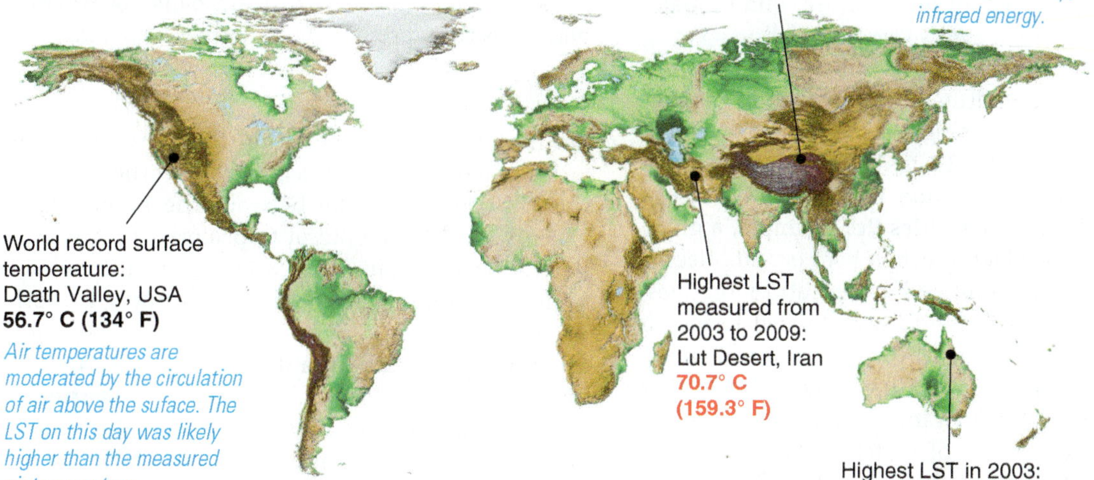

5.2 EFFECT OF ALBEDO AND LAND COVER ON LST

These images of China's Turpan Depression show how spatial variations in albedo affect LST. Notice that the lowest LSTs are in areas with agriculture and other plant cover. [NASA]

Light-colored rock and sediment with a high albedo reflect more solar radiation and are therefore cooler.

Dark-colored sand dunes have a lower albedo and absorb more solar radiation, heating the surface.

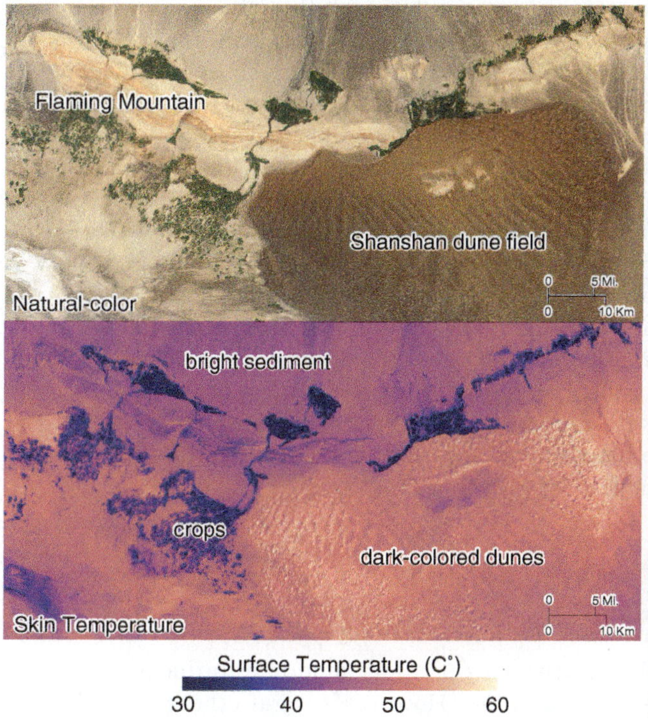

What does it mean to be hot?

MasteringGeography™

Visit the Study Area in MasteringGeography™ to explore temperature extremes.
Visualize: Study a geoscience animation of temperature extremes.
Assess: Demonstrate understanding of temperature extremes (if assigned by instructor).

GEOquiz

1. **Explain:** Why are the land surface temperatures in Figure 5.1 hotter than Death Valley's air temperature?
2. **Analyze:** Based on the two images in Figure 5.3, explain how and why LST varies across the Shanshan dune field.

rate of temperature rise or fall, and the temperature variation over time throughout the day and night. In 1992, the World Meteorological Organization and other international climate organizations established the Global Climate Observing System to coordinate the reading and recording of temperature and other climate factors among countries worldwide. One goal is to establish a reference network of one station per 250,000 km^2 (95,800 mi^2) across the globe. (For an overview of temperature- and climate-observing stations, go to http://www.wmo.int/pages/prog/gcos/index.php?name=ObservingSystemsandData.)

NASA's Goddard Institute for Space Studies constantly refines the methodology of collecting average-temperature measurements. To maintain a quality data set, scientists assess each station for the possibility of effects caused by human activities (for example, a given city location may be subject to urban heat island effects). For stations with human-altered temperature regimes, the long-term temperature trends are adjusted to match the average conditions of nearby rural stations where such impacts are absent. When using absolute (rather than long-term average) temperatures in a data set, scientists do not make these kinds of adjustments.

Satellites do not measure air temperature in the same way as thermometers; instead they measure *land-surface temperature* (LST), or land "skin" temperature, which is the heating of the land surface and is often much hotter than air temperature. You have felt this difference when walking barefoot across hot sand or pavement—the surface under your feet is much hotter than the air around your body above. Land skin temperatures record the heating of the ground from insolation and other heat flows; LSTs tend to be highest in dry environments with clear skies and surfaces with low albedo that absorb solar radiation.

Three expressions of temperature are common: The *daily mean temperature* is an average of hourly readings taken over a 24-hour day, but may also be the average of the daily minimum–maximum readings. The *monthly mean temperature* is the total of daily mean temperatures for the month divided by the number of days in the month. An *annual temperature range* expresses the difference between the lowest and highest monthly mean temperatures for a given year.

Principal Temperature Controls

Insolation is the single most important influence on temperature variations. However, several other physical controls interact with it to produce Earth's temperature patterns. These include latitude, altitude and elevation, cloud cover, and land–water heating differences.

The effects of human activity are altering some of these natural controls on temperature. Increasing amounts of greenhouse gases and human-made aerosols affect temperatures, as discussed in previous chapters. Recent evidence suggests that soot, such as the black carbon emitted from shipping and other industrial practices, may be a more significant cause of warming temperatures than previously thought.

Latitude

We learned in Chapter 2 that the subsolar point is the latitude where the Sun is directly overhead at noon, and that this point migrates between the Tropic of Cancer at 23.5° N and the Tropic of Capricorn at 23.5° S latitude. Between the tropics, insolation is more intense than at higher latitudes where the Sun is never directly overhead (at a 90° angle) during the year. The intensity of incoming solar radiation decreases away from the equator and toward the poles. Daylength also varies with latitude during the year, influencing the duration of insolation exposure. Variations in these two factors—Sun angle and daylength—throughout the year drive the seasonal effect of latitude on temperature.

Temperature patterns throughout the year for the five cities in Figure 5.5 demonstrate the effects of latitudinal position. Note the range from near-constant warm temperatures at Salvador, Brazil, near the equator, to wide-ranging seasonal temperature variation at Barrow, Alaska, at 71° N latitude. From equator to poles, Earth ranges from continually warm, to seasonally variable, to continually cold.

Altitude and Elevation

From Chapter 3, remember that within the troposphere, temperatures decrease with increasing altitude above Earth's surface. (Recall from Figure 3.3 that the *normal lapse rate* of temperature change with altitude is 6.4 C°/1000 m, or 3.5 F°/1000 ft.) The density of the atmosphere also diminishes with increasing altitude. In fact, the density of the atmosphere at an elevation of 5500 m (18,000 ft) is about half that at sea level. As the atmosphere thins, it contains less sensible heat. Thus, worldwide, mountainous areas experience lower temperatures than do regions nearer sea level, even at similar latitudes.

Two terms, altitude and elevation, are commonly used to refer to heights on or above Earth's surface. *Altitude* refers to airborne objects or heights *above* Earth's

GEOreport 5.2 Alaska and Montana hold U.S. records for low temperatures

The lowest temperature ever recorded in the United States is −62°C (−80°F), which occurred on January 23, 1971, at Prospect Creek, Alaska, a small settlement located about 290 km (180 mi) northwest of Fairbanks at an elevation of 335 m (1100 ft). In the contiguous United States (the "lower 48"), a record cold temperature of −57°C (−70°F) occurred at Rogers Pass, Montana, on January 20, 1954, at an elevation of 1667 m (5470 ft).

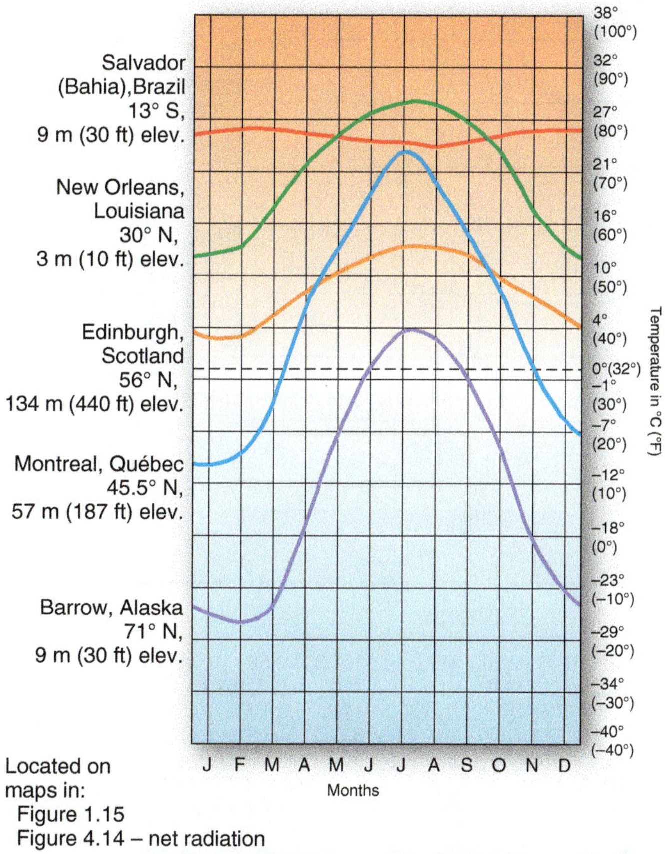

▲Figure 5.5 **Latitudinal effects on temperatures.** A comparison of five cities from near the equator to north of the Arctic Circle demonstrates changing seasonality and increasing differences between average minimum and maximum temperatures with increasing latitude.

surface. *Elevation* usually refers to the height of a point *on* Earth's surface above some plane of reference, such as elevation above sea level. Therefore, the height of a flying jet is expressed as altitude, whereas the height of a mountain ski resort is expressed as elevation.

In the thinner atmosphere at high elevations in mountainous regions or on high plateaus, surfaces gain and lose energy rapidly to the atmosphere. The result is that average air temperatures are lower, nighttime cooling is greater, and the temperature range between day and night is greater than at low elevations. The temperature difference between areas of sunlight and shadow is greater at higher elevation than at sea level, and temperatures drop rapidly after sunset. You may have felt this difference if you live in or have visited the mountains.

The *snow line* seen on mountain slopes is the lower limit of permanent snow and indicates where winter snowfall exceeds the amount of snow lost through summer melting and evaporation. The snow line's location is a function both of latitude and of elevation, and to a lesser extent, is related to local microclimatic conditions. Even at low latitudes, permanent ice fields and glaciers exist on mountain summits, such as in the Andes and East Africa. In equatorial mountains, the snow line occurs at approximately 5000 m (16,400 ft). With increasing latitude, snow lines gradually lower in elevation from 2700 m (8850 ft) in the midlatitudes to lower than 900 m (2950 ft) in southern Greenland.

The effects of latitude and elevation combine to create temperature characteristics at many locations. In Bolivia, two cities, both at about 16° S latitude, located less than 800 km (497 mi) apart, have quite different climates (Figure 5.6). Concepción is a low-elevation city at 490 m (1608 ft), with a warm, moist climate that is typical for many low-latitude locations with consistent Sun angle and daylength throughout the year. This city has an annual average temperature of 24°C (75°F).

La Paz, at an elevation of 4103 m (13,461 ft) above sea level, is situated on a high plateau with a cool, dry climate. La Paz has moderate annual temperatures, averaging about 11°C (51.8°F). Despite its high elevation, people living around La Paz are able to grow wheat, barley, and potatoes—crops characteristic of the midlatitudes—in the fertile highland soils. Compared to the harsh climates at similar elevations in the United States, such as the 4301-m (14,111-ft) summit of Pikes Peak at 38° N in Colorado or the 4392-m (14,410-ft) top of Mount Rainier at 46° N in Washington, La Paz has a mild and hospitable climate resulting from its low-latitude location.

Cloud Cover

At any given moment, approximately 50% of Earth is covered by clouds. Examine the extent of cloud cover over Earth on the two images inside the front cover. In Chapter 4, we learned that clouds affect the Earth–atmosphere energy balance by reflecting and absorbing radiation, and that their effects vary with cloud type, height, and density.

The presence of cloud cover at night has a moderating effect on temperature; you may have experienced the colder temperatures outside on a clear night, especially before dawn, the coldest time of the day. At night, clouds act as an insulating layer that reradiates longwave energy back to Earth, preventing rapid energy loss to space. Thus, in general, the presence of clouds raises minimum nighttime temperatures. During the day, clouds reflect insolation, lowering daily maximum temperatures; this is the familiar shading effect you feel when clouds move in on a hot summer day. Clouds also reduce seasonal temperature differences as a result of these moderating effects.

Clouds are the most variable factor influencing Earth's radiation budget, and studies are ongoing as to their effects on Earth's temperatures. For more information, read about the Global Energy and Water Cycle Experiment (GEWEX) at http://www.gewex.org/gewex_overview.html#foci, or check out NASA's Clouds and the Earth Radiant Energy System (CERES) experiment at http://ceres.larc.nasa.gov/.

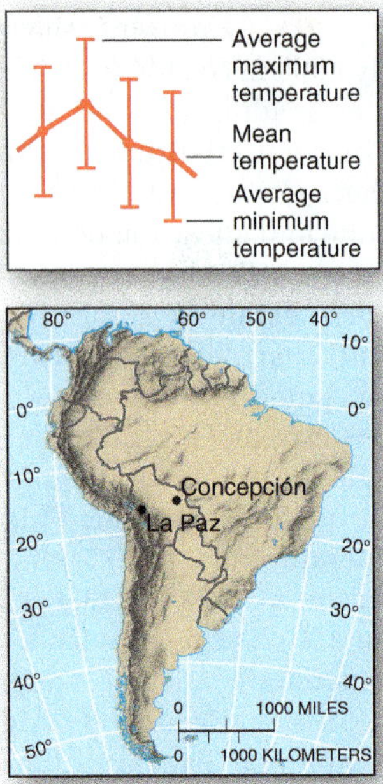

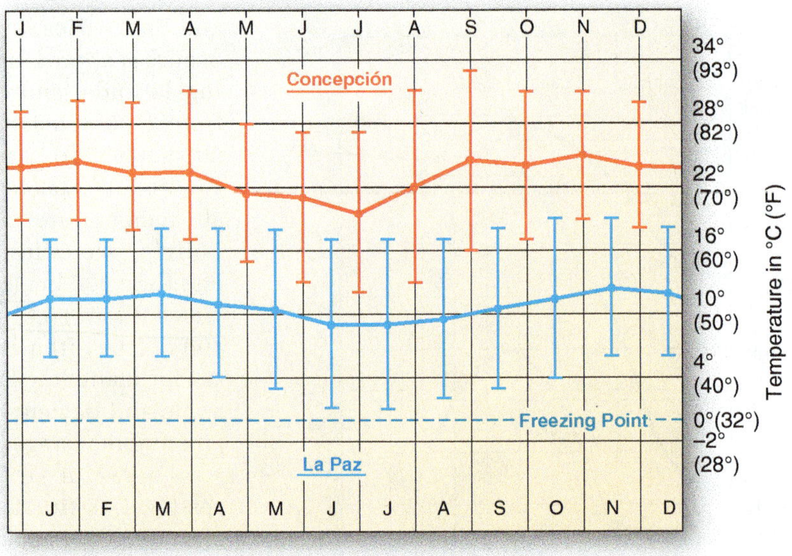

(a) Comparison of temperatures for two Bolivian cities.

	Station: Concepción, Bolivia	Station: La Paz, Bolivia
Latitude/longitude	16° 15′ S 62° 03′ W	16° 30′ S 68° 10′ W
Elevation	**490 m (1608 ft)**	**4103 m (13,461 ft)**
Avg. ann. temperature	23°C (73.4°F)	11°C (51.8°F)
Ann. temperature range	6.5 C° (11.7 F°)	3.5 C° (6.3 F°)
Ann. precipitation	121.2 cm (47.7 in.)	55.5 cm (21.9 in.)
Population	10,000	810,300 (administrative division 1.6 million)

(b) Tropical dry forests cover the lower elevations of east-central Bolivia near Concepción; some forests have been cleared for farmland and ranching.

(c) High-elevation villages near La Paz are in view of permanent ice-covered peaks of the Bolivian Cordillera Real in the Andes Mountains.

▲**Figure 5.6 Effects of latitude and elevation.** [(b) Peter Langer/Design Pics/Corbis. (c) Seux Paule/AGE Fotostock America, Inc.]

Land–Water Heating Differences

An important control over temperature is the difference in the ways land and water surfaces respond to insolation. On Earth, these two surfaces occur in an irregular arrangement of continents and oceans. Land and water absorb and store energy differently, with the result that water bodies tend to have more moderate temperature patterns, whereas continental interiors have more temperature extremes.

The physical differences between land (rock and soil) and water (oceans, seas, and lakes) are the reasons for **land–water heating differences**, the most basic of which is that land heats and cools faster than water. Figure 5.7 summarizes these differences, which relate to the principles and processes of evaporation, transparency, specific heat, and movement. We include ocean currents and sea-surface temperatures in this section because of their effects on temperatures in coastal locations.

Evaporation The process of *evaporation* dissipates significant amounts of the energy arriving at the ocean's surface, much more than over land surfaces where less water is available. An estimated 84% of all evaporation on Earth is from the oceans. When water evaporates, it

Figure 5.7 Land–water heating differences. The differential heating of land and water produces contrasting marine (more moderate) and continental (more extreme) temperature regimes.

changes from liquid to vapor, absorbing heat energy in the process and storing it as latent heat.

You experience the cooling effect of evaporative heat loss by wetting the back of your hand and then blowing on the moist skin. Sensible heat energy is drawn from your skin to supply some of the energy for evaporation, and you feel the cooling as a result. As surface water evaporates, it absorbs energy from the immediate environment, resulting in a lowering of temperatures. (Remember that the water and vapor remain the same temperature throughout the process; the vapor stores the absorbed energy as latent heat.) Land temperatures are affected less by evaporative cooling than are temperatures over water.

Transparency Soil and water differ in their transmission of light: Solid ground is opaque; water is transparent. Light striking a soil surface does not pass through, but is absorbed, heating the ground surface. That energy is accumulated during times of sunlight exposure and is rapidly lost at night or when shaded.

Maximum and minimum daily temperatures for soil surfaces generally occur at the ground surface level. Below the surface, even at shallow depths, temperatures remain about the same throughout the day. You encounter this at a beach, where surface sand may be painfully hot to your feet, but as you dig in your toes and feel the sand a few centimeters below the surface, it is cooler, offering relief. On the *MasteringGeography* website, you will find a profile of daily temperatures for a column of soil and the air above it, showing the pattern just described.

In contrast, when light reaches a body of water, it penetrates the surface because of water's **transparency**—water is clear, and light passes through it to an average depth of 60 m (200 ft) in the ocean. This illuminated zone occurs in some ocean waters to depths of 300 m (1000 ft). The transparency of water results in the distribution of available heat energy over a much greater depth and volume, forming a larger reservoir of energy storage than that which occurs on land.

Specific Heat The energy needed to increase the temperature of water is greater than for an equal volume of land. Overall, water can hold more heat than can soil or rock. The heat capacity of a substance is **specific heat.** On average, the specific heat of water is about four times that of soil. Therefore, a given volume of water represents a more substantial energy reservoir than does the same volume of soil or rock and consequently heats and cools more slowly. For this reason, day-to-day temperatures near large water bodies tend to be moderated rather than having large extremes.

Movement In contrast to the solid, rigid characteristics of land, water is fluid and capable of movement. The movement of currents results in a mixing of cooler and warmer waters, and that mixing spreads the available energy over an even greater volume than if the water were still. Surface water and deeper waters mix, redistributing energy in a vertical direction as well. Both ocean and land surfaces radiate longwave radiation at night, but land loses its energy more rapidly than does the moving reservoir of oceanic energy, with its more extensive distribution of heat.

Ocean Currents and Sea-Surface Temperatures

Although our full discussion of ocean circulation is in Chapter 6, we include a brief discussion of currents here because they influence temperature in coastal locations. Ocean currents affect land temperatures in different ways, depending on whether the currents are warm or cold. Along midlatitude and subtropical west coasts of continents, cool ocean currents flowing toward the equator moderate air temperatures on land. An example is the effect of the cold Humboldt Current flowing offshore from

Figure 5.8 The Gulf Stream. Satellite instruments sensitive to thermal infrared wavelengths imaged the Gulf Stream. Temperature differences are noted by computer-enhanced false colors: reds and oranges = 25° to 29°C (76° to 84°F), yellows and greens = 17° to 24°C (63° to 75°F); blues = 10° to 16°C (50° to 61°F); purples = 2° to 9°C (36° to 48°F). [Imagery by RSMAS, University of Miami.]

Lima, Peru, which has a cooler climate than might be expected at that latitude. When conditions in these regions are warm and moist, fog frequently forms in the chilled air over the cooler currents.

The warm current known as the **Gulf Stream** moves northward off the east coast of North America, carrying warm water far into the North Atlantic (Figure 5.8). As a result, the southern third of Iceland experiences much milder temperatures than would be expected for a latitude of 65° N, just south of the Arctic Circle (66.5° N). In Reykjavík, on the southwestern coast of Iceland, monthly temperatures average above freezing during all months of the year. The Gulf Stream also moderates temperatures in coastal Scandinavia and northwestern Europe. In the western Pacific Ocean, the warm Kuroshio, or Japan Current, functions much the same as the Gulf Stream, having a warming effect on temperatures in Japan, the Aleutians, and along the northwestern margin of North America.

Across the globe, ocean water is rarely found warmer than 31°C (88°F), although in 2005, Hurricanes Katrina, Rita, and Wilma intensified as they moved over 33°C (92°F) sea-surface temperatures in the Gulf of Mexico. Higher ocean temperatures produce higher evaporation rates, and more energy is dissipated from the ocean as latent heat. As the water vapor content of the overlying air increases, the ability of that air to absorb longwave radiation also increases, leading to warming. The warmer the air and the ocean become, the more evaporation that occurs, increasing the amount of water vapor entering the air. More water vapor leads to cloud formation, which reflects insolation and produces lower temperatures. Lower temperatures of air and ocean reduce evaporation rates and the ability of the air mass to absorb water vapor—an interesting *negative feedback* mechanism.

Ocean temperatures are typically measured at the surface and recorded as the *sea-surface temperature*, or SST. Maps of global average SSTs measured from satellites, such as those in Figure 5.9, reveal that the region with the highest average ocean temperatures in the world is the *Western Pacific Warm Pool* in the southwestern Pacific Ocean, where temperatures are often above 30°C (86°F). Although the difference in SSTs between the equator and the poles is apparent on both maps, note the seasonal changes in ocean temperatures, such as the northward shifting of the Western Pacific Warm Pool in July. The warm Gulf Stream is apparent off the coast of Florida in both images; cooler currents occur off the west coasts of North and South America, Europe, and Africa.

Following the same recent trends as global air temperatures, average annual SSTs increased steadily from 1982 through 2010 to record-high levels—2010 breaking records for both ocean and land temperatures. Increasing warmth is measured at depths to 1000 m (3200 ft), and in 2004, scientists reported slight increases even in the temperature of deep bottom water. These data suggest that the ocean's ability to absorb excess heat energy from the atmosphere may be nearing its capacity.

Examples of Marine Effects and Continental Effects

The land–water heating differences that affect temperature regimes worldwide can be summarized in terms of continental and marine effects. The **marine effect**, or *maritime effect,* refers to the moderating influences of the ocean and usually occurs in locations along coastlines or on islands. The **continental effect**, or condition of *continentality,* refers to the greater range between maximum and minimum temperatures on both a daily and a yearly basis that occurs in areas that are inland from the ocean or distant from other large water bodies.

The monthly temperatures of San Francisco, California, and Wichita, Kansas, both at approximately 37° 40' N latitude, illustrate these effects (Figure 5.10). In San Francisco, only a few days a year have summer maximums that exceed 32.2°C (90°F). Winter minimums rarely drop below freezing.

In contrast, Wichita, Kansas, is susceptible to freezing temperatures from late October to mid-April—the record low temperature is –30°C (–22°F). Wichita's temperature reaches 32.2°C (90°F) or higher more than 65 days each year, with 46°C (114°F) as a record high. During the summer of 2012, temperatures exceeded 38°C (100°F) for a record 53 nonconsecutive days.

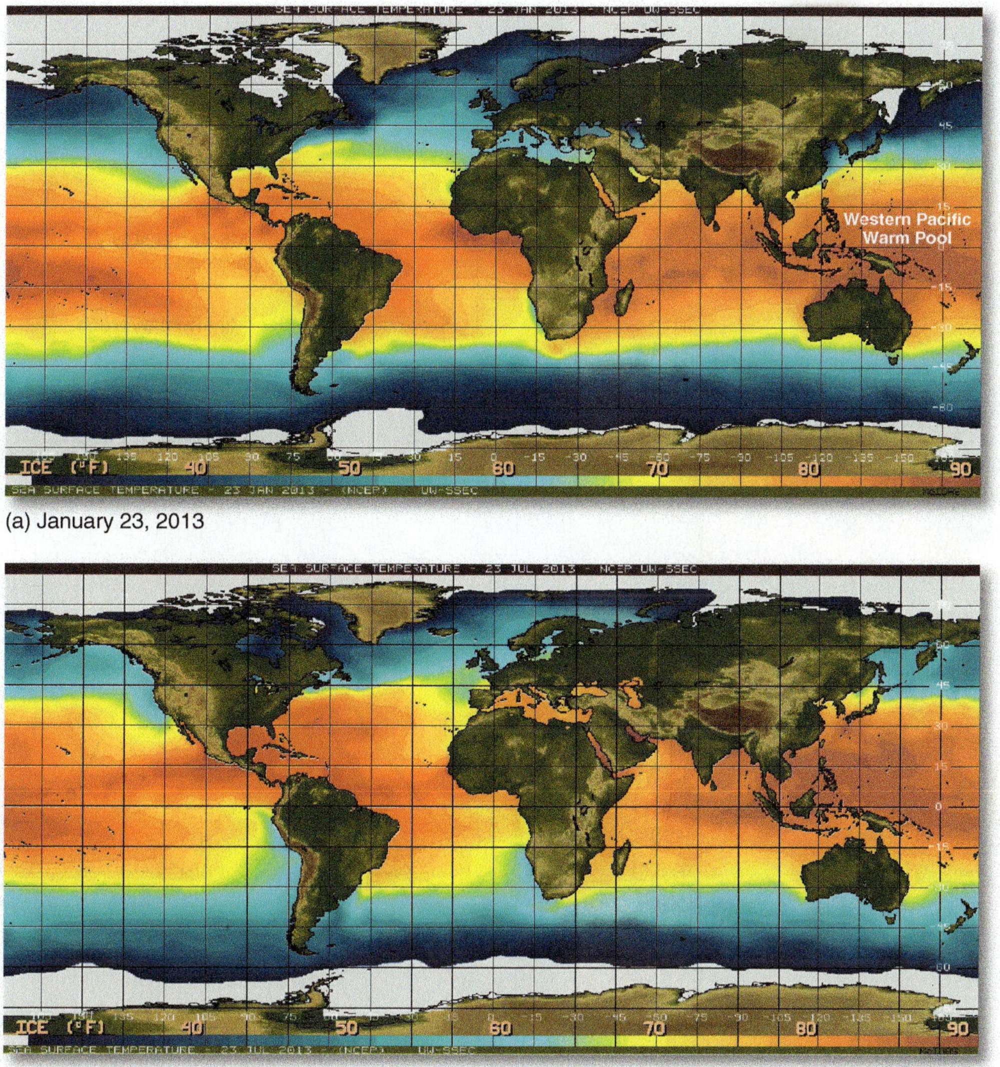

(a) January 23, 2013

(b) July 23, 2013

▲**Figure 5.9 Average monthly sea-surface temperatures for January and July.** [Satellite data courtesy of Space Science and Engineering Center, University of Wisconsin, Madison.]

Satellite
Global Sea-Surface Temperatures

In Eurasia, similar differences exist for cities in marine versus continental locations. Figure 5.11 shows temperature data from stations in Trondheim, Norway, and Verkhoyansk, Russia, at similar latitudes and elevations. Trondheim's coastal location moderates its annual temperature regime. January minimum and maximum temperatures range between −17°C and +8°C (+1.4°F and +46°F), and July minimum and maximum temperatures range between +5°C and +27°C (+41°F and +81°F). The lowest minimum and highest maximum temperatures ever recorded in Trondheim are −30°C and +35°C (−22°F and +95°F). In contrast, Verkhoyansk has a continental location and a 63 C° (113 F°) range in average annual temperatures during an average year. This location has 7 months of temperatures below freezing, including at least 4 months below −34°C (−30°F). Temperature extremes reflect continental effects: Verkhoyansk recorded a minimum temperature of −68°C (−90°F) in January, and a maximum temperature of +37°C (+98°F) occurred in July—an incredible 105 C° (189 F°) minimum–maximum range for the record temperature extremes! Verkhoyansk has a population of 1400 and has been occupied continuously since 1638.

(text continued on page 120)

CRITICALthinking 5.1
Compare and Explain Coastal and Inland Temperatures

Using the map, graphs, and other data in Figure 5.10, explain the effects of San Francisco's marine location on its average monthly temperatures. Why are summer temperatures higher in Wichita relative to those in San Francisco? Why does San Francisco's average monthly temperature peak occur later in the summer than that of Wichita? Is your location subject to marine or continental effects on temperature? (Find the explanation at the end of the Key Learning Concept Review.) ●

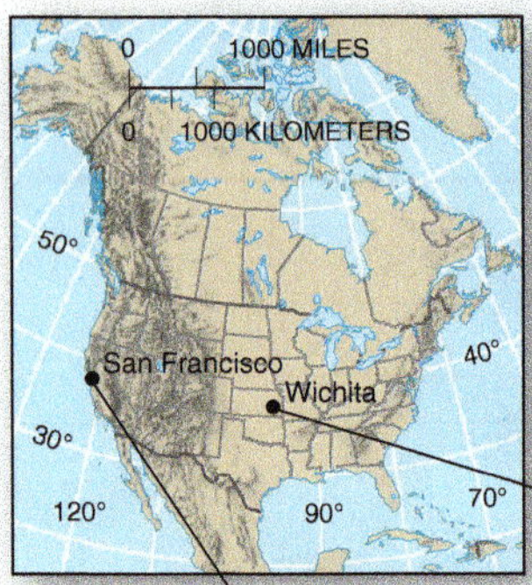

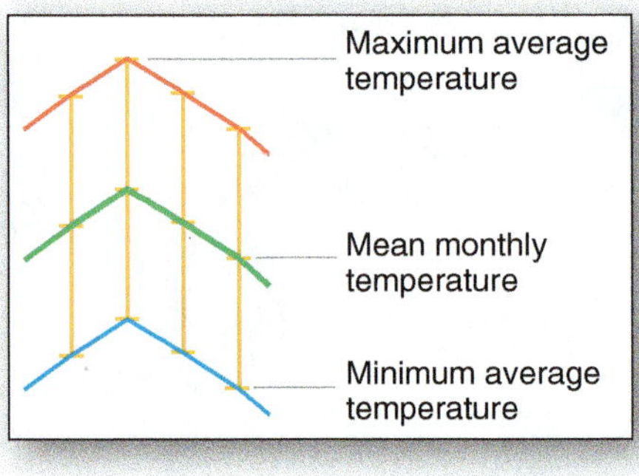

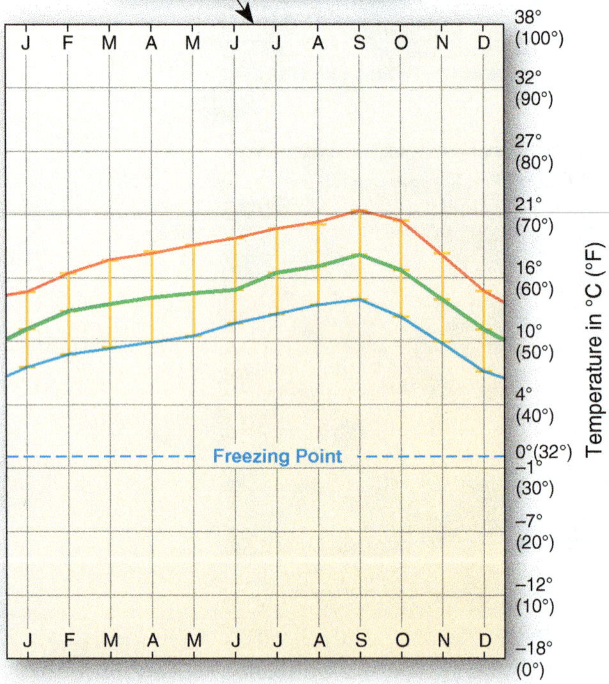

Station: San Francisco, California
Lat/long: 37° 46′ N 122° 23′ W
Avg. ann. temp.: 14.6°C (58.3°F)
Total ann. precip.: 56.6 cm (22.3 in.)
Elevation: 5 m (16.4 ft)
Population: 777,000
Ann. temp. range: 11.4 C° (20.5 F°)

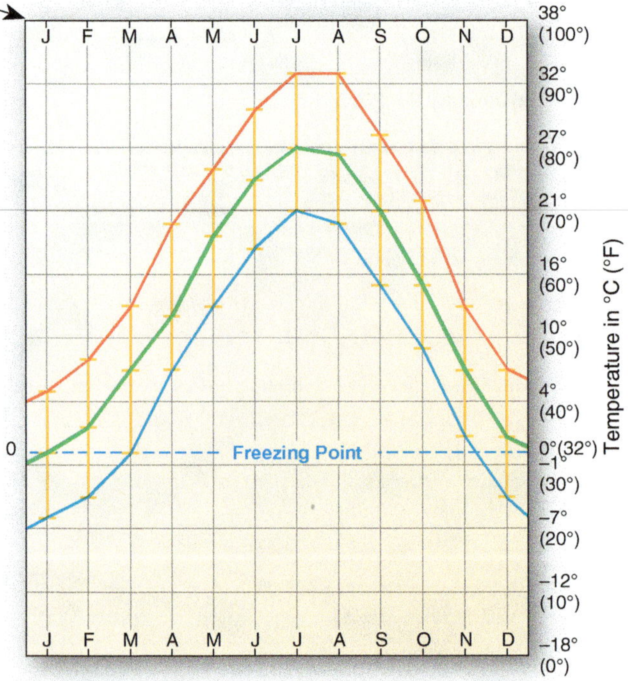

Station: Wichita, Kansas
Lat/long: 37° 39′ N 97° 25′ W
Avg. ann. temp.: 13.7°C (56.6°F)
Total ann. precip.: 72.2 cm (28.4 in.)
Elevation: 402.6 m (1321 ft)
Population: 327,000
Ann. temp. range: 27 C° (48.6 F°)

▲**Figure 5.10 Marine and continental cities—United States.** Compare temperatures in coastal San Francisco, California, with those of continental Wichita, Kansas. [Bobbé Christopherson.]

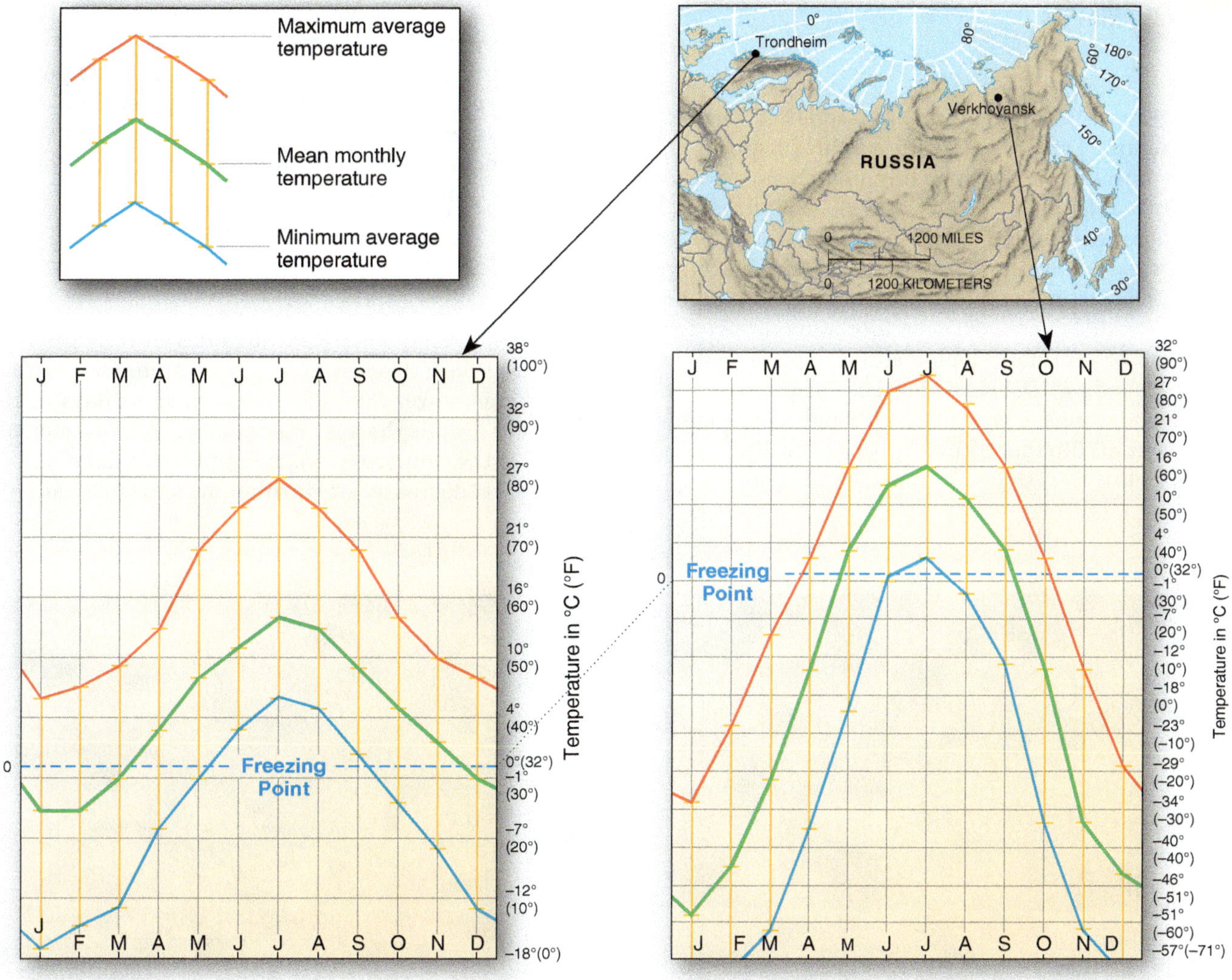

Station: Trondheim, Norway
Lat/long: 63° 25′ N 10° 27′ E
Avg. ann. temp.: 5°C (41°F)
Total ann. precip.: 85.7 cm (33.7 in.)
Elevation: 115 m (377.3 ft)
Population: 139,000
Ann. temp. range: 17 C° (30.6 F°)

Station: Verkhoyansk, Russia
Lat/long: 67° 33′ N 133° 24′ E
Avg. ann. temp.: −15°C (5°F)
Total ann. precip.: 15.5 cm (6.1 in.)
Elevation: 107 m (351 ft)
Population: 1400
Ann. temp. range: 63 C° (113.4 F°)

▲**Figure 5.11 Marine and continental cities—Eurasia.** Compare temperatures in coastal Trondheim, Norway, with continental Verkhoyansk, Russia. Note that the freezing levels on the two graphs are positioned differently to accommodate the contrasting data. [Bobbé Christopherson; Martin Hartley/Image Bank/Getty Images.]

Earth's Temperature Patterns

Figures 5.12 through 5.16 are a series of maps to help us visualize Earth's temperature patterns: global mean temperatures for January and July, polar region mean temperatures for January and July, and global annual temperature ranges (differences between averages of the coolest and warmest months). Maps are for January and July instead of the solstice months of December and June because of the lag that occurs between insolation received and maximum or minimum temperatures experienced (explained in Chapter 4). Data used for these maps are representative of conditions since 1950, although some temperatures are from ship reports dating back to 1850 and land reports to 1890.

The lines on temperature maps are known as *isotherms*. An **isotherm** is an isoline—a line along which there is a constant value—that connects points of equal temperature to portray the temperature pattern, just as a contour line on a topographic map illustrates points of equal elevation. Isotherms are useful for the spatial analysis of temperatures.

January and July Global Temperature Maps

In January, high Sun altitudes and longer days in the Southern Hemisphere cause summer weather conditions; lower Sun altitudes and short days in the Northern Hemisphere are associated with winter. Isotherms on the January average-temperature map mark the general decrease in insolation and net radiation with

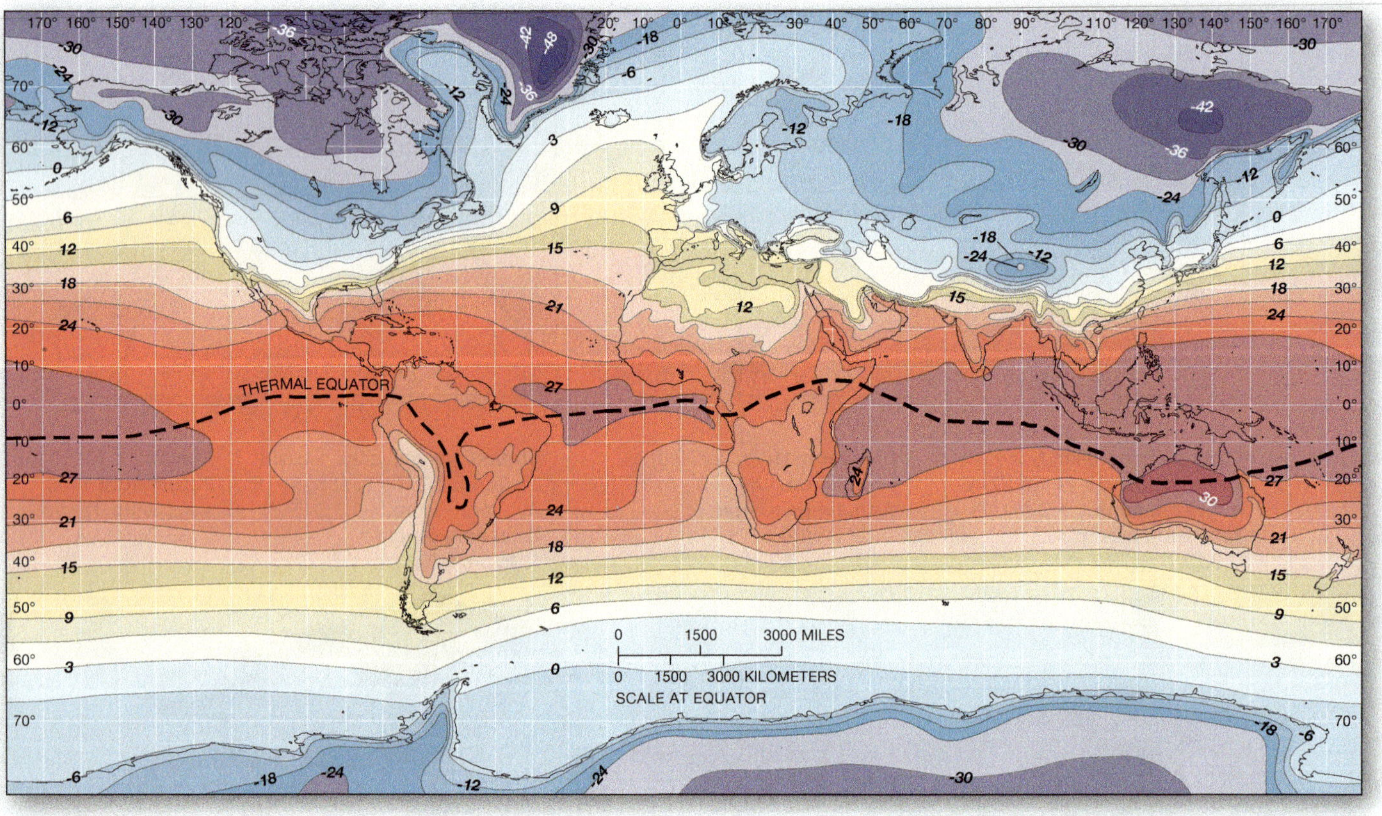

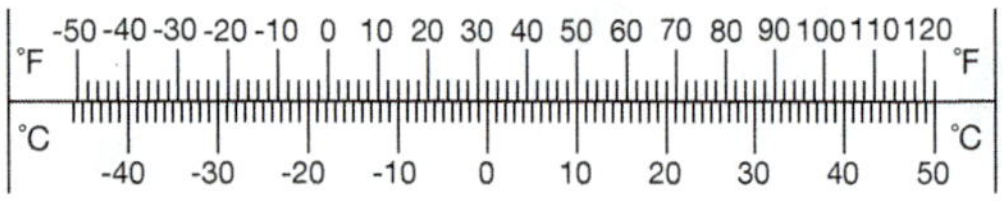

▲**Figure 5.12 Global average temperatures for January.** Temperatures are in Celsius (convertible to Fahrenheit by means of the scale) as taken from separate air-temperature databases for ocean and land. Note the inset map of North America and the equatorward-trending isotherms in the interior. [Adapted by author and redrawn from National Climatic Data Center, *Monthly Climatic Data for the World*, 47 (January 1994), and WMO and NOAA.]

MG™ Satellite
Global Surface Temperatures, Land and Ocean

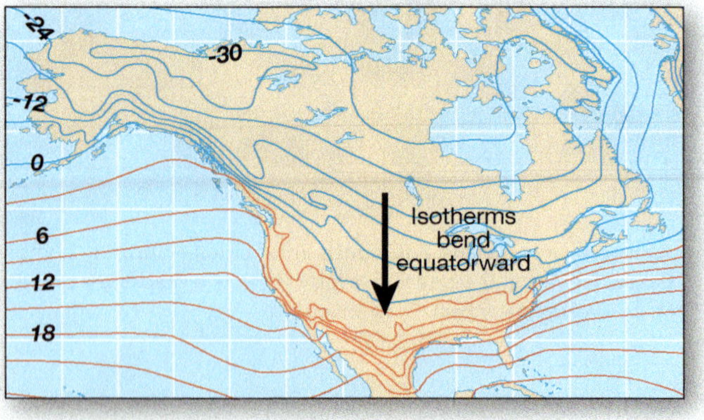

distance from the equator (Figure 5.12). Isotherms generally trend east–west, are parallel to the equator, and are interrupted by the presence of landmasses. This interruption is caused by the differential heating of land and water discussed earlier.

The **thermal equator** is an isotherm connecting all points of highest mean temperature, roughly 27°C (80°F); it trends southward into the interior of South America and Africa, indicating higher temperatures over the interiors of landmasses. In the Northern Hemisphere, isotherms shift toward the equator as cold air chills the continental interiors. More moderate temperatures occur over oceans, with warmer conditions extending farther north than over land at comparable latitudes.

To see the temperature differences over land and water, follow along 50° N latitude (the 50th parallel) in Figure 5.12 and compare isotherms: 3°C to 6°C in the North Pacific and 3°C to 9°C in the North Atlantic, as contrasted with −18°C in the interior of North America and −24°C to −30°C in central Asia. Also, note the orientation of isotherms over areas with mountain ranges and how they illustrate the cooling effects of elevation—check the South American Andes as an example.

For a continental region other than Antarctica, Russia—and specifically, northeastern Siberia—is the coldest area (Figure 5.12). The intense cold results from winter conditions of consistently clear, dry, calm air; small insolation input; and an inland location far from moderating maritime effects. Prevailing global winds prevent moderating effects from the Pacific Ocean to the east.

Figure 5.13 maps average July global temperatures for comparison. The longer days of summer and higher

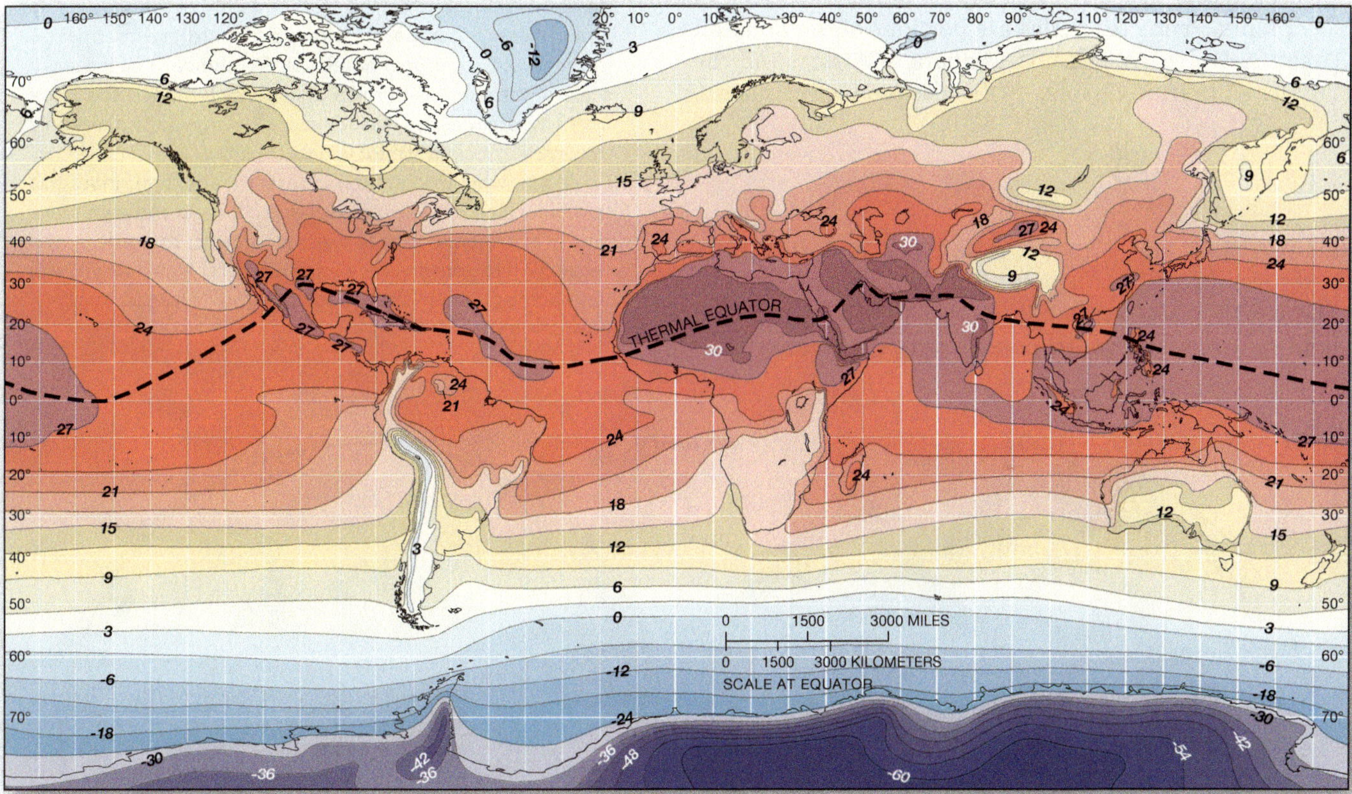

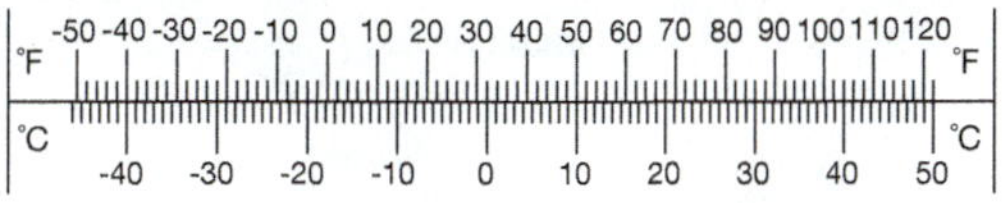

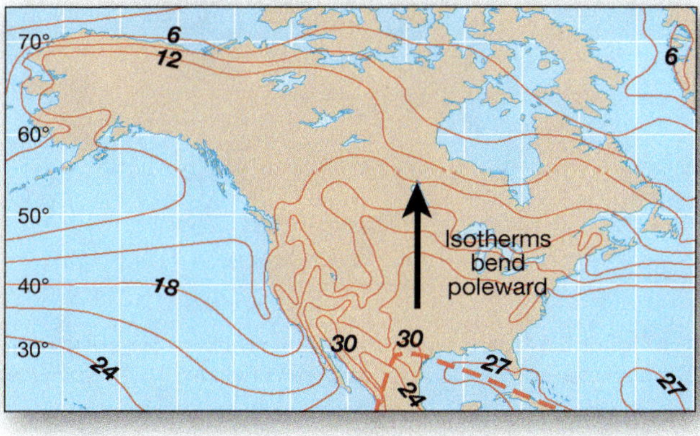

▲**Figure 5.13 Global average temperatures for July.** Temperatures are in Celsius (convertible to Fahrenheit by means of the scale) as taken from separate air-temperature databases for ocean and land. Note the inset map of North America and the poleward-trending isotherms in the interior. [Adapted by author and redrawn from National Climatic Data Center, *Monthly Climatic Data for the World*, 47 (July 1994), and WMO and NOAA.]

MG

Satellite
Global Surface Temperatures, Land and Ocean

Sun altitude are in the Northern Hemisphere. Winter dominates the Southern Hemisphere, although it is milder than winters north of the equator because continental landmasses, with their greater temperature ranges are smaller. The *thermal equator* shifts northward with the high summer Sun and reaches the Persian Gulf–Pakistan–Iran area. The Persian Gulf is the site of the highest recorded sea-surface temperature—an astounding 36°C (96°F).

During July in the Northern Hemisphere, isotherms shift toward the poles over land as higher temperatures dominate continental interiors. July temperatures in Verkhoyansk, Russia, average more than 13°C (56°F), which represents a 63 C° (113 F°) seasonal range between January and July averages. The Verkhoyansk region of Siberia is probably Earth's most dramatic example of continental effects on temperature.

The hottest places on Earth occur in Northern Hemisphere deserts during July, caused by clear skies, strong surface heating, virtually no surface water, and few plants. Prime examples are portions of the Sonoran Desert of North America, the Sahara of Africa, and the Lut Desert in Iran (Chapter 10 discusses desert climates).

January and July Polar-Region Temperature Maps

Figures 5.14 and 5.15 show January and July temperatures in the north polar and south polar regions. The north polar region is an ocean surrounded by land, whereas the south polar region is the enormous Antarctic ice sheet (covering the Antarctic continent) surrounded by ocean. The island of Greenland, holding Earth's second largest ice sheet, has a maximum elevation of 3240 m (10,630 ft), the highest elevation north of the Arctic Circle. Two-thirds of the island is north of the Arctic Circle, and its north shore is only 800 km (500 mi) from the North Pole. The combination of high latitude and interior high elevation on this ice sheet produces cold midwinter temperatures in January (Figure 5.14a).

In Antarctica (Figure 5.14b), "summer" is in December and January. This is Earth's coldest and highest

CRITICALthinking 5.2
Begin a Full Physical Geography Profile of Your Area

With each temperature map (Figures 5.12, 5.13, and 5.16), begin by finding your own city or town and noting the temperatures indicated by the isotherms for January and July and the annual temperature range. Record the information from these maps in your notebook. Remember that the small scale of these maps permits only generalizations about actual temperatures at specific sites, allowing you to make only a rough approximation for your location.

As you work through the different maps throughout this text, add such data as atmospheric pressure and winds, annual precipitation, climate type, landforms, soil orders, and vegetation to your profile. By the end of the course, you will have a complete physical geography profile for your regional environment. ●

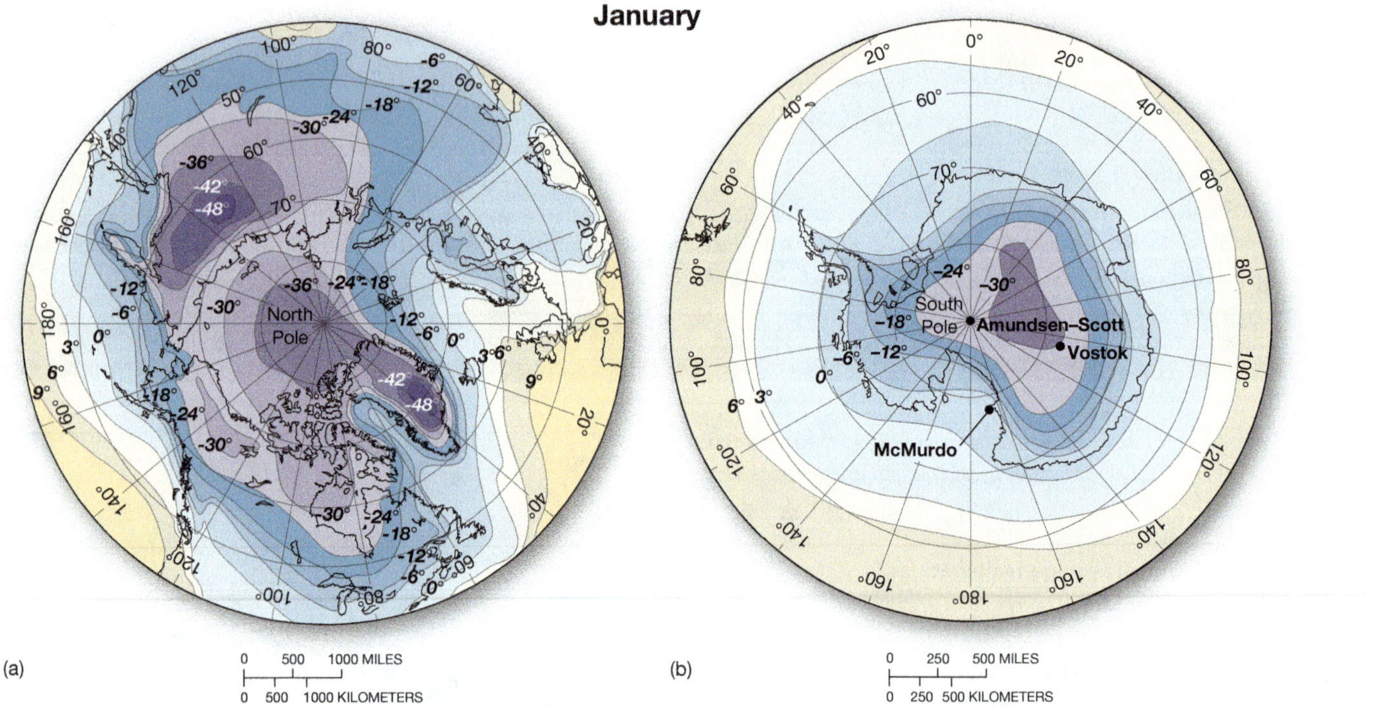

▲**Figure 5.14 January average temperatures for polar regions.** January temperatures in Celsius for (a) north polar region and (b) south polar region. Use temperature conversions in Figure 5.12. Note that each map is at a different scale. [Author-prepared maps using same sources as Figure 5.12.]

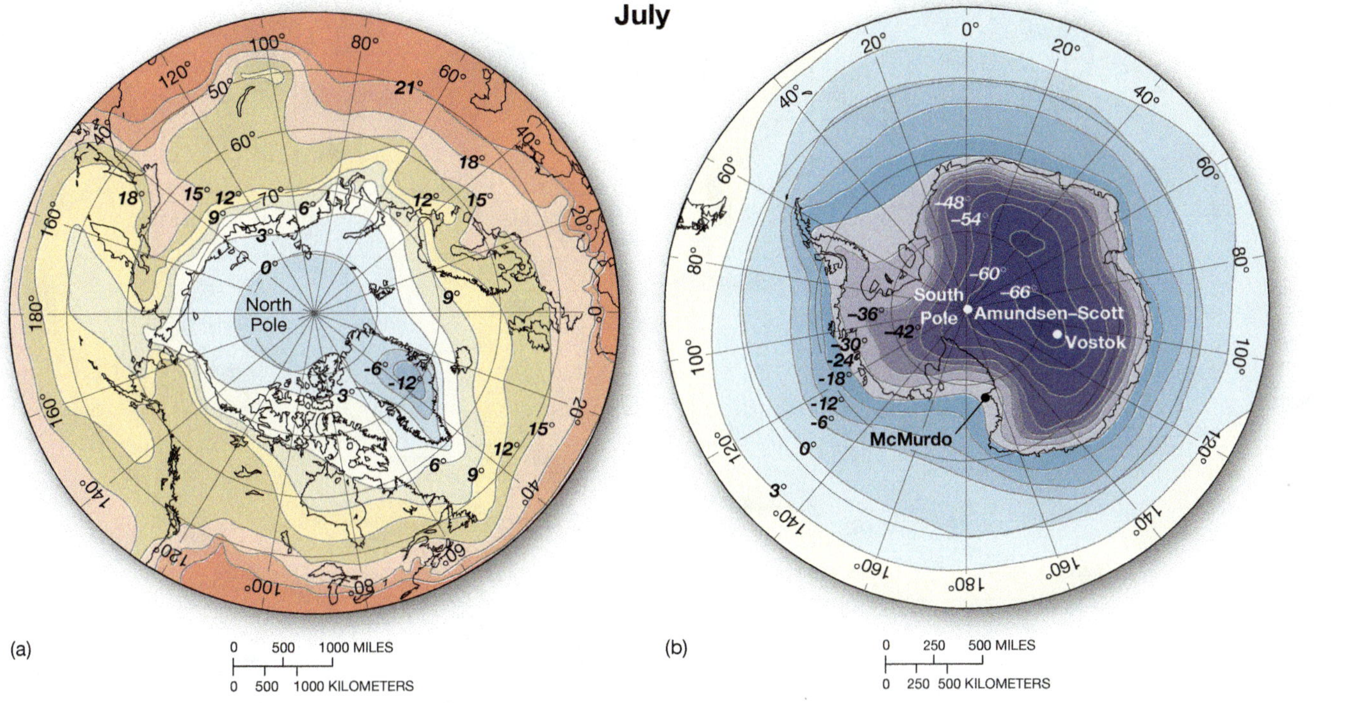

▲**Figure 5.15 July average temperatures for polar regions.** July temperatures in Celsius for (a) north polar region and (b) south polar region. Use temperature conversions in Figure 5.12. Note that each map is at a different scale. [Author-prepared maps using same sources as Figure 5.13.]

landmass (in terms of average elevation). January average temperatures for the three scientific bases noted on the map are −3°C (26.6°F) at McMurdo Station (on the coast); −28°C (−18.4°F) at the Amundsen–Scott Station at the South Pole (elevation 2835 m, 9301 ft); and −32°C (−25.6°F) at the Russian Vostok Station (elevation 3420 m, 11,220 ft), the most continental location of the three.

Figure 5.15 maps July average temperatures in the north and south polar regions. July is "summer" in the Arctic Ocean (Figure 5.15a), where rising air and ocean temperatures are causing sea-ice melting, as discussed in the Chapter 4 *Geosystems Now*. Open cracks and water channels through ice, or "leads," stretched all the way to the North Pole over the past several years.

In July, nights in Antarctica are 24 hours long. This lack of insolation results in the lowest natural temperatures reported on Earth; the record is a frigid −89.2°C (−128.6°F) recorded on July 21, 1983, at the Russian Vostok Station. This temperature is 11 C° (19.8 F°) colder than the freezing point of dry ice (solid carbon dioxide). If the concentration of carbon dioxide were large enough, such a cold temperature would theoretically freeze tiny carbon dioxide dry-ice particles out of the sky.

The average July temperature around the Vostok Station is −68°C (−90.4°F). For comparison, average temperatures are −60°C (−76°F) and −26°C (−14.8°F) at the Amundsen–Scott and McMurdo Stations, respectively (Figure 5.15b). Note that the coldest temperatures in Antarctica are usually in August, not July, at the end of the long polar night just before the equinox sunrise in September.

GEOreport 5.3 Polar regions show greatest rates of warming

Climate change is affecting the higher latitudes at a pace exceeding that in the middle and lower latitudes. Since 1978, warming has increased in the Arctic region to a rate of 1.2 C° (2.2 F°) per decade, which means the last 20 years warmed at nearly seven times the rate of the last 100 years. Since 1970, nearly 60% of the Arctic sea ice has disappeared in response to increasing air and ocean temperatures, with record low levels of ice since 2007. The term *Arctic amplification* refers to the tendency for far northern latitudes to experience enhanced warming relative to the rest of the Northern Hemisphere. This phenomenon is related to the presence of snow and ice, and the positive feedback loops triggered by snow and ice melt, as discussed in previous chapters (review Figure 1.8 and *Geosystems Now* 4). Similar warming trends are affecting the Antarctic Peninsula and the West Antarctic Ice Sheet, as ice shelves collapse and retreat along the coast.

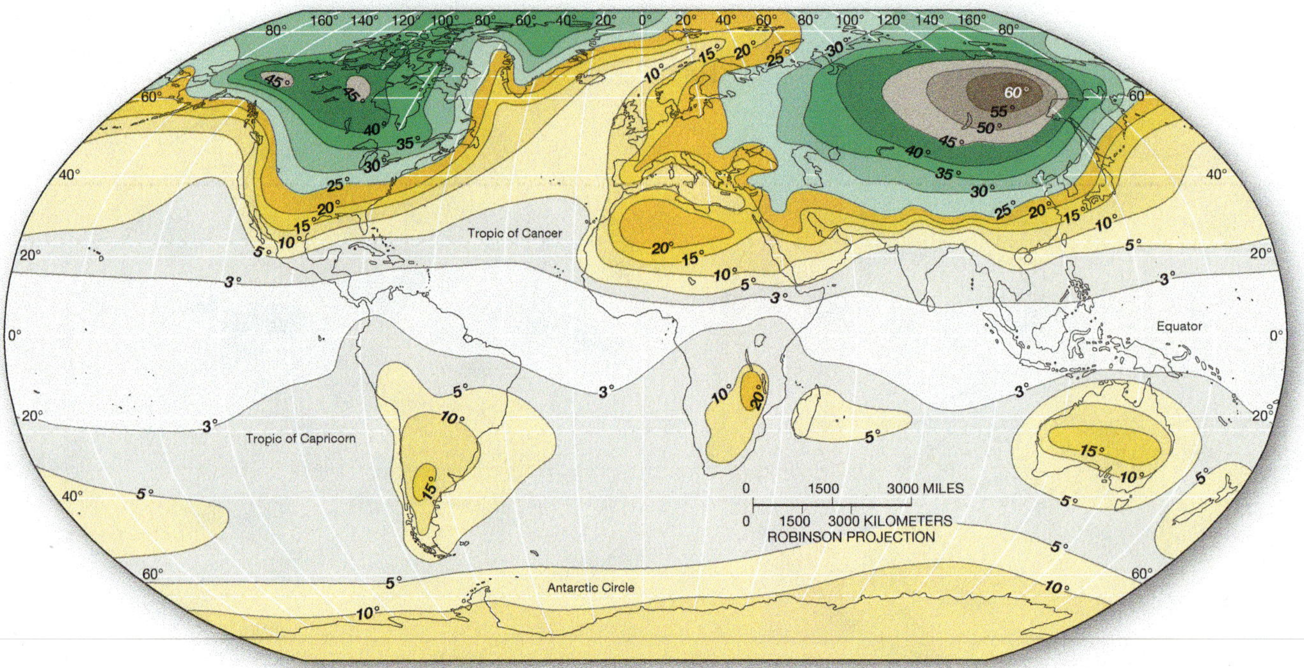

F°	5	9	18	27	36	45	54	63	72	81	90	99	108	F°
C°	3	5	10	15	20	25	30	35	40	45	50	55	60	C°

▲Figure 5.16 **Global annual temperature ranges.** The annual ranges of global temperatures in Celsius degrees, C°, with conversions to Fahrenheit degrees, F°, shown on scale. The mapped data show the difference between average January and July temperature maps.

MG™ Animation
Global Warming, Climate Change

MG™ Notebook
GISS Surface Temperature Analysis, 1891 to 2006

Annual Temperature Range Map

The largest average annual temperature ranges occur at subpolar locations within the continental interiors of North America and Asia (Figure 5.16), where average ranges of 64 C° (115 F°) are recorded (see the dark brown area on the map). Smaller temperature ranges in the Southern Hemisphere indicate less seasonal temperature variation owing to the lack of large landmasses and the vast expanses of water to moderate temperature extremes. Thus, continental effects dominate in the Northern Hemisphere, and marine effects dominate in the Southern Hemisphere.

However, temperatures in interior regions in the Southern Hemisphere show some continental characteristics. For example, in January (Figure 5.12), Australia is dominated by isotherms of 20°–30°C (68°–86°F), whereas in July (Figure 5.13), Australia is crossed by the 12°C (54°F) isotherm. The Northern Hemisphere, with greater land area overall, registers a slightly higher average surface temperature than does the Southern Hemisphere.

Recent Temperature Trends and Human Response

Studies of past climates, discussed at length in Chapter 17, indicate that present temperatures are higher than at any time during the past 125,000 years. Global temperatures rose an average of 0.17 C° (0.3 F°) per decade since 1970, and this rate is accelerating. Heat waves are also on the increase; summer temperatures in the United States and Australia rose to record levels during heat waves in 2012 and 2013. Persistent high temperatures present challenges to people, especially those living in cities in humid environments.

Record Temperatures and Greenhouse Warming

One way scientists assess and map temperature patterns is using global temperature anomalies (Figure 5.17). A temperature *anomaly* is a difference, or irregularity, found by comparing recorded average annual temperatures against the long-term average annual temperature for a time period selected as the *baseline*, or base period (the basis for comparison).

The maps of surface-temperature anomalies by decade in Figure 5.17 reveal the differences between the temperatures of each specified decade and the temperatures of the baseline period 1951 to 1980. Colors on the map indicate either positive (warmer, shown by oranges and reds) or negative (cooler, shown by blues) anomalies. The overall trend toward warming temperatures is apparent by the 2000s, especially over the Northern Hemisphere.

On the *MasteringGeography* website, you will find an animation of yearly global temperature anomalies

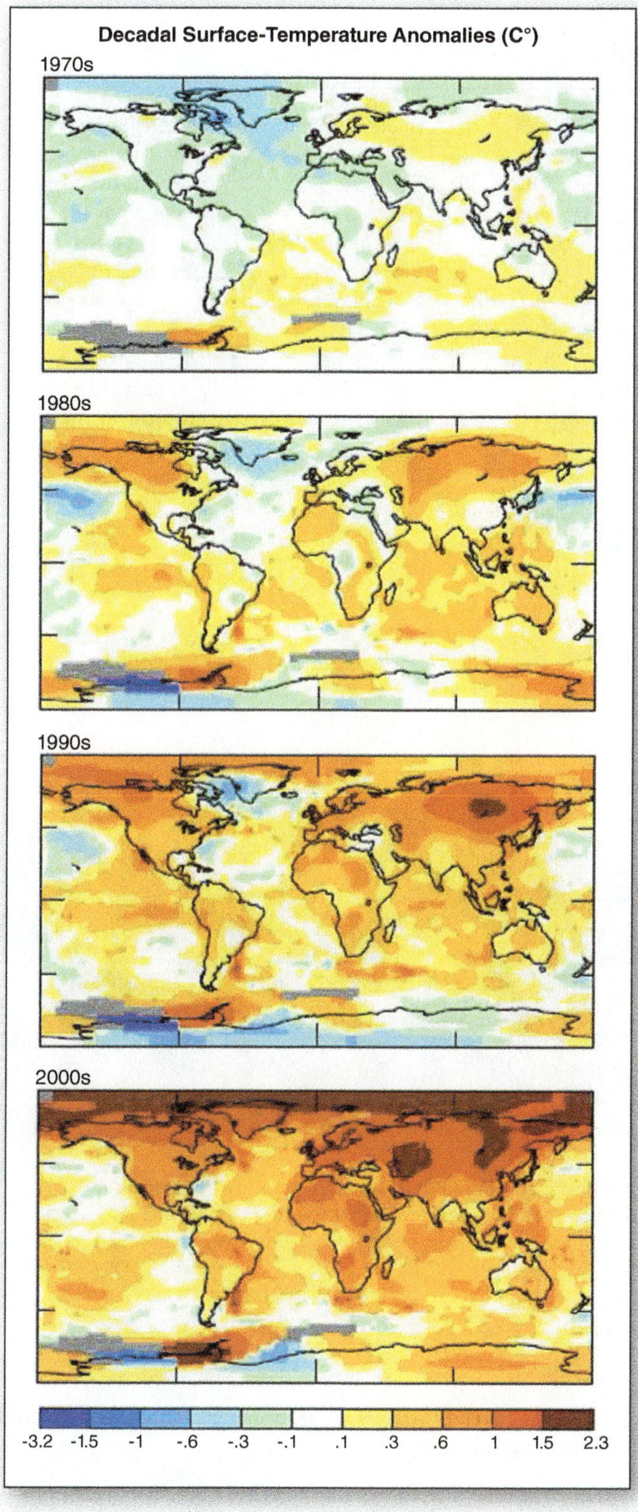

▲**Figure 5.17 Surface-temperature anomalies by decade, compared to 1951–1980 baseline.** Reds and oranges indicate positive temperature anomalies; blues indicate negative anomalies. The temperature-anomaly maps are based on data from 6300 stations. [GISS/NASA.]

from 1884 to 2012 (as compared to the baseline average global temperature from 1951 to 1980). This visualization clearly shows the trend toward above-average temperatures, especially in the past decade. The data for these maps, from over 1000 meteorological stations, as well as satellite observations of sea-surface temperature and Antarctic research-station measurements, are in close agreement with other global data sets, such as those from NCDC in the United States and Met Office Hadley Centre in the United Kingdom.

The last 15 years feature the warmest years in the climate record since 1880. According to NOAA's National Climatic Data Center (NCDC), the year 2012 was the hottest year ever in the contiguous United States. Globally, 2012 was the tenth highest on record (the warmest years were 2005 and 2010, part of the warmest decade on record.)

To get an idea of the magnitude of this temperature trend, consider these statistics: In 2012, the average global temperature was 14.6°C (58.3°F). This is 0.6 C° (1.0 F°) warmer than the baseline temperature in the mid-1900s. It is also 0.8 C° (1.4 F°) warmer than any other average yearly global temperature since recordkeeping began.

According to the U.S. National Research Council of the National Academy of Sciences, for each Celsius degree of global temperature increase, we can expect 5%–10% changes in precipitation in many regions; 3%–10% increases in the amount of rain falling during the heaviest precipitation events; 5%–10% changes in flows of streams and rivers (either up or down); 25% decreases in the extent of Arctic summer sea ice; 5%–15% reductions in crop yields (as currently grown); and 200%–400% increases in the area burned by wildfire in some areas of the western United States. Thus, a 1-C° change can have far-reaching effects on Earth systems.

This so-called global warming is related to complex changes now under way in the lower atmosphere. Scientists agree that human activities, principally the burning of fossil fuels, are increasing atmospheric greenhouse gases that absorb longwave radiation, delaying losses of heat energy to space. The bottom line is that human actions are enhancing Earth's natural greenhouse effect and forcing climate change.

However, "global warming" is not the same thing as global climate change, and the two terms should not be considered interchangeable. Climate change encompasses all the effects of atmospheric warming—these effects vary with location and relate to humidity, precipitation, sea-surface temperatures, severe storms, and many other Earth processes. An example already discussed is the positive feedback loop created by Arctic sea-ice melting and temperature rise in Chapter 4. Other impacts will be discussed in chapters ahead—the effects of climate change link to almost all Earth systems.

The long-term climate records discussed in Chapter 10B show that climates have varied over the last 2 million years. In essence, climate is always changing. However, many of the changes occurring today are beyond what can be explained by the natural variability of Earth's climate patterns. The scientific consensus on human-forced climate change is now overwhelming. In

Focus Study 5.1 Climate Change

Heat Waves

Heat waves are most deadly in midlatitude regions, where extremes of temperature and humidity become concentrated over stretches of days or weeks during the warmer months. Heat waves are often most severe in urban areas, where urban heat island effects worsen weather conditions, and indoor air-conditioning systems may be unaffordable for the poor. The oppressive heat in an urban environment during such an event can result in many deaths; those most susceptible to heat-related illness are the young, the elderly, and those with pre-existing medical conditions. The deadly effects of high temperatures, especially in cities, result from extreme daily temperature maximums combined with lack of nighttime cooling.

The Chicago heat wave of July 1995, which caused over 700 deaths in the city core, is an example. The combination of high temperatures with the persistence of a stable, unmoving air mass and moist air from the Gulf of Mexico produced stifling conditions, particularly affecting the sick and elderly. The record temperatures logged at Midway Airport (41°C; 106°F) were exceeded by heat-index temperatures of 54°C (130°F) in some apartments without air conditioning. For nearly a week the heat-index values for those dwellings signaled "extreme danger."

Heat waves are a major cause of weather-related deaths. A heat wave paralyzed much of Europe during the summer of 2003 when temperatures topped 40°C (104°F) in June, July, and August. An estimated 40,000 people died in six western European countries, with the highest number in France.

Heat waves are often associated with increased wildfires, as in Russia during the summer of 2010 (Figure 5.1.1). The Russian heat wave brought persistent high temperatures throughout eastern Europe that were the most destructive in 130 years, causing an estimated 55,000 heat-related deaths, massive crop losses, over 1 million hectares (2.5 million acres) of land burned by wildfires, and an overall estimated economic cost of US$15 billion.

The record-breaking Australian heat wave of January 2013 lasted not days but weeks, with temperatures

▲**Figure 5.1.1 Heat wave in Russia, summer 2010.** American tourists in Moscow wear face masks to filter smoke from nearby forest fires as temperatures top 38°C (100°F). [Pavel Golovkin, AP.]

2007, the Intergovernmental Panel on Climate Change (IPCC), the leading body of climate change scientists in the world, concluded,

> Warming of the climate system is unequivocal, as is now evident from observations of increases in global average air and ocean temperatures, widespread melting of snow and ice, and rising global average sea level.

In 2010, the U.S. National Research Council of the National Academy of Sciences, in its report *Advancing the Science of Climate Change,* stated,

> Climate change is occurring, is caused largely by human activities, and poses significant risks for—and in many cases is already affecting—a broad range of human and natural systems.

The Association of American Geographers, composed of geographers and related professionals working in the public, private, and academic sectors, is one of hundreds of professional organizations and national academies of sciences that are now on record as supporting action to slow rates of climate change. In the news on almost a daily basis, climate change has become one of the most complicated yet vital issues facing world leaders and human society in this century.

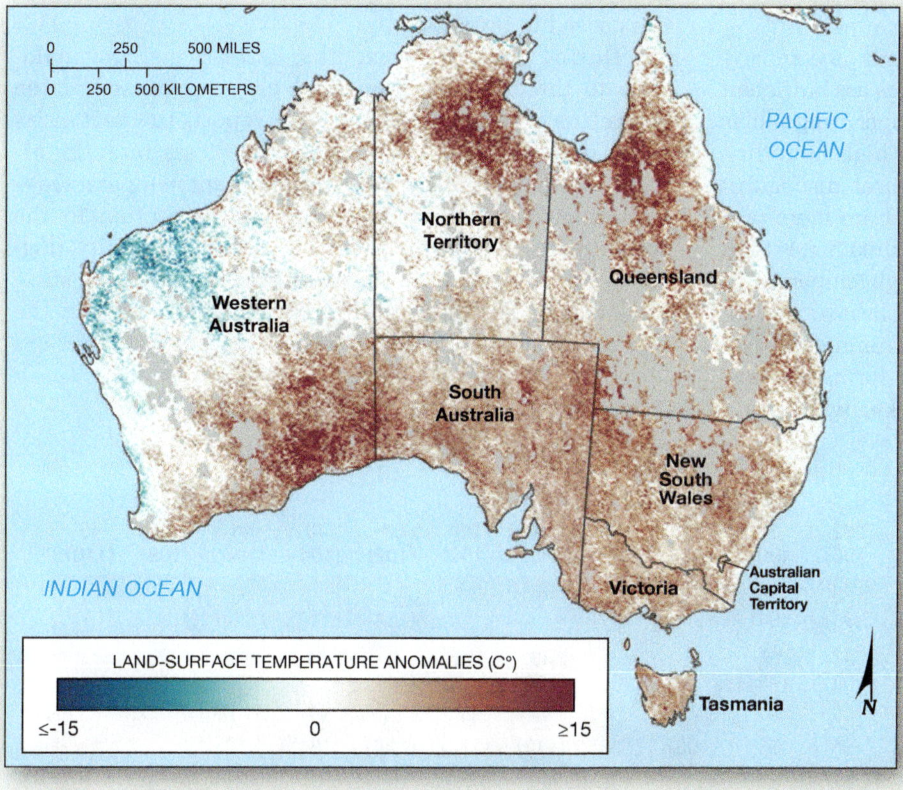

▲**Figure 5.1.2 Record-setting heat wave in Australia, 2013.** Map shows land-surface temperature differences recorded from satellite data in early January. Above-average values (positive anomalies) are shown by red colors; below-average values (negative anomalies) are shown by blue colors. The baseline temperatures for comparison are from the same week during 2005–2012. [Data from *Aqua* MODIS, NASA.]

regularly exceeding 45°C (113°F) at several locations across the country. According to the Australian Bureau of Meteorology, the heat wave registered the highest average temperature ever recorded across the country (40.3°C, or 104.6°F) during the hottest two-day period (January 7–8) in Australia's history. Figure 5.1.2 shows the pattern of above-average land-surface temperatures across the continent during this time. In mid-January, weather forecasters added two new colors to their heat maps to extend the range of temperatures beyond 50°C (122°F) to 54°C (129°F). This heat wave continued a trend of four consecutive months of record-breaking temperatures in Australia.

Although heat waves in Australia are nothing new, the extent of heat across the country and the persistence of the heat wave over time represent new conditions and suggest a continuing trend of temperature increase for the country. According to the chairperson of the International Panel on Climate Change (IPCC), this heat wave, and others, relates to recent trends of global climate change.

In the United States, 2011 and 2012 brought the warmest two consecutive summers on record at that time, and the second and third warmest summers ever as of 2012 (the warmest occurred in 1936). The 2011 heat wave was worst in the southern Plains states, centering on Texas and Oklahoma. In 2012, high temperatures were spread over the entire country, with one-third of the nation's population exposed to 10 or more days over 38°C (100°F). (For animations of surface land and air temperatures during the July 2011 heat wave in the United States, see http://photojournal.jpl.nasa.gov/catalog/PIA14480.)

Heat Stress and the Heat Index

One of the challenges humans are facing in adapting to the effects of climate change is an increase in the frequency of heat waves, putting more people at risk from the effects of prolonged high temperatures during the summer season. By definition, a **heat wave** is a prolonged period of abnormally high temperatures, usually, but not always, in association with humid weather. Focus Study 5.1 discusses recent heat waves, such as Australia's record-breaking summer of 2013.

Through several complex mechanisms, the human body maintains an average internal temperature ranging within a degree of 36.8°C (98.2°F), slightly lower in the morning or in cold weather and slightly higher at emotional times or during exercise and work.* When exposed to extreme heat and humidity, the human body reacts in various ways (perspiration is one such response) to maintain this core temperature and to protect the brain at all cost.

Humidity is the presence of water vapor in the air and is commonly expressed as relative humidity (see full

*The traditional value for "normal" body temperature, 37°C (98.6°F), was set in 1868 using old methods of measurement. According to Dr. Philip Mackowiak of the University of Maryland School of Medicine, a more accurate modern assessment places normal at 36.8°C (98.2°F), with a range of 2.7 C° (4.8 F°) for the human population (*Journal of the American Medical Association*, September 23– 0, 1992).

discussion in Chapter 7); the higher the amount of water vapor, the higher the relative humidity. Under humid conditions, the air cannot absorb as much moisture, so perspiration is not as effective a cooling mechanism as in dry environments.

When low humidity and strong winds accompany high temperatures, evaporative cooling rates are sufficient to keep body temperature within the proper range. It is the combination of high air temperatures, high humidity, and low winds that produces the most heat discomfort for humans. This is why the effects of heat are more pronounced in the hot, humid southeastern United States than in the dry environments of Arizona; although temperatures in deserts may be more extreme, the risk of heat-related illness, or *heat stress*, is higher in humid environments.

Heat stress in humans takes such forms as heat cramps, heat exhaustion, and heat stroke, which is a life-threatening condition. A person with heat stroke has overheated to the point where the body is unable to cool itself—at this point, internal temperature may have risen to as high as 41°C (106°F), and the sweating mechanism has ceased to function.

During appropriate months, using a method analogous to that for reporting wind chill (discussed at the beginning of the chapter), the NWS reports the *heat index* in its daily weather summaries to indicate how the air feels to an average person—its apparent temperature—and gauge the human body's probable reaction to the combined effects of air temperature and humidity (Figure 5.18). Canada uses the *humidex*, based on a similar formula. For more information and forecasts in the United States, see http://www.nws.noaa.gov/om/heat/index.shtml; in Canada, go to http://www.ec.gc.ca/meteo-weather/default.asp?lang=En&n=86C0425B-1#h2).

Temperature in °C (°F)

Relative Humidity (%)	27 (80)	28 (82)	29 (84)	30 (86)	31 (88)	32 (90)	33 (90)	34 (94)	36 (96)	37 (98)	38 (100)	39 (102)	40 (104)	41 (106)	42 (108)	43 (110)
40	27 (80)	27 (81)	28 (83)	29 (85)	31 (88)	33 (91)	34 (94)	36 (97)	38 (101)	41 (105)	43 (109)	46 (114)	48 (124)	51 (106)	54 (130)	58 (136)
45	27 (80)	28 (82)	29 (84)	31 (87)	32 (89)	34 (93)	36 (96)	38 (100)	40 (104)	43 (109)	46 (114)	48 (119)	51 (124)	54 (130)	58 (137)	
50	27 (81)	28 (83)	29 (85)	31 (88)	33 (91)	35 (95)	37 (99)	39 (103)	42 (108)	45 (113)	48 (118)	51 (124)	55 (131)	58 (137)		
55	27 (81)	29 (84)	30 (86)	32 (89)	34 (93)	36 (97)	38 (101)	41 (106)	44 (112)	47 (117)	51 (124)	54 (129)	58 (137)			
60	28 (82)	29 (84)	31 (88)	33 (91)	35 (95)	38 (100)	41 (105)	43 (110)	47 (116)	51 (123)	54 (129)	58 (137)				
65	28 (82)	29 (85)	32 (89)	34 (93)	37 (98)	39 (103)	42 (108)	46 (114)	49 (121)	53 (128)	58 (136)					
70	28 (83)	30 (86)	32 (90)	35 (95)	38 (100)	41 (105)	44 (112)	48 (119)	52 (126)	57 (134)						
75	29 (84)	31 (88)	33 (92)	36 (97)	39 (103)	43 (109)	47 (116)	51 (124)	56 (132)							
80	29 (84)	32 (89)	34 (94)	38 (100)	41 (106)	45 (113)	49 (121)	54 (129)								
85	29 (85)	32 (90)	36 (96)	39 (102)	43 (110)	47 (117)	52 (126)	57 (135)								
90	30 (86)	33 (91)	37 (98)	41 (105)	45 (113)	50 (122)	55 (131)									
95	30 (86)	34 (93)	38 (100)	42 (108)	47 (117)	53 (127)										
100	31 (87)	35 (95)	39 (103)	44 (112)	49 (121)	56 (132)										

Likelihood of heat disorders with prolonged exposure or strenuous activity

Caution	Fatigue and heat cramps possible
Extreme Caution	Heat cramps, heat exhaustion, or heat stroke possible
Danger	Heat cramps, heat exhaustion likely; heat stroke probable
Extreme Danger	Heat stroke imminent

▲**FIGURE 5.18** **Heat index for various temperatures and relative humidity levels.**

GEOreport 5.4 Record-breaking heat hits China in 2013.

Temperatures soared across Asia in July and August, 2013, setting records and causing heat-related fatalities. In Shanghai, China, with a population of over 23 million, a 3-week heat wave—including a new high temperature of 40.8°C (105.4°F) recorded on August 7—prompted officials to issue the country's first ever weather warning for heat and caused at least 40 reported deaths, although the actual fatalities may be much higher. See http://earthobservatory.nasa.gov/IOTD/view.php?id=81870 for maps, images, and more on this heat wave.

TEMPERATURE ⇒ HUMANS

- Temperature patterns drive Earth systems, making the planet habitable for humans and other life.
- Heat and cold determine individual comfort levels.

HUMANS ⇒ TEMPERATURE

- Humans produce atmospheric gases and aerosols that affect clouds and the Earth–atmosphere energy budget, which in turn affects temperature and climate.
- Rising sea-surface temperatures alter evaporation rates, affecting clouds and rainfall, with associated effects on the biosphere.

5a

Measurements taken in 1909 and in 2004 indicate that the McCarty Glacier in Alaska retreated over 15 km (9.8 mi). Accelerated loss of ice is happening to glaciers in high latitude and mountain regions owing to climate change. (More on glaciers in Chapter 17.) [1909 by G. S. Grant, USGS; 2004 by Bruce F. Molnia, USGS.]

5b

A severe heat wave hit India during the summer of 2010, bringing the hottest temperatures in over 50 years and causing hundreds of casualties. [Bappa Majumdar/Reuters.]

5d

During 2012, the warmest summer on record in the United States up to that time, large fires in Idaho and Montana contributed to the over 3.6 million hectares (9 million acres) burned across the country, with the increase in wildfires tied to climate change impacts. [NASA/Jeff Schmaltz/LANCE MODIS Rapid Response.]

5c

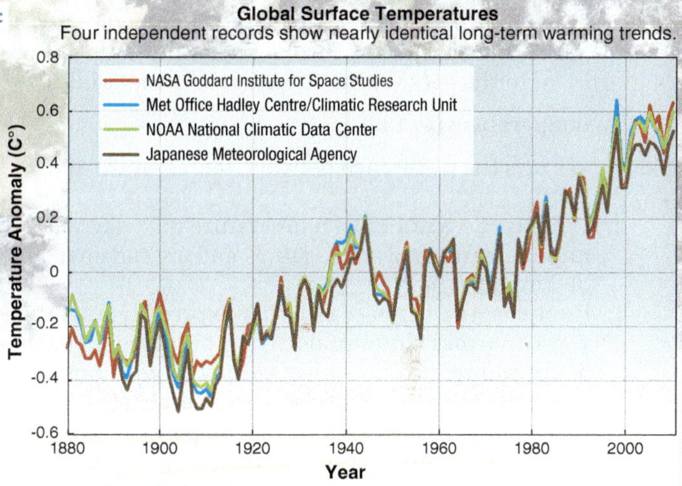

Records from several international scientific agencies are in agreement as to recent climate change impacts on global temperatures. [NASA Earth Observatory/Robert Simmon.]

ISSUES FOR THE 21ST CENTURY

- Rising average global temperatures will enhance climate change effects across the globe.
- Changes in sea-surface temperatures will have far-ranging effects on Earth systems.

GEOSYSTEMS**connection**

Global temperature patterns are a significant output of the energy–atmosphere system. We examined the complex interactions of several factors that produce these patterns and studied maps of their distributions. In the next chapter, we shift to another output of this energy–atmosphere system, that of global circulation of winds and ocean currents. We look at the forces that interact to produce these movements of air and water over Earth. We also examine multiyear fluctuations in global circulation patterns, such as the phenomena of El Niño and La Niña patterns in the Pacific Ocean and their far-reaching effects on world climates.

KEY LEARNING concepts review

■ ***Define*** **the concept of temperature, and** ***distinguish*** **between Kelvin, Celsius, and Fahrenheit temperature scales and how they are measured.**

Temperature is a measure of the average kinetic energy, or molecular motion, of individual molecules in matter. Heat transfer occurs from object to object when there is a temperature difference between them. The *wind-chill factor* indicates the enhanced rate at which body heat is lost to the air under conditions of cold temperatures and wind. As wind speeds increase, heat loss from the skin increases, decreasing the *apparent temperature*, or the temperature that we perceive.

Temperature scales include

- Kelvin scale: 100 units between the melting point of ice (273 K) and the boiling point of water (373 K).
- Celsius scale: 100 degrees between the melting point of ice (0°C) and the boiling point of water (100°C).
- Fahrenheit scale: 180 degrees between the melting point of ice (32°F) and the boiling point of water (212°F).

Scientists use the Kelvin scale because temperature readings on that scale start at absolute zero and thus are proportional to the actual kinetic energy in a material.

temperature (p. 109)

1. What is the difference between temperature and heat?
2. What is the wind-chill temperature on a day with an air temperature of –12°C (10°F) and a wind speed of 32 kmph (20 mph)?
3. Compare the three scales that express temperature. Find "normal" human body temperature on each scale and record the three values in your notes.
4. What is your source of daily temperature information? Describe the highest temperature and the lowest temperature you have experienced. From what we discussed in this chapter, can you identify the factors that contributed to these temperatures?

■ ***Explain*** **the effects of latitude, altitude and elevation, and cloud cover on global temperature patterns.**

Principal controls and influences on temperature patterns include latitude (the distance north or south of the equator), altitude and elevation, and cloud cover (reflection, absorption, and radiation of energy). *Altitude* describes the height of an object above Earth's surface, whereas *elevation* relates to a position on Earth's surface relative to sea level. Latitude and elevation work in combination to determine temperature patterns in a given location.

5. Explain the effects of altitude and elevation on air temperature. Why is air at higher altitude lower in temperature? Why does it feel cooler standing in shadows at higher elevation than at lower elevation?
6. What noticeable effect does air density have on the absorption and radiation of energy? What role does elevation play in that process?
7. Why is it possible to grow moderate-climate-type crops such as wheat, barley, and potatoes at an elevation of 4103 m (13,460 ft) near La Paz, Bolivia?
8. Describe the effect of cloud cover with regard to Earth's temperature patterns. From the last chapter, review the effects of different cloud types on temperature, and relate the concepts with a simple sketch.

■ ***Review*** **the differences in heating of land versus water that produce continental effects and marine effects on temperatures, and** ***utilize*** **a pair of cities to illustrate these differences.**

Differences in the physical characteristics of land (rock and soil) compared to water (oceans, seas, and lakes) lead to **land–water heating differences** that have an important effect on temperatures. These physical differences, related to *evaporation, transparency, specific heat,* and *movement*, cause land surfaces to heat and cool faster than water surfaces.

Because of water's **transparency**, light passes through it to an average depth of 60 m (200 ft) in the ocean. This penetration distributes available heat energy through a much greater volume than is possible through land, which is opaque. At the same time, water has a higher **specific heat**, or heat capacity, requiring far more energy to increase its temperature than does an equal volume of land.

Ocean currents and *sea-surface temperatures* also affect land temperature. An example of the effect of ocean currents is the **Gulf Stream**, which moves northward off the east coast of North America, carrying warm water far into the North Atlantic. As a result, the southern third of Iceland experiences much milder temperatures than would be expected for a latitude of 65° N, just below the Arctic Circle (66.5°).

Moderate temperature patterns occur in locations near water bodies, and more extreme temperatures occur inland. The **marine effect**, or maritime effect, usually seen along coastlines or on islands, is the moderating influence of the ocean. In contrast, the **continental effect** occurs in areas that are less affected by the sea and therefore experience a greater range between maximum and minimum temperatures on a daily and yearly basis.

land–water heating differences (p. 114)
transparency (p. 115)
specific heat (p. 115)
Gulf Stream (p. 116)
marine effect (p. 116)
continental effect (p. 116)

9. List the physical characteristics of land and water that produce their different responses to heating

from absorption of insolation. What is the specific effect of transparency in a medium?

10. What is specific heat? Compare the specific heat of water and soil.
11. Describe the pattern of sea-surface temperatures (SSTs) as determined by satellite remote sensing. Where is the warmest ocean region on Earth?
12. What effect does sea-surface temperature have on air temperature? Describe the negative feedback mechanism created by higher sea-surface temperatures and evaporation rates.
13. Differentiate between temperatures at marine versus continental locations. Give an example of each from the text discussion.

■ *Interpret* the pattern of Earth's temperatures from their portrayal on January and July temperature maps and on a map of annual temperature ranges.

Maps for January and July instead of the solstice months of December and June are used for temperature comparison because of the natural lag that occurs between insolation received and maximum or minimum temperatures experienced. Each line on these temperature maps is an **isotherm**, an isoline that connects points of equal temperature. Isotherms portray temperature patterns.

Isotherms generally trend east–west, parallel to the equator, marking the general decrease in insolation and net radiation with distance from the equator. The **thermal equator** (isoline connecting all points of highest mean temperature) trends southward in January and shifts northward with the high summer Sun in July. In January, it extends farther south into the interior of South America and Africa, indicating higher temperatures over landmasses.

In the Northern Hemisphere in January, isotherms shift equatorward as cold air chills the continental interiors. The coldest area on the map is in Russia—specifically, northeastern Siberia. The intense cold experienced there results from winter conditions of consistently clear, dry, calm air; small insolation input; and an inland location far from any moderating maritime effects.

isotherm (p. 120)
thermal equator (p. 121)

14. What is the thermal equator? Describe its location in January and in July. Explain why it shifts position annually.
15. Observe trends in the pattern of isolines over North America as seen on the January average-temperature map compared with the July map. Why do the patterns shift?
16. Describe and explain the extreme temperature range experienced in north-central Siberia between January and July.
17. Where are the hottest places on Earth? Are they near the equator or elsewhere? Explain. Where is the coldest place on Earth?
18. Compare the maps in Figures 5.14 and 5.15: (a) Describe what you find in central Greenland in January and July; (b) look at the south polar region and describe seasonal changes there. Characterize conditions along the Antarctic Peninsula in January and July (around 60° W longitude).

■ *Discuss* heat waves and the heat index as a measure of human heat response.

Global climate change is presenting challenges to people across Earth. Recent **heat waves**, prolonged periods of high temperatures lasting days or weeks, have caused fatalities and billions of dollars in economic losses. The *heat index* indicates the human body's reaction to air temperature and water vapor. The level of humidity in the air affects our natural ability to cool through evaporation from skin.

heat wave (p. 127)

19. On a day when temperature reaches 37.8°C (100°F), how does a relative humidity reading of 50% affect apparent temperature?
20. Discuss recent heat waves in Australia and the United States. How do these events differ from previous heat waves in these regions?

Answer to Critical Thinking 5.1: San Francisco exhibits marine effects on temperature because the cooling waters of the Pacific Ocean and San Francisco Bay surround the city on three sides. Summer fog is common and helps delay the warmest summer month in San Francisco until September. Wichita has a continental location and therefore a greater annual temperature range. Note that Wichita is also slightly higher in elevation, which has a small effect on daily temperature variations.

MasteringGeography™

Looking for additional review and test prep materials? Visit the Study Area in *MasteringGeography*™ to enhance your geographic literacy, spatial reasoning skills, and understanding of this chapter's content by accessing a variety of resources, including MapMaster interactive maps, geoscience animations, satellite loops, author notebooks, videos, RSS feeds, flashcards, web links, self-study quizzes, and an eText version of *Geosystems*.

6 Atmospheric and Oceanic Circulations

Wind turbines line the horizon near Wasco in eastern Oregon, a state that ranks fifth out of 39 states contributing to overall U.S. wind energy production. Worldwide, wind energy is growing at about 10% every year. In the United States, installed wind power capacity increased over 500% between 2005 and 2012. Wind is a primary driver of Earth's atmospheric and oceanic circulation patterns, discussed in this chapter. [Darrell Gulin/Getty Images.]

KEY LEARNING concepts

After reading the chapter, you should be able to:

- ***Define*** the concept of air pressure, and ***describe*** instruments used to measure air pressure.
- ***Define*** wind, and ***explain*** how wind is measured, how wind direction is determined, and how winds are named.
- ***Explain*** the four driving forces within the atmosphere—gravity, pressure gradient force, Coriolis force, and friction force—and ***locate*** the primary high- and low-pressure areas and principal winds.
- ***Describe*** upper-air circulation, and ***define*** the jet streams.
- ***Explain*** the regional monsoons and several types of local winds.
- ***Sketch*** the basic pattern of Earth's major surface ocean currents and deep thermohaline circulation.
- ***Summarize*** several multiyear oscillations of air temperature, air pressure, and circulation associated with the Arctic, Atlantic, and Pacific Oceans.

Ocean Currents Bring Invasive Species

As human civilization discharges ever more refuse and chemicals into the oceans, the garbage travels in currents that sweep across the globe. A dramatic episode of such ocean-current transport happened in the South Atlantic in 2006. We begin the story here and continue with its ecological impacts in Chapter 20.

The South Atlantic Gyre Across Earth, winds and ocean currents move in distinct patterns that we examine in this chapter. In the South Atlantic Ocean, a counterclockwise flow drives the South Equatorial Current westward and the West Wind Drift eastward (Figure GN 6.1). Along the African coast, the cool Benguela Current flows northward, while the warm Brazil Current moves south along the coast of South America. This is the prevailing circulation of the South Atlantic.

Along the southeastern portion of this prevailing circular flow, called a *gyre*, is a remote island group, the Tristan da Cunha archipelago, comprising four islands some 2775 km (1725 mi) from Africa and 3355 km (2085 mi) from South America (see location in Figure GN 6.1).

Isolated Tristan Only 298 people live on Tristan. This unique small society does subsistence farming, collectively grows and exports potatoes, and depends on rich marine life that the people manage carefully. The Tristan rock lobster (a crawfish) is harvested and quick-frozen at a community factory and exported around the world. Although several ships a year transport the island's commodities, Tristan has no pier or port, and also lacks an airport. However, in 2006, ocean currents brought to this isolated island a grim reminder of the outside world.

▼**Figure GN 6.1 Principal ocean currents of the South Atlantic.** Note the location of Tristan da Cunha and the probable route of the oil-drilling platform.

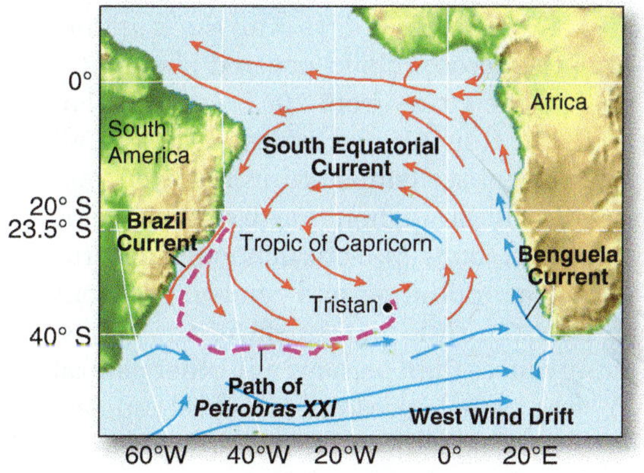

▲**Figure GN 6.2 *Petrobras XXI* aground in Trypot Bay, Tristan.** [Photo by biologist Sue Scott. All rights reserved.]

Path of the Drilling Rig Oil-drilling platforms are like small ships that can be towed to a desired site for drilling undersea wells. Petrobras, Brazil's state-controlled oil company, had such a platform, the *Petrobras XXI*. The oil company towed the 80-m × 67-m × 34-m-high (264-ft × 220-ft × 113-ft-high) platform from Macaé, Brazil, on March 5, 2006—destination Singapore. The company chartered a tug, the *Mighty Deliverer*, to do the job. This tug was actually built for inland work as a "pusher" tug, like ones you might see on the Mississippi River or in the Great Lakes; yet, it was contracted to haul the drilling rig across the Southern Ocean, easily the most treacherous seas on Earth. The *Deliverer* and *Petrobras XXI* headed south, encountering rough seas after a couple of weeks. Conditions forced the tug's crew to cut the platform loose on April 30, and after a few days, it was lost.

The *Petrobras XXI* was missing in rough seas, caught in the wind system of the westerlies and the currents of the West Wind Drift. Sometime in late May, after almost a month adrift, the oil-drilling platform ran aground on Tristan in Trypot Bay, where it was discovered by Tristanian fishermen on June 7 (Figure GN 6.2).

Invasive Species Land in Tristan The owners of the drilling platform had neglected to clean it in preparation for towing. (Clean rigs move through the water with less friction, require less fuel, and result in lower labor costs.) Because of this oversight, *Petrobras XXI* carried some 62 nonnative species of marine life, including free-swimming silver porgy and blenny, fish that tagged along with the rig as it drifted in the currents.

What the marine scientists found when they surveyed the grounded rig was an unprecedented species invasion of Tristan. Exclusive photos by one of the scientist divers, interviewed for this report, are displayed in Chapter 20. Here in Chapter 6, we learn about wind and ocean circulation, the forces that brought the *Petrobras XXI* to Tristan.

GEOSYSTEMS NOW ONLINE Go to Chapter 6 on the *MasteringGeography* website (www.masteringgeography.com) for more on this story and the effects of the spilled materials. In March 2011, a vessel carrying soybeans en route between Brazil and Singapore ran aground on Nightingale Island, just south of Tristan da Cunha. Two days later, the vessel broke up, leaking oil into the waters around this World Heritage site and spilling soya onto lobster beds, thus endangering local fisheries. For dramatic photos of the wreck and the local effort to clean oil-coated Northern Rockhopper Penguins, go to http://www.tristandc.com/newsmsoliva.php. MG

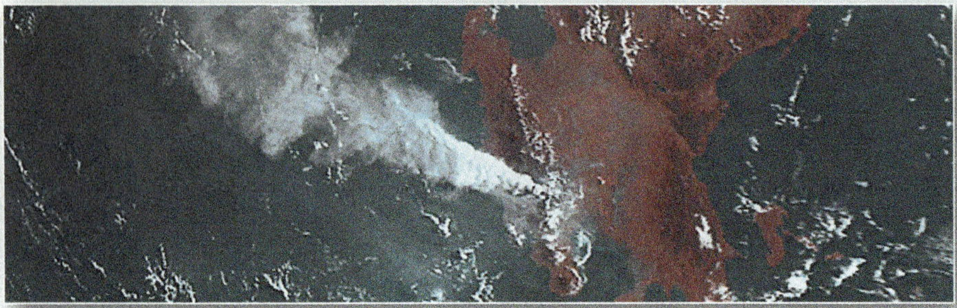

(a) Mount Pinatubo eruption, June 15, 1991. Images below track the movement of aerosols across the globe.

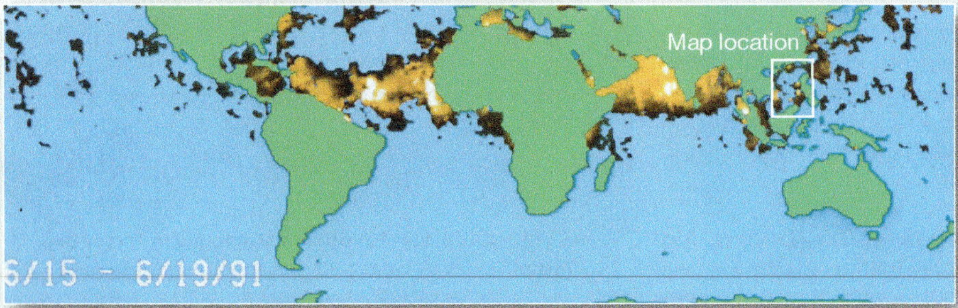

(b) False-color images show aerosol optical thickness: White has the highest concentration of aerosols; yellow shows medium values; and brown lowest values. Note dust, smoke from fires, and haze in the atmosphere at the time of the eruption.

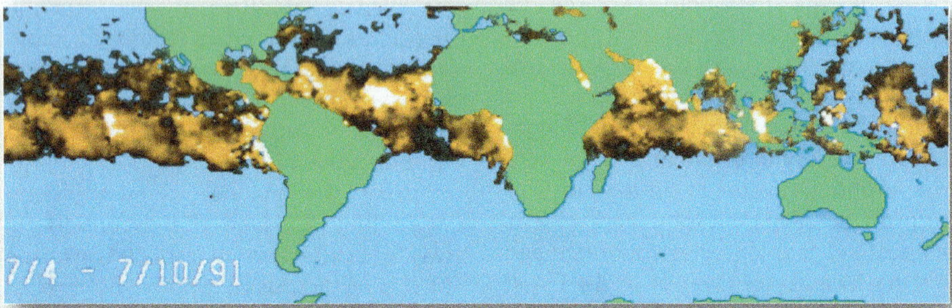

(c) Aerosol layer circles the entire globe 21 days after the eruption.

(d) The effects of the eruption cover 42% of the globe after 2 months.

▲**Figure 6.1 Atmospheric effects of the Mount Pinatubo volcanic eruption and global winds.** [(a) USGS. (b–d) AOT images from the advanced very high resolution radiometer (AVHRR) instrument aboard *NOAA-11;* NESDIS/NOAA.]

The 1991 eruption in the Philippines of Mount Pinatubo, after 635 years of dormancy, provided a unique opportunity to assess the dynamics of atmospheric circulation using satellite monitoring and tracking of contaminants from the volcanic explosion (Figure 6.1a). The Mount Pinatubo event had tremendous impact, lofting 15–20 million tons of ash, dust, and sulfur dioxide (SO_2) into the atmosphere. As the sulfur dioxide rose into the stratosphere, it quickly formed sulfuric acid (H_2SO_4) aerosols, which concentrated at an altitude of 16–25 km (10–15.5 mi).

In Figure 6.1b, which shows atmospheric aerosols in the first few days after the eruption, we see some of the millions of tons of dust from African soils that cross the Atlantic each year, borne by the winds of atmospheric circulation. We also see the effects of smoke from Kuwaiti oil well fires set during the first Persian Gulf War, as well as smoke from forest fires in Siberia and haze off the East Coast of the United States.

Figures 6.1c and d show Mount Pinatubo's aerosol's mixed with airborne debris from dust storms, fires, and industrial haze as global winds swept them around Earth. This debris increased atmospheric albedo about 1.5%. Some 60 days after the eruption, the aerosol cloud covered about 42% of the globe, from 20° S to 30° N. For almost 2 years, colorful sunrises and sunsets and a small temporary lowering of average temperatures followed.

Aerosols move freely throughout Earth's atmosphere, unconfined by political borders. International concerns about transboundary air pollution and nuclear weapons testing illustrate how the fluid movement of the atmosphere links humanity more than perhaps any other natural or cultural factor. The global spread of low-level radioactive contamination from Japan's nuclear disaster associated with the earthquake and tsunami in 2011 is another example of this linkage. More than any other Earth system, our atmosphere is shared by all humanity—one person's or country's exhalation is another's inhalation.

In this chapter: We begin with a discussion of wind essentials, including air pressure and the measurement of wind. We examine the driving forces that produce and determine the speed and direction of surface winds: pressure gradients, the Coriolis force, and friction. We then look at the circulation of Earth's atmosphere, including the principal pressure systems, patterns of global surface winds, upper-atmosphere winds, monsoons, and local winds. Finally, we consider Earth's wind-driven oceanic currents and explain multiyear oscillations in atmospheric and oceanic flows. The energy driving all this movement comes from one source: the Sun.

Wind Essentials

The large-scale circulation of winds across Earth has fascinated explorers, sailors, and scientists for centuries, although only in the modern era is a clear picture emerging of global winds. Driven by the imbalance between equatorial energy surpluses and polar energy deficits, Earth's atmospheric circulation transfers both energy and mass on a grand scale, determining Earth's weather patterns and the flow of ocean currents. The atmosphere is the dominant medium for redistributing energy from about 35° latitude to the poles in each hemisphere, whereas ocean currents redistribute more heat in a zone straddling the equator between the 17th parallels in each hemisphere. Atmospheric circulation also spreads air pollutants, whether natural or human-caused, worldwide, far from their point of origin.

Air Pressure

Air pressure—the weight of the atmosphere described as force per unit area—is key to understanding wind. The molecules that constitute air create **air pressure** through their motion, size, and number, and this pressure is exerted on all surfaces in contact with air. As we saw in Chapter 3, the number of molecules and their motion are also the factors that determine the density and temperature of the air.

Pressure Relationships As Chapter 3 discussed, both pressure and density decrease with altitude in the atmosphere. The low density in the upper atmosphere means the molecules are far apart, making collisions between them less frequent and thereby reducing pressure (review Figure 3.2). However, differences in air pressure are noticeable even between sea level and the summits of Earth's highest mountains.

The subjective experience of "thin air" at altitude is caused by the smaller amount of oxygen available to inhale (fewer air molecules means less oxygen). Mountaineers feel the effects of thin air as headaches, shortness of breath, and disorientation as less oxygen reaches their brain—these are the symptoms of *acute mountain sickness.* Near the summits of the highest Himalayan peaks, some climbers use oxygen tanks to counteract these effects, which are worsened by ascending quickly without giving the body time to acclimatize, or adapt, to the decrease in oxygen.

Remember from Chapter 5 that temperature is a measure of the average kinetic energy of molecular motion. When air in the atmosphere is heated, molecular activity increases and temperature rises. With increased activity, the spacing between molecules increases so that density is reduced and air pressure decreases. Therefore, warmer air is less dense, or lighter, than colder air, and exerts less pressure.

The amount of water vapor in the air also affects its density. Moist air is lighter because the molecular weight of water is less than that of the molecules making up dry air. If the same total number of molecules has a higher percentage of water vapor, mass will be less than if the air were dry (that is, than if it were made up entirely of oxygen and nitrogen molecules). As water vapor in the air increases, density decreases, so humid air exerts less pressure than dry air.

The end result over Earth's surface is that warm, humid air is associated with low pressure and cold, dry air is associated with high pressure. These relationships between pressure, density, temperature, and moisture are important to the discussion ahead.

Air Pressure Measurement In 1643, work by Evangelista Torricelli, a pupil of Galileo, on a mine-drainage problem led to the first method for measuring air pressure (Figure 6.2a). Torricelli knew that pumps in the mine were able to "pull" water upward about 10 m (33 ft) but no higher, and that this level fluctuated from day to day. Careful observation revealed that the limitation was not the fault of the pumps but a property of the atmosphere itself. He figured out that air pressure, the weight of the air, varies with weather conditions and that this weight determined the height of the water in the pipe.

To simulate the problem at the mine, Torricelli devised an instrument using a much denser fluid than water—mercury (Hg)—and a glass tube 1 m (3.3 ft) high. He sealed the glass tube at one end, filled it with mercury, and inverted it into a dish containing mercury, at which point a small space containing a vacuum was formed in the tube's closed end (Figure 6.2b). Torricelli found that the average height of the column of mercury remaining in the tube was 760 mm (29.92 in.), depending

GEOreport 6.1 Blowing in the wind

Dust originating in Africa sometimes increases the iron content of the waters off Florida, promoting the toxic algal blooms (*Karenia brevis*) known as "red tides." In the Amazon, soil samples bear the dust print of these former African soils that crossed the Atlantic. Active research on such dust is part of the U.S. Navy's Aerosol Analysis and Prediction System; see links and the latest navy research at http://www.nrlmry.navy.mil/7544.html.

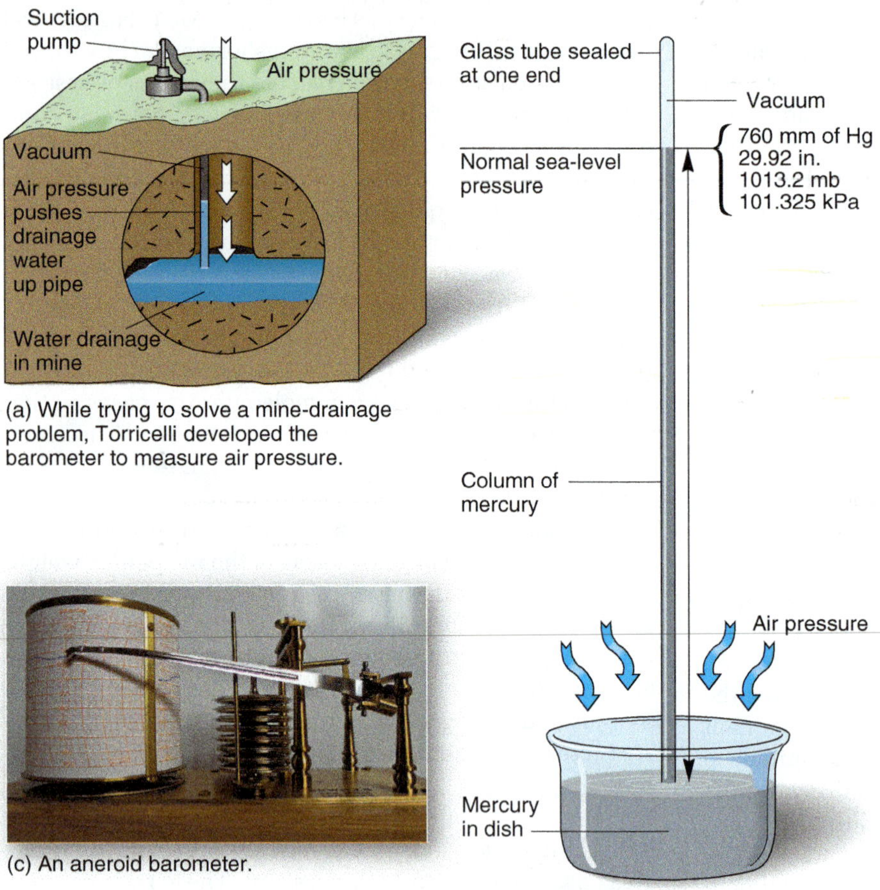

(a) While trying to solve a mine-drainage problem, Torricelli developed the barometer to measure air pressure.

(b) Idealized sketch of a mercury barometer.

(c) An aneroid barometer.

◀**Figure 6.2 Developing the barometer.** Have you used a barometer? If so, what type was it? Did you try to reset it using a local weather information source? [(c) Stuart Aylmer/Alamy]

on the weather. He concluded that the mass of surrounding air was exerting pressure on the mercury in the dish and thus counterbalancing the weight of the column of mercury in the tube.

Any instrument that measures air pressure is a barometer (from the Greek *baros*, meaning "weight"). Torricelli developed a **mercury barometer**. A more compact barometer design, which works without a meter-long tube of mercury, is the **aneroid barometer** (Figure 6.2c). Aneroid means "using no liquid." The aneroid barometer principle is simple: Imagine a small chamber, partially emptied of air, which is sealed and connected to a mechanism attached to a needle on a dial. As the air pressure outside the chamber increases, it presses inward on the chamber; as the outside air pressure decreases, it relieves the pressure on the chamber—in both cases causing changes in the chamber that move the needle. An aircraft altimeter is a type of aneroid barometer.

Today, atmospheric pressure is measured at weather stations by electronic sensors that provide continuous measurement over time using millibars (mb, which express force per square meter of a surface area) or hectopascals (1 millibar = 1 hectopascal). To compare pressure conditions from one place to another, pressure measurements are adjusted to a standard of normal sea-level pressure, which is 1013.2 mb or 29.92 in. of mercury (Hg). In Canada and certain other countries, normal sea-level pressure is expressed as 101.32 kilopascal, or kPa (1 kPa = 10 mb). The adjusted pressure is known as *barometric pressure*.

Figure 6.3 shows comparative scales in millibars and inches of mercury for air pressure. Note that the normal range of Earth's atmospheric pressure from strong high pressure to deep low pressure is about 1050 to 980 mb (31.00 to 29.00 in.). The figure also indicates pressure extremes recorded for the United States and worldwide.

Wind: Description and Measurement

Simply stated, **wind** is generally the horizontal motion of air across Earth's surface. Within the boundary layer at the surface, turbulence adds wind updrafts and downdrafts and thus a vertical component to this definition. Differences in air pressure between one location and another produce wind.

Wind's two principal properties are speed and direction, and instruments measure each. An **anemometer** measures wind speed in kilometers per hour (kmph), miles per hour (mph), meters per second (mps), or knots. (A knot is a nautical mile per hour, covering 1 minute of Earth's arc in an hour, equivalent to 1.85 kmph, or 1.15 mph.) A **wind vane** determines wind direction; the

GEOreport 6.2 Pressure changes in an airplane cabin

As an airplane descends to the landing strip and air pressure in the cabin returns to normal, passengers commonly feel a "popping" in their ears. This same effect can happen in an elevator, or even during a drive down a steep mountain road. The "popping" generally follows a plugged-up sensation produced when a change of air pressure affects the pressure balance of the middle and outer ear, distorting the eardrum and causing sound to be muffled. Actions such as yawning, swallowing, or chewing gum that open the eustachian tube connecting the middle ear to the upper throat can equalize the pressure in the middle ear and "pop" the eardrum back to its normal shape. (See http://www.mayoclinic.com/health/airplane-ear/DS00472 for more.)

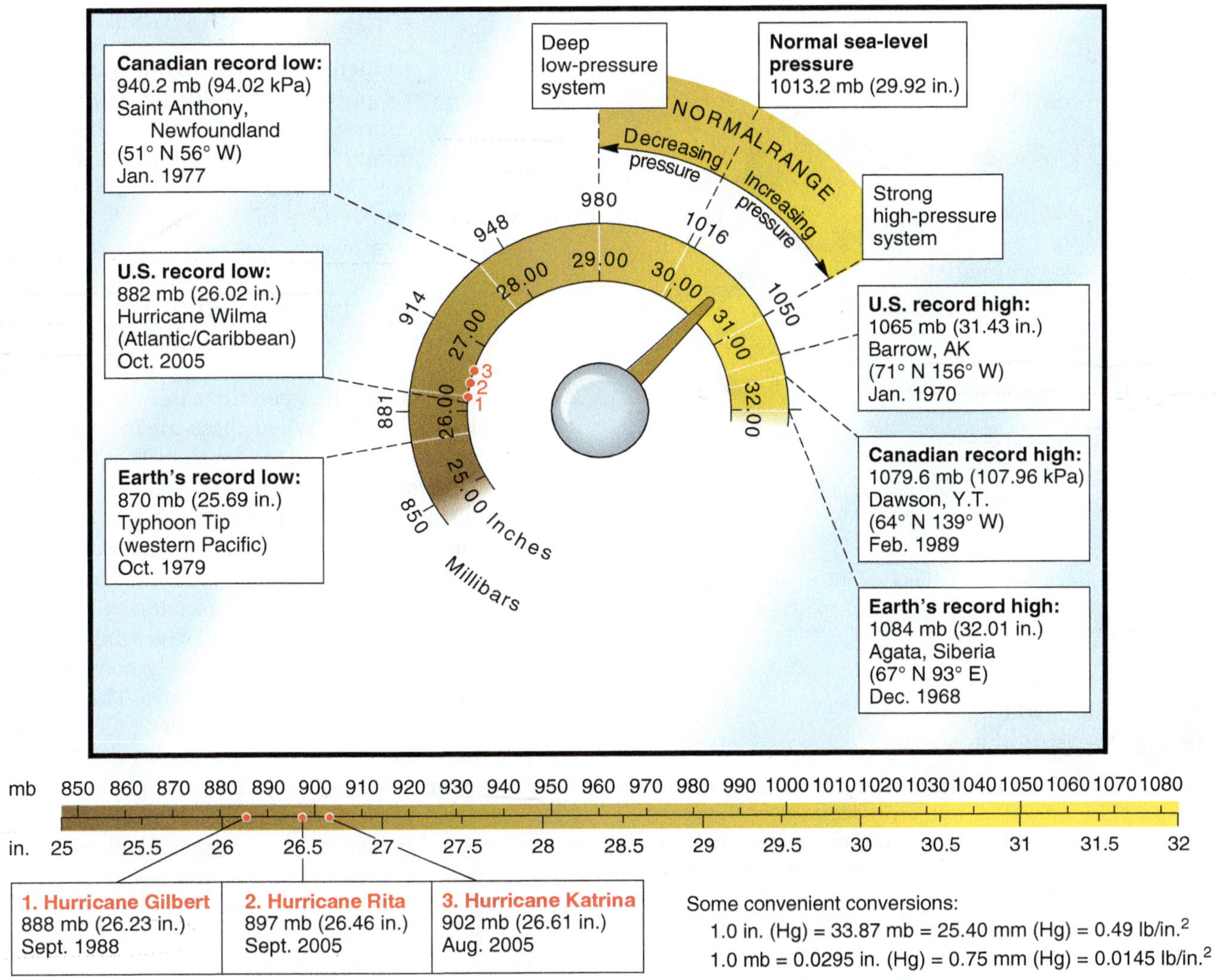

▲**Figure 6.3 Air pressure readings and conversions.** Scales express barometric air pressure in millibars and inches of mercury (Hg), with average air pressure values and recorded pressure extremes. Note the positions of Hurricanes Gilbert, Rita, and Katrina on the pressure dial (numbers 1–3).

standard measurement is taken 10 m (33 ft) above the ground to reduce the effects of local topography on wind direction (Figure 6.4).

Winds are named for the direction from which they originate. For example, a wind from the west is a westerly wind (it blows eastward); a wind out of the south is a southerly wind (it blows northward). Figure 6.5 illustrates a simple wind compass, naming 16 principal wind directions used by meteorologists.

The traditional Beaufort wind scale (named after Admiral Beaufort of the British Navy, who introduced the scale in 1806) is a descriptive scale useful in visually estimating wind speed and is posted on our *MasteringGeography* website. Ocean charts still reference the scale (see http://www.srh.noaa.gov/jetstream/ocean/beaufort_max.htm), enabling estimation of wind speed without instruments, even though most ships today use sophisticated equipment to perform such measurements.

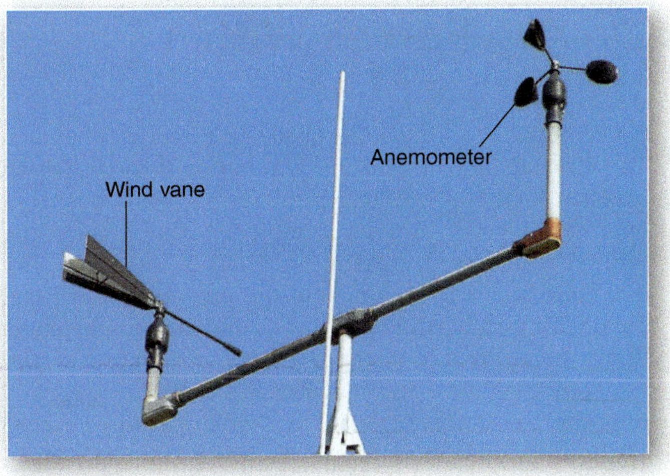

▲**Figure 6.4 Wind vane and anemometer.** Instruments used to measure wind direction and speed at a weather station installation. [NOAA Photo Library.]

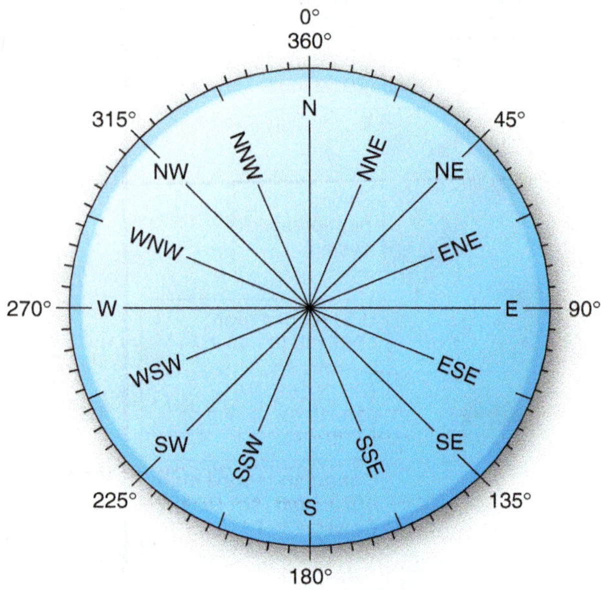

▲Figure 6.5 **Sixteen wind directions identified on a wind compass.** Winds are named for the direction from which they originate. For example, a wind from the west is a westerly wind.

CRITICALthinking 6.1
Measure the Wind

Estimate wind speed and direction as you walk across campus on a day with wind. You might access the URL given above or go to *MasteringGeography* to view the Beaufort scale to assist you. Record your estimates at least twice during the day—more often if you note changing wind patterns. For wind direction, moisten your finger, hold it in the air, and sense evaporative cooling on one side of your finger to indicate from which direction the wind is blowing. Check the Internet to find a weather station on your campus or at a nearby location, and compare your measurements to actual data. What changes in wind speed and direction do you notice over several days? How did these changes relate to the weather you experienced? ●

Driving Forces Within the Atmosphere

Four forces determine both speed and direction of winds. The first of these is Earth's *gravitational force*, which exerts a virtually uniform pressure on the atmosphere over all of Earth. Gravity compresses the atmosphere, with the density decreasing as altitude increases. The gravitational force counteracts the outward centrifugal force acting on Earth's spinning surface and atmosphere. (Centrifugal force is the apparent force drawing a rotating body away from the center of rotation; it is equal and opposite to the centripetal, or "center-seeking" force.) Without gravity, there would be no atmospheric pressure—or atmosphere, for that matter.

The other forces affecting winds are the pressure gradient force, Coriolis force, and friction force. All of these forces operate on moving air and ocean currents at Earth's surface and influence global wind-circulation patterns.

Pressure Gradient Force

The **pressure gradient force** drives air from areas of higher barometric pressure (more-dense air) to areas of lower barometric pressure (less-dense air), thereby causing winds. A *gradient* is the rate of change in some property over distance. Without a pressure gradient force, there would be no wind.

High- and low-pressure areas exist in the atmosphere principally because Earth's surface is unequally heated. For example, cold, dry, dense air at the poles exerts greater pressure than warm, humid, less-dense air along the equator. On a regional scale, high- and low-pressure areas are associated with specific masses of air that have varying characteristics. When these air masses are near each other, a pressure gradient develops that leads to horizontal air movement.

In addition, vertical air movement can create pressure gradients. This happens when air descends from the upper atmosphere and diverges at the surface or when air converges at the surface and ascends into the upper atmosphere. Strongly subsiding and diverging air is associated with high pressure, and strongly converging and rising air is associated with low pressure. These horizontal and vertical pressure differences establish a pressure gradient force that is a causal factor for winds.

An **isobar** is an isoline (a line along which there is a constant value) plotted on a weather map to connect points of equal pressure. The pattern of isobars provides a portrait of the pressure gradient between an area of higher pressure and one of lower pressure. The spacing between isobars indicates the intensity of the pressure difference, or pressure gradient.

Just as closer contour lines on a topographic map indicate a steeper slope on land and closer isotherms on a temperature map indicate more extreme temperature gradients, so closer isobars denote steepness in the pressure gradient. In Figure 6.6a, note the spacing of the isobars. A steep gradient causes faster air movement from a high-pressure area to a low-pressure area. Isobars spaced wider apart from one another mark a more gradual pressure gradient, one that creates a slower airflow. Along a horizontal surface, a pressure gradient force that is acting alone (uncombined with other forces) produces movement at right angles to the isobars, so wind blows across the isobars from high to low pressure. Note the location of steep ("strong winds") and gradual ("light winds") pressure gradients and their relationship to wind intensity on the weather map in Figure 6.6b.

Coriolis Force

The **Coriolis force** is a deflective force that makes wind traveling in a straight path appear to be deflected in relation to Earth's rotating surface. This force is an effect of Earth's rotation. On a nonrotating Earth, surface winds would move in a straight line from areas of higher pressure to areas of lower pressure. But on our rotating planet, the Coriolis force deflects anything that flies or flows across Earth's surface—wind, an airplane, or ocean

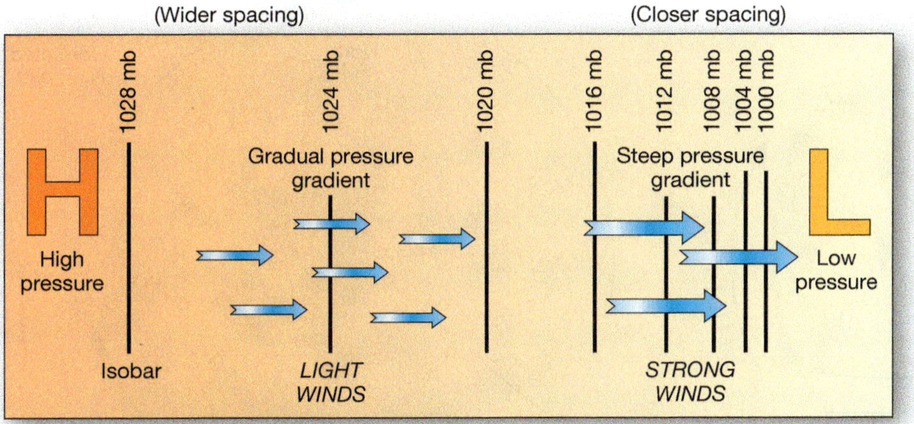

(a) Pressure gradient, isobars, and wind strength.

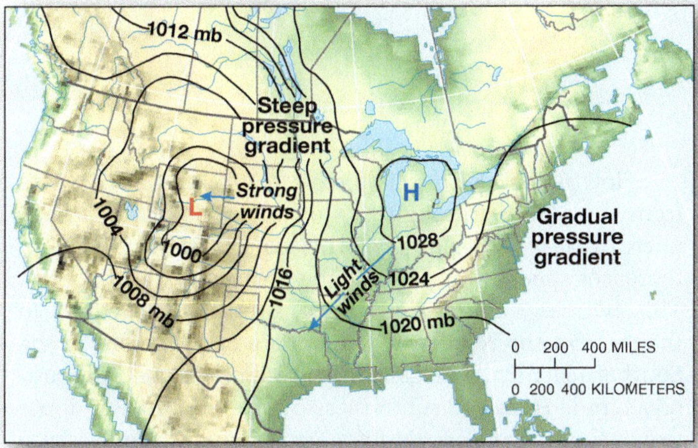

(b) Pressure gradient and wind strength portrayed on a weather map.

▲**Figure 6.6 Effect of pressure gradient on wind speed.**

currents—from a straight path. Because Earth rotates eastward, such objects appear to curve to the right in the Northern Hemisphere and to the left in the Southern Hemisphere. Because the speed of Earth rotation varies with latitude, the strength of this deflection varies, being weakest at the equator and strongest at the poles.

Note that we call Coriolis a *force*. This label is appropriate because, as the physicist Sir Isaac Newton (1643–1727) stated, when something is accelerating over a distance, a force is in operation (mass times acceleration). This apparent force (in classical mechanics, an inertial force) exerts an effect on moving objects. It is named for Gaspard Coriolis (1792–1843), a French mathematician and researcher of applied mechanics, who first described the force in 1831. For deeper insight into the physics of this phenomenon, go to http://www.real-world-physics-problems.com/coriolis-force.html.

Coriolis Force Example A simple example of an airplane helps explain this subtle but significant force affecting moving objects on Earth. From the viewpoint of an airplane that is passing over Earth's surface, the surface is seen to rotate slowly below. But, looking from the surface at the airplane, the surface seems stationary, and the airplane appears to curve off course. The airplane does not actually deviate from a straight path, but it appears to do so because we are standing on Earth's rotating surface beneath the airplane. Because of this apparent deflection, the airplane must make constant corrections in flight path to maintain its "straight" heading relative to a rotating Earth (see Figure 6.7a and b).

A pilot leaves the North Pole and flies due south toward Quito, Ecuador. If Earth were not rotating, the aircraft would simply travel along a meridian of longitude and arrive at Quito. But Earth is rotating eastward beneath the aircraft's flight path. As the plane travels toward the equator, the speed of Earth's rotation increases from about 838 kmph (521 mph) at 60° N to about 1675 kmph (1041 mph) at 0°. If the pilot does not allow for this increase in rotational speed, the plane will reach the equator over the ocean along an apparently curved path, far to the west of the intended destination (Figure 6.7a). On the return flight northward, if the pilot does not make corrections, the plane will end up to the east of the pole, in a right-hand deflection.

This effect also occurs if the plane is flying in an east–west direction. During an eastward flight from California to New York, in the same direction as Earth's rotation, the centrifugal force pulling outward on the plane in flight (Earth rotation speed + plane speed) becomes so great that it cannot be balanced by the gravitational force pulling toward Earth's axis. Therefore, the plane experiences an overall movement away from Earth's axis, observed as a right-hand deflection toward the equator. Unless the pilot corrects for this deflective force, the flight will end up somewhere in North Carolina (Figure 6.7b). In contrast, flying westward on a return flight opposite Earth's rotation direction decreases the centrifugal force (Earth rotation speed − plane speed) so that it is less than the gravitational force. In this case, the plane experiences an overall movement toward Earth's axis, observed in the Northern Hemisphere as a right-hand deflection toward the pole.

Distribution and Significance Figure 6.7c summarizes the distribution of the effects of the Coriolis force on Earth. Note the deflection to the right in the Northern Hemisphere and to the left in the Southern Hemisphere.

Several factors contribute to the Coriolis force on Earth. First, the strength of this deflection varies with the speed of Earth's rotation, which varies with latitude (please review Table 2.1). Remember that rotational speed is 0 kmph at the poles, where Earth's surface is closest to its axis, and 1675 kmph (1041 mph) at the equator, where Earth's surface is farthest from its axis. Thus, deflection is zero along the equator, increases to half the maximum deflection at 30° N and 30° S latitude, and reaches maximum deflection for objects moving away from the poles. Second, the deflection occurs regardless of the direction in which the object is moving, as illustrated in Figure 6.7, and does not change the speed of the moving object. Third, the deflection increases as the speed of the moving object increases; thus, the faster the wind speed, the greater its

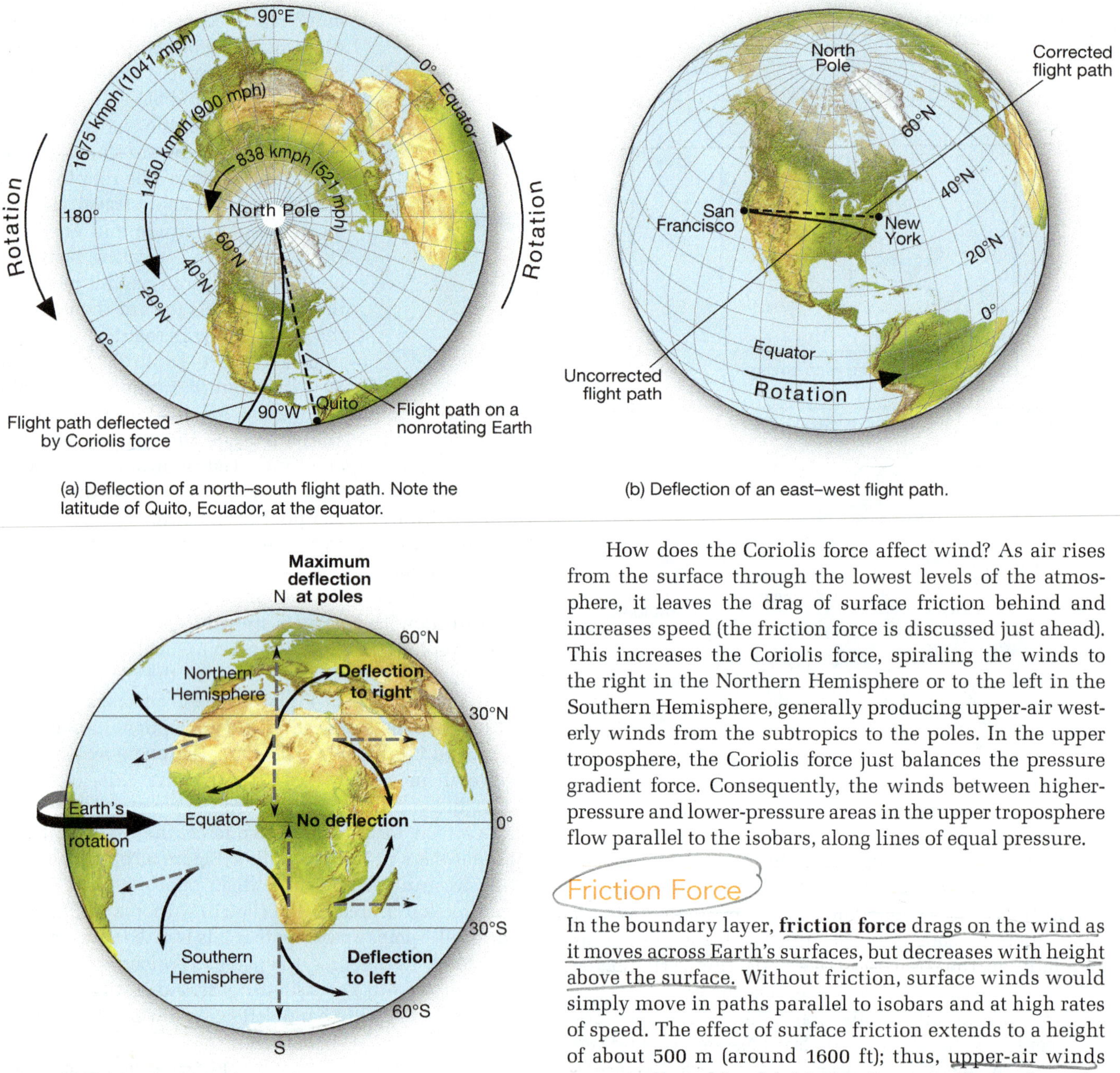

(a) Deflection of a north–south flight path. Note the latitude of Quito, Ecuador, at the equator.

(b) Deflection of an east–west flight path.

(c) Distribution of the Coriolis force on Earth. Apparent deflection is to the right in the Northern Hemisphere and to the left in the Southern Hemisphere; dashed lines show intended route and solid lines show actual movement.

▲Figure 6.7 The Coriolis force—an apparent deflection.

MG™ Animation Coriolis Force

apparent deflection. Although the Coriolis force affects all moving objects on Earth to some degree, its effects are negligible for small-scale motions that cover insignificant distance and time, such as a Frisbee or an arrow.

How does the Coriolis force affect wind? As air rises from the surface through the lowest levels of the atmosphere, it leaves the drag of surface friction behind and increases speed (the friction force is discussed just ahead). This increases the Coriolis force, spiraling the winds to the right in the Northern Hemisphere or to the left in the Southern Hemisphere, generally producing upper-air westerly winds from the subtropics to the poles. In the upper troposphere, the Coriolis force just balances the pressure gradient force. Consequently, the winds between higher-pressure and lower-pressure areas in the upper troposphere flow parallel to the isobars, along lines of equal pressure.

Friction Force

In the boundary layer, **friction force** drags on the wind as it moves across Earth's surfaces, but decreases with height above the surface. Without friction, surface winds would simply move in paths parallel to isobars and at high rates of speed. The effect of surface friction extends to a height of about 500 m (around 1600 ft); thus, upper-air winds are not affected by the friction force. At the surface, the effect of friction varies with surface texture, wind speed, time of day and year, and atmospheric conditions. In general, rougher surfaces produce more friction.

Summary of Physical Forces on Winds

Winds are a result of the combination of these physical forces (Figure 6.8). When the pressure gradient acts alone, shown in Figure 6.8a, winds flow from areas of high pressure to areas of low pressure. Note the descending, diverging air associated with high pressure and the

GEOreport 6.3 Coriolis: Not a force on sinks or toilets

A common misconception about the Coriolis force is that it affects water draining out of a sink, tub, or toilet. Moving water or air must cover some distance across space and time before the Coriolis force noticeably deflects it. Long-range artillery shells and guided missiles do exhibit small amounts of deflection that must be corrected for accuracy. But water movements down a drain are too small in spatial extent to be noticeably affected by this force.

ascending, converging air associated with low pressure in the side view.

Figure 6.8b illustrates the combined effect of the pressure gradient force and the Coriolis force on air currents in the upper atmosphere, above about 1000 m (3300 ft). Together, they produce winds that do not flow directly from high to low, but that flow around the pressure areas, remaining parallel to the isobars. Such winds are **geostrophic winds** and are characteristic of upper tropospheric circulation. (The suffix *-strophic* means "to turn.") Geostrophic winds produce the characteristic pattern shown on the upper-air weather map just ahead in Figure 6.12.

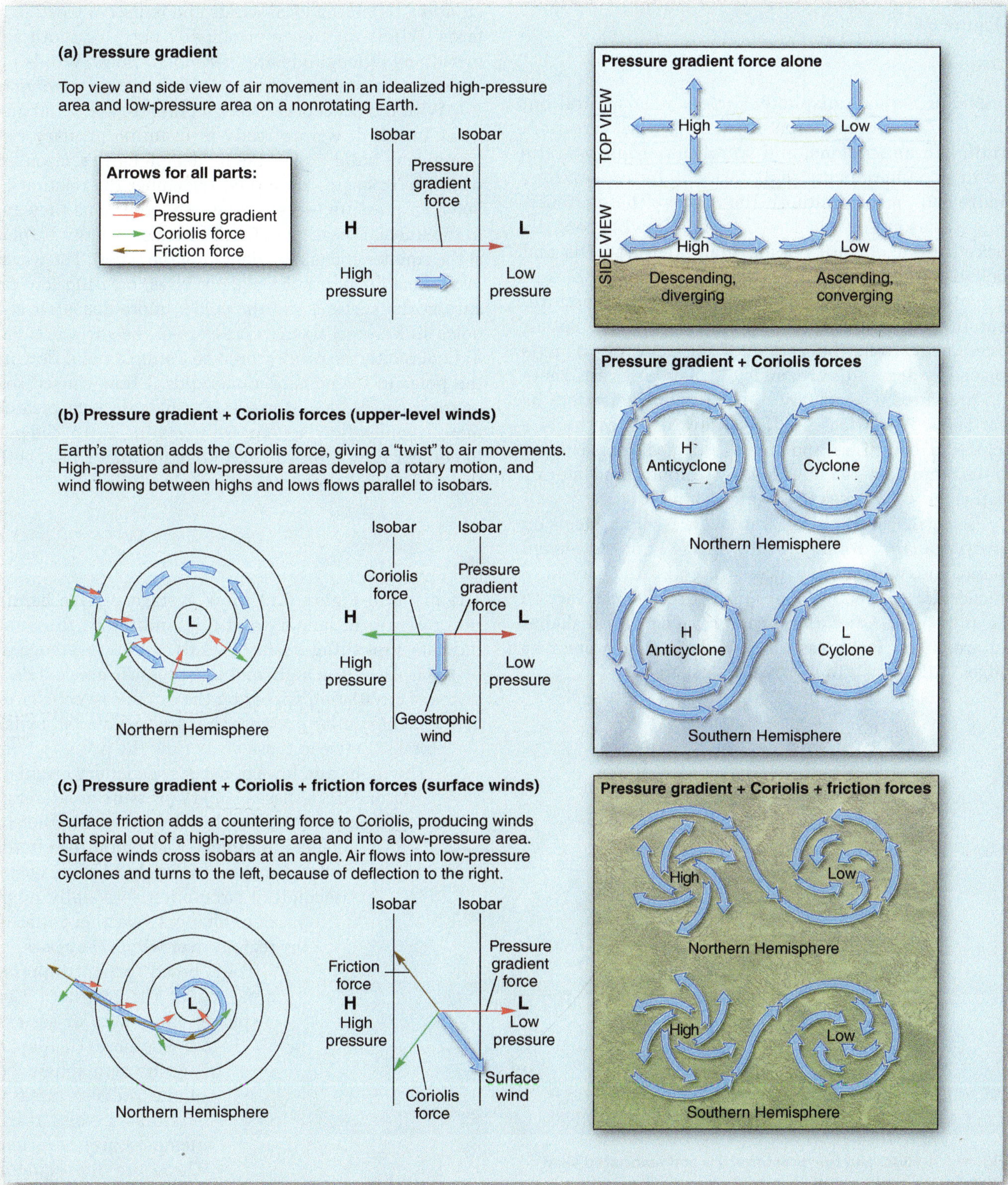

▲**Figure 6.8 Three physical forces that produce winds.** Three physical forces interact to produce wind patterns at the surface and in the upper atmosphere: (a) pressure gradient force; (b) Coriolis force, which counters the pressure gradient force, producing a geostrophic wind flow in the upper atmosphere; and (c) friction force, which, combined with the other two forces, produces characteristic surface winds.

MG Animation Wind Pattern Development

Near the surface, friction prevents the equilibrium between the pressure gradient and Coriolis forces that results in geostrophic wind flows in the upper atmosphere (Figure 6.8c). Because surface friction decreases wind speed, it reduces the effect of the Coriolis force and causes winds to move across isobars at an angle. Thus, wind flows around pressure centers form enclosed areas called *pressure systems*, or *pressure cells*, as illustrated in Figure 6.8c.

High- and Low-Pressure Systems

In the Northern Hemisphere, surface winds spiral out from a *high-pressure area* in a clockwise direction, forming an **anticyclone**, and spiral into a *low-pressure area* in a counterclockwise direction, forming a **cyclone** (Figure 6.8). In the Southern Hemisphere these circulation patterns are reversed, with winds flowing counterclockwise out of anticyclonic high-pressure cells and clockwise into cyclonic low-pressure cells.

Anticyclones and cyclones have vertical air movement in addition to these horizontal patterns. As air moves away from the center of an anticyclone, it is replaced by descending, or subsiding (sinking), air. These high-pressure systems are typically characterized by clear skies. As surface air flows toward the center of a cyclone, it converges and moves upward. These rising motions promote the formation of cloudy and stormy weather, as we will see in Chapters 7 and 8.

Figure 6.9 shows high- and low-pressure systems on a weather map, with a side view of the wind movement around and within each pressure cell. You may have noticed that on weather maps, pressure systems vary in size and shape. Often these cells have elongated shapes and are called low-pressure "troughs" or high-pressure "ridges" (illustrated in Figure 6.12 just ahead).

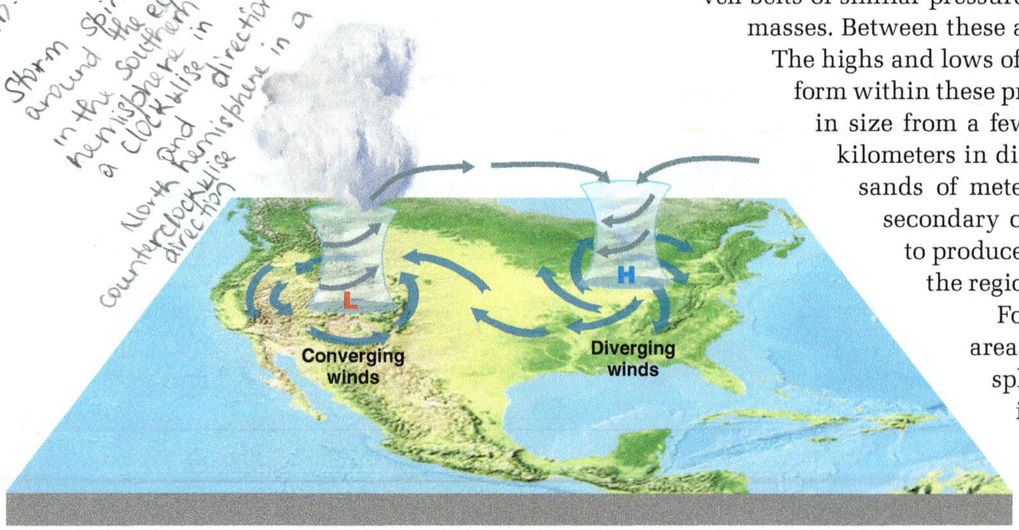

▲**Figure 6.9 High- and low-pressure cells and associated wind movement.** Side view of high- and low-pressure cells over the United States. Note surface winds spiraling clockwise out of the high-pressure area toward the low pressure, where winds spiral counterclockwise into the low.

Atmospheric Patterns of Motion

Atmospheric circulation is categorized at three levels: *primary circulation*, consisting of general worldwide circulation; *secondary circulation*, consisting of migratory high-pressure and low-pressure systems; and *tertiary circulation*, including local winds and temporal weather patterns. Winds that move principally north or south along meridians of longitude are *meridional flows*. Winds moving east or west along parallels of latitude are *zonal flows*.

With the concepts related to pressure and wind movement in mind, we are ready to examine primary circulation and build a general model of Earth's circulation patterns. To begin, we should remember the relationships between pressure, density, and temperature as they apply to the unequal heating of Earth's surface (energy surpluses at the equator and energy deficits at the poles). The warmer, less-dense air along the equator rises, creating low pressure at the surface, and the colder, more-dense air at the poles sinks, creating high pressure at the surface. If Earth did not rotate, the result would be a simple wind flow from the poles to the equator, a meridional flow caused solely by pressure gradient. However, Earth does rotate, creating a more complex flow system. On a rotating Earth, the poles-to-equator flow is broken up into latitudinal zones, both at the surface and aloft in the upper-air winds.

Primary Pressure Areas and Associated Winds

The maps in Figure 6.10 show average surface barometric pressures in January and July. Indirectly, these maps indicate prevailing surface winds, which are suggested by the isobars. The high- and low-pressure areas of Earth's primary circulation appear on these maps as cells or uneven belts of similar pressure that are interrupted by landmasses. Between these areas flow the primary winds. The highs and lows of Earth's secondary circulation form within these primary pressure areas, ranging in size from a few hundred to a few thousand kilometers in diameter and hundreds to thousands of meters in height. The systems of secondary circulation seasonally migrate to produce changing weather patterns in the regions over which they pass.

Four broad primary pressure areas cover the Northern Hemisphere, and a similar set exists in the Southern Hemisphere. In each hemisphere, two of the pressure areas are stimulated by *thermal* (temperature) factors. These are the **equatorial low** (marked by the ITCZ line on the maps) and the weak **polar highs** at the North and South poles (not shown, as the maps are cut off at 80° N and 80° S). Remember from

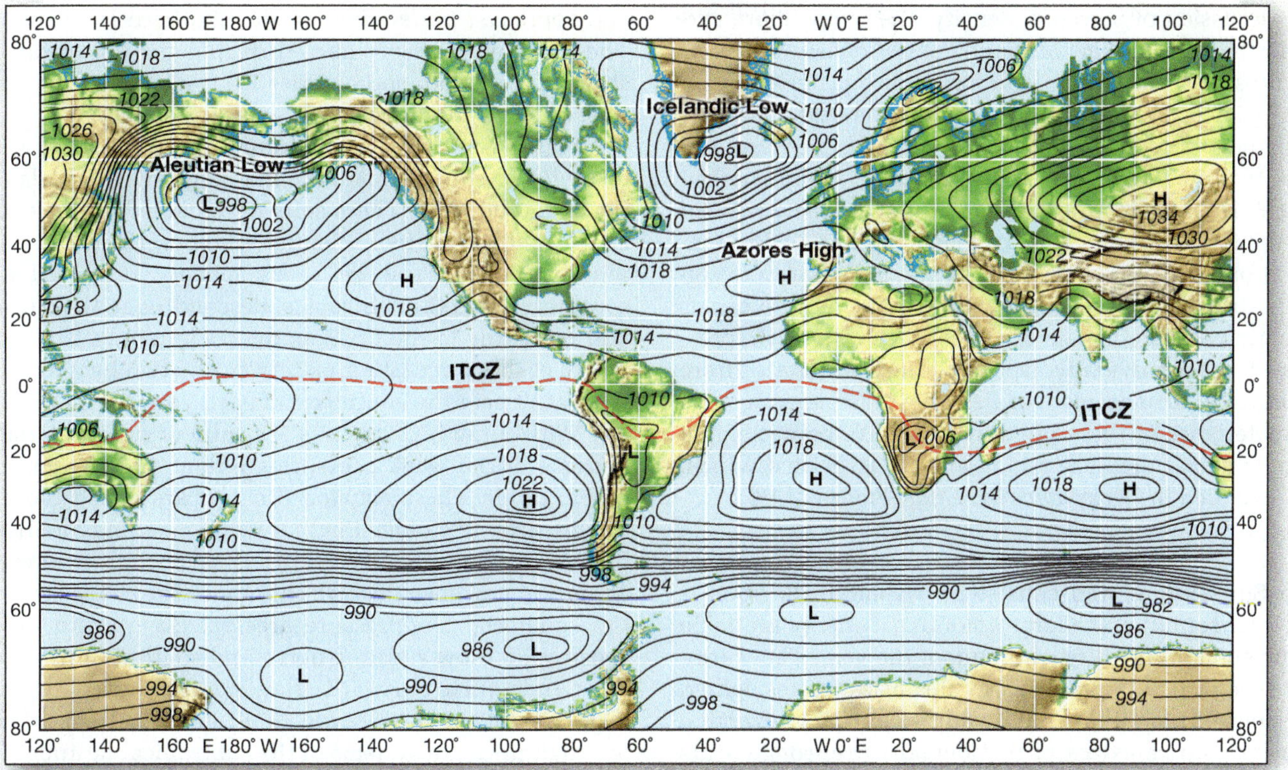

(a) January average surface barometric pressures (millibars); dashed line marks the general location of the intertropical convergence zone (ITCZ).

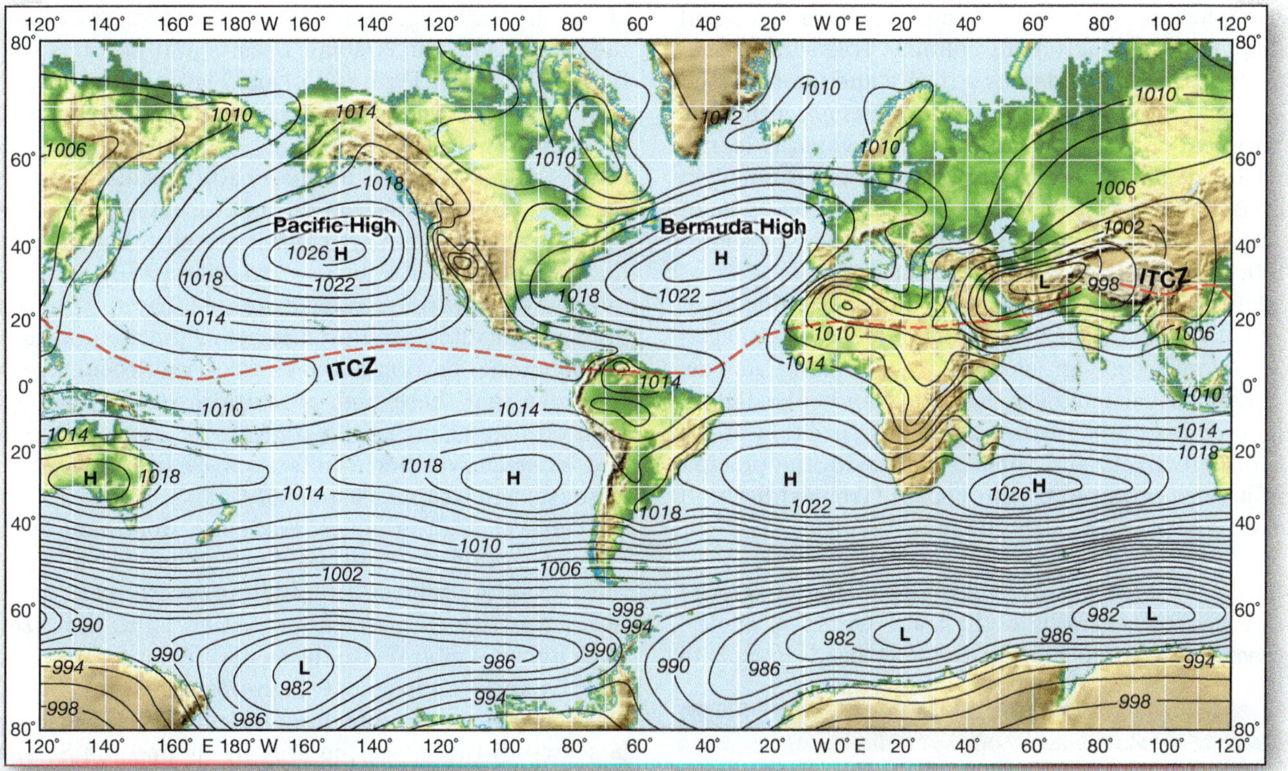

(b) July average surface barometric pressures. Compare pressures in the North Pacific, the North Atlantic, and the central Asian landmass with the January map above.

▲Figure 6.10 Global barometric pressures for January and July. [Adapted by author and redrawn from National Climatic Data Center, Monthly Climatic Data for the World, 46 (January and July 1993), and WMO and NOAA.]

MG™ Animation Global Patterns of Pressure

MG™ Satellite Global Sea-Level Pressure

our discussion of pressure, density, and temperature earlier in the chapter that warmer air is less dense and exerts less pressure. The warm, light air in the equatorial region is associated with low pressure, while the cold, dense air in the polar regions is associated with high pressure. The other two pressure areas—the **subtropical highs** (marked with an H on the map) and **subpolar lows** (marked with an L)—are formed by *dynamic* (mechanical) factors. Remember in our discussion of pressure gradients that converging, rising air is associated with low pressure, whereas subsiding, diverging air is associated with high pressure—these are dynamic factors because they result from the physical displacement of air. Table 6.1 summarizes the characteristics of these pressure areas. We now examine each principal pressure region and its associated winds, all illustrated in *Geosystems In Action*, Figure GIA 6.

Equatorial Low or ITCZ: Warm and Rainy Constant high Sun altitude and consistent daylength (12 hours a day, year-round) make large amounts of energy available in the equatorial region throughout the year. The warming associated with these energy surpluses creates lighter, less-dense, ascending air, with surface winds converging along the entire extent of the low-pressure trough. This converging air is extremely moist and full of latent heat energy. As it rises, the air expands and cools, producing condensation; consequently, rainfall is heavy throughout this zone (condensation and precipitation are discussed in Chapter 7). Vertical cloud columns frequently reach the tropopause, in thunderous strength and intensity.

The equatorial low, or *equatorial trough*, forms the **intertropical convergence zone (ITCZ)**, which is identified by bands of clouds along the equator and is noted on Figure 6.10 and Figure GIA 6.1a as a dashed line. In January, the zone crosses northern Australia and dips southward in eastern Africa and South America.

Figure GIA 6.2 shows the band of precipitation associated with the ITCZ on January and July satellite images; precipitation forms an elongated, undulating narrow band that is consistent over the oceans and only slightly interrupted over land surfaces. Note the position of the ITCZ in Figure 6.10 and compare this with the precipitation pattern captured by the TRMM (Tropical Rainfall Measuring Mission) sensors in Figure GIA 6.2.

Trade Winds The winds converging at the equatorial low are known generally as the **trade winds**, or trades. *Northeast trade winds* blow in the Northern Hemisphere and *southeast trade winds* in the Southern Hemisphere. The trade winds were named during the era of sailing ships that carried merchandise for trade across the seas. These are the most consistent winds on Earth.

Figure GIA 6.1b shows circulation cells, called *Hadley cells*, in each hemisphere that begin with winds rising along the ITCZ. These cells were named for the eighteenth-century English scientist who described the trade winds. Within these cells, air moves northward and southward into the subtropics, descending to the surface and returning to the ITCZ as the trade winds. The symmetry of this circulation pattern in the two hemispheres is greatest near the equinoxes of each year.

Within the ITCZ, winds are calm or mildly variable because of the weak pressure gradient and the vertical ascent of air. These equatorial calms are called the *doldrums* (from an older English word meaning "foolish") because of the difficulty sailing ships encountered when attempting to move through this zone. The rising air from the equatorial low-pressure area spirals upward into a geostrophic flow running north and south. These upper-air winds turn eastward, flowing from west to east, beginning at about 20° N and 20° S, and then descend into the high-pressure systems of the subtropical latitudes.

Subtropical Highs: Hot and Dry Between 20° and 35° latitude in both hemispheres, a broad high-pressure zone of hot, dry air brings clear, frequently cloudless skies over the Sahara and the Arabian Deserts and portions of the Indian Ocean (see Figures 6.10 and Figure GIA 6.2 and the world physical map on the inside back cover of the book).

These subtropical anticyclones generally form as air above the subtropics is mechanically pushed downward and heats by compression on its descent to the surface. Warmer air has a greater capacity to absorb water vapor than does cooler air, making this descending warm air relatively dry (discussed in Chapter 7). The air is also dry because moisture is removed as heavy precipitation along the equatorial portion of the circulation. Recent research indicates that these high-pressure areas may intensify with climate change, with impacts on regional climates and extreme weather events such as tropical cyclones (discussed in Chapters 8 and 10).

Several high-pressure areas are dominant in the subtropics (Figure 6.10). In the Northern Hemisphere, the Atlantic subtropical high-pressure cell is the **Bermuda High** (in the western Atlantic) or the **Azores High** (when it migrates to the eastern Atlantic in winter). The Atlantic subtropical high-pressure area features clear, warm waters and large quantities of *Sargassum* (a seaweed) that gives the area its name—the Sargasso Sea. The **Pacific High**, or *Hawaiian High*, dominates the Pacific in July, retreating southward in January. In the Southern Hemisphere, three large high-pressure centers dominate the Pacific, Atlantic, and Indian Oceans, especially in January, and tend to move along parallels of latitude in shifting zonal positions.

Because the subtropical belts are near 25° N and 25° S latitude, these areas sometimes are known as the "calms

TABLE 6.1 Four Hemispheric Pressure Areas

Name	Cause	Location	Air Temperature/ Moisture
Polar high	Thermal	90° N, 90° S	Cold/dry
Subpolar low	Dynamic	60° N, 60° S	Cool/wet
Subtropical high	Dynamic	20°–35° N, 20°–35° S	Hot/dry
Equatorial low	Thermal	10° N to 10° S	Warm/wet

of Cancer" and the "calms of Capricorn." These zones of windless, hot, dry desert air, so deadly in the era of sailing ships, earned the name *horse latitudes*. Although the term's true origin is uncertain, it is popularly attributed to becalmed and stranded sailing crews of past centuries who destroyed the horses on board, not wanting to share food or water with the livestock.

The entire high-pressure system migrates with the summer high Sun, fluctuating about 5°–10° in latitude. The eastern sides of these anticyclonic systems are drier and more stable (exhibit less convective activity) and are associated with cooler ocean currents. These drier eastern sides influence climate along subtropical and midlatitude west coasts (discussed in Chapter 10 and shown in Figure 6.11). In fact, Earth's major deserts generally occur within the subtropical belt and extend to the west coast of each continent except Antarctica. In Figures 6.11 and 6.18, note that the desert regions of Africa come right to the shore in both hemispheres, with the cool, southward-flowing *Canaries Current* offshore in the north and the cool, northward-flowing *Benguela Current* offshore in the south.

Westerlies Surface air diverging within the subtropical high-pressure cells generates Earth's principal surface winds: the trade winds that flow toward the equator, and the **westerlies**, which are the dominant winds flowing from the subtropics toward higher latitudes. The westerlies diminish somewhat in summer and are stronger in winter in both hemispheres. These winds are less consistent than the trade winds, with variability resulting from midlatitude migratory pressure systems and topographic barriers that can change wind direction.

Subpolar Lows: Cool and Moist In January, two low-pressure cyclonic cells exist over the oceans around 60° N latitude, near their namesake islands: the North Pacific **Aleutian Low** and the North Atlantic **Icelandic Low** (see Figure 6.10a). Both cells are dominant in winter and weaken or disappear in summer with the strengthening of high-pressure systems in the subtropics. The area of contrast between cold air from higher latitudes and warm air from lower latitudes forms the **polar front**, where masses of air with different characteristics meet (air masses and weather are the subjects of Chapter 8). This front encircles Earth, focused in these low-pressure areas.

Figure GIA 6 illustrates the polar front, where warm, moist air from the westerlies meets cold, dry air from the polar and Arctic regions. Warm air is displaced upward above the cool air at this front, leading to condensation and precipitation (frontal precipitation is discussed in Chapter 8). Low-pressure cyclonic storms migrate out of the Aleutian and Icelandic frontal areas and may produce precipitation in North America and Europe, respectively. Northwestern sections of North America and Europe generally are cool and moist as a result of the passage of these cyclonic systems onshore—consider the weather in British Columbia, Washington, Oregon, Ireland, and the United Kingdom. In the Southern Hemisphere, a discontinuous belt of subpolar low-pressure systems surrounds Antarctica.

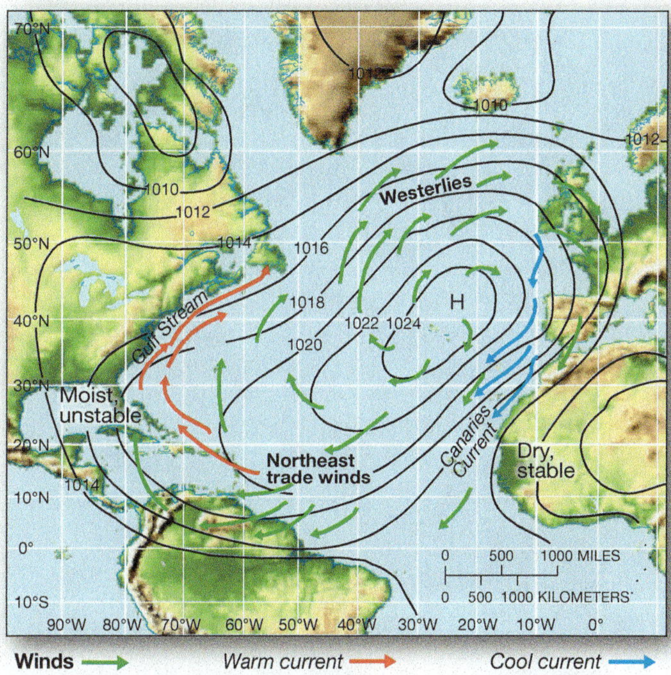

▲**Figure 6.11 Subtropical high-pressure system in the Atlantic.** Characteristic circulation in the Northern Hemisphere. Note deserts extending to the shores of Africa with offshore cool currents, whereas the southeastern United States is moist and humid, with offshore warm currents.

Polar Highs: Frigid and Dry Polar high-pressure cells are weak. The polar atmospheric mass is small, receiving little energy from the Sun to put it into motion. Variable winds, cold and dry, move away from the polar region in an anticyclonic direction. They descend and diverge clockwise in the Northern Hemisphere (counterclockwise in the Southern Hemisphere) and form weak, variable winds of the **polar easterlies** (shown in Figure GIA 6).

Of the two polar regions, the Antarctic has the stronger and more persistent high-pressure system, the **Antarctic High**, forming over the Antarctic landmass. Less pronounced is a polar high-pressure cell over the Arctic Ocean. When it does form, it tends to locate over the colder northern continental areas in winter (Canadian and Siberian Highs) rather than directly over the relatively warmer Arctic Ocean.

Upper Atmospheric Circulation

Circulation in the middle and upper troposphere is an important component of the atmosphere's general circulation. For surface-pressure maps, we plot air pressure using the fixed elevation of sea level as a reference datum—a *constant height surface*. For upper-atmosphere pressure maps, we use a fixed pressure value of 500 mb as a reference datum and plot its elevation above sea level throughout the map to produce a **constant isobaric surface**.

Figure 6.12a and b illustrate the undulating surface elevations of a 500-mb constant isobaric surface for an April day. Similar to surface-pressure maps, closer spacing of the height contours indicates faster winds; wider spacing indicates slower winds. On this map, altitude variations in the isobaric surface are ridges for high

(text and figures continued on page 149)

Earth's atmospheric circulation transfers thermal energy from the equator toward the poles. The overall pattern of the atmospheric circulation (GIA 6.1) arises from the distribution of high- and low- pressure regions, which determines patterns of precipitation (GIA 6.2) as well as winds.

6.1a GENERAL ATMOSPHERIC CIRCULATION MODEL

In both the Northern and Southern Hemispheres, zones of unstable, rising air (lows) and stable, sinking air (highs) divide the troposphere into *circulation cells*, which are symmetrical on both sides of the equator.

Subpolar Low-Pressure Cells Persistent lows (cyclones) over the North Pacific and North Atlantic cause cool, moist conditions. Cold, northern air masses clash with warmer air masses to the south, forming the *polar front*. Cause: *Dynamic*

Polar High-Pressure Cells A small atmospheric polar mass is cold and dry, with weak anticyclonic high pressure. Limited solar energy results in weak, variable winds called the *polar easterlies*. Cause: *Thermal*

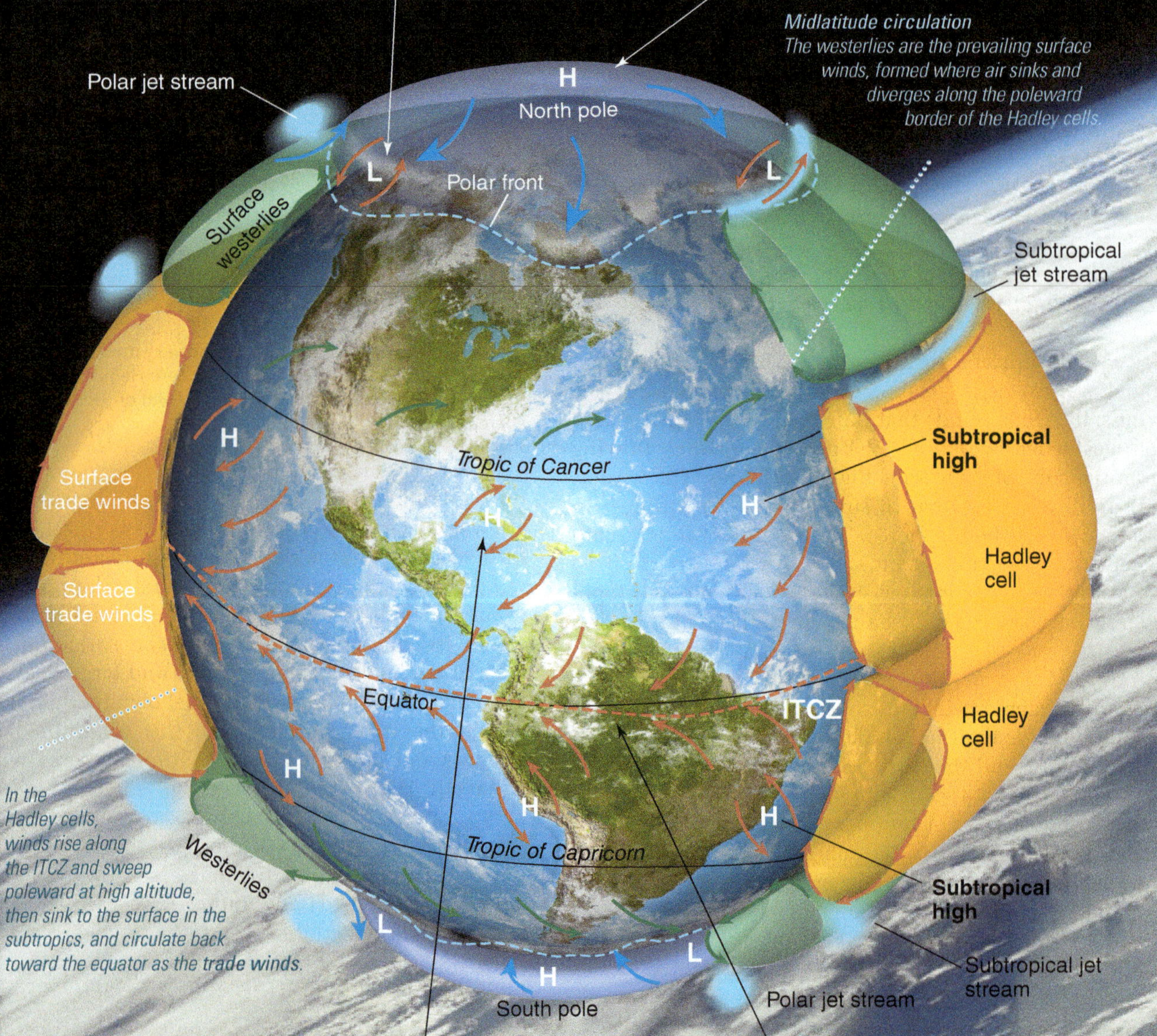

Infer: In what directions would the westerlies and northeast trade winds blow if there were no Coriolis force?

Subtropical High-Pressure Cells Persistent highs (anticyclones) produce regions where air is mechanically pushed downward, compressed, and warmed. Earth's major deserts form beneath these cells. Cause: *Dynamic*

Intertropical Convergence Zone (ITCZ) Lying along the equator, the ITCZ is a trough of low pressure and light or calm winds—the doldrums. Moist, unstable air rises in the ITCZ, causing heavy precipitation year-round. Cause: *Thermal*

Animation
Global Atmospheric Circulation

MasteringGeography™

Visit the Study Area in MasteringGeography™ to explore atmospheric circulation.

Visualize: Study geoscience animations of atmospheric circulation patterns.

Assess: Demonstrate understanding of Earth's atmospheric circulation (if assigned by instructor).

6.1b **Cross Section of Atmospheric Circulation** The cross section shows the relationship between pressure cells (rising or sinking air) and the circulation cells and winds.

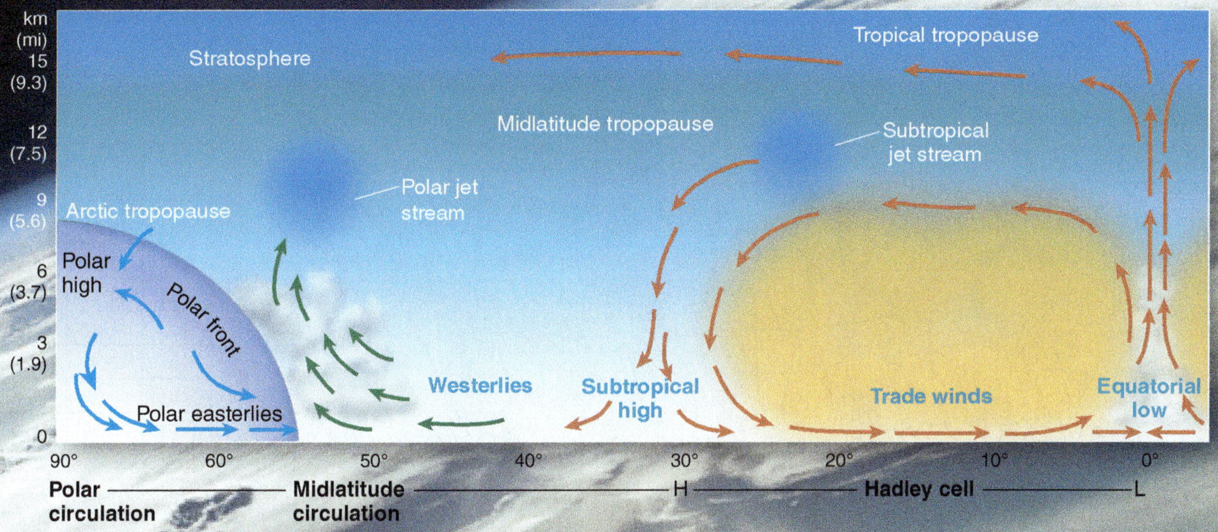

6.2 PRECIPITATION PATTERNS AND ATMOSPHERIC CIRCULATION

Areas of higher precipitation in green, yellow, and orange are zones of low pressure and moist, rising air. Areas of lower precipitation, shown in white on the maps, are zones of high pressure, where air sinks and dries out. Notice the band of heavy rainfall along the ITCZ, and how areas of dryness and moisture vary seasonally on both maps [GSFC/NASA].

[Arne Huckelheim.]

Explain: What causes the difference in precipitation between the dry and rainy seasons in the Western Ghats mountain range of India, pictured to the right?

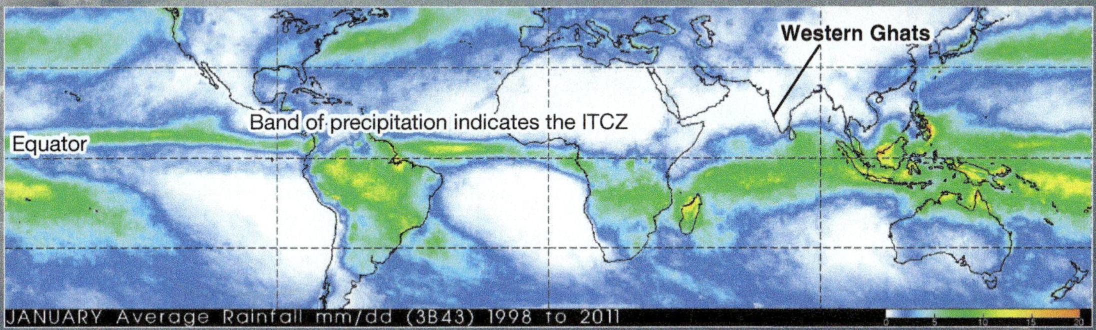

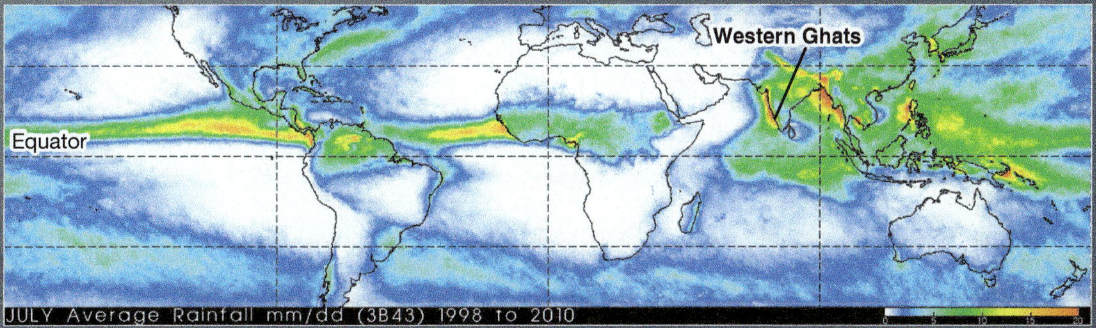

Thunderstorms on the Brazilian horizon as seen from the International Space Station [NASA].

GEOquiz

1. **Explain:** The position of subtropical highs and subpolar lows shifts with the seasons. Explain how this shift affects climate patterns of the midlatitudes.

2. **Compare:** Describe surface air movements where a Hadley cell meets a midlatitude cell and where two Hadley cells meet the ITCZ. How do these movements explain the climate patterns along these boundaries?

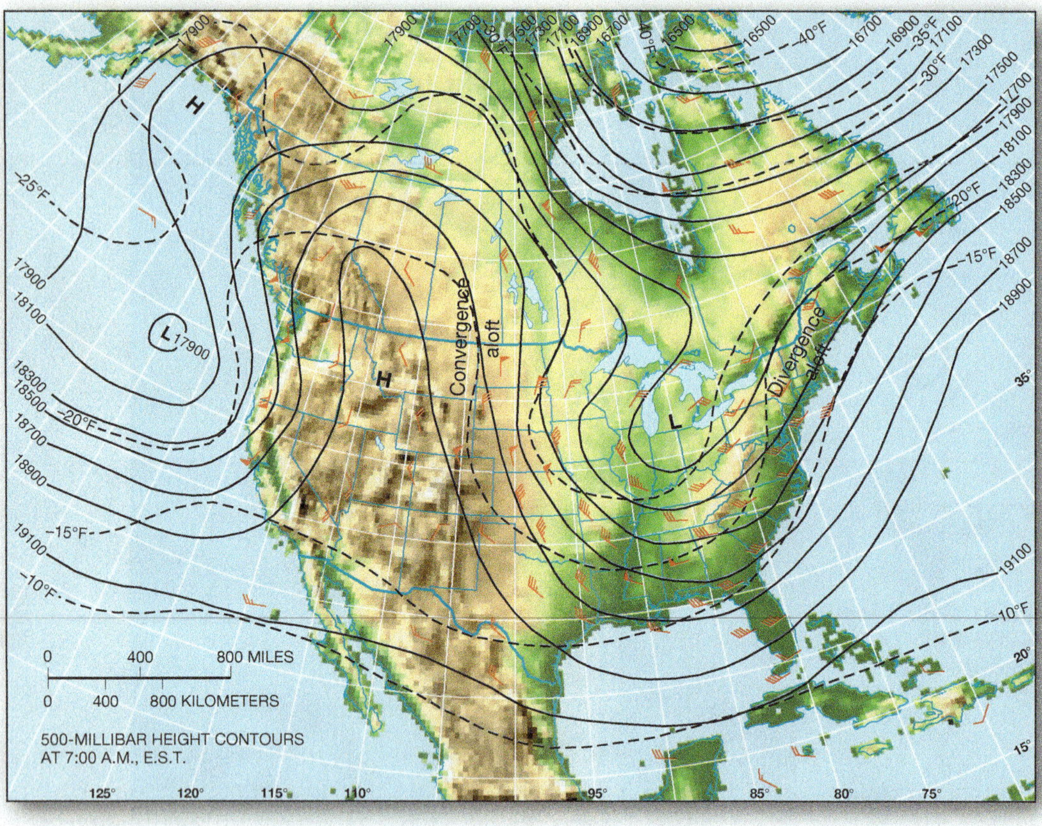

Wind Speed Symbol	Miles (statute) per Hour	Knots
◎	Calm	Calm
	1–2	1–2
	3–8	3–7
	9–14	8–12
	15–20	13–17
	21–25	18–22
	26–31	23–27
	32–37	28–32
	38–43	33–37
	44–49	38–42
	50–54	43–47
	55–60	48–52
	61–66	53–57
	67–71	58–62
	72–77	63–67
	78–83	68–72
	84–89	73–77
	119–123	103–107

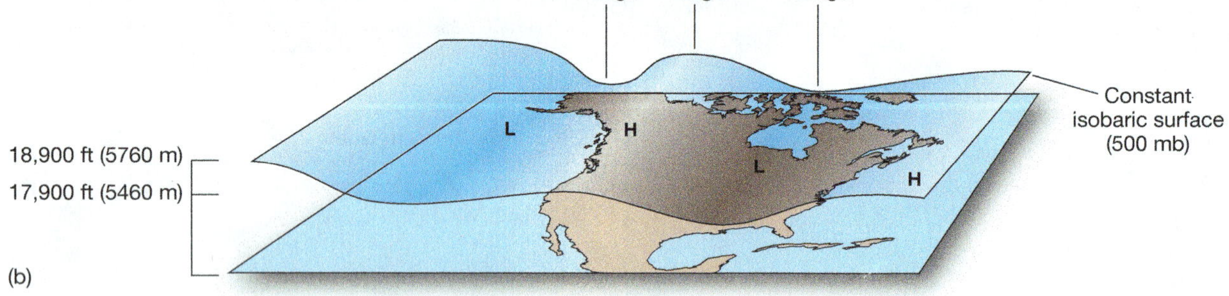

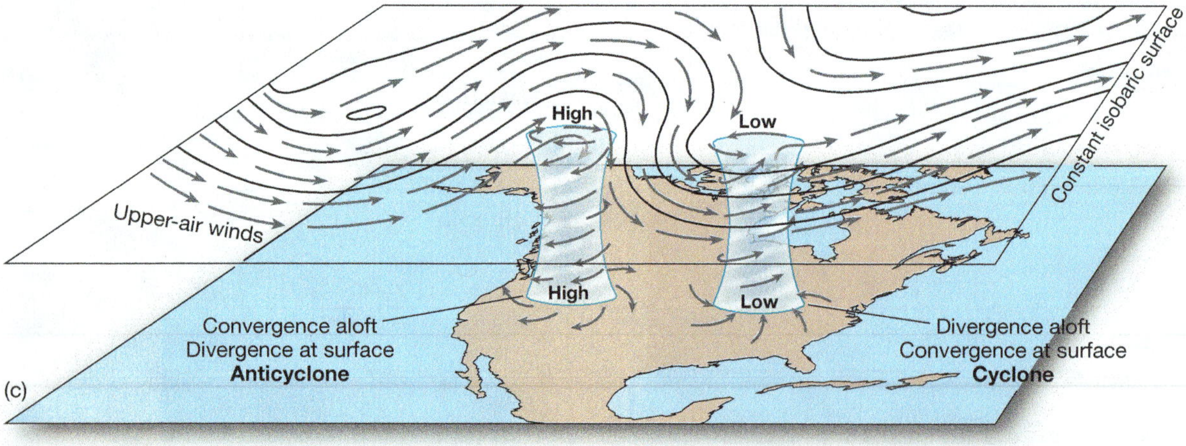

▲**Figure 6.12 Analysis of a constant isobaric surface for an April day.** (a) Contours show elevation (in feet) at which 500-mb pressure occurs—a constant isobaric surface. The pattern of contours reveals geostrophic wind patterns in the troposphere ranging from 16,500 to 19,100 ft in elevation. (b) Note on the map and in the sketch beneath the chart the "ridge" of high pressure over the Intermountain West, at an altitude of 5760 m, (18,900 ft) and the "trough" of low pressure over the Great Lakes region and off the Pacific Coast, at an altitude of 5460 m, (17,900 ft). (c) Note areas of convergence aloft (corresponding to surface divergence) and divergence aloft (corresponding to surface convergence).

MG Animation
Cyclones and Anticyclones

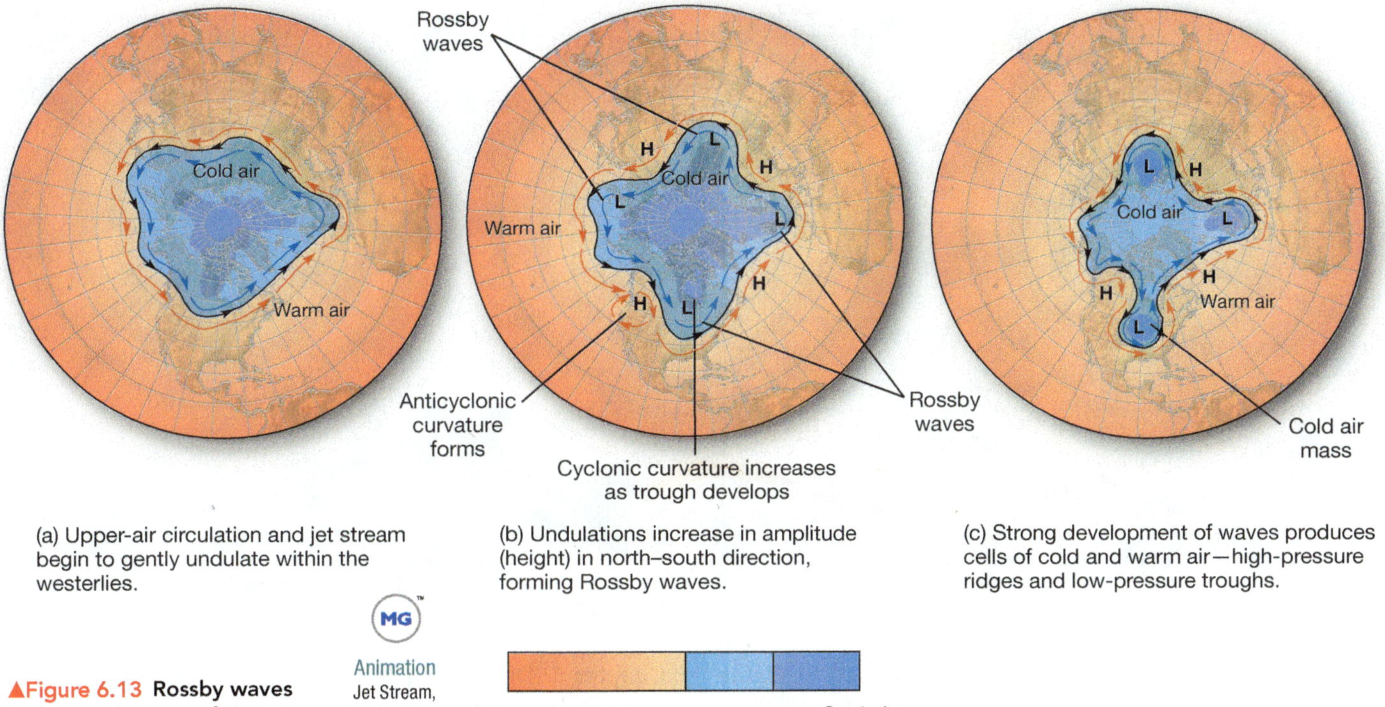

(a) Upper-air circulation and jet stream begin to gently undulate within the westerlies.

(b) Undulations increase in amplitude (height) in north–south direction, forming Rossby waves.

(c) Strong development of waves produces cells of cold and warm air—high-pressure ridges and low-pressure troughs.

▲**Figure 6.13 Rossby waves in the upper atmosphere.**

MG Animation Jet Stream, Rossby Waves

pressure (with height contours on the map bending poleward) and troughs for low pressure (with height contours on the map bending equatorward).

The pattern of ridges and troughs in the upper-air wind flow is important in sustaining surface cyclonic (low-pressure) and anticyclonic (high-pressure) circulation. Along ridges, winds slow and converge (pile up); along troughs, winds accelerate and diverge (spread out). Note the wind-speed indicators and labels in Figure 6.12a near the ridge (over Alberta, Saskatchewan, Montana, and Wyoming), and compare them with the wind-speed indicators around the trough (over Kentucky, West Virginia, the New England states, and the Maritimes). Also, note the wind relationships off the Pacific Coast.

Figure 6.12c shows convergence and divergence in the upper-air flow. Divergence aloft is important to cyclonic circulation at the surface because it creates an outflow of air aloft that stimulates an inflow of air into the low-pressure cyclone (like what happens when you open an upstairs window to create an upward draft). Similarly, convergence aloft is important to anticyclonic circulation at the surface, driving descending airflows and causing airflow to diverge from high-pressure anticyclones.

Rossby Waves Within the westerly flow of geostrophic winds are great waving undulations, the **Rossby waves**, named for meteorologist Carl G. Rossby, who first described them mathematically in 1938. Rossby waves occur along the polar front, where colder air meets warmer air, and bring tongues of cold air southward, with warmer tropical air moving northward. The development of Rossby waves begins with undulations that then increase in amplitude to form waves (Figure 6.13). As these disturbances mature, circulation patterns form in which warmer air and colder air mix along distinct fronts. These wave-and-eddy formations and upper-air divergences support cyclonic storm systems at the surface. Rossby waves develop along the flow axis of a jet stream.

Jet Streams The most prominent movement in the upper-level westerly geostrophic wind flows are the **jet streams**, irregular, concentrated bands of wind occurring at several different locations that influence surface weather systems (Figure GIA 6.1 shows the location of four jet streams). The jet streams normally are 160–480 km (100–300 mi) wide by 900–2150 m (3000–7000 ft) thick, with core speeds that can exceed 300 kmph (190 mph). Jet streams in each hemisphere tend to weaken during the hemisphere's summer and strengthen during its winter as the streams shift closer to the equator. The pattern of high-pressure ridges and low-pressure troughs in the meandering jet streams causes variation in jet-stream speeds.

The *polar jet stream* meanders between 30° and 70° N latitude, at the tropopause along the polar front, at altitudes

GEOreport 6.4 Icelandic ash caught in the jet stream

Although smaller than either the Mount St. Helens 1980 eruption or the Mount Pinatubo 1991 eruption, the 2010 eruption of Eyjafjallajökull volcano in Iceland injected about a tenth of a cubic kilometer of volcanic debris into the jet-stream flow. The ash cloud from Iceland was swept toward the European mainland and the United Kingdom. Aircraft cannot risk ingesting volcanic ash into jet engines; therefore, airports were shut down and thousands of flights canceled. People's attention was focused on the guiding jet stream as it impacted their flight schedules and lives. See the ash cloud path in the satellite image in Figure 12.22.

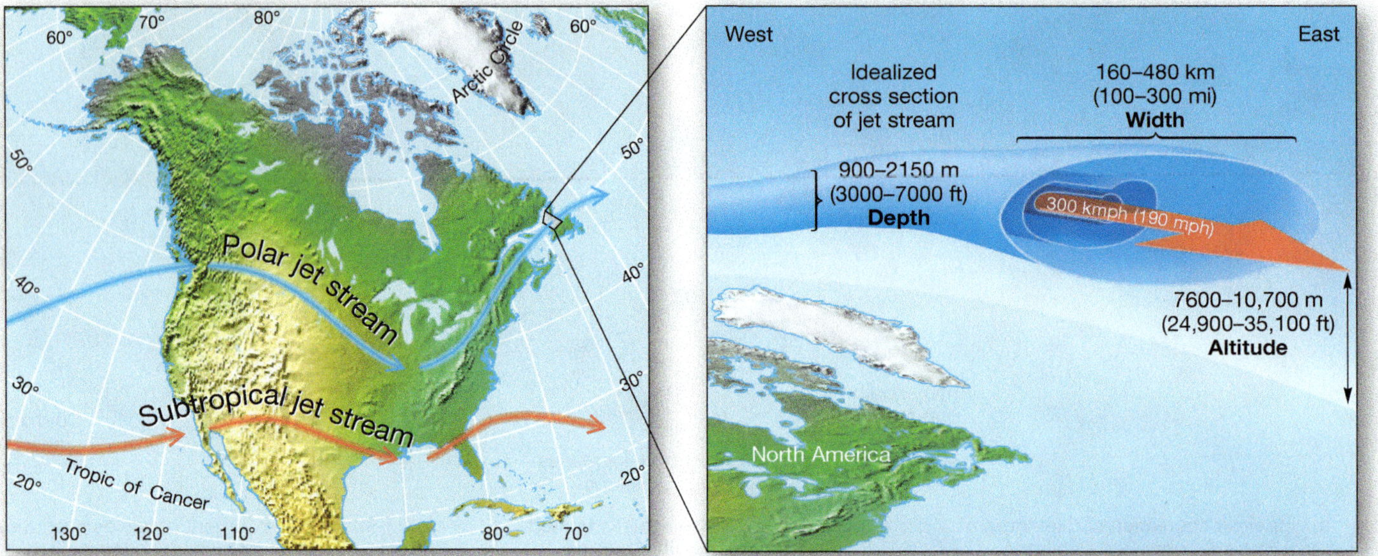

(a) Average locations of the two jet streams over North America.

(b) Width, depth, and altitude and core speed of an idealized polar jet stream.

▲Figure 6.14 **Jet streams.**

between 7600 and 10,700 m (24,900 and 35,100 ft; Figure 6.14a). The polar jet stream can migrate as far south as Texas, steering colder air masses into North America and influencing surface storm paths traveling eastward. In the summer, the polar jet stream remains at higher latitudes and exerts less influence on midlatitude storms. Figure 6.14b shows the cross-sectional view of a polar jet stream.

In subtropical latitudes, near the boundary between tropical and midlatitude air, the *subtropical jet stream* flows near the tropopause. The subtropical jet stream meanders from 20° to 50° latitude and may occur over North America simultaneously with the polar jet stream—sometimes the two will actually merge for brief episodes.

Monsoonal Winds

Regional winds are part of Earth's secondary atmospheric circulation. A number of regional wind systems change direction on a seasonal basis. Such regional wind flows occur in the tropics over Southeast Asia, Indonesia, India, northern Australia, and equatorial Africa; a mild regional flow also occurs in the extreme southwestern United States in Arizona. These seasonally shifting wind systems are **monsoons** (from the Arabic word *mausim*, meaning "season") and involve an annual cycle of returning precipitation with the summer Sun. Note the changes in precipitation between January and July visible on the *TRMM* images in Figure GIA 6.2, page 147.

The Asian Monsoon Pattern The unequal heating between the Asian landmass and the Indian Ocean drives the monsoons of southern and eastern Asia (Figure 6.15). This process is heavily influenced by the shifting migration of the ITCZ during the year, which brings moisture-laden air northward during the Northern Hemisphere summer.

A large difference is seen between summer and winter temperatures over the large Asian landmass—a result of the continental effect on temperature discussed in Chapter 5. During the Northern Hemisphere winter, an intense high-pressure cell dominates this continental landmass (see Figure 6.10a and Figure 6.15a). At the same time, the ITCZ is present over the central area of the Indian Ocean. The pressure gradient from about November to March between land and water produces cold, dry winds from the Asian interior that flow over the Himalayas and southward across India. These winds desiccate, or dry out, the landscape, especially in combination with hot temperatures from March through May.

During the Northern Hemisphere summer, the ITCZ shifts northward over southern Asia, and the Asian continental interior develops low pressure associated with high average temperatures (remember the summer warmth in Verkhoyansk, Siberia, from Chapter 5). Meanwhile, subtropical high pressure dominates over the Indian Ocean, causing warming of sea-surface temperatures (Figure 6.15b). Therefore, the pressure gradient is reversed from the winter pattern. As a result, hot subtropical air sweeps over the warm ocean toward India, producing extremely high evaporation rates.

By the time this air reaches India, the air is laden with moisture and clouds, which produce the monsoonal rains from about June to September (Figure 6.15b). These rains are welcome relief from the dust, heat, and parched land of Asia's springtime. World-record rainfalls occur in this region: Cherrapunji, India, holds the record for both the second highest average annual rainfall (1143 cm, or 450 in.) and the highest single-year rainfall (2647 cm, or 1042 in.) on Earth. In the Himalayas, the monsoon brings snowfall.

Human Influences on the Asian Monsoon In Chapter 4, we discussed the effects of increased aerosols from air pollution over the Indian Ocean, which appear to be causing a reduction in monsoon precipitation. Both air pollution and the warming of atmospheric and oceanic temperatures associated with climate change affect monsoon circulation. Research shows complicated interactions between

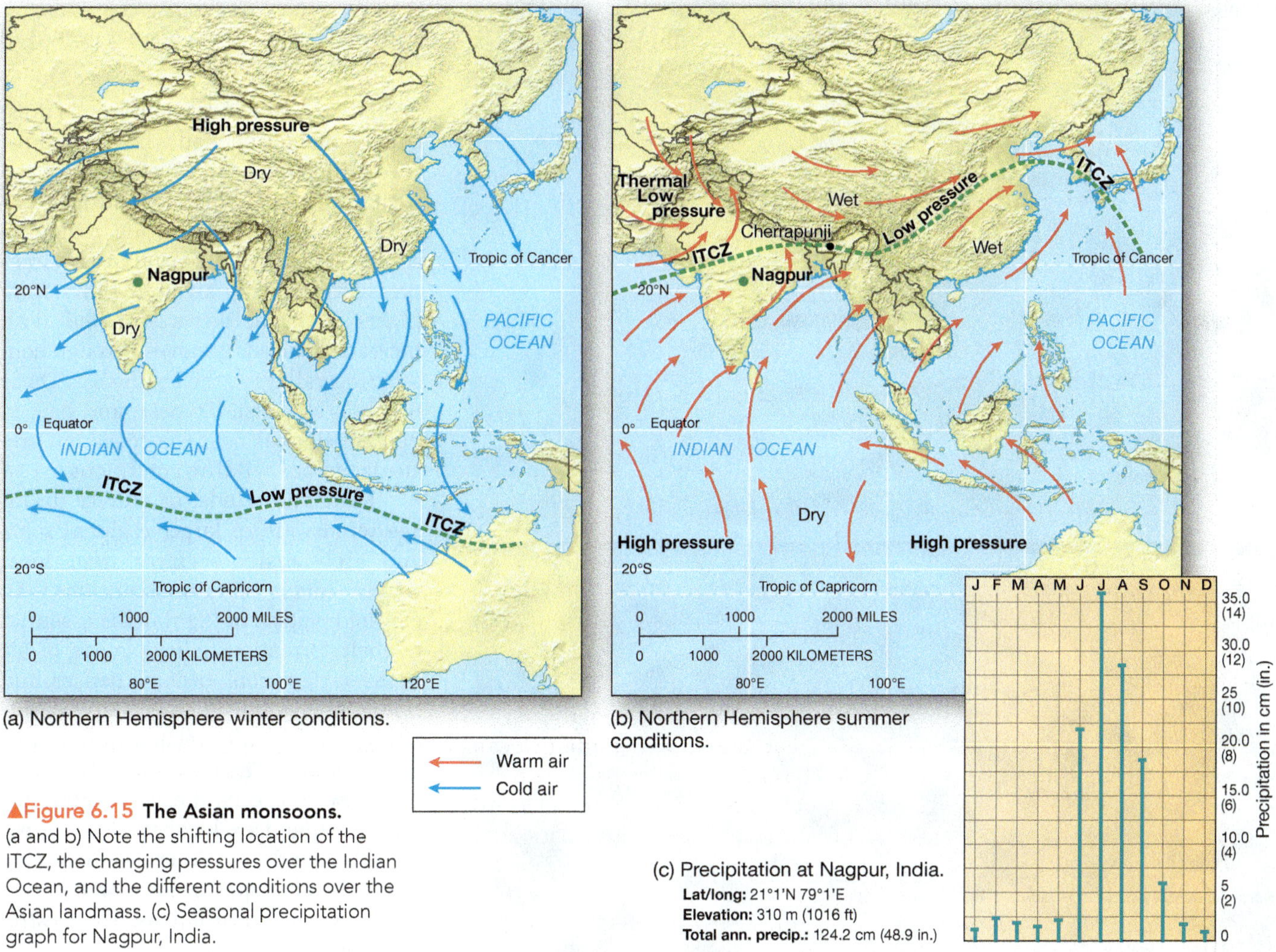

(a) Northern Hemisphere winter conditions.

(b) Northern Hemisphere summer conditions.

(c) Precipitation at Nagpur, India.
Lat/long: 21°1'N 79°1'E
Elevation: 310 m (1016 ft)
Total ann. precip.: 124.2 cm (48.9 in.)

▲**Figure 6.15 The Asian monsoons.** (a and b) Note the shifting location of the ITCZ, the changing pressures over the Indian Ocean, and the different conditions over the Asian landmass. (c) Seasonal precipitation graph for Nagpur, India.

these factors. New studies indicate that warmer temperatures caused by rising greenhouse gas concentrations have increased monsoon precipitation in the Northern Hemisphere over the past few decades. However, other research suggests that rising concentrations of aerosols—principally sulfur compounds and black carbon—cause an overall drop in monsoon precipitation. Air pollution reduces surface heating and therefore decreases the pressure differences at the heart of monsoonal flows.

A further complicating factor is that these influences are occurring at a time of some unusually heavy precipitation events, such as the monsoonal deluge on July 27, 2005, in Mumbai, India, that produced 94.2 cm (37.1 in.) of rain in only a few hours, causing widespread flooding. Again in August 2007, India experienced extensive flooding from an intense monsoon, and in 2010 Pakistan was devastated by record-breaking monsoon rains.

Further study is critical to understand these complex relationships, as well as the role of natural oscillations in global circulation (such as El Niño, discussed later in the chapter) in affecting monsoonal flows. Considering that 70% of the annual precipitation for the entire south Asian region comes during the wet monsoon, alterations in precipitation would have important impacts on water resources.

CRITICALthinking 6.2
What Causes the North Australian Monsoon?

Using your knowledge about global pressure and wind patterns, and the maps provided in this chapter and on the back inside cover of the book, sketch a map of the seasonal changes that cause the monsoonal winds over northern Australia. Begin by examining the pressure patterns and associated winds over this continent. How do they change throughout the year? Sketch the patterns for January and July on your map. Where is the position of the ITCZ? Finally, during which months do you expect a rainy season related to monsoonal activity to occur in this region? (Find the answers at the end of the chapter.) ●

Local Winds

Local winds, which occur on a smaller scale than the global and regional patterns just discussed, belong to the tertiary category of atmospheric circulation. **Land and sea breezes** are local winds produced along most coastlines (Figure 6.16). The different heating characteristics of land and water surfaces create these breezes. Land gains heat energy and warms faster than the water offshore during the day. Because warm air is less dense,

Figure 6.16 Temperature and pressure patterns for daytime sea breezes and nighttime land breezes.

it rises and triggers an onshore flow of cooler marine air to replace the rising warm air—the flow is usually strongest in the afternoon, forming a sea breeze. At night, land cools, by radiating heat energy, faster than offshore waters do. As a result, the cooler air over the land subsides (sinks) and flows offshore toward the lower-pressure area over the warmer water, where the air is lifted. This nighttime land-breeze pattern reverses the process that developed during the day.

Mountain and valley breezes are local winds resulting, respectively, when mountain air cools rapidly at night and when valley air gains heat energy rapidly during the day (Figure 6.17). Valley slopes are heated sooner during the day than valley floors. As the slopes heat up and warm the air above, this warm, less-dense air rises and creates an area of low pressure. By the afternoon, winds blow out of the valley in an upslope direction along this slight pressure gradient, forming a valley breeze. At night, heat is lost from the slopes, and the cooler air then subsides downslope in a mountain breeze.

Santa Ana winds result from a pressure gradient generated when high pressure builds over the Great Basin of the western United States. A strong, dry wind flows out across this region to southern California coastal areas. Compression heats the air as it flows from higher to lower elevations, and with increasing speed, it moves through constricting valleys to the southwest. These winds bring dust, dryness, and heat to populated areas near the coast and create dangerous wildfire conditions.

Katabatic winds, or gravity drainage winds, are of larger regional scale and are usually stronger than local winds, under certain conditions. An elevated plateau or highland is essential to their formation, where layers of air at the surface cool, become denser, and flow downslope. Such gravity winds are not specifically related to the pressure gradient. The ferocious winds that can blow off the ice sheets of Antarctica and Greenland are katabatic in nature.

Worldwide, various terrains produce distinct types of local winds that are known by local names. The *mistral* of the Rhône Valley in southern France is a cold north wind that can cause frost damage to vineyards as it moves over the region on its way to the Gulf of Lions and the Mediterranean Sea. The frequently stronger *bora*, driven by the cold air of winter high-pressure systems occurring inland over the Balkans and southeastern Europe, flows across the Adriatic Coast to the west and south. In Alaska, such winds are called the *taku*. In the U.S. West, *chinook winds* are dry, warm downslope winds occurring on the leeward side of mountain ranges such as the Cascades in Washington or the Rockies in Montana. These winds are known for their

CRITICALthinking 6.3
Construct Your Own Wind-Power Assessment Report

Go to http://www.awea.org/ and http://www.ewea.org/, the websites of the American Wind Energy Association and the European Wind Energy Association, respectively. Sample the materials presented and make your own assessment of the potential for wind-generated electricity, the reasons for delays in development and implementation, and the economic pros and cons. Propose a brief action plan for the future of this resource. •

Figure 6.17 Conditions for daytime valley breezes and nighttime mountain breezes.

ability to melt snow rapidly by sublimation, as discussed in Chapter 7.

Regionally, wind represents a significant and increasingly important source of renewable energy. Focus Study 6.1 briefly explores the wind-power resource.

Oceanic Currents

The atmospheric and oceanic systems are intimately connected in that the driving force for ocean currents is the frictional drag of the winds. Also important in shaping ocean currents is the interplay of the Coriolis force, density differences caused by temperature and salinity, the configuration of the continents and ocean floor, and the astronomical forces that cause tides.

Surface Currents

Figure 6.18 portrays the general patterns of major ocean currents. Because ocean currents flow over long distances, the Coriolis force deflects them. However, their pattern of deflection is not as tightly circular as that of the atmosphere. Compare this ocean-current map with the map showing Earth's pressure systems (see Figure 6.10), and you can see that ocean currents are driven by the atmospheric circulation around subtropical high-pressure cells in both hemispheres. The oceanic circulation systems are known as *gyres* and generally appear to be offset toward the western side of each ocean basin. Remember, in the Northern Hemisphere, winds and ocean currents move clockwise about high-pressure cells; in the Southern Hemisphere, circulation is counterclockwise, as shown on the map. You saw in *Geosystems Now* how these currents guided an oil-drilling platform to Tristan da Cunha, where it ran aground.

Examples of Gyre Circulation In 1992, a child at Dana Point, California (33.5° N), a small seaside community south of Los Angeles, placed a letter in a glass juice bottle and tossed it into the waves, where it entered the vast clockwise-circulating gyre around the Pacific High (Figure 6.19). Three years passed as ocean currents carried the message in a bottle to the white sands of Mogmog, a small island in Micronesia (7° N). Imagine the journey of that note from California—traveling through storms and calms, clear moonlit nights and typhoons.

In January 1994, a powerful storm ravaged a container ship from Hong Kong loaded with toys and other goods. One of the containers on board split apart in the wind off the coast of Japan, dumping nearly 30,000 rubber ducks, turtles, and frogs into the North Pacific. Westerly winds and the North Pacific Current swept this floating cargo at up to 29 km (18 mi) a day across the ocean to the coast of Alaska, Canada, Oregon, and California. Other toys drifted through the Bering Sea and into the Arctic Ocean (see dashed red line in Figure 6.19).

In August 2006, Tropical Storm Ioke formed about 1285 km (800 mi) south of Hawai'i. The storm's track moved westward across the Pacific Ocean as it became the strongest super typhoon in recorded history—a category 5, as we discuss in Chapter 8. Turning northward before reaching Japan, the storm moved into higher latitudes around the Pacific gyre. Typhoon Ioke remnants eventually crossed over the Aleutian Islands, reaching 55° N as an extratropical depression. The storm roughly followed the path around the Pacific Gyre and along the track of the rubber duckies.

Marine debris circulating in the Pacific Gyre is the subject of ongoing scientific study. Although plastics, especially small-sized plastic fragments, are predominant, this debris also consists of metals, fishing gear, and abandoned vessels, some of which remains in circulation within the gyre and some of which makes landfall. Debris from the 2011 Japan tsunami added more material into the North Pacific; computer models based on winds and currents estimate the extent of the debris, some of which has already washed up on the U.S. coastlines (Figure 6.19). (Go to http://marinedebris.noaa.gov/tsunamidebris for more on the tsunami debris.)

Equatorial Currents Trade winds drive the ocean surface waters westward in a concentrated channel along the equator (Figure 6.18). These equatorial currents remain near the equator because of the weakness of the Coriolis force, which diminishes to zero at that latitude. As these surface currents approach the western margins of the oceans, the water actually piles up against the eastern shores of the continents. The average height of this pileup is 15 cm (6 in.). This phenomenon is the **western intensification.**

The piled-up ocean water then goes where it can, spilling northward and southward in strong currents, flowing in tight channels along the eastern shorelines. In the Northern Hemisphere, the Gulf Stream and the Kuroshio (a current east of Japan) move forcefully northward as a result of western intensification. Their speed and depth increase with the constriction of the area they occupy. The warm, deep, clear water of the ribbon-like Gulf Stream (Figure 5.8) usually is 50–80 km (30–50 mi) wide and 1.5–2.0 km (0.9–1.2 mi) deep, moving at 3–10 kmph (1.8–6.2 mph). In 24 hours, ocean water can move 70–240 km (40–150 mi) in the Gulf Stream.

Upwelling and Downwelling Flows Where surface water is swept away from a coast, either by surface divergence (induced by the Coriolis force) or by offshore winds, an **upwelling current** occurs. This cool water generally is rich in nutrients and rises from great depths to replace the vacating water. Such cold upwelling currents exist off the Pacific coasts of North and South America and the subtropical and midlatitude west coast of Africa. These areas are some of Earth's prime fishing regions.

In other regions with an accumulation of water—such as at the western end of an equatorial current, or in the Labrador Sea, or along the margins of Antarctica—the excess water gravitates downward in a **downwelling current.** These are the deep currents that flow vertically and along the ocean floor and travel the full extent of the ocean basins, redistributing heat energy and salinity over the globe.

Thermohaline Circulation—The Deep Currents

Differences in temperatures and salinity (the amount of salts dissolved in water) produce density differences important to the flow of deep currents on Earth known as **thermohaline circulation**, or THC (*thermo-* refers to temperature and *-haline* refers to salinity). Traveling at slower speeds than wind-driven surface currents, the thermohaline circulation hauls larger volumes of water. (Figure 16.4 illustrates the ocean's physical structure and profiles of temperature, salinity, and dissolved gases; note that temperature and salinity vary with depth.)

To picture the THC, imagine a continuous channel of water beginning with the flow of the Gulf Stream and the North Atlantic Drift (Figure 6.18). When this warm, salty water mixes with the cold water of the Arctic Ocean, it cools, increases in density, and sinks. The cold water downwelling in the North Atlantic, on either side of Greenland, produces the deep current that then flows southward. Downwelling also occurs in the high southern latitudes as warm equatorial surface currents meet cold

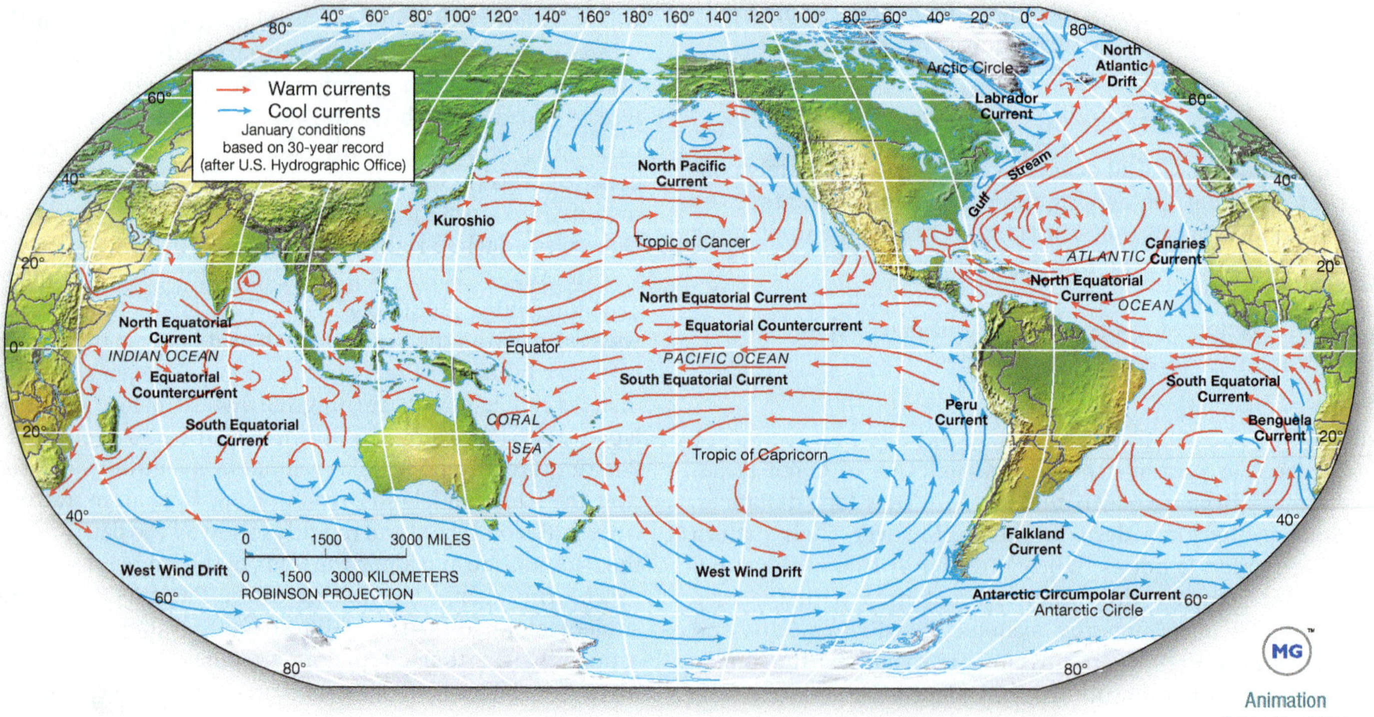

MG Animation Ocean Circulation

▲Figure 6.18 **Major ocean currents.** [After the U.S. Naval Oceanographic Office.]

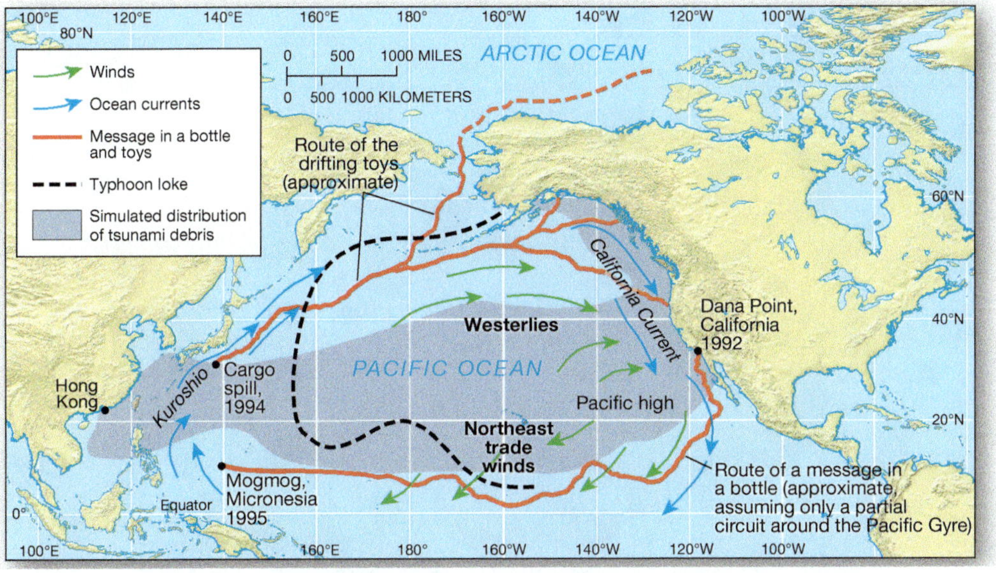

▲**Figure 6.19 Transport of marine debris by Pacific Ocean currents.** The paths of a message in a bottle, toy rubber duckies, and Typhoon Ioke show the movement of currents around the Pacific Gyre. The distribution of debris from the 2011 Japan tsunami is a computer simulation based on expected winds and currents through January 7, 2012. [NOAA.]

Antarctic waters (Figure 6.20). As water then moves northward, it warms; areas of upwelling occur in the Indian Ocean and North Pacific. A complete circuit of these surface and subsurface currents may require 1000 years.

Ocean surface waters undergo "freshening" in the polar regions because water releases salt when frozen (the salt is essentially squeezed out of the ice structure), and is then salt-free when it melts. This ocean freshening through the melting of sea ice is currently being accelerated by climate change. Increased rates of glacial and ice-sheet melting are producing fresh, lower-density surface waters that ride on top of the denser saline water. In theory a large input of fresh water into the North Atlantic could reduce the density of seawater enough that downwelling would no longer occur there—effectively shutting down the THC.

Ongoing scientific research shows the effects of climate change in the Arctic: rising temperatures, melting sea ice, thawing permafrost, melting glaciers, increased runoff in rivers, increased rainfall, all adding to an overall increase in the amount of freshwater entering the Arctic Ocean. Current models suggest that a weakening of the THC is possible by the end of the 21st century. For more information on this vital research frontier, check http://sio.ucsd.edu/ or http://www.whoi.edu/ for updates.

▼**Figure 6.20 Deep-ocean thermohaline circulation.** This vast conveyor belt of water draws heat energy from warm, shallow currents and transports it to higher latitudes for release in the depths of the ocean basins in cold, deep, salty currents. Four blue areas at high latitudes are where surface water cools, sinks, and feeds the deep circulation.

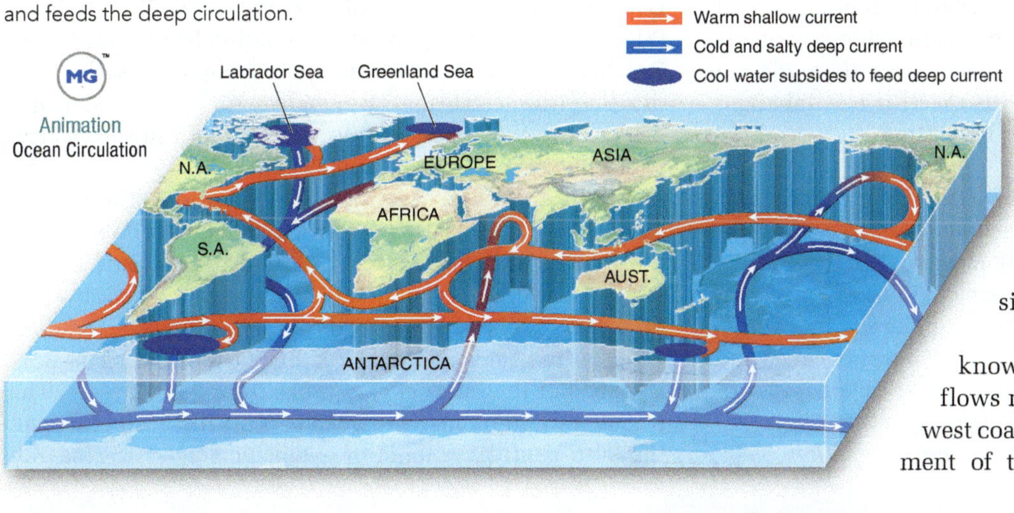

Natural Oscillations in Global Circulation

Several system fluctuations that occur in multiyear or shorter periods are important in the global circulation picture. Multiyear oscillations affect temperatures and air pressure patterns and thus affect global winds and climates. The most famous of these is the El Niño–Southern Oscillation (ENSO) phenomenon, which affects temperature and precipitation on a global scale. Here we describe ENSO and briefly introduce three other hemisphere-scale oscillations.

El Niño–Southern Oscillation

Climate is the consistent behavior of weather over time, but normal weather conditions can include extremes that depart from the average in a given region. The **El Niño–Southern Oscillation (ENSO)** in the Pacific Ocean forces the greatest interannual variability of temperature and precipitation on a global scale. Peruvians coined the name El Niño ("the boy child") because these episodes seem to occur around the time of the traditional December celebration of Christ's birth. Actually, El Niños can occur as early as spring and summer and persist through the year.

The cold Peru Current (also known as the Humbodt Current) flows northward off South America's west coast, joining the westward movement of the South Equatorial Current

Focus Study 6.1 Sustainable Resources

Wind Power: An Energy Resource for the Present and Future

The principles of wind power are ancient, but the technology is modern and the benefits are substantial. Scientists estimate that wind as a resource could potentially produce many times more energy than is currently in demand on a global scale. Yet, despite the available technology, wind-power development continues to be slowed, mainly by the changing politics of renewable energy.

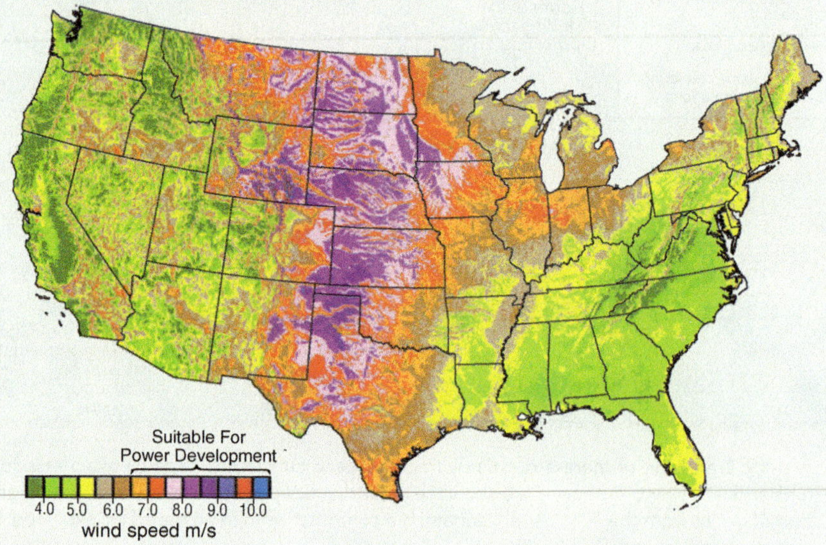

▲**Figure 6.1.1 Wind-speed map of the contiguous United States.** Map shows predicted average wind speeds at a height of 80 m (164 ft) above the ground. Areas with wind speeds greater than 6.5 m/s (21 ft/s) are considered suitable for power development. Map has a spatial resolution of 2.5 km (1.6 mi). [NREL and AWS Truepower.]

The Nature of Wind Energy

Power generation from wind depends on site-specific characteristics of the wind resource. Favorable settings for consistent wind are areas (1) along coastlines influenced by trade winds and westerly winds; (2) where mountain passes constrict air flow and interior valleys develop thermal low-pressure areas, thus drawing air across the landscape; and (3) where localized winds occur, such as an expanse of relatively flat prairies, or areas with katabatic or monsoonal winds. Many developing countries are located in areas blessed by such steady winds, such as the trade winds across the tropics.

Where winds are sufficient, electricity is generated by groups of wind turbines (in wind farms) or by individual installations. If winds are reliable less than 25%–30% of the time, only small-scale use of wind power is economically feasible.

The potential of wind power in the United States is enormous (Figure 6.1.1). In the Midwest, power from the winds of North and South Dakota and Texas alone could meet all U.S. electrical needs. In the California Coast Ranges, land and sea breezes blow between the Pacific and Central Valley, peaking in intensity from April to October, which happens to match peak electrical demands for air conditioning during the hot summer months.

On the eastern shore of Lake Erie sits a closed Bethlehem Steel mill, contaminated with industrial waste until the site was redeveloped with eight 2.5-MW wind-power turbines in 2007 (Figure 6.1.2). Six more turbines were added in 2012, making this a 35-MW electrical generation facility, the largest urban installation in the country. The former "brownfield" site now supplies enough electricity to power 15,000 homes in western New York. A proposed expansion would add 500 MW from some 167 turbines to be installed offshore in Lake

near the equator (Figure 6.18). The Peru Current is part of the normal counterclockwise circulation of winds and surface ocean currents around the subtropical high-pressure cell dominating the eastern Pacific in the Southern Hemisphere. As a result, a location such as Guayaquil, Ecuador, normally receives 91.4 cm (36 in.) of precipitation each year under dominant high pressure, whereas islands in the Indonesian archipelago receive more than 254 cm (100 in.) under dominant low pressure. This normal alignment of pressure is shown in Figure 6.21a.

El Niño—ENSO's Warm Phase Occasionally, for unexplained reasons, pressure patterns and surface ocean temperatures shift from their usual locations in the Pacific. Higher pressure than normal develops over the western Pacific, and lower pressure develops over the eastern Pacific. Trade winds normally moving from east to west weaken and can be reduced or even replaced by an eastward (west-to-east) flow. The shifting of atmospheric pressure and wind patterns across the Pacific is the *Southern Oscillation.*

Sea-surface temperatures may increase to more than 8 C° (14 F°) above normal in the central and eastern Pacific during an ENSO, replacing the normally cold, nutrient-rich water along Peru's coastline. Such ocean-surface warming, creating the "warm pool," may extend to the International Date Line. This surface pool of warm water is the El Niño (Figure 6.21b), leading to the designation ENSO—El Niño–Southern Oscillation. During El Niño conditions, the *thermocline* (the transition layer between surface water and colder deep-water beneath) lowers in depth in the eastern Pacific Ocean, blocking upwelling. The change in wind direction and the warmer surface water slow the normal upwelling currents that control nutrient availability off the South American coast. This loss of nutrients affects the phytoplankton and food chain, depriving fish, marine mammals, and predatory birds of nourishment.

The expected interval for ENSO recurrence is 3 to 5 years, but the interval may range from 2 to 12 years. The frequency and intensity of ENSO events increased through the 20th century, a topic of extensive scientific research

Erie. With the slogan "Turning the Rust Belt into the Wind Belt," this former steel town is using wind power to lift itself out of an economic depression.

Although most U.S. wind power is land-based, offshore wind development has high potential. The proposed Cape Wind Farm near Cape Cod, Massachusetts, was recently approved as the nation's first offshore project. Proponents hope that despite the additional expense of installation, offshore production will increase, especially along the eastern seaboard, where population centers are close together. At least 12 offshore projects are currently under consideration, most of them on the East Coast.

The appeal of land-based wind-power development is enhanced by the income it brings. Farmers in Iowa and Minnesota receive about $2000 in annual income from electrical production by a leased turbine and about $20,000 a year from electrical production by an owned turbine—requiring only one-quarter acre to site the wind machine. The Midwest is on the brink of an economic boom if this wind-energy potential is developed and transmission-line capacity installed. In fact, Iowa now ranks third for installed wind-power capacity in the United States.

Wind-Power Status and Benefits

Wind-generated energy resources are the fastest-growing energy technology—capacity has risen worldwide in a continuing trend of doubling every 3 years. Total world capacity approached 282,000 MW (megawatts), or 282 GW (gigawatts), by the end of 2012 from installations in over 81 countries, including sub-Saharan Africa's first commercial wind farm in Ethiopia; this is an increase of 19% over 2011.

In the United States, installed wind capacity exceeded 60 GW through 2012, a 517% increase in 7 years. Installations are operating in 39 states, with the greatest wind-power capacity in Texas, California, Iowa, Illinois, and Oregon. This puts U.S. installed wind capacity second highest on a global level behind China's 77 GW, with Germany third.

The European Wind Energy Association announced installed capacity exceeding 105,600 MW at the end of 2012, enough to meet 7% of its electricity needs. Germany has the most, followed by Spain, the United Kingdom, Italy, and France. The European Union has a goal of 20% of all energy from renewable sources by 2020.

The economic and social benefits from using wind resources are numerous. With all costs considered, wind energy is cost-competitive and actually cheaper than oil, coal, natural gas, and nuclear power. Wind power is renewable and does not cause adverse human health effects or environmental degradation. The main challenges of wind-generated power are the high initial financial investment required to build the turbines and the cost of building transmission lines to bring electricity from rural wind farms to urban locations.

To put numbers in meaningful perspective, every 10,000 MW of wind-generation capacity reduces carbon dioxide emissions by 33 million metric tons if it replaces coal or by 21 million metric tons if it replaces mixed fossil fuels. If countries rally and create a proposed $600 billion industry by installing 1,250,000 MW of wind capacity by 2020, that would supply 12% of global electrical needs. By the middle of this century, wind-generated electricity, along with other renewable energy sources, could be routine.

▲**Figure 6.1.2 Wind turbines at a former industrial site on Lake Erie.** The "Steel Winds" project began in 2007 and in 2012 achieved a total generating capacity of 35 MW, enough to power about 15,000 homes in western New York. [Ken JP Stucynski.]

looking for a link to global climate change. Although recent studies suggest that this phenomenon might be more responsive to global change than previously thought, scientists have found no definitive connection.

The two strongest ENSO events in 120 years occurred in 1982–1983 and 1997–1998. The latest El Niño subsided in May 2010 (Figure 6.21d). Although the pattern began to build again in late summer 2012, it resulted in a weak El Niño that ended in early 2013.

La Niña—ENSO's Cool Phase When surface waters in the central and eastern Pacific cool to below normal by 0.4 C° (0.7 F°) or more, the condition is dubbed *La Niña*, Spanish for "the girl." This condition is weaker and less consistent than El Niño; otherwise, there is no correlation in strength or weakness between the two phases. For instance, following the record 1997–1998 ENSO event, the subsequent La Niña was not as strong as predicted and shared the Pacific with lingering warm water (Figure 6.21c).

In contrast, the 2010–2011 La Niña was one of the strongest on record, according to atmospheric indicators (and correlated with sea-surface height anomalies, Figure 6.21e). This event corresponded with the wettest December in history in Queensland and across eastern Australia, where heavy rainfall led to the country's worst flooding in 50 years. Weeks of precipitation caused the Fitzroy River to overflow its banks, inundating an area the size of France and Germany combined and causing the evacuation of thousands of people (see Figure 6.22 at the end of the chapter). This La Niña intensified again in late 2011, lasting into 2012.

Global Effects Related to ENSO Weather and short-term climate effects related to ENSO occur worldwide, but vary depending on location. El Niño is tied to droughts in South Africa, southern India, Australia, and the Philippines; strong hurricanes in the Pacific, including Tahiti and French Polynesia; and flooding in the United States (in the southwestern and mountain states), Bolivia, Cuba, Ecuador, and Peru. In India, every drought for more than 400 years

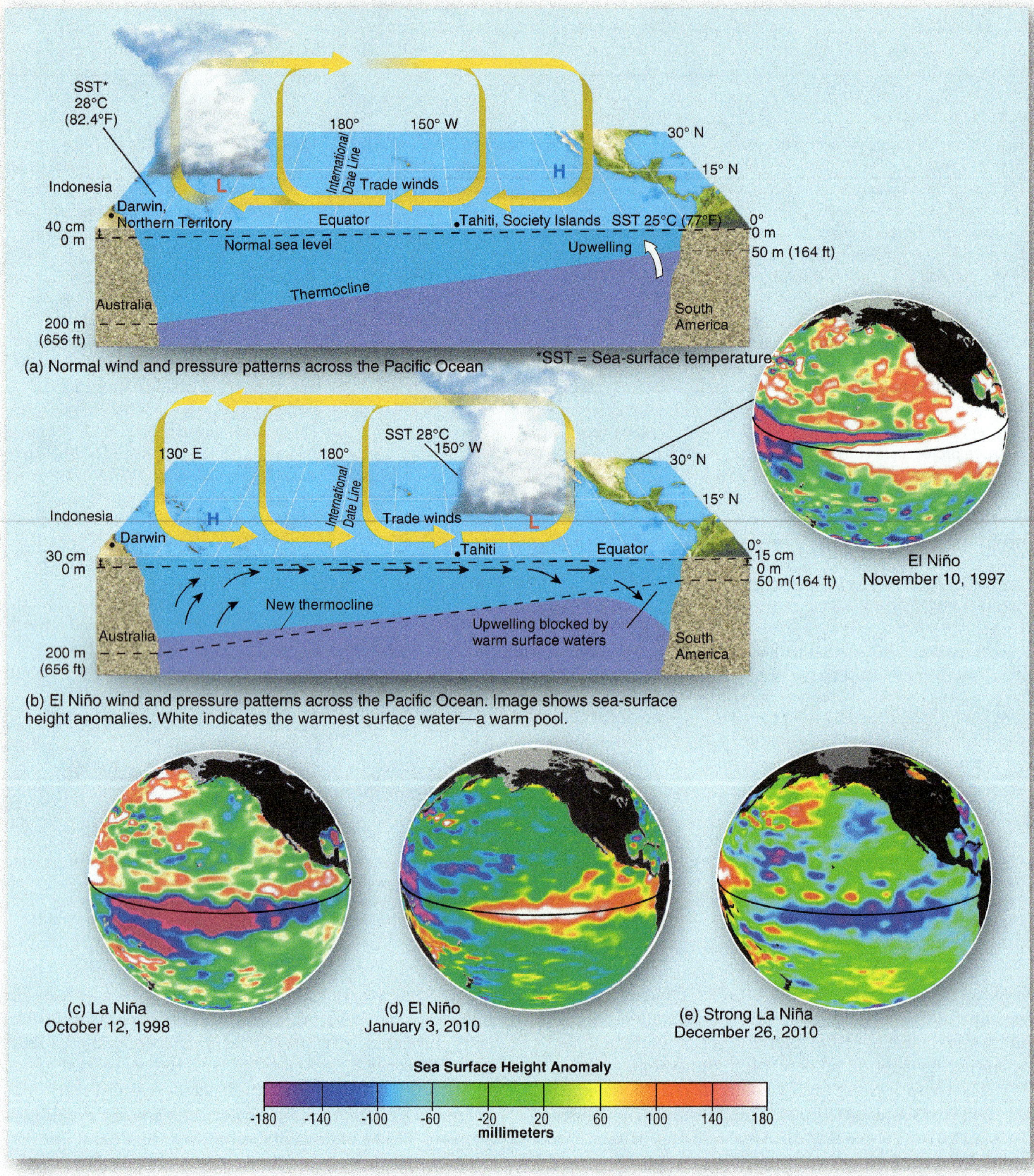

▲**Figure 6.21 Normal, El Niño, and La Niña conditions in the Pacific.** [(a) and (b) adapted and corrected from C. S. Ramage, "El Niño." © 1986 by *Scientific American, Inc.* (b image)–(c) *TOPEX/Poseidon*, (d) *Jason–1*, (e) *OSTM-Jason–2* images courtesy of Jet Propulsion Laboratory, NASA.]

seems linked to this warm phase of ENSO. La Niña often brings wetter conditions throughout Indonesia, the South Pacific, and northern Brazil. The Atlantic hurricane season weakens during El Niño years and strengthens during La Niña years. (See http://www.esrl.noaa.gov/psd/enso/ for more on ENSO, or go to NOAA's El Niño Theme Page at http://www.pmel.noaa.gov/toga-tao/el-nino/nino-home.html.)

Pacific Decadal Oscillation

The *Pacific Decadal Oscillation (PDO)* is a pattern of sea-surface temperatures, air pressure, and winds that shifts between the northern and tropical western Pacific (off the coast of Asia) and the eastern tropical Pacific (along the U.S. West Coast). The PDO, lasting 20 to 30 years, is longer-lived than the 2- to 12-year variation in the ENSO.

The PDO is strongest in the North Pacific, rather than in the tropical Pacific, another distinction from the ENSO.

The PDO negative phase, or cool phase, occurs when higher-than-normal temperatures dominate in the northern and tropical regions of the western Pacific and lower temperatures occur in the eastern tropical region; such conditions occurred from 1947 to 1977.

A switch to a positive phase, or warm phase, in the PDO ran from 1977 to the 1990s, when lower-than-normal temperatures were found in the northern and western Pacific and higher-than-normal temperatures dominated the eastern tropical region. This coincided with a time of more intense ENSO events. In 1999, a negative phase began for 4 years, a mild positive phase continued for 3 years, and now the PDO has been in a negative phase since 2008. This PDO negative phase can mean a decade or more of drier conditions in the U.S. Southwest, as well as cooler and wetter winters in the U.S. Northwest.

The PDO affects fisheries along the U.S. Pacific coast, with more productive regions shifting northward toward Alaska during PDO warm phases and southward along the California coast during cool phases. Causes of the PDO and its cyclic variability over time are unknown. For more information, see http://www.nc-climate.ncsu.edu/climate/patterns/PDO.html.

North Atlantic and Arctic Oscillations

A north–south fluctuation of atmospheric variability marks the *North Atlantic Oscillation (NAO)*, as pressure differences between the Icelandic Low and the Azores High in the Atlantic alternate from a weak to a strong pressure gradient. The NAO is in its positive phase when a strong pressure gradient is formed by a lower-than-normal Icelandic low-pressure system and a higher-than-normal Azores high-pressure cell (review their location in Figure 6.10). Under this scenario, strong westerly winds and jet streams cross the eastern Atlantic. In the eastern United States, winters tend to be less severe in contrast to the strong, warm, wet storms hitting northern Europe; however, the Mediterranean region is dry.

In its negative phase, the NAO features a weaker pressure gradient than normal between the Azores and Iceland and reduced westerlies and jet streams. Storm tracks shift southward in Europe, bringing moist conditions to the Mediterranean and cold, dry winters to northern Europe. The eastern United States experiences cold, snowy winters as Arctic air masses plunge to lower latitudes.

The NAO flips unpredictably between positive and negative phases, sometimes changing from week to week. From 1980 to 2008, the NAO was more strongly positive; however, through 2009 and into early 2013, the NAO was more strongly negative (see http://www.ncdc.noaa.gov/teleconnections/nao/).

Variable fluctuations between middle- and high-latitude air mass conditions over the Northern Hemisphere produce the *Arctic Oscillation (AO)*. The AO is associated with the NAO, especially in winter, and their phases correlate. In the AO positive, or warm, phase (positive NAO), the pressure gradient is affected by lower pressure than normal over the North Pole region and relatively higher pressures at lower latitudes. This sets up stronger westerly winds and a consistently strong jet stream as well as the flow of warmer Atlantic water currents into the Arctic Ocean. Cold air masses do not migrate as far south in winter, whereas winters are colder than normal in Greenland.

In the AO negative, or cold, phase (negative NAO), the pattern reverses. Higher-than-normal pressure is over the polar region, and relatively lower pressure is over the central Atlantic. The weaker zonal wind flow during the winter allows cold air masses into the eastern United States, northern Europe, and Asia, and sea ice in the Arctic Ocean becomes a bit thicker. Greenland, Siberia, northern Alaska, and the Canadian Archipelago are all warmer than normal.

The Northern Hemisphere winter of 2009–2010 featured unusual flows of cold air into the midlatitudes. As temperatures dropped in the U.S. Northeast and Midwest, the circumpolar region was experiencing conditions 15 to 20 C° (27 to 36 F°) above average. In December 2009, the AO was in its most negative phase since 1970; it was even more negative in February 2010 (see http://nsidc.org/arcticmet/patterns/arctic_oscillation.html).

Recent research shows that melting sea ice in the Arctic region may be forcing a negative AO. As sea ice melts, the open ocean retains heat, which it releases to the atmosphere in the fall, warming the Arctic air. This reduces the temperature difference between the poles and the midlatitudes, and reduces the strength of the polar vortex, strong wind patterns that trap arctic air masses at the poles. Weakening of the polar vortex causes a weaker jet stream, which then meanders in a north–south direction. This forces a negative AO, leading to mostly colder winters over the Northern Hemisphere and the presence of a blocking high-pressure system over Greenland. Figure HD 6, *The Human Denominator*, illustrates this phenomenon, as well as other examples of interactions between humans and Earth's circulation patterns.

GEOreport 6.5 2010–2011 La Niña breaks records

The *Southern Oscillation Index (SOI)* measures the difference in air pressure between Tahiti and Darwin, Australia, and is one of several atmospheric indexes used to monitor ENSO. In general, the SOI is negative during an El Niño and positive during a La Niña. This index corresponds with sea-surface temperature (SST) changes across the Pacific: Negative SOI values coincide with warm SSTs (El Niño), and positive SOI values correlate with cold SSTs (La Niña). During the 2010–2011 La Niña, the SOI reached record levels, coinciding with dramatic weather events in Australia.

THE**human**DENOMINATOR 6 Global Circulation

ATMOSPHERIC AND OCEANIC CIRCULATION ⇨ HUMANS

- Wind and pressure contribute to Earth's general atmospheric circulation, which drives weather systems and spreads natural and anthropogenic pollution across the globe.
- Natural oscillations in global circulation, such as ENSO, affect global weather.
- Ocean currents carry human debris and non-native species into remote areas and spread oil spills across the globe.

HUMANS ⇨ ATMOSPHERIC AND OCEANIC CIRCULATION

- Climate change may be altering patterns of atmospheric circulation, especially in relation to Arctic sea-ice melting and the jet stream, as well as possible intensification of subtropical high-pressure cells.
- Air pollution in Asia affects monsoonal wind flow; weaker flow could reduce rainfall and affect water availability.

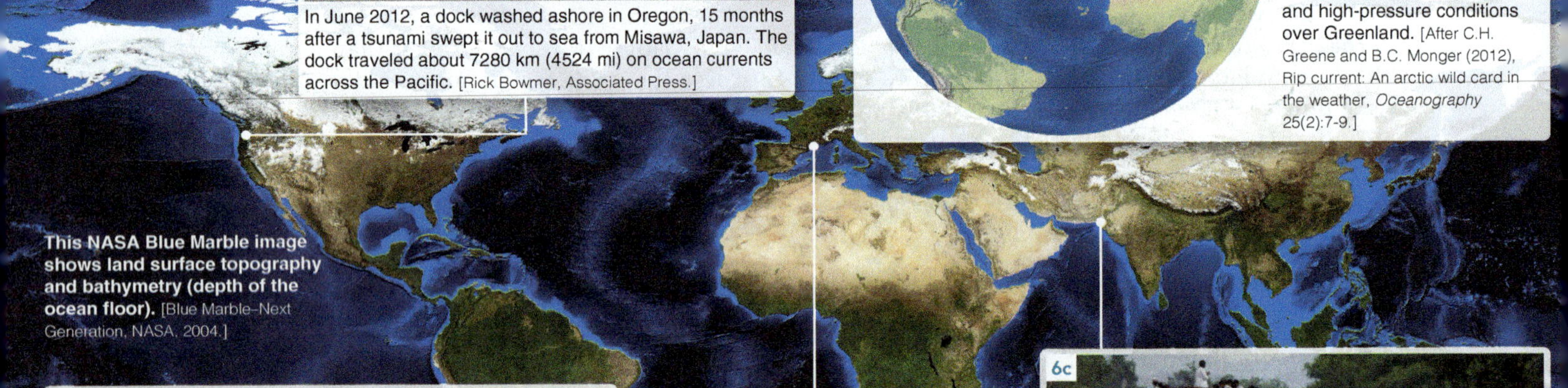

This NASA Blue Marble image shows land surface topography and bathymetry (depth of the ocean floor). [Blue Marble–Next Generation, NASA, 2004.]

6a

In June 2012, a dock washed ashore in Oregon, 15 months after a tsunami swept it out to sea from Misawa, Japan. The dock traveled about 7280 km (4524 mi) on ocean currents across the Pacific. [Rick Bowmer, Associated Press.]

6b

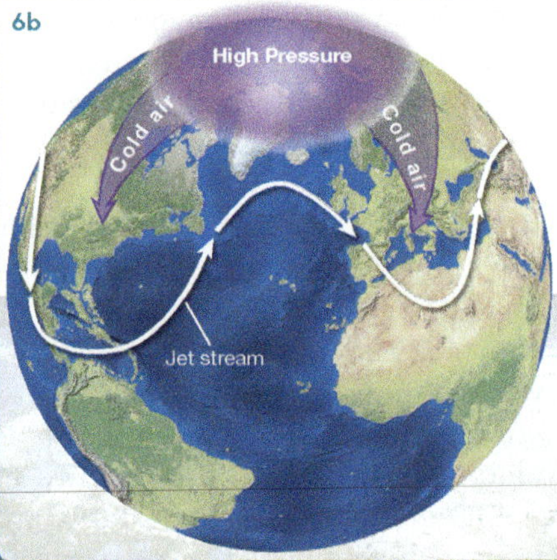

Scientists think that melting sea ice in the Arctic owing to climate change is altering the temperature and pressure balance between polar and midlatitude regions. This weakens the jet stream (which steers weather systems from west to east around the globe), creating large meanders that bring colder conditions to the United States and Europe, and high-pressure conditions over Greenland. [After C.H. Greene and B.C. Monger (2012), Rip current: An arctic wild card in the weather, *Oceanography* 25(2):7-9.]

6d

Windsurfers enjoy the effects of the mistral winds off the coast of southern France. These cold, dry winds driven by pressure gradients blow southward across Europe through the Rhône River valley. [Gardel Bertrand/Hemis/Alamy.]

6c

In August 2010, monsoon rainfall caused flooding in Pakistan that affected 20 million people and led to over 2000 fatalities. The rains came from an unusually strong monsoonal flow, worsened by La Niña conditions. See Chapter 14 for satellite images of the Indus River during this event. [Andrees Latif/Reuters.]

ISSUES FOR THE 21ST CENTURY

- Wind energy is a renewable resource that is expanding in use.
- Ongoing climate change may affect ocean currents, including the thermohaline circulation, as well as natural oscillations in circulation, such as the AO and the PDO.

GEOSYSTEMS**connection**

With this chapter, you have completed the inputs–actions–outputs pathway through PART I, "The Energy–Atmosphere System." In the process, you examined the many ways these Earth systems affect our lives and the many ways human society is impacting these systems. From this foundation, we now move on to PART II, "The Water, Weather, and Climate Systems." Water and ice cover 71% of Earth's surface by area, making Earth the water planet. This water is driven by Earth–atmosphere energy systems to produce weather, water resources, and climate.

KEY LEARNING
concepts review

Define the concept of air pressure, and *describe* instruments used to measure air pressure.

The weight of the atmosphere in terms of force per unit area is **air pressure**, created by motion, size, and number of molecules. A **mercury barometer** measures air pressure at the surface (mercury in a tube—closed at one end and open at the other, with the open end placed in a vessel of mercury—that changes level in response to pressure changes), as does an **aneroid barometer** (a closed cell, partially evacuated of air, that detects changes in pressure).

air pressure (p. 135)
mercury barometer (p. 136)
aneroid barometer (p. 136)

1. How does air exert pressure? Describe the basic instrument used to measure air pressure. Compare the operation of two different types of instruments discussed.
2. What is the relationship between air pressure and density, and between air pressure and temperature?
3. What is normal sea-level pressure in millimeters? Millibars? Inches? Kilopascals?

Define wind, and *explain* how wind is measured, how wind direction is determined, and how winds are named.

Wind is the horizontal movement of air across Earth's surface; turbulence adds updrafts and downdrafts, and thus a vertical component, to the definition. Wind speed is measured with an **anemometer** (a device with cups that are pushed by the wind) and its direction with a **wind vane** (a flat blade or surface that is directed by the wind).

wind (p. 136)
anemometer (p. 136)
wind vane (p. 136)

4. What is a possible explanation for the beautiful sunrises and sunsets during the summer of 1992 in North America? Relate your answer to global circulation.
5. Explain this statement: "The atmosphere socializes humanity, making the world a spatially linked society." Illustrate your answer with some examples.
6. Define wind. How is it measured? How is its direction determined?

Explain the four driving forces within the atmosphere—gravity, pressure gradient force, Coriolis force, and friction force—and *locate* the primary high- and low-pressure areas and principal winds.

The pressure that Earth's *gravitational force* exerts on the atmosphere is virtually uniform worldwide. The **pressure gradient force** drives winds, as air moves from areas of high pressure to areas of low pressure. Maps portray air pressure patterns using the **isobar**—an isoline that connects points of equal pressure. The **Coriolis force** causes an apparent deflection in the path of winds or ocean currents, owing to the rotation of Earth. This force deflects objects to the right in the Northern Hemisphere and to the left in the Southern Hemisphere. The **friction force** drags winds along Earth's varied surfaces in opposition to the pressure gradient. The pressure gradient and Coriolis force in combination (absent the friction force) produce **geostrophic winds**, which move parallel to isobars, characteristic of winds above the surface frictional layer.

In a high-pressure system, or **anticyclone**, winds descend and diverge, spiraling outward in a clockwise direction in the Northern Hemisphere. In a low-pressure system, or **cyclone**, winds converge and ascend, spiraling upward in a counterclockwise direction in the Northern Hemisphere. (The rotational directions are reversed for each in the Southern Hemisphere.)

The pattern of high and low pressures on Earth in generalized belts in each hemisphere produces the distribution of specific wind systems. These primary pressure regions are the **equatorial low**, the weak **polar highs** (at both the North and the South Poles), the **subtropical highs**, and the **subpolar lows**.

All along the equator, winds converge into the equatorial low, creating the **intertropical convergence zone (ITCZ)**. Air rises in this zone and descends in the subtropics in each hemisphere. The winds returning to the ITCZ from the northeast in the Northern Hemisphere and from the southeast in the Southern Hemisphere produce the **trade winds**.

The subtropical high-pressure cells on Earth are generally between 20° and 35° in each hemisphere. In the Northern Hemisphere, they include the **Bermuda High**, **Azores High**, and **Pacific High**. Winds flowing out of the subtropics to higher latitudes produce the **westerlies** in each hemisphere.

In January, two low-pressure cells known as, the **Aleutian Low** and **Icelandic Low** dominate the North Pacific and Atlantic, respectively. The region of contrast between cold polar air and the warmer air toward the equator is the **polar front**. The weak and variable **polar easterlies** diverge from the high-pressure cells at each pole, the stronger of which is the **Antarctic High**.

pressure gradient force (p. 138)
isobar (p. 138)
Coriolis force (p. 138)
friction force (p. 140)
geostrophic wind (p. 141)
anticyclone (p. 142)
cyclone (p. 142)
equatorial low (p. 142)
polar high (p. 142)
subtropical high (p. 144)
subpolar low (p. 144)
intertropical convergence zone (ITCZ) (p. 144)
trade winds (p. 144)
Bermuda High (p. 144)
Azores High (p. 144)
Pacific High (p. 144)
westerlies (p. 145)
Aleutian Low (p. 145)
Icelandic Low (p. 145)
polar front (p. 145)
polar easterlies (p. 145)
Antarctic High (p. 145)

7. What does an isobaric map of surface air pressure portray? Contrast pressures over North America for January and July.
8. Describe the effect of the Coriolis force. Explain how it appears to deflect atmospheric and oceanic circulations.
9. What are geostrophic winds, and where are they encountered in the atmosphere?
10. Describe the horizontal and vertical air motions in a high-pressure anticyclone and in a low-pressure cyclone.
11. Construct a simple diagram of Earth's general circulation; begin by labeling the four principal pressure belts or zones, and then add arrows between these pressure systems to denote the three principal wind systems.
12. How is the intertropical convergence zone (ITCZ) related to the equatorial low? How does the ITCZ appear on the satellite images of accumulated precipitation for January and July in GIA 6.2?
13. Characterize the belt of subtropical high pressure on Earth: Name several specific cells. Describe the generation of westerlies and trade winds and their effects on sailing conditions.
14. What is the relationship among the Aleutian Low, the Icelandic Low, and migratory low-pressure cyclonic storms in North America? In Europe?

Describe upper-air circulation, and *define* the jet streams.

A **constant isobaric surface**—a surface that varies in altitude from place to place according to where a given air pressure, such as 500 mb, occurs—is useful for visualizing geostrophic wind patterns in the middle and upper troposphere. The variations in altitude of this surface show the ridges and troughs around high- and low-pressure systems. Areas of converging upper-air winds sustain surface highs, and areas of diverging upper-air winds sustain surface lows.

Vast wave motions in the upper-air westerlies are known as **Rossby waves**. Prominent streams of high-speed westerly winds in the upper-level troposphere are the **jet streams**. Depending on their latitudinal position in either hemisphere, they are termed the *polar jet stream* or the *subtropical jet stream*.

constant isobaric surface (p. 145)
Rossby waves (p. 149)
jet stream (p. 149)

15. What is the relation between wind speed and the spacing of isobars?
16. How is the constant isobaric surface, especially the ridges and troughs, related to surface-pressure systems? To divergence aloft and surface lows? To convergence aloft and surface highs?
17. Relate the jet-stream phenomenon to general upper-air circulation. How is the presence of this circulation related to airline schedules for the trip from New York to San Francisco and for the return trip to New York?

Explain the regional monsoons and several types of local winds.

Intense, seasonally shifting wind systems occur in the tropics over Southeast Asia, Indonesia, India, northern Australia, equatorial Africa, and southern Arizona. These winds are associated with an annual cycle of returning precipitation with the summer Sun and named using the Arabic word for season, *mausim*, or **monsoon**. The location and size of the Asian landmass and its proximity to the seasonally shifting ITCZ over the Indian Ocean drive the monsoons of southern and eastern Asia.

The difference in the heating characteristics of land and water surfaces creates **land and sea breezes**. Temperature differences during the day and evening between valleys and mountain summits cause **mountain and valley breezes**. **Katabatic winds**, or gravity drainage winds, are of larger regional scale and are usually stronger than valley and mountain breezes, under certain conditions. An elevated plateau or highland is essential, where layers of air at the surface cool, become denser, and flow downslope.

monsoon (p. 150)
land and sea breezes (p. 151)
mountain and valley breezes (p. 152)
katabatic wind (p. 152)

18. Describe the seasonal pressure patterns that produce the Asian monsoonal wind and precipitation patterns. Contrast January and July conditions.
19. People living along coastlines generally experience variations in winds from day to night. Explain the factors that produce these changing wind patterns.
20. The arrangement of mountains and nearby valleys produces local wind patterns. Explain the day and night winds that might develop.
21. This chapter presents wind-power technology as well developed and cost effective. Given the information presented and your additional critical thinking work, what conclusions have you reached?

Sketch the basic pattern of Earth's major surface ocean currents and deep thermohaline circulation.

Ocean currents are primarily caused by the frictional drag of wind and occur worldwide at varying intensities, temperatures, and speeds, both along the surface and at great depths in the oceanic basins. The circulation around subtropical high-pressure cells in both hemispheres is discernible on the ocean-circulation map—these gyres are usually offset toward the western side of each ocean basin.

The trade winds converge along the ITCZ and push enormous quantities of water that pile up along the eastern shore of continents in a process known as the **western intensification**. Where surface water is swept away from a coast, either by surface divergence (induced by the Coriolis force) or by offshore winds, an **upwelling current** occurs. This cool water generally is rich with nutrients and rises from great depths to replace the vacating water. In other oceanic regions where water accumulates, the excess water gravitates downward in a **downwelling current**. These currents generate vertical mixing of heat energy and salinity.

Differences in temperatures and salinity produce density differences important to the flow of deep, sometimes vertical, currents; this is Earth's **thermohaline circulation**. Traveling at slower speeds than wind-driven surface currents, the thermohaline circulation hauls larger volumes of water. Scientists are concerned that increased surface temperatures in the ocean and atmosphere, coupled with climate-related changes in salinity, can alter the rate of thermohaline circulation in the oceans.

western intensification (p. 154)
upwelling current (p. 154)
downwelling current (p. 154)
thermohaline circulation (p. 154)

22. Define the western intensification. How is it related to the Gulf Stream and the Kuroshio Current?
23. Where on Earth are upwelling currents experienced? What is the nature of these currents? Where are the four areas of downwelling that feed these dense bottom currents?
24. What is meant by deep-ocean thermohaline circulation? At what rates do these currents flow? How might this circulation be related to the Gulf Stream in the western Atlantic Ocean?
25. Relative to Question 24, what effects might climate change have on these deep currents?

■ *Summarize* several multiyear oscillations of air temperature, air pressure, and circulation associated with the Arctic, Atlantic, and Pacific Oceans.

Several system fluctuations that occur in multiyear or shorter periods are important in the global circulation picture. The most famous of these is the **El Niño–Southern Oscillation (ENSO)** in the Pacific Ocean, which affects interannual variability in climate on a global scale.

The *Pacific Decadal Oscillation (PDO)* is a pattern in which sea-surface temperatures and related air pressure vary back and forth between two regions of the Pacific Ocean: (1) the northern and tropical western Pacific and (2) the eastern tropical Pacific, along the U.S. West Coast. The PDO switches between positive and negative phases in 20- to 30-year cycles.

A north–south fluctuation of atmospheric variability marks the *North Atlantic Oscillation (NAO)*, in which pressure differences between the Icelandic Low and the Azores High in the Atlantic alternate between weaker and stronger pressure gradients. The *Arctic Oscillation (AO)* is the variable fluctuation between middle- and high-latitude air mass conditions over the Northern Hemisphere. The AO is associated with the NAO, especially in winter. During the winter of 2009–2010, the AO was at its most strongly negative phase since 1970.

El Niño–Southern Oscillation (ENSO) (p. 155)

26. Describe the changes in sea-surface temperatures and atmospheric pressure that occur during El Niño and La Niña, the warm and cool phases of the ENSO. What are some of the climatic effects that occur worldwide?
27. What is the relationship between the PDO and the strength of El Niño events? Between PDO phases and climate in the western United States?
28. What phases are identified for the NAO and AO? What winter weather conditions generally affect the eastern United States during each phase? What happened during the 2009–2010 winter season in the Northern Hemisphere?

Answer for Critical Thinking 6.2: The dry season in northern Australia is from about May through October, during the Southern Hemisphere winter. The southeast trades blow dry air from the Australian continent northward over the western Pacific toward Indonesia. High pressure is over Australia at this time (see the July pressure in Figure 6.10). The wet season is from about November to April, during the Australian summer (see the January rainfall pattern in Figure GIA 6.2). The ITCZ brings moist, warm air over northern Australia during this time. For more information, go to http://www.environment.gov.au/soe/2001/publications/theme-reports/atmosphere/atmosphere02-1.html and scroll down to "monsoon."

MasteringGeography™

Looking for additional review and test prep materials? Visit the Study Area in *MasteringGeography*™ to enhance your geographic literacy, spatial reasoning skills, and understanding of this chapter's content by accessing a variety of resources, including MapMaster interactive maps, geoscience animations, satellite loops, author notebooks, videos, RSS feeds, flashcards, web links, self-study quizzes, and an eText version of *Geosystems*.

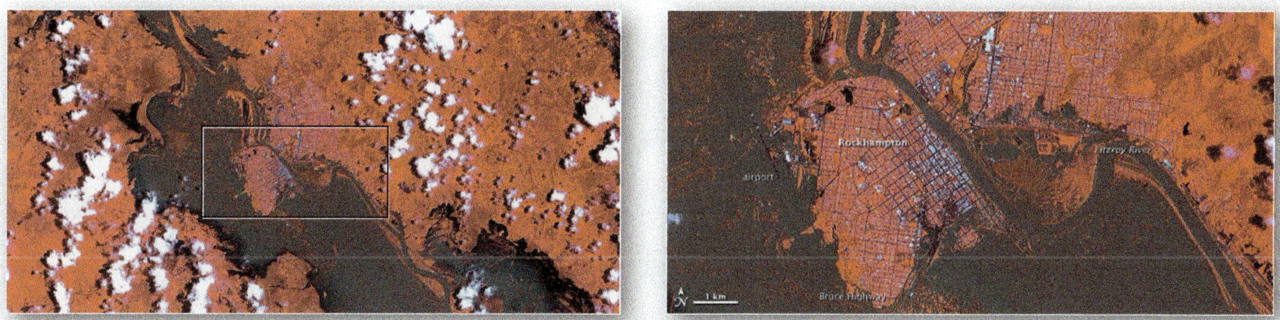

(a) The Fitzroy River surrounds the west portion of Rockhampton; population over 60,000.

(b) A closer view shows the city itself; the river flooded over 300 homes.

▲Figure 6.22 **Flooding on the Fitzroy River, Queensland, Australia during La Niña.** In January 2011, strong La Niña conditions on the Pacific caused weeks of heavy rains across Australia. These images combine infrared and visible light to increase the contrast between muddy floodwater and other surfaces. Buildings are white; vegetation is red. More on floods and river systems is in Chapter 14. [Terra-ASTER, NASA/GSFC/METI/ERSDAC/JAROS and US/Japan ASTER Science Team.]

II The Water, Weather, and Climate Systems

Earth is the water planet. Chapter 7 describes the remarkable qualities and properties water possesses. It also examines the daily dynamics of the atmosphere—the powerful interaction of moisture and energy in the form of latent heat, the resulting stability and instability of atmospheric conditions, and the varieties of cloud forms—all important to understanding weather. Chapter 8 examines weather and its causes. Topics include the interactions of air masses, features of the daily weather map, and analysis of violent phenomena and recent trends in thunderstorms, tornadoes, and hurricanes.

▼ Ocean waves at sunset. [Photoff/Shutterstock.]

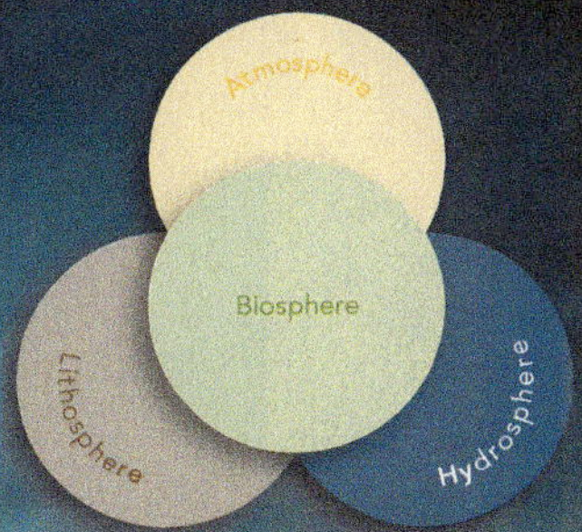

Chapter 9 describes the distribution of water on Earth, including water circulation in the hydrologic cycle; these water resources are an output of the water–weather system. We examine the water-budget concept, which is useful in understanding soil-moisture and water-resource relationships on global, regional, and local scales. Potable (drinking) water is emerging as the critical global political issue this century. In Chapter 10, we see the spatial implications over time of the energy–atmosphere and water–weather systems and the output of Earth's climatic patterns. These observations interconnect all the system elements from Chapters 2 through 9. PART II closes with a discussion of global climate change science, a look at present conditions, and a forecast of future climate trends.

7 Water and Atmospheric Moisture

Coastal fog covers part of a redwood forest in Humboldt Redwoods State Park, California. Coast redwoods (*Sequoia sempervirens*) depend on summer fog as moisture source. As sea surface temperatures warm and circulation patterns shift, summer fog is declining along the California coast, with potentially negative effects on tree health. *Geosystems Now* discusses fog and redwoods, part of the focus on water and atmospheric moisture in this chapter. [Michael Nichols/National Geographic.]

KEY LEARNING concepts

After reading the chapter, you should be able to:

- ***Describe*** the heat properties of water, and ***identify*** the traits of its three phases: solid, liquid, and gas.
- ***Define*** humidity and relative humidity, and ***explain*** dew-point temperature and saturated conditions in the atmosphere.
- ***Define*** atmospheric stability, and ***relate*** it to a parcel of air that is ascending or descending.
- ***Illustrate*** three atmospheric conditions—unstable, conditionally unstable, and stable—with a simple graph that relates the environmental lapse rate (ELR) to the dry adiabatic rate (DAR) and moist adiabatic rate (MAR).
- ***Identify*** the requirements for cloud formation, and ***explain*** the major cloud classes and types, including fog.

GEOSYSTEMS now

Summer Fog Protects the World's Tallest Trees

Along the Pacific coast of California and extreme southern Oregon live the world's tallest trees, the coast redwoods (*Sequoia sempervirens*), which can reach over 91 m (300 ft) in height. These trees occupy a coastal habitat that is often enshrouded in summer fog, as low clouds form over the cold waters flowing offshore in the California Current (Figure GN 7.1). Local winds carry this moist air toward the warmer continental interior, pushing it across the Coast Ranges where the trees live. This moist air provides a cool, humid summer climate for the coast redwoods, which have evolved to depend on this seasonal moisture.

Decline in Summer Fog Recently, University of California, Berkeley, researchers examined a variety of factors affecting summer fog along California's coast. The scientists found, first, that the average number of hours of daily fog in summer has dropped by about 14% since 1901, which amounts to about 3 hours per day less fog. Second, they found that frequent coastal fog was almost always associated with a large temperature gradient between coastal and inland areas, and that this gradient has decreased since 1950, not just in California but along the entire coast from Seattle to San Diego. One comparison between temperature stations showed that a coastal–inland temperature difference of 9.5 C° (17 F°) at the beginning of this century had dropped to a 6.1 C° (11 F°) difference in 2008. Overall, the study found a 33% reduction in fog frequency within the range of the coast redwoods since the early 20th century.

Decreased coastal fog is linked to warmer sea-surface temperatures, which in turn are related to the Pacific Decadal Oscillation, a pattern of shifting pressure and wind over the northern Pacific Ocean (discussed in Chapter 6). In addition, the scientists linked fog frequency to the presence of a temperature inversion that traps fog between the coast and the coastal mountains during the summer—a stronger inversion keeps the fog from passing across the Coast Ranges into the interior. This temperature inversion has weakened with the decline in the coastal–inland temperature gradient, causing warmer and drier coastal conditions and allowing cooler air to move inland.

▲**Figure GN 7.1** **Coast redwood distribution.** [USGS.]

Fog Drip and Redwood Ecology Because coast redwoods occupy a narrow range of coastal habitat about 8 to 56 km (5 to 35 mi) in width with consistent climatic conditions, these trees have specific temperature and humidity requirements. As fog moves into the forest from the coast, water droplets accumulate on branches and needles, eventually falling onto the vegetation and ground beneath (Figure GN 7.2). This "fog drip" reduces temperature and raises humidity in the forest, an important process during periods when temperatures are higher and rainfall scant (as during California's Mediterranean dry-summer climate, discussed in Chapter 10). Redwood roots, which may spread 24 m (80 ft) out from the tree's base, absorb about 35% of the fog drip, drawing moisture up into the tree's internal water system.

▲**Figure GN 7.2** **California coast redwoods in fog.** [NPS.]

Fog Helps Redwoods Conserve Water Redwoods depend on humid conditions to properly regulate their water use. Redwood needle leaves, like other plant leaves, have special openings called stomata through which the tree loses water to the atmosphere by *transpiration* (a process that is similar to evaporation). Stomata also provide an entryway for carbon dioxide, which is converted to food energy by photosynthesis during the day. Redwoods are unusual in that their stomata remain open at night as well as during the day, making them vulnerable to excessive transpiration during hot, dry conditions.

As in previous studies, the Berkeley scientists found significantly reduced redwood transpiration rates in the presence of summer fog, indicating the importance of fog as an atmospheric mechanism for water conservation in these trees. A decline in this summer moisture increases drought stress, with unknown consequences for tree health.

The intricate relationship between coast redwood trees and summer fog decline is one example of the effects of "Water and Atmospheric Moisture" on Earth systems, the subject of Chapter 7.

GEOSYSTEMS NOW ONLINE Go to Chapter 7 on the *MasteringGeography* website (www.masteringgeography.com) for more information about Pacific coastal fog and its environmental effects. For more about coast redwood tree ecology, go to http://www.nps.gov/redw/naturescience/about-the-trees.htm.

Water is everywhere in the atmosphere, in visible forms (clouds, fog, and precipitation) and in microscopic forms (water vapor). Out of all the water present in Earth systems, less than 0.03%, some 12,900 km^3 (3095 mi^3), is stored in the atmosphere. If this amount were to fall to Earth as rain, it would cover the surface to a depth of only 2.5 cm (1.0 in). However, the atmosphere is a key pathway for the movement of water around the globe; in fact, some 495,000 km^3 (118,757 mi^3) are cycled through the atmosphere each year.

Water is an extraordinary compound. It is the only common substance that naturally occurs in all three states of matter: liquid, solid, and vapor. When water changes from one state of matter to another (as from liquid to gas or solid), the heat energy absorbed or released helps power the general circulation of the atmosphere, which drives daily weather patterns.

This chapter begins our study of the *hydrologic cycle*, or *water cycle*. This cycle includes the movement of water throughout the atmosphere, hydrosphere, lithosphere, and biosphere. We discuss the surface and subsurface components of the cycle in Chapter 9, and they are summarized in Figure 9.1. Although the hydrologic cycle forms a continuous loop, its description often begins with the movement of water through the atmosphere as evaporation from oceans, and includes the formation of clouds and precipitation over land and water. These are the processes that power weather systems on Earth.

Water vapor in the air affects humans in numerous ways. We examined the heat index, a summation of the effects of heat and humidity that cause heat stress, in Chapter 5. Water vapor forms clouds, which affect the Earth's energy balance. As discussed in Chapter 4, clouds have a warming effect, seen by comparing humid and dry regions over the world. In humid areas, such as the equatorial regions, clouds reradiate outgoing longwave energy so that nighttime cooling is less. In dry regions with few clouds, such as in the areas of the subtropical highs, nighttime cooling is more significant.

In this chapter: We examine the dynamics of atmospheric moisture, beginning with the properties of water in all its states—frozen ice, liquid water, and vapor in the air. We discuss humidity, its daily patterns, and the instruments that measure it. We then study adiabatic processes and atmospheric conditions of stability and instability, and relate them to cloud development. We end the chapter with a look at the processes that form clouds, perhaps our most beautiful indicators of atmospheric conditions, and fog.

▲**Figure 7.1 Surface tension of water.** Surface tension causes water to form beads on a plant leaf. [céline kriébus/Fotolia.]

Water's Unique Properties

Earth's distance from the Sun places it within a most remarkable temperate zone when compared with the locations of the other planets. This temperate location allows all three states of water—ice, liquid, and vapor—to occur naturally on Earth. Two atoms of hydrogen and one of oxygen, which readily bond, make up each water molecule. Once the hydrogen and oxygen atoms join in covalent bonds, they are difficult to separate, thereby producing a water molecule that remains stable in Earth's environment.

The nature of the hydrogen–oxygen bond gives the hydrogen side of a water molecule a positive charge and the oxygen side a negative charge. As a result of this *polarity*, water molecules attract each other: The positive (hydrogen) side of a water molecule attracts the negative (oxygen) side of another—an interaction called *hydrogen bonding*. The polarity of water molecules also explains why water is able to dissolve many substances; pure water is rare in nature because of its ability to dissolve other substances within it.

The effects of hydrogen bonding in water are observable in everyday life, creating the *surface tension* that allows a steel needle to float lengthwise on the surface of water, even though steel is much denser than water. This *surface tension* allows you to slightly overfill a glass with water; webs of millions of hydrogen bonds hold the water slightly above the rim (Figure 7.1).

Hydrogen bonding is also the cause of *capillarity*, which you observe when you "dry" something with a paper towel. The towel draws water through its fibers because hydrogen bonds make each molecule pull on its neighbor. In chemistry laboratory classes, students observe the concave *meniscus*, or inwardly curved surface of the water, which forms in a cylinder or a test tube because hydrogen bonding allows the water to slightly "climb" the glass sides. *Capillary action* is an important component of soil-moisture processes, discussed in Chapter 18. Without hydrogen bonding to make the molecules in water and ice attract each other, water would be a gas at normal surface temperatures.

Phase Changes and Heat Exchange

For water to change from one state to another, heat energy must be added to it or released from it. The amount of heat energy absorbed or released must be sufficient to affect the hydrogen bonds between molecules. This relation between water and heat energy is important to atmospheric processes. In fact, the heat exchanged between physical states of water provides more than 30% of the energy that powers the general circulation of the atmosphere.

Figure 7.2 presents the three states of water and the terms describing each change between states, known as a **phase change**. *Melting* and *freezing* describe the familiar phase changes between solid and liquid. *Condensation* is the process through which water vapor in the air becomes liquid water—this is the process that forms clouds. *Evaporation* is the process through which liquid water becomes water vapor—the cooling process discussed in Chapter 4. This phase change is called *vaporization* when water is at boiling temperature.

The phase changes between solid ice and gaseous water vapor may be less familiar. *Deposition* is the process through which water vapor attaches directly to an ice crystal, leading to the formation of *frost*. You may have seen this on your windows or car windshield on a cold morning. It also occurs inside your freezer. **Sublimation** is the process by which ice changes directly to water vapor. A classic sublimation example is the water-vapor clouds associated with the vaporization of dry ice (frozen carbon dioxide) when it is exposed to air. Sublimation is an important contributor to the shrinking of snowpacks in dry, windy environments. The warm chinook winds that blow downslope on the lee side of the Rocky Mountains in the western United States, as well as the *fóhn*, or *foehn*, winds in Europe, are known as "snow-eaters" for their ability to vaporize snow rapidly by sublimation (Figure 7.3).

Ice, the Solid Phase As water cools from room temperature, it behaves like most compounds and contracts in volume. At the same time, it increases in density, as the same number of molecules now occupy a smaller space. When other liquids cool, they congeal into the solid state by the time they reach their greatest density. However, when water has cooled to the point of greatest density, at 4°C (39°F), it is still in a liquid state. Below a temperature of 4°C (39°F), water behaves differently from other compounds. Continued cooling makes it expand as more hydrogen bonds form among the slowing molecules, creating the hexagonal (six-sided) crystalline structure characteristic of ice (see Figure 7.2c).

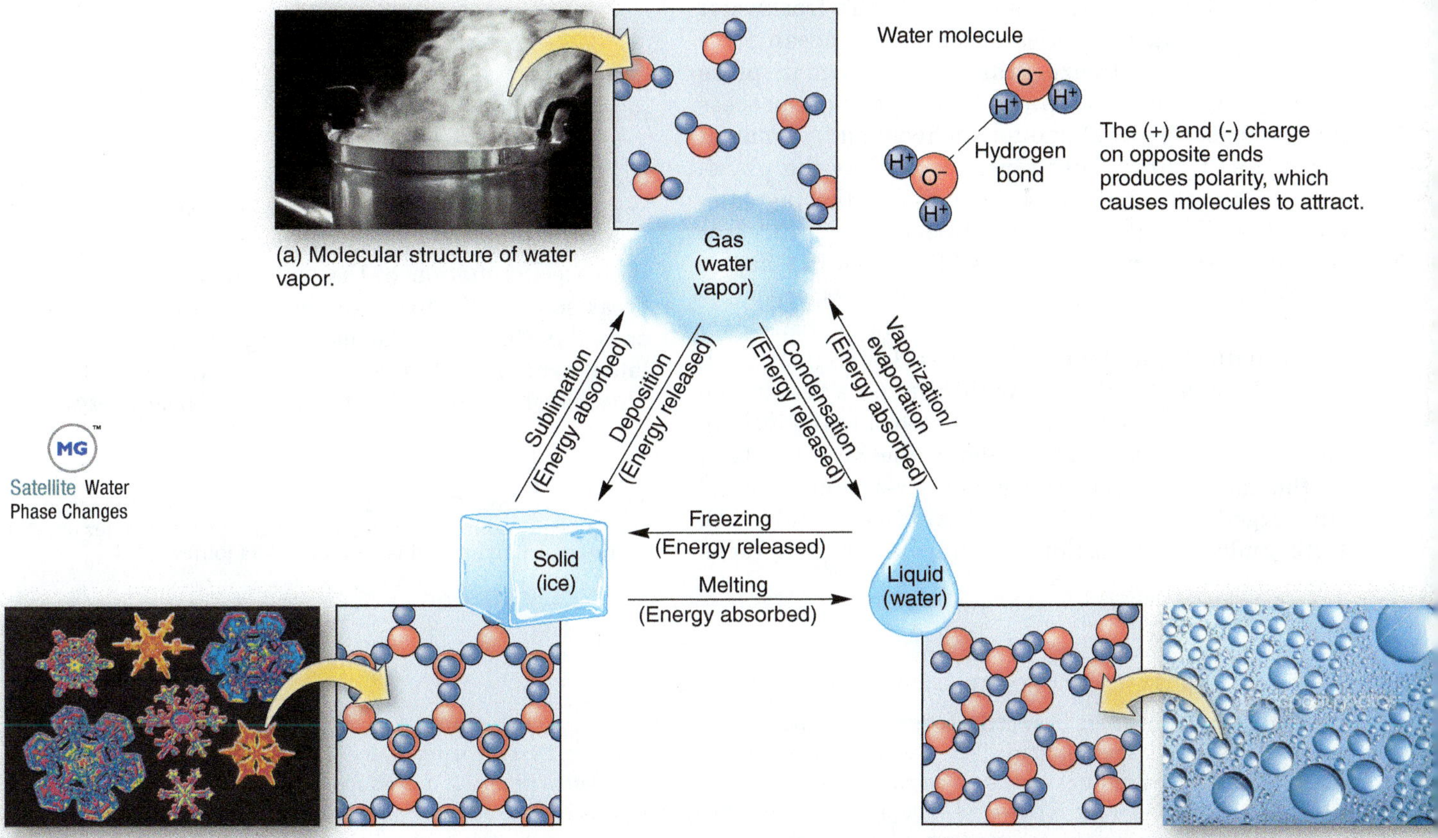

MG™ Satellite Water Phase Changes

▲**Figure 7.2 Three physical states of water and phase changes between them.** [(a) toa555/Fotolia. (b) Olga Miltsova/Shutterstock. (c) Photo enhancement © Scott Camazine/Photo Researchers, Inc., after W. A. Bentley.]

▲Figure 7.3 **Sublimation of a mountain snowpack.** Blowing snow enhances sublimation in the Canadian Rockies near Lake Louise, Banff National Park, Alberta. [Mick House/Alamy]

▲Figure 7.4 **The buoyancy of ice.** A bergy bit (small iceberg) off the coast of Antarctica illustrates the reduced density of ice compared to cold water. [Bobbé Christopherson.]

This six-sided preference applies to ice crystals of all shapes: plates, columns, needles, and dendrites (branching or treelike forms). Ice crystals demonstrate a unique interaction of chaos (all ice crystals are different) and the determinism of physical principles (all have a six-sided structure).

As temperatures descend further below freezing, ice continues to expand in volume and decrease in density to a temperature of −29°C (−20°F)—up to a 9% increase in volume is possible. Pure ice has 0.91 times the density of water, so it floats. Without this unusual pattern of density change, much of Earth's freshwater would be bound in masses of ice on the ocean floor (the water would freeze, sink, and remain in place forever). At the same time, the expansion process just described is to blame for highway and pavement damage and burst water pipes, as well as the physical breakdown of rocks known as weathering (discussed in Chapter 12) and the freeze–thaw processes that affect soils in cold regions (discussed in Chapter 17). For more on ice crystals and snowflakes, see http://www.its.caltech.edu/~atomic/snowcrystals/primer/primer.htm.

In nature, the density of ice varies slightly with age and the air contained within it. As a result, the amount of water that is displaced by a floating iceberg varies, with an average of about one-seventh (14%) of the mass exposed and about six-sevenths (86%) submerged beneath the ocean's surface (Figure 7.4). With underwater portions melting faster than those above water, icebergs are inherently unstable and will overturn.

CRITICALthinking 7.1 Iceberg Analysis

Examine the iceberg in Antarctic waters in Figure 7.4. In a clear glass almost filled with water, place an ice cube. Then approximate the amount of ice above the water's surface compared to the amount below water. Pure ice is 0.91 the density of water; however, because ice usually contains air bubbles, most icebergs are about 0.86 the density of water. How do your measurements compare with this range of density? ●

Water, the Liquid Phase Water, as a liquid, is a noncompressible fluid that assumes the shape of its container. For ice to change to water, heat energy must increase the motion of the water molecules enough to break some of the hydrogen bonds (Figure 7.2b). As discussed in Chapter 4, the heat energy of a phase change is **latent heat** and is hidden within the structure of water's physical state. In total, 80 calories* of heat energy must

*Remember, from Chapter 2, that a calorie (cal) is the amount of energy required to raise the temperature of 1 g of water (at 15°C) by 1 degree Celsius and is equal to 4.184 joules.

GEOreport 7.1 Breaking roads and pipes

Road crews are busy in the summer repairing winter damage to streets and freeways in regions where winters are cold. Rainwater seeps into roadway cracks and then expands as it freezes, thus breaking up the pavement. Perhaps you have noticed that bridges suffer the greatest damage; cold air can circulate beneath a bridge and produce more freeze–thaw cycles. The expansion of freezing water is powerful enough to crack plumbing or an automobile radiator or engine block. Wrapping water pipes with insulation to avoid damage is a common winter task in many places. Historically, this physical property of water was put to use in quarrying rock for building materials. Holes were drilled and filled with water before winter so that, when cold weather arrived, the water would freeze and expand, cracking the rock into manageable shapes.

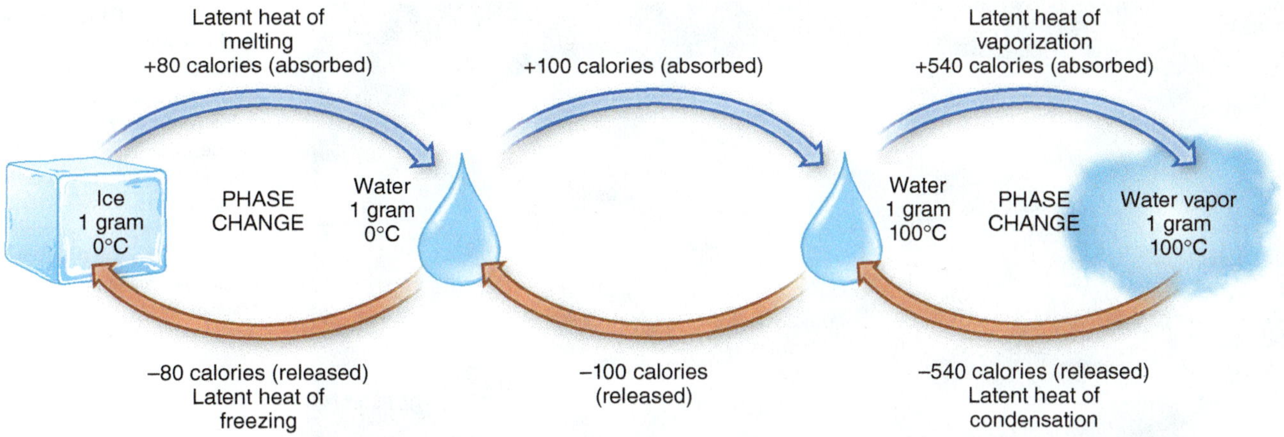

(a) Latent heat absorbed or released in phase changes between ice and water and water vapor. To transform 1 g of ice at 0° C to 1 g of water vapor at 100° C requires 720 cal: 80 + 100 + 540.

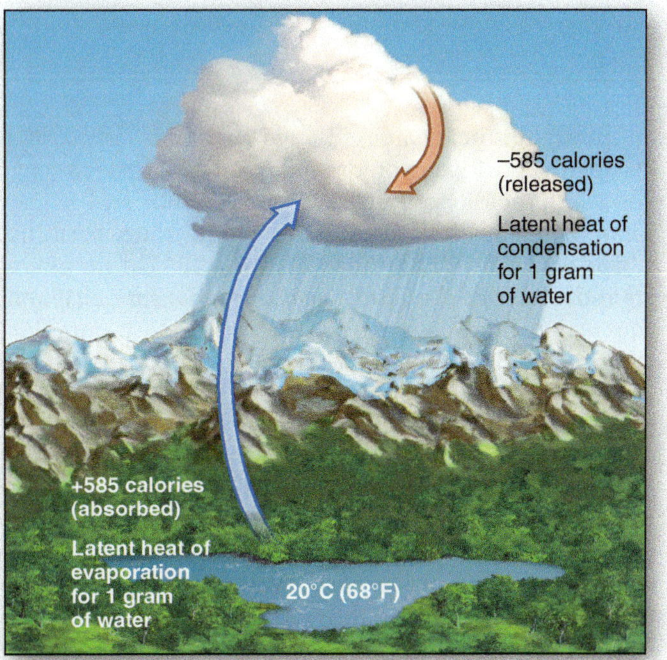

(b) Latent heat exchange between water in a lake at 20° C and water vapor in the atmosphere, under typical conditions.

▲**Figure 7.5 Water's heat-energy characteristics.**

MG Animation Water Phase Changes

be absorbed for the phase change of 1 g of ice melting to 1 g of water—this latent heat transfer occurs despite the fact that the sensible temperature remains the same: both ice and water measure 0°C (32°F; Figure 7.5a). When the phase change is reversed and a gram of water freezes, latent heat is released rather than absorbed. The *latent heat of melting* and the *latent heat of freezing* are each 80 cal/g.

To raise the temperature of 1 g of water at 0°C (32°F) to boiling at 100°C (212°F), we must add 100 calories (an increase of 1 C°, or 1.8 F°, for each calorie added). No phase change is involved in this temperature gain.

Water Vapor, the Gas Phase Water vapor is an invisible and compressible gas in which each molecule moves independently of the others (Figure 7.2a). When the phase change from liquid to vapor is induced by boiling, it requires the addition of 540 cal for each gram, under normal sea-level pressure; this amount of energy is the **latent heat of vaporization** (Figure 7.5). When water vapor condenses to a liquid, each gram gives up its hidden 540 cal as the **latent heat of condensation**. We see water vapor in the atmosphere after condensation has occurred, in the form of clouds, fog, and steam. Perhaps you have felt the release of the latent heat of condensation on your skin from steam when you drained steamed vegetables or pasta or filled a hot teakettle.

In summary, the changing of 1 g of ice at 0°C to water and then to water vapor at 100°C—from a solid to a liquid to a gas—*absorbs* 720 cal (80 cal + 100 cal + 540 cal). Reversing the process, or changing the phase of 1 g of water vapor at 100°C to water and then to ice at 0°C, *releases* 720 cal into the surrounding environment. Go to the *MasteringGeography* website for an excellent animation illustrating these concepts.

The **latent heat of sublimation** absorbs 680 cal as a gram of ice transforms into vapor. Water vapor freezing directly to ice releases a comparable amount of energy.

Latent Heat Transfer Under Natural Conditions

In a lake or stream or in soil water, at 20°C (68°F), every gram of water that breaks away from the surface through evaporation must absorb from the environment approximately 585 cal as the *latent heat of evaporation* (see Figure 7.5b). This is slightly more energy than would be required if the water were at a higher temperature (if the water is boiling, 540 cal are required). You can feel this absorption of latent heat as evaporative cooling on your skin when it is wet. This latent heat exchange is the dominant cooling process in Earth's energy budget. (Remember from Chapter 4 that the latent heat of evaporation is

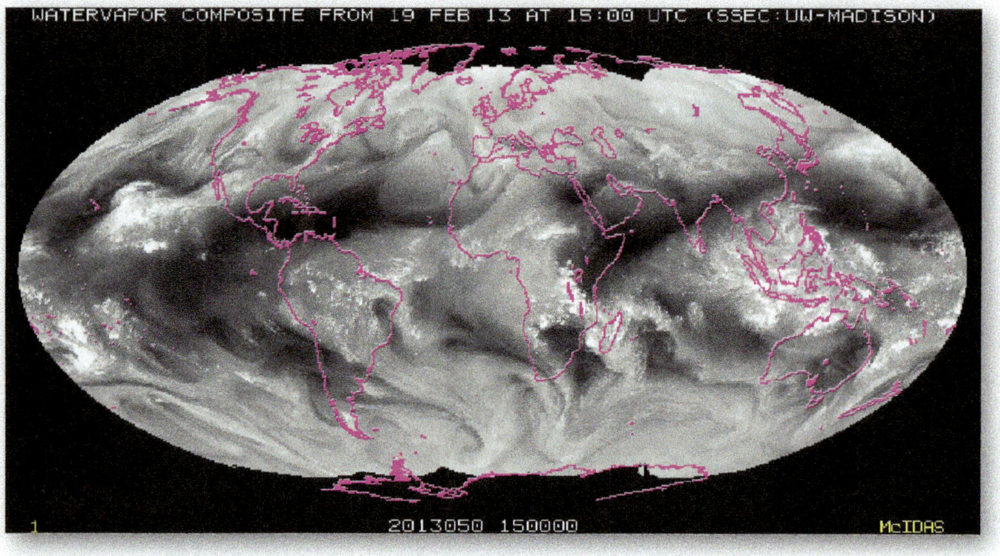

◀**Figure 7.6 Global water vapor in the atmosphere.** High water-vapor content is lighter and lower water-vapor content is darker in this February 19, 2013, composite image from the *GOES* (United States), *Meteosat* (European Space Agency), and *MTSAT* (Japan) satellites. [Satellite data courtesy of Space Science and Engineering Center, University of Wisconsin, Madison.]

the most significant nonradiative heat transfer process dissipating positive net radiation at Earth's surface.)

The process reverses when air cools and water vapor condenses back into the liquid state, forming moisture droplets and thus liberating 585 cal for every gram of water as the *latent heat of condensation*. When you realize that a small, puffy, fair-weather cumulus cloud holds 500–1000 tons of moisture droplets, think of the tremendous latent heat released when water vapor condenses to droplets.

Satellites using infrared sensors now routinely monitor water vapor in the lower atmosphere. Water vapor absorbs long wavelengths (infrared), making it possible to distinguish areas of relatively high water vapor from areas of low water vapor (Figure 7.6). This technology is important to weather forecasting because it shows available moisture in the atmosphere and therefore the available latent heat energy and precipitation potential.

Water vapor is also an important greenhouse gas. However, it is different from other greenhouse gases in that its concentration is tied to temperature. As global temperatures rise, evaporation increases from lakes, oceans, soils, and plants, amplifying the concentration of water vapor in the atmosphere and strengthening the greenhouse effect over Earth. Scientists have already observed an increase in average global water vapor in recent decades and project a 7% increase for every 1 C° of warming in the future. As water vapor increases, precipitation patterns will change and the amount of rainfall will likely increase during the heaviest precipitation events.

Humidity

The amount of water vapor in the air is **humidity**. The capacity of air for water vapor is primarily a function of the temperatures of both the air and the water vapor, which are usually the same.

As discussed in Chapter 5, humidity and air temperatures determine our sense of comfort. North Americans spend billions of dollars a year to adjust the humidity in buildings, either with air conditioners, which remove water vapor as they cool building interiors, or with air humidifiers, which add water vapor to lessen the drying effects of cold temperatures and dry climates. We also saw the relationship between heat and humidity, and its effects on humans, in our discussion of the heat index in Chapter 5.

Relative Humidity

The most common measure of humidity in weather reports is **relative humidity**, a ratio (expressed as a percentage) of the amount of water vapor that is actually in the air compared to the maximum water vapor possible in the air at a given temperature.

Relative humidity varies because of water vapor or temperature changes in the air. The formula to calculate

GEOreport 7.2 Katrina had the power

Meteorologists estimated that the moisture in Hurricane Katrina (2005) weighed more than 30 trillion tons at its maximum power and mass. With about 585 cal released for every gram as the latent heat of condensation, a weather event such as a hurricane involves a staggering amount of energy. Do the quick math in English units (585 calories times 28 g to an ounce, times 16 ounces to the pound, times 2000 pounds to the ton, times 30 trillion tons).

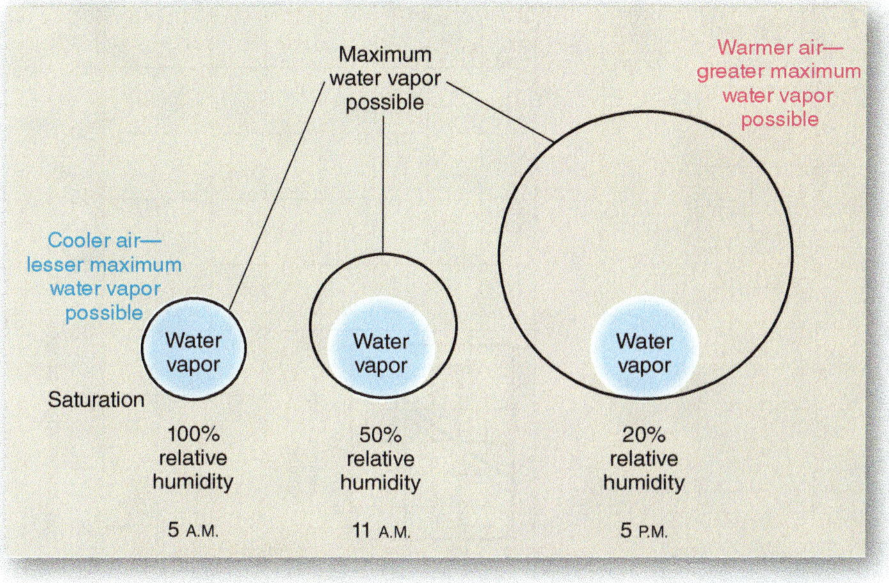

▲Figure 7.7 **Water vapor, temperature, and relative humidity.** The maximum water vapor possible in warm air is greater (net evaporation more likely) than that possible in cold air (net condensation more likely), so relative humidity changes with temperature, even though in this example the actual water vapor present in the air stays the same during the day.

the relative humidity ratio and express it as a percentage places actual water vapor in the air as the numerator and water vapor possible in the air at that temperature as the denominator:

$$\text{Relative humidity} = \frac{\text{Actual water vapor in the air}}{\text{Maximum water vapor possible in the air at that temperature}} \times 100$$

Warmer air increases the evaporation rate from water surfaces, whereas cooler air tends to increase the condensation rate of water vapor onto water surfaces. Because there is a maximum amount of water vapor that can exist in a volume of air at a given temperature, the rates of evaporation and condensation can reach equilibrium at some point; the air is then saturated, and the balance is *saturation equilibrium.*

Figure 7.7 shows changes in relative humidity throughout a typical day. At 5 A.M., in the cool morning air, saturation equilibrium exists, and any further cooling or addition of water vapor produces net condensation. When the air is saturated with maximum water vapor for its temperature, the relative humidity is 100%. At 11 A.M., the air temperature is rising, so the evaporation rate exceeds the condensation rate; as a result, the same volume of water vapor now occupies only 50% of the maximum possible capacity. At 5 P.M., the air temperature is just past its daily peak, so the evaporation rate exceeds condensation by an even greater amount, and relative humidity is at 20%.

Saturation and Dew Point

Relative humidity tells us how near the air is to saturation and is an expression of an ongoing process of water molecules moving between air and moist surfaces. At **saturation**, or 100% relative humidity, any further addition of water vapor or any decrease in temperature that reduces the evaporation rate results in active condensation (forming clouds, fog, or precipitation).

The temperature at which a given sample of vapor-containing air becomes saturated and net condensation begins to form water droplets is the **dew-point temperature.** The air is saturated when the dew-point temperature and the air temperature are the same. When temperatures are below freezing, the *frost point* is the temperature at which the air becomes saturated, leading to the formation of frost (ice) on exposed surfaces.

A cold drink in a glass provides a familiar example of these conditions (Figure 7.8a). The air near the glass chills to the dew-point temperature

(a) When the air reaches the dew-point temperature, water vapor condenses out of the air and onto the glass as dew.

(b) Cold air above the rain-soaked rocks is at the dew point and is saturated. Water evaporates from the rock into the air and condenses in a changing veil of clouds.

▲Figure 7.8 **Dew-point temperature examples.** [Author.]

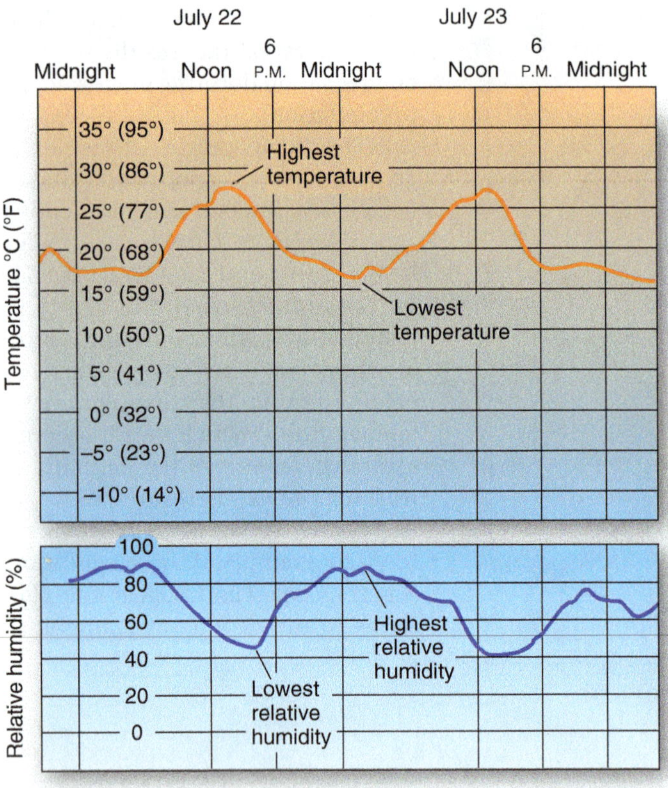

▲Figure 7.9 **Daily relative humidity patterns.** Typical daily variations demonstrate temperature and relative humidity relations.

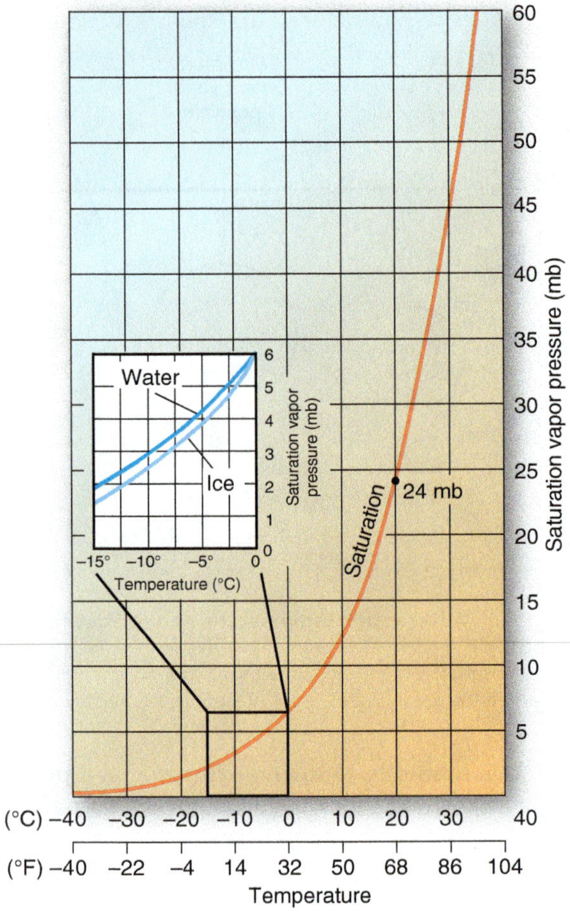

▲Figure 7.10 **Saturation vapor pressure at various temperatures.** Saturation vapor pressure is the maximum possible water vapor, as measured by the pressure it exerts (mb). Inset compares saturation vapor pressures over water surfaces with those over surfaces at subfreezing temperatures. Note the point indicating 24 mb, discussed in the text.

and becomes saturated, causing the water vapor in that cooled air to form water droplets on the outside of the glass. Figure 7.8b shows active condensation in the saturated air above a cool, wet rock surface. As you walk to classes on some cool mornings, you perhaps notice damp lawns or dew on windshields, an indication of dew-point temperature conditions.

Daily and Seasonal Relative Humidity Patterns An inverse relation occurs during a typical day between air temperature and relative humidity—as temperature rises, relative humidity falls (Figure 7.9). Relative humidity is highest at dawn, when air temperature is lowest. If you park outdoors, you know about the wetness of the dew that condenses on your car or bicycle overnight. You have probably also noticed that the morning dew on windows, cars, and lawns evaporates by late morning as net evaporation increases with air temperature.

Relative humidity is lowest in the late afternoon, when higher temperatures increase the rate of evaporation. As shown in Figure 7.7, the actual water vapor present in the air may remain the same throughout the day. However, relative humidity changes because the temperature, and therefore the rate of evaporation, varies from morning to afternoon. Seasonally, January readings are higher than July readings because air temperatures are lower overall in winter. Humidity data from most weather stations demonstrate this seasonal relationship.

Specialized Expressions of Humidity

There are several specific ways to express humidity and relative humidity. Each has its own utility and application. Two examples are vapor pressure and specific humidity.

Vapor Pressure As free water molecules evaporate from surfaces into the atmosphere, they become water vapor. Now part of the air, water-vapor molecules exert a portion of the air pressure along with nitrogen and oxygen molecules. The share of air pressure that is made up of water-vapor molecules is **vapor pressure**, expressed in millibars (mb).

Air that contains as much water vapor as possible at a given temperature is at *saturation vapor pressure*. Any temperature increase or decrease will change the saturation vapor pressure.

Figure 7.10 graphs the saturation vapor pressure at various air temperatures. For every temperature increase of 10 C° (18 F°), the saturation vapor pressure in air nearly doubles. This relationship explains why warm tropical air over the ocean can contain so much water vapor, thus providing much latent heat to power tropical storms. It also explains why cold air is "dry" and why cold air toward the poles does not produce a lot of precipitation (it contains too little water vapor, even though it is near the dew-point temperature).

As marked on the graph, air at 20°C (68°F) has a saturation vapor pressure of 24 mb; that is, the air is saturated if the water-vapor portion of the air pressure also is at 24 mb. Thus, if the water vapor actually present is exerting a vapor pressure of only 12 mb in 20°C air, the relative humidity is 50% (12 mb ÷ 24 mb = 0.50 × 100 = 50%). The inset in Figure 7.10 compares saturation vapor pressure over water and over ice surfaces at subfreezing temperatures. You can see that saturation vapor pressure is greater above a water surface than over an ice surface—that is, it takes more water-vapor molecules to saturate air above water than it does above ice. This fact is important to condensation processes and rain-droplet formation, both of which are described later in this chapter.

Specific Humidity A useful humidity measure is one that remains constant as temperature and pressure change. **Specific humidity** is the mass of water vapor (in grams) per mass of air (in kilograms) at any specified temperature. Because it is measured in mass, specific humidity is not affected by changes in temperature or pressure, as occur when air rises to higher elevations. Specific humidity stays constant despite volume changes.*

The maximum mass of water vapor possible in a kilogram of air at any specified temperature is the *maximum specific humidity*, plotted in Figure 7.11. Noted on the graph is that a kilogram of air could hold a maximum specific humidity of 47 g of water vapor at 40°C (104°F), 15 g at 20°C (68°F), and about 4 g at 0°C (32°F). Therefore, if a kilogram of air at 40°C has a specific humidity of 12 g, its relative humidity is 25.5% (12 g ÷ 47 g = 0.255 × 100 × 25.5%). Specific humidity is useful in describing the moisture content of large air masses that are interacting in a weather system and provides information necessary for weather forecasting.

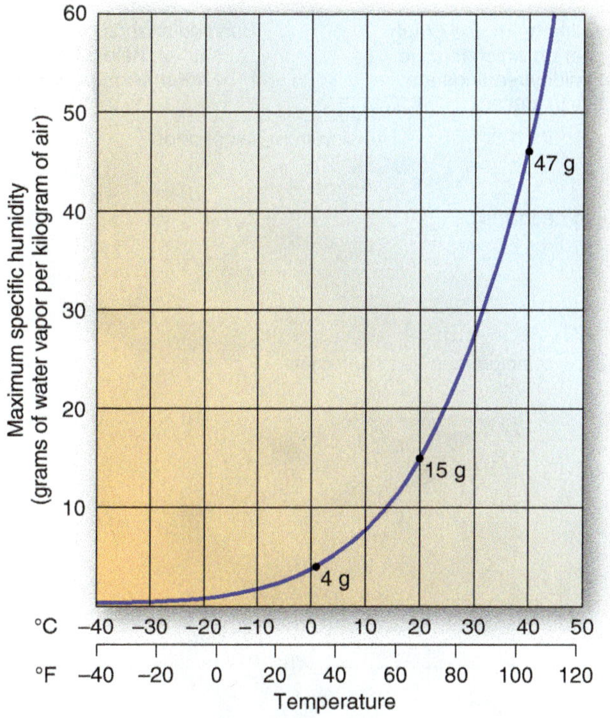

▲**Figure 7.11 Maximum specific humidity at various temperatures.** Maximum specific humidity is the maximum possible water vapor in a mass of water vapor per unit mass of air (g/kg). Note the points corresponding to 47 g, 15 g, and 4 g mentioned in the text discussion.

Instruments for Measuring Humidity

Various instruments measure relative humidity. The **hair hygrometer** uses the principle that human hair changes as much as 4% in length between 0% and 100% relative humidity. The instrument connects a standardized bundle of human hair through a mechanism to a gauge. As the hair absorbs or loses water in the air, it changes length, indicating relative humidity (Figure 7.12a).

Another instrument used to measure relative humidity is a **sling psychrometer**, which has two thermometers mounted side by side on a holder (Figure 7.12b). One is the *dry-bulb thermometer*; it simply records the ambient (surrounding) air temperature. The other thermometer is the *wet-bulb thermometer*; it is set lower in the holder, and the bulb is covered by a moistened cloth wick. The psychrometer is then spun, or "slung," by its handle or placed where a fan forces air over the wet-bulb wick. After several minutes of spinning, the temperatures on each bulb are compared with a relative humidity (psychrometric) chart to find the relative humidity.

The rate at which water evaporates from the wick depends on the relative saturation of the surrounding air. If the air is dry, water evaporates quickly from the wet-bulb thermometer and its wick, cooling the thermometer and causing its temperature to lower (the *wet-bulb depression*). In conditions of high humidity, little water evaporates from the wick; in low humidity, more water evaporates.

CRITICALthinking 7.2
Using Relative Humidity and Dew-Point Maps

Please refer to our discussion of relative humidity, air temperature, dew-point temperature, and saturation on the *MasteringGeography* website. Examine the dew-point temperature map in relation to the map of air temperatures. As you compare and contrast the maps, describe in general terms what relative humidity conditions you find in the Northwest. How about along the Gulf Coast? How does this contrast with the Southeast and with the Southwest? Remember, the closer the air temperature is to the dew-point temperature, then the closer the air is to saturation and the possibility of condensation. ●

*Another similar measure used for relative humidity that approximates specific humidity is the *mixing ratio*—that is, the ratio of the mass of water vapor (grams) per mass of dry air (kilograms), as in g/kg.

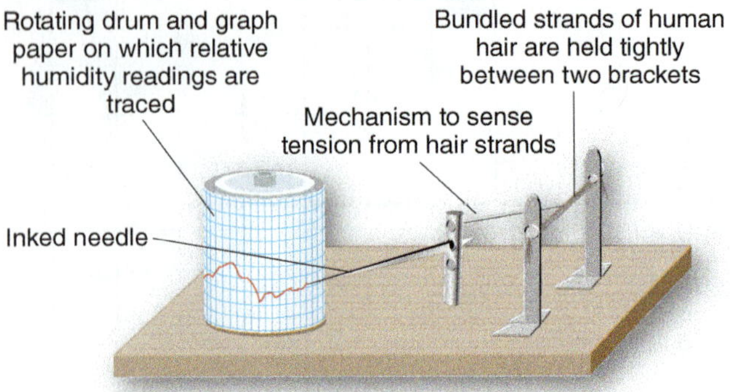

(a) The principle of a hair hygrometer.

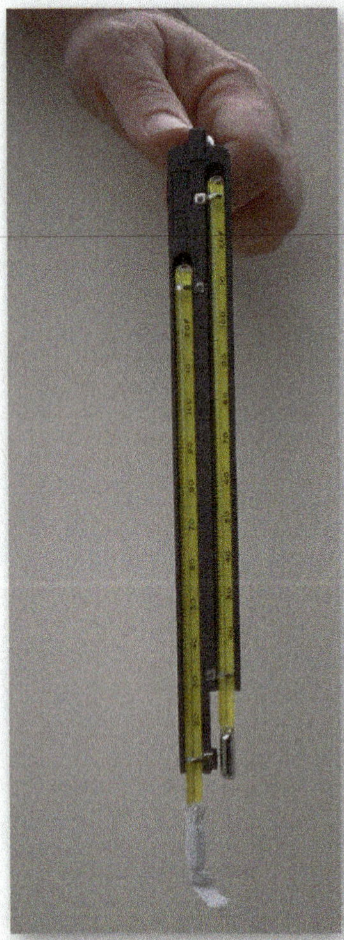

(b) Sling psychrometer with wet and dry bulbs.

▲**Figure 7.12 Instruments that measure relative humidity.** [Bobbé Christopherson.]

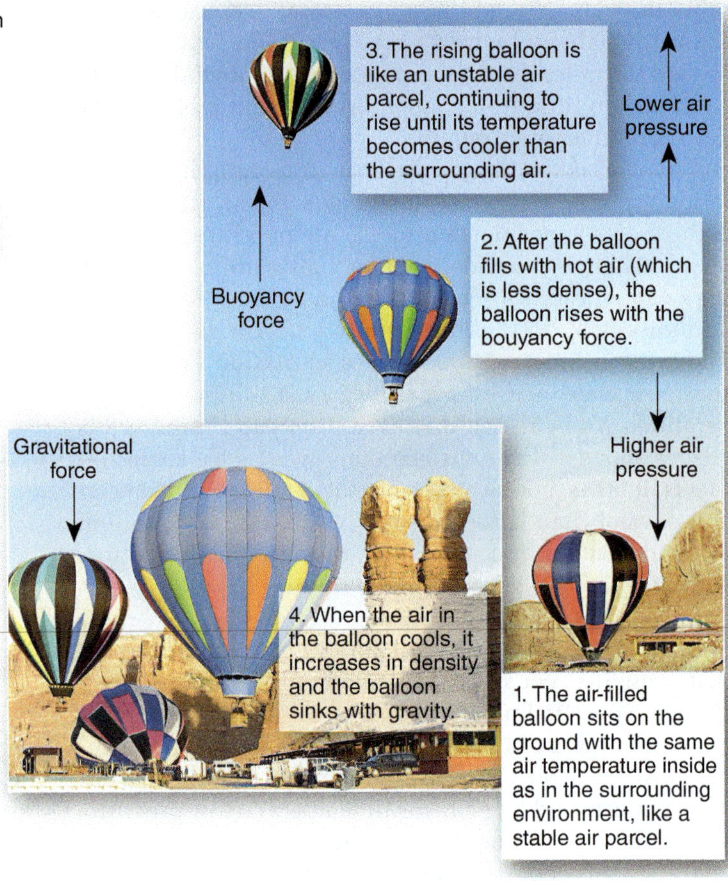

▲**Figure 7.13 Principles of air stability and balloon launches.** Hot-air balloons being launched in southern Utah illustrate the principles of stability. Why do you think these balloon launches are taking place in the early morning? [Steven K. Huhtala.]

MG Animation Atmospheric Stability

Atmospheric Stability

Meteorologists use the term *parcel* to describe a body of air that has specific temperature and humidity characteristics. Think of an air parcel as a volume of air that is about 300 m (1000 ft) in diameter or more.

Two opposing forces decide the vertical position of a parcel of air: an upward *buoyant force* and a downward *gravitational force*. A parcel of lower density than the surrounding air is buoyant, so it rises; a rising parcel expands as external pressure decreases. In contrast, a parcel of higher density descends under the force of gravity because it is not buoyant; a falling parcel compresses as external pressure increases. The temperature of the volume of air determines the density of the air parcel—warm air has lower density; cold air has higher density. Therefore, buoyancy depends on density, and density depends on temperature. (Look ahead to *Geosystems In Action*, Figure GIA 7 on page 178, for an illustration.)

Stability refers to the tendency of an air parcel either to remain in place or to change vertical position by ascending (rising) or descending (falling). An air parcel is *stable* if it resists displacement upward or, when disturbed, tends to return to its starting place. An air parcel is *unstable* if it continues to rise until it reaches an altitude where the surrounding air has a density and temperature similar to its own. The behavior of a hot-air balloon illustrates these concepts (Figure 7.13). The relative stability of air parcels in the atmosphere is an indicator of weather conditions.

Adiabatic Processes

The stability or instability of an air parcel depends on two temperatures: the temperature inside the parcel and the temperature of the air surrounding the parcel. The difference between these two temperatures determines stability. Such temperature measurements are made daily with *radiosonde* instrument packages carried aloft by helium-filled balloons at thousands of weather stations (see Figure HD 3e on page 79 for an example from Antarctica.)

The *normal lapse rate*, introduced in Chapter 3, is the average decrease in temperature with increasing altitude, a value of 6.4 C°/1000 m (3.5 F°/1000 ft). This rate of temperature change is for still, calm air, and it can vary greatly under different weather conditions. In contrast, the *environmental lapse rate (ELR)* is the actual lapse rate at a particular place and time. It can vary by several degrees per thousand meters.

Two generalizations predict the warming or cooling of an ascending or descending parcel of air. *An ascending parcel of air tends to cool by expansion*, responding to the reduced pressure at higher altitudes. In contrast, *descending air tends to heat by compression*. These mechanisms of cooling and heating are adiabatic. *Diabatic* means occurring with an exchange of heat; **adiabatic** means occurring without a loss or gain of heat—that is, without any heat exchange between the surrounding environment and the vertically moving parcel of air. Adiabatic temperature changes are measured with one of two specific rates, depending on moisture conditions in the parcel: dry adiabatic rate (DAR) and moist adiabatic rate (MAR). These processes are illustrated in Figure GIA 7.

Dry Adiabatic Rate The **dry adiabatic rate (DAR)** is the rate at which "dry" air cools by expansion as it rises or heats by compression as it falls. "Dry" refers to air that is less than saturated (relative humidity is less than 100%). The average DAR is 10 C°/1000 m (5.5 F°/1000 ft).

To see how a specific example of dry air behaves, consider an unsaturated parcel of air at the surface with a temperature of 27°C (81°F), shown in Figure GIA 7.1b. It rises, expands, and cools adiabatically at the DAR, reaching an altitude of 2500 m (approximately 8000 ft). What happens to the temperature of the parcel? Calculate the temperature change in the parcel, using the dry adiabatic rate:

$$(10\ \text{C}°/1000\ \text{m}) \times 2500\ \text{m} = 25\ \text{C}°\ \text{of total cooling}$$

$$(5.5\ \text{F}°/1000\ \text{ft}) \times 8000\ \text{ft} = 44\ \text{F}°\ \text{of total cooling}$$

Subtracting the 25 C° (44 F°) of adiabatic cooling from the starting temperature of 27°C (81°F) gives the temperature in the air parcel at 2500 m as 2°C (36°F).

In Figure GIA 7.2b, assume that an unsaturated air parcel with a temperature of –20°C at 3000 m (24°F at 9800 ft) descends to the surface, heating adiabatically. Using the dry adiabatic lapse rate, we determine the temperature of the air parcel when it arrives at the surface:

$$(10\ \text{C}°/1000\ \text{m}) \times 3000\ \text{m} = 30\ \text{C}°\ \text{of total warming}$$

$$(5.5\ \text{F}°/1000\ \text{ft}) \times 9800\ \text{ft} = 54\ \text{F}°\ \text{of total warming}$$

Adding the 30 C° of adiabatic warming to the starting temperature of –20°C gives the temperature in the air parcel at the surface as 10°C (50°F).

Moist Adiabatic Rate The **moist adiabatic rate (MAR)** is the rate at which an ascending air parcel that is moist, or saturated, cools by expansion. The average MAR is 6 C°/1000 m (3.3 F°/1000 ft). This is roughly 4 C° (2 F°) less than the dry adiabatic rate. From this average, the MAR varies with moisture content and temperature and can range from 4 C° to 10 C° per 1000 m (2 F° to 5.5 F° per 1000 ft). (Note that a descending parcel of saturated air warms at the MAR as well, because the evaporation of liquid droplets, absorbing sensible heat, offsets the rate of compressional warming.)

The cause of this variability, and the reason that the MAR is lower than the DAR, is the latent heat of condensation. As water vapor condenses in the saturated air, latent heat is liberated, becoming sensible heat, thus decreasing the adiabatic rate. The release of latent heat may vary with temperature and water-vapor content. The MAR is much lower than the DAR in warm air, whereas the two rates are more similar in cold air.

Stable and Unstable Atmospheric Conditions

The relationship of the DAR and MAR to the environmental lapse rate, or ELR, at a given time and place determines the stability of the atmosphere over an area. In turn, atmospheric stability affects cloud formation and precipitation patterns, some of the essential elements of weather.

Temperature relationships in the atmosphere produce three conditions in the lower atmosphere: unstable, conditionally unstable, and stable. For the sake of illustration, the three examples in Figure 7.14 begin with an air parcel at the surface at 25°C (77°F). In each example, compare the temperatures of the air parcel and the surrounding environment. Assume that a lifting mechanism, such as surface heating, a mountain range, or weather fronts, is present to get the parcel started (we examine lifting mechanisms in Chapter 8).

Given unstable conditions in Figure 7.14a, the air parcel continues to rise through the atmosphere because it is warmer (less dense and more buoyant) than the surrounding environment. Note that the environmental lapse rate in this example is 12 C°/1000 m (6.6 F°/1000 ft). That is, the air surrounding the air parcel is cooler by 12 C° for every 1000-m increase in altitude. By 1000 m (about 3300 ft), the rising air parcel has cooled adiabatically by expansion at the DAR from 25° to 15°C, while the surrounding air cooled from 25°C at the surface to 13°C. By comparing the temperature in the air parcel and the surrounding environment, you see that the temperature in the parcel is 2 C° (3.6 F°) warmer than the surrounding air at 1000 m. *Unstable* describes this condition because the less-dense air parcel will continue to lift.

Eventually, as the air parcel continues rising and cooling, it may achieve the dew-point temperature, saturation, and active condensation. This point where saturation begins is the *lifting condensation level* that you see in the sky as the flat bottoms of clouds.

Two forces act on a parcel of air: an upward buoyant force and a downward gravitational force. A parcel's temperature and density determine its buoyancy and whether it will rise, sink, or remain in place. Because air pressure decreases with altitude, air expands as it rises (GIA 7.1a) and is compressed as it sinks (GIA 7.2a). At the same time, its temperature changes due to adiabatic cooling or heating. In an adiabatic process, there is no loss or gain of heat. The temperature change occurs when air rises, expands, and cools (GIA 7.1b) or when air sinks, is compressed, and warms (GIA 7.2b).

7.1a COOLING BY EXPANSION

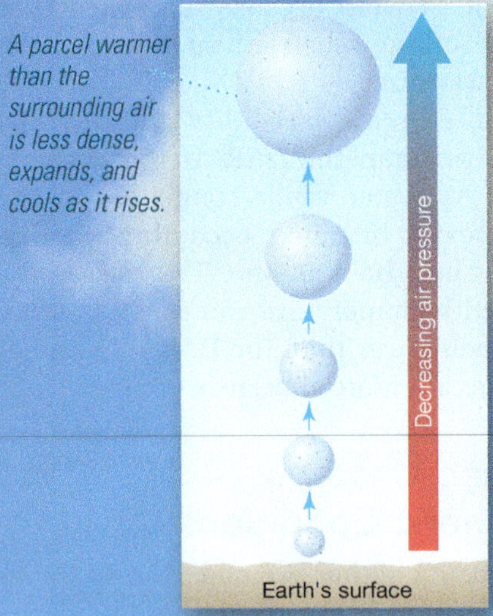

7.1b ADIABATIC COOLING

Temperature cools as pressure falls and altitude increases. The temperature change depends on the relative humidity of the parcel. Rising air that is "dry" (relative humidity less than 100 percent) cools at the *dry adiabatic rate* (DAR) of about 10 C° per 1000 m (5.5 F° per 1000 ft).

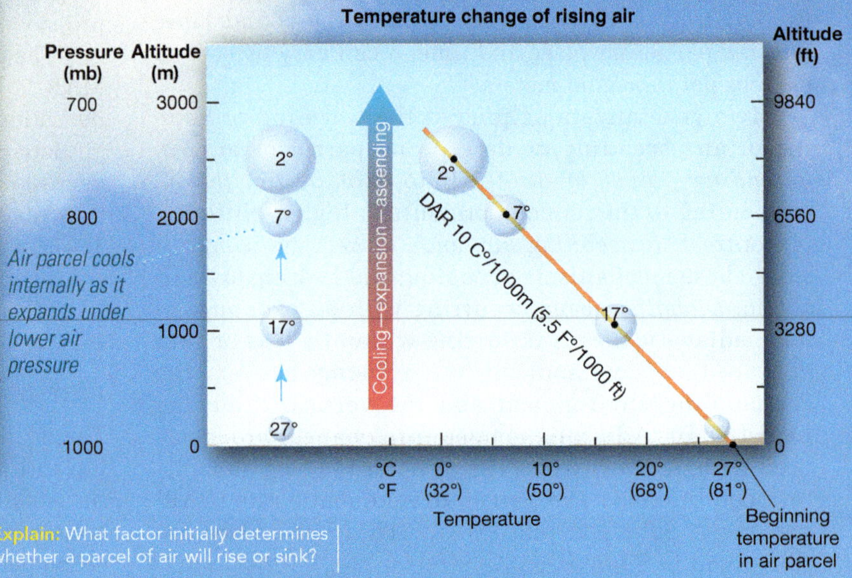

Explain: What factor initially determines whether a parcel of air will rise or sink?

7.2a HEATING BY COMPRESSION

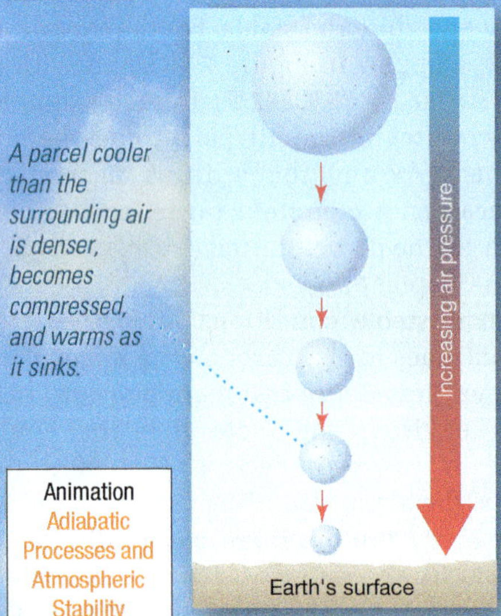

7.2b ADIABATIC HEATING

Temperature warms as pressure increases and altitude decreases. Sinking air that is "dry" warms at the dry adiabatic rate.

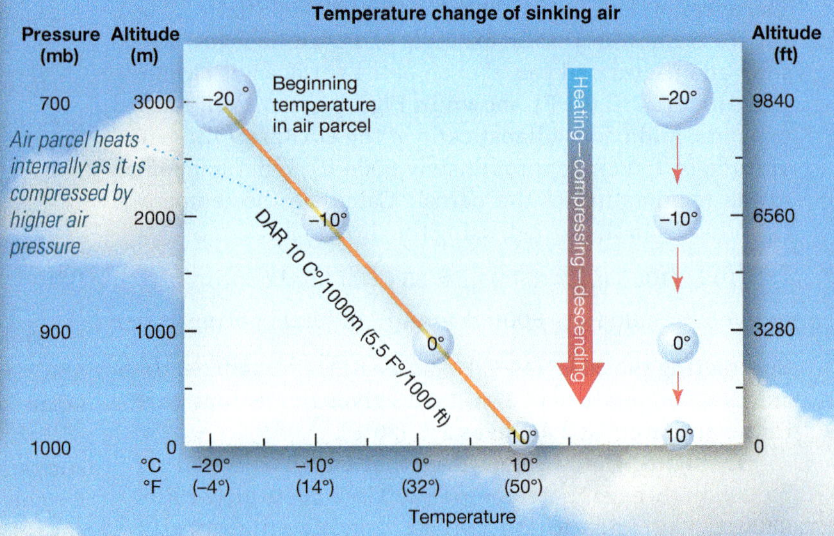

Animation
Adiabatic Processes and Atmospheric Stability

MasteringGeography™

Visit the Study Area in MasteringGeography™ to explore Earth systems.

Visualize: Study geoscience animations of adiabatic cooling and heating.

Assess: Demonstrate understanding of adiabatic processes and atmospheric stability/instability (if assigned by instructor).

GEOquiz

1. Explain: Why does the temperature of a parcel of air change as it rises or sinks?

2. Calculate: A parcel of air at the surface has a temperature of 25° C (77° F). If the parcel rises and cools at the dry adiabatic rate, what is its temperature at 730 m (2400 ft)?

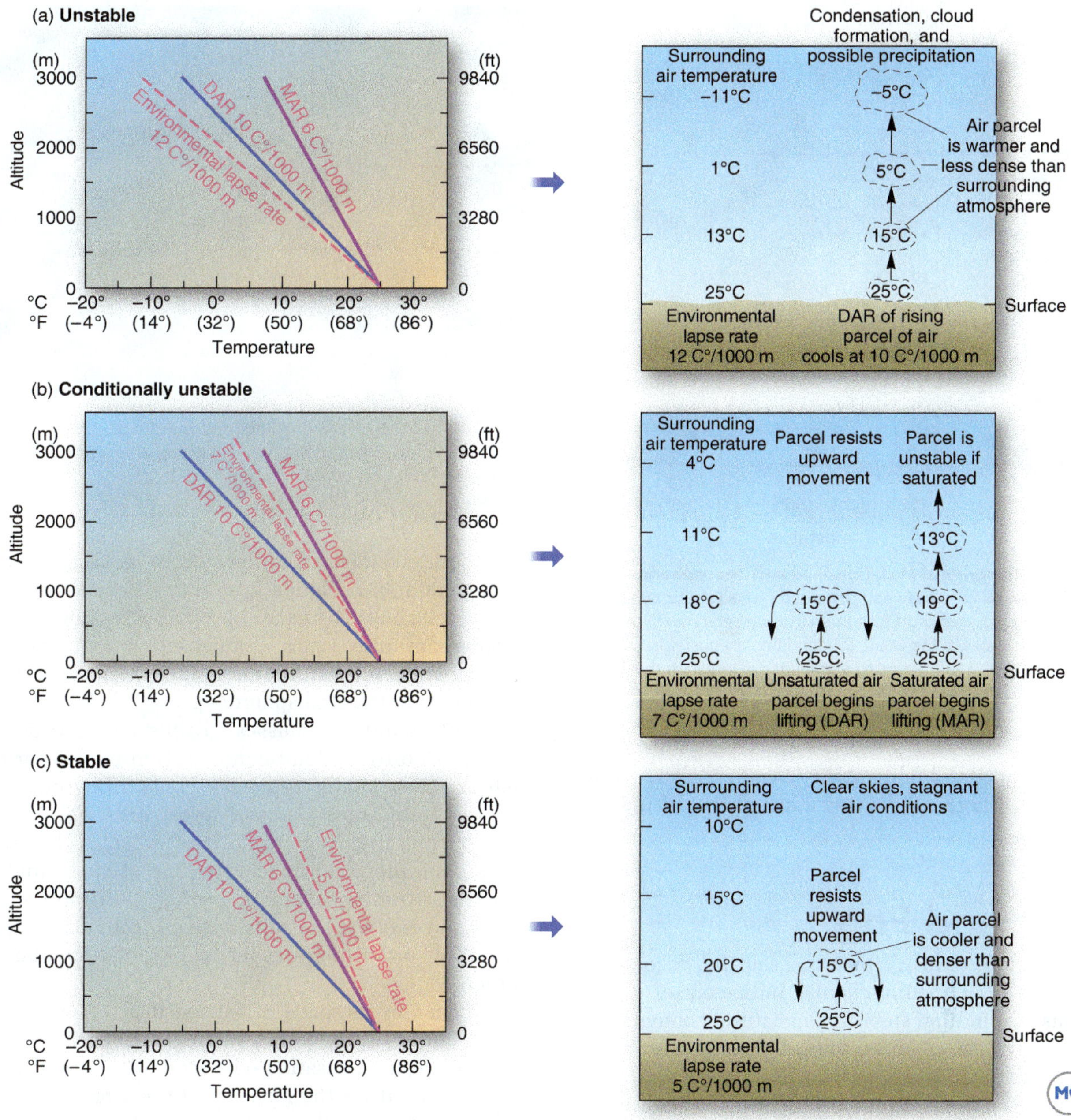

▲Figure 7.14 Stability—three examples. Specific examples of (a) unstable, (b) conditionally unstable, and (c) stable conditions in the lower atmosphere. Note the response to these three conditions in the air parcel on the right side of each diagram.

MG Animation Atmospheric Stability

The example in Figure 7.14c shows stable conditions resulting when the ELR is only 5 C°/1000 m (3 F°/1000 ft). An ELR of 5 C°/1000 m is less than both the DAR and the MAR, a condition in which the air parcel has a lower temperature (is more dense and less buoyant) than the surrounding environment. The relatively cooler air parcel tends to settle back to its original position—it is *stable*. The denser air parcel resists lifting, unless forced by updrafts or a barrier, and the sky remains generally cloud-free. If clouds form, they tend to be stratiform (flat clouds) or cirroform (wispy), lacking vertical development. In regions experiencing air pollution, stable conditions in the atmosphere worsen the pollution by slowing exchanges in the surface air.

If the ELR is somewhere between the DAR and the MAR, conditions are neither unstable nor stable. In Figure 7.14b, the ELR is measured at 7 C°/1000 m. Under these conditions, the air parcel resists upward movement, unless forced, if it is less than saturated. But if the air parcel becomes saturated and cools at the MAR, it acts unstable and continues to rise.

One example of such conditionally unstable air occurs when stable air is forced to lift as it passes over a mountain range. As the air parcel lifts and cools to the dew point, the air becomes saturated and condensation begins. Now the MAR is in effect, and the air parcel behaves in an unstable manner. The sky may be clear and without a cloud, yet huge clouds may develop over a nearby mountain range.

The overall relationships between the dry and moist adiabatic rates and environmental lapse rates that produce conditions of stability, instability, and conditional instability are summarized in Figure 7.15. We will work

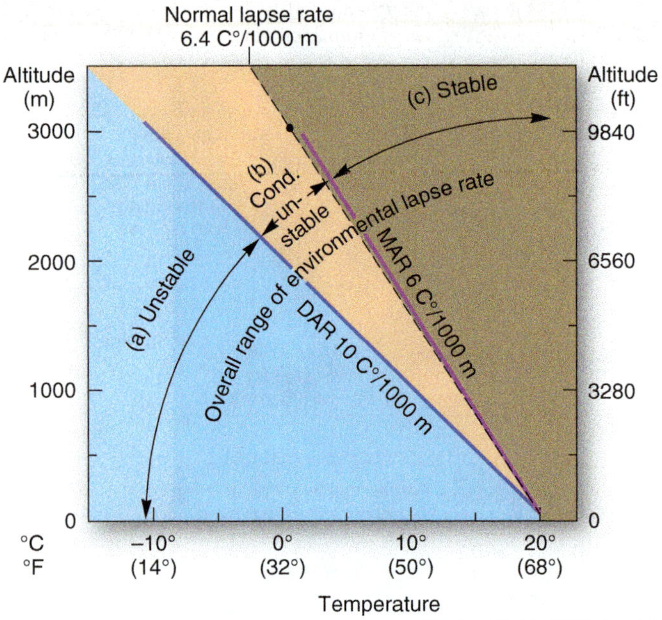

▲Figure 7.15 **Temperature relationships and atmospheric stability.** The relationship between dry and moist adiabatic rates and environmental lapse rates produces three atmospheric conditions: (a) unstable (ELR exceeds the DAR), (b) conditionally unstable (ELR is between the DAR and MAR), and (c) stable (ELR is less than the DAR and MAR).

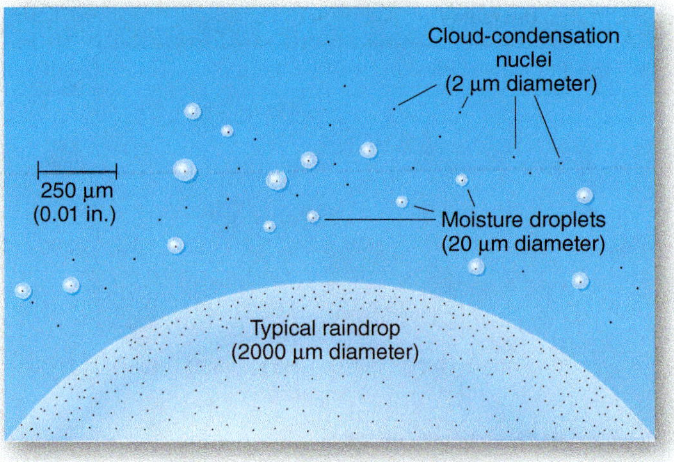

▲Figure 7.16 **Moisture droplets and raindrops.** Cloud-condensation nuclei, moisture droplets, and a raindrop enlarged many times—compared at roughly the same scale.

more with lapse rates and adiabatic cooling and heating in Chapter 8, where we discuss atmospheric lifting and precipitation.

Clouds and Fog

Clouds are more than whimsical, beautiful decorations in the sky; they are fundamental indicators of overall conditions, including stability, moisture content, and weather. They form as air becomes saturated with water. Clouds are the subject of much scientific inquiry, especially regarding their effect on net radiation patterns, as discussed in Chapters 4 and 5. With a little knowledge and practice, you can learn to "read" the atmosphere from its signature clouds.

Cloud Formation Processes

A **cloud** is an aggregation of tiny moisture droplets and ice crystals that are suspended in air and are great enough in volume and concentration to be visible. Fog, discussed later in the chapter, is simply a cloud in contact with the ground. Clouds may contain raindrops, but not initially. At the outset, clouds are a great mass of moisture droplets, each invisible without magnification. A **moisture droplet** is approximately 20 μm (micrometers) in diameter (0.002 cm, or 0.0008 in.). It takes a million or more such droplets to form an average raindrop with a diameter of 2000 μm (0.2 cm, or 0.078 in.), as shown in Figure 7.16.

As an air parcel rises, it may cool to the dew-point temperature and 100% relative humidity. (Under certain conditions, condensation may occur at slightly less or more than 100% relative humidity.) More lifting of the air parcel cools it further, producing condensation of water vapor into water. Condensation requires **cloud-condensation nuclei**, microscopic particles that always are present in the atmosphere.

Continental air masses, discussed in Chapter 8, average 10 billion cloud-condensation nuclei per cubic meter. These nuclei typically come from dust, soot, and ash from volcanoes and forest fires, and particles from burned fuel, such as sulfate aerosols. The air over cities contains great concentrations of such nuclei. In maritime air masses, nuclei average 1 billion per cubic meter and include sea salts derived from ocean sprays. The lower atmosphere never lacks cloud-condensation nuclei.

Given the presence of saturated air, cloud-condensation nuclei, and cooling (lifting) mechanisms in the atmosphere, condensation will occur. Two principal processes account for the majority of the world's raindrops and snowflakes: the *collision–coalescence process*, involving warmer clouds and falling coalescing droplets, and the *Bergeron ice-crystal process*, in which supercooled water droplets evaporate and are absorbed by ice crystals that grow in mass and fall.

Cloud Types and Identification

In 1803, English biologist and amateur meteorologist Luke Howard established a classification system for clouds and coined Latin names for them that we still use. A sampling of cloud types according to this system is presented in Table 7.1 and Figure 7.17.

Altitude and *shape* are key to cloud classification. Clouds occur in three basic forms—flat, puffy, and wispy—and in four primary altitude classes. Flat and layered clouds with horizontal development are classed as *stratiform*. Puffy and globular clouds with vertical development are *cumuliform*. Wispy clouds, usually quite high in altitude and made of ice crystals, are

TABLE 7.1 Cloud Classes and Types

Cloud Class, Altitude, and Midlatitude Composition	Cloud Type	Description
Low clouds (C_L) • Up to 2000 m (6500 ft) • Water	Stratus (St)	Uniform, featureless, gray clouds that look like high fog.
	Stratocumulus (Sc)	Soft, gray, globular cloud masses in lines, groups, or waves.
	Nimbostratus (Ns)	Gray, dark, low clouds with drizzling rain.
Middle clouds (C_M) • 2000–6000 m (6500–20,000 ft) • Ice and water	Altostratus (As)	Thin to thick clouds, with no halos. Sun's outline just visible through clouds on a gray day.
	Altocumulus (Ac)	Clouds like patches of cotton balls, dappled, and arranged in lines or groups.
High clouds (C_H) • 6000–13,000 m (20,000–43,000 ft) • Ice	Cirrus (Ci)	"Mares' tails" clouds—wispy, feathery, with delicate fibers, streaks, or plumes.
	Cirrostratus (Cs)	Clouds like veils, formed from fused sheets of ice crystals, having a milky look, with Sun and Moon halos.
	Cirrocumulus (Cc)	Dappled clouds in small white flakes or tufts. Occur in lines or groups, sometimes in ripples, forming a "mackerel sky."
Vertically developed clouds • Near surface to 13,000 m (43,000 ft) • Water below, ice above	Cumulus (Cu)	Sharply outlined, puffy, billowy, flat-based clouds with swelling tops. Associated with fair weather.
	Cumulonimbus (Cb)	Dense, heavy, massive clouds associated with dark thunderstorms, hard showers, and great vertical development, with towering, cirrus-topped plume blown into anvil-shaped head.

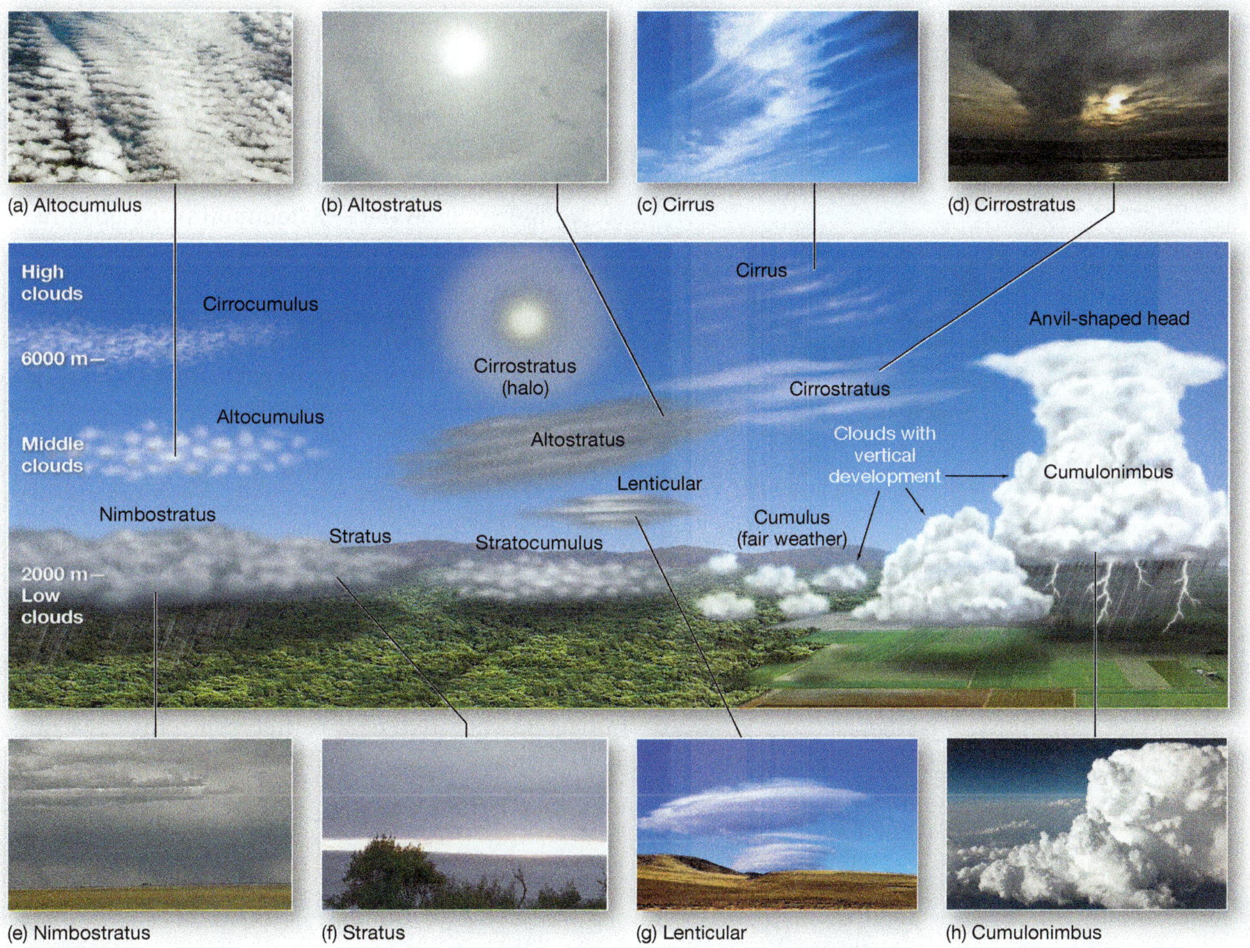

▲**Figure 7.17 Principal cloud types and special cloud forms.** Cloud types according to form and altitude (low, middle, high, and vertically developed). [(a), (b), (c), and (h) by Bobbé Christopherson; (d), (e), and (f) by author; (g) by Judy A. Mosby.]

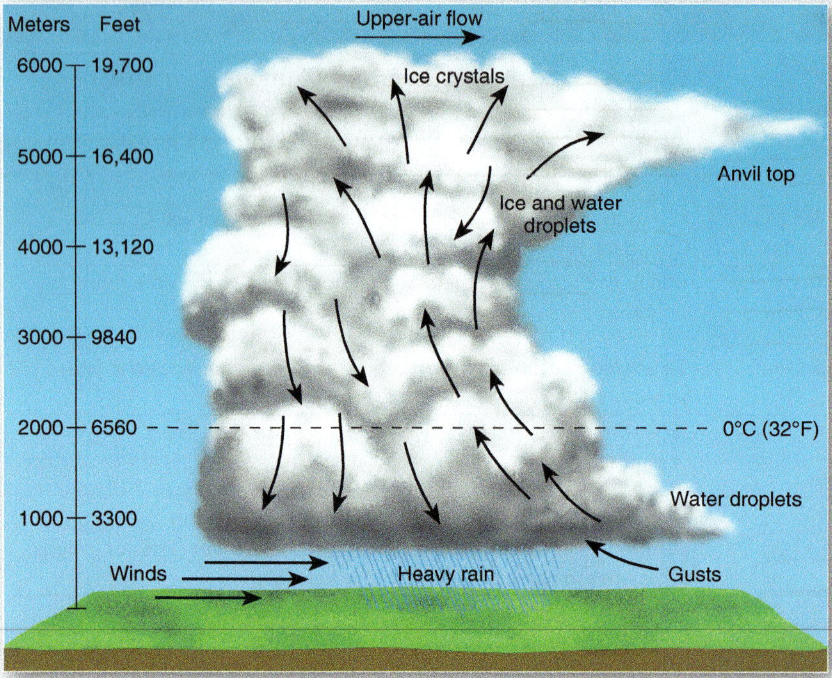

(a) Structure and form of a cumulonimbus cloud. Violent updrafts and downdrafts mark the circulation within the cloud. Blustery wind gusts occur along the ground.

(b) A dramatic cumulonimbus thunderhead over Africa at 13.5° N latitude near the Senegal–Mali border.

◄**Figure 7.18 Cumulonimbus thunderhead.** [(b) ISS Astronaut photograph, NASA.]

cirroform. The four altitudinal classes are low, middle, high, and clouds vertically developed through the troposphere. Combinations of shape and altitude result in 10 basic cloud types.

Low clouds, ranging from the surface up to 2000 m (6500 ft) in the middle latitudes, are simply *stratus* or *cumulus* (Latin for "layer" and "heap," respectively). **Stratus** clouds appear dull, gray, and featureless. When they yield precipitation, they become **nimbostratus** (*nimbo-* denotes "stormy" or "rainy"), and their showers typically fall as drizzling rain (Figure 7.17e).

Cumulus clouds appear bright and puffy, like cotton balls. When they do not cover the sky, they float by in infinitely varied shapes. Vertically developed cumulus clouds are in a separate class in Table 7.1 because further vertical development can produce cumulus clouds that extend beyond low altitudes into middle and high altitudes (illustrated at the far right in Figure 7.17 and the photo in Figure 7.17h).

Sometimes near the end of the day **stratocumulus** may fill the sky in patches of lumpy, grayish, low-level clouds. Near sunset, these spreading, puffy, stratiform remnants may catch and filter the Sun's rays, sometimes indicating clearing weather.

The prefix *alto-* (meaning "high") denotes middle-level clouds. They are made of water droplets, mixed, when temperatures are cold enough, with ice crystals. **Altocumulus** clouds, in particular, represent a broad category that includes many different styles: patchy rows, wave patterns, a "mackerel sky," or lens-shaped (lenticular) clouds.

Ice crystals in thin concentrations compose clouds occurring above 6000 m (20,000 ft). These wispy filaments, usually white except when colored by sunrise or sunset, are **cirrus** clouds (Latin for "curl of hair"), sometimes dubbed "mares' tails." Cirrus clouds look as though an artist took a brush and made delicate feathery strokes high in the sky. Cirrus clouds can indicate an oncoming storm, especially if they thicken and lower in elevation. The prefix *cirro-*, as in *cirrostratus* and *cirrocumulus*, indicates other high clouds that form a thin veil or have a puffy appearance, respectively.

GEOreport 7.3 Lenticular clouds signal mountain weather

Along the mountains of the world, lenticular clouds may warn of high-speed winds at altitude, sometimes signaling the onset of severe weather. Lenticular, or lens-shaped, clouds often form on the lee side of mountain ranges, where winds passing over the terrain develop a wave pattern. These clouds often appear stationary, but they are not. The flow of moist air continuously resupplies the cloud on the windward side as air evaporates from the cloud on the leeward side.

▲**Figure 7.19 Winter radiation fog.** This December 2005 image shows fog in California's Central Valley, trapped by the topographic barriers of the Cascades to the north, the Sierra Nevada to the east, and the Coastal Ranges to the west. Coastal areas of higher elevation are the coast redwood forests discussed in this chapter's *Geosystems Now*. [*Terra* MODIS image, NASA.]

▲**Figure 7.20 Advection fog.** San Francisco's Golden Gate Bridge is shrouded by an invading advection fog characteristic of summer conditions along a western coast. [Brad Perks Lightscapes/Alamy.]

A cumulus cloud can develop into a towering giant called **cumulonimbus** (again, *nimbus* in Latin denotes "rain storm" or "thundercloud"; Figure 7.18). Such clouds are known as *thunderheads* because of their shape and associated lightning and thunder. Note the surface wind gusts, updrafts and downdrafts, heavy rain, and ice crystals present at the top of the rising cloud column. High-altitude winds may then shear the top of the cloud into the characteristic anvil shape of the mature thunderhead.

Processes That Form Fog

By international definition, **fog** is a cloud layer on the ground, with visibility restricted to less than 1 km (3300 ft). The presence of fog tells us that the air temperature and the dew-point temperature at ground level are nearly identical, indicating saturated conditions. A temperature-inversion layer generally caps a fog layer (warmer temperatures above and cooler temperatures below the inversion altitude), with as much as 22 C° (40 F°) difference in air temperature between the cooler ground under the fog and the warmer, sunny skies above.

Almost all fog is warm—that is, its moisture droplets are above freezing. Supercooled fog, in which the moisture droplets are below freezing, is special because it can be dispersed by means of artificial seeding with ice crystals or other crystals that mimic ice, following the principles of the Bergeron process mentioned earlier.

Radiation When radiative cooling of a surface chills the air layer directly above that surface to the dew-point temperature, creating saturated conditions, a **radiation fog** forms. This fog occurs over moist ground, especially on clear nights; it does not occur over water because water does not cool appreciably overnight.

Winter radiation fog is typical in the Central Valley of California, encompassing the Sacramento Valley in the north to the San Joaquin Valley in the south, and is locally known as a *tule fog* (pronounced "toolee") because of its association with the tule (bulrush) plants that line the low-elevation islands and marshes of the Sacramento River and San Joaquin River delta regions (Figure 7.19). Tule fog can reduce visibility to 3 m (10 ft) or less, and is a leading cause of weather-related traffic accidents in California.

▲Figure 7.21 **Valley fog.** Cold air settles in the Saar River valley, Saarland, Germany, chilling the air to the dew point and forming a valley fog. [Hans-Peter Merten/Getty Images.]

▲Figure 7.22 **Evaporation fog.** Evaporation fog, or sea smoke, highlights dawn on a cold morning at Donner Lake, California. Later that morning as air temperatures rose, what do you think happened to the evaporation fog? [Bobbé Christopherson.]

Advection When air in one place migrates to another place where conditions are right for saturation, an **advection fog** forms. For example, when warm, moist air moves over cooler ocean currents, lake surfaces, or snow masses, the layer of migrating air directly above the surface becomes chilled to the dew point and fog develops. Off all subtropical west coasts in the world, summer fog forms in the manner just described (Figure 7.20).

One type of advection fog forms when moist air flows to higher elevations along a hill or mountain. This upslope lifting leads to adiabatic cooling by expansion as the air rises. The resulting **upslope fog** forms a stratus cloud at the condensation level of saturation. Along the Appalachians and the eastern slopes of the Rockies such fog is common in winter and spring.

Another advection fog associated with topography is **valley fog**. Because cool air is denser than warm air, it settles in low-lying areas, producing a fog in the chilled, saturated layer near the ground in the valley (Figure 7.21).

Evaporation Another type of fog that is related to both advection and evaporation forms when cold air lies over the warm water of a lake, an ocean surface, or even a swimming pool. This wispy **evaporation fog**, or *steam fog*, may form as water molecules evaporate from the water surface into the cold overlying air, effectively humidifying the air to saturation, followed by

CRITICALthinking 7.3
Identify Two Kinds of Fog

In Figure CT 7.3.1, can you tell which two kinds of fog are pictured? These questions may help: How does the temperature of the river water compare with that of the overlying air, especially beyond the bend in the river? Could the temperature of the moist farmlands have changed overnight, and if so, how? Might that contribute to fog formation? Can you see any evidence of air movement, such as a light breeze? How might this affect the presence of fog? •

▲Figure CT 7.3.1 **Two kinds of fog.** [Bobbé Christopherson.]

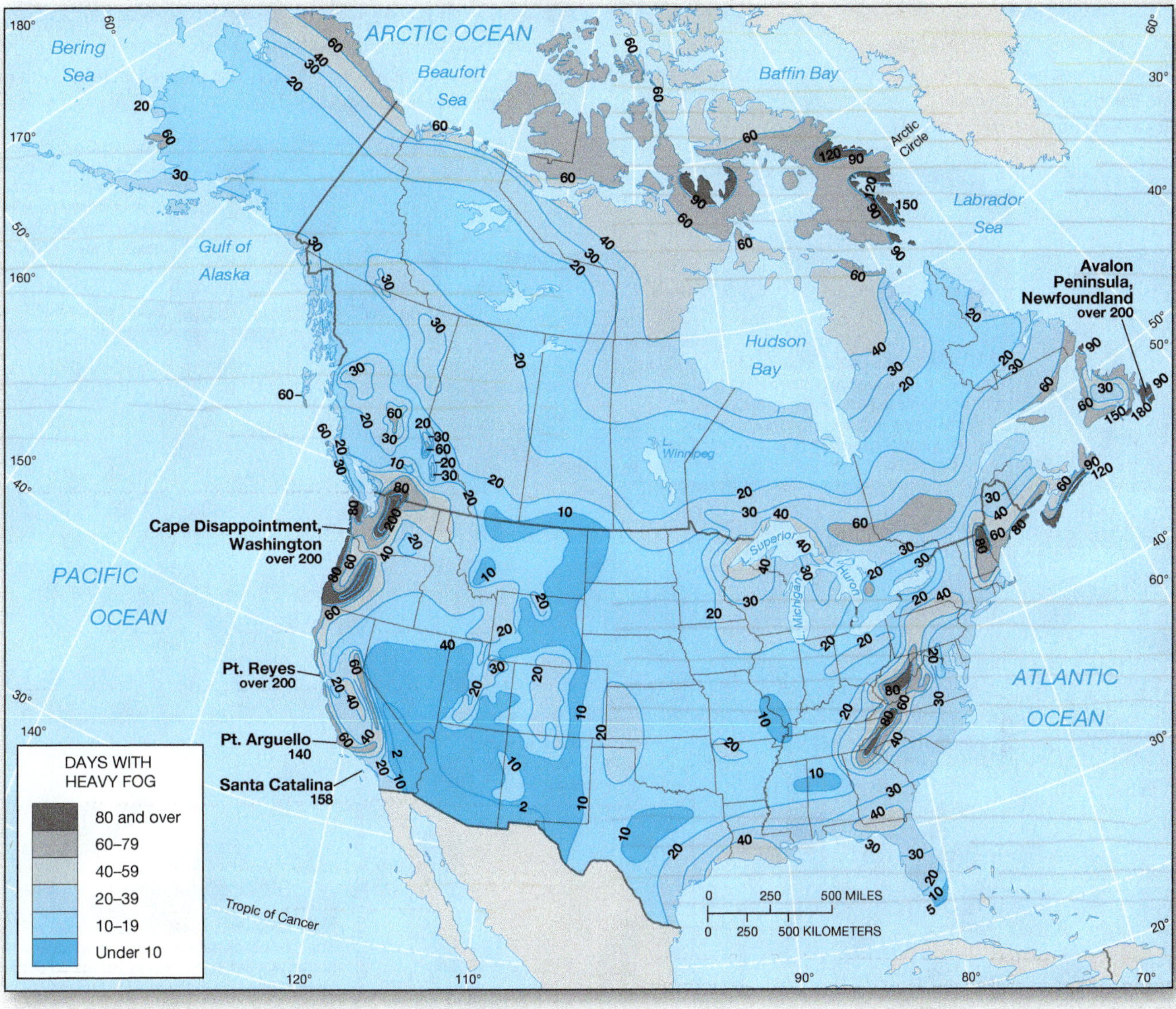

▲**Figure 7.23 Mean annual number of days with heavy fog in the United States and Canada.** The foggiest spot in the United States is the mouth of the Columbia River, where it enters the Pacific Ocean at Cape Disappointment, Washington. One of the foggiest places in the world is Newfoundland's Avalon Peninsula, specifically Argentina and Belle Isle, which regularly exceed 200 days of fog each year. [Data courtesy of NWS; *Climatic Atlas of Canada*, Atmospheric Environment Service Canada; and *The Climates of Canada*, Environment Canada, 1990.]

condensation to form fog (Figure 7.22). When evaporation fog happens at sea, it is a shipping hazard called *sea smoke*.

The prevalence of fog throughout the United States and Canada is shown in Figure 7.23. Every year the media carry stories of multicar pileups on stretches of highway where vehicles drive at high speed in foggy conditions. These crash scenes can involve dozens of cars and trucks. Fog is a hazard to drivers, pilots, sailors, pedestrians, and cyclists, even though its conditions of formation are quite predictable. The distribution of regional fog occurrence should be a planning consideration for any proposed airport, harbor facility, or highway.

Fog is an important moisture source for many organisms, as discussed in *Geosystems Now* for the coast redwood trees of California. Humans are increasingly using fog as a water resource in some regions. Throughout history, people have harvested water from fog, perhaps taking a cue from the remarkable adaptations of insects. Sand beetles in the Namib Desert in extreme southwestern Africa harvest water from fog by holding up their wings so condensation collects and runs down to their mouths. As the day's heat arrives, they burrow into the sand, not emerging until the next night or morning when the advection fog brings in more water for harvesting. For centuries, coastal villages in the deserts of Oman collected water drips deposited on trees by coastal fogs. In the Atacama Desert of Chile and Peru, residents stretch large nets to intercept advection fog; moisture condenses on the netting, drips into trays, and then flows through pipes to a 100,000-L (26,000-gal.) reservoir. At least 30 countries across the globe experience foggy conditions suitable for harvesting water. (See http://www.oas.org/dsd/publications/unit/oea59e/ch33.htm, and *The Human Denominator,* Figure HD 7d.)

THE**human**DENOMINATOR 7 Atmospheric Moisture

ATMOSPHERIC MOISTURE ⇨ HUMANS

- Water in the atmosphere provides energy that drives global weather systems and affects Earth's energy balance.
- Humidity affects human comfort levels.

HUMANS ⇨ ATMOSPHERIC MOISTURE

- Water vapor is a greenhouse gas that will increase with rising temperatures from climate change.

7a

To reduce the frequency of deadly fog-related traffic accidents in parts of California's Central Valley, electronic freeway signs warn drivers when potentially hazardous road conditions exist. [Jeremy Walker/Getty Images.]

7b

Morning Glory clouds are rare, tubular clouds that form in only several locations on Earth. Northern Australia's Gulf of Carpentaria is the only place where they are observed on a regular basis. Gliders flock to this area to ride the associated winds. Learn more at http://www.morninggloryaustralia.com/. [Mick Pertroff/NASA.]

7c

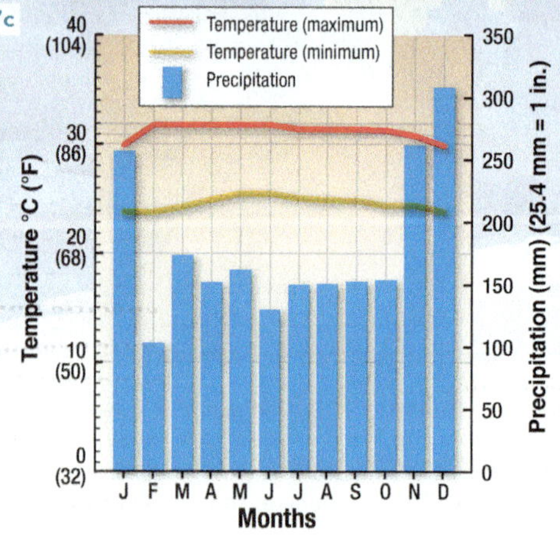

The tropical climate of Malaysia is hot and humid all year. In Singapore, at the tip of the Malay peninsula, daily relative humidity averages 84% and often exceeds 90% at night. Rain falls on most days, and dramatic afternoon thunderstorms can occur throughout the year. [ak_phuong/Flickr Open/Getty Images. Climate data from WeatherWise Singapore.]

7d

A net captures moisture from advection fog that forms over the cold Humboldt Current in the Pacific Ocean and creeps toward the Andes Mountains of Chile and Peru. Nets such as this can collect up to 5 L/m² of water, which translates to 200 L/day per net, a significant amount in this dry region. [Mariana Bazo/Reuters.]

ISSUES FOR THE 21ST CENTURY

- Fog will continue to be used to supplement drinking water supplies in some regions of the world.
- Rising temperatures affect evaporation and condensation rates, in turn affecting cloud formation. Ongoing studies indicate that cloud cover will be less in a warming world, creating a positive feedback that accelerates warming.

GEOSYSTEMS**connection**

We explored the remarkable physical properties of water and began our discussion of the role of water in the atmosphere. We looked at humidity concepts, which underlie the dynamics of atmospheric stability, and at the formation of clouds and fog. With this foundation we move to Chapter 8 and the identification of air masses and the conditions that force these to lift, cool, condense, and produce weather phenomena. Also, we look at violent weather events, such as thunderstorms, strong winds, tornadoes, and tropical cyclones.

KEY LEARNING concepts review

■ *Describe* the heat properties of water, and *identify* the traits of its three phases: solid, liquid, and gas.

Water is the most common compound on the surface of Earth, and it possesses unusual solvent and heat characteristics. Owing to Earth's temperate position relative to the Sun, water exists here naturally in all three states—solid, liquid, and gas. A change from one state to another is a **phase change**. The change from liquid to solid is freezing; from solid to liquid, melting; from vapor to liquid, condensation; from liquid to vapor, vaporization or evaporation; from vapor to solid, deposition; and from solid to vapor, **sublimation**.

The heat energy required for water to change phase is **latent heat** because, once absorbed, it is hidden within the structure of the water, ice, or water vapor. For 1 g of water to become 1 g of water vapor by boiling requires the addition of 540 cal, or the **latent heat of vaporization**. When this 1 g of water vapor condenses, the same amount of heat energy, 540 calories, is liberated and is the **latent heat of condensation**. The **latent heat of sublimation** is the energy exchanged in the phase change from ice to vapor and vapor to ice. Weather is powered by the tremendous amount of latent heat energy involved in the phase changes among the three states of water.

phase change (p. 169)
sublimation (p. 169)
latent heat (p. 170)
latent heat of vaporization (p. 171)
latent heat of condensation (p. 171)
latent heat of sublimation (p. 171)

1. Describe the three states of matter as they apply to ice, water, and water vapor.
2. What happens to the physical structure of water as it cools below 4°C (39°F)? What are some visible indications of these physical changes?
3. What is latent heat? How is it involved in the phase changes of water?
4. Take 1 g of water at 0°C and follow the changes it undergoes to become 1 g of water vapor at 100°C, describing what happens along the way. What amounts of energy are involved in the changes that take place?

■ *Define* humidity and relative humidity, and *explain* dew-point temperature and saturated conditions in the atmosphere.

The amount of water vapor in the atmosphere is **humidity**. The maximum water vapor possible in air is principally a function of the temperature of the air and of the water vapor (usually these temperatures are the same). Warmer air produces higher net evaporation rates and maximum possible water vapor, whereas cooler air can produce net condensation and lower the possible water vapor.

Relative humidity is a ratio of the amount of water vapor actually in the air to the maximum amount possible at a given temperature. Relative humidity tells us how near the air is to saturation. Relatively dry air has a lower relative humidity value; relatively moist air has a higher relative humidity percentage. Air is said to be at **saturation** when the rate of evaporation and the rate of condensation reach equilibrium; any further addition of water vapor or temperature lowering will result in active condensation (100% relative humidity). The temperature at which air achieves saturation is the **dew-point temperature**.

Among the various ways to express humidity and relative humidity are vapor pressure and specific humidity. **Vapor pressure** is that portion of the atmospheric pressure produced by the presence of water vapor. A comparison of vapor pressure with the saturation vapor pressure at any moment yields a relative humidity percentage. **Specific humidity** is the mass of water vapor (in grams) per mass of air (in kilograms) at any specified temperature. Because it is measured as a mass, specific humidity does not change as temperature or pressure changes, making it a valuable measurement in weather forecasting. A comparison of specific humidity with the maximum specific humidity at any moment produces a relative humidity percentage.

Two instruments measure relative humidity, and indirectly the actual humidity content of the air; they are the **hair hygrometer** and the **sling psychrometer**.

humidity (p. 172)
relative humidity (p. 172)
saturation (p. 173)
dew-point temperature (p. 173)
vapor pressure (p. 174)
specific humidity (p. 175)
hair hygrometer (p. 175)
sling psychrometer (p. 175)

5. What is humidity? How is it related to the energy present in the atmosphere? To our personal comfort and how we perceive apparent temperatures?
6. Define relative humidity. What does the concept represent? What is meant by the terms *saturation* and *dew-point temperature*?
7. Using Figures 7.10 and 7.11, derive relative humidity values (vapor pressure/saturation vapor pressure; specific humidity/ maximum specific humidity) for levels of humidity in the air different from the ones presented as examples in the chapter discussion.
8. How do the two instruments described in this chapter measure relative humidity?
9. How does the daily trend in relative humidity values compare with the daily trend in air temperature?

■ *Define* atmospheric stability, and *relate* it to a parcel of air that is ascending or descending.

In meteorology, a *parcel* of air is a volume (on the order of 300 m, or 1000 ft, in diameter) that is homogenous in temperature and humidity. The temperature of the volume of air determines the density of the air parcel. Warm air has a lower density in a given volume of air; cold air has a higher density.

Stability refers to the tendency of an air parcel, with its water-vapor cargo, either to remain in place or to change vertical position by ascending (rising) or descending (falling). An air parcel is *stable* if it resists displacement upward or, when disturbed, it tends to return to its starting place. An air parcel is *unstable* if it continues to rise until it reaches an altitude where the surrounding air has a density (air temperature) similar to its own.

stability (p. 176)

10. Differentiate between stability and instability of a parcel of air lifted vertically in the atmosphere.
11. What are the forces acting on a vertically moving parcel of air? How are they affected by the density of the air parcel?

■ *Illustrate* three atmospheric conditions—unstable, conditionally unstable, and stable—with a simple graph that relates the environmental lapse rate (ELR) to the dry adiabatic rate (DAR) and moist adiabatic rate (MAR).

An ascending (rising) parcel of air cools by expansion, responding to the reduced air pressure at higher altitudes. A descending (falling) parcel heats by compression. Temperature changes in ascending and descending air parcels are **adiabatic**, meaning they occur as a result of expansion or compression, without any significant heat exchange between the surrounding environment and the vertically moving parcel of air.

The **dry adiabatic rate (DAR)** is the rate at which "dry" air cools by expansion (if ascending) or heats by compression (if descending). The term *dry* is used when air is less than saturated (relative humidity is less than 100%). The DAR is 10 C°/1000 m (5.5 F°/1000 ft). The **moist adiabatic rate (MAR)** is the average rate at which moist (saturated) air cools by expansion on ascent or warms by compression on descent. The average MAR is 6 C°/1000 m (3.3 F°/1000 ft). This is roughly 4 C° (2 F°) less than the dry rate. The MAR, however, varies with moisture content and temperature and can range from 4 to 10 C° per 1000 m (2 to 5.5 F° per 1000 ft).

A simple comparison of the DAR and MAR in a vertically moving parcel of air with the *environmental lapse rate* (ELR) in the surrounding air reveals the atmosphere's stability—whether it is unstable (air parcel continues lifting), stable (air parcel resists vertical displacement), or conditionally unstable (air parcel behaves as though unstable if the MAR is in operation and stable otherwise).

adiabatic (p. 177)
dry adiabatic rate (DAR) (p. 177)
moist adiabatic rate (MAR) (p. 177)

12. How do the adiabatic rates of heating or cooling in a vertically displaced air parcel differ from the normal lapse rate and environmental lapse rate?
13. Why is there a difference between the dry adiabatic rate (DAR) and the moist adiabatic rate (MAR)?
14. What atmospheric temperature and moisture conditions would you expect on a day when the weather is unstable? When it is stable? Relate in your answer what you would experience if you were outside watching.
15. Use the "Atmospheric Stability" animation in the Mastering Geography™ Study Area. Try different temperature settings on the sliders to produce stable and unstable conditions.

■ *Identify* the requirements for cloud formation, and *explain* the major cloud classes and types, including fog.

A **cloud** is an aggregation of tiny moisture droplets and ice crystals suspended in the air. Clouds are a constant reminder of the powerful heat-exchange system in the environment. **Moisture droplets** in a cloud form when saturation and the presence of **cloud-condensation nuclei** in air combine to cause *condensation*. Raindrops are formed from moisture droplets through either the *collision–coalescence process* or the *Bergeron ice-crystal process*.

Low clouds, ranging from surface levels up to 2000 m (6500 ft) in the middle latitudes, are **stratus** (flat clouds, in layers) or **cumulus** (puffy clouds, in heaps). When stratus clouds yield precipitation, they are **nimbostratus**. Sometimes near the end of the day, lumpy, grayish, low-level clouds called **stratocumulus** may fill the sky in patches. Middle-level clouds are denoted by the prefix *alto-*. **Altocumulus** clouds, in particular, represent a broad category that includes many different types. Clouds at high altitude, principally composed of ice crystals, are called **cirrus**. A cumulus cloud can develop into a towering giant **cumulonimbus** cloud (-*nimbus* in Latin denotes "rain storm" or "thundercloud"). Such clouds are called *thunderheads* because of their shape and their associated lightning, thunder, surface wind gusts, updrafts and downdrafts, heavy rain, and hail.

Fog is a cloud that occurs at ground level. Radiative cooling of a surface that chills the air layer directly above the surface to the dew-point temperature creates saturated conditions and a **radiation fog**. **Advection fog** forms when air in one place migrates to another place where conditions exist that can cause saturation—for example, when warm, moist air moves over cooler ocean currents. **Upslope fog** is produced when moist air is forced to higher elevations along a hill or mountain. Another fog caused by topography is **valley fog**, formed because cool, denser air settles in low-lying areas, producing fog in the chilled, saturated layer near the ground. Another type of fog resulting from evaporation and advection forms when cold air flows over the warm water of a lake, ocean surface, or swimming pool. This **evaporation fog**, or steam fog, may form as the water molecules evaporate from the water surface into the cold overlying air.

cloud (p. 180)
moisture droplet (p. 180)
cloud-condensation nuclei (p. 180)
stratus (p. 182)
nimbostratus (p. 182)
cumulus (p. 182)
stratocumulus (p. 182)
altocumulus (p. 182)
cirrus (p. 182)
cumulonimbus (p. 183)
fog (p. 183)
radiation fog (p. 183)
advection fog (p. 184)
upslope fog (p. 184)
valley fog (p. 184)
evaporation fog (p. 184)

16. Specifically, what is a cloud? Describe the droplets that form a cloud.
17. Explain the condensation process: What are the requirements? What two principal processes are discussed in this chapter?
18. What are the basic forms of clouds? Using Table 7.1, describe how the basic cloud forms vary with altitude.
19. Explain how clouds might be used as indicators of the conditions of the atmosphere and of expected weather.
20. What type of cloud is fog? List and define the principal types of fog.
21. Describe the occurrence of fog in the United States and Canada. Where are the regions of highest incidence?

Answer to Critical Thinking 7.3: The river water is warmer than the cold overlying air, producing an evaporation fog, especially beyond the bend in the river. The moist farmlands have radiatively cooled overnight, chilling the air along the surface to the dew point, resulting in active condensation. Wisps of radiation fog reveal the flow of light air movements from right to left in the photo.

MasteringGeography™

Looking for additional review and test prep materials? Visit the Study Area in *MasteringGeography*™ to enhance your geographic literacy, spatial reasoning skills, and understanding of this chapter's content by accessing a variety of resources, including MapMaster interactive maps, geoscience animations, satellite loops, author notebooks, videos, RSS feeds, flashcards, web links, self-study quizzes, and an eText version of *Geosystems*.

▲Figure 7.24 **What type of fog is this?** The scene is at about 80° north latitude, and in the background is the Austfonna Ice Cap on Nordaustlandet Island, part of the Svalbard Archipelago. Walrus moms and pups gather in a haul out on a rocky unmapped islet, exposed by recent ice retreat. As air colder than the 0°C (32°F) ocean descends off the ice cap near the coast, evaporation humidifies this air layer. Away from the coast, conditions are clear. [Bobbé Christopherson.]

8 Weather

KEY LEARNING **concepts**

After reading the chapter, you should be able to:

- ***Describe*** air masses that affect North America, and ***relate*** their qualities to source regions.
- ***Identify*** and ***describe*** four types of atmospheric lifting mechanisms, and give an example of each.
- ***Explain*** the formation of orographic precipitation, and ***review*** an example of orographic effects in North America.
- ***Describe*** the life cycle of a midlatitude cyclonic storm system, and ***relate*** this to its portrayal on weather maps.
- ***List*** the measurable elements that contribute to modern weather forecasting, and ***describe*** the technology and methods employed.
- ***Identify*** various forms of violent weather by their characteristics, and ***review*** several examples of each.

On June 8, 2006, this supercell thunderstorm developed near the Crow Reservation in south-central Montana along the Wyoming border. Although these severe thunderstorms can develop the counterclockwise rotating circulation that forms a tornado, this one did not. The storm produced heavy rain, lightning, and small hail, all common thunderstorm by-products. The spreading clouds aloft indicate a large amount of convective circulation, characteristic of the cumulonimbus clouds discussed in Chapter 7. In this chapter, we explore weather systems, including the violent weather events that can cause damage and human casualties. [Eric Nguyen/Corbis.]

On the Front Lines of Intense Weather

We frequently see images of violent weather and floods that evoke horror, sadness, and a range of other emotions. Yet most of us are distant from these events and cannot fully imagine what people face on the front lines. As these storms become more intense and destructive, better warning systems reduce human fatalities. However, as population grows in vulnerable areas, recovery becomes increasingly difficult, costly, and demoralizing.

An Oklahoma Tornado On the afternoon of May 20, 2013, as tornado warnings hit media outlets—including Internet weather apps, television, and continuously-broadcasting weather radio—the people of Moore, Oklahoma, south of Oklahoma City, retreated into their basements and storm shelters, while school children huddled in unprotected hallways. Just after 3 p.m., an EF-5–intensity tornado roared in from the southwest with winds over 322 kmph (200 mph).

The tornado was 2.1 km (1.3 mi) wide as it moved through Moore, devastating a 27-km (17-mile) swath of the city that included two elementary schools. With hundreds at first missing and feared dead, people searched the rubble for friends and family. Days later, officials reported the final death toll at 23 people, with over 200 injured, including 70 children (Figure GN 8.1).

The U.S. Federal Emergency Management Agency (FEMA) declared the area a federal disaster and mobilized search and rescue personnel as the Oklahoma governor deployed the National Guard. In a few days politicians began their tours. People began the process of acquiring disaster assistance, not anticipating the weeks of delay that lay ahead. Residents and volunteers started clearing debris from over 1200 homes destroyed, and another 12,000 damaged.

Residents told stories that aired nationwide; but as time passed, the outsiders left, and months rolled by with little media attention. People were left with the daily tasks of cleaning up and beginning to rebuild as former neighborhoods of Moore lay piled in a hastily made dump outside of town. How do people on disaster front lines physically and mentally deal with such changes in their lives?

Aftermath of Hurricane Sandy In Staten Island, New York, three months after Hurricane Sandy made landfall along the Atlantic seaboard in October 2012, over 300 buildings were still without power or heat, and another 200 were uninhabitable, still awaiting repair or demolition. In these communities, some of the hardest hit by the storm, a coalition of volunteers continued to check door-to-door on residents, many of them still waiting for financial and other assistance.

▲**Figure GN 8.2** Sand and debris line the road in Seaside Heights, New Jersey, 3 months after Hurricane Sandy, as the cleanup process continues. [Julio Cortez/AP.]

Similar stories played out all along the East Coast. In Seaside Heights, New Jersey, one of several towns built on a 32-km-long (20-mi-long) barrier island, Sandy's winds and storm surge damaged 80%–90% of the buildings, at an estimated cost of $1 billion (see Figure GN 8.2 and Focus Study 8.1 in this chapter). Residents worried that debris from the storm, still washing up on shore, would threaten the region's multi-billion-dollar summer tourist industry.

The aftereffects of traumatic events such as these are still reverberating in New Orleans years after Hurricanes Katrina, Rita, Gustav, and Cindy; across Central Europe since the 2013 floods along the Danube River and its tributaries; and in Joplin, Missouri, hit by an EF-5 tornado in 2011 (pictured in Figure 8.17). Contemplate the human dimension as you work through the weather chapter.

▼**Figure GN 8.1** Damage from the May 2013 tornado in Moore, Oklahoma. [Major Geoff Legler/Oklahoma National Guard.]

GEOSYSTEMS NOW ONLINE Go to Chapter 8 on the *MasteringGeography* website (www.masteringgeography.com) for more about tornadoes and severe weather. For a links to and information about extreme weather events in the United States and their costs, see http://www.ncdc.noaa.gov/oa/climate/severeweather/extremes.html or http:/www.ncdc.noaa.gov/billions/. MG

Water has a leading role in the vast drama played out daily on Earth's stage. It affects the stability of air masses and their interactions and produces powerful and beautiful effects in the lower atmosphere. Air masses conflict; they move and shift, dominating first one region and then another, varying in strength and characteristics. Think of the weather as a play, Earth's continents and oceans as the stage, air masses as actors of varying ability, and water as the lead.

Weather is the short-term, day-to-day condition of the atmosphere, contrasted with *climate,* which is the long-term average (over decades) of weather conditions and extremes in a region. Weather is both a "snapshot" of atmospheric conditions and a technical status report of the Earth–atmosphere heat-energy budget. Important elements that contribute to weather are temperature, air pressure, relative humidity, wind speed and direction, and seasonal factors, such as insolation receipt, related to day length and Sun angle.

Meteorology is the scientific study of the atmosphere. (*Meteor* means "heavenly" or "of the atmosphere.") Meteorologists study the atmosphere's physical characteristics and motions; related chemical, physical, and geologic processes; the complex linkages of atmospheric systems; and weather forecasting. Computers handle the volumes of data from ground instruments, aircraft, and satellites used to accurately forecast near-term weather and to study trends in long-term weather, climates, and climatic change.

The cost of weather-related destruction can be staggering—Hurricane Katrina alone produced $146 billion (adjusted to current dollars) in damage losses, part of a total $210 billion for weather-related losses in 2005. The cost of damage related to droughts, floods, hail, tornadoes, derechos, tropical storms, storm surges, blizzards and ice storms, and wildfires has risen more than 500% since 1975. Since 1980, 144 weather events exceeded $1 billion each in cost (see http://www.ncdc.noaa.gov/billions/events.pdf). Studies estimate that annual weather-related damage losses could exceed $1 trillion by 2040. Global climate change is driving this increasing total, at a pace that is accelerating.

In this chapter: We follow huge air masses across North America, observe powerful lifting mechanisms in the atmosphere, revisit the concepts of stable and unstable conditions, and examine migrating cyclonic systems with attendant cold and warm fronts. We conclude with a portrait of violent and dramatic weather so often in the news in recent years.

Water, with its ability to absorb and release vast quantities of heat energy, drives this daily drama in the atmosphere. The spatial implications of these weather phenomena and their relationship to human activities strongly link meteorology and weather forecasting to the concerns of physical geography.

Air Masses

Each area of Earth's surface imparts its temperature and moisture characteristics to overlying air. The effect of a location's surface on the air creates a homogenous mix of temperature, humidity, and stability that may extend through the lower half of the atmosphere. Such a distinctive body of air is an **air mass**, and it initially reflects the characteristics of its *source region*. Examples include the "cold Canadian air mass" and "moist tropical air mass" often referred to in weather forecasts. The various masses of air over Earth's surface interact to produce weather patterns.

Air Masses Affecting North America

We classify air masses according to the general moisture and temperature characteristics of their source regions: *Moisture* is designated **m** for maritime (wet) or **c** for continental (dry). *Temperature* is directly related to latitude and is designated **A** for arctic, **P** for polar, **T** for tropical, **E** for equatorial, and **AA** for Antarctic. Figure 8.1 shows the principal air masses that affect North America in winter and summer.

Continental polar (cP) air masses form only in the Northern Hemisphere and are most developed in winter and cold-weather conditions. These cP air masses are major players in middle- and high-latitude weather, as their cold, dense air displaces moist, warm air in their path, lifting and cooling the warm air and causing its vapor to condense. An area covered by cP air in winter experiences cold, stable air; clear skies; high pressure; and anticyclonic wind flow. The Southern Hemisphere lacks the necessary continental landmasses at high latitudes to create such a cP air mass.

Maritime polar (mP) air masses in the Northern Hemisphere sit over the northern oceans. Within them, cool, moist, unstable conditions prevail throughout the year. The Aleutian and Icelandic subpolar low-pressure cells reside within these mP air masses, especially in their well-developed winter pattern (see the January isobaric pressure map in Figure 6.10a).

Two *maritime tropical* (mT) air masses—the mT Gulf/Atlantic and the mT Pacific—influence North America. The humidity experienced in the North American East and Midwest is created by the mT Gulf/Atlantic air mass, which is particularly unstable and active from late spring to early fall. In contrast, the mT Pacific is stable to conditionally unstable and generally lower in moisture content and available energy. As a result, the western United States, influenced by this weaker Pacific air mass, receives lower average precipitation than the rest of the country. Please review Figure 6.11 and the discussion of subtropical high-pressure cells.

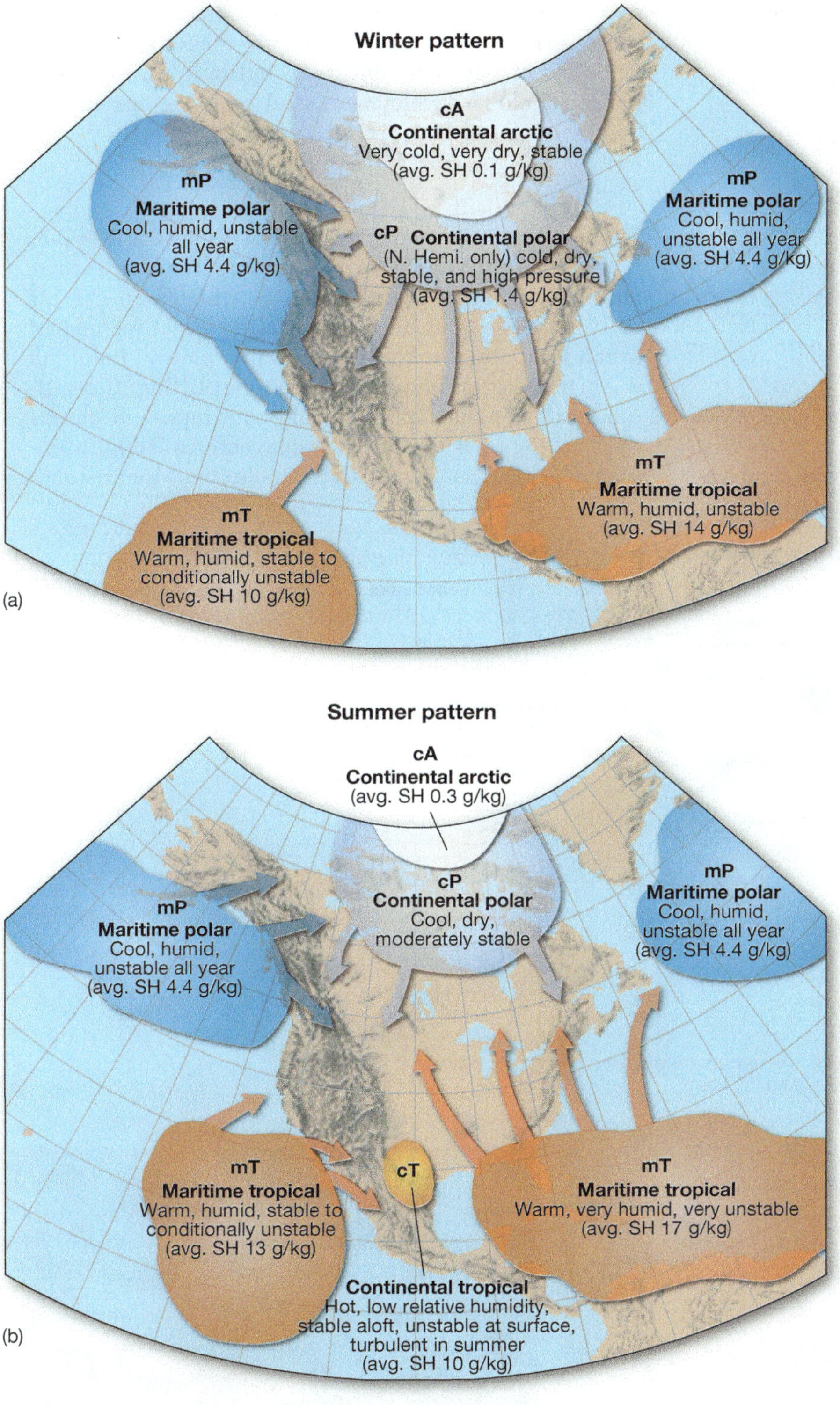

▲**Figure 8.1 Principal air masses affecting North America.** Air masses and their source regions influencing North America during (a) winter and (b) summer. (SH = specific humidity.)

Air Mass Modification

The longer an air mass remains stationary over a region, the more definite its physical attributes become. As air masses migrate from source regions, their temperature and moisture characteristics slowly change to the characteristics of the land over which they pass. For example, an mT Gulf/Atlantic air mass may carry humidity to Chicago and on to Winnipeg, but it gradually loses its initial high humidity and warmth with each day's passage northward.

Similarly, below-freezing temperatures occasionally reach into southern Texas and Florida, brought by an invading winter cP air mass from the north. However, that air mass warms to above the −50°C (−58°F) of its winter source region in central Canada, especially after it leaves areas covered by snow.

Modification of cP air as it moves south and east produces snowbelts that lie to the east of each of the Great Lakes. As below-freezing cP air passes over the warmer Great Lakes, it absorbs heat energy and moisture from the lake surfaces and becomes *humidified*. In what is called the *lake effect,* this enhancement produces heavy snowfall downwind of the lakes, into Ontario, Québec, Michigan, northern Pennsylvania, and New York—some areas receiving in excess of 250 cm (100 in.) in average snowfall a year (Figure 8.2).

The severity of the lake effect also depends on the presence of a low-pressure system positioned north of the Great Lakes, with counterclockwise winds pushing air across the lakes. Climate change is expected to enhance lake-effect snowfall over the next several decades, since warmer air can absorb more water vapor. In contrast, some climate models show that later in this century, lake-effect snowfall will decrease as temperatures rise, but that rainfall totals will continue to increase over the regions leeward of the Great Lakes. Research is ongoing as to the effects of climate change on precipitation in this region.

Atmospheric Lifting Mechanisms

When an air mass is lifted, it cools adiabatically (by expansion). When the cooling reaches the dew-point temperature, moisture in the saturated air can condense, forming clouds and perhaps precipitation. Four principal lifting mechanisms, illustrated in Figure 8.3, operate in the atmosphere:

- Convergent lifting results when air flows toward an area of low pressure.
- Convectional lifting happens when air is stimulated by local surface heating.
- Orographic lifting occurs when air is forced over a barrier such as a mountain range.
- Frontal lifting occurs as air is displaced upward along the leading edges of contrasting air masses.

Convergent Lifting

Air flowing from different directions into the same low-pressure area is converging, displacing air upward in **convergent lifting** (Figure 8.3a). All along the equatorial region, the southeast and northeast trade winds converge, forming the intertropical convergence zone (ITCZ) and areas of extensive convergent uplift, towering cumulonimbus cloud development, and high average annual precipitation (see Figure 6.10).

Convectional Lifting

When an air mass passes from a maritime source region to a warmer continental region, heating from the warmer land surface causes lifting and convection in the air mass. Other sources of surface heating might include an urban heat island or the dark soil in a plowed field; the warmer surfaces produce **convectional lifting**. If conditions are unstable, initial lifting continues and clouds develop. Figure 8.3b illustrates convectional action stimulated by local heating, with unstable conditions present

(a) Heavy local snowfall is associated with the lee side of each Great Lake; storms come from the west or northwest.

Figure 8.2 Lake-effect snowbelts of the Great Lakes. [(a) *Climatic Atlas of the United States*, p. 53. (c) *Terra* MODIS image, NASA/GSFC. (d) David Duprey/AP.]

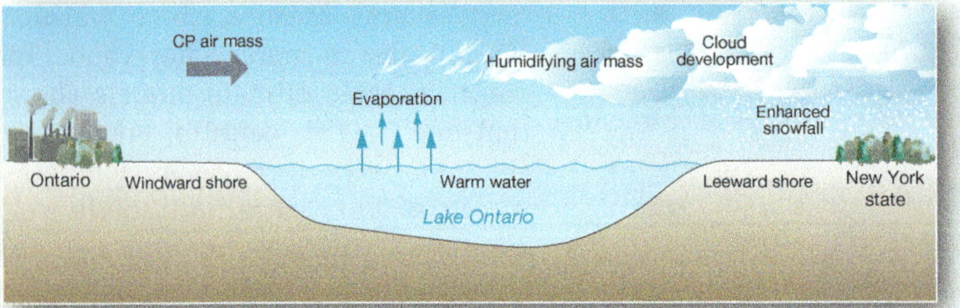

(b) Processes causing lake-effect snowfall are generally limited to about 50 km (30 mi) to 100 km (60 mi) inland.

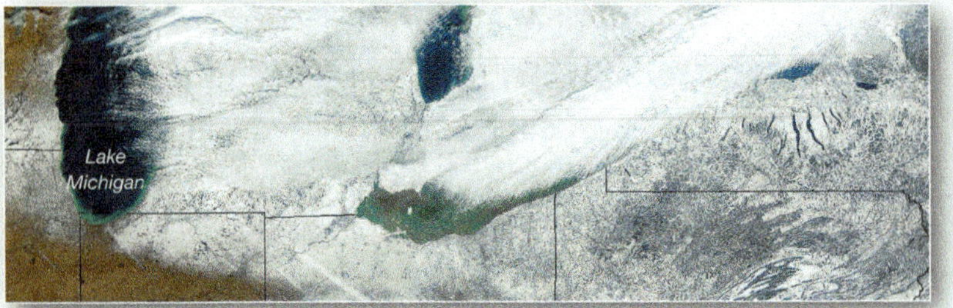

(c) Satellite image shows lake-effect weather in December.

(d) Heavy snowfall buries Buffalo, New York, in December 2010, closing the New York State Thruway.

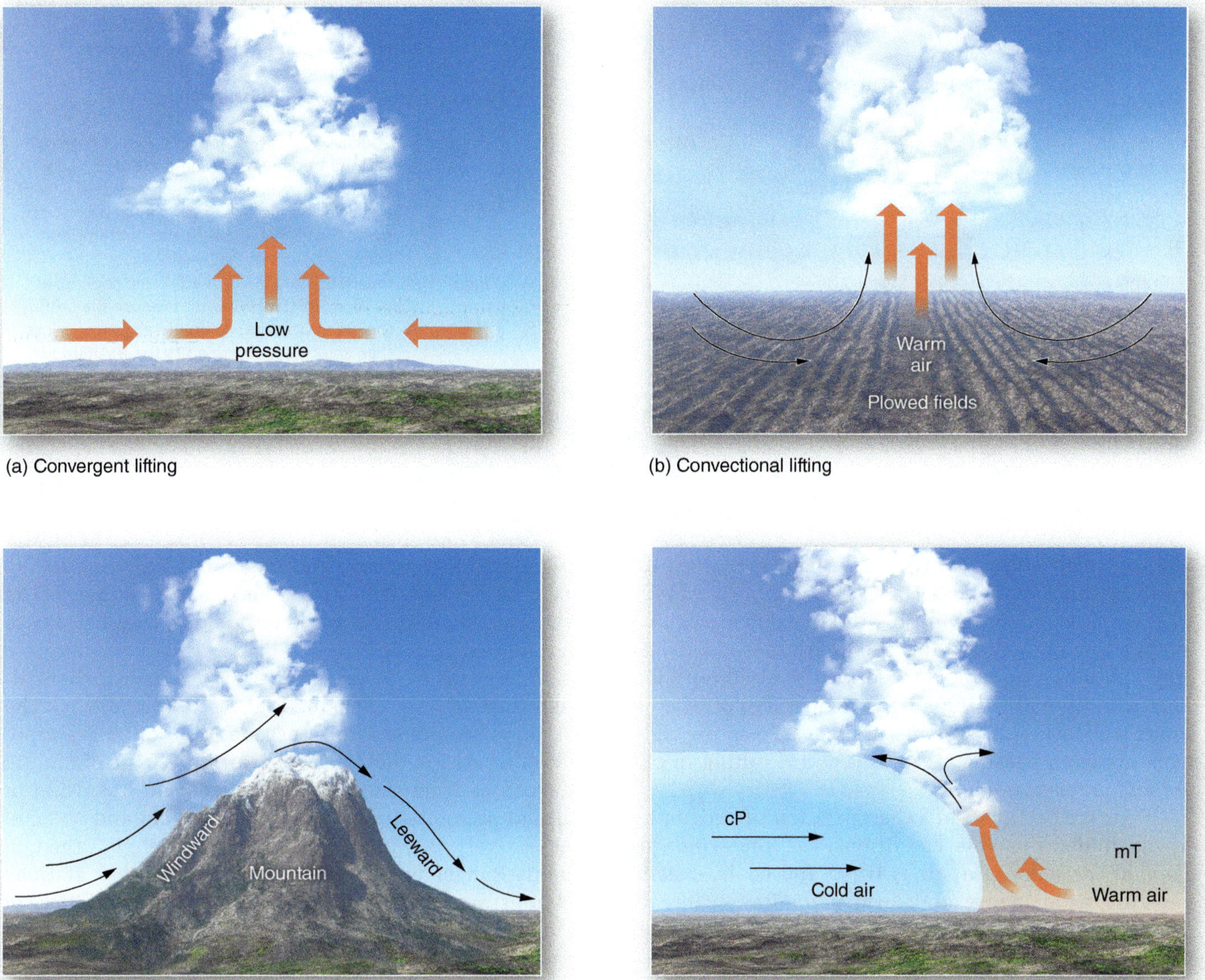

▲**Figure 8.3 Four atmospheric lifting mechanisms.**

MG™ Animation
Atmospheric Stability

in the atmosphere. The rising parcel of air continues its ascent because it is warmer and therefore less dense than the surrounding environment (Figure 8.4).

Florida's precipitation generally illustrates both convergent and convectional lifting mechanisms. Heating of the land produces convergence of onshore winds from the Atlantic and the Gulf of Mexico. As an example of local heating and convectional lifting, Figure 8.5 depicts a day on which the landmass of Florida was warmer than the surrounding Gulf of Mexico and Atlantic Ocean. Because the Sun's radiation gradually heats the land throughout the day and warms the air above it, convectional showers tend to form in the afternoon and early evening. Thus, Florida has the highest frequency of days with thunderstorms in the United States.

Orographic Lifting

The physical presence of a mountain acts as a topographic barrier to migrating air masses. **Orographic lifting** (*oro* means "mountain") occurs when air is forcibly lifted upslope as it is pushed against a mountain (Figure 8.3c).

GEOreport 8.1 Lake-effect snow makes an avalanche of a storm

In October 2006, a lake-effect storm dubbed the "surprise storm" dropped 0.6 m (2 ft) of wet snow in the Buffalo, New York, area. This was the earliest date for such a storm in 137 years of records and remarkable because the ratio of snow to water content was 6:1. Four months later, in February 2007, another lake-effect storm—an "avalanche of a storm," as the media described it—hit upstate New York with 2 m (6.5 ft) of snow in 1 day and over 3.7 m (12 ft) over 7 days.

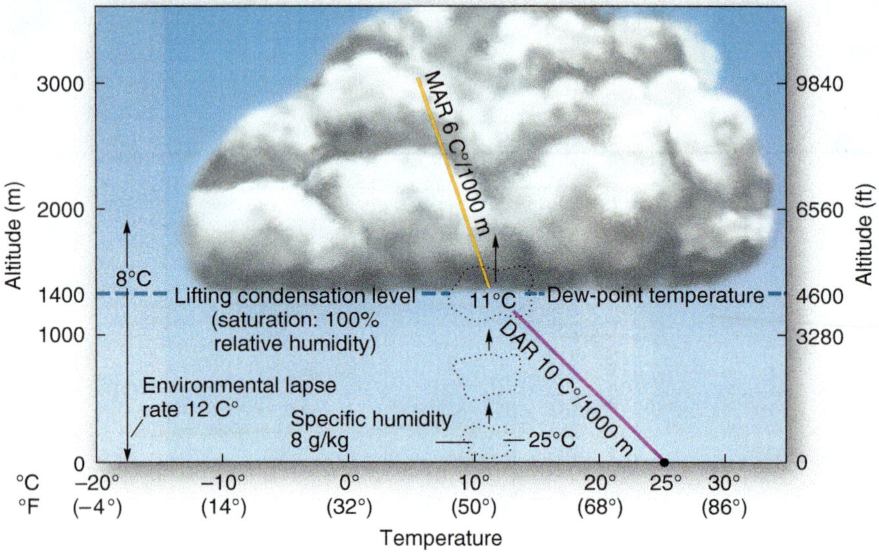

◀Figure 8.4 Convectional activity in unstable conditions. Looking back to Figure 7.15, you can see that these are unstable atmospheric conditions. The environmental lapse rate is 12 C°/1000 m. Specific humidity of the air parcel is 8 g/kg, and the beginning temperature is 25°C. In Figure 7.11, you find on the graph that air with a specific humidity of 8 g/kg must be cooled to 11°C to achieve the dew-point temperature. Here you can see that the dew point is reached, after 14 C° of adiabatic cooling, at 1400 m (4600 ft). Note that we use the DAR (dry adiabatic rate) when the air parcel is less than saturated, changing to the MAR (moist adiabatic rate) above the lifting condensation level at 1400 m.

The lifting air cools adiabatically. Stable air forced upward in this manner may produce stratiform clouds, whereas unstable or conditionally unstable air usually forms a line of cumulus and cumulonimbus clouds. An orographic barrier enhances convectional activity and causes additional lifting during the passage of weather fronts and cyclonic systems, thereby extracting more moisture from passing air masses and resulting in *orographic precipitation*.

Figure 8.6a illustrates the operation of orographic lifting under unstable conditions. On the *windward slope* of the mountain, air is lifted and cools, causing moisture to condense and form precipitation; on the *leeward slope*, the descending air mass heats by compression, and any remaining water in the air evaporates (Figure 8.6b). Thus, air beginning its ascent up a mountain can be warm and moist, but finishing its descent on the leeward slope, it becomes hot and dry. The term **rain shadow** is applied to this dry, leeward side of mountains.

The state of Washington provides an excellent illustration of this concept (Figures 8.6c and 8.7). The Olympic Mountains and Cascade Mountains are topographic barriers that cause orographic lifting of the mP air masses from the North Pacific Ocean, resulting in annual precipitation of more than 500 cm and 400 cm, respectively (200 in. and 160 in.). In Figure 8.7, the Quinault Ranger and Rainier Paradise weather stations demonstrate windward-slope rainfall. The cities of Sequim, in the Puget Trough, and Yakima, in the Columbia Basin, are in the rain shadows (on the leeward sides) of these mountain ranges and have characteristically low annual rainfall. Data for each station include monthly precipitation (portrayed as bars) and temperature (as a line), presented together on a *climograph*, a specialized graph that shows basic climatic characteristics for a certain location (Chapter 10 includes many such graphs).

In North America, **chinook winds** (called *föhn* or *foehn* winds in Europe) are the warm, downslope airflows characteristic of the leeward side of mountains. Such winds can bring a 20-C° (36-F°) jump in temperature and greatly reduce relative humidity.

◀Figure 8.5 Convectional activity over the Florida peninsula. Cumulus clouds cover the land, with several cells developing into cumulonimbus thunderheads. [*Terra* MODIS image, NASA/GSFC.]

Animation
Convectional Activity over the Florida Peninsula

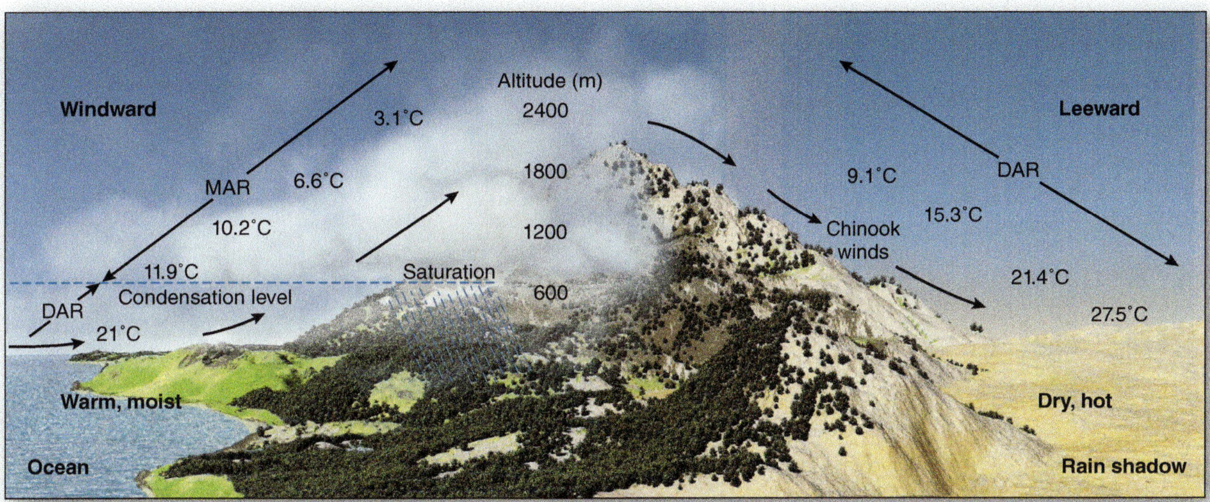

(a) Prevailing winds force warm, moist air upward against a mountain range, producing adiabatic cooling, eventual saturation and net condensation, cloud formation, and precipitation. On the leeward slope, as the "dried" air descends, compressional heating warms it and evaporation dominates, creating the hot, relatively dry rain shadow.

Leeward ← Windward

(b) Rain shadow produced by descending, warming air contrasts with the clouds of the windward side of the Sierra Nevada. Dust is stirred up by downslope winds.

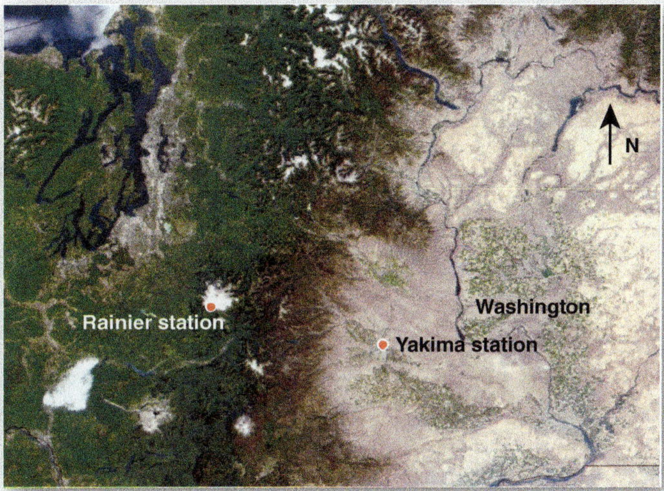

Windward → Leeward

(c) The wetter windward slopes are in contrast to the drier leeward landscapes in Washington (check the map in Figure 8.7).

▲**Figure 8.6 Orographic precipitation, unstable conditions assumed.** [(b) Author. (c) *Terra* MODIS image, NASA/GSFC.]

In the United States, rain-shadow conditions also occur east of the Sierra Nevada and Rocky Mountains. The precipitation pattern of windward and leeward slopes is seen worldwide, as confirmed by the precipitation maps for North America (Figure 9.7) and the world (Figure 10.1).

Frontal Lifting (Cold and Warm Fronts)

The leading edge of an advancing air mass is its *front.* Vilhelm Bjerknes (1862–1951) first applied the term while working with a team of meteorologists in Norway during World War I. Weather systems seemed to them to be

GEOreport 8.2 Mountains cause record rains

Mount Waialeale, on the island of Kaua'i, Hawai'i, rises 1569 m (5147 ft) above sea level. On its windward slope, rainfall averaged 1234 cm (486 in., or 40.5 ft) a year for the years 1941–1992. In contrast, the rain-shadow side of Kaua'i receives only 50 cm (20 in.) of rain annually. If no islands existed at this location, this portion of the Pacific Ocean would receive only an average 63.5 cm (25 in.) of precipitation a year. (These statistics are from established weather stations with a consistent record of weather data; several stations claim higher rainfall values but do not have dependable measurement records.)

Cherrapunji, India, is 1313 m (4309 ft) above sea level at 25° N latitude, in the Assam Hills south of the Himalayas. Summer monsoons pour in from the Indian Ocean and the Bay of Bengal, producing 930 cm (366 in., or 30.5 ft) of rainfall in 1 month. Not surprisingly, Cherrapunji is the all-time precipitation record holder for a single year, 2647 cm (1042 in., or 86.8 ft), and for every other time interval from 15 days to 2 years. The average annual precipitation there is 1143 cm (450 in., 37.5 ft), placing it second only to Mount Waialeale.

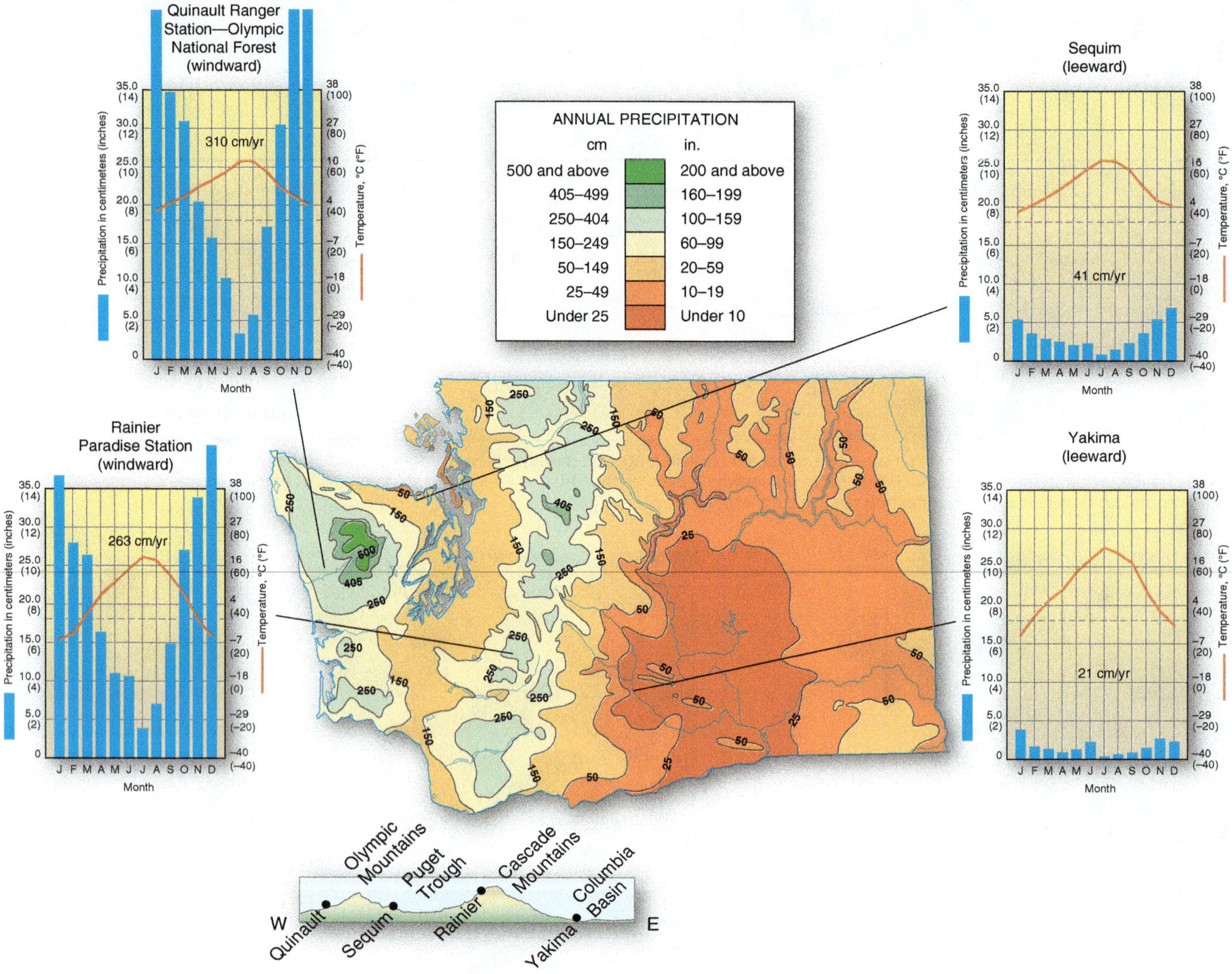

▲Figure 8.7 **Orographic patterns in Washington State.** Four stations in Washington provide examples of orographic effects: windward precipitation and leeward rain shadows. Isohyets (isolines of equal precipitation amounts) on the map indicate rainfall (in inches). Note each station on the landscape profile at bottom. [Data from J. W. Scott and others, *Washington: A Centennial Atlas* (Bellingham, WA: Center for Pacific Northwest Studies, Western Washington University, 1989), p. 3.]

migrating air mass "armies" doing battle along fronts. A front is a place of atmospheric discontinuity, a narrow zone forming a line of conflict between two air masses of different temperature, pressure, humidity, wind direction and speed, and cloud development. The leading edge of a cold air mass is a **cold front**, whereas the leading edge of a warm air mass is a **warm front** (Figures 8.8 and 8.9).

Cold Front

The steep face of an advancing cold air mass reflects the ground-hugging nature of cold air, caused by its greater density and more uniform characteristics compared to the warmer air mass it displaces (Figure 8.8a). Warm, moist air in advance of the cold front lifts upward abruptly and experiences the same adiabatic rates of cooling and factors of stability or instability that pertain to all lifting air parcels.

A day or two ahead of a cold front's arrival, high cirrus clouds appear. Shifting winds, dropping temperature, and lowering barometric pressure mark the front's advance due to lifting of the displaced warmer air along the front's leading edge. At the line of most intense lifting, usually traveling just ahead of the front itself, air pressure drops to a local low. Clouds may build along the cold front into characteristic cumulonimbus form and may appear as an advancing wall of clouds. Precipitation usually is heavy, containing large droplets, and can be accompanied by hail, lightning, and thunder.

The aftermath of a cold front's passage usually brings northerly winds in the Northern Hemisphere and southerly winds in the Southern Hemisphere as anticyclonic high pressure advances. Temperatures drop and air pressure rises in response to the cooler, denser air; cloud cover breaks and clears.

On weather maps, such as the example in Critical Thinking 8.1 just ahead, a cold front is depicted as a line with triangular spikes that point in the direction of frontal movement along an advancing air mass. The

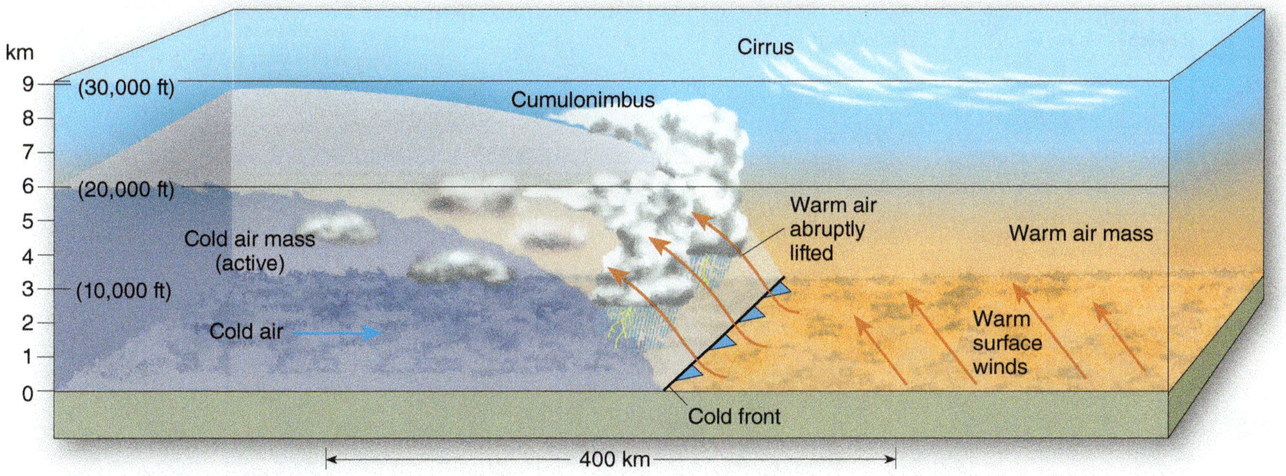

(a) Denser, advancing cold air forces warm, moist air to lift abruptly. As the air is lifted, it cools by expansion at the DAR, rising to a level of condensation and cloud formation, where it cools to the dew-point temperature.

(b) A sharp line of cumulonimbus clouds near the Texas coast and Gulf of Mexico marks a cold front and squall line. The cloud formation rises to 17,000 m (56,000 ft). The passage of such a frontal system over land often produces strong winds, cumulonimbus clouds, large raindrops, heavy showers, lightning and thunder, hail, and the possibility of tornadoes.

MG™ Animation Cold and Warm Fronts

▲**Figure 8.8 A typical cold front.** [(b) NASA.]

particular shape and size of the North American landmass and its latitudinal position present conditions where cP and mT air masses are best developed and have the most direct access to each other. The resulting contrast can lead to dramatic weather, particularly in late spring, with sizable temperature differences from one side of a cold front to the other.

A fast-advancing cold front can cause violent lifting, creating a zone known as a **squall line** right along or slightly ahead of the front. (A *squall* is a sudden episode of high winds that is generally associated with bands of thunderstorms.) Along a squall line, such as the one in the Gulf of Mexico shown in Figure 8.8b, wind patterns are turbulent and wildly changing, and precipitation is intense. The well-defined frontal clouds in the photograph rise abruptly, feeding the formation of new thunderstorms along the front. Tornadoes also may develop along such a squall line.

Warm Front Warm air masses can be carried by the jet stream into regions with colder air, such as when an airflow called the "Pineapple Express" carries warm, moist air from Hawai'i and the Pacific into California. Warm air also advances during monsoonal flow conditions, such as when mT air moves from the Pacific Ocean into the low-pressure areas over the heated landmasses of the U.S. Southwest during July and August. The leading edge of an advancing warm air mass is unable to displace cooler, passive air, which is denser along the surface. Instead, the warm air tends to push the cooler, underlying air into a characteristic wedge shape, with the warmer air sliding up over the cooler air. Thus, in the cooler-air region, a temperature inversion is present, sometimes causing poor air drainage and stagnation.

Figure 8.9 illustrates a typical warm front, in which gentle lifting of mT air leads to stratiform cloud development and characteristic nimbostratus clouds as well as drizzly precipitation. A warm front presents a progression of cloud development to an observer: High cirrus and cirrostratus clouds announce the advancing frontal system; then come lower and thicker altostratus clouds; and finally, still lower and thicker stratus clouds appear within several hundred kilometers of the front. A line with semicircles facing in the direction of frontal movement denotes a warm front on weather maps (see the map in Figure GIA 8.1 on page 202).

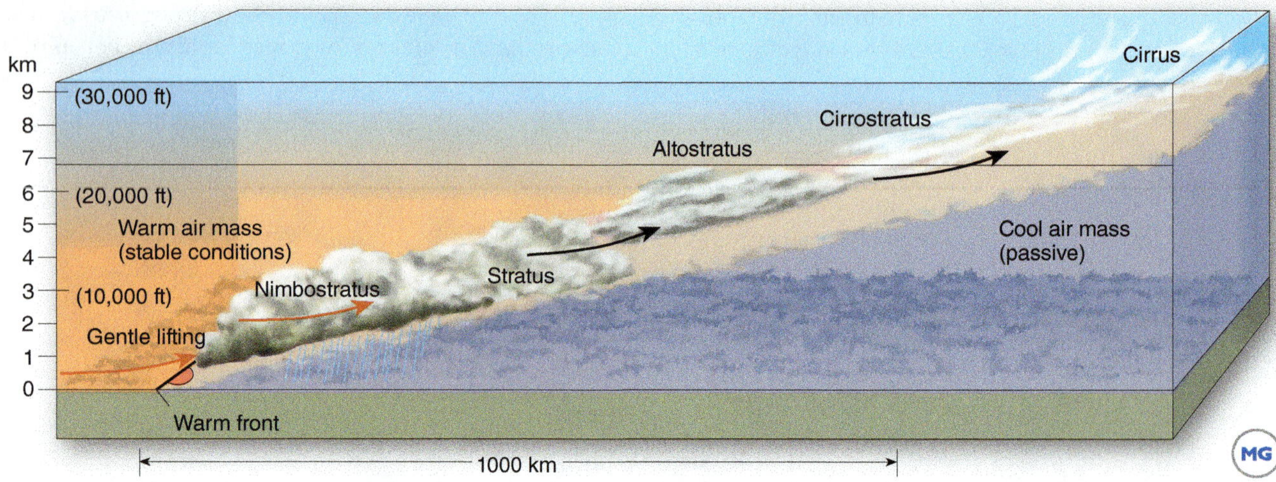

▲Figure 8.9 **A typical warm front.** Note the sequence of cloud development as the warm front approaches. Warm air slides upward over a wedge of cooler, passive air near the ground. Gentle lifting of the warm, moist air produces nimbostratus and stratus clouds and drizzly rain showers, in contrast to the more dramatic cold-front precipitation.

MG Animation Cold and Warm Fronts

Midlatitude Cyclonic Systems

The conflict between contrasting air masses can develop a **midlatitude cyclone**, also known as a **wave cyclone** or *extratropical cyclone*. Midlatitude cyclones are migrating low-pressure weather systems that occur in the middle latitudes, outside the tropics. They have a low-pressure center with converging, ascending air spiraling inward counterclockwise in the Northern Hemisphere and inward clockwise in the Southern Hemisphere, owing to the combined influences of the *pressure gradient force*, *Coriolis force*, and *surface friction* (see discussion in Chapter 6). Because of the undulating nature of frontal boundaries and of the jet streams that steer these cyclones across continents, the term *wave* is appropriate. An emerging model of this interactive system characterizes air mass flows as being like "conveyor belts" as described in Figure GIA 8.2, just ahead.

Wave cyclones, which can be 1600 km (1000 mi) wide, dominate weather patterns in the middle and higher latitudes of both the Northern and the Southern Hemispheres. A midlatitude cyclone can originate along the polar front, particularly in the region of the Icelandic and Aleutian subpolar low-pressure cells in the Northern Hemisphere. Certain other areas are associated with cyclone development and intensification: the eastern slope of the Rocky Mountains, home of the Colorado Low in the United States and of the Alberta Clipper in Canada; the Gulf Coast, home of the Gulf Low; and the eastern seaboard, home of the nor'easter and Hatteras Low (Figure 8.10).

Nor'easters are notorious for bringing heavy snows to the northeastern United States. The February 2013 nor'easter resulted from the merging of two low-pressure areas off the northeast coast on February 8, producing record snowfall in Portland, Maine (81 cm, or 32 in), and a maximum snowfall of 100 cm (40 in) in Hamden, Connecticut. This precipitation arrived with pressure readings around 968 mb, wind gusts up to 164 kmph (102 mph), and storm surges up to 1.3 m (4.2 ft).

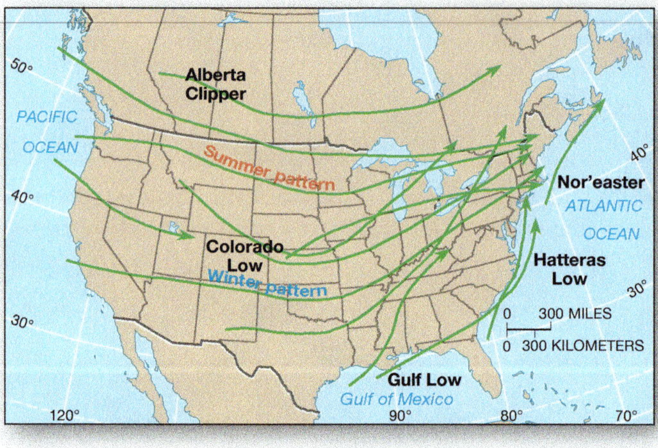

▲Figure 8.10 **Typical cyclonic storm tracks over North America.** Cyclonic storm tracks vary seasonally. Note the regional names reflecting locations of cyclogenesis.

The intense high-speed winds of the jet streams guide cyclonic systems, with their attendant air masses, across the continent (review jet streams in Figures 6.13 and 6.14) along **storm tracks** that shift in latitude with the Sun and the seasons. Typical storm tracks crossing North America are farther northward in summer and farther southward in winter. As the storm tracks begin to shift northward in the spring, cP and mT air masses are in their clearest conflict. This is the time of strongest frontal activity, featuring thunderstorms and tornadoes.

Life Cycle of a Midlatitude Cyclone

shows the birth, maturity, and death of a typical midlatitude cyclone in several stages, along with an idealized weather map. On average, a midlatitude cyclonic system takes 3–10 days to progress through this life cycle from the area where it develops to the area where it finally dissolves. However, every day's weather map departs from this model in some manner.

Cyclogenesis The first stage is **cyclogenesis**, the atmospheric process in which low-pressure wave cyclones develop and strengthen. This process usually begins along the polar front, where cold and warm air masses converge and are drawn into conflict, creating potentially unstable conditions. For a wave cyclone to form along the polar front, a compensating area of divergence aloft must match a surface point of air convergence. Even a slight disturbance along the polar front, perhaps a small change in the path of the jet stream, can initiate the converging, ascending flow of air and thus a surface low-pressure system in Figure GIA 8.1, Stage 1.

In addition to the polar front, certain other areas are associated with wave cyclone development and intensification: the eastern slope of the Rockies and other north–south mountain barriers, the Gulf Coast, and the east coasts of North America and Asia.

Open Stage In the open stage, to the east of the developing low-pressure center of a Northern Hemisphere midlatitude cyclone, warm air begins to move northward along an advancing front, while cold air advances southward to the west of the center. See this organization in Figure GIA 8.1, Stage 2 and around the center of low pressure, located over western Nebraska, in the GI A map. As the midlatitude cyclone matures, the counterclockwise flow draws the cold air mass from the north and west and the warm air mass from the south. In the cross section, you can see the profiles of both a cold front and a warm front and each air mass segment.

Occluded Stage Next is the occluded stage (in Figure GIA 8.1, Stage 3.). Remember the relation between air temperature and the density of an air mass. The colder cP air mass is denser than the warmer mT air mass. This cooler, more unified air mass, acting like a bulldozer blade, moves faster than the warm front. Cold fronts can travel at an average 40 kmph (25 mph), whereas warm fronts average roughly half that, 16–24 kmph (10–15 mph). Thus, a cold front often overtakes the cyclonic warm front and wedges beneath it, producing an **occluded front** (*occlude* means "to close").

On the idealized weather map in GI A, an occluded front stretches south from the center of low pressure in Virginia to the border between the Carolinas. Precipitation may be moderate to heavy initially and then taper off as the warmer air wedge is lifted higher by the advancing cold air mass. Note the still active warm front in the extreme Southeast and the flow of mT air. What conditions do you observe in Tallahassee, Florida? What might it be like there in the next 12 hours if the cold front passes south of the city?

When there is a stalemate between cooler and warmer air masses such that airflow on either side is almost parallel to the front, although in opposite directions, a **stationary front** results. Some gentle lifting might produce light to moderate precipitation. Eventually, the stationary front will begin to move, as one of the air masses assumes dominance, evolving into a warm or a cold front.

Dissolving Stage Finally, the dissolving stage of the midlatitude cyclone occurs when its lifting mechanism is completely cut off from the warm air mass, which was its source of energy and moisture. Remnants of the cyclonic system then dissipate in the atmosphere, perhaps after passage across the country in Figure GIA 8.1, Stage 4.

Although the actual patterns of cyclonic passage over North America are widely varied in shape and duration, you can apply this general model of the stages of a midlatitude cyclone, along with your understanding of warm and cold fronts, to reading the daily weather map. Examples of the standard weather symbols used on maps are shown below the map in Figure GIA 8.1.

Weather Maps and Forecasting

Synoptic analysis is the evaluation of weather data collected at a specific time. Building a database of wind, pressure, temperature, and moisture conditions is key to *numerical*, or computer-based, *weather prediction* and the development of weather-forecasting models. Development of numerical models is a great challenge because the atmosphere operates as a nonlinear system, tending toward chaotic behavior. Slight variations in input data or slight changes in the model's basic assumptions can produce widely varying forecasts. The accuracy of forecasts continues to improve with technological advancements in instruments and software and with our increasing knowledge of the atmospheric interactions that produce weather.

Weather data necessary for the preparation of a synoptic map and forecast include the following:

- Barometric pressure (sea level and altimeter setting)
- Pressure tendency (steady, rising, falling)
- Surface air temperature
- Dew-point temperature
- Wind speed, direction, and character (gusts, squalls)
- Type and movement of clouds
- Current weather
- State of the sky (current sky conditions)
- Visibility; vision obstruction (fog, haze)
- Precipitation since last observation

Environmental satellites are one of the key tools in forecasting weather and analyzing climate. Massive computers handle volumes of data from surface, aircraft, and orbital platforms for accurate forecasting of near-term weather. These data are also used for assessing climatic change. In the United States, the National Weather Service (NWS) provides weather forecasts and current satellite images (see http://www.nws.noaa.gov/). In Canada, the Meteorological Service of Canada provides forecasts at http://www.weatheroffice.gc.ca/canada_e.html.

(text continued on page 204)

A midlatitude cyclone is a low-pressure system that forms when a cool air mass (cP) collides with a warm, moist air mass (mT). Steered by the jet stream, these storms typically migrate from west to east. A cyclonic system of air mass interactions is represented in the established Norwegian model (GIA 8.1), or in a new conceptualization depicting air mass flows in a conveyor belt model (GIA 8.2). Remote sensing from satellite platforms is essential for analyzing cyclonic structure (GIA 8.3).

8.1 AIR-MASS MODEL

The life-cycle of a midlatitude cyclone has four stages that unfold as contrasting cold and warm air masses meet, driven by flows of conflicting air.

Stage 1: Cyclogenesis
A disturbance develops along the polar front or in certain other areas. Warm air converges near the surface and begins to rise, creating instability.

Stage 2: Open stage
Cyclonic, counterclockwise flow pulls warm, moist air from the south into the low-pressure center while cold air advances southward west of the center.

Stage 3: Occluded stage
The faster-moving cold front overtakes the slower warm front and wedges beneath it. This forms an occluded front, along which cold air pushes warm air upward, causing precipitation.

Stage 4: Dissolving stage
The midlatitude cyclone dissolves when the cold air mass completely cuts off the warm air mass from its source of energy and moisture.

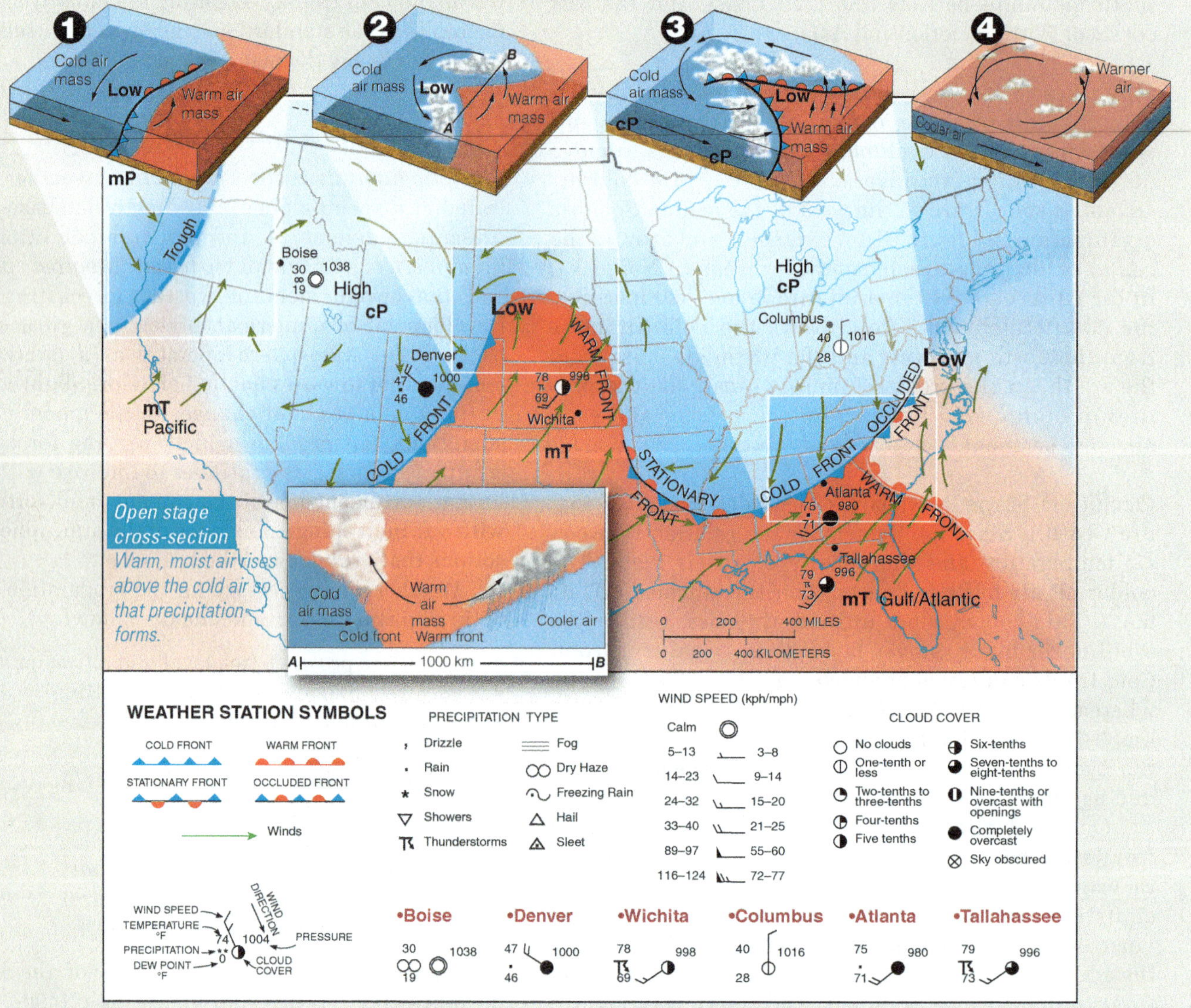

Describe: Based on the map, what is the weather in Denver and Wichita? How will their weather change as the storm moves east?

Animation
Midlatitude Cyclones

MasteringGeography™

Visit the Study Area in MasteringGeography™ to explore midlatitude cyclones.

Visualize: Study geosciences animations of a midlatitude cyclone.

Assess: Demonstrate understanding of midlatitude cyclones (if assigned by instructor).

8.2 CONVEYOR BELT MODEL

As you can see from Figure GIA 8.1, air mass flows and cyclonic circulation involve interactions in a physical, fluid system. A way of viewing these channels of air, in a three-dimensional perspective, is to think of an air mass flow as a conveyor belt, although lacking a return flow component. In this illustration, you see three conveyors of air and moisture, one aloft and two initially along the surface, that interact to produce a midlatitude cyclone and sustain it as a dynamic system.

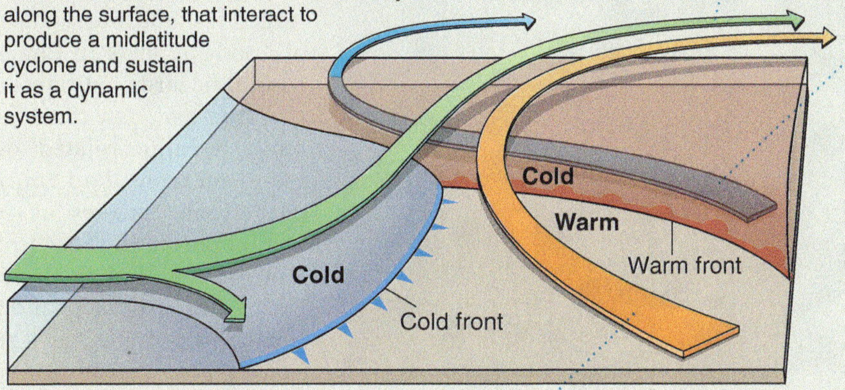

Dry conveyor belt
Dry, cold air aloft flows in the cyclonic circulation from the west, with some descending behind the cold front as clear, cold air. Another branch of this flow moves cyclonically toward the low. This generally cloud-free dry sector can form a "dry slot," separating warm and cold cloud bands, clearly visible on satellite images.

Cold conveyor belt
Cold surface air flows westward beneath the less dense air. Approaching convergence with the low forces lifting, with one lifted stream turning counterclockwise around the low. Another stream moves clockwise to join the westerly flow aloft. As the cold conveyor passes beneath the warm channel it picks up moisture, and becomes saturated as it rises. Thus, this cold conveyor can be an important snow producer northwest of the low; an area labeled in GIA 8.3a as a "comma head."

Warm conveyor belt
Warm, moist air moves as a surface flow into the system, riding upward over cooler air to the north. A warm front structure results with gentle lifting and stratus clouds. During its passage, moisture is delivered ahead of the cold front, subjected to abrupt lifting, condensation, and cumulonimbus clouds. This flow is the principal moisture source for the frontal systems. Eventually, the warm air conveyor turns eastward and joins the westerly flow aloft.

Explain: How do the conveyor belts interact to produce precipitation north of the warm front?

8.3 OBSERVING A MIDLATITUDE CYCLONE

Satellite images reveal the flow of moist and dry air that drives a midlatitude cyclone, as well as how the storm changes over time.

(a) September 26, 2011- Occluded stage. This satellite image shows a midlatitude cyclone over the midwestern U.S. [NASA.]

(b) Water vapor image. The cold conveyor belt delivers cold, dry air (in yellow) that will soon cut off the storm's supply of warm, moist air. [NOAA]

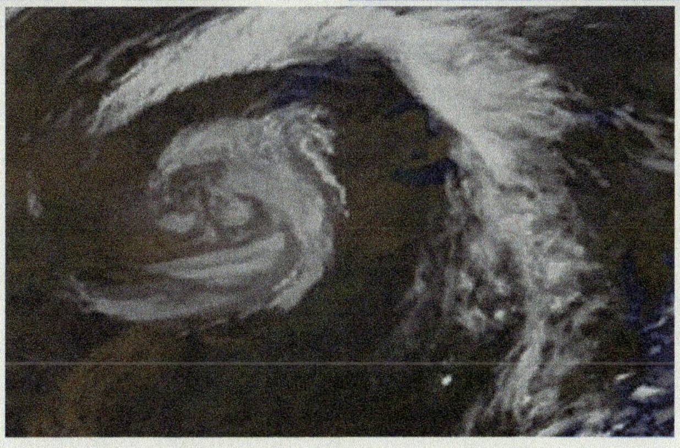

(c) September 27, 2011– Dissolving stage. Without a source of warm, moist air, the storm begins to dissolve [NOAA].

GEOquiz

1. **Summarize:** In your own words, summarize the life cycle of a midlatitude cyclone.

2. **Analyze:** Which of the three conveyor belts is most critical in maintaining a midlatitude cyclone? Explain.

3. From the sources given in this chapter, find several satellite images of midlatitude cyclones and identify the basic elements described here. List the dates you found.

(a) A radar antenna is within the dome structure of the NWS weather installation at the Indianapolis International Airport.

(b) ASOS weather-instrument station.

Figure 8.11 NWS weather installation and ASOS weather instruments. [Bobbé Christopherson.]

MG Animation Midlatitude Cyclones

Internationally, the World Meteorological Organization coordinates weather information (see http://www.wmo.ch/).

An essential element of weather forecasting is Doppler radar. Using backscatter from two radar pulses, it detects the direction of moisture droplets toward or away from the radar source indicating wind direction and speed (Figure 8.11). This information is critical to making accurate severe storm warnings. As part of the Next Generation Weather Radar (NEXRAD) program, the NWS operates 159 Doppler radar systems, mainly in the United States (http://radar.weather.gov/); installations also exist in Japan, Guam, South Korea, and the Azores. For links to weather maps, current forecasts, satellite images, and the latest radar, please go to the *MasteringGeography* website.

Weather information in the United States comes mainly from the Automated Surface Observing System (ASOS), which is installed at over 900 airports across the country. An ASOS instrument array is made up of numerous sensors that supply continuous on-the-ground data concerning weather elements, helping the NWS increase the timeliness and accuracy of its forecasts. Because weather data come from a large number of sources, NOAA developed the Advanced Weather Interactive Processing System (AWIPS), which consists of high-tech computer workstations that assemble the various types of information, thereby improving forecast accuracy and severe weather warnings. AWIPS is able to display and integrate a wide range of data—for example, air pressure, water vapor, humidity, Doppler radar, lightning strikes in real time, and wind profiles.

Violent Weather

Weather is a continuous reminder that the flow of energy across the latitudes can at times set into motion destructive, violent events. We focus in this chapter on ice storms, thunderstorms, derechos, tornadoes, and hurricanes; coverage of floods appears in Chapter 14 and coastal hazards in Chapter 16.

Weather is often front-page news. Weather-related destruction has risen more than 500% over the past three decades as population has increased in areas prone to violent weather and as climate change intensifies weather anomalies. In the United States, government research and monitoring of violent weather is centered at NOAA's National Severe Storms Laboratory and Storm Prediction Center—see http://www.nssl.noaa.gov/ and http://www.spc.noaa.gov/; consult these sites for each of the topics that follow.

Ice Storms and Blizzards

Ice storms and blizzards are types of violent weather that are generally confined to the mid- to high-latitude regions of the world. According to the National Weather Service, an *ice storm* is a winter storm in which at least 6.4 mm (0.25 in.) of ice accumulates on exposed surfaces. Ice storms occur when a layer of warm air is between two layers of cold air. When precipitation falls through the warm layer into a below-freezing layer of air nearer the ground layer, it may form **sleet**, which is composed of freezing rain, ice glaze, and ice pellets (see *The Human Denominator*, Figure HD 8, at the end of the chapter). During the ice storm of January 1998, the 700,000 residents of a large region of Canada and the United States were without power for weeks as ice-coated power lines and tree limbs collapsed under the added weight. Freezing rain and drizzle continued for over 80 hours, more than double the typical ice-storm duration. In Montreal, over 100 mm (4 in.) of ice accumulated. Hypothermia claimed 25 lives.

Blizzards are snowstorms with frequent gusts or sustained winds greater than 56 kmph (35 mph) for a period of time longer than 3 hours and blowing snow that reduces visibility to 400 m (0.25 mi) or less. These storms often result in large snowfall and can paralyze regional transportation both during the storm and for days afterward.

Thunderstorms

By definition, a *thunderstorm* is a type of turbulent weather accompanied by lightning and thunder. Such storms are characterized by a buildup of giant cumulonimbus clouds that can be associated with squall lines of heavy rain, including sleet, blustery winds, hail, and tornadoes. Thunderstorms may develop within an air mass, in a line along a front (particularly a cold front), or where mountain slopes cause orographic lifting.

Thousands of thunderstorms occur on Earth at any given moment. Equatorial regions and the ITCZ experience many of them, exemplified by the city of Kampala, Uganda, in East Africa (north of Lake Victoria), which sits virtually on the equator and averages a record 242 days a year with thunderstorms. In North America, most thunderstorms occur in areas dominated by mT air masses (Figure 8.12).

A thunderstorm is fueled by the rapid upward movement of warm, moist air. As the air rises, cools, and condenses to form clouds and precipitation, tremendous energy is liberated by the condensation of large quantities of water vapor. This process locally heats the air, causing violent updrafts and downdrafts as rising parcels of air pull surrounding air into the column and as the frictional drag of raindrops pulls air toward the ground (review the illustration of a cumulus cloud in Figure 7.18).

CRITICALthinking 8.1
Analyzing a Weather Map

The weather map in Figure CT 8.1.1 shows a classic open-stage midlatitude cyclone from March 31, 2007, 7 A.M. EST, along with a *GOES-12* satellite image of water vapor at that time. Referring to Figure GIA 8.1, p. 202 for the meaning of weather map symbols, briefly analyze this map: Find the center of low pressure, note the counterclockwise winds, compare temperatures on either side of the cold front, note air temperatures and dew-point temperatures, determine why the fronts are positioned where they are, and locate the center of high pressure.

You can see that the air and dew-point temperatures in New Orleans are 68°F and 63°F, respectively; yet, in St. George, Utah (southwest corner of state), you see 27°F and 22°F, respectively. To become saturated, the warm, moist air in New Orleans needs to cool to 63°F, whereas the dry, cold air in St. George needs to cool to 22°F for saturation. In the far north, find Churchill on the western shore of Hudson Bay. The air and dew-point temperatures are 6°F and 3°F, respectively, and the state of the sky is clear. If you were there, what would you experience at this time? Why would you be reaching for the lip balm?

Describe the pattern of air masses on this weather map. How are these air masses interacting, and what kind of frontal activity do you see? What does the pattern of isobars tell you about high- and low-pressure areas? •

▼Figure CT 8.1.1 Weather map and water-vapor image, March 31, 2007. [Map courtesy of Hydrometeorological Prediction Center, NCEP, NWS.]

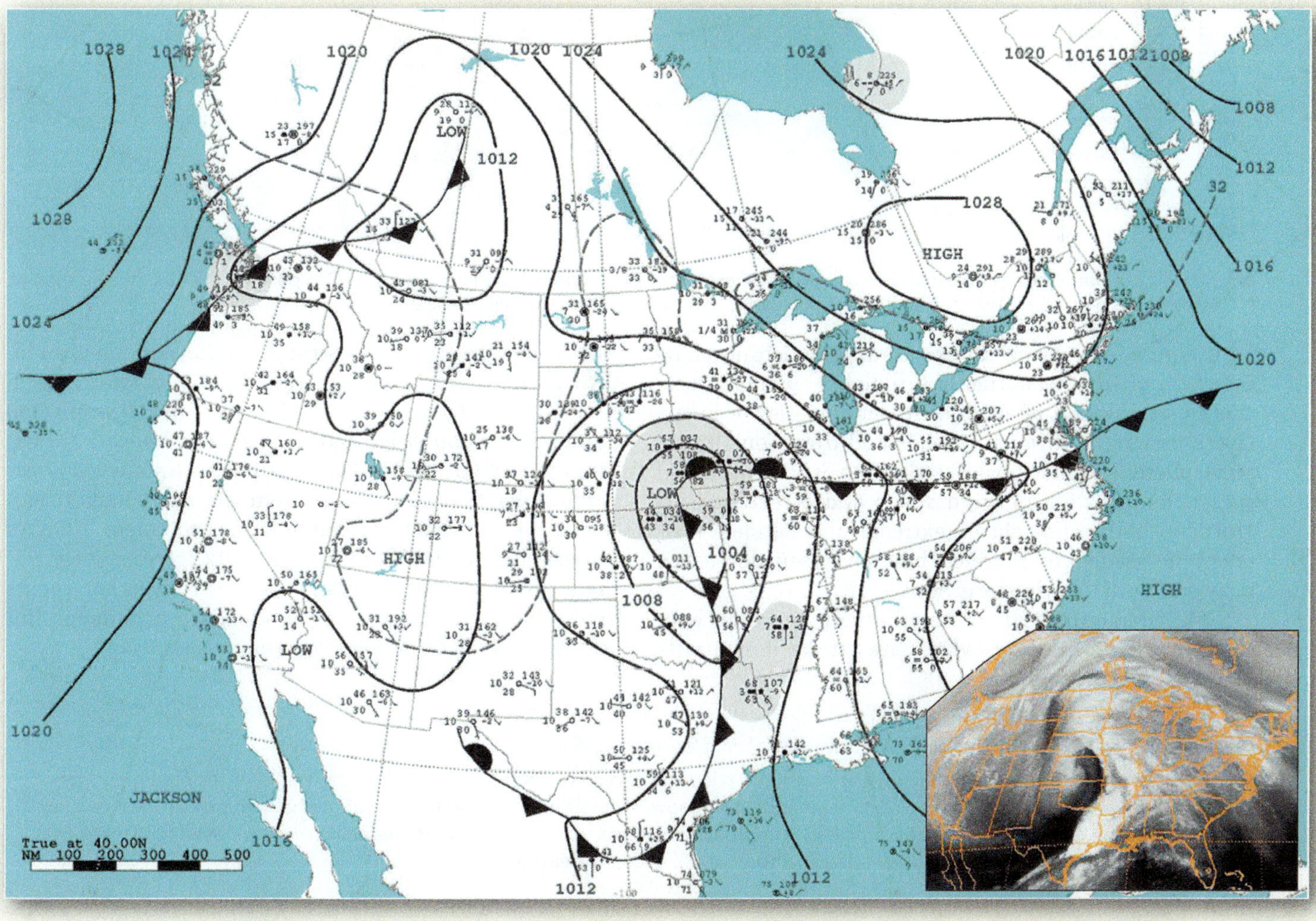

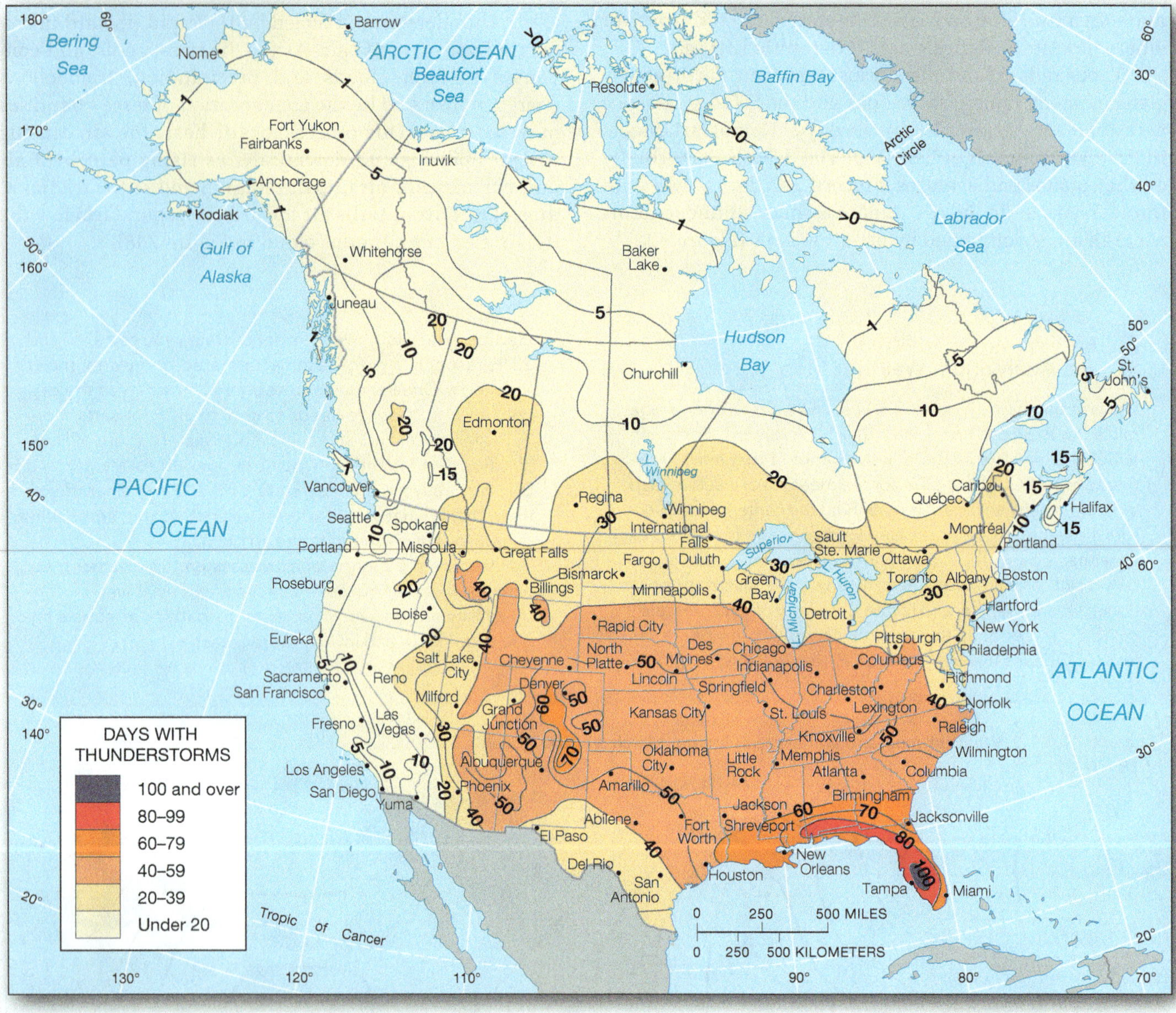

▲Figure 8.12 Thunderstorm occurrence. Average annual number of days experiencing thunderstorms. Compare this map with the location of mT air masses in Figure 8.1. [Data courtesy of NWS; Map Series 3; *Climatic Atlas of Canada*, Atmospheric Environment Service, Canada.]

Turbulence and Wind Shear A distinguishing characteristic of thunderstorms is turbulence, which is created by the mixing of air of different densities or by air layers moving at different speeds and directions in the atmosphere. Thunderstorm activity also depends on *wind shear*, the variation of wind speed and direction with altitude—high wind shear (extreme and sudden variation) is needed to produce hail and tornadoes, two by-products of thunderstorm activity.

Thunderstorms can produce severe turbulence in the form of *downbursts*, which are strong downdrafts that cause exceptionally strong winds near the ground. Downbursts are classified by size: A *macroburst* is at least 4.0 km (2.5 mi) wide and in excess of 210 kmph (130 mph); a *microburst* is smaller in size and speed. Downbursts are characterized by the dreaded high-wind-shear conditions that can bring down aircraft. Such turbulence events are short-lived and hard to detect. NOAA's forecasting model, launched in 2012, gives hourly updates to improve predictions for severe weather events and aviation hazards such as clear-air turbulence (for more information, see http://rapidrefresh.noaa.gov/).

GEOreport 8.3 Kentucky ice storm causes record power losses

In late January 2009, an ice storm hit southern Indiana and central Kentucky. Over a 3-day period, precipitation began as freezing rain, later changing to sleet and finally snow. Across Kentucky, the ice-related damage to power lines resulted in 609,000 homes and businesses without power; outages continued for up to 10 days, and area schools were closed for a week. Ice storms of similar intensity blanketed the Midwest, upper South, and Northeast in February 2011.

Supercells The strongest thunderstorms are known as supercell thunderstorms, or *supercells*, and give rise to some of the world's most severe and costly weather events (such as hailstorms and tornadoes). Supercells often contain a deep, persistently rotating updraft called a **mesocyclone**, a spinning, cyclonic, rising column of air associated with a convective storm and ranging up to 10 km (6 mi) in diameter. A well-developed mesocyclone will produce heavy rain, large hail, blustery winds, and lightning; some mature mesocyclones will generate tornado activity (as discussed later in this chapter).

The conditions conducive to forming thunderstorms and more intense supercells—lots of warm, moist air and strong convective activity—are enhanced by climate change. However, wind shear, another important factor in thunderstorm and supercell formation, will likely lessen in the midlatitudes as Arctic warming reduces overall temperature differences across the globe. Research is ongoing as to which of these effects will be more important for determining severe thunderstorm frequency in different regions of the world.

Lightning and Thunder An estimated 8 million lightning strikes occur each day on Earth. **Lightning** is the term for flashes of light caused by enormous electrical discharges—tens of millions to hundreds of millions of volts—that briefly superheat the air to temperatures of 15,000°–30,000°C (27,000°–54,000°F). A buildup of electrical-energy polarity between areas within a cumulonimbus cloud or between the cloud and the ground creates lightning. The violent expansion of this abruptly heated air sends shock waves through the atmosphere as the sonic bang of **thunder**.

Lightning poses a hazard to aircraft, people, animals, trees, and structures, causing nearly 200 deaths and thousands of injuries each year in North America. When lightning is imminent, the NWS issues *severe storm warnings* and cautions people to remain indoors. People caught outdoors as a lightning charge builds should not seek shelter beneath a tree, as trees are good conductors of electricity and often are hit by lightning. Data from NASA's Lightning Imaging Sensor (LIS) show that about 90% of all strikes occur over land in response to increased convection over relatively warmer continental surfaces (Figure 8.13; see http://thunder.msfc.nasa.gov/data/data_nldn.html).

Hail Ice pellets larger than 0.5 cm (0.20 in) that form within a cumulonimbus cloud are known as **hail**—or *hailstones*, after they fall to the ground. During hail formation, raindrops circulate repeatedly above and below the freezing level in the cloud, adding layers of ice until the circulation in the cloud can no longer support their weight. Hail may also grow from the addition of moisture on a snow pellet.

Pea-sized hail (0.63 cm, or 0.25 in., in diameter) is common, although hail can range from the size of quarters (2.54 cm, or 1.00 in.) to softballs (11.43 cm, or 4.50 in.). Baseball-sized (6.98 cm, or 2.75 in.) hail fell a half dozen times in 2010 in the United States alone. For larger hail to form, the frozen pellets must stay aloft for longer periods. The largest authenticated hailstone in the world fell from a thunderstorm supercell in Aurora, Nebraska, June 22, 2003, measuring 47.62 cm (18.75 in.) in circumference. However, the largest hailstone by diameter and weight fell in Vivian, South Dakota, in July 2010 (Figure 8.14).

▼Figure 8.13 Global lightning strikes, 1998 to 2012. [(a) Lightning Imaging Sensor-Optical Transient Detector (LIS-OTD) global lightning image obtained from the NASA EOSDIS Global Hydrology Resource Center DAAC, Huntsville, AL. Reprinted by permission of Richard Blakeslee. (b) Keith Kent/Photo Researchers.]

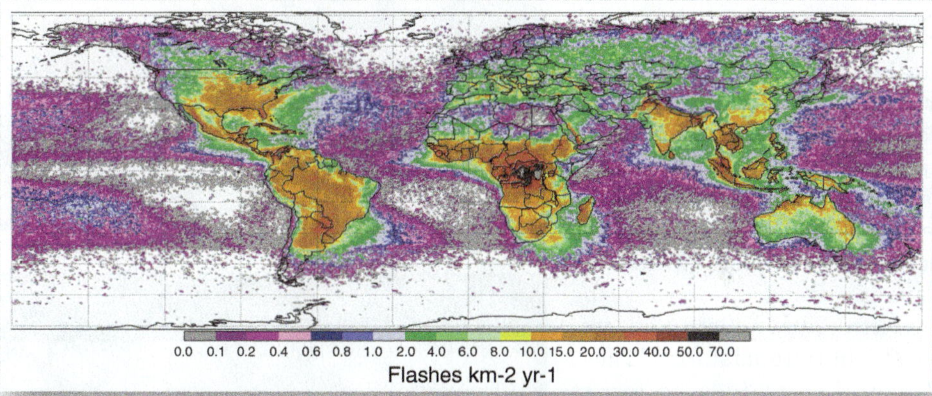

(a) Map of total annual lightning strikes (flashes) from January 1998 to February 2012. The lightning imaging sensor aboard the *TRMM* satellite combines optical and electronic elements that can detect lightning within individual storms, day or night, between 35° N and 35° S latitudes.

(b) Multiple lightning strikes in southern Arizona captured in a time-lapse photo.

◀**Figure 8.14 Largest-diameter hailstone ever recorded.** This hailstone measuring 20.32 cm (8.0 in.) in diameter fell in South Dakota during a supercell thunderstorm with winds exceeding 129 kmph (80 mph). Its circumference measured 47.307 cm (18.625 in), just under the world record. Read more about this hail event at http://www.crh.noaa.gov/abr/?n=stormdamagetemplate. [NOAA.]

Hail is common in the United States and Canada, although somewhat infrequent at any given place. Hail occurs perhaps every 1 or 2 years in the highest-frequency areas. Annual hail damage in the United States tops $800 million.

Derechos

Straight-line winds associated with fast-moving, severe thunderstorms can cause significant damage to urban areas, as well as crop losses in agricultural regions. Such winds, known as **derechos**—or *plow winds,* in Canada—are produced by the powerful downbursts characteristic of thunderstorms. These strong, linear winds, in excess of 26 m/s (58 mph), tend to blast in straight paths fanning out along curved-wind fronts over a wide swath of land. The name, coined by physicist G. Hinrichs in 1888, derives from a Spanish word meaning "direct" or "straight ahead."

Because downbursts occur in irregular clusters, the winds associated with a derecho may vary considerably in intensity along its path. One of the strongest derecho wind gusts ever measured occurred in 1998 in eastern Wisconsin, exceeding 57 m/s (128 mph). In August 2007, a series of derechos across northern Illinois reached this same intensity. Researchers identified 377 derechos between 1986 and 2003, an average of about 21 per year.

Derechos pose distinct hazards to summer outdoor activities by overturning boats, hurling flying objects, and breaking tree trunks and limbs. Their highest frequency (about 70%) is from May to August in the region stretching from Iowa, across Illinois, and into the Ohio River Valley in the upper Midwest. In June 2012, a derecho originated near Chicago and traveled southeastward to hit Washington, D.C., causing damage and power outages for millions of people, as well as several fatalities (Figure 8.15). By September and on through to

▼**Figure 8.15 The June 2012 derecho event.** [(a) Courtesy of Kevin Gould, NOAA. (b) Base image by G. Carbin, NOAA Storm Prediction Center.]

(a) A shelf cloud over LaPorte, Indiana, indicates the strong winds of the advancing derecho.

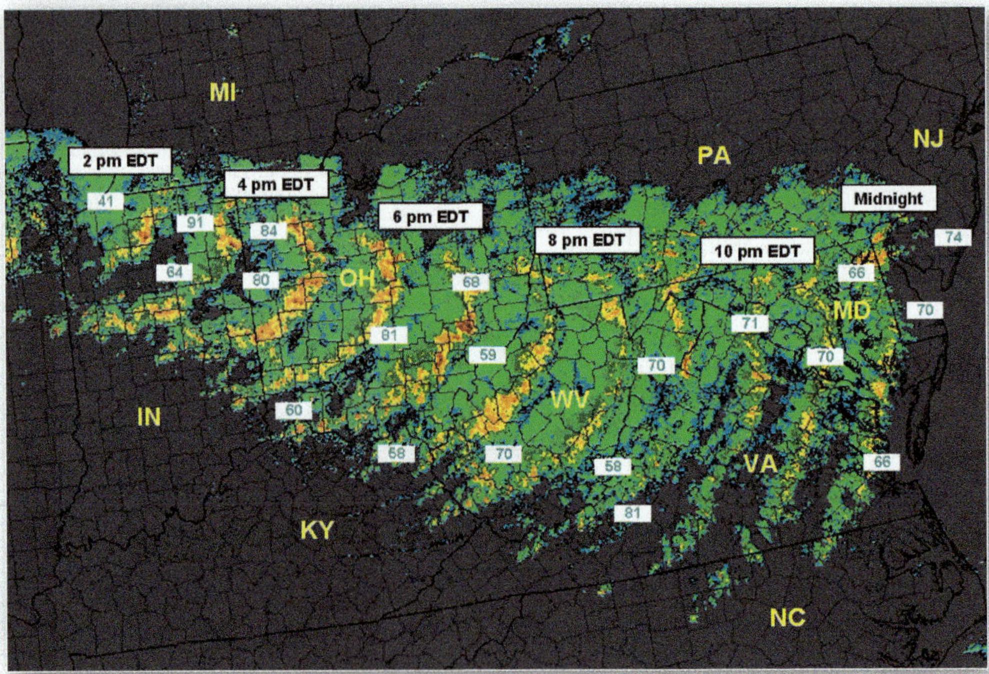

(b) Composite of hourly radar reflectivity imagery shows the development of the June 29, 2012, derecho event, including selected wind gusts (mph), beginning at 2 p.m. Eastern Daylight Time (far left) and ending at midnight (far right).

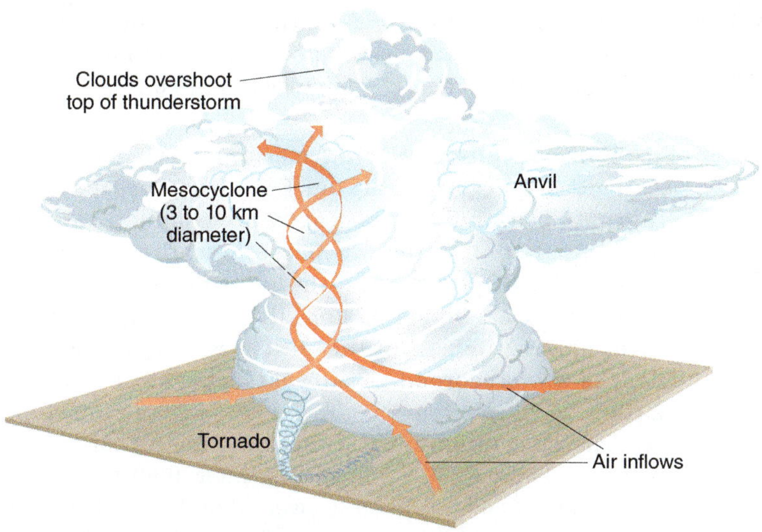

(a) Strong wind aloft establishes spinning, and updraft from thunderstorm development tilts the roataing air, causing a mesocyclone to form as a rotating updraft within the thunderstorm. If one forms, a tornado will descend from the lower portion of the mesocyclone.

(b) A tornado descends from the base of a supercell cloud near Spearman, Texas. Strong hail is falling to the left of the tornado.

▲**Figure 8.16 Mesocyclone and tornado formation.**
[(b) Photo by Howard Bluestein, all rights reserved.]

MG™ Animation
Tornado Wind Patterns

April, areas of derecho activity shift southward to eastern Texas through Alabama.

In May 2009, a fierce windstorm crossed Kansas, Missouri, and Illinois, reaching wind speeds of 100 mph and creating a path of destruction 100 miles wide. Scientists coined the term *super derecho* to describe this new phenomenon: a derecho with an eyelike structure similar to a hurricane that spun off 18 tornadoes as it moved across Kansas. Scientists were able to forecast this wind event 24 hours in advance and are using their forecasting models to investigate the development of this weather phenomenon.

Reported derecho wind events are on the increase since 2000 and may continue to increase with climate change. For more information, see http://www.spc.noaa.gov/misc/AbtDerechos/derechofacts.htm.

Tornadoes

A **tornado** is a violently rotating column of air in contact with the ground surface, usually visible as spining vortex of clouds and debris. A tornado can range from a few meters to more than a kilometer in diameter and can last anywhere from a few moments to tens of minutes.

The updrafts associated with thunderstorm squall lines and supercells are the beginning stages of tornado development (however, fewer than half of supercells produce tornadoes). As moisture-laden air is drawn up into the circulation of a mesocyclone, energy is liberated by condensation, and the rotation of air increases speed (Figure 8.16a). The narrower the mesocyclone, the faster the spin of converging parcels of air being sucked into the rotation; this movement may form a smaller, dark gray **funnel cloud** that pulses from the bottom side of the supercell cloud. This funnel cloud may lower to Earth, resulting in a tornado (Figure 8.16b). When tornado circulation occurs over water, surface water is drawn some 3–5 m (10–16 ft) up into the funnel, forming a **waterspout**.

Tornado Measurement Pressures inside a tornado usually are about 10% less than those in the surrounding

GEOreport 8.4 Storm causes Hawai'i hailstorm and tornado

Although the conditions necessary to form supercell thunderstorms and large hail are rare in Hawai'i, a March 2012 storm produced a grapefruit-sized hailstone on the windward side of Oahu—measured at 10.8 cm (4.25 in.) in diameter. The same storm spawned a waterspout offshore that turned into a small tornado after it hit land, both rare occurrences in Hawai'i. The NWS confirmed this event as an EF-0 tornado, with wind speeds reaching 97–113 kmph (60–70 mph).

TABLE 8.1 The Enhanced Fujita Scale

EF-Number	3-Second-Gust Wind Speed; Damage
EF-0 Gale	105–137 kmph, 65–85 mph; *light damage*: branches broken, chimneys damaged.
EF-1 Weak	138–177 kmph, 86–110 mph; *moderate damage*: beginning of hurricane wind-speed designation, roof coverings peeled off, mobile homes pushed off foundations.
EF-2 Strong	178–217 kmph, 111–135 mph; *considerable damage*: roofs torn off frame houses, large trees uprooted or snapped, boxcars pushed over, small missiles generated.
EF-3 Severe	218–266 kmph, 136–165 mph; *severe damage*: roofs torn off well-constructed houses, trains overturned, trees uprooted, cars thrown.
EF-4 Devastating	267–322 kmph, 166–200 mph; *devastating damage*: well-built houses leveled, cars thrown, large missiles generated.
EF-5 Incredible	More than 322 kmph, >200 mph; *incredible damage*: houses lifted and carried distance to disintegration, car-sized missiles fly farther than 100 m, bark removed from trees.

Note: See http://www.depts.ttu.edu/weweb/Pubs/fscale/EFScale.pdf for details.

air. The inrushing convergence created by such a horizontal pressure gradient causes high wind speeds. The late Theodore Fujita, a noted meteorologist from the University of Chicago, designed the Fujita Scale, which classifies tornadoes according to wind speed as indicated by related property damage. A refinement of this 1971 scale, adopted in February 2007, is the Enhanced Fujita Scale, or EF Scale (Table 8.1). The revision met the need to better assess damage, correlate wind speed to damage caused, and account for structural construction quality. To assist with wind estimates, the EF Scale includes Damage Indicators representing types of structures and vegetation affected, along with Degree of Damage ratings, both of which are listed at the URL cited in the table note.

The Storm Prediction Center in Kansas City, Missouri, provides short-term forecasting for thunderstorms and tornadoes to the public and to NWS field offices. Warning times of from 12 to 30 minutes are possible with current technology. From 1950 to 2010 in the United States, tornadoes caused over 5000 deaths (about 85 deaths per year), more than 80,000 injuries, and property damage of over $28 billion. The yearly average cost of tornado damage is rising each year. In 2011, Joplin, Missouri, was hit by an EF-5 tornado that killed at least 159 people and cost an estimated $3 billion in damages, the most expensive in U.S. history (Figure 8.17).

Tornado Frequency North America experiences more tornadoes than anywhere on Earth, because its latitudinal position and topography are conducive to the meeting of contrasting air masses and the formation of frontal precipitation and thunderstorms. Tornadoes have struck all 50 states and all the Canadian provinces and territories. In the United States, 54,030 tornadoes were recorded in the years from 1950 through 2010.

The long-term annual average number of tornadoes before 1990 was 787. Interestingly, after 1990 this long-term average per year rose to over 1000. Since 1990, the years with the highest number of tornadoes were 2004 (1820 tornadoes), 2011 (1691 tornadoes), and 1998 (1270

▼Figure 8.17 The 2011 Joplin, Missouri, tornado. [(a) Tom Uhlenbrock/UPI/Newscom. (b) NOAA.]

(a) An EF-5 tornado struck Joplin, Missouri, just before 6 p.m. (CDT) on May 22, 2011.

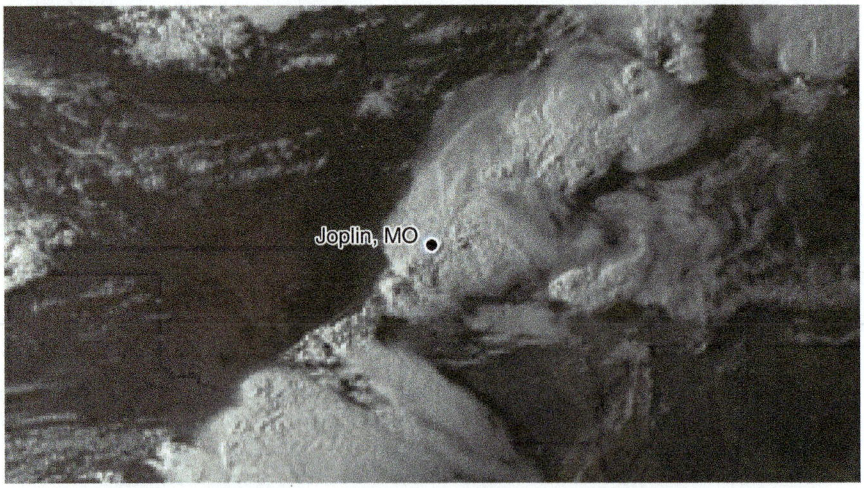

(b) *GOES East* satellite image shows the storm system over the central United States, moments before spawning the tornado.

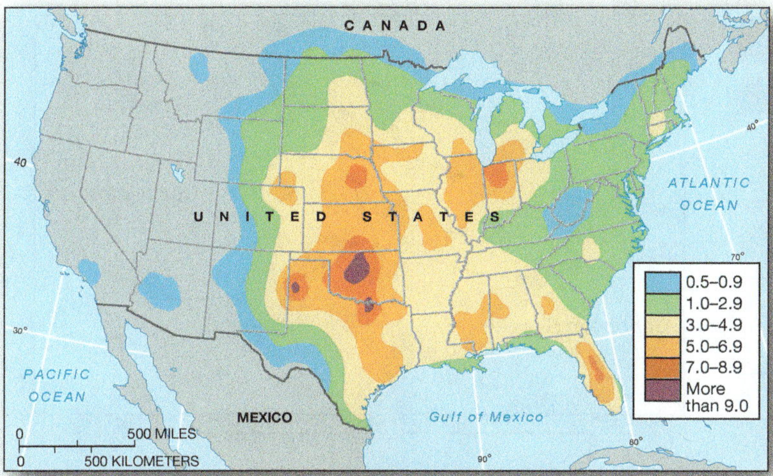

(a) Average number of tornadoes per 26,000 km² (10,000 mi²). Tornado numbers in Alaska and Hawai'i are negligible.

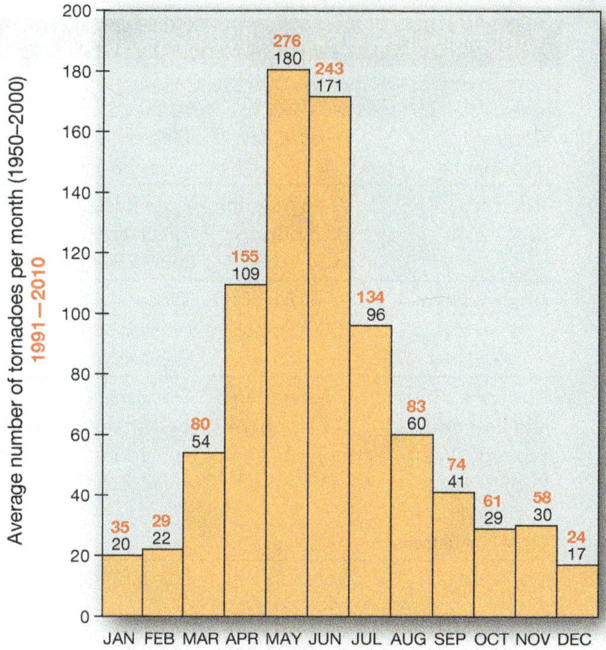

(b) Average number of tornadoes per month from 1950 to 2000, with additional updated statistics from 1991 to 2010 (in red).

▲Figure 8.18 **Trends in tornado occurrence in the United States.** [Data courtesy of the Storm Prediction Center, NWS, and NOAA sources.]

tornadoes). On April 27, 2011, 319 tornadoes were sighted, the third highest number ever recorded on a single day. Figure 8.18a shows areas of tornado occurrence in the United States, which is highest in Texas and Oklahoma (the southern part of the region known as "tornado alley"), Indiana, and Florida.

According to records since 1950, May and June are the peak months for tornadoes in the United States. The graph in Figure 8.18b shows average monthly tornadoes per month for 1950 to 2000, and for 1991 to 2010. Although the trend is toward increasing tornadoes for the most recent 20-year period, scientists agree that these data are unreliable indicators of actual changes in tornado occurrence because they correspond with more people being in the right place to see and photograph tornadoes and improved communication regarding tornado activity. Any actual increases in tornado occurrence may in part relate to rising sea-surface temperatures. Warmer oceans increase evaporation rates, which increase the availability of moisture in the mT air masses, thus producing more intense thunderstorm activity over certain areas of the United States. Other factors in tornado development, such as wind shear, are not as well understood, are difficult to model, and cannot yet be definitively linked to climate change.

Canada experiences an average of 80 observed tornadoes per year, but those in sparsely populated rural areas can go unreported. In the United Kingdom, observers report about 50 per year, all classified less than EF-3. In Europe overall, 330 tornadoes are reported each year, although some experts estimate that as many as 700 per year occur. In Australia, observers report about 16 tornadoes every year. Other continents experience a small number of tornadoes annually.

Tropical Cyclones

Originating entirely within tropical air masses, **tropical cyclones** are powerful manifestations of the Earth–atmosphere energy budget. (Remember that the tropics extend from the Tropic of Cancer at 23.5° N latitude to the Tropic of Capricorn at 23.5° S latitude.)

Tropical cyclones are classified according to wind speed; the most powerful are **hurricanes**, **typhoons**, or *cyclones*, which are different regional names for the same type of tropical storm. The three names are based on location: Hurricanes occur around North America, typhoons in the western Pacific (mainly in Japan and the Philippines), and cyclones in Indonesia, Bangladesh, and India. A full-fledged hurricane, typhoon, or cyclone has wind speeds greater than 119 kmph (74 mph, or 64 knots); the wind-speed criteria for tropical storms, depressions, and disturbances are listed in Table 8.2. For coverage and reporting, see the National Hurricane Center at http://www.nhc.noaa.gov/ or the Joint Typhoon Warning Center at http://www.usno.navy.mil/JTWC.

In the western Pacific, a strong tropical cyclone is designated a *super typhoon* when winds speeds reach 241 kmph (150 mph, or 130 knots). In November 2013, Super Typhoon Haiyan, hit the Philippines with sustained winds at 306–314 kmph (190–195 mph), the strongest ever recorded for a tropical cyclone at landfall (see the satellite image of Haiyan on the title-facing page of this book).

Storm Development Cyclonic systems forming in the tropics are quite different from midlatitude cyclones because the air of the tropics is essentially homogeneous, with no fronts or conflicting air masses of differing temperatures. In addition, the warm air and warm

TABLE 8.2 Tropical Cyclone Classification

Designation	Winds	Features
Tropical disturbance	Variable, low	Definite area of surface low pressure; patches of clouds
Tropical depression	Up to 63 kmph (38 mph); up to 34 knots	Gale force, organizing circulation; light to moderate rain
Tropical storm	63–118 kmph (39–73 mph); 35–63 knots	Closed isobars; definite circular organization; heavy rain; assigned a name
Hurricane (Atlantic and East Pacific) Typhoon (West Pacific) Cyclone (Indian Ocean, Australia)	Greater than 119 kmph (74 mph); 64 knots	Circular, closed isobars; heavy rain, storm surges; tornadoes in right-front quadrant

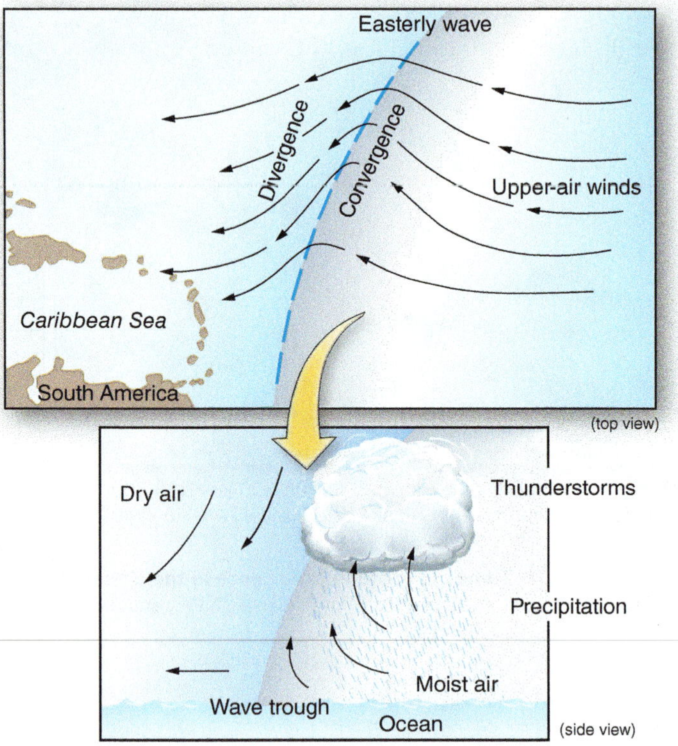

▲**Figure 8.19 Easterly wave in the tropics.** A low-pressure center develops along an easterly (westward-moving) wave. Moist air rises in an area of convergence at the surface to the east of the wave trough. Wind flows bend and converge before the trough and diverge downwind from the trough.

seas ensure abundant water vapor and thus the necessary latent heat to fuel these storms. Tropical cyclones convert heat energy from the ocean into mechanical energy in the wind—the warmer the ocean and atmosphere, the more intense the conversion and powerful the storm.

What triggers the start of a tropical cyclone? Cyclonic motion begins with slow-moving easterly waves of low pressure in the trade-wind belt of the tropics (Figure 8.19). If the sea-surface temperatures exceed approximately 26°C (79°F), a tropical cyclone may form along the eastern (leeward) side of one of these migrating troughs of low pressure, a place of convergence and rainfall. Surface airflow then converges into the low-pressure area, ascends, and flows outward aloft. This important divergence aloft acts as a chimney, pulling more moisture-laden air into the developing system. To maintain and strengthen this vertical convective circulation, there must be little or no wind shear to interrupt or block the vertical airflow.

Physical Structure Tropical cyclones have steep pressure gradients that generate inward-spiraling winds toward the center of low pressure—lower central pressure causes stronger pressure gradients, which in turn cause stronger winds. However, other factors come into play so that the storms with lowest central pressure are not always the strongest or cause the most damage. The lowest central pressure for a storm in the Atlantic is 882 mb (26.02 in), recorded for Hurricane Wilma in 2005.

As winds rush toward the center of a tropical cyclone, they turn upward, forming a wall of dense rain bands called the *eyewall*—this is the zone of most intense precipitation. The central area is designated the *eye* of the storm, where wind and precipitation subside; this is the warmest area of the storm, and although clear skies can appear here they may not always be present, as commonly believed. The structure of the rain bands, eyewall, and central eye are clearly visible for Hurricane Gilbert in Figure 8.20, from a top view, an oblique view (from an artist's perspective), and a side-radar view.

Tropical cyclones range in diameter from a compact 160 km (100 mi)–1000 km (600 mi) to the 1300–1600 km (800–1000 mi) attained by some western

GEOreport 8.5 Research aircraft dissect Hurricane Karl

In 2010, scientists launched the Genesis and Rapid Intensification Processes (GRIP) mission to study hurricane development using satellites, aircraft, and unmanned aerial vehicles. In September, an aircraft flew into category 3 Hurricane Karl as it made landfall along the Mexican coast. At 37,000 ft, the plane collected data using nine instruments and launched dropsondes, which record measurements as they fall through the atmosphere to the ocean surface. Another aircraft flew at 56,000 ft, using specialized radiometers to measure rain-cloud systems and surface winds. Meanwhile, the remotely piloted *Global Hawk* flew over the storm for more than 15 hours, sampling the upper reaches of a hurricane for the first time ever. The *Global Hawk* aircraft is being used in the new 5-year Hurricane and Severe Storm Sentinel (HS3) mission launched in 2012. More information is at http://www.nasa.gov/mission_pages/hurricanes/missions/grip/main/index.html.

(a) *GOES-7* satellite image of Hurricane Gilbert, September 13, 1988. For the Western Hemisphere, this hurricane attained the record size (1600 km, or 1000 mi, in diameter) and second lowest barometric pressure (888 mb, or 26.22 in.). Gilbert's sustained winds reached 298 kmph (185 mph), with peaks exceeding 320 kmph (200 mph).

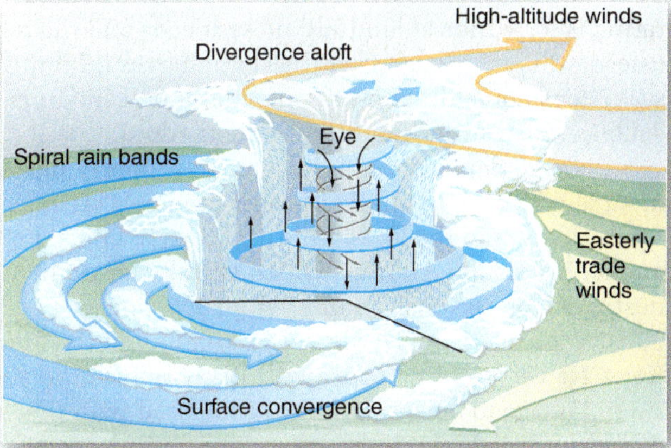

(b) A stylized portrait of a mature hurricane, drawn from an oblique perspective (cutaway view shows the eye, rain bands, and wind-flow patterns).

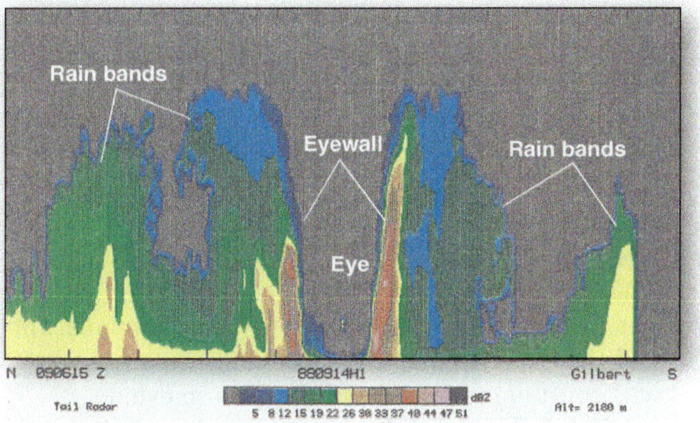

(c) *SLAR* (side-looking airborne radar) image from an aircraft flying through the center of Hurricane Gilbert. Rain bands of greater cloud density are false-colored in yellows and reds. Note the clear sky in the central eye.

▲**Figure 8.20 Profile of a hurricane.** [(a) and (c) NOAA and the NHC.]

MG Animation Hurricane Wind Patterns

MG Satellite Hurricane Georges

Pacific super typhoons. Vertically, a tropical cyclone dominates the full height of the troposphere. These storms move along over water at about 16–40 kmph (10–25 mph). The strongest winds are usually recorded in the right-front quadrant (relative to the storm's directional path) of the storm. At **landfall**, where the eye moves ashore, dozens of fully developed tornadoes may be located in this high-wind sector. For example, Hurricane Camille in 1969 had up to 100 tornadoes embedded in its right-front quadrant.

Damage Potential When you hear meteorologists speak of a "category 4" hurricane, they are using the Saffir–Simpson Hurricane Wind Scale to estimate possible damage from hurricane-force winds. Table 8.3 presents this scale, which uses sustained wind speed to rank hurricanes and typhoons in five categories, from smaller category 1 storms to extremely dangerous category 5 storms. The

TABLE 8.3 Saffir–Simpson Hurricane Wind Scale

Category	Wind Speed	Types of Damage	Recent Atlantic Examples (landfall rating)
1	119–153 kmph (74–95 mph; 65–82 knots)	Some damage to homes	2012 Isaac
2	154–177 kmph (96–110 mph; 83–95 knots)	Extensive damage to homes; major roof and siding damage	2003 Isabel (was a cat. 5), Juan; 2004 Francis; 2008 Dolly; 2010 Alex
3	178–208 kmph (111–129 mph; 96–112 knots)	Devastating damage; removal of roofs and gables	2004 Ivan (was a cat. 5), Jeanne; 2005 Dennis, Katrina, Rita, Wilma (were cat. 5); 2008 Gustav, Ike (were cat. 4); 2011 Irene; 2012 Sandy
4	209–251 kmph (130–156 mph; 113–136 knots)	Catastrophic damage; severe damage to roofs and walls	2004 Charley; 2005 Emily (was a cat. 5)
5	>252 kmph (>157 mph; >137 knots)	Catastrophic damage; total roof failure and wall collapse on high percentage of homes	2004 Ivan; 2007 Dean, Felix Other Notable: 1935 No. 2; 1938 No. 4; 1969 Camille; 1971 Edith; 1979 David; 1988 Gilbert, Mitch; 1989 Hugo; 1992 Andrew

rating is for winds at landfall; these speeds and the rating category may decrease after the storm moves inland. This scale does not address other potential hurricane impacts, such as storm surge, flooding, and tornadoes.

Damage depends on the degree of property development at a storm's landfall site, how prepared citizens are for the blow, and the local building codes in effect. For example, Hurricane Andrew, which struck Florida in 1992, destroyed or seriously damaged 70,000 homes and left 200,000 people homeless between Miami and the Florida Keys. Newer building codes will likely reduce the structural damage to newer buildings compared with that predicted by this scale, although local political pressure to weaken standards is active.

When a tropical cyclone makes landfall, additional hazards arise from storm surge and the flooding associated with heavy rainfall. In fact, the impact of hurricanes on human population centers is as much connected to storm surge as it is to damaging winds. **Storm surge** is the seawater that is pushed inland during a hurricane and can combine with the normal tide to create a *storm tide* of 4.5 m (15 ft) or more in height. The landfall of Hurricane Sandy in 2012 coincided with high tide to create record storm surge in New York City—as high as 4.2 m (13.9 ft) at the southern tip of Manhattan (see Focus Study 8.1). As sea levels rise from melting ice and the expansion of seawater as it warms, storm tides will continue to increase during storm events.

Rainfall and flooding can cause devastation and loss of life, making even a weak, slow-moving tropical cyclone quite dangerous and destructive. In 2011, heavy rainfall from Hurricane Irene caused flooding from North Carolina's outer banks northward to New England. In 2012, rainfall totals associated with Sandy were greater than 180 mm (7.0 in) over some coastal areas from South Carolina to New Jersey, and snowfall topped 76 cm (30 in.) in parts of Tennessee and West Virginia.

Formation Areas and Storm Tracks The map in Figure 8.21a shows the seven primary formation areas, or "basins," for hurricanes, typhoons, and cyclones, and the months during which these storms are most likely to form. Figure 8.21b shows the pattern of actual tracks and intensities for tropical cyclones between 1856 and 2006.

In the Atlantic basin, tropical depressions (low-pressure areas) tend to intensify into tropical storms as they cross the Atlantic toward North and Central America. If tropical storms mature early along their track, before reaching approximately 40° W longitude, they tend to curve northward toward the north Atlantic and miss the United States. If a tropical storm matures after it reaches the longitude of the Dominican Republic (70° W), then it has a higher probability of hitting the United States.

The number of Atlantic hurricanes broke records in 2005, including the most named tropical storms in a single year, totaling 27 (the average until then was 10); the most hurricanes, totaling 15 (the average was 5); and the highest number of intense hurricanes, category 3 or higher, totaling 7 (the average was 2). The 2005 season was also the first time that three category 5 storms (Katrina, Rita, and Wilma) occurred in the Gulf of Mexico (see image and wind speed map of Katrina in Focus Study 8.1). The 2005 season also recorded the greatest damage total in one year, at more than $130 billion. On our *MasteringGeography* website, a map, table, and satellite-image movie detail the storms of this remarkable season.

In the Southern Hemisphere, no hurricane was ever observed turning from the equator into the south Atlantic until Hurricane Catarina made landfall in Brazil in March 2004 (see Figure 8.21b). The satellite image of Catarina in Figure 8.21c clearly shows an organized hurricane, with characteristic central eye and rain bands. Note the Coriolis force in action, clockwise in the Southern Hemisphere. (Compare this image with the Northern Hemisphere circulation patterns of Hurricane Gilbert in Figure 8.20a.)

Although tropical cyclones are rare in Europe, in October 2005, the remnants of Tropical Storm Vince became the first Atlantic tropical cyclone on record to strike Spain. Likewise, Super Cyclone Gonu in 2007 became the strongest tropical cyclone on record occurring in the Arabian Sea, eventually hitting Oman and the Arabian Peninsula. In January 2011, an unusually strong category 5 tropical cyclone hit northeastern Australia in the same area that underwent severe flooding a few months before.

Coastal Flooding from Hurricane Katrina The flooding in New Orleans after Hurricane Katrina in 2005 resulted more from human engineering and construction errors than from the storm itself, which was actually downgraded to a category 3 hurricane when it made landfall. About half of the city of New Orleans is below sea level in elevation, the result of years of draining wetlands, compacting soils, and overall land subsidence. In addition, a system of canals, built throughout the 20th century for drainage and navigation, runs through the city. To address the ever-present flood danger from coastal storms, the U.S. Army Corps of Engineers reinforced the canals with concrete floodwalls and *levees*, earthen embankments constructed along banks of waterways to prevent overflow of the channel.

Hurricane Katrina's landfall, just south of the city, was accompanied by high rainfall and storm surges that moved into the city through canals. As the level in Lake Pontchartrain rose from rainfall and storm surge, four levee breaks and at least four dozen levee breaches (in which water flows over the embankment) permitted the inundation of this major city (Figure 8.22). Some neighborhoods were submerged up to 6.1 m (20 ft); the polluted water remained for weeks.

The poor performance of the New Orleans waterway infrastructures during Hurricane Katrina will long stand as one of the greatest engineering failures of the government. In the wake of the storm, six investigations by civil engineers, scientists, and political bodies agreed that in concept, design, construction, and maintenance, the flood "protection" system was flawed. Many of the structures failed before they reached design-failure limits.

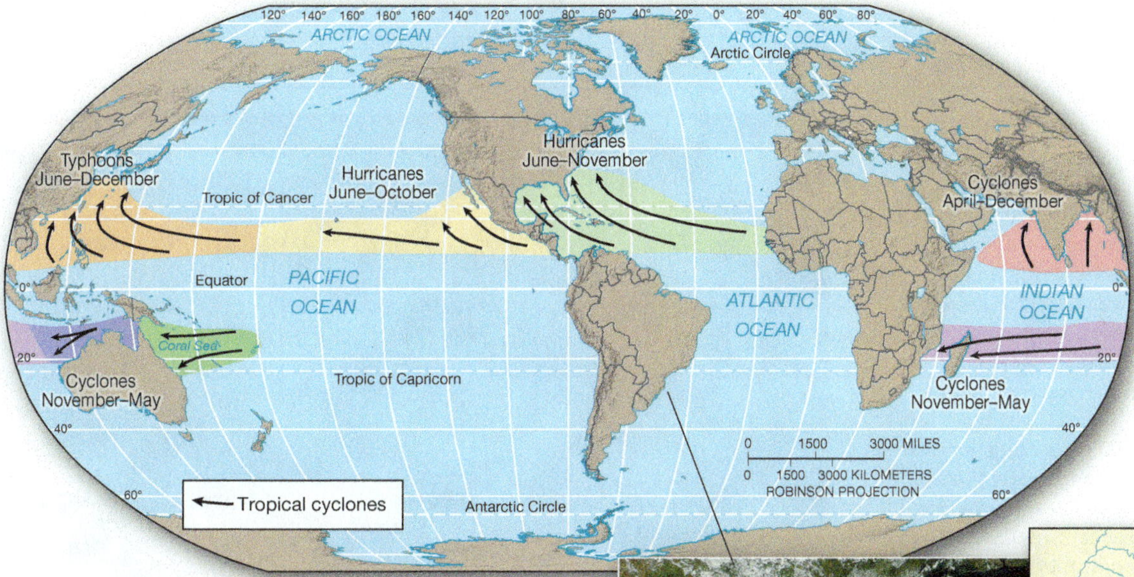

(a) Seven primary areas of tropical cyclone formation, with regional names and principal months of occurrence. Note that the general time period is the same for the Southern Hemisphere cyclones.

MG Satellite Hurricane Isabel 9/6–9/19/04

MG Satellite Florida Hurricanes Satellite Coverage

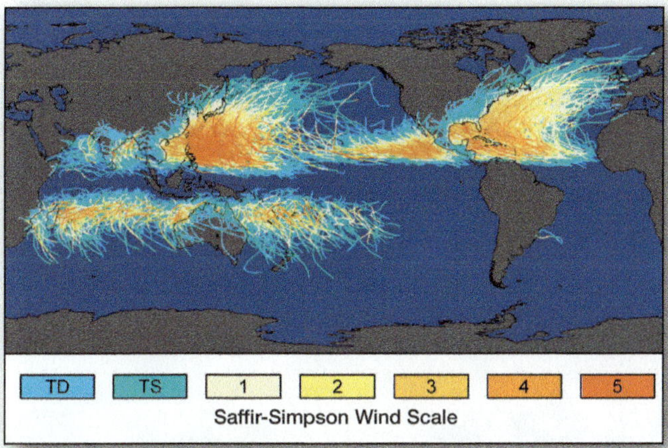

(b) Global tropical cyclone tracks from 1856 to 2006. Note the track of Hurricane Catarina in the south Atlantic.

(c) Hurricane Catarina approaches Brazil's southeastern coast, March 27, 2004—a unique occurrence in recorded history.

▲**Figure 8.21 Worldwide pattern of the most intense tropical cyclones.** [(b) R. Rohde/NASA/GSFC. (c) *Terra* MODIS image, NASA/GSFC.]

In addition, the composition of some levee fill materials throughout the city was determined after Katrina to be too high in sand—a recipe for future problems.

Repair and recovery work in New Orleans is ongoing. Since Katrina, the Army Corps of Engineers has strengthened levees, floodwalls, floodgates, and pumping stations in preparation for future hurricanes, at a cost of over $12 billion.

An Avoidable Cycle Tropical cyclones are potentially the most destructive storms experienced by humans, claiming thousands of lives each year worldwide. The tropical cyclone that struck Bangladesh in 1970 killed an estimated 300,000 people, and the one in 1991 claimed over 200,000. In Central and North America, death tolls are much lower, but still significant. The Galveston, Texas, hurricane of 1900 killed 6000. Hurricane Mitch (October 26–November 4, 1998) was the deadliest Atlantic hurricane in two centuries, killing more than 12,000 people, mainly in Honduras and Nicaragua. Hurricane Katrina and the associated engineering failures killed more than 1830 people in Louisiana, Mississippi, and Alabama in 2005.

Despite these statistics, the risk of human fatalities is decreasing in most parts of the world owing to better warning and rescue systems, and ongoing improvements in the forecasting of these storms. At the same time, the damage caused by tropical cyclones is increasing substantially as more and more development occurs along susceptible coastlines.

The history of major hurricanes in the United States reveals a recurrent, yet avoidable, cycle—construction, devastation, reconstruction, devastation—especially in the Gulf Coast, beginning with Hurricane Camille over 40 years ago. The same Gulf Coast towns that Camille washed

(text continued on page 218)

Focus Study 8.1 Natural Hazards

Hurricanes Katrina and Sandy: Storm Development and Links to Climate Change

Hurricane Katrina, devastating the Gulf Coast in 2005, and Hurricane Sandy, slamming the mid-Atlantic coast in 2012, rank as the two most expensive hurricanes in U.S. history. Although both are memorable for destructiveness and human loss, the storms themselves differed greatly. And in the 7-year interval between them, public awareness of the impacts of climate change on global weather events has increased.

On August 28, 2005, Hurricane Katrina reached category 5 strength over the Gulf of Mexico (Figure 8.1.1). The next day, the storm made landfall near New Orleans as a strong category 3 hurricane, with recorded sustained wind speeds over 200 kmph (125 mph) along the Louisiana coast. Katrina was a textbook tropical cyclone, developed from warm waters of the tropical Atlantic with a compact low-pressure center and a symmetrical wind field (Figure 8.1.2a).

▲**Figure 8.1.1 Hurricane Katrina.** View of Katrina from the *GOES* satellite on August 28, 2005, 20:45 UTC. [NASA/NOAA as processed by SSEC CIMSS, University of Wisconsin–Madison.]

MG Notebook
Record-breaking 2005 Atlantic hurricane season

MG Satellite
2005: 27 Storms, Arlene to Zeta

Sandy's Development and Effects

As the 2012 Atlantic hurricane season neared its end, Hurricane Sandy began as tropical depression number 18 in the Caribbean Sea, reaching hurricane strength on October 23. Sandy moved northward over Cuba and the Bahamas toward the mid-Atlantic coast, where it became wedged between a stationary cold front over the Appalachian Mountains and a high-pressure air mass over Canada. These systems blocked the storm from moving north or curving east, as it would normally have done, and finally drove it toward the coast.

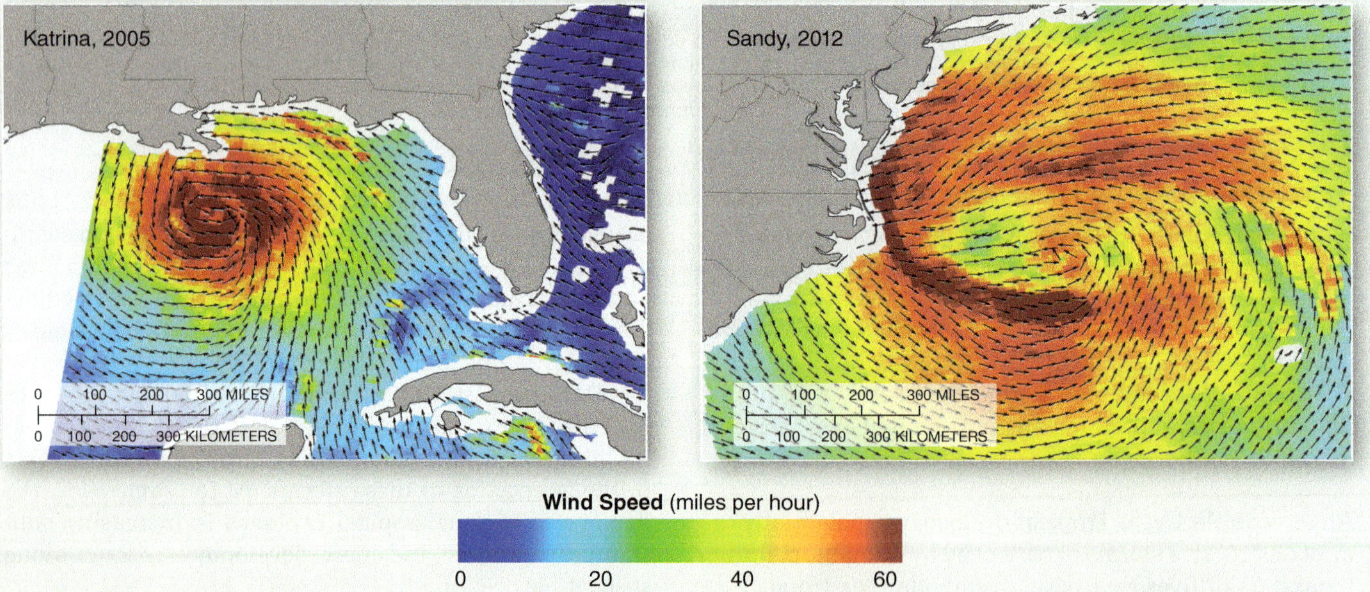

(a) Ocean surface wind map shows Katrina's stronger wind speeds east of the eye, amidst the eyewall, a condition typical of Northern Hemisphere tropical cyclones.

(b) Similar map for Sandy shows weaker winds to the east, indicating the influence of nearby air masses and associated pressure systems.

▲**Figure 8.1.2 Wind-speed maps for Katrina and Sandy.** [(a) *OceanSat-2*, Indian Space Research Organization. (b) *QuickSat*, NASA/JPL.]

The center of the hurricane hit the New Jersey shoreline just after 11 pm (EDT) on October 29 (Figure 8.1.3)

According to the criteria for storm classification by the National Hurricane Center, Hurricane Sandy transitioned into a *post-tropical*, or *extratropical*, storm just before making landfall. At this point, Sandy departed from the classic tropical cyclone pattern and was instead gathering energy from sharp temperature contrasts between air masses. The storm now exhibited characteristics more closely aligned with nor'easters, midlatitude cyclonic winter storms that typically cover a large area, with strong winds and precipitation far from the center of the storm. Just before landfall, Sandy's wind patterns were asymmetrical, with a broad wind and cloud field shaped like a comma (rather than a circle), yet retaining strong, hurricane-force winds (Figure 8.1.2b).

Moving inland, Sandy brought rainfall to low-elevation areas and blizzard conditions to the mountains of West Virginia, North Carolina, and Tennessee. This huge and unusual storm system affected an estimated 20% of the U.S. population, resulted in over 100 fatalities, and cost $75 billion.

Sandy's storm surges broke records along the New York and New Jersey coastlines. It knocked out power for millions of people, destroyed homes, eroded coastlines, and flooded lower Manhattan (see *Geosystems Now* in Chapter 16). A full moon made tides higher than average; at the height of the storm, a buoy in New York harbor measured a record-setting 10-m (32.5-ft) wave, 2 m (6.5 ft) taller than the 7.6-m (25-ft) wave recorded during Hurricane Irene in 2011.

▲Figure 8.1.3 **Hurricane Sandy just before landfall.** Sandy's circulation, October 29 at 1:35 p.m. (EDT), with the center southeast of Atlantic City, New Jersey, as the storm moved northward with maximum sustained winds of 150 kmph (90 mph). The storm covered 1.8 million square miles, from the mid-Atlantic coast to the Ohio Valley and Great Lakes, and north into Canada. [VIIRS instrument, *Suomi NPP*, NASA.]

Hurricanes and Climate Change

Perhaps the most certain causal link between recent increases in hurricane damage and climate change is the effect of rising sea level on storm surge. Sandy's destruction was worsened by recent sea-level rise—as much as 2 mm (0.08 in) per year since 1950—along the coast from North Carolina to Massachusetts, reflecting both global sea-level rise and shifting ocean currents along the mid-Atlantic shoreline.

Higher sea-surface temperatures caused by climate change are another certain causal link. Research has correlated longer tropical storm lifetimes and greater intensity with rising SSTs. As oceans warm, the energy available to fuel tropical cyclones is increasing. This connection is established for the Atlantic basin, where the number of hurricanes has increased over the past 20 years. Models suggest that the number of category 4 and 5 storms in this basin may double by the end of this century (see http://www.gfdl.noaa.gov/21st-century-projections-of-intense-hurricanes).

Continued ocean warming combined with the current trend of increasing coastal population will likely result in substantial hurricane-related property losses. Eventually, though, the more-intense storms and rising sea level could lead to shifts in population along U.S. coasts, and even the abandonment of some coastal resort communities, such as those along North Carolina's outer banks and the New Jersey shore.

away in 1969—Waveland, Bay Saint Louis, Pass Christian, Long Beach, and Gulfport, among others—were obliterated again by Hurricane Katrina in 2005. This time, the below-sea-level sections of New Orleans may never recover; nevertheless, the mantra repeated after each storm—that "the Gulf Coast will rebuild bigger and better"—was heard once again.

No matter how accurate storm forecasts become, coastal and lowland property damage will continue to increase until better hazard zoning and development restrictions are in place. The property insurance industry appears to be taking action to promote these improvements, requiring tougher building standards to qualify for coverage—or, in some cases, refusing to insure property along vulnerable coastal lowlands. Given rising sea level and the increased intensity of tropical storms, the public, politicians, and business interests must respond to somehow mitigate this hazardous predicament.

Weather has many consequences for human society, especially as anthropogenic climate change worsens severe weather events across the globe. The human denominator summarizes some of the interactions between weather events and humans, with a few examples of severe weather events across the globe.

CRITICALthinking 8.2
Hazard Perception and Planning: What Seems to be Missing?

Along coasts subject to extreme tropical weather, the cycle of "construction, devastation, reconstruction, devastation" means a cycle of ever-increasing dollar losses to property from tropical storms and hurricanes, even though improved forecasts have resulted in a significant reduction in loss of life. Given rising sea levels along coastlines and the increase in total power dissipation in these tropical storms since 1970, in your opinion, what is the solution to halting this cycle of destruction and increasing losses? How would you implement your ideas? •

Figure 8.22 New Orleans before and after the Katrina disaster. [(a) and (b) USGS *Landsat* Image Gallery. (d) AP Photo/U.S. Coast Guard, Petty Officer 2nd Class Kyle Niemi.]

(a) Part of New Orleans on April 24, 2005, before Hurricane Katrina.

(b) Flooded portions of the city on August 30, after Hurricane Katrina.

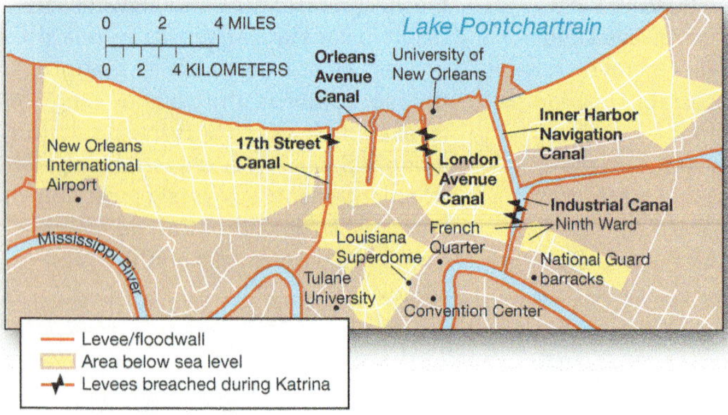

(c) Location of levee breaks.

(d) New Orleans on August 29, looking toward Lake Pontchartrain with Interstate 10 at West End Boulevard in the foreground. The 17th Street Canal is just to the left, off the photo.

THE**human**DENOMINATOR 8 Weather

WEATHER ⇨ HUMANS

- Frontal activity and midlatitude cyclones bring severe weather that affects transportation systems and daily life.
- Severe weather events such as ice storms, derechos, tornadoes, and tropical cyclones cause destruction and human casualties.

HUMANS ⇨ WEATHER

- Rising temperatures with climate change have caused shrinking spring snow cover in the Northern Hemisphere.
- Sea-level rise is increasing hurricane storm surge on the U.S. east coast.

An ice-covered car sits beside Lake Geneva in Versoix, Switzerland, during a February 2012 arctic cold snap that brought freezing temperatures as far south as North Africa, claiming 300 lives. Ice storms occur when freezing rain and sleet (illustrated at right) cause at least 6.4 mm (0.25 in.) of ice to accumulate on exposed surfaces. [Fabrice Coffrini/AFP/Getty Images; NOAA.]

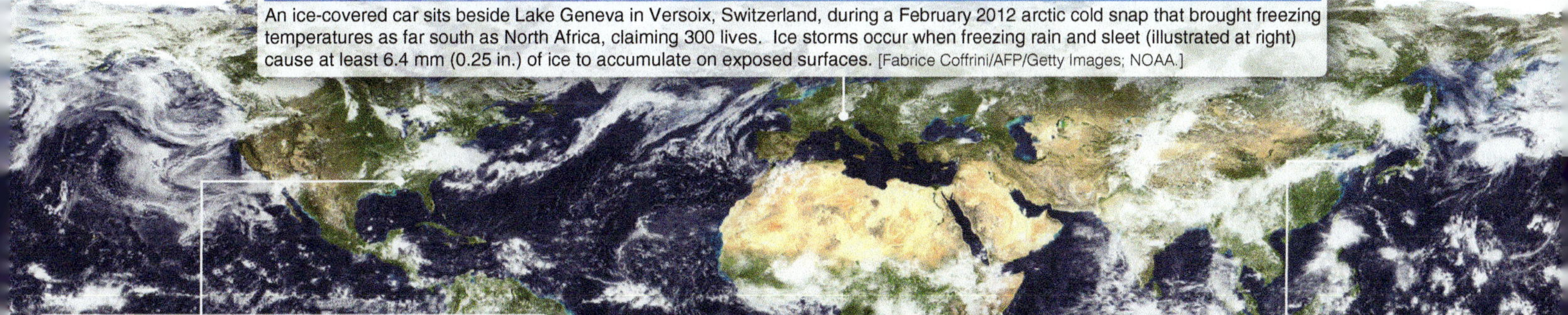

An EF-5 tornado, almost 2 km (1.2 mi) wide at its base, tore across Alabama in April 2011. The tornado hit Tuscaloosa (pictured here) near the University of Alabama, where 44 people died, and continued on to hit the suburbs of Birmingham. As thunderstorms intensify with climate change, tornado frequency may increase. [x77/press/Newscom; David Mabel/Alamy.]

In February 2011, 100 cm of snow fell on parts of South Korea's east coast over a 2-day period, the heaviest since record keeping began in 1911. The unusually cold weather may be driven in part by the Arctic Oscillation and in part by the trend toward more extreme snowfall events associated with climate change. [Yu Hyung-jae/AP.]

ISSUES FOR THE 21ST CENTURY

- Global snowfall will decrease, with less snow falling during a shorter winter season; however, extreme snowfall events (blizzards) will increase in intensity. Lake-effect snowfall, transitioning to rainfall, will increase owing to increased lake temperatures.
- Increasing ocean temperatures with climate change will strengthen the intensity and frequency of tropical cyclones by the end of the century.

GEOSYSTEMS**connection**

Weather is the expression of the interactions of energy, water, water vapor, and the atmosphere at any given moment. The patterns of precipitation produced across the globe form the input to surface water supplies in lakes, rivers, glaciers, and groundwater—the final components in our study of the hydrologic cycle. In the next chapter, we examine these surface water resources and the inputs and outputs of the water-balance model. Water quality and quantity and the availability of potable water loom as major issues for the global society.

KEY LEARNING concepts review

Weather is the short-term condition of the atmosphere; **meteorology** is the scientific study of the atmosphere. The spatial implications of atmospheric phenomena and their relationship to human activities strongly link meteorology to physical geography.

weather (p. 192)
meteorology (p. 192)

■ *Describe* air masses that affect North America, and *relate* their qualities to source regions.

An **air mass** is a regional volume of air that is homogenous in humidity, stability, and cloud coverage and that may extend through the lower half of the troposphere. Air masses are categorized by their moisture content—**m** for maritime (wetter) and **c** for continental (drier)—and their temperature, a function of latitude—designated **A** (arctic), **P** (polar), **T** (tropical), **E** (equatorial), and **AA** (Antarctic). Air masses take on the characteristics of their source region and carry those characteristics to new regions as they migrate. As air masses move, their characteristics change to reflect their underlying regions.

air mass (p. 192)

1. How does a source region influence the type of air mass that forms over it? Give specific examples of each basic classification.
2. Of all the air mass types, which are of greatest significance to the United States and Canada? What happens to them as they migrate to locations different from their source regions? Give an example of air mass modification.

■ *Identify* and *describe* four types of atmospheric lifting mechanisms, and give an example of each.

Air masses can rise through **convergent lifting** (airflows conflict, forcing some of the air to lift); **convectional lifting** (air passing over warm surfaces gains buoyancy); **orographic lifting** (air passes over a topographic barrier); and *frontal lifting*. Orographic lifting creates wetter windward slopes and drier leeward slopes situated in the **rain shadow** of the mountain. In North America, **chinook winds** (called föhn or foehn winds in Europe) are the warm, downslope airflows characteristic of the leeward side of mountains. Conflicting air masses may produce a **cold front** (and sometimes a zone of strong wind and rain) or a **warm front**. A zone right along or slightly ahead of the front, called a **squall line**, is characterized by turbulent and wildly changing wind patterns and intense precipitation.

convergent lifting (p. 194)
convectional lifting (p. 194)
orographic lifting (p. 195)
rain shadow (p. 196)
chinook wind (p. 196)
cold front (p. 198)
warm front (p. 198)
squall line (p. 199)

3. Explain why it is necessary for an air mass to be lifted if there is to be saturation, condensation, and precipitation.
4. What are the four principal lifting mechanisms that cause air masses to ascend, cool, condense, form clouds, and perhaps produce precipitation? Briefly describe each.
5. Differentiate between the structure of a cold front and that of a warm front.

■ *Explain* the formation of orographic precipitation, and *review* an example of orographic effects in North America.

The physical presence of a mountain acts as a topographic barrier to migrating air masses. *Orographic lifting* (*oro-* means "mountain") occurs when air is forcibly lifted upslope as it is pushed against a mountain. It cools adiabatically. An orographic barrier enhances convectional activity. The precipitation pattern of windward and leeward slopes is seen worldwide.

6. When an air mass passes across a mountain range, many things happen to it. Describe each aspect of a moist air mass crossing a mountain. What is the pattern of precipitation that results?
7. Explain how the distribution of precipitation in the state of Washington is influenced by the principles of orographic lifting.

■ *Describe* the life cycle of a midlatitude cyclonic storm system, and *relate* this to its portrayal on weather maps.

A **midlatitude cyclone,** or **wave cyclone,** is a vast low-pressure system that migrates across a continent, pulling air masses into conflict along fronts. These systems are guided by the jet streams of the upper troposphere along seasonally shifting **storm tracks**. **Cyclogenesis**, the birth of the low-pressure circulation, can occur off the west coast of North America, along the polar front, along the lee slopes of the Rockies, in the Gulf of Mexico, and along the East Coast. A midlatitude cyclone can be thought of as having a life cycle of birth, maturity, old age, and dissolution. An **occluded front** is produced when a cold front overtakes a warm front in the maturing cyclone. Sometimes a **stationary front** develops between conflicting air masses, where airflow is parallel to the front on both sides.

midlatitude cyclone (p. 200)
wave cyclone (p. 200)
storm track (p. 200)
cyclogenesis (p. 201)
occluded front (p. 201)
stationary front (p. 201)

8. Differentiate between frontal lifting at an advancing cold front and at an advancing warm front, and describe what you would experience with each one.
9. How does a midlatitude cyclone act as a catalyst for conflict between air masses? How do flows of warm, cold, and dry "conveyors" of air interact in such a system?
10. What is meant by cyclogenesis? In what areas does it occur and why? What is the role of upper-tropospheric circulation in the formation of a surface low?

11. Diagram a midlatitude cyclonic storm during its open stage. Label each of the components in your illustration, and add arrows to indicate wind patterns in the system.

■ *List* the measurable elements that contribute to modern weather forecasting, and *describe* the technology and methods employed.

Synoptic analysis involves the collection of weather data at a specific time. Computer-based weather prediction and the development of weather-forecasting models rely on data such as barometric pressure, surface air temperatures, dew-point temperatures, wind speed and direction, clouds, sky conditions, and visibility. On weather maps, these elements appear as specific weather station symbols.

12. What is your principal source of weather data, information, and forecasts? Where does your source obtain its data?

■ *Identify* various forms of violent weather by their characteristics, and *review* several examples of each.

The violent power of some weather phenomena poses a hazard to society. Severe ice storms involve **sleet** (freezing rain, ice glaze, and ice pellets), snow blizzards, and crippling ice coatings on roads, power lines, and crops. Thunderstorms are fueled by rapid upward movement of warm, moist air and are characterized by turbulence and wind shear. In strong thunderstorms known as *supercells*, a cyclonic updraft—a **mesocyclone**—may form within a cumulonimbus cloud, sometimes rising to the mid-troposphere. Thunderstorms produce **lightning** (electrical discharges in the atmosphere), **thunder** (sonic bangs produced by the rapid expansion of air after intense heating by lightning), and **hail** (ice pellets formed within cumulonimbus clouds). Strong linear winds in excess of 26 m/s (58 mph), known as **derechos**, are associated with thunderstorms and bands of showers crossing a region. These straight-line winds can cause significant damage and crop losses.

A **tornado** is a violently rotating column of air in contact with the ground surface, usually visible as a dark gray **funnel cloud** pulsing from the bottom side of the parent cloud. A **waterspout** forms when a tornado circulation occurs over water.

Within tropical air masses, large low-pressure centers can form along easterly wave troughs. Under the right conditions, a tropical cyclone is produced. A **tropical cyclone** becomes a **hurricane**, **typhoon**, or *cyclone* when winds exceed 65 knots (119 kmph, 74 mph). As forecasting of weather-related hazards improves, loss of life decreases, although property damage continues to increase. Great damage occurs to occupied coastal lands when hurricanes make **landfall** and when winds drive ocean water inland in **storm surges**.

sleet (p. 204)
mesocyclone (p. 207)
lightning (p. 207)
thunder (p. 207)
hail (p. 207)
derecho (p. 208)
tornado (p. 209)
funnel cloud (p. 209)
waterspout (p. 209)
tropical cyclone (p. 211)
hurricane (p. 211)
typhoon (p. 211)
landfall (p. 213)
storm surge (p. 214)

13. What constitutes a thunderstorm? What type of cloud is involved? What type of air mass would you expect in an area of thunderstorms in North America?
14. Lightning and thunder are powerful phenomena in nature. Briefly describe how they develop.
15. Describe the formation process of a mesocyclone. How is this development associated with that of a tornado?
16. Evaluate the pattern of tornado activity in the United States. What generalizations can you make about the distribution and timing of tornadoes? Do you perceive a trend in tornado occurrences in the United States? Explain.
17. What are the different classifications for tropical cyclones? List the various names used worldwide for hurricanes. Have any hurricanes ever occurred in the south Atlantic?
18. Why have damage figures associated with hurricanes increased even though loss of life has decreased over the past 30 years?
19. How did the effects of Hurricane Katrina in part relate to engineering and flood control structures in New Orleans?
20. Explain several differences between Hurricanes Katrina and Sandy. How is present climate change affecting hurricane intensity and damage costs?

MasteringGeography™

Looking for additional review and test prep materials? Visit the Study Area in *MasteringGeography*™ to enhance your geographic literacy, spatial reasoning skills, and understanding of this chapter's content by accessing a variety of resources, including MapMaster interactive maps, geoscience animations, satellite loops, author notebooks, videos, RSS feeds, flashcards, web links, self-study quizzes, and an eText version of *Geosystems*.

9 Water Resources

KEY LEARNING **concepts**

After reading the chapter, you should be able to:

- ***Describe*** the origin of Earth's waters, ***report*** the quantity of water that exists today, and ***list*** the locations of Earth's freshwater supply.
- ***Illustrate*** the hydrologic cycle with a simple sketch, and ***label*** it with definitions for each water pathway.
- ***Construct*** the water-budget equation, ***define*** each of the components, and ***explain*** its use.
- ***Discuss*** water storage in lakes and wetlands, and ***describe*** several large water projects involving hydroelectric power production.
- ***Describe*** the nature of groundwater, and ***define*** the elements of the groundwater environment.
- ***Evaluate*** the U.S. water budget, and ***identify*** critical aspects of present and future freshwater supplies.

The valley of Oued Todgha (Todgha River) forms an oasis that supports palm plantations and other agriculture near the city of Tinghir in the desert of Morocco in northwest Africa. The Todgha flows southward from the Atlas Mountains—a range that extends across Morocco, Algeria, and Tunisia, reaching elevations over 4000 m (13,100 ft)—carrying snowmelt runoff in surface and subsurface flows until finally disappearing permanently into the ground during most of the year. Water in this arid region is a precious resource; extensive canal systems direct surface flows to irrigate fields, and pumping systems extract subsurface water for agriculture and other uses. Dissolved salts from farm fields now threaten groundwater quality throughout Morocco. This chapter examines Earth's water supply and distribution, and their importance for human populations. [Ignacio Palacios/age fotostock.]

Earth's Largest Lake Warms with Changing Climate

Lake Baikal, located between 52° and 56° north latitude in south-central Siberia, is the world's largest (by volume), deepest, oldest, and most biologically diverse lake. Holding 23,600 km^3 (5662 mi^3) of water, almost as much as all five of the U.S. Great Lakes combined, the lake contains 20% of the world's liquid freshwater, earning it the label "well of the planet." At its deepest point, the lake descends 1642 m (5387 ft) and is deeper than the Grand Canyon. At nearly 644 km (400 mi) long with a surface area of 31,500 km^2 (12,200 mi^2), this enormous crescent-shaped lake is easily seen in satellite images or on a world map (Figure GN 9.1).

Changes in Lake Temperature The region around Lake Baikal is currently experiencing some of the most rapid climate change on Earth. During the past century, annual air temperatures increased 1.2 C° (2.1 F°), twice the global average, with a greater temperature increase occurring in winter than summer. Water temperatures warmed rapidly to a depth of 25 m (82 ft) during the past 60 years. By the end of this century, scientists predict Baikal's winters will be warmer by 4.3 C° (7.7 F°) and much wetter, with 26% more winter precipitation than present averages. As air temperatures warm and winter snow increases, several changes occur in the lake environment: The timing of ice cover is altered, reducing duration and ice thickness; at the same time, wind patterns change, clearing less snow away and diminishing the transparency of the ice. These changes are already affecting Lake Baikal's ecosystems (Figure GN 9.2).

Effects on Lake Stratification Warmer water interrupts vertical mixing of the lake water and nutrient cycling. Normally, most lakes exhibit stratification in the summer: water at the surface, which is warmer, fails to mix with the cooler water down below. This leads to a depletion of nutrients in the shallows by late summer. In the fall, however, as the surface cools, water sinks, creating a turnover that replenishes surface nutrients throughout the lake. In Lake Baikal, wind assists such mixing. But now, as regional temperatures warm, the summerlike stratification appears earlier in the year and persists later in the fall, and mixing slows or stops. Scientists suggest that prolonged stratification could adversely affect Baikal's phytoplankton, microscopic algae that form the basis of the lake's food web.

Effects on Biota The lake freezes for 4 to 5 months every year, and most of its flora and fauna have adapted to life under the ice. Lake Baikal is the only lake in the world in which both the dominant primary producers (phytoplankton) and the top predator (the Baikal seal, locally known as the nerpa) require ice for reproduction.

Phytoplankton are microscopic organisms that live in fresh- or saltwater environments. Baikal's phytoplankton include green algae, which can grow explosively in "blooms" that may last days or weeks. Ice thickness and transparency determine the amount of light reaching the water, a critical factor for phytoplankton growth. Because these unique algae have adapted to under-ice conditions, present changes in the ice have decreased algae growth rates and slowed spring algal blooms. The effects of this decrease then move up the food chain, from the enormous quantities of tiny crustaceans that eat the algae to the fish that eat the crustaceans to the seals that depend on fish as their main food source.

The Baikal seal (*Pusa sibirica*), smallest of the world's seals and the only species exclusively living in freshwater, mates and gives birth on the lake ice. The seals require ice in early spring for shelter. If ice melt occurs early, the seals are forced into the water, and the extra energy expended affects female fertility and nurturing ability.

From a systems perspective, these changes in water temperature and in the timing and thickness of Lake Baikal's ice cover exemplify how changes in the atmosphere link to changes in the hydrosphere and biosphere. The lake's unique living ecosystem is already feeling the effects of changing climate, and many of Earth's other major lakes are experiencing similar warming trends. In this chapter, we examine lakes and other parts of the hydrologic cycle as we assess Earth's water resources.

▲**Figure GN 9.2 Forest ecosystem surrounding the lake.** [Debra Sharkey.]

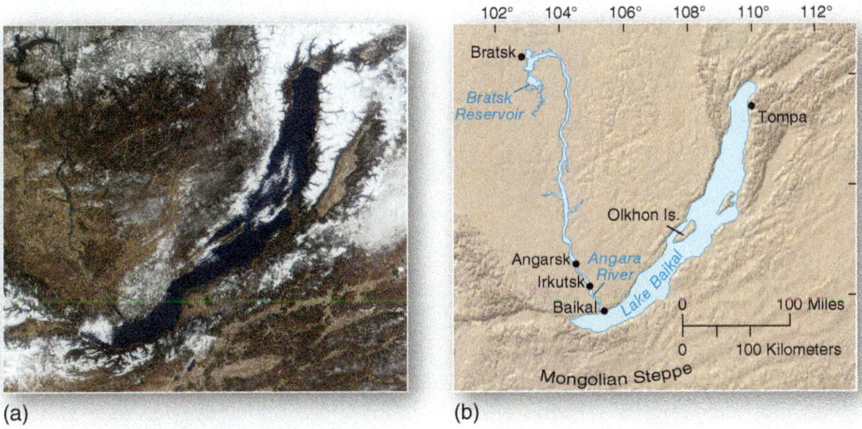

▲**Figure GN 9.1 Lake Baikal.** (a) A satellite view of the lake in October, as snowpack accumulates in the mountains. (b) Some 300 rivers flow into the lake, but only the Angara River flows out. [(a)*Terra* MODIS, NASA/GSFC.]

GEOSYSTEMS NOW ONLINE Go to Chapter 9 on the *MasteringGeography* website (www.masteringgeography.com) for more information about the effects of changing climate on Earth's lakes. To read about the effects of warming temperatures on Lake Superior in the United States, go to http://www.scientificamerican.com/article.cfm?id=lake-superior-a-natural-global-warming-gauge-is-running-a-fever. MG

Earth's physical processes are dependent on water, which is the essence of all life. Humans are about 70% water, as are plants and other animals. We use water to cook, bathe, wash clothes, dilute wastes, and run industrial processes. We use water to produce food, from the scale of small gardens to vast agricultural tracts. Water is the most critical resource supplied by Earth systems.

In the Solar System, water occurs in significant quantities only on our planet, covering 71% of Earth by area. Yet water is not always naturally available where and when we want it. Consequently, we rearrange water resources to suit our needs. We drill wells to tap groundwater and dam and divert streams to redirect surface water, either spatially (geographically, from one area to another) or temporally (over time, from one part of the calendar to another). All of this activity constitutes water-resource management.

Fortunately, water is a renewable resource, constantly cycling through the environment in the hydrologic cycle. Even so, some 1.1 billion people lack safe drinking water. People in 80 countries face impending water shortages, either in quantity or in quality, or both. Approximately 2.4 billion people lack adequate sanitary facilities—80% of these in Africa and 13% in Asia. This translates to some 2 million deaths a year due to lack of water and 5 million deaths a year from waterborne infections and disease. Investment in safe drinking water, sanitation, and hygiene could decrease these numbers. During the first half of this century, water availability per person will drop by 74% as population increases and adequate water decreases.

A 2011 report from the World Water Assessment Program of UNESCO (http://www.unesco.org/water/wwap/) states,

> As water demand and availability become more uncertain, all societies become more vulnerable to a wide range of risks associated with inadequate water supply, including hunger and thirst, high rates of disease and death, lost productivity and economic crises, and degraded ecosystems. These impacts elevate water to a crisis of global concern . . . [A]ll water users are—for better or worse and knowingly or unknowingly—agents of change who affect and are affected by, and connected through, the water cycle.

In this chapter: We begin with the origin and distribution of water on Earth. We then examine the hydrologic cycle, which gives us a model for understanding the global water balance. Next we introduce a water-budget approach to look at water resources. Similar in many ways to a money budget, it focuses attention on water "receipts" and "expenses" at specific locations. This budget approach can be applied at any scale, from a small garden to a farm to a regional landscape such as the High Plains Aquifer or Lake Baikal in Asia, discussed in *Geosystems Now.*

We also examine the various types of surface and groundwater resources, and discuss issues concerning water quantity and quality on a national and global basis. The chapter concludes by considering water supply in North America—specifically, the U.S. water budget made up of withdrawn and consumed water for irrigation, and industrial, and municipal uses. For many parts of the world, the question of water quantity and quality looms as the most important resource issue in this century.

Water on Earth

Earth's hydrosphere contains about 1.36 billion cubic kilometers of water (more specifically, 1,359,208,000 km^3, or 326,074,000 mi^3). Much of Earth's water originated from icy comets and from hydrogen- and oxygen-laden debris within the planetesimals that coalesced to form the planet. In 2007, the orbiting Spitzer Space Telescope observed for the first time the presence of water vapor and ice during the formation of new planets in a system 1000 light-years from Earth. Such discoveries prove water to be abundant throughout the Universe. As a planet forms, water from within migrates to its surface and outgasses.

Outgassing on Earth is a continuing process in which water and water vapor emerge from layers deep within and below the crust, 25 km (15.5 mi) or more below the surface, and are released in the form of gas (Figure 9.1). In the early atmosphere, massive quantities of outgassed water vapor condensed and then fell to Earth in torrential rains. For water to remain on Earth's surface, land temperatures had to drop below the boiling point of 100°C (212°F), something that occurred about 3.8 billion years ago. The lowest places across the face of Earth then began to fill with water—first forming ponds, then lakes and seas, and eventually ocean-sized bodies of water. Massive flows of water washed over the landscape, carrying both dissolved and solid materials to these early seas and oceans. Outgassing of water has continued ever since and

GEOreport 9.1 The water we use

In terms of direct water usage, urban dwellers in the United States average 680 L (180 gal) of water per person per day, whereas rural populations average only 265 L (70 gal) or less per person per day. (Take a moment and speculate about why there is a difference.) When calculated in overall terms, however, individual water averages are actually much higher than these figures because of indirect use, such as consuming food and beverages made with water-intensive practices. See http://www.worldwater.org/data.html, World's Water 2008–2009, Table 19, for more on the amount of water used for everyday foods and other products.

▲**Figure 9.1 Water outgassing from the crust.** Outgassing of water from Earth's crust occurs in geothermal areas such as southern Iceland west of where the Eyjafjallajökull volcano erupted in 2010. [Bobbé Christopherson.]

is visible in volcanic eruptions, geysers, and seepage to the surface.

Worldwide Equilibrium

Today, water is the most common compound on the surface of Earth. The present volume of water circulating throughout Earth's surface systems was attained approximately 2 billion years ago, and this quantity has remained relatively constant even though water is continuously gained and lost. Gains occur as pristine water not previously at the surface emerges from within Earth's crust. Losses occur when water dissociates into hydrogen and oxygen and the hydrogen escapes Earth's gravity to space or when it breaks down and forms new compounds with other elements. The net result of these water inputs and outputs is that Earth's hydrosphere is in a steady-state equilibrium in terms of quantity.

Within this overall balance, the amount of water stored in glaciers and ice sheets varies, leading to periodic global changes in sea level (discussed further in Chapter 16). The term **eustasy** refers to changes in global sea level caused by changes in the volume of water in the oceans. Such sea-level changes caused specifically by glacial ice melt are *glacio-eustatic* factors (see Chapter 17). During cooler climatic conditions, when more water is bound up in glaciers (at high latitudes and at high elevations worldwide) and in ice sheets (on Greenland and Antarctica), sea level lowers. During warmer periods, less water is stored as ice, so sea level rises. Today, sea level is rising worldwide at an accelerating pace as higher temperatures melt more ice and, in addition, cause ocean water to thermally expand.

Distribution of Earth's Water Today

From a geographic point of view, ocean and land surfaces are distributed unevenly on Earth. If you examine a globe, it is obvious that most of Earth's continental land is in the Northern Hemisphere, whereas water dominates the surface in the Southern Hemisphere. In fact, when you look at Earth from certain angles, it appears to have an *oceanic hemisphere* and a *land hemisphere* (Figure 9.2).

The present distribution of all of Earth's water between the liquid and frozen states and between fresh and saline, surface and underground, is shown in Figure 9.3. The oceans contain 97.22% of all water, with about 48% of that water in the Pacific Ocean (as measured by ocean surface area; Figure 9.3b). The remaining 2.78% is freshwater (nonoceanic) and is either surface or subsurface water, as detailed in the middle pie chart in the figure. Ice sheets and glaciers contain the greatest amount of Earth's freshwater. Groundwater, either shallow or deep, is the second largest amount. The remaining freshwater, which resides in lakes, rivers, and streams, actually represents less than 1% of all water.

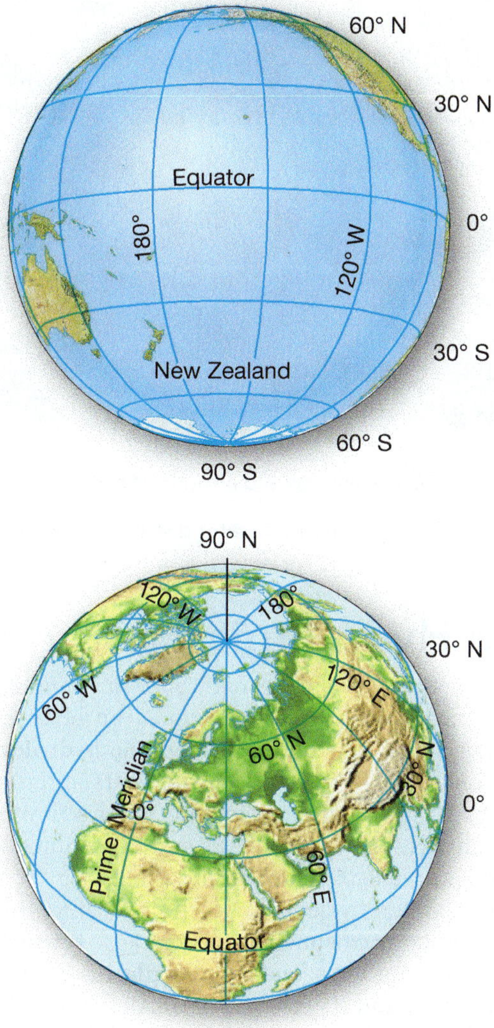

▲**Figure 9.2 Land and water hemispheres.** Two perspectives that roughly illustrate Earth's ocean hemisphere and land hemisphere.

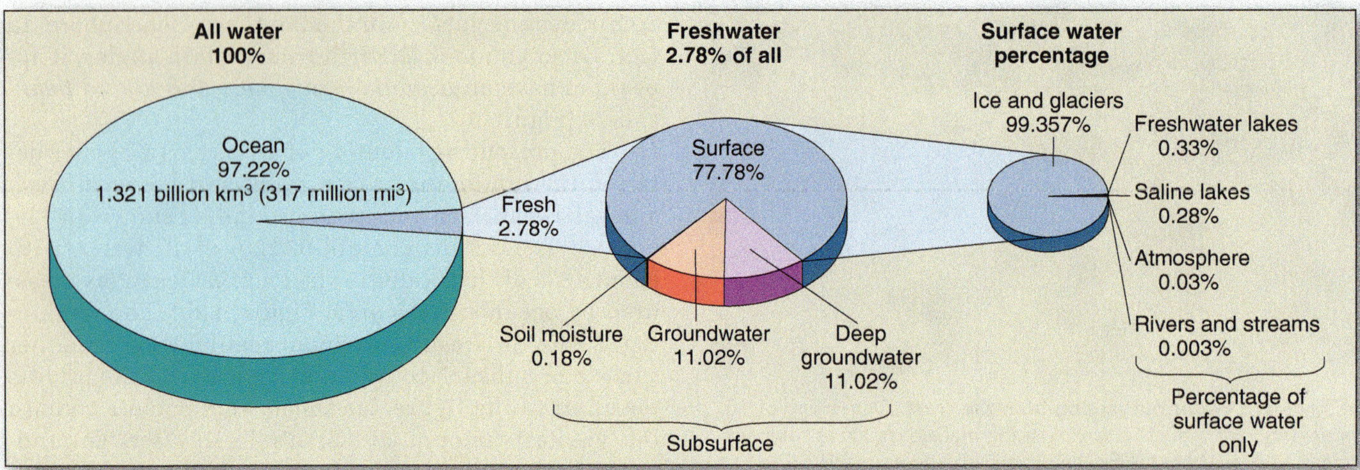

(a) The location and percentages of all water on Earth, with detail of the freshwater portion (surface and subsurface) and a breakdown of the surface water component.

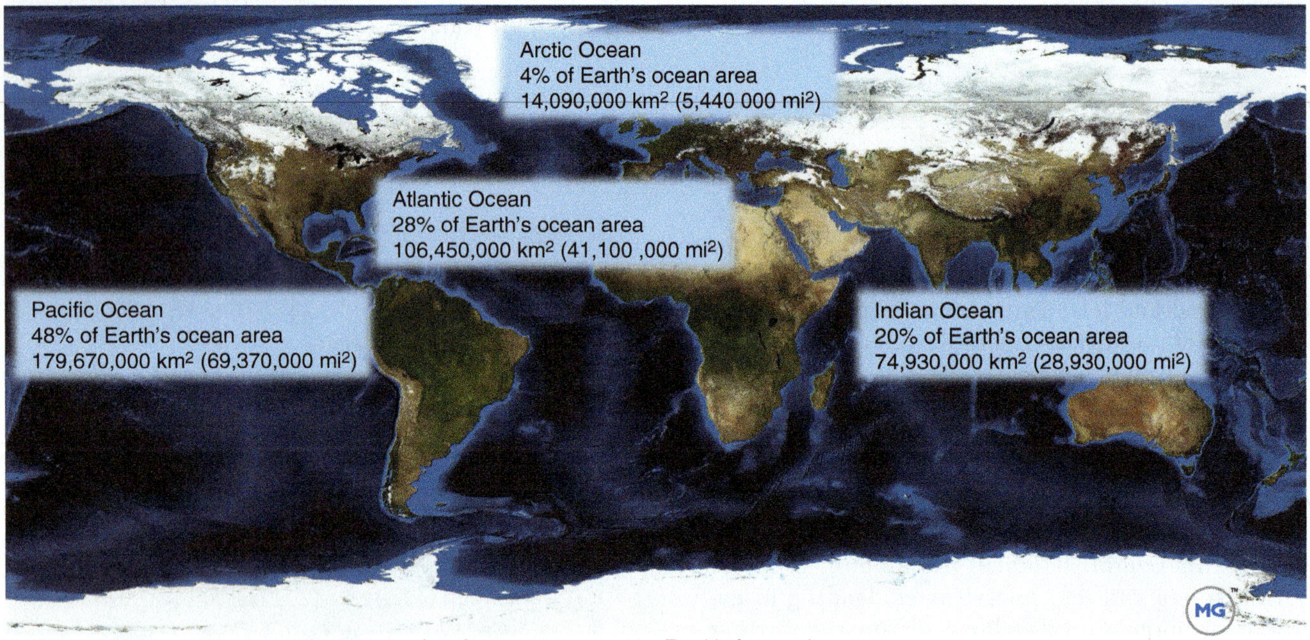

(b) The relative percentages and amounts of surface area covered by Earth's four major oceans.

▲Figure 9.3 **Ocean and freshwater distribution on Earth.**

MG Animation
Earth's Water and the Hydrologic Cycle

The Hydrologic Cycle

Vast currents of water, water vapor, ice, and associated energy are flowing continuously in an elaborate, open, global system. Together, they form the **hydrologic cycle**, which has operated for billions of years, circulating and transforming water throughout Earth's lower atmosphere, hydrosphere, biosphere, and lithosphere to several kilometers beneath the surface.

The water cycle can be divided into three main components: atmosphere, surface, and subsurface. The residence time for a water molecule in any component of the cycle, and its effect on climate, is variable. Water has a short residence time in the atmosphere—an average of 10 days—where it plays a role in temporary fluctuations in regional weather patterns. Water has longer residence times in deep-ocean circulation groundwater, and glacial ice (as long as 3000–10,000 years), where it acts to moderate temperature and climatic changes. These slower parts of the hydrologic cycle, the parts where water is stored and released over long periods, can have a "buffering" effect during periods of water shortage.

Water in the Atmosphere

Figure 9.4 is a simplified model of the hydrologic system, with estimates of the volume of water involved in the main pathways (in thousands of cubic kilometers). We use the ocean as a starting point for our discussion, although we could jump into the model at any point. More than 97% of Earth's water is in the oceans, and it is over these water bodies that 86% of Earth's evaporation occurs. As discussed in previous chapters, **evaporation** is the net movement of free water molecules away from a wet surface into air that is less than saturated.

Water also moves into the atmosphere from land environments, including water moving from the soil into plant

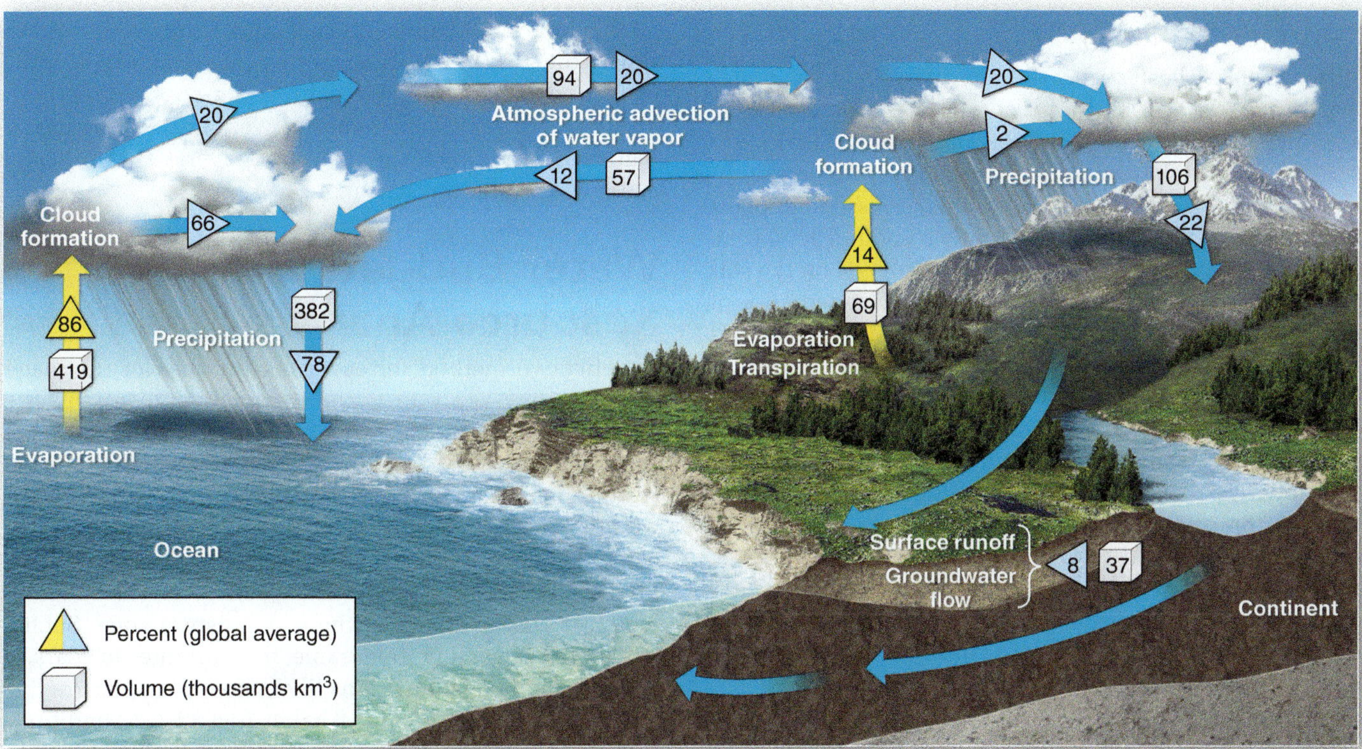

▲**Figure 9.4 The hydrologic cycle model.** Water travels endlessly through the hydrosphere, atmosphere, lithosphere, and biosphere. The triangles show global average values as percentages. Note that all evaporation (86% + 14% = 100%) equals all precipitation (78% + 22% = 100%), and advection in the atmosphere is balanced by surface runoff and subsurface groundwater flow when all of Earth is considered. (To convert volume, 1 $km^3 \times 0.24 = 1\ mi^3$.)

MG Animation Earth's Water and the Hydrologic Cycle

roots and passing through their leaves to the air. This process is transpiration, introduced in the Chapter 7 *Geosystems Now* case study about coastal redwoods and summer fog. During **transpiration**, plants release water to the atmosphere through small openings called stomata in their leaves. Transpiration is partially regulated by the plants themselves, as control cells around the stomata conserve or release water. On a hot day, a single tree can transpire hundreds of liters of water; a forest, millions of liters. Evaporation and transpiration from Earth's land surfaces together make up **evapotranspiration**, which represents 14% of the water entering Earth's atmosphere in Figure 9.4.

Figure 9.4 shows that of the 86% of evaporation rising from the oceans, 66% combines with 12% advected (moved horizontally) from the land to produce the 78% of all precipitation that falls over the oceans. The remaining 20% of moisture evaporated from the ocean, plus 2% of land-derived moisture, produces the 22% of all precipitation that falls over land. Clearly, the bulk of continental precipitation comes from the oceanic portion of the cycle. The different parts of the cycle vary over different regions on Earth, creating imbalances that, depending on the local climate, lead to water surpluses in one place and water shortages in another.

Water at the Surface

Precipitation that reaches Earth's surface as rain follows two basic pathways: It either flows overland or soaks into the soil. Along the way, **interception** also occurs, in which precipitation lands on vegetation or other ground cover before reaching the surface. Intercepted water that drains across plant leaves and down their stems to the ground is known as *stem flow*. Precipitation that falls directly to the ground, including drips from vegetation that are not stem flow, is *throughfall*. Precipitation that reaches Earth's surface as snow may accumulate for a period of hours or days before melting, or it may accumulate as part of the snowpack that remains throughout winter and melts in the spring.

After reaching the ground surface as rain, or after snowmelt, water may soak into the subsurface through **infiltration**, or penetration of the soil surface (Figure 9.5). If the ground surface is impermeable (does not permit the passage of liquids), then the water will begin to flow downslope as **overland flow**, also known as **surface runoff**. Overland flow will also occur if the soil has been infiltrated to full capacity and is saturated. Excess water may remain in place on the surface in puddles or ponds, or may flow until it forms channels—at this point it becomes *streamflow*, a term that describes surface water flow in streams, rivers, and other channels.

Figure 9.4 shows that 8% of the water in the cycle is moving on or through land. Most of this movement—about 95%—comes from surface waters that wash across land as overland flow and streamflow. Only 5% of water movement is slow-moving subsurface groundwater. Although only a small percentage of water is in rivers and streams, this portion is dynamic and fast-moving compared to its sluggish, subsurface counterpart.

Water in the Subsurface

Water that infiltrates into the subsurface moves downward into soil or rock by **percolation**, the slow passage of water through a porous substance (shown in Figure 9.5). The **soil-moisture zone** contains the volume of subsurface water stored in the soil that is accessible to plant roots. Within this zone, some water is bound to soil so that it is not available to plants—this depends on the soil texture (discussed in Chapter 18). An estimated 76% of precipitation over land infiltrates the subsurface, and about 85% of this water returns to the atmosphere either by evaporation from soil or transpiration from plants.

If the soil is saturated, then any water surplus within the soil body becomes *gravitational water*, percolating downward into the deeper groundwater. The latter defines the *zone of saturation*, where the soil spaces are completely filled with water. The top of this zone is known as the *water table*. At the point where the water table intersects a stream channel, water naturally discharges at the surface, producing **base flow**, which refers to the portion of streamflow that consists of groundwater.

Under natural conditions, streams and groundwater ultimately flow into oceans, thus continuing movement through the hydrologic cycle. In some cases, streams flow into closed lake basins, where water evaporates or soaks into the ground. Many streams flow into reservoirs behind dams, where water is stored until it evaporates or is released into the channel downstream. Groundwater flows slowly toward the sea, intersecting the surface or seeping from underground after reaching the coast, sometimes mixing with seawater in coastal wetlands and estuaries (bodies of water near the mouths of rivers). Groundwater is discussed later in the chapter.

▼Figure 9.5 Pathways for precipitation on Earth's surface. The principal pathways for precipitation include interception by plants; throughfall to the ground; collection on the surface and overland flow to streams; transpiration and evaporation from plants; evaporation from land and water; and gravitational water moving to subsurface groundwater.

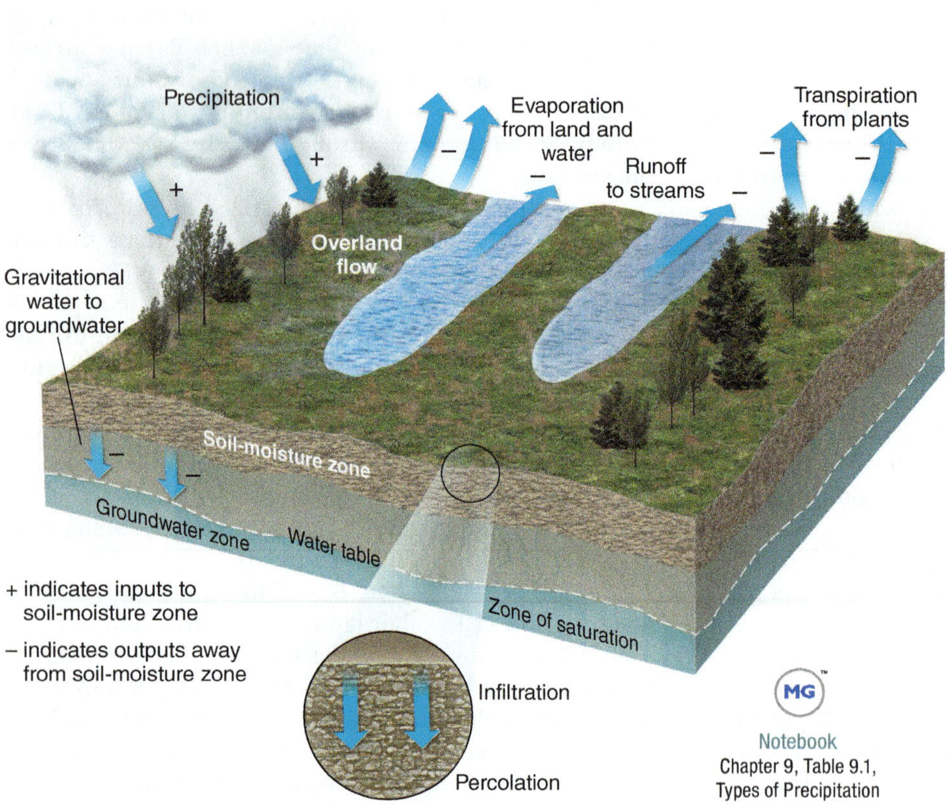

MG Notebook
Chapter 9, Table 9.1, Types of Precipitation

Water Budgets and Resource Analysis

An effective method for assessing portions of the water cycle as they apply to water resources is to establish a **water budget** for any area of Earth's surface—a continent, country, region, field, or front yard. A water budget is derived from measuring the input of precipitation and its distribution and the outputs of evapotranspiration, including evaporation from ground surfaces and transpiration from plants, and surface runoff. Also included in this budget is moisture that is stored in the soil-moisture zone. Such a budget can cover any time frame, from minutes to years.

A water budget functions like a money budget: Precipitation is the income that must balance against expenditures for evaporation, transpiration, and runoff. Soil-moisture storage acts as a savings account, accepting deposits and yielding withdrawals of water. Sometimes all expenditure demands are met, and any extra water results in a **surplus**. At other times, precipitation and soil-moisture savings are inadequate to meet demands, and a **deficit**, or water shortage, results.

Geographer Charles W. Thornthwaite (1899–1963) pioneered applied water-resource analysis using water budgets, working with others to develop a methodology for solving real-world problems related to irrigation and water use for maximizing crop yields. Thornthwaite also recognized the important relationship between water supply and water demand as it varies with climate.

Components of the Water Budget

To understand Thornthwaite's water-budget methodology and "accounting" procedures, we must first define certain terms and concepts. We begin by discussing water supply, demand, and storage as components of the water-budget equation.

Water Supply: Precipitation The moisture supply to Earth's surface is **precipitation** (PRECIP) in all its forms, such as rain, sleet, snow, or hail. A summary table of the different types of precipitation is on the *MasteringGeography* website. As you read through this table, recall the forms of precipitation you have experienced.

One way to measure precipitation is with a **rain gauge**, essentially a large measuring cup that collects rainfall and snowfall so the water can be measured by depth, weight, or volume (Figure 9.6). Wind causes underestimation because the drops or snowflakes are not falling vertically; the wind shield reduces the undercatch by catching raindrops that arrive at an angle.

Regular precipitation measurements are made at more than 100,000 locations worldwide. A global map of annual precipitation averages is displayed in Chapter 10. Figure 9.7 shows precipitation patterns in the United States and Canada, which relate to the air masses and lifting mechanisms presented in Chapter 8.

▲Figure 9.6 **A rain gauge.** A funnel guides water into a bucket sitting on an electronic weighing device. The gauge minimizes evaporation, which would cause low readings. The wind shield around the top of the gauge minimizes the undercatch produced by wind. [Bobbé Christopherson.]

Water Demand: Potential Evapotranspiration Evapotranspiration is an actual expenditure of water to the atmosphere. In contrast, **potential evapotranspiration** (POTET) is the amount of water that would evaporate and transpire under optimum moisture conditions when adequate precipitation and soil moisture are present.

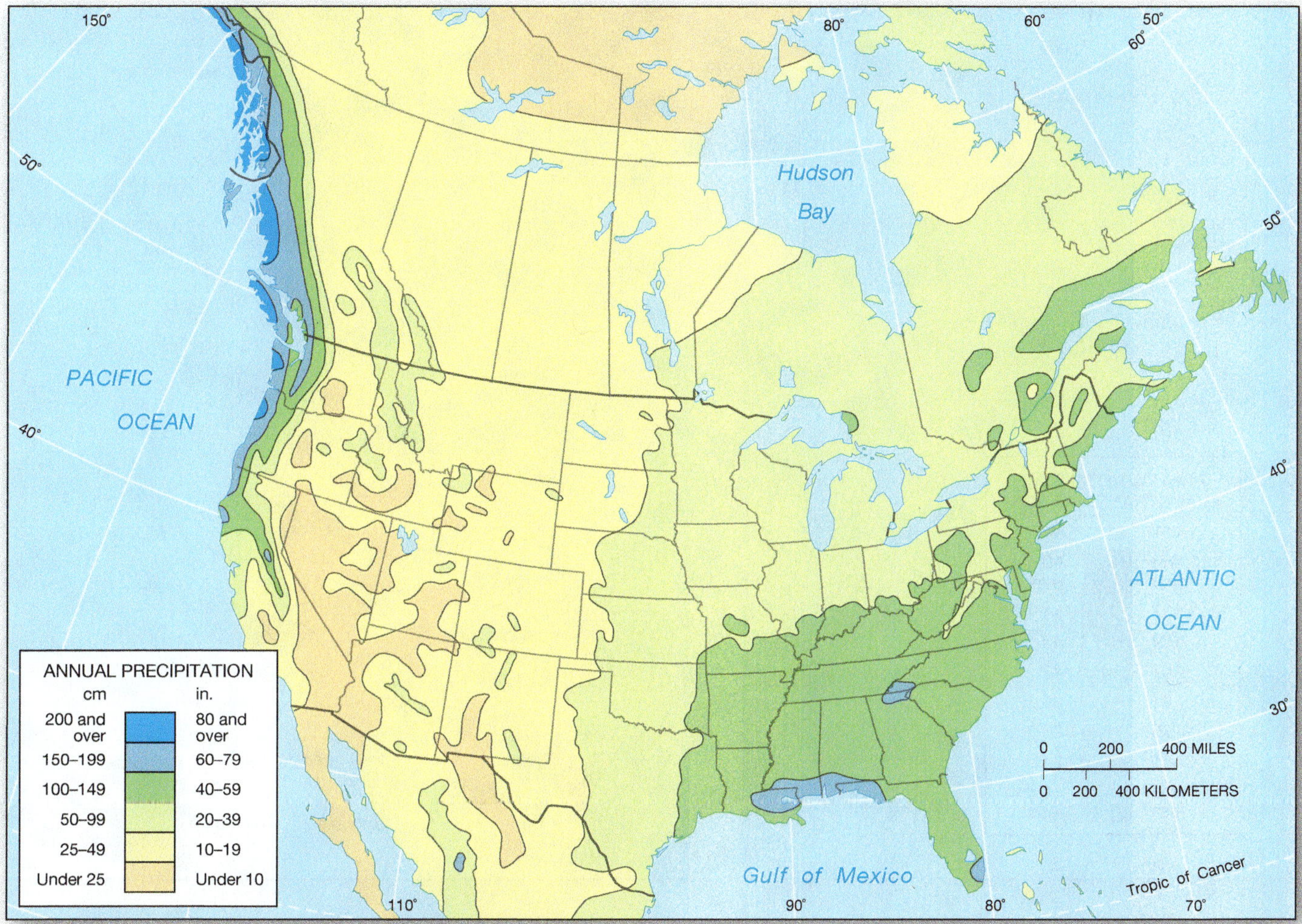

▲Figure 9.7 **Precipitation in North America—the water supply.** [Based on NWS, U.S. Department of Agriculture, and Environment Canada.]

MG™ Satellite **Global Water Balance Components**

Filling a bowl with water and letting the water evaporate illustrates this concept: When the bowl becomes dry, some degree of evaporation demand remains. If the bowl could be constantly replenished with water, the amount of water that would evaporate given this constant supply is the POTET—the total water demand. If the bowl dries out, the amount of POTET that is not met is the water deficit. If we subtract the deficit from the potential evapotranspiration, we derive what actually happened—**actual evapotranspiration** (ACTET).

Precise measurement of evapotranspiration is difficult. One method employs an *evaporation pan*, or evaporimeter. As evaporation occurs, water in measured amounts is automatically replaced in the pan, equaling the amount that evaporated. A more elaborate measurement device is a *lysimeter*, which isolates a representative volume of soil, subsoil, and plant cover to allow measurement of the moisture moving through the sampled area (an illustration is on the *MasteringGeography* website). A rain gauge next to the lysimeter measures the precipitation input.

Thornthwaite developed an easy and fairly accurate indirect method of estimating POTET for most midlatitude locations using mean air temperature and daylength to approximate POTET. (Remember from Chapter 2 that daylength is a function of a station's latitude.) His method works well for large-scale regional applications using data from hourly, daily, monthly, or annual time frames, and works better in some climates than in others.

Figure 9.8 presents POTET values for the United States and Canada derived using Thornthwaite's approach. Note that higher values occur in the South, with highest readings in the Southwest, where higher average air temperature and lower relative humidity exist. Lower POTET values are found at higher latitudes and elevations, which have lower average temperatures.

Compare Figures 9.7 and 9.8, the maps of water supply in the form of precipitation and water demand in the form of potential evapotranspiration. Can you identify regions where PRECIP is greater than POTET (for example, the eastern United States)? Or where POTET is greater than PRECIP (for example, the southwestern United States)? Where you live, is the water demand usually met by the precipitation supply? Or does your area experience a natural shortage? How might you find out?

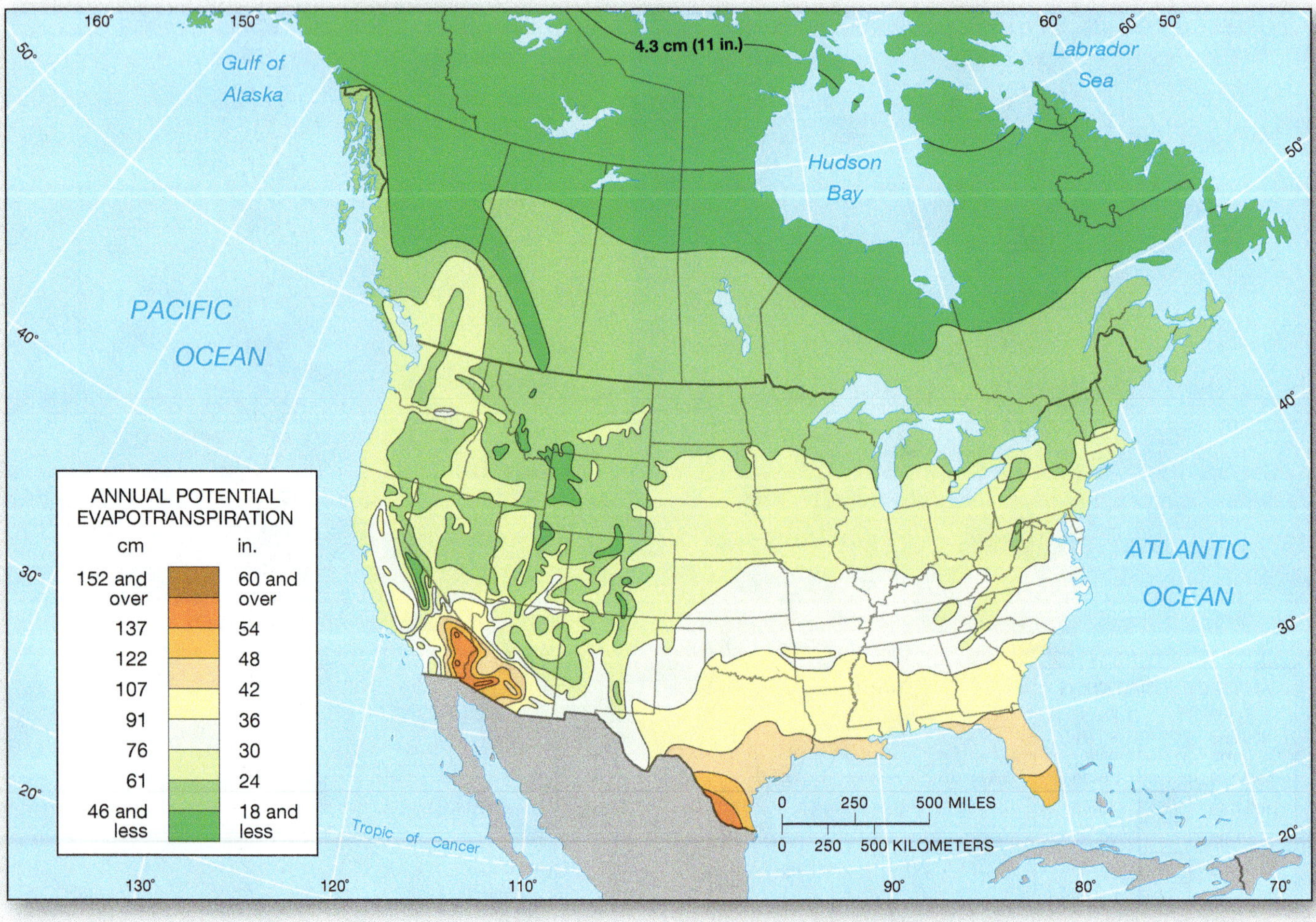

▲**Figure 9.8 Potential evapotranspiration for the United States and Canada—the water demand.** [From C. W. Thornthwaite, "An approach toward a rational classification of climate," *Geographical Review* 38 (1948): 64, © American Geographical Society. Canadian data adapted from M. Sanderson, "The climates of Canada according to the new Thornthwaite classification," *Scientific Agriculture* 28 (1948): 501–517.]

MG Satellite
Global Water Balance Component

Figure 9.9 Types and availability of soil moisture. [(a) After D. Steila, The Geography of Soils, © 1976, p. 45, © Pearson Prentice Hall, Inc. (b) After U.S. Department of Agriculture, 1955 Yearbook of Agriculture—Water, p. 120.]

MG Satellite Global Water Balance Components

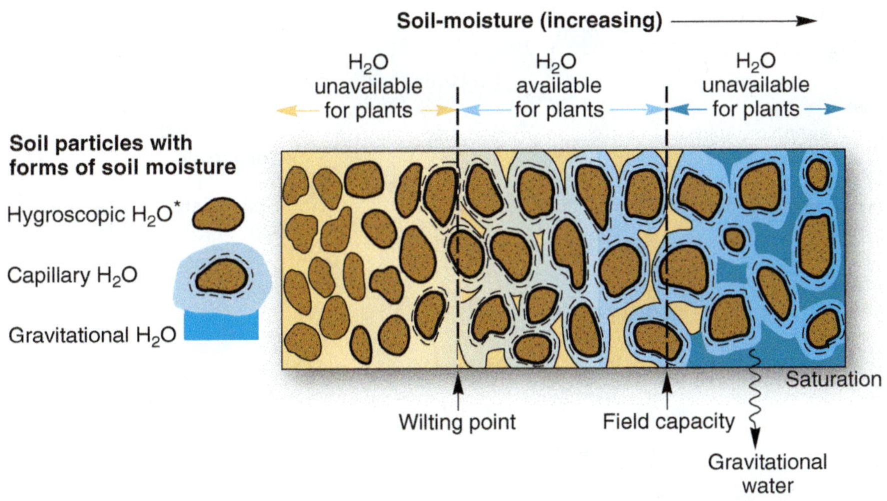

(a) Hygroscopic water bound to soil particles and gravitational water draining through the soil moisture zone are not available to plant roots.

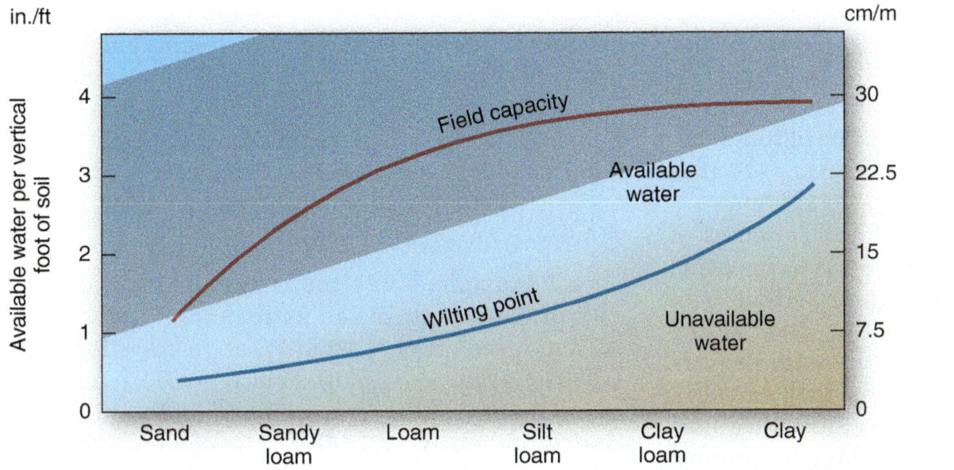

(b) The relationship between soil-moisture availability and soil texture determines the distance between the two curves that show field capacity and wilting point. A loam soil (one-third each of sand, silt, and clay) has roughly the most available water per vertical foot of soil exposed to plant roots.

*Note: Some capillary water is bound to hygroscopic water on soil particles and is also unavailable.

Water Storage: Soil Moisture As part of the water budget, the volume of water in the subsurface soil-moisture zone that is accessible to plant roots is **soil-moisture storage** (STRGE). This is the savings account of water that receives deposits (or recharge) and provides for withdrawals (or utilization). The soil-moisture environment includes three categories of water—hygroscopic, capillary, and gravitational. Only hygroscopic and capillary water remain in the soil-moisture zone; gravitational water fills the soil pore spaces and then drains downward under the force of gravity. Of the two types of water that remain, only capillary water is accessible to plants (Figure 9.9a).

When only a small amount of soil moisture is present, it may be unavailable to plants. **Hygroscopic water** is inaccessible to plants because it is a molecule-thin layer that is tightly bound to each soil particle by the hydrogen bonding of water molecules. Hygroscopic water exists in all climates, even in deserts, but it is unavailable to meet POTET demands. Soil moisture is at the **wilting point** for plants when all that remains is this inaccessible water; plants wilt and eventually die after a prolonged period at this degree of moisture stress.

Capillary water is generally accessible to plant roots because it is held in the soil, against the pull of gravity, by hydrogen bonds between water molecules (that is, by surface tension), and by hydrogen bonding between water molecules and the soil. Most capillary water is *available water* in soil-moisture storage. After some water drains from the larger pore spaces, the amount of available water remaining for plants is termed **field capacity**, or storage capacity. This water can meet POTET demands through the action of plant roots and surface evaporation. Field capacity is specific to each soil type; the texture and the structure of the soil dictate available pore spaces, or *porosity* (discussed in Chapter 18).

Gravitational water is the water surplus in the soil body after the soil becomes saturated during a precipitation event. This water is unavailable to plants, as it percolates downward to the deeper groundwater zone. Once the soil-moisture zone reaches saturation, the pore spaces are filled with water, leaving no room for oxygen or gas exchange by plant roots until the soil drains.

Figure 9.9b shows the relation of soil texture to soil-moisture content. Different plant species send roots to different depths and therefore reach different amounts of soil moisture. A soil blend that maximizes available water is best for plants (see the discussion of soil texture in Chapter 18).

When water demand exceeds the precipitation supply, **soil-moisture utilization**—usage by plants of the available moisture in the soil—occurs. As water is removed from the soil, plants have increased difficulty extracting the amount of moisture they need. Eventually, even though a small amount of water may remain in the

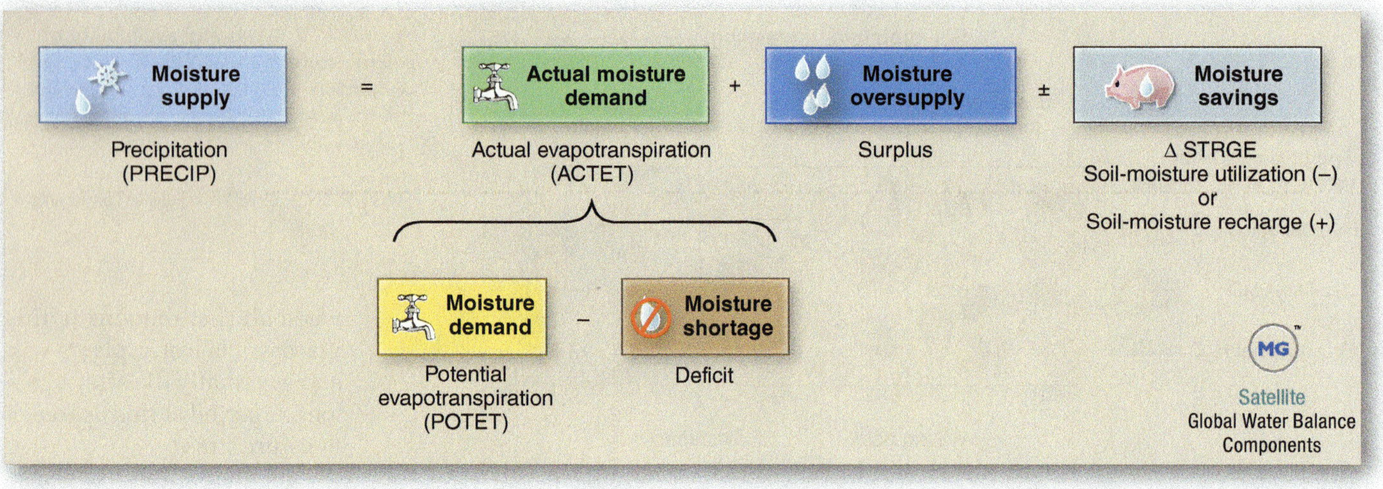

▲**Figure 9.10 The water-budget equation explained.**

soil, plants may be unable to use it. In agriculture, farmers use irrigation to avoid a deficit and enhance plant growth with adequate amounts of available water.

When water infiltrates the soil and replenishes available water, whether from natural precipitation or artificial irrigation, **soil-moisture recharge** occurs. The property of the soil that determines the rate of soil-moisture recharge is its **permeability**, which depends on particle sizes and the shape and packing of soil grains.

Water infiltration is rapid in the first minutes of precipitation and slows as the upper soil layers become saturated, even though the deeper soil may still be dry. Agricultural practices such as plowing and adding sand or manure to loosen soil structure can improve both soil permeability and the depth to which moisture can efficiently penetrate to recharge soil-moisture storage. You may have found yourself working to improve soil permeability for a houseplant or garden—that is, working the soil to increase the rate of soil-moisture recharge.

Water Deficit and Surplus

A deficit, or moisture shortage, occurs at a given location when the POTET demand cannot be satisfied by precipitation inputs, by moisture stored in the soil, or through additional inputs of water by artificial irrigation. Deficits cause drought conditions (defined and discussed just ahead).

A surplus occurs where additional water exists after POTET is satisfied by precipitation inputs; this surplus often becomes runoff, feeding surface streams and lakes and recharging groundwater. Under ideal conditions for plants, potential and actual amounts of evapotranspiration are about the same, so plants do not experience a water shortage.

The Water-Budget Equation

The water-budget equation explained in Figure 9.10 states that, for a certain location or portion of the hydrologic cycle, water inputs are equal to the water outputs plus or minus the change in water storage. The delta symbol, Δ, means "change"—in this case, the change in soil-moisture storage, which includes both recharge and utilization.

In summary, precipitation (mostly rain and snow) provides the moisture input. This supply is distributed as actual water used for evaporation and plant transpiration, extra water running into streams and subsurface groundwater, and water that moves in and out of soil-moisture storage. As in all equations, the two sides must balance; that is, the precipitation input (left side) must equal the outputs (right side).

Sample Water Budgets

As an example, study the water-budget graph for the city of Kingsport, in the extreme northeastern corner of Tennessee (36.6° N, 82.5° W) at an elevation of 390 m (1280 ft). Figure 9.11 plots PRECIP, ACTET, and POTET, using monthly averages, which smooth out the actual daily and hourly variability. The cooler time from October to May shows a net surplus (blue areas), as precipitation is higher than potential evapotranspiration. The warm days from June to September create a net water demand. If we assume a soil-moisture storage capacity of 100 mm (4.0 in.), typical of shallow-rooted plants, this water demand is satisfied through soil-moisture utilization (green area), with a small summer soil-moisture deficit (orange area).

Kingsport experiences water supply-and-demand patterns typical of a humid continental region. In other climatic regimes, the relationships between water-budget components are different. Figure 9.12 presents water-budget graphs for the cities of Berkeley, California, which has a summer minimum in precipitation, and Phoenix,

CRITICALthinking 9.1
Your Local Water Budget

Select an area of interest, such as your campus, yard, or even a houseplant container, and apply the water-budget concepts to it. Where does the water supply originate—its source? Estimate the ultimate water supply and demand for the area you selected. For a general idea of PRECIP and POTET, find your locale on Figures 9.7 and 9.8. Consider the seasonal timing of this supply and demand, and estimate water needs and how they vary as components of the water-budget change. ●

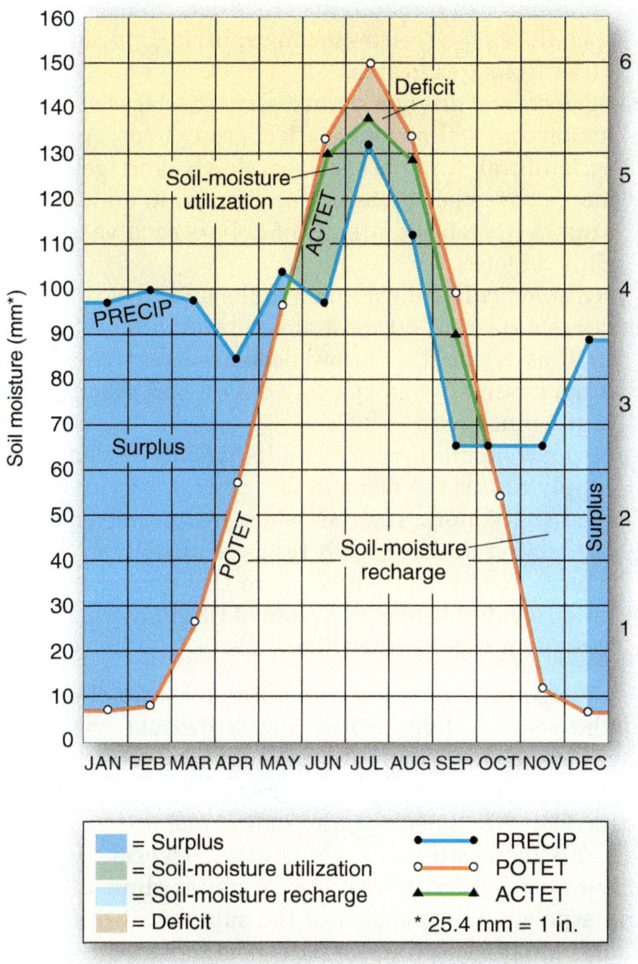

▲Figure 9.11 **Sample water budget for Kingsport, Tennessee.** Compare the average plots for precipitation inputs and potential evapotranspiration outputs to determine the condition of the soil-moisture environment. A typical pattern for Kingsport exhibits a spring surplus, summer soil-moisture utilization, a small summer deficit, autumn soil-moisture recharge, and a year-ending surplus.

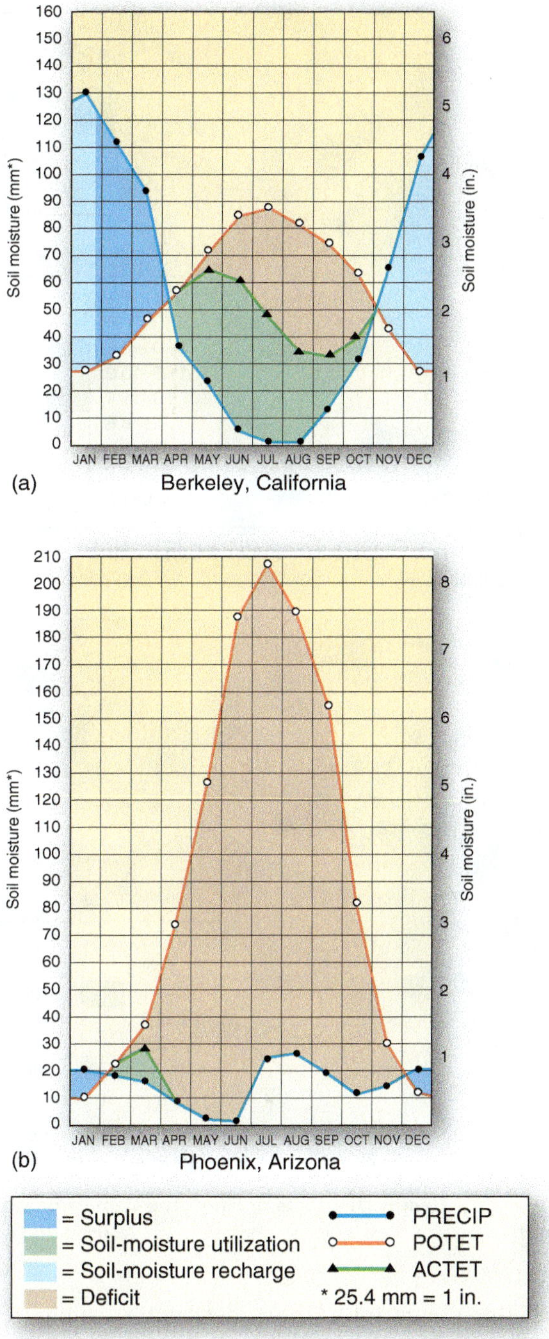

▲Figure 9.12 **Sample water budgets for stations near two U.S. cities.**

Arizona, which has low precipitation throughout the year. Compare the size and timing of the water deficit at these locations with the Kingsport graph.

Water-Budget Application: Hurricane Camille

Hurricane Camille, in 1969, one of the most devastating hurricanes of the 20th century, provides an interesting application of water-budget analysis. Ironically, Camille was significant not only for its 256 fatalities and $1.5 billion in damage, but also for its generally positive effects on regional water resources.

After Camille made landfall on the evening of August 17 along the Gulf Coast, the hurricane-force winds sharply diminished (Figure 9.13a), leaving a vast rainstorm that traveled through Mississippi, western Tennessee, and Kentucky and into central Virginia. These rains ended a yearlong drought along major portions of Camille's storm track. Torrential rains produced record flooding near the point of landfall and in the James River basin of Virginia.

Figure 9.13a maps precipitation along the storm's track from landfall in Mississippi to the Virginia and Delaware coasts. We can analyze the impact of the storm by comparing actual water budgets with simulated water budgets for the same 3 days in which we remove Camille's rainfall totals. Figures 9.13b maps the degree to which moisture shortages were reduced because of Camille's rains—an outcome described as *deficit abatement*. Over vast portions of the affected area, Camille reduced dry-soil conditions, invigorated agriculture, and raised low reservoir levels. The monetary benefits outweighed the damage by an estimated 2-to-1 ratio. (Of course, the tragic loss of life does not fit into a financial equation.)

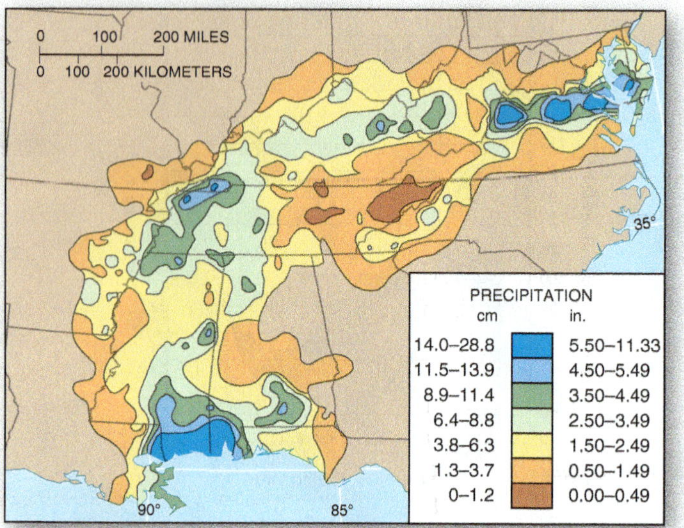

(a) Precipitation attributable to Hurricane Camille (the moisture supply).

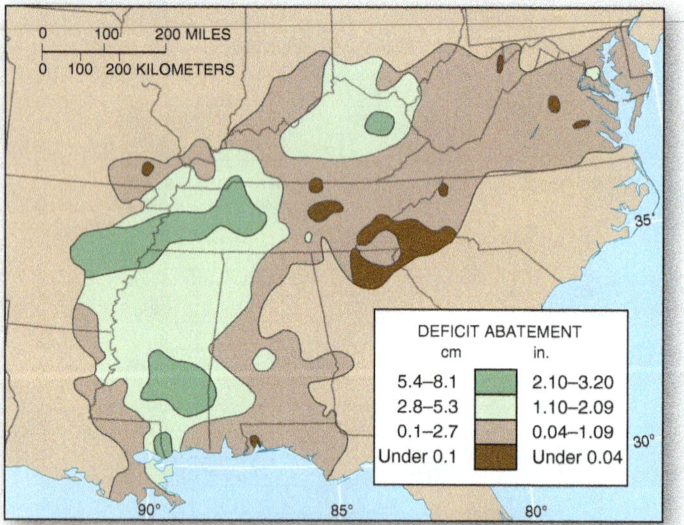

(b) Resulting deficit abatement (avoided moisture shortages).

▲**Figure 9.13 Beneficial effects of Hurricane Camille on water budgets.** [Data and maps by Robert Christopherson. All rights reserved.]

Thus, along with their destructive potential for coastal lowlands, hurricanes also contribute to the precipitation regimes of these areas. According to weather records, about one-third of all hurricanes making landfall in the United States provide beneficial precipitation inputs to local water budgets.

Drought: The Water Deficit

Drought is a commonly used term, and might seem simple to define: Less precipitation and higher temperatures make for drier conditions over an extended period of time. However, scientists and resource managers use four distinct technical definitions for **drought** based on not only precipitation, temperature, and soil-moisture content, but also the water-resource demand.

- *Meteorological drought* is defined by the degree of dryness, as compared to a regional average, and the duration of dry conditions. This definition is region-specific, since it relates to atmospheric conditions that differ from area to area.
- *Agricultural drought* occurs when shortages of precipitation and soil moisture affect crop yields. Although agricultural drought evolves slowly and gets little media coverage, losses can be significant and are running in the tens of billions of dollars each year in the United States.
- *Hydrological drought* refers to the effects of precipitation shortages (both rain and snow) on water supply, such as when streamflow decreases, reservoir levels drop, mountain snowpack declines, and groundwater mining increases.
- *Socioeconomic drought* results when reduced water supply causes the demand for goods or services to exceed the supply, such as when hydroelectric power production declines with reservoir depletion. This is a more comprehensive measure that considers water rationing, wildfire events, loss of life, and other widespread impacts of water shortfalls.

Drought is a natural and recurrent feature of climate. In the southwestern United States, drought conditions have existed since early 2000, one of several such droughts evident in the region's climatic record over the last 1000 years. However, scientists are finding mounting evidence that this increased aridity, or climatic dryness, links not only to natural factors but also to global climate change and a poleward expansion of the subtropical dry zones. Thus, human-caused warming is combining with natural climate variability to create a trend toward lasting drought.

Studies suggest that droughts such as those occurring previously in the U.S. Southwest, which were related to sea-surface temperature changes in the distant tropical Pacific Ocean, will still occur, but they will be worsened by climate change and the further expansion of the hot, dry, subtropical high-pressure system and the summertime continental tropical (cT) air mass into the region. The spatial implications of such shifting of Earth's primary circulation systems and semi-permanent drought into a region of steady population growth and urbanization are serious and suggest the immediate need for water-resource planning. Focus Study 9.1 discusses this drought and its effects on the Colorado River watershed.

Droughts of various intensities occurred on every continent throughout 2012. Australia is in a decade-long drought, its worst in 110 years, correlating with record high temperatures, as discussed in Chapter 5. In 2010 and 2011, the Horn of Africa (including Somalia and Ethiopia) experienced the worst drought in 60 years and an associated food crisis. Research suggests that ongoing drought patterns in east Africa are tied to sea-surface temperature patterns in the Indian Ocean.

In the United States, the region of western Oklahoma and Texas is in an intense drought, as evidenced by 4 million acres burned by wildfires between November 2010 and November 2011, including almost 3000 homes destroyed. Towns in the Oklahoma Panhandle endured more than a year with little or no precipitation (never more than 6 mm, 0.25 in., of rain in a day). In July of 2012, the U.S. Department of Agriculture declared almost one-third of all U.S. counties federal disaster areas owing to drought conditions—the largest natural disaster area ever declared. According to the National Climatic Data Center, 16 of the droughts that occurred from 1980 to 2011 cost over $1 billion each, making drought one of the costliest U.S. weather events. (See http://www.drought.unl.edu/ for a weekly Drought Monitor map and other resources for assessing and mitigating drought.)

Surface Water Resources

Water distribution over Earth's surface is uneven over space and time. Because humans require a steady water supply, we increasingly rely on large-scale management projects intended to redistribute water resources either geographically, by moving water from one place to another, or through time, by storing water until it is needed. In this way, deficits are reduced, surpluses are held for later release, and water availability is improved to satisfy natural and human demands.

The freshwater on Earth's surface is found primarily in snow and ice, rivers, lakes, and wetlands. Surface water is also stored in reservoirs, artifical lakes formed by dams on rivers. Figure 9.14 shows the world's major rivers, lakes, reservoirs, and wetlands, all discussed in this section.

Snow and Ice

The largest amount of surface freshwater on Earth is stored in glaciers, permafrost, and polar ice (review Figure 9.3). Seasonal melting from glaciers and the annual snowpack in temperate regions feeds streamflow, contributing to water supplies. Snowpack melting captured in reservoirs behind dams is a primary water source for humans in many parts of the world.

Glaciers provide a semi-permanent form of water storage, although rising temperatures associated with recent climate change are causing accelerated rates of glacial melting. The residence time of water in glaciers can range between decades and centuries, and the relatively small but continuous meltwater from glaciers can sustain streamflow throughout the year (Chapter 17 discusses glaciers and ice sheets).

Rising temperatures have caused glacier melt rates to accelerate, and some scientists estimate that most glaciers will be gone by 2035 if present melt rates continue. On Asia's Tibetan Plateau, the world's largest and highest plateau at 3350 km (11,000 ft) elevation, climate change is causing mountain glaciers to recede at rates faster than anywhere else in the world. More than 1000 lakes store water on this plateau, forming the headwaters for several of the world's longest rivers. Almost half the world's population lives within the watersheds of these rivers; the Yangtze and Yellow Rivers (both flowing westward through China) alone supply water to approximately 520 million people in China. Even as glacial melting has increased streamflows, the worsening drought in western China is causing these flows to evaporate or infiltrate into the ground before reaching the largest population centers. Although disappearing glaciers will not significantly change water availability in the lower-elevation regions, which depend on monsoonal precipitation and snowmelt, these changes will affect high-elevation water supplies, especially during the dry season.

Rivers and Lakes

Surface runoff and base flow from groundwater move across Earth's surface in rivers and streams, forming vast arterial networks that drain the continents (Figure 9.14). Freshwater lakes are fed by precipitation, streamflow, and groundwater, and store about 125,000 km^3 (30,000 mi^3), or about 0.33%, of the freshwater on Earth's surface. About 80% of this volume is in just 40 of the largest lakes, and about 50% is contained in just 7 lakes (Figure 9.14).

The greatest single volume of lake water resides in 25-million-year-old Lake Baikal in Siberian Russia, discussed in *Geosystems Now*. This lake contains almost as much water as all five North American Great Lakes combined. Africa's Lake Tanganyika contains the next largest volume, followed by the five Great Lakes. About one-fourth of global freshwater lake storage is in small lakes. More than 3 million lakes exist in Alaska alone; Canada has at least that many in number and has more total surface area of lakes than any country in the world.

Not connected to the ocean are saline lakes and salty inland seas, containing about 104,000 km^3 (25,000 mi^3) of water. They usually exist in regions of interior river drainage (no outlet to the ocean), which allows salts resulting from evaporation over time to become concentrated. Examples of such lakes include Utah's Great Salt

GEOreport 9.2 How is water measured?

In most of the United States, hydrologists measure streamflow in cubic feet per second (ft^3/s); Canadians use cubic meters per second (m^3/s). For large-scale assessments, water managers in the eastern United States use millions of gallons a day (MGD), billions of gallons a day (BGD), or billions of liters per day (BLD). In the western United States, where irrigated agriculture is so important, total annual streamflows are frequently measured in acre-feet per year. One acre-foot is an acre of water, 1 ft deep, equivalent to 325,872 gal (43,560 ft^3, or 1234 m^3, or 1,233,429 L). An acre is an area that is about 208 ft on a side and is 0 .4047 hectares. For global measurements, 1 km^3 = 1 billion m^3 = 810 million acre-feet; 1000 m^3 = 264,200 gal = 0.81 acre-feet. For smaller measures, 1 m^3 = 1000 L = 264.2 gal.

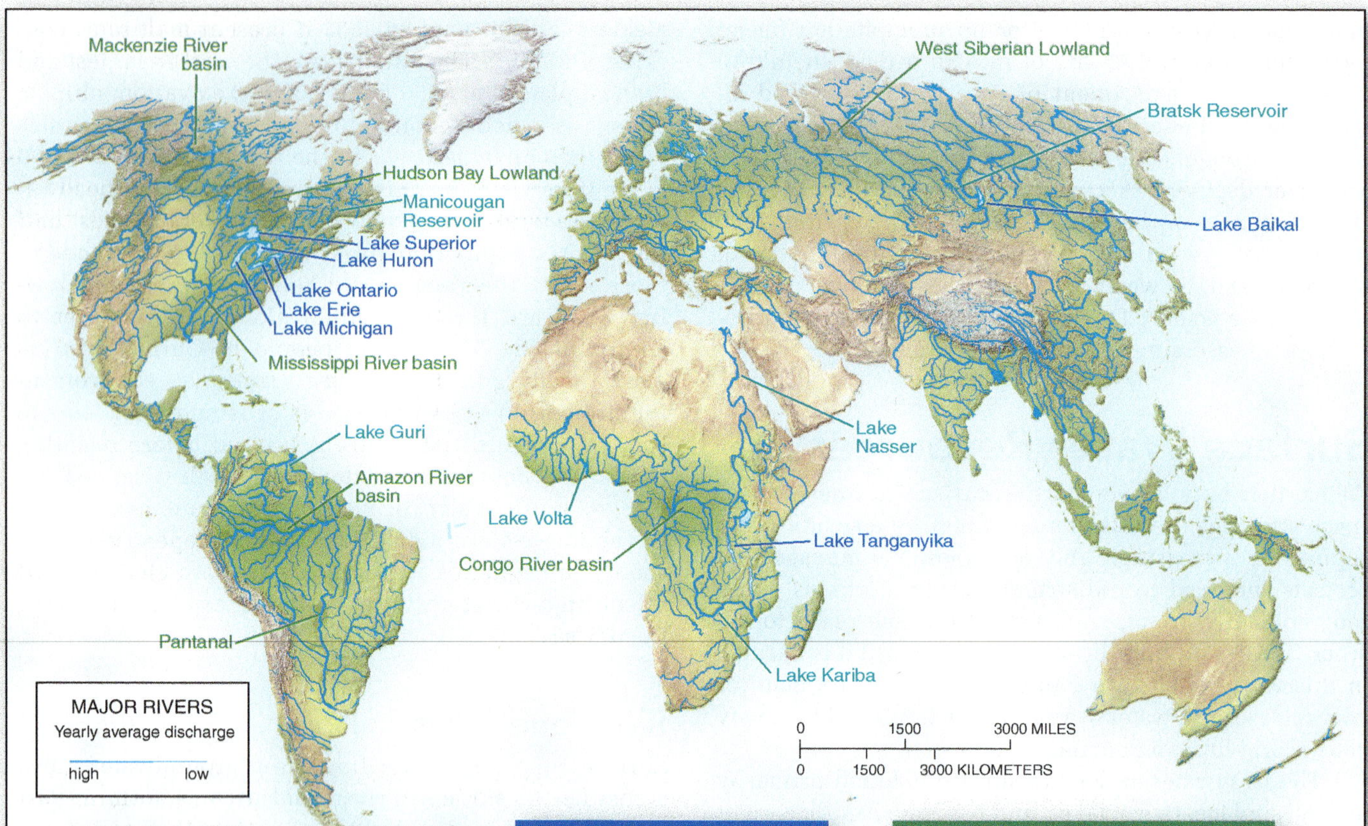

Major Reservoirs	Volume in km³ (mi³)
Lake Kariba, Zambia/Zimbabwe	181 (43.4)
Bratsk Reservoir, Russia	169 (40.5)
Lake Nasser, Egypt/Sudan	157 (37.7)
Lake Volta, Ghana	150 (36.0)
Manicougan Reservoir, Canada	142 (34.0)
Lake Guri, Venezuela	135 (32.4)

Major Freshwater Lakes	Volume in km³ (mi³)
Lake Baikal, Russia	22,000 (5280)
Lake Tanganyika	18,750 (4500)
Lake Superior	12,500 (3000)
Lake Michigan	4920 (1180)
Lake Huron	3545 (850)
Lake Ontario	1640 (395)
Lake Erie	485 (115)

Major Freshwater Wetlands	Area in km² (mi²)
West Siberian Lowland	2,745,000 (1,059,850)
Amazon River basin	1,738,000 (671,045)
Hudson Bay Lowland	374,000 (144,402)
Congo River basin	189,000 (72,973)
Mackenzie River basin	166,000 (64,093)
Pantanal	138,000 (53,282)
Mississippi River basin	108,000 (41,699)

▲Figure 9.14 **The world's major rivers, lakes, and wetlands.** Earth's largest streamflow volumes occur within and just adjacent to the tropics, reflecting the continual rainfall associated with the Intertropical Convergence Zone (ITCZ). Regions of lower streamflow coincide with Earth's subtropical deserts, rain-shadow areas, and continental interiors, particularly in Asia. [Adapted from William E. McNulty, National Geographic Society, based on data from USGS; World Wildlife Fund; State Hydrological Institute, Russia; University of Kassel Center for Environmental Systems Research, Germany.]

Lake, California's Mono Lake and Salton Sea (see Figure 9.19 on page 255), Southwest Asia's Caspian and Aral Seas, and the Dead Sea between Israel and Jordan.

Effects of Climate Change on Lakes Increasing air temperatures are affecting lakes throughout the world. Some lake levels are rising in response to the melting of glacial ice; others fall as a result of drought and high evaporation rates. Longer, warmer summers change the thermal structure of a lake, blocking the normal mixing between deep and surface waters. Lake Tahoe in the Sierra Nevada mountain range along the California–Nevada border is warming at about 1.3 C° (2.3 F°) per decade. The rate of warming is highest in the upper 10 m (33 ft), and mixing has slowed. Non-native, invasive species such as large-mouth bass, carp, and Asian clam are on the rise in warming lakes, while cold-water species decline.

In East Africa, Lake Tanganyika is surrounded by an estimated 10 million people, with most depending on its fish stocks, especially freshwater sardines, for food. Present water temperatures have risen to 26°C (79°F), the highest in a 1500-year climate record exposed by lake-sediment cores. The mixing of surface and deep waters is necessary to replenish nutrients in the upper 200 m (656 ft) of the lake, where the sardines reside. As mixing slows or stops, scientists fear that fish stocks will continue to decline.

Hydroelectric Power Human-made lakes are generally called *reservoirs*, although the term *lake* often appears in their name. Dams built on rivers cause surface water reservoirs to form upstream; the total volume worldwide of such reservoirs is estimated at 5000 km³ (1200 mi³). The largest reservoir in the world by volume is Lake Kariba in Africa, impounded by the Kariba Dam on the Zambezi River on the border between Zambia and Zimbabwe. The third and fourth largest are also in Africa (Figure 9.14). The largest U.S. reservoir is Lake Mead on the Colorado River, discussed in Focus Study 9.1.

Although flood control and water-supply storage are two primary purposes for dam construction, an associated benefit is power production. Hydroelectric power, or **hydropower**, is electricity generated using the power of moving water. Currently, hydropower supplies almost one-fifth of the world's electricity and is the most widely used source of renewable energy. However, because it depends on precipitation, hydropower is highly variable from month to month and year to year.

China is the leading world hydropower producer. The Three Gorges Dam on the Yangtze River in China is 2.3 km (1.5 mi) long and 185 m (607 ft) high, making it the largest dam in the world in overall size, including all related construction at the dam site (Figure 9.15). Entire cities were relocated (including more than 1.2 million people) to make room for the 600-km-long (370-mi-long) reservoir upstream from the dam. The immense scale of environmental, historical, and cultural losses associated with the project was the subject of great controversy. Benefits from the project include flood control, water storage for redistribution, and electrical power production, with capacity at 22,000 MW. In October 2012, the reservoir was filled to its design capacity; however, water pollution and landslides along the reservoir banks may force the relocation of an additional 300,000 people.

(text continued on page 240)

(a) The $25 billion Three Gorges Dam on the Yangtze River, China.

(b) Satellite view shows the dam and shipping channel.

MG Satellite
Global Water Balance Components

▲**Figure 9.15 Major dam in China.** [(a) John Henshall/Alamy (b) *Terra* ASTER, NASA/GSFC/MITI/ERSDAC/JAROS.]

Focus Study 9.1 Climate Change

The Colorado River: A System Out of Balance

The most important water resource in the U.S. Southwest is the Colorado River and its tributaries. This *exotic stream*—so designated because even though its headwaters are in a region of water surpluses, it flows mostly through arid lands—travels almost 2317 km (1440 mi) southwestward to the sea. Orographic precipitation totaling 102 cm (40 in.) per year falls mostly as snow in the Rockies, feeding the Colorado River headwaters in Rocky Mountain National Park (Figure 9.1.1a). But at Yuma, Arizona, near the river's end, annual precipitation is a scant 8.9 cm (3.5 in.), an extremely small amount when compared with the high annual potential evapotranspiration demand of 140 cm (55 in.) in the Yuma region.

From its wetter source region, the Colorado River spills out into the dry climates of western Colorado and eastern Utah (Figure 9.1.1b), where the river flows onward to Lake Powell, the reservoir behind Glen Canyon Dam (Figure 9.1.1c), and then through the Grand Canyon. The river then turns southward, marking its final 644 km (400 mi) as the Arizona–California border. Along this stretch sits Hoover Dam and the major storage reservoir Lake Mead, just east of Las Vegas (Figure 9.1.1d); Davis Dam, built to control the releases from Hoover Dam (Figure 9.1.1e); Parker Dam, for the water needs of Los Angeles; three more dams for irrigation water (Palo Verde, Imperial, and Laguna; Figure 9.1.1f); and finally, Morelos Dam at the Mexican border. Mexico owns the end of the river and whatever water is left. Hydrologically exhausted, the river rarely flows into Mexico, instead ending as a trickle kilometers short of its former mouth in the Gulf of California (Figure 9.1.1g).

Figure 9.1.1h shows the annual streamflow and suspended sediment load for the Colorado River at Yuma, Arizona, from 1905 to 1964. The completion of Hoover Dam in the 1930s dramatically reduced water and sediment at Yuma; additional dams in years following reduced both streamflow and sediment to nearly zero by the mid-1960s (the effects of sediment reduction on rivers below dams are discussed in Chapter 14).

Overall, the river basin encompasses 641,025 km^2 (247,500 mi^2), including parts of seven states and two countries. From 2011 to 2012, Utah, Nevada, Colorado, and Arizona were ranked the fourth through eighth fastest growing U.S. states in terms of population. The rapidly expanding urban areas of Las Vegas, Phoenix, Tucson, Denver, San Diego, and Albuquerque all depend on Colorado River water, which supplies approximately 30 million people in seven states. The Colorado River has played a crucial part in the history of the Southwest and will have a defining role in the future of this drought-plagued region.

Dividing the Colorado's Dammed Water

John Wesley Powell (1834–1902) was the first Euro-American of record to successfully navigate the Colorado River through the Grand Canyon. Lake Powell is named after him. Powell perceived that the challenge of settling the West was too great for individual efforts and believed that solutions to problems such as water availability could be met only through private cooperative efforts. His 1878 study (reprinted in 1962), *Report of the Lands of the Arid Region of the United States*, is a conservation landmark.

Today, Powell probably would be skeptical of the intervention by government agencies in building large-scale reclamation projects. An anecdote in Wallace Stegner's book *Beyond the Hundredth Meridian* tells of an 1893 international irrigation conference, held in Los Angeles, where development-minded delegates bragged that the entire West could be conquered and reclaimed from nature and that "rain will certainly follow the plow." Powell spoke against that sentiment: "I tell you, gentlemen, you are piling up a heritage of conflict and litigation over water rights, for there is not sufficient water to supply the land."* He was booed from the hall, but history has shown Powell to be correct.

The Colorado River Compact was signed by six of the seven basin states in 1923. (The seventh, Arizona, signed in 1944, the same year that the United States signed the Mexican Water Treaty.) With this compact, the Colorado River basin was divided into an upper basin and a lower basin, arbitrarily separated for administrative purposes at Lees Ferry near the Utah–Arizona border (Figure 9.1.1). In 1928, Congress authorized Hoover Dam as the first major reclamation project and the All-American Canal to carry water for irrigation into the Imperial Valley, which required the Imperial Dam to divert water into the canal. Los Angeles then began its project to bring Colorado River water 390 km (240 mi) from still another dam and reservoir on the river to the city.

After construction of Hoover Dam, establishing flood protection for downstream resources, the other projects were quickly completed. Eight major dams and numerous canals and diversions are now in place. The last effort to redistribute Colorado River water was the Central Arizona Project, which carries water to the Phoenix area (see Figure 9.16).

Highly Variable River Flows

The most notable flaw in long-range planning and water distribution in the Colorado River system relates to the river's highly variable flows, typical of an exotic stream. For example, in 1917, the annual flow measured at Lees Ferry totaled 24 million acre-feet (maf), whereas in 1934 it dropped by nearly 80%, to only 5.03 maf. In 1977, the flow dropped again, to 5.02 maf, but in 1984, it rose to an all-time high of 24.5 maf. Although flow measurements at this site also reflect upstream water withdrawals for human use, the overall connection between flow variability and climate is apparent.

The period from 2000 to 2010 marked the lowest 10-year-average streamflow in the 103-year record, with 2002 dropping to an all-time record low of 3.1 maf. By September 2010, the level of Lake Mead behind Hoover Dam dropped to 330 m (1083 ft), far below the 375-m (1225-ft) full capacity of the reservoir reached in July 1983.

The government based the terms of the Colorado River Compact on the river's average annual streamflows from 1914 up to the compact signing in 1923—an exceptionally high 18.8 maf. That amount was perceived as more than enough to supply the upper and lower basins with 7.5 maf each, and, later, to allow Mexico to receive 1.5 maf (as determined by the 1944 Mexican Water Treaty).

Should long-range planning be based on data from such a short time period on a river with highly variable flows? Climate reconstructions for about the past 1000 years suggest that the only other time

*W. Stegner, *Beyond the Hundredth Meridian* (Boston, MA: Houghton Mifflin, 1954), p. 343.

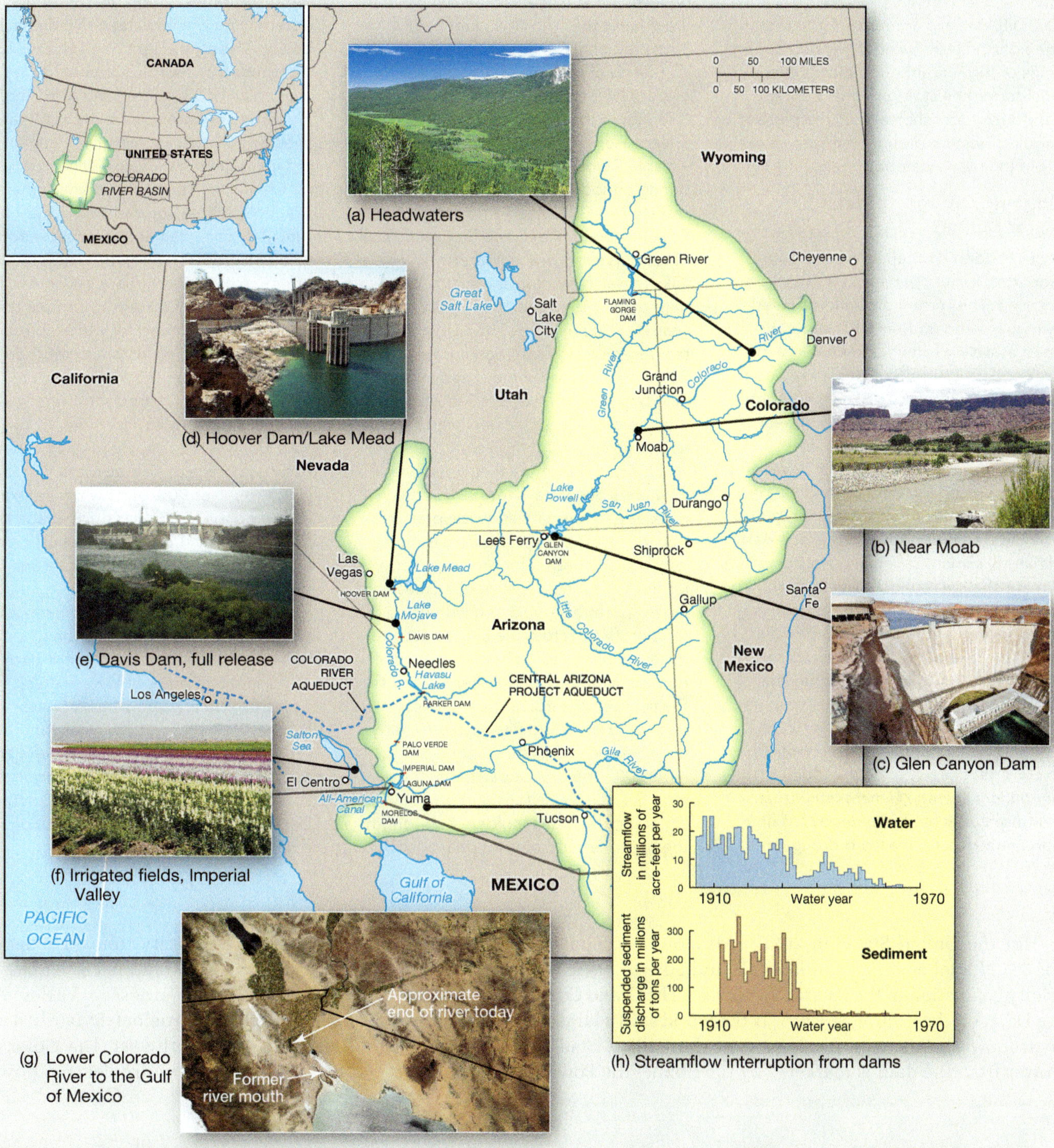

▲**Figure 9.1.1 The Colorado River drainage basin.** The Colorado River basin is divided into upper and lower basins at Lees Ferry in northern Arizona. [(a), (c), (d), (f) Bobbé Christopherson. (b), (e) Author. (g) *Terra* image, NASA/GSFC. (h) Data from USGS, *National Water Summary 1984*, Water Supply Paper 2275, p. 55.]

Colorado River streamflows were at the high 1914–1923 level was between A.D. 1606 and 1625. The overall trends show that the period 1900–2000 had consistently higher flows than any other time period during the past 500 years. Since then, planners have regularly overestimated river flows, a situation that has created shortfalls as demand exceeds supply in the water budget.

The seven states want rights to Colorado River water that total as much as 25.0 maf per year. When added to the guarantee for Mexico, this comes to 26.5 maf of yearly demand (more by far

(continued)

Focus Study 9.1 *(continued)*

than the river contains). Additionally, a 2012 U.S.–Mexico agreement grants storage rights in Lake Mead to Mexico during times of surplus, although the country must forfeit the storage rights during periods of water shortage. In the present ongoing western drought, no surplus exists in the river's water budget.

Ongoing Drought and the Water Budget

Drought conditions affecting the Colorado River system began in 2000 and are ongoing. The current drought is caused by higher temperatures and evaporation rates, as well as reduced mountain snowpack and earlier spring melt. According to regional tree-ring records, nine droughts have affected the American Southwest since A.D. 1226. However, the present drought is the first to occur with increasing human demand for water, as population growth continues throughout the region.

Whether this ongoing drought is part of natural climate variability or a sign of climate change is under investigation. Meteorological data and computer models suggest that the drought is directly linked to the expansion of subtropical high-pressure systems into higher latitudes with climate change. The temperature and moisture changes associated with this climatic shift has led scientists to consider this a change the region's *permanent average climate* rather than a drought, as it has been described until now. In the future, the word *drought* may be reserved strictly for water deficits that are greater than the new, drier base state of the regional climate.

Effects of prolonged drought are evident in declining reservoirs levels throughout the Colorado River system. These reservoirs store water to offset variable Colorado River flows, especially during periods of drought. The largest reservoirs—Lake Mead and Lake Powell—together account for over 80% of total system storage. Water inflow to Lake Powell is running at less than one-third of long-term averages (in the lowest year, 2002, it ran at just 25%). Levels in system reservoirs were at record lows by mid-2010 and then rose somewhat over the next two years. However, scientists agree that the overall trend of declining snowfall and ongoing drought will continue.

Engineers estimate that it could take 13 winters of normal snowpack in the Rockies and normal basinwide precipitation to "reset" the system. This is much longer than needed for previous recoveries from drought, because of increasing evaporation associated with rising temperatures across the Southwest and because of increasing human water use. Forecasts for the period after 2016 put the chances of surplus streamflow occurring in the Colorado River system in any given year at only 20%.

In 2007, the U.S. Department of the Interior signed a historic agreement with Colorado River stakeholders throughout the basin, setting rules for dealing with water shortages during the drought and determining where water deliveries will be cut. This agreement also set rules for distributing any surpluses, should they ever again occur, and encouraged implementation of water conservation initiatives (see http://www.usbr.gov/lc/region/programs/strategies/RecordofDecision.pdf).

Clearly, current water-management practices in the Colorado River system are not sustainable. Conservation (use of less water) and efficiency (more effective use of water) are only now being discussed as ways to reduce the tremendous demand for water across the Southwest. For instance, southern Nevada has launched a campaign to replace lawns with drought-tolerant xeriscaping (desert landscaping), but with 150,000 hotel rooms, we might question the choice of strategy. Las Vegas, which receives 90% of its water from the Colorado River, could reduce usage by 30% simply by requiring water-efficient fixtures and appliances in hotels and residences.

Limiting or halting further metropolitan construction and population growth in the Colorado River basin is not part of present strategies to lower water demand. The focus remains on water supply. We might wonder what John Wesley Powell would think if he were alive today to witness such errant attempts to control the mighty and variable Colorado River. He foretold such a "heritage of conflict and litigation."

In the United States in 2011, hydropower accounted for 8% of total electricity production and 50% of the electricity generation in the Pacific Northwest. The Grand Coulee Dam, on the Columbia River, is the world's fifth largest hydropower plant and the largest in the United States. Power from the dam is marketed by the Bonneville Power Administration, which sells electricity from 31 federally operated hydropower plants in the Columbia River basin.

In the U.S. Southeast, the Tennessee Valley Authority operates 29 dams and hydroelectric facilities in the Tennessee Valley. In the Southwest, the Colorado River is an important source of hydroelectric power;

GEOreport 9.3 Satellite *GRACE* enables groundwater measurements

Groundwater is a difficult resource to study and measure since it lies hidden beneath Earth's surface. In 2002, NASA launched its Gravity Recovery and Climate Experiment (*GRACE*) satellites to measure Earth's gravity field by noting the tiniest changes associated with changes in Earth's mass. Scientists are now using *GRACE* data to study changes in groundwater storage on land. Recent analyses reveal water table declines in northern India as large as 33 cm (1 ft) between 2002 and 2008, almost entirely caused by human use, which is depleting the resource more quickly than it can recharge. In the Middle East, *GRACE* data show that reservoir volumes have declined precipitously, and scientists attribute the cause to groundwater pumping by humans. *GRACE* has given scientists a "scale in the sky" with which to track groundwater changes—yielding information that may prove critical for initiating action on water conservation in these regions.

Focus Study 9.1 examines the impacts of numerous water projects, many of them hydropower facilities, on this river system that is being stressed by climate change and increasing water demands.

In the United States, dams have already been built on the best multipurpose dam sites, with numerous detrimental consequences for river environments. Many of the largest hydropower projects are old, and overall production of hydropower is declining. Chapter 14 examines the environmental effects of dams and reservoirs on river ecosystems and reports on recent dam removals. Worldwide, however, hydropower is increasing; several large projects are proposed and under construction in Brazil alone.

Water Transfer Projects The transfer of water over long distances in pipelines and aqueducts is especially important in dry regions where the most dependable water resources are far away from population centers. The California Water Project is a system of storage reservoirs, aqueducts, and pumping plants that rearranges the water budget in the state: Water distribution over time is altered by holding back winter runoff for release in summer, and water distribution over space is altered by pumping water from the northern to the southern parts of the state. Completed in 1971, the 1207-km-long (750-mi-long) California Aqueduct is a "river" flowing from the Sacramento River delta to the Los Angeles region, servicing irrigated agriculture in the San Joaquin Valley along the way. The Central Arizona Project, another system of aqueducts, moves water from the Colorado River to the cities of Phoenix and Tucson (Figure 9.16).

China is currently constructing a massive infrastructure of water transfer canals to bring water to the northern, industrialized regions of the country. With a goal of completion by 2050, the project will construct three main diversions that link four major rivers and will displace hundreds of thousands of people.

▲**Figure 9.16 The Central Arizona Project.** A Central Arizona Project canal transports water through the desert west of Phoenix. [Central Arizona Project.]

Wetlands

A **wetland** is an area that is permanently or seasonally saturated with water and characterized by vegetation adapted to *hydric* soils (soils saturated long enough to develop anaerobic, or "oxygen-free," conditions). The water found in wetlands can be freshwater or saltwater (Chapter 16 discusses saltwater wetlands). Marshes, swamps, bogs, and peatlands (bog areas composed of peat, or partly decayed vegetation) are types of freshwater wetlands that occur worldwide along river channels and lakeshores, in surface depressions such as prairie potholes in the U.S Great Plains region, and in the cool, lowland, high-latitude regions of Canada, Alaska, and Siberia. Figure 9.14 shows the global distribution of some major wetlands.

Large wetlands are important sources of freshwater and recharge groundwater supplies. When rivers flow over their banks, wetlands absorb and spread out the floodwaters. For example, the Amazon River floodplain is a major wetland that stores water and mitigates flooding within the river system—the river and its associated wetlands provide about one-fifth of the freshwater flowing into the world's oceans. Wetlands are also significant for improving water quality by trapping sediment and removing nutrients and pollutants. In fact, constructed wetlands are increasingly used globally for water purification. (See Chapter 19 for more on freshwater wetlands.)

Groundwater Resources

Although **groundwater** lies beneath the surface, beyond the soil-moisture zone and the reach of most plant roots, it is an important part of the hydrologic cycle. In fact, it is the largest potential freshwater source on Earth—larger than all surface lakes and streams combined. In the region from the soil-moisture zone to a depth of 4 km (13,000 ft) worldwide is an amount of water totaling some 8,340,000 km^3 (2,000,000 mi^3), a volume comparable to 70 times all the freshwater lakes in the world. Groundwater is not an independent source of water; it is tied to surface supplies for recharge through pores in soil and rock. An important consideration in many regions is that groundwater accumulation occurred over millions of years, so care must be taken not to deplete this long-term buildup with excessive short-term demands.

Groundwater provides about 80% of the world's irrigation water for agriculture and nearly half the world's drinking water. It is generally free of sediment, color, and disease organisms, although polluted groundwater conditions are considered irreversible. Where groundwater pollution does occur, it threatens water quality. Overconsumption is another problem, depleting groundwater volume in quantities beyond natural replenishment rates and thus threatening global food security.

About 50% of the U.S. population derives a portion of its freshwater from groundwater sources. In some states, such as Nebraska, groundwater supplies 85% of water needs, with that figure as high as 100% in rural areas. Between 1950 and 2000, annual groundwater withdrawal in the United States and Canada increased more than 150%. Figure 9.17 shows potential groundwater resources in both countries. For the latest research on U.S. groundwater resources, go to http://water.usgs.gov/ogw/gwrp/; for maps and data, see http://groundwaterwatch.usgs.gov/.

The Groundwater Environment

Geosystems in Action, Figure GIA 9.1 brings together many groundwater concepts in a single illustration. Follow the 13 numbers on the GIA illustration as you read about each part of the groundwater environment.

Precipitation is the main source of groundwater, percolating downward as gravitational water from the soil-moisture zone. This water moves through the **zone of aeration**, where soil and rock are less than saturated (some pore spaces contain air), an area also known as the *unsaturated zone* (GIA number 1).

Eventually, gravitational water accumulates in the **zone of saturation**, where soil pore spaces are completely filled with water (GIA number 2). Like a hard sponge made of sand, gravel, and rock, the zone of saturation stores water in its countless pores and voids. It is bounded at the bottom by an impermeable layer of rock that obstructs further downward movement of water. The upper limit of the zone of saturation is the **water table**, the point of transition between the zone of aeration and the zone of saturation (note the white dashed lines across Figure GIA 9.1). The slope of the water

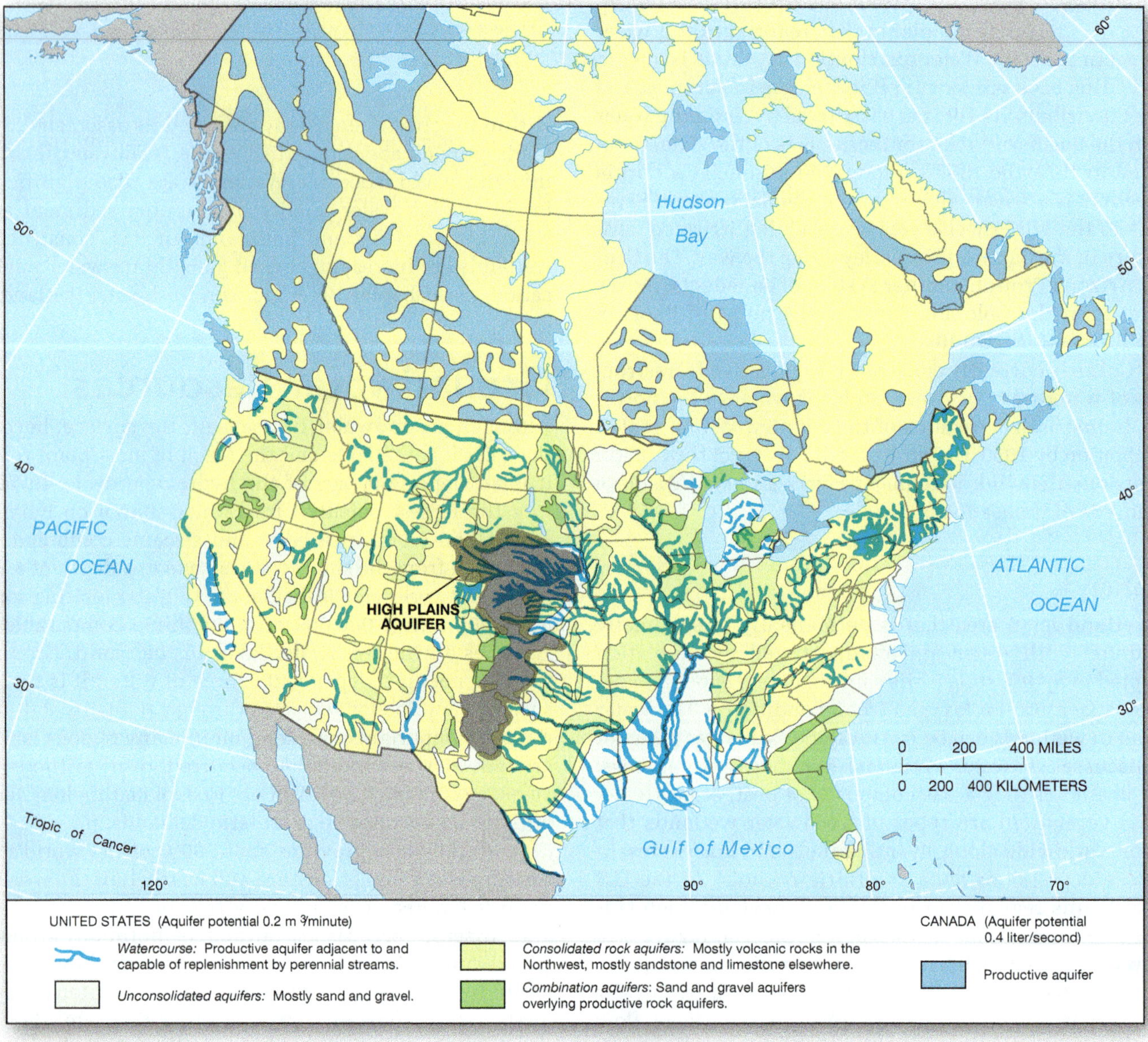

▲Figure 9.17 **Groundwater resource potential for the United States and Canada.** Areas of the United States indicated in dark blue are underlain by productive aquifers capable of yielding freshwater to wells at 0.2 m^3/min or more (for Canada, 0.4 L/s). [Courtesy of Water Resources Council for the United States and the Inquiry on Federal Water Policy for Canada.]

table, which generally follows the contours of the land surface, drives groundwater movement toward areas of lower elevation and lower pressure (GIA number 3).

Aquifers and Wells As discussed earlier, permeable rock or materials conduct water readily, while impermeable rock obstructs water flow. An **aquifer** is a subsurface layer of permeable rock or unconsolidated materials (silt, sand, or gravels) through which groundwater can flow in amounts adequate for wells and springs. The blue underground area in Figure GIA 9.1 is an unconfined aquifer; note the water wells on the left-hand side. An **unconfined aquifer** has a permeable layer above, which allows water to pass through, and an impermeable one beneath (GIA number 4). A **confined aquifer** is bounded above and below by impermeable layers of rock or unconsolidated materials (GIA number 9). The solid, impermeable layer that forms such a boundary is known as an *aquiclude*. An *aquitard* is a layer that has low permeability but cannot conduct water in usable amounts. The zone of saturation may include the saturated portion of the aquifer and a part of the underlying aquiclude (GIA number 7).

Humans commonly extract groundwater using wells that are drilled downward into the ground until they penetrate the water table. Shallow drilling results in a "dry well" (GIA number 6); drilling too deeply will punch through the aquifer and into the impermeable layer below, also yielding little water. The water in a well drilled into an unconfined aquifer is not under pressure and so must be pumped to rise above the water table (GIA number 5). In contrast, the water in a confined aquifer is under the pressure of its own weight, creating a pressure level called the **potentiometric surface** to which the water can rise on its own.

The potentiometric surface can be above ground level (GIA number 10). Under this condition, **artesian water**, or groundwater confined under pressure, may rise in a well and even flow at the surface without pumping if the top of the well is lower than the potentiometric surface (GIA number 11). (These wells are called *artesian* for the Artois area in France, where they are common.) In other wells, however, pressure may be inadequate, and the artesian water must be pumped the remaining distance to the surface.

The size of the *aquifer recharge area*, where surface water accumulates and percolates downward, differs for unconfined and confined aquifers. For an unconfined aquifer, the recharge area generally extends above the entire aquifer; the water simply percolates down to the water table. But in a confined aquifer, the recharge area is far more restricted. Pollution of this limited area causes groundwater contamination; note in Figure GIA 9.1, number 12, the pollution caused by leakage from the disposal pond on the aquifer recharge area, contaminating the nearby well.

Groundwater at the Surface Where the water table intersects the ground surface (GIA number 8), water flows outward in the form of springs, streams, lakes, and wetlands. Springs are common in karst environments, in which water dissolves rock (primarily limestone) by chemical processes and flows underground until it finds a surface outlet (karst discussion is in Chapter 13). Hot springs are common in volcanic environments where water is heated underground before emerging under pressure at the surface. In the southwestern United States, a ciénega (the Spanish term for *spring*) is a marsh where groundwater seeps to the surface.

Groundwater interacts with streamflow to provide base flow during dry periods when runoff does not occur. Conversely, streamflow supplements groundwater during periods of water surplus. Geosystems in Action 9.2 illustrates the relationship between groundwater and surface streams in two different climatic settings. In humid climates, the water table is higher in elevation than the stream channel and generally supplies a continuous base flow to a stream. In this environment, the stream is *effluent* because it receives the water flowing out from the surrounding ground. The Mississippi River is a classic example, among many other humid-region streams. In drier climates, the water table is lower than the stream, causing *influent* conditions in which streamflow feeds groundwater, sustaining deep-rooted vegetation along the stream. The Colorado and Rio Grande rivers of the American West are examples of influent streams.

When a water table declines so that the bottom of a streambed is no longer in contact with it, streamflow seeps into the aquifer. In Central and Western Kansas, the Arkansas River channel is now dry as a result of overuse of the High Plains Aquifer (see the photo in Figure GIA 9.2).

Overuse of Groundwater

As water is pumped from a well, the surrounding water table within an unconfined aquifer might experience **drawdown**, or become lowered. Drawdown occurs if the pumping rate exceeds the replenishment flow of water into the aquifer or the horizontal flow around the well. The resultant lowering of the water table around the well is a **cone of depression** (Figure GIA 9.1, left-hand side).

An additional problem arises when aquifers are overpumped near the ocean or seacoast. Along a coastline, fresh groundwater and salty seawater establish a natural interface, or *contact surface*, with the less-dense freshwater flowing on top. But excessive withdrawal of freshwater can cause this interface to migrate inland. As a result, wells near the shore become contaminated with saltwater, and the aquifer becomes useless as a freshwater source (GIA number 13). Pumping freshwater back into the aquifer may halt seawater intrusion, but once contaminated, the aquifer is difficult to reclaim.

Groundwater Mining The utilization of aquifers beyond their flow and recharge capacities is known as **groundwater mining.** In the United States, chronic groundwater overdrafts occur in the Midwest, West, lower Mississippi

(text continued on page 246)

G*roundwater* forms when rainfall and snowmelt seep down through the soil and accumulate in the pore spaces in fractured bedrock or sediment. GIA 9.1 shows the structure of groundwater deposits, called *aquifers*, and threats to this resource from pollution and overuse. GIA 9.2 shows how groundwater helps to maintain streamflow.

9.1 THE WATER TABLE AND AQUIFERS

The *water table* is a boundary between the zones of aeration and saturation. Beneath the zone of saturation, an impermeable layer of rock blocks further downward movement of water. An *aquifer* contains groundwater stored in the zone of saturation.

(1) Zone of aeration
In this layer, some pore spaces contain air.

(2) Zone of saturation
In this layer, water fills the spaces between particles of sand, gravel, and rock.

(3) Slope and flow
The water table follows the slope of the land surface above it. The water in an aquifer flows toward areas of lower elevation and lower pressure. A plume of water pollution from septic systems or landfills can flow through an aquifer, contaminating wells.

(4) Unconfined aquifer
An ***unconfined aquifer*** *has a permeable layer above, and an impermeable layer beneath.*

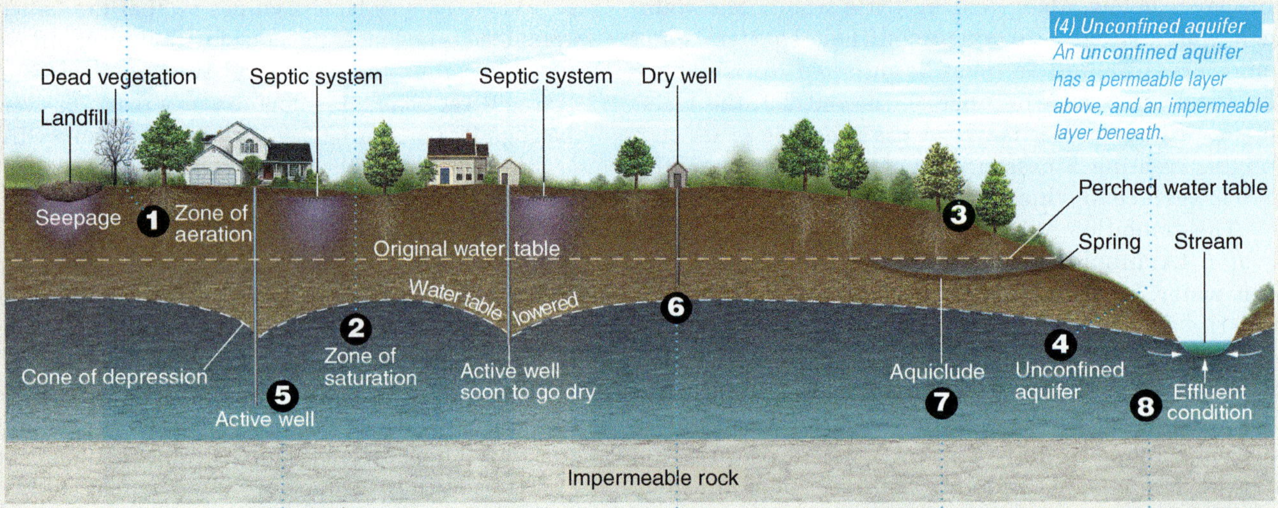

(5) Wells in an unconfined aquifer
The water in an unconfined aquifer is not under pressure and must be pumped to the surface.

(6) Dry wells and aquifer overuse
Wells pump groundwater to the surface, lowering the water table. Overuse, or groundwater mining, occurs when drawdown exceeds an aquifer's recharge capacity. A dry well results if a well is not drilled deep enough or if the water table falls below the depth of the well.

(7) Aquicludes and Springs
An ***aquiclude*** *is a layer of impermeable rock or unconsolidated material that prevents water from seeping farther down.* ***Springs*** *form where the perched water table intersects the surface.*

(8) Water table at the surface
Streams, lakes, and wetlands form where the water table intersects the surface.

Infer: In which direction does the water flow in this aquifer? How can you tell?

Animation
The Water Table

MasteringGeography™

Visit the Study Area in MasteringGeography™ to explore groundwater.

Visualize: Study a geoscience animation of groundwater, the water table, and aquifers.

Assess: Demonstrate understanding of groundwater (if assigned by instructor).

Animation
Groundwater Cone of Depression

9.2 GROUNDWATER INTERACTION WITH STREAMFLOW

Runoff and groundwater together supply the water to keep streams flowing. Groundwater can maintain streamflow when runoff does not occur. The diagrams show the relationship of the water table and streamflow in humid and dry climates.

[Bobbé Christopherson]

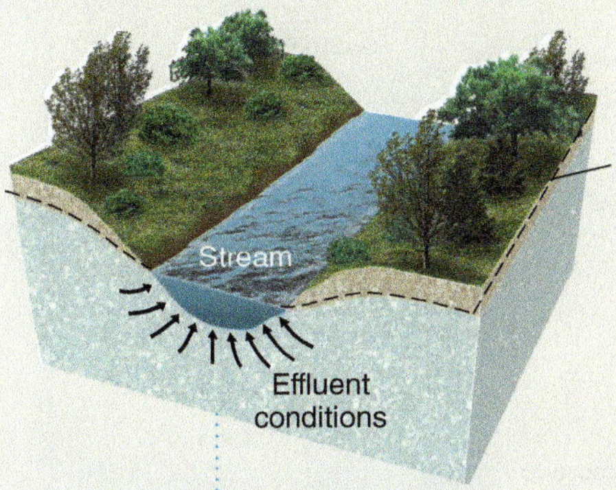

Humid climate– effluent conditions

The water table is higher than the stream channel, so water flows out from the surrounding ground into the stream.

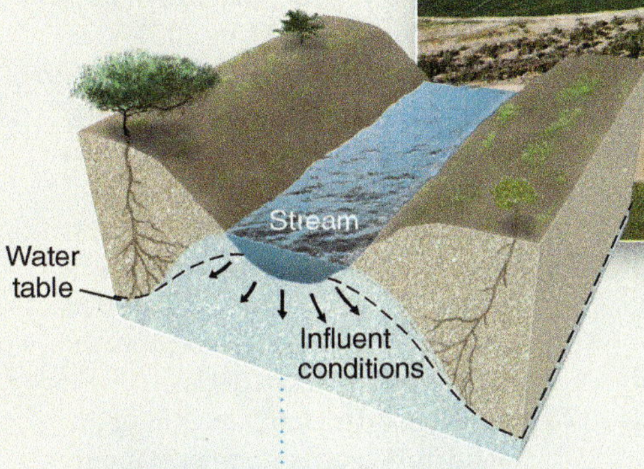

Dry climate– influent conditions

The water table is lower than the stream channel, so the stream's water flows into groundwater.

(9) Confined aquifer

A ***confined aquifer*** *is bounded above and below by impermeable layers (aquicludes)*

(10) Potentiometric surface

The ***potentiometric surface*** *is the level to which groundwater under pressure can rise on its own, and can be above ground level.*

(11) Artesian wells

Artesian water *is groundwater in a sloping, confined aquifer where groundwater is under pressure. If the top of a well is lower than the potentiometric surface, artesian water may rise in the well and flow at the surface without pumping.*

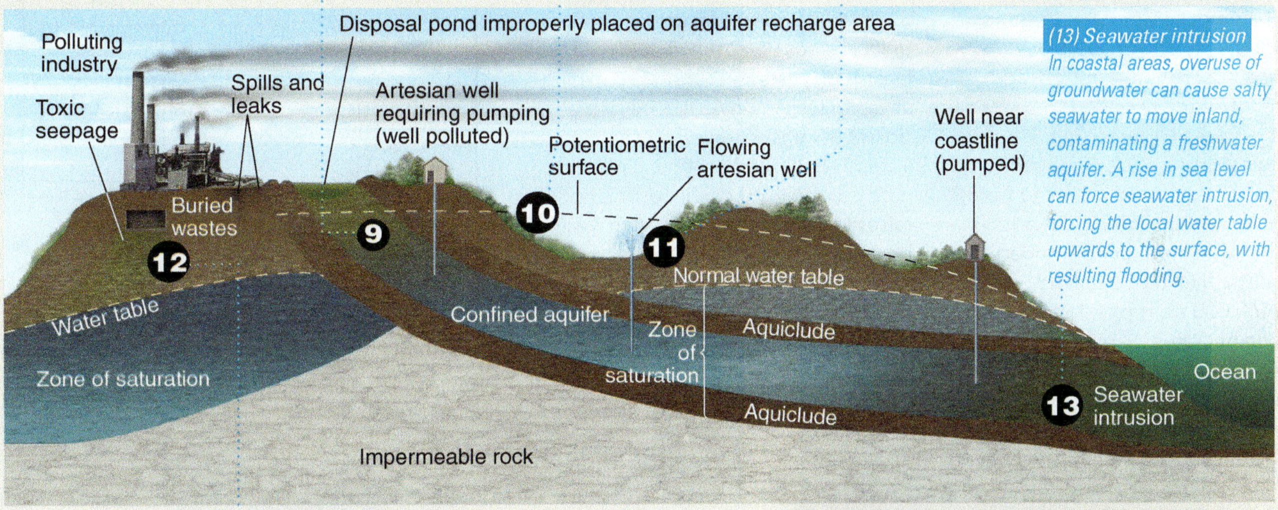

(13) Seawater intrusion

In coastal areas, overuse of groundwater can cause salty seawater to move inland, contaminating a freshwater aquifer. A rise in sea level can force seawater intrusion, forcing the local water table upwards to the surface, with resulting flooding.

(12) Pollution of aquifer

In industrial areas, spills, leaks, and improper disposal of wastes can pollute groundwater. Notice the improper placement of a disposal pond in the aquifer recharge area.

Explain: Suggest steps to prevent such groundwater pollution.

Springs occur where the water table meets the ground surface (see number 8 in GIA 9.1). [holbox/Shutterstock]

GEOquiz

1. Apply Concepts: Where in Figure 9.1 could you drill a well that was free of pollutants and saltwater contamination? Explain.

2. Predict: What would happen to a stream with influent conditions if runoff were greatly reduced?

Focus Study 9.2 Sustainable Resources

High Plains Aquifer Overdraft

North America's largest known aquifer system is the High Plains Aquifer, which underlies a 450,600-km^2 (174,000-mi^2) area shared by eight states and extending from southern South Dakota to Texas. Also known as the Ogallala Aquifer, for the principle geologic unit forming the aquifer system, it is composed mainly of sand and gravel, with some silt and clay deposits. The average thickness of the saturated parts of the aquifer is highest in Nebraska, southwestern Kansas, and the Oklahoma Panhandle (Figure 9.2.1). Throughout the region, groundwater flows generally from east to west, discharging at the surface into streams and springs. Precipitation, which varies widely over the region, is the main source of recharge; annual average precipitation ranges from about 30 cm in the southwest to 60 cm in the northeast (12 to 24 in.). Drought conditions have prevailed throughout this region since 2000.

Heavy mining of High Plains groundwater for irrigation began about 70 years ago, intensifying after World War II with the introduction of center-pivot irrigation, in which large, circular devices provide water for wheat, sorghums, cotton, corn, and about 40% of the grain fed to cattle in the United States (Figure 9.2.2). The U.S. Geological Survey (USGS) began monitoring this groundwater mining from a sample of more than 7000 wells in 1988.

The High Plains Aquifer now irrigates about one-fifth of all U.S. cropland, with more than 160,000 wells providing water for 5.7 million hectares (14 million acres). The aquifer also supplies drinking water for nearly 2 million people. In 1980, water was pumped from the aquifer at the rate of 26 billion cubic meters (21 million acre-feet) a year, an increase of more than 300% since 1950. By 2000, withdrawals had decreased slightly due to declining well yields and increasing pumping costs, which led to the abandonment of thousands of wells.

The overall effect of groundwater withdrawals has been a drop in the water table of more than 30 m (100 ft) in most of the region. Throughout the 1980s, the water table declined an average of 2 m (6 ft) each year. During the period from predevelopment (about 1950) to 2011, the level of the water table declined more than 45 m (150 ft) in parts of northern Texas, where the saturated thickness of the aquifer is least, and in western Kansas (Figure 9.2.3). Rising water levels are noted in small areas of Nebraska and Texas due to recharge from surface irrigation, a short period of above-normal-precipitation years, and downward percolation from canals and reservoirs. (See http://ne.water.usgs.gov/ogw/hpwlms/.)

The USGS estimates that recovery of the High Plains Aquifer (those portions that have not collapsed) would take at least 1000 years if groundwater mining stopped today. Obviously, billions of dollars of agricultural activity cannot be abruptly halted, but neither can profligate water mining continue. This issue raises tough questions: How do we best manage cropland? Can extensive irrigation continue? Can the region continue to meet the demand to produce commodities for export? Should we continue high-volume farming of certain crops that are in chronic oversupply?

Scientists now suggest that irrigated agriculture is unsustainable on

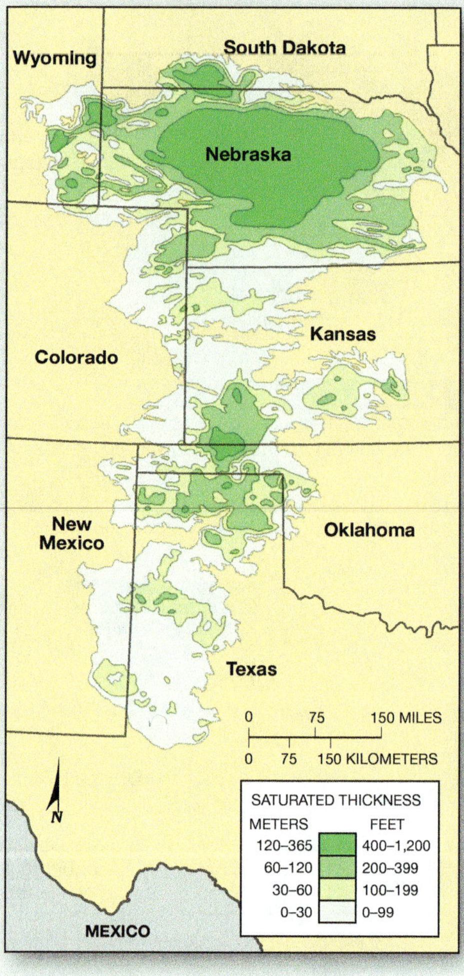

▲Figure 9.2.1 **Average saturated thickness of the High Plains Aquifer.** [After D. E. Kromm and S. E. White, "Interstate groundwater management preference differences: The High Plains region," *Journal of Geography* 86, no. 1 (January–February 1987): 5.]

Valley, and Florida and in the intensely farmed Palouse region of eastern Washington. In many places, the water table or artesian water level has declined more than 12 m (40 ft). Groundwater mining is of special concern for the massive High Plains Aquifer, discussed in Focus Study 9.2.

About half of India's irrigation water needs and half of its industrial and urban water needs are met by the groundwater reserve. In rural areas, groundwater supplies 80% of domestic water from some 3 million hand-pumped bore holes. In approximately 20% of India's agricultural districts, groundwater mining through more than 17 million wells is beyond recharge rates.

In the Middle East, groundwater overuse is even more severe. The groundwater resources beneath Saudi Arabia accumulated over tens of thousands of years, forming "fossil aquifers," so named because they receive little or no recharge in the desert climate that exists in the region today. Thus, increasing withdrawals in Saudi Arabia at present are not being naturally recharged—in essence, groundwater has become a nonrenewable resource. Libya's water supply comes mainly

(a) Center-pivot irrigation system waters a wheat field.

(b) A pattern of quarter-section circular fields results from center-pivot irrigation systems near Dalhart, Texas. In each field, a sprinkler arm pivots around a center, delivering about 3 cm (1.18 in.) of High Plains Aquifer water per revolution.

▲**Figure 9.2.2 Center-pivot irrigation.** [(a) Gene Alexander, USDA/NRCS. (b) USDA National Agricultural Imagery Program, 2010.]

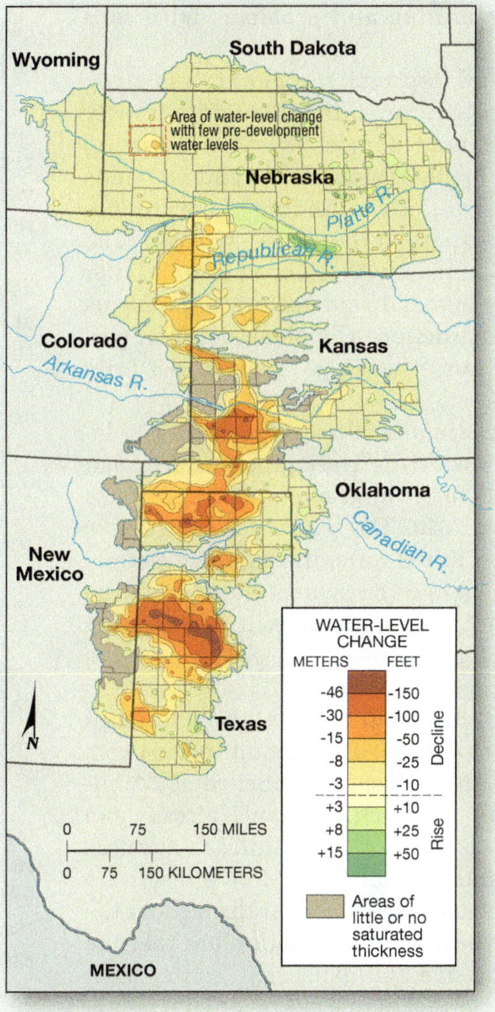

◀**Figure 9.2.3 Water-level changes in the High Plains Aquifer, 1950 to 2011.** The color scale indicates widespread declines and a few areas of water-level rise. [Adapted from "Water-level and storage changes in the High Plains aquifer, predevelopment to 2011 and 2009–2011," by V. L. McGuire, USGS Scientific Investigations Report 2012–5291, 2013, Fig. 1; available at http://ne.water.usgs.gov/ogw/hpwlms/.]

the southern High Plains. Present irrigation practices, if continued, will deplete about half of the High Plains Aquifer (and two-thirds of the Texas portion) by 2020. Eventually, farmers will be forced to switch to non-irrigated crops, such as sorghum, and these are more vulnerable to drought conditions (and will also yield smaller economic returns). Add to this the approximate 10% loss of soil moisture due to increased evapotranspiration demand caused by climatic warming for the region by 2050, as forecast by computer models, and we have a portrait of a major regional water problem.

from fossil aquifers, some of them 75,000 years old; an elaborate system of pipes and storage reservoirs in place since the 1980s makes this water available for the country's population. Some researchers suggest that groundwater in the region will be depleted in a decade, although water-quality problems are already apparent and worsening. In Yemen, the largest fossil aquifer is down to its last years of extractable water.

Desalination In the Middle East and other areas with declining groundwater reserves, **desalination** of seawater is an increasingly important method for obtaining freshwater. Desalination processes remove organic compounds, debris, and salinity from seawater, brackish (slightly saline) water found along coastlines, and saline groundwater, yielding potable water for domestic uses. More than 14,000 desalination plants are now in operation worldwide; the volume of freshwater produced by desalination worldwide is projected to nearly double between 2010 and 2020. Approximately 50% of all desalination plants are in the Middle East; the world's largest is the Jebel Ali Desalination Plant in the United Arab Emirates, capable of producing 300 million m^3/year of water. In Saudi Arabia, 30 desalination plants currently supply 70% of the country's drinking water needs, providing an alternative to further groundwater mining and problems with saltwater intrusion.

In the United States, especially in Florida and along the coast of southern California, desalination is slowly increasing in use. Around Tampa Bay–St. Petersburg, Florida, water levels in the region's lakes and wetlands are declining and land surfaces subsiding as the groundwater drawdown for export to the cities increases. The Tampa Bay desalination plant was finished in 2008, after being plagued with financial and technical problems during its 10-year construction. The plant met final performance criteria for operation in 2013, making it fully operational.

The Carlsbad Desalination Project, including a desalination plant and a water-delivery pipeline, near San Diego, California, will be the largest in the United States when it is finished in 2016. The plant will use *reverse osmosis*, a process that forces water through semipermeable membranes to separate the solutes (salts) from the solvent (water), in effect removing the salt and creating freshwater (see http://carlsbaddesal.com/pipeline for more information).

Desalination of saline aquifers has also increased in Texas over the past decade. The amount of brackish groundwater beneath the surface is estimated at 2.7 billion acre-feet, and 44 desalination plants already tap those reserves (none of the plants desalinizes seawater). The largest of the plants is in El Paso. More projects are in the works as the recent severe drought throughout the state continues. Drawbacks to desalination are that the process is energy-intensive and expensive, and concentrated salts must be disposed of in a manner that does not contaminate freshwater supplies.

Collapsing Aquifers A possible effect of removing water from an aquifer is that the ground will lose internal support and collapse as a result (remember that aquifers are layers of rock or unconsolidated material). Water in the pore spaces is not compressible, so it adds structural strength to the rock or other material. If the water is removed through overpumping, air infiltrates the pores. Air is readily compressible, and the tremendous weight of overlying rock may crush the aquifer. On the surface, the visible result may be land subsidence, cracks in foundations, and changes in surface drainage.

In Houston, Texas, the removal of groundwater and crude oil caused land throughout an 80-km (50-mi) radius to subside more than 3 m (10 ft) over the years. In the Fresno area of California's San Joaquin Valley, after years of intensive pumping of groundwater for irrigation, land levels dropped almost 10 m (33 ft) because of a combination of water removal and soil compaction from agricultural activity.

Pollution of Groundwater

If surface water is polluted, groundwater inevitably becomes contaminated during recharge. Whereas pollution in surface water flushes downstream, slow-moving groundwater, once contaminated, remains polluted virtually forever.

Pollution can enter groundwater from many sources: industrial injection wells (which pump waste into the ground), septic tank outflows, seepage from hazardous-waste disposal sites, industrial toxic waste, agricultural residues (pesticides, herbicides, fertilizers), and urban solid-waste landfills. An example is the suspected leakage from some 10,000 underground gasoline storage tanks at U.S. gasoline stations, thought to be contaminating thousands of local water supplies with cancer-causing gasoline additives. About 35% of groundwater pollution comes from *point sources*, such as a gasoline tank or septic tank; 65% is categorized as *nonpoint source*, coming from a broad area, such as runoff from an agricultural field or urban community. Regardless of the spatial nature of the source, pollution can spread over a great distance.

An controversial source of groundwater contamination is shale gas extraction, discussed in the Chapter 1 *Geosystems Now*. The process of hydraulic fracturing, or fracking, requires large quantities of water and chemicals to be pumped under high pressure into subsurface rock, fracturing it to release natural gas. The wastewater produced is contaminated with chemicals used as lubricants in the fracking process and is often held in wells or containment ponds. Leaks or spills send toxic wastewater into surface water supplies and groundwater. More than 500,000 such wells already exist in the United States, more than 50,000 of these in Pennsylvania alone, and numbers are increasing. Natural gas can also leak into groundwater, causing contamination and methane buildup in drinking water wells—several studies have attributed the presence of methane in wells to fracking operations. However, further research is needed, especially since the effects of hydraulic fracturing on groundwater relate to regional geology and other site-specific conditions.

As discussed in Geo Report 9.3, scientists are using satellite data to estimate the overall volume of the groundwater resource. Assessing groundwater quality remains problematic, however, since aquifers are generally inaccessible to measurement and analysis.

Our Water Supply

Human thirst for adequate water supplies, both in quantity and in quality, will be a major issue in this century. Internationally, increases in per capita water use are double the rate of population growth. Since we are so dependent on water, it seems that humans should cluster where good water is plentiful. But accessible water supplies are not well correlated with population distribution or the regions where population growth is greatest.

TABLE 9.1 Regional Comparison of Factors Influencing Global Water Supply

Region	2012 Population (in millions)	Share of Global Population	Land Area in Thousands of km² (mi²)	Share of Global Land Area	Mean Annual Streamflow (BGD)	Share of Annual Streamflow	2050 Population as a Multiple of 2012	2006 CO_2 Emissions per capita (metric tons)
Africa	1072	15%	30,600 (11,800)	23%	3060	11%	2.2	0.9
Asia	4260	60%	44,600 (17,200)	33%	9540	34%	1.2	3.0
Australia–Oceania	37	0.5%	8,420 (3,250)	6%	1420	5%	1.6	19.0 (Aust.)
Europe	740	10.5%	9,770 (3,770)	7%	2280	8%	1.0	8.4
North America*	465	6.6%	22,100 (8,510)	16%	4310	15%	1.4	18.4**
Central and South America	483	6.8%	17,800 (6,880)	13%	7510	27%	1.3	2.5
Global (excluding Antarctica)	7058	—	134,000 (51,600)	—	28,100	—	1.4	4.1

*Includes Canada, Mexico, and the United States.
Note: BGD, billion gallons per day. Population data from 2012 World Population Data Sheet (Washington, DC: Population Reference Bureau, 2012). CO_2 data from PRB 2009.
**CO_2 data for U.S. and Canada; Mexico per capita emissions are 4.0 metric tons.

Table 9.1 provides statistics that, taken together, indicate the unevenness of Earth's water supply. These data include population, land area, annual streamflow, and projected population change for six world regions. Also note 2006 data on carbon dioxide emissions per person for each region. The adequacy of Earth's water supply is tied, first, to climatic variability and, second, to water usage, which is, in turn, tied to level of development, affluence, and per capita consumption.

For example, North America's mean annual streamflow is 4310 BGD and Asia's is 9540 BGD. However, North America has only 6.6% of the world's population, whereas Asia has 60%, with a population doubling time less than half that of North America. In northern China, 550 million people living in approximately 500 cities lack adequate water supplies. For comparison, consider that the 1990 floods cost China $10 billion, whereas water shortages are running at more than $35 billion a year in costs to the Chinese economy.

In Africa, 56 countries draw from a varied water-resource base; these countries share more than 50 river and lake watersheds. Population growth that is ever more concentrated in urban areas and the need for increased irrigation during periods of drought are enhancing the water demand. The condition of water stress (where people have less than 1700 m³ of water per person per year) presently occurs in 12 African countries; water scarcity (less than 1000 m³ per person per year) occurs in 14 countries. Recent research indicates, however, that the total volume of groundwater on the African continent is much larger than previously thought, with substantial reserves below the dry northern countries of Libya, Algeria, and Chad. But with ongoing drought in this region, even these reserves could be quickly depleted.

Water resources are different from other resources in that *there is no alternative substance to water.* Water shortages increase the probability for international conflict, endanger public health, reduce agricultural productivity, and damage life-supporting ecological systems. Water-resource stress related to decreasing quantity and quality will dominate future political agendas. In this century, we must transition away from large, centralized water-development projects toward decentralized, community-based strategies for more efficient technologies and increased water conservation.

CRITICALthinking 9.2
Calculate Your Water Footprint

How much water have you used today? From showers to tooth brushing to cooking and dish cleanup to quenching our thirst, our households have a water "footprint" that relates to affluence and technology. Just as you calculated your carbon footprint in Chapter 1, CT 1.1, you can calculate your water footprint at http://www.gracelinks.org/1408/water-footprint-calculator. How does your individual water use compare to the average American's? Can you think of ways to reduce your water footprint? •

GEOreport 9.4 The water it takes for food and necessities

Simply providing the foods we enjoy requires voluminous water. For example, 77 g (2.7 oz) of broccoli requires 42 L (11 gal) of water to grow and process; producing 250 mL (8 oz) of milk requires 182 L (48 gal) of water; producing 28 g (1 oz) of cheese requires 212 L (56 gal); producing 1 egg requires 238 L (63 gal); and producing a 113-g (4-oz) beef patty requires 2314 L (616 gal). And then there are our toilets, the majority of which still flush approximately 4 gal of water. Imagine the spatial complexity of servicing the desert city of Las Vegas, with 150,000 hotel rooms times the number of toilet flushes per day; of providing all support services for 38 million visitors per year, plus 38 golf courses. One hotel confirms that it washes 14,000 pillowcases a day.

Water Supply in the United States

The U.S. water supply (excluding Alaska and Hawai'i) derives from surface and groundwater sources that are fed by an average precipitation of 4200 BGD (billion gallons a day). In terms of space, this input is unevenly distributed across the 48 contiguous states, and in terms of time, it is unevenly distributed throughout the year. (Annual average precipitation across the contiguous U.S states is 76.2 cm, or 30 in.; review the map in Figure 9.7.) For example, New England's water supply is so abundant that only about 1% of available water is consumed each year.

Viewed daily, the national water budget has two general outputs: 71% actual evapotranspiration and 29% surplus (Figure 9.18). The 71% actual evapotranspiration occurs over land with nonagricultural vegetation, agricultural crops and pastures, and forest and browse (shrubs and woody vegetation eaten by wildlife). The remaining 29% surplus is what we directly use.

Note that groundwater is considered part of surface freshwater (remember that it is linked to surface supplies). In 2005, about 67% of groundwater withdrawals were for irrigation and 18% for public supply water needs. Just six states accounted for 50% of withdrawals: California, Texas, Nebraska, Arkansas, Idaho, and Florida, where more than half of the public supply of water is from groundwater.

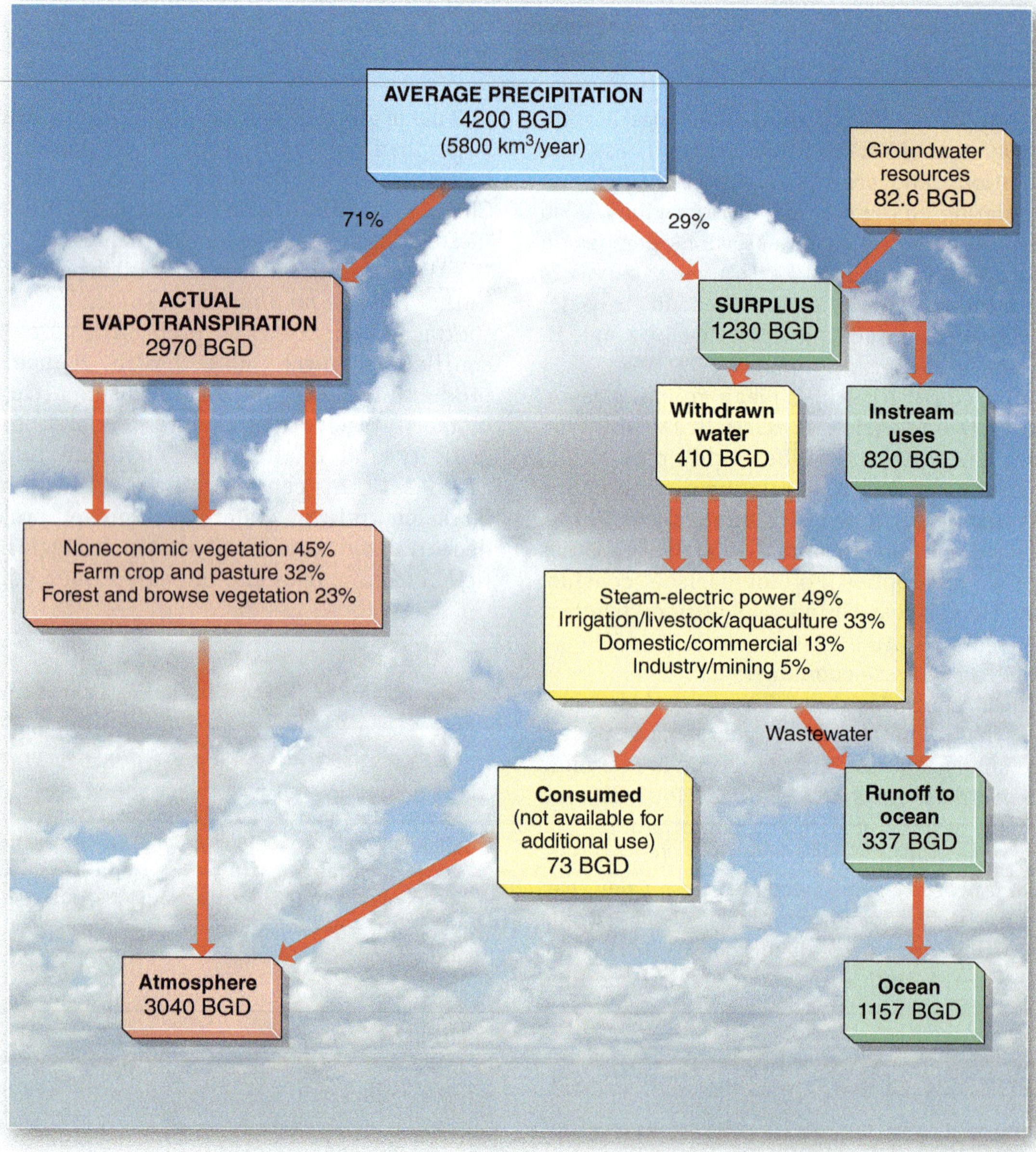

▲**Figure 9.18 Daily water budget for the contiguous 48 states in billions of gallons a day (BGD).** [Data from J. Kenny, N. Barber, S. Hutson, K. Linsey, J. Lovelace, and M. Maupin, "Estimated Use of Water in the United States in 2005," USGS Circular 1344, 2005—the latest year for which data are available.] [Bobbé Christopherson.]

Water Withdrawal and Consumption

Rivers and streams represent only a tiny percentage (0.003%) of Earth's overall surface water (review Figure 9.3). In terms of volume, they represent 1250 km^3 (300 mi^3), the smallest of any of the freshwater categories. Yet streamflow represents about four-fifths of all the water making up the surplus 1230 BGD that is available for withdrawal, consumption, and various instream uses.

- **Water withdrawal**, sometimes called *nonconsumptive use* or *offstream use*, refers to the removal or diversion of water from surface or groundwater supplies followed by the subsequent return of that water to the same supply. Examples include water use by industry, agriculture, and municipalities and in steam-electric power generation. A portion of the water withdrawn may be consumed.
- **Consumptive use** refers to the permanent removal of water from the immediate water environment. This water is not returned and so is not available for a second or third use. Examples include water lost to evapotranspiration, consumed by humans or livestock, or used in manufacturing.
- *Instream use* refers to uses of streamflow while it remains in the channel, without being removed. Examples include transportation, waste dilution and removal, hydroelectric power production, fishing, recreation, and ecosystem maintenance, such as sustaining wildlife.

Water withdrawal provides the opportunity to extend the water resource through reuse. The four main uses of withdrawn water in the United States in 2005 were (1) steam-electric power (49%, about 30% of which is met with saline ocean and coastal water); (2) irrigation, livestock, and aquaculture (33%); (3) domestic and commercial use (13%); and (4) industry and mining (5%; see Figure 9.18). However, when water returns to the stream system, its quality usually is altered—contaminated chemically with pollutants or waste or contaminated thermally with heat energy.

Contaminated or not, returned water becomes a part of all water systems downstream as runoff moves into oceans. In New Orleans, the last city to withdraw municipal water from the Mississippi River, streamflows include diluted and mixed contaminants from the entire Missouri–Ohio–Mississippi River system. This includes the effluent from chemical plants, agricultural runoff from fields treated with fertilizer and pesticides, treated and untreated sewage, pollution from oil spills and gasoline leaks, wastewater from thousands of industries, runoff from urban streets and storm drains, and particulates from countless construction sites and from mining, farming, and logging activities. Abnormally high cancer rates among citizens living along the Mississippi River between Baton Rouge and New Orleans have led to the ominous label "Cancer Alley" for the region. The infamous "muck" that covered the region when the levees broke after Hurricane Katrina in 2005 contained some of these contaminants in its brew.

The estimated U.S. withdrawal of water for 2005 was 410 BGD, double the usage rate in 1950 but down 5% from the peak usage year in 1980. Increased water prices helped produce the recent downward trend. For studies of water use in the United States and updated USGS Estimated Use of Water reports (available for download), see http://water.usgs.gov/public/watuse/.

CRITICALthinking 9.3
That Next Glass of Water

You have no doubt had several glasses of water so far today. Where did this water originate? Obtain the name of the water company or agency, determine whether it is using surface or groundwater to meet demands, and check how the water is metered and billed. If your state or province requires water-quality reporting, obtain a copy from the water supplier of an analysis of your tap water. Read this report, for you always want to know about that "next glass of water." •

Future Considerations

When water supply and demand are examined in terms of water budgets, the limits of water resources become apparent. In the water-budget equation in Figure 9.10, any change in one side (such as increased demand for surplus water) must be balanced by an adjustment in the other side (such as an increase in precipitation). How can we satisfy the growing demand for water? Water availability per person declines as population increases, and individual demand increases with economic development, affluence, and technology. World population growth since 1970 reduced per capita water supplies by a third. In addition, pollution limits the water-resource base, so that even before quantity constraints are felt, quality problems may limit the health and growth of a region.

Since surface and groundwater resources do not respect political boundaries, nations must share water resources, a situation that inevitably creates problems. For example, 145 countries of the world share a river basin with at least one other country. Streamflow truly represents a global commons.

The human denominator lists some of the interactions between humans and water resources, and some of the water-related issues for the century that lies ahead. One of the greatest challenges for humans will be the projected water shortages connected with global warming and climate change.

Clearly, international cooperation is needed, yet we continue toward a water crisis without a concept of a world water economy as a frame of reference. One of the most important questions is, When will more international coordination begin, and which country or group of countries will lead the way to sustain future water resources? The water-budget approach detailed in this chapter is a place to start.

THE**human**DENOMINATOR 9 Water Use

WATER RESOURCES ⇨ HUMANS

- Freshwater, stored in lakes, rivers, and groundwater, is a critical resource for human society and life on Earth.
- Drought results in water deficits, decreasing regional water supplies and causing declines in agriculture.

HUMANS ⇨ WATER RESOURCES

- Climate change affects lake depth, thermal structure, and associated organisms.
- Water projects (dams and diversions) redistribute water over space and time.
- Groundwater overuse and pollution depletes and degrades the resource, with side effects such as collapsed aquifers and saltwater contamination.

Desalination is an important supplement to water supplies in regions with large variations in rainfall throughout the year and declining groundwater reserves. This plant in Andalucía, Spain, uses the process of reverse osmosis to remove salts and impurities. [Jerónimo Alba/Alamy.]

The third largest reservoir in the world, Lake Nasser is impounded by the Aswan High Dam on the Nile River in Egypt. Its water is used for agricultural, industrial, and domestic purposes, as well as for hydropower. [WitR/Shutterstock.]

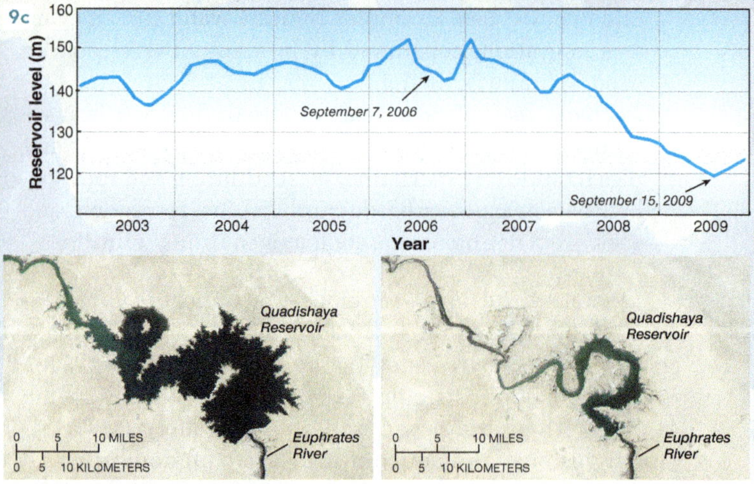

Data from *GRACE* reveal a rapid decline in reservoir levels from 2003 to 2009 along the Euphrates River in the Middle East; Quadishaya Reservoir is an example. The graph shows the surface-level decline, with dates of the images marked. About 60% of the volume loss is attributed to groundwater withdrawals in the region. [Image: Landsat-5, NASA. Graph based on data from UC Center for Hydrologic Monitoring.]

The Itaipu Dam and power plant on the Paraná River bordering Brazil and Paraguay produces more electricity annually than the Three Gorges Dam in China. Itaipu Reservoir displaced over 10,000 people and submerged Guaira Falls, formerly the world's largest waterfall by volume. [Mike Goldwater/Alamy.]

ISSUES FOR THE 21ST CENTURY

- Maintaining adequate water quantity and quality will be a major issue. Desalination will increase to augment freshwater supplies.
- Hydropower is a renewable energy resource; however, drought-related streamflow declines and drops in reservoir storage interfere with production.
- Drought in some regions will intensify, with related pressure on groundwater and surface water supplies.
- In the next 50 years water availability per person will drop as population increases, and continuing economic development will increase water demand.

GEOSYSTEMS**connection**

In this chapter, we looked at water resources through a water-budget approach. Such a systems view, considering both water supply and water demand, is the best method to understand the water resource. The ongoing drought in the western United States brings the need for such a water-budget strategy to the forefront. Water resources being the ultimate output of the water–weather system, we now shift our attention to climate. In Chapter 10, we examine world climates and in Chapter 10B, we discuss natural and human-caused climate change.

KEY LEARNING
concepts review

■ *Describe* the origin of Earth's waters, *report* the quantity of water that exists today, and *list* the locations of Earth's freshwater supply.

Water molecules came from within Earth over a period of billions of years in the **outgassing** process. Thus began endless cycling of water through the hydrologic system of evaporation–condensation–precipitation. Water covers about 71% of Earth. Approximately 97% of it is salty seawater, and the remaining 3% is freshwater—most of it frozen.

The present volume of water on Earth is estimated at 1.36 billion km^3 (326 million mi^3), an amount achieved roughly 2 billion years ago. This overall steady-state equilibrium might seem in conflict with the many changes in sea level that have occurred over Earth's history, but it is not. **Eustasy** refers to worldwide changes in sea level and relates to changes in volume of water in the oceans. The amount of water stored in glaciers and ice sheets explains these changes as *glacio-eustatic* factors. At present, sea level is rising because of increases in the temperature of the oceans and the record melting of glacial ice.

outgassing (p. 224)
eustasy (p. 225)

1. Approximately where and when did Earth's water originate?
2. If the quantity of water on Earth has been quite constant in volume for at least 2 billion years, how can sea level have fluctuated? Explain.
3. Describe the locations of Earth's water, both oceanic and fresh. What is the largest repository of freshwater at this time? In what ways is this distribution of water significant to modern society?
4. Why would climate change be a concern, given this distribution of water?
5. Why might you describe Earth as the water planet? Explain.

■ *Illustrate* the hydrologic cycle with a simple sketch, and *label* it with definitions for each water pathway.

The **hydrologic cycle** is a model of Earth's water system, which has operated for billions of years from the lower atmosphere to several kilometers beneath Earth's surface. **Evaporation** is the net movement of free water molecules away from a wet surface into air. **Transpiration** is the movement of water through plants and back into the atmosphere; it is a cooling mechanism for plants. Evaporation and transpiration are combined into one term—**evapotranspiration**.

Interception occurs when precipitation strikes vegetation or other ground cover. Water soaks into the subsurface through **infiltration**, or penetration of the soil surface. Water may puddle on the surface or flow across the surface toward stream channels. This **overland flow**, also called **surface runoff**, may become *streamflow* as it moves into channels on the surface.

Surface water becomes groundwater when it permeates soil or rock through vertical downward movement called **percolation**. The volume of subsurface water stored in the soil that is accessible to plant roots is contained in the **soil-moisture zone**. Groundwater is the largest potential freshwater source in the hydrologic cycle and is tied to surface supplies. The portion of streamflow that discharges naturally at the surface from groundwater is the **base flow**.

hydrologic cycle (p. 226)
evaporation (p. 226)
transpiration (p. 227)
evapotranspiration (p. 227)
interception (p. 227)
infiltration (p. 227)
overland flow (p. 227)
surface runoff (p. 227)
percolation (p. 228)
soil-moisture zone (p. 228)
base flow (p. 228)

6. Sketch and explain a simplified model of the complex flows of water on Earth—the hydrologic cycle.
7. What are the possible routes that a raindrop may take on its way to and into the soil surface?
8. Compare precipitation and evaporation volumes from the ocean with those over land. Describe advection flows of moisture and the countering flows of surface and subsurface runoff.

■ *Construct* the water-budget equation, *define* each of the components, and *explain* its use.

A **water budget** can be established for any area of Earth's surface by measuring the precipitation input and the output of various water demands in the area considered. If demands are met and extra water remains, a **surplus** occurs. If demand exceeds supply, a **deficit**, or water shortage results. Understanding both the supply of the water resource and the natural demands on the resource is essential to sustainable human interaction with the hydrologic cycle. **Precipitation** (PRECIP) is the moisture supply to Earth's surface, arriving as rain, sleet, snow, and hail and measured with the **rain gauge**. The ultimate demand for moisture is **potential evapotranspiration** (POTET), the amount of water that would evaporate and transpire under optimum moisture conditions (adequate precipitation and adequate soil moisture). If we subtract the deficit from the POTET, we determine **actual evapotranspiration**, or ACTET. Evapotranspiration is measured with an *evaporation pan* (evaporimeter) or the more elaborate *lysimeter*.

The volume of water stored in the soil that is accessible to plant roots is the **soil-moisture storage** (STRGE). This is the "savings account" of water that receives deposits and provides withdrawals as water-balance conditions change. In soil, **hygroscopic water** is inaccessible to plants because it is a molecule-thin layer that is tightly bound to each soil particle by hydrogen bonding. As available water is utilized, soil reaches the **wilting point** (all that remains is unextractable water). **Capillary water** is generally accessible to plant roots because it is held in the soil by surface tension and hydrogen bonding between water and soil. Almost all capillary water is available water in soil-moisture storage. After water drains from the larger pore spaces, the available water remaining for plants is termed **field capacity**, or storage capacity. When soil is saturated after a precipitation event, surplus water in the soil becomes **gravitational water** and percolates to groundwater. **Soil-moisture utilization** removes soil water, whereas **soil-moisture recharge** is the rate at which needed moisture enters the soil. The texture and the structure of the soil dictate available pore spaces, or *porosity*. The soil's **permeability** is the degree to which water can flow through it. Permeability depends on particle sizes and the shape and packing of soil grains.

Unsatisfied POTET is a deficit. If POTET is satisfied and the soil is full of moisture, then additional water input becomes a surplus. **Drought** can occur in at least four forms: meteorological drought, agricultural drought, hydrologic drought, and/or socioeconomic drought.

water budget (p. 228)
surplus (p. 228)
deficit (p. 228)
precipitation (p. 228)
rain gauge (p. 229)
potential evapotranspiration (p. 229)
actual evapotranspiration (p. 230)
soil-moisture storage (p. 231)
hygroscopic water (p. 231)
wilting point (p. 231)
capillary water (p. 231)
field capacity (p. 231)
gravitational water (p. 231)
soil-moisture utilization (p. 231)
soil-moisture recharge (p. 232)
permeability (p. 232)
drought (p. 232)

9. What are the components of the water-balance equation? Construct the equation, and place each term's definition below its abbreviation in the equation.
10. Explain how to derive actual evapotranspiration (ACTET) in the water-balance equation.
11. What is potential evapotranspiration (POTET)? How do we go about estimating this potential rate? What factors did Thornthwaite use to determine this value?
12. Explain the difference between soil-moisture utilization and soil-moisture recharge. Include discussion of capillary water and the field capacity and wilting point concepts.
13. In the case of silt loam soil from Figure 9.9, roughly what is the available water capacity? How is this value derived?
14. Use the water-balance equation to explain the changing relation of PRECIP to POTET in the annual water-balance chart for Kingsport, Tennessee.
15. How does the water-budget concept help your understanding of the hydrologic cycle, water resources, and soil moisture for a specific location? Give a specific example.
16. Describe the four definitions of drought. Describe current scientific thinking about the causes of recent drought conditions in the U.S. Southwest.

■ *Discuss* water storage in lakes and wetlands, and *describe* several large water projects involving hydroelectric power production.

Surface water is transferred in canals and pipelines for redistribution over space and stored in reservoirs for redistribution over time to meet water demand. Hydroelectric power, or **hydropower**, provides 20% of the world's electricity, and many large projects have changed river systems and affected human populations.

Lakes and wetlands are important freshwater storage areas. A **wetland** is an area that is permanently or seasonally saturated with water and that is characterized by vegetation adapted to *hydric* soils (soils saturated for a long enough period to develop anaerobic, or "oxygen-free," conditions).

hydropower (p. 236)
wetland (p. 241)

17. What changes occur along rivers as a result of the construction of large hydropower facilities?
18. Define wetland, and discuss the distribution of lakes and wetlands on Earth.

■ *Describe* the nature of groundwater, and *define* the elements of the groundwater environment.

Groundwater lies beneath the surface beyond the soil-moisture root zone, and its replenishment is tied to surface surpluses. Excess surface water moves through the **zone of aeration**, where soil and rock are less than saturated. Eventually, the water reaches the **zone of saturation**, where the pores are completely filled with water. The upper limit of the water that collects in the zone of saturation is the **water table**, forming the contact surface between the zones of saturation and aeration.

The permeability of subsurface rocks depends on whether they conduct water readily (higher permeability) or tend to obstruct its flow (lower permeability). They can even be impermeable. An **aquifer** is a rock layer that is permeable to groundwater flow in usable amounts. An **unconfined aquifer** has a permeable layer on top and an impermeable one beneath. A **confined aquifer** is bounded above and below by impermeable layers of rock or unconsolidated material. An *aquiclude* is a solid, impermeable layer that forms such a boundary, while an *aquitard* has low permeability but cannot conduct water in usable amounts.

Water in a confined aquifer is under the pressure of its own weight, creating a pressure level to which the water can rise on its own. This **potentiometric surface** can be above ground level. Groundwater confined under pressure is **artesian water**; it may rise up in wells and even flow out at the surface without pumping if the head of the well is below the potentiometric surface.

As water is pumped from a well, the surrounding water table within an unconfined aquifer will experience **drawdown**, or become lower, if the rate of pumping exceeds the horizontal flow of water in the aquifer around the well. This excessive pumping causes a **cone of depression**. Aquifers frequently are pumped beyond their flow and recharge capacities, a condition known as **groundwater mining**.

In many areas, especially where groundwater levels are declining, desalination is becoming an increasingly important method for meeting water demands. **Desalination** of seawater and saline groundwater involves the removal of organics, debris, and salinity through distillation or reverse osmosis. This processing yields potable water for domestic uses.

groundwater (p. 241)
zone of aeration (p. 242)
zone of saturation (p. 242)
water table (p. 242)
aquifer (p. 243)
unconfined aquifer (p. 243)
confined aquifer (p. 243)
potentiometric surface (p. 243)
artesian water (p. 243)
drawdown (p. 243)
cone of depression (p. 243)
groundwater mining (p. 243)
desalination (p. 247)

19. Are groundwater resources independent of surface supplies, or are the two interrelated? Explain your answer.
20. Make a simple sketch of the subsurface environment, labeling zones of aeration and saturation and the water table in an unconfined aquifer. Then add a confined aquifer to the sketch.
21. At what point does groundwater utilization become groundwater mining? Use the High Plains Aquifer example to explain your answer.
22. What is the nature of groundwater pollution? Can contaminated groundwater be cleaned up easily? What is fracking and how does it impact groundwater? Revisit Geosystems Now in Chapter 1 to assist with your answer.

■ *Evaluate* the U.S. water budget, and *identify* critical aspects of present and future freshwater supplies.

The world's water supply is distributed unevenly over Earth's surface. In the United States, water surpluses are used in several ways. **Water withdrawal**, also known as *nonconsumptive use* or *offstream use*, temporarily removes water from the supply, returning it later. **Consumptive use** permanently removes water from a stream. *Instream* use leaves water in the stream channel; examples are recreation and hydropower production. Americans in the 48 contiguous states withdraw approximately one-third of the available water surplus for irrigation, industry, and municipal uses.

water withdrawal (p. 251)
consumptive use (p. 251)

23. Describe the principal pathways involved in the water budget of the contiguous 48 states. What is the difference between withdrawal and consumptive use of water resources? Compare these with instream uses.
24. Briefly assess the status of world water resources. What challenges exist in meeting the future needs of an expanding population and growing economies?
25. If wars in the 21st century are predicted to be about water availability in the needed quantity and quality, what action could we take to understand the issues and avoid the conflicts?

MasteringGeography™

Looking for additional review and test prep materials? Visit the Study Area in *MasteringGeography™* to enhance your geographic literacy, spatial reasoning skills, and understanding of this chapter's content by accessing a variety of resources, including MapMaster interactive maps, geoscience animations, satellite loops, author notebooks, videos, RSS feeds, flashcards, web links, self-study quizzes, and an eText version of *Geosystems*.

▶**Figure 9.19 The Salton Sea.** California's largest lake is the Salton Sea (located on the map in Figure 9.1.1 on page 239). This lake bed, which has filled and dried out with changing climate over thousands of years, formed most recently in 1905 when irrigation canals from the Colorado River basin were breached by floodwaters. The lake, which has no outlet to the sea, has since become increasingly saline with agricultural runoff. This March 2013 image shows the irrigated agriculture of the Coachella Valley to the northwest, and the Imperial Valley to the southeast. [LDCM image by Matthew Montanaro and Robert Simmon using the *Landsat-8* instrument, NASA/USGS.]

10 Global Climate Systems

KEY LEARNING **concepts**

After reading the chapter, you should be able to:

- ***Define*** climate and climatology, and ***review*** the principal components of Earth's climate system.
- ***Describe*** climate classification systems, ***list*** the main categories of world climates, and ***locate*** the regions characterized by each climate type on a world map.
- ***Discuss*** the subcategories of the six world climate groups, including their causal factors.
- ***Explain*** the precipitation and moisture-efficiency criteria used to classify the arid and semiarid climates.

Most of Indonesia lies between 5° N and 10° S latitude, and receives rain throughout the year, making it part of the tropical rain forest climate region. The Maluku Islands of eastern Indonesia, seen here with the volcanic island of Ternate in the distance, once were the world's largest producer of cloves. In 2011, Ternate's Mount Gamalama erupted, closing the airport and causing thousands of residents to evacuate, but producing no fatalities. [Fadil/Corbis.]

A Large-Scale Look at Puerto Rico's Climate

Puerto Rico, an official territory of the United States, is centered between the Greater and Lesser Antilles island chains, separating the Atlantic Ocean and the Caribbean Sea. The island is nearly rectangular, stretching 65 km (40 mi) north and south and 180 km (112 mi) east and west. Puerto Rico's climate, which is determined by the overall pattern of weather over many years, is "tropical" in nature because all months have warm temperatures (averaging above 18°C, or 64.4°F) and moisture is abundant.

Small-scale maps of world climates, such as Figure 10.2, show generalizations. Large-scale maps of particular regions, such as Figure GN 10.1, show local detail—for example, the pattern of precipitation and temperatures as it relates to differences in elevation and orientation to the trade winds on the island of Puerto Rico. These local climatic variations are reflected in the vegetation types across the island.

Several factors interact to determine Puerto Rico's climate zones. The 18th parallel runs along the island's south coast, positioning it to receive trade winds from the northeast. The mountains of the Cordillera Central, rising to 1338 m (4389 ft), divide Puerto Rico lengthwise, and the Sierra de Luquillo rise to 650 m (2133 ft) in the east. These mountains add an orographic component to the island's climate patterns, increasing annual precipitation as the northern and eastern slopes intercept moisture brought by the trade winds; note the annual rainfall totals in these areas. In contrast, the slopes and coastal plain of the southern region are in the rain shadow of these mountains.

Tropical rain forest climates, where precipitation exceeds 6 cm (2.4 in.) every month, dominate most of the island. A tropical monsoon climate, where precipitation is less than 6 cm for 3 to 5 months during the year, is present on the leeward southern mountain slopes. A tropical savanna climate, where rainfall for 6 months is less than 6 cm (2.4 in.), occurs on the southern flatlands near Ponce and along the coast. This region is characterized by acacia trees and grasslands, commonly called savannas. As we examine the global distribution of Earth's climates in this chapter, remember the local variation that occurs when climates are studied using a larger map scale.

▲**Figure GN 10.1 Local climates of Puerto Rico.** Climate types, with annual rainfall totals for each photo location. [Bobbé Christopherson.]

GEOSYSTEMS NOW ONLINE Go to Chapter 10 on the *MasteringGeography* website (www.masteringgeography.com) for resources and activities. Another example of climatic variation over a small region is New Zealand, whose two islands can be classified into more than eight climate zones. For a closer look, go to www.niwa.co.nz/education-and-training/schools/resources/climate/overview. MG

The climate where you live may be humid with distinct seasons, or dry with consistent warmth, or moist and cool—almost any combination is possible. Some places have rainfall totaling more than 20 cm (8 in.) each month, with monthly average temperatures remaining above 27°C (80°F) year-round. Other places may be rainless for a decade at a time. A climate may have temperatures that average above freezing every month, yet still threaten severe frost problems for agriculture. Students reading *Geosystems* in Singapore experience precipitation every month, totaling 228.1 cm (89.8 in.) during an average year, whereas students at the university in Karachi, Pakistan, measure only 20.4 cm (8 in.) of annual rainfall.

Climate is the collective pattern of weather over many years. As we have seen, Earth experiences an almost infinite variety of *weather* at any given time and place. But, if we consider a longer time scale, and the variability and extremes of weather over such a time scale, a pattern emerges that constitutes climate. For a given region, this pattern is dynamic rather than static; that is, climate changes over time (we examine this in Chapter 10B).

Climatology is the study of climate and its variability, including long-term weather patterns over time and space and the controls that produce Earth's diverse climatic conditions. No two places on Earth's surface experience exactly the same climatic conditions; in fact, Earth is a vast collection of microclimates. However, broad similarities among local climates permit their grouping into **climatic regions**, which are areas with similarity in weather statistics. As you will see in Chapter 10B, the climate designations we study in this chapter are shifting as temperatures rise over the globe.

In this chapter: Many of the physical systems studied in the first nine chapters of this text interact to explain climates. Here we survey the patterns of climate using a series of sample cities and towns. *Geosystems* uses a simplified classification system based on physical factors that help answer the question "Why are climates in certain locations?" Though imperfect, this method is easily understood and is based on a widely used classification system devised by climatologist Wladimir Köppen.

Review of Earth's Climate System

Several important components of the energy–atmosphere system work together to determine climatic conditions on Earth. Simply combining the two principal climatic components—temperature and precipitation—reveals general climate types, sometimes called *climate regimes*, such as tropical deserts (hot

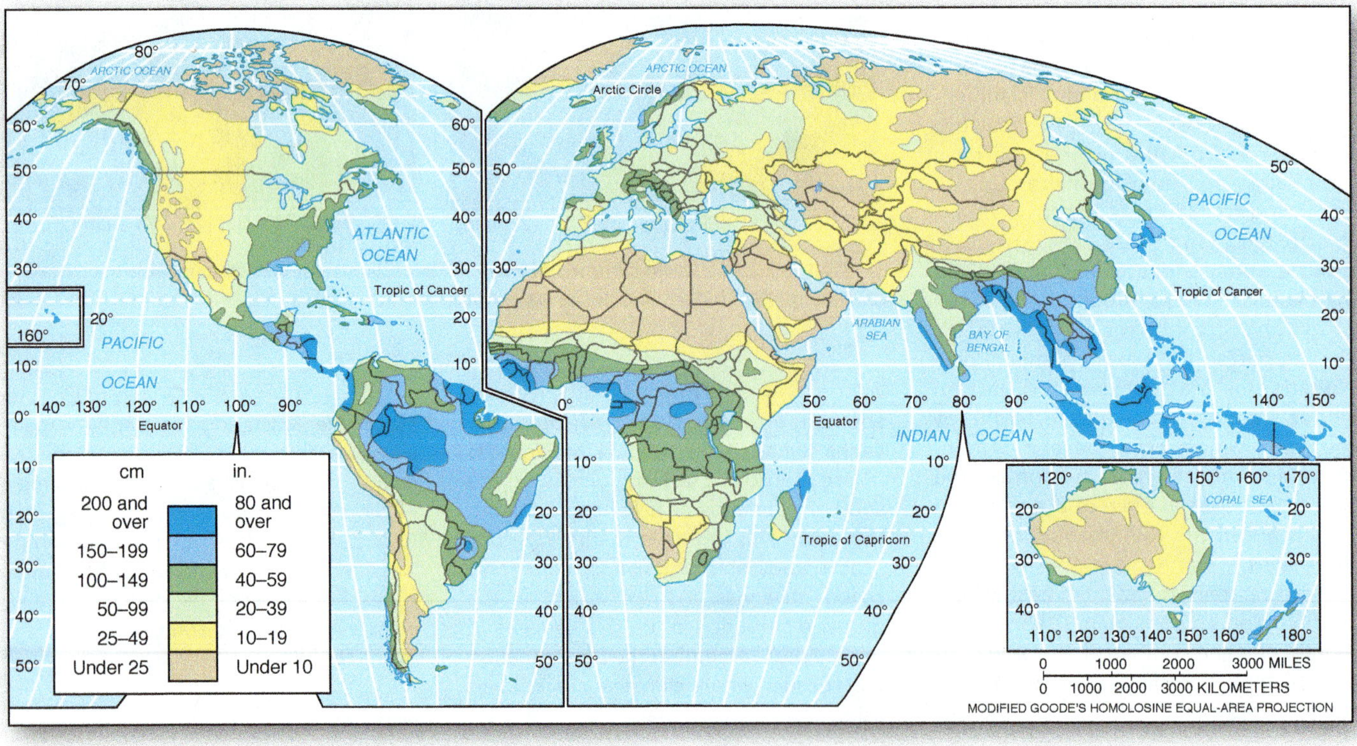

▲**Figure 10.1 Worldwide average annual precipitation.**

MG™ Animation Global Patterns of Precipitation

and dry), polar ice sheets (cold and dry), and equatorial rain forests (hot and wet).

Figure 10.1 maps the worldwide distribution of precipitation. These patterns reflect the interplay of numerous factors that should now be familiar to you, including temperature and pressure distributions; air mass types; convergent, convectional, orographic, and frontal lifting mechanisms; and the general energy availability that decreases toward the poles. The principal components of Earth's climate system are summarized in the Geosystems in Action feature on pages 260–261.

Classifying Earth's Climates

Classification is the ordering or grouping of data or phenomena into categories of varying generality. Such generalizations are important organizational tools in science and are especially useful for the spatial analysis of climatic regions. Observed patterns confined to specific regions are at the core of climate classification. When using classifications, we must remember that the boundaries of these regions are *transition zones*, or areas of gradual change. The placement of climate boundaries depends on overall climate patterns rather than precise locations where classifications change.

A climate classification based on *causative* factors—for example, the interaction of air masses—is a **genetic classification**. This approach explores "why" a certain mix of climatic ingredients in certain locations occurs. A climate classification based on *statistics* or other data determined by measurement of observed effects is an **empirical classification**.

Climate classifications based on temperature and precipitation are examples of the empirical approach. One empirical classification system, published by C. W. Thornthwaite, identified climate regions according to moisture, using aspects of the water-budget approach (discussed in Chapter 9) and vegetation types. Another empirical system is the widely recognized Köppen climate classification, designed by Wladimir Köppen (pronounced KUR-pen; 1846–1940), a German climatologist and botanist. His classification work began with an article on heat zones in 1884 and continued throughout his career. The first wall map showing world climates, coauthored with his student Rudolph Geiger, was introduced in 1928 and soon was widely adopted. Köppen continued to refine it until his death. In Appendix B, you find a description of his system and the detailed criteria he used to distinguish climatic regions and their boundaries.

The classification system used in *Geosystems* is a compromise between genetic and empirical systems. It focuses on temperature and precipitation measurements (and for the desert areas, moisture efficiency) and also on causal factors that produce the climates. Six basic climate categories provide the structure for our chapter discussion.

- Tropical climates: tropical latitudes, winterless
- Mesothermal climates: midlatitudes, mild winters
- Microthermal climates: mid- and high latitudes, cold winters
- Polar climates: high latitudes and polar regions
- Highland climates: high elevations at all latitudes
- Dry climates: permanent moisture deficits at all latitudes

Each of these climates is divided into subcategories, presented in the world climate map in Figure 10.2 and described in the following sections. The upcoming discussions also include at least one **climograph** for each climate subcategory, showing monthly temperature and precipitation for a weather station at a selected city. Listed along the top of each climograph are the dominant weather features that influence that climate's characteristics. A location map and selected statistics—including location coordinates, average annual temperature, total annual precipitation, and elevation—complete the information for each station. For each main climate category, a text box introduces the climate characteristics and causal elements, and includes a world map showing the general distribution of that climate type and in most cases the location of representative weather stations.

Climates greatly influence *ecosystems*, the natural, self-regulating communities formed by plants and animals in their nonliving environment. On land, the basic climatic regions determine to a large extent the location of the world's major ecosystems. These broad regions, with their associated soil, plant, and animal communities, are called *biomes*; examples include forest, grassland, savanna, tundra, and desert. Discussions of the major terrestrial biomes that fully integrate these global climate patterns appear in Part IV of this text (see Table 20.1). In this chapter, we mention the ecosystems and biomes associated with each climate type.

CRITICALthinking 10.1
Finding Your Climate

Using Figure 10.2, locate your campus and your birthplace, and the associated climate description. Then obtain monthly precipitation and temperature data for those places. Briefly describe the information sources you used: library, Internet, teacher, or phone calls to state or provincial climatologists. Next, refer to Appendix B to refine your assessment of climate for the two locations. Briefly show how you worked through the Köppen climate criteria in the appendix to establish the climate classification for your two cities. ●

(text continued on page 264)

Earth's climate system is the result of interactions among several components. These include the input and transfer of energy from the Sun (GIA 10.1 and 10.2); the resulting changes in atmospheric temperature and pressure (GIA 10.3 and 10.4); the movements and interactions of air masses (GIA 10.5); and the transfer of water—as vapor, liquid, or solid— throughout the system (GIA 10.6).

10.1 INSOLATION

Incoming solar radiation is the energy input for the climate system. Insolation varies by latitude, as well as on a daily and seasonal basis with changing day length and Sun angle. (*Chapter 2; review Figures 2.8, 2.10, and GIA 2*)

North Cape Midnight Sun, Nordkapp, Norway [marcokenya/Shutterstock]

10.2 EARTH'S ENERGY BALANCE

The imbalance created by energy surpluses at the equator and energy deficits at the poles causes the global circulation patterns of winds and ocean currents that drive weather systems. (*Chapter 4; review Figure 4.10*)

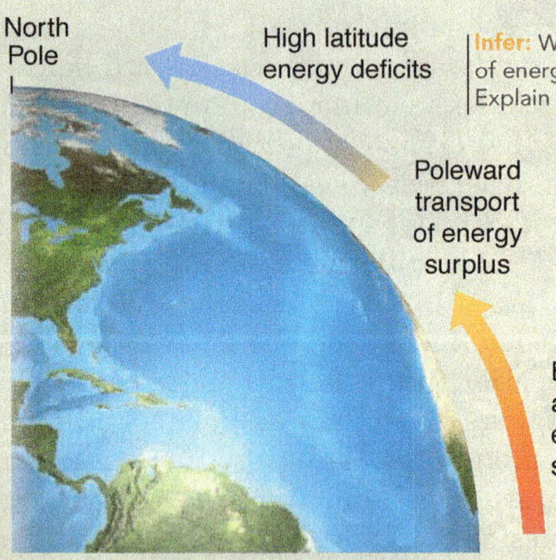

Infer: What is the general pattern of energy flow in the atmosphere? Explain your answer.

10.3 TEMPERATURE

Primary temperature controls are latitude, elevation, cloud cover, and land–water heating differences. The pattern of world temperatures is affected by global winds, ocean currents, and air masses. (*Chapter 5; review Figures 5.12 through 5.15*)

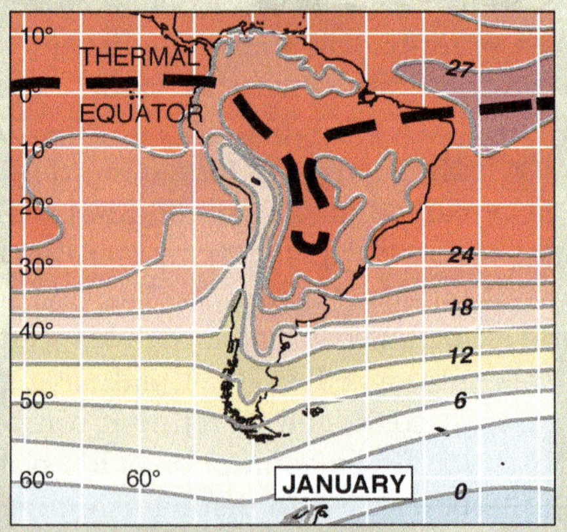

Video
Supercomputing the Climate

MasteringGeography™

Visit the Study Area in MasteringGeography™ to explore Earth's climate system.

Visualize: Study a NASA video of modeling Earth's climate.

Assess: Demonstrate understanding of Earth's climate system (if assigned by instructor).

10.4 AIR PRESSURE

Winds flow from areas of high pressure to areas of low pressure. The equatorial low creates a belt of wet climates. Subtropical highs create areas of dry climates. Pressure patterns influence atmospheric circulation and movement of air masses. Oceanic circulation and multi-year oscillations in pressure and temperature patterns over the oceans also affect weather and climate. (*Chapter 6; review Figures 6.10 and GIA 6*)

Explain: Where are the trade winds on this view of the globe? Explain and locate areas of warm, wet climates and hot, dry climates in relation to the Hadley cells.

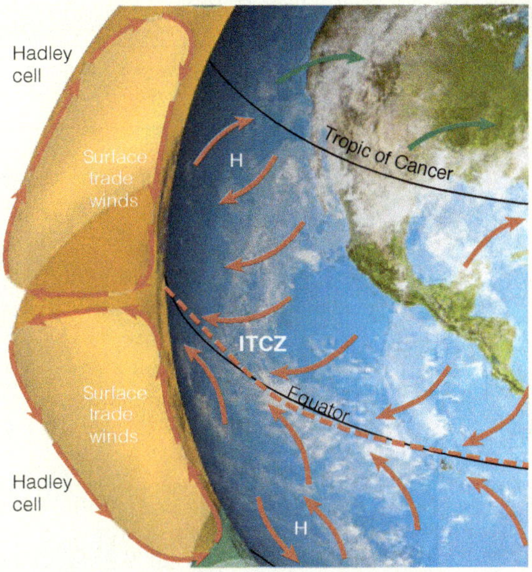

10.5 AIR MASSES

Vast bodies of homogeneous air form over oceanic and continental source regions, taking on the characteristics of their source region. As these air masses migrate, they carry their temperature and moisture conditions to new regions. At fronts, where air masses meet, precipitation or violent weather may occur. (*Chapter 8; review Figure 8.1*)

The warm, tropical mT air mass brings humid, unstable air to the U.S. East and Midwest. Here, a storm approaches Miami Beach.
[Jorg Hackemann/Shutterstock]

10.6 ATMOSPHERIC MOISTURE

The hydrologic cycle transfers moisture through Earth's climate system. Within this cycle, the processes of evaporation, transpiration, condensation, and precipitation are essential components of weather. (*Chapters 7 and 9; review Figure 9.4*)

[cpphotoimages/Shutterstock]

GEOquiz

1. Analyze: How do these six components of Earth's climate system interact to produce the world climates in Figure 10.2?

2. Discuss: Describe the role that each component of the climate system plays in causing precipitation patterns on Earth.

Animation
Global Climate Maps, World Map References

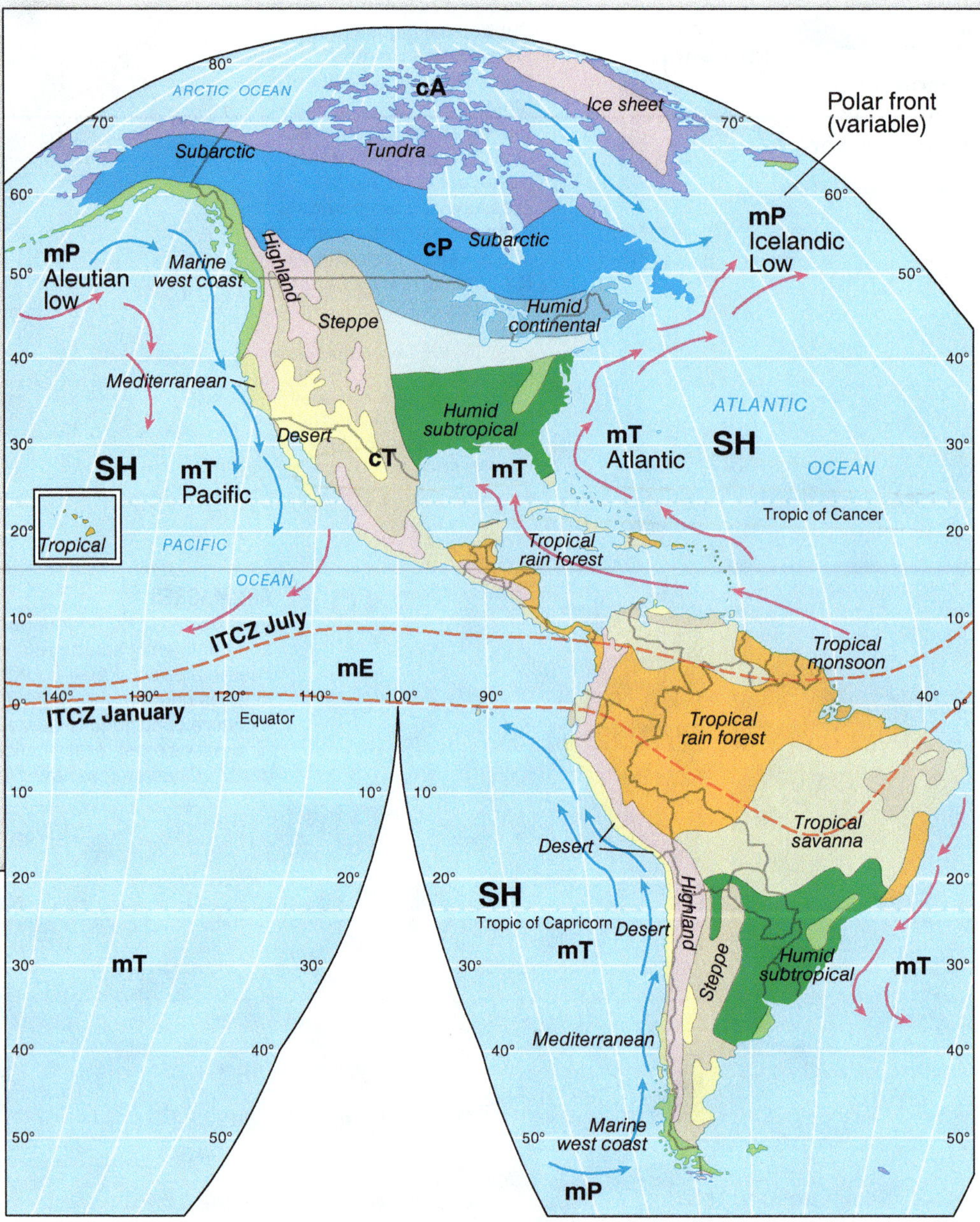

OCEAN CURRENTS (Fig. 6.18)

- Warm current
- Cool current

INTERTROPICAL CONVERGENCE ZONE

Figs. 6.10, GIA 6

- ITCZ July
- ITCZ January

AIR PRESSURE SYSTEMS (Fig. 6.10)

SH Subtropical high

AIR MASSES (Fig. 8.1)

mP Maritime polar (cool, humid)
cP Continental polar (cool, cold, dry)
mT Maritime tropical (warm, humid)
cA Continental arctic (very cold, dry)
cT Continental tropical (hot, dry summer only)
mE Maritime equatorial (warm, wet)

TROPICAL CLIMATES

- Tropical rain forest
  Rainy all year
- Tropical monsoon
  6 to 12 months rainy
- Tropical savanna
  Less than 6 months rainy

MESOTHERMAL CLIMATES

- Humid subtropical
  Moist all year, hot summer
- Humid subtropical
  Winter-dry, hot to warm summers
- Marine west coast
  Moist all year, warm to cool summers
- Mediterranean
  Summer-dry, hot to warm summers

▲**Figure 10.2 World climate classification.** Annotated on this map are selected air masses, nearshore ocean currents, pressure systems, and the January and July locations of the ITCZ. Use the colors in the legend to locate various climate types; some labels of the climate names appear in italics on the map to guide you.

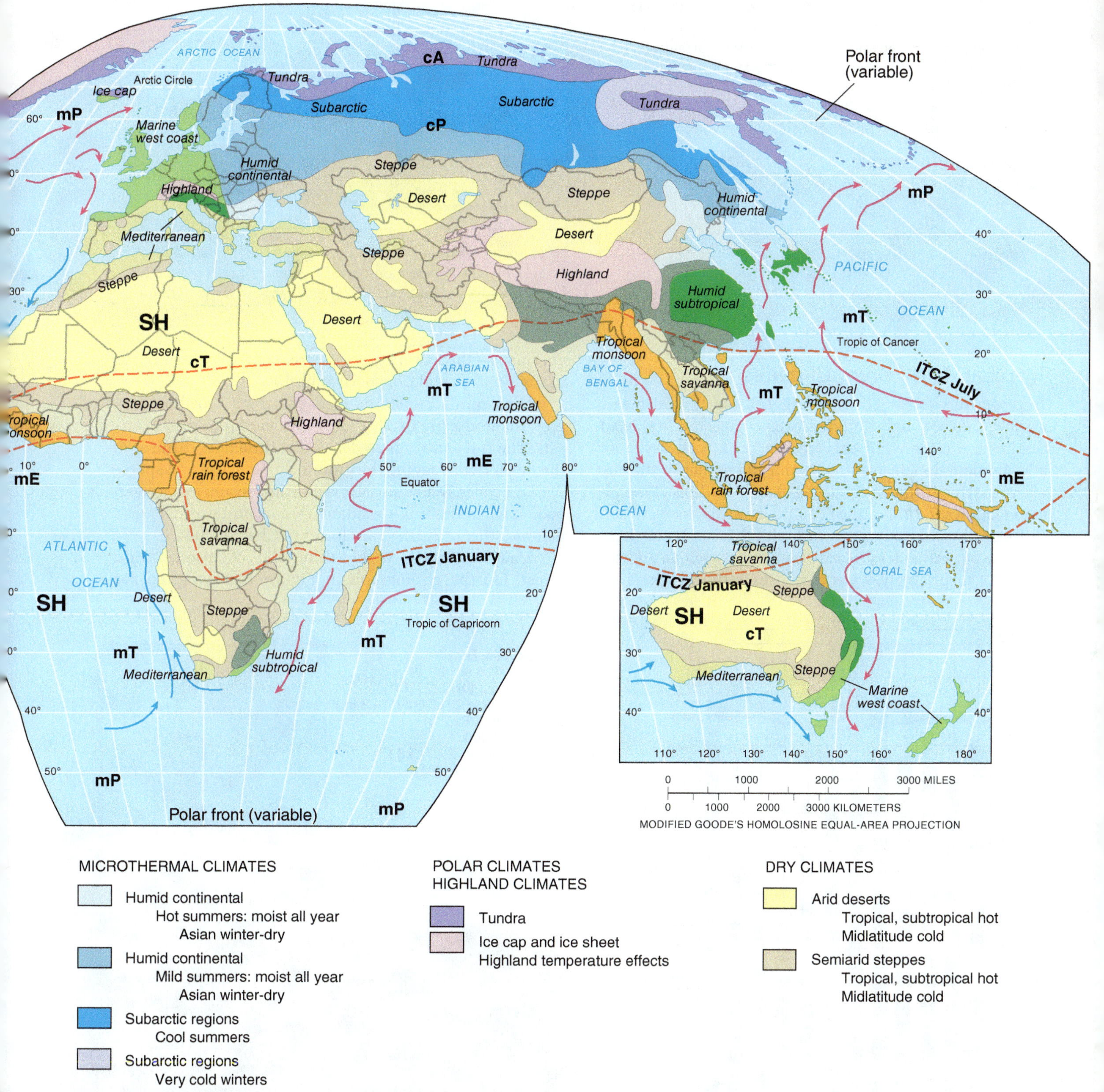
ARCTIC OCEAN
cA
Tundra
Polar front (variable)
Arctic Circle
Ice cap
mP
Marine west coast
Subarctic
cP
Humid continental
Steppe
Desert
Highland
Mediterranean
SH
cT
PACIFIC
OCEAN
mT
Humid subtropical
Tropic of Cancer
ITCZ July
ARABIAN SEA
BAY OF BENGAL
Tropical monsoon
Tropical savanna
Tropical rain forest
mE
Equator
INDIAN
OCEAN
ATLANTIC
OCEAN
ITCZ January
Tropic of Capricorn
CORAL SEA
Marine west coast
Polar front (variable)
0 1000 2000 3000 MILES
0 1000 2000 3000 KILOMETERS
MODIFIED GOODE'S HOMOLOSINE EQUAL-AREA PROJECTION
MICROTHERMAL CLIMATES
Humid continental
Hot summers: moist all year
Asian winter-dry
Humid continental
Mild summers: moist all year
Asian winter-dry
Subarctic regions
Cool summers
Subarctic regions
Very cold winters
POLAR CLIMATES
HIGHLAND CLIMATES
Tundra
Ice cap and ice sheet
Highland temperature effects
DRY CLIMATES
Arid deserts
Tropical, subtropical hot
Midlatitude cold
Semiarid steppes
Tropical, subtropical hot
Midlatitude cold

Tropical Climates (tropical latitudes)

Tropical climates are the most extensive, occupying about 36% of Earth's surface, including both ocean and land areas. The tropical climates straddle the equator from about 20° N to 20° S, roughly between the Tropics of Cancer and Capricorn—thus, the name. Tropical climates stretch northward to the tip of Florida and to south-central Mexico, central India, and Southeast Asia and southward to northern Australia, Madagascar, central Africa, and southern Brazil. These climates truly are winterless. Important causal elements include:

- Consistent daylength and insolation, which produce consistently warm temperatures;
- Effects of the intertropical convergence zone (ITCZ), which brings rains as it shifts seasonally with the high Sun;
- Warm ocean temperatures and unstable maritime air masses.

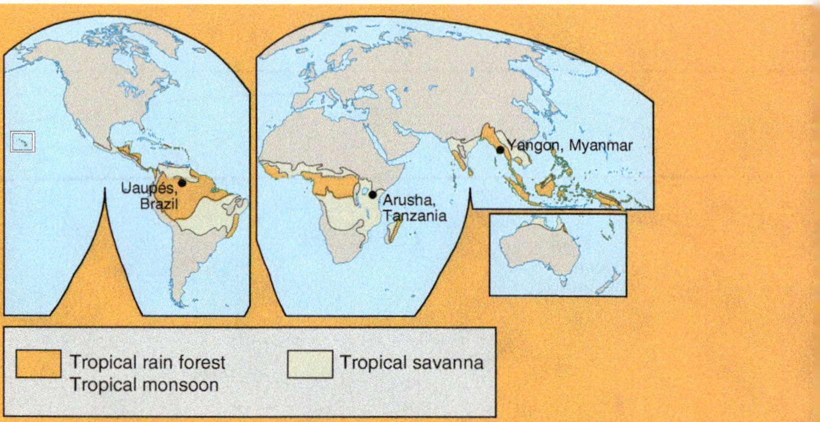

Tropical climates have three distinct regimes: *tropical rain forest* (ITCZ present all year), *tropical monsoon* (ITCZ present 6 to 12 months annually), and *tropical savanna* (ITCZ present less than 6 months).

Tropical Rain Forest Climates

Tropical rain forest climates are constantly moist and warm. Convectional thunderstorms, triggered by local heating and trade-wind convergence, peak each day from midafternoon to late evening inland and earlier in the day where marine influence is strong along coastlines. Precipitation follows the migrating ITCZ (review Chapter 6), which shifts northward and southward with the Sun throughout the year but influences tropical rain forest regions all year long. Not surprisingly, water surpluses in these regions are enormous—the world's greatest streamflow volumes occur in the Amazon and Congo River basins.

High rainfall sustains lush evergreen broadleaf tree growth, producing Earth's equatorial and tropical rain forests. The leaf canopy is so dense that little light diffuses to the forest floor, leaving the ground surface dim and sparse in plant cover. Dense surface vegetation occurs along riverbanks, where light is abundant. (We examine widespread deforestation of Earth's rain forest in Chapter 20.)

Uaupés, Brazil, is characteristic of tropical rain forest. On the climograph in Figure 10.3, you can see

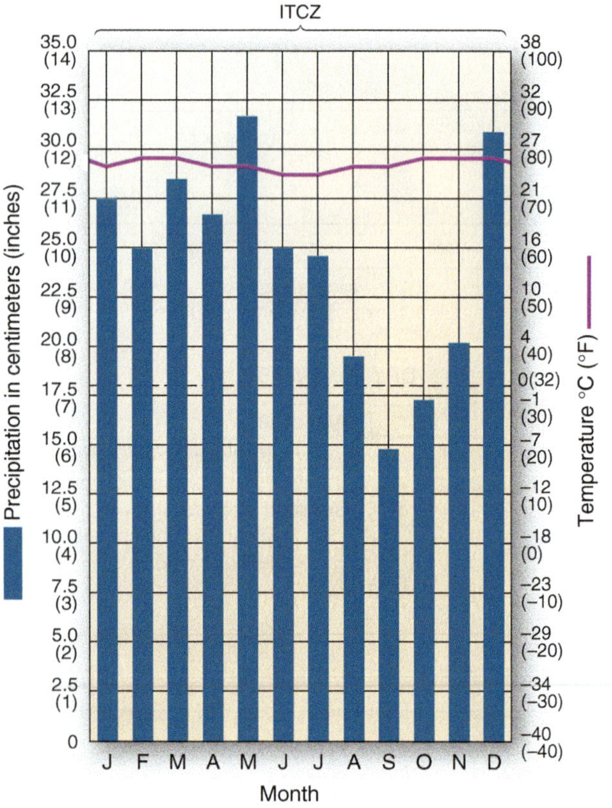

(a) Climograph for Uaupés, Brazil.

Station: Uaupés, Brazil
Lat/long: 0° 06′ S 67° 02′ W
Avg. Ann. Temp.: 25°C (77°F)
Total Ann. Precip.: 291.7 cm (114.8 in.)
Elevation: 86 m (282.2 ft)
Population: 10,000
Ann. Temp. Range: 2 C° (3.6 F°)
Ann. Hr of Sunshine: 2018

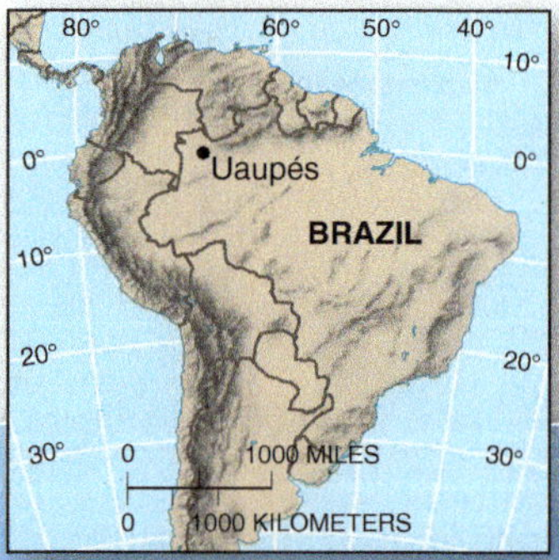

(b) The rain forest along a tributary of the Rio Negro River, Amazonas state, Brazil.

▲**Figure 10.3 Tropical rain forest climate.** [(b) Sue Cunnigham Photographic/Alamy.]

▼**Figure 10.4 Tropical monsoon climate.** [(b) shaileshnanal/Shuttershock.]

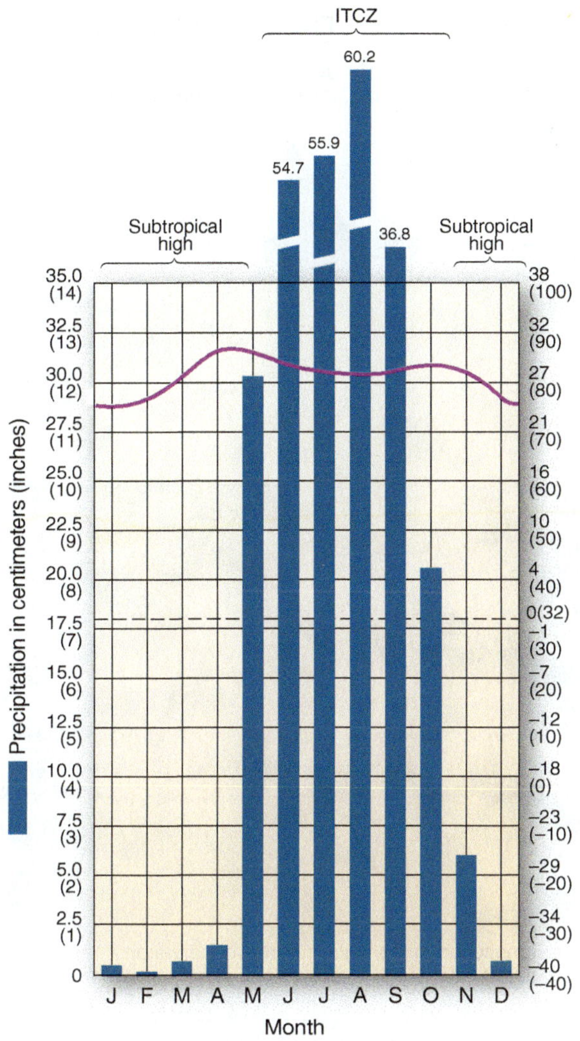

(a) Climograph for Yangon, Myanmar (formerly Rangoon, Burma); city of Sittwe also noted on map.

Station: Yangon, Myanmar*
Lat/long: 16° 47′ N 96° 10′ E
Avg. Ann. Temp.: 27.3°C (81.1°F)
Total Ann. Precip.: 268.8 cm (105.8 in.)
Elevation: 23 m (76 ft)
Population: 6,000,000
Ann. Temp. Range: 5.5 C° (9.9 F°)
*(Formerly Rangoon, Burma)

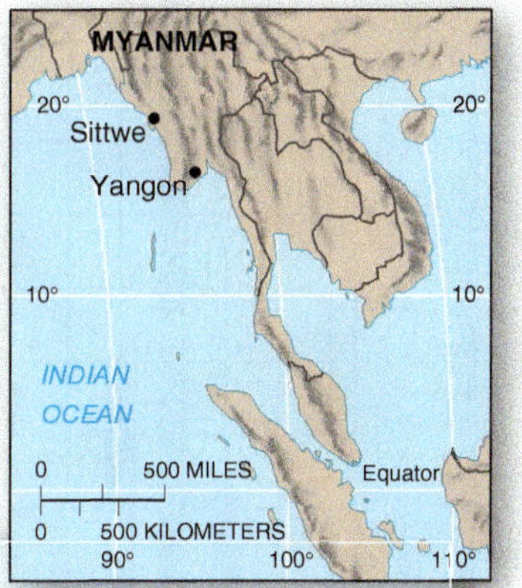

(b) Mixed monsoonal forest and scrub characteristic of the region in eastern India.

that the month of lowest precipitation receives nearly 15 cm (6 in.) and the annual temperature range is barely 2 C° (3.6 F°). In all such climates, the diurnal (day-to-night) temperature range exceeds the annual average minimum–maximum (coolest to warmest) range: Day–night differences can range more than 11 C° (20 F°), more than five times the annual monthly average range.

The only interruption in the distribution of tropical rain forest climates across the equatorial region is in the highlands of the South American Andes and in East Africa (see Figure 10.2). There, higher elevations produce lower temperatures; Mount Kilimanjaro is less than 4° south of the equator, but at 5895 m (19,340 ft), it has permanent glacial ice on its summit (although this ice has now nearly disappeared due to increasing air temperatures ; see Figure HD 17c, page 548). Such mountainous sites fall within the *highland* climate category.

Tropical Monsoon Climates

Tropical monsoon climates feature a dry season that lasts 1 or more months. Rainfall brought by the ITCZ falls in these areas from 6 to 12 months of the year. (Remember, the ITCZ affects the tropical rain forest climate region throughout the year.) The dry season occurs when the ITCZ has moved away so that the convergence effects are not present. Yangon, Myanmar (formerly Rangoon, Burma), is an example of this climate type (Figure 10.4). Mountains prevent cold air masses from central Asia from moving over Yangon, resulting in its high average annual temperatures.

About 480 km (300 mi) north in another coastal city, Sittwe (Akyab), Myanmar, on the Bay of Bengal, annual precipitation rises to 515 cm (203 in.), considerably higher than Yangon's 269 cm (106 in.). Therefore, Yangon is a drier area than that farther north along the coast, but it still exceeds the 250-cm annual precipitation criterion in use for the tropical monsoon classification.

Tropical monsoon climates lie principally along coastal areas within the tropical rain forest climatic realm and experience seasonal variation of wind and precipitation. Vegetation in this climate type typically consists of evergreen trees grading into thorn forests on the drier margins near the adjoining tropical savanna climates.

▼Figure 10.5 Tropical savanna climate.
[(b) Blaine Harrington III/Corbis.]

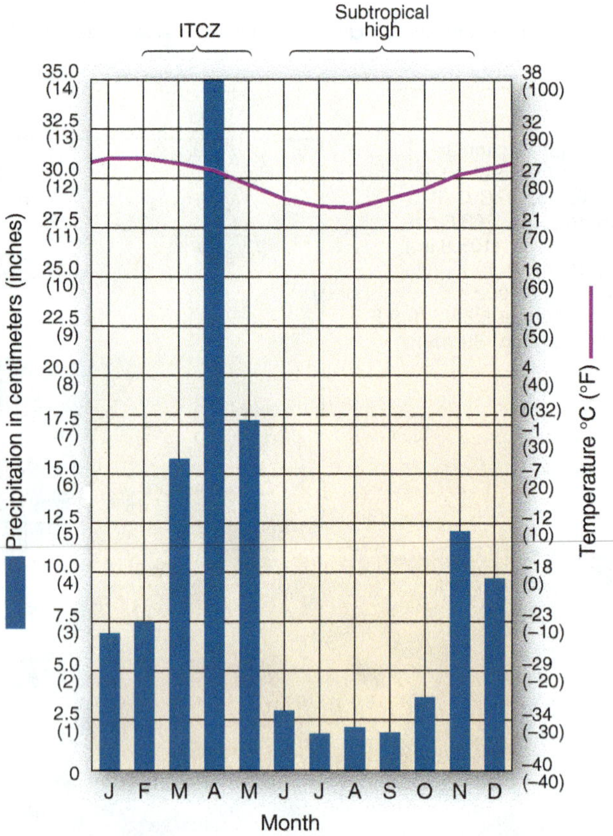

(a) Climograph for Arusha, Tanzania; note the intense dry period.

Station: Arusha, Tanzania
Lat/long: 3° 24′ S 36° 42′ E
Avg. Ann. Temp.: 26.5°C (79.7°F)
Total Ann. Precip.: 119 cm (46.9 in.)
Elevation: 1387 m (4550 ft)
Population: 1,368,000
Ann. Temp. Range: 4.1 C° (7.4 F°)
Ann. Hr of Sunshine: 2600

(b) Characteristic landscape in the Ngorongoro Conservation Area, Tanzania, near Arusha with plants adapted to seasonally dry water budgets.

Tropical Savanna Climates

Tropical savanna climates exist poleward of the tropical rain forest climates. The *Geosystems Now* photo shows that Ponce, Puerto Rico, is such a region. The ITCZ reaches these climate regions for about 6 months or less of the year as it migrates with the summer Sun. Summers are wetter than winters because convectional rains accompany the shifting ITCZ when it is overhead. In contrast, when the ITCZ is farthest away and high pressure dominates, conditions are notably dry. Thus, POTET (natural moisture demand) exceeds PRECIP (natural moisture supply) in winter, causing water-budget deficits.

Temperatures vary more in tropical savanna climates than in tropical rain forest regions. The tropical savanna regime can have two temperature maximums during the year because the Sun's direct rays are overhead twice—before and after the summer solstice in each hemisphere as the Sun moves between the equator and the tropics. Grasslands with scattered trees, drought resistant to cope with the highly variable precipitation, dominate the tropical savanna regions.

The climate of Arusha, Tanzania, represents tropical savanna conditions (Figure 10.5). This metropolitan area is near the grassy plains of the Serengeti, a heavily visited national park that hosts one of the largest annual mammal migrations in the world. Temperatures are consistent with tropical climates, despite the elevation (1387 m, 4550 ft) of the station. On the climograph, note the marked dryness from June to October, which indicates changing dominant pressure systems rather than annual changes in temperature. This region is near the transition to the drier *desert hot steppe* climates to the northeast (discussed later in the chapter).

GEOreport 10.1 Tropical climate zones advance to higher latitudes

The belt of tropical climates that straddles the equator is getting wider. Recent research suggests that this zone has widened more than 2° of latitude since 1979, with an overall advance of 0.7° of latitude per decade. Evidence indicates that the southward advance of this climate boundary is affected by stratospheric ozone depletion over the Antarctic region, while the northward advance relates to increases in black carbon aerosols and tropospheric (ground layer) ozone caused by fossil fuel burning in the Northern Hemisphere—these pollutants absorb sunlight and warm the atmosphere. As the tropical climates move poleward, the dry subtropical regions are becoming drier, with more frequent droughts.

Mesothermal Climates (midlatitudes, mild winters)

Mesothermal, meaning "middle temperature," describes these warm and temperate climates, where true seasonality begins. More than half the world's population resides in mesothermal climates, which occupy about 27% of Earth's land and sea surface—the second largest percentage behind the tropical climates.

The mesothermal climates, and nearby portions of the microthermal climates (cold winters), are regions of great weather variability, for these are the latitudes of greatest air mass interaction. Causal elements include:

- Latitudinal effects on insolation and temperature, as summers transition from hot to warm to cool moving poleward from the tropics;
- Shifting of maritime and continental air masses, as they are guided by upper-air westerly winds;
- Migration of cyclonic (low-pressure) and anticyclonic (high-pressure) systems, bringing changeable weather conditions and air mass conflicts;
- Effects of sea-surface temperatures on air mass strength: Cooler temperatures along west coasts weaken air masses, and warmer temperatures along east coasts strengthen air masses.

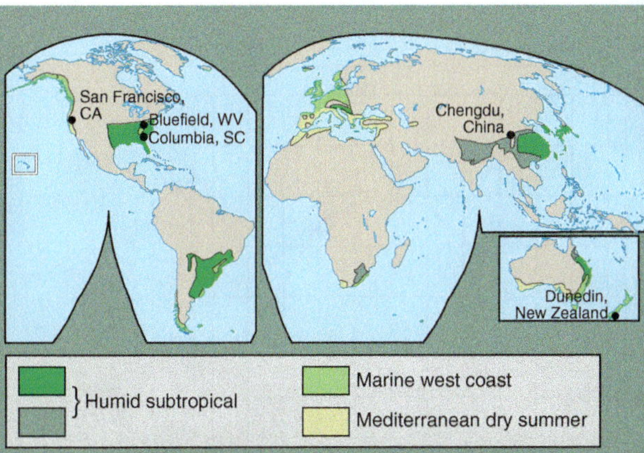

Mesothermal climates are humid, except where subtropical high pressure produces dry-summer conditions. Their four distinct regimes based on precipitation variability are *humid subtropical hot-summer* (moist all year), *humid subtropical winter-dry* (hot to warm summers, in Asia), *marine west coast* (warm to cool summers, moist all year), and *Mediterranean dry-summer* (warm to hot summers).

Humid Subtropical Hot-Summer Climates

Humid subtropical hot-summer climates either are moist all year or have a pronounced winter-dry period, as occurs in eastern and southern Asia. Maritime tropical air masses generated over warm waters off eastern coasts influence these climates during summer. This warm, moist, unstable air produces convectional showers over land. In fall, winter, and spring, maritime tropical and continental polar air masses interact, generating frontal activity and frequent midlatitude cyclonic storms. These two mechanisms produce year-round precipitation, which averages 100–200 cm (40–80 in.) a year.

In North America, humid subtropical hot-summer climates are found across the southeastern United States. Columbia, South Carolina, is a representative station (Figure 10.6), with characteristic winter precipitation from cyclonic storm activity (other examples are Atlanta, Memphis, and New Orleans). Nagasaki, Japan, is characteristic of an Asian humid subtropical hot-summer station (Figure 10.7), where winter precipitation is less because of the effects of the East Asian monsoon. However, the lower precipitation of winter is not quite dry enough to change the climate category to *humid subtropical winter-dry*. Nagasaki receives more overall annual precipitation (196 cm, 77 in.) than similar climates in the United States, owing to the monsoonal flow pattern.

Humid Subtropical Winter-Dry Climates

Humid subtropical winter-dry climates are related to the winter-dry, seasonal pulse of the monsoons. They extend poleward from tropical savanna climates and have a summer month that receives 10 times more precipitation than their driest winter month. Chengdu, China, is a representative station in Asia. Figure 10.8 demonstrates the strong correlation between precipitation and the high-summer Sun.

Large numbers of people live in the humid subtropical hot-summer and humid subtropical winter-dry climates, demonstrated by the large populations of north-central India, southeastern China, and the southeastern United States. Although these climates are relatively habitable for humans, natural hazards exist; for example, the intense summer rains of the Asian monsoon cause flooding in India and Bangladesh that affects millions of people. In the U.S. Southeast, dramatic thunderstorms are common, often spawning tornadoes, and rainfall associated with hurricanes can cause seasonal flooding events.

Marine West Coast Climates

Marine west coast climates, featuring mild winters and cool summers, are characteristic of Europe and other middle- to high-latitude west coasts (see Figure 10.2). In the United States, these climates, with their cooler summers, are in contrast to the humid subtropical hot-summer climate of the Southeast.

Maritime polar air masses—cool, moist, unstable—dominate marine west coast climates. Weather systems forming along the polar front and maritime polar air masses move into these regions throughout the year, making weather quite unpredictable. Coastal fog, annually totaling 30 to 60 days, is a part of the moderating marine influence. Frosts are possible and tend to shorten the growing season.

Marine west coast climates are unusually mild for their latitude. They extend along the coastal margins of the Aleutian Islands in the North Pacific, cover the southern third of Iceland in the North Atlantic and coastal Scandinavia, and dominate the British Isles. Many of us might find it hard to imagine that such high-latitude

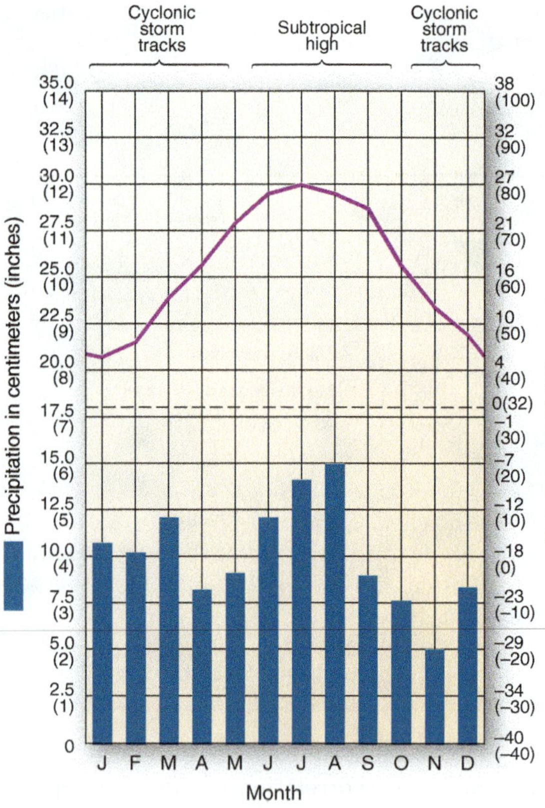

(a) Climograph for Columbia, South Carolina.

Station: Columbia, South Carolina
Lat/long: 34° N 81° W
Avg. Ann. Temp.: 17.3°C (63.1°F)
Total Ann. Precip.: 126.5 cm (49.8 in.)
Elevation: 96 m (315 ft)
Population: 116,000
Ann. Temp. Range: 20.7 C° (37.3 F°)
Ann. Hr of Sunshine: 2800

(b) Water lilies and mixed evergreen forest of cypress and pine in southern Georgia.

▲Figure 10.6 Humid subtropical hot-summer climate, American region. [(b) Bobbé Christopherson.]

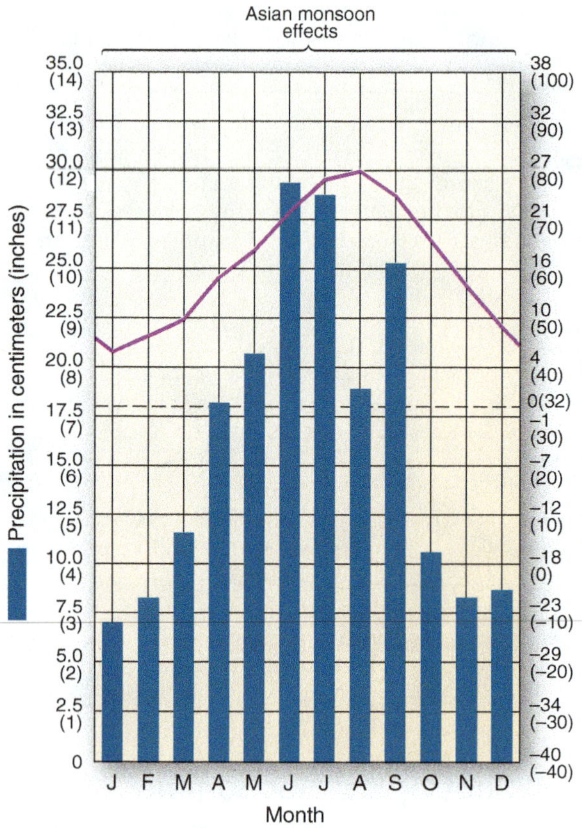

(a) Climograph for Nagasaki, Japan.

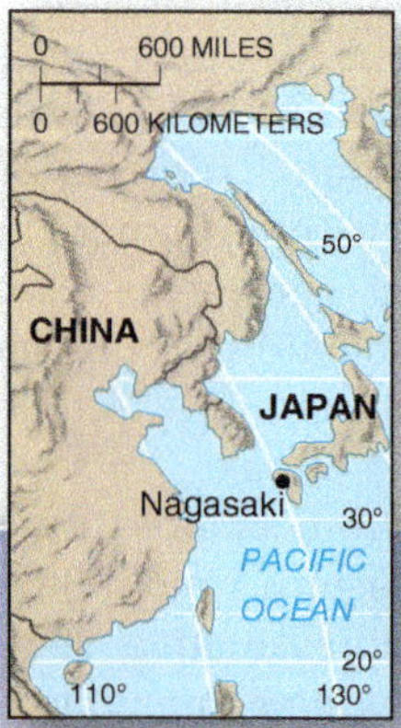

Station: Nagasaki, Japan
Lat/long: 32° 44′ N 129° 52′ E
Avg. Ann. Temp.: 16°C (60.8°F)
Total Ann. Precip.: 195.7 cm (77 in.)
Elevation: 27 m (88.6 ft)
Population: 1,585,000
Ann. Temp. Range: 21 C° (37.8 F°)
Ann. Hr of Sunshine: 2131

(b) Landscape on Kitakyujukuri Island, near Nagasaki and Sasebo, Japan, in spring season.

▲Figure 10.7 Humid subtropical hot-summer climate, Asian region. [(b) JTB Photo/photolibrary.com.]

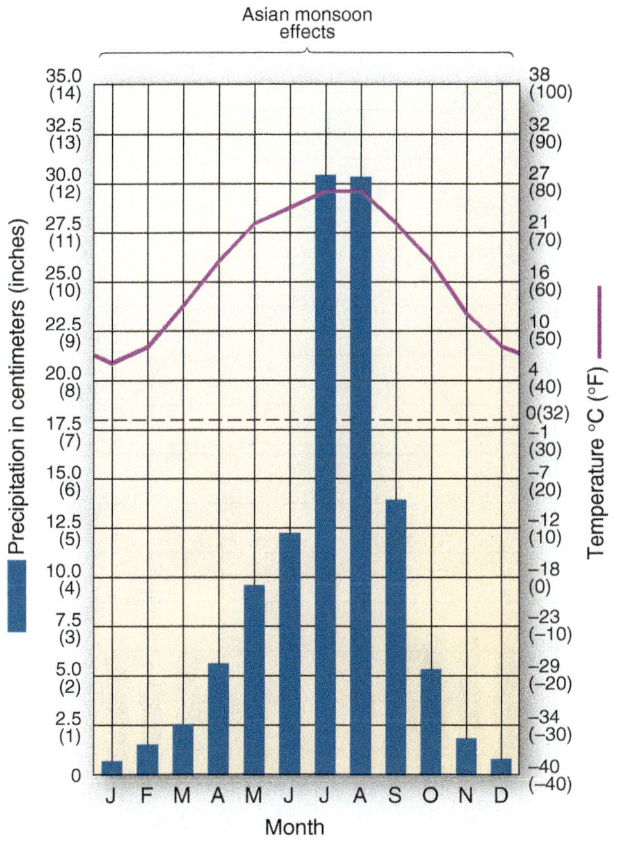

(a) Climograph for Chengdu, China. Note the summer-wet monsoonal precipitation.

Station: Chengdu, China
Lat/long: 30° 40′ N 104° 04′ E
Avg. Ann. Temp.: 17°C (62.6°F)
Total Ann. Precip.: 114.6 cm (45.1 in.)
Elevation: 498 m (1633.9 ft)
Population: 2,500,000
Ann. Temp. Range: 20 C° (36 F°)
Ann. Hr of Sunshine: 1058

(b) Agricultural fields near Chengdu, Sichuaun, China.

▲**Figure 10.8 Humid subtropical winter-dry climate.**
[(b) TAO Images Limited/Alamy.]

locations can have average monthly temperatures above freezing throughout the year. Unlike Europe, where the marine west coast regions extend quite far inland, mountains in Canada, Alaska, Chile, and Australia restrict this climate to relatively narrow coastal environs. In the Southern Hemisphere country of New Zealand, the marine west coast climate extends across the country. The climograph for Dunedin, New Zealand, demonstrates the moderate temperature patterns and the annual temperature range for this climate type (Figure 10.9).

An interesting anomaly occurs in the eastern United States. In portions of the Appalachian highlands, which is in the humid subtropical hot-summer climate region of the continent, increased elevation affects temperatures, producing a cooler summer and an isolated area of marine west coast climate. The climograph for Bluefield, West Virginia (Figure 10.10), reveals marine west coast temperature and precipitation patterns, despite its continental location in the East. Vegetation similarities between the Appalachians and the Pacific Northwest have enticed many emigrants from the East to settle in these climatically familiar environments in the Northwest.

Mediterranean Dry-Summer Climates

The *Mediterranean dry-summer* climate designation specifies that at least 70% of annual precipitation occurs during the winter months. This is in contrast to climates in most of the rest of the world, which exhibit summer-maximum precipitation. Across narrow bands of the planet during summer months, shifting cells of subtropical high pressure block moisture-bearing winds from adjacent regions. This shifting of stable, warm to hot, dry air over an area in summer and away from that area in winter creates a pronounced dry-summer and wet-winter pattern. For example, in summer the continental tropical air mass over the Sahara in Africa shifts northward over the Mediterranean region and blocks maritime air masses and cyclonic storm tracks.

Worldwide, cool offshore ocean currents (the California Current, Canary Current, Peru Current, Benguela Current, and West Australian Current) produce stability in overlying air masses along west coasts, poleward of subtropical high pressure. The world climate map in Figure 10.2 shows Mediterranean dry-summer climates along the western margins of North America, central Chile, and the southwestern tip of Africa as well as across southern Australia and the Mediterranean Basin—the climate's namesake region. Examine the offshore currents along each of these regions on the map.

Figure 10.11 compares the climographs of the Mediterranean dry-summer cities of San Francisco, California, and Sevilla (Seville), Spain. Coastal maritime effects moderate San Francisco's climate, producing a cooler summer. The transition to a hot summer occurs no more than 24–32 km (15–20 mi) inland from San Francisco.

The Mediterranean dry-summer climate brings summer water-balance deficits. Winter precipitation recharges soil moisture, but water use usually exhausts

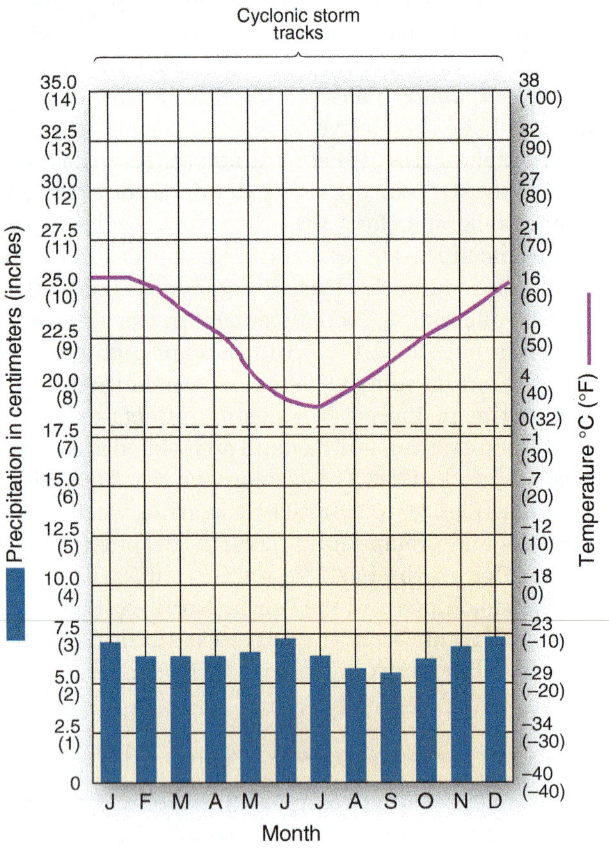

(a) Climograph for Dunedin, New Zealand.

Station: Dunedin, New Zealand
Lat/long: 45° 54′ S 170° 31′ E
Avg. Ann. Temp.: 10.2°C (50.3°F)
Total Ann. Precip.: 78.7 cm (31.0 in.)
Elevation: 1.5 m (5 ft)
Population: 120,000
Ann. Temp. Range: 14.2 C° (25.5 F°)

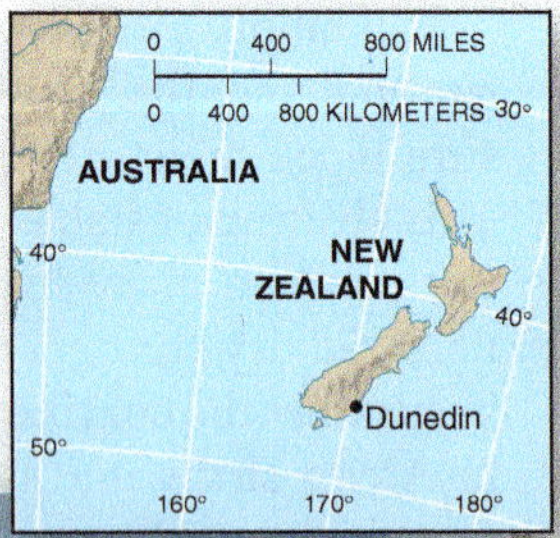

(b) Meadow, forest, and mountains on South Island, New Zealand.

▲Figure 10.9 A Southern Hemisphere marine west coast climate. [(b) Brian Enting/Photo Researchers, Inc.]

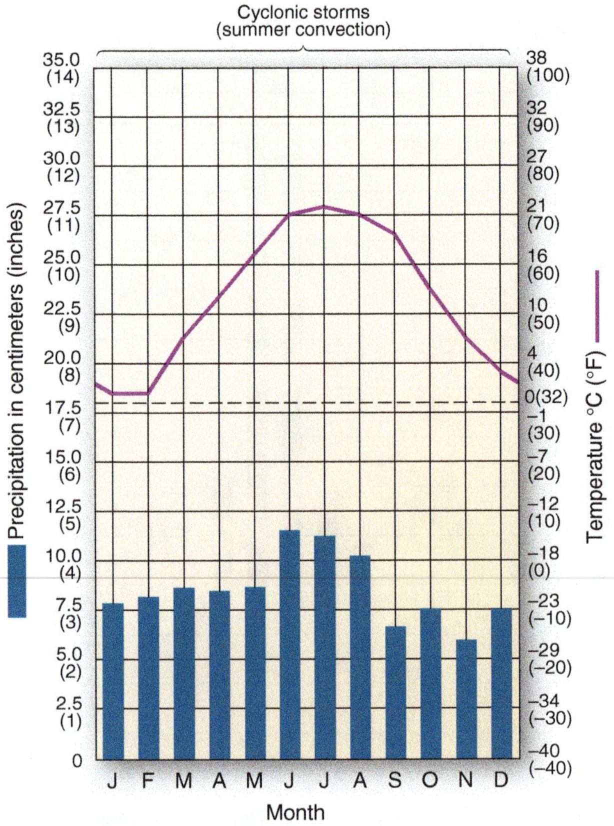

(a) Climograph for Bluefield, West Virginia.

Station: Bluefield, West Virginia
Lat/long: 37° 16′ N 81° 13′ W
Avg. Ann. Temp.: 12°C (53.6°F)
Total Ann. Precip.: 101.9 cm (40.1 in.)
Elevation: 780 m (2559 ft)
Population: 11,000
Ann. Temp. Range: 21 C° (37.8 F°)

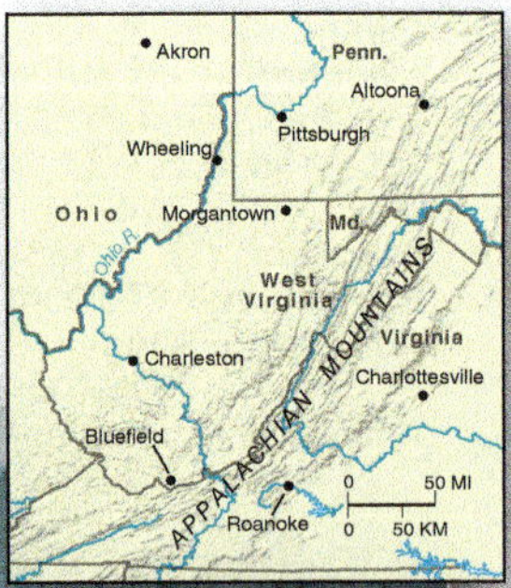

(b) Characteristic mixed forest in Appalachia in winter.

▲Figure 10.10 Marine west coast climate in the Appalachians of the eastern United States. [(b) Author.]

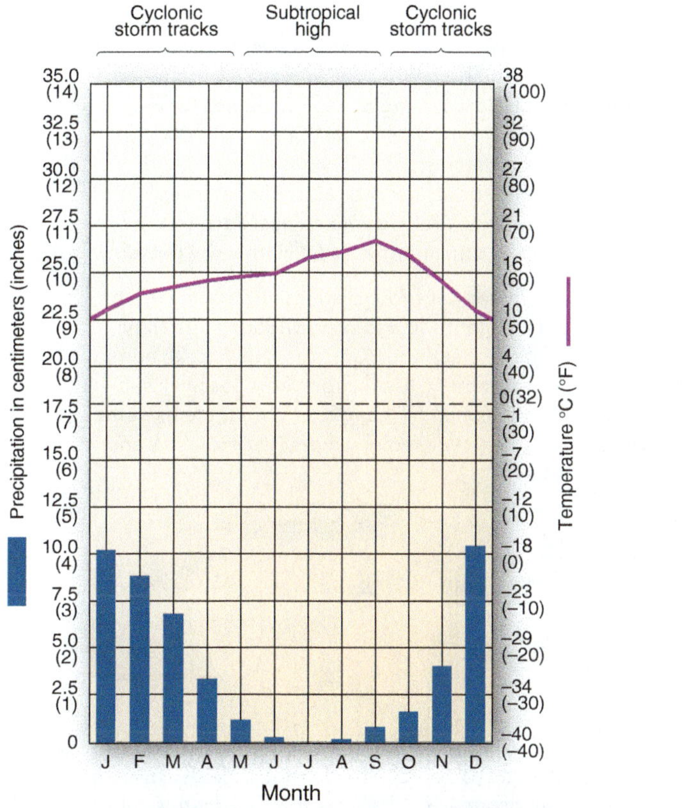

(a) Climograph for San Francisco, California, with its cooler dry summer.

Station: San Francisco, California
Lat/long: 37° 37′ N 122° 23′ W
Avg. Ann. Temp.: 14.6°C (57.2°F)
Total Ann. Precip.: 56.6 cm (22.3 in.)
Elevation: 5 m (16.4 ft)
Population: 777,000
Ann. Temp. Range: 11.4 C° (20.5 F°)
Ann. Hr of Sunshine: 2975

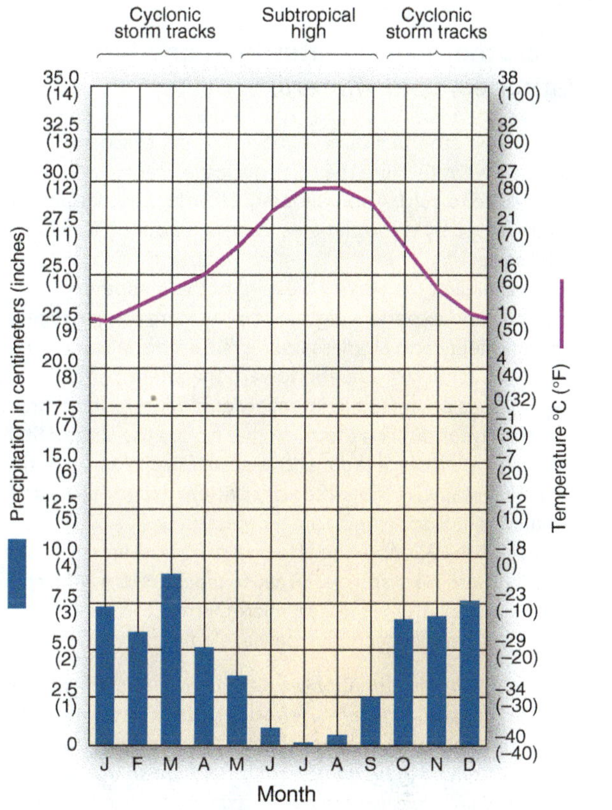

(b) Climograph for Sevilla, Spain with its hotter dry summer.

Station: Sevilla, Spain
Lat/long: 37° 22′ N 6° W
Avg. Ann. Temp.: 18°C (64.4°F)
Total Ann. Precip.: 55.9 cm (22 in.)
Elevation: 13 m (42.6 ft)
Population: 1,764,000
Ann. Temp. Range: 16 C° (28.8 F°)
Ann. Hr of Sunshine: 2862

(c) Central California Mediterranean landscape of oak savanna.

(d) Sevilla, Spain, the El Peñon Mountains in the distance.

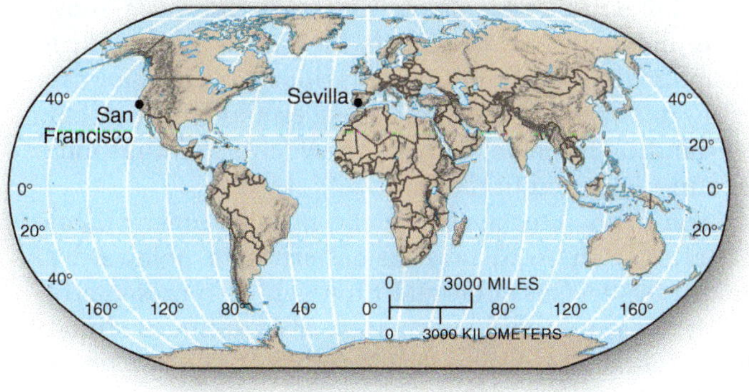

▲**Figure 10.11 Mediterranean climates, California and Spain.** [(c) Bobbé Christopherson. (d) Michael Thornton/DesignPics/Corbis.]

soil moisture by late spring. Large-scale agriculture in this climate requires irrigation, although some subtropical fruits, nuts, and vegetables are uniquely suited to these conditions. Hard-leafed, drought-resistant vegetation, known locally as *chaparral* in the western United States, is common. (Chapter 20 discusses this type of vegetation in other parts of the world.)

Microthermal Climates (mid and high latitudes, cold winters)

Humid microthermal climates have a winter season with some summer warmth. Here the term *microthermal* means cool temperate to cold. Approximately 21% of Earth's land surface is influenced by these climates, equaling about 7% of Earth's total surface.

These climates occur poleward of the mesothermal climates and experience great temperature ranges related to continentality and air mass conflicts. Temperatures decrease with increasing latitude and toward the interior of continental landmasses and result in intensely cold winters. In contrast to moist-all-year regions (northern tier across the United States; the lower part of Canada; and eastern Europe through the Ural Mountains) is the winter-dry pattern associated with the Asian dry monsoon and cold air masses.

In Figure 10.2, note the absence of microthermal climates in the Southern Hemisphere. Because the Southern Hemisphere lacks substantial landmasses, microthermal climates develop only in highlands. Important causal elements include:

- Increasing seasonality (daylength and Sun altitude) and greater temperature ranges (daily and annually);
- Latitudinal effects on insolation and temperature: summers become cool moving northward, with winters becoming cold to very cold;
- Upper-air westerly winds and undulating Rossby waves, which bring warmer air northward and colder air southward for cyclonic activity; and convectional thunderstorms from maritime tropical air masses in summer;
- Continental interiors serving as source regions for intense continental polar air masses that dominate winter, blocking cyclonic storms;
- Continental high pressure and related air masses, increasing from the Ural Mountains eastward to the Pacific Ocean, producing the Asian winter-dry pattern.

Microthermal climates have four distinct regimes based on increasing cold with latitude and precipitation variability: humid continental hot-summer (Chicago, New York); humid continental mild-summer (Duluth, Toronto, Moscow); subarctic cool-summer (Churchill); and the formidable extremes of frigid subarctic with very cold winters (Verkhoyansk and northern Siberia).

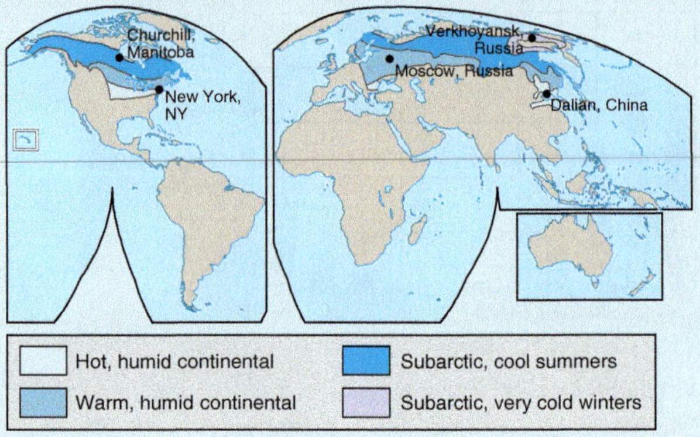

Humid Continental Hot-Summer Climates

Humid continental hot-summer climates have the warmest summer temperatures of the microthermal category. In the summer, maritime tropical air masses influence precipitation, which may be consistent throughout the year or have a distinct winter-dry period. In North America, frequent weather activity is possible between conflicting air masses—maritime tropical and continental polar—especially in winter. New York City and Dalian, China (Figure 10.12), exemplify the two types of hot-summer microthermal climates—*moist-all-year* and *winter-dry*. The Dalian climograph demonstrates a dry-winter tendency caused by the intruding cold continental airflow that forces dry monsoon conditions.

Before European settlement, forests covered the humid continental hot-summer climatic region of the United States as far west as the Indiana–Illinois border. Beyond that approximate line, tall-grass prairies extended westward to about the 98th meridian (98° W, in central Kansas) and the approximate location of the 51-cm (20-in.) *isohyet* (a line of equal precipitation). Farther west, the presence of shortgrass prairies reflected lower precipitation receipts.

Deep sod made farming difficult for the first settlers of the American prairies; however, domesticated crops such as wheat and barley soon replaced native grasses. Various inventions (barbed wire, the self-scouring steel plow, well-drilling techniques, windmills, railroads, and the six-shooter) aided the expansion of farming and ranching into the region. In the United States today, the humid continental hot-summer region is the location of corn, soybean, hog, feed crop, dairy, and cattle production (Figure 10.13).

Humid Continental Mild-Summer Climates

Located farther toward the poles, *humid continental mild-summer* climates are slightly cooler. Figure 10.14 presents a climograph for Moscow, Russia, which is at 55° N, or about the same latitude as the southern shore of Hudson Bay, in Canada. In the United States, a characteristic city having this mild-summer climate is Duluth, Minnesota.

Agricultural activity remains important in the cooler microthermal climates and includes dairy, poultry, flax, sunflower, sugar beet, wheat, and potato production. Frost-free periods range from fewer than 90 days in the northern portions of these regions to as many as 225 days in the southern parts. Overall, precipitation is less than in the hot-summer regions to the south; however, snowfall is notably heavier, and its melting is important to soil-moisture recharge. Among various strategies for capturing this snow are the use of fences and of tall stubble left standing in fields to create snowdrifts and thus more moisture retention in the soil.

The dry-winter aspect of the mild-summer climate occurs only in Asia, in a far-eastern area poleward of the winter-dry mesothermal climates. A representative of this type of humid continental mild-summer climate along Russia's east coast is Vladivostok, usually one of only two ice-free ports in that country.

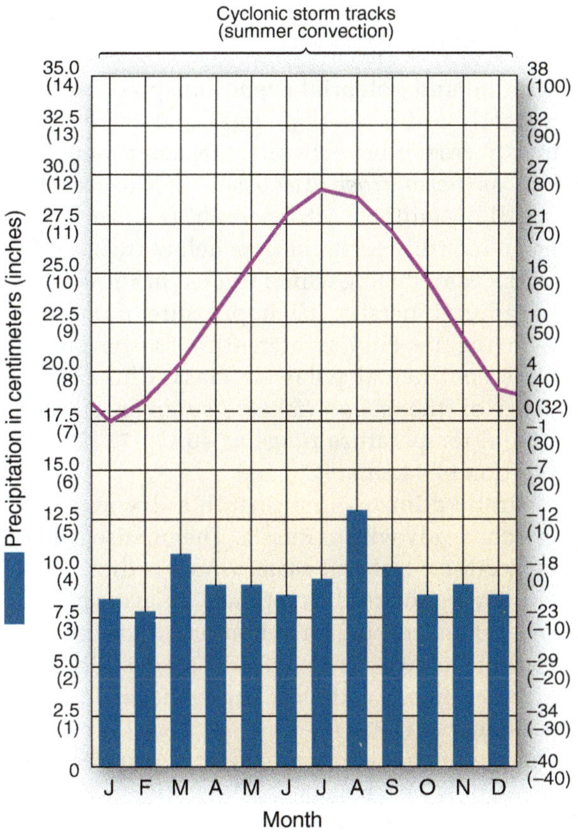

(a) Climograph for New York City (humid continental hot-summer, moist all year).

Station: New York, New York
Lat/long: 40° 46′ N 74° 01′ W
Avg. Ann. Temp.: 13°C (55.4°F)
Total Ann. Precip.: 112.3 cm (44.2 in.)
Elevation: 16 m (52.5 ft)
Population: 8,092,000
Ann. Temp. Range: 24 C° (43.2 F°)
Ann. Hr of Sunshine: 2564

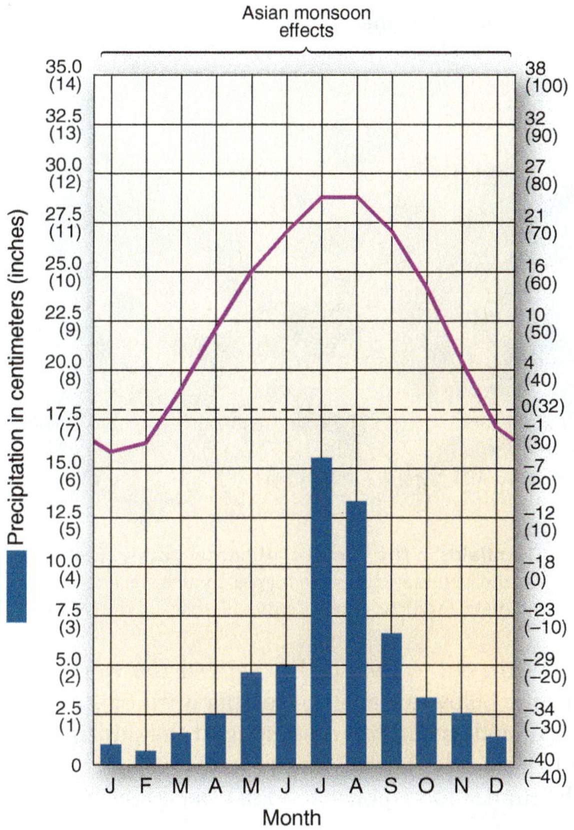

(b) Climograph for Dalian, China (humid continental hot-summer, winter-dry).

Station: Dalian, China
Lat/long: 38° 54′ N 121° 54′ E
Avg. Ann. Temp.: 10°C (50°F)
Total Ann. Precip.: 57.8 cm (22.8 in.)
Elevation: 96 m (314.9 ft)
Population: 5,550,000
Ann. Temp. Range: 29 C° (52.2 F°)
Ann. Hr of Sunshine: 2762

(c) Belvedere Castle, built in 1872, in New York's Central Park; location of weather station from 1919 to 1960.

(d) Dalian, China, cityscape and park in summer.

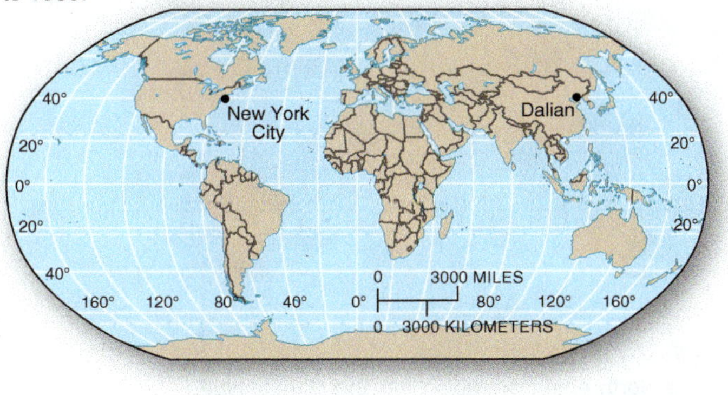

▲**Figure 10.12 Humid continental hot-summer climates, New York and China.** [(c) Bobbé Christopherson. (d) Paul Louis Collection.]

Subarctic Climates

Farther poleward, seasonal change becomes greater. The short growing season is more intense during long summer days. The *subarctic* climates include vast stretches of Alaska, Canada, and northern Scandinavia, with their cool summers, and Siberian Russia, with its very cold winters (Figure 10.2).

Areas that receive 25 cm (10 in.) or more of precipitation a year on the northern continental margins and are covered by the so-called snow forests of fir, spruce, larch, and birch are the *boreal forests* of Canada and the *taiga* of Russia. These forests are in transition to the more open northern woodlands and to the tundra region of the far

▲**Figure 10.13 Cornfields in the humid continental prairie.** East of Minneapolis near the boundary between humid continental hot- and mild-summer climatic regions. [Bobbé Christopherson.]

north. Forests thin out to the north wherever the warmest month drops below an average temperature of 10°C (50°F). Climate models and forecasts suggest that, during the decades ahead, the boreal forests will shift northward into the tundra in response to higher temperatures.

Precipitation and potential evapotranspiration both are low, so soils are generally moist and either partially or totally frozen beneath the surface, a phenomenon known as *permafrost* (discussed in Chapter 17). The Churchill, Manitoba, climograph (Figure 10.15) shows average monthly temperatures below freezing for 7 months of the year, during which time light snow cover and frozen ground persist. High pressure dominates Churchill during its cold winter—this is the source region for the continental polar air mass. Churchill is representative of the *subarctic cool-summer* climate, with an annual temperature range of 40 C° (72 F°) and low precipitation of 44.3 cm (17.4 in.).

The subarctic climates that feature a dry and very cold winter occur only within Russia. The intense cold of Siberia and north-central and eastern Asia is difficult to comprehend, for these areas experience an average temperature lower than freezing for 7 months, and minimum temperatures of below −68°C (−90°F), as described in Chapter 5. Yet summer-maximum temperatures in these same areas can exceed 37°C (98°F).

An example of this extreme *subarctic climate with very cold winters* is Verkhoyansk, Siberia (Figure 10.16).

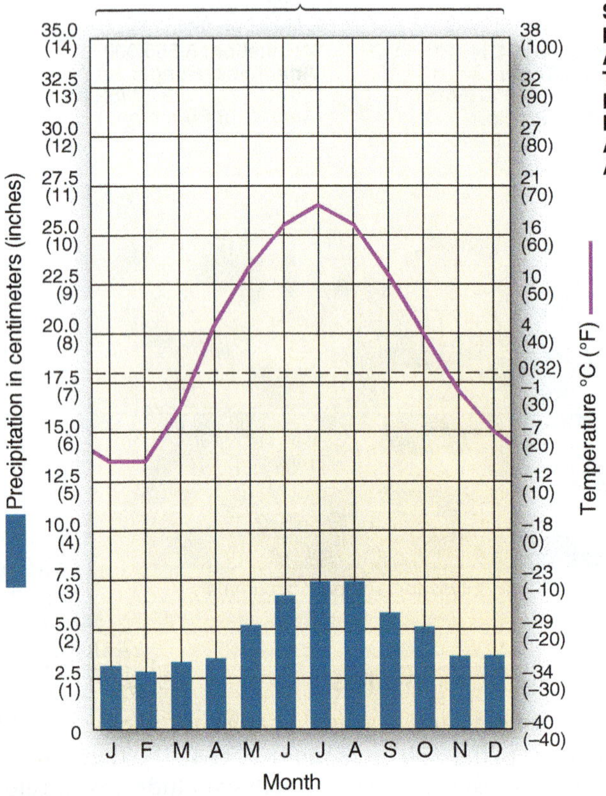

(a) Climograph for Moscow, Russia.

Station: Moscow, Russia
Lat/long: 55° 45′ N 37° 34′ E
Avg. Ann. Temp.: 4°C (39.2°F)
Total Ann. Precip.: 57.5 cm (22.6 in.)
Elevation: 156 m (511.8 ft)
Population: 11,460,000
Ann. Temp. Range: 29 C° (52.2 F°)
Ann. Hr of Sunshine: 1597

(b) Landscape between Moscow and St. Petersburg along the Volga River.

(c) Winter scene of mixed forest near Brunswick, Maine, in the North American region with this climate type.

▲**Figure 10.14 Humid continental mild-summer climate.** [(b) Dave G Houser/Corbis. (c) Bobbé Christopherson.]

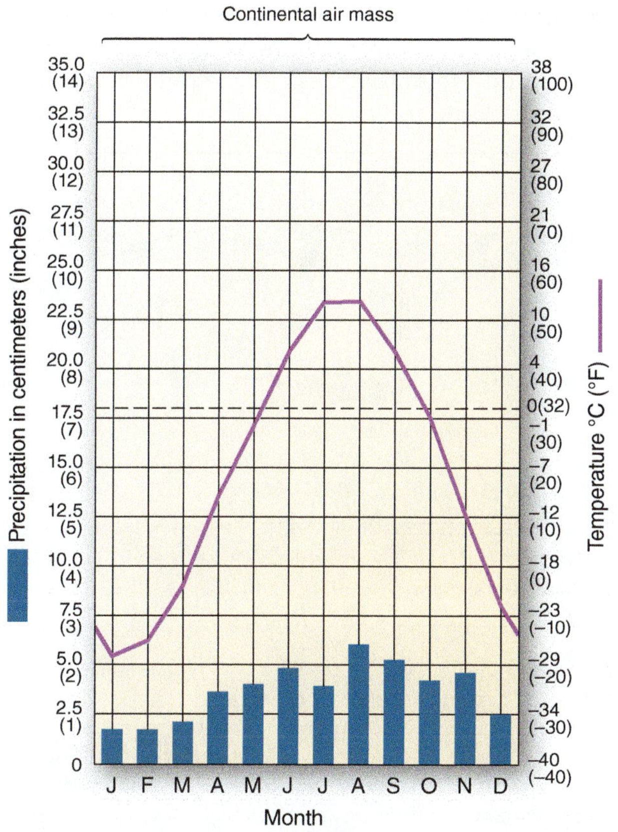

(a) Climograph for Churchill, Manitoba.

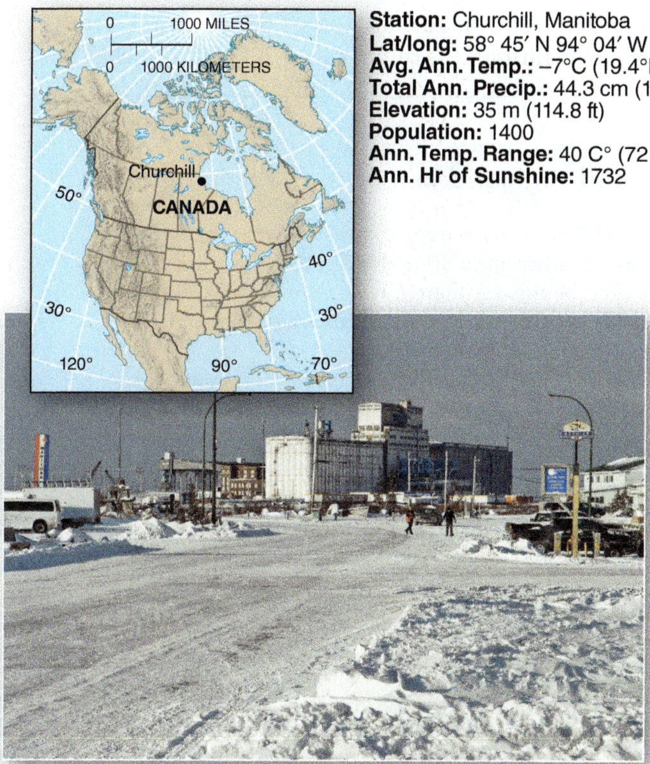

Station: Churchill, Manitoba
Lat/long: 58° 45′ N 94° 04′ W
Avg. Ann. Temp.: −7°C (19.4°F)
Total Ann. Precip.: 44.3 cm (17.4 in.)
Elevation: 35 m (114.8 ft)
Population: 1400
Ann. Temp. Range: 40 C° (72 F°)
Ann. Hr of Sunshine: 1732

(b) This and other port facilities on Hudson Bay may expand with renewed interest in mineral and petroleum reserves in subarctic regions.

▲**Figure 10.15 Subarctic cool-summer climate.** [(b) Bobbé Christopherson.]

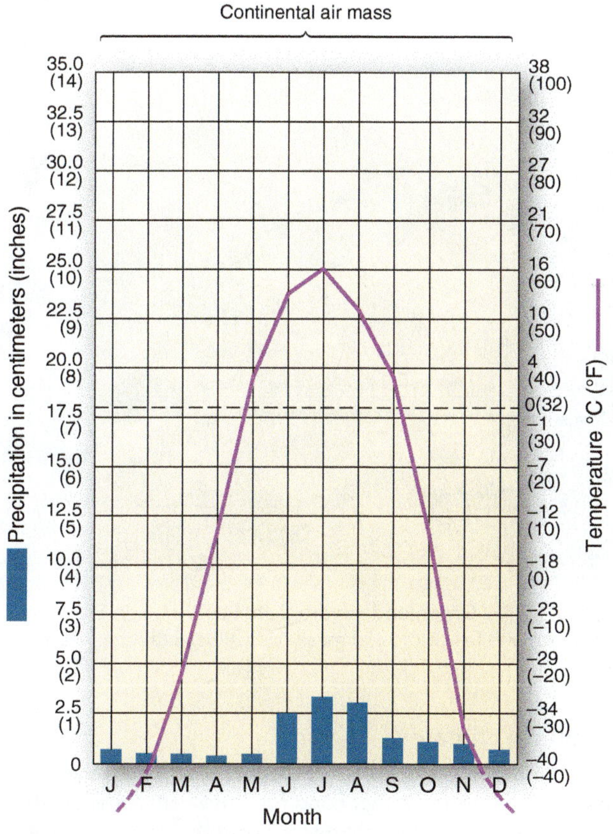

(a) Climograph for Verkhoyansk, Russia.

Station: Verkhoyansk, Russia
Lat/long: 67° 35′ N 133° 27′ E
Avg. Ann. Temp.: −15°C (5°F)
Total Ann. Precip.: 15.5 cm (6.1 in.)
Elevation: 137 m (449.5 ft)
Population: 1500
Ann. Temp. Range: 63 C° (113.4 F°)

(b) A summer scene shows one of many ponds created by thawing permafrost.

▲**Figure 10.16 Extreme subarctic cold-winter climate.** [(b) Dean Conger/Corbis.]

For 4 months of the year, average temperatures fall below −34°C (−30°F). Verkhoyansk has probably the world's greatest annual temperature range from winter to summer: a remarkable 63 C° (113.4 F°). In Verkhoyansk, metals and plastics are brittle in winter; people install triple-thick windowpanes to withstand temperatures that render straight antifreeze a solid.

Polar and Highland Climates

The polar climates have no true summer like that in lower latitudes. Overlying the South Pole is the Antarctic continent, surrounded by the Southern Ocean, whereas the North Pole region is covered by the Arctic Ocean, surrounded by the continents of North America and Eurasia. Poleward of the Arctic and Antarctic Circles, daylength increases in summer until daylight becomes continuous, yet average monthly temperatures never rise above 10°C (50°F). These temperature conditions do not allow tree growth. (Please review the polar region temperature maps for January and July in Figures 5.14 and 5.15.) Important causal elements of polar climates include:

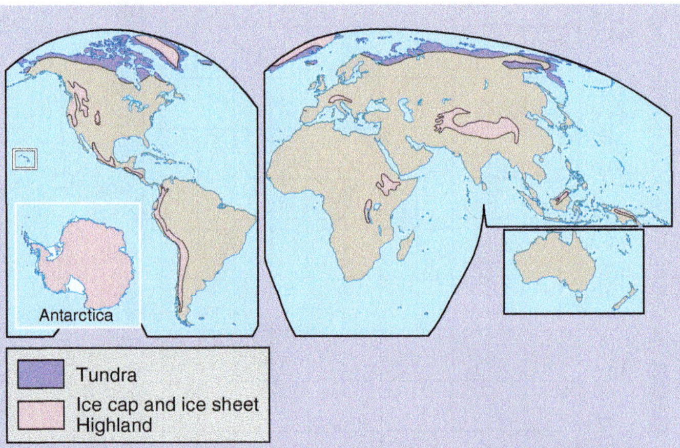

- Low Sun altitude even during the long summer days, which is the principal climatic factor;
- Extremes of daylength between winter and summer, which determine the amount of insolation received;
- Extremely low humidity, producing low precipitation amounts—these regions are Earth's frozen deserts;
- Surface albedo impacts, as light-colored surfaces of ice and snow reflect substantial energy away from the ground, thus reducing net radiation.

Polar climates have three regimes: *tundra* (at high latitude or high elevation); *ice-cap* and *ice-sheet* (perpetually frozen); and *polar marine* (with an oceanic association and slight moderation of extreme cold).

Also in this climate category are *highland* climates, in which tundra and polar conditions occur at non-polar latitudes because of the effects of elevation. Glaciers on tropical mountain summits attest to the cooling effects of altitude. Highland climates on the map follow the pattern of Earth's mountain ranges.

Tundra Climates

The term *tundra* refers to the characteristic vegetation of high latitudes and high elevations, where plant growth is restricted by cold temperatures and a short growing season. In *tundra* climates, land is under some snow cover for 8–10 months, with the warmest month above 0°C (32°F), yet never warming above 10°C (50°F). These climates occur only in the Northern Hemisphere, except for elevated mountain locations in the Southern Hemisphere and a portion of the Antarctic Peninsula. Because of its elevation, the summit of Mount Washington in New Hampshire (1914 m, or 6280 ft) statistically qualifies as a highland tundra climate despite its limited areal extent. In contrast, approximately 410,500 km^2 (158,475 mi^2) of Greenland make up an area of tundra and rock about the size of California.

In spring when the snow melts, numerous plants appear—stunted sedges, mosses, dwarf shrubs, flowering plants, and lichens—and persist through the short summer (Figure 10.17). Some of the dwarf willows (7.5 cm, or

▲**Figure 10.17 Greenland tundra.** Late September in East Greenland, with fall colors and musk oxen. [Bobbé Christopherson.]

GEOreport 10.2 Boundary considerations and shifting climates

The boundary between mesothermal and microthermal climates is sometimes placed along the isotherm where the coldest month is −3°C (26.6°F) or lower. That might be a suitable criterion for Europe, but for conditions in North America, the 0°C (32°F) isotherm is considered more appropriate. The distance between the 0°C and −3°C isotherms is about the width of the state of Ohio. In Figure 10.2, you see the 0°C boundary used.

Scientists estimate that the poleward shift of climate regions will be 150- to 550-km (90- to 350-mi) in the midlatitudes during this century. As you examine North America in Figure 10.2, use the graphic scale to get an idea of the magnitude of these potential shifts.

(a) In the Antarctic Sound, between Bransfield Strait and the Weddell Sea, mountains rise from the mist behind a flank of tabular icebergs.

(b) Glaciers move toward the ocean in West Greenland, with the ice sheet on the horizon.

▲**Figure 10.18 Earth's ice sheets—Antarctica and Greenland.** [Bobbé Christopherson.]

3 in., tall) can exceed 300 years in age. Much of the area experiences permafrost and ground ice conditions; these are Earth's periglacial regions, discussed in Chapter 17.

Ice-Cap and Ice-Sheet Climates

An *ice sheet* is a continuous layer of ice covering an extensive continental region. Earth's two ice sheets cover the Antarctic continent and most of the island of Greenland (Figure 10.18). An *ice cap* is smaller in extent, roughly less than 50,000 km^2 (19,300 mi^2), but completely buries the landscape like an ice sheet. The Vatnajökull ice cap in southeastern Iceland is an example (see the NASA image in Figure 17.5).

Most of Antarctica and central Greenland fall within the *ice-cap and ice-sheet* climate category, as does the North Pole, with all months averaging below freezing (the area of the North Pole is actually a sea covered by ice, rather than a continental landmass). These regions are dominated by dry, frigid air masses, with vast expanses that never warm above freezing. In fact, winter minimum temperatures in central Antarctica (July) frequently drop below the temperature of solid carbon dioxide, or "dry ice" (–78°C, or –109°F). Antarctica is constantly snow-covered, but receives less than 8 cm (3 in.) of precipitation each year. However, Antarctic ice has accumulated to several kilometers deep and is the largest repository of freshwater on Earth.

Polar Marine Climates

Polar marine climates are more moderate than other polar climates in winter, with no month below –7°C (20°F), yet overall they are not as warm as tundra climates. Because of marine influences, annual temperature ranges are low. This climate exists along the Bering Sea, on the southern tip of Greenland, and in northern Iceland and northern Norway; in the Southern Hemisphere, it generally occurs over oceans between 50° S and 60° S. Macquarie Island at 54° S in the Southern Ocean, south of New Zealand, is polar marine.

South Georgia Island, made famous as the place where Ernest Shackleton sought rescue help for himself and his men in 1916 after their failed Antarctic expedition, exemplifies a polar marine climate (Figure 10.19). Although the island is in the Southern Ocean and part of Antarctica, the annual temperature range is only

▲**Figure 10.19 South Georgia Island, a polar marine climate.** This abandoned whaling station at Grytviken, South Georgia, processed more than 50,000 whales between 1904 and 1964, about a third of all whales processed on the island during this period. The whales of the Southern Ocean were driven almost to extinction. [Bobbé Christopherson.]

GEOreport 10.3 Tundra climates respond to warming

Global warming is bringing dramatic changes to the tundra climate regions, where temperatures in the Arctic are warming at a rate twice that of the global average increase. In parts of Canada and Alaska, near-record temperatures as much as 5 to 10 C° (9 to 18 F°) above average are a regular occurrence. As organic peat deposits in the tundra thaw, vast stores of carbon and methane are released to the atmosphere, further adding to the greenhouse gas problem (more discussion is in Chapters 10B, 17, and 18).

8.5 C° (15.3 F°) between the seasons (the averages are 7°C in January and –1.5°C in July; 44.6°F and 29.3°F), with 7 months averaging slightly above freezing. Ocean temperatures, ranging between 0°C and 4°C (32°F and 39°F), help to moderate the climate so that temperatures are warmer than expected at its 54°-S-latitude location. Average precipitation is 150 cm (59 in.), and it can snow during any month.

Dry Climates (permanent moisture deficits)

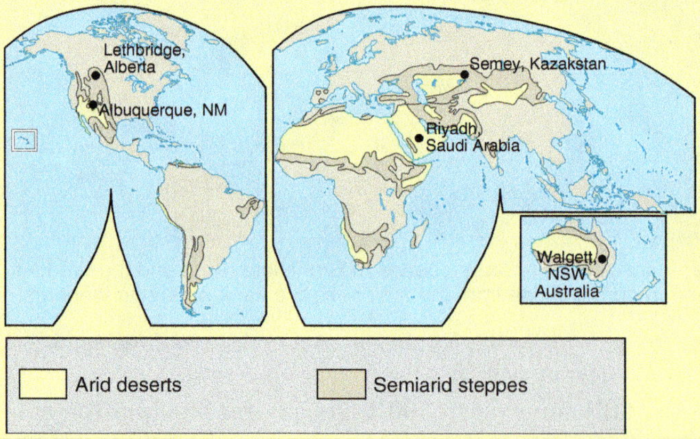

For understanding the dry climates, we consider moisture efficiency (both timing and quantity of moisture) along with temperature. These dry regions occupy more than 35% of Earth's land area and are by far the most extensive climate over land. Sparse vegetation leaves the landscape bare; water demand exceeds the precipitation water supply throughout, creating permanent water deficits. The extent of these deficits distinguishes two types of dry climatic regions: *arid deserts*, where the precipitation supply is roughly less than one-half of the natural moisture demand; and *semiarid steppes*, where the precipitation supply is roughly more than one-half of natural moisture demand. (Review pressure systems in Chapter 6 and temperature controls, including the highest recorded temperatures, in Chapter 5. Desert environments are discussed in Chapter 20.)

Important causal elements in these dry lands include:

- The dominant presence of dry, subsiding air in subtropical high-pressure systems;
- Location in the rain shadow (or leeward side) of mountains, where dry air subsides after moisture is intercepted on the windward slopes;
- Location in continental interiors, particularly central Asia, which are far from moisture-bearing air masses;
- Location along western continental margins with cool, stabilizing ocean currents;
- Shifting subtropical high-pressure systems, which produce semiarid steppes around the periphery of arid deserts.

Dry climates fall into four distinct regimes, according to latitude and to the amount of moisture deficit: arid climates include *tropical, subtropical hot desert* and *midlatitude cold desert* regimes; semiarid climates include *tropical, subtropical hot steppe* and *midlatitude cold steppe* regimes.

Characteristics of Dry Climates

Dry climates are subdivided into deserts and steppes according to moisture—deserts have greater moisture deficits than do steppes, but both have permanent water shortages. **Steppe** is a regional term referring to the vast semiarid grassland biome of Eastern Europe and Asia (the equivalent biome in North America is shortgrass prairie, and in Africa, the savanna; see Chapter 20). In this chapter, we use steppe in a climatic context; a *steppe climate* is considered too dry to support forest, but too moist to be a desert.

The timing of precipitation (winter rains with dry summers, summer rains with dry winters, or even distribution throughout the year) affects moisture availability in these dry lands. Winter rains are most effective because they fall at a time of lower moisture demand. Relative to temperature, the lower-latitude deserts and steppes tend to be hotter with less seasonal change than the midlatitude deserts and steppes, where mean annual temperatures are below 18°C (64.4°F) and freezing winter temperatures are possible.

Earth's dry climates cover broad regions between 15° and 30° latitude in the Northern and Southern Hemispheres, where subtropical high-pressure cells predominate, with subsiding, stable air and low relative humidity. Under generally cloudless skies, these subtropical deserts extend to western continental margins, where cool, stabilizing ocean currents operate offshore and summer advection fog forms. The Atacama Desert of Chile, the Namib Desert of Namibia, the Western Sahara of Morocco, and the Australian Desert each lie adjacent to such a coastline (Figure 10.20).

However, dry regions also extend into higher latitudes. Deserts and steppes occur as a result of orographic lifting over mountain ranges, which intercepts moisture-bearing weather systems to create rain shadows, especially in North and South America (Figure 10.20). The isolated interior of Asia, far distant from any moisture-bearing air masses, also falls within the dry climate classification.

The world's largest desert, as defined by moisture criteria, is the Antarctic region. The largest non-polar deserts in surface area are the Sahara (3.5 million mi²), the Arabian, the Gobi in China and Mongolia, the Patagonian in Argentina, the Great Victoria in Australia, the Kalahari in South Africa, and the Great Basin of the western United States.

▼**Figure 10.20 Important deserts and steppes.** Worldwide distribution of arid and semiarid climates, with the world's major deserts and steppes labeled. [(a) Bobbé Christopherson. (b) Jacques Jangoux/Photo Researchers, Inc.]

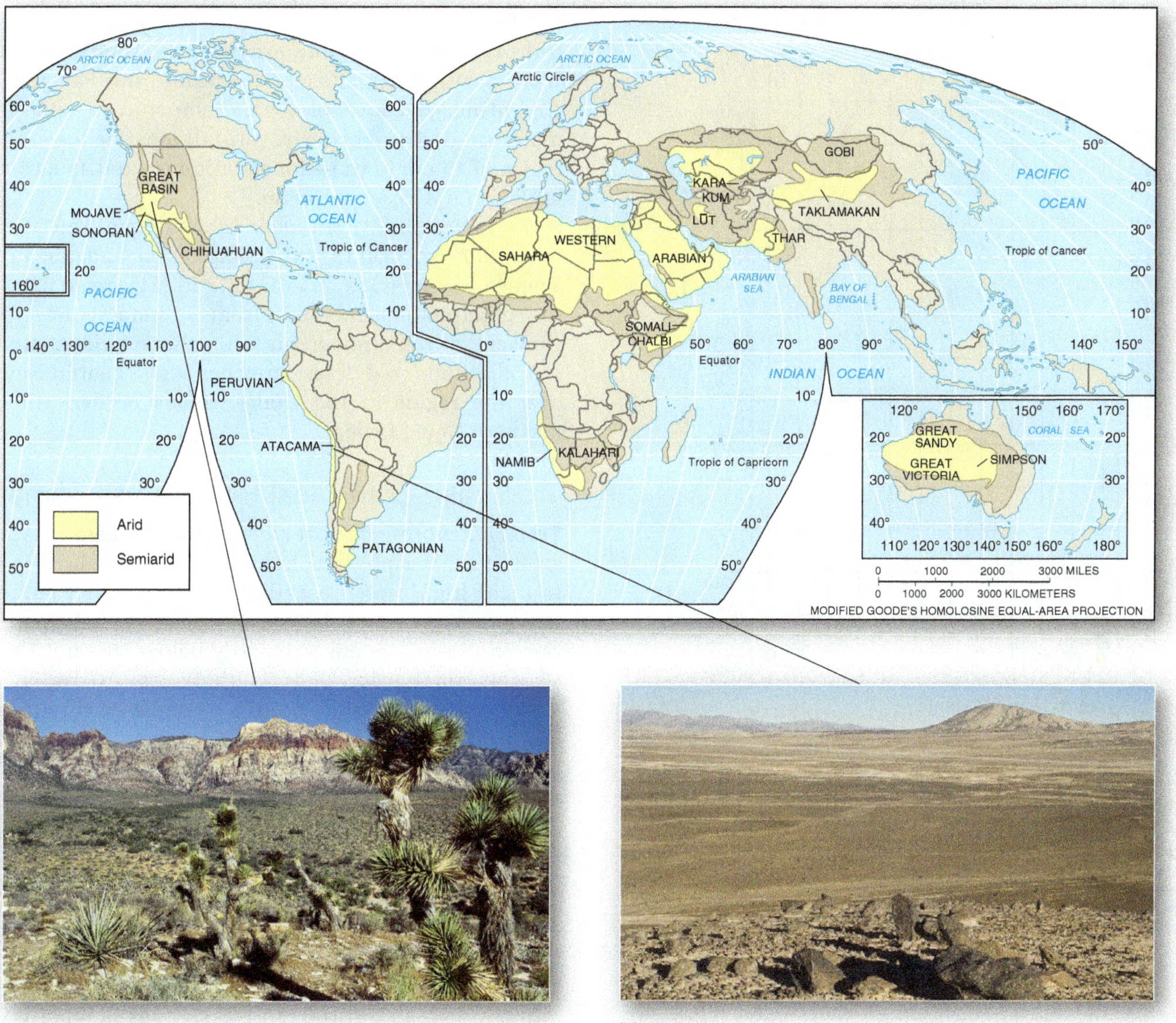

(a) Mojave Desert

(b) Atacama

Tropical, Subtropical Hot Desert Climates

Tropical, subtropical hot desert climates are Earth's true tropical and subtropical deserts and feature annual average temperatures above 18°C (64.4°F). They generally are found on the western sides of continents, although Egypt, Somalia, and Saudi Arabia also fall within this classification. Rainfall is from local summer convectional showers. Some regions receive almost no rainfall, whereas others may receive up to 35 cm (14 in.) of precipitation a year. A representative subtropical hot desert city is Riyadh, Saudi Arabia (Figure 10.21).

Along the Sahara Desert's southern margin in Africa is a drought-prone region called the Sahel, where human populations suffer great hardship as desert conditions gradually expand over their homelands. *Geosystems Now* in Chapter 18 examines the process of desertification (expanding desert conditions), an ongoing problem in many dry regions of the world.

In California, Death Valley holds the record for highest temperature ever recorded—57°C (134°F) during July 1913. Extremely hot summer temperatures occur in other hot desert climates, such as around Baghdad, Iraq, where air temperatures regularly reach 50°C (122°F) and higher in the city. Baghdad records zero precipitation from May to September, as it is dominated by an intense subtropical high-pressure system. In January, averages for Death Valley (11°C; 52°F) and Baghdad (9.4°C; 49°F) are comparable. Death Valley is drier, with 5.9 cm of precipitation, compared to 14 cm in Baghdad (2.33 in.; 5.5 in.), both low amounts.

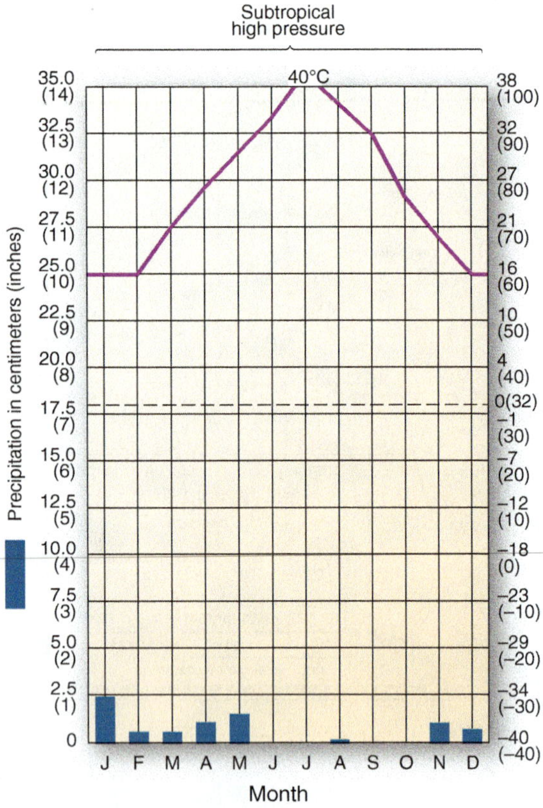

(a) Climograph for Riyadh, Saudi Arabia.

Station: Riyadh, Saudi Arabia
Lat/long: 24° 42′ N 46° 43′ E
Avg. Ann. Temp.: 26°C (78.8°F)
Total Ann. Precip.: 8.2 cm (3.2 in.)
Elevation: 609 m (1998 ft)
Population: 5,024,000
Ann. Temp. Range: 24 C° (43.2 F°)

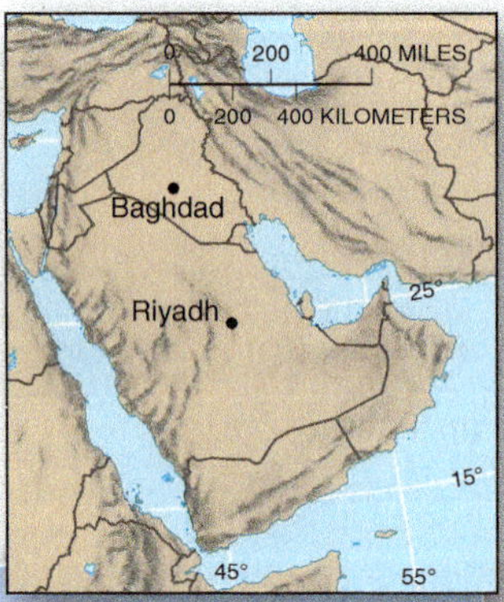

(b) The Arabian desert landscape of Red Sands near Riyadh.

▲Figure 10.21 Tropical, subtropical hot desert climate.
[(b) Andreas Wolf/agefotostock.]

Midlatitude Cold Desert Climates

Midlatitude cold desert climates cover only a small area: the countries along the southern border of Russia, the Taklamakan Desert, and Mongolia in Asia; the central third of Nevada and areas of the American Southwest, particularly at high elevations; and Patagonia in Argentina. Because of lower temperature and lower moisture-demand criteria, rainfall must be low—in the realm of 15 cm (6 in.)—for a station to qualify as a midlatitude cold desert climate.

A representative station is Albuquerque, New Mexico, with 20.7 cm (8.1 in.) of precipitation and an annual average temperature of 14°C (57.2°F) (Figure 10.22). Note the precipitation increase from summer convectional showers on the climograph. This characteristic expanse of midlatitude cold desert stretches across central Nevada, over the region of the Utah–Arizona border, and into northern New Mexico.

Tropical, Subtropical Hot Steppe Climates

Tropical, subtropical hot steppe climates generally exist around the periphery of hot deserts, where shifting subtropical high-pressure cells create a distinct summer-dry and winter-wet pattern. Average annual precipitation in these climates is usually below 60 cm (23.6 in.). Walgett, in interior New South Wales, Australia, provides a Southern Hemisphere example of this climate (Figure 10.23). This climate is seen around the Sahara's periphery and in the Iran, Afghanistan, Turkmenistan, and Kazakhstan region.

Midlatitude Cold Steppe Climates

The *midlatitude cold steppe* climates occur poleward of about 30° latitude and of the *midlatitude cold desert* climates. Such midlatitude steppes are not generally found in the Southern Hemisphere. As with other dry climate regions, rainfall in the steppes is widely variable and undependable, ranging from 20 to 40 cm (7.9 to 15.7 in.). Not all rainfall is convectional, for cyclonic storm tracks penetrate the continents; however, most storms produce little precipitation.

Figure 10.24, page 282, presents a comparison between Asian and North American midlatitude cold steppe climates. Consider Semey (Semipalatinsk) in Kazakhstan, with its greater temperature range and more evenly distributed precipitation, and Lethbridge, Alberta, with its lesser temperature range and summer-maximum convectional precipitation.

Climate Regions and Climate Change

The boundaries of climate regions are changing worldwide. The current expansion of tropical climates to higher latitudes means that subtropical high-pressure

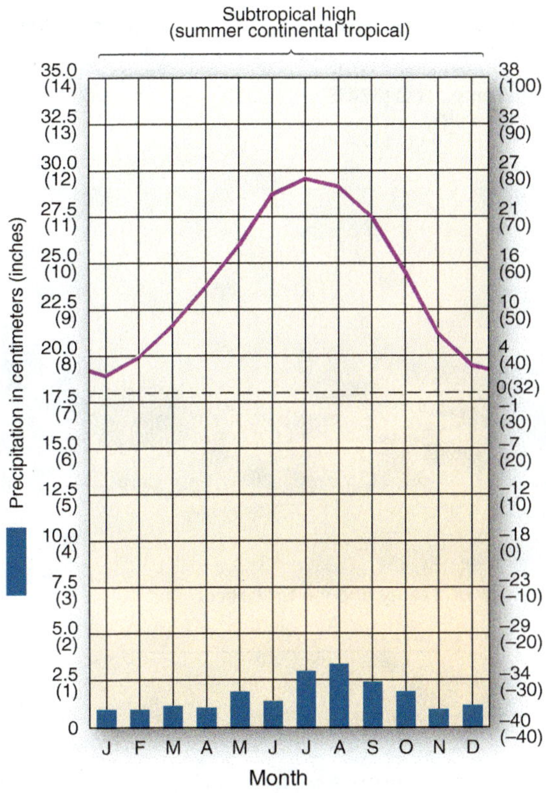

(a) Climograph for Albuquerque, New Mexico.

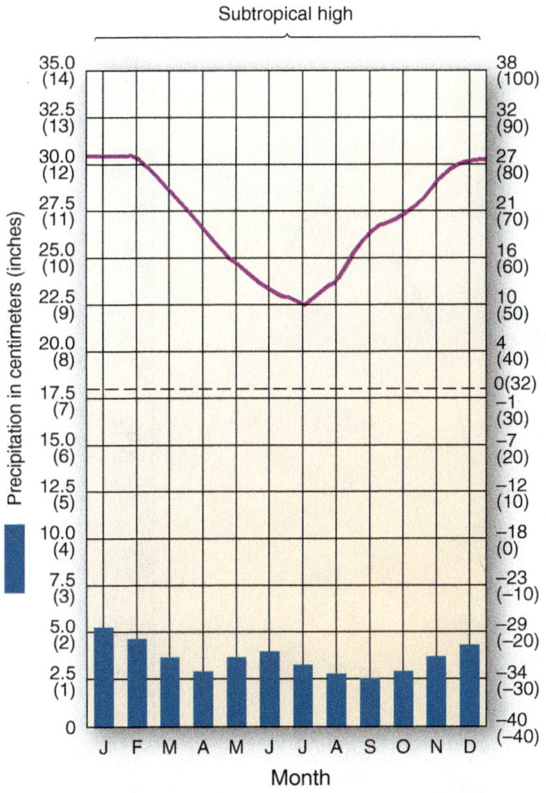

(a) Climograph for Walgett, New South Wales, Australia

Station: Albuquerque, New Mexico
Lat/long: 35° 03′ N 106° 37′ W
Avg. Ann. Temp.: 14°C (57.2°F)
Total Ann. Precip.: 20.7 cm (8.1 in.)
Elevation: 1620 m (5315 ft)
Population: 522,000
Ann. Temp. Range: 24 C° (43.2 F°)
Ann. Hr of Sunshine: 3420

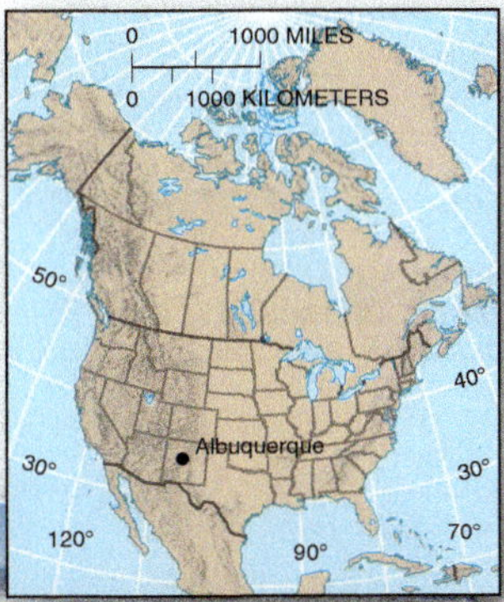

(b) A winter scene near the New Mexico – Arizona border.

▲Figure 10.22 Midlatitude cold desert climate.
[(b) Bobbé Christopherson.]

Station: Walgett, New South Wales, Australia
Lat/long: 30° S 148° 07′ E
Avg. Ann. Temp.: 20°C (68°F)
Total Ann. Precip.: 45.0 cm (17.7 in.)
Elevation: 133 m (436 ft)
Population: 8200
Ann. Temp. Range: 17 C° (31 F°)

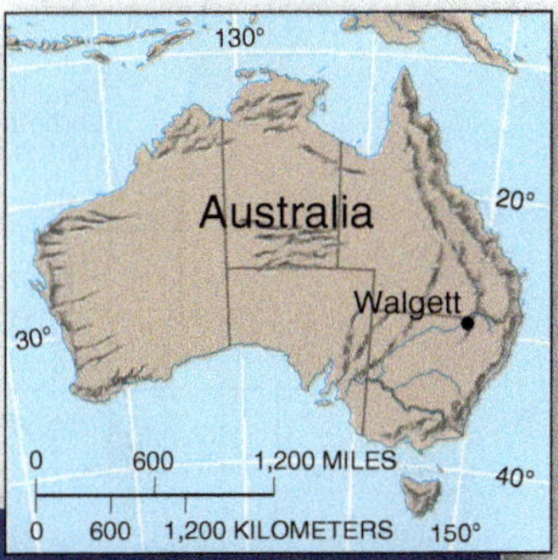

(b) Vast plains characteristic of north-central New South Wales.

▲Figure 10.23 Tropical, subtropical hot steppe climate.
[(b) Prisma/SuperStock.]

areas and dry conditions are also moving to higher latitudes. In addition, warming temperatures are making these areas more prone to drought. At the same time, storm systems are being pushed further into the midlatitudes. In many cases, the evidence for shifting climate regions comes from changes in associated ecosystems; for example, the growth of trees in tundra climate regions, or the expanding range of animals to higher latitudes or higher elevations on mountains. More discussion of ecosystems, ranges, and biomes is in Chapters 19 and 20.

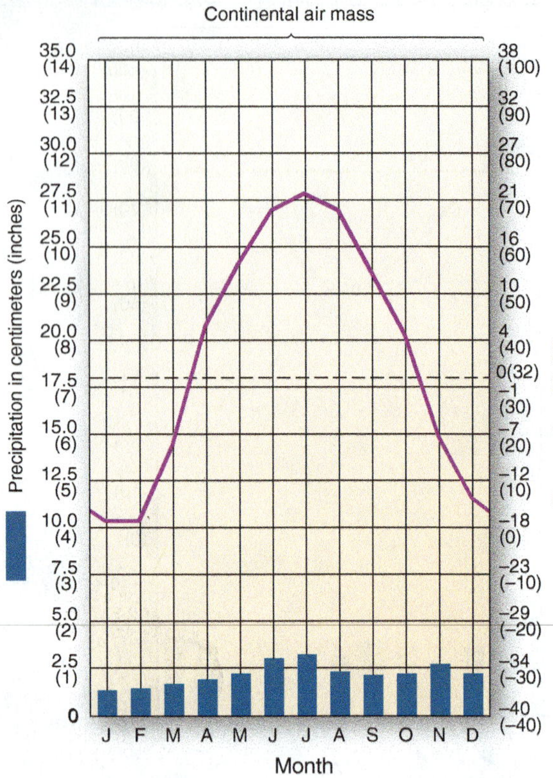

(a) Climograph for Semey (Semipalatinsk), Kazakhstan.

Station: Semey (Semipalatinsk), Kazakhstan
Lat/long: 50° 21′ N 80° 15′ E
Avg. Ann. Temp.: 3°C (37.4°F)
Total Ann. Precip.: 26.4 cm (10.4 in.)
Elevation: 206 m (675.9 ft)
Population: 270,500
Ann. Temp. Range: 39 C° (70.2 F°)

(b) Aerial view of Semey and the Ertis River.

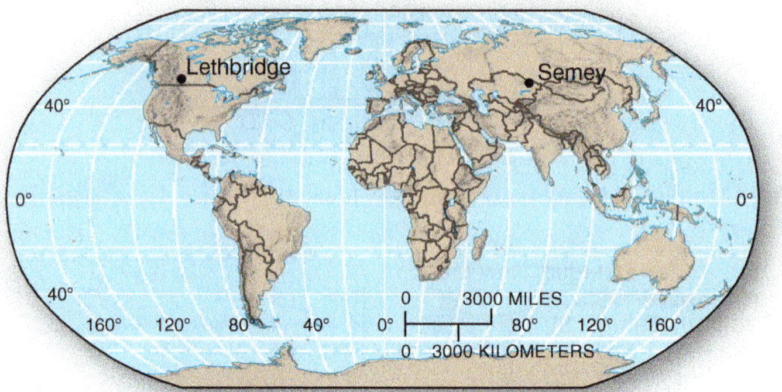

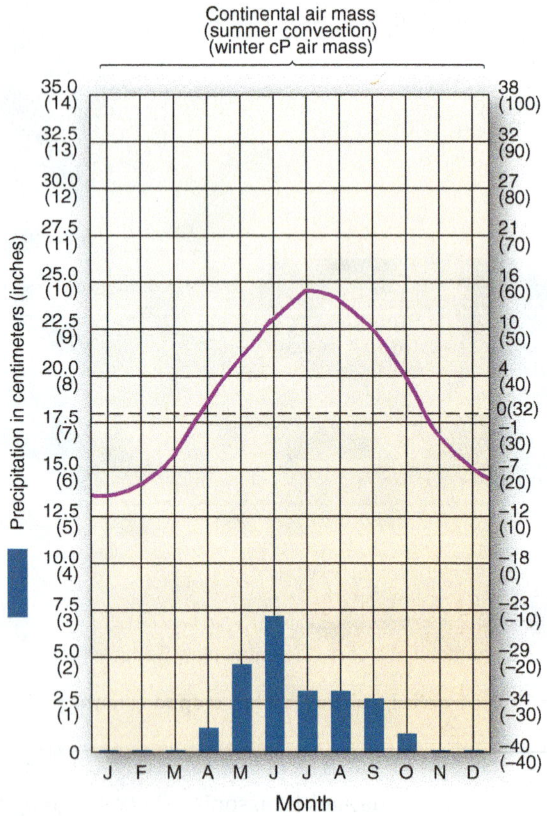

(c) Climograph for Lethbridge, Alberta.

Station: Lethbridge, Alberta
Lat/long: 49° 42′ N 110° 50′ W
Avg. Ann. Temp.: 2.9°C (37.3°F)
Total Ann. Precip.: 25.8 cm (10.2 in.)
Elevation: 910 m (2985 ft)
Population: 73,000
Ann. Temp. Range: 24.3 C° (43.7 F°)

(d) Grain elevators highlight an Alberta, Canada, landscape.

▲**Figure 10.24 Midlatitude cold steppe climates, Kazakhstan and Canada.** [(b) Dinara Sagatova. (d) Design Pics RF/Getty Images.]

CLIMATES ⇨ HUMANS

- Climates affect many facets of human society, including agriculture, water availability, and natural hazards such as floods, droughts and heat waves.

HUMANS ⇨ CLIMATES

- Anthropogenic climate change is altering Earth systems that affec temperature and moisture, and therefore climate.

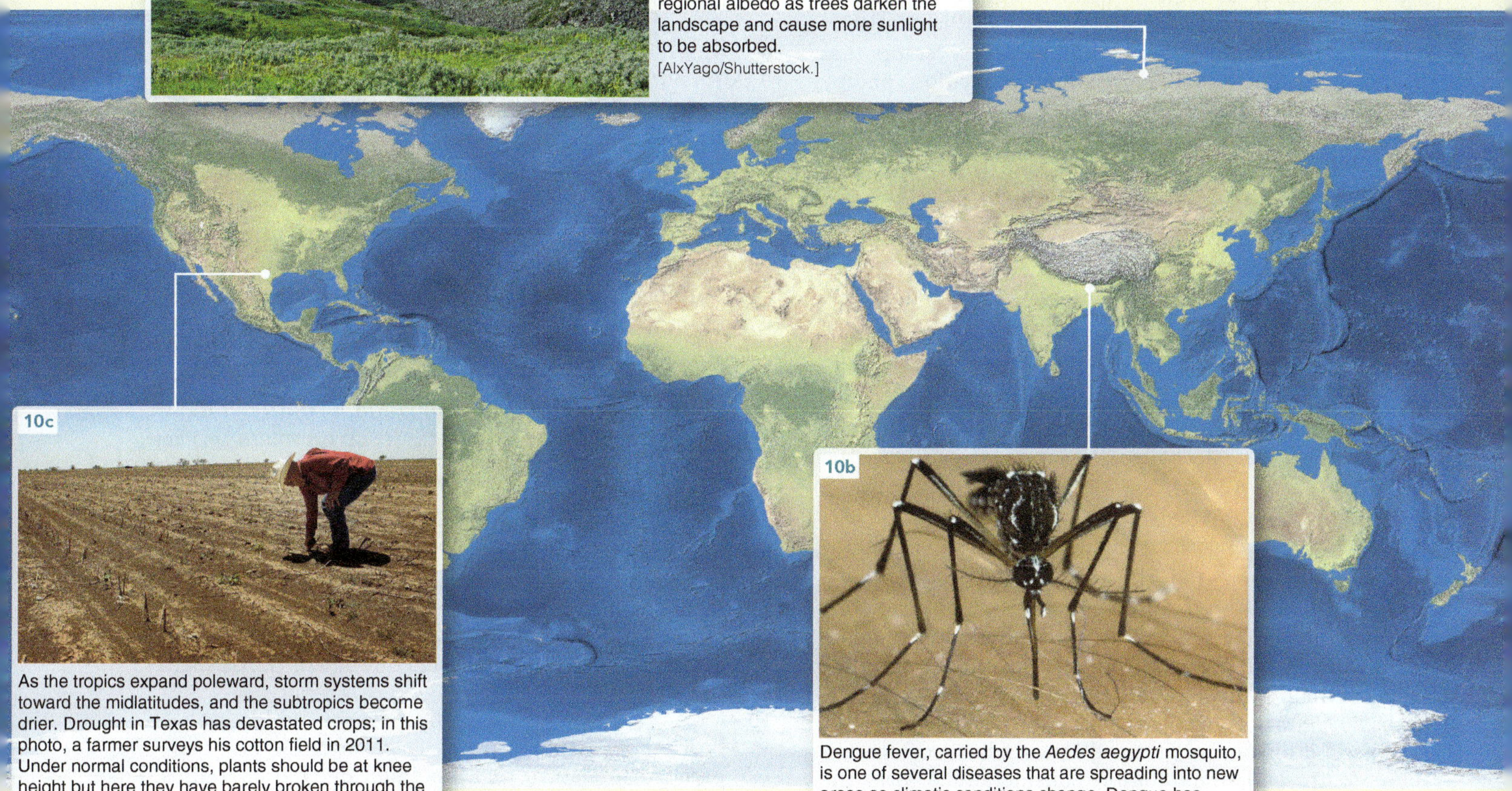

10a In the Eurasian Arctic tundra, 30 years of warming temperatures have allowed willow and alder shrubs to grow into small trees. New research from the southern parts of this region shows that the shrubs, formerly about 1 m (3.3 ft) in height, now reach over 2 m (6.6 ft). This trend may lead to changes in regional albedo as trees darken the landscape and cause more sunlight to be absorbed. [AlxYago/Shutterstock.]

10c As the tropics expand poleward, storm systems shift toward the midlatitudes, and the subtropics become drier. Drought in Texas has devastated crops; in this photo, a farmer surveys his cotton field in 2011. Under normal conditions, plants should be at knee height but here they have barely broken through the soil surface. The drought has also affected beef production and lowered reservoir levels, cutting off irrigation water for rice crops–with millions of dollars lost. [Scott Olsen/Getty Images.]

10b Dengue fever, carried by the *Aedes aegypti* mosquito, is one of several diseases that are spreading into new areas as climatic conditions change. Dengue has spread into previously unaffected parts of India, and into Nepal and Bhutan. In the United States, dengue is still uncommon, but reported cases are rising. [Nigel Cattlin/Alamy.]

ISSUES FOR THE 21ST CENTURY

- Human-caused global warming is driving a poleward shift in the boundaries of climate regions.

GEOSYSTEMS**connection**

Collectively, the climate regions presented in Chapter 10 provide a synthesis of the systems studied in Chapters 2 through 9 of *Geosystems*. Climate classification categories are portraits of the interplay between solar insolation, moisture, and weather conditions that determine climates, and their associated ecosystems. Next, we move to the last chapter of Part II, where we survey and explain the many facets of climate change now underway. We examine the method scientists use to assess past climates and natural climate variability. We also look at present changes in atmospheric composition that are raising global temperatures, with associated effects on all Earth systems.

s review

■ matology, and *review* the princi- rth's climate system.

mate is a synthesis of weather phenomena at many scales, from planetary to local, in contrast to weather, which is the condition of the atmosphere at any given time and place. Earth experiences a wide variety of climatic conditions that can be grouped by general similarities into climatic regions. **Climatology** is the study of climate and attempts to discern similar weather statistics and identify **climatic regions**. The principal factors that influence climates on Earth include insolation, energy imbalances between the equator and the poles, temperature, air pressure, air masses, and atmospheric moisture (including humidity and the supply of moisture from precipitation).

climate (p. 258)
climatology (p. 258)
climatic region (p. 258)

1. Define climate and compare it with weather. What is climatology?

■ *Describe* climate classification systems, *list* the main categories of world climates, and *locate* the regions characterized by each climate type on a world map.

Classification is the ordering or grouping of data or phenomena into categories. A **genetic classification** is one based on causative factors, such as the interaction of air masses. An **empirical classification** is one based on statistical data, such as temperature or precipitation. This text analyzes climate using aspects of both approaches. Temperature and precipitation data are measurable aspects of climate and are plotted on **climographs** to display the basic characteristics that determine climate regions.

World climates are grouped into six basic categories. Temperature and precipitation considerations form the basis of five climate categories and their regional types:

- Tropical (tropical latitudes)
- Mesothermal (midlatitudes, mild winters)
- Microthermal (mid and high latitudes, cold winters)
- Polar (high latitudes and polar regions)
- Highland (high elevations at all latitudes)

Only one climate category is based on moisture efficiency as well as temperature:

- Dry (permanent moisture deficits)

classification (p. 259)
genetic classification (p. 259)
empirical classification (p. 259)
climograph (p. 259)

2. What are the differences between a genetic and an empirical classification system?
3. What are some of the climatological elements used in classifying climates? Why is each of these used? Use the approach on the climate classification map in Figure 10.2 in preparing your answer.
4. List and discuss each of the principal climate categories. In which one of these general types do you live? Which category is the only type associated with the annual distribution and amount of precipitation?
5. What is a climograph, and how is it used to display climatic information?
6. Which of the major climate types occupies the most land and ocean area on Earth?
7. How do radiation receipts, temperature, air pressure inputs, and precipitation patterns interact to produce climate types? Give an example from a humid environment and one from an arid environment.

■ *Discuss* the subcategories of the six world climate groups, including their causal factors.

Tropical climates include *tropical rain forest* (rainy all year), *tropical monsoon* (6 to 12 months rainy), and *tropical savanna* (less than 6 months rainy). The shifting ITCZ is a major causal factor for seasonal moisture in these climates. Mesothermal climates include *humid subtropical* (hot to warm summers), *marine west coast* (warm to cool summers), and *Mediterranean* (dry summers). These warm, temperate climates are humid, except where high pressure produces dry-summer conditions. Microthermal climates have cold winters, the severity of which depends on latitude; subcategories include *humid continental* (hot or mild summers) and *subarctic* (cool summers to very cold winters). Polar climates have no true summer and include *tundra* (high latitude or high elevation), *ice-cap and ice-sheet* (perpetually frozen), and *polar marine* (moderate polar) climates.

8. Characterize the tropical climates in terms of temperature, moisture, and location.
9. Using Africa's tropical climates as an example, characterize the climates produced by the seasonal shifting of the ITCZ with the high Sun.
10. Mesothermal climates occupy the second largest portion of Earth's entire surface. Describe their temperature, moisture, and precipitation characteristics.
11. Explain the distribution of the *humid subtropical hot-summer* and *Mediterranean dry-summer* climates at similar latitudes and the difference in precipitation patterns between the two types. Describe the difference in vegetation associated with these two climate types.
12. Which climates are characteristic of the Asian monsoon region?
13. Explain how a *marine west coast* climate can occur in the Appalachian region of the eastern United States.
14. What role do offshore ocean currents play in the distribution of the *marine west coast* climates? What type of fog is formed in these regions?
15. Discuss the climatic conditions for the coldest places on Earth outside the poles.

■ *Explain* the precipitation and moisture-efficiency criteria used to classify the arid and semiarid climates.

The dry climates of the tropics and midlatitudes consist of arid deserts and semiarid steppes. In arid deserts, precipitation (the natural water supply) is less than one-half of natural water demand. In *semiarid steppes*, precipitation, though insufficient, is more than one-half of natural water demand. A **steppe** is a regional term referring to the vast semiarid grassland biome of Eastern Europe and Asia. The dry climates are subdivided into *tropical, subtropical hot deserts; midlatitude cold deserts; tropical, subtropical hot steppes;* and *midlatitude cold steppes.*

steppe (p. 278)

16. In general terms, what are the differences among the four desert classifications? How are moisture and temperature distributions used to differentiate these subtypes?

17. Describe the factors that contribute to the location of arid and semiarid climates in the western United States. What explains the presence of these climates in northern Africa?

MasteringGeography™

Looking for additional review and test prep materials? Visit the Study Area in *MasteringGeography*™ to enhance your geographic literacy, spatial reasoning skills, and understanding of this chapter's content by accessing a variety of resources, including **MapMaster** interactive maps, geoscience animations, satellite loops, author notebooks, videos, RSS feeds, flashcards, web links, self-study quizzes, and an eText version of *Geosystems*.

◄**Figure 10.25 Waimea Canyon, Kaua'i, Hawai'i.** The cover photo of this textbook shows Waimea Canyon, known as "the Grand Canyon of the Pacific," on the western side of Kaua'i. The Waimea River and its tributaries carved this 900-m (3000-ft) deep canyon, carrying runoff from the slopes of Mount Wai'ale'ale (1569 m, 5147 ft), one of the wettest locations in the world. (Please review GeoReport 8.2, page 197, which describes orographic precipitation on the island of Kaua'i.)

In this aerial photo, looking northeast from 1060 m (3000 ft) altitude, Po'amau Stream flows through layers of ancient lava flows. Despite the large amounts of precipitation on the windward side of the island, annual rainfall here in the rain shadow of Wai'ale'ale ranges from 76 cm to 152 cm (30 in. to 60 in.) per year, with a distinct summer-dry pattern. Temperatures are typical of the tropics, with all months averaging above 18°C (64.4°F). Mount Wai'ale'ale has a tropical rainforest climate, while only 18 km (11 mi) to the west, Waimea Canyon is characterized by a tropical summer-dry climate—rare among the tropical climate types and small in spatial extent on Earth. [Bobbé Christopherson.]

10B Climate Change

KEY LEARNING concepts

After reading the chapter, you should be able to:

- ***Describe*** scientific tools used to study paleoclimatology.
- ***Discuss*** several natural factors that influence Earth's climate, and ***describe*** climate feedbacks, using examples.
- ***List*** the key lines of evidence for present global climate change, and ***summarize*** the scientific evidence for anthropogenic forcing of climate.
- ***Discuss*** climate models, and ***summarize*** several climate projections.
- ***Describe*** several mitigation measures to slow rates of climate change.

In March 2013, scientists began the fifth year of Operation IceBridge, NASA's airborne, multi-instrument survey of Earth's rapidly changing polar ice. This view of Saunders Island and Wolstenholme Fjord in northwest Greenland in April 2013 shows Arctic sea ice as air and ocean temperature warm. Thinner seasonal ice appears clearer in the foreground; thicker multiyear ice appears whiter in the distance. Much of the Arctic Ocean is now dominated by seasonal ice, which melts rapidly every summer. Ice melt in the polar regions and at high altitudes is an important indicator of Earth's changing climate, the subject of this chapter [NASA/Michael Studinger.]

Greenhouse Gases Awaken in the Arctic

In the subarctic and tundra climate regions of the Northern Hemisphere, perennially frozen soils and sediment, known as permafrost, cover about 24% of the land area. With Arctic air temperatures currently rising at a rate more than two times that of the midlatitudes, ground temperatures are increasing, causing permafrost thaw. This results in changes to land surfaces, primarily sinking and slumping, that damage buildings, forests, and coastlines (Figure GN 10B.1). Permafrost thaw also leads to the decay of soil material, a process that releases vast amounts of carbon, in the form of the greenhouse gases carbon dioxide (CO_2) and methane (CH_4), into the atmosphere.

Carbon in Permafrost Soils Permafrost is, by definition, soil and sediment that have remained frozen for two or more consecutive years. The "active layer" is the seasonally frozen ground on top of subsurface permafrost. This thin layer of soil and sediment thaws every summer, providing substrate for seasonal grasses and other plants that absorb CO_2 from the atmosphere. In winter, the active layer freezes, trapping plant and animal material before it can decompose completely. Over hundreds of thousands of years, this carbon-rich material has become incorporated into permafrost and now makes up roughly half of all the organic matter stored in Earth's soils—twice the amount of carbon that is stored in the atmosphere. In terms of real numbers, the latest estimate of the amount of carbon stored in Arctic permafrost soils is 1700 gigatonnes (or 1700 billion tons).

▲Figure GN 10B.1 **Blocks of melting permafrost collapse into the Beaufort Sea, Alaska.** [USGS Alaska Science Center.]

A Positive Feedback Loop As summers become warmer in the Arctic, heat radiating through the ground thaws the permafrost layers. Microbial activity in these layers increases, enhancing the breakdown of organic matter. As this occurs, bacteria and other organisms release CO_2 into the atmosphere in a process known as *microbial respiration*. In anaerobic (oxygen-free) environments, such as lakes and wetlands, the process releases methane. Studies show that thousands of methane seeps can develop under a single lake, a huge amount when multiplied by hundreds of thousands of lakes across the northern latitudes (Figure GN 10B.2).

Carbon dioxide and methane are major greenhouse gases, which absorb outgoing longwave radiation and radiate it back toward Earth, enhancing the greenhouse effect and leading to atmospheric warming. Methane is especially important because, although its relative percentage is small in the atmosphere, it is over 20 times more effective than CO_2 at trapping atmospheric heat. Thus, a positive feedback loop forms: As temperatures rise, permafrost thaws, causing a release of CO_2 and CH_4 into the atmosphere, which causes more warming, leading to more permafrost thaw.

▲Figure GN 10B.2 **Methane lies under arctic lakebeds, and like natural gas, is highly flammable.** [Todd Paris/AP Images.]

Melting Ground Ice In addition to frozen soil and sediment, permafrost also contains ground ice, which melts as the permafrost thaws. When the supporting structure provided by the ice is removed, land surfaces collapse and slump. Subsurface soils are then exposed to sunlight, which speeds up microbial processes, and to water erosion, which moves organic carbon into streams and lakes, where it is mobilized into the atmosphere. Research suggests that this process may release bursts of CO_2 and CH_4 into the atmosphere, in contrast to the slower top-down melting of permafrost.

Permafrost soils are now warming at a rate faster than Arctic air temperatures, releasing vast amounts of "ancient" carbon into the atmosphere. Scientists are actively researching the locations and amounts of vulnerable permafrost, the current and projected rates of thaw, and the potential impacts to the permafrost–carbon positive feedback. The thawing Arctic is one of many immediate concerns we discuss in this chapter regarding the causes and impacts of changing climate on Earth systems.

GEOSYSTEMS NOW ONLINE 10B the permafrost thaw and climate change. To learn about NASA's Carbon in Arctic Reservoirs Vulnerability Experiment (CARVE), which measures CO_2 and CH_4 gas emissions in permafrost regions, go to http://science1.nasa.gov/missions/carve/ (the mission website) or http://www.nasa.gov/topics/earth/features/earth20130610.html#.UhwYVj_pxXJ (mission background and early results). MG

Everything we have learned in *Geosystems* up to this point sets the stage for our exploration of **climate change science**—the interdisciplinary study of the causes and consequences of changing climate for all Earth systems and the sustainability of human societies. Climate change is one of the most critical issues facing humankind in the twenty-first century, and is today an integral part of physical geography and Earth systems science. Three key elements of climate change science are the study of past climates, the measurement of current climatic changes, and the modeling and projection of future climate scenarios—all of which are discussed in the chapter ahead. We will revisit some of the principles of the scientific method as we explore global climate change, and by doing so address some of the confusion that complicates the public discussion of this topic.

The physical evidence for global climate change is extensive and is observable by scientists and nonscientists alike—record-breaking global average temperatures for air, land surfaces, lakes, and oceans; ice losses from mountain glaciers and from the Greenland and Antarctic ice sheets; declining soil-moisture conditions and resultant effects on crop yields; changing distributions of plants and animals; and the pervasive impact of global sea-level rise, which threatens coastal populations and development worldwide (Figure 10B.1). These are only a few of the many complex and far-reaching issues that climate change science must address to understand and mitigate the environmental changes ahead.

Although climate has fluctuated naturally over Earth's long history, most scientists now agree that present climate change is resulting from human activities that produce greenhouse gases. The consensus is overwhelming and includes the vast majority of professional scientific societies, associations, and councils in the United States and throughout the world. Scientists agree that natural climate variability, which we discuss in the chapter ahead, cannot explain the present warming trend.

▲Figure 10B.1 Sea-level rise along the world's coastlines. From 1993 to 2010, sea-level rise occurred at a rate of 3.2 mm/year (0.13 in./year), causing high tides and storm waves to encroach on developed areas and hastening coastal erosion processes (discussed in this chapter and in Chapter 16). Pictured here are houses near Cape Hatteras, North Carolina, in 2011. [Archive Image/Alamy]

The observed changes in the global climate over recent decades are happening at a pace much faster than seen in the historical climate records or in the climate reconstructions that now extend millions of years into the past.

In this chapter: We examine techniques used to study past climates, including oxygen isotope analysis of sediment cores extracted from ocean-floor and ice cores, carbon isotope analysis, and dating methods using tree rings, speleothems, and corals. We examine long-term climate trends and discuss mechanisms of natural climate fluctuation, including Milankovitch cycles, solar variability, tectonics, and atmospheric factors. We survey the evidence of accelerating climate change now underway as measured by record-high ocean, land, and atmospheric temperatures; increasing acidity of the oceans; declining glacial ice and ice-sheet losses worldwide and record losses of Arctic sea ice; accelerating rates of sea-level rise; and the occurrence of severe weather events. We then examine the human causes of contemporary climate change, and the climate models that provide evidence and scenarios for future trends. The chapter concludes with a look at the path ahead and the actions that people can take now, on an individual, national, and global level.

Population Growth and Fossil Fuels—The Setting for Climate Change

As discussed in earlier chapters, carbon dioxide (CO_2) produced from human activities is amplifying Earth's natural greenhouse effect. It is released into the atmosphere naturally from outgassing (discussed in Chapter 9) and from microbial and plant respiration and decomposition on land and in the world's oceans (discussed in Chapter 19). These natural sources have contributed to atmospheric CO_2 for over a billion years, unaffected by the presence of humans. In recent times, however, the growing human population on Earth has produced significant quantities of atmospheric CO_2. The primary anthropogenic source is the burning of fossil fuels (coal, oil, and natural gas), which has increased dramatically in the last few centuries and added to greenhouse gas concentrations. To illustrate the increase, Figure 10B.2 shows carbon dioxide (CO_2) levels for the last 800,000 years, including the steadily rising CO_2 trend since the Industrial Revolution began in the 1800s.

During the twentieth century, population increased from about 1.6 billion people to about 6.1 billion people (please review Chapter 1, Figure 1.5, and see Figure 10B.3). At the same time, CO_2 emissions increased by a factor of 10 or more. The burning of fossil fuels as an energy source has contributed to most of this increase, with secondary effects from the clearing and burning of land for development and agriculture. Notice in Figure 10B.3 that the majority of population growth is now occurring in the less-developed countries (LDCs). It is especially rapid in China and India. Although more-developed countries (MDCs) currently

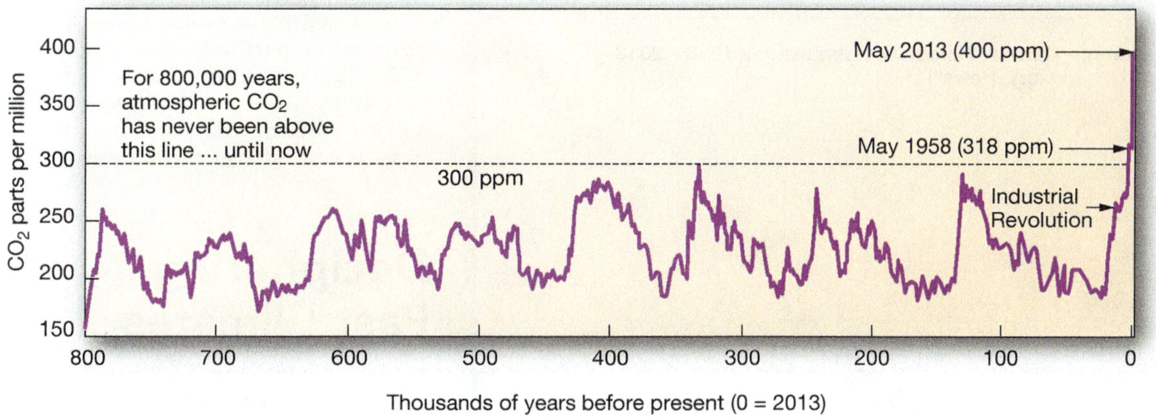

MG Animation Global Warming, Climate Change

◀**Figure 10B.2 Carbon dioxide concentrations for the last 800,000 years.** Data are from atmospheric samples contained in ice cores and direct CO_2 measurements. Note the rise in CO_2 over the past several hundred years since the Industrial Revolution. [NOAA.]

emit the greatest share of total greenhouse gases and lead in per capita emissions, this portion is changing. According to some projections, LDCs will contribute over 50% of global greenhouse gas emissions by 2050 (Figure 10B.4).

The rate of population growth increased dramatically after 1950 (see Figure 10B.3), and this trend correlates with a dramatic rise in atmospheric CO_2 since that time. In 1953, Charles David Keeling of the Scripps Institute of Oceanography began collecting detailed measurements of atmospheric carbon dioxide (CO_2) in California. In 1958, he began CO_2 measurements in Hawai'i, resulting in what is considered by many scientists to be the single most important environmental data set of the twentieth century. Figure 10B.5 shows the graph, known as the "Keeling Curve," of monthly average carbon dioxide (CO_2) concentrations from 1958 to the present as recorded at the Mauna Loa Observatory in Hawai'i.

The uneven line on the graph in Figure 10B.5 shows fluctuations in CO_2 that occur throughout the year, with May and October usually being the highest and lowest months, respectively, for CO_2 readings. This annual fluctuation between spring and fall reflects seasonal changes in vegetation cover in the higher latitudes of the northern hemisphere. Vegetation is dormant during the Northern Hemisphere winter, allowing CO_2 to build up in the atmosphere; in spring, when plant growth resumes, vegetation takes in CO_2 for photosynthesis (discussed in Chapter 19), causing a decline in atmospheric CO_2. More important than these yearly fluctuations, however, is the overall trend, which is that from 1992 to 2012, atmospheric CO_2 increased 16%, and in May 2013, CO_2 concentrations crossed the 400-ppm threshold—a level that is unprecedented during the last 800,000 years, and perhaps longer. These data match records from hundreds of other stations across the globe.

The Mauna Loa CO_2 record is now one of the best-known graphs in modern science and is an iconic symbol of the effects of humans on Earth systems. The corresponding increases in human population, fossil-fuel usage, and atmospheric CO_2 are the basic ingredients for understanding the causes of present-day climate change and mitigating its effects.

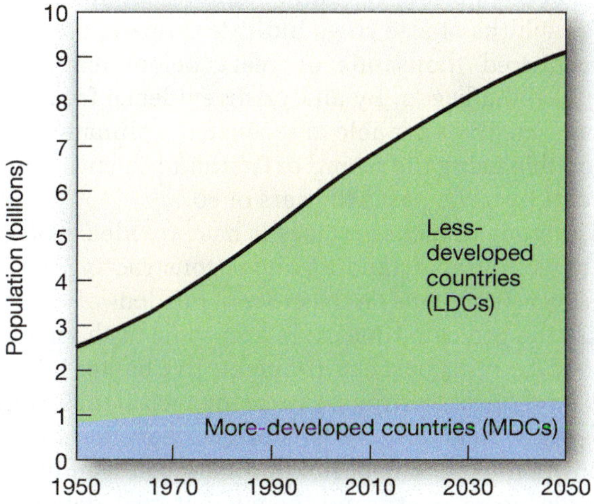

▲**Figure 10B.3 Human population growth since 1950, and projected to 2050.** Since 1950, population has increased in LDCs far more than in MDCs, a trend that is expected to increase to 2050. [Reprinted by permission of the Population Reference Bureau from http://www.prb.org/Publications/Datasheets/2012/world-population-data-sheet/fact-sheet-world-population.aspx]

▲**Figure 10B.4 Industrial growth and rising CO_2 emissions in India.** Farmers plow a field next to one of India's many industrial plants. India's population now makes up about 18% of Earth's total, and the country's rapidly growing economy relies heavily on the use of fossil fuels for energy. [Tim Graham/Alamy.]

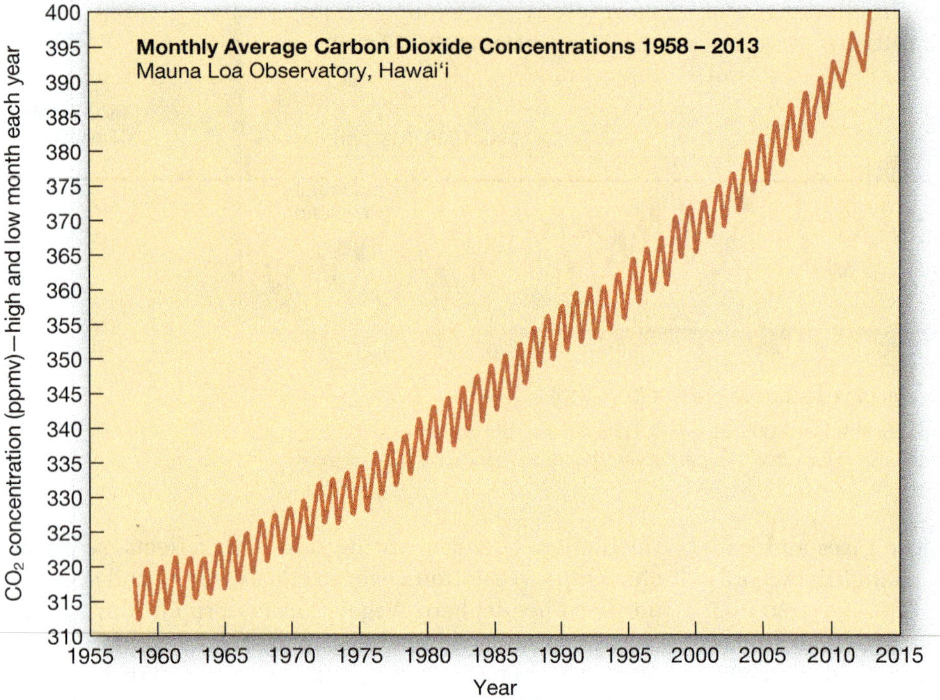

(a) Fifty-five years of CO_2 data measured through May 2013. The highest and lowest average months, usually May and October, respectively, are plotted for each year on the graph in parts per million by volume (ppmv).

(b) The Mauna Loa Observatory is located on the north slope of Mauna Loa volcano at 3397 m (11,141 ft) on Hawai'i's Big Island. The remote location minimizes pollution and the effects of vegetation, and is above the atmospheric inversion layer, thus making air quality data representative of global values.

Figure 10B.5 The Keeling Curve. Increase in carbon dioxide concentration, 1958–2013, measured at the Mauna Loa Observatory. [(a) Data from the NOAA posted at ftp://ftp.cmdl.noaa.gov/ccg/co2/trends/co2_mm_mlo.txt. (b) Jonathan Kingston/Getty Images.]

CRITICAL thinking 10B.1
Crossing the 450-ppm Threshold for Carbon Dioxide

Using the accelerating trend in atmospheric CO_2 concentrations for the past 10 years, shown in Figure 10B.5, calculate the year in which we might reach the 450-ppm threshold for atmospheric carbon dioxide. To help stabilize climate change effects, scientists think that we should implement policies to return the level of CO_2 to 350 ppm. By what annual percentage would we need to reduce atmospheric CO_2 concentrations to achieve that goal by 2020? •

Deciphering Past Climates

To understand present climatic changes, we begin with a discussion of the climates of the past; specifically, how scientists reconstruct climates that occurred before human recordkeeping began. Clues to past climates are stored in a variety of environments on Earth. Among these climatic indicators are gas bubbles in glacial ice, fossil plankton in ocean-bottom sediments, fossil pollen from ancient plants, and growth rings in trees, speleothems (mineral formations in caves), and corals. Scientists access these environmental indicators by extracting cores from deep within the various materials, which are then analyzed by various methods to determine age and climate-related characteristics. In this way, scientists can establish a chronology of environmental conditions over time periods of thousands or millions of years.

The study of Earth's past climates is the science of **paleoclimatology**, which tells us that Earth's climate has fluctuated over hundreds of millions of years. To learn about past climates, scientists use **proxy methods** instead of direct measurements. A *climate proxy* is a piece of information from the natural environment that can be used to reconstruct climates that extend back further than our present instrumentation allows. For example, the widths of tree rings indicate climatic conditions that occurred thousands of years before temperature recordkeeping began. By analyzing evidence from proxy sources, scientists are able to reconstruct climate in ways not possible using the record of firsthand scientific measurements over the past 140 years or so.

The study of rocks and fossils has provided geologists with tools for understanding and reconstructing past climates over time spans on the order of millions of years (we discuss the geologic time scale and some of the methods for reconstructing past environments in Chapter 11). Geologists study past environments using techniques ranging from simple field observations of the characteristics and composition of rock deposits (for example, whether they are made up of dust, sand, or ancient coal deposits) to complex and costly laboratory analyses of rock samples (for example, mass spectrometry analyzes the chemical makeup of single molecules from rock and other materials). Fossils of animal and plant material preserved within rock layers also provide important climate clues; for example, fossils of tropical plants indicate warmer climate conditions; fossils of ocean-dwelling creatures indicate ancient marine environments. The beds of coal that we rely on today for energy were formed of organic matter from plants that grew in warm, wet tropical and temperate climate conditions about 325 million years ago.

Climate reconstructions spanning millions of years show that Earth's climate has cycled between periods that were colder and warmer than today. An extended period of cold (not a single brief cold spell), in some cases lasting several million years, is known as an *ice age*, or *glacial age*. An ice age is a time of generally cold climate that includes one or more *glacials* (glacial periods, characterized by glacial advance) interrupted by brief warm periods known as *interglacials*. The most recent ice age is known as the Pleistocene Epoch, discussed in Chapter 17, from about 2.5 million years ago to about 11,700 years ago. The geologic time scale, which places the Pleistocene within the context of Earth's 4.6-billion-year history, is presented in Chapter 11, Figure 11.1, on page 327.

Methods for Long-Term Climate Reconstruction

Some paleoclimatic techniques yield long-term records that span hundreds of thousands to millions of years. Such records come from cores drilled into ocean-bottom sediments or into the thickest ice sheets on Earth. Once cores are extracted, layers containing fossils, air bubbles, particulates, and other materials provide information about past climates.

The basis for long-term climate reconstruction is **isotope analysis**, a technique that uses the atomic structure of chemical elements, specifically the relative amounts of their isotopes, to identify the chemical composition of past oceans and ice masses. Using this knowledge, scientists can reconstruct temperature conditions. Remember that the nuclei of atoms of a given chemical element, such as oxygen, always contain the same number of protons but can differ in the number of neutrons. Each number of neutrons found in the nucleus represents a different *isotope* of that element. Different isotopes have slightly different masses and therefore slightly different physical properties.

Oxygen Isotope Analysis

An oxygen atom contains 8 protons, but may have 8, 9, or 10 neutrons. The atomic weight of oxygen, which is approximately equal to the number of protons and neutrons combined, may therefore vary from 16 atomic mass units ("light" oxygen) to 18 ("heavy" oxygen). Oxygen-16, or ^{16}O, is the most common isotope found in nature, making up 99.76% of all oxygen atoms. Oxygen-18, or ^{18}O, comprises only about 0.20% of all oxygen atoms.

We learned in Chapter 7 that water, H_2O, is made up of two hydrogen atoms and one oxygen atom. Both the ^{16}O and ^{18}O isotopes occur in water molecules. If the water contains "light" oxygen (^{16}O), it evaporates more easily but condenses less easily. The opposite is true for water containing "heavy" oxygen (^{18}O), which evaporates less easily, but condenses more easily. These property differences affect where each of the isotopes is more likely to accumulate within Earth's vast water cycle. As a result, the relative amount, or *ratio*, of heavy to light oxygen isotopes ($^{18}O/^{16}O$) in water varies with climate; in particular, with temperature. By comparing the isotope ratio with an accepted standard, scientists can determine to what degree the water is enriched or depleted in ^{18}O relative to ^{16}O.

Since ^{16}O evaporates more easily, over time the atmosphere becomes relatively rich in "light" oxygen. As this water vapor moves toward the poles, enrichment with ^{16}O continues, and eventually this water vapor condenses and falls to the ground as snow, accumulating in glaciers and ice sheets (Figure 10B.6). At the same time, the oceans become relatively rich in ^{18}O—partly as a result of ^{16}O evaporating at a greater rate and partly from ^{18}O condensing and precipitating at a greater rate once it enters the atmosphere.

During periods of colder temperatures, when "light" oxygen is locked up in snow and ice in the polar regions, "heavy" oxygen concentrations are highest in the oceans (Figure 10B.7a). During warmer periods, when snow and ice melt returns ^{16}O to the oceans, the concentration of ^{18}O in the oceans becomes relatively less—the isotope ratio is essentially in balance (Figure 10B.7b). The result is that higher levels of "heavy" oxygen (a higher ratio of $^{18}O/^{16}O$) in ocean water indicates colder climates (more water is tied up in snow and ice), whereas lower levels of "heavy" oxygen (a lower ratio of $^{18}O/^{16}O$) in the oceans indicates warmer climates (melting glaciers and ice sheets).

Ocean Sediment Cores

Oxygen isotopes are found not only in water molecules but also in calcium carbonate ($CaCO_3$), the primary component of the exoskeletons, or shells, of marine microorganisms called *foraminifera*. Today, floating (planktonic) or bottom-dwelling (benthic) foraminifera are some of the world's most abundant shelled marine organisms, living in a variety of environments from the equator to the poles. Upon the death of the organism, foraminifera shells accumulate on the ocean bottom and build up in layers of sediment. By extracting a core of these ocean-floor sediments and comparing the ratio of oxygen isotopes in the $CaCO_3$ shells, scientists can determine the isotope ratio of seawater at the time the shells were formed. Foraminifera shells with a high $^{18}O/^{16}O$ ratio were formed during cold periods; those with low ratios were formed during warm periods. In an ocean sediment core, shells accumulate in layers that reflect these temperature conditions.

Specialized drilling ships have powerful rotary drills able to bore into ocean-bottom rock and sediment, extracting a cylinder of material—a *core sample*—within a hollow metal pipe. Such a core may contain dust, minerals, and fossils that have accumulated in layers over long periods of time on the ocean floor (Figure 10B.8a). Over the past 50 years, the Integrated Ocean Drilling Program

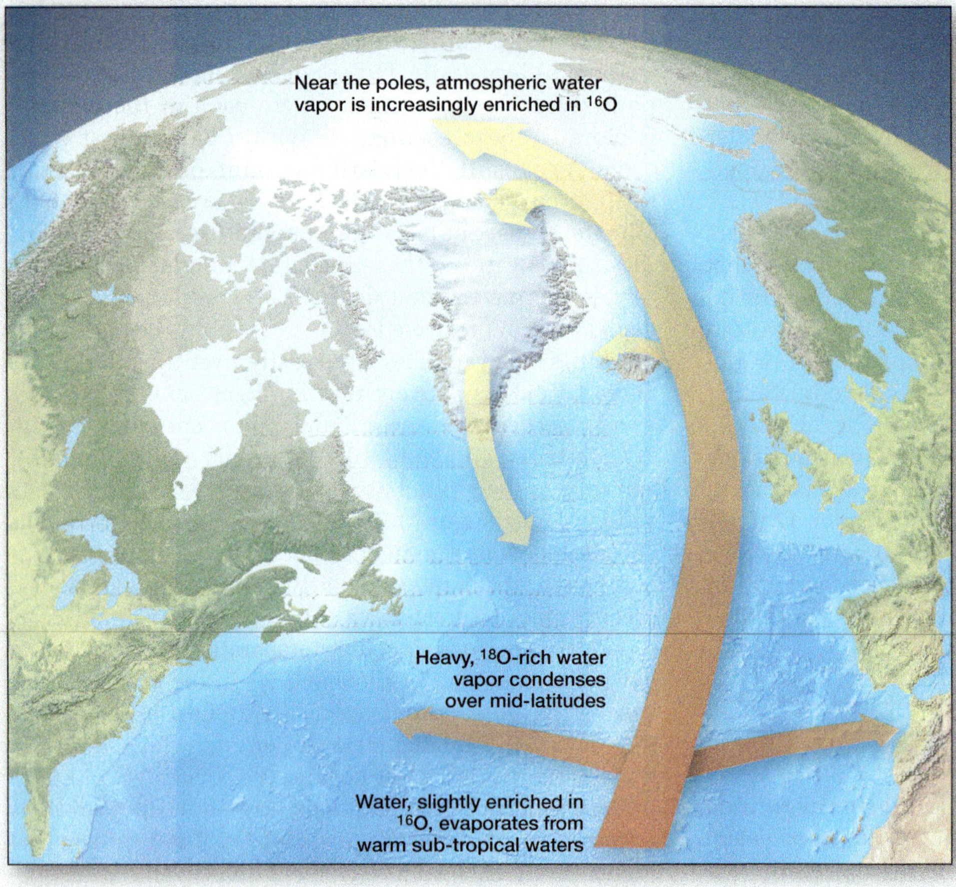

▲**Figure 10B.6 Changing concentrations of ^{16}O (light oxygen) and ^{18}O (heavy oxygen) in water vapor between equator and North Pole.** [Modified from illustration by Robert Simmon, NASA GSFC.]

completed about 2000 cores in the ocean floor, yielding more than 35,000 samples for researchers (see http://www.oceandrilling.org/). The international program includes two drilling ships: The U.S. *JOIDES Resolution*, in operation since 1985 (Figure 10B.8b), and Japan's *Chikyu*, operating since 2007. *Chikyu* set a new record in 2012 for the deepest hole drilled into the ocean floor—2466 m (8090 ft)—and is capable of drilling 10,000 m (32,800 ft) below sea level and yielding undisturbed core samples. Recent improvements in both isotope analysis techniques and the quality of ocean core samples have led to improved resolution of climate records for the past 70 million years (see Figure 10B.10 on page 294).

Ice Cores In the cold regions of the world, snow accumulates seasonally in layers, and in regions where snow is permanent on the landscape, these layers of snow eventually form glacial ice (Figure 10B.9a). Scientists have extracted cores of such glacial ice to reconstruct climate. The world's largest accumulations of glacial ice occur in Greenland and Antarctica. Scientists have extracted cores drilled thousands of meters deep into the thickest part of these ice sheets to provide a shorter but more detailed record of climate than ocean sediment cores, presently pushing the record back 800,000 years.

Extracted from areas where the ice is undisturbed, ice cores are about 13 cm (5 in.) in diameter and are composed

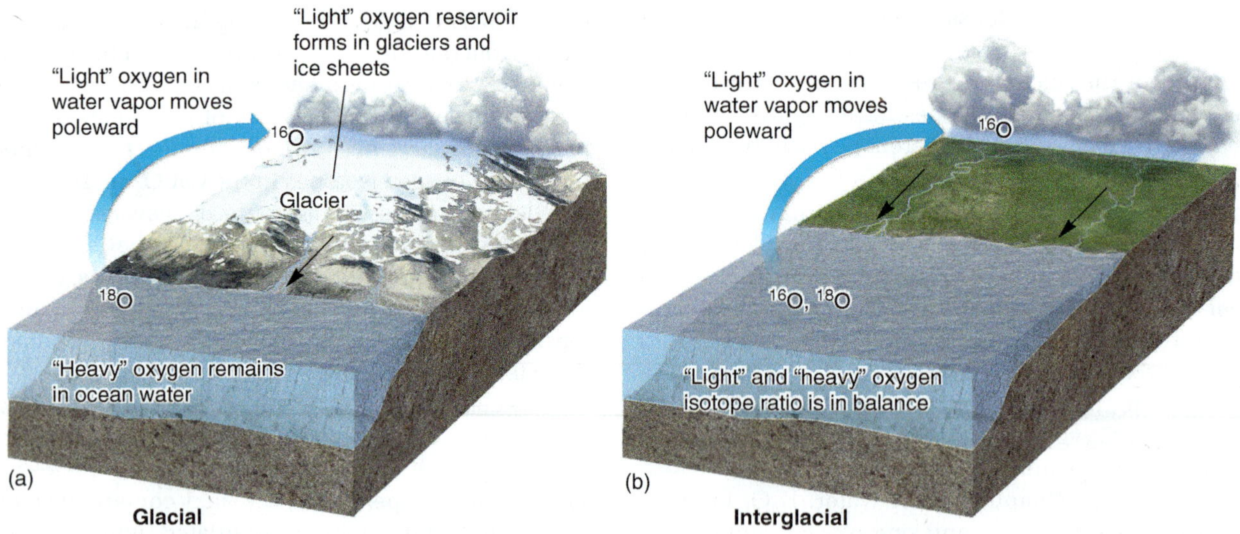

▲**Figure 10B.7 Relative oceanic concentrations of ^{16}O and ^{18}O during colder (glacial) and warmer (interglacial) periods.** [Based on *Analysis of Vostok Ice Core Data*, Global Change, available at http://www.globalchange.umich.edu/globalchange1/current/labs/Lab10_Vostok/Vostok.htm.]

(a) Core samples of ocean sediments are split open for analysis.

(b) The U.S. *JOIDES Resolution* drilling ship.

▲**Figure 10B.8 Ocean-bottom core sample and ocean drilling ship.** [IODP.]

of distinct layers of younger ice at the top and less-defined layering of older ice beneath (Figure 10B.9b). At the bottom of the core, the oldest layers are deformed from the weight of ice above. In this part of the core, layers can be defined based on horizons of dust and volcanic ash that landed on the ice surface and mark specific time periods. For example, scientists can identify the exact beginning of the Bronze Age—about 3000 B.C.—owing to concentrations of ash and smoke particulates associated with copper smelting by the ancient Greeks and Romans at the time.

Within a core, any given year's accumulation consists of a layer of winter and a layer of summer ice, each differing in chemistry and texture. Scientists use oxygen isotope ratios to correlate these layers with environmental temperature conditions. Earlier, we discussed oxygen isotope ratios in ocean water and their application to the analysis of ocean sediment cores. We now discuss oxygen isotope ratios in ice, which have a different relationship to climate.

In ice cores, a *lower* $^{18}O/^{16}O$ ratio (less "heavy" oxygen in the ice) suggests colder climates, where more ^{18}O is tied up in the oceans and more light oxygen is locked into glaciers and ice sheets. Conversely, a *higher* $^{18}O/^{16}O$ ratio (more "heavy" oxygen in the ice) indicates a warmer climate during which more ^{18}O evaporates and precipitates onto ice-sheet surfaces. Therefore, the oxygen isotopes in ice cores are a proxy for air temperature.

Ice cores also reveal information about past atmospheric composition. Within the ice layers, trapped air bubbles reveal concentrations of gases—mainly carbon dioxide and methane—indicative of past environmental conditions at the time the bubble was sealed into the ice (Figure 10B.9c and d).

Several ice-core projects in Greenland have produced data spanning more than 250,000 years. In Antarctica, the Dome C ice core (part of the European Project for Ice Coring in Antarctica, or EPICA), completed in 2004, reached a depth of 3270.20 m (10,729 ft), producing the longest ice-core record at that time: 800,000 years of Earth's past climate history. This record was correlated with a core record of 400,000 years from the nearby Vostok Station and matched with ocean sediment core records to provide scientists with a sound reconstruction of climate changes throughout this time period. In 2011, American scientists extracted an ice core from the West Antarctic Ice Sheet (WAIS) that will reveal 30,000 years of annual climate history and 68,000 years at resolutions from annual to decadal—a higher time resolution than previous coring projects. (For more information, see the Byrd Polar Research Center at Ohio State University at http://bprc.osu.edu/Icecore/ and the WAIS Divide Ice Core site at http://www.waisdivide.unh.edu/news/.)

Earth's Long-Term Climate History

Climatic reconstructions using fossils and deep-ocean sediment cores reveal long-term changes in Earth's climate, shown on two different time scales in Figure 10B.10. Over the span of 70 million years, we see that Earth's climate was much warmer in the distant past, during which time tropical conditions extended to higher latitudes than today. Since the warmer times of about 50 million years ago, climate has generally cooled (Figure 10B.10a).

A distinct short period of rapid warming occurred about 56 million years ago (known as the Paleocene–Eocene Thermal Maximum, or PETM; see the geologic time scale in Figure 11.1). Scientists think that this temperature maximum was caused by a sudden increase in atmospheric carbon, the cause of which is still uncertain. One prominent hypothesis is that a massive carbon release in the form of methane occurred from the melting of methane hydrates, ice-like chemical compounds—each composed of one methane molecule surrounded by a cage of water molecules—that are stable when frozen under conditions of cold temperature and high pressure. If some warming event (such as abrupt ocean warming from a sudden change in ocean circulation) caused large amounts of hydrates to melt, the release of large quantities of methane, a short-lived but potent greenhouse gas, could drastically alter Earth's temperature. (Focus Study 10B.1, later in the chapter, discusses methane hydrates and their potential impacts on present-day warming.)

(a) A scientist stands in a snowpit at the West Antarctic Ice Sheet, where layers of snow and ice revealing individual snowfall events are backlit by a neighboring snowpit.

(b) Scientists inspect an ice-core segment at Dome C. A quarter-section of each ice core is kept on site in case an accident occurs during transport to labs in Europe.

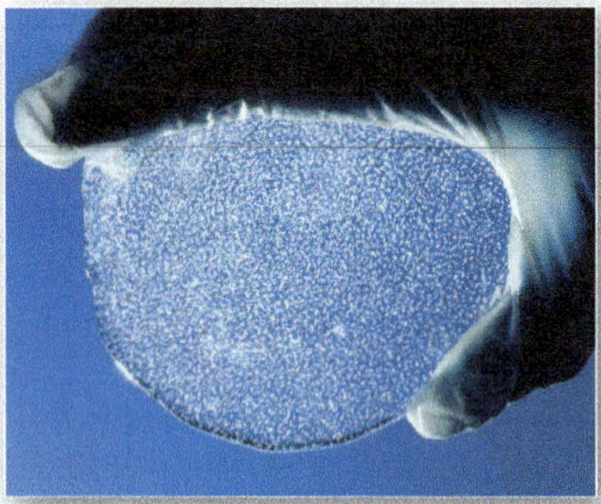

(c) Light shines through a thin section from the ice core, revealing air bubbles trapped within the ice.

(d) Air bubbles in the glacial ice indicate the composition of past atmospheres.

▲**Figure 10B.9 Ice-core analysis.** [(a) NASA LIMA. (b) and (c) British Antarctic Survey, http://www.antarctica.ac.uk. (d) Bobbé Christopherson.]

During the PETM, the rise in atmospheric carbon probably happened over a period of about 20,000 years or less—a "sudden" increase in terms of the vast scale of geologic time. Today's accelerating concentrations of atmospheric CO_2 are building at a more rapid pace. The amount of carbon that entered the atmosphere

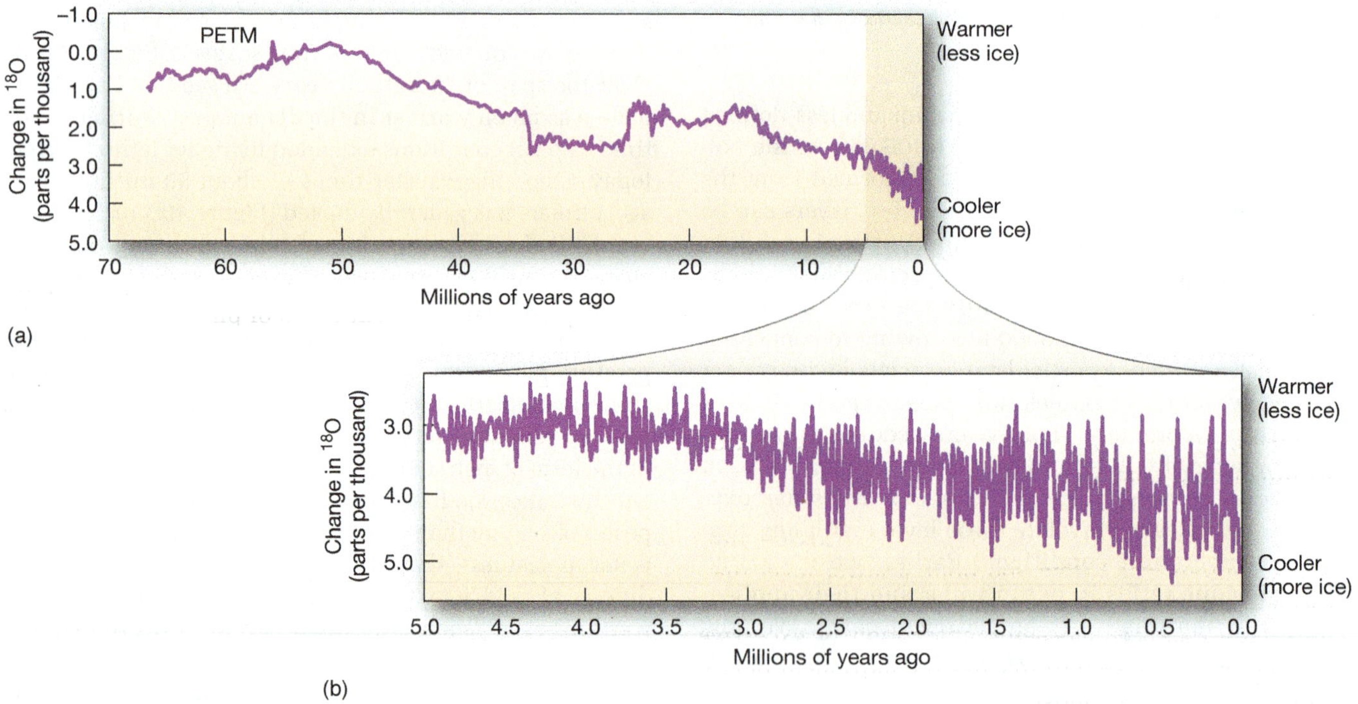

▲**Figure 10B.10 Climate reconstructions using oxygen isotopes (^{18}O) over two time scales.** The vertical axis (y-axis) shows the change in ^{18}O parts per thousand, indicating warmer and colder periods over the past 70 million years (top graph) and the past 5 miilion years (bottom graph). Note the brief, distinct rise in temperature about 56 million years ago during the Paleocene-Eocene Thermal Maximum (PETM). The bottom graph shows alrenating periods of warmer and colder temperatures within the 5 million year time span. [After Edward Aguado and James Burt, *Understanding Weather and Climate*, 6th edition, © 2013 by Pearson Education, Inc. Reprinted and electronically reproduced by permission of Pearson Education, Inc., Upper Saddle River, New Jersey.]

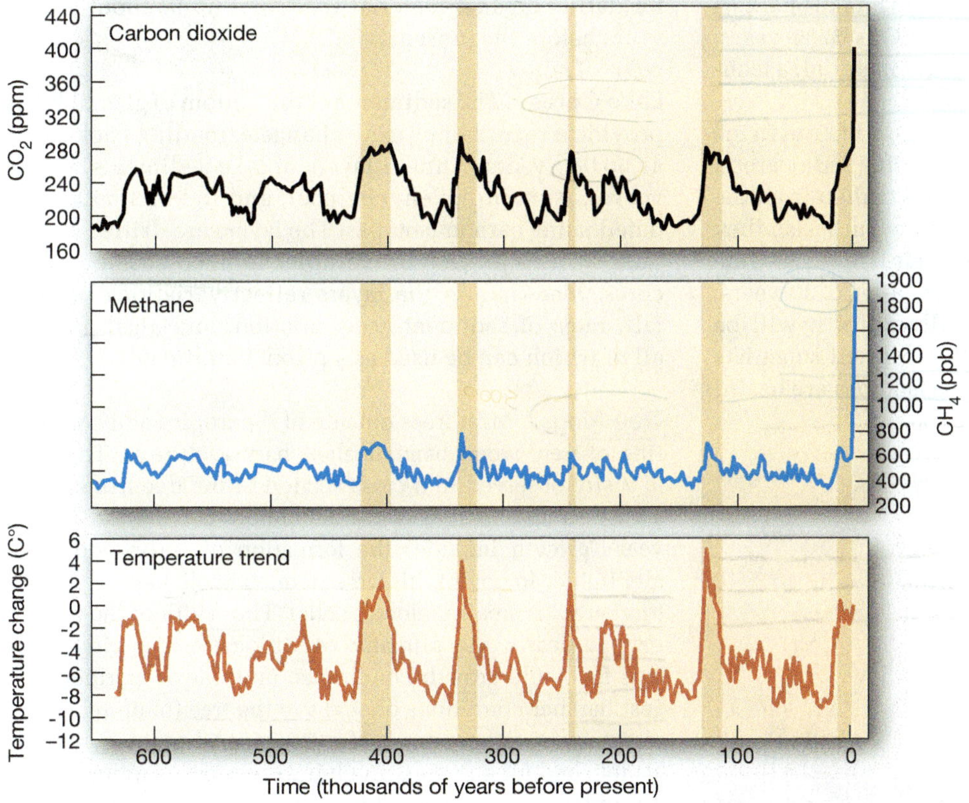

▲**Figure 10B.11 The 650,000-year record for carbon dioxide (CO_2), methane (CH_4), and temperature from ice-core data.** The shaded bands are interglacials, periods of elevated temperature and greenhouse gas concentrations. [Adapted from "Climate Change 2007: Working Group I: The Physical Science Basis," IPCC, available at http://www.ipcc.ch/publications_and_data/ar4/wg1/en/tssts-2-1-1.html]

during the PETM is estimated to be similar to the amount of carbon that human activity would release to the atmosphere with the burning of all Earth's fossil-fuel reserves.

Over the span of the last 5 million years, high-resolution climatic reconstructions using foraminifera from deep-ocean sediment cores reveal a series of cooler and warmer periods (Figure 10B.10b). These periods are known as marine isotope stages (MIS), or oxygen isotope stages; many have been assigned numbers that correspond to a specific chronology of cold and warm periods during this time span. These ocean core data are correlated with ice-core records, which show nearly identical trends.

Within this stretch of years, the last time temperatures were similar to the present-day interglacial period was during the Eemian interglacial about 125,000 years ago, during which time temperatures were warmer than at present. Notably, atmospheric carbon dioxide during the Eemian was below 300 ppm, a lower level than expected and which scientists interpret as resulting from the buffering effect of oceanic absorption of excessive atmospheric CO_2. (We will discuss the movement of CO_2 within Earth's carbon budget later in the chapter.)

As discussed earlier, ice cores also provide data on atmospheric composition, specifically for concentrations of CO_2 and methane, as measured from air bubbles trapped in the ice. Figure 10B.11 shows the changing concentrations of those two greenhouse gases, as well as changing temperature, during the last 650,000 years. Note the close correlation between the two gas concentrations, and between the gases and temperature on the graphs. Analyses have shown that the greenhouse gas concentrations lag behind the temperature changes, generally by about 1000 years. This interesting relationship suggests the presence and importance of climate feedbacks, discussed later in the chapter.

Methods for Short-Term Climate Reconstruction

Based on the paleoclimatic evidence just discussed, scientists know that Earth has undergone long-term climate cycles that have included conditions that were warmer and colder than today. Using a different set of indicators, they have also determined and verified climatic trends on shorter time scales, on the order of hundreds or thousands of years. The tools for short-term climate analysis consist mainly of radiocarbon dating and the analysis of growth rings of trees, speleothems, and corals.

Carbon Isotope Analysis Like oxygen, carbon is an element with several stable isotopes. Scientists use ^{12}C (carbon-12) and ^{13}C (carbon-13) to decipher past environmental conditions by analyzing the $^{13}C/^{12}C$ ratio in a similar manner to oxygen isotope analysis. In converting light energy from the Sun to food energy for growth, different plants use different types of photosynthesis, each of which produces a different carbon isotope ratio in the plant products. Thus, scientists can use the carbon isotope ratio of dead plant material to determine past vegetation assemblages and their associated rainfall and temperature conditions.

Up to this point, we have discussed "stable" isotopes of oxygen and carbon, in which protons and neutrons remain together in an atom's nucleus. However, certain isotopes are "unstable" because the number of neutrons compared to protons is large enough to cause the isotope to decay, or break down, into a different element. During this process, the nucleus emits radiation. This type of unstable isotope is a **radioactive isotope**.

Atmospheric carbon includes the unstable isotope ^{14}C, or carbon-14. The additional neutrons compared to protons in this isotope cause it to decay into a different atom, ^{14}N, or nitrogen-14. The rate of decay is constant and is measured as a *half-life*, or the time it takes for half of a sample to decay. The half-life of ^{14}C is 5730 years. This decay rate can be used to date plant material, a technique known as *radiocarbon dating*.

As an example, pollen is a plant material found in ice and lake sediments and is often dated using radiocarbon dating techniques. Since land plants use carbon from the air (specifically, carbon dioxide in photosynthesis), they contain ^{14}C in some amount, as does their pollen. As time passes, this radioactive carbon decays: after 5730 years, half of it will be gone, and eventually all of it will be gone. The amount of ^{14}C in the pollen can tell scientists how long ago it was alive. Radioactive isotopes are useful for dating organic material with ages up to about 50,000 years before the present.

Lake Cores The sediments at the bottom of glacial lakes provide a record of climate change extending back as far as 50,000 years. Annual layers of lake sediments, called *varves*, contain pollen, charcoal, and fossils that can be dated using carbon isotopes. The layers are drilled to produce lake sediment cores similar to deep-ocean and ice cores. Materials in the layers reflect variations in rainfall, rates of sediment accumulation, and algal growth, all of which can be used as a proxy for climate.

Tree Rings Most trees outside of the tropics add a growth ring of new wood beneath their bark each year. This ring is easily observed in a cross section of the tree trunk, or in a core sample analyzed in a laboratory (Figure 10B.12). A year's growth includes the formation of earlywood (usually lighter in color with large-diameter cells) and latewood (darker with small-diameter cells). The width of the growth ring indicates the climatic conditions: wider rings suggest favorable growth conditions, and narrower rings suggest harsher conditions or stress to the tree (often related to moisture or temperature). If a tree-ring chronology can be established for a region, involving cross-correlations among a number of trees, then this technique can be effective for assessing climatic conditions in the recent past. The dating of tree rings by these methods is *dendrochronology*; the study of past climates using tree rings is **dendroclimatology**.

To use tree rings as a climate proxy, dendroclimatologists compare tree-ring chronologies with local climate records. These correlations are then used to estimate relationships between tree growth and climate, which in some cases can yield a continuous record over hundreds or even thousands of years. Long-lived species are the most useful for dendroclimatological studies. For example, bristlecone pine in the western United States are some of the oldest organisms

(a) A cross section of a tree trunk shows growth rings; trees generally add one ring each year.

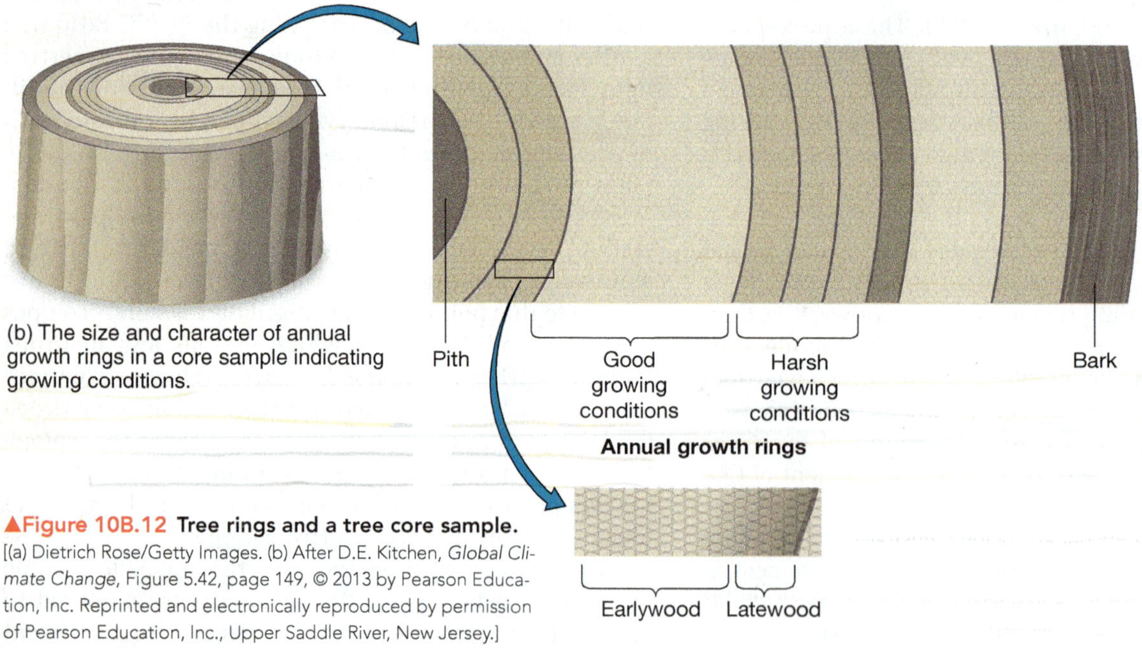

(b) The size and character of annual growth rings in a core sample indicating growing conditions.

▲**Figure 10B.12 Tree rings and a tree core sample.** [(a) Dietrich Rose/Getty Images. (b) After D.E. Kitchen, *Global Climate Change*, Figure 5.42, page 149, © 2013 by Pearson Education, Inc. Reprinted and electronically reproduced by permission of Pearson Education, Inc., Upper Saddle River, New Jersey.]

on Earth, reaching up to 5000 years in age. Evidence from tree rings in the U.S. Southwest is particularly important for assessing the magnitude of the present drought, as discussed for the Colorado River basin in Chapter 9, Focus Study 9.1, on pages 238-240.

Speleothems Limestone is a sedimentary rock that is easily dissolved by water (rocks and minerals are discussed in Chapter 11). Natural chemical processes at work on limestone surfaces often form caves and underground rivers, producing a landscape of *karst topography* (discussed in Chapter 13). Within caves and caverns are calcium carbonate ($CaCO_3$) mineral deposits called **speleothems** that take thousands of years to form. Speleothems include *stalactites*, which grow downward from a cave roof, and *stalagmites*, which grow upward from the cave floor. Speleothems form as water drips or seeps from the rock and subsequently evaporates, leaving behind a residue of $CaCO_3$ that builds up over time (Figure 10B.13).

The rate of growth of speleothems depends on several environmental factors, including the amount of rainwater percolating through the rocks that form the cave, its acidity, and the temperature and humidity conditions in the cave. Like trees, speleothems have growth rings whose size and properties reflect the environmental conditions present when they formed, and that can be dated using uranium isotopes. These growth rings also contain isotopes of oxygen and carbon, whose ratios indicate temperature and the amount of rainfall.

Scientists have correlated speleothem ring chronologies with temperature patterns in New Zealand and Russia (especially in Siberia), and with temperature and precipitation in the U.S. Southwest, among many other places. Some speleothem chronologies date to 350,000 years ago and are often combined with other paleoclimatic data to corroborate evidence of climate change.

Corals Like the shelled marine organisms found in ocean sediment cores, corals are marine invertebrates with a body called a *polyp* that extracts calcium carbonate from seawater and then excretes it to form a calcium carbonate exoskeleton. These skeletons accumulate over time in warm, tropical oceans, forming coral reefs (see discussion in Chapter 16). X-rays of core samples extracted from coral reefs reveal seasonal growth bands similar to those of trees, yielding information as to the water chemistry at the time the exoskeletons were formed (Figure 10B.14). Climatic data covering hundreds of years can be obtained this way. Although the process damages polyps living at the surface of the drill site, it does not damage the reef, and drill holes are recolonized by polyps within a few years.

Earth's Short-Term Climate History

The Pleistocene Epoch, Earth's most recent period of repeated glaciations, began 2.5 million years ago. The last glacial period lasted from about 110,000 years ago to about 11,700 years ago, with the *last glacial maximum* (LGM), the time when ice extent in the last glacial period was greatest, occurring about 20,000 years ago. Chapter 17 discusses changes to Earth's landscapes during this time, and Figure 17.25 shows the extent of glaciation during this prolonged cold period. The climate record for the past 20,000 years reveals the period of cold temperatures

(a) Speloethems in Royal Cave Buchanan, Victoria, Australia.

(b) Growth bands in a speleothem cross section.

▲**Figure 10B.13 Speleothems in a cavern and in cross section.** [(a) Chris Howes/Wild Places Photography/Alamy. (b) Pauline Treble, Australian Nuclear Science and Technology Organisation.]

(a) Scientists drill into corals on the Clipperton Atoll, an uninhabited coral reef in the Pacific Ocean off the west coast of Central America.

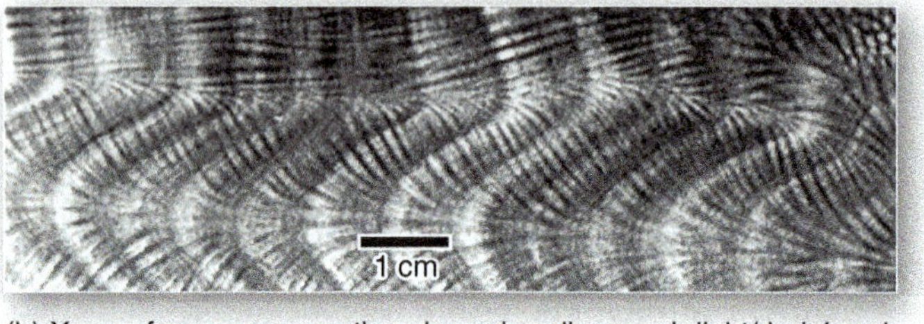

(b) X-ray of core cross section shows banding; each light/dark band indicates one year of growth.

◀**Figure 10B.14 Extraction and cross section of coral core samples.** [(a) Maris Kazmers, NOAA Paleoclimatology Program (b) Thomas Felis, Research Center Ocean Margins, Bremen/NASA.]

and little snow acccumulation that occurred from the LGM to about 15,000 years ago. (Figure 10B.15).

About 14,000 years ago, average temperatures abruptly increased for several thousand years, and then dropped again during the colder period known as the *Younger Dryas*. The abrupt warming about 11,700 years ago marked the end of the Pleistocene Epoch. Note in Figure 10B.15 that snow accumulation is less during colder glacial periods. As we learned in Chapters 7 and 8, the capacity of cold air to absorb water vapor is less than for warm air, resulting in decreased snowfall during glacial periods even though a greater volume of ice is present over Earth's surface.

From A.D. 800 to 1200, a number of climate proxies (tree rings, corals, and ice cores) show a mild climatic episode, now known as the *Medieval Climate Anomaly* (a period during which the Vikings settled Iceland and coastal areas of Greenland). During this time, warmer temperatures—as warm or warmer than today—occurred in some regions whereas cooling occurred in other regions. The warmth over the North Atlantic region allowed a variety of crops to grow at higher latitudes in Europe, shifting settlement patterns northward. Scientists think that the cooling in some places during this time is linked to the cool La Niña phase of the ENSO phenomenon over the tropical Pacific.

From approximately A.D. 1250 through about 1850, temperatures cooled globally during a period known as the *Little Ice Age*. Winter ice was more extensive in the North Atlantic Ocean, and expanding glaciers in western Europe blocked many key mountain passes. During the coldest years, snowlines in Europe lowered about 200 m (650 ft) in elevation. This was a 600-year span of somewhat inconsistent colder temperatures, a period that included many rapid, short-term climate fluctuations that lasted only decades and are probably related to volcanic activity and multiyear oscillations in global circulation patterns, specifically the North Atlantic Oscillation (NAO) and Arctic Oscillation (AO) (see discussion in Chapter 6).

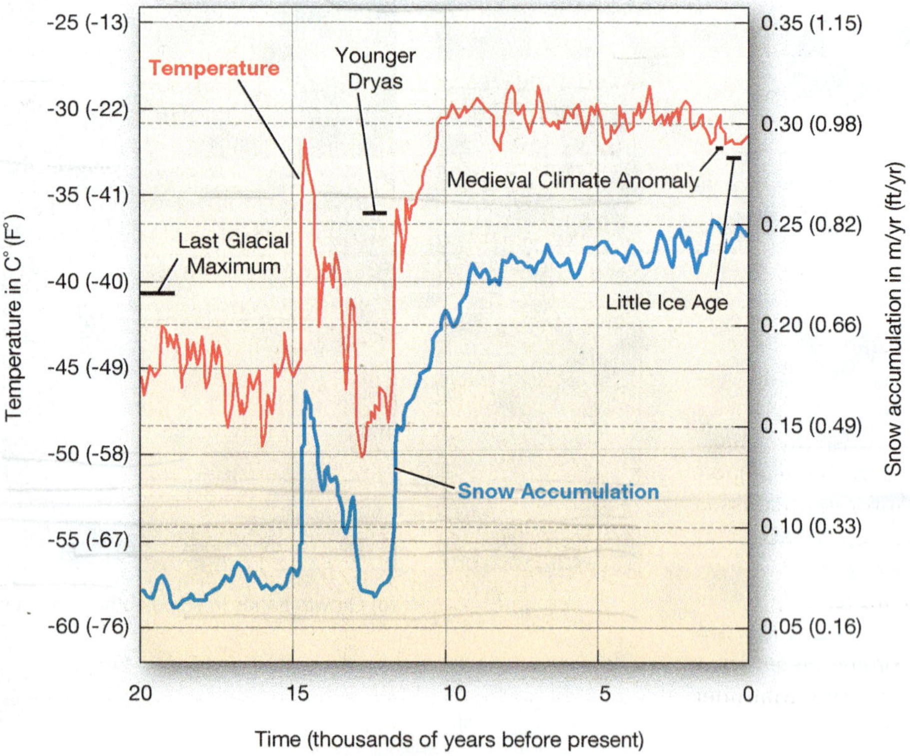

◀**Figure 10B.15 The past 20,000 years of temperature and snow accumulation.** Evidence from Greenland ice cores shows periods of colder temperatures occurring during the last glacial maximum and the *Younger Dryas*, and an abrupt temperature rise occurring about 14,000 years ago and again about 12,000 years ago at the end of the *Younger Dryas*. Although this graph uses ice-core data, these temperature trends correlate with other climate proxy records. [From R.B. Alley, "The Younger Dryas cold interval as viewed from central Greenland," *Quaternary Science Reviews* 19 (January 2000): 213-226; available at http://www.ncdc.noaa.gov/paleo/pubs/alley2000/.]

After the *Little Ice Age* temperatures steadily warmed, and with growing human population and the onset of the Industrial Revolution, this warming has continued—a trend that is accelerating today.

Mechanisms of Natural Climate Fluctuation

In reviewing climate records on various scales, we have seen that Earth's climate cycles between warmer and colder periods. When temperature is viewed over certain time scales, such as over periods of about 650,000 years (illustrated in Figure 10B.11), patterns are apparent that appear to follow cycles of about 100,000 years, 40,000 years, and 20,000 years. Scientists have evaluated a number of natural mechanisms that affect Earth's climate and might cause these long-term cyclical climate variations.

Solar Variability

As we learned in earlier chapters, energy from the Sun is the most important driver of the Earth–atmosphere climate system. The Sun's output of energy toward Earth, known as *solar irradiance*, varies over several timescales, and these natural variations can affect climate. Over billions of years, solar output has generally increased; overall, it has increased by about one-third since the formation of the solar system. Within this time frame, variations on the scale of thousands of years are linked to changes in the solar magnetic field. Over recent decades, scientists have measured slight variations in the amount of radiation received at the top of the atmosphere using satellite data and have correlated these variations to sunspot activity.

As discussed in Chapter 2, the number of sunspots varies over an 11-year solar cycle. When sunspot abundance is high, solar activity and output increases; when sunspot abundance is low, solar output decreases. Scientists have determined that these relationships are reflected in climatic indicators such as temperature. For example, the record of sunspot occurrences shows a prolonged solar minimum (a period with little sunspot activity) from about 1645 to 1715, during one of the coldest periods of the Little Ice Age. Known as the **Maunder Minimum**, this 70-year period would suggest a causal effect between decreased sunspot abundance and cooling in the North Atlantic region. However, recent temperature increases have occurred during a prolonged solar minimum (from 2005 to 2010), which corresponds with a period of reduced solar irradiance. Thus, the causal effect is not definitive. The IPCC 5th Assessment Report includes solar irradiance as a climate forcing (discussed later in the chapter), however scientists agree that solar irradiance does not appear to be the primary driver of recent global warming trends (as an example, see http://www.giss.nasa.gov/research/news/20120130b/).

To explain the apparent correlation between solar output and cooler temperatures during the Maunder Minimum, many scientists think that reduced solar output, while not the cause of cooling, did serve to reinforce the colder temperatures through feedback mechanisms such as the ice–albedo feedback (discussed ahead). Recent research suggests that solar output may have some effects on regional climate, such as occurred during the Maunder Minimum, without affecting overall global climate trends (see http://science.nasa.gov/science-news/science-at-nasa/2013/08jan_sunclimate/).

Earth's Orbital Cycles

If the overall amount of energy from the Sun does not drive climate change, then another logical hypothesis is that Earth–Sun relationships affect climate, a reasonable connection given that orbital relationships affect energy receipts and seasonality on Earth. These relationships include Earth's distance from the Sun, which varies within its orbital path, and Earth's orientation to the Sun, which varies as a result of the "wobble" of Earth on its axis, and because of Earth's varying axial tilt (please review Chapter 2, where we discussed Earth–Sun relations and the seasons).

Milutin Milankovitch (1879–1958), a Serbian astronomer, studied the irregularities in Earth's orbit around the Sun, its rotation on its axis, and its axial tilt, and identified regular cycles that relate to climatic patterns (Figure 10B.16).

- Earth's elliptical orbit about the Sun known, as *eccentricity*, is not constant and changes in cycles over several time scales. The most prominent is a 100,000-year cycle in which the shape of the ellipse varies by more than 17.7 million kilometers (11 million miles), from a

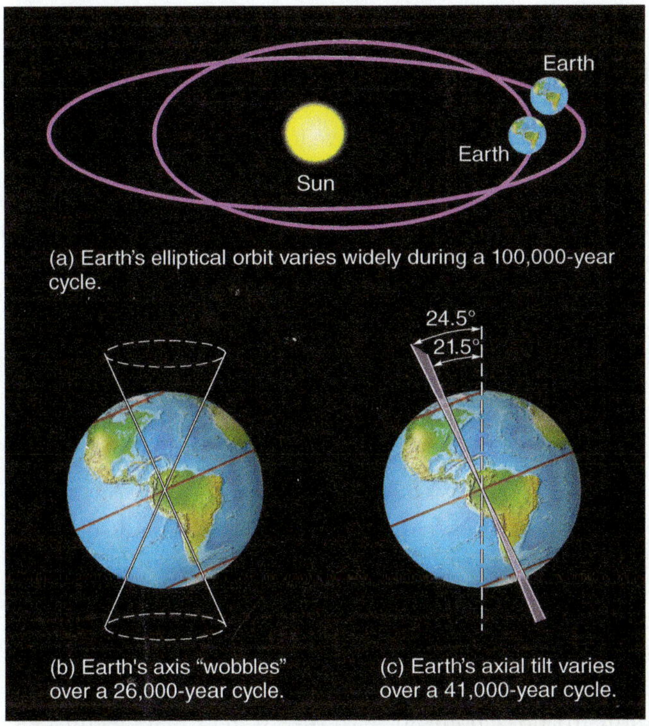

▲**Figure 10B.16 Astronomical factors that may affect broad climatic cycles.** Figures are an exaggeration of actual orbital paths, axis wobble, and axial tilt.

shape that is nearly circular to one that is more elliptical (Figure 10B.16a).

- Earth's axis "wobbles" through a 26,000-year cycle, in a movement much like that of a spinning top winding down (Figure 10B.16b). Earth's wobble, known as *precession*, changes the orientation of hemispheres and landmasses to the Sun.
- Earth's axial tilt, at present about 23.5°, varies from 21.5° to 24.5° during a 41,000-year period (Figure 10B.16c).

Although many of his ideas were rejected by the scientific community at the time, these consistent orbital cycles, now called **Milankovitch cycles**, are today accepted as a causal factor for long-term climatic fluctuations, though their role is still being investigated. Scientific evidence in ice cores from Greenland and Antarctica and in the accumulated sediments of Lake Baikal in Russia have confirmed a roughly 100,000-year climatic cycle; other evidence supports the effect of shorter-term cycles of roughly 40,000 and 20,000 years on climate. Milankovitch cycles appear to be an important cause of glacial–interglacial cycles, although other factors probably amplify the effects (such as changes in the North Atlantic Ocean, albedo effects from gains or losses of Arctic sea ice, and variations in greenhouse gas concentrations, discussed just ahead).

Continental Position and Topography

In Chapters 11 and 12, we discuss plate tectonics and the movement of the continents over the past 400 million years. Because Earth's lithosphere is composed of moving plates, continental rearrangement has occurred throughout geologic history (look ahead to Figure 11.13). This movement affects climate, since landmasses have strong effects on the general circulation of the atmosphere. As discussed in previous chapters, the relative proportions of land and ocean area affect surface albedo, as does the position of landmasses relative to the poles or equator. The position of the continents also impacts ocean currents, which are critical for redistributing heat throughout the world's oceans. Finally, the movement of continental plates causes episodes of mountain building and spreading of the seafloor, discussed in Chapter 12. These processes affect Earth's climate system as high-elevation mountain ranges accumulate snow and ice during glacial periods, and as CO_2 from volcanic outgassing enters the atmosphere (from volcanoes) and the ocean (from spreading of the seafloor).

Atmospheric Gases and Aerosols

Natural processes can release gases and aerosols into Earth's atmosphere with varying impacts on climate. Natural outgassing from Earth's interior through volcanoes and vents in the ocean floor is the primary natural source of CO_2 emissions to the atmosphere. Water vapor is a natural greenhouse gas present in Earth's atmosphere and is discussed ahead with regard to climate feedbacks. Over long time scales, higher levels of greenhouse gases generally correlate with warmer interglacials, and lower levels correlate with colder glacials. As greenhouse gas concentrations change, Earth's surface heats or cools in response. Further, these gases can amplify the climatic trends (discussed ahead with climate feedbacks). In cases where huge amounts of greenhouse gases are released into the atmosphere, such as may have happened during the PETM, these gases can potentially drive climatic change.

In addition to outgassing, volcanic eruptions also produce aerosols that scientists definitively have linked to climatic cooling. Accumulations of aerosols ejected into the stratosphere can create a layer of particulates that increases albedo so that more insolation is reflected and less solar energy reaches Earth's surface. Studies of twentieth-century eruptions have shown that sulfur aerosol accumulations affect temperatures on timescales of months to years. As examples, the aerosol cloud from the 1982 El Chichón eruption in Mexico lowered temperatures worldwide for several months, and the 1991 Mount Pinatubo eruption lowered temperatures for 2 years (discussed in Chapter 6, Figure 6.1; also see Figure 10B.17 ahead). Scientific evidence also suggests that a series of large volcanic eruptions may have initiated the colder temperatures of the Little Ice Age in the second half of the thirteenth century.

Climate Feedbacks and the Carbon Budget

Earth's climate system is subject to a number of feedback mechanisms. As discussed in Chapter 1, systems can produce outputs that sometimes influence their own operations via positive or negative feedback loops. Positive feedback amplifies system changes and tends to destabilize the system; negative feedback inhibits system changes and tends to stabilize the system. **Climate feedbacks** are processes that either amplify or reduce climatic trends, toward either warming or cooling.

A good example of a positive climate feedback is the *ice–albedo feedback* introduced in Chapter 1 (see Figure 1.8) and discussed in the Chapter 4 *Geosystems Now*. This feedback is accelerating the current climatic trend toward global warming. However, ice–albedo feedback can also amplify global cooling, because lower temperatures lead to more snow and ice cover, which increases albedo, or reflectivity, and causes less sunlight to be absorbed by Earth's surface. Scientists think that the ice–albedo feedback may have amplified global cooling following the volcanic eruptions at the start of the Little Ice Age. As atmospheric aerosols increased and temperatures decreased, more ice formed, further increasing albedo, leading to further cooling and more ice formation. These conditions persisted until the recent influx of anthropogenic greenhouse gases into the atmosphere at the onset of the Industrial Revolution in the 1800s.

Earth's Carbon Budget

Many climate feedbacks involve the movement of carbon through Earth systems and the balance of carbon over time within these systems. The carbon on Earth cycles through atmospheric, oceanic, terrestrial, and living systems in a biogeochemical cycle known as the *carbon cycle* (discussed and illustrated in Chapter 19). Areas of carbon release are carbon sources; areas of storage are called **carbon sinks**, or carbon reservoirs. The overall exchange of carbon between the different systems on Earth is the **global carbon budget**, which should naturally remain balanced as carbon moves between sources and sinks. *Geosystems in Action*, Figure GIA 10B, illustrates the components, both natural and anthropogenic, of Earth's carbon budget.

Several areas on Earth are important carbon sinks. The ocean is a major carbon storage area, taking up CO_2 by chemical processes as it dissolves in seawater and by biological processes through photosynthesis in microscopic marine organisms called phytoplankton. Rocks, another carbon sink, contain "ancient" carbon from dead organic matter that was solidified by heat and pressure, including the shells of ancient marine organisms that lithified to become limestone (discussed in Chapter 11). Forests and soils, where carbon is stored in both living and dead organic matter, are also important carbon sinks. Finally, the atmosphere is perhaps the most critical area of carbon storage today, as human activities cause emissions that lead to increasing concentrations of CO_2 into the atmosphere, changing the balance of carbon on Earth.

Humans have impacted Earth's carbon budget for thousands of years, beginning with the clearing of forests for agriculture, which reduces the areal extent of one of Earth's natural carbon sinks (forests) and transfers carbon to the atmosphere. With the onset of the Industrial Revolution, around 1850, the burning of fossil fuels became a large source of atmospheric CO_2 and began the depletion of fossilized carbon stored in rock. These activities have transferred solid carbon stored in plants and rock to gaseous carbon in the atmosphere.

Given the large concentrations of CO_2 currently being released by human activities, scientists have for several decades wondered why the amount of CO_2 in Earth's atmosphere is not higher. Where is the missing carbon? Studies suggest that uptake of carbon by the oceans is offsetting some of the atmospheric increase. When dissolved CO_2 mixes with seawater, carbonic acid (H_2CO_3) forms, in a process of *ocean acidification*. The increased acidity affects seawater chemistry and harms marine organisms, such as corals and some types of plankton, that build shells and other external structures from calcium carbonate (discussed in Chapter 16). Scientists estimate that the oceans have absorbed some 50% of the rising concentrations of atmospheric carbon, slowing the warming of the atmosphere. However, as the oceans increase in temperature, their ability to dissolve CO_2 is lessened. Thus, as global air and ocean temperatures warm, more CO_2 will likely remain in the atmosphere, with related impacts on Earth's climate.

Uptake of excess carbon is also occurring in Earth's terrestrial environment, as increased CO_2 levels in the atmosphere enhance photosynthesis in plants. Research suggests that this produces a "greening" effect as plants produce more leaves in some regions of the world (see Chapter 19, Figure HD 19c on page 614).

Water-Vapor Feedback

Water-vapor is the most abundant natural greenhouse gas in the Earth–atmosphere system. Water-vapor feedback is a function of the effect of air temperature on the amount of water vapor that air can absorb, a subject discussed in Chapters 7 and 8. As air temperature rises, evaporation increases, because the capacity to absorb water vapor is greater for warm air than for cooler air. Thus, more water enters the atmosphere from land and ocean surfaces, humidity increases, and greenhouse warming accelerates. As temperatures increase further, more water vapor enters the atmosphere, greenhouse warming further increases, and the positive feedback continues.

The water-vapor climate feedback is still not well understood, in large part because measurements of global water vapor are limited, especially when compared to the relatively strong data sets for other greenhouse gases, such as CO_2 and methane. Another complicating factor is the role of clouds in Earth's energy budget. As atmospheric water vapor increases, higher rates of condensation will lead to more cloud formation. Remember from Chapter 4 (Figure 4.8, page 91) that low, thick cloud cover increases the albedo of the atmosphere and has a cooling effect on Earth (cloud albedo forcing). In contrast, the effect of high, thin clouds can cause warming (called cloud greenhouse forcing).

Carbon–Climate Feedbacks

We saw earlier that over long time periods, CO_2 and methane concentrations track temperature trends, with a slight lag time (Figure 10B.11). One hypothesis for this relationship is that warming temperatures caused by changes in Earth's orbital configuration may trigger the release of greenhouse gases (both CO_2 and methane), which then act as a positive feedback mechanism: Initial warming leads to increases in gas concentrations; elevated gas concentrations then amplify warming; and so on.

In this chapter's *Geosystems Now*, we discussed this carbon–climate feedback as it occurs in permafrost areas. Figure GIA 10B illustrates the *permafrost–carbon feedback* now underway in the Arctic. This process occurs as warming temperatures lead to permafrost thaw. Rising atmospheric CO_2 leads to increased plant growth and microbial activity, and eventually leads to a greater amount of carbon emitted to the atmosphere rather than stored in the ground—a positive feedback that accelerates warming.

CO_2–Weathering Feedback

Not all climate feedback loops act on short time scales and not all have a positive, or amplifying, effect on climatic trends. Some feedback is negative, acting to slow the warming or cooling trend. For example, increasing CO_2 stored in the atmosphere increases global warming, which increases the amount of water vapor present in the atmosphere (warm air masses absorb more moisture). Greater atmospheric moisture in a warmer climate generally leads to greater precipitation. With increasing rainfall comes an increase in the breakdown of exposed rock on Earth's

(text continued on page 304)

Several processes transfer carbon between the atmosphere, hydrosphere, lithosphere, and biosphere. (GIA 10B.1). Over time, the flow of carbon among the spheres—Earth's carbon budget—has remained roughly in balance. Today, human activities, primarily the burning of fossil fuels (GIA 10B.2) and the removal of forests (GIA 10B.3), are increasing the atmospheric concentration of carbon dioxide, altering the carbon budget, and affecting Earth's climate. As global temperatures rise, Arctic permafrost thaws, releasing additional carbon into the atmosphere (GIA 10B.4).

10B.1 COMPONENTS OF THE CARBON BUDGET

Carbon is taken up by plants through photosynthesis or becomes dissolved in the oceans. On land, carbon is stored in vegetation and soils. Fossil carbon is stored within sedimentary rock strata beneath Earth's surface. On the ocean floor, the carbon-containing shells of marine organisms build up in thick layers that form limestone. Carbon flows in and out of Earth's main areas of carbon storage.

Gases in the atmosphere: (800 Gt)

Organic material: Soils, plants, animals (2,850 Gt)

Rock in Earth's crust: Primarily limestone and shale (16,000 Gt)

CO_2 dissolved in water: (38,000 Gt)

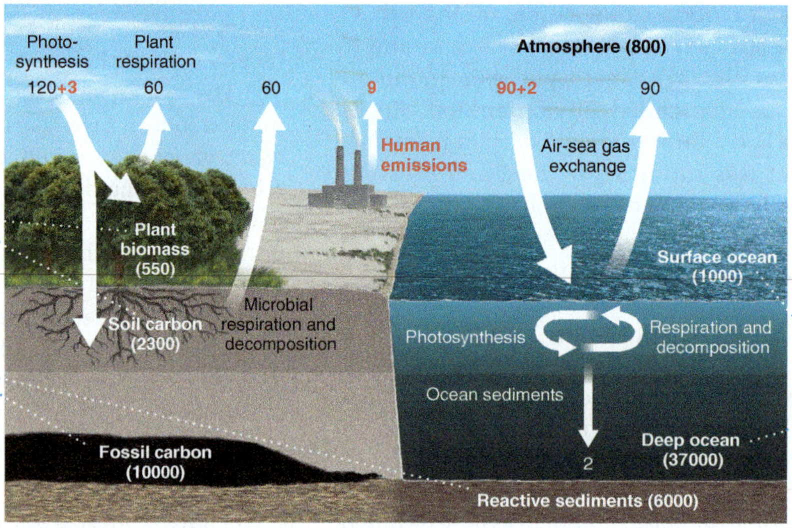

Numbers are in Gt (gigatonnes, or billions of tons) of carbon per year. White numbers are natural flows; red are human contributions; bold numbers are carbon sinks, or areas of carbon storage. [After U.S. DOE and NASA.]

Analyze: How much carbon leaves the atmosphere each year through processes on land? Explain.

10B.2 CARBON EMISSIONS FROM FOSSIL-FUEL BURNING

Carbon emissions caused by human activities rose throughout the twentieth century and continue to rise (GIA 10B.2a). A major contributor to this increase is the burning of fossil fuels, which releases carbon dioxide into the atmosphere. As GIA 10B.2b shows, carbon dioxide emissions are projected to continue to rise in coming decades. (values are in Gt, or gigatonnes).

Calculate: How many times more carbon emissions do humans produce annually today than in 1950?

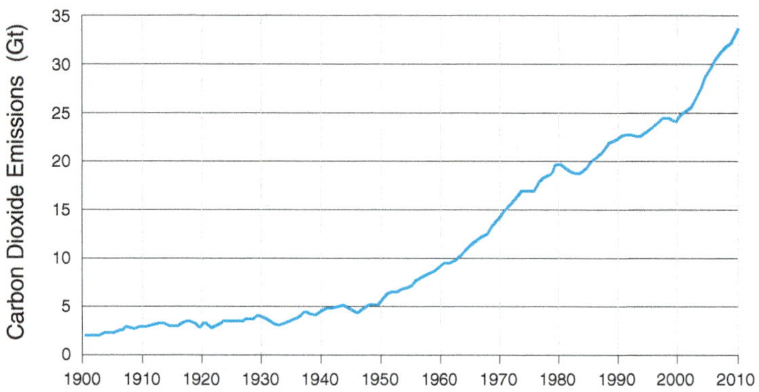

10B.2a World carbon dioxide emissions, 1900-2010 [Carbon Dioxide Information Analysis Center, U.S. DOE, 2013.]

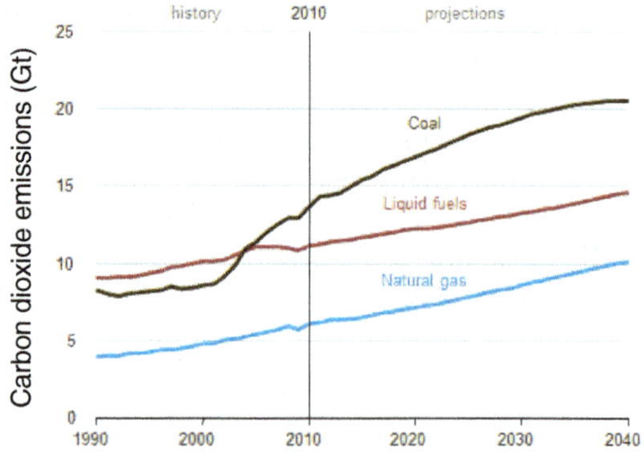

10B.2b World energy-related carbon dioxide emissions by fuel type, 1990-2040 [U.S. EIA, 2013.]

Video
Keeping Up with Carbon

MasteringGeography™

Visit the Study Area in MasteringGeography™ to explore the carbon budget.

Visualize: Study videos of the carbon cycle and climate modeling.

Assess: Demonstrate understanding of the carbon budget (if assigned by instructor).

Video
Taking Earth's Temperature

10B.3 CARBON EMISSIONS FROM DEFORESTATION

Earth's forests are a major carbon sink, storing large amounts of carbon in their wood and leaves. Today, forests worldwide are threatened as they are cut for wood products and as forest lands are converted to other types of land use. Deforestation releases about one Gt of carbon into the atmosphere each year, amplifying climate change [Climate Change and Environmental Risk Atlas, 2012, Maplecroft.]

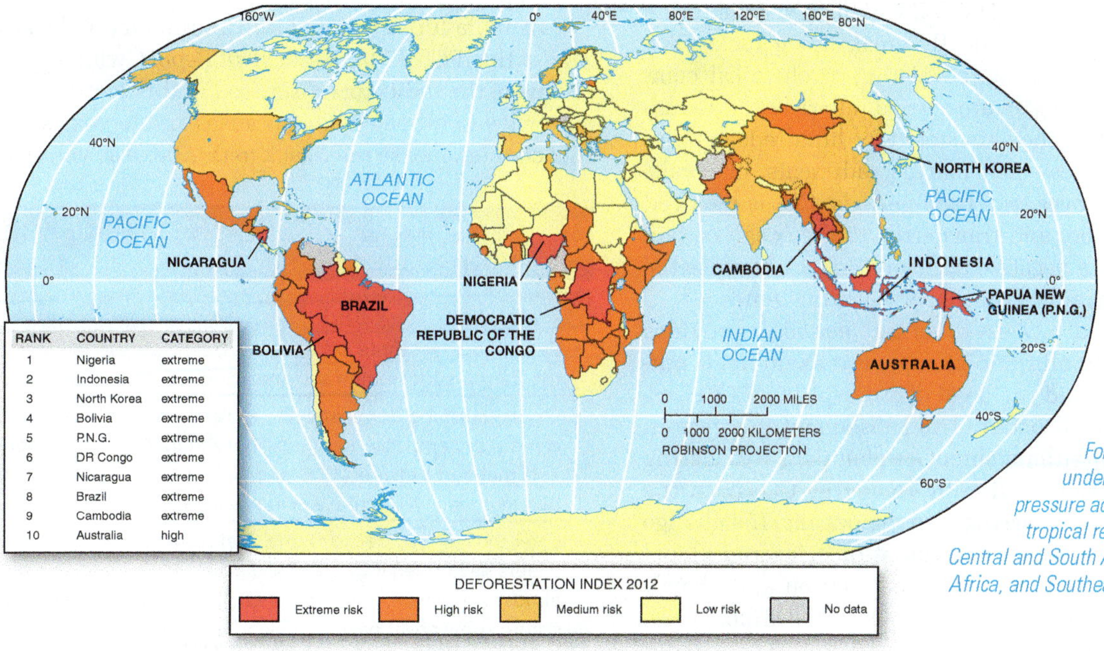

RANK	COUNTRY	CATEGORY
1	Nigeria	extreme
2	Indonesia	extreme
3	North Korea	extreme
4	Bolivia	extreme
5	P.N.G.	extreme
6	DR Congo	extreme
7	Nicaragua	extreme
8	Brazil	extreme
9	Cambodia	extreme
10	Australia	high

Forests are under intense pressure across the tropical regions of Central and South America, Africa, and Southeast Asia.

Explain: On which continent do forests seem most threatened? Least threatened? What factors do you think might account for the difference?

10B.4 CARBON EMISSIONS FROM ARCTIC PERMAFROST THAW

When permafrost remains intact, the carbon budget remains in balance. Because soil organic matter decays slowly (if at all) in cold temperatures, Arctic soils absorb and release only a small amount of carbon (GIA 10B.4a). Warming temperatures cause vegetation and soils to take up more carbon (GIA 10B.4b), and as permafrost thaws more carbon from tundra soils is released than is absorbed (GIA 10B.4c) [After Zina Deretsky, NSF, based on research by Ted Schuur, University of Florida.]

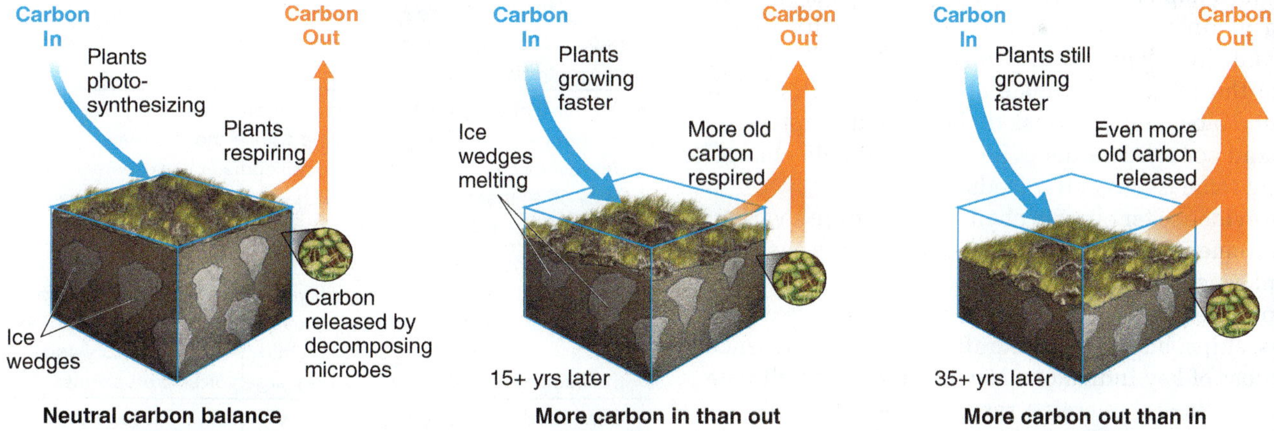

10B.4a *Permafrost Intact*

Arctic vegetation absorbs carbon through photosynthesis during the warm summer months. At the same time, vegetation and microbial activity release carbon to the atmosphere by respiration.

10B.4b *Starting to thaw*

As temperatures warm, the Arctic growing season lengthens, and carbon uptake by vegetation and soils increases. Higher concentrations of atmospheric CO_2 can accelerate plant growth, causing plants to take up even more carbon. Soil microbes that decompose organic material become more abundant and release more stored carbon.

10B.4c *More thawed*

As warming continues, thawing of permafrost and soil allows microbes to flourish and decompose even more organic material, thus releasing larger amounts of carbon.

GEOquiz

1. Infer: Refer to GIA 10B.1. Some of the 9 Gt of carbon from human emissions is taken up by plants and some is dissolved in the oceans. How much is left to increase CO_2 levels in the atmosphere?

2. Analyze: What are the inputs, actions, and outputs in the Arctic permafrost system shown in GIA 10B.4? Explain.

surface by chemical weathering processes (discussed in Chapter 13). This process occurs over long time periods as CO_2 in the atmosphere dissolves in rainwater to form a weak acid (carbonic acid, H_2CO_3), which then falls to the ground and works to chemically break down rocks. The outputs of this weathering, consisting of chemical compounds dissolved in water, are carried into rivers and eventually oceans, where they are stored for hundreds of thousands of years in seawater, marine sediments, or corals. The result is that, over long time periods, CO_2 is removed from the atmosphere and transferred to the ocean carbon sink. This creates a negative climate feedback because the overall effect is to reduce the global warming trend.

This so-called *CO_2–weathering feedback* provides a natural buffer to climatic change. For example, if outgassing increases and the level of atmospheric CO_2 rises, global warming will occur. The subsequent increase in precipitation and enhancement of chemical weathering then acts to buffer warming by removing CO_2 from the atmosphere and transferring it to the oceans. If a change in orbital cycles or oceanic circulation triggered global cooling, then the reduction of precipitation and chemical weathering would leave more CO_2 in the atmosphere, where it works to increase temperatures. Throughout Earth's history, this natural buffer has helped prevent Earth's climate from becoming too warm or too cold. The rapid pace of present climate change, however, is evidently beyond the ability of natural systems to moderate.

Evidence for Present Climate Change

In previous chapters, we discussed many aspects of contemporary climate change as we explored Earth's atmosphere and hydrosphere. In subsequent chapters, we examine the effects of climate change on Earth's lithosphere and biosphere. The task of the present chapter is to review and consolidate the evidence for climate change, revisiting some issues and introducing others.

The evidence for climate change comes from a variety of measurements showing global trends over the past century, and especially over the past two decades. Data gathered from weather stations, orbiting satellites, weather balloons, ships, buoys, and aircraft confirm the presence of a number of key indicators. New evidence and climate change reports are emerging regularly; good sources for the latest updates on climate change science are the U.S. Global Change Research Program, at http://www.globalchange.gov/home, and the Intergovernmental Panel on Climate Change (IPCC), operating since 1988 and issuing its *Fifth Assessment Report* between September 2013 and April 2014 (see http://www.ipcc.ch/).

In this section, we present and discuss the measurable indicators that unequivocally show climatic warming:

- Increasing temperatures over land and ocean surfaces, and in the troposphere
- Increasing sea-surface temperatures and ocean heat content
- Melting glacial ice and sea ice
- Rising sea level
- Increasing humidity

In Table 10B.1, you will find a summary of the key findings of the IPCC *Fifth Assessment Report*, which includes a review of the indicators discussed in this section as well as a preview of climate change causes and forecasts ahead in this chapter. Please refer back to this table as you read.

TABLE 10B.1 Highlights of the 2013–2014 IPCC *Fifth Assessment Report*

Key Points of the Fifth Assessment Summary*
• Each of the last three decades has been successively warmer at the Earth's surface than any preceding decade since 1850.
• Ocean warming dominates the increase in energy stored in the climate system, accounting for more than 90% of the energy accumulated between 1971 and 2010. Further uptake of carbon by the ocean will increase ocean acidification.
• Over the last two decades, the Greenland and Antarctic ice sheets have been losing mass, glaciers have continued to shrink almost worldwide, and Arctic sea ice and Northern Hemisphere spring snow cover have continued to decrease in extent.
• The rate of sea-level rise since the mid-nineteenth century has been larger than the mean rate during the previous two millennia. Over the period 1901–2010, global mean sea level rose by 0.19 m (0.6 ft).
• The atmospheric concentrations of carbon dioxide (CO_2), methane, and nitrous oxide have increased to levels unprecedented in at least the last 800,000 years, primarily from fossil-fuel emissions and secondarily from net land-use-change emissions. The ocean has absorbed about 30% of the emitted anthropogenic carbon dioxide, causing ocean acidification.
• Total radiative forcing is positive and has led to an uptake of energy by the climate system. The largest contribution to total radiative forcing is caused by the increase in the atmospheric concentration of CO_2 since 1750.
• Warming of the climate system is unequivocal. Many of the temperature changes observed since the 1950s are unprecedented over decades to millennia. It is *extremely likely* (95%–99%) that human influence has been the dominant cause of the observed warming since the mid-twentieth century.
• Climate models have improved since the *Fourth Assessment Report*. Models reproduce observed continental-scale surface-temperature patterns and trends over many decades, including the more rapid warming since the mid-twentieth century and the cooling immediately following large volcanic eruptions.
• Continued emissions of greenhouse gases will cause further warming and changes in all components of the climate system. Limiting climate change will require substantial and sustained reductions of greenhouse gas emissions.
• Changes in the global water cycle will not be uniform. The contrast in precipitation between wet and dry regions and between wet and dry seasons will increase.
• Global mean sea level will continue to rise. The rate of sea-level rise will *very likely* exceed that observed during 1971–2010, due to increased ocean warming and increased loss of mass from glaciers and ice sheets.

*Source: *Climate Change 2013, The Physical Science Basis*, Summary for Policy Makers (SPM), Working Group I, Contribution to the Fifth Assessment Report of the IPCC.

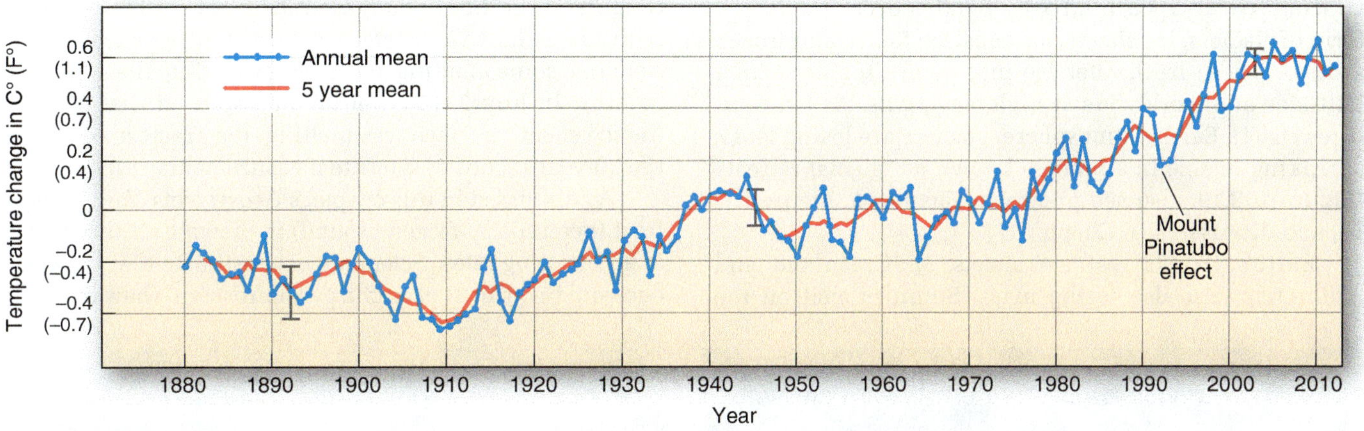

▲**Figure 10B.17 Global land–ocean temperature trends, 1880–2010.** The graph shows change in global surface temperatures relative to the 1951–1980 global average. The gray bars represent uncertainty in the measurements. Note the inclusion of both annual average temperature anomalies and 5-year mean temperature anomalies; together they give a sense of overall trends. [Based on data from NASA/GISS; available at http://climate.nasa.gov/key_indicators#co2.]

Temperature

In previous chapters, we discussed the rise in atmospheric temperatures during this century. In Chapter 5, *The Human Denominator*, Figure HD 5c, presents an important graph of data from four independent surface-temperature records showing a warming trend since 1880. These records, each collected and analyzed using slightly different techniques, show remarkable agreement. Figure 10B.17 plots the NASA temperature data of global mean annual surface air temperature anomalies (as compared to the 1951–1980 temperature-average base line) and 5-year mean temperatures from 1880 through 2012. (Remember from Chapter 5 that temperature anomalies are the variations from the mean temperature during some period of record.)

The temperature data unmistakably show a warming trend. Since 1880, in the Northern Hemisphere, the years with the warmest land-surface temperatures were 2005 and 2010 (a statistical tie). For the Southern Hemisphere, 2009 was the warmest in the modern record. In Chapter 5, we saw that the period from 2000 to 2010 was the warmest decade since 1880 (look back to Figure 5.17 on page 125). The data from long-term climate reconstructions of temperature point to the present time as the warmest in the last 120,000 years (see Figure 10B.11). These reconstructions also suggest that the increase in temperature during the twentieth century is *extremely likely* (within a confidence interval of greater than 95%–100%*) the largest to occur in any century over the past 1000 years.

Record-setting summer daytime temperatures are being recorded in many countries (Figure 5.18). For example, in August 2013 temperatures in western Japan topped 40°C (104°F) for 4 days, and on August 12th, the temperature in Kochi Prefecture reached 41°C (106°F), the highest ever recorded in that country (Focus Study 5.1 on page 126 discusses record temperatures in Australia in 2013). In the United States, the number of unusually hot summer days has been rising since 1990 (see http://www.epa.gov/climatechange/science/indicators/weather-climate/high-low-temps.html). The year 2012 broke all U.S records (since 1895) by 0.7 C° (1.4 F°), with an average temperature for the year of 12.9°C (55.3°F).

Ocean temperatures are also rising. As discussed in Chapter 5, sea surface temperatures increased at an average rate of 0.07 C° (0.13 F°) per year from 1901 to 2012 as oceans absorbed atmospheric heat. This rise is reflected in measurements of upper-ocean heat content, which includes the upper 700 m (2296 ft) of ocean (http://www.ncdc.noaa.gov/indicators/, click on "warming climate"). This increasing heat content is consistent with sea-level rise resulting from the thermal expansion of seawater (discussed ahead).

Ice Melt

Heating of Earth's atmosphere and oceans is causing land ice and sea ice to melt. Chapter 17 discusses the character and distribution of snow and ice in Earth's cryosphere.

▲**Figure 10B.18 Heat wave hits the United Kingdom in 2013.** Temperatures topped 32°C (89.6°F) for over a week in southern England in July 2013, the highest recordings in seven years. [Luke MacGregor/Reuters.]

*The Intergovernmental Panel on Climate Change uses the following as standard references to indicate levels of confidence in predictions concerning climate change: *virtually certain* >99% probability of occurrence; *extremely likely* >95%; *very likely* >90%; *likely* >66%; *more likely than not* >50%; *unlikely* < 33%; *very unlikely* < 10%; and *extremely unlikely* < 5%.

Glacial Ice and Permafrost Land ice occurs in the form of glaciers, ice sheets, ice caps, ice fields, and frozen ground. These freshwater ice masses are found at high latitudes and worldwide at high elevations. As temperatures rise in Earth's atmosphere, glaciers are losing mass, shrinking in size in a process known as "glacial retreat" (Figure 10B.19; also see photos in Chapter 5, Figure HD 5a, and discussion in Chapter 17).

Earth's two largest ice sheets, in Greenland and Antarctica, are also losing mass. Summer melt on the Greenland Ice Sheet increased 30% from 1979 to 2006, with about half the surface area of the ice sheet experiencing some melting on average during the summer months. In July 2012, satellite data showed that 97% of the ice sheet's surface was melting, the greatest extent in the 30-year record of satellite measurements.

As discussed in this chapter's *Geosystems Now*, permafrost (perennially frozen ground) is thawing in the Arctic at accelerating rates. Scientists now estimate that between one- and two-thirds of Arctic permafrost will thaw over the

(a) Alaska's Muir Glacier, August 13, 1941.

(b) Muir Glacier, August 31, 2004

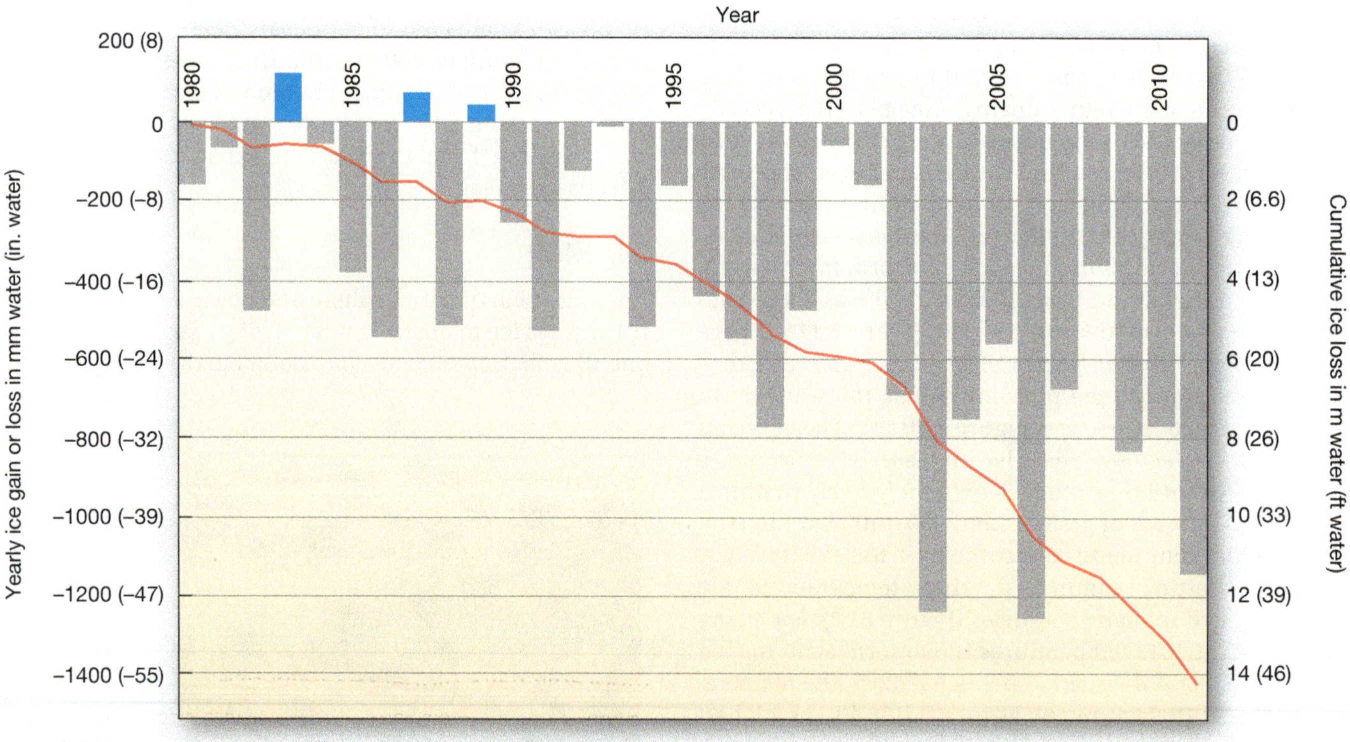

(c) Data from glaciers worldwide show negative mass balance every year since 1990 (indicated by gray bars). The red line shows the cumulative annual balance.

▲**Figure 10B.19** Muir Glacier and annual glacial mass balance (the gain of snow minus the melt losses) show net losses of glacial ice. Glaciers grow when annual winter snowfall exceeds annual summer melt losses; when snowfall and melting are equal, glacial mass balance is zero (see discussion in Chapter 17). [NOAA graph adapted from *State of the Climate in 2012*, Bulletin of the American Meteorological Society report.]

next 200 years, if not sooner; these permafrost reserves took tens of thousands of years to form. Warming land and ocean temperatures may also cause the thaw of methane hydrates stored in permafrost and in deep-ocean sediments on the seafloor (discussed in Focus Study 10B.1 on page 312).

Sea Ice In earlier chapters, we discussed the effects of Arctic sea ice on global temperatures. Sea ice is composed of frozen seawater, which forms over the ocean (sea ice does not include ice shelves and icebergs, which are made up of freshwater originating on land). Arctic sea ice, also called *pack ice*, is especially important for global climate owing to its effects on global albedo; remember that the Arctic region is an ocean surrounded by landmasses, and sea ice in this region helps cool the planet by reflecting sunlight.

The extent of Arctic sea ice varies over the course of a year, as shown in Chapter 4, *The Human Denominator*, Figure HD 4d. Every summer, some amount of sea ice thaws; in winter, the ice refreezes. Satellite data show that the summer sea-ice minimum extent (occurring in September) and winter sea-ice maximum extent (occurring in February or early March) have declined since 1979 (Figure 10B.20a). September sea ice is declining at a rate of 11% per decade (as compared to the 1979–2000 average) and reached its lowest extent in the modern record in 2012 (Figure 10B.20b). The accelerating decline of summer sea ice, in association with record losses of sea ice in 2007 and 2012, suggests that summer sea ice may disappear sooner than predicted by most models; some scientists estimate an ice-free summer Arctic Ocean within the next few decades.

As evidence of accelerating losses of Arctic sea ice, scientists have recently noted a decline in *multiyear ice*, the oldest and thickest ice, having survived through two or more summers. Younger, thinner, *seasonal ice* forms over one winter and typically melts rapidly the following summer. Figure 10B.21 shows the decline in multiyear ice extent from 1980 to 2012. During that time, the average winter seasonal ice extent also declined, but at a much smaller rate. Scientists think that the decline of multiyear ice causes an overall thinning of the Arctic pack ice that thus becomes vulnerable for further, accelerating melt. In addition, as the season for ice formation becomes shorter, multiyear ice cannot be replaced. Only a persistent cold spell would enable multiyear ice to form and reverse the current trend.

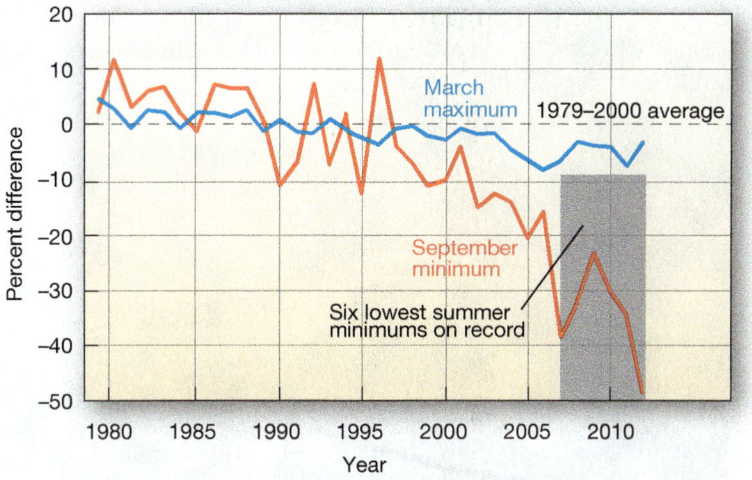

(a) Percent difference between annual March winter maximum (blue line) and September summer minimum (red line) as compared to the 1979–2000 average (dashed line). Note the rapid decrease during the last decade.

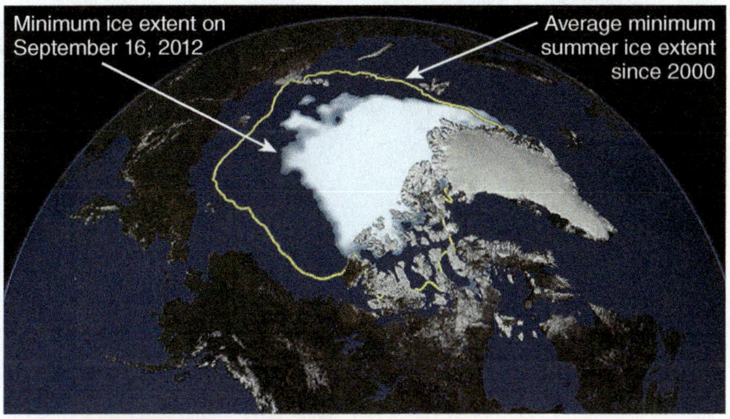

(b) Summer sea ice dropped to a record low in 2012.

▲**Figure 10B.20 Recent changes in annual Arctic sea-ice extents.** [(a) Based on *Sea ice extents since 1979*, available at http://www.climate.gov/sites/default/files/seaice1979-2012_final.png. (b) Based on *Arctic Sea Ice Hits Smallest Extent In Satellite Era*, NASA, available at http://www.nasa.gov/topics/earth/features/2012-seaicemin.html.]

Sea-Level Rise

Sea level is rising more quickly than predicted by most climate models, and the rate appears to be accelerating. During the last century, sea level rose 17–21 cm (6.7–8.3 in.),

GEOreport 10B.1 Rainfall over Australia temporarily halts global sea-level rise

The discovery of complicated interactions and unexpected system responses within Earth's climate system is common in climate change science. For example, for an 18-month period beginning in 2010, global sea level dropped by about 7 mm (0.3 in.), offsetting the consistent annual rise in recent decades. New research shows that heavy rainfall caused vast amounts of water to collect on the Australian continent during 2010 and 2011, temporarily slowing global sea-level rise. The unique topography and soils of the Australian continent enable water to collect on the continental interior, where it eventually evaporates or infiltrates into the soil rather than running off to the ocean. The heavy rainfall over Australia was generated by the unusual convergence of distinct atmospheric patterns over the Indian and Pacific Oceans. Since that time, rain over the tropical oceans has returned to the normal, heavy patterns, and sea level is rising once again.

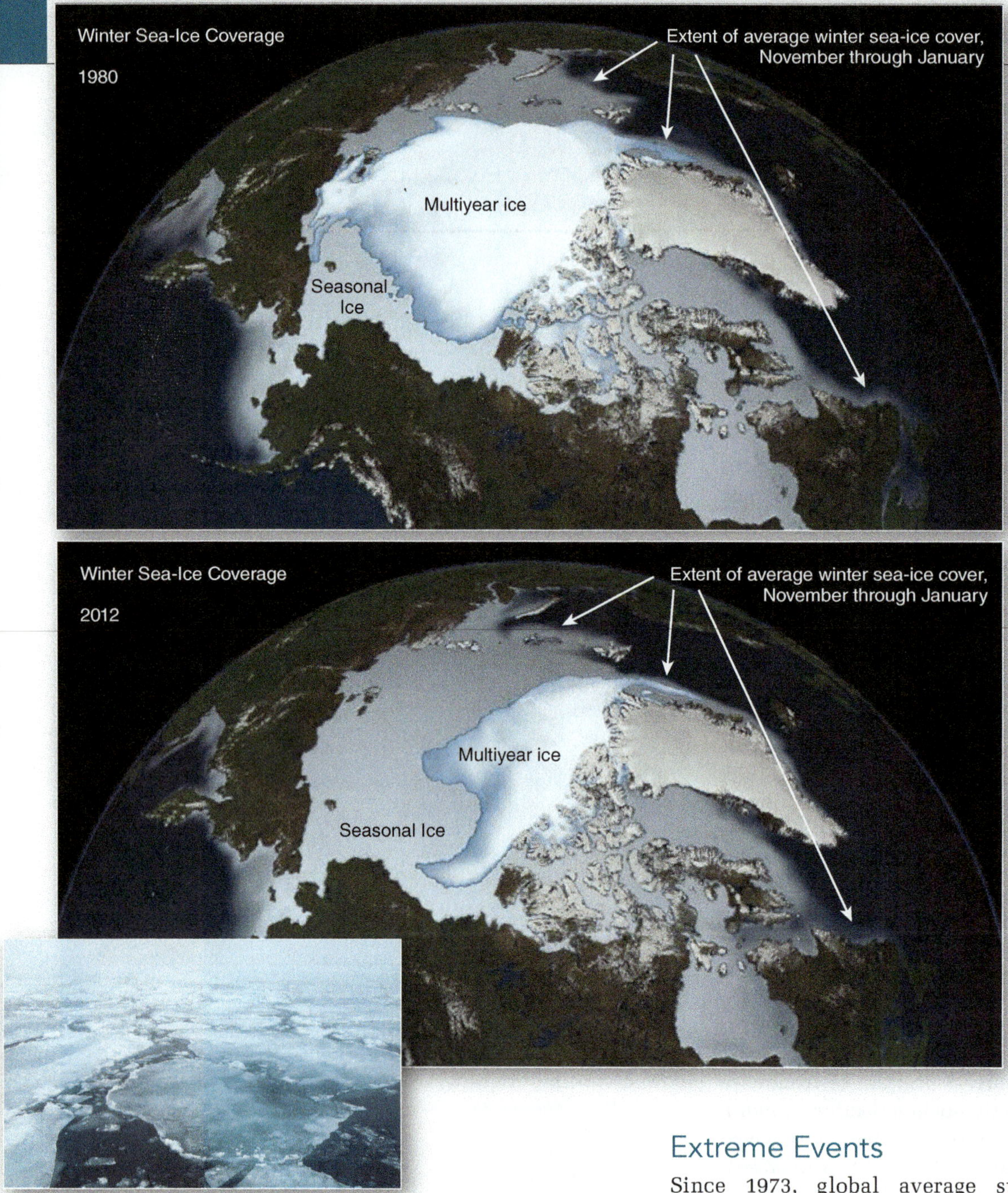

◀Figure 10B.21 Comparison of Arctic winter sea ice in 1980 and 2012, showing multiyear and seasonal ice. Photo shows the edge of the pack ice above 82° north latitude in the Arctic Ocean in 2013, where seasonal ice is now dominant rather than the multiyear ice dominant just a few years earlier. [Images NASA; photo Bobbé Christopherson.]

a greater rise in some areas (such as the U.S. Atlantic coast) than at any time during the past 2000 years. From 1901 to 2010, tidal gauge records show that sea level rose at a rate of 1.7 mm (0.07 in.) per year. From 1993 to 2013, satellite data show that sea level rose 3.16 mm (0.12 in.) per year.

Two primary factors are presently contributing to sea-level rise. Roughly two-thirds of the rise comes from the melting of land ice in the form of glaciers and ice sheets. The other third comes from the thermal expansion of seawater that occurs as oceans absorb heat from the atmosphere and expand in volume. Chapter 16 discusses sea level and its measurement; Chapter 17 provides further discussion of ice losses.

Extreme Events

Since 1973, global average specific humidity has increased by about 0.1 g of water vapor per kilogram of air per decade. This change is consistent with rising air temperatures, since warm air has a greater capacity to absorb water vapor. A greater amount of water vapor in the atmosphere affects weather in a number of ways and can lead to "extreme" events involving temperature, precipitation, and storm intensity. The Annual Climate Extremes Index (CEI) for the United States, which tracks extreme events since 1900, shows such an increase during the past four decades (see the data at http://www.ncdc.noaa.gov/extremes/cei/graph/cei/01-12). According to the World Meteorological Organization, the decade from 2001 to 2010 showed evidence of a worldwide increase in extreme events, notably heat waves, increased precipitation, and floods. However, to assess extreme weather trends and definitively link these events to climate change requires data for a longer timeframe than is now available.

Causes of Present Climate Change

Both the long-term proxy records and the short-term instrumental records show that fluctuations in the average surface temperature of the Earth correlate with fluctuations in the concentration of CO_2 in the atmosphere. Furthermore, as previously discussed, studies indicate that high levels of atmospheric carbon can act as a climate feedback that amplifies short-term temperature changes. At present, data show that record-high global temperatures have dominated the past two decades—for both land and ocean and for both day and night—in step with record-high levels of CO_2 in the atmosphere.

Scientists agree that rising concentrations of greenhouse gases in the atmosphere are the primary cause of recent worldwide temperature increases. As discussed earlier, CO_2 emissions associated with the burning of fossil fuels, primarily coal, oil, and natural gas, have increased with growing human population and rising standard of living. But how can scientists be sure that the primary source of atmospheric CO_2 over the past 150 years is from human rather than natural factors?

Scientists track the amount of carbon burned by humans over periods of years or decades by estimating the amount of CO_2 released by different activities. For example, the amount of CO_2 emitted at a particular facility can be computed by estimating the amount of fuel burned multiplied by the amount of carbon in the fuel. Although some large power plants now have devices on smoke stacks to measure exact emission rates, estimation is the more common practice and is now standardized worldwide.

Recently, scientists have used carbon isotopes to more accurately determine the atmospheric CO_2 emitted from fossil fuels. The basis for this method is that fossil carbon—formed millions of years ago from organic matter that now lies deeply buried within layers of rock—contains low amounts of the carbon isotope ^{13}C (carbon-13) and none of the radioactive isotope ^{14}C (carbon-14), with its half-life of 5730 years. This means that as the concentration of CO_2 produced from fossil fuels rises, the proportions of ^{13}C and ^{14}C drop measurably. Scientists first discovered the decreasing proportion of ^{14}C within atmospheric CO_2 in the 1950s; regular measurement of this carbon isotope began in 2003. These data show that most of the CO_2 increase comes from the burning of fossil fuels.

Contributions of Greenhouse Gases

We discussed in Chapter 4 the effect of greenhouse gases on Earth's energy balance. Increasing concentrations of greenhouse gases absorb longwave radiation, delaying losses of heat energy to space and resulting in a warming trend in the atmosphere. Global warming causes complex changes in the lower atmosphere that drive the shifts in climate discussed earlier and examined in the chapters ahead. Today, CO_2 levels far exceed the natural range that has been the norm for hundreds of thousands of years.

The contribution of each greenhouse gas toward warming the atmosphere varies depending on which wavelengths of energy the gas absorbs and on the gas's *residence time*, the length of time that it resides in the atmosphere. The primary greenhouse gases in Earth's atmosphere are water vapor (H_2O), carbon dioxide (CO_2), methane (CH_4), nitrous oxide (N_2O), and halogenated gases. Of these, water vapor is the most abundant. However, as discussed in Chapter 7, water vapor has a short residence time in the atmosphere (about 90 days) and is subject to phase changes at certain temperatures. Carbon dioxide, in contrast, has a longer residence time in the atmosphere and remains in a gaseous state at a larger range of temperatures.

Carbon Dioxide As discussed earlier and shown in Figures 10B.2, 10B.5, and 10B.11, the present concentration of CO_2 in Earth's atmosphere is higher than at any time in the last 800,000 years, and perhaps longer. The record shows CO_2 ranging between 100 ppm and 300 ppm over that time period, while never changing 30 ppm upward or downward in any span of less than 1000 years. Yet in May 2013, atmospheric CO_2 reached 400 ppm after rising over 30 ppm in only the last 13 years (having hit 370 in May 2000).

Carbon dioxide has a residence time of 50 to 200 years in the atmosphere; however, rates of uptake vary for different removal processes. For example, the uptake of atmospheric CO_2 into long-term carbon sinks such as marine sediments can take tens of thousands of years. As mentioned earlier, CO_2 emissions come from a number of sources: the combustion of fossil fuels; biomass burning (such as the burning of solid waste for fuel); removal of forests; industrial agriculture; and cement production. (Cement is used to make concrete, which is used globally for construction activities, accounting for about 5% of total CO_2 emissions). Fossil-fuel burning accounts for over 70% of the total. Overall, from 1990 to 1999, CO_2 emissions rose an average of 1.1% per year; from 2000 to 2010 that rate increased to 2.7% per year.

Methane After CO_2, methane is the second most prevalent greenhouse gas produced by human activities. Today, atmospheric methane concentrations are increasing at a rate even faster than carbon dioxide. Reconstructions of the past 800,000 years show that methane levels never topped 750 parts per billion (ppb) until relatively modern times, yet in Figure 10B.22a we see present levels at 1890 ppb.

GEOreport 10B.2 China leads the world in overall CO_2 emissions

Over the past several years, China took the lead in overall carbon dioxide emissions (29%), with the United States second (16%), and the European Union third (11%). On a per capita basis in 2011, China, with 19.5% of global population, produced 7.2 tonnes per person, on par with the European Union (at 7.5 tonnes per person). The United States, with 4.5% of world population, produced 17.3 tons of CO_2 emissions per person. Among the world's major industrialized countries, Australia had the highest per capita CO_2 output at 18.3 tonnes per person in 2011.

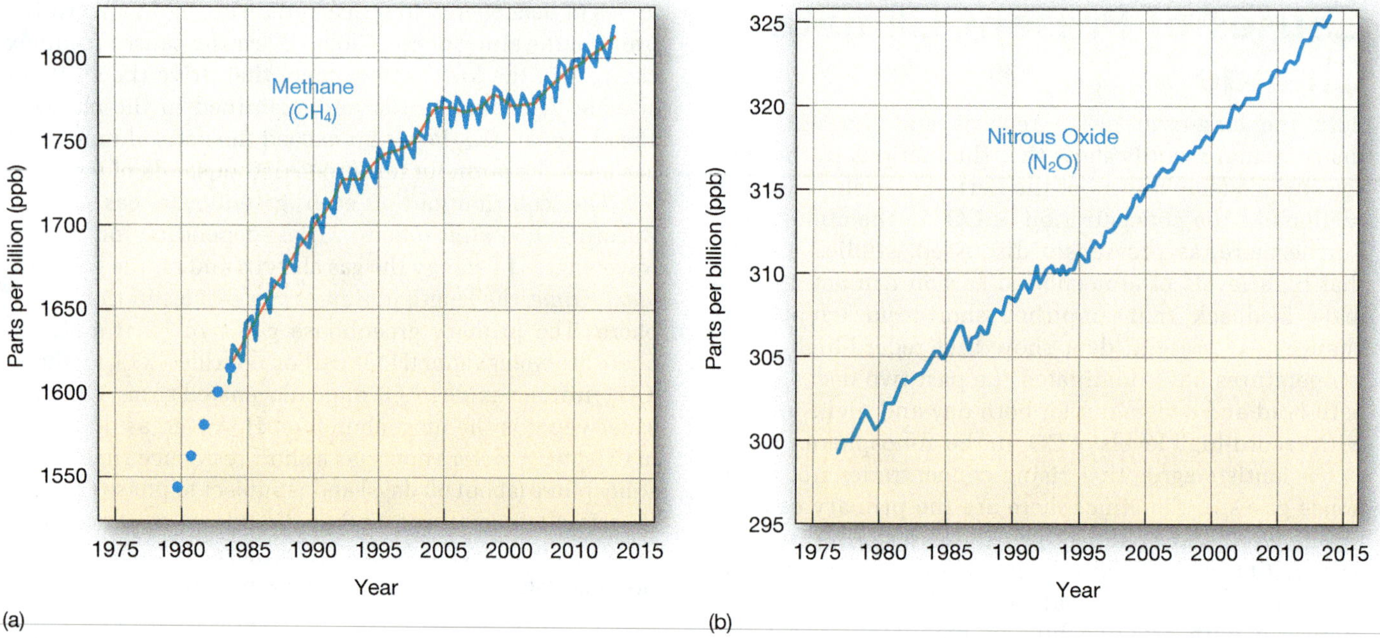

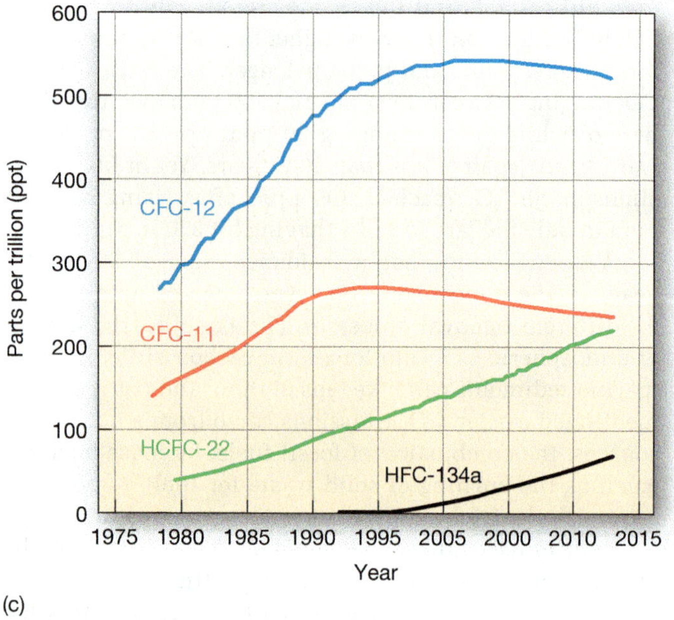

▲**Figure 10B.22 Concentrations of methane, nitrous oxide, and flourinated gases since 1978.** Gas concentrations are in parts per billion (ppb) or parts per trillion (ppt), indicating the number of molecules of each gas per billion or trillion molecules of air. [From *Greenhouse gases continue climbing; 2012 a record year*, NOAA, August 2013, available at http://research.noaa.gov/News/NewsArchive/LatestNews/TabId/684/ArtMID/1768/ArticleID/10216/Greenhouse-gases-continue-climbing-2012-a-recordyear.aspx.]

Methane has a residence time of about 12 years in the atmosphere, much shorter than that of CO_2. However, methane is more efficient at trapping longwave radiation. Over a 100-year timescale, methane is 25 times more effective at trapping atmospheric heat than CO_2, making its global warming potential higher. On a shorter timescale of 20 years, methane is 72 times more effective than CO_2. After about a decade, methane oxidizes to CO_2 in the atmosphere.

The largest sources of atmospheric methane are anthropogenic, accounting for about two-thirds of the total. Of the anthropogenic methane released, about 20% is from livestock (from waste and from bacterial activity in the animals' intestinal tracts); about 20% is from the mining of coal, oil, and natural gas, including shale gas extraction (discussed in Chapter 1, *Geosystems Now*); about 12% is from anaerobic ("without oxygen") processes in flooded fields, associated with rice cultivation; and about 8% is from the burning of vegetation in fires. Natural sources include methane released from wetlands (associated with natural anaerobic processes, some of which occur in areas of melting permafrost, as described in *Geosystems Now*) and bacterial action inside the digestive systems of termite populations. Finally, scientists suggest that methane is released from permafrost areas and along continental shelves in the Arctic as methane hydrates thaw, potentially a significant source. Focus Study 10B.1 discusses methane hydrates (also called gas hydrates), a controversial potential energy source that could have serious consequences for global climates if large amounts of methane are extracted.

Nitrous Oxide The third most important greenhouse gas produced by human activity is nitrous oxide (N_2O), which increased 19% in atmospheric concentration since 1750 and is now higher than at any time in the past 10,000 years. Nitrous oxide has a lifetime in the atmosphere of about 120 years—giving it a high global warming potential.

Although it is produced naturally as part of Earth's nitrogen cycle (discussed in Chapter 19), human activities, primarily the use of fertilizer in agriculture, but also wastewater management, fossil-fuel burning, and some industrial practices, also release N_2O to the atmosphere. Scientists attribute the recent rise in atmospheric concentrations mainly to emissions associated with agricultural activities (Figure 10B.22b).

Halogenated Gases Containing fluorine, chlorine, or bromine, halogenated gases are produced only by human activities. These gases have high global warming potential; even small quantities can accelerate greenhouse warming. Of this group, *fluorinated gases*, sometimes called *F-gases*, comprise a large portion. The most important of these are chlorofluorocarbons (CFCs), especially CFC-12 and CFC-11, and hydrochlorofluorocarbons (HCFCs), especially HCFC-22. Although CFC-12 and CFC-11 account for a small portion of the rising greenhouse gas accumulations since 1979, their concentrations have decreased in recent years owing to regulations in the Montreal Protocol (Figure 10B.22c; also see discussion of stratospheric ozone in Focus Study 3.1). However, hydrofluorocarbons (HFCs), fluorinated gases that are used as substitutes for CFCs and other ozone-depleting substances, have been increasing since the early 1990s. In general, fluorinated gases are the most potent greenhouse gases with the longest atmospheric residence times.

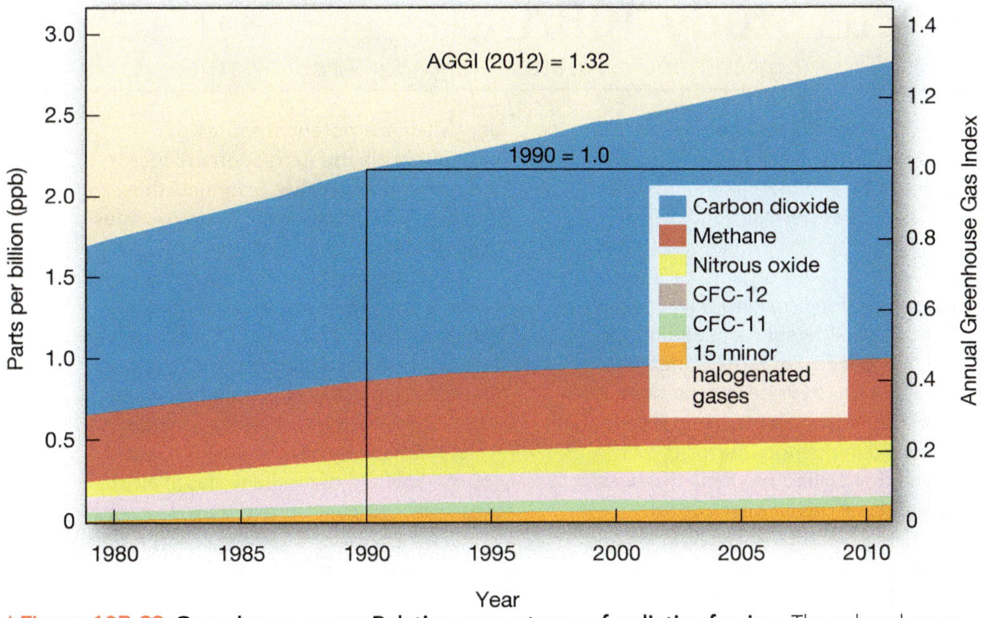

▲**Figure 10B.23 Greenhouse gases: Relative percentages of radiative forcing.** The colored areas indicate the amount of radiative forcing accounted for by each gas, based on the concentrations present in Earth's atmosphere. Note that CO_2 accounts for the largest amount of radiative forcing. The right side of the graph shows radiative forcing converted to the Annual Greenhouse Gas Index (AGGI), set to a value of 1.0 in 1990. In 2012, the AGGI was 1.32, an increase of 30% in 22 years. [From *the NOAA Annual Greenhouse Gas Index (AGGI)*, NOAA, updated summer 2013, available at http://www.esrl.noaa.gov/gmd/aggi/aggi.html]

Sources of Radiative Forcing

We learned in Chapter 4 that Earth's energy balance is theoretically zero, meaning that the amount of energy arriving at Earth's surface is equal to the amount of energy eventually radiated back to space. However, Earth's climate has cycled through periods where this balance is not achieved and Earth systems are either gaining or losing heat. The term **radiative forcing**, also called *climate forcing*, describes the amount by which some perturbation causes Earth's energy balance to deviate from zero; a positive forcing indicates a warming condition, a negative forcing indicates cooling.

Anthropogenic Greenhouse Gases

Scientists have measured the radiative forcing, quantified in watts of energy per square meter of Earth's surface (W/m²), of greenhouse gases on Earth's energy budget since 1979. Figure 10B.23, which compares the radiative forcing (RF) exerted by 20 greenhouse gases, shows that CO_2 is the dominant gas

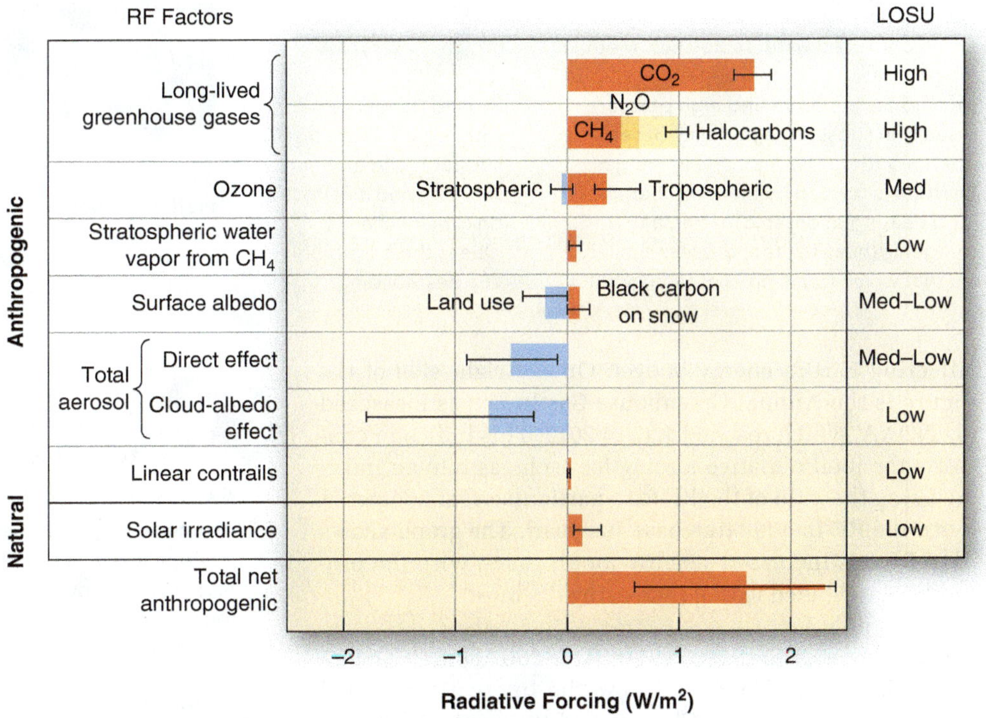

▲**Figure 10B.24 Analysis of radiative forcing of temperatures.** The top group of items in the table (for example, long-lived greenhouse gases) represent anthropogenic radiative forcing. The only long-term natural climate forcing included is solar irradiance. Note that the overall direction of forcing is positive (raising temperatures), caused mainly by atmospheric CO_2. In the preliminary 2013 IPCC *Fifth Assessment Report from* Working Group I, total net anthropogenic forcing is increased to 2.3 W/m² from the 1.6 W/m² calculated in 2006 and shown above. [IPCC, Working Group I, *Fourth Assessment Report: Climate Change 2007: The Physical Science Basis*, Figure SPM-2, p. 4, and Figure 2.20, p. 203.]

Focus Study 10B.1 Climate Change

Thawing Methane Hydrates—Another Arctic Methane Concern

In this chapter's *Geosystems Now*, we discussed the release of greenhouse gases—CO_2 and methane—into the atmosphere as permafrost thaws in Arctic regions. This process is biogenic, produced by living organisms, a result of bacterial action breaking down organic matter in shallow soils and sediments. Meanwhile, another form of methane exists as a gas hydrate, stored deep beneath permafrost deposits on land in the Arctic and in offshore deposits on the ocean floor. Called *methane hydrates*, these natural gas deposits consist of methane molecules encased in ice and become destabilized with warming temperatures. The process of melting breaks down the crystalline hydrate structures, releasing a burst of methane—Earth's most potent greenhouse gas—into the oceans and atmosphere. If these isolated bursts multiply over a large area, atmospheric methane concentrations could increase enough to accelerate global warming.

Methane Hydrate Essentials

Methane hydrate is a solid, icy compound containing a central methane molecule surrounded by a structure, or "cage," of interconnected water molecules. Methane hydrates exist only under conditions of cold temperature and high pressure, usually in subsurface deposits of sedimentary rock. The methane in the gas hydrate is formed by the deep burial and heating of organic matter, a thermogenic (heat-related) process similar to that which forms oil. Methane hydrate deposits are a potential source of energy, although the extraction process is difficult and expensive. Scientists think that over 90% of the world's gas hydrates occur in deep-ocean settings.

As temperatures rise in the ground and oceans, methane hydrate deposits are at risk of dissociation, or melt, a process that would release high concentrations of methane gas to the atmosphere (Figure 10B.1.1). For example, melting 1 m^3 (35 ft^3) of methane hydrate releases about 160 m^3 (5650 ft^3) of methane gas. If all of these gas hydrates were to thaw, the result would be a pulse of methane so large that it would almost certainly trigger abrupt climate change. Scientists think that a massive release of carbon observed in climate records from ocean sediment cores about 55 million years ago may be related to a gas hydrate dissolution event (see the temperature spike called the PETM in Figure 10B.10).

Causes and Effects of Methane Hydrate Thaw

Several dangers exist with regard to the dissociation of methane hydrates and its effect on climate change. One is the possibility of a destabilizing event causing a sudden release of enough methane to accelerate global warming. For example, the dissociation of large gas hydrate deposits—breakdown of the solids into liquids and gases—can destabilize seafloor sediments, causing a loss of structural support that can lead to subsidence and collapse in the form of submarine landslides. This type of major landslide could release large quantities of methane. A second danger is the potential for massive methane release as a by-product of energy extraction.

Methane hydrates are now thought to be the world's largest reserve of carbon-based fuel—scientists think that 10,000 Gt

▲Figure 10B.1.1 **The "ice that burns" is extracted from seafloor deposits.** Solid methane hydrate extracted from about 6 m (20 ft) beneath the seafloor near Vancouver Island, Canada. As methane hydrate warms, it releases enough methane to sustain a flame. [USGS.]

affecting Earth's energy budget. On the right side of the figure is the Annual Greenhouse Gas Index, as measured by NOAA, which reached 1.32 in 2012. This indicator converts the total radiative forcing for each gas into an index by using the ratio of the RF for a particular year compared with the RF in 1990 (the baseline year). The graph shows that RF has increased steadily for all gases, with the proportion attributed to CO_2 increasing the most.

Comparison of RF Factors In their 2007 *Fourth Assessment Report on Climate Change*, the International Panel on Climate Change (IPCC) estimated the amount of radiative forcing for a number of natural and anthropogenic factors (Figure 10B.24). The factors included in the analysis were long-lived greenhouse gases, stratospheric and tropospheric ozone, stratospheric water vapor, changes in surface albedo related to land use and pollution, atmospheric

CRITICALthinking 10B.2 Thinking Through an Action Plan to Reduce Human Climate Forcing

Figure 10B.24 illustrates some of the many factors that force climate. Let us consider how these variables might inform decision making regarding climate change policy and mitigation. Assume you are a policy maker with a goal of reducing the rate of climate change—that is, reducing positive radiative forcing of the climate system. What strategies do you suggest to alter the extent of radiative forcing or adjust the mix of elements that cause temperature increases? Assign priorities to each suggested strategy to denote the most to least effective in moderating climate change. Try brainstorming and discussing your strategies with others. In your opinion, how should the jump to 2.3 W/m^2 in human climate forcing over the past 6 years influence policy and action strategies?

of methane is trapped in gas hydrates worldwide, an amount that exceeds the energy available in coal, oil, and other natural gas reserves combined. These deposits occur at depths greater than about 900 m (3000 ft) under seafloor sediments. Several countries are exploring the use of methane hydrates as an energy source. In March 2013, Japan's deep-water ocean drilling rig, *Chikyu*, successfully extracted gas hydrates from a depth of 1000 m (3280 ft) in the Pacific Ocean. The country's goal is to achieve commercial production of methane hydrates within 6 years. However, the possibility of accidental uncontrolled methane releases during extraction is an important concern.

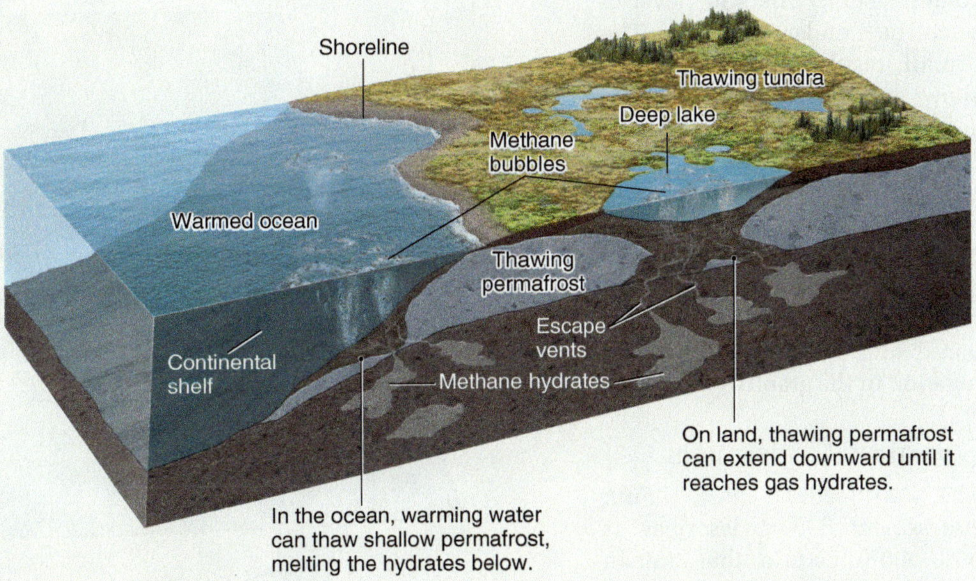

▲**Figure 10B.1.2 Methane hydrate deposits in arctic permafrost and under continental shelves.** Two theoretical pathways for methane hydrate thaw: On land, thawing permafrost can extent downward until it reaches gas hydrates; in the ocean, warming water can thaw shallow permafrost, melting the hydrates below. [From Walter Anthony, K. 2009. Methane: A menace surfaces. *Scientific American* 301: 68-75 doi:10.1038/scientificamerican1209-68.]

Under today's conditions of atmospheric warming, scientists think that the thawing of methane hydrates from two of its general source regions can potentially affect atmospheric methane concentrations (Figure 10B.1.2). On land, at high latitudes in the Arctic, permafrost could thaw to depths of 180 m (600 ft), causing methane hydrates in rock to dissociate. In the Arctic Ocean, below subsea permafrost at shallow depths along continental shelves, rising ocean and land temperatures could also thaw permafrost to depths that would compromise hydrate structures. In the East Siberian Sea, off the shore of northern Russia, scientists think that an estimated 50 billion tons of methane stored in the form of hydrates is already beginning to dissociate, producing rising plumes of methane that reach the atmosphere. The process, they think, is triggered by changes in summer sea ice, whose extent has declined so much above the Siberian shelf that ocean surface temperatures in ice-free areas have warmed as much as 7 C° (3.9 F°), according to satellite data. The warming extends downward about 50 m (164 ft) to the shallow seafloor, melting the frozen sediments. In areas where the seabed is deeper along continental slopes, gas hydrates may dissociate if ocean warming continues, but scientists do not yet know whether the methane released would reach the atmosphere.

The overall processes and environmental effects of methane hydrate thaw are a focus of ongoing research. For more information, see the article "Good Gas, Bad Gas" at http://ngm.nationalgeographic.com/2012/12/methane/lavelle-text and the U.S. Geological Survey Gas Hydrates Project page at http://woodshole.er.usgs.gov/project-pages/hydrates/.

aerosols from both human and natural sources, linear contrails, and solar irradiance (the output of energy from the Sun, discussed earlier).

In Figure 10B.24, the factors that warm the atmosphere, causing a positive radiative forcing, are in red, orange, or yellow; those that cool the atmosphere, causing a negative forcing, are in blue. The estimates of radiative forcing (RF) in watts-per-square-meter units are given on the *x*-axis (horizontal axis). The horizontal black lines superimposed on the colored bars represent the uncertainty

GEOreport 10B.3 Causes of extreme weather events in a changing climate

According to the 2013 NOAA report "Explaining Extreme Events of 2012 from a Climatic Perspective," *Bulletin of the American Meteorological Society*, Volume 94 (9), scientific analyses of 12 extreme weather and climate events in 2012 found that anthropogenic climate change was a contributing factor to half the events—either to their occurrence or outcomes—and that the magnitude and likelihood of each were boosted by climate change. Also important for these extreme events was the role of natural climate and weather fluctuations, such as El Niño–Southern Oscillation and other global circulation patterns, factors that may be affected by global warming. The new, developing science of "event attribution" seeks to find the causes of extreme weather and climate events and has important applications for risk management, preparing for future events, and overall mitigation of climate change effects.

range for each factor (for example, the cloud-albedo effect has a large uncertainty range). In the far-right column, LOSU refers to "level of scientific understanding." The overall results of this analysis show that the RF of greenhouse gases far surpasses the RF of other factors, whether natural or anthropogenic in origin. In the 2013 IPCC *Fifth Assessment Report*, Working Group I states that total net anthropogenic forcing increased to 2.3 W/m^2, compared to 1.6 W/m^2 in 2006 (see notation in the graph).

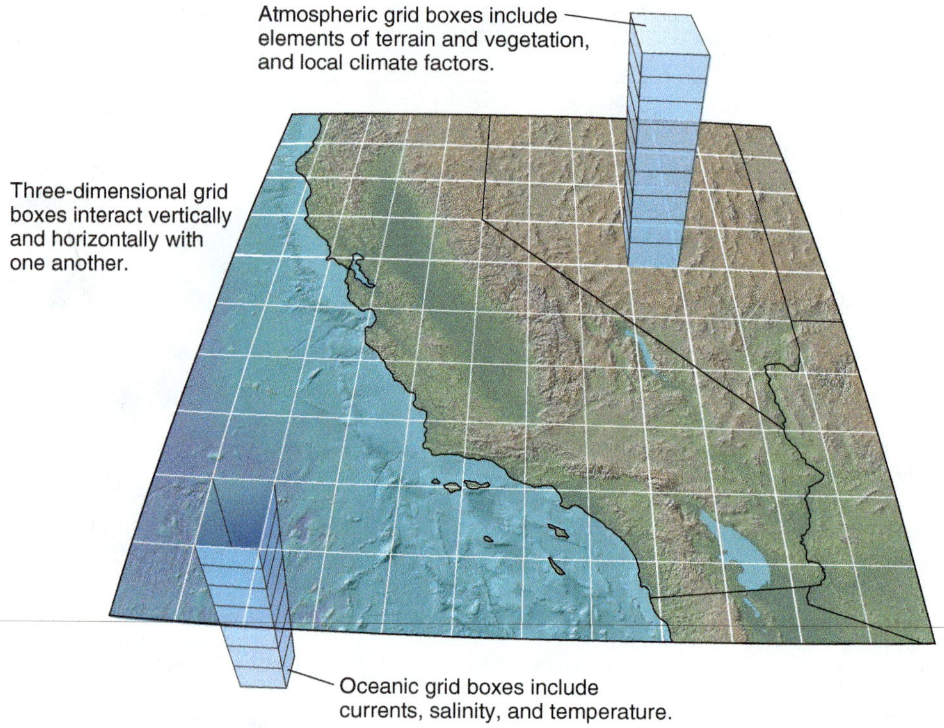

▲**Figure 10B.25** **Grid boxes and layers in a general circulation model.**

Scientific Consensus

The 2013–2014 IPCC *Fifth Assessment Report* describes as 95%–100% certain that human activities are the primary cause of present climate change. This assertion represents consensus from the world's leading climate scientists and is confirmed throughout the scientific community. The *Fifth Assessment Report* also concludes with 95% certainty that humans are responsible for the temperature increase from 1951 to 2010—in other words, the IPCC scientists are 95% certain that the observed warming during this period matches the estimated human contribution to warming. The *Fifth Assessment Report* is available at http://www.ipcc.ch/index.htm#.UiDyVj_pxfk.

The IPCC, formed in 1988 and operating under sponsorship of the United Nations Environment Programme (UNEP) and the World Meteorological Organization (WMO), is the international scientific organization coordinating global climate change research, climate forecasts, and policy formulation—truly a global collaboration of scientists and policy experts from many disciplines. Its reports represent peer-reviewed, consensus opinions among experts in the scientific community concerning the causes of climate change as well as the uncertainties and areas where further research is needed. In 2007, the IPCC shared the Nobel Peace Prize for its two decades of work raising understanding and awareness of global climate change science.

Climate Models and Forecasts

In addition to using paleoclimatic records and actual measurements of present-day climatic elements, scientists use computer models of climate to assess past trends and forecast future changes. A climate model is a mathematical representation of the interacting factors that make up Earth's climate systems, including the atmosphere and oceans, and all land and ice. Some of the most complex computer climate models are **general circulation models (GCMs)**, based on mathematical models originally established for forecasting weather.

The starting point for such a climate model is a three-dimensional "grid box" for a particular location on or above Earth's surface (Figure 10B.25). The atmosphere is divided up vertically and horizontally into such boxes, each having distinct characteristics regarding the movement of energy, air, and water. Within each grid box, physical, chemical, geological, and biological characteristics are represented in equations based on physical laws. All these climatic components are translated into computer codes so that they can "talk" to each other, as well as interact with the components of grid boxes on all sides.

GCMs incorporate all climatic components, including climate forcings, to calculate the three-dimensional motions of Earth–atmosphere systems. They can be programed to model the effects of linkages between specific climatic components over different time frames and at various scales. Submodel programs for the atmosphere, ocean, land surface, cryosphere, and biosphere may be used within the GCMs. The most sophisticated models couple atmosphere and ocean submodels and are known as **Atmosphere–Ocean General Circulation Models (AOGCMs)**. At least a dozen established GCMs are now in operation around the world.

Radiative Forcing Scenarios

Scientists can use GCMs to determine the relative effects of various climate forcings on temperature (remember that a climate forcing is a perturbation in Earth's radiation

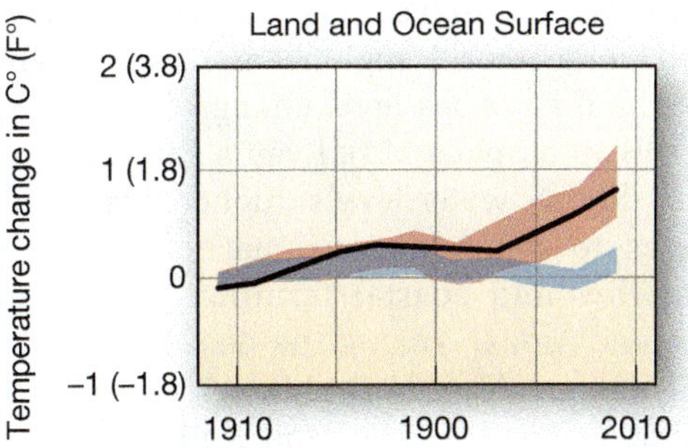

▲**Figure 10B.26 Climate models showing relative effects of natural and anthropogenic forcing.** Computer models track the agreement between observed temperature anomalies (black line) with two forcing scenarios: combined natural and anthropogenic forcings (pink shading) and natural forcing only (blue shading). The natural forcing factors include solar activity and volcanic activity, which alone do not explain the temperature increases. [*Climate Change 2013: The Physical Science Basis*, Working Group I, IPCC *Fifth Assessment Report*, September 27, 2013: Fig. SPM-6, p. 32.]

budget that causes warming or cooling). One question that scientists have sought to answer is, "Does positive radiative forcing have natural or anthropogenic causes?"

Figure 10B.26 compares results from two sets of climate simulations as compared to actual global average temperature observations for land and ocean (black line) made from 1906 to 2010. In the graph, the actual temperature data are compared with simulations in which both natural and anthropogenic forcings were included (shaded pink area and black line). These data are compared with simulations that included natural forcing only, modeled from solar variability and volcanic output alone (shaded blue area), with no anthropogenic forcing.

The model using both natural and anthropogenic forcings (including greenhouse gas concentrations) was the one that produced the closer match to the actual observed temperature averages. Natural forcings alone do not match the increasing temperature trend. These models are consistent with the hypothesis regarding the effect of human activities on climate; the primary anthropogenic input to Earth's climate system is increased amounts of greenhouse gases.

Future Temperature Scenarios

GCMs do not predict specific temperatures, but they do offer various future scenarios of global warming. GCM-generated maps correlate well with the observed global warming patterns experienced since 1990, and a variety of AOGCM forecast scenarios were used by the IPCC to predict temperature change during this century. Figure 10B.27 depicts several of these scenarios, each based on different combinations of economic factors, population, degree of global cooperation, and level of greenhouse gas emissions. The "A" scenarios (A1FI, A1T, and A1B) represent a future of rapid economic growth and the introduction of new technologies: the A1FI scenario represents continued use of fossil fuels; A1T represents an emphasis on non-fossil energy sources; and A1B represents a balance between the two types of energy use. The A2 scenario represents continuous population growth and a lower level of economic growth and technological change.

The "B" scenarios represent emphasis on environmental sustainability: the B1 scenario is similar to A1, but includes the use of greater amounts of new and clean energy technologies and a focus on global solutions; B2 represents lower levels of economic growth and technological change, with an emphasis on local solutions. The orange line is the simulation experiment where greenhouse gas emissions are held at 2000 values with no increases. The gray bars to the right of the graph are the best estimate of surface warming (solid bar within the gray area) and likely range (gray area) for each scenario. The B1 scenario predicts the lowest levels of warming; the A1FI scenario predicts the highest levels. Thus, according to these models, the

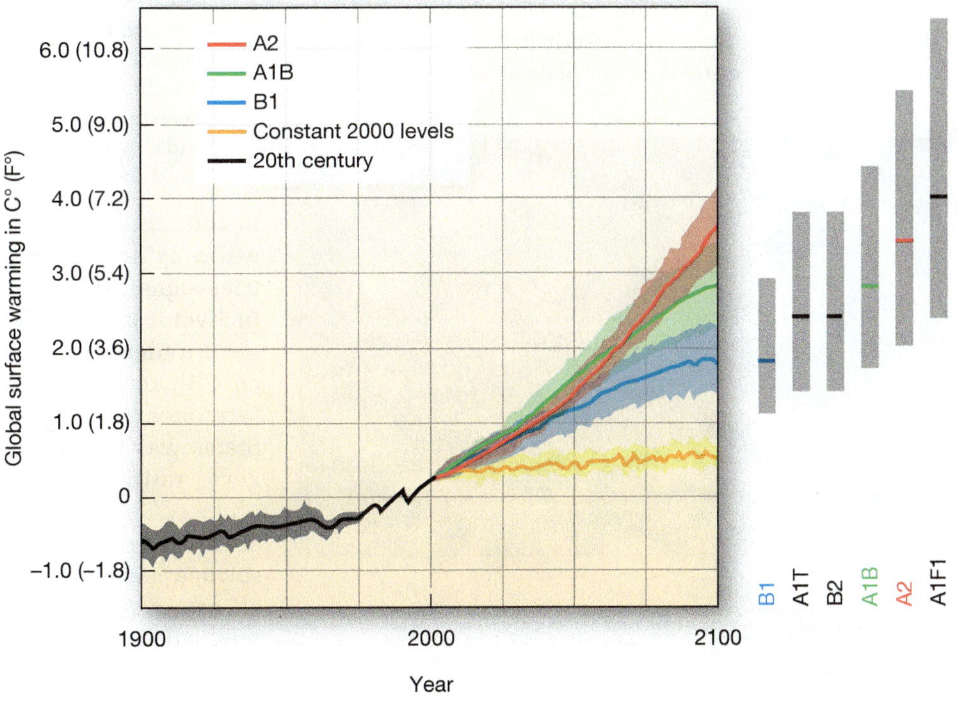

▲**Figure 10B.27 AOGCM scenarios for surface warming during this century.** The lowest temperature impacts occur when atmospheric greenhouse gases are held constant at 2000 levels. The lowest–CO_2-emission scenarios are B1 and A1T, and correspond to the lowest amounts of global warming. The highest-emission scenarios are A1F1 and A2, corresponding to the greatest temperature increases. [IPCC, Working Group I, *Fourth Assessment Report: Climate Change 2007: The Physical Science Basis*, Figure SPM-5, page 14.]

TABLE 10B.2 Global Sea-Level-Rise Scenarios*

Scenario	Sea-Level Rise by 2100, m (ft)
High	2.0 (6.6)
Intermediate–high	1.2 (3.9)
Intermediate–low	0.5 (1.6)
Low	0.2 (0.7)

*Data from NOAA, *Global Sea-Level Rise Scenarios for the United States, National Climate Assessment*, Technical Report OAR CPO-1, 2012. (Available at http://cpo.noaa.gov/sites/cpo/Reports/2012/NOAA_SLR_r3.pdf.)

continued use of fossil fuels in a rapidly growing economy causes the greatest amount of warming over the twenty-first century. The 2013–2014 IPCC *Fifth Assessment Report* confirms these scenarios and analyses from the previous IPCC report.

Sea-Level Projections

In 2012, NOAA scientists developed scenarios for sea-level rise based on present ice-sheet losses coupled with losses from mountain glaciers and ice caps worldwide (Table 10B.2). These models projected a sea-level rise of 2.0 m (6.6 ft) by 2100 at the high end of the range, with 1.2 m (4.0 ft) as the intermediate projection. For perspective, a 0.3-m (1.0-ft) rise in sea level would produce a shoreline retreat of 30 m (98 ft) in some places; a 1.0-m (3.2-ft) rise would displace an estimated 130 million people.

In the United States, sea level is rising more quickly on the east coast than on the west coast. Figure 10B.28 shows the area affected by sea-level increases along the U.S. East and Gulf coasts. The most severe effects are around the low-elevation cities of Miami and New Orleans. Many states, including California, presently use a 1.4-m sea-level rise this century as a standard for planning purposes.

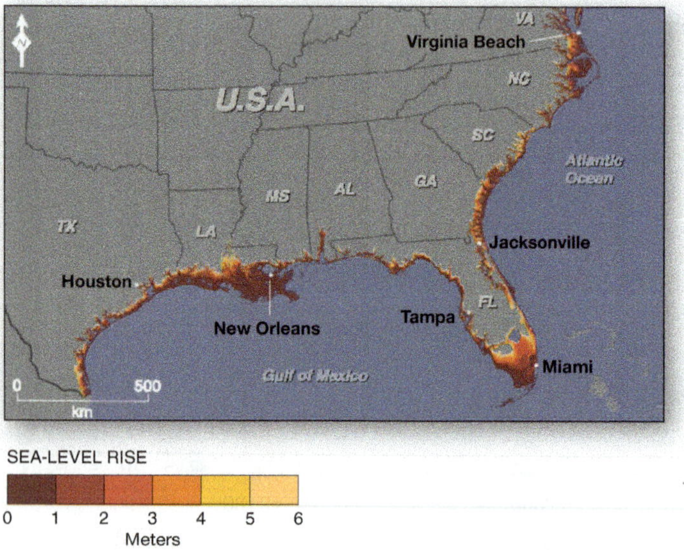

▲**Figure 10B.28 U.S. coastal areas that would be inundated with a 1–6 m (3.2 to 19.6 ft) rise in sea level.** Labeled cities in vulnerable areas have populations greater than 300,000 (as of the 2000 census). [Adapted by permission from J. Wiess, J. Overpeck, and B. Strauss, 2011, Implications of recent sea level rise science for low-elevation areas in coastal cities of the conterminous U.S.A., *Climatic Change* 105: 635–645.]

The effects of sea-level rise will vary from place to place, with some world regions more vulnerable than others (see the map of sea level change from 1992 to 2013 in Figure 10B.30 on page 321). Even a small sea-level rise would bring higher water levels, higher tides, and higher storm surges to many regions, particularly impacting river deltas, lowland coastal farming valleys, and low-lying mainland areas. Among the densely populated cities most at risk are New York, Miami, and New Orleans in the United States, Mumbai and Kolkata in India, and Shanghai in China, where large numbers of people will have to leave coastal areas. The social and economic consequences will especially affect small, low-elevation island states. For example, in Malé, the capital city of the Maldives archipelago in the Indian Ocean, over 100,000 people reside behind a sea wall at elevations of about 2.0 to 2.4 m (6.7 to 8.0 ft). If the high sea-level rise scenarios occur by 2100, major portions of this island will be inundated. National and international migration—a flood of environmental refugees driven by climate change—would be expected to continue for decades. Sea-level increases will continue beyond 2100, even if greenhouse gas concentrations were to be stabilized today. To see an interactive map of the effects of rising sea level on different areas of the world, go to http://geology.com/sea-level-rise/.

The Path Ahead

Over 25 years ago, in an April 1988 article in *Scientific American*, climatologists described the climatic condition:

> The world is warming. Climatic zones are shifting. Glaciers are melting. Sea level is rising. These are not hypothetical events from a science fiction movie; these changes and others are already taking place, and we expect them to accelerate over the next years as the amounts of carbon dioxide, methane, and other trace gases accumulating in the atmosphere through human activities increase.

In 1997, under the evolving threat of accelerating global warming and associated changes in climate, 84 countries signed the *Kyoto Protocol* at a climate conference in Kyoto, Japan, a legally binding international agreement under the United Nations Framework Convention on Climate Change (UNFCCC) that set specific targets to reduce emissions of greenhouse gases. The United States was among the signatory countries; however, it never ratified the treaty. (See http://unfccc.int/kyoto_protocol/items/2830.php.) Since the time of these initial climate warnings and policy actions, climate change science has advanced, and the issue of human-caused climate change has moved from one of scientific and public debate to one of general scientific agreement.

Taking a Position on Climate Change

Despite the consensus among scientists, considerable controversy still surrounds the topic among the public at large. The disagreement takes two forms: first, disagreement about whether climate change is occurring, and second, disagreement about whether its cause is anthropogenic. The fuel for the continuing "debate" on this topic appears to come, at least partly, from media coverage that at times is biased, alarmist, or factually incorrect. The bias often reflects the influence of special-interest groups, and some errors come from simple misinterpretation of the facts. Media misinformation often comes from corporate interests whose financial gains are at stake if climate change solutions are imposed. Another source of misinformation are the growing number of blogs and other social media that often present results not yet evaluated by other scientists. Such information is sometimes inaccurate and may be sensationalized. The bottom line is that having an informed position on climate change requires an understanding of Earth's physical laws and system operations, and an awareness of the scientific evidence and ongoing research.

In Chapter 1, we discussed the scientific process, which encourages peer evaluation, criticism, and cautious skepticism through the scientific method. Many would argue that skepticism concerning climate change is simply part of this process. However, climate scientists have generally reached consensus: The case for anthropogenic climate change has become more convincing as scientists gather new data and complete more research, and as we witness actual events in the environment. However, as new information becomes available, scientists will constantly need to reevaluate evidence and formulate new hypotheses.

When considering the facts behind climate change, several key questions can help guide you to an informed position based on a scientific approach:

- *Does increasing atmospheric carbon dioxide in the atmosphere cause warming temperatures?*

  Yes. Scientists know that CO_2 acts as a greenhouse gas and that increased concentrations produce warming in the lower atmosphere. Scientists have understood the physical processes related to atmospheric CO_2 for almost 100 years, since well before the effects of global warming became apparent to the scientific community or to the public in general.
- *Does the rise of global temperatures cause global climate change?*

  Yes. *Global warming* is an unusually rapid increase in Earth's average surface temperature. Scientists know, based on physical laws and empirical evidence, that global warming affects overall climate; for example, it changes precipitation patterns, causes ice melt, lengthens growing seasons, impacts ecosystems, and leads to rising sea level and coastal erosion.
- *Have human activities increased the amount of greenhouse gases in the atmosphere?*

  Yes. As discussed earlier, scientists use radioactive carbon isotopes to measure the amount of atmospheric CO_2 that originates from fossil-fuel burning and other human activities. They now know that human sources account for a large and growing percentage of these CO_2 concentrations.
- *If climate change on Earth has occurred in the past, then why are the present conditions problematic?*

  Carbon dioxide concentrations are today rising more quickly than is seen throughout most of the long-term climate record. This rate of change puts Earth systems in uncharted territory for assessing impacts, at a time when Earth's human population exceeds 7.3 billion.
- *Can scientists definitively attribute the changes we are seeing in climate (including extreme events and weather anomalies) to anthropogenic causes alone?*

  Not with 100% certainty, although an increasing amount of research indicates this to be the case. One issue needing further study is the effect of multiyear oscillations in global circulation patterns (such as ENSO) on short-term changes in climate. Scientists do not yet know the extent to which global warming is driving changes in the intensity of ENSO and other oscillations, or whether the changes result in part from natural variability.

Action Now Means "No Regrets"

Given this scientific knowledge base, taking action on climate change must focus on lowering atmospheric CO_2. As emphasized in the 2007 and 2013–2014 IPCC reports and many others, opportunities to reduce carbon dioxide emissions are available and have benefits that far offset their costs. These are sometimes called "no regrets" opportunities, because societies have little to lose and much to gain by employing these strategies. For example, the benefits from reducing greenhouse gas emissions include improved air quality, with related benefits to human health, reduced oil import costs and related oil tanker spills, and an increase in renewable and sustainable energy development, with related business opportunities. These benefits exist separately from the fact that all of them will decrease the rate of CO_2 accumulation in the atmosphere, and thus slow climate change.

For Europe, scientists determined that carbon emissions could be reduced to less than half the 1990 level by 2030, at no cost to society. A study by the U.S. Department of Energy found that the Unites States could meet the carbon-emission reduction targets called for in the Kyoto Protocol with a cash savings ranging from \$7 billion to \$34 billion a year.

A comprehensive economic analysis prepared by economist Nicholas Stern for the British government in 2007 (*The Economics of Climate Change, The Stern Review*, Cambridge University Press, 2007) reached several conclusions concerning action on climate change. Foremost was that the worst impacts of climate change can be avoided if we take strong action now; delaying action is a dangerous and more costly alternative. Stern stressed the need for action on a global level, an international response across both wealthy and poor countries, and that a range of options already exists to cut emissions.

Mitigating Climate Change: What Can You Do?

We often hear people say, "I am only one person; what can I do?" Our answer is that actions taken on an individual level by millions of people can slow the pace of climate change for us and for future generations. The principal way we can do this—as individuals, as a country, and as an international community—is to reduce carbon emissions, especially in our burning of fossil fuels. Perhaps you can begin by examining the sustainability of your daily practices and weeding out wasteful lifestyle choices. For example, placing dishes in a dishwasher without pre-washing saves 6500 gallons of water a year. Driving a vehicle that gets 30 miles per gallon saves 2.9 annual tons of CO_2 over one that gets 20 miles per gallon.

These efforts are "no regrets" opportunities that save money at the same time that they save resources, energy, and materials. For example, CFL (compact fluorescent light) and LED bulbs can save homeowners thousands of dollars over their lifetime, as energy use is reduced. Similarly, driving a gas-efficient motor vehicle and staying near the speed limit saves money while lowering emissions.

In Chapter 1, we discussed the concept of an individual's "footprint," whether an ecological footprint, carbon footprint, or lifestyle footprint. Although most footprint assessments are grossly simplified, they can give you an idea of your impact and even an estimate of how many planets it would take to sustain that lifestyle and economy if everyone lived like you. Actions that reduce our footprint can make a difference; for example, decisions about the type and origin of the food we eat (vegetables or red meat; from local growers or from a major supermarket chain). Every action we take, every decision we make, has a consequence. Everything in Earth's environment is linked, and every human action affects something else.

On a collective level as a human society, we can work to improve certain land-use practices, especially those that remove forests and other vegetation, and we can plant trees and expand green spaces in urban areas to take up CO_2 and moderate higher temperatures. We can alter our agricultural practices so that more carbon is retained in the soil, and so that crop varieties are bred to withstand heat, drought, or flood inundation. We can call for and support government regulations to increase motor vehicle fuel-efficiency standards. In coastal areas, we can alter development practices to utilize natural shoreline protection, such as sand dunes, and allow for shifting of natural features during storms. We can also promote efficient water use, especially in areas prone to drought. These are only a few examples among many for mitigating the causes and effects of climate change (for links and more information, see http://www.unep.org/climatechange/mitigation/ and http://climate.nasa.gov/solutions).

Present greenhouse gas concentrations will remain in the atmosphere for many decades to come, but the time for action is now. Scientists describe 450 ppm as a possible climatic threshold near which Earth systems would transition into a chaotic mode; with accelerating CO_2 emissions, this level of CO_2 could occur in the decade of the 2020s. The goal of avoiding this threshold is a practical start to alleviating the worst consequences of our current trajectory. In fact, as stated in Critical Thinking 10B.1, scientists suggest that we aim for reducing atmospheric levels of CO_2 to 350 ppm*. Raising public awareness of this goal is one climate change mitigation strategy (see *The Human Denominator*, Figure HD 10B). The technologies necessary for reducing atmospheric CO_2 to the target goal of 350 ppm are already established. The information presented in this chapter is offered in the hope of providing motivation and empowerment—personally, locally, regionally, nationally, and globally.

*Among many sources, see J. Hansen et al., "Target atmospheric CO_2: Where should humanity aim?" *Open Atmospheric Science Journal* 2 (March 2008): 217–231; and K. Zickfeld, M. Eby, H. D. Matthews, and A. J. Weaver, "Setting cumulative emissions targets to reduce the risk of dangerous climate change," *Proceedings of the National Academy of Sciences* 106, 38 (September 22, 2009): 16129–16134.

◄**Figure 10B.29 Climate change mitigation strategies in cities.** Lights in the United States reflect population patterns, and correlate to many of the large urban areas working to reduce greenhouse gas emissions. According to the non-profit Carbon Disclosure Project (CDP), 110 cities worldwide reported in 2013 that actions to mitigate climate change have co-benefits that save money, attract new businesses, and improve the health of residents. See the CDP report at https://www.cdproject.net/CDPResults/CDP-Cities-2013-Global-Report.pdf. [*Suomi NPP*, VIIRS instrument, NASA.]

THEhumanDENOMINATOR 10B Taking Action on Climate Cha

CLIMATE CHANGE ⇨ HUMANS

- Climate change affects all Earth systems.
- Climate change drives weather and extreme events, such as drought, heat waves, storm surge, and sea-level encroachment, which cause human hardship and fatalities.

HUMANS ⇨ CLIMATE CHANGE

- Anthropogenic activities produce greenhouse gases that alter Earth's radiation balance and drive climate change.

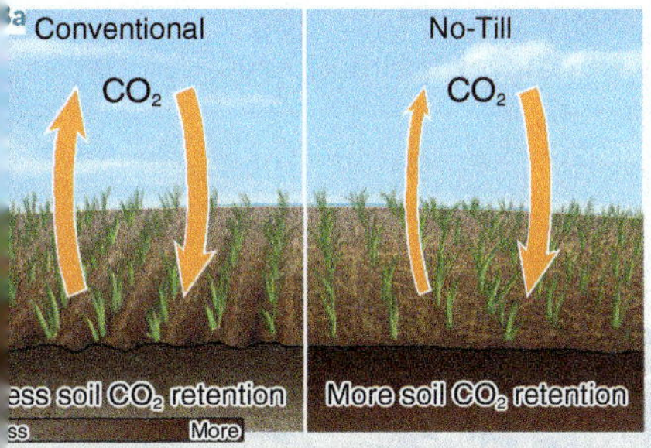

More carbon is retained in soils with the practice of no-till agriculture, in which farmers do not plow fields after a harvest but instead leave crop residue on the fields. (See discussion in Chapter 18.) In the photo, a farmer uses no-till agricultural practices in New York by planting corn into a cover crop of barley. [After U.S. Department of Energy Pacific Northwest National Laboratory. Photo by NRCS.]

10Bb

Parade participants in different costumes march during the Global Day of Action Against Climate Change on December 8, 2007, in suburban Quezon City, north of Manila, Philippines. [Pat Roque/AP Photo.]

10Bd

Through the International Small Group and Tree Planting Program (TIST), subsistence farmers have been planting trees to reverse deforestation and combat climate change. Farmers earn greenhouse gas credits for each tree planted based on growth and the amount of carbon stored; these credits translate into small cash stipends. Scientists think that planting trees is one of the simplest and most effective ways to combat climate change. [Charles Sturge/Alamy.]

10Bc

Climate-awareness advocates holding umbrellas form the number "350" on the steps of the Sydney Opera House in downtown Sydney, Australia, in 2009. The "350" signifies the concentration (in ppm) of atmospheric carbon dioxide that scientists have determined to be sustainable for Earth's climate system. [David Hill/Getty Images.]

ISSUES FOR THE 21ST CENTURY

How will human society curb greenhouse gas emissions and mitigate change effects? Some examples:

- Using agricultural practices that help soils retain carbon.
- Planting trees to help sequester carbon in terrestrial ecosystems.
- Supporting renewable energy, and changing lifestyles to use fewer resources.
- Protecting and restoring natural ecosystems that are carbon reservoirs.

GEOSYSTEMSconnection

Chapter 10B presents a synthesis of climate change evidence and its human causes, ending Part II and the focus on Earth's atmosphere. Next we move to Part III, Chapters 11 through 17, "The Earth–Atmosphere Interface"—where systems meet at the lithosphere. We begin with the endogenic system in Chapters 11 and 12, describing how energy wells up within Earth with enough power to drag and push portions of Earth's crust as continents migrate. We also study rock-forming processes and the nature of crustal deformation, as well as the dramatic events of earthquakes and volcanic eruptions.

ARNING ncepts review

■ *Describe* scientific tools used to study paleoclimatology.

The study of the causes and consequences of changing climate on Earth systems is **climate change science**. Growing human population on Earth has led to an accelerating use of natural resources, increasing the release of greenhouse gases—most notably CO_2—into the atmosphere.

The study of natural climatic variability over the span of Earth's history is the science of **paleoclimatology**. Since scientists do not have direct measurements for past climates, they use **proxy methods**, or *climate proxies*—information about past environments that represent changes in climate. Climate reconstructions spanning millions of years show that Earth's climate has cycled between periods both colder and warmer than today. One tool for long-term climatic reconstruction is **isotope analysis**, a technique that uses relative amounts of the isotopes of chemical elements to identify the composition of past oceans and ice masses. **Radioactive isotopes**, such as ^{14}C (carbon-14), are unstable and decay at a constant rate measured as a *half-life* (the time it takes half the sample to break down). The science of using tree growth rings to study past climates is **dendroclimatology**. Analysis of mineral deposits in caves that form **speleothems** and the growth rings of ocean corals can also identify past environmental conditions.

climate change science (p. 288)
paleoclimatology (p. 290)
proxy method (p. 290)
isotope analysis (p. 291)
radioactive isotope (p. 295)
dendroclimatology (p. 296)
speleothem (p. 297)

1. Describe the change in atmospheric CO_2 over the past 800,000 years. What is the Keeling Curve? Where do its measurements put us today in relation to the past 50 years?
2. Describe an example of a climate proxy used in the study of paleoclimatology.
3. Explain how oxygen isotopes can identify glacials and interglacials.
4. What climatic data do scientists obtain from ice cores? Where on Earth have scientists drilled the longest ice cores?
5. How can pollen be used in radiocarbon dating?
6. Describe how scientists use tree rings, corals, and speleothems to determine past climates.

■ *Discuss* several natural factors that influence Earth's climate, and *describe* climate feedbacks, using examples.

Several natural mechanisms can potentially cause climatic fluctuations. The Sun's output varies over time, but this variation has not been definitely linked to climate change. The **Maunder Minimum**, a solar minimum from about 1645 to 1715, corresponded with one of the coldest periods of the Little Ice Age. However, other solar minimums do not correlate with colder periods. Earth's orbital cycles and Earth-Sun relationships, called **Milankovitch cycles**, appear to affect Earth's climate—especially glacial and interglacial cycles—although their role is still under study. Continental position and atmospheric aerosols, such as those produced by volcanic eruptions, are other natural factors that affect climate.

Climate feedbacks are processes that either amplify or reduce climatic trends toward warming or cooling. Many climate feedbacks involve the movement of carbon through Earth systems, between carbon sources, areas where carbon is released, and **carbon sinks**, areas where carbon is stored (carbon reservoirs)—the overall exchange between sources and sinks is the **global carbon budget**. The *permafrost–carbon feedback* and *CO_2–weathering feedback* involve the movement of CO_2 within Earth's carbon budget.

Maunder Minimum (p. 299)
Milankovitch cycles (p. 300)
climate feedback (p. 300)
carbon sink (p. 300)
global carbon budget (p. 301)

7. What is the connection between sunspots and solar output? What happened to sunspot activity during the Maunder Minimum? What was the status of solar activity in 2005 to 2010?
8. What are the three time periods of cyclical variation in Milankovitch cycles?
9. Describe the effect of volcanic aerosols on climate.
10. Name several of the most important carbon sinks in the global carbon budget. What are the most important carbon sources?
11. Define a climate feedback, and sketch an example as a feedback loop.
12. Does the CO_2–weathering feedback work in a positive or negative direction? Explain.

■ *List* the key lines of evidence for present global climate change, and *summarize* the scientific evidence for anthropogenic forcing of climate.

Several indicators provide strong evidence of climate warming: increasing air temperatures over land and oceans, increasing sea-surface temperatures and ocean heat content, melting of glacial ice and sea ice, rising global sea level, and increasing specific humidity. The scientific consensus is that present climate change is caused primarily by increased concentrations of atmospheric greenhouse gases resulting from human activities. The primary greenhouse gases produced by human activities are carbon dioxide, methane, nitrous oxide, and halogenated gases, such as chlorofluorocarbons (CFCs) and hydrofluorocarbons (HFCs). The increasing presence of these gases is causing a positive **radiative forcing** (or climate forcing), the amount that some perturbation causes the Earth–atmosphere energy balance to deviate from zero. Studies show that CO_2 has the largest radiative forcing among greenhouse gases and that this forcing surpasses other natural and anthropogenic factors that force climate.

radiative forcing (p. 311)

13. What is the role of multiyear ice in overall global sea-ice losses? What is the status of this ice today? (Check some websites such as that of the National Snow and Ice Data Center.)
14. What are the two most significant factors currently contributing to global sea-level rise?
15. What are the main sources of carbon dioxide? What are the main sources of atmospheric methane? Why is methane considered to be a more potent greenhouse gas than CO_2?
16. What are methane hydrates, and how can they potentially affect atmospheric greenhouse gas concentrations?

■ *Discuss* climate models, and *summarize* several climate projections.

A **general circulation model (GCM)** is a complex computerized climate model used to assess past climatic trends and their causes, and project future changes in climate. The most sophisticated atmosphere and ocean submodels are known as **Atmosphere–Ocean General Circulation Models (AOGCMs)**. Climate models show that positive radiative forcing is caused by anthropogenic greenhouse gases rather than natural factors such as solar variability and volcanic aerosols.

general circulation model (GCM) (p. 314)
Atmosphere–Ocean General Circulation Model (AOGCM) (p. 314)

17. What do climate models tell us about radiative forcing and future temperature scenarios?
18. How might we alter future scenarios by changing national policies regarding fossil-fuel usage?

■ *Describe* several mitigation measures to slow rates of climate change.

Actions taken on an individual level by millions of people can slow the pace of climate change for us and for future generations. The principal way we can do this—as individuals, as a country, and as an international community—is to reduce carbon emissions, especially in our burning of fossil fuels.

19. What are the actions being taken at present to delay the effects of global climate change? What is the Kyoto Protocol?
20. Take a moment and reflect on possible personal, local, regional, national, and international mitigation actions to reduce climate-change impacts.

MasteringGeography™

Looking for additional review and test prep materials? Visit the Study Area in *MasteringGeography*™ to enhance your geographic literacy, spatial reasoning skills, and understanding of this chapter's content by accessing a variety of resources, including MapMaster interactive maps, geoscience animations, satellite loops, author notebooks, videos, RSS feeds, flashcards, web links, self-study quizzes, and an eText version of *Geosystems.*

▼**Figure 10B.30 Rate of global mean sea-level change, 1992–2013.** Sea level trends as measured by *TOPEX/Poseidon*, *Jason*-1, and *Jason*-2 satellites vary with geographic location. [Laboratory for Satellite Altimetry/NOAA.]

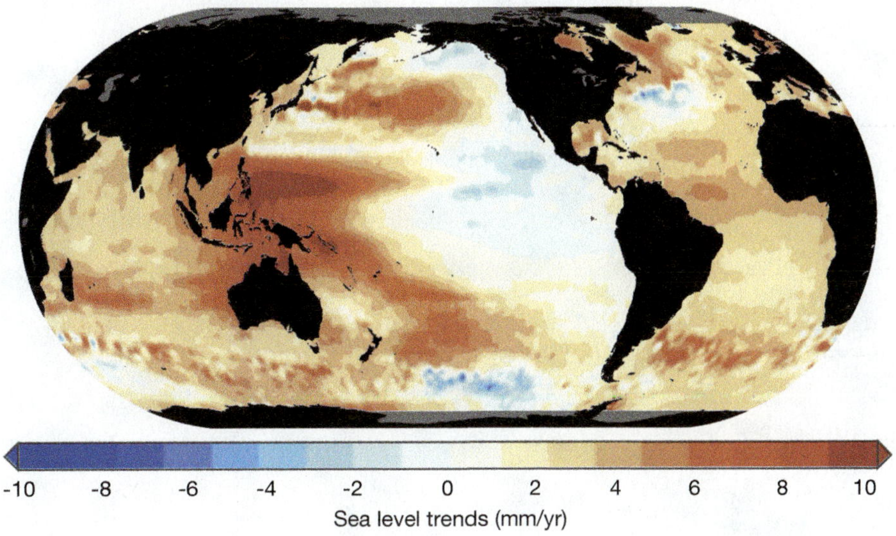

III The Earth–Atmosphere Interface

Earth is a dynamic planet whose surface is shaped by active physical agents of change. Two broad systems—endogenic and exogenic—organize these agents in Part III. The *endogenic system* (Chapters 11 and 12) encompasses internal processes that produce flows of heat and material from deep below Earth's crust. Radioactive decay is the principal source of power for these processes. The

▼ The Colorado River near Moab, Utah. [Scott Prokop/Shutterstock.]

materials involved constitute the solid realm of Earth. Earth's surface responds by moving, warping, and breaking, sometimes in dramatic episodes of earthquakes and volcanic eruptions, constructing the crust.

The *exogenic system* (Chapters 13–14, 16–17) consists of external processes at Earth's surface that set into motion air, water, and ice, all powered by solar energy. These media carve, shape, and wear down the landscape. One such process, *weathering*, breaks up and dissolves the crust. *Erosion* picks up these materials; transports them in rivers, coastal waves, winds, and flowing glaciers; and deposits them in new locations. Thus, Earth's surface is the interface between two vast open systems: one that builds the landscape and creates topographic relief, and one that tears the landscape down into relatively low elevation plains of sedimentary deposits.

11 The Dynamic Planet

KEY LEARNING **concepts**

After reading the chapter, you should be able to:

- ***Distinguish*** between the endogenic and exogenic systems that shape Earth, and ***name*** the driving force for each.
- ***Explain*** the principle of uniformitarianism, and ***discuss*** the time spans into which Earth's geologic history is divided.
- ***Depict*** Earth's interior in cross section, and ***describe*** each distinct layer.
- ***Describe*** the three main groups of rock, and ***diagram*** the rock cycle.
- ***Describe*** Pangaea and its breakup, and ***explain*** the physical evidence that crustal drifting is continuing today.
- ***Draw*** the pattern of Earth's major plates on a world map, and ***relate*** this pattern to the occurrence of earthquakes, volcanic activity, and hot spots.

The volcanic landscapes of Iceland are a surface expression of Earth's endogenic, or internal, system. The ash-covered environment and vapor cloud results from the activity of the Eyjafjallajökull volcano, which erupted in 2010. (See the ash plume produced by this eruption in Figure 12.21.) This chapter examines the materials and processes that shape Earth's surface, laying the foundation for our study of the Earth–atmosphere interface. [Kersten Langenberger/Imagebroker/Corbis.]

Earth's Migrating Magnetic Poles

In Chapter 1, we discussed the north geographic pole—the axial pole centered where the meridians of longitude converge. This is *true north* and is a fixed point. Another "north pole" also exists, this one in association with Earth's magnetic field. This is the pole toward which a compass needle points, known as the *North Magnetic Pole* (NMP). Before the Global Positioning System (GPS) was in widespread use, people relied on the compass to find direction, making the location of this pole critical for navigation. Since the location of this pole changes, the NMP must be periodically detected and pinpointed by magnetic surveys.

The deep interior of Earth—the core—has a solid inner region and a fluid outer region. Earth's magnetic field is principally generated by motions in the fluid material of the outer core. Like a bar magnet, the magnetic field has poles with opposite charges. At the NMP, the pull of the magnetic field is directed vertically downward; imagine this pull as an arrow pointing downward, intersecting Earth's surface at the north and south magnetic poles. The movement of the magnetic poles results from changes in Earth's magnetic field.

Magnetic Declination Today, a compass needle does not point to true north. The angular distance in degrees between the direction of the compass needle and the line of longitude at a given location is known as the *magnetic declination*. Since the NMP is constantly on the move, knowledge of its present location is essential for determining the magnetic declination. Then, to calculate true north from a compass reading, you add or subtract (depending on your longitude relative to the NMP) the magnetic declination appropriate to your location. For example, in 2008, north of San Francisco, California, the declination was 15° west. However, at the junction of the Iowa, Missouri, and Illinois borders, the declination was 0°—at this location, the NMP and true north were aligned. At the same time, in Boston, the declination was 15° east.

Movement of the Poles During the past century, the NMP moved 1100 km (685 mi) across the Canadian Arctic. Presently the NMP is moving northwest toward Siberia at approximately 55–60 km (34–37 mi) per year. The observed positions for 1831–2012 are mapped in Figure GN 11.1. Each day, the actual magnetic pole migrates in a small oval pattern around the average locations given on the map. The Geological Survey of Canada (GSC) tracks the location and movement of the NMP (see http://geomag.nrcan.gc.ca/index-eng.php).

In 2012, the NMP was near 85.9° N by 147° W in the Canadian Arctic. Its *antipode*, or opposite pole, the South Magnetic Pole (SMP), lay off the coast of Wilkes Land, Antarctica. The SMP moves separately from the NMP and is presently headed northwest at just 5 km (3.1 mi) per year. As we explore Earth's interior in this chapter, we discuss the changing intensity of Earth's magnetic field and how the field periodically reverses polarity. We discuss in later chapters the effects of the magnetic field on animal migration and how birds and turtles, among others, can read magnetic declination.

GEOSYSTEMS NOW ONLINE Go to Chapter 11 on the *MasteringGeography* website (www.masteringgeography.com/) for more information about Earth's magnetic field and the movement of the magnetic poles. For more about magnetic declination, go to http://geomag.nrcan.gc.ca/mag_fld/magdec-eng.php, scroll down to the declination calculator, and determine the declination for your current location. MG

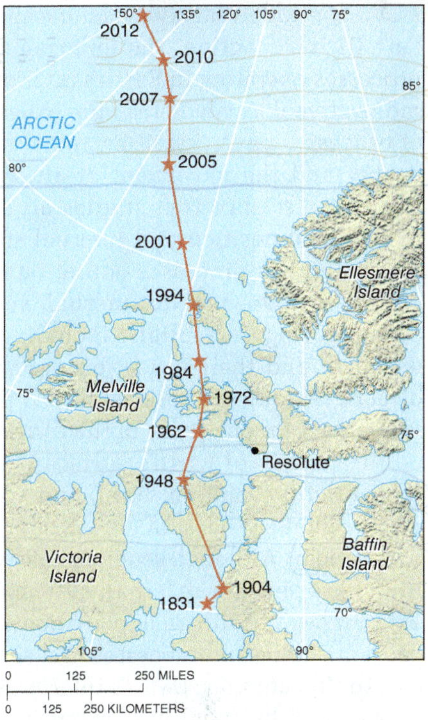

▲**Figure GN 11.1 North Magnetic Pole Movement, 1831 to 2012.**

TABLE GN 11.1 Approximate North Magnetic Pole Coordinates, 2003 to 2015

Year	Latitude (°N)	Longitude (°W)
2003	82.0	112.4
2004	82.3	113.4
2005	82.7	114.4
2006	83.9	119.9
2007	84.4	121.7
2008	84.2	124.9
2009	84.9	131.0
2010	85.0	132.6
2011	85.1	134.0
2012	85.9	147.0
2013	85.9	148.0
2014	85.9	149.0
2015 (predicted)	86.1	153.0

The Earth–atmosphere interface is the meeting place of internal and external processes that build up and wear away landscapes. As stated in the Part III introduction, the **endogenic system** consists of processes operating in Earth's interior, driven by heat and radioactive decay, whereas the **exogenic system** consists of processes operating at Earth's surface, driven by solar energy and the movement of air, water, and ice. *Geology* is the science that studies all aspects of Earth—its history, composition and internal structure, surface features, and the processes acting on them. Our overview of Earth's *endogenic*, or internal, system in this chapter covers geology essentials, including the types of rocks on Earth and their formation processes and the theory of plate tectonics. These essentials provide a conceptual framework for our study of the lithosphere in this book.

The study of Earth's surface landforms, specifically their origin, evolution, form, and spatial distribution, is **geomorphology**—a subfield of both physical geography and geology. Although geomorphology is primarily related to the *exogenic*, or external, system, our study of Earth's exterior begins with an explanation of Earth's interior, the basic materials and processes that shape Earth's surface.

In this chapter: Earth's interior is organized as a core surrounded by roughly concentric shells of material. It is unevenly heated by the radioactive decay of unstable elements. A rock cycle produces three classes of rocks through igneous, sedimentary, and metamorphic processes. A tectonic cycle moves vast sections of Earth's crust, called plates, accompanied by the spreading of the ocean floor. Collisions of these plate produce irregular surface fractures and mountain ranges both on land and on the ocean floor. This movement of crustal material results from endogenic forces within Earth; the surface expressions of these forces include earthquakes and volcanic events.

The Pace of Change

In Chapter 10B, we discussed paleoclimatic techniques that establish chronologies of past environments, enabling scientists to reconstruct the age and character of past climates. An assumption of these reconstructions is that the movements, systems, and cycles that occur today also operated in the past. This guiding principle of Earth science, called **uniformitarianism**, presupposes that the same physical processes now active in the environment were operating throughout Earth's history. The phrase "the present is the key to the past" describes the principle. For example, the processes by which streams carve valleys at present are assumed to be the same as those that carved valleys 500 million years ago. Evidence from the geologic record, preserved in layers of rock that formed over millennia, supports this concept, which was first hypothesized by geologist James Hutton in the 18th century and later amplified by Charles Lyell in his seminal book *Principles of Geology* (1830).

Although the principle of uniformitarianism applies mainly to the gradual processes of geologic change, it also includes sudden, catastrophic events such as massive landslides, earthquakes, volcanic episodes, and asteroid impacts. These events have geological importance and may occur as small interruptions in the generally uniform processes that shape the slowly evolving landscape. Thus, uniformitarianism means that the natural laws that govern geologic processes have not changed throughout geologic time even though the rate at which these processes operate is variable.

The full scope of Earth's history can be represented in a summary timeline known as the **geologic time scale** (Figure 11.1). The scale breaks the past 4.6 billion years down into *eons*, the largest time span (although some refer to the Precambrian as a *supereon*), and then into increasingly shorter time spans of *eras*, *periods*, and *epochs*. Major events in Earth's history determine the boundaries between these intervals, which are not equal in length. Examples are the six major extinctions of life forms in Earth history, labeled in Figure 11.1. The timing of these events ranges from 440 million years ago (m.y.a.) to the ongoing present-day extinction episode caused by modern civilization (discussed in Chapter 19; for more on the geologic timescale, see http://www.ucmp.berkeley.edu/exhibit/geology.html).

Geologists assign ages to events or specific rocks, structures, or landscapes using this time scale, based on either relative time (what happened in what order) or numerical time (the actual number of years before the present). *Relative age* refers to the age of one feature with respect to another within a sequence of events and is deduced from the relative positions of rock strata above or below each other. *Numerical age* (sometimes called absolute age) is today determined most often using isotopic dating techniques (introduced in Chapter 10B).

Determinations of relative age are based on the general principle of *superposition*, which states that rock and unconsolidated particles are arranged with the youngest layers "superposed" toward the top of a rock formation and the oldest at the base. This principle holds true as long as the materials have remained undisturbed. The horizontally arranged rock layers of the Grand Canyon and many other canyons of the U.S. Southwest are an example. The scientific study of these sequences is **stratigraphy**. Important time clues—for example, *fossils*, the remains of ancient plants and animals—lie embedded within these strata. Since approximately 4.0 billion years ago, life has left its evolving imprint in the rocks.

Numerical age is determined by scientific methods such as *radiometric dating*, which uses the rate of decay for different unstable isotopes to provide a steady time clock to pinpoint the ages of Earth materials. Precise knowledge of radioactive decay rates allows scientists to determine the date a rock formed by comparing the amount of original isotope in the sample with the amount of decayed end product in the sample. Numerical ages

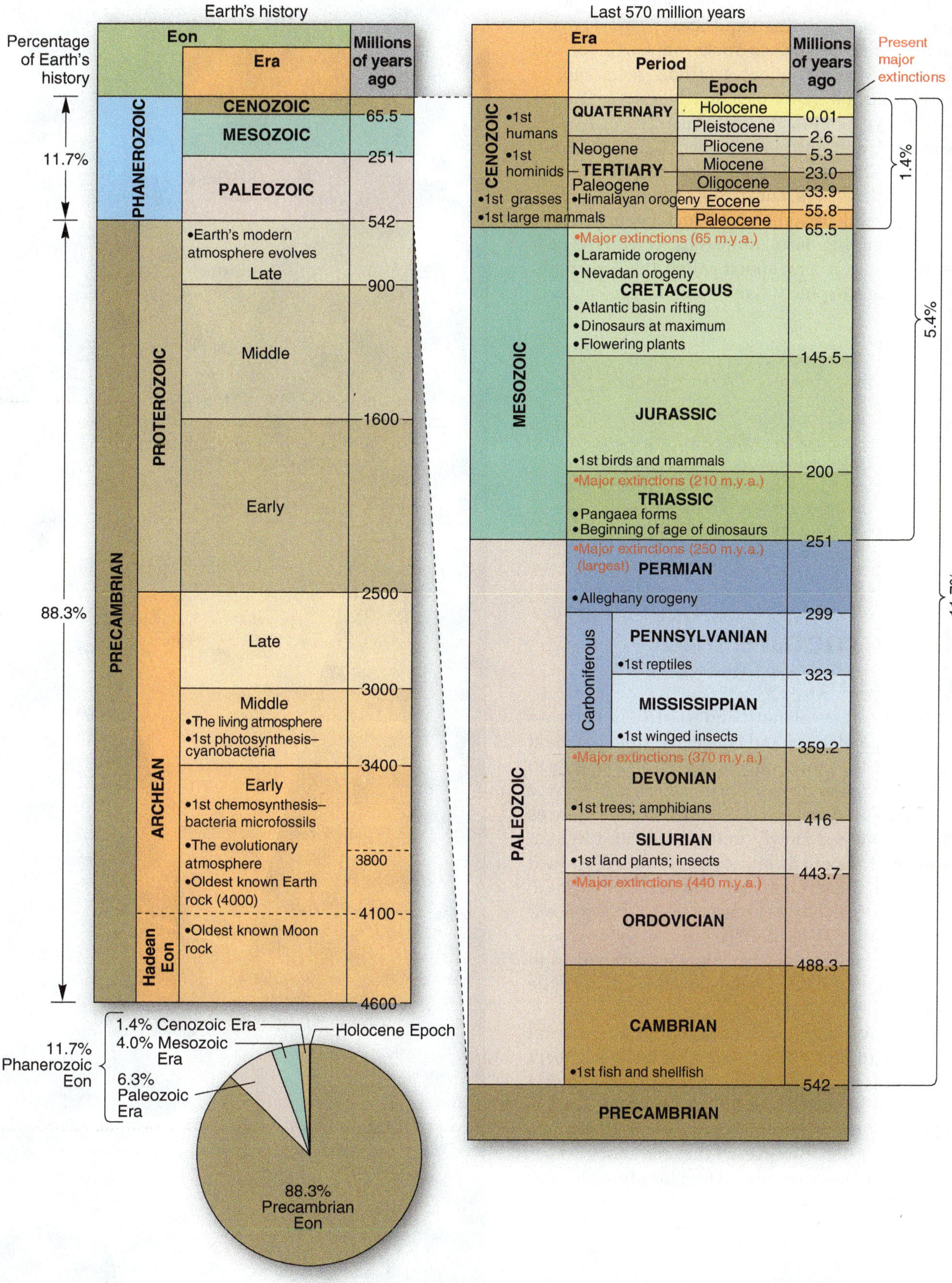

▲**Figure 11.1 Geologic time scale, showing highlights of Earth's history.** Dates appear in m.y.a (millions of years ago). The scale uses currently accepted names of time intervals, except for the Hadean Eon, which, while not an official designation, is the proposed name for the period before the Archean. The six major extinctions or depletions of life forms are shown in red. In the column to the left, note that the Precambrian Eon spans 88.3% of geologic time. [Data from Geological Society of America and *Nature 429* (May 13, 2004): 124–125.]

MG Animation Applying Relative Dating Principles

permit scientists to refine the geologic scale and improve the accuracy of relative dating sequences.

The oldest known surface rocks on Earth formed during the Archean Eon, about 4 billion years ago, in Greenland (3.8 billion years old), northwestern Canada (about 4 billion years old), Western Australia (4.2 to 4.4 billion years old), and northern Quebec, Canada (4.3 billion years old). The most recent epoch in the geologic time scale is the *Holocene*, consisting of the 11,500 years since the last glacial period. As the impacts of humans on Earth systems increase, numerous scientists now agree that we are in a new epoch called the Anthropocene (discussed in Chapter 1, Geo Report 1.1).

CRITICALthinking 11.1
Thoughts about an "Anthropocene Epoch"

Take a moment to explore the idea of naming our current epoch in the geologic time scale for humans. Develop some arguments for doing so that consider landscape alteration, deforestation, and climate change. If we were to designate the late Holocene as the Anthropocene, what do you think the criteria for the beginning date should be? ●

Earth's Structure and Internal Energy

Along with the other planets and the Sun, Earth is thought to have condensed and congealed from a nebula of dust, gas, and icy comets about 4.6 billion years ago (discussed in Chapter 2). As Earth solidified, gravity sorted materials by density. Heavier, denser substances such as iron gravitated slowly to its center, and lighter, less-dense elements such as silica slowly welled upward to the surface and became concentrated in the outer shell. Consequently, Earth's interior consists of roughly concentric layers (Figure 11.2), each distinct in either composition or temperature. Heat energy migrates outward from the center by conduction, as well as by convection in the *plastic*, or fluid, layers.

Scientists have direct evidence of Earth's internal structure down to about 2 km (1.2 mi), from sediment cores drilled into Earth's outer surface layer. Below this region, scientific knowledge of Earth's internal layers is acquired entirely through indirect evidence. In the late 19th century, scientists discovered that the shock waves created by earthquakes were useful for identifying Earth's

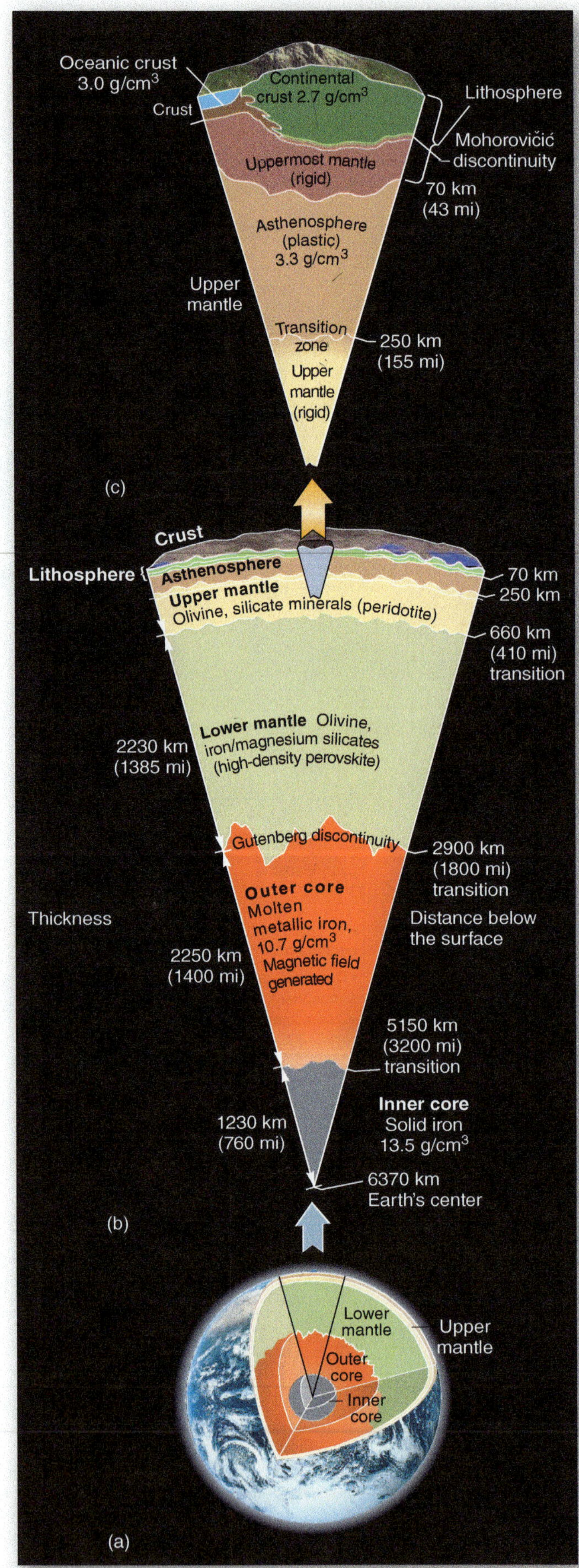

▶**Figure 11.2 Earth in cross section.** (a) Cutaway showing Earth's interior. (b) Earth's interior in cross section, from the inner core to the crust. (c) Detail of the structure of the lithosphere and its relation to the asthenosphere. (For comparison to the densities noted, the density of water is 1.0 g/cm³, and mercury, a liquid metal, is 13.0 g/cm³.)

internal materials. Earthquakes are the surface vibrations felt when rocks near the surface suddenly fracture, or break (discussed at length in Chapter 12). These fractures generate **seismic waves**, or shock waves, that travel throughout the planet. The speed of the waves varies as it passes through different materials—cooler, more rigid areas transmit seismic waves at a higher velocity than do the hotter, more fluid areas. Plastic zones do not transmit some seismic waves; they absorb them. Waves may also be refracted (bent) or reflected, depending on the density of the material. Thus, scientists are now able to identify the boundaries between different layers within Earth by measuring the depths of changes in seismic wave velocity and direction. This is the science of *seismic tomography*; for animations and information, go to http://www.iris.edu/hq/programs/education_and_outreach/animations/7.

Earth's Core and Mantle

A third of Earth's entire mass, but only a sixth of its volume, lies in its dense core. The **core** is differentiated into two regions—*inner core* and *outer core*—divided by a transition zone several hundred kilometers wide (see Figure 11.2b). Scientists think that the inner core formed before the outer core, shortly after Earth condensed. The inner core is solid iron that is well above the melting temperature of iron at the surface but remains solid because of tremendous pressure. The iron is not pure, but probably is combined with silicon and possibly oxygen and sulfur. The outer core is molten, metallic iron with lighter densities than the inner core. The high temperatures keep the outer core in a liquid state, and the flow of this material generates Earth's magnetic field (discussed just ahead).

Earth's outer core is separated from the mantle by a transition zone several hundred kilometers wide at an average depth of about 2900 km (1800 mi; see Figure 11.2b). This zone is a *discontinuity*, or a place where physical differences occur between adjoining regions in Earth's interior. By studying the seismic waves of more than 25,000 earthquakes, scientists determined that this transition area, the *Gutenberg discontinuity*, is uneven, with ragged peak-and-valley-like formations.

Together, the lower and upper **mantle** represent about 80% of Earth's total volume. The mantle is rich in iron and magnesium oxides and in silicates, which are dense and tightly packed at depth, grading to lesser densities toward the surface. Temperatures are highest at depth and decrease toward the surface; materials are thicker at depth, with higher viscosity, due to increased pressure. A broad transition zone of several hundred kilometers, centered between 410 and 660 km (255–410 mi) below the surface, separates the lower mantle from the upper mantle. Rocks in the lower mantle are at high enough temperature that they become soft and are able to flow slowly, deforming over time scales of millions of years.

The boundary between the uppermost mantle and the crust above is another discontinuity, known as the **Mohorovičić discontinuity**, or **Moho** for short. It is named for the Yugoslavian seismologist who determined that seismic waves change at this depth due to sharp contrasts in material composition and density.

Earth's Crust

Above the Moho is Earth's outer layer, the **crust**, which makes up only a fraction of Earth's overall mass. The crust also makes up only a small portion of the overall distance from Earth's center to its surface (Figure 11.3). The distances through Earth's interior compared with surface distances in North America provide a sense of size and scale: An airplane flying from Anchorage, Alaska, to Fort Lauderdale, Florida, would travel the same distance as that from Earth's center to its surface. The last 30 km (18.6 mi) of that journey represent the thickness of Earth's crust.

The thickness of Earth's crust varies over the extent of the planet. Crustal areas beneath mountain masses are thicker, extending to about 50–60 km (31–37 mi), whereas the crust beneath continental interiors averages about 30 km (19 mi) in thickness. Oceanic crust averages only 5 km (3 mi) in thickness. Drilling through the crust and Moho discontinuity (the crust–mantle boundary) into the uppermost mantle remains an elusive scientific goal.

Just eight natural elements make up over 98% of Earth's crust by weight, and just two of these—oxygen and silicon—together account for 74.3% (Table 11.1). Oxygen is the most reactive gas in the lower atmosphere, readily combining with other elements. For this reason, the percentage of oxygen is higher in the crust (at 46%) than in the atmosphere, where it makes up 21%. The internal differentiation process, in which less-dense elements are nearer the surface, explains the relatively large percentages of elements such as silicon and aluminum in the crust.

Continental crust differs greatly from oceanic crust in composition and texture, and the difference has a bearing on the dynamics of plate tectonics and continental drift discussed later in the chapter.

- Continental crust is relatively low in density, averaging 2.7 g/cm^3 (or 2700 kg/m^3), and is composed mainly of *granite*. It is crystalline and high in silica, aluminum, potassium, calcium, and sodium. Sometimes continental crust is called *sial*, shorthand for the dominant elements of *si*lica and *al*uminum.

GEOreport 11.1 Radioactive elements drive Earth's internal heat

The internal heat that fuels endogenic processes beneath Earth's surface comes from residual heat left over from the planet's formation and from the decay of radioactive elements—specifically, the radioactive decay of the isotopes potassium-40 (^{40}K), uranium-238 and 235 (^{238}U and ^{235}U), and thorium-232 (^{232}Th).

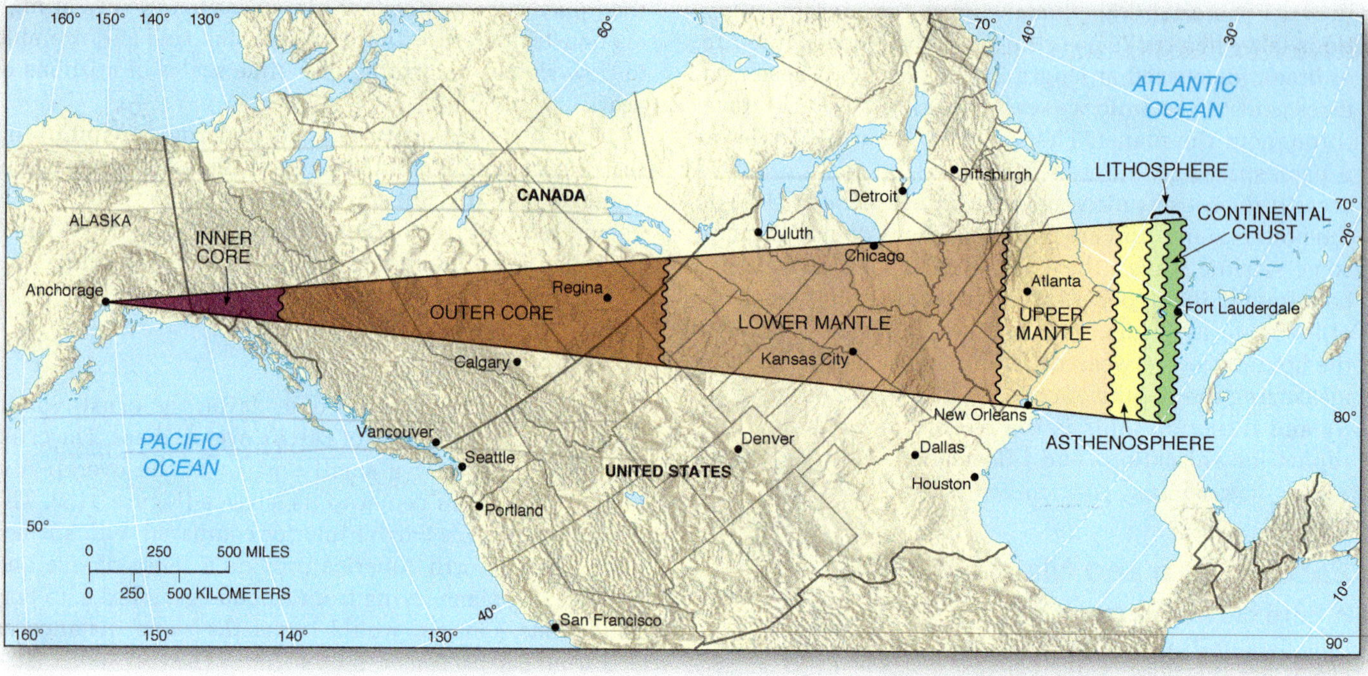

▲Figure 11.3 **Distances from core to crust.** The distance from Anchorage, Alaska, to Fort Lauderdale, Florida, is the same as the distance from Earth's center to its outer crust. The distance from Boca Raton, Florida, to Fort Lauderdale (30 km, or 18.6 mi) represents the thickness of the continental crust.

- Oceanic crust is denser than continental crust, averaging 3.0 g/cm^3 (or 3000 kg/m^3), and is composed of *basalt.* It is granular and high in silica, magnesium, and iron. Sometimes oceanic crust is called *sima*, shorthand for the dominant elements of *si*lica and *ma*gnesium.

The Asthenosphere and Lithosphere

The interior layers of core, mantle, and crust are differentiated by chemical composition. Another way to distinguish layers within the Earth is by their rigid or plastic character. A rigid layer will not flow when a force acts on it; instead it will bend or break. A plastic layer will slowly flow when a force is present. Using this criterion, scientists divide the outer part of Earth into two layers: the **lithosphere**, or rigid layer (from the Greek *lithos*, or "rocky"), and the **asthenosphere**, or plastic layer (from the Greek *asthenos*, meaning "weak").

The lithosphere includes the crust and the uppermost mantle, to about 70 km (43 mi) in depth, and forms the rigid, cooler layer at Earth's surface (Figure 11.2c). Note that the terms *lithosphere* and *crust* are not the same; the crust makes up the upper portion of the lithosphere.

The asthenosphere lies within the mantle from about 70 km to 250 km (43 mi to 155 mi) in depth. This is the hottest region of the mantle—about 10% of the asthenosphere is molten in uneven patterns. The movement of convection currents in this zone in part causes the shifting of lithospheric plates, discussed later in the chapter.

Adjustments in the Crust

We discussed the *buoyancy force* in Chapter 7 with regard to parcels of air—if a parcel of air is less dense than the surrounding air, it is buoyant and will rise. In essence, buoyancy is the principle that something less dense, such as wood, floats in something denser, such as water. The balance between the buoyancy and gravitational forces is the principle of **isostasy**, which explains the elevations of continents and the depths of ocean floors as determined by vertical movements of Earth's crust.

Earth's lithosphere floats on the denser layers beneath, much as a boat floats on water. If a load is placed on the surface, such as the weight of a glacier, a mountain range, or an area of sediment accumulation (rock material that has been transported by exogenic processes), the lithosphere tends to sink, or ride lower in the asthenosphere (Figure 11.4, page 332). When this happens, the rigid lithosphere bends, and the plastic asthenosphere flows out of the way. If the load is removed, such as when a glacier melts, the crust rides higher and the asthenosphere flows back toward the region of uplifting lithosphere. The uplift after removal

GEOreport 11.2 Earth on the scales
How much does our planet weigh? A revised estimate of Earth's mass, or weight, calculated in 2000 set its weight at 5.972 sextillion metric tons (5972 followed by 18 zeros).

TABLE 11.1 Common Elements in Earth's Crust

Element	Percentage of Earth's Crust by Weight
Oxygen (O)	46.6
Silicon (Si)	27.7
Aluminum (Al)	8.1
Iron (Fe)	5.0
Calcium (Ca)	3.6
Sodium (Na)	2.8
Potassium (K)	2.6
Magnesium (Mg)	2.1
All others	1.5
Total	100.00

[Quartz: Stefano Cavoretto/Shutterstock.]

A quartz crystal (SiO_2) consists of Earth's two most abundant elements, silicon (Si) and oxygen (O).

of surface load is known as *isostatic rebound*. The entire crust is in a constant state of isostatic adjustment, slowly rising and sinking in response to weight at the surface.

In southeast Alaska, the retreat of glacial ice following the last ice age about 10,000 years ago removed weight from the crust. Researchers today, using an array of GPS receivers to measure isostatic rebound, expected to find a slowed rate of crustal rebound occurring now in southeastern Alaska, as compared to the more rapid response in the distant past when the ice first retreated. Instead, they detected some of the most rapid vertical motion on Earth, averaging about 36 mm (1.42 in.) per year. Scientists attribute this isostatic rebound to the loss of modern glaciers in the region, especially in the areas from Yakutat Bay and the Saint Elias Mountains in the north, through Glacier Bay and Juneau in the south. This rapid rebound is attributable to climate change—causing glacial melt and retreat over the past 150 years—and correlates with record warmth across Alaska.

Earth's Magnetism

As mentioned earlier, Earth's fluid outer core generates most (at least 90%) of Earth's magnetic field and the magnetosphere that surrounds and protects Earth from solar wind and cosmic radiation. One hypothesis explains that circulation in the outer core converts thermal and gravitational energy into magnetic energy, thus producing the magnetic field. The locations of the north and south magnetic poles are surface expressions of Earth's magnetic field and migrate as plotted on the map in *Geosystems Now* for this chapter.

At various times in Earth's history, the magnetic field has faded to zero and then returned to full strength with the polarity reversed (meaning that the north magnetic pole then lies near the south geographic pole, so that a compass needle would point south). In the process, the field does not blink on and off but instead diminishes slowly to low intensity, perhaps 25% strength, and then rapidly regains full power. This **geomagnetic reversal** has taken place nine times during the past 4 million years and hundreds of times over Earth's history. During the transition interval of low strength, Earth's surface receives higher levels of cosmic radiation and solar particles, but not to such an extent as to cause species extinctions. Life on Earth has weathered many of these transitions.

The average period of a magnetic reversal is about 500,000 years, ranging from as short as 20,000 years to as long as 50 million years. The last reversal was 790,000 years ago, preceded by a transition ranging from 2000 years in length along the equator to 10,000 years in the midlatitudes. Given rates of magnetic field decay measured over the last 150 years, we are perhaps 1000 years away from entering the next phase of field changes, although there is no expected pattern to forecast the timing.

The reasons for these magnetic reversals are unknown. However, they have become a key tool in understanding the evolution of landmasses and the movements of the continents. As rocks cool and solidify from molten material (lava) at Earth's surface, the small magnetic particles in the material align according to the orientation of the magnetic poles at that time, and this alignment locks in place.

GEOreport 11.3 Deep-drilling the continental crust

The Kola Borehole in Russia, north of the Arctic Circle, is 12.23 km deep (7.6 mi, or 40,128 ft), drilled over a 20-year period purely for exploration and science. This is the deepest drilling attempt for scientific purposes in continental crust; drilling for oil wells has gone deeper—the record is 12,289 m (40,318 ft) in the Al Shaheen oil field in Qatar. The Kola Borehole reached rock 1.4 billion years old at 180°C (356°F) in the Earth's crust, and for two decades was the deepest borehole ever drilled (see http://www.icdp-online.org/front_content.php?idcat=695).

All across Earth, rocks of the same age bear identical alignments of magnetic materials, such as iron particles. Across the ocean floor, scientists have found a record of magnetic reversals in the form of measurable rock "stripes" indicating periods of normal polarity and reversed polarity recorded in newly formed rock. These stripes illustrate global patterns of changing magnetism. Later in this chapter, we see the importance of these magnetic reversals.

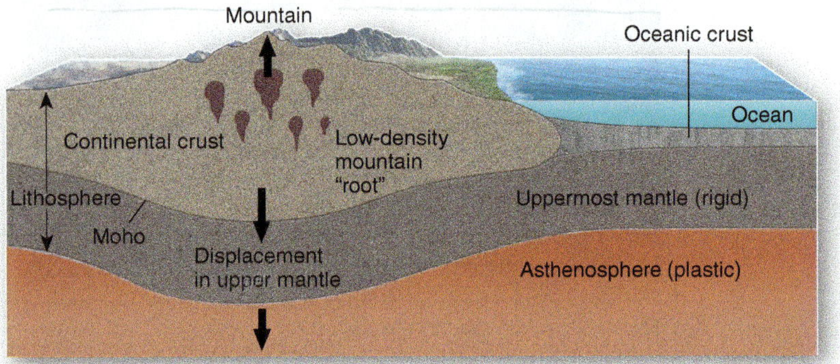

(a) The mountain mass slowly sinks, displacing mantle material.

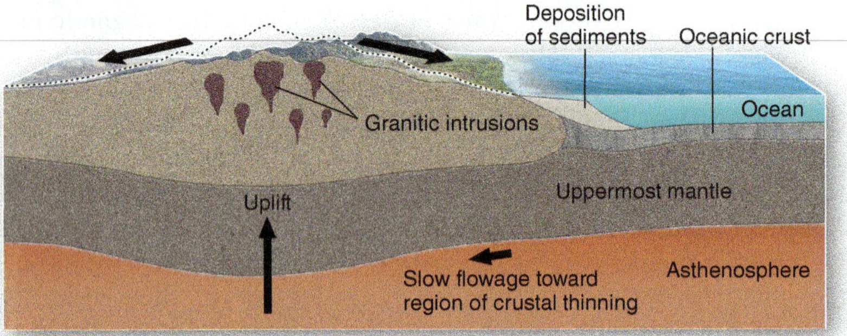

(b) Weathering and erosion transport sediment from land into oceans; as land loses mass, the crust isostatically adjusts upward.

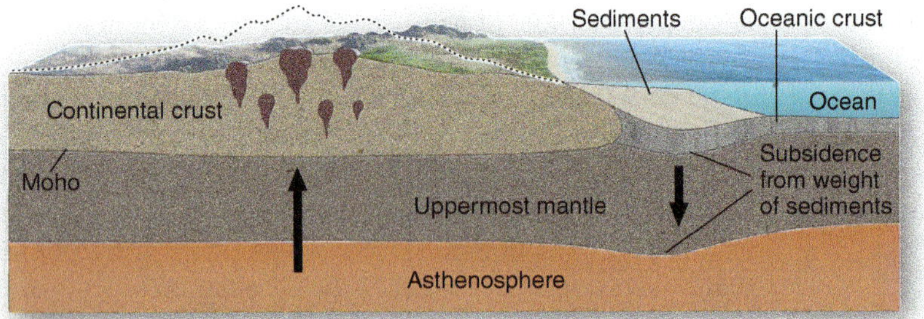

(c) As the continental crust erodes, the heavy sediment load offshore deforms the lithosphere beneath the ocean.

(d) The melting of ice from the last ice age and losses of overlying sediments are thought to produce an ongoing isostatic uplift of portions of the Sierra Nevada batholith.

▲**Figure 11.4 Isostatic adjustment of the crust.** Earth's entire crust is in a constant state of adjustment, as suggested by these three sequential stages. [(d) Bobbé Christopherson.]

Earth Materials and the Rock Cycle

We have already mentioned several types of rocks, such as granite and basalt, and described processes involving rocks and their formation. To understand and classify rocks from a scientific viewpoint, we must begin with minerals, which are the building blocks of rocks. A **mineral** is an inorganic, or nonliving, natural solid compound having a specific chemical formula and usually possessing a crystalline structure. Each mineral has its own characteristic color, texture, crystal shape, and density, among other unique properties. For example, the common mineral *quartz* is silicon dioxide, SiO_2, and has a distinctive six-sided crystal. Ice fits the definition of a mineral, although water does not.

Mineralogy is the study of the composition, properties, and classification of minerals (see http://www.mindat.org/). Of the more than 4200 minerals known, about 30 are the most common components of rocks. Roughly 95% of Earth's crust is made up of *silicates*, one of the most widespread mineral families—not surprising considering the percentages of silicon and oxygen on Earth and their readiness to combine with each other and with other elements. This mineral family includes quartz, feldspar, clay minerals, and numerous gemstones. Several other groups of minerals are also important: *Oxides* are minerals in which oxygen combines with metallic elements; *sulfides* and *sulfates* are minerals in which sulfur compounds combine with metallic elements; and *carbonates* feature carbon in combination with oxygen and other elements such as calcium, magnesium, and potassium.

A **rock** is an assemblage of minerals bound together (such as granite, a rock containing three minerals); or a mass composed of a single mineral (such as rock salt); or of undifferentiated material (such as the noncrystalline glassy obsidian); or even solid organic material (such as coal). Scientists have identified thousands of different rocks, all of which can be sorted according to three types that depend on the processes that formed them: *igneous* (formed from

(a) Granite shaped by weathering processes near Joshua Tree National Park, California, the southern region of the extensive California batholith (we examine weathering in Chapter 13.)

(b) Basaltic lava flows on Hawai`i. The glowing opening is a skylight into an active tube where molten lava flows; the shiny surface is where lava recently flowed out of the skylight.

▲Figure 11.5 **Examples of intrusive granite and extrusive basalt.** [(a) Egon Bömsch/Imagebroker/SuperStock. (b) Bobbé Christopherson.]

molten material), *sedimentary* (formed from compaction or chemical processes), and *metamorphic* (altered by heat and pressure). The movement of material through these processes is known as the *rock cycle* and is summarized at the end of this section.

Igneous Processes

An **igneous rock** is one that solidifies and crystallizes from a molten state. Igneous rocks form from **magma**, which is molten rock beneath Earth's surface (hence the name *igneous*, which means "fire-formed" in Latin). When magma emerges at the surface, it is **lava**, although it retains its molten characteristics. Overall, igneous rocks make up approximately 90% of Earth's crust, although sedimentary rocks, soil, or oceans frequently cover them.

Igneous Environments

Magma is fluid, highly gaseous, and under tremendous pressure. The result is that it either *intrudes* into crustal rocks, cooling and hardening below the surface to form **intrusive igneous rock**, or it *extrudes* onto the surface as lava and cools to form **extrusive igneous rock**. Extrusive igneous rocks result from volcanic eruptions and flows, and are discussed in additional detail in Chapter 12.

The location and rate of cooling determine the crystalline texture of a rock, that is, whether it is made of coarser (larger) or finer (smaller) materials. Thus, the texture indicates the environment in which the rock formed. The slower cooling of magma beneath the surface allows more time for crystals to form, resulting in coarse-grained rocks such as **granite**. Even though this rock cooled below Earth's surface, subsequent uplift of the landscape has exposed granitic rocks (Figure 11.5a), some of which form the world's most famous cliff faces and rock climbing destinations—El Capitan and Half Dome in Yosemite Valley, California, and the Great Trango Tower in Pakistan are examples. The faster cooling of lava at the surface forms finer-grained rocks, such as **basalt**, the most common extrusive igneous rock. As discussed later in the chapter, basalt makes up the bulk of the ocean floor, accounting for 71% of Earth's surface. Basalt is actively forming on the Big Island of Hawai'i, where lava flows are a major tourist destination (Figure 11.5b).

If cooling is so rapid that crystals cannot form, the result is a glassy rock such as *obsidian*, or volcanic glass. *Pumice* is another glassy rock (one that does not have a crystal structure), which forms when the bubbles from escaping gases create a frothy texture in the lava. Pumice is full of small holes, is light in weight, and is low enough in density to float in water.

Igneous Rock Classification

Scientists classify the many types of igneous rocks according to their texture and composition (Table 11.2). The same magma that produces coarse-grained granite when it cools beneath the surface can form fine-grained basalt when it cools above the surface. The mineral composition of a rock, especially the relative amount of silica (SiO_2), provides information about the source of the magma that formed it and affects its physical characteristics. *Felsic* igneous rocks, such as granite, are high in silicate minerals, such as feldspar and quartz (pure silica), and have low melting points. The category name is derived from *fel*dspar and *sil*ica. Rocks formed from felsic minerals generally are lighter in color and less dense than those from mafic minerals. *Mafic* igneous rocks, such as basalt, are derived from *ma*gnesium and iron (the Latin word for iron is *ferrum*). Mafic rocks are lower in silica and higher in magnesium and iron and have high melting points. Rocks formed from mafic minerals are darker in color and of greater density than those from felsic minerals. Ultramafic rocks

TABLE 11.2 Classification of Igneous Rocks

	Felsic →		Mafic (Ultramafic)
Silica content of magma	High	Medium	Low
Chemical composition	High in potassium and sodium		High in magnesium and iron
Mineral content	Quartz Feldspars	Mica Amphibole	Pyroxine (Olivine)
Coloration	Lighter →		Darker
Melting temperature	Lower →		Higher
Resistance to weathering	Higher →		Lower
Intrusive rock type (coarse-grained)	Granite	Diorite	Gabbro
Extrusive rock type (fine-grained)	Rhyolite	Andesite	Basalt

[Granite: Givaga/Shutterstock. Diorite and basalt: Tyler Boyes/Shutterstock. Gabbro: Siim Sepp/Shutterstock. Andesite: Susan E. Degginger/Alamy.]

have the lowest silica content; an example is peridotite (less than 45% silica).

Igneous Landforms If igneous rocks are uplifted by exogenic processes, the work of air, water, and ice then sculpts them into unique landforms. Figure 11.6 illustrates the formation environments of several intrusive and extrusive igneous rock types and landforms.

Intrusive igneous rock that cools slowly in the crust forms a **pluton**, the general term for any intrusive igneous body, regardless of size or shape. (The Roman god of the underworld, Pluto, is the namesake.) The largest plutonic form is a **batholith**, defined as an irregular-shaped mass with an exposed surface greater than 100 km^2 (40 mi^2). Batholiths form the mass of many large mountain ranges—for example, the Sierra Nevada batholith in California (see Figure 11.4d), the Idaho batholith, and the Coast Range batholith of British Columbia and Washington State.

Smaller plutons include the magma conduits of ancient volcanoes that have cooled and hardened. Those that form parallel to layers of sedimentary rock are *sills*; those that cross layers of the rock they invade are *dikes* (Figure 11.6). Magma also can bulge between rock strata and produce a lens-shaped body called a *laccolith*, a type of sill. In addition, magma conduits may solidify in roughly cylindrical forms that stand starkly above the landscape when finally exposed by weathering and erosion. Shiprock volcanic neck in New Mexico is such a feature, rising 518 m (1700 ft) above the surrounding plain; note the radiating dikes in the aerial photo in Figure 11.6.

Sedimentary Processes

Solar energy and gravity drive the processes that form **sedimentary rock**, in which loose *clasts* (grains or fragments) are cemented together (Figure 11.7). The clasts that become solid rock are derived from several sources: the weathering and erosion of existing rock (origin of the sand that forms sandstone), the accumulation of shells on the ocean floor (which make up one form of limestone), the accumulation of organic matter from ancient plants (which form coal), and the precipitation of minerals from water solution (origin of the calcium carbonate, $CaCO_3$, that forms chemical limestone). Sedimentary rocks are divided into several categories—clastic, biochemical, organic, and chemical—based on their origin.

Clastic Sedimentary Rocks The formation of clastic sedimentary rock involves several processes. *Weathering and erosion*, discussed in detail in Chapter 13, disintegrate and dissolve existing rock into clasts. *Transportation* by gravity, water, wind, and ice then carries these rock particles across landscapes; at this point the moving material is called **sediment**. Transport occurs from "higher-energy" sites, where the carrying medium has the energy to pick up and move the material, to "lower-energy" sites, where the sediment is deposited. *Deposition* is the process whereby sediment settles out of the transporting medium and results in material dropped along river channels, on beaches, and on ocean bottoms, where it is eventually buried.

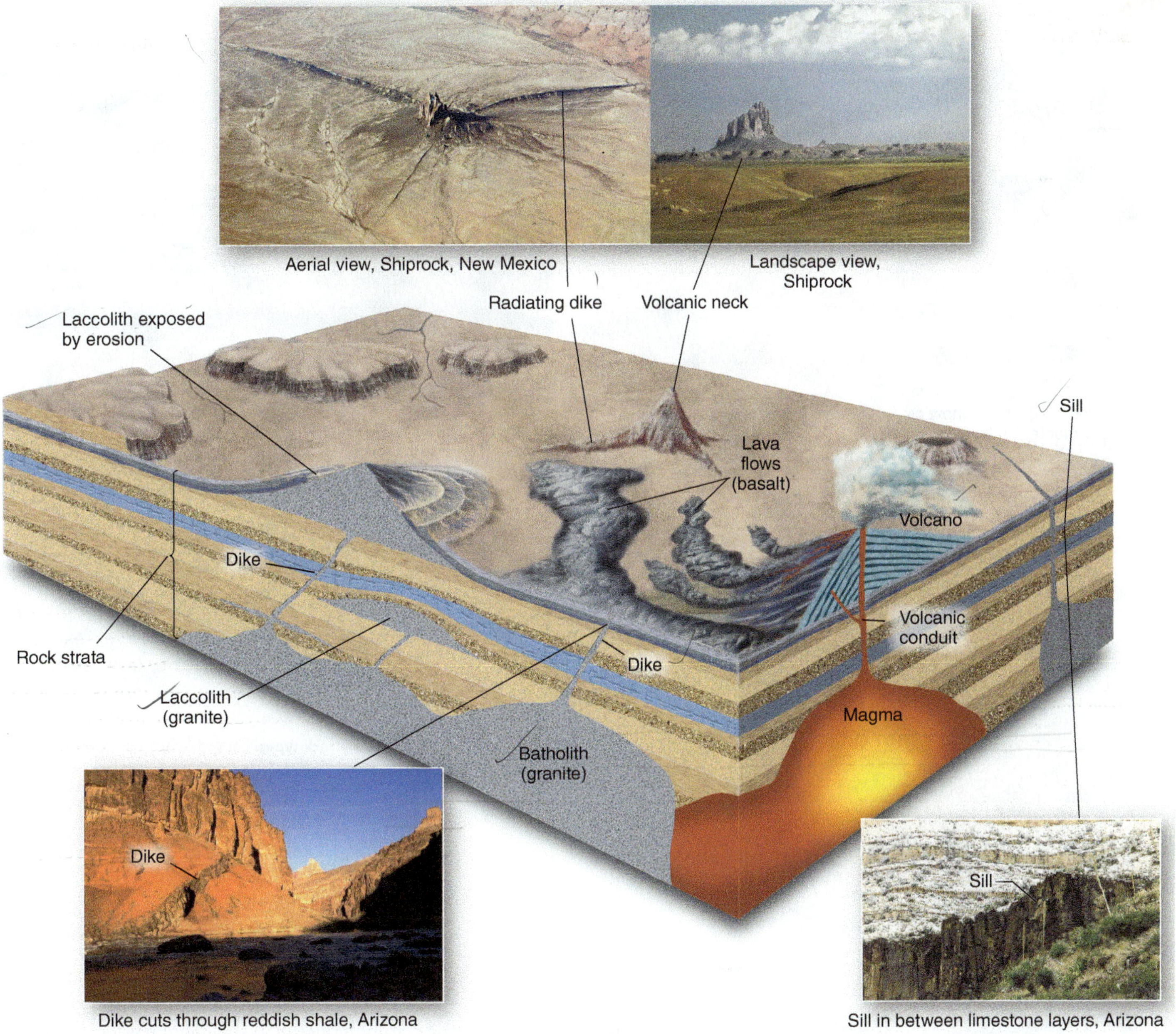

▲**Figure 11.6 Igneous landforms.** Varieties of igneous rocks, both intrusive (below the surface) and extrusive (on the surface), and associated landforms. [Aerial view: Bobbé Christopherson; landscape view: Author. Dike: NPS. Sill: Larry Fellows, Arizona Geological Survey.]

Lithification occurs as loose sediment is hardened into solid rock. This process involves *compaction* of buried sediments as the weight of overlying material squeezes out the water and air between clasts and *cementation* by minerals, which fill any remaining spaces and fuse the clasts—principally quartz, feldspar, and clay minerals—into a coherent mass. The type of cement varies with different environments. Calcium carbonate ($CaCO_3$) is the most common cement, followed by iron oxides and silica. Drying (dehydration) and heating can also unite particles.

The different sediments that make up sedimentary rock range in size from boulders to gravel to sand to microscopic clay particles (Table 11.3). After lithification, these size classes, combined with their composition, sorting, and cement characteristics, determine the common sedimentary rock types. For example, pebbles and gravels become conglomerate, silt-sized particles become siltstone or mudstone, and clay-sized particles become shale.

Chemical Sedimentary Rocks Some sedimentary rocks are formed not from pieces of broken rock, but instead from the shells of organisms that contain calcium carbonate (a biochemical process) or from dissolved minerals that precipitate out of water solutions (a chemical process) and build up to form rock. *Chemical precipitation* is the formation of a separate solid substance from

(a) Sandstone strata show cross-bedding from ancient sand dune environments, Utah.

(b) Chemical limestone in south-central Indiana; inset samples show biochemical limestone with shells and clasts cemented together.

(c) Ancient sediment deposition is evident on Mars, in the western Arabia Terra.

Sandstone

Limestone

▲**Figure 11.7** **Sedimentary rock types.** [(a) and (b) Bobbé Christopherson. (c) *Mars Global Surveyor*, NASA/JPL/Malin Space Science Systems. Sandstone: michal812/Shutterstock.]

TABLE 11.3 Clast Sizes and Related Sedimentary Rocks

Clast Size	Sediment Type	Rock Type
>80 mm (very coarse)	Boulders, cobbles	Conglomerate (breccia, if pieces are angular)
>2 mm (coarse)	Pebbles, gravel	Conglomerate
0.5–2.0 mm (coarse to medium)	Sand	Sandstone
0.062–0.5 mm (medium to fine)	Sand	Sandstone
0.0039–0.062 mm (fine)	Silt	Siltstone (mudstone)
<0.0039 mm (very fine)	Clay	Shale

a solution, such as when water evaporates and leaves behind a residue of salts. These processes are especially important in oceanic environments, as well as in areas of karst topography (discussed in Chapter 13).

The most common chemical sedimentary rock is **limestone**, and the most common form of limestone is biochemical limestone from marine organic origins. As discussed in Chapter 10B, many organisms extract dissolved $CaCO_3$ from seawater to construct solid shells. When these organisms die, the solid shell material builds up on the ocean floor and is then lithified to become limestone (Figure 11.7b).

Limestone is also formed from a chemical process in which $CaCO_3$ in solution is chemically precipitated out of groundwater that has seeped to the surface. This process forms *travertine*, a mineral deposit that commonly forms terraces or mounds near springs (Figure 11.8). The precipitation of carbonates from the water of these natural springs is driven in part by "degassing": Carbon dioxide bubbles out of solution at the surface, making the remaining solutes more likely to precipitate. Cave features such as speleothems, discussed in Chapter 10B, are another type of travertine deposit.

Hydrothermal deposits, consisting of metallic minerals accumulated by chemical precipitation from hot

▲Figure 11.8 **Chemical sedimentary rock.** Travertine, a chemical limestone composed of calcium carbonate, at the natural hot springs of Pamukkale in southwestern Turkey. [funkyfood London-Paul Williams/ Alamy.]

▲Figure 11.9 **Chemical sediments at a hydrothermal vent.** Black smokers and associated mineral deposits along the Mid-Atlantic Ridge. [Scripps Institution of Oceanography, University of California, San Diego.]

water, often are found near vents in the ocean floor—often along mid-ocean ridges created by spreading of the seafloor (discussed later in the chapter). As water seeps into the magma below the crust, it becomes superheated and then gushes out of the ocean floor at high speed. Such hydrothermal vents, called "black smokers," belch dark clouds of hydrogen sulfides, minerals, and metals that the hot water (in excess of 380°C, or 716°F) leached from the basalt (Figure 11.9). These materials may build up to form towers around the vents that support life forms uniquely suited to the chemical conditions of the vent fluids. The deposits are caused by mineral precipitation but are closely associated with igneous activity within newly forming crust.

Salt deposits that precipitate when water evaporates can build up to form another type of chemical sedimentary rock. Examples of these *evaporites* are found in Utah on the Bonneville Salt Flats, created when an ancient salt lake evaporated, and across the dry landscapes of the American Southwest. The pair of photographs in Figure 11.10 dramatically demonstrates this process; the first photo was taken in Death Valley National Park the day after a record 2.57-cm (1.01-in.) rainfall, and the second photo was taken a month later at

(a) The day after a record rainfall, several square kilometers are covered by water only a few centimeters deep.

(b) A month later, the water has evaporated and the valley floor is coated with evaporites (inset photo shows close-up view of crystallized salt deposits.

▲Figure 11.10 **Death Valley evaporites after a rain.** [Author. Inset by Andrea Paggiaro/Shutterstock.]

the exact same spot, after the water had evaporated and the valley was covered in evaporites.

Both clastic and chemical sedimentary rocks are deposited in layered strata that form an important record of past ages. Using the principle of superposition discussed earlier, scientists use the stratigraphy (the ordering of layers), thickness, and spatial distribution of strata to determine the relative age and origin of the rocks. The Part III opening photo shows the stratigraphy along the Colorado River near Moab, Utah. Note the different strata, which form cliffs or slopes depending on the resistance of the rock to exogenic forces such as weathering and erosion. The strata also indicate the region's climatic history, since each layer was formed under different environmental conditions.

Metamorphic Processes

Any igneous or sedimentary rock may be transformed into a **metamorphic rock** by going through profound physical or chemical changes under pressure and increased temperature. (The name *metamorphic* comes from a Greek word meaning "to change form.") Metamorphic rocks generally are more compact than the original rock and therefore are harder and more resistant to weathering and erosion (Figure 11.11).

The four processes that can cause metamorphism are heating, pressure, heating and pressure together, and compression and shear. When heat is applied to rock, the atoms within the minerals may break their chemical bonds, move, and form new bonds, leading to new mineral assemblages that develop into solid rock. When pressure is applied to rock, mineral structure may change as atoms become packed more closely. When rock is subject to both heat and pressure at depth, the original mineral assemblage becomes unstable and changes. Finally, rocks may be compressed by overlying weight and subject to shear when one part of the mass moves sideways relative to another part. These processes change the shape of the rock, leading to changes in the mineral alignments within.

▼Figure 11.11 Metamorphic rocks. A metamorphic rock outcrop in Greenland, the Amitsoq Gneiss, at 3.8 billion years old, one of the oldest rock formations on Earth. [Kevin Schafer/Photolibrary/Peter Arnold, Inc.]

MG™ Animation Foliation (Metamorphic Rock)

Metamorphic rock may form from igneous rocks as the lithospheric plates shift, especially when one plate is thrust beneath another (discussed with plate tectonics, just ahead). *Contact metamorphism* occurs when magma rising within the crust "cooks" adjacent rock; this type of metamorphism occurs adjacent to igneous intrusions and results from heat alone. *Regional metamorphism* occurs when a large areal extent of rock is subject to metamorphism. This can occur when sediments collect in broad depressions in Earth's crust and, because of their own weight, create enough pressure in the bottommost layers to transform the sediments into metamorphic rock. Regional metamorphism also occurs as lithospheric plates collide and mountain building occurs (discussed in Chapter 12).

Metamorphic rocks have textures that are foliated or nonfoliated, depending on the arrangement of minerals after metamorphism (Table 11.4). *Foliated* rock has a banded or layered appearance, demonstrating the alignment of minerals, which may appear as wavy striations (streaks or lines) in the new rock. *Nonfoliated* rocks do not exhibit this alignment.

On the Isle of Lewis in the island group known as the Outer Hebrides northwest of the Scottish coast, people constructed the Standing Stones of Calanais (Callanish) beginning approximately 5000 years ago (Table 11.4, far-right photo). These ancient people used the 3.1-billion-year-old metamorphic Lewisian gneiss for their monument, arranging the stones so that the foliations aligned vertically. The standing stone in the photo is about 3.5 m (11.5 ft) tall.

The Rock Cycle

Although rocks appear stable and unchanging, they are not. The **rock cycle** is the name for the continuous alteration of Earth materials from one rock type to another (Figure 11.12). For example, igneous rock formed from magma may break down into sediment by weathering and erosion and then lithify into sedimentary rock. This rock may subsequently become buried and exposed to pressure and heat deep within Earth, forming metamorphic rock. This may, in turn, break down and become sedimentary rock. Igneous rock may also take a shortcut through that cycle by directly becoming metamorphic rock. As the arrows indicate, there are many pathways through the rock cycle.

Two cyclic systems drive the rock cycle. At and above Earth's surface, the hydrologic cycle, fueled by solar energy, drives the exogenic processes. Below Earth's surface and within the crust, the tectonic cycle,

TABLE 11.4 Metamorphic Rocks

Parent Rock	Metamorphic Equivalent	Texture
Shale (clay minerals)	Slate	Foliated
Granite, slate, shale	Gneiss	Foliated
Basalt, shale, peridotite	Schist	Foliated
Limestone, dolomite	Marble	Nonfoliated
Sandstone	Quartzite	Nonfoliated

Gneiss

Marble

Slate

[Slate and Lewisian gneiss: Bobbé Christopherson; marble and gneiss: *Laboratory Manual in Physical Geology*, 3rd ed., R. M. Busch, ed. © 1993 by Macmillan Publishing Co.]

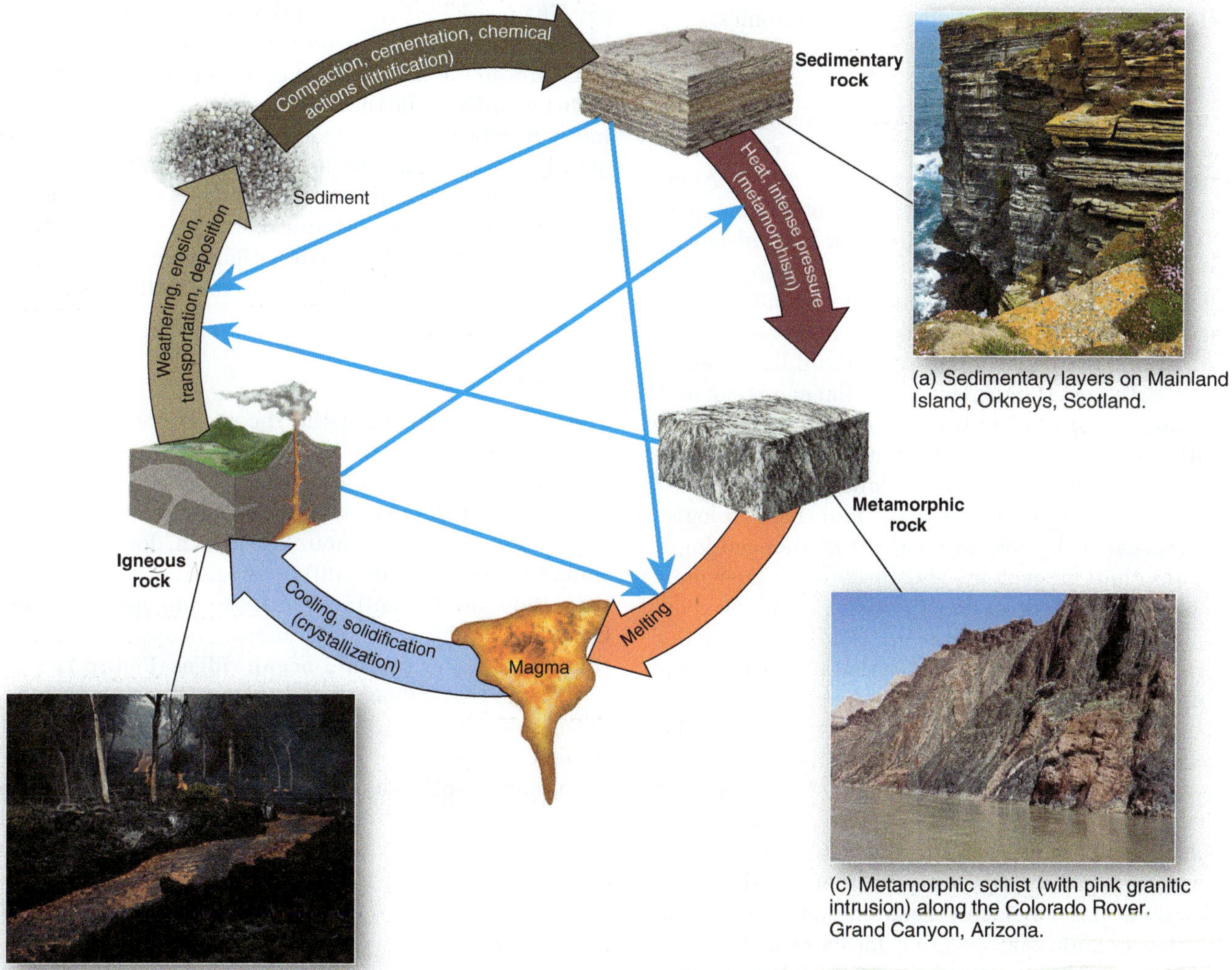

(a) Sedimentary layers on Mainland Island, Orkneys, Scotland.

(b) Active lava flow and surrounding basalt, Big Island, Hawai`i.

(c) Metamorphic schist (with pink granitic intrusion) along the Colorado Rover. Grand Canyon, Arizona.

▲**Figure 11.12 The rock cycle.** In this schematic of the relations among igneous, sedimentary, and metamorphic processes, the thin blue arrows indicate "shortcuts"—such as when igneous rock is melted and becomes metamorphic rock without first going through a sedimentary stage. [(a) Bobbé Christopherson. (b) USGS. (c) NPS.

MG™
Animation
The Rock Cycle

powered by internal heat, drives the endogenic processes. We now discuss plate tectonics theory and the tectonic cycle.

Plate Tectonics

Have you ever looked at a world map and noticed that a few of the continental landmasses appear to have matching shapes like pieces of a jigsaw puzzle—particularly South America and Africa? The reality is that the continental pieces once did fit together. Continental landmasses not only migrated to their present locations, but are continuing to move today at speeds up to 6 cm (2.4 in.) per year. We say that the continents are *adrift* because convection currents in the asthenosphere and upper mantle provide upwelling and downwelling forces that push and pull portions of the lithosphere. Thus, the arrangement of continents and oceans we see today is not permanent, but is in a continuing state of change. This theory was at first controversial, but it is now accepted and is the underlying foundation for much of Earth systems science. Let us trace the discoveries that led to the theory we now know as plate tectonics.

Continental Drift

As early mapping gained accuracy, some observers noticed the agreement in the profiles of the continents, particularly those of South America and Africa. Abraham Ortelius (1527–1598), a geographer, noted the apparent fit of some continental coastlines in his *Thesaurus Geographicus* (1596). In 1620, English philosopher Sir Francis Bacon noted gross similarities between the edges of Africa and South America (although he did not suggest that they had drifted apart). Benjamin Franklin wrote in 1780 that Earth's crust must be a shell that can break and shift by movements of fluid below. Others wrote—unscientifically—about such apparent relationships, but it was not until much later that a valid explanation emerged.

In 1912, German geophysicist and meteorologist Alfred Wegener presented an idea that challenged long-held assumptions in geology, and three years later published his book *Origin of the Continents and Oceans*. After studying the geologic record represented in the rock strata, Wegener found evidence that the rock assemblages on the east coast of South America were the same as those on the west coast of Africa, suggesting that the continents were at one time connected. The fossil record provided further evidence, as did the climatic record found in sedimentary rocks. Wegener hypothesized that the coal deposits found today in the midlatitudes exist because these regions were at one time nearer the equator and were covered by lush vegetation that later became the lithified organic material that forms coal. Wegener concluded that all landmasses migrate, and that approximately 225 m.y.a. they formed one supercontinent that he named **Pangaea**, meaning "all Earth."

Today, scientists regard Wegener as the father of plate tectonics, which he first called **continental drift**. However, scientists at the time were unreceptive to Wegener's revolutionary proposal. A great debate began with Wegener's book and lasted almost 50 years. As modern scientific capabilities led to discoveries that built the case for continental drift, the 1950s and 1960s saw a revival of interest in Wegener's concepts and, finally, confirmation. Although his initial model kept the landmasses together too long and his proposal included an incorrect driving mechanism for the moving continents, Wegener's arrangement of Pangaea and its breakup was correct.

Figure 11.13 shows the changing arrangement of the continents, beginning with the pre-Pangaea configuration of 465 m.y.a. (during the Middle Ordovician Period; Figure 11.13a) and moving to an updated version of Wegener's Pangaea, 225–200 m.y.a. (Triassic–Jurassic Periods; Figure 11.13b); followed by the configuration that occurred by 135 m.y.a., with the continents of Gondwana and Laurasia (the beginning of the Cretaceous Period; Figure 11.13c); the arrangement 65 m.y.a. (shortly after the beginning of the Tertiary Period; Figure 11.13d); and finally the present arrangement in modern geologic time (the late Cenozoic Era; Figure 11.13e).

The word *Tectonic*, from the Greek *tektonikùs*, meaning "building" or "construction," refers to changes in the configuration of Earth's crust as a result of internal forces. **Plate tectonics** is the theory that the lithosphere is divided into a number of plates that float independently over the mantle and along whose boundaries occur the formation of new crust, the building of mountains, and the seismic activity that causes earthquakes. Plate tectonics theory describes the motion of Earth's lithosphere; we discuss the various principles of this theory ahead in this section and in Chapter 12.

Seafloor Spreading

The key to establishing the theory of continental drift was a better understanding of the seafloor crust. As scientists acquired information about the bathymetry (depth variations) of the ocean floor, they discovered an interconnected worldwide mountain chain, forming a ridge some 64,000 km (40,000 mi) in extent and averaging more than 1000 km (600 mi) in width (see the opening map in Chapter 12). The underwater mountain systems that form this chain are termed **mid-ocean ridges** (Figure 11.14).

In the early 1960s, geophysicist Harry H. Hess proposed that these mid-ocean ridges are new ocean floor formed by upwelling flows of magma from hot areas in the upper mantle and asthenosphere and perhaps from the deeper lower mantle. As upwelling occurs, the new seafloor then moves outward from the ridge as plates pull apart and new crust is formed. This process, now called **seafloor spreading**, is the mechanism that builds mid-ocean ridges and drives continental movement.

Hess and other geologists then faced a new problem: If seafloor spreading and the creation of new crust are ongoing, then old ocean crust must somewhere be consumed; otherwise, Earth would be expanding. Harry Hess and another geologist, Robert S. Dietz, proposed that old seafloor sinks back into the Earth's mantle at deep-ocean

▼Figure 11.13 Continents adrift, from 465 m.y.a. to the present. Observe the formation and breakup of Pangaea and the types of motions occurring at plate boundaries. [(a) From R. K. Bambach, "Before Pangaea: The geography of the Paleozoic world," *American Scientist 68* (1980): 26–38, reprinted by permission. (b–e) From R. S. Dietz and J. C. Holden, *Journal of Geophysical Research 75*, no. 26 (September 10, 1970): 4939–4956, © The American Geophysical Union.

MG™ Animation
Plate Motions Through Time

(a)

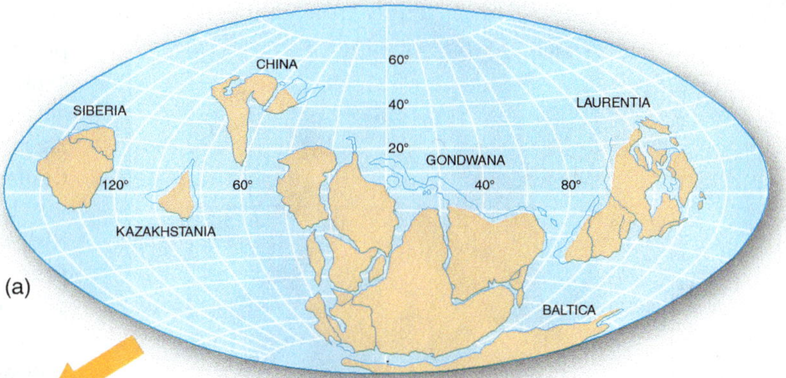

465 million years ago

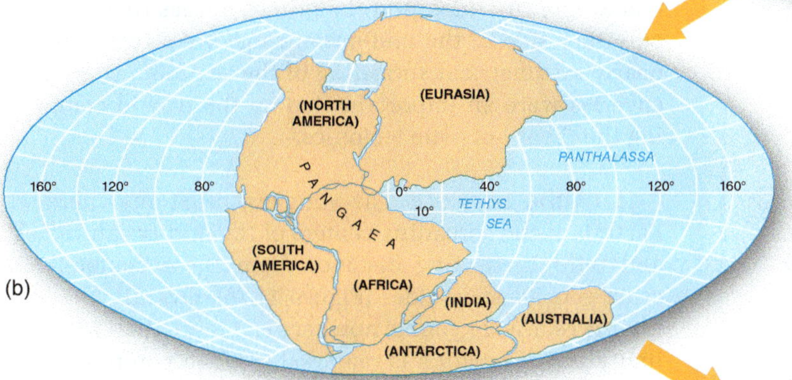

(b)

225 million years ago

Pangaea all Earth, Panthalassa all seas—no Atlantic Ocean. The Appalachian Mountains in the eastern U.S. and the Lesser Atlas Mountains of northwestern Africa are portions of this older range; Africa connected to North and South Americas.

New seafloor is highlighted (purple tint). An active spreading center rifted North America away from landmasses to the east. India moved toward its collision with Asia, a spreading center to the south and a subduction zone to the north.

(c)

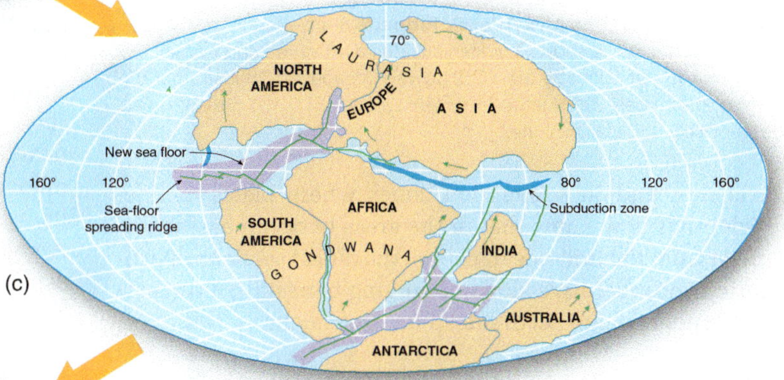

135 million years ago

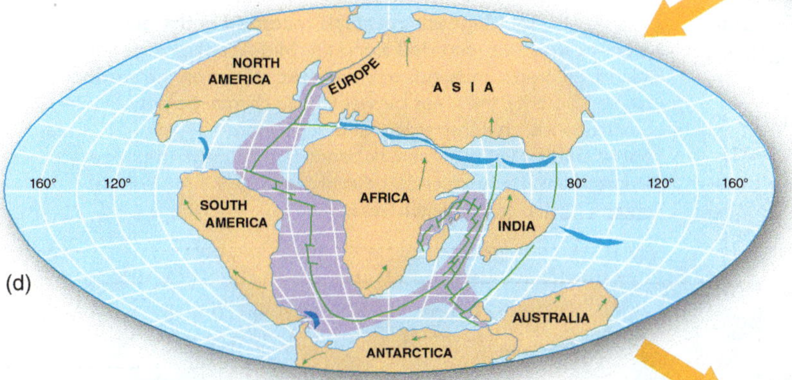

(d)

65 million years ago

Along the Mid-Atlantic Ridge, seafloor spread some 3000 km (almost 1900 mi) in 70 million years. The rifting along where the Red Sea would form began. Of all the major plates, India traveled the farthest—almost 10,000 km.

From 65 m.y.a. until the present, more than half of the ocean floor was renewed. The northern reaches of the India plate underthrust the southern mass of Asia through subduction, forming the Himalayas. Plate motions continue to this day.

(e)

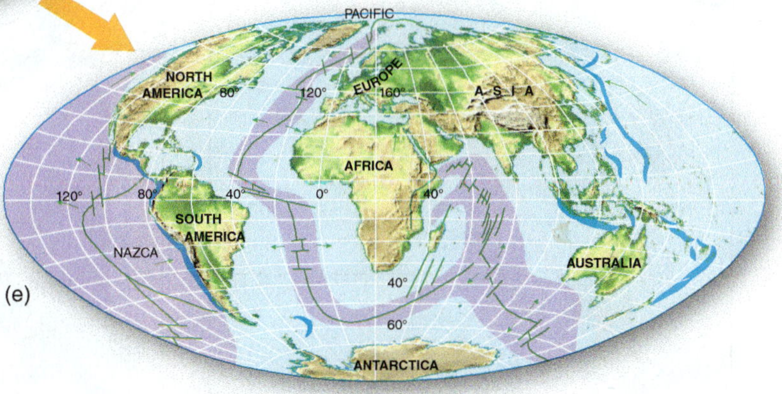

Today

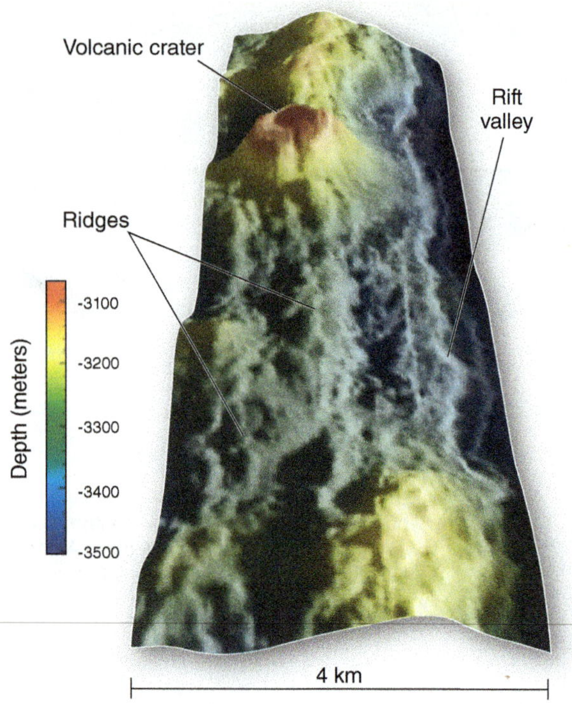

▲Figure 11.14 **The Mid-Atlantic Ridge.** A 4-km-wide image of the Mid-Atlantic Ridge shows a volcanic crater, a rift valley, and ridges. The image was taken by the *TOBI* (towed ocean-bottom instrument) at approximately 29° N latitude. [Image courtesy of D. K. Smith, Woods Hole Oceanographic Institute.]

trenches and in subduction zones where plates collide. Scientists now know that in the areas of ocean basins farthest from the mid-ocean ridges, the oldest sections of oceanic lithosphere are slowly plunging beneath continental lithosphere along Earth's deep-ocean trenches.

Magnetic Reversals Earlier we discussed the history of reversals in Earth's magnetic field. As seafloor spreading occurs and magma emerges at the surface, magnetic particles in the lava orient with the magnetic field in force at the time it cools and hardens. The particles become locked in this alignment as part of the new seafloor, creating an ongoing magnetic record of Earth's polarity. Using isotopic dating methods, scientists have established a chronology for these reversals—that is, the actual years in which the polarity reversal occurred. Ages for materials on the ocean floor proved to be a fundamental piece of the plate tectonics puzzle.

Figure 11.15 shows a record from the Mid-Atlantic Ridge south of Iceland, illustrating the magnetic stripes preserved in the minerals—the colored bands are areas of reversed polarity, the areas between them have normal polarity. The relative ages of the rocks increase with distance from the ridge, and the mirror images that develop on either side of the mid-ocean ridge are a result of the nearly symmetrical spreading of the seafloor. These discoveries were an important step on the way to completing the theory of plate tectonics.

Age of the Seafloor The youngest crust anywhere on Earth is at the spreading centers of the mid-ocean ridges, and with increasing distance from these centers, the crust gets steadily older (Figure 11.16). Overall, the seafloor is relatively young; nowhere is it more than 280 million years old, which is remarkable when you remember that Earth is 4.6 billion years old. In the Atlantic Ocean, the oldest large-scale area of seafloor is along the continental margins, farthest from the Mid-Atlantic Ridge. In the Pacific, the oldest seafloor is in the western region near Japan (dating to the Jurassic Period). Note on the map the distance between this part of the basin and its spreading center in the South Pacific, west of South America. Parts of the Mediterranean Sea contain the oldest seafloor remnants, which may have been part of the Tethys Sea, dating to about 280 m.y.a (see Figure 11.13).

▼Figure 11.15 **Magnetic reversals recorded in the ocean floor.** Magnetic reversals recorded along the Mid-Atlantic Ridge south of Iceland; colored bands indicate magnetic stripes on the seafloor with reversed polarity, the areas between have normal polarity. Similar colored bands on either side of the ridge indicate symmetrical seafloor spreading with oldest rock bands farthest from the ridge. [Adapted from J. R. Heirtzler, S. Le Pichon, and J. G. Baron, *Deep-Sea Research 13*, © 1966, Pergamon Press, p. 247.]

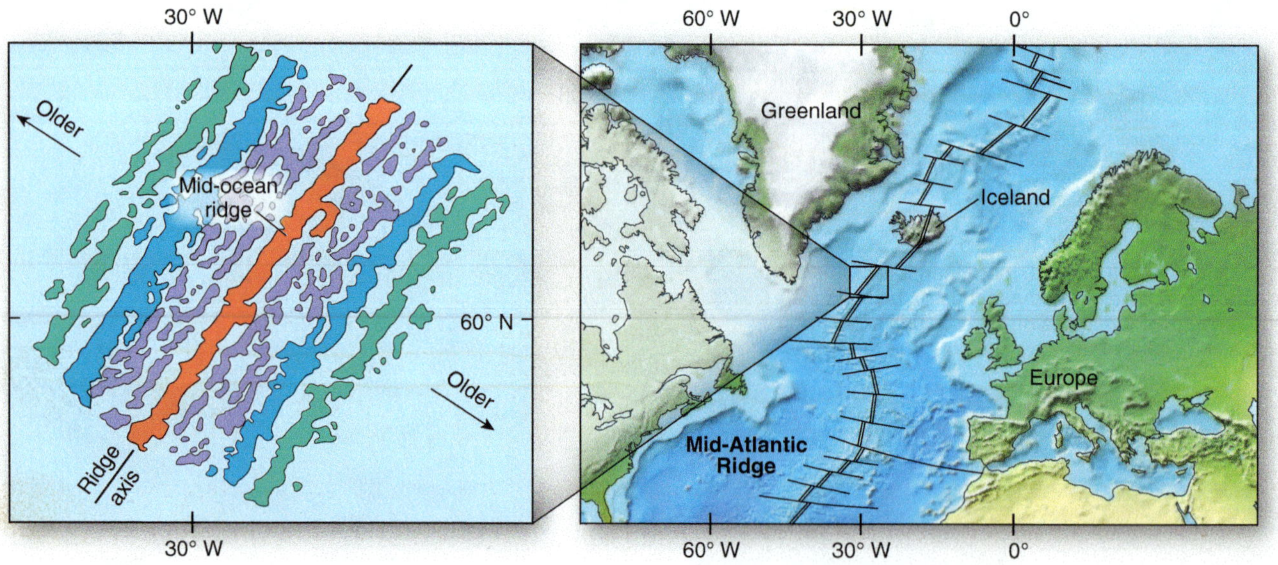

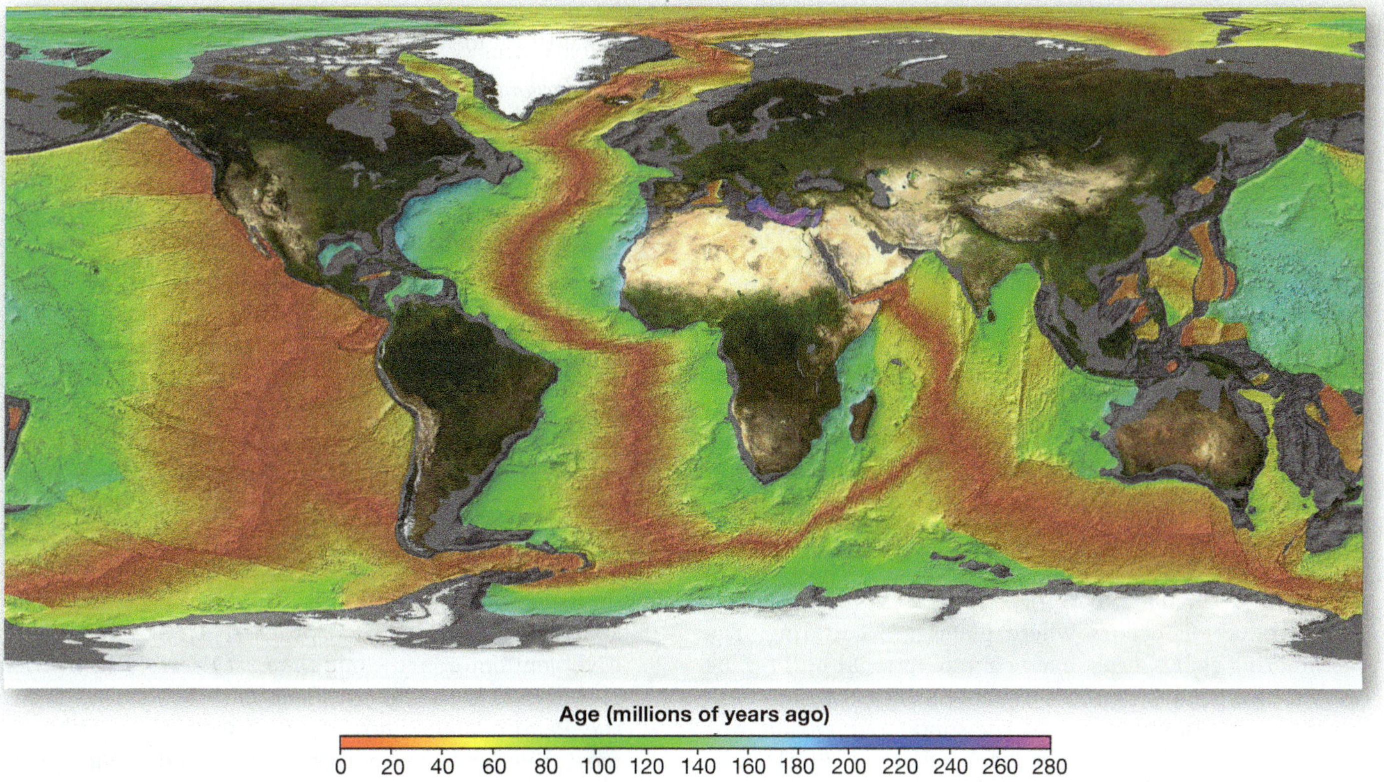

▲**Figure 11.16 Relative ages of oceanic crust.** Compare the width of the red color (young crust) near the East Pacific Rise in the eastern Pacific Ocean with the width of the red color along the Mid-Atlantic Ridge. What does the difference tell you about the rates of plate motion in the two locations? [Image by Elliot Lim, CIRES.]

Today, scientists know that mid-ocean ridges occur where plates are moving apart (Figure 11.17). As the seafloor spreads, magma rises and accumulates in magma chambers beneath the centerline of the ridge. Some of the magma rises and erupts through fractures and small volcanoes along the ridge, forming new oceanic crust. Scientists now think that this upward movement of material beneath an ocean ridge is a consequence of seafloor spreading rather than the cause. As the plates continue to move apart, more magma rises from below to fill the gaps.

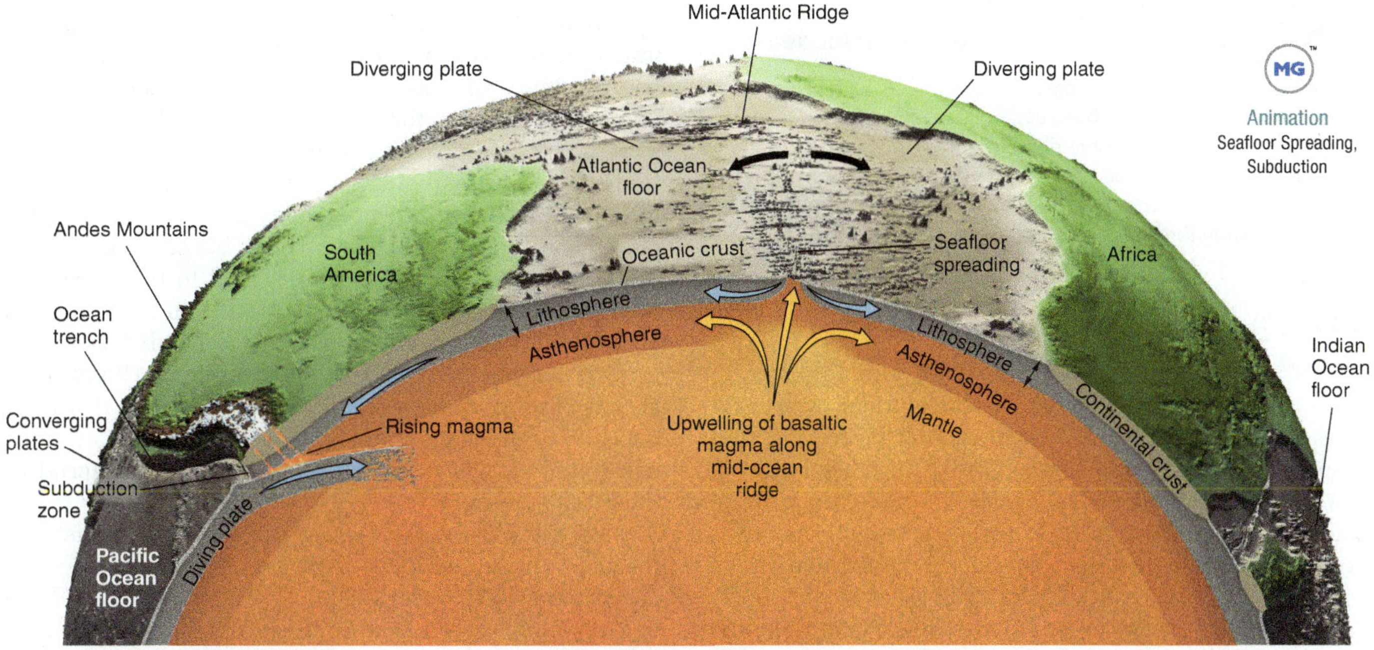

MG Animation Seafloor Spreading, Subduction

▲**Figure 11.17 Plate movements.** Seafloor spreading, upwelling currents, subduction, and plate movements, shown in cross section. Arrows indicate the direction of the spreading.

Subduction

When one portion of the lithosphere descends beneath another and dives downward into the mantle, the process is called *subduction* and the area is a **subduction zone**. As discussed earlier, the basaltic ocean crust has an average density of 3.0 g/cm^3, whereas continental crust averages a lighter 2.7 g/cm^3. As a result, when continental crust and oceanic crust slowly collide, the denser ocean floor will grind beneath the lighter continental crust, thus forming a subduction zone.

The world's deep-ocean trenches coincide with these subduction zones and are the lowest features on Earth's surface. The Mariana Trench near Guam is the deepest, descending below sea level to –11,030 m (–36,198 ft). The Tonga Trench, also in the Pacific, is the next deepest, dropping to –10,882 m (–35,702 ft). For comparison, in the Atlantic Ocean, the Puerto Rico Trench drops to –8605 m (–28,224 ft), and in the Indian Ocean, the Java Trench drops to –7125 m (–23,376 ft).

Subduction occurs where plates are colliding. The subducting slab of crust exerts a gravitational pull on the rest of the plate—a pull now known to be an important driving force in plate motion. The subducted portion travels down into the asthenosphere, where it remelts and eventually is recycled as magma, rising again toward the surface through deep fissures and cracks in crustal rock (left side of Figure 11.17). Volcanic mountains such as the Andes in South America and the Cascade Range from northern California to the Canadian border form inland of these subduction zones as a result of rising plumes of magma. Sometimes the diving plate remains intact for hundreds of kilometers, whereas at other times it can break into large pieces, thought to be the case under the Cascade Range with its fragmented distribution of volcanoes.

CRITICALthinking 11.2
Tracking Your Location Since Pangaea

Using the maps in this chapter, determine your present location relative to Earth's crustal plates. Now, using Figure 11.13b, identify approximately where your present location was 225 m.y.a.; express it in a rough estimate using the equator and the longitudes noted on the map. Can you track your location throughout parts c and d of Figure 11.13? •

Plate Boundaries

Earth's present crust is divided into at least 14 plates, of which about half are major and half are minor in terms of area (Figure 11.18). Hundreds of smaller pieces and perhaps dozens of microplates migrating together make up these broad, moving plates. Arrows in the figure indicate the direction in which each plate is presently moving, and the length of the arrows suggests the relative rate of movement during the past 20 million years.

The boundaries where plates meet are dynamic places when considered over geologic time scales, although slow-moving within human time frames. The block diagrams in Figure 11.18 show the three general types of boundaries and the interacting movements of plates at those locations.

- *Convergent boundaries* occur in areas of crustal collision and subduction. As discussed earlier, where areas of continental and oceanic lithosphere meet, crust is compressed and lost in a destructional process as it moves downward into the mantle. Convergent boundaries form subduction zones, such as off the west coast of South and Central America, along the Aleutian Island trenches (see Figure 11.21 ahead), and along the east coast of Japan, where a magnitude 9.0 earthquake struck in 2011. Convergent boundaries also occur where two plates of continental crust collide, such as the collision zone between India and Asia, and where oceanic plates collide, such as along the deep trenches in the western Pacific Ocean.
- *Divergent boundaries* occur in areas of seafloor spreading, where upwelling material from the mantle forms new seafloor and lithospheric plates spread apart in a constructional process. An example is the divergent boundary along the East Pacific Rise, which gives birth to the Nazca plate (moving eastward) and the Pacific plate (moving northwestward). Whereas most divergent boundaries occur at mid-ocean ridges, a few occur within continents themselves. An example is the Great Rift Valley of East Africa, where continental crust is rifting apart.
- *Transform boundaries* occur where plates slide past one another, usually at right angles to a seafloor spreading center. These are the fractures stretching across the mid-ocean ridge system worldwide, first described in 1965 by University of Toronto geophysicist Tuzo Wilson. As plates move past each other horizontally, they form a type of *fault*, or fracture, in Earth's crust, called a **transform fault**.

Along these fracture zones that intersect ridges, a transform fault occurs only along the fault section that lies *between* two segments of the fragmented mid-ocean ridge (Figure 11.18). Along the fracture zone outside of the transform fault, the crust moves in the same direction (away from the mid-ocean ridge) as the spreading plates. The movement along transform faults is that of horizontal displacement—no new crust is formed or old crust subducted.

The name *transform* was assigned to these features because of the apparent transformation in the direction of fault movement—these faults can be distinguished from other horizontal faults (discussed in Chapter 12) because the movement along one side of the fault line is opposite to movement along the other side. This unique movement results from the creation of new material as the seafloor spreads.

All the seafloor spreading centers on Earth feature these fractures, which are perpendicular to the mid-ocean ridges (Figure 11.19). Some are a few hundred kilometers long; others, such as those along the East Pacific Rise, stretch out 1000 km or more (over 600 mi).

Transform boundaries are associated with earthquake activity, especially where they cut across portions

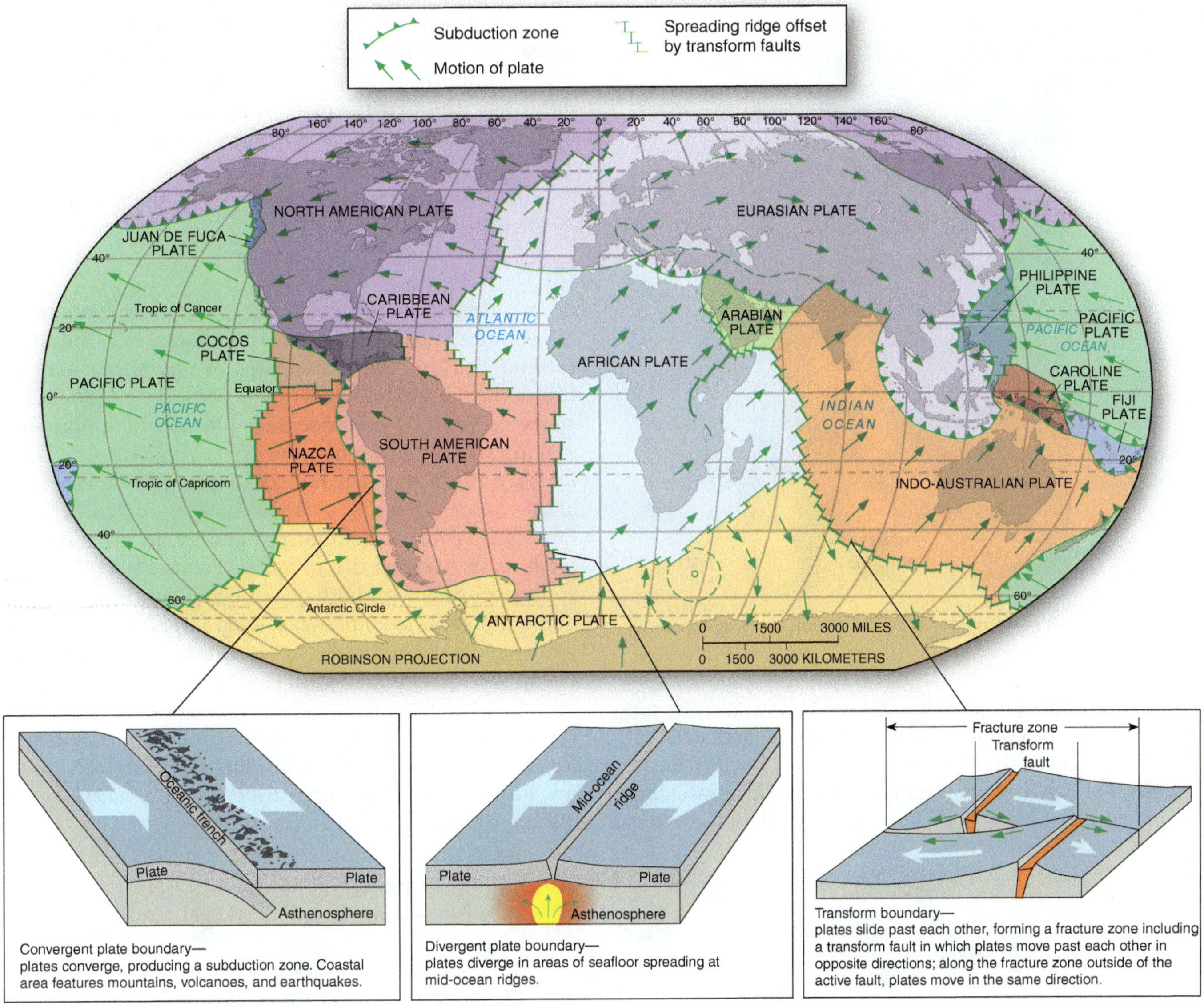

▲**Figure 11.18 Earth's major lithospheric plates and their movements.** Each arrow represents 20 million years of movement. The longer arrows indicate that the Pacific and Nazca plates are moving more rapidly than are the Atlantic plates. Compare the length of these arrows with the purple areas on Figure 11.13. [Adapted from U.S. Geodynamics Committee, National Academy of Sciences and National Academy of Engineering.]

MG™ Animation Motions at Plate Boundaries

MG™ Notebook Plate Boundaries

MG™ Animation Correlating Processes and Plate Boundaries

MG™ Animation Forming a Divergent Boundary

MG™ Animation India Collision with Asia

MG™ Notebook Plate Boundaries

of continental crust, such as along the San Andreas fault in California where the Pacific and North American plates meet, and along the Alpine fault in New Zealand, the boundary between the Indo-Australian and Pacific plates. The San Andreas, running through several metropolitan areas of California, is perhaps the most famous transform fault in the world and is discussed in Chapter 12.

Earthquake and Volcanic Activity

Plate boundaries are the primary locations of earthquake and volcanic activity, and the correlation of these phenomena is an important aspect of plate tectonics. The massive earthquakes that hit Haiti, Chile, New Zealand, and Japan in 2010 and 2011, as well as the 2010 volcanic eruption in Iceland, focused world attention on these plate boundaries

GEOreport 11.4 Spreading along the East Pacific Rise

The fastest rate of seafloor spreading on Earth occurs along the East Pacific Rise, which runs roughly north–south along the eastern edge of the Pacific plate from near Antarctica to North America. Spreading is occurring at a rate of 6 to 16 cm/year (2.4 to 6.3 in./year), depending on location. For perspective, human fingernails grow at a rate of about 4 cm/year (1.6 in/yr).

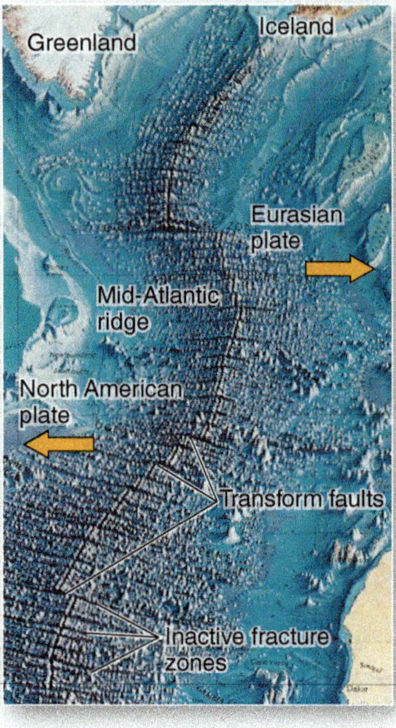

▲**Figure 11.19 Transform faults.** [Office of Naval Research.]

MG Animation Transform Faults, Plate Margins

and the principles of plate tectonics. The next chapter discusses earthquakes and volcanic activity in more detail.

The "ring of fire" surrounding the Pacific Basin, named for the frequent incidence of volcanoes along its margin, is evident on the map of earthquake zones, volcanic sites, hot spots, and plate motion in Figure 11.20. The features that form this "ring" are caused by the subducting edge of the Pacific plate as it thrusts deep into the crust and mantle and produces molten material that makes its way back toward the surface. The upwelling magma forms active volcanoes along the Pacific Rim. Such processes occur at similar subduction zones throughout the world.

Hot Spots

As mentioned, volcanic activity is often associated with plate boundaries. However, scientists have found an estimated 50 to 100 active sites of upwelling material that exist independent of plate boundaries. These **hot spots** (or hot-spot volcanoes) are places where plumes of magma rise from the mantle, producing volcanic activity as well as thermal effects in the groundwater and crust. Some of these sites produce enough heat from Earth's interior, or **geothermal energy**, to be developed for human uses, as discussed in Focus Study 11.1. Hot spots occur beneath both oceanic and continental crust. Some hot spots are anchored deep in the lower mantle, tending to remain fixed relative to migrating plates; others appear to be above plumes that move by themselves or shift with plate motion. In the case of a fixed hot spot, the area of a plate that passes above it is locally heated for the brief geologic time it remains above that spot (a few hundred thousand or million years), sometimes producing a chain of volcanic features.

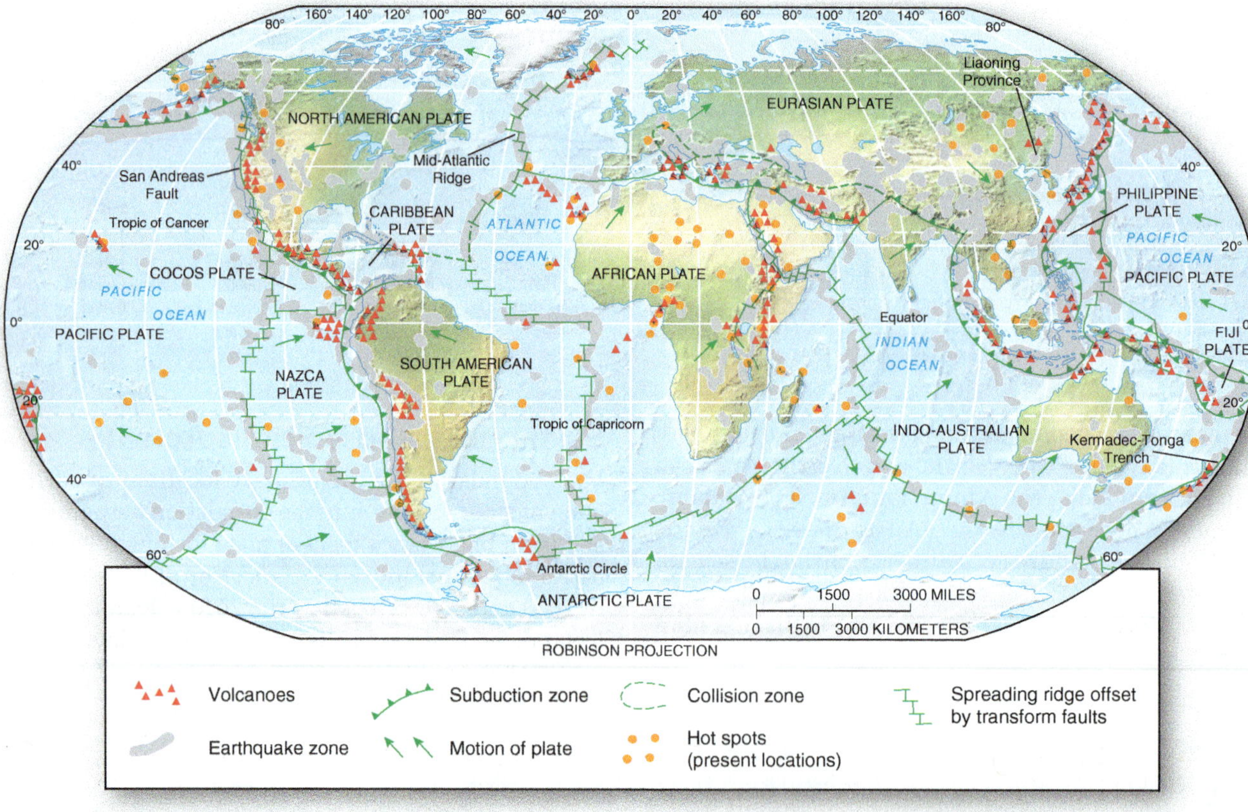

▲**Figure 11.20 Locations of earthquake and volcanic activity.** Earthquake and volcanic activity in relation to major tectonic plate boundaries and principal hot spots. [Earthquake, volcano, and hot-spot data adapted from U.S. Geological Survey.]

Hydrothermal Features Groundwater that is heated by pockets of magma and other hot portions of the crust may emerge as a hot spring or erupt explosively from the ground as a *geyser*—a spring characterized by intermittent discharge of water and steam. A hot spot below Yellowstone National Park in Montana and Wyoming produces several types of hydrothermal features, including geysers, mud pots, and fumaroles (see *The Human Denominator*, Figure HD 11a, on page 351). *Mud pots* are bubbling pools that are highly acidic with limited water supply, and thus produce mainly gases that are broken down by microorganisms to release sulfuric acid, which in turn breaks down rock into clays that form mud. As gases escape through the mud, they produce the bubbling action that makes these features distinct. *Fumaroles* are steam vents that emit gases that may contain sulfur dioxide, hydrogen chloride, or hydrogen sulfide—some areas of Yellowstone are known for the characteristic odor of "rotten eggs" caused by hydrogen sulfide gas.

Hot-Spot Island Chains A hot spot in the Pacific Ocean produced, and continues to form, the Hawaiian–Emperor Islands chain, including numerous seamounts, which are submarine mountains that do not reach the surface (Figure 11.21). The Pacific plate moved across this hot, upward-erupting plume over the last 80 million years, creating a string of volcanic islands and seamounts with ages increasing northwestward away from the hot spot. The oldest island

(text continued on page 350)

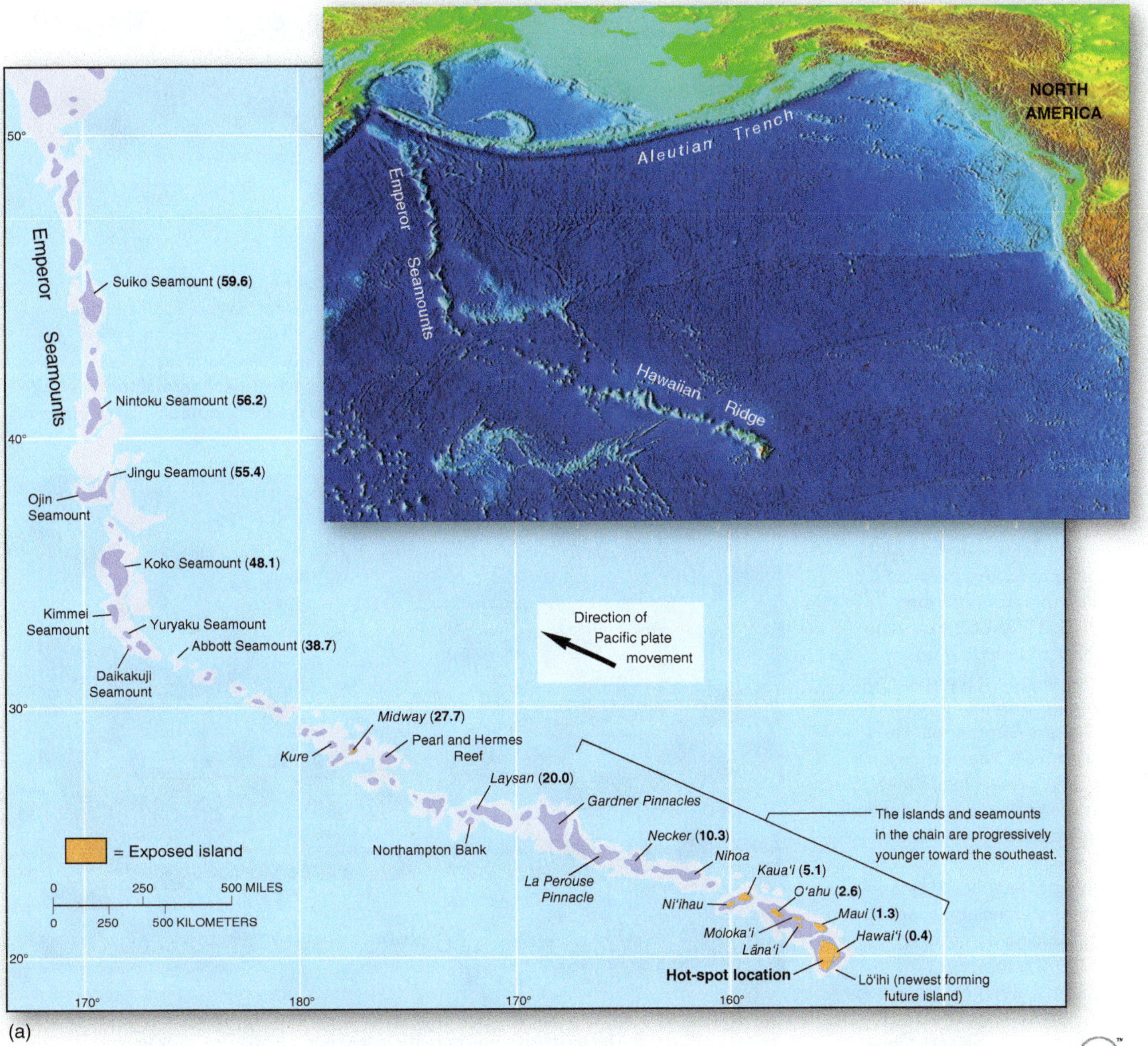

▲Figure 11.21 Hot-spot tracks across the North Pacific. Hawai'i and the linear volcanic chain of islands known as the Emperor Seamounts. (a) The islands and seamounts in the chain are progressively younger toward the southeast. Ages, in m.y.a., are shown in parentheses. Note that Midway Island is 27.7 million years old, meaning that the site was over the plume 27.7 m.y.a. [(a) After D. A. Clague, "Petrology and K–Ar (Potassium–Argon) ages of dredged volcanic rocks from the western Hawaiian ridge and the southern Emperor seamount chain," *Geological Society of America Bulletin 86* (1975): 991; inset map courtesy of NOAA.]

MG Animation
Hot-Spot Volcano Tracks

Focus Study 11.1 Sustainable Resources

Heat from Earth—Geothermal Energy and Power

Geothermal energy, the tremendous amount of endogenic heat within Earth's interior, can in some places be harnessed for heating and power production by means of wells and pipes that transmit heated water or steam to the surface. Underground reservoirs of hot water at varying depths and temperatures are among the geothermal resources that can be tapped and brought to the surface. Resources of this kind provide *direct geothermal heating,* which uses water at low-to-moderate temperatures (20°C to 150°C, or 68°F to 302°F) in heat-exchange systems in buildings, commercial greenhouses, and fish farms, among many examples. Where available, direct geothermal heating provides energy that is inexpensive and clean.

In Boise, Idaho, which has used geothermal resources for decades, the state capitol building uses direct geothermal heating, and most city locations return used geothermal water to the aquifer through injection wells. In Reykjavík, Iceland, the majority of space heating systems (more than 87%) are geothermal.

▲**Figure 11.1.1 Geysers Geothermal Field, California.** [James P. Blair/National Geographic.]

Geothermal Power Production

Geothermal electricity is produced using steam from a natural underground reservoir to drive a turbine that powers a generator. Because the steam comes directly from Earth's interior (without burning fuel to produce it), geothermal electricity is a relatively clean energy. Ideally, groundwater for this purpose should have a temperature of from 180°C to 350°C (355°F to 600°F) and be moving through rock of high porosity and permeability (allowing the water to move freely through connecting pore spaces). The Geysers Geothermal Field in northern California (so named despite the absence of any geysers in the area) is the largest geothermal power production plant using this method in the world (Figure 11.1.1).

Today, geothermal applications are in use in 70 countries and include over 200 power plants producing a total output of about 11,000 MW. The top countries for installed geothermal electrical generation are the United States, Philippines, Indonesia, Mexico, Italy, New Zealand, Iceland, and Japan. In the Philippines, almost 27% of total electrical production is generated with geothermal energy; in Iceland, the percentage is 30% (Figure 11.1.2).

In the United States, the geothermal industry grew by 5% during 2012, with power now totaling 3300 MW operating in Alaska, California, Hawai'i, Idaho, Nevada, Oregon, Utah, and Wyoming. For more information, see http://geology.com/geothermal/ and http://www1.eere.energy.gov/geothermal/.

The newest geothermal technology seeks to create conditions for geothermal

▲**Figure 11.1.2 Surface geothermal activity in Iceland.** On the Reykjavík Peninsula, Iceland, the Svartsengi geothermal power plant produces electricity and provides hot water for homes and businesses. Nearby is the Blue Lagoon Spa filled with warm geothermal waters. [Eco Images/Universal Images Group/Getty.]

power production at locations where underground rock temperatures are high but where water or permeability is lacking. In an *enhanced geothermal system* (EGS), cold water is pumped underground into hot rock, causing the rock to fracture and become permeable to water flow; the cold water, in turn, is heated to steam as it flows through the high-temperature rock.

Several EGS projects are either operational or under development worldwide; the largest of these is in Australia's Cooper Basin. The potential for EGS, as with other geothermal technologies, is highest in the western United States and in areas of the world along plate boundaries that produce upwelling pockets of magma and volcanic activity (Figure 11.1.3).

Geothermal as a Sustainable Resource

Geothermal power production has many advantages, including minimal production of carbon dioxide (CO_2 is released during the extraction of steam, but in far smaller amounts than the CO_2 emitted to the atmosphere from the use of fossil fuels). Geothermal power can be produced 24 hours a day, an advantage when compared with solar or wind energy, in which the timing of production is linked to daylight or other natural variations in the resource.

Although geothermal is billed as renewable and self-sustaining, research shows that some geysers and geothermal fields are being depleted as the extraction rate exceeds the rate of recharge. In addition, the drilling of wells or the injection or removal of water in connection with geothermal energy projects may pose some risk of *induced seismicity* (that is, minor earthquakes and tremors produced in association with human activity, discussed further in Chapter 12). In Switzerland, geothermal power developers are working on technology to reduce seismic hazards after several geothermal projects there have produced small earthquakes. In the United States, scientists have correlated earthquake activity with geothermal development in several locations. Research is ongoing to address and mitigate the seismicity issues. Despite these problems, geothermal remains a promising source of clean energy for the future.

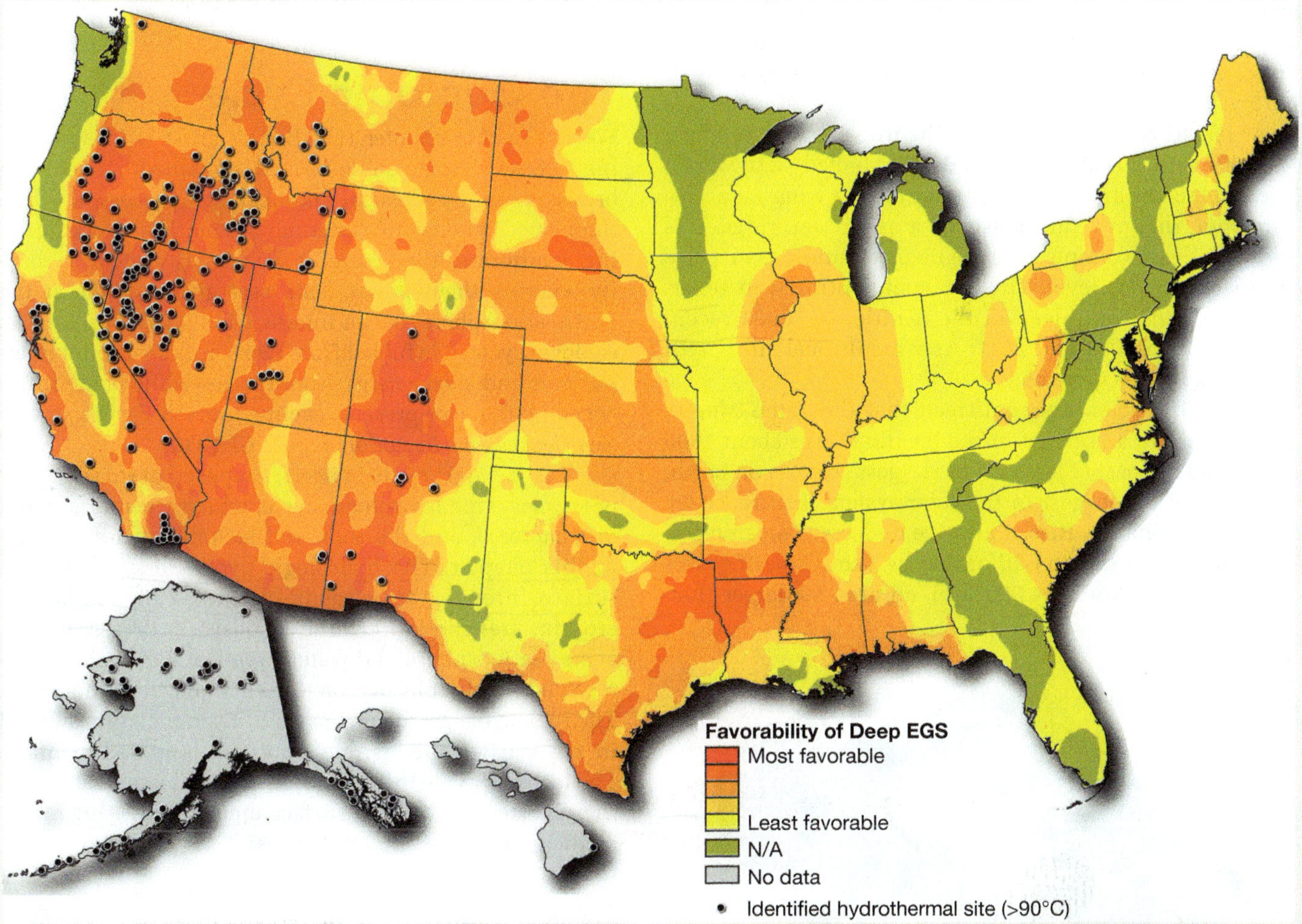

▲Figure 11.1.3 **Geothermal resource potential in the United States.** Identified hydrothermal resources (black circles) and potential enhanced geothermal resources (most favorable to least favorable) in the United States. [NREL.]

in the Hawaiian part of the chain is Kaua'i, approximately 5 million years old; today, it is weathered and eroded into the deep canyons and valleys pictured on the cover of this book.

To the northwest of Hawai'i, the island of Midway rises as a part of the same system. From there, the Emperor seamounts spread northwestward and are progressively older until they reach about 40 million years in age. At that point, this linear island chain shifts direction northward. This bend in the chain is now thought to be from both movement in the plume and a possible change in the plate motion itself, a revision of past thinking that all hot-spot plumes remain fixed relative to the migrating plate. At the northernmost extreme, the seamounts that formed about 80 m.y.a. are now approaching the Aleutian Trench, where they eventually will be subducted beneath the Eurasian plate.

The big island of Hawai'i, the youngest in the island chain, actually took less than 1 million years to build to its present stature. The island is a huge mound of lava, formed from magma from several seafloor fissures and volcanoes and rising from the seafloor 5800 m (19,000 ft) to the ocean surface. From sea level, its highest peak, Mauna Kea, reaches an elevation of 4205 m (13,796 ft). This total height of almost 10,000 m (32,800 ft) represents the highest mountain on Earth if measured from the ocean floor.

A new addition to the Hawaiian chain is still a seamount. It rises 3350 m (11,000 ft) from its base, but is still 975 m (3200 ft) beneath the ocean surface. Even though this new island will not experience the tropical Sun for about 10,000 years, it is already named Lo'ihi (noted on the map).

Iceland is another island that results from an active hot spot, this one sitting astride a mid-ocean ridge. It is an excellent example of a segment of mid-ocean ridge rising above sea level. This hot spot continues to generate eruptions from deep in the mantle, the most recent occurring in 2010 and 2011. As a result, Iceland is still growing in area and volume.

In 2013, scientists confirmed that the Tamu Massif, located above a hot spot in the Pacific Ocean about 1600 km (1000 mi) east of Japan, is the largest volcano on Earth, and one of the biggest in the solar system (Figure 11.22). Part of the Shatsky Rise, a massive undersea volcanic plateau similar in size to California, the Tamu Massif covers about 310,800 km^2 (120,000 mi^2), an area far larger than Mauna Loa on Hawai'i, covering only 5200 km^2 (2000 mi^2). Until recently, scientists thought Tamu Massif was a composite of smaller volcanic structures, formed in a similar manner as the island of Hawai'i. However, recent research determined that this feature is composed of related materials, formed some 145 million years ago above a hot spot that coincides with the boundaries of three tectonic plates. The form and processes of this undersea feature are further evidence that, indeed, Earth is a dynamic planet.

CRITICALthinking 11.3
How Fast is the Pacific Plate Moving?

Tracing the motion of the Pacific plate shown in Figure 11.21 (note the map's graphic scale in the lower-left corner) reveals that the island of Midway formed 27.7 m.y.a. over the hot spot that today is active under the southeast coast of the big island of Hawai'i. Use the scale of the map to roughly determine the average annual speed of the Pacific plate in centimeters per year for Midway to have traveled this distance. Given your calculation for the plate speed and assuming the directions of movement remain the same, approximately how many years will it take the remnants of Midway to reach 50° N in the upper-left corner of the map? •

The Geologic Cycle

We see in this chapter that Earth's crust is in an ongoing state of change, being formed, deformed, moved, and broken down by physical and chemical processes. While the planet's endogenic (internal) system is at work building landforms, the exogenic (external) system is busily wearing them down. This vast give-and-take at the Earth–atmosphere–ocean interface is summarized in the **geologic cycle**. It is fueled from two sources—Earth's internal heat and solar energy from space—while being influenced by the ever-present leveling force of Earth's gravity (see *Geosystems In Action*, Figure GIA 11.).

The geologic cycle is itself composed of three principal cycles—the hydrologic cycle, which we summarized in Chapter 9, and the rock and tectonic cycles covered in this chapter. The hydrologic cycle works on Earth's surface through the exogenic processes of weathering, erosion, transportation, and deposition driven by the energy–atmosphere and water–weather systems and represented by the physical action of water, ice, and wind. The rock cycle produces the three basic rock types found in the crust—igneous, metamorphic, and sedimentary. The tectonic cycle brings heat energy and new material to the surface and recycles surface material, creating movement and deformation of the crust.

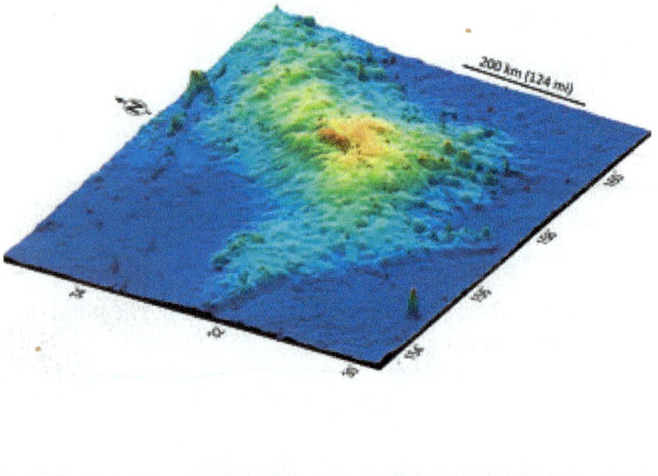

Figure 11.22 Tamu Massif, the largest volcano on Earth. This image of the seafloor shows the shape and size of Tamu Massif in the northern Pacific Ocean. Mauna Loa in Hawai'i is higher than Tamu Massif above the seafloor, but is much smaller in overall area. [Image courtesy of Will Sager.]

THE**human**DENOMINATOR 11 Earth Materials and Plate Tectonics

ENDOGENIC PROCESSES ⇨ HUMANS

- Endogenic processes cause natural hazards such as earthquakes and volcanic events that affect humans and ecosystems.
- Rocks provide materials for human use; geothermal power is a renewable resource.

HUMANS ⇨ ENDOGENIC PROCESSES

- Wells drilled into Earth's crust in association with oil and gas drilling and Enhanced Geothermal Systems may cause earthquakes.

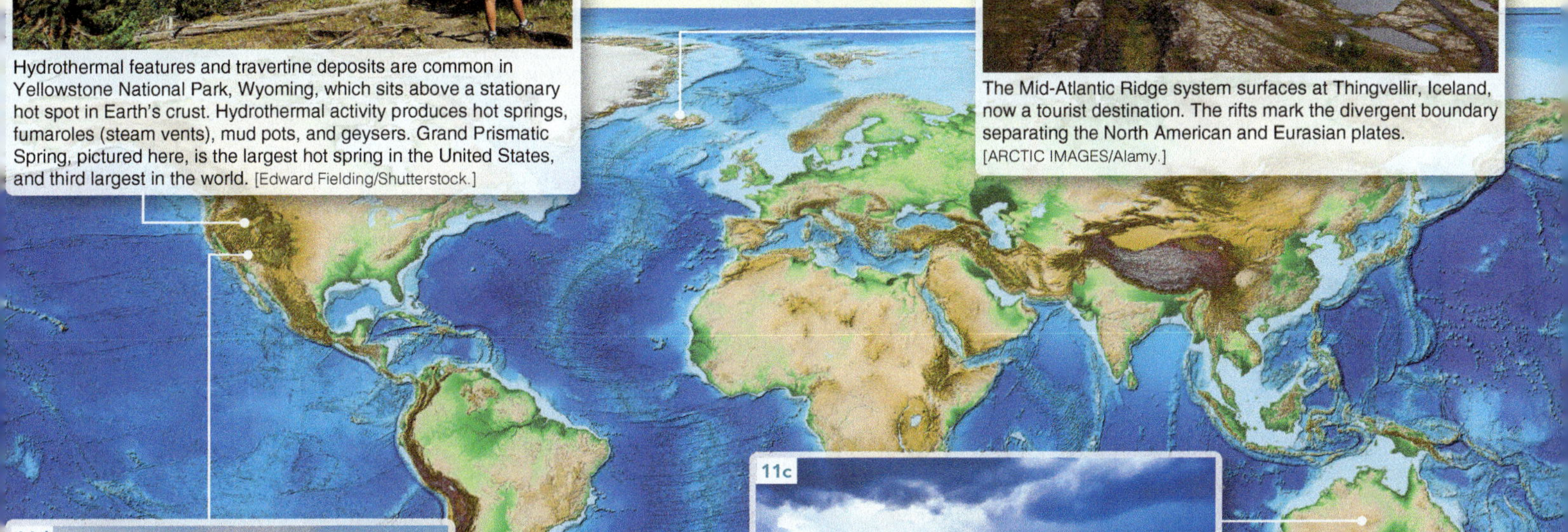

11a

Hydrothermal features and travertine deposits are common in Yellowstone National Park, Wyoming, which sits above a stationary hot spot in Earth's crust. Hydrothermal activity produces hot springs, fumaroles (steam vents), mud pots, and geysers. Grand Prismatic Spring, pictured here, is the largest hot spring in the United States, and third largest in the world. [Edward Fielding/Shutterstock.]

11b

The Mid-Atlantic Ridge system surfaces at Thingvellir, Iceland, now a tourist destination. The rifts mark the divergent boundary separating the North American and Eurasian plates. [ARCTIC IMAGES/Alamy.]

11d

In April 2013, the Nevada Desert Peak Enhanced Geothermal System (EGS) became the first U.S. enhanced geothermal project to supply electricity to the power grid. [Inga Spence/Alamy.]

11c

Uluru, also known as Ayers Rock, is probably Australia's best known landmark. This steep-sided isolated sandstone feature, about 3.5 km long and 1.9 km (1.2 mi) wide, was formed from endogenic and exogenic processes, and has cultural significance for the Aboriginal peoples. [Penny Tweedie/Alamy.]

[NOAA/NGDC.]

ISSUES FOR THE 21ST CENTURY

- Geothermal capacity will continue to be explored as an alternative energy source to fossil fuels.
- Mapping of tectonically active regions will continue to inform policy actions with regard to seismic hazards.

GEOSYSTEMS**connection**

We surveyed the internal structure of Earth and discussed the internal energy flow. Movement in Earth's crust results from these internal dynamics. Plate tectonics is the unifying theory that describes the lithosphere in terms of continent-sized migrating pieces of crust that can collide with other plates. Earth's present surface map is the result of these vast forces and motions. In Chapter 12, we focus more closely on the surface expressions of all this energy and matter in motion: the stress and strain of folding, faulting, and deformation; the building of mountains; and the sometimes dramatic activity of earthquakes and volcanoes.

The geologic cycle is a model made up of the hydrologic, rock, and tectonic cycles (GIA 11.1). Earth's exogenic (external) and endogenic (internal) and systems, driven by solar energy and Earth's internal heat, interact within the geologic cycle (GIA 11.2). The processes of the geologic cycle create distinctive landscapes (GIA 11.3).

11.1 INTERACTIONS WITHIN THE GEOLOGIC CYCLE

The cycles that make up the geologic cycle influence each other. For example, over millions of years, the tectonic cycle slowly leads to the building of mountains, which affects global precipitation patterns, one aspect of the hydrologic cycle.

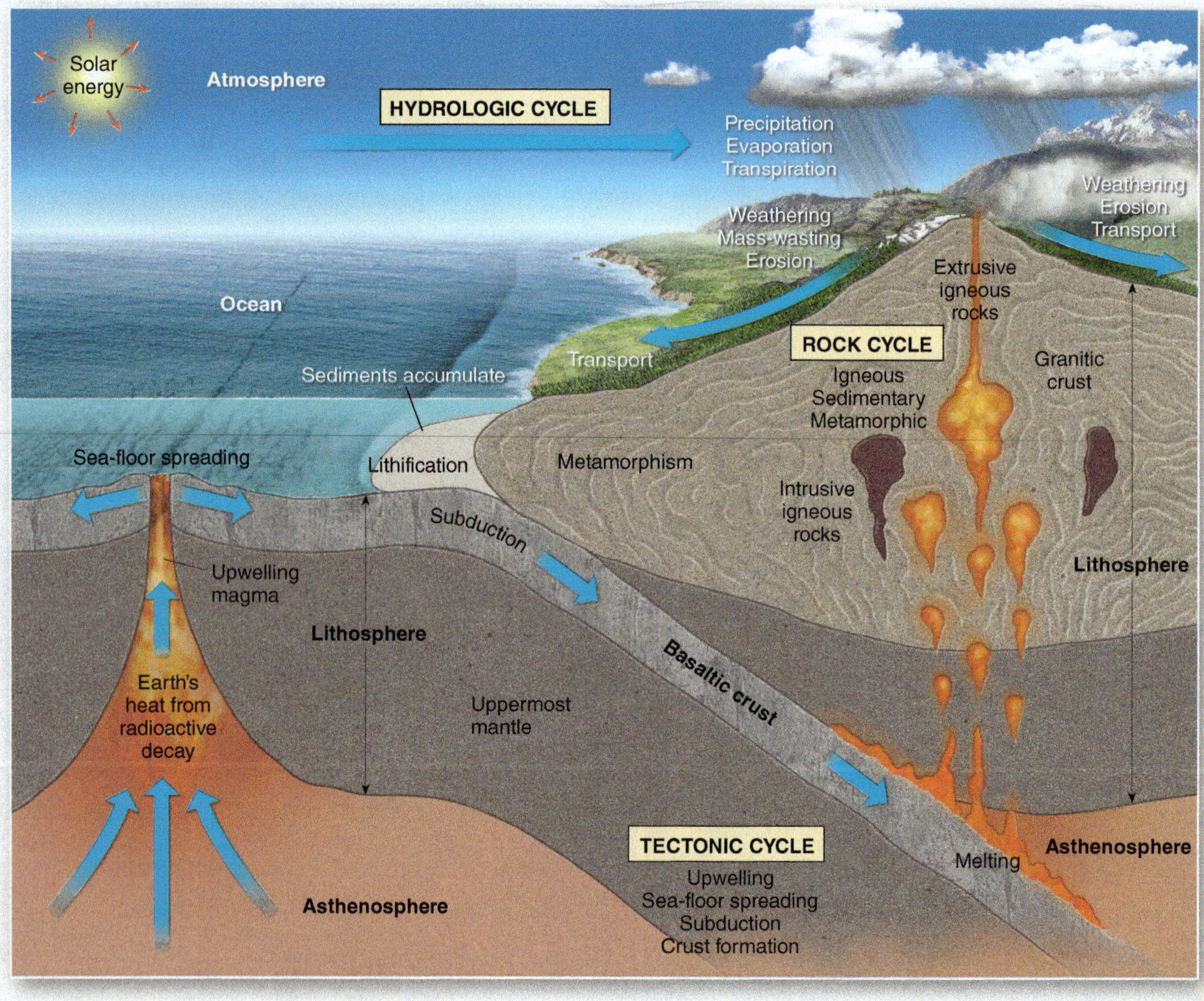

11.2 ROLE OF EXOGENIC AND ENDOGENIC SYSTEMS

Exogenic processes drive the hydrologic cycle; both endogenic and exogenic processes contribute to the rock cycle; and endogenic processes drive the tectonic cycle.

Give examples: List two additional examples of ways in which the cycles within the geologic cycle could affect each other.

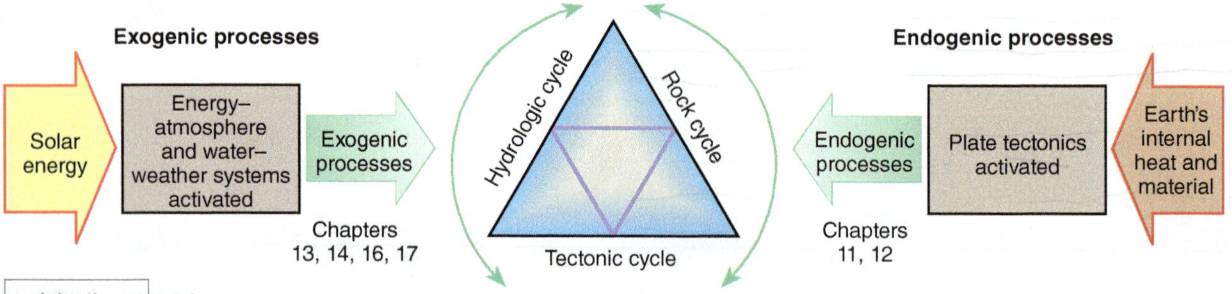

Animation
Convection and Plate Tectonics

MasteringGeography™

Visit the Study Area in MasteringGeography™ to explore the geologic cycle.

Visualize: Study a geosciences animation of convection and plate tectonics.

Assess: Demonstrate understanding of the geologic cycle (if assigned by instructor).

11.3 PROCESSES AND LANDSCAPES OF THE GEOLOGIC CYCLE

During the monsoon in India, atmospheric moisture streams inland from the Indian Ocean and falls as heavy rains. [Annie Owen/Getty Images]

11.3a

HYDROLOGIC CYCLE

Earth's water cycles continuously among the atmosphere, hydrosphere, geosphere, and biosphere.

Sources of Energy:
Solar energy provides the heat necessary for water to evaporate. Gravity causes precipitation to fall.

Example of systems interaction:
Water weathers, erodes, and deposits sediments—processes that are part of the rock cycle.

These layers of sandstone in Canyonlands National Park, Utah formed from particles of other rock, pressed and cemented together. [Tom Grundy/Shutterstock]

11.3b

ROCK CYCLE

The processes that form igneous, sedimentary, and metamorphic rocks operate so that each rock can enter the cycle and be transformed into other rock types.

Sources of Energy:
Solar energy drives the processes involved in weathering, eroding, and transporting sediment. Gravity causes sediment deposition and compaction. Earth's internal heat causes melting of rock.

Example of systems interaction:
Igneous rock forms from the cooling and hardening of molten rock, a process that is part of the tectonic cycle.

11.3c

TECTONIC CYCLE

Earth's plates diverge, collide, subduct, and slide past each other, changing the continents and ocean basins and causing earthquakes, volcanoes, and mountain building.

Sources of Energy:
Earth's internal heat provides energy that powers plate motions and melts rock to form magma. Gravity drives the subduction of lithospheric plates.

Example of systems interaction:
Volcanic activity releases gases and particles that change the atmosphere, and thus affect the hydrologic cycle.

Tectonic processes produced Indonesia's many active volcanoes, including these in Bromo National Park, Indonesia. [Somchai Buddha/Shutterstock]

Explain: How could processes in the tectonic cycle contribute to the formation of metamorphic rock in the rock cycle?

GEOquiz

1. Infer: How could mountain building (tectonic cycle) affect weather and climate (hydrologic cycle)?

2. Apply concepts: Describe the role of exogenic and endogenic processes in the rock cycle.

...ts review

■ ...sh between the endogenic and exogenic ...ns that shape Earth, and *name* the driving force ... each.

...he Earth–atmosphere interface is where the **endogenic system** (internal), powered by heat energy from within the planet, interacts with the **exogenic system** (external), powered by insolation and influenced by gravity. These systems work together to produce Earth's diverse landscape. **Geomorphology** is the subfield within physical geography that studies the development and spatial distribution of landforms. Knowledge of Earth's endogenic processes helps us understand these surface features.

endogenic system (p. 326)
exogenic system (p. 326)
geomorphology (p. 326)

1. Define the endogenic and the exogenic systems. Describe the driving forces that energize these systems.

■ *Explain* the principle of uniformitarianism, and *discuss* the time spans into which Earth's geologic history is divided.

The most fundamental principle of Earth science is **uniformitarianism**, which assumes that the same physical processes active in the environment today have been operating throughout geologic time. Dramatic, catastrophic events such as massive landslides or volcanic eruptions can interrupt the long-term processes that slowly shape Earth's surface. The **geologic time scale** is an effective device for organizing the vast span of geologic time. Geologists assign *relative age* based on the age of one feature relative to another in sequence, or *numerical age* acquired from isotopic or other dating techniques. **Stratigraphy** is the study of layered rock strata, including its sequence (superposition), thickness, and spatial distribution, which yield clues to the age and origin of the rocks.

uniformitarianism (p. 326)
geologic time scale (p. 326)
stratigraphy (p. 326)

2. Explain the principle of uniformitarianism in the Earth sciences.
3. How is the geologic time scale organized? What era, period, and epoch are we living in today? What is the difference between the relative and numerical ages of rocks?

■ *Depict* Earth's interior in cross section, and *describe* each distinct layer.

We have learned about Earth's interior indirectly, from the way its various layers transmit **seismic waves**. The **core** is differentiated into an inner core and an outer core, divided by a transition zone. Above Earth's core lies the **mantle**, differentiated into lower mantle and upper mantle. It experiences a gradual temperature increase with depth and flows slowly over time at depth, where it is hot and pressure is greatest. The boundary between the uppermost mantle and the crust is the **Mohorovičić discontinuity**, or **Moho**. The outer layer is the **crust**.

The outer part of Earth's crust is divided into two layers. The uppermost mantle, along with the crust, makes up the **lithosphere**. Below the lithosphere is the **asthenosphere**, or plastic layer. It contains pockets of increased heat from radioactive decay and is susceptible to slow convective currents in these hotter materials. The principles of buoyancy and balance produce the important principle of isostasy. **Isostasy** explains certain vertical movements of Earth's crust, such as isostatic rebound when the weight of ice is removed.

Earth's magnetic field is generated almost entirely within Earth's outer core. Polarity reversals in Earth's magnetism are recorded in cooling magma that contains iron minerals. The patterns of **geomagnetic reversal** in rock help scientists piece together the history of Earth's mobile crust.

seismic wave (p. 329)
core (p. 329)
mantle (p. 329)
Mohorovičić discontinuity (Moho) (p. 329)
crust (p. 329)
lithosphere (p. 330)
asthenosphere (p. 330)
isostasy (p. 330)
geomagnetic reversal (p. 331)

4. Make a simple sketch of Earth's interior, label each layer, and list the physical characteristics, temperature, composition, and depth of each on your drawing.
5. How does Earth generate its magnetic field? Is the magnetic field constant or does it change? Explain the implications of your answer.
6. Describe the asthenosphere. Why is it also known as the plastic layer? What are the consequences of its convection currents?
7. What is a discontinuity? Describe the principal discontinuities within Earth.
8. Define isostasy and isostatic rebound, and explain the crustal equilibrium concept.
9. Diagram the uppermost mantle and crust. Label the density of the layers in grams per cubic centimeter. What two types of crust were described in the text in terms of rock composition?

■ *Describe* the three main groups of rock, and *diagram* the rock cycle.

A **mineral** is an inorganic natural compound having a specific chemical formula and possessing a crystalline structure. A **rock** is an assemblage of minerals bound together (such as granite, a rock containing three minerals), or it may be a mass of a single mineral (such as rock salt).

Igneous rock forms from **magma**, which is molten rock beneath the surface. **Lava** is the name for magma once it has emerged onto the surface. Magma either intrudes into crustal rocks, cools, and hardens, forming **intrusive igneous rock**, or extrudes onto the surface, forming **extrusive igneous rock**. The crystalline texture of igneous rock is related to the rate of cooling. **Granite** is a coarse-grained intrusive igneous rock; it is crystalline and high in silica, aluminum, potassium, calcium, and sodium. **Basalt** is a fine-grained extrusive igneous rock; it is granular and high in silica, magnesium, and iron. Intrusive igneous rock that cools slowly in the crust forms a **pluton**. The largest pluton form is a **batholith**.

Sedimentary rock is formed when loose *clasts* (grains or fragments) derived from several sources are compacted and cemented together in the process of **lithification**. Clastic sedimentary rocks are derived from the fragments of weathered and eroded rocks and the material that is transported and deposited as **sediment**. Chemical sedimentary rocks are formed by either biochemical processes or from the chemical dissolution of minerals into solution; the most common is **limestone**, which is lithified calcium carbonate, $CaCO_3$.

Any igneous or sedimentary rock may be transformed into **metamorphic rock** by going through profound physical or chemical changes under pressure and increased temperature. The **rock cycle** describes the three principal rock-forming processes and the rocks they produce.

mineral (p. 332)
rock (p. 332)
igneous rock (p. 333)
magma (p. 333)
lava (p. 333)
intrusive igneous rock (p. 333)
extrusive igneous rock (p. 333)
granite (p. 333)
basalt (p. 333)
pluton (p. 334)
batholith (p. 334)
sedimentary rock (p. 334)
sediment (p. 334)
lithification (p. 335)
limestone (p. 336)
metamorphic rock (p. 338)
rock cycle (p. 338)

10. What is a mineral? A mineral family? Name the most common minerals on Earth. What is a rock?
11. Describe igneous processes. What is the difference between intrusive and extrusive types of igneous rocks?
12. Explain what coarse- and fine-grained textures say about the cooling history of a rock.
13. Briefly describe sedimentary processes and lithification. Describe the sources and particle sizes of sedimentary rocks.
14. What is metamorphism, and how are metamorphic rocks produced? Name some original parent rocks and their metamorphic equivalents.

■ *Describe* Pangaea and its breakup, and *explain* the physical evidence that crustal drifting is continuing today.

The present configuration of the ocean basins and continents is the result of tectonic processes involving Earth's interior dynamics and crust. **Pangaea** was the name Alfred Wegner gave to a single assemblage of continental crust existing some 225 m.y.a. that subsequently broke apart. Wegener coined the phrase **continental drift** to describe his idea that the crust is moved by vast forces within the planet. The theory of **plate tectonics** is that Earth's lithosphere is fractured into huge slabs or plates, each moving in response to gravitational pull and to flowing currents in the mantle that create frictional drag on the plate. Geomagnetic reversals along **mid-ocean ridges** on the ocean floor provide evidence of **seafloor spreading** which accompanies the movement of plates toward the continental margins of ocean basins. At some plate boundaries, denser oceanic crust dives beneath lighter continental crust along **subduction zones.**

Pangaea (p. 340)
continental drift (p. 340)
plate tectonics (p. 340)
mid-ocean ridge (p. 340)
seafloor spreading (p. 340)
subduction zone (p. 344)

15. Briefly review the history of the theory of plate tectonics, including the concepts of continental drift and seafloor spreading. What was Alfred Wegener's role?
16. What was Pangaea? What happened to it during the past 225 million years?
17. Describe the process of upwelling as it refers to magma under the ocean floor. Define subduction, and explain that process.
18. Characterize the three types of plate boundaries and the actions associated with each type.

■ *Draw* the pattern of Earth's major plates on a world map, and *relate* this pattern to the occurrence of earthquakes, volcanic activity, and hot spots.

Earth's lithosphere is made up of 14 large plates, and many smaller ones, that move and interact to form three types of plate boundaries: divergent, convergent, and transform. Along the offset portions of mid-ocean ridges, horizontal motions produce **transform faults**. Earthquakes and volcanoes often correlate with plate boundaries. As many as 50 to 100 **hot spots** exist across Earth's surface, where plumes of magma—some anchored in the lower mantle, others originating from shallow sources in the upper mantle—generate an upward flow. Some hot spots produce **geothermal energy**, or heat from Earth's interior, which may be used for direct geothermal heating or geothermal power. The **geologic cycle** is a model of the internal and external interactions that shape the crust—including the hydrologic, rock, and tectonic cycles.

transform fault (p. 344)
hot spot (p. 346)
geothermal energy (p. 346)
geologic cycle (p. 350)

19. What is the relation between plate boundaries and volcanic and earthquake activity?
20. What is the nature of motion along a transform fault?
21. What is geothermal energy?
22. Illustrate the geologic cycle, and define each component: rock cycle, tectonic cycle, and hydrologic cycle.

12 Tectonics, Earthquakes, and Volcanism

KEY LEARNING concepts

After reading the chapter, you should be able to:

- ***Describe*** first, second, and third orders of relief, and ***list*** Earth's six major topographic regions.
- ***Describe*** the formation of continental crust, and ***define*** displaced terranes.
- ***Explain*** the process of folding, and ***describe*** the principal types of faults and their characteristic landforms.
- ***List*** the three types of plate collisions associated with orogenesis, and ***identify*** specific examples of each.
- ***Explain*** earthquake characteristics and measurement, ***describe*** earthquake fault mechanics, and ***discuss*** the status of earthquake forecasting.
- ***Describe*** volcanic landforms, and ***distinguish*** between an effusive and an explosive volcanic eruption.

In 1977, geologists and oceanographers Marie Tharp and Bruce Heezen, aided by artist Heinrich Berann, published this comprehensive map of the ocean basins, allowing scientists their first look at the global ocean floor. Tharp and Heezen's work in the 1950s processing thousands of sonar depth measurements led to the first identification of the vast Mid-Atlantic Ridge, providing solid evidence for plate tectonics theory. On the map, seafloor spreading centers are marked by oceanic ridges that stretch over 64,000 km (40,000 mi), subduction zones are indicated by deep oceanic trenches, and transform faults slice across oceanic ridges. Critical Thinking 12.2 at the end of the chapter revisits this map to apply some of the concepts covered in Chapters 11 and 12. [Office of Naval Research.]

The San Jacinto Fault Connection

In southern California, people live with earthquakes and with the ever-present question, "When will the 'Big One' occur?" The hundreds of faults that make up the San Andreas fault system in that region are produced by the relative horizontal motions, approximately 5 cm (2 in.) per year, of the northwestward-moving Pacific plate against the southeastward-moving North American plate. The San Jacinto fault, running from San Bernardino southward to the Mexican border, is an active part of this system. The San Jacinto runs somewhat parallel to the Elsinore fault, which runs northward into Orange County, and the Whittier fault, which crosses into Los Angeles. Numerous other related faults lace through the region with this same alignment (Figure GN 12.1).

Potential for a Major Earthquake In 1994, the Northridge earthquake in the San Fernando Valley north of Los Angeles caused 66 fatalities and set the record, which still stands today, for earthquake-related property damage in the United States at $30 billion. The U.S. Geological Survey (USGS) regards the San Jacinto fault as capable of an M 7.5 quake ("M" is the abbreviation for moment magnitude and refers to an earthquake rating scale discussed in this chapter). An M 7.5 would release almost 30 times more energy than that produced by the M 6.7 Northridge earthquake—thus, an M 7.5 is a major quake as compared to an M 6.7 strong quake.

Scientific evidence regarding the nature of faulting and crustal movement suggests that a major earthquake may be preceded by a number of small earthquakes that occur in a "wavelike pattern" as rupturing spreads along the fault. Recent earthquake activity along the San Jacinto and Elsinore faults may be following this pattern.

Records show that at least six quakes have occurred along the San Jacinto fault in the last 50 years: an M 5.8 and M 6.5 in 1968, an M 5.3 in 1980, an M 5.0 in 2005, and an M 4.1 in 2010, south of Palm Springs. Then in April 2010, an earthquake occurred along a nearby system of small faults that transferred strain to the San Jacinto; the M 7.2 El Mayor–Cucapah quake, located southeast of El Centro, in Baja, Mexico, was the largest to affect the region since 1994 (Figure GN 12.2). This temblor occurred along a complex, previously unknown fault system connecting the Gulf of California (see Figure 12.12a) to the Elsinore fault, and caused large clusters of aftershocks toward the north along the San Jacinto fault structure. The M 5.4 Borrego quake occurred several months later, with aftershocks again spreading to the northwest. This pattern indicates that strain is moving northward along the system—a cause for concern that this sequence of seismic events may be leading to the "Big One."

▲**Figure GN 12.1 Southern California seismic map.** [Based on data from USGS and the California Geological Survey.]

Planning for Earthquakes The Global Positioning System is now an essential part of earthquake forecasting, and a network of 100 GPS stations monitors crustal change throughout southern California. Given scientific advancements in earthquake analysis and forecasting, a major quake along the San Jacinto and related faults should not catch the region by surprise, assuming that proper planning, zoning, and preparation are put in place now. This chapter examines earthquakes and other tectonic processes that determine Earth's surface topography.

GEOSYSTEMS NOW ONLINE Go to Chapter 12 on the *MasteringGeography* website (www.masteringgeography.com/) for more on southern California earthquakes. For information and links regarding the Northridge quake of 1994, see http://earthquake.usgs.gov/earthquakes/states/events/1994_01_17.php, and for the 2010 El Mayor–Cucapah earthquake, see http://www.scsn.org/2010sierraelmayor.html. MG

▲**Figure GN 12.2 Land-surface shift of up to 3.0 m (9.8 ft) along the Borrego Fault caused by the El Major–Cucapah quake.** [John Fletcher, CICESE/NASA.]

The arrangement of continents and oceans, the topography of the land and the seafloor, the uplift and erosion of mountain ranges, and global patterns of earthquake and volcanic activity are all evidence of our dynamic Earth. Earth's endogenic systems send flows of heat and material toward the surface to form crust. These ongoing processes alter continental landscapes and produce oceanic seafloor crust, sometimes in dramatic events that make headline news.

Natural hazards such as earthquakes and volcanic eruptions near population centers pose threats to lives and property. The world was riveted to images of massive earthquakes in Haiti and Chile in 2010 and in New Zealand and Japan in 2011, as unstable plate boundaries snapped into new positions. (More on these events is in Focus Study 12.1 in this chapter.) Earth systems science is now providing analysis and warnings of seismic and volcanic activity to affected populations as never before.

In this chapter: We examine processes that create the various landforms and crustal features that make up Earth's surface. Continental crust has been forming throughout most of Earth's 4.6-billion-year existence. Tectonic processes deform Earth's crust, which is then weathered and eroded into recognizable landforms: mountains, basins, faults and folds, and volcanoes. Such processes sometimes occur suddenly, as in the case of earthquakes, but more often the movements that shape the landscape are gradual, such as regional uplift and folding.

We begin our look at tectonics and volcanism on the ocean floor, where they are hidden from direct view. The map that opens this chapter and highlights the details of the ocean floor is a striking representation of the concepts learned in Chapter 11, which lays the foundation for this and subsequent chapters. Try to correlate this seafloor illustration with the maps of crustal plates and plate boundaries shown in Figure 11.18.

Earth's Surface Relief

Landscapes across Earth occur at varying elevations; think of the low coastal plains of India and Bangladesh, the middle-elevation foothills and mountains of India and Nepal, and the high elevations of the Himalayas in Nepal and Tibet. These vertical elevation differences upon a surface are known as **relief**. The general term for the undulations and other variations in the shape of Earth's surface, including its relief, is **topography**—the lay of the land.

The relief and topography of Earth's landforms have played a vital role in human history: High mountain passes both protected and isolated societies, ridges and valleys dictated transportation routes, and vast plains encouraged the development of faster methods of communication and travel. Earth's topography has stimulated human invention and spurred adaptation.

Studying Earth's Topography

To study Earth's relief and topography, scientists use satellite radar, LiDAR systems (which use laser scanners), and tools such as GPS (which report location and elevation; please review Chapter 1 for discussion of these tools.). During an 11-day period in 2000, the Shuttle Radar Topography Mission (SRTM) instruments on board the Space Shuttle *Endeavor* surveyed almost 80% of Earth's land surface; Figure 12.1 shows examples of SRTM images (see the image gallery at http://www2.jpl.nasa.gov/srtm). The data acquired from this mission is the most complete high-resolution topographic dataset available and is commonly used with a satellite image overlay for topographic display.

A recent advance in the study of topography is the development of digital elevation models (DEMs) to display elevation data in digital form. LiDAR, which provides the highest resolution for mapping Earth's surface, is often used in conjunction with DEMs for scientific purposes. The USGS, known for its comprehensive topographic mapping program for the United States, is currently collaborating with NASA and the National Park Service to acquire LiDAR data for mapping U.S. National Parks (see http://ngom.usgs.gov/dsp/mapping/lidar_topographic_mapping.php).

(a) Second order of relief on the Kamchatka Peninsula, Russia.

(b) Third order of relief demonstrated by the local landscape near San Jose, Costa Rica, with the volcanoes Irazu (3401 m; 11,161 ft) and Turrialba (3330 m; 10,925 ft) in the distance.

▲**Figure 12.1 Second and third orders of relief.** [Shuttle Radar Topography Mission, courtesy of JPL/NGA/NASA–CalTech.]

In another USGS project, topography and geology are combined in a shaded relief map of the United States called "A Tapestry of Time and Terrain." This map is made up of 12 million spot elevations, with less than a kilometer between any two, offering a detailed composite color illustration of land surfaces and the ages of underlying rock formations (see http://tapestry.usgs.gov/).

Orders of Relief

For convenience of description, geographers group landscapes into three *orders of relief.* These orders classify landscapes by scale, from vast ocean basins and continents down to local hills and valleys. The first order of relief is the coarsest level of landforms, consisting of the continents and oceans. **Continental landmasses** are those portions of crust that reside above or near sea level, including the undersea continental shelves along the coastlines. **Ocean basins**, featured in the chapter-opening map illustration, are portions of the crust that are entirely below sea level. Approximately 71% of Earth is covered by ocean.

The second order of relief is the intermediate level of landforms, both on continents and in ocean basins (Figure 12.1a). Continental features in the second order of relief include mountain masses, plains, and lowlands. A few examples are the Alps, Canadian and American Rockies, west Siberian lowland, and Tibetan Plateau. This order includes the great rock "shields" that form the heart of each continental mass. In the ocean basins, the second order of relief includes continental rises and slopes, flat plains (called *abyssal plains*), mid-ocean ridges, submarine canyons, and oceanic trenches (subduction zones)—all visible in the seafloor illustration that opens this chapter.

The third and most detailed order of relief includes individual mountains, cliffs, valleys, hills, and other landforms of smaller scale (Figure 12.1b). These features characterize local landscapes.

Earth's Hypsometry

Hypsometry (from the Greek *hypsos*, meaning "height") is the measurement of land elevation relative to sea level (the measurement of underwater elevations is bathymetry, mentioned in Chapter 11). Figure 12.2 is a hypsographic curve that shows the distribution of Earth's surface area according to elevation above and depth below sea level. Relative to Earth's diameter of 12,756 km (7926 mi), the surface has low relief—only about 20 km (12.5 mi) from highest peak to lowest oceanic trench. For perspective, Mount Everest is 8.8 km (5.5 mi) above sea level, and the Mariana Trench is 11 km (6.8 mi) below sea level.

The average elevation of Earth's solid surface is actually under water: –2070 m (–6790 ft) below mean sea level. The average elevation for exposed land is only 875 m (2870 ft). For the ocean depths, the average elevation is –3800 m (–12,470 ft). From this description you can

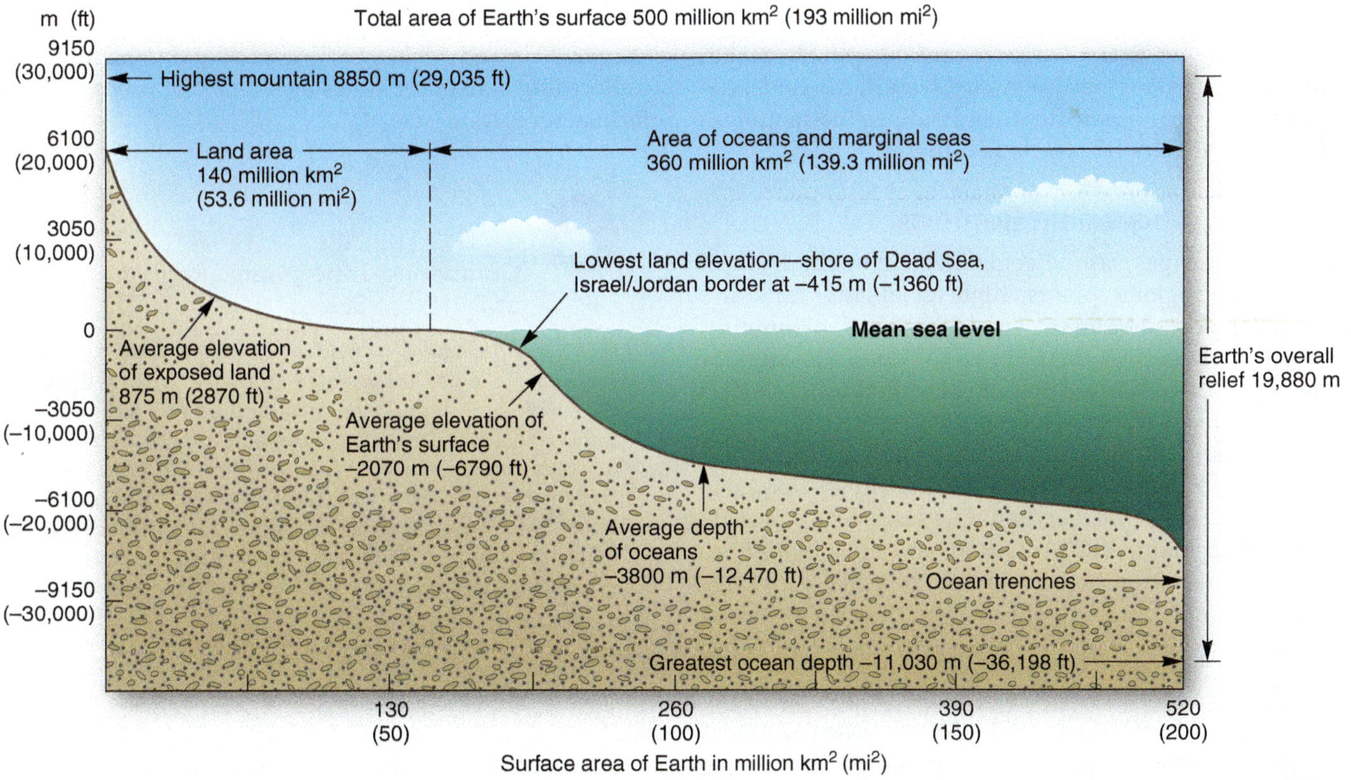

▲**Figure 12.2 Earth's hypsometry.** Hypsographic curve of Earth's surface area and elevation as related to mean sea level. From the highest point above sea level (Mount Everest) to the deepest oceanic trench (Mariana Trench), Earth's overall relief is almost 20 km (12.5 mi).

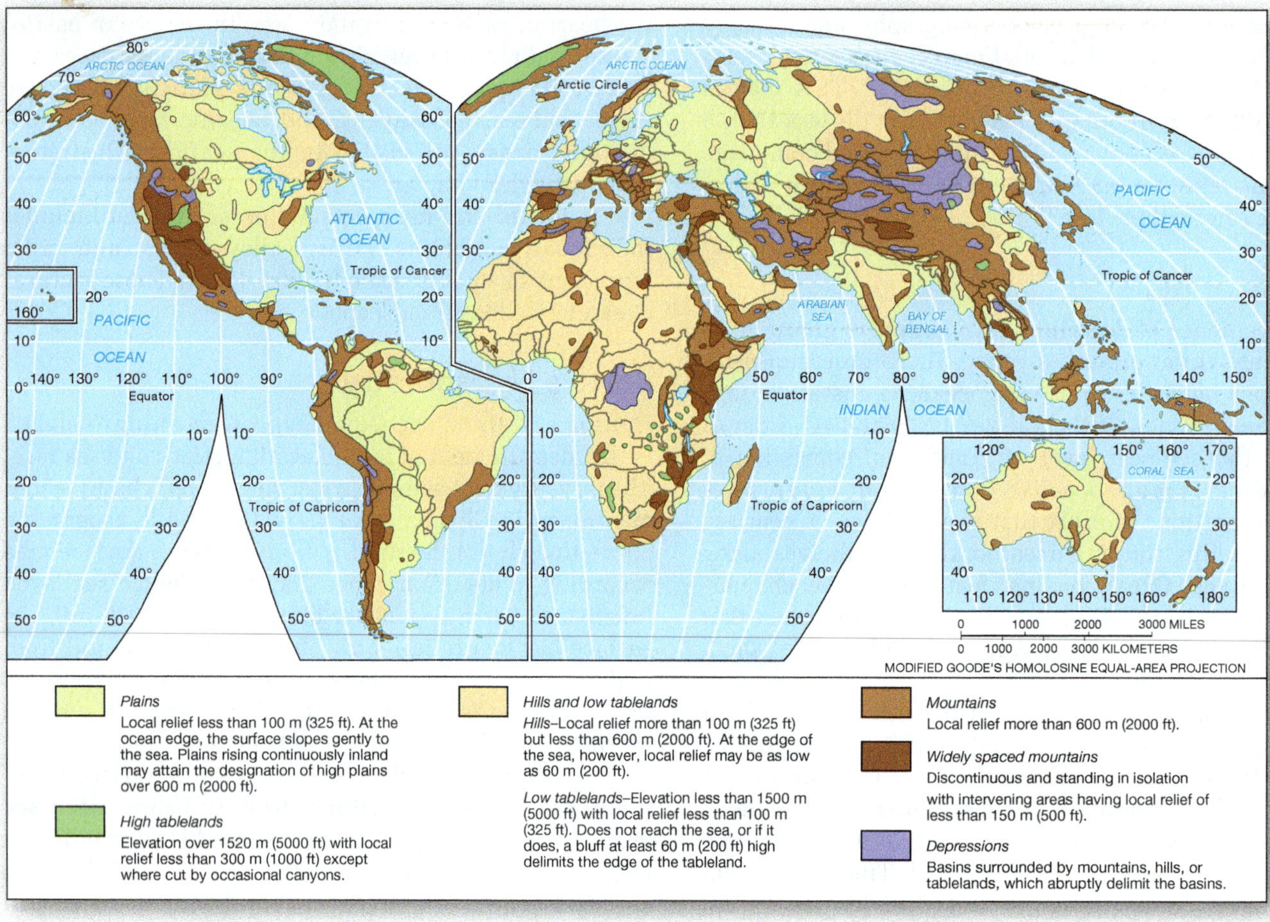

▲**Figure 12.3 Earth's topographic regions.** Compare the map and legend to the world physical map inside the back cover of this text. [R. E. Murphy, "Landforms of the world," *Annals of the Association of American Geographers* 58, 1 (March 1968). Adapted by permission.]

see that, on the average, the oceans are much deeper than continental regions are high. Overall, the underwater ocean basins, ocean floor, and submarine mountain ranges form Earth's largest "landscape."

Earth's Topographic Regions

Earth's landscapes can be generalized into six types of topographic regions: plains, high tablelands, hills and low tablelands, mountains, widely spaced mountains, and depressions (Figure 12.3). An arbitrary elevation or other descriptive limit in common use defines each type of topography (see the map legend).

North and South America, Asia, and Australia possess extensive *plains*. *Hills* and *low tablelands* dominate Africa and part of Europe and Australia. The Colorado Plateau, Greenland, and Antarctica are notable *high tablelands* (the latter two composed of ice). *Mountains* occur on each continent; *depressions*, or basins, occur only in Asia and Africa. Earth's relief and topography are undergoing constant change as a result of processes that form and rearrange crust.

CRITICALthinking 12.1 Comparing Topographic Regions at Different Scales

Using the definitions in Figure 12.3, identify the type of topographic region represented by the area within 100 km (62 mi) of your campus, and then do the same for the area within 1000 km (620 mi) of your campus. Use Google Earth™ or consult maps and atlases to describe the topographic character and the variety of relief within these two regional scales. Do you perceive that the type of topographic region influences lifestyles? Economic activities? Transportation? Was topography influential in the history of the region? ●

GEOreport 12.1 Mount Everest measured by GPS

In 1999, climbers determined the height of Mount Everest using direct GPS placement on the mountain's icy summit. The newly measured elevation of 8850 m (29,035 ft) replaced the previous official measure of 8848 m (29,028 ft) made in 1954 by the Survey of India. The 1999 GPS readings also indicated that the Himalayas are moving northeastward at a rate of up to 6 mm (0.25 in.) per year, as the mountain range is being driven farther into Asia by the continuing collision of the Indian and Asian landmasses.

Crustal Formation

How did Earth's continental crust form? What gave rise to the three orders of relief just discussed? Earth's surface is a battleground of opposing processes: On the one hand, tectonic activity, driven by our planet's internal energy, builds crust; on the other hand, the exogenic processes of weathering and erosion, powered by the Sun and occurring through the actions of air, water, waves, and ice, tear down crust.

Tectonic activity generally is slow, requiring millions of years. Endogenic processes result in gradual uplift and new landforms, with major mountain building occurring along plate boundaries. The uplifted crustal regions are quite varied but can be grouped into three general categories:

- Residual mountains and stable continental cratons, consisting of inactive remnants of ancient tectonic activity
- Tectonic mountains and landforms, produced by active folding, faulting, and crustal movements
- Volcanic landforms, formed by the surface accumulation of molten rock from eruptions of subsurface materials

The various processes mentioned in these descriptions operate in concert to produce the continental crust we see around us.

Continental Shields

All continents have a nucleus, called a *craton*, consisting of ancient crystalline rock on which the continent "grows" through the addition of crustal fragments and sediments. Cratons are generally old and stable masses of continental crust that have been eroded to a low elevation and relief. Most date to the Precambrian Eon and can be more than 2 billion years old. The lack of basaltic components in these cratons offers a clue to their stability. The lithosphere underlying a craton is often thicker than that underlying younger portions of continents and oceanic crust.

A **continental shield** is a large region where a craton is exposed at the surface (Figure 12.4). Layers of younger sedimentary rock—called continental platforms—surround

(b) Canadian shield landscape in northern Québec, stable for hundreds of millions of years, stripped by past glaciations and marked by intrusive igneous dikes (magmatic intrusions).

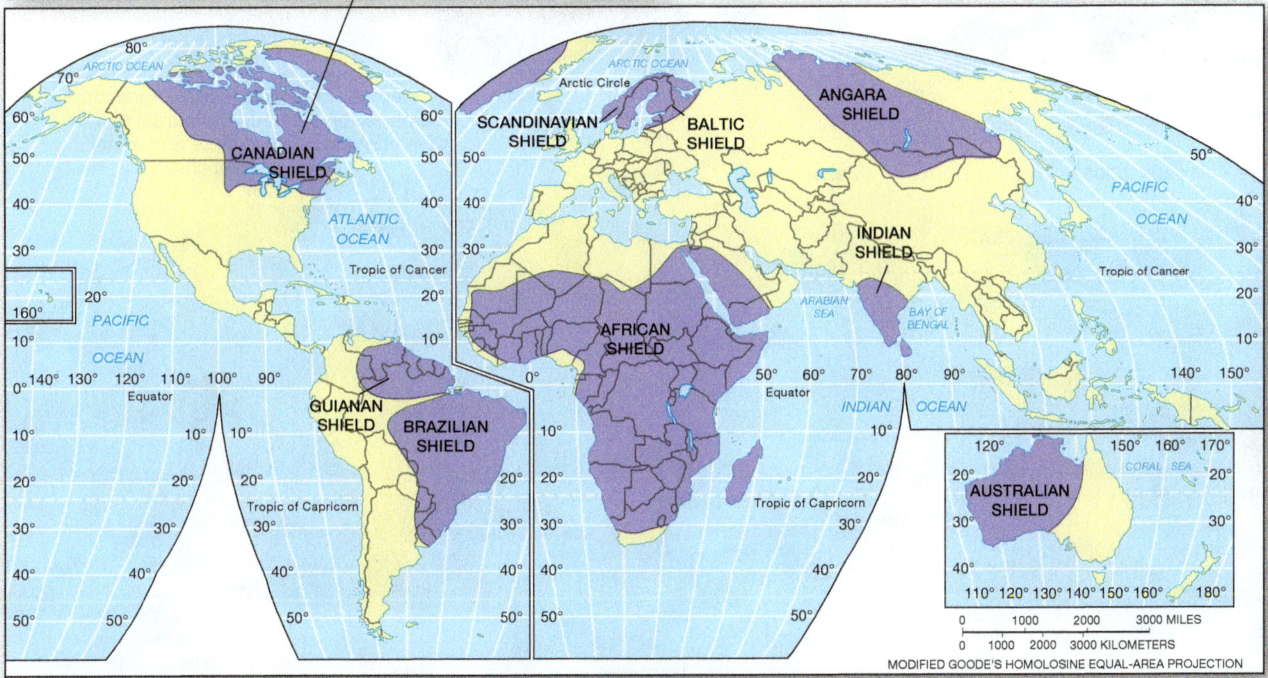

(a) Earth's major continental shields exposed by erosion. Adjacent portions of these shields remain covered by younger sedimentary layers.

▲**Figure 12.4 Continental shields.** [(a) R. E. Murphy, "Landforms of the world," *Annals of the Association of American Geographers* 58, 1 (March 1968). Adapted by permission. (b) Bobbé Christopherson.]

MG Notebook
World Structural Regions and Major Mountain Systems

these shields and appear quite stable over time. Examples of such stable platforms include the region from east of the Rockies to the Appalachians and northward into central and eastern Canada, a large portion of China, eastern Europe to the Ural Mountains, and portions of Siberia.

On the *MasteringGeography* website, you will find a map of seven world structural regions, including shields and their surrounding sedimentary deposits. Various mountain chains, rifted regions, and isolated volcanic areas are also noted on the map, which you can refer back to as you read through this chapter.

Building Continental Crust and Accretion of Terranes

The formation of continental crust is complex and takes hundreds of millions of years. It involves the entire sequence of seafloor spreading and formation of oceanic crust, eventual subduction and remelting of that oceanic crust, and the subsequent rise of remelted material as new magma, all summarized in Figure 12.5. In this process of crustal formation, you can literally follow the cycling of materials through the tectonic cycle.

To understand this process, study Figure 12.5 and the inset photos. Begin with the magma that originates in the asthenosphere and wells up along the mid-ocean ridges. Basaltic magma is formed from minerals in the upper mantle that are rich in iron and magnesium. Such magma contains less than 50% silica and has a low-viscosity (thin) texture—it tends to flow. This mafic material rises to erupt at spreading centers and cools to form new basaltic seafloor, which spreads outward to collide with continental crust along its far edges. The oceanic crust, being denser, plunges beneath the lighter continental crust, into the mantle, where it remelts. The new magma then rises and cools, forming more continental crust in the form of intrusive granitic igneous rock.

As the subducting oceanic plate works its way under a continental plate, it takes with it trapped seawater and sediment from eroded continental crust. The remelting incorporates the seawater, sediments, and surrounding crust into the mixture. As a result, the magma, generally called a *melt*, migrating upward from a subducted plate,

▲**Figure 12.5 Crustal formation.** Material from the asthenosphere upwells along seafloor spreading centers. Basaltic ocean floor is subducted beneath lighter continental crust, where it melts, along with its cargo of sediments, water, and minerals. This melting generates magma, which makes its way up through the continental crust to form igneous intrusions and extrusive eruptions. [Photos by Bobbé Christopherson.]

(a) Wrangell terranes occur in four segments, highlighted in red among the outlines of other terranes along the western margin of North America (light blue shaded area).

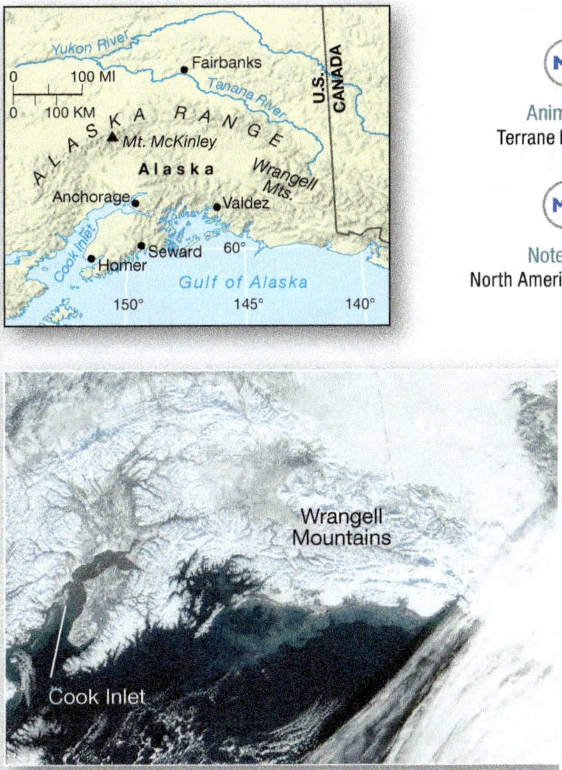

(b) The snow-covered Wrangell Mountains stretch from east-central Alaska across the Canadian border. Mount McKinley (Denali), the highest elevation in North America at 6194 m (20,322 ft), is a granite batholith surrounded by accreted terranes of varying age. McKinley is relatively young by comparison and is still gaining height as a result of plate convergence.

MG Animation Terrane Formation

MG Notebook North American Terranes

▲**Figure 12.6 North American terranes.** [(a) Based on data from USGS. (b) Terra MODIS image, NASA/GSFC.]

contains 50%–75% silica and aluminum. These give the melt a high-viscosity (thick) texture, and therefore it tends to block and plug conduits to the surface.

Bodies of such silica-rich magma may reach the surface in explosive volcanic eruptions, or they may stop short and become subsurface intrusive bodies in the crust, cooling slowly to form granitic crystalline plutons such as batholiths (see Figure 12.6). As noted earlier, their composition is quite different from that of the magma that rises directly from the asthenosphere at seafloor spreading centers.

Each of Earth's major lithospheric plates actually is a collage of many crustal pieces acquired from a variety of sources. Over time, slowly migrating fragments of ocean floor, curving chains (or arcs) of volcanic islands, and pieces of crust from other continents all have been forced against the edges of continental shields and platforms. These varied crustal pieces that have become attached, or accreted, to the plates are called **terranes** (not to be confused with *terrain*, which refers to the topography of a tract of land). Such displaced terranes, sometimes called *microplate*, or *exotic, terranes*, have histories different from those of the continents that capture them. They are usually framed by fault-zone fractures and differ in rock composition and structure from their new continental homes.

In the regions surrounding the Pacific Ocean, accreted terranes are particularly prevalent. At least 25% of the growth of western North America can be attributed to the accretion of at least 50 terranes since the early Jurassic Period (190 million years ago). A good example is the Wrangell Mountains, which lie just east of Prince William Sound and the city of Valdez, Alaska. The *Wrangellia terranes*—a former volcanic island arc and associated marine sediments from near the equator—migrated approximately 10,000 km (6200 mi) to form the Wrangell Mountains and three other distinct formations along the western margin of the continent (Figure 12.6).

The Appalachian Mountains, extending from Alabama to the Maritime Provinces of Canada, possess bits of land once attached to ancient Europe, Africa, South America, Antarctica, and various oceanic islands. The discovery of terranes, which occurred as recently as the 1980s, revealed one of the ways continents are assembled.

Crustal Deformation

Rocks, whether igneous, sedimentary, or metamorphic, are subjected to powerful stress by tectonic forces, gravity, and the weight of overlying rocks. *Stress* is any force that affects an object, measured as force per unit area;

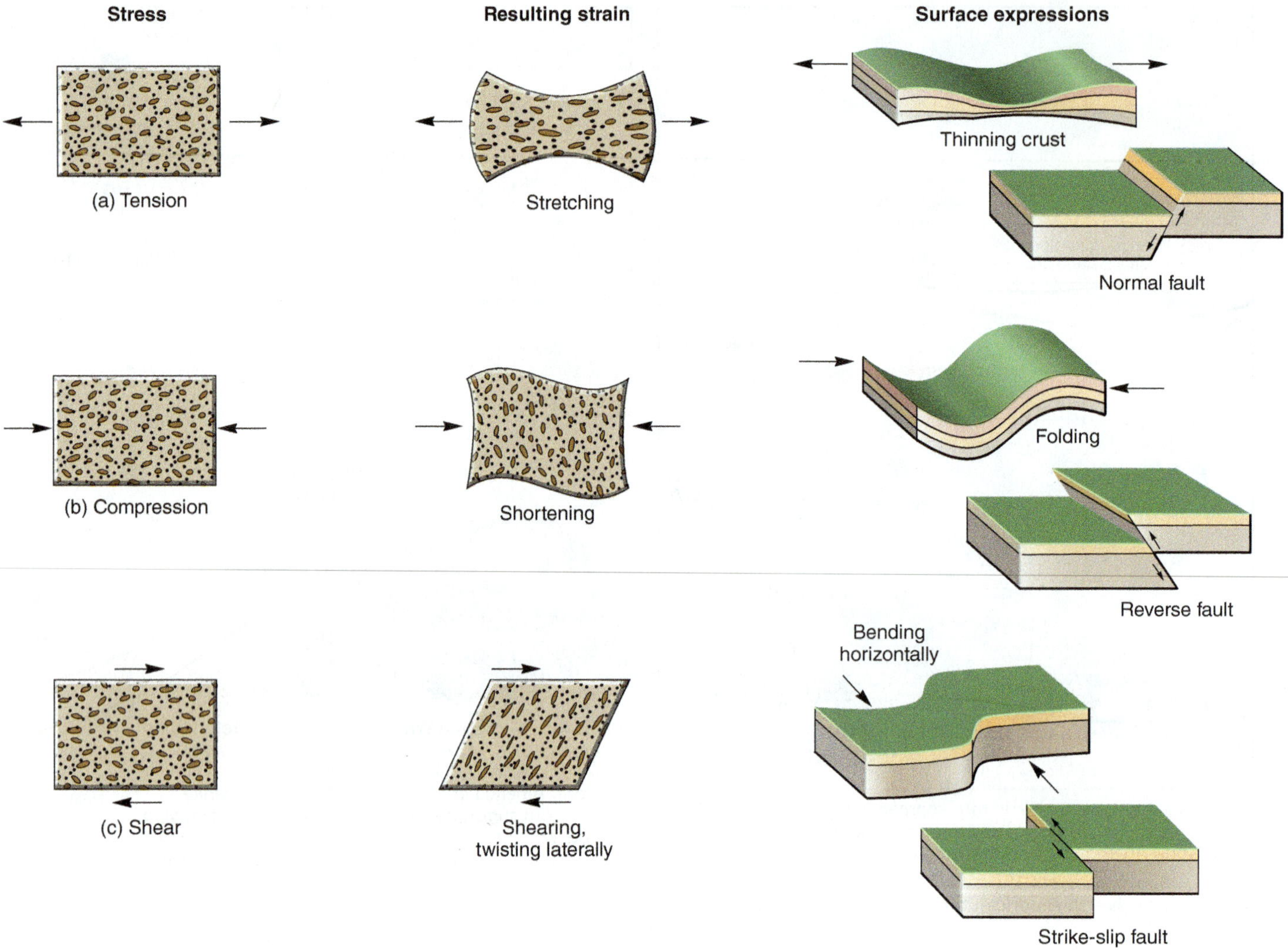

▲Figure 12.7 **Three kinds of stress and strain and the resulting surface expressions on Earth's crust.**

note that these units are the same as for pressure (defined in Chapter 3). Three types of stress are important for crustal deformation: *tension*, which causes stretching; *compression*, which causes shortening; and *shear*, which causes twisting or tearing as objects slide parallel to one another (Figure 12.7).

Although stress is an important force in shaping Earth's crust, the landforms we see result from strain, which is how rocks respond to stress. *Strain* is, by definition, a dimensionless measure of the amount of deformation undergone by an object. Strain is the stretching, shortening, and twisting that results from stress and is expressed in rocks by *folding* (bending) or *faulting* (breaking). Whether a rock bends or breaks depends on several factors, including its composition and the amount of pressure it is undergoing. Figure 12.7 illustrates each type of stress and the resulting strain and surface expressions that develop.

Folding and Broad Warping

When rock strata that are layered horizontally are subjected to compressional forces, they become deformed (Figure 12.8). **Folding** occurs when rocks are deformed as a result of compressional stress and shortening. We can visualize this process by stacking sections of thick fabric on a table and slowly pushing on opposite ends of the stack. The cloth layers will bend and rumple into folds similar to those shown in the landscape of Figure 12.8a, with some folds forming arches (upward folds) and some folds forming troughs (downward folds).

An arch-shaped upward fold is an **anticline**; the rock strata slope downward away from an imaginary center axis that divides the fold into two parts. A trough-shaped downward fold is a **syncline**; the strata slope upward away from the center axis. The erosion of a syncline may form a *synclinal ridge*, produced when the different rock strata offer different degrees of resistance to weathering processes (Figure 12.8b).

The *hinge* is the horizontal line that defines the part of the fold with the sharpest curvature. If the hinge is not horizontal, meaning that it is not "level" (parallel with Earth's surface), the fold is *plunging*, or dipped down (inclined) at an angle. If the *axial plane* of the fold, an imaginary surface that parallels the hinge but descends downward through each layer, is inclined from vertical,

MG
Animation
Folds, Anticlines, and Synclines

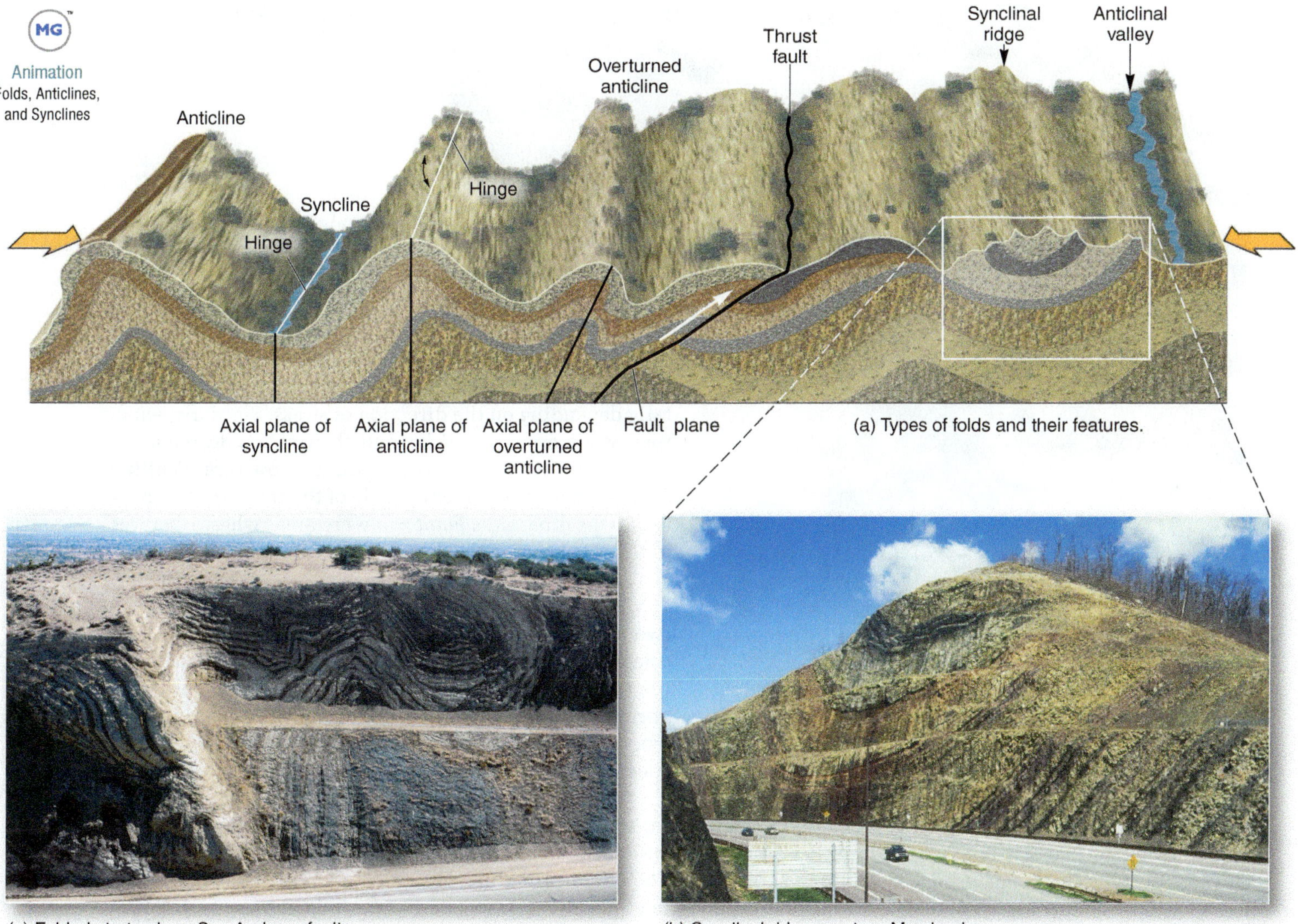

(c) Folded strata along San Andreas fault

(b) Synclinal ridge, western Maryland

▲**Figure 12.8 Folded landscapes.** [(b) Mike Boroff/Photri-Microstock. (c) Bobbé Christopherson.]

the resulting configuration is an *overturned anticline,* in which folds have been compressed so much that they overturn upon their own strata. Further stress eventually fractures the rock strata along distinct lines, a process that forms a thrust fault (see Figure 12.8; faults are discussed ahead); some overturned folds are thrust upward, causing a considerable shortening of the original strata. Areas of intense stress, compressional folding, and faulting are visible along the San Andreas fault in southern California (Figure 12.8c).

Knowledge of folding and stratigraphy are important for the petroleum industry. For example, petroleum geologists know that oil and natural gas collect in the upper portions of anticlinal folds in permeable rock layers such as sandstone.

Over time, folded structures can erode to produce interesting landforms (Figure 12.9). An example is a *dome,* which is an area of uplifted rock strata resembling an anticline that has been heavily eroded over time (Figure 12.9a). Since an anticline is a fold that is convex in an upward direction, erosion exposes the oldest rocks in the center of a dome, which often have a circular pattern that resembles a bull's eye when viewed from the air—the Richat Structure in Mauritania is an example. The Black Hills of South Dakota are another dome structure. A *basin* forms when an area resembling a syncline is uplifted and then erodes over time; in this structure, the oldest rock strata are at the outside of the circular structure (Figure 12.9b). Since a syncline is concave when considered from above, erosion exposes the youngest rocks in the center of the structure.

Mountain ranges in North America, such as the Canadian Rockies and the Appalachian Mountains, and in the Middle East exhibit the complexity that folding can produce. The area north of the Persian Gulf in the Zagros Mountains of Iran, for example, was a dispersed terrane that separated from the Eurasian plate. However, the northward push of the Arabian plate is now shoving this terrane back into Eurasia and forming an active plate margin known as the Zagros crush zone, a continuing

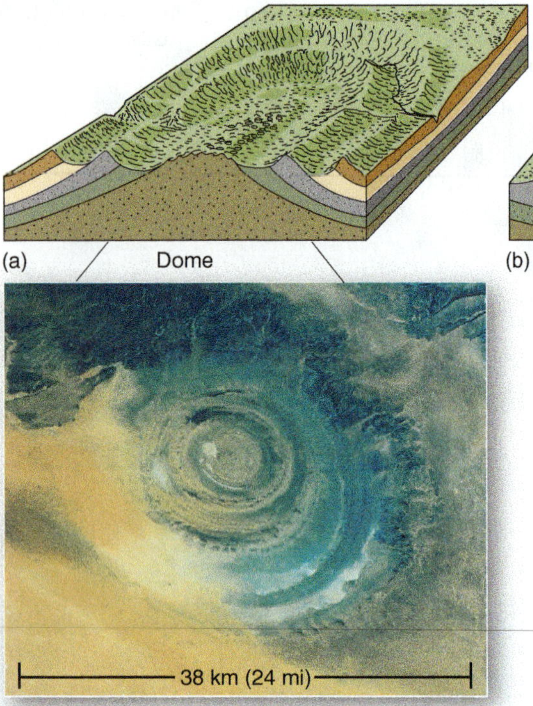

(c) The Richat dome, Mauritania; notice the sand dunes moving toward the dome from the southeast.

Figure 12.9 Domes and basins. [(c) *Terra* ASTER image, NASA/GSFC/MITI.]

collision more than 400 km (250 mi) wide. In the satellite image of this zone in Figure 12.10, anticlines form the parallel ridges; active weathering and erosion processes are exposing the underlying strata.

In addition to the types of folding discussed above, broad warping actions are another cause of bending in continental crust. The bends produced by these actions, however, are far greater in extent than the folds produced by compression. Forces responsible for such large-scale warping include mantle convection, isostatic adjustment (such as that caused by the weight of previous ice loads across northern Canada), and crustal swelling above an underlying hot spot.

Faulting

A freshly poured concrete sidewalk is smooth and strong. Adding stress to the sidewalk by driving heavy equipment over it may result in strain that causes a fracture. Pieces on either side of the fracture may move up, down, or horizontally depending on the direction of stress. Similarly, when rock strata are stressed beyond their ability to remain a solid unit, they express the strain as a fracture. **Faulting** occurs when rocks on either side of the fracture shift relative to the other side. *Fault zones* are areas where fractures in the rock demonstrate crustal movement.

Types of Faults The fracture surface along which the two sides of a fault move is the *fault plane*; the tilt and orientation of this plane is the basis for differentiating the three main types of faults introduced in Figure 12.7: normal, reverse, and strike-slip, caused, respectively, by tensional stress, by compressional stress, and by lateral-shearing stress.

When forces pull rocks apart, the tensional stress causes a **normal fault**, in which rock on one side moves vertically along an inclined fault plane (Figure 12.11a). The downward-shifting side is the *hanging wall;* it drops relative to the *footwall block*. An exposed fault plane sometimes is visible along the base of faulted mountains, where individual ridges are truncated by the movements of the fault and end in triangular facets. A displacement of the ground surface caused by faulting is commonly called a *fault scarp*, or *escarpment*.

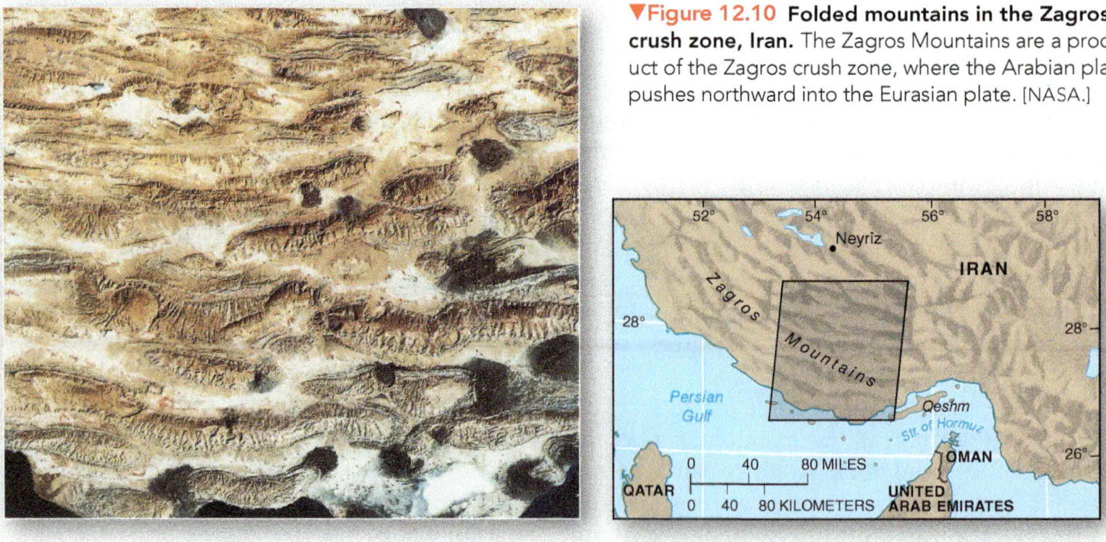

Figure 12.10 Folded mountains in the Zagros crush zone, Iran. The Zagros Mountains are a product of the Zagros crush zone, where the Arabian plate pushes northward into the Eurasian plate. [NASA.]

When forces push rocks together, such as when plates converge, the compression causes a **reverse fault**, in which rocks move upward along the fault plane (Figure 12.11b). On the surface, it appears similar to a normal fault, although more collapse and landslides may occur from the hanging-wall component. In England, when miners worked along a reverse fault, they would stand on the lower side (footwall) and hang their lanterns on the upper side (hanging wall), giving rise to these terms.

A **thrust fault**, or *overthrust fault*, occurs when the fault plane forms a low angle relative to the horizontal, so that the overlying block has shifted far over the underlying

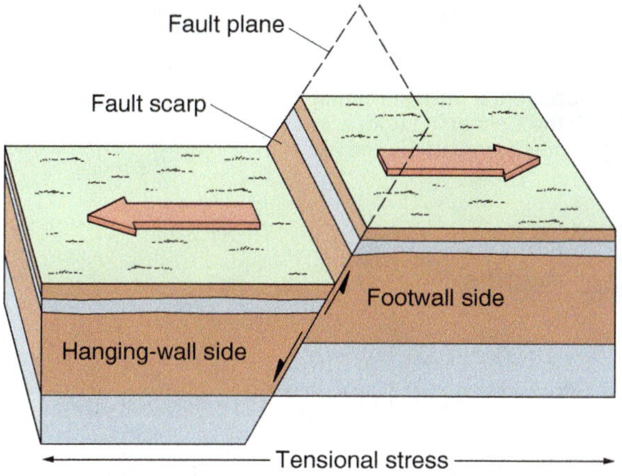

(a) Normal fault (tension)

(a) A normal fault visible along the edge of mountain ranges in California and Utah.

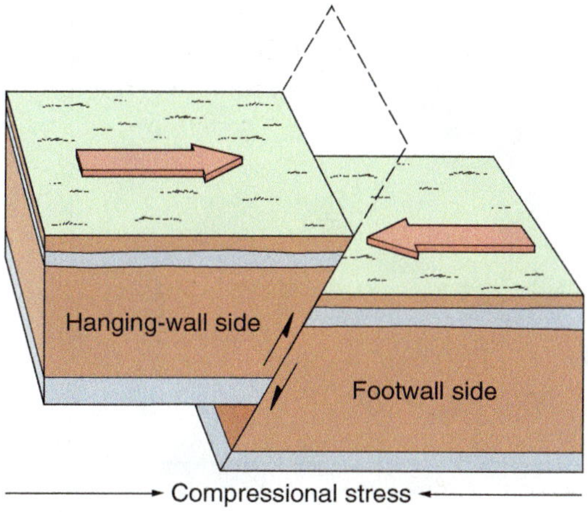

(b) Thrust or reverse fault (compression)

(b) A thrust, or reverse, fault visible in these offset strata in coal seams and volcanic ash in British Columbia.

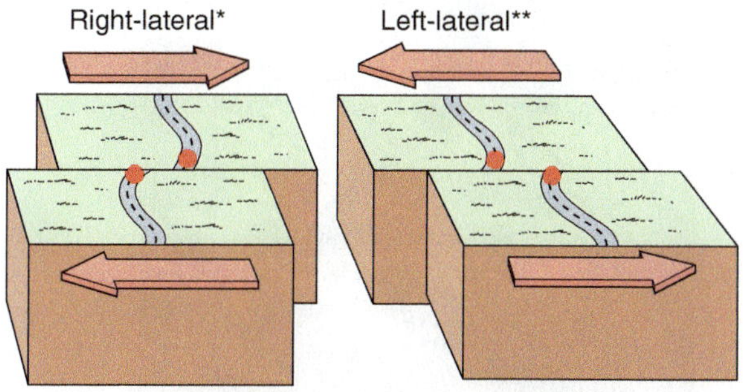

(c) Strike-slip fault (lateral shearing)

* Viewed from either dot on each road, movement of opposite side is *to the right*.
** Viewed from either dot on each road, movement of opposite side is *to the left*.

(c) Aerial view of a right-lateral strike-slip fault in southern Nevada.

▲**Figure 12.11 Types of faults.** [Photographs by (a) Bobbé Christopherson. (b) Fletcher and Baylis/Photo Researchers, Inc. (c) Marli Miller.]

MG™
Animation
Fault Types, Transform Faults, Plate Margins

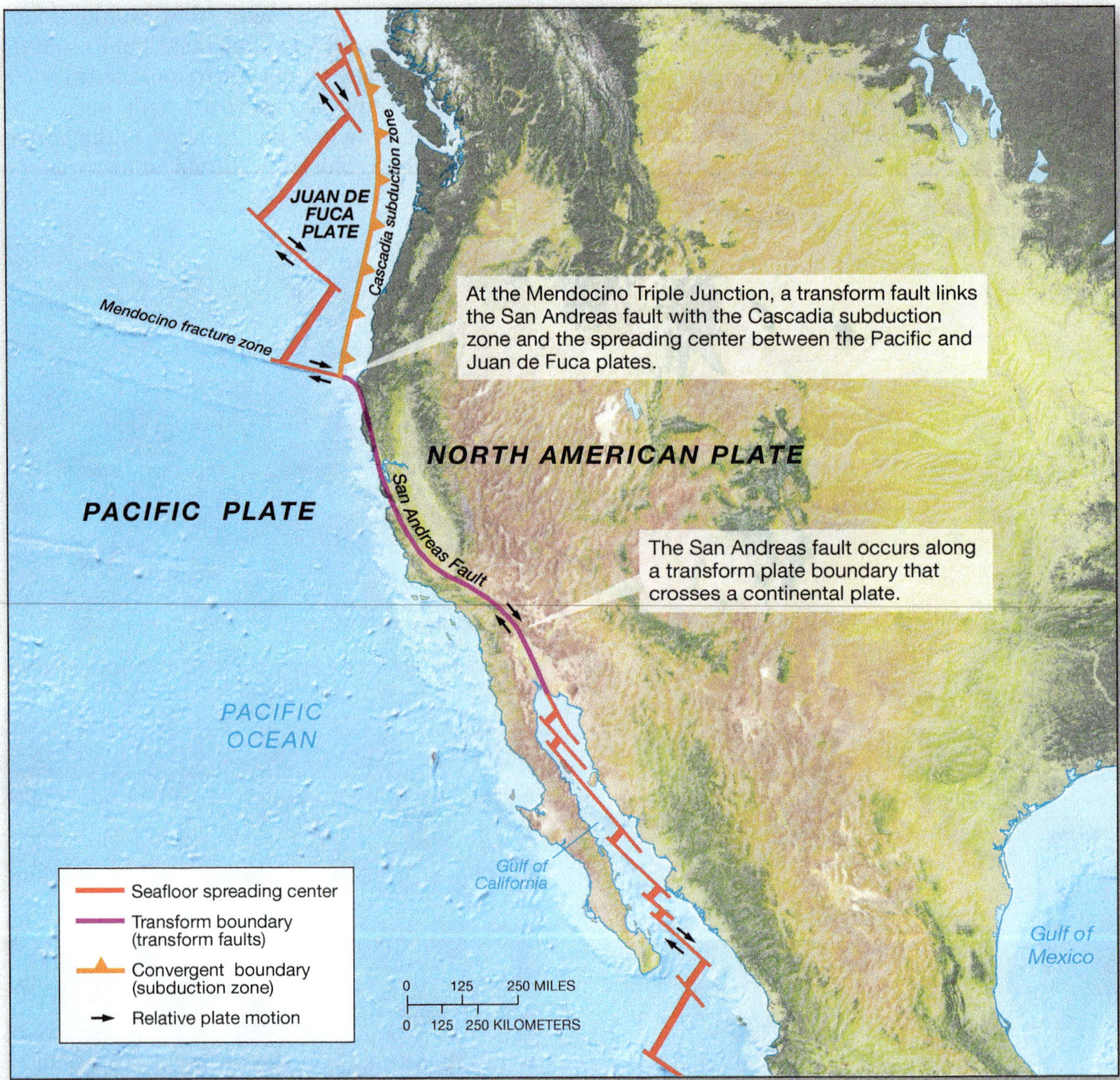

(a) The western margin of North America is the meeting point of three plates with different types of boundaries. Between the Juan de Fuca and Pacific plates is a spreading center with transform faults linking mid-ocean ridges. Between the Juan de Fuca and North American plates is the Cascadia subduction zone. The San Andreas transform fault separates the Pacific and North American plates. The meeting point of three plates is a "triple junction."

(b) The San Andreas fault, San Luis Obispo County, California.

▲**Figure 12.12 Plate boundaries of western North America and the San Andreas fault.** [(a, b) Lloyd Cluff/Corbis]

block (see Figure 12.8). Place your hands palms-down on your desk, with fingertips together, and slide one hand up over the other—this is the motion of a low-angle thrust fault, with one side pushing over the other.

In the Alps, several such overthrusts have resulted from compressional forces of the ongoing collision between the African and Eurasian plates. Beneath the Los Angeles Basin, overthrust faults caused numerous earthquakes in the twentieth century, including the $30 billion 1994 Northridge earthquake. Many of the faults beneath the Los Angeles region are "blind thrust faults," meaning that no evidence of rupture exists at the surface. Such faults are essentially buried under the crust but remain a major earthquake threat.

When lateral shear causes horizontal movement along a fault plane, such as produced along a transform plate boundary and the associated transform faults, the fault is called a **strike-slip fault** (Figure 12.11c). The movement is right lateral or left lateral, depending on the direction of motion an observer on one side of the fault sees occurring on the other side.

The San Andreas fault system is a famous example of both a transform and strike-slip fault. Recall from Chapter 11 that transform faults occur along transform plate boundaries; most are found in the ocean basins along mid-ocean ridges, but some cross continental crust. The crust on either side of a transform fault moves parallel to the fault itself; thus, transform faults are a type of strike-slip fault. However, since some strike-slip faults occur at locations away from plate boundaries, not all strike-slip faults are transform.

Along the western margin of North America, plate boundaries take several forms. The San Andreas transform fault is the product of a continental plate (the North American plate) overriding an oceanic transform plate boundary. The San Andreas fault meets a seafloor spreading center and a subduction zone at the Mendocino Triple Junction (a *triple junction* is where three plates intersect). Each of these plate boundaries is illustrated in Figure 12.12a.

Strike-slip faults often create linear valleys, or troughs, along the fracture zone, such as those along the San Andreas (Figure 12.12b). Offset streams are another landform associated with strike-slip faults, characterized by an abrupt bend in the course of the stream as it crosses the fault.

Faulted Landscapes Across some landscapes, pairs of faults act in concert to form distinctive terrain. The term **horst** applies to upward-faulted blocks; **graben** refers to downward-faulted blocks (Figure 12.13). The Great Rift Valley of East Africa, associated with crustal spreading, is an example of such a horst-and-graben landscape. This rift extends northward to the Red Sea, which fills the rift formed by parallel normal faults. Lake Baikal in Siberia, discussed in the Chapter 9 *Geosystems Now*, is also a graben. This lake basin is the deepest continental "rift valley" on Earth and continues to widen at an average rate of about 2.5 cm (1 in.) per year. Another example, among many, is the Rhine graben, through which the Rhine River flows in Europe. Go to the *MasteringGeography* website for photos and images of these landscapes.

A large region that is identified by several geologic or topographic traits is known as a *physiographic province*. In the U.S. interior west, the **Basin and Range Province** is a physiographic province recognized for its north-and-south-trending basins and mountains—the basins are low-elevation areas that dip downward toward the center; the ranges are interconnected mountains of varying elevations above the basins. These roughly parallel mountains and valleys (known as *basin-and-range topography*) are aligned pairs of normal faults, an example of a horst-and-graben landscape (Figure 12.14a and b).

The driving force for the formation of this landscape is the westward movement of the North American plate; the faulting results from tensional forces caused by the uplifting and thinning of the crust. Basin-and-range relief is abrupt, and its rock structures are angular and rugged. As the ranges erode, transported materials accumulate to great depths in the basins, gradually producing extensive plains. Basin elevations average roughly 1200–1500 m (4000–5000 ft) above sea level, with mountain crests rising higher by another 900–1500 m (3000–5000 ft).

Death Valley, California, is the lowest of these basins, with an elevation of –86 m (–282 ft). Directly to the west, the Panamint Range rises to 3368 m (11,050 ft) at Telescope Peak, producing almost 3.5 vertical kilometers (2.2 mi) of relief from the desert valley to the mountain peak.

Several other landforms are associated with basin-and-range topography. The area composed of slopes and basin between the crests of two adjacent ridges in a dry region of internal drainage is a **bolson** (Figure 12.14b). A *playa* is a dry lakebed characterized by an area of salt crust left behind by evaporation of water in a bolson or valley (Figure 12.14c and d). Chapter 14 discusses other landforms associated with running water in this dry region.

Orogenesis (Mountain Building)

The geologic term for mountain building is **orogenesis**, literally meaning the birth of mountains (*oros* comes from the Greek for "mountain"). An *orogeny* is a mountain-building episode, occurring over millions of years, usually caused by large-scale deformation and uplift of the crust. An orogeny may begin with the capture of migrating exotic terranes and their accretion to the continental margins, or with the intrusion of granitic magmas to form plutons. The net result of this accumulating material is a thickening of the crust. The next event in the orogenic cycle is uplift, which is followed by the work of weathering and erosion, exposing granite plutons and creating rugged mountain topography.

The locations of Earth's major chains of folded and faulted mountains, called *orogens*, are remarkably well correlated with the plate tectonics model. Two major examples are the Cordilleran mountain system in North and South America and the Eurasian–Himalayan system, which stretches from the Alps across Asia to the Himalayas. Go to the *Mastering Geography* website to view a map of world structural regions, including these mountain systems.

No orogeny is a simple event; many involve previous developmental stages dating back far into Earth's past, and the processes are ongoing today. Major mountain ranges, and the latest related orogenies that caused them, include the following (review Figure 12.1 for orogeny dates within the context of the geologic time scale):

- Appalachian Mountains and the folded Ridge and Valley Province of the eastern United States and Canadian

Figure 12.13 Faulted landscapes. [(b) Airphoto-Jim Wark.]

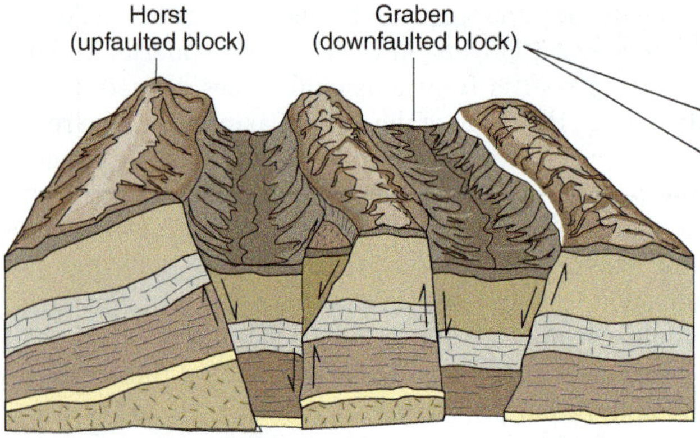

(a) Pairs of faults produce a horst-and-graben landscape.

(b) Horsts and grabens result from normal faults in Canyonlands National Park, Utah.

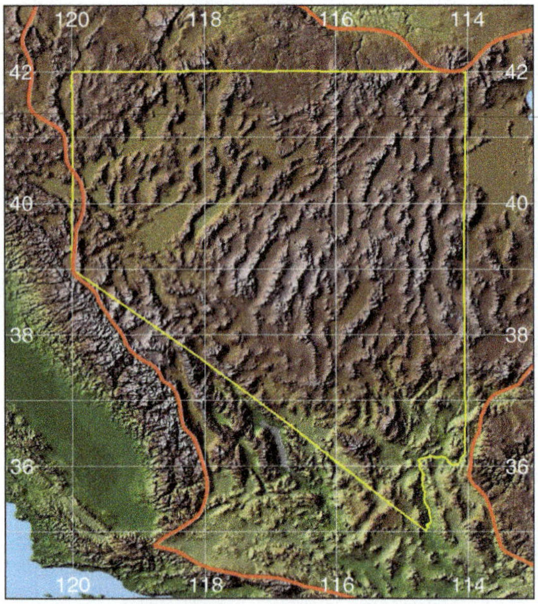

(a) DEM shaded relief map with the Basin and Range Province within the red outline.

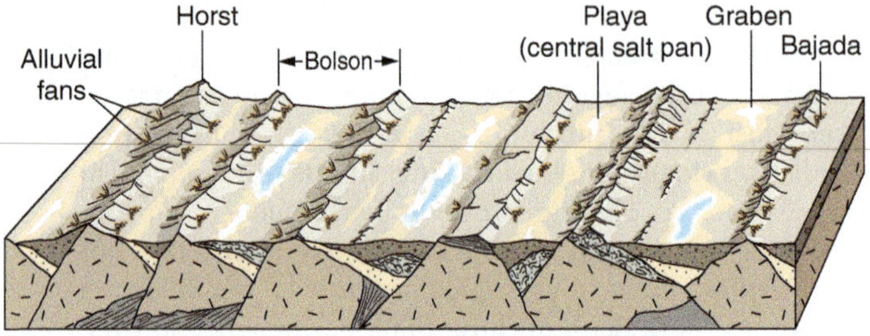

(b) Parallel faults produce a series of ranges and basins. Infrequent precipitation events leave alluvial fan deposits at the mouths of canyons; coalesced alluvial fans join to form a bajada.

(c) A playa in Death Valley, California

(d) Parallel mountain ranges, bolsons, and playas in Nevada.

Figure 12.14 Landforms of the Basin and Range Province. [(a) GSFC/NASA. (c) Author. (d) Bobbé Christopherson.]

Maritime Provinces—formed during the *Alleghany orogeny*, 250–300 million years ago (m.y.a.), associated with the collision of Africa and North America

- Rocky Mountains of North America—formed mainly during the *Laramide orogeny*, 40–80 m.y.a., but also during several earlier orogenies, beginning 170 m.y.a. (including the *Sevier orogeny*)
- Sierra Nevada and Klamath Mountains of California—formed during the *Nevadan orogeny*, with faulting 29–35 m.y.a. (older batholithic intrusions date to 80–180 million m.y.a.)
- Alps of Europe—formed during the *Alpine orogeny*, 2–66 m.y.a., principally in the Tertiary Period and continuing to the present across southern Europe and the Mediterranean, with many earlier episodes (see Figure 12.15)
- Himalayas of Asia—Himalayan orogeny, 45–54 m.y.a., beginning with the collision of the India and Eurasia plates and continuing to the present.

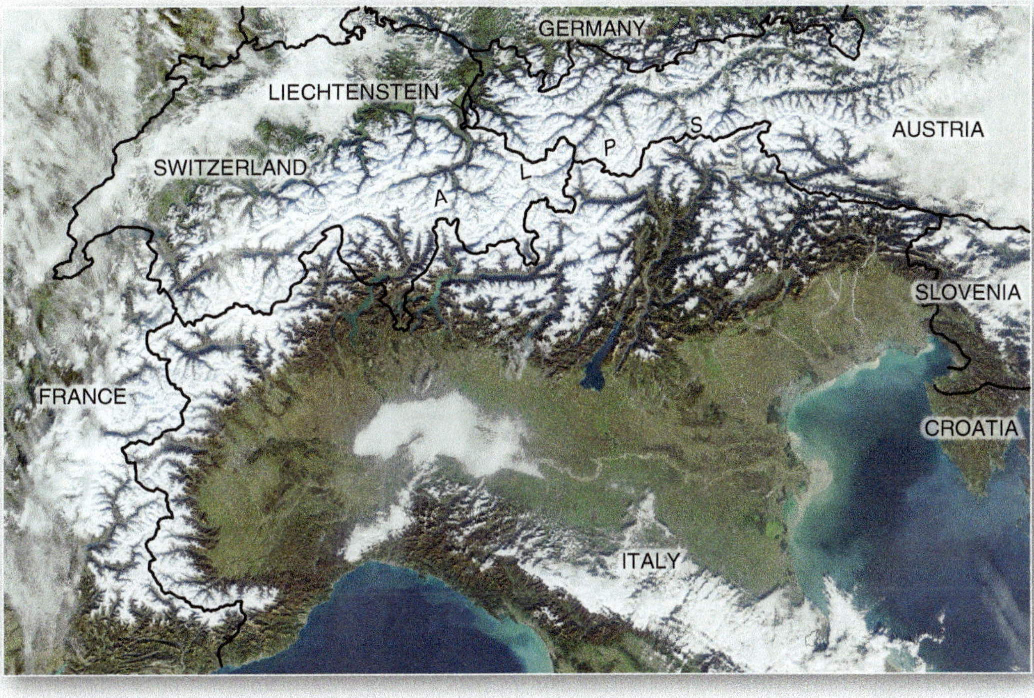

▲Figure 12.15 **European Alps.** The Alps are some 1200 km (750 mi) in length, occupying a crescent of 207,000 km² (80,000 mi²) with western (France), central (Italy), and eastern (Austria) segments. Note the snow coverage in this December image. [*Terra* MODIS image, NASA/GSFC.]

Types of Orogenesis

Three types of tectonic activity cause mountain building along convergent plate margins (illustrated in *Geosystems In Action*, Figure GIA 12.2a). As discussed in Chapter 11, an *oceanic plate–continental plate collision* produces a subduction zone as the denser oceanic plate dives beneath the continental plate. This convergence creates magma below Earth's surface that is forced upward to become magma intrusions, resulting in granitic plutons and sometimes volcanic activity at the surface. Compressional forces cause the crust to uplift and buckle. This type of convergence is now occurring along the Cordilleran mountain system that follows the Pacific coast of the Americas and has formed the Andes, the Sierra Madre of Central America, and the Rockies (Figure GIA 12.1).

An *oceanic plate–oceanic plate collision* can produce curving belts of mountains called *island arcs* that rise from the ocean floor. When the plates collide, one is forced beneath the other, creating an oceanic trench. Magma forms at depth and rises upward, erupting as it reaches the ocean bottom and beginning the construction of a volcanic island. As the process continues along the trench, the eruptions and accumulation of volcanic material form a volcanic island arc (Figure GIA 12.2b). These processes formed the chains of island arcs and volcanoes that range from the southwestern Pacific into the western Pacific, the Philippines, the Kurils, and on through portions of the Aleutians. Some of the arcs are complex, such as Indonesia and Japan, which exhibit surface rock deformation and metamorphism of rocks and granitic intrusions.

These two types of plate collisions are active around the Pacific Rim, and each is part thermal in nature because the diving plate melts and migrates back toward the surface as molten rock. The region of active volcanoes and earthquakes around the Pacific is known as the **circum-Pacific belt** or, more popularly, the **Ring of Fire**.

The third type of orogenesis occurs during a *continental plate–continental plate collision*. This process is mainly mechanical, as large masses of continental crust are subjected to intense folding, overthrusting, faulting, and uplifting (Figure 12.2c). The converging plates crush and deform both marine sediments and basaltic oceanic crust. The European Alps are a result of such compression forces and exhibit considerable crustal shortening in conjunction with great overturned folds, called *nappes*.

The collision of India with the Eurasian landmass, producing the Himalayas, is estimated to have shortened the overall continental crust by as much as 1000 km (about 600 mi) and to have produced telescoping sequences of thrust faults at depths of 40 km (25 mi). The Himalayas feature the tallest above-sea-level mountains on Earth, including all 10 of Earth's highest peaks.

The Tetons and the Sierra Nevada

Mountain landscapes can be altered when a normal fault along one side of a range produces a tilted linear landscape with dramatic relief known as a *tilted-fault-block* mountain range. The Tetons of Wyoming and the Sierra Nevada of California are recent examples of this stage of mountain building. In both these ranges, mountain building first began hundreds of millions of years ago with magma intrusions that cooled to form granitic cores of coarsely crystalline rock. Tectonic uplift followed, with

(text continued on page 374)

Orogenesis, or mountain building, is the result of plate interactions and related processes that thicken and uplift the crust, such as folding, faulting, and volcanism. Combined with weathering, erosion, and isostatic adjustment, these processes produce the striking landscapes of Earth's mountain ranges (GIA 12.1). Collisions of Earth's plates produce three distinct kinds of orogenesis (GIA 12.2).

12.1 MAJOR MOUNTAIN RANGES AND OROGENIES

The mountain ranges we see today have roots deep in geologic time—some, such as the Appalachians, have repeatedly been formed, eroded away, and uplifted again as Earth's plates interacted over hundreds of millions of years.

Mount Katahdin, Maine [WIN-Initiative/Getty Images.]

Appalachian Mountains: *Formed during the* ***Alleghany orogeny****, 250–300 million years ago (m.y.a.), when Africa and North America collided. Includes folded Ridge and Valley Province of the eastern United States and extends into Canada's Maritime Provinces.*

French Alps [Calin Tatu/Shutterstock.]

Alps: *Formed during the* ***Alpine orogeny****, 2–66 m.y.a., and continuing to the present across southern Europe and the Mediterranean, with many earlier episodes (see Figure 12.15)*

Three Sisters, Canadian Rockies, Alberta [Bradley L. Grant/Getty Images.]

Rocky Mountains: *Formed mainly during the* ***Laramide orogeny****, 40–80 m.y.a., but also during several earlier orogenies, beginning 170 m.y.a. (including the* ***Sevier orogeny****)*

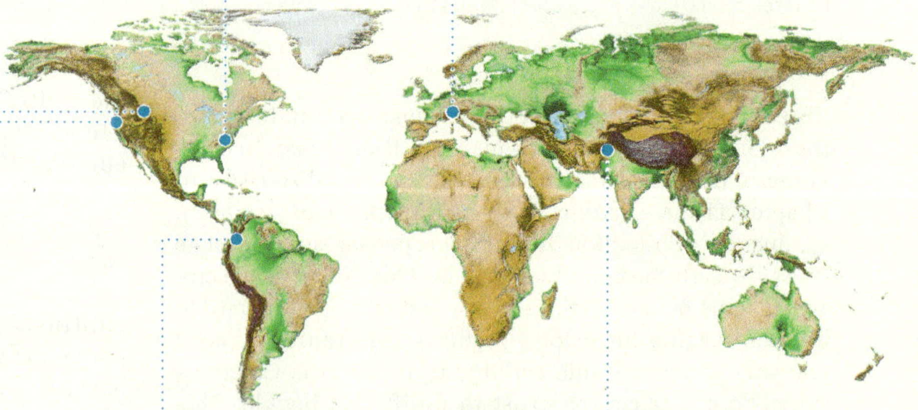

Klamath Mountains, Oregon [Spring Images/Alamy.]

Sierra Nevada and Klamath Mountains: *Formed during the* ***Nevadan orogeny****, with faulting 29–35 m.y.a. (older batholithic intrusions date to 80 to 180 million m.y.a.)*

Chimborazo, Ecuador [Michael Mellinger/Getty Images.]

Andes: *Formed during the* ***Andean orogeny*** *over the past 65 million years, the Andes are the South American segment of a vast north–south belt of mountains running along the western margin of the Americas from Tierra del Fuego to Alaska.*

Himalayas, Pakistan [Microstock Man/Shutterstock.]

Himalayas: *Formed during the* ***Himalayan orogeny****, 45–54 m.y.a., beginning with the collision of the Indian and Eurasian plates and continuing to the present.*

Identify: What geologic event triggered the Alleghany orogeny, which formed the Appalachians?

Animation Subduction Zones

MasteringGeography™

Visit the Study Area in MasteringGeography™ to explore mountain building.

Visualize: Study geosciences animations of subduction zones and plate boundaries.

Assess: Demonstrate understanding of mountain building (if assigned by instructor).

Animation Plate Boundaries

12.2 MOUNTAIN BUILDING AT CONVERGENT BOUNDARIES

Three different types of lithospheric plate collisions result in mountain building:

- oceanic plate–continental plate
- oceanic plate–oceanic plate
- continental plate–continental plate

Each plate interaction leads to a different kind of orogenesis.

Oceanic plate–continental plate

Where a dense oceanic plate collides with a less-dense continental plate, a subduction zone forms and the oceanic plate is subducted. Magma forms above the descending plate. Where the magma erupts to the surface through the continental plate, volcanic mountains form. Magma may also harden beneath the surface, forming batholiths.
Example: *The Andes of South America formed as a result of the subduction of the oceanic Nazca plate beneath the continental South American plate.*

Andes, Peru: Ancient Incan city of Machu Picchu [vdLee/Shutterstock.]

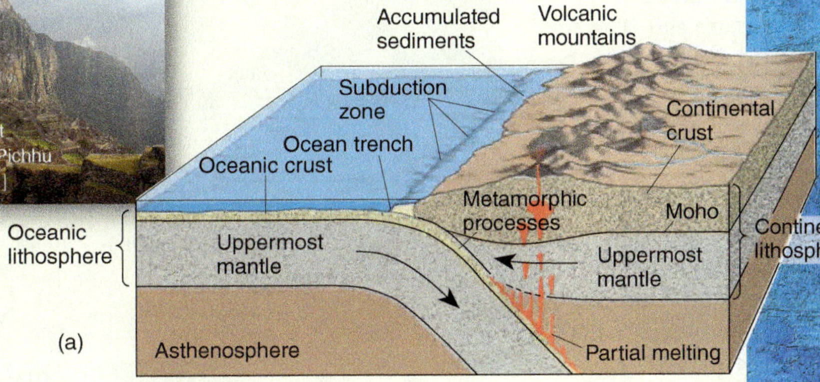

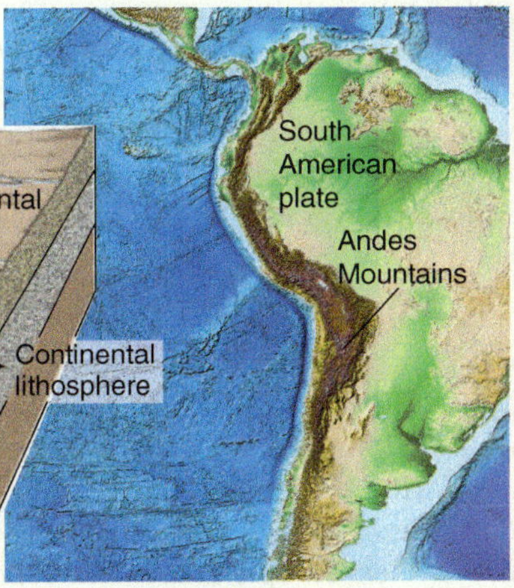

Oceanic plate–oceanic plate

Where two oceanic plates collide, one plate is subducted beneath the other. Magma forms above the descending plate, giving rise to a volcanic island arc.
Example: *As part of the "Ring of Fire," there are many volcanic island arcs where the Pacific plate interacts with other oceanic plates. These arcs extend from the southwestern Pacific (shown here) through Indonesia, the Philippines, and Japan to the Aleutians.*

Mayon volcano, Philippines [suronin/Shutterstock.]

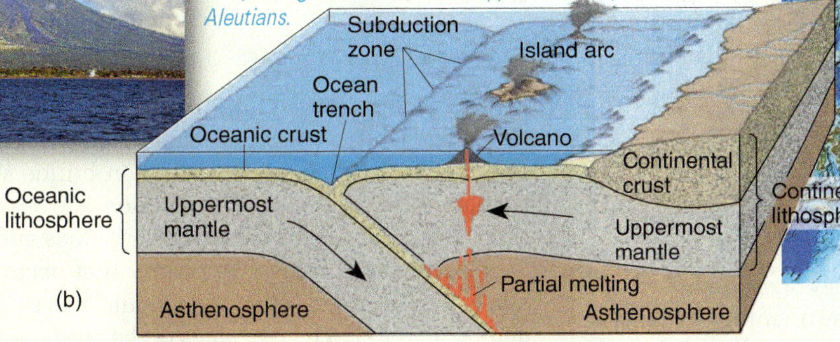

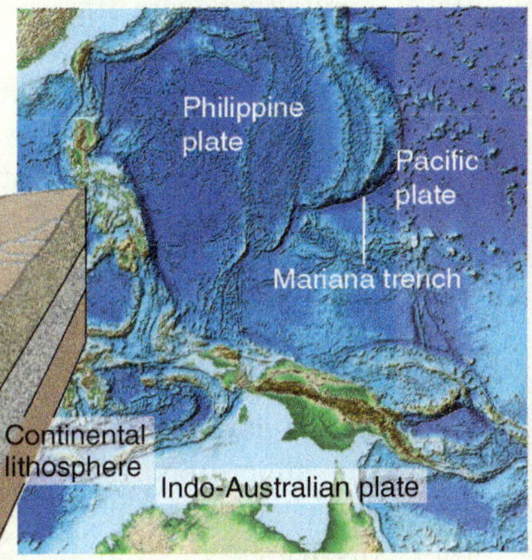

Himalayas, Nepal [Purepix/Alamy.]

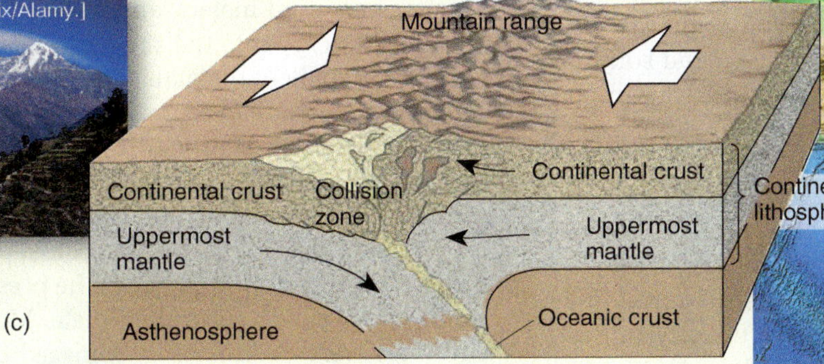

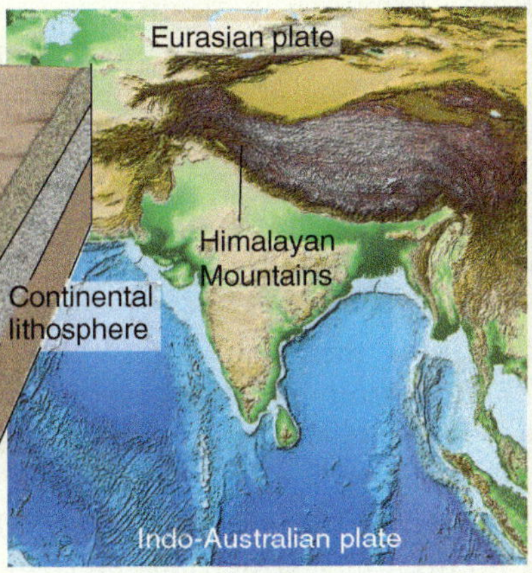

Continental plate–continental plate

Where two continental plates collide, neither plate is subducted. Instead, the collision subjects the plates to powerful compression forces that fold, fault, and uplift the crust, pushing up huge mountain ranges.
Example: *The Himalayas formed as a result of the ongoing collision of the Indian and Eurasian plates. The Himalayas are part of a long east–west mountain belt, stretching from Europe across Asia, formed by similar collisional processes.*

Infer: Why doesn't subduction occur when two continental plates collide?

GEOquiz

1. **Predict:** Will the Himalayas keep growing higher and higher indefinitely? Explain your answer.

2. **Compare:** In terms of orogenesis and plate tectonics, how are the Alps and the Andes similar? How are they different?

subsequent faulting and tilting occurring less than 10 m.y.a. The removal of overlying material through weathering, erosion, and transport then exposed the granitic masses, leaving the rugged topography and steep vertical relief we see today (Figure 12.16).

Recent research in the Sierra Nevada disclosed that some of the uplift in this range was isostatic, in response to the erosion of overlying material and the melting of ice following the last ice age some 18,000 years ago. The accumulation of sediments in the adjoining valley to the west depressed the crust there, further enhancing the topographic relief.

The Appalachian Mountains

The origin of the Appalachian Mountains of the eastern United States and southeastern Canada dates to the formation of Pangaea, and the collision of Africa with North America (250–300 m.y.a.). The complexity of the *Alleghany orogeny* derives from at least two earlier orogenic cycles of uplift and the accretion of several terranes.

Mountain ranges that were linked during the time of Pangaea but that are now separated by the Atlantic Ocean are similar in both structure and composition. Such similarities indicate that the Lesser (or Anti-) Atlas Mountains of Mauritania and northwestern Africa were connected to the Appalachians in the past (Figure 12.13). Subsequent plate movement created active plate boundaries and several orogenic periods that gave rise to the mountains of today, stretching from Newfoundland, Canada, to central Alabama in the United States.

The central Appalachian Mountain region includes several landscape subregions. The Appalachian Plateau on the western edge is a plateau of eroded sedimentary rock; the Ridge and Valley Province consists of elongated sequences of folded sedimentary strata; the Blue Ridge Province is an area principally of crystalline rock, highest where North Carolina, Virginia, and Tennessee converge; the Piedmont is a region of hilly to gentle terrain along most of the eastern and southern margins of the mountains; and the coastal plain consists of gentle hills descending to flat plains that extend to the Atlantic coast (Figure 12.17).

In the Ridge and Valley Province, folded anticlines and synclines form prominent linear ridges. Rivers such as the Susquehanna in east-central Pennsylvania have carved *water gaps* where the watercourse flows through a mountain ridge. Water gaps usually indicate that the river is older than the mountain ridge through which it cuts, suggesting that the river had established its position prior to the uplift of the landscape and thus was able to erode through the uplift and maintain its course.

Water gaps in the rugged Appalachians were important breaks in topography that greatly influenced human migration, settlement patterns, and the diffusion of cultural traits in the 1700s. The initial flow of people, goods, and ideas toward the west was guided by these geomorphic features.

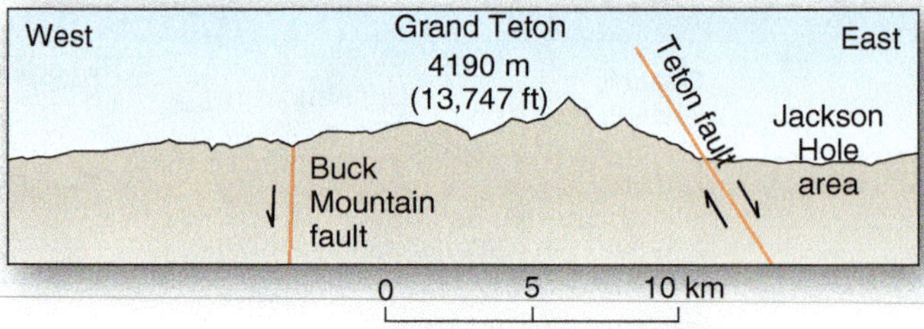

▲**Figure 12.16 Tilted fault block.** The Tetons in Wyoming are an example of a tilted-fault-block mountain range, featuring 2130 m (7000 ft) of relief between the valley of Jackson Hole and the mountain summits. Erosion of the upthrown footwall side of the Teton fault (a normal fault) causes the rugged topography and vertical relief. [Author.]

Earthquakes

Crustal plates do not glide smoothly past one another. Instead, tremendous friction exists along plate boundaries. The stress, or force, of plate motion builds strain, or deformation, in the rocks until friction is overcome and the sides along plate boundaries or fault lines suddenly break loose. The sharp release of energy that occurs at the moment of fracture, producing seismic waves, is an **earthquake**, or *quake*. The two sides of the fault plane then lurch into new positions, moving distances ranging from centimeters to several meters, and release enormous amounts of seismic energy into the surrounding crust. This energy radiates throughout the planet, diminishing with distance.

The plate collision in which India is moving northeastward at about 2 m (6.6 ft) a year creates disruption that has reached far under China, causing frequent earthquakes. As evidence of this ongoing strain, the January 2001 quake in Gujarat, India, occurred along a shallow thrust fault that gave way under the pressure of the northward-pushing Indo-Australian plate. More than 1 million buildings were destroyed or damaged during this event. In October 2005, a 40-km (25-mi) segment of the 2500-km (1553-mi) fault zone that marks the plate boundary snapped, hitting the Kashmir region of Pakistan with an M 7.6 quake that killed more than 83,000 people.

Tectonic earthquakes are those quakes associated with faulting. Earthquakes can also occur in association with volcanic activity. Recent research suggests that the injection of wastewater from oil and gas drilling into subsurface areas is another cause, leading to episodes of *induced seismicity* (discussed with fault mechanics ahead).

(a) Landscape subregions of the central Appalachian Mountains. The folded Ridge and Valley Province extends from Pennsylvania south through Maryland, Virginia, and West Virginia.

(b) Astronaut photograph of the central Appalachians from Pennsylvania to West Virginia.

(c) Fall colors highlight the ridges of the Susquehanna River Valley, Pennsylvania.

▲**Figure 12.17 The folded Ridge and Valley Province of the Appalachian Mountains.** [(a) USGS. (b) *ISS* Crew Earth Observations Experiment and Image Science & Analysis Laboratory, Johnson Space Center, NASA. (c) *Landsat-7*, NASA.]

Earthquake Anatomy

The subsurface area where the motion of seismic waves is initiated along a fault plane is the *focus*, or hypocenter, of an earthquake (see example in Figure 12.18a). The area at the surface directly above the focus is the *epicenter*. Shock waves produced by an earthquake radiate outward through the crust from the focus and epicenter. As discussed earlier, scientists use the seismic wave patterns and the nature of their transmission to learn about the deep layers of Earth's interior.

A *foreshock* is a quake that precedes the main shock. The pattern of foreshocks is now regarded as an important consideration in earthquake forecasting. An *aftershock* occurs after the main shock, sharing the same general area of the epicenter; some aftershocks rival the main tremor in magnitude. For instance, on the South Island of New Zealand an M 7.1 quake struck in September 2010 with an epicenter 45 km (30 mi) from Christchurch, causing USD $2.7 billion in damage and no deaths. Seventeen days later, a 6.3 aftershock with an epicenter just 6 km (3 mi) from the city produced building collapses, 350 deaths, and more than USD $15 billion in damage. These events show that the distance from the epicenter is an important factor in determining overall earthquake effects on population centers.

In 1989, the Loma Prieta earthquake hit south of San Francisco, California, east of Santa Cruz. Damage totaled $8 billion, 14,000 people were displaced from their homes, 4000 were injured, and 63 were killed. Unlike previous earthquakes—such as the San Francisco quake of 1906, when the plates shifted a maximum of 6.4 m (21 ft) relative to each other—no fault plane or rifting was evident at the surface in the Loma Prieta quake. Instead, the Pacific and North American plates moved horizontally approximately 2 m (6 ft) past each other deep below the surface, with the Pacific plate thrusting 1.3 m (4.3 ft) upward (Figure 12.18). This vertical motion is unusual for the San Andreas fault and indicates that this portion of the San Andreas system is more complex than previously thought.

Earthquake Intensity and Magnitude

A **seismometer** (also called a seismograph) is an instrument used to detect and record the ground motion that occurs during an earthquake. This instrument records motion in only a single direction, so scientists use a combination of vertical-motion and horizontal-motion

GEOreport 12.2 Ongoing earthquake activity in Sumatra, Indonesia

In 2009, an M 7.6 earthquake caused over 1000 fatalities in southern Sumatra, Indonesia, near the same plate boundary where six quakes greater than M 7.9 had already occurred since 1998. In 2010, two more quakes hit in the region just to the north. At this boundary, the Indo-Australian plate moves northeast, subducting beneath the Sunda plate at a relative speed of 6.6 cm/year (2.6 in./year). This is where the M 9.1 Sumatra–Andaman earthquake struck on December 26, 2004, triggering the devastating Indian Ocean tsunami (more on tsunami in Chapter 16); the quake and seismic sea wave took 228,000 lives. For complete listings of these and other earthquakes, go to http://earthquake.usgs.gov/earthquakes.

seismometers to determine the source and strength of seismic waves. The instrument detects body waves (traveling through Earth's interior) first, followed by surface waves, recording both on a graphic plot called a *seismogram*. Scientists use a worldwide network of more than 4000 seismometers to sense earthquakes, and these quakes are then classified on the basis of either damage intensity or the magnitude of energy released.

Before the invention of modern earthquake instrumentation, damage to terrain and structures and severity of shaking were used to assess the size of earthquakes. These surface effects are a measure of earthquake *intensity*. The *Modified Mercalli Intensity (MMI) scale* is a Roman-numeral scale from I to XII that ranges from earthquakes that are "barely felt" (lower numbers) to those that cause "catastrophic total destruction" (higher numbers; see Table 12.1). The scale was designed in 1902 and modified in 1931.

Earthquake *magnitude* is a measure of the energy released and provides a way to compare earthquake size. In 1935, Charles Richter designed a system to estimate earthquake magnitude based on measurement of maximum wave amplitude on a seismometer. *Amplitude* is the height of a seismic wave and is directly related to the amount of ground movement. The size and timing of maximum seismic wave height can be plotted on a chart called the **Richter scale**, which then provides a number for earthquake magnitude in relation to a station located 100 km (62 mi)

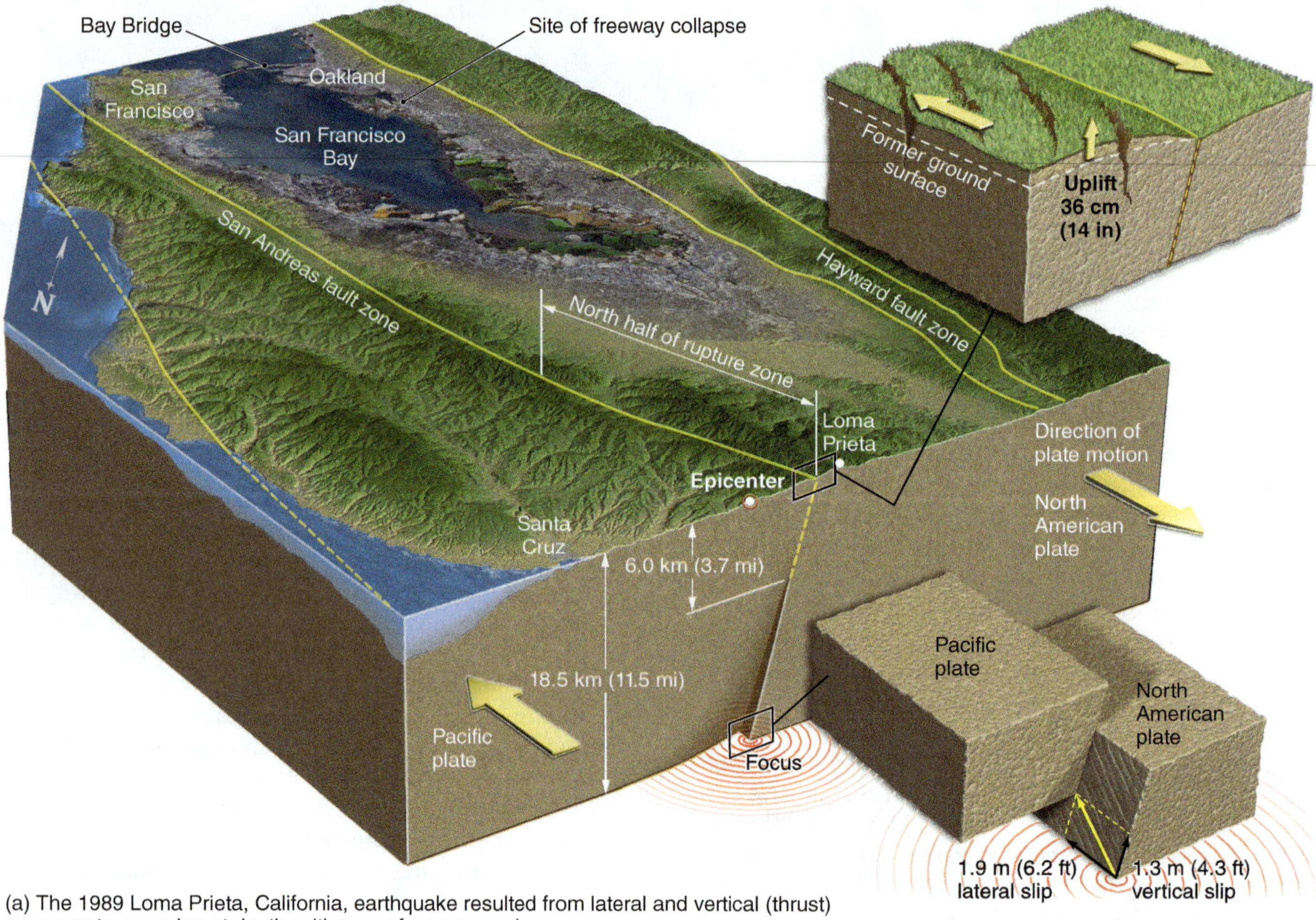

(a) The 1989 Loma Prieta, California, earthquake resulted from lateral and vertical (thrust) movements occurring at depth, with no surface expression.

(b) More than 2 km (1.2 mi) of the Route 880 Cypress Freeway collapsed, in part due to liquefaction of the soft muds on which the freeway was built.

▲**Figure 12.18 Anatomy of an earthquake.** [(a) USGS. (b) Courtesy of California Department of Transportation.]

MG™ Animation Seismograph, How It Works

MG™ Animation P- and S- Waves, Seismology

TABLE 12.1 Intensity, Magnitude, and Expected Frequency of Earthquakes

Description	Effects on Populated Areas	Modified Mercalli Scale	Moment Magnitude Scale	Number Expected per Year*
Great	Damage nearly total	XII	8.0 and higher	1
Major	Great damage	X–XI	7–7.9	17
Strong	Considerable-to-serious damage to buildings; railroad tracks bent	VIII–IX	6–6.9	134
Moderate	Felt-by-all, with slight building damage	V–VII	5–5.9	1319
Light	Felt-by-some to felt-by-many	III–IV	4–4.9	13,000 (estimated)
Minor	Slight, some feel it	I–II	3–3.9	130,000 (estimated)
Very minor	Not felt, but recorded	None to I	2–2.9	1,300,000 (estimated)

Based on observations since 1990.
Source: USGS Earthquake Information Center.

from the epicenter of the quake. (For more on the Richter scale, see http://earthquake.usgs.gov/learn/topics/richter.php.)

The Richter scale is logarithmic: Each whole number on it represents a 10-fold increase in the measured wave amplitude. Translated into energy, each whole number signifies a 31.5-fold increase in energy released. Thus, a magnitude of 3.0 on the Richter scale represents 31.5 times more energy than a 2.0 and 992 times more energy than a 1.0. Although useful for measuring shallow earthquakes that are in close proximity to seismic stations, the Richter scale does not properly measure or differentiate between quakes of high intensity.

The **moment magnitude (M) scale**, in use since 1993, is more accurate for large earthquakes than is Richter's amplitude magnitude scale. Moment magnitude considers the amount of fault slippage produced by the earthquake, the size of the surface (or subsurface) area that ruptured, and the nature of the materials that faulted, including how resistant they were to failure. Technically, the M is equal to the rigidity of Earth multiplied by the average amount of slip on the fault and its area. This scale considers extreme ground acceleration (movement upward), which the Richter amplitude magnitude method underestimates.

Table 12.1 shows the MMI and M scales and the expected number of quakes in each category in a year. In 1960, an M 9.5 earthquake in Chile produced Mercalli XII damage—this was the strongest earthquake in recorded history.

A reassessment of pre-1993 quakes using the moment magnitude scale has increased the rating of some and decreased that of others. As an example, the 1964 earthquake at Prince William Sound in Alaska had an amplitude magnitude of 8.6 (on the Richter scale), but on the moment magnitude scale, it increased to an M 9.2.

On the *MasteringGeography* website, you will find a sampling of significant earthquakes with their MMI and M ratings. For a range of information about earthquakes, see http://earthquake.usgs.gov/regional/neic/ or http://www.ngdc.noaa.gov/hazard/earthqk.shtml. Focus Study 12.1 offers more details on the quakes in Haiti, Chile, and Japan in 2010 and 2011.

Fault Mechanics

Earlier in the chapter, we described faulting, types of faults, and direction of faulting motions. The specific mechanics of how a fault breaks, however, remain under study. **Elastic-rebound theory** describes the basic process. Generally, two sides along a fault appear to be locked by friction, resisting any movement despite the powerful forces acting on the adjoining pieces of crust. Stress continues to build strain along the fault-plane surfaces, storing elastic energy like a wound-up spring. When the strain buildup finally exceeds the frictional lock, both sides of the fault abruptly move to a condition of less strain, releasing a burst of mechanical energy.

Think of the fault plane as a surface with irregularities that act as sticking points to prevent movement, similar to two pieces of wood held together by drops of glue of different sizes rather than an even coating of glue. These small areas are points of high strain, known as *asperities*—when these sticking points break, they release the sides of the fault.

If the fracture along the fault line is isolated to a small asperity break, the quake will be small in magnitude. As some asperities break (perhaps recorded as small foreshocks), the strain increases on surrounding asperities that remain intact (Figure 12.19). Thus, small earthquakes in an area may be precursors to a major quake. However, if the break involves the release of strain along several asperities, the quake will be greater in extent and will involve the shifting of massive amounts of crust. The latest evidence suggests that movement along the fault occurs in a wave-like pattern, as rupturing spreads along the fault plane, rather than the entire fault surface giving way at once.

Human activities can exacerbate the natural processes occurring along faults in a process called *induced seismicity*. For example, if the pressure on the pores and fractures in rocks is increased by the addition of fluids (called fluid injection), then earthquake activity may accelerate. The extraction of fluids, especially at a rapid rate, can also induce seismicity by causing subsidence of the ground and enhancing slippage along faults.

(text continued on page 380)

Focus Study 12.1 Natural Hazards

Earthquakes in Haiti, Chile, and Japan: A Comparative Analysis

In 2010 and 2011, three quakes struck areas near major population centers, causing massive destruction and fatalities. These earthquakes—in the countries of Haiti, Chile, and Japan—all occurred at plate boundaries and ranged in magnitude from M 7.0 to M 9.0 (Figure 12.1.1 and Table 12.1.1).

The Human Dimension

The 2010 Haiti earthquake hit an impoverished country where little of the infrastructure was built to withstand earthquakes. Over 2 million people live in the capital city of Port-au-Prince, which has been destroyed by earthquakes several times, mostly notably in 1751 and 1770. The total damage there from the 2010 quake exceeded the country's $14 billion gross domestic product (GDP). In developing countries such as Haiti, earthquake damage is worsened by inadequate construction, lack of enforced building codes, and the difficulties of getting food, water, and medical help to those in need.

The Maule, Chile, earthquake, which occurred just 6 weeks later, caused only minimal damage, in large part due to the fact that the country enacted strict building codes in 1985. The result was a fraction of the human cost compared to the Haiti earthquake.

The Japan quake resulted in an enormous and tragic human fatality count, mainly due to the massive Pacific Ocean tsunami (defined as a set of seismic sea waves; discussed in Chapter 16). When an area of ocean floor some 338 km (N–S) by

(a) Destruction in Port-au-Prince, Haiti, in 2010. The quake epicenter was along multiple surface faults and a previously unknown subsurface thrust fault.

(b) A collapsed bridge in Santiago, Chile, after the M8.8 earthquake hit Maule, 95 km (60 mi) away. The epicenter was on a convergent plate boundary between the Nazca and South American plates.

(c) Honshu Island, Japan, after the quake and tsunami. The epicenter was on a convergent plate boundary between the Pacific and North American plates.

(d) Tsunami moves ashore, Iwanuma, Japan. Iwanuma is 20 km (12.4 mi) south of Sendai, the city closest to the epicenter.

▲**Figure 12.1.1 The Haiti, Chile, and Japan earthquakes and the Japan tsunami.** [(a) Julie Jacobson/AP Images. (b) Martin Bernetti/Getty Images. (c) and (d) Kyodo/Reuters.]

TABLE 12.1.1 Summary of Three Major Earthquakes in 2010 and 2011

Location, date, and local time	Moment magnitude*	Focus depth	Epicenter distance to nearest city	Human dimension	Damage cost (USD)
Port-au-Prince, Haiti, Jan. 12, 2010, 4:53 P.M.	M 7.0	13 km (8 mi) Ocean	15 km (10 mi) southwest from Port-au-Prince	222,570 killed, 300,000 injured, 1.3 million displaced	$25 billion
Maule, Chile, Feb. 27, 2010, 3:34 A.M.	M 8.8	35 km (22 mi) Ocean	95 km (60 mi) from Santiago	521 killed, 12,000 injured, 800,000 displaced	$30 billion
Tohoku, Honshu, Japan, Mar. 11, 2011 2:46 P.M.	M 9.0	32 km (20 mi) Ocean	129 km (80 mi) from Sendai	15,856 deaths, 6000 injured, 3000 missing (as of April 2012)	~$325 billion (maybe as high as $500 billion)

** Reported by USGS.*

150 km (210 mi by 93 mi) snapped and was abruptly lifted as much as 80 m (260 ft), the ocean was displaced above it. This disturbance caused the tsunami, in which the largest wave averaged 10 m (33 ft) along the coast of Honshu (Figure 12.1.1d). Where it entered narrow harbors and embayments, wave height reached nearly 30 m (98 ft). Although Japan's tsunami warning system sent out immediate alerts, there was not enough time for evacuation. This event illustrates the damage and human cost associated with an earthquake and tsunami, even in a country with strict and extensive earthquake preparedness standards.

Faulting and Plate Interactions

The Chile and Japan quakes both occurred along subduction zones. Along the coast of Chile, the Nazca plate is moving eastward beneath the westward-moving South American plate at a relative speed of 7 to 8 cm (2.7 to 3.1 in.) per year. This is the same subduction zone that produced the M 9.6 that hit Chile in 1960, the largest earthquake of the twentieth century. Off the coast of Japan, the Japan Trench defines the subduction zone in which the westward-moving Pacific plate is pulled beneath the North American plate at a rate averaging 8.3 cm/year (3.5 in./year). Several microplates form this plate boundary, visible on the ocean-floor map that begins this chapter, and in Figure 11.18.

The Haiti earthquake involved more complex fault interactions. Scientists first thought that this earthquake occurred along a 50-km section of the Enriquillo–Plantain Garden strike-slip fault, where the Caribbean plate moves eastward relative to the North American plate's westward shift. After extensive analysis, experts now think that the quake resulted from slip along multiple faults, primarily along a previously unknown subsurface thrust fault.

Key data for the analysis came from radar *interferograms*, remotely sensed images produced by comparing radar topography measurements before and after earthquake events. The Haiti interferogram shows surface deformation in the area of the fault rupture; the narrow rings of color represent contours of ground motion (Figure 12.1.2). Overall, the earthquake displaced Léogâne upward about 0.5 m (20 in.). Scientists found no evidence of surface rupture after a field survey along the Enriquillo fault. Because the slip was not near the surface, scientists believe that strain is continuing to accumulate, making future rupture at the surface likely.

For a listing and details of the 10 biggest earthquakes in history, including the Chile (ranked 6th) and Japan (ranked 4th) events, go to http://earthquake.usgs.gov/earthquakes/world/10_largest_world.php.

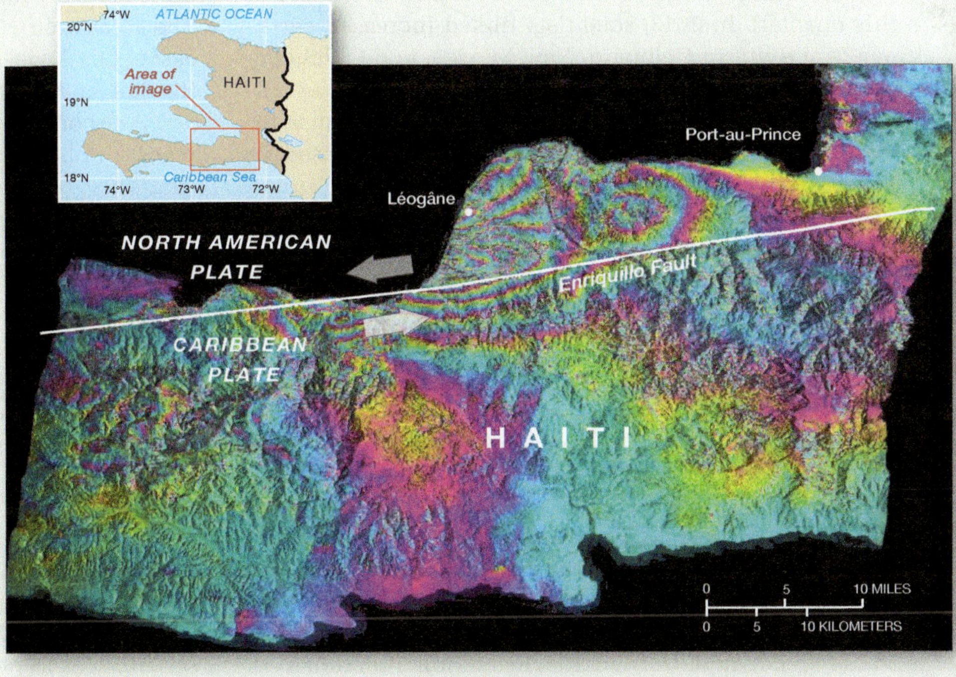

▲Figure 12.1.2 **Radar image of Haiti earthquake faults.** Synthetic-aperture radar image shows ground deformation near Léogâne, west of Port-au-Prince; narrow bands of color are contours, each representing 11.8 cm, or 4.6 in., of ground motion). This radar interferogram combines data from topographic surveys before and after the earthquake. [NASA/JPL/JAXA/METI.]

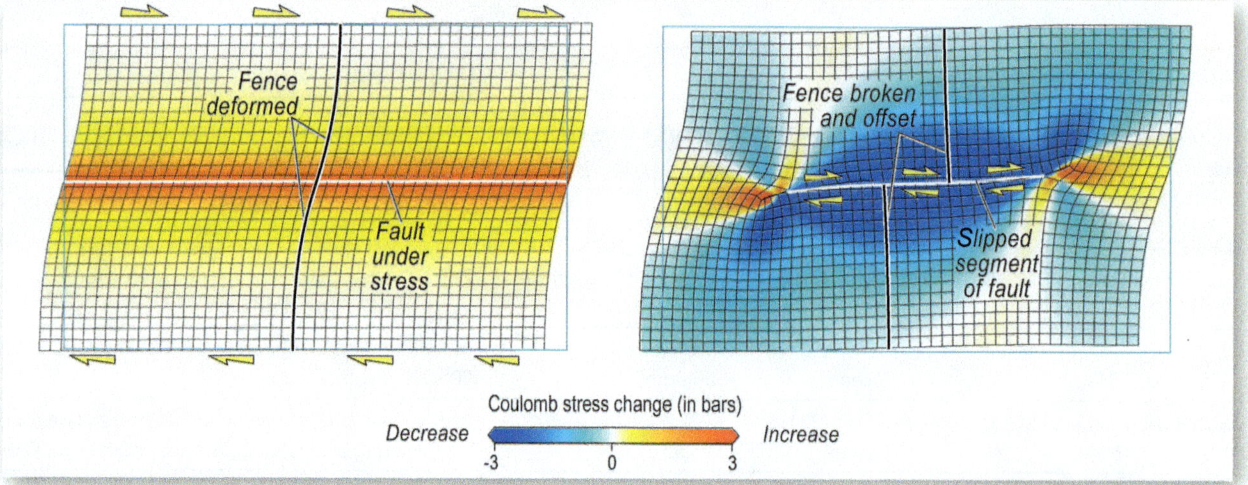

(a) Note the line denoting a fence that crosses the fault line. When stress on the fault has built to the break point, the two sides snap into new positions, strain is released, and the fence no longer forms a continuous line.

(b) The segments of the fault that did not move past each other, and instead remain locked, now have added strain.

▲**Figure 12.19 Buildup and release of stress and strain along a fault system.** A coulomb is a unit of shear-stress change on a fault, measured in bars (units of pressure; 1 bar is approximately equal to atmospheric pressure at sea level); the color scale ranges from blue (decreased stress) to red (increased stress). [Illustrations courtesy of Serkan Bozkurt, USGS.]

MG Animation Elastic Rebound

Both fluid injection and fluid extraction are common activities associated with oil and natural gas drilling and with geothermal energy production. Hydraulic fracturing, or fracking, associated with shale gas extraction injects large quantities of fluid to break up subsurface rock, making this process a probable cause for induced seismicity (please review the Chapter 1 *Geosystems Now*, and see *The Human Denominator*, Figure HD 12, at the end of this chapter). In 2013, scientists linked increased earthquake activity in Colorado, New Mexico, and Oklahoma to fluid injection associated with fracking. In the Los Angeles region, fracking near active fault zones is cause for concern. Enhanced Geothermal Systems, discussed in Focus Study 12.1, also use fracking, which has resulted in seismic activity at several locations, including the Geysers in California. A 2012 National Research Council report found only a minimal risk of seismicity associated with energy technologies.

Earthquake Forecasting

The probability of earthquakes varies across the United States, with some regions having a particularly high risk (such as California and the Yellowstone area of Wyoming, Idaho, and Montana). The map in Figure 12.20 plots the earthquake hazard for the United States in terms of the probability of ground motion; data for this map is from the most recent seismic, geologic, and geodetic information regarding earthquakes and ground shaking. The National Earthquake Hazards Reduction Program is a multi-agency program that includes the Advanced National Seismic System, which provides key data for the hazard maps (see http://earthquake.usgs.gov/monitoring/anss/).

A major challenge for scientists is to predict earthquake occurrences. One approach to earthquake forecasting is the science of *paleoseismology*, which studies the history of plate boundaries and the frequency of past earthquakes. Paleoseismologists construct maps that estimate expected earthquake activity based on past performance. An area that is quiet and overdue for an earthquake is a *seismic gap*; such an area possesses accumulated strain. The area along the Aleutian Trench subduction zone had three such gaps until the great 1964 Alaskan earthquake filled one of them. The areas around San Francisco and northeast and southeast of Los Angeles represent other such seismic gaps along the San Andreas fault system.

A second approach to forecasting is to observe and measure phenomena that might precede an earthquake. *Dilatancy* refers to the slight increase in rock volume produced by small cracks that form under stress and accumulated strain. One indication of dilatancy is a tilting and swelling in the affected region in response to strain, as measured by instruments called *tiltmeters*. Another indicator of dilatancy is an increase in the amount of radon (a naturally occurring, slightly radioactive gas) dissolved in groundwater. At present, earthquake hazard zones have thousands of radon monitors taking samples in test wells.

Both tiltmeters and gas-monitoring wells are used for earthquake monitoring at the most intensely studied seismic area in the world, along the San Andreas fault outside Parkfield, California. The area also features a newly completed drill hole, called the San Andreas Fault Observatory at Depth, that allows instruments to be placed deep into the fault (between 2 km and 3 km, or 1.2 mi and 1.9 mi, down) with the goal of measuring the physical and chemical processes occurring within an active fault. Research sites such as this provide critical data for predicting future seismic events.

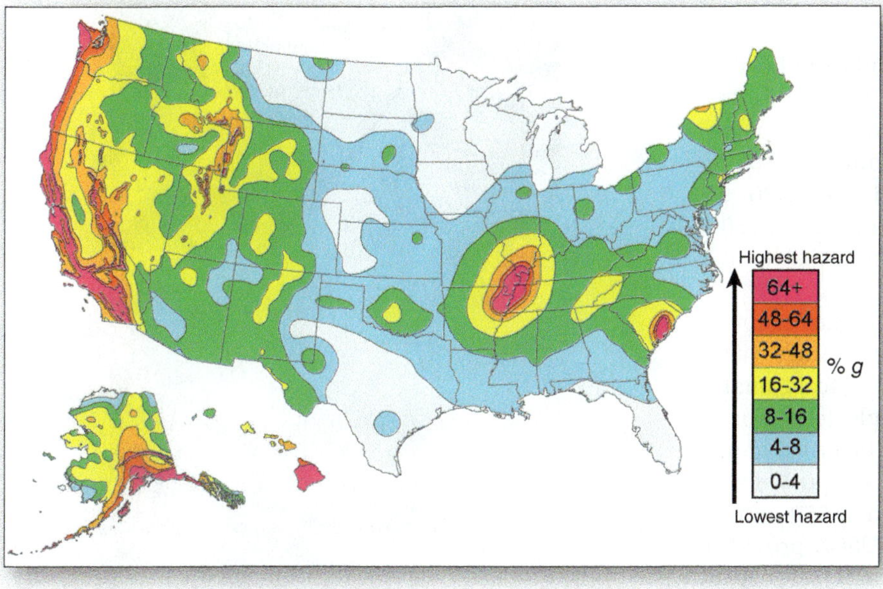

▲**Figure 12.20 Earthquake hazard map for the United States.** Colors on the map show the levels of horizontal shaking that have a 2-in-100 chance of being exceeded during a 50-year period. Shaking is expressed as a percentage of *g* (the acceleration of a falling object with gravity), with red being the highest shaking. Active seismic regions include the West Coast, the Wasatch Front of Utah northward into Canada, the central Mississippi Valley, the southern Appalachians, and portions of South Carolina, upstate New York, and Ontario. [USGS; see http://earthquake.usgs.gov/hazards/products/.]

Although scientists are not yet able to accurately forecast earthquakes, they have made strides in predicting earthquake probabilities over periods of decades. For example, in 2008, a working group of scientists and engineers reported that the probability of an M 6.7 earthquake along the San Andreas fault system in the San Francisco Bay Area between 2007 and 2030 is 63%. Adding to the risk presented by such a quake is the extent of landfill in the Bay Area, where about half the original bay is now filled and occupied with buildings. In an earthquake, this type of landfill fails in a process of *liquefaction*, in which shaking brings water to the surface and liquefies the soil.

In 2013, researchers reported that the probability of an M 8.0 earthquake in Japan was higher for the northern region than for the south. These results, based on new methodology combining data of various kinds into a mathematical probability model, directly contradict Japan's national seismic hazard map, which puts higher quake risk in the south. Once refined, the ability to predict earthquakes over long time scales will be useful for hazard planning.

In the meantime, earthquake warning systems have been successfully implemented in some countries. A Mexico City warning system provides 70-second notice of arriving seismic waves. The system was effective in March 2012, when a senate hearing at the capitol was interrupted by the sirens indicating an imminent earthquake. Shaking began about a minute later. In Japan, a warning system was activated during the Tohoku earthquake, sending alerts to televisions and cell phones and automatically shutting down some transportation and industrial services. (The tsunami warning system in the Pacific is discussed in Chapter 16.)

In the United States, an earthquake early warning system is still in development. The Southern California Seismographic Network monitors 350 instruments for immediate earthquake location analysis and disaster coordination (see http://www.scsn.org/ for more information).

Earthquake Planning

Someday accurate earthquake forecasting may be a reality, but several questions remain: How will humans respond to a forecast? Can a major metropolitan region be evacuated for short periods of time? Can a city relocate after a disaster to an area of lower risk?

Actual implementation of an action plan to reduce death, injury, and property damage from earthquakes is difficult to achieve. For example, such a plan is likely to be unpopular politically, since it involves large expenditures of money before a quake has even hit. Moreover, the negative image created by the idea of the possibility of earthquakes in a given municipality is not likely to be welcomed by banks, real estate agents, politicians, or the chamber of commerce. These factors work against the adoption of effective prediction methods and planning.

A valid and applicable generalization is that *humans and their institutions seem unable or unwilling to perceive hazards in a familiar environment.* In other words, people tend to feel secure in their homes, even in communities known to be sitting on a quiet fault zone. Such an axiom of human behavior certainly helps explain why large populations continue to live and work in earthquake-prone settings. Similar statements also can be made about populations in areas vulnerable to floods, hurricanes, and other natural hazards. (See the Natural Hazards Center at the University of Colorado at http://www.colorado.edu/hazards/index.html.)

GEOreport 12.3 Large earthquakes affect Earth's axial tilt

Scientific evidence is mounting that Earth's largest earthquake events have a global influence. Both the 2004 Sumatran–Andaman quake and the 2011 Tohoku quake in Japan caused Earth's axial tilt to shift several centimeters. NASA scientists estimate that the redistribution of mass in each quake shortened daylength by 6.8 millionths of a second for the 2004 event and 1.8 millionths of a second for the 2011 event.

Volcanism

Volcanic eruptions across the globe remind us of Earth's tremendous internal energy and of the dynamic forces shaping the planet's surface. The distribution of ongoing volcanic activity matches the distribution of plate boundaries, as shown on the map in Figure 12.20, as well as indicating the location of hot spots. Over 1300 identifiable volcanic cones and mountains exist on Earth, although fewer than 600 are active.

An *active* volcano is defined as one that has erupted at least once in recorded history. In an average year, about 50 volcanoes erupt worldwide, varying from small-scale venting of lava or fumes to major explosions. North America has about 70 volcanoes (mostly inactive) along the western margin of the continent. The Global Volcanism Program lists information for more than 8500 eruptions at http://www.volcano.si.edu/, and the USGS provides extensive information about current volcanic activity at http://volcanoes.usgs.gov/.

Eruptions in remote locations and at depths on the seafloor go largely unnoticed, but the occasional eruption of great magnitude near a population center makes headlines. Even a distant eruption has global atmospheric effects. For example, in April 2010, the eruption of the Eyjafjallajökull volcano in southern Iceland garnered world attention for its effects on air transportation (Figure 12.21). The first eruption of this volcano since the 1820s produced an ash cloud that rose to 10,660 m (35,000 ft) and quickly dispersed toward Europe and commercial airline corridors. Airspace was closed for 5 days, canceling more than 100,000 flights. When flights resumed, routes were changed to avoid lingering ash. Iceland's Meteorological Office monitored the eruption with its 56-station seismic network.

▲Figure 12.21 **The Eyjafjallajökull eruption in Iceland, May 10, 2010.** The ash plume is seen rising to between 5 and 6 km (3.1 and 3.7 mi). [*Aqua*, MODIS sensor NASA/GSFC.]

Settings for Volcanic Activity

Volcanic activity occurs in three settings, listed below with representative examples and illustrated in Figure 12.22:

- Along *subduction boundaries* at continental plate–oceanic plate convergence (Mount St. Helens; Kliuchevskoi, Siberia) or oceanic plate–oceanic plate convergence (the Philippines; Japan)

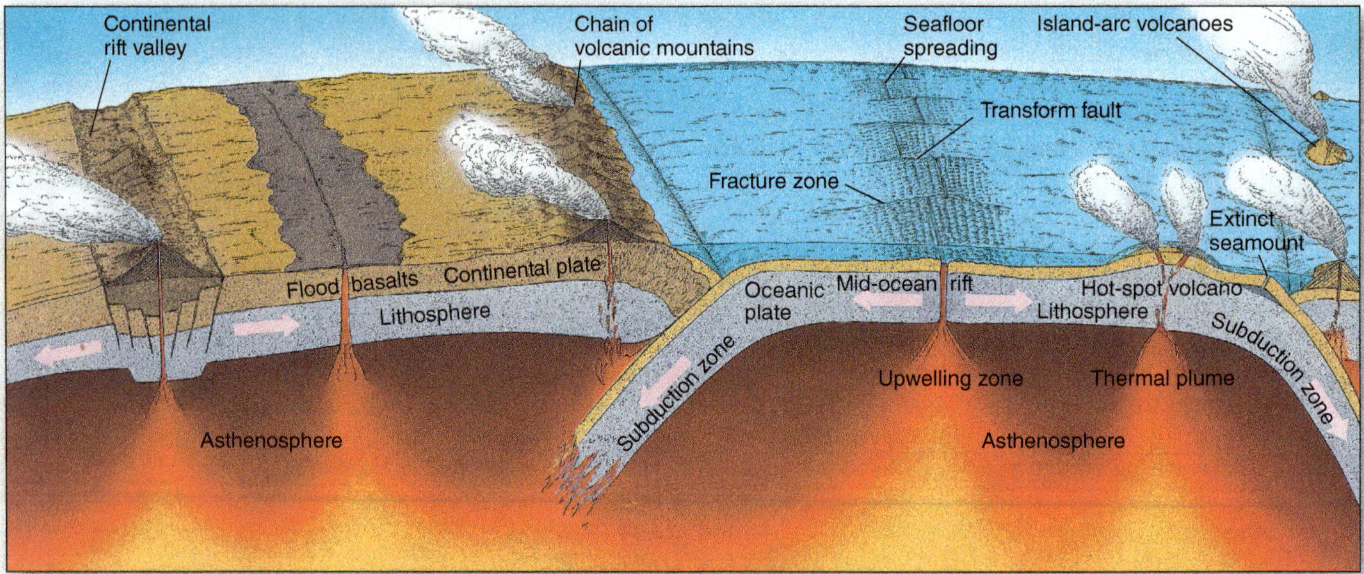

▲Figure 12.22 **Tectonic settings of volcanic activity.** Magma rises and lava erupts from rifts, through crust above subduction zones, and where thermal plumes at hot spots break through the crust. [Adapted from U.S. Geological Survey, *The Dynamic Planet* (Washington, DC: Government Printing Office, 1989).]

MG Animation
Forming Types of Volcanoes

- Along *sea-floor spreading centers* on the ocean floor (Iceland, on the Mid-Atlantic Ridge; off the coast of Oregon and Washington) and along areas of rifting on continental plates (the rift zone in East Africa)
- At *hot spots*, where individual plumes of magma rise to the crust (Hawai'i; Yellowstone National Park)

Volcanic Materials

A **volcano** is the structure in the Earth's crust containing an opening at the end of a central vent or pipe through which magma rises from the asthenosphere and upper mantle. Magma rises and collects in a magma chamber deep below the volcano until conditions are right for an eruption. This subsurface magma emits tremendous heat; in some areas, it boils groundwater, as seen in the thermal springs and geysers of Yellowstone National Park and at other locations with surface expressions of geothermal energy.

Various materials pass through the central vent to the surface to build volcanic landforms, including lava (magma that has cooled to form rock), gases, and **pyroclastics**—pulverized rock and clastic materials of various sizes ejected violently during an eruption (also called *tephra*). These materials may emerge explosively, or they may emerge effusively (flowing gently) from the vent (eruption types are discussed ahead).

As discussed in Chapter 11, solidified magma forms igneous rock. When magma emerges at the surface, it is *lava*. The chemistry of lava determines its behavior (whether it is thin and liquid or is thick and forms a plug).

Geologists classify lava as felsic, intermediate, or mafic, depending on its chemical composition (review Table 11.2). Mafic, or basaltic, lava has two principal forms, both known by Hawaiian names (Figure 12.23). The composition of both these forms of lava is the same; the texture difference results from the manner in which the lava flows while it cools. Rough and jagged basalt with sharp edges is **aa**; it forms as a thick skin over the surface of a slowing lava flow, cracking and breaking as it cools and solidifies. Shiny and smooth basalt that resembles coiled, twisted rope is **pahoehoe**; it forms as a thin crust that develops folds as the lava cools. Both forms can come from the same eruption, and sometimes pahoehoe will become aa as the flow progresses. Other types of basaltic magma are described later in this section.

During a single eruption, a volcano may behave in several different ways, which depend primarily on the chemistry and gas content of the lava. These factors determine the lava's *viscosity*, or resistance to flow. Viscosity can range from low (very fluid) to high (thick and flowing slowly). For example, pahoehoe has lower viscosity than aa.

(a) Aa is a rough, sharp-edged lava said to get its name from the sounds people make if they attempt to walk on it.

(b) Pahoehoe forms ropy cords in twisted folds.

▲**Figure 12.23 Two types of basaltic lava—Hawaiian examples, close up.** [Bobbé Christopherson.]

Volcanic Landforms

Volcanic eruptions result in structures that range among several forms, such as hill, cone, or mountain. In a volcanic mountain, a **crater**, or circular surface depression, is usually found at or near the summit (see Figures 12.29b and 12.30b).

A **cinder cone** is a small, cone-shaped hill usually less than 450 m (1500 ft) high, with a truncated top formed from cinders that accumulate during moderately explosive eruptions. Cinder cones are made of pyroclastic material and *scoria* (cindery rock, full of air bubbles). Several notable cinder cones are located on the San Francisco volcanic field of northern Arizona. On Ascension Island at about 8° S latitude in the Atlantic Ocean, over 100 cinder cones and craters mark this island's volcanic history (Figure 12.24).

A **caldera** (Spanish for "kettle") is a large, basin-shaped depression that forms when summit material on a volcanic mountain collapses inward after an eruption or other loss of magma. A caldera may fill with rainwater to form a lake, such as Crater Lake in southern Oregon.

▲**Figure 12.24 Cinder cones on Ascension Island.** Ascension Island in the south Atlantic Ocean is a massive composite volcano that rises over 3000 m (9842 ft) from the ocean floor, with 858 m above sea level. [Bobbe' Christopherson.]

(a) The Long Valley Caldera in eastern California formed over 700,000 years ago.

▼Figure 12.25 Long Valley Caldera and forest death from CO_2 emission, California. [(a) NASA/GSFC. (b) Bobbé Christopherson.]

(b) The effects of CO_2 gassing from subsurface magma and related volcanic activity have killed one person to date.

The Long Valley Caldera (Figure 12.25), near the California–Nevada border, was formed by a powerful volcanic eruption 760,000 years ago. The area is characterized today by hydrothermal activity such as hot springs and fumaroles that fuel the Casa Diablo geothermal plant, which supplies power to about 40,000 homes. About 1200 tons of carbon dioxide are coming up through the soil in the caldera each day, killing acres of forests in six different areas of the caldera floor. The source is active and moving magma at a depth of some 3 km (1.86 mi). These gas emissions signal volcanic activity and are useful indicators of potential eruptions. For updates, see http://volcanoes.usgs.gov/volcanoes/long_valley/.

Effusive Eruptions

Effusive eruptions are outpourings of low-viscosity magma that produce enormous volumes of lava annually on the seafloor and in places such as Hawai'i and Iceland. These eruptions flow directly from the asthenosphere and upper mantle, releasing fluid magma that cools to form a dark, basaltic rock low in silica (less than 50%) and rich in iron and magnesium. Gases readily escape from this magma because of its low viscosity. Effusive eruptions pour out on the surface with relatively small explosions and few pyroclastics. However, dramatic fountains of basaltic lava sometimes shoot upward, powered by jets of rapidly expanding gases.

An effusive eruption may come from a single vent or from the flank of a volcano through a side vent. If such vent openings are linear in form, they are *fissures*, which sometimes erupt in a dramatic "curtain of fire" as sheets of molten rock spray into the air.

On the island of Hawai'i, the continuing Kīlauea eruption is the longest in recorded history—active since January 3, 1983 (Figure 12.26). Although this eruption is located on the slopes of the massive Mauna Loa volcano, scientists have determined that Kīlauea has its own magma system extending down some 60 km (37 mi) into Earth. To date, the active crater on Kīlauea (called Pu'u O'o) has produced more lava than any other in recorded history—some 3.1 km^3 (0.7 mi^3). To see a series of photographs showing the actively changing Pu'u O'o crater from 1999 to 2011, go to Chapter 12 on the *MasteringGeography* website.

A typical mountain landform built from effusive eruptions is gently sloped, gradually rising from the surrounding landscape to a summit crater. The shape is similar in outline to a shield of armor lying face up

▲Figure 12.26 Kīlauea landscape. The newest land on the planet is produced by the massive flows of basaltic lavas from the Kīlauea volcano, Hawai'i Volcanoes National Park. In the distance, ocean water meets 1200°C (2200°F) lava, producing steam and hydrochloride mist. [Bobbé Christopherson.]

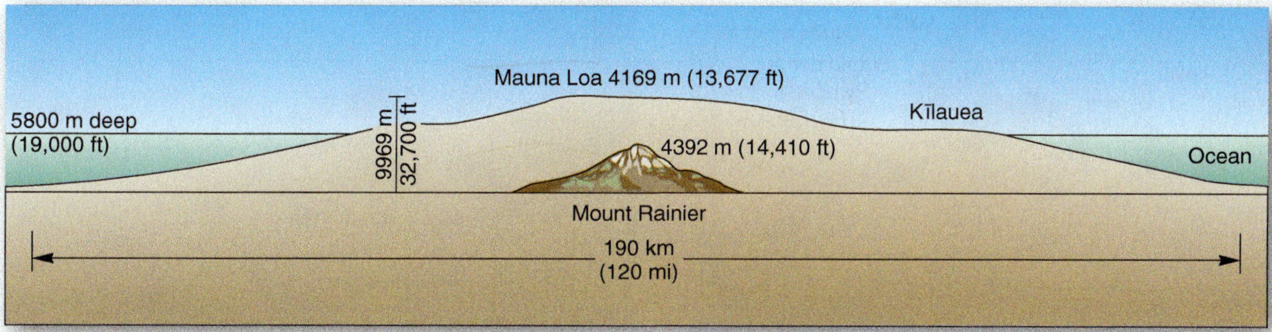

(a) Comparison of Mauna Loa in Hawai'i, a shield volcano, and Mount Rainier in Washington State, a composite volcano. Their strikingly different profiles reveal their different tectonic origins.

(b) Mauna Loa's gently sloped shield shape dominates the horizon.

▲**Figure 12.27 Shield and composite volcanoes compared.** [(a) After USGS, Eruption of Hawaiian Volcanoes, 1986. (b) Bobbé Christopherson.]

on the ground and therefore is called a **shield volcano**. Mauna Loa is one of five shield volcanoes that make up the island of Hawai'i. The height of the Mauna Loa shield is the result of successive eruptions, flowing one on top of another. At least 1 million years were needed to accumulate this shield volcano, forming the most massive single mountain on Earth (although Mauna Kea, also located on Hawai'i, is slightly taller). The shield shape and size of Mauna Loa are distinctive when compared with Mount Rainier in Washington, which is a different type of volcano (explained shortly) and the largest in the Cascade Range (Figure 12.27).

In volcanic settings above hot spots and in continental rift valleys, effusive eruptions send material out through elongated fissures, forming extensive sheets of basaltic lava on the surface (see Figure 12.22). The Columbia Plateau of the northwestern United States, some 2–3 km thick (1.2–1.8 mi), is the result of the eruption of these **flood basalts** (sometimes called *plateau basalts*).

Flood basalts cover several large regions on Earth, sometimes referred to as *igneous provinces* (Figure 12.28). The Deccan Traps is an igneous province in west-central India that is more than double the size of the Columbian

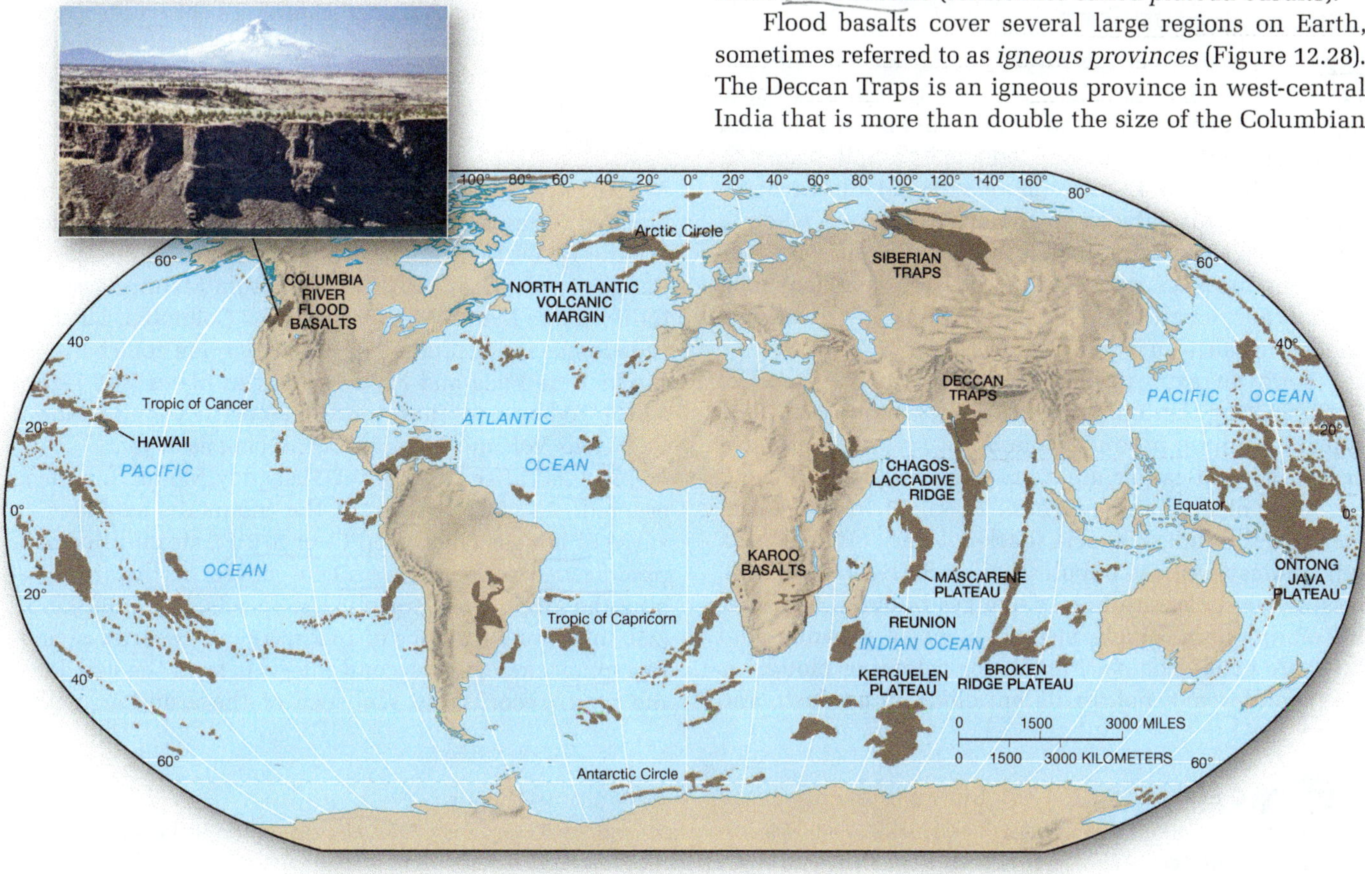

▲**Figure 12.28 Earth's igneous provinces.** Flood basalt provinces include regions of ancient volcanic activity as well as active flows. Inset photo shows flood basalts on the Columbia Plateau in Oregon in the foreground and Mount Hood, a composite volcano, in the background. [(a) After M. F. Coffin and O. Eldholm, "Large igneous provinces," *Scientific American* (October 1993): 42–43. © Scientific American, Inc. (b) Author.]

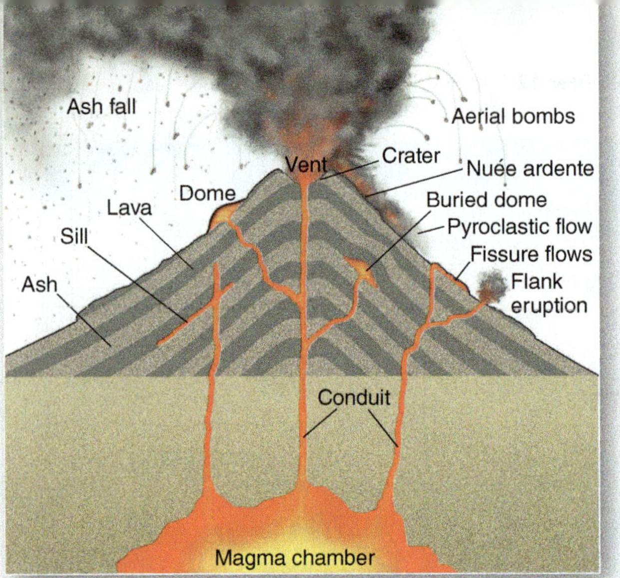

(a) A typical composite volcano with its cone-shaped form in an explosive eruption.

(b) Mount Redoubt, Alaska, erupts in 2009.

▲**Figure 12.29 A composite volcano.** [(b) Game McGimsey, USGS.]

Plateau. *Trap* is Dutch for "staircase," referring to the typical steplike form of the eroded flood basalts. The Siberian Traps is more than twice the area of the igneous province in India and is exceeded in size only by the Ontong Java Plateau, which covers an extensive area of the seafloor in the Pacific. None of the presently active sites forming flood basalts (such as on Mauna Loa) come close in size to the largest of the extinct igneous provinces, some of which formed more than 200 million years ago.

Explosive Eruptions

Explosive eruptions are violent explosions of magma, gas, and pyroclastics driven by the buildup of pressure in a magma conduit. This buildup occurs because magma produced by the melting of subducted oceanic plate and other materials is thicker (more viscous) than magma that forms effusive volcanoes. Consequently, it tends to block the magma conduit by forming a plug near the surface. The blockage traps and compresses gases (so much so that they remain liquefied) until their pressure is great enough to cause an explosive eruption.

Such an explosion is equivalent to megatons of TNT blasting the top and sides off the mountain. This type of eruption produces much less lava than effusive eruptions, but larger amounts of pyroclastics, which include volcanic ash (<2 mm, or <0.08 in., in diameter), dust, cinders, scoria (dark-colored, cindery rock with holes from gas bubbles), pumice (lighter-colored, less-dense rock with holes from gas bubbles), and *aerial bombs* (explosively ejected blobs of incandescent lava). A *nuée ardente*, French for "glowing cloud," is an incandescent, hot, turbulent cloud of gas, ash, and pyroclastic that can jet across the landscape in these kinds of eruptions (Figure 12.29).

A mountain produced by a series of explosive eruptions is a **composite volcano**, formed by multiple layers of lava, ash, rock, and pyroclastics. These landforms are sometimes called *stratovolcanoes* to describe the alternating layers of ash, rock, and lava, but shield volcanoes also can exhibit a stratified structure, so composite is the preferred term. Composite volcanoes tend to have steep sides and a distinct conical shape, and therefore are also known as *composite cones*. If a single summit vent erupts repeatedly, a remarkable symmetry may develop as the mountain grows in size, as demonstrated by Popocatépetl in Mexico (see Figure HD 12c, on page 389) and the pre-1980-eruption shape of Mount St. Helens in Washington (Figure 12.30a).

Mount St. Helens Probably the most studied and photographed composite volcano on Earth is Mount St. Helens (Figure 12.30), the youngest and most active of the Cascade Range of volcanoes, which extend from Mount Lassen in California to Mount Meager in British Columbia. The Cascade Range is the product of the subduction zone between the Juan de Fuca and North American plates (see Figure 12.12). Today, more than 1 million tourists visit the Mount St. Helens Volcanic National Monument each year.

In 1980, after 123 years of dormancy, Mount St. Helens erupted. As the contents of the mountain exploded, a surge of hot gas (about 300°C, or 570°F), steam-filled ash, pyroclastics, and a nuée ardente moved northward, hugging the ground and traveling at speeds up to 400 kmph (250 mph) for a distance of 28 km (17 mi). A series of photographs, taken at 10-second intervals from the east looking west, records the sequence of the eruption, which

GEOreport 12.4 Slow slip events across Kīlauea's south flank

The entire south flank of the Kīlauea volcano is in a long-term motion toward the ocean, along a low-angle fault at a rate of about 7 cm (2.75 in.) per year. GPS units, tiltmeters, and small earthquakes mark the slow movement. In February 2010, in one 36-hour period, a slow slip event of 3 cm (1.18 in.) occurred, accompanied by a flurry of small earthquakes. A sizeable collapse or failure of this new basaltic landscape into the sea is possible, but the eventual outcome of these slow slip events is uncertain. In 2011, lava flows to the ocean changed course as new eruptions began west of Pu'u O'o.

(a) Mount St. Helens prior to the 1980 eruption.

(b) After the eruption, the scorched earth and tree blow-down area covered some 38,950 hectares (96,000 acres).

▲**Figure 12.30 Mount St. Helens, before and after the 1980 eruption.** [(a) Pat and Tom Lesson/Photo Researchers. (b) Krafft-Explorer/Photo Researchers, Inc.]

continued with intensity for 9 hours and then blasted new material intermittently for days (Figure 12.31).

The slumping that occurred along the north face of the mountain produced the greatest landslide witnessed in recorded history; about 2.75 km^3 (0.67 mi^3) of rock, ice, and trapped air, all fluidized with steam, surged at speeds approaching 250 kmph (155 mph). Landslide materials traveled for 21 km (13 mi) into the valley, blanketing the forest, covering a lake, and filling the rivers below.

As destructive as such eruptions are, they also are constructive, for this is the way in which a volcano eventually builds its height. Before the eruption, Mount St. Helens was 2950 m (9677 ft) tall; the eruption blew away 418 m (1370 ft). Today, Mount St. Helens is building a lava dome within its crater. The thick lava rapidly and repeatedly plugs and breaks in a series of lesser dome eruptions that may continue for several decades. The buildup of the lava dome is now more than 300 m (1000 ft) high, so a new mountain is being born from the eruption of the old.

Mount Pinatubo The 1991 eruption of Mount Pinatubo in the Philippines was the second largest of that century (the largest was Novarupta on the Katmai Peninsula in Alaska in 1912), and the largest to affect a densely populated area. The eruption produced 15 to 20 million tons of ash and 12 km^3 (3.0 mi^3) of magma, ash, and pyroclastics—about 12 times the volume of material produced by Mount St. Helens. The loss of this vast amount of material caused the summit of the volcano to collapse, forming a caldera 2.5 km (1.6 mi) in diameter. The eruption killed 800 people and devastated many surrounding villages. However, accurate prediction of this event saved many lives, as approximately 60,000 evacuated their homes prior to the eruption.

Although volcanoes are local events, they have global effects. As discussed in Chapters 1 and 6, the Mount Pinatubo eruption affected Earth's climate, releasing an aerosol cloud that changed atmospheric albedo, impacted atmospheric absorption of insolation, and altered net radiation at Earth's surface (please review Figures 1.11 and 6.1).

Volcano Forecasting and Planning

After 23,000 people died in the 1985 eruption of Nevado del Ruiz in Colombia, the USGS and the Office of Foreign Disaster Assistance established the Volcano Disaster Assistance Program (VDAP; see http://vulcan.wr.usgs.gov/Vdap/). In the United States, VDAP helps local scientists forecast eruptions by setting up mobile volcano-monitoring systems at sites with activity or with vulnerable populations. A number of "volcano cams" are positioned around the world for 24-hour surveillance (go to http://vulcan.wr.usgs.gov/Photo/volcano_cams.html).

At Mount St. Helens, a dozen survey benchmarks and several tiltmeters have been placed within the

CRITICALthinking 12.2
Ocean-Floor Tectonics Tour

Using the chapter-opening map, follow the ocean ridge and spreading center known as the East Pacific Rise northward as it trends beneath the west coast of the North American plate, disappearing under earthquake-prone California. Locate continents, offshore submerged continental shelves, and expanses of sediment-covered abyssal plain on the map.

On the floor of the Indian Ocean, you see the wide track along which the Indian plate traveled northward to its collision with the Eurasian plate. Vast deposits of sediment cover the Indian Ocean floor, south of the Ganges River and to the east of India. Sediments derived from the Himalayan Range blanket the floor of the Bay of Bengal (south of Bangladesh) to a depth of 20 km (12.4 mi). These sediments result from centuries of soil erosion in the land of the monsoons.

In the Pacific, a chain of islands and seamounts marks the hot-spot track on the Pacific plate from Hawai'i to the Aleutians. Visible as dark trenches are subduction zones south and east of Alaska and Japan and along the western coast of South and Central America. Follow the Mid-Atlantic Ridge spreading center the full length of the Atlantic to where it passes through Iceland, sitting astride the ridge. Take time with this map and find other examples in it of the many concepts discussed in Chapters 11 and 12. ●

crater to monitor the building lava dome. The monitoring, accompanied by intensive scientific research, has paid off, as every eruption since 1980 was successfully forecasted from days to as long as 3 weeks in advance (with the exception of one small eruption in 1984). Swarms of minor earthquakes along the north flank of the mountain occurred in 2001, and dome eruptions of varying intensities occurred through 2007.

Early warning systems for volcanic activity are now possible through integrated seismographic networks and monitoring. In addition, satellite remote sensing allows scientists to monitor eruption cloud dynamics and the climatic effects of volcano emissions and to estimate volcanic hazard potential. In 2005, the USGS identified 57 priority volcanoes in the United States in need of improved monitoring as part of the National Volcano Early Warning System.

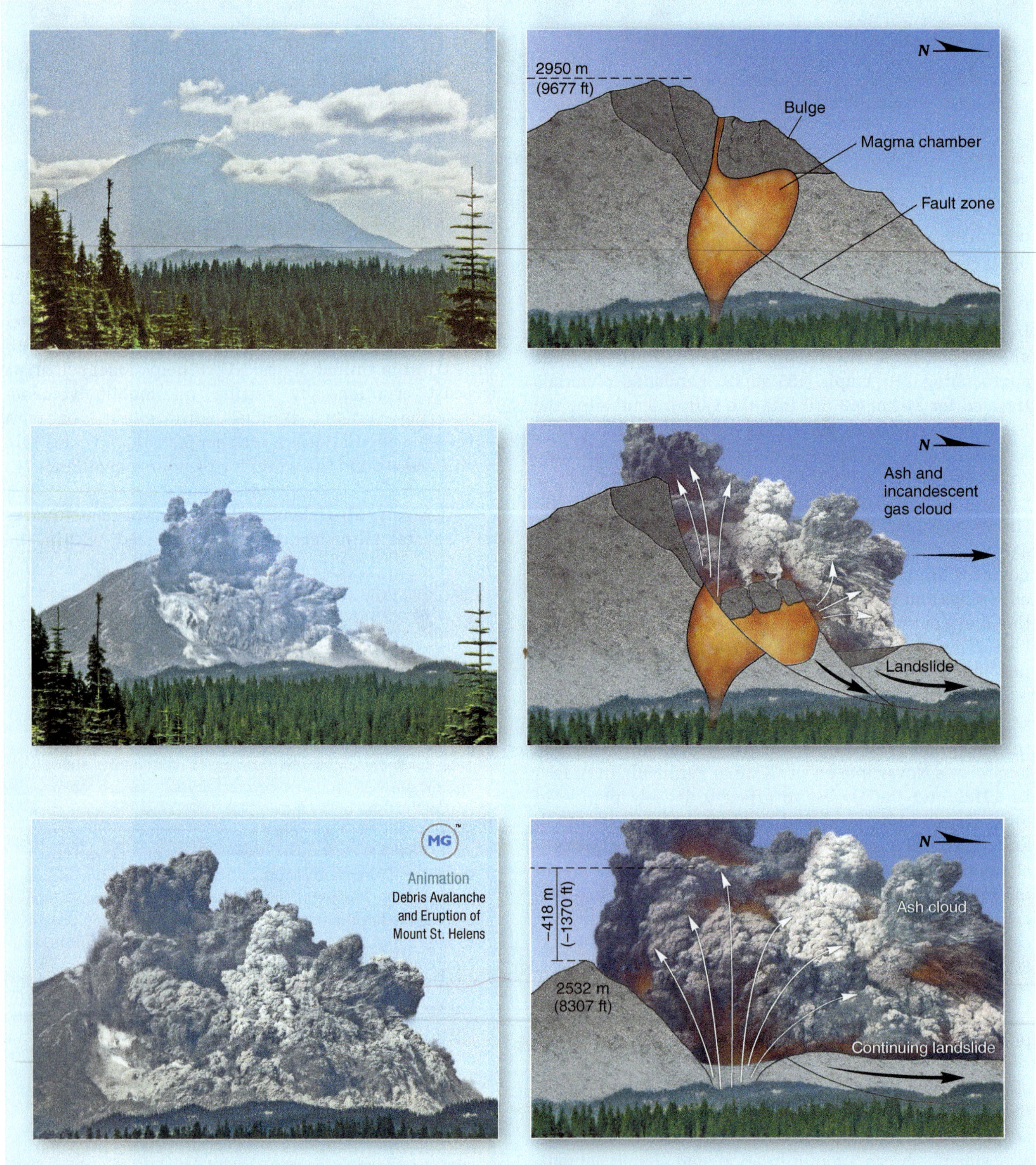

▲**Figure 12.31 The Mount St. Helens eruption sequence and corresponding schematics.** [Photo sequence by Keith Ronnholm. All rights reserved.]

THE**human**DENOMINATOR 12 Tectonics

TECTONIC PROCESSES ⇨ HUMANS

- Earthquakes cause damage and human casualties; destruction is amplified in developing countries, such as Haiti.
- Volcanic eruptions can devastate human population centers, disrupt human transportation, and affect global climate.

HUMANS ⇨ TECTONIC PROCESSES

- Human activity such as subsurface fluid injections associated with gas drilling can cause earthquakes.

12a

In 1959, an M 7.5 earthquake hit the area near West Yellowstone, Montana, fracturing highways along the shores of Hebgen Lake. This quake caused a landslide that dammed the Madison River, discussed in Chapter 13. [USGS.]

12b

In 2012, the government approved exploratory shale gas drilling in Great Britain, despite environmental concerns. Recent research has linked oil and gas injection wells to increased earthquake activity in several U.S. states. [Christopher Furlong/Getty Images.]

12c

Mexico's Popocatépetl is an active composite volcano located about 70 km (45 mi) from Mexico City. Ongoing activity since 1994 includes growth of the lava dome within the crater, episodic steaming and ash emission, and ejection of material. Evacuation routes and shelters are ready for another eruption, which could affect millions of people living nearby. [Violeta Schmidt/Reuters.]

12d

A scientist at the Hawaiian Volcano Observatory takes a sample of lava at Kīlauea, part of an effort to understand changes in lava chemistry on Mauna Loa. [USGS.]

ISSUES FOR THE 21ST CENTURY

- Growing human population centers in regions prone to seismic activity and near active volcanoes will increase the hazard.
- Scientific research is needed for earthquake prediction and volcano forecasting.

GEOSYSTEMS**connection**

Chapter 12 completes our study of Earth's endogenic system with a look at the processes that build continental crust and create Earth's topography. Mountain building, earthquakes, and volcanism are the outputs of this system, which is powered by radioactive decay in the planet's interior. We focus in the next four chapters on the exogenic system, where solar energy and gravity empower weathering, erosion, and transportation of material by gravity, water, wind, waves, and ice—these are the processes that reduce landscape relief.

KEY LEARNING concepts review

■ *Describe* first, second, and third orders of relief, and *list* Earth's six major topographic regions.

Tectonic forces generated within the planet dramatically shape Earth's surface. **Relief** is the vertical elevation difference in a local landscape. The variations in the physical surface of Earth, including relief, are **topography**. *Orders of relief* are convenient descriptive categories for landforms; the coarsest level includes the **continental landmasses** (portions of crust that reside above or near sea level) and **ocean basins** (portions of the crust that are entirely below sea level), and the finest level comprises local hills and valleys.

relief (p. 358)
topography (p. 358)
continental landmass (p. 359)
ocean basin (p. 359)

1. How does the map of the ocean floor in the chapter-opening illustration exhibit the principles of plate tectonics? Briefly analyze.
2. What is meant by an order of relief? Give an example from each order.
3. Explain the difference between relief and topography.

■ *Describe* the formation of continental crust, and *define* displaced terranes.

A continent has a nucleus of ancient crystalline rock called a *craton*. A region where a craton is exposed is a **continental shield**. As continental crust forms, it is enlarged through accretion of dispersed **terranes**. An example is the Wrangellia terrane of the Pacific Northwest and Alaska.

continental shield (p. 361)
terrane (p. 363)

4. What is a craton? Describe the relationship of cratons to continental shields and platforms, and describe these regions in North America.
5. What is an accreted terrane, and how does it add to the formation of continental landmasses? Briefly describe the journey and current location of the Wrangellia terrane.

■ *Explain* the process of folding, and *describe* the principal types of faults and their characteristic landforms.

Folding, broad warping, and faulting deform the crust and produce characteristic landforms. Compression causes rocks to deform in a process known as **folding**, during which rock strata bend and may overturn. Along the ridge of a fold, layers slope downward away from the axis, forming an **anticline**. In the trough of a fold, however, layers slope downward toward the axis; this is a **syncline**.

When rock strata are stressed beyond their ability to remain a solid unit, they express the strain as a fracture. Rocks on either side of the fracture are displaced relative to the other side in a process known as **faulting**. Thus, fault zones are areas where fractures in the rock demonstrate crustal movement.

When forces pull rocks apart, the tension causes a **normal fault**, sometimes visible on the landscape as a scarp, or escarpment. Compressional forces associated with converging plates force rocks to move upward, producing a **reverse fault**. A low-angle fault plane is referred to as a **thrust fault**. Horizontal movement along a fault plane, often producing a linear rift valley, is a **strike-slip fault**. The term **horst** is applied to upward-faulted blocks; **graben** refers to downward-faulted blocks. In the U.S. interior west, the **Basin and Range Province** is an example of aligned pairs of normal faults and a distinctive horst-and-graben landscape. A **bolson** is the slope-and-basin area between mountain ridges in this type of arid region.

folding (p. 364)
anticline (p. 364)
syncline (p. 364)
faulting (p. 366)
normal fault (p. 366)
reverse fault (p. 367)
thrust fault (p. 367)
strike-slip fault (p. 369)
horst (p. 369)
graben (p. 369)
Basin and Range Province (p. 369)
bolson (p. 369)

6. Diagram a simple folded landscape in cross section, and identify the features created by the folded strata.
7. Define the four basic types of faults. How are faults related to earthquakes and seismic activity?
8. How did the Basin and Range Province evolve in the western United States? What other examples exist of this type of landscape?

■ *List* the three types of plate collisions associated with orogenesis, and *identify* specific examples of each.

Orogenesis is the birth of mountains. An *orogeny* is a mountain-building episode, occurring over millions of years, that thickens continental crust. It can occur through large-scale deformation and uplift of the crust. It also may include the capture and cementation of migrating terranes to the continental margins and the intrusion of granitic magmas to form plutons.

Three types of tectonic activity cause mountain building along convergent plate margins. *Oceanic plate–continental plate collisions* are now occurring along the Pacific coast of the Americas, forming the Andes, the Sierra Madre of Central America, the Rockies, and other western mountains. *Oceanic plate–oceanic plate collisions* produce volcanic island arcs such as Japan, the Philippines, the Kurils, and portions of the Aleutians. The region around the Pacific contains expressions of each type of collision in the **circum-Pacific belt**, or the **Ring of Fire**. In a *continental plate–continental plate collision*, large masses of continental crust, such as the Himalayan Range, are subjected to intense folding, overthrusting, faulting, and uplifting.

orogenesis (p. 369)
circum-Pacific belt (p. 371)
Ring of Fire (p. 371)

9. Define *orogenesis*. What is meant by the birth of mountain chains?

10. Name some significant orogenies.
11. Identify on a map several of Earth's mountain chains. What processes contributed to their development?
12. How are plate boundaries related to episodes of mountain building? Explain how different types of plate boundaries produce differing orogenic episodes and different landscapes.
13. Relate tectonic processes to the formation of the Appalachians and the Alleghany orogeny.

■ *Explain* earthquake characteristics and measurement, *describe* earthquake fault mechanics, and *discuss* the status of earthquake forecasting.

An **earthquake** is the release of energy that occurs at the moment of fracture along a fault in the crust, producing seismic waves. Earthquakes generally occur along plate boundaries. Seismic motions are measured with a **seismometer**, also called a seismograph.

Scientists measure earthquake magnitude using the **moment magnitude scale**, a more precise and quantitative scale than the **Richter scale**, which was mainly an effective measure for small-magnitude quakes. The **elastic-rebound theory** describes the basic process of how a fault breaks, although the specific details are still under study. The small areas that are sticking points along a fault are points of high strain, known as *asperities*—when these sticking points break, they release the sides of the fault. When the elastic energy is released abruptly as the rock breaks, both sides of the fault return to a condition of less strain. Earthquake forecasting remains a major challenge for scientists.

earthquake (p. 374)
seismometer (p. 375)
Richter scale (p. 376)
moment magnitude (M) scale (p. 377)
elastic-rebound theory (p. 377)

14. What is the relationship between an epicenter and the focus of an earthquake? Give an example from the Loma Prieta, California, earthquake.
15. Differentiate among the Mercalli, moment magnitude (M), and amplitude magnitude (Richter) scales. How are these used to describe an earthquake? Reference some recent quakes in your discussion.
16. How do the elastic-rebound theory and asperities help explain the nature of faulting? In your explanation, relate the concepts of stress (force) and strain (deformation) along a fault. How does this lead to rupture and earthquake?
17. Describe the San Andreas fault and its relationship to ancient seafloor spreading movements along transform faults.
18. How are paleoseismology and the seismic gap concept related to expected earthquake occurrences?
19. What do you see as the biggest barrier to effective earthquake prediction?

■ *Describe* volcanic landforms, and *distinguish* between an effusive and an explosive volcanic eruption.

A **volcano** forms at the end of a central vent or pipe that rises from the asthenosphere through the crust. Eruptions produce lava (molten rock), gases, and **pyroclastics** (pulverized rock and clastic materials ejected violently during an eruption) that pass through the vent to openings and fissures at the surface and build volcanic landforms. Basaltic lava flows occur in two principal textures: **aa**, rough and sharp-edged lava, and **pahoehoe**, smooth, ropy folds of lava.

Landforms produced by volcanic activity include **craters**, or circular surface depressions, usually formed at the summit of a volcanic mountain; **cinder cones**, which are small conical-shaped hills; and **calderas**, large basin-shaped depressions sometimes caused by the collapse of a volcano's summit.

Volcanoes are of two general types, based on the chemistry and gas content of the magma involved. An **effusive eruption** produces a **shield volcano** (such as Kīlauea in Hawai'i) and extensive deposits of **flood basalts**, or *plateau basalts*. **Explosive eruptions** (such as Mount Pinatubo in the Philippines) produce a **composite volcano**. Volcanic activity has produced some destructive moments in history, but constantly creates new seafloor, land, and soils.

volcano (p. 383)
pyroclastics (p. 383)
aa (p. 383)
pahoehoe (p. 383)
crater (p. 383)
cinder cone (p. 383)
caldera (p. 383)
effusive eruption (p. 384)
shield volcano (p. 385)
flood basalt (p. 385)
explosive eruption (p. 386)
composite volcano (p. 386)

20. What is a volcano? In general terms, describe some related features.
21. Where do you expect to find volcanic activity in the world? Why?
22. Compare effusive and explosive eruptions. Why are they different? What distinct landforms are produced by each type? Give examples of each.
23. Describe several recent volcanic eruptions, such as in Hawai'i and Iceland. What is the present status in each place? Specifically, what changes are occurring in Hawai'i?

MasteringGeography™

Looking for additional review and test prep materials? Visit the Study Area in *MasteringGeography*™ to enhance your geographic literacy, spatial reasoning skills, and understanding of this chapter's content by accessing a variety of resources, including MapMaster interactive maps, geoscience animations, satellite loops, author notebooks, videos, RSS feeds, flashcards, web links, self-study quizzes, and an eText version of *Geosystems*.

13 Weathering, Karst Landscapes, and Mass Movement

KEY LEARNING concepts

After reading the chapter, you should be able to:

- ***Describe*** the dynamic equilibrium approach to the study of landforms, and ***illustrate*** the forces at work on materials residing on a slope.
- ***Define*** weathering, and ***explain*** the importance of parent rock and joints and fractures in rock.
- ***Describe*** the physical weathering processes of frost action, salt-crystal growth, and pressure-release jointing.
- ***Explain*** the chemical weathering processes of hydration, hydrolysis, oxidation, carbonation, and dissolution.
- ***Review*** the processes and features associated with karst topography.
- ***Categorize*** the various types of mass movements, and ***identify*** examples of each by moisture content and speed of movement.

In March 2011, weeks of heavy rains triggered landslides that devastated several neighborhoods in La Paz, Bolivia, crumpling roads and flattening hundreds of homes. No fatalities occurred from this event, though landslides kill 8,000 people on average every year worldwide. Landslides represent one type of mass movement, or mass wasting, process in which large bodies of earth materials are carried downslope in sudden events. [Aizar Raldes, AFP/Getty Images.]

Human-Caused Mass Movement at the Kingston Steam Plant, Tennessee

During the early morning hours of December 22, 2008, a catastrophic displacement of soil and sediment occurred at the containment ponds of the Kingston coal-fired power plant in Tennessee. This single event, in which the embankment of a fly-ash holding pond was breached, released 4.13 million m^3 (5.4 million yd^3) of toxic ash laced with an array of potentially harmful pollutants into the Emory River and over the surrounding area.

The term *mass movement,* or *mass wasting,* is used to describe the movement of a mass of earth materials with gravity; examples include landslides and mudflows (which with other kinds of mass movement are covered in this chapter). In areas where humans have disturbed the landscape, mass movement may occur more frequently, with disastrous consequences. Such was the case with this fly-ash spill at the Kingston Steam Plant (KSP).

The KSP is located in Roane County near the town of Kingston. Built by the Tennessee Valley Authority (TVA) in the 1950s to provide power for the atomic energy installations at nearby Oak Ridge, this coal-fired 1456-MW power plant burns 14,000 tons of pulverized coal a day. Before the exhaust smoke leaves the 305-m (1000-ft) smokestacks, fly ash is extracted from it (to reduce air pollution), fluidized with water, and moved as slurry to ash ponds (Figure GN 13.1a). In these ponds, the solids settle and are gradually collected in dredge holding cells constructed along the Emory River floodplain.

Embankments surrounding the dredge cells (labeled in the aerial photo) consisted originally of clay and later were built up with materials derived from the ash-waste product, without the benefit of engineering design specifications. As more waste ash was added, the walls eventually rose to 18 m (59 ft) in height. Over the years, seepage and leaking were reported.

During the first three weeks of December 2008, above-normal rainfall of more than 16 cm (6.5 in.) fell in the region. On December 22 the entire northwest corner and sides of the ash-pond embankment failed, triggering a mass-wasting event that released fly-ash slurry onto the landscape, destroying homes and infrastructure. The volume of the spill was equal to 1.1 billion gallons, or about 100 times the 1989 *Exxon–Valdez* oil spill in Alaska (Figure GN 13.1b). Worse, the slurry contained harmful metals and other pollutants, such as arsenic, copper, barium, cadmium, chromium, lead, mercury, nickel, and thallium. These contaminants pose a significant risk to human health.

Almost a year after the disaster, the TVA had removed about two-thirds of the ash in the Emory River, moving it by train to a landfill in Alabama. The next cleanup phase—in which the TVA began to permanently store on-site all of the ash released into the embayment, thus reducing the hazards of transport—was completed in 2010. The final phase, now in progress, includes ash excavation from the embayment to an on-site landfill and restoration of the aquatic and riparian ecosystems; it is expected to be finished in 2015, bringing the total cost to about $1.1 billion.

This human-caused mass-movement event illustrates the hazards of ignoring scientific and engineering principles, in both natural and human-modified environments.

GEOSYSTEMS NOW ONLINE Go to Chapter 13 on the *MasteringGeography* website (www.masteringgeography.com) for resources and activities. In 2010, a similar event occurred in western Hungary when a retaining wall around a toxic holding pond broke at an aluminum oxide plant, sending a torrent of chemical waste over land and into waterways (see http://www.npr.org/blogs/thetwo-way/2010/10/05/130351938/red-sludge-from-hungarian-aluminum-plant-spill-an-ecological-disaster). What are the differences between these two events? MG

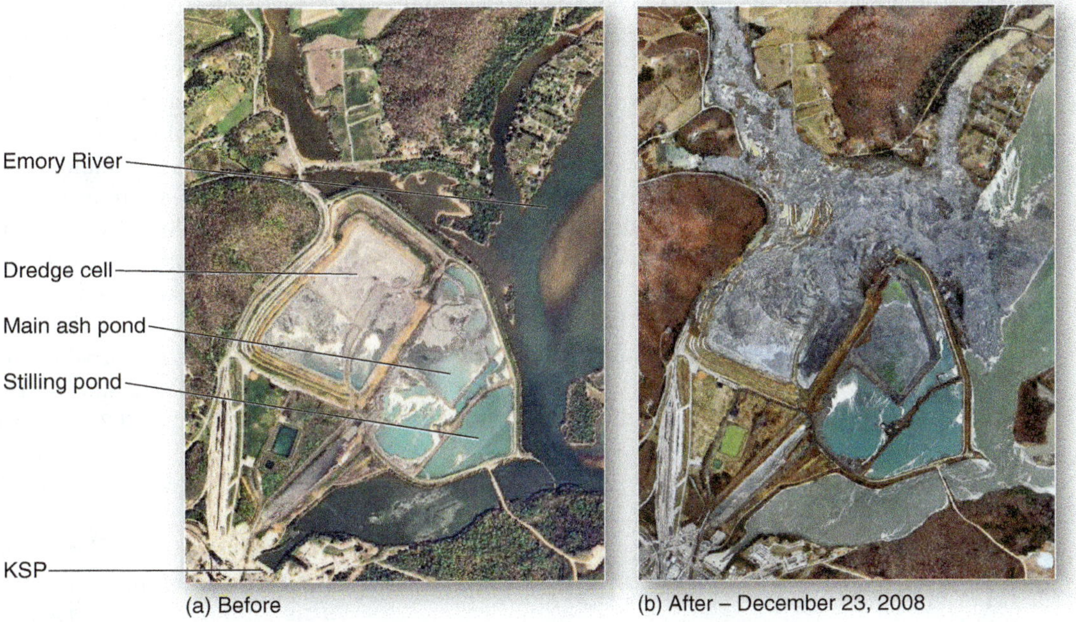

▲**Figure GN 13.1 Before and after the TVA ash slurry spill disaster.** (a) The Emory River enters from the north and joins the Clinch River near the center of this aerial photo. The dredge cells, main ash pond, and stilling pond are in the river floodplain. The power plant itself is off the photo's lower edge. (b) December 23, the day after the spill, the extent of the contaminated wastage is clearly visible. [Courtesy of TVA.]

As mentioned in earlier chapters, the exogenic processes at work on Earth's landscapes include weathering, erosion, transportation, and deposition of materials. In this chapter and the four chapters that follow, we look at exogenic agents and their handiwork—weathering and mass-movement processes, river systems and their landforms, landscapes shaped by waves and wind, and landforms worked by ice and glaciers. All of these are subjects of geomorphology, the science of the origin, development, and spatial distribution of landforms. Whether your preference is for hiking in the mountains or wandering along a river, for visiting sand dunes in the desert or catching waves along a coastline—or perhaps you live in a place where glaciers once carved the land—there is something of interest for you in these chapters.

We begin our study of Earth's exogenic systems with weathering, the process that breaks down rock by disintegrating it into mineral particles or dissolving it into water. Weathering produces an overall weakening of surface rock, which makes it more susceptible to other exogenic processes. The difference between weathering and erosion is important: *Weathering* is the breakdown of materials, whereas *erosion* includes the transport of weathered materials to different locations.

Along with the earthquakes and volcanoes that were the focus of Chapter 12, many events related to exogenic processes are often in the news: for example, a debris avalanche in Austria or Pakistan, a landslide in China or Turkey, a massive rockfall in the Utah canyon country, or a mudflow down a hill in a southern California suburb. In 2008, the world watched as Haiti was deluged by three hurricanes that caused massive landslides and mudflows from deforested mountain slopes. Then in 2010, Haiti suffered more landslides, caused by the earthquake described in Focus Study 12.1.

In this chapter: We look at physical (mechanical) and chemical weathering processes that break up, dissolve, and generally reduce the landscape. Such weathering releases essential minerals from bedrock for soil formation and enrichment. In limestone regions, chemical weathering produces sinkholes, caves, and caverns. In these karst environments, water has dissolved enormous underground areas that are still being discovered by scientists and explorers. In addition, we examine types of mass movements and discuss the processes that cause them.

Landmass Denudation

Denudation is any process that wears away or rearranges landforms. The principal denudation processes affecting surface materials include *weathering, mass movement, erosion, transportation,* and *deposition,* as produced by moving water, air, waves, and ice—all influenced by the pull of gravity.

▲**Figure 13.1 Delicate Arch, Arches National Park, Utah.** Resistant rock strata at the top of the structure helped preserve the arch beneath as surrounding rock eroded away. Note the person standing at the base in the inset photo for a sense of scale. In the distance are the snow-covered La Sal Mountains, an example of a laccolith (a type of igneous intrusion) exposed by erosion. [Bobbé Christopherson.]

Interactions between the structural elements of the land and the processes of denudation are complex. They represent an ongoing opposition between the forces of weathering and erosion and the resistance of Earth materials.

The iconic 15-story-tall Delicate Arch in Utah is dramatic evidence of this conflict (Figure 13.1). An assortment of weathering processes have worked in combination with the differing resistances of the rocks to produce this delicate sculpture—an example of **differential weathering**, where a more resistant cap rock protects supporting strata below.

The buttes, pinnacles, and mesas of the U.S. Southwest are other examples of resistant horizontal rock strata that have eroded differentially. Removal of strata that are less resistant produces unusual desert sculptures—arches, windows, pedestals, and delicately balanced rocks. For an idea of the quantity of material removed by weathering and erosion to produce these landscapes, imagine a line intersecting the tops of the Mitten Buttes shown in Figure 13.2. These buttes exceed 300 m (1000 ft) in height, which is similar to the Chrysler Building in New York City or the U.S. Bank Tower (Library Tower) in Los Angeles.

As stated in earlier chapters, endogenic processes, such as tectonic uplift and volcanic activity, build landforms into *initial landscapes,* whereas exogenic processes tear landforms down, developing *sequential landscapes* characterized by lower relief, gradual change, and stability. However, these countering sets of processes happen simultaneously. Scientists have proposed several hypotheses to model denudation processes and to account for the appearance of the landscape.

▲**Figure 13.2 Monument Valley landscape.** (a) Mitten Buttes, Merrick Butte, and rainbow in Monument Valley, Navajo Tribal Park, which have appeared as a backdrop in many movies, along the Utah–Arizona border. (b) A schematic of the tremendous removal of material by weathering, erosion, and transport. [Author.]

Dynamic Equilibrium Approach to Understanding Landforms

A landscape is an open system, with highly variable inputs of energy and materials. The Sun provides radiant energy that converts into *heat energy* that drives the hydrologic cycle and other Earth systems. The hydrologic cycle imparts *kinetic energy* through the mechanical motion of moving air and water. *Chemical energy* is available from the atmosphere and various reactions within the crust. In addition, uplift of the land by tectonic processes creates *potential energy of position* as land rises above sea level. Remember from Chapter 4 that potential energy is stored energy that has the capacity to do work under the right conditions, such as the pull of gravity down a hillslope.

As landscapes and the forces acting on them change, the surface constantly responds in search of equilibrium. Every change produces compensating actions and reactions. Tectonic uplift creates disequilibrium, an imbalance, between relief and the energy required to maintain stability. The idea of landscape formation as a balancing act between uplift and reduction by weathering and erosion is the **dynamic equilibrium model**. Landscapes in a dynamic equilibrium show ongoing adaptations to the ever-changing conditions of local relief, rock structure, and climate.

Endogenic events, such as faulting or a volcanic eruption, or exogenic events, such as a heavy rainfall or a forest fire, may change the relationships between landscape elements and within landscape systems. During or following a destabilizing event, a landform system sometimes arrives at a **geomorphic threshold**, or tipping point, where the system lurches to a new operational level. This threshold is reached when a geomorphic system moves from the slow accumulation of small adjustments (as occurs in a steady-state equilibrium) to a point of abrupt change that takes it to a new system state (as occurs in a dynamic equilibrium)—such as when a flood establishes a new river channel or a hillslope adjusts after a landslide. Such a threshold can also occur at the precise moment when force overcomes resistance within a system; for example, when slope stability fails and movement ensues in a downhill direction (as during a landslide). After crossing this threshold, the system establishes a new set of equilibrium relationships. (Please review Figure 1.9 in Chapter 1.)

The dynamic equilibrium model encompasses a series of steps that usually follow a sequence over time. First is equilibrium stability, in which the system fluctuates around some average. Next is a destabilizing event, followed by a period of adjustment. Last is the development of a new and different condition of equilibrium stability. Slow, continuous-change events, such as soil development and erosion, tend to maintain a near-equilibrium condition in the system. Dramatic events such as a major landslide require longer recovery times before equilibrium is reestablished. Figure GIA 13.1 provides an example, in which the failure of saturated slopes caused a landslide that brought sediment and debris into a river and introduced a disequilibrium condition.

Slopes

Material loosened by weathering is susceptible to erosion and transportation. However, for material to move downslope, the forces of erosion must overcome other forces: friction, inertia (the resistance to movement), and the cohesion of particles to one another (GIA). If the angle is steep enough for gravity to overcome frictional forces, or if the impact of raindrops or moving animals or even wind dislodges material, then erosion of particles and transport downslope can occur.

Slopes, or *hillslopes*, are curved, inclined surfaces that form the boundaries of landforms. The basic components of a slope, illustrated in *Geosystems In Action*, Figure GIA 13, vary with conditions of rock structure and climate. Slopes generally feature an upper *waxing slope* near the top (*waxing* means increasing). This convex surface curves downward and may grade into

(text continued on page 398)

A slope, like the recently disturbed slope in GIA 13.1, is an open system that tends toward dynamic equilibrium. If the forces acting on slope materials —shown in GIA 13.2—are balanced, the slope remains stable. If the forces become unbalanced, the slope changes until a new equilibrium is reached. Over time, this equilibrium-seeking process gives slopes a characteristic structure, or "anatomy" (GIA 13.3).

13.1 A SLOPE IN DISEQUILIBRIUM
Unstable, saturated soils gave way on this hillslope, leaving a debris dam partly blocking the river. The hillslope, river, and forest ecosystem are in disequilibrium as adjustments to new conditions proceed. [Author.]

13.2 FORCES ON A SLOPE
Directional forces (noted by arrows) act on materials along an inclined slope. If the forces opposing motion fall below the force of gravity, the slope is destabilized and material moves downhill. A variety of events can destabilize a slope, including heavy rain, a wildfire that destroys protective plant cover, or an earthquake.

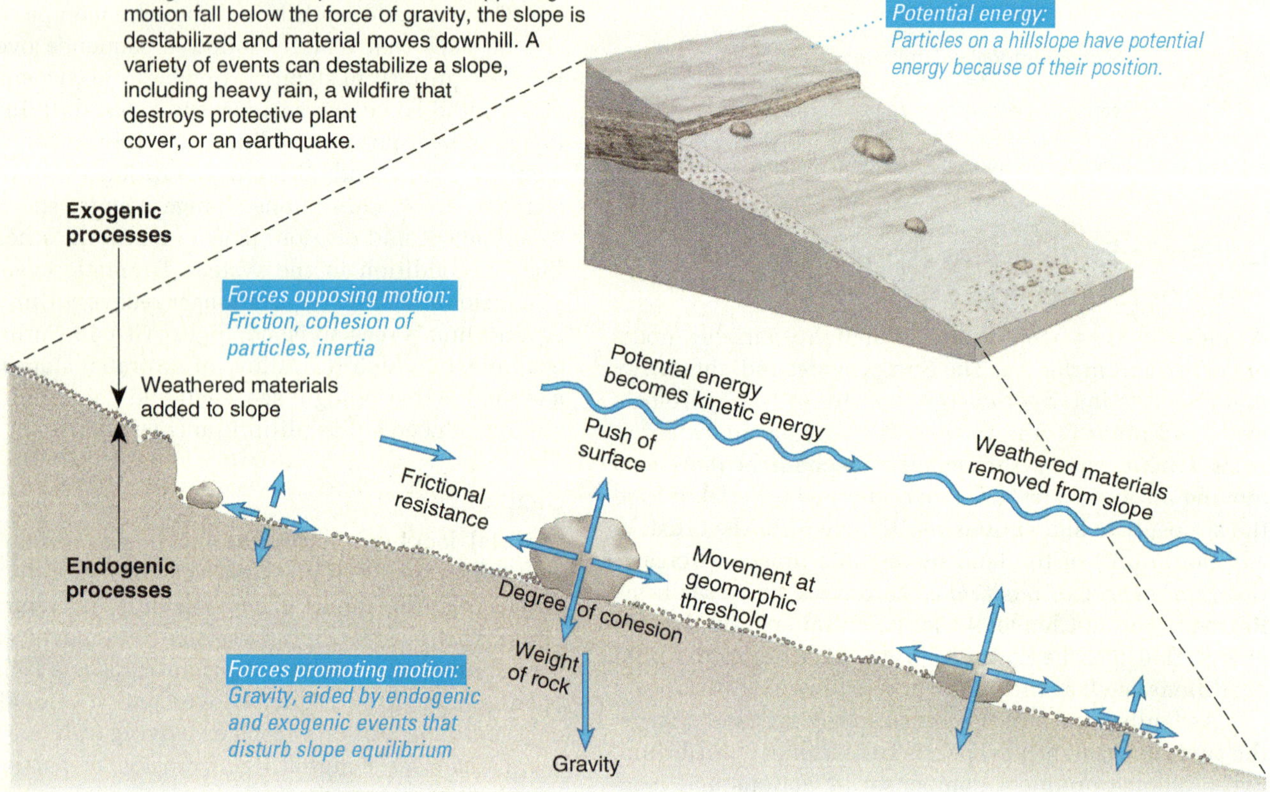

Infer: What events or processes could reduce the degree of cohesion of particles on a slope?

Animation
Mass Movement

MasteringGeography™

Visit the Study Area in MasteringGeography™ to explore slopes and the dynamic equilibrium model.

Visualize: Study a geosciences animation of mass movement.

Assess: Demonstrate understanding of slopes and the dynamic equilibrium model (if assigned by instructor).

13.3 ANATOMY OF A SLOPE

Hillslopes typically develop a structure made up of several elements: a convex *waxing slope*, rock outcrop, debris slope, and concave *waning slope*. One main process predominates on each part of a hillslope: physical and chemical weathering on the upper slope, transportation on the debris slope, and deposition on the lower, waning slope.

Rock outcrop (free face): *The rock outcrop interrupts the slope. Frost wedging loosens rock fragments from the outcrop to form the debris slope.*

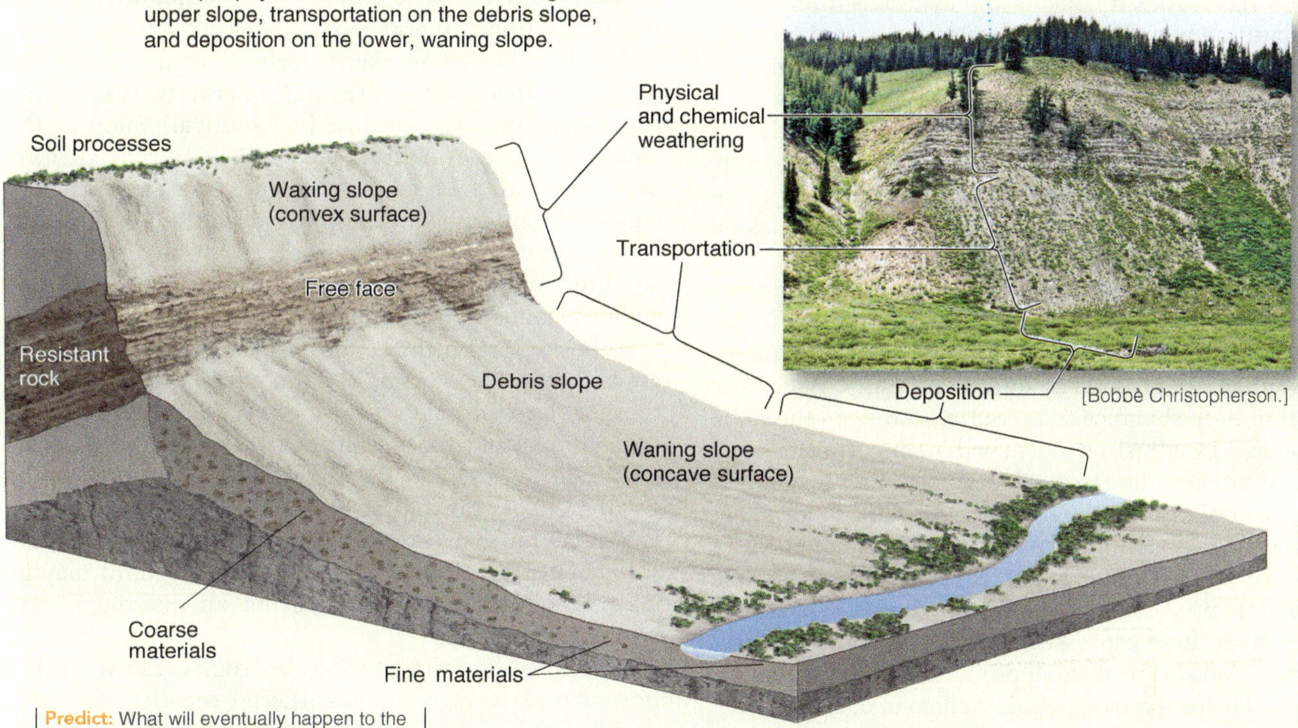

[Bobbè Christopherson.]

Predict: What will eventually happen to the coarse materials on the debris slope?

Many slope elements are visible in the Gros Ventre Range near Jackson Hole Airport, Jackson, Wyoming. Locate some of the named features in the illustration above and in GIA 13.2. [Bobbè Christopherson.]

GEOquiz

1. Explain: Explain how an exogenic event results in a rock particle's movement down a hillslope. Refer to the forces and energy types shown in GIA 13.2 in your answer.

2. Apply concepts: Is the slope in GIA 13.3 at its angle of equilibrium? Explain your answer, referring to specific parts of the slope and the processes that affect them.

a *free face* below, a steep scarp or cliff whose presence indicates an outcrop of resistant rock.

Downslope from the free face is a *debris slope,* which receives rock fragments and materials from above. The condition of a debris slope reflects the local climate. In humid climates, continually moving water carries material away, lowering the angle of the debris slope. But in arid climates, debris slopes accumulate material. A debris slope grades into a *waning slope,* a concave surface along the base of the slope. You can identify these slope components and conditions on the actual hillslope shown in GIA.

Slopes are open systems and seek an *angle of equilibrium* among the forces described here. Conflicting forces work simultaneously on slopes to establish a compromise incline that balances these forces optimally. When any condition in the balance is altered, all forces on the slope compensate by adjusting to a new dynamic equilibrium.

In summary, the rates of weathering and breakup of slope materials, coupled with the rates of mass movement and material erosion, determine the shape and stability of the slope. A slope is *stable* if its strength exceeds these denudation processes and *unstable* if its materials are weaker than these processes. Why are hillslopes shaped in certain ways? How does slope anatomy evolve? How do hillslopes behave during rapid, moderate, or slow uplift? These are topics of active scientific study and research.

CRITICALthinking 13.1
Find a Slope; Apply the Concepts

Locate a slope, possibly near campus, your home, or exposed in a local road cut. Using GIA, can you identify the different parts of the hillslope? What forces act on the hillslope, and what is the evidence of their activity? How would you go about assessing the stability of the slope? Do you see evidence of slope instability near your campus or the region in which you are located—perhaps at a construction site or other disturbed area? ●

Weathering Processes

Weathering is the process that breaks down rock at Earth's surface and slightly below, either disintegrating rocks into mineral particles or dissolving it into water. Weathering weakens surface rock, making the rock more susceptible to the pull of gravity. Weathering processes are both physical (mechanical), such as the wedging action of frost in the cracks of a rock surface, and chemical, such as the dissolution of minerals into water. The interplay of these two broad types of weathering is complex; in many cases, the suite of processes combine synergistically to produce unique landforms such as Delicate Arch in Figure 13.1.

On a typical hillside, loose surface material such as gravel, sand, clay, or soil overlies consolidated, or solid, rock called **bedrock**. In most areas, the upper surface of bedrock undergoes continual weathering, creating broken-up rock called **regolith**. As regolith continues to weather, or is transported and deposited, the loose surface material that results becomes the basis for soil development (Figure 13.3). In some areas, regolith may be missing or undeveloped, exposing an outcrop of unweathered bedrock.

As a result of this process, bedrock is known as the *parent rock* from which weathered regolith and soils develop. Wherever a soil is relatively young, its parent rock is traceable through similarities in composition. For example, in the canyon country of the U.S. Southwest, sediments derive their color and character from the parent rock that is the substance of the cliffs seen in Figure 13.4a, just as the sediments on Mars derive their characteristics from weathered parent material seen in Figure 13.4b. **Parent material** is the consolidated or unconsolidated material from which soils develop, ranging from unconsolidated sediments and weathered rock (the fragments in Figure GIA 13.3) to bedrock (the cliff in Figure 13.4a). We discuss soils in Chapter 18.

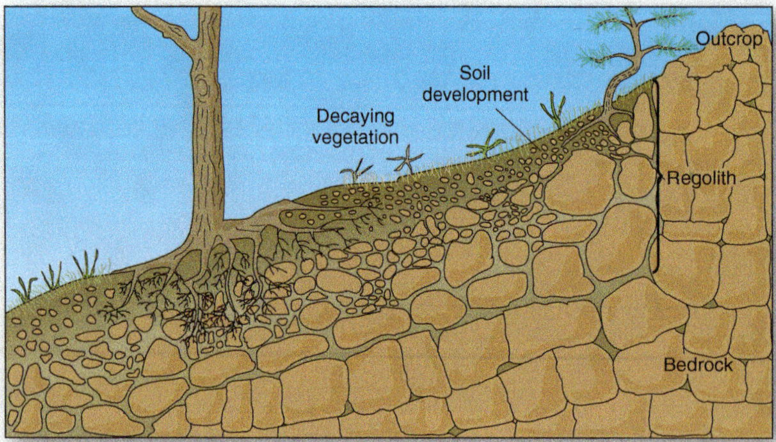

(a) A cross section of a typical hillside.

(b) A cliff exposes hillside components.

▲**Figure 13.3 Regolith and soil.** [Author.]

(a) Reddish-colored dunes in the Navajo Tribal Park, near the Utah–Arizona border, derive their color from the red-sandstone parent materials in the background.

(b) Weathered rocks and windblown sand (in approximate true colors) on Mars, looking toward the Columbia Hills, imaged by the Mars Exploration Rover *Spirit* in 2004.

▲Figure 13.4 **Parent materials.** [(a) Author. (b) Mars image courtesy of NASA/JPL and Cornell University.]

Factors Influencing Weathering Processes

A number of factors influence weathering processes.

- Rock composition and structure (jointing). The character of the bedrock (hard or soft, soluble or insoluble, broken or unbroken) and its mineral composition (different minerals weather at different rates) influence the rate of weathering. **Joints** are fractures or separations in rock that occur without displacement of the rock on either side (in contrast with faulting). Jointing increases the surface area of rock exposed to both physical and chemical weathering.
- Climate (precipitation and temperature). Wetter, warmer environments speed up chemical weathering processes; colder environments have freeze–thaw cycles that cause physical weathering. Rocks that weather rapidly in warm, humid climates may be resistant to weathering in dry climates (an example is limestone).
- Slope orientation. Whether a slope faces north, south, east, or west controls the slope's exposure to Sun, wind, and precipitation. Slopes facing away from the Sun's rays tend to be cooler, moister, and more vegetated than are slopes in direct sunlight. This effect of orientation is especially noticeable in the middle and higher latitudes.
- Subsurface water. The position of the water table and water movement within soil and rock structures influence weathering.
- Vegetation. Although vegetative cover can protect rock by shielding it from raindrop impact and providing roots to stabilize soil, it also produces organic acids, from the partial decay of organic matter, that contribute to chemical weathering. Moreover, plant roots can enter crevices and mechanically break up a rock, exerting enough pressure to force rock segments apart, thereby exposing greater surface area to other weathering processes (Figure 13.5). You may have observed how tree roots can heave the sections of a sidewalk or driveway sufficiently to raise and crack concrete.

Weathering processes occur on micro- as well as macroscopic scales. In particular, research at *microscale* levels reveals a more complex relationship between

▲Figure 13.5 **Physical weathering by tree roots.** [Bobbé Christopherson.]

MG Animation Physical Weathering

climate and weathering than previously thought. At the small scale of actual reaction sites on the rock surface, both physical and chemical weathering processes can occur over a wide range of climate types. At this scale, soil moisture (hygroscopic water and capillary water) activates chemical weathering processes even in the driest landscape. (Review types of soil moisture in Figure 9.9.) Similarly, the role of bacteria in weathering is an important new area of research, with these organisms potentially affecting physical processes as they colonize rock surfaces and chemical processes as they metabolize certain minerals and secrete acids.

Keep in mind that, in the complexity of nature, all these factors influencing weathering rates are operating in concert, and that physical and chemical weathering processes usually operate together. *Time* is the final critical factor affecting weathering, for these processes require long periods. Usually, the longer the duration of exposure for a particular surface, the more it will be weathered.

▲Figure 13.6 **Physical weathering by frost wedging.** Ice expansion involved in freeze–thaw processes broke this marble (a metamorphic rock) apart. [Bobbé Christopherson.]

Physical Weathering Processes

Physical weathering, or *mechanical weathering*, is the disintegration of rock without any chemical alteration. By breaking up rock, physical weathering produces more surface area on which all weathering may operate. For example, breaking a single stone into eight pieces exposes double the surface area susceptible to weathering processes. Physical weathering occurs primarily by frost action, salt-crystal growth, and exfoliation.

Frost Wedging

When water freezes, its volume expands as much as 9% (see Chapter 7). This expansion produces a powerful mechanical force that can overcome the tensional strength of rock. Repeated freezing (expanding) and thawing (contracting) of water is *frost action*, or *freeze–thaw*, which breaks rocks apart in the process of **frost wedging** (Figure 13.6).

The work of ice begins in small openings along existing joints and fractures, gradually expanding them and cracking or splitting them in varied shapes, depending on the rock structure. Sometimes frost wedging results in blocks of rock, or *joint-block separation* (Figure 13.7).

Frost wedging is an important weathering process in the humid microthermal climates (*humid continental* and *subarctic*) and the polar climates, and in the highland climates at high elevations in mountains worldwide. At high latitudes, frost action is important in soils affected by permafrost (discussed in further detail in Chapters 17 and 18).

As winter ends and temperatures warm in mountainous terrain, the falling of rocks from cliff faces occurs more frequently. The rising temperatures that melt the winter's ice cause newly fractured pieces to fall without warning and sometimes start rockslides. Many such incidents are reported in the European Alps, and they appear to be on the increase. The falling rock pieces shatter on impact—another form of physical weathering (Figure 13.8).

Salt-Crystal Growth (Salt Weathering)

Especially in arid climates where heating is intense, evaporation draws moisture to the surface of rocks, leaving behind previously dissolved minerals in the form of crystals (a process called crystallization). Over time, as the crystals accumulate and grow, they exert a force great enough to separate the grains making up the rock and begin breaking the rock to pieces, a process known as *salt-crystal growth*, or *salt weathering*.

In many areas of the U.S. Southwest, groundwater that meets an impermeable rock layer, such as shale, within sandstone rock strata will flow laterally until it emerges at a surface. The water then evaporates and leaves salt crystals that loosen the sand grains within the rock. Subsequent erosion of the grains by water and wind complete the sculpting process, forming alcoves at the base of sandstone cliffs. More than 1000 years

GEOreport 13.1 Rockfalls in Yosemite

Records indicate that in the past 150 years, at least 600 rockfalls have occurred in Yosemite Valley, located in the Sierra Nevada of California. In July 1996, a 162,000-ton granite slab dropped 670 m at 260 kmph (2200 ft at 160 mph), felling more than 500 trees before pulverizing into a light dust that covered 50 acres. In 1999, another rockfall occurred, taking a life. More recently, large rockfalls occurred on Half Dome (see Figure 13.10b) in 2006, Glacier Point in 2008, and Ahwiyah Point in 2009. A National Park Service (NPS) map of rockfalls from 1857 to 2011 is at http://www.nps.gov/yose/naturescience/rockfall.htm. Note from the NPS map that winter and spring events dominate the record, although large rockfalls can occur during the warmer months as well.

(a) Physical weathering along joints in sandstone produces discrete blocks in Canyonlands National Park, Utah. Note the differential weathering, as the softer supporting rock underneath the slabs has already weathered and eroded.

(b) Joint-block separation in slate at Alkenhornet, Isfjord, on Spitsbergen Island in the Arctic Ocean, where freezing is intense.

▲**Figure 13.7 Physical weathering along joints.** [(a) Author. (b) Bobbé Christopherson.]

▲**Figure 13.8 Rockfall.** Shattered rock debris from a large rockfall in Yosemite National Park. Freshly exposed, light-colored rock shows where the rockfall originated. [Author.]

ago, Native Americans built entire villages in these weathered niches, as in Mesa Verde in Colorado and Arizona's Canyon de Chelly (pronounced "canyon duh shay"; Figure 13.9).

Exfoliation The process whereby rock peels or slips off in sheets instead of breaking up into grains is **exfoliation**, a term that generally refers to the removal or shedding of an outer layer. This process is also known as *sheeting*. Exfoliation creates arch-shaped and dome-shaped features on the exposed landscape. These *exfoliation domes* are probably the largest weathering features, in areal extent, on Earth (Figure 13.10).

Exfoliation is thought to occur as pressure is released from the removal of overlying rock. Recall from Chapter 11 how magma that rises into the crust and then remains deeply buried under high pressure forms intrusive igneous rocks called plutons. These plutons cool slowly into coarse-grained, crystalline, granitic rocks that may then be uplifted and subjected to weathering and erosion. As the tremendous weight of overlying material is removed from a granite pluton, the pressure of deep burial is relieved. Over millions of years, the granite slowly responds with an enormous physical heave, initiating a process known as *pressure-release jointing,* in which the rock cracks into joints. Exfoliation is the mechanical weathering that separates the joints into layers resembling curved slabs or plates, often thinner at the top of the rock structure and thicker at the sides. Recent research suggests that exfoliation may also result from the force of gravity working a curved surface, creating tension beneath a dome, that augments pressure-release processes.

Chemical Weathering Processes

Chemical weathering refers to the chemical breakdown, always in the presence of water, of the constituent minerals in rock. The chemical decomposition and decay become more intense as both temperature and precipitation increase. Although individual minerals vary in susceptibility, all rock-forming minerals are responsive to some degree of chemical weathering.

A familiar example of chemical weathering is the eating away of cathedral façades and the etching of tombstones by acid precipitation. In Europe, where increasingly acidic rains resulted from the burning of coal, chemical weathering processes are visible on many buildings. An example is Saint Magnus Cathedral, in the Orkney Islands north of Scotland, which was built of red and yellow sandstone beginning around A.D. 1137. Almost nine centuries of chemical

(b) Water and an impervious rock layer helped concentrate weathering processes in a niche in the overlying sandstone.

(a) Ancient cliff dwelling in a niche formed partially by salt weathering, Canyon de Chelly, Arizona. The dark streaks on the cliff are thin coatings of desert varnish, composed of manganese that is taken up and metabolized by microbes and transformed into oxide minerals.

▲**Figure 13.9 Physical weathering in sandstone.** [Bobbé Christopherson.]

weathering have dissolved cementing materials in the sandstone, breaking the rock down and making the building's intricate decorative sculpture look "melted" and "out-of-focus" (Figure 13.11).

Spheroidal weathering is chemical weathering that softens and rounds the sharp edges and corners of jointed rock (thus the name *spheroidal*) as water penetrates the joints and dissolves weaker minerals or cementing materials (Figure 13.12). A boulder can be attacked from all sides by such weathering, shedding spherical shells of decayed rock, like the layers of an onion. Spheroidal weathering of rock resembles exfoliation, but it does not result from pressure-release jointing.

Hydration and Hydrolysis Chemical decomposition of rock by water can occur by the simple combination of water with a mineral, in the process of *hydration*, and by the chemical reaction of water with a mineral, in the process of *hydrolysis*. **Hydration**, meaning "combination with water," involves little chemical change (it does not form new chemical compounds) but does involve a change in structure. Water becomes part of the chemical composition of the mineral, forming a hydrate. One such hydrate is gypsum, which is hydrous calcium sulfate ($CaSO_4 \cdot 2H_2O$).

When some minerals undergo hydration, they expand, creating a strong mechanical wedging effect that stresses the rock, forcing grains apart in a physical weathering process. A cycle of hydration and dehydration can lead to granular disintegration and further susceptibility of the rock to chemical weathering. Hydration also works with other processes to convert feldspar, a common mineral in many rocks, into clay minerals. The hydration process is also at work on the sandstone niches shown in Figure 13.9.

Hydrolysis is the decomposition of a chemical compound by reaction with water. In geomorphology, hydrolysis is of interest as a process that breaks down silicate minerals in rocks. In contrast with hydration, in which water merely combines with minerals in the rock, hydrolysis chemically breaks down a mineral, thereby producing a different mineral through the chemical reaction.

For example, the weathering of feldspar minerals in granite can occur by reaction with the normal mild acids dissolved in precipitation:

feldspar (K, Al, Si, O) + carbonic acid and water →
residual clays + dissolved minerals + silica

The products of chemical weathering of feldspar in granite include clay (such as kaolinite) and silica. The particles of quartz (silica, or SiO_2) formed in this process are resistant to further chemical breakdown and may wash downstream, eventually becoming sand on some distant beach. Clay minerals become a major component in soil and in shale, a common sedimentary rock.

When minerals in rock are changed by hydrolysis, the interlocking crystal network consolidating the rock breaks down and *granular disintegration* takes place. Such disintegration in granite may make the rock appear corroded and even crumbly (Figure 13.12b).

In Table 11.2, Classification of Igneous Rocks, the sixth line compares the various rocks' resistance to chemical weathering. It shows that the "ultramafic" (low-silica) minerals pyroxene and olivine (on the far right side of the table) are most susceptible to chemical weathering.

(a) Exfoliated granite, White Mountains, New Hampshire.

(b) Exfoliation forms characteristic granite domes such as Half Dome in Yosemite, California.

(c) Rock sheeting exposed along Beverly Sund, Nordaustlandet Island, in the Arctic Ocean.

▲Figure 13.10 **Exfoliation in granite.** Exfoliation loosens slabs of rock, freeing them for further weathering and downslope movement. [(a) and (c) Bobbé Christopherson. (b) Author.]

High-silica minerals such as feldspar and quartz are more resistant. The chemical properties of the minerals determine the resistance of the rock to weathering; for example, basalt (a mafic rock) weathers faster chemically than does granite (a felsic rock).

Oxidation Another type of chemical weathering occurs when certain metallic elements combine with oxygen to form oxides. This process is known as **oxidation**. Perhaps the most familiar oxidation is the "rusting" of iron to produce iron oxide (Fe_2O_3). You see the result of this oxidation after leaving a tool or nails outside only to find them, weeks later, coated with a crumbly reddish-brown substance. Its rusty color is visible on the surfaces of rock and in heavily oxidized soils such as those in the humid southeastern United States, the arid U.S. Southwest, and the tropics (Figure 13.13). Here is a simple oxidation reaction in iron:

$$\text{iron (Fe)} + \text{oxygen } (O_2) \rightarrow \text{iron oxide (hematite; } Fe_2O_3)$$

When oxidation reactions remove iron from the minerals in a rock, the disruption of the crystal structures makes the rock more susceptible to further chemical weathering and disintegration.

▼Figure 13.11 **Chemical weathering of sandstone.** The Saint Magnus Cathedral, in Kirkwall, Scotland, shows the signs of almost nine centuries of chemical weathering. [Bobbé Christopherson.]

(a) Chemical weathering processes act on the joints in granite to dissolve weaker minerals, leading to a rounding of the edges of the cracks in the Alabama Hills. Mount Whitney is visible on the crest of the Sierra Nevada in the background.

(b) Rounded granite outcrop demonstrates spheroidal weathering and the disintegration of rock. The surface is actually crumbly.

◀Figure 13.12 **Chemical weathering and spheroidal weathering.** [Bobbé Christopherson.]

Dissolution of Carbonates Chemical weathering also occurs when a mineral dissolves into solution—for example, when sodium chloride (common table salt) dissolves in water. Remember, water is called the universal solvent because it is capable of dissolving at least 57 of the natural elements and many of their compounds.

Water vapor readily dissolves carbon dioxide, thereby yielding precipitation containing carbonic acid (H_2CO_3). This acid is strong enough to dissolve many minerals, especially limestone, by a type of reaction called **carbonation**. This type of chemical weathering breaks down minerals that contain calcium, magnesium, potassium, or sodium. When rainwater attacks formations of limestone (mainly calcium carbonate, $CaCO_3$), the constituent minerals dissolve and wash away with the mildly acidic rainwater:

$$\text{calcium carbonate} + \text{carbonic acid and water} \rightarrow \text{calcium bicarbonate } (Ca_2{}^{2+}CO_2H_2O)$$

The dissolution of marble, a metamorphic form of limestone, is apparent on tombstones in many cemeteries

(a) Oxidation of iron minerals produces these colors in the tropical rain forest soils of eastern Puerto Rico.

(b) Soils produced by warm, moist conditions in Sumter County, Georgia, bear the color of iron and aluminum oxides. The crop is peanuts.

▲Figure 13.13 **Oxidation processes in soil.** [Bobbé Christopherson.]

GEOreport 13.2 Weathering on bridges in Central Park, NYC

In New York City's Central Park, 36 bridges are built out of various rock types from sources across the U.S. Northeast and Canada. For more than 140 years, physical and chemical processes have weathered these bridges. As air pollution from the burning of fossil fuels has increased, acidity in rain and snow has hastened weathering rates, a problem compounded by the use in winter of salt on the roads crossing the bridges. Decorative design elements on the bridges are now disappearing as weathering tears at the surface rock. Some heavily weathered sandstone blocks have had to be replaced by cast concrete.

▲**Figure 13.14 Dissolution of limestone.** A marble tombstone is chemically weathered beyond recognition in a Scottish churchyard. Marble is a metamorphic form of limestone. Readable dates on surrounding tombstones suggest that this one is about 228 years old. [Bobbé Christopherson.]

(Figure 13.14). In environments where adequate water is available for dissolution, weathered limestone and marble take on a pitted and worn appearance. Acid precipitation also enhances carbonation processes (see Focus Study 3.2, Acid Deposition: Damaging to Ecosystems).

Karst Topography

In certain areas of the world with extensive limestone formations, chemical weathering involving dissolution of carbonates dominates entire landscapes (Figure 13.15). These areas are characterized by pitted, bumpy surface topography, poor surface drainage, and well-developed *solution channels* (dissolved openings and conduits) underground. In landscapes of this type, weathering and erosion caused by groundwater may result in remarkable mazes of underworld caverns.

These are the hallmark features and landforms of **karst topography**, named for the Krš Plateau in Slovenia (formerly Yugoslavia), where karst processes were first studied. Approximately 15% of Earth's land area has some karst features, with outstanding examples found in southern China, Japan, Puerto Rico, Jamaica, the Yucatán of Mexico, Kentucky, Indiana, New Mexico, and Florida.

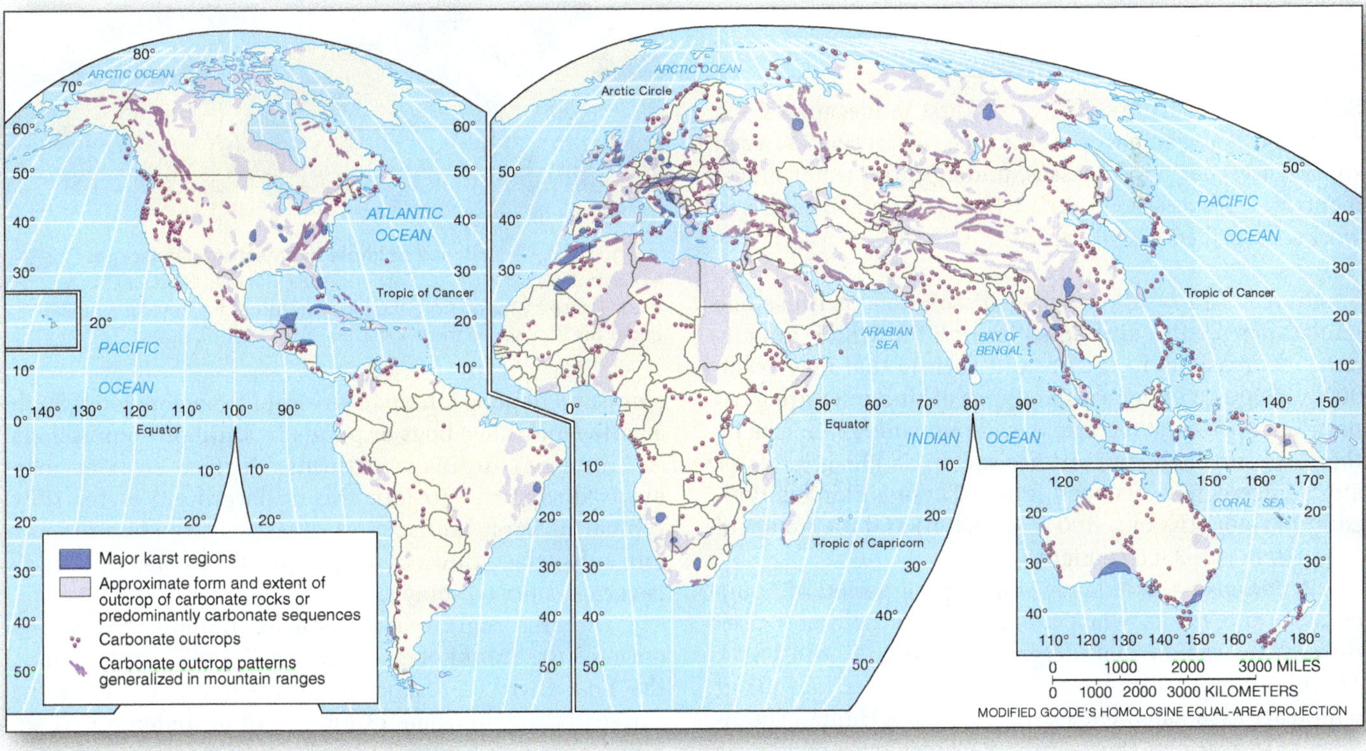

▲**Figure 13.15 Karst landscapes and limestone regions.** Major karst regions exist on every continent except Antarctica. The outcrops of carbonate rocks or predominantly carbonate sequences are limestone and dolomite (calcium magnesium carbonate) but may contain other carbonate rocks. [Map adapted by Pam Schaus, after USGS sources, and D. C. Ford and P. Williams, *Karst Geomorphology and Hydrology*, p. 601. © 1989 by Kluwer Academic Publishers. Adapted by permission.]

Formation of Karst

For a limestone landscape to develop into karst topography, several conditions are necessary:

- The limestone formation must contain 80% or more calcium carbonate for dissolution processes to proceed effectively.
- Complex patterns of joints in the otherwise impermeable limestone are needed for water to form routes to subsurface drainage channels.
- An aerated (air-containing) zone must exist between the ground surface and the water table.
- Vegetation cover is needed to supply varying amounts of organic acids that enhance the dissolution process.

The role of climate in providing optimum conditions for karst processes remains under debate, although the amount and distribution of rainfall appear important. The karst features found today in arid regions were formed during past climatic conditions of greater humidity. Karst is rare in the Arctic and Antarctic regions because subsurface water, although present there, is generally frozen.

As with all weathering processes, time is an important factor. Early in the twentieth century, scientists proposed that karst landscapes progress through identifiable stages of development, from youth to old age. Evidence has not supported this idea, and today karst landscapes are thought to be locally unique, a result of site-specific conditions. Nonetheless, mature karst landscapes display certain characteristic forms.

▲Figure 13.16 **Buraco das Araras, a solution sinkhole near Goiás, Brazil.** [J. Carvalho/Getty Images.]

▲Figure 13.17 **Collapse sinkhole in limestone landscape.** Collapse of this sinkhole near Frederick, Maryland, in 2003 was likely made worse by runoff from the highway (note the storm drain pipe under the highway in the photo). [Randall Orndorff, USGS.]

Features of Karst Landscapes

Several landforms are typical of karst landscapes. Each form results to some extent from the interaction between surface weathering processes, underground water movement, and processes occurring in subterranean cave networks, described just ahead.

Sinkholes The weathering by dissolution of limestone landscapes creates **sinkholes**, or *dolines*, which are circular depressions in the ground surface that may reach 600 m (2000 ft) in depth. Two types of sinkholes are most prominent in karst terrain. A *solution sinkhole* forms by the slow subsidence of surface materials along joints or at an intersection between joints. These sinkholes typically have depths of 2–100 m (7–330 ft) and diameters of 10–1000 m (33–3300 ft; Figure 13.16).

A *collapse sinkhole* develops over a period of hours or days and forms when a solution sinkhole collapses through the roof of an underground cavern (Figure 13.17). These sinkholes can have dramatic features, not all of which are associated with karst processes. Human activities cause many of these sinkhole subsidence events, as described in Geo Report 13.3.

Karst Valleys Through continuing dissolution and collapse, sinkholes may coalesce to form a *karst valley*—an elongated depression up to several kilometers long. Such a valley may have bogs or ponds in sinkhole depressions and unusual drainage patterns. Surface streams may even "disappear" to join the underground water flow typical of karst landscapes; disappearing streams may join subsurface flows by way of joints or holes linking to cavern systems or may flow directly into caves.

The area southwest of Orleans, Indiana, has an average of 1022 sinkholes per 2.6 km^2 (1 mi^2). In this area, the Lost River, a disappearing stream, flows from the surface into more than 13 km (8 mi) of underground solution channels before it resurfaces in a spring, or "rise." The Lost River's dry channel can be seen on the lower left of the topographic map in Figure 13.18d. The Orangeville Rise near Orleans is the second largest spring in Indiana (Figure 13.18e).

▶**Figure 13.18 Features of karst topography in Indiana.** [(d) Mitchell, Indiana quadrangle, USGS. (b), (c), and (e) Bobbé Christopherson.]

(b) Rolling karst landscape and cornfields near Orleans, Indiana.

(c) Pond in a sinkhole depression near Palmyra, Indiana

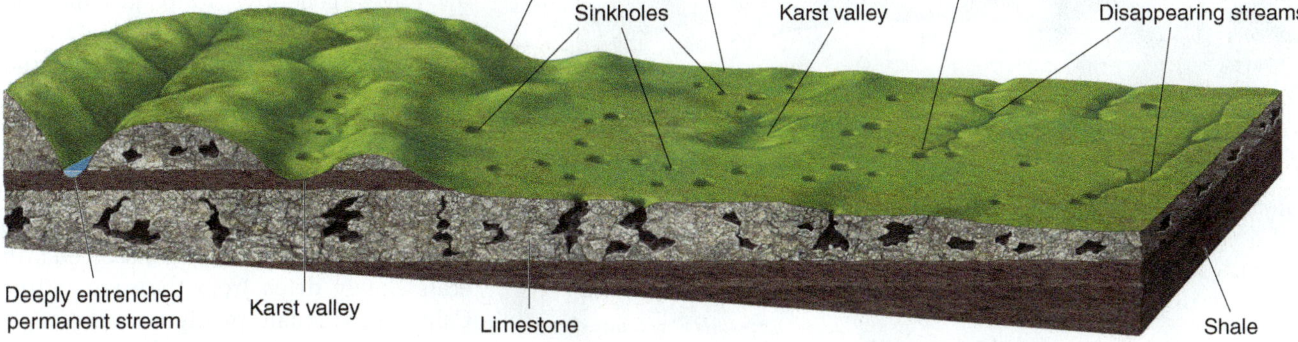

(a) Idealized features of karst topography in southern Indiana.

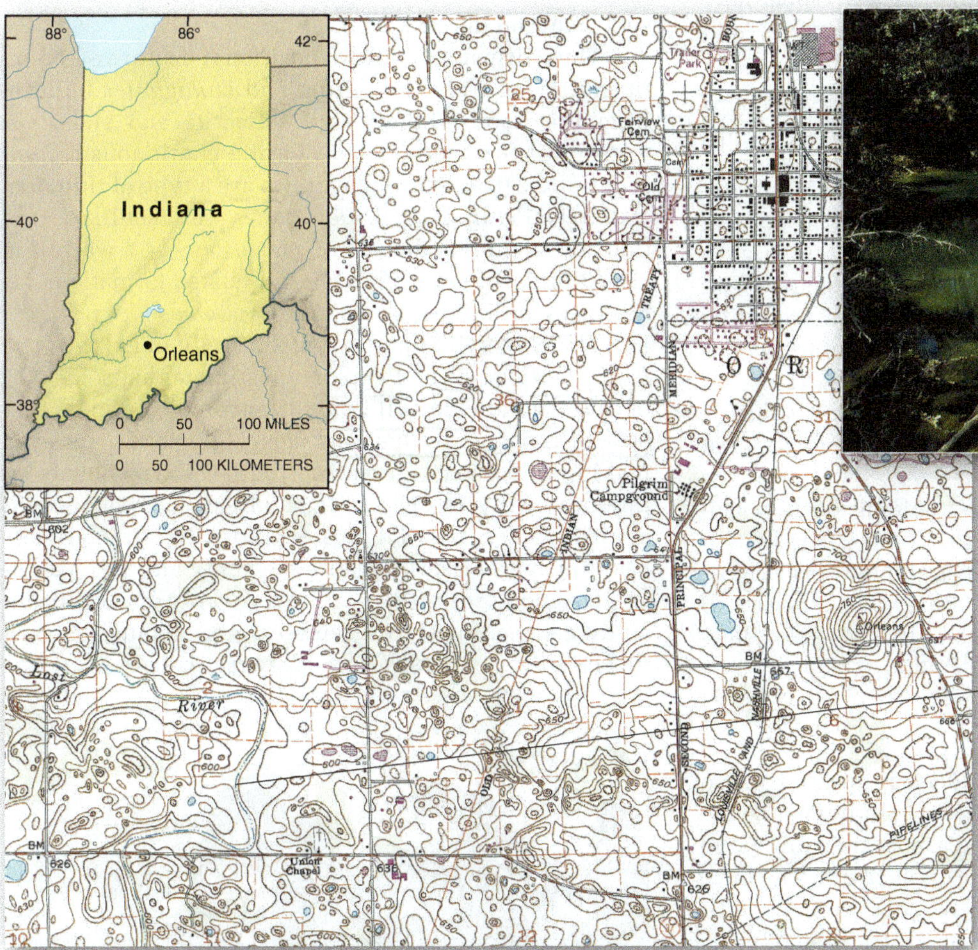

(e) Orangeville Rise, a spring just north of the Lost River Rise.

(d) Karst topography near Orleans, Indiana; note the contour lines and the depressions, indicated with small hachures (tick marks) on the downslope side of the contour lines.

MG Notebook
Karst Farm Park, Indiana

GEOreport 13.3 Sinkholes caused by human activities

Along Bushkill Creek, downstream from the Hercules mining quarry enterprise, several dozen sinkholes have collapsed since 2000, taking down one of the bridges from State Route 33 and threatening neighborhoods in Stockertown, Northampton County, Pennsylvania. Sinkhole development correlated with groundwater pumping at the nearby quarry, which lowered water tables. In Chicago in 2013, a sinkhole opened in the middle of a street, swallowing three cars. A burst water main caused the collapse, which led to only minor injuries. (See the *MasteringGeography* website for other examples of sinkhole collapse from groundwater removal.)

◄**Figure 13.19 Deep-space research using cockpit karst topography.** Cockpit karst topography near Arecibo, Puerto Rico, is the setting for Earth's largest radio telescope. The discoloration of the dish does not affect telescope reception. The suspended movable receivers where the signals focus are 168 m (550 ft) above the dish. [Bobbé Christopherson; inset Cornell University.]

Tropical Karst In tropical climates, karst topography includes two characteristic landforms—cockpits and cones—with prominent examples found in the Caribbean region (Puerto Rico, Jamaica) and southeast Asia (China, Vietnam, and Thailand). Weathering in these wet climates, where thick beds of limestone are deeply jointed (exposing a large surface area for dissolution processes), forms a complex topography called *cockpit karst*, resembling the shape of an egg-carton (Figure 13.19). The "cockpits" are steep-sided, star-shaped hollows in the landscape with water drainage occurring by percolation from the bottom of the cockpit to the underground water flow. Sinkholes may form in the cockpit bottoms, and according to some theories, solution sinkhole collapse is an important cause of cockpit karst topography.

Dissolution weathering in the tropics also leaves isolated resistant limestone blocks that form cones known as *tower karst*. These resistant cones and towers are most remarkable in several areas of China, where towers up to 200 m (660 ft) high interrupt an otherwise flat, low-elevation plain (Figure 13.20).

Caves and Caverns

Caves are defined as natural underground areas large enough for humans to enter. Caves form in limestone because it is so easily dissolved by carbonation; any large cave formed by chemical processes is a *cavern*. The largest limestone caverns in the United States are Mammoth Cave in Kentucky (also the longest surveyed cave in the world at 560 km, or 350 mi), Carlsbad Caverns in New Mexico, and Lehman Cave in Nevada.

Carlsbad Caverns are in 200-million-year-old limestone formations deposited when shallow seas covered the area. Regional uplifts associated with the building of the Rockies (the Laramide orogeny, 40–80 million years ago) elevated the region above sea level, subsequently leading to active cave formation.

Caves generally form just beneath the water table, where later lowering of the water level exposes them to further development (Figure 13.21). As discussed in Chapter 10B, speleothems are formations consisting of mineral deposits inside caves and occur in various characteristic shapes. *Dripstones* are speleothems formed as water containing dissolved minerals slowly drips from the cave ceiling. Calcium carbonate precipitates out of the evaporating solution, literally one molecular layer at a time, and accumulates on a spot below on the cave floor.

Thus dripstones are depositional features—*stalactites* growing from the ceiling and *stalagmites* building from the floor. Sometimes a stalactite and stalagmite grow until they connect and form a continuous *column* (Figure 13.21b and c). Soda straws are a type of thin, long stalactite (Figure 13.21e). *Flowstones* are sheet-like formations of calcium carbonate on cave floors and walls (Figure 13.21d). For more on caves and related formations, see http://www.goodearthgraphics.com/virtcave/virtcave.html.

The exploration and scientific study of caves is *speleology*. Scientists and explorers estimate that some 90% of caves worldwide still lie undiscovered, and more than 90% of known caves have not been biologically surveyed, making this a major research frontier. Cave habitats are unique, nearly closed, self-contained ecosystems with simple food chains and great stability. In total darkness, bacteria synthesize inorganic elements and produce organic compounds that sustain many types of cave life, including algae, small invertebrates, amphibians, and fish. *Biospeleology* is the study of cave organisms.

▲**Figure 13.20 Tower karst, Li River valley, China.** [Keren Su/Corbis]

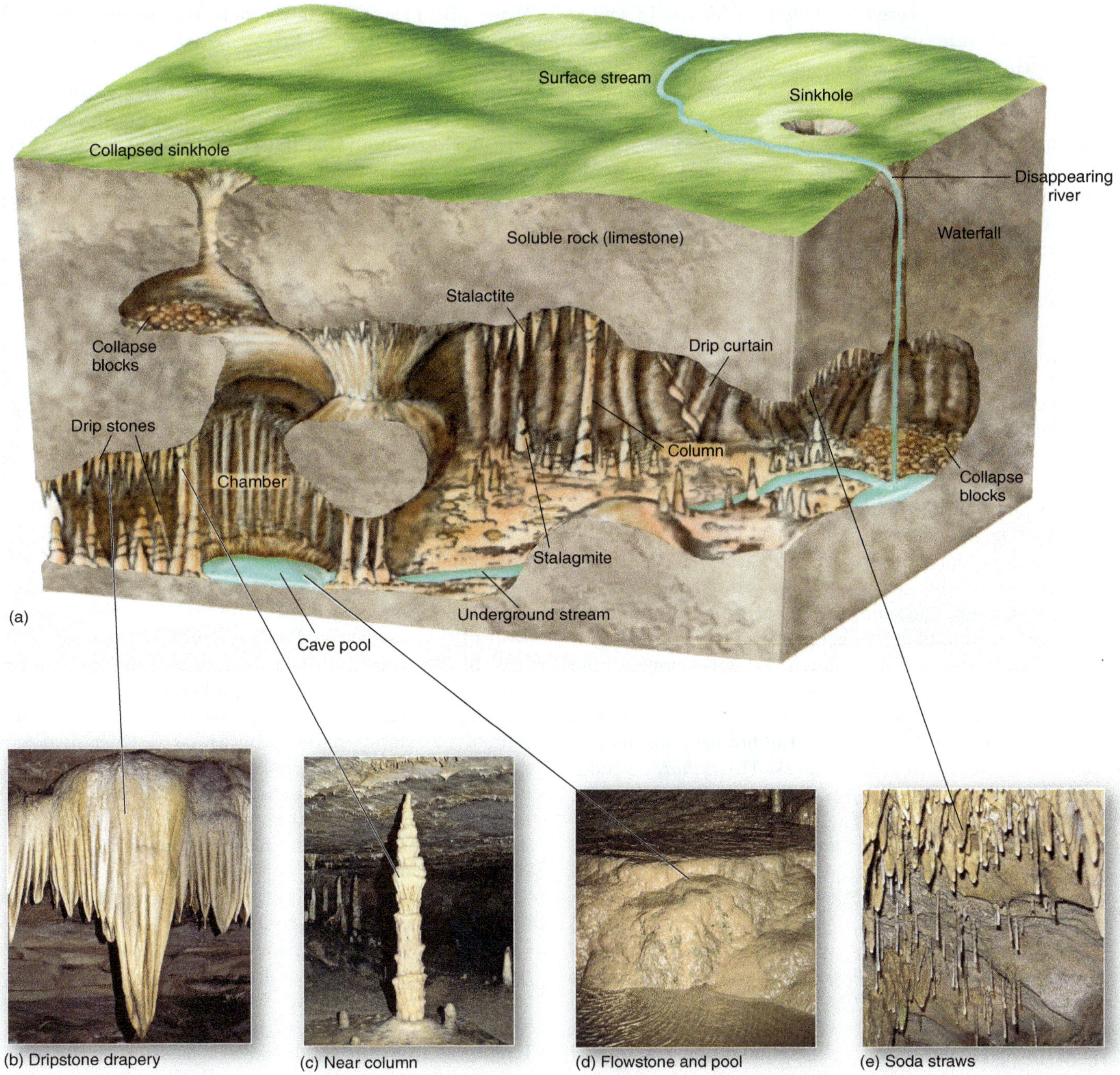

▲Figure 13.21 **An underground cavern and related forms in limestone.** A column is created when stalactites from the ceiling and stalagmites from the floor connect. All cave photos from Marengo Caves, Marengo, Indiana (see Geo Report 13.4).
[All photos by Bobbé Christopherson.]

GEOreport 13.4 Amateurs make cave discoveries

In the early 1940s near Bedford, Indiana, a man awoke to find his farm pond at the bottom of a deep collapsed sinkhole. This sinkhole became the entrance to the extensive Marengo cave system, which includes a subterranean navigable stream. Several years ago in Viet Nam, a local farmer reported the entrance to a large, previously unknown cave in the remote jungle. In 2010, this farmer led a joint British–Vietnamese expedition into what is now believed to be the largest single cave passage in the world. Clearly, in addition to the ongoing activities of scientists in this field, local residents and amateur cavers—"spelunkers"—have made many important discoveries, a fact that keeps this popular sport very much alive. For nearly a thousand worldwide links and information, see http://www.cbel.com/speleology.

In a cave discovered in 1986, near Movile in southeastern Romania, cave-adapted invertebrates were discovered after millions of years of sunless isolation. Thirty-one of these organisms were previously unknown. Without sunlight, the ecosystem in Movile is sustained on sulfur-metabolizing bacteria that synthesize organic matter using energy from oxidation processes. These chemosynthetic bacteria feed other bacteria and fungi that, in turn, support cave animals. The sulfur bacteria produce sulfuric acid compounds that may prove to be important in the chemical weathering of some caves.

Mass-Movement Processes

In the South American country of Colombia, Nevado del Ruiz is the northernmost of two dozen dormant (not extinct, sometimes active) volcanic peaks in the Cordillera Central. This volcano erupted six times during the past 3000 years, killing 1000 people during its last eruption, in 1845. On November 13, 1985, at 11 P.M., after a year of earthquakes and harmonic tremors (seismic energy releases associated with volcanoes), a growing bulge on its northeast flank, and months of small summit eruptions, Nevado del Ruiz violently erupted in a lateral explosion and triggered a mudflow down its slopes toward the sleeping city and villages below.

The mudflow was a mixture of liquefied mud and volcanic ash that developed as the hot eruption melted ice on the mountain's snowy peak. This *lahar,* an Indonesian word referring to mudflows of volcanic origin, moved rapidly down the Lagunilla River toward the villages below. The wall of mud was at least 40 m (130 ft) high as it approached Armero, a regional center with a population of 25,000. The lahar buried the sleeping city: 23,000 people were killed; thousands were injured; 60,000 were left homeless across the region. The debris flow generated by the 1980 eruption of Mount St. Helens was also a lahar.

Landslides are another type of mass movement that poses a major hazard, causing an average of 25 fatalities every year in the United States. For more information, see http://landslides.usgs.gov/ or the American Geophysical Union's landslide blog at http://blogs.agu.org/landslideblog/.

Mass-Movement Mechanics

Mass movement, also called **mass wasting**, is the downslope movement of a body of material made up of soil, sediment, or rock propelled by the force of gravity. Mass movements can occur on land, or they can occur beneath the ocean as submarine landslides.

Slope Angle and Forces

All mass movements occur on slopes under the influence of gravitational stress. If we pile dry sand on a beach, the grains will flow downslope until equilibrium is achieved. The steepness of the resulting slope, called the **angle of repose**, depends on the size and texture of the grains. This angle represents a balance of the driving force (gravity) and resisting force (friction and shear). The angle of repose for various materials ranges between 33° and 37° (from horizontal) and between 30° and 50° for snow avalanche slopes.

As noted, the *driving force* in mass movement is gravity. It works in conjunction with the weight, size, and shape of the surface material; the degree to which the slope is oversteepened (how far it exceeds the angle of repose); and the amount and form of moisture available (frozen or fluid). The greater the slope angle, the more susceptible the surface material is to mass-wasting processes.

The *resisting force* is the shear strength of the slope material—that is, its cohesiveness and internal friction, which work against gravity and mass wasting. To reduce shear strength is to increase shear stress, which eventually reaches the point at which gravity overcomes friction, initiating slope failure.

Conditions for Slope Failure

Several conditions can lead to the slope failure that causes mass movement. Failure can occur when a slope becomes saturated by a heavy rainfall; when a slope becomes oversteepened (40° to 60° slope angle), such as when river or ocean waves erode the base; when a volcanic eruption melts snow and ice, as happened on Nevado del Ruiz and Mount St. Helens; or when an earthquake shakes debris loose or fractures the rock that stabilizes an oversteepened slope.

Water content is an important factor for slope stability; an increase in water content may cause rock or regolith to begin to flow. Clay surfaces are highly susceptible to hydration (physical swelling in response to the presence of water). When clay surfaces are wet, they deform slowly in the direction of movement; when saturated, they form a viscous fluid that fails easily with overlying weight. The 1995 La Conchita mudslide in California (see Figure 13.23a) occurred during an unusually wet year. The same slope failed again in 2005, after a two-week period of near-record rainfall.

The shocks and vibrations associated with earthquakes often cause mass movement, as happened in the Madison River Canyon near West Yellowstone, Montana. Around midnight on August 17, 1959, an M 7.5 earthquake broke a dolomite (a type of limestone) block along the foot of a deeply weathered and oversteepened slope (white area in Figure 13.22), releasing 32 million m^3 (1.13 billion ft^3) of mountainside. The material moved downslope at 95 kmph (60 mph), causing gale-force winds through the canyon. Momentum carried the

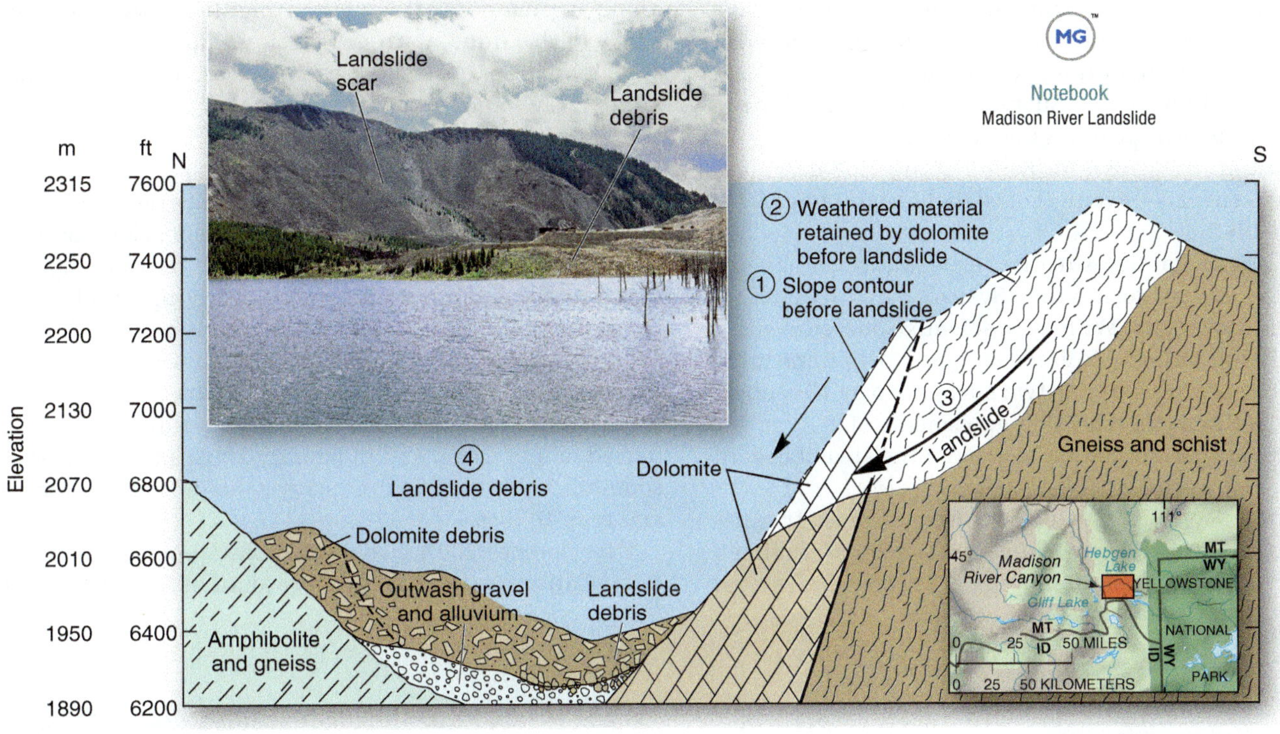

▲**Figure 13.22 Madison River landslide.** Cross section showing geologic structure of the Madison River Canyon in Montana, where an earthquake triggered a landslide in 1959. The landslide debris blocked the canyon and dammed the Madison River, visible in the inset photo. [USGS Professional Paper 435-K, August 1959, p. 115. Inset photo by Bobbé Christopherson.]

material more than 120 m (about 400 ft) up the opposite side of the canyon, trapping several hundred campers with about 80 m (260 ft) of rock and killing 28 people.

The mass of material that dammed the Madison River as a result of this event created a new lake, dubbed Quake Lake. To prevent overflow and associated erosion and flooding downstream, the U.S. Army Corps of Engineers excavated a channel through which the lake could drain.

The M 8.0 earthquake in the Sichuan Province of China in 2008 caused thousands of landslides throughout the region, many of which created dams on rivers and earthquake lakes. At the largest of these landslide dams on the Qianjiang River, channel dredging successfully prevented overtopping and downstream flooding. In 2010 in northern Pakistan, a massive landslide dammed the Hunza River. In this case, dredging was impossible due to muddy conditions at the site. The dam survived repeated overtoppings; then, in 2012, a spillway was blasted to reduce the water level of the lake.

Classes of Mass Movements

In any mass movement, gravity pulls on a mass of material until the critical shear-failure point is reached—a geomorphic threshold. The material then can *fall, slide, flow,* or *creep*—the four classes of mass movement. These classes range in volume of material (small to massive), moisture content (dry to wet), and rate of movement (rapid free-falling rock to slow-moving creep). Figure 13.23 displays the specific types of mass movement discussed ahead according to the moisture and speed categories.

Rockfalls and Debris Avalanches Rockfalls and debris avalanches are types of mass movement that occur at faster rates and in materials that have little to intermediate water content. A **rockfall** is simply a volume of rock that falls through the air and hits a surface (Figure 13.8). During a rockfall, individual pieces fall independently and characteristically form cone-shaped piles of irregular broken rocks known as talus cones that coalesce in a **talus slope** at the base of a steep incline (Figure 13.24 on page 414).

A **debris avalanche** is a mass of falling and tumbling rock, debris, and soil traveling at high velocity owing to the presence of ice and water that fluidize the debris. The extreme danger of a debris avalanche results from its tremendous speed. In 1962 and again in 1970, debris avalanches roared down the west face of Nevado Huascarán, the highest peak in the Peruvian Andes. An earthquake initiated the 1970 event in which upward of 100 million m^3 (3.53 billion ft^3) of debris traveling at 300 kmph (185 mph) buried the city of Yungay, killing 18,000 people (Figure 13.25 on page 414).

Landslides A sudden rapid movement of a cohesive mass of regolith or bedrock that is not saturated with moisture is a **landslide**—a large amount of material failing simultaneously. Surprise creates the danger, for the downward pull of gravity wins the struggle for equilibrium in an instant. Focus Study 13.1 describes one such surprise event that struck near Longarone, Italy, in 1963.

To eliminate the surprise element, scientists are using the Global Positioning System (GPS) to monitor landslide movement. With GPS, scientists measure slight land shifts in vulnerable areas for clues to an impending danger of mass wasting. At two sites in Japan, GPS effectively identified pre-landslide movements of 2–5 cm per year, providing information to expand the area of hazard concern and warning.

Slides occur in one of two basic forms: translational or rotational (see Figure 13.23 for an idealized view of each). *Translational slides* involve movement along a planar (flat) surface roughly parallel to the angle of the slope, with no rotation. The Madison Canyon landslide described earlier was a translational slide. Flow and creep patterns also are considered translational in nature.

Rotational slides, also called *slumps*, occur when surface material moves along a concave surface. Frequently, underlying clay presents an impervious barrier to percolating water. As a result, water flows along the clay's surface, undermining the overlying block. The overlying material may rotate as a single unit, or it may acquire a stepped appearance. Continuing rotational mudslides in response to heavy rainfall plague La Conchita, California (Figure 13.23a). In 1995, a slump landslide buried homes there, and in January 2005, a mudslide episode following a period of heavy rains buried 30 homes and took 10 lives.

In October 2007, a rotational slump occurred in La Jolla, California, damaging more than 100 homes along

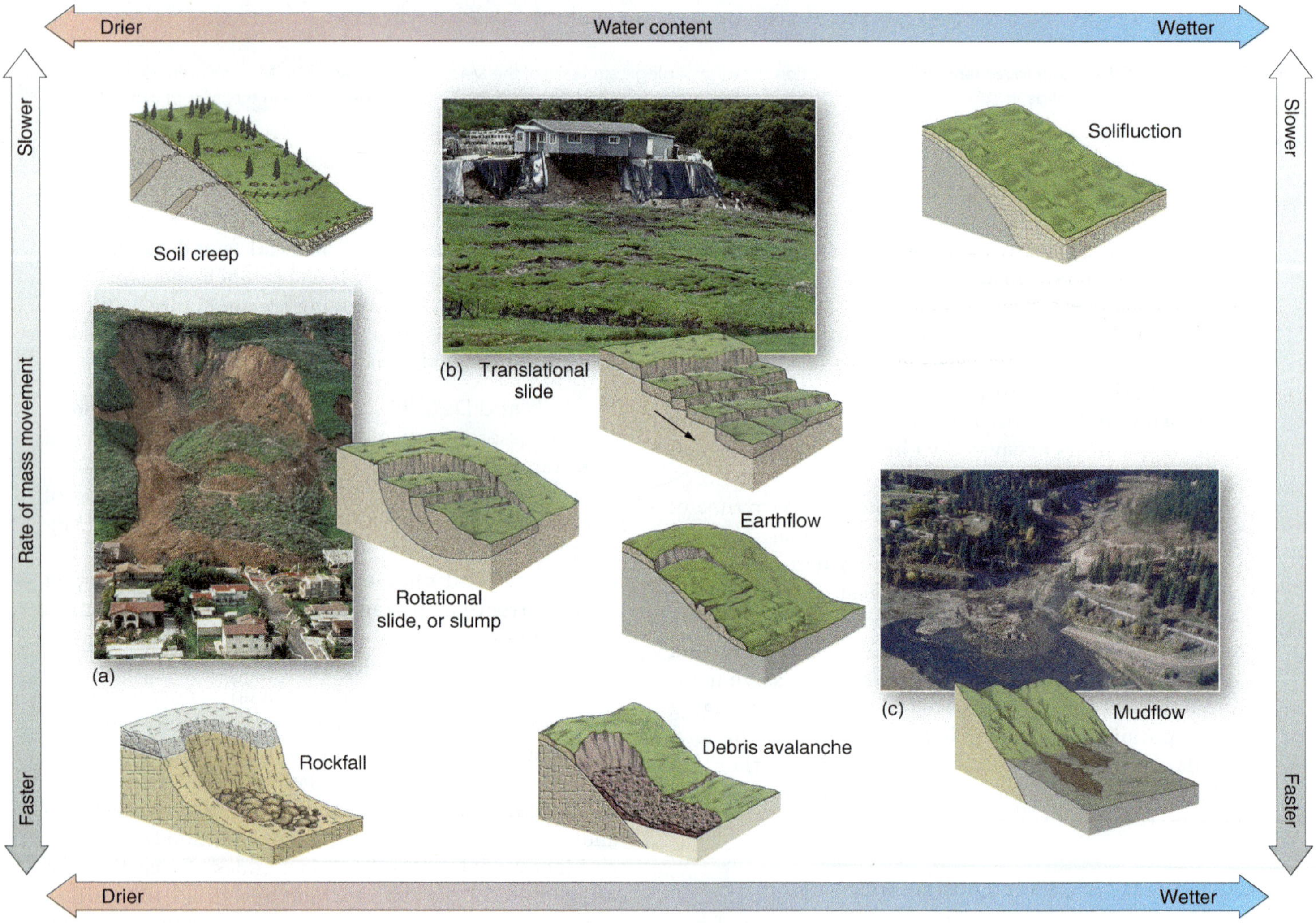

▲**Figure 13.23 Mass-movement classification.** Principal types of mass movement produced by variations in water content and rates of movement. (a) A 1995 slide in La Conchita, California, also the site of a 2005 mudslide event. (b) Saturated hillsides fail, Santa Cruz County, California. (c) A 1998 mudflow buried part of a railroad track and highway near Bonner's Ferry in northern Idaho. [(a) Robert L. Schuster/USGS. (b) Alexander Lowry/Photo Researchers, Inc. (c) NOAA/NGDC, D. Krammer, Disaster Services Boundary County, Idaho.]

MG Animation Mass Movements

Focus Study 13.1 Natural Hazards

Vaiont Reservoir Landslide Disaster

In the early 1960s, one of the worst landslide and dam disasters in history occurred at Vaiont Canyon in the Italian Alps of northeastern Italy. The landslide occurred during the filling of a reservoir behind a hydropower dam. The 262-m (860-ft) high Vaiont Dam, built in a narrow canyon opening to populated lowlands to the west, was nearing completion in February 1960 when filling of the reservoir began. During this phase, as scientists had warned in their preconstruction report, slopes on either side of the dam became unstable.

Preconstruction Geologic Analysis

An analysis of the area's geology and hillslope stability at the damsite yielded several important points. First, the steep canyon walls were composed of interbedded limestone and shale, with open fractures in the shale inclined toward the future reservoir body. The steepness of the canyon walls exacerbated the strong driving force of gravity at work on the rock structures.

The preconstruction analysis also indicated a high potential for bank storage (water absorption by canyon walls) into the groundwater system that would increase water pressure on all rocks in contact with the reservoir. The nature of the shale beds was such that cohesion would be reduced as their clay minerals became saturated. Finally, the analysis indicated evidence of ancient rockfalls and landslides on the north side of the canyon and evidence of creep activity along the south side.

Despite the potential for mass wasting identified by the analysis, the authorities decided to begin design and construction of an experimental thin-arch dam at this site.

Events During Construction and Filling

Large volumes of concrete were injected into the bedrock as "dental work," that is, for the purpose of reinforcing the fractured rock. Despite such precautions, during reservoir filling in 1960, 700,000 m^3 (2.5 million ft^3) of rock and soil slid into the reservoir from the south side. A slow creep of slope materials along the entire south side began shortly thereafter, increasing to about 1 cm per week by January 1963. Creep rate increased as the water level in the reservoir rose. By mid-September 1963, the creep rate exceeded 40 cm (16 in.) a day.

This sequence of events set the stage for disaster. Heavy rains began on September 28, 1963. Alarmed at the runoff volume into the reservoir and the increase in creep rate along the south wall, engineers opened the outlet tunnels on October 8 in an attempt to bring down the reservoir level—but it was too late.

The next evening it took only 30 seconds for a landslide of 240 million m^3 (8.5 billion ft^3) to crash into the reservoir, shaking seismographs across Europe. A slab of mountainside 150 m thick and 2 km by 1.6 km in area (500 ft thick and 1.2 mi by 1 mi in area) gave way, sending shock waves of wind and water up the canyon 2 km (1.2 mi) and splashing a 100-m (330-ft) wave over the dam. The former reservoir was effectively filled with bedrock, regolith, and soil that almost entirely displaced its water content (Figure 13.1.1). Amazingly, the experimental dam design held.

Downstream, in the unsuspecting town of Longarone, near the mouth of Vaiont Canyon on the Piave River, people heard the distant rumble, followed quickly by a 69-m (226-ft) wave of water that came out of the canyon. That night 3000 people perished. In 2008, the United Nations named this tragedy one of five cautionary events caused by failures of geologists and engineers.

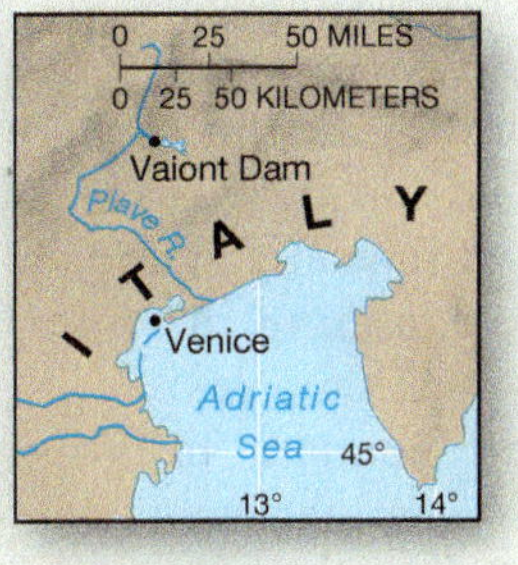

(a) Vaiont dam and landslide debris, October 1963.

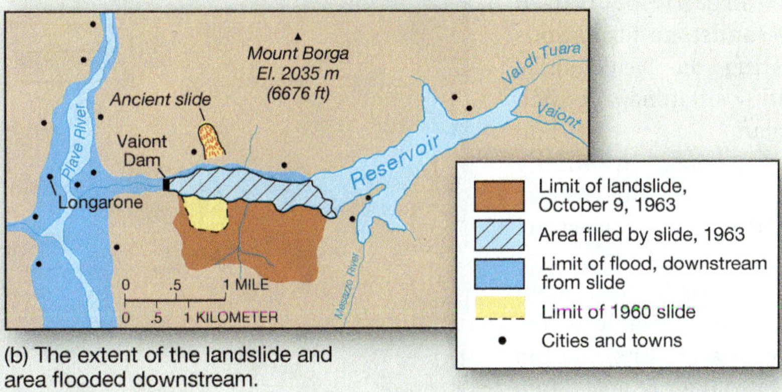

(b) The extent of the landslide and area flooded downstream.

▲**Figure 13.1.1 Vaiont Reservoir disaster.** [(a) Archivio Cameraphoto Epoche/Getty Images. (b) After G. A. Kiersch, "The Vaiont Reservoir disaster," *Mineral Information Service* (California Division of Mines and Geology), 18, 7 (1964): 129–138.]

▲**Figure 13.24 Talus slope.** Rockfall and talus deposits at the base of a steep slope along Duve Fjord, Nordaustlandet Island. Can you see the lighter rock strata above that are the source for the three talus cones? [Bobbé Christopherson.]

▲**Figure 13.25 Debris avalanche, Peru.** A 1970 debris avalanche falls more than 4100 m (2.5 mi) down the west face of Nevado Huascarán, burying the city of Yungay, Peru. The same area was devastated by a similar avalanche in 1962 and by others in pre-Columbian times. The towns in the valley remain at great risk from possible future mass movements. [George Plafker, USGS.]

Soledad Mountain Road. The area is inherently unstable, experiencing three other collapse events since 1961. Early warning signs of the 2007 occurrence began in July. The cause is attributed to unrestricted irrigation of lawns and gardens, feeding drainage runoff water into subsurface strata and, as a result, increasing instability and lubricating slides.

Flows When the moisture content of moving material is high, the suffix *-flow* is used, as in *earthflows* and more fluid **mudflows**. Heavy rains can saturate barren mountain slopes and set them moving, as was the case in the Gros Ventre River valley east of Jackson Hole, Wyoming, in the spring of 1925. About 37 million m^3 (1.3 billion ft^3) of wet soil and rock moved down one side of the canyon and surged 30 m (100 ft) up the other side, damming the river and forming a lake. The water content of this mass-wasting event was great enough to classify it as an earthflow, caused by sandstone formations resting on weak shale and siltstone, which became moistened and soft, and eventually failed with the weight of the overlying strata (Figure 13.26).

In 1925 equipment was not available to excavate a channel, so the new lake filled. Two years later the lake water broke through the temporary earthflow dam,

▲**Figure 13.26 The Gros Ventre earthflow near Jackson, Wyoming.** Evidence of the 1925 earthflow is still visible after more than 89 years. [Steven K. Huhtala.]

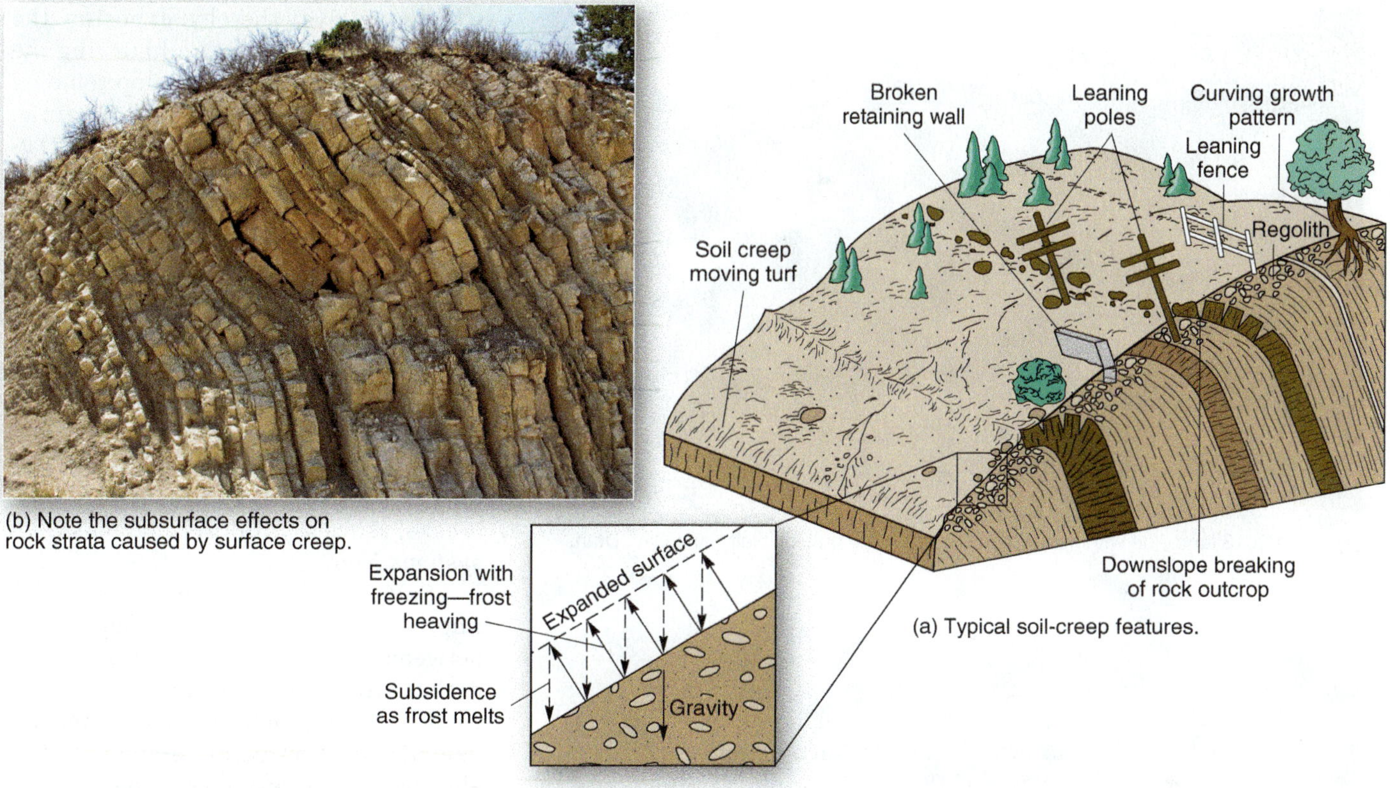

(b) Note the subsurface effects on rock strata caused by surface creep.

(a) Typical soil-creep features.

▲**Figure 13.27 Soil creep and its effects.** [(b) Bobbé Christopherson.]

transporting a tremendous quantity of debris over the region downstream.

Creep A persistent, gradual mass movement of surface soil is **soil creep**. In creep, individual soil particles are lifted and disturbed, whether by the expansion of soil moisture as it freezes, by cycles of moistness and dryness, by diurnal temperature variations, or by grazing livestock or digging animals.

In the freeze–thaw cycle, particles are lifted at right angles to the slope by freezing soil moisture, as shown in Figure 13.27. When the ice melts, however, the particles fall straight downward in response to gravity. As the process repeats, the surface soil gradually creeps its way downslope.

The overall wasting of a creeping slope may cover a wide area and may cause fence posts, utility poles, and even trees to lean downslope. Various strategies are used to arrest the mass movement of slope material—grading the terrain, building terraces and retaining walls, planting ground cover—but the persistence of creep often renders these strategies ineffective.

In polar regions and at high elevations, freeze–thaw processes are critical for mass wasting. During the summer when the upper layers of soil thaw and become saturated, slow downslope movement occurs, called *solifluction* (shown in Figure 13.23 and discussed with periglacial environments in Chapter 17).

Humans as a Geomorphic Agent

Any human disturbance of a slope—highway road cutting, surface mining, or construction of a shopping mall, housing development, or home—can hasten mass wasting. Large open-pit surface mines—such as the Bingham Canyon Copper Mine west of Salt Lake City; the abandoned Berkeley Pit in Butte, Montana; several iron mines in the Brazilian rain forest; and numerous large coal surface mines in the eastern and western United States (such as Black Mesa, Arizona)—are examples of human impacts that move sediment, soil, and rock material, a process known as **scarification**.

At the Bingham Canyon Copper Mine, a mountain literally was removed since mining began in 1906, forming a pit 4 km wide and 1.2 km deep, or 2.5 mi wide and

GEOreport 13.5 Open pit mining in the Amazon region

The Carajás Mine in northern Brazil is the world's largest iron ore mining complex. Since the 1980s, open pit mining operations at Carajás have destroyed large tracts of rain forest, causing runoff of sediment and pollutants into streams, and triggered intense conflict over land with indigenous communities. According to 2009 estimates, the region contains 7.2 billion metric tons of iron ore reserves. For a satellite image of the pit, go to earthobservatory.nasa.gov/IOTD/view.php?id=39581.

(a) Large 2013 landslide within Bingham Canyon Mine outside Salt Lake City, Utah.

(b) Mountaintop removal on Kayford Mountain, West Virginia.

▲**Figure 13.28 Scarification.** [(a) Ravell Call/AP Photo (b) Dr. Chris Mayda, Eastern Michigan University, all rights reserved.]

0.75 mi deep. This is easily the largest human-made excavation on Earth. In April 2013, a large landslide occurred within the open pit on an unstable slope that was being closely monitored for safety reasons (Figure 13.28a). No fatalities occurred.

The disposal of tailings (mined ore of little value) and waste material is a significant problem at any mine. Such large excavations produce tailing piles that are unstable and susceptible to further weathering, mass wasting, or wind dispersal. Additionally, the leaching of toxic materials from tailings and waste piles poses an ever-increasing problem for streams, aquifers, and public health across the country.

Where underground mining is common, particularly for coal in the Appalachians, land subsidence and collapse may produce mass movement on hillslopes. Homes, highways, streams, wells, and property values are severely affected. A controversial form of mining called *mountaintop removal* is done by removing ridges and summits and dumping the debris into stream valleys, thereby exposing the coal seams for mining and burying the stream channels. Mountaintop removal on Kayford Mountain, West Virginia, and elsewhere in the region has flattened more than 500 mountains, removing an estimated 1.2 million acres and filling some 2000 km (1245 mi) of streams with tailings (Figure 13.28b). These valley fills affect downstream water quality with concentrations of potentially toxic nickel, lead, cadmium, iron, and selenium that generally exceed government standards. Valley-fill collapses have caused floods, as have coal- and ash-sludge containment breaks, as we saw in this chapter's *Geosystems Now.*

Scientists have made informal, but impressive, quantitative comparisons between scarification and natural denudation processes. Geologist R. L. Hooke used estimates of U.S. excavations for new housing, mineral production (including the three largest types—stone, sand and gravel, and coal), and highway construction. He then prorated these quantities of moved earth for all countries, based on their gross domestic product, energy consumption, and agriculture's effect on river sediment loads. From these, he calculated a global estimate for human earth moving. Later researchers confirmed and expanded on these findings.

Hooke estimated for the early 1990s that humans, as a geomorphic agent, annually moved 40–45 billion tons (40–45 Gt) of the planet's surface. Compare this quantity with the natural movement of river sediment (14 Gt/year), or the movement due to wave action and erosion along coastlines (1.25 Gt/year), or deep-ocean sedimentation (7 Gt/year). In 2005, geologist Bruce Wilkinson corroborated these measurements, concluding that humans are 10 times more active in shaping the landscape than natural processes. As Hooke concluded about humans,

> *Homo sapiens* has become an impressive geomorphic agent. Coupling our earth-moving prowess with our inadvertent adding of sediment load to rivers and the visual impact of our activities on the landscape, one is compelled to acknowledge that, for better or for worse, this biogeomorphic agent may be the premier geomorphic agent of our time.*

*R. L. Hooke, "On the efficacy of humans as geomorphic agents," *GSA Today* (The Geological Society of America), 4, 9 (September 1994): 217–226.

THEhumanDENOMINATOR 13 Weathering, Karst, and Hillslopes

GEOMORPHIC PROCESSES ⇨ HUMANS

- Chemical weathering processes break down carvings made by humans in rock, such as tombstones, cathedral facades, and bridges.
- Sudden sinkhole formation in populated areas can cause damage and human casualties.
- Mass movements cause human casualties and sometimes catastrophic damage, burying cities, damming rivers, and sending flood waves downstream.

HUMANS ⇨ GEOMORPHIC PROCESSES

- Mining causes scarification, often moving contaminated sediments into surface water systems and groundwater.
- Removal of vegetation on hillslopes may lead to slope failure, destabilizing streams and associated ecosystems.
- Lowering of water tables from groundwater pumping causes sinkhole collapse in population centers.

13a The 71-m-tall (233-ft-tall) Grand Buddha at Leshan in the Sichuan province of southern China is an example of chemical weathering accelerated by air pollution. Carved over 1000 years ago, the statue is now being corroded by acid rain from nearby industrial development. [Bennett Dean/Corbis.]

13c Blue holes are typical karst sinkholes located in what are now offshore areas but that formed during times when sea level was lower. The Great Blue Hole near Belize is part of the Belize Barrier Reef Reserve System designated as a World Heritage site by the United Nations. [Schafer & Hill/Getty Images.]

13b In April 2010, a massive landslide covered parts of a highway near Taipei, Taiwan. The cause of the translational slide is uncertain, as it was apparently not related to earthquake activity or excessive rainfall. [Patrick Lin/Getty Images.]

ISSUES FOR THE 21ST CENTURY

- Global climate change will affect forest health; declining forests (from disease or drought) will increase slope instability and mass movement events.
- Open-pit mining worldwide will continue to move massive amounts of Earth materials, with associated impacts on ecosystems and water quality. Reclamation will help mitigate long-term mining impacts.
- Improved engineering of containment ponds holding industrial by-products will prevent the spread of toxic materials.

GEOSYSTEMSconnection

We began our study of Earth's exogenic processes with the basic principles of landmass denudation and slope morphology. We then moved on to the processes that deliver materials for erosion and transport: the physical and chemical weathering of rock, dissolution of limestone landscapes, and mass-movement processes. In the next chapter, we examine river systems and the landforms that result from fluvial processes. The erosional and depositional activities of running water, wind, waves and coastal actions, and ice flow through the next four chapters.

KEY LEARNING concepts review

■ *Describe* the dynamic equilibrium approach to the study of landforms, and *illustrate* the forces at work on materials residing on a slope.

Geomorphology is the science that analyzes and describes the origin, evolution, form, and spatial distribution of landforms. Earth's exogenic system, powered by solar energy and gravity, tears down the landscape through processes of landmass **denudation** involving weathering, mass movement, erosion, transportation, and deposition. Different rocks offer differing resistance to these weathering processes and produce a pattern on the landscape of **differential weathering**.

Agents of change include moving air, water, waves, and ice. Since the 1960s, research and understanding of the processes of denudation have moved toward the **dynamic equilibrium model,** which considers slope and landform stability to be consequences of the resistance of rock materials to the attack of denudation processes. When a destabilizing event occurs, a landform or landform system may reach a **geomorphic threshold**, where force overcomes resistance and the system moves to a new level and toward a new equilibrium state.

Slopes are shaped by the relation between the rate of weathering and breakup of slope materials and the rate of mass movement and erosion of those materials. Slopes that form the boundaries of landforms have several general components: *waxing slope*, *free face*, *debris slope*, and *waning slope*. Slopes seek an *angle of equilibrium* among the operating forces.

denudation (p. 394)
differential weathering (p. 394)
dynamic equilibrium model (p. 395)
geomorphic threshold (p. 395)
slope (p. 395)

1. Define *landmass denudation*. What processes are included in the concept?
2. What is the interplay between the resistance of rock structures and differential weathering?
3. Describe what is at work to produce the landform in Figure 13.1.
4. What are the principal considerations in the dynamic equilibrium model?
5. Describe conditions on a hillslope that is right at the geomorphic threshold. What factors might push the slope beyond this point?
6. Given all the interacting variables, do you think a landscape ever reaches a stable, old-age condition? Explain.
7. What are the general components of an idealized slope?
8. Relative to slopes, what is meant by an *angle of equilibrium*? Can you apply this concept to the photograph in Figure GIA 13.3?

■ *Define* weathering, and *explain* the importance of parent rock and joints and fractures in rock.

Weathering processes disintegrate both surface and subsurface rock into mineral particles or dissolve them in water. On a typical hillside, loose surface material overlies consolidated, or solid, rock called **bedrock**. In most areas, the upper surface of bedrock undergoes continual weathering, creating broken-up rock called **regolith**. The unconsolidated, fragmented material that is carried across landscapes by erosion, transportation, and deposition is sediment, which along with weathered rock forms the **parent material** from which soil evolves.

Important in weathering processes are **joints,** the fractures and separations in the rock. Jointing opens up rock surfaces on which weathering processes operate. Factors that influence weathering include the character of the bedrock (hard or soft, soluble or insoluble, broken or unbroken), climatic elements (temperature, precipitation, freeze–thaw cycles), position of the water table, slope orientation, surface vegetation and its subsurface roots, and time.

weathering (p. 398)
bedrock (p. 398)
regolith (p. 398)
parent material (p. 398)
joint (p. 399)

9. Describe weathering processes operating on an open expanse of bedrock. How does regolith develop? How is sediment derived?
10. Describe the relationship between climate and weathering at microscale levels.
11. What is the relationship between parent rock, parent material, regolith, and soil?
12. What role do joints play in the weathering process? Give an example from this chapter.

■ *Describe* the physical weathering processes of frost action, salt-crystal growth, and pressure-release jointing.

Physical weathering, or mechanical weathering, refers to the breakup of rock into smaller pieces with no alteration of mineral identity. The physical action of water when it freezes (expands) and thaws (contracts) causes rock to break apart in the process of **frost wedging**. Working in joints, expanded ice can produce *joint-block separation* through this process. Another physical weathering process is *salt-crystal growth* (*salt weathering*); as crystals in rock grow and enlarge over time by crystallization, they force apart mineral grains and break up rock.

Removal of overburden from a granitic batholith relieves the pressure of deep burial, producing joints. **Exfoliation**, or *sheeting*, occurs as mechanical forces enlarge the joints, separating the rock into layers of curved slabs or plates (rather than granular disintegration that occurs with many weathering processes). The resulting arch-shaped or dome-shaped feature is an *exfoliation dome*.

physical weathering (p. 400)
frost action (p. 400)
exfoliation (p. 401)

13. What is physical weathering? Give an example.
14. Why is freezing water such an effective physical weathering agent?
15. What weathering processes produce a granite dome? Describe the sequence of events.

■ *Explain* the chemical weathering processes of hydration, hydrolysis, oxidation, carbonation, and dissolution.

Chemical weathering is the chemical decomposition of minerals in rock. It can cause **spheroidal weathering,** in which chemical weathering that occurs in cracks in the rock removes cementing and binding materials, so that the sharp edges and corners of rock disintegrate and become rounded.

Hydration occurs when a mineral absorbs water and expands, thus changing the mineral structure. This process also creates a strong mechanical (physical weathering) force that stresses rocks. **Hydrolysis** breaks down silicate minerals in rock through reaction with water, as in the chemical weathering of feldspar into clays and silica. **Oxidation** is a chemical weathering process in which oxygen reacts with certain metallic elements, the most familiar example being the rusting of iron to produce iron oxide. The *dissolution* of materials into solution is also considered chemical weathering. An important type of dissolution is **carbonation,** resulting when carbonic acid in rainwater reacts to break down certain minerals, such as those containing calcium, magnesium, potassium, or sodium.

chemical weathering (p. 401)
spheroidal weathering (p. 402)
hydration (p. 402)
hydrolysis (p. 402)
oxidation (p. 403)
carbonation (p. 404)

16. What is chemical weathering? Contrast this set of processes to physical weathering.
17. What is meant by the term *spheroidal weathering*? How does spheroidal weathering occur?
18. What is hydration? What is hydrolysis? Differentiate between these processes. How do they affect rocks?
19. Iron minerals in rock are susceptible to which form of chemical weathering? What characteristic color is associated with this type of weathering?
20. With what kind of minerals does carbonic acid react, and what circumstances bring this type of reaction about? What is this weathering process called?

■ *Review* the processes and features associated with karst topography.

Karst topography refers to distinctively pitted and weathered limestone landscapes. **Sinkholes** are circular surface depressions that may be *solution sinkholes* formed by slow subsidence or *collapse sinkholes* formed in a sudden collapse through the roof of an underground cavern below. In tropical climates, karst landforms include *cockpit karst* and *tower karst*. The creation of caverns is a result of karst processes and groundwater erosion. Limestone caves feature many unique erosional and depositional features.

karst topography (p. 405)
sinkhole (p. 406)

21. Describe the development of limestone topography. What is the name applied to such landscapes? From what area was this name derived?
22. Explain and differentiate among the formation of sinkholes, karst valleys, cockpit karst, and tower karst. Which forms are found in the tropics?
23. In general, how would you characterize the region southwest of Orleans, Indiana?
24. What are some of the characteristic erosional and depositional features you find in a limestone cavern?

■ *Categorize* the various types of mass movements, and *identify* examples of each by moisture content and speed of movement.

Any movement of a body of material, propelled and controlled by gravity, is **mass movement,** or **mass wasting.** The **angle of repose** of loose sediment grains represents a balance of driving and resisting forces on a slope. Mass movement of Earth's surface produces some dramatic incidents, including **rockfalls** (volumes of falling rocks), which can form a **talus slope** of loose rock along the base of the cliff; **debris avalanches** (masses of tumbling, falling rock, debris, and soil moving at high speed); **landslides** (large amounts of material failing simultaneously); **mudflows** (material in motion with a high moisture content); and **soil creep** (persistent movement of individual soil particles that are lifted by the expansion of soil moisture as it freezes, by cycles of wetness and dryness, by temperature variations, or by the impact of grazing animals). In addition, human mining and construction activities have created massive **scarification** of landscapes.

mass movement (p. 410)
mass wasting (p. 410)
angle of repose (p. 410)
rockfall (p. 411)
talus slope (p. 411)
debris avalanche (p. 411)
landslide (p. 412)
mudflow (p. 414)
soil creep (p. 415)
scarification (p. 415)

25. Define the role of slopes in mass movements, using the terms *angle of repose, driving force, resisting force,* and *geomorphic threshold.*
26. What events occurred in the Madison River Canyon in 1959?
27. What are the classes of mass movement? Describe each briefly and differentiate among these classes.
28. Name and describe the type of mudflow associated with a volcanic eruption.
29. Describe the difference between a landslide and what happened on the slopes of Nevado Huascarán.
30. What is scarification, and how does it relate to mass movement? Give several examples of scarification. Why are humans a significant geomorphic agent?

MasteringGeography™

Looking for additional review and test prep materials? Visit the Study Area in *MasteringGeography*™ to enhance your geographic literacy, spatial reasoning skills, and understanding of this chapter's content by accessing a variety of resources, including MapMaster interactive maps, geoscience animations, satellite loops, author notebooks, videos, RSS feeds, flashcards, web links, self-study quizzes, and an eText version of Geosystems.

14 River Systems

Carving through the volcanic strata of the rugged Owyhee plateau, the Owyhee River flows from northern Nevada, through Idaho and southeast Oregon, and eventually into the Snake River on the Idaho-Oregon border. The remote, high-desert canyons of the Owyhee and its tributaries are considered by many to be some of the last remaining truly wild country in the American West. Over 560 km (350 mi) of the Owyhee, in several segments, are protected as part of the U.S. National Wild and Scenic Rivers System, which mandates preservation of each river segment's free-flowing condition. [Michael Melford/Getty Images.]

KEY LEARNING concepts

After reading the chapter, you should be able to:

- ***Sketch*** a basic drainage basin model, and ***identify*** different types of drainage patterns by visual examination.
- ***Explain*** the concepts of stream gradient and base level, and ***describe*** the relationship between stream velocity, depth, width, and discharge.
- ***Explain*** the processes involved in fluvial erosion and sediment transport.
- ***Describe*** common stream channel patterns, and ***explain*** the concept of a graded stream.
- ***Describe*** the depositional landforms associated with floodplains and alluvial fan environments.
- ***List*** and ***describe*** several types of river deltas, and ***explain*** flood probability estimates.

Environmental Effects of Dams on the Nu River in China

In the United States, the razing of dams and restoration of rivers has become a multibillion dollar industry, discussed in Focus Study 14.1 in this chapter. However, in China the trend is quite different. There, and in certain other countries, dam building and hydropower development continue in order to meet increasing energy demands. In 2013, the Chinese government announced plans to build a series of large dams on the Nu River—Southeast Asia's longest free-flowing watercourse.

The Nu River, known as the Salween in Thailand and Myanmar, flows from the mountains of Tibet, through the Yunnan province of southwestern China, and into the Andaman Sea (Figure GN 14.1). The river's headwaters lie adjacent to those of the Mekong and Yangtze Rivers within the Three Parallel Rivers World Heritage Site, famed for its biological diversity. The upper watershed is home to people from 13 ethnic groups, many relying on subsistence farming as a way of life (Figure GN 14.2). The Chinese government is proposing to relocate some 60,000 people from these communities to make way for the reservoirs that are part of the hydropower development. In some areas, the relocation has already begun as dam sites are being prepared for construction.

▲**Figure GN 14.1 Map of the Nu/Salween River system, Southeast Asia.**

Interruptions to Water and Sediment Flows Rivers move both water and sediment, a process that builds fertile soils on valley floors and in the flatlands where rivers meet the sea. Dams interrupt this process, starving downstream areas of sediment and nutrients that naturally replenish farmland in the cycle of annual spring floods. Farmers in the Nu River's lower reaches stand to suffer soil degradation and crop losses if the dams move forward as planned.

Sediment is also critical for maintaining stream habitat, as well as coastal wetlands near the river's mouth. When sediment is trapped behind a dam, excessive erosion often results downstream, leading to degradation of banks and the riverbed. Alterations to water and sediment movement also have detrimental effects on fisheries, which are critical food resources for populations in the downstream portions of the river system.

Induced Seismicity Recently, scientists have warned that reservoirs may be connected to earthquake activity in tectonically active southwest China. Although the idea of "reservoir-induced seismicity" (earthquakes induced by the location of large reservoirs along fault lines) is not new, the proposed dams on the Nu River have prompted scientists to take a fresh look at this phenomenon. The mass of a large reservoir creates pressure that forces water into cracks and fissures in the ground beneath, increasing instability and possibly lubricating fault zones. Seismic events induced by large reservoirs may damage the dams that impound them, increasing the risk of catastrophic flooding in the event of dam failure.

▲**Figure GN 14.2 Locals crossing a suspension bridge spanning the Nu River, Yunnan, China.** [Redlink/Corbis.]

First proposed almost 10 years ago, the Nu River dams were delayed in light of environmental concerns, primary among which were the high numbers of sensitive species in the watershed and the potential earthquake hazard. However, development along the river has been underway since 2006 in preparation for dam construction.

China already suffers from poor air quality owing to industrial pollution; as power demand increases, the government continues to develop hydropower on its rivers despite the environmental risks. This chapter examines stream and river systems—their natural processes and the role of humans in altering them.

GEOSYSTEMS NOW ONLINE Go to Chapter 14 on the *MasteringGeography* website (www.masteringgeography.com) for more on the downstream effects of dams on river systems. For a recent study on the impacts of small dams in southwestern China, see http://www.agu.org/news/press/pr_archives/2013/2013-22.shtml. MG

Earth's rivers and waterways form vast arterial networks that drain the continents, and at any moment, approximately 1250 km^3 (300 mi^3) of water is flowing through them. Even though this volume is only 0.003% of all freshwater, the work performed by this energetic flow makes it a dominant natural agent of landmass denudation. Rivers shape the landscape by removing the products of weathering, mass movement, and erosion and transporting them downstream. Rivers also serve society in many ways. They not only provide us with essential water supplies, but also receive, dilute, and transport wastes, provide critical cooling water for industry, and form critical transportation networks.

Of the world's rivers, those with the greatest *discharge* (the streamflow volume past a point in a given unit of time, discussed in this chapter) are the Amazon of South America, the Congo of Africa, the Yangtze (Chàng Jiang) of Asia, and the Orinoco of South America. In North America, the greatest discharges are from the Missouri–Ohio–Mississippi, Saint Lawrence, and Mackenzie River systems (Table 14.1).

Hydrology is the science of water and its global circulation, distribution, and properties—focusing on water at and below Earth's surface. Processes that are related expressly to streams and rivers are termed **fluvial** (from the Latin *fluvius,* meaning "river"). There is some overlap in usage between the terms *river* and *stream.* Specifically, the term *river* is applied to the trunk, or main stream of the network of tributaries forming a *river system. Stream* is a more general term for water flowing in a channel and is not necessarily related to size. Fluvial systems, like all natural systems, have characteristic processes and produce recognizable landforms, yet also can behave with randomness and seeming disorder.

In this chapter: We begin by examining the organization of river systems into drainage basins and the types of drainage patterns. With this foundation, we examine gradient, base level, and stream discharge. We then discuss factors that affect flow characteristics and the work performed by flowing water, including erosion and transport. We also examine the effects of urbanization on stream hydrology and the impacts of dams on sediment regimes. Depositional features are illustrated with a detailed look at floodplains and the Mississippi River delta. Finally, we examine floods, floodplain management, and river restoration.

Drainage Basins and Drainage Patterns

Streams, which come together to form river systems, lie within drainage basins, the portions of landscape from which they receive their water. Every stream has its own **drainage basin**, or *watershed,* ranging in size from tiny to vast. A major drainage basin system is made up of many smaller drainage basins, each of which gathers and delivers its runoff and sediment to a larger basin, eventually concentrating the volume into the main stream. Figure 14.1 illustrates the drainage basin of the Amazon River, from headwaters to the river's mouth (where the river meets the ocean). The Amazon has an average flow greater than 175,000 m^3/s (6.2 million cfs, or cubic feet per second) and moves millions of tons

TABLE 14.1 Largest Rivers on Earth Ranked by Discharge

Rank by Volume	Average Discharge at Mouth in Thousands of m^3/s (cfs)	River (Major Tributaries)	Outflow (Location)	Length, km (mi)	Rank by Length
1	180 (6350)	Amazon (Ucayali, Tambo, Ene, Apurimac)	Atlantic Ocean (Amapá-Pará, Brazil)	6800 (4225)	1*
2	41 (1460)	Congo (Lualaba)	Atlantic Ocean (Angola, Congo)	4630 (2880)	10
3	34 (1201)	Yangtze or Chàng Jiang	East China Sea (Kiangsu, China)	6300 (3915)	3
4	30 (1060)	Orinoco	Atlantic Ocean (Venezuela)	2737 (1700)	27
5	21.8 (779)	Rio de la Plata (Paraná, Uruguay)	Atlantic Ocean (Argentina)	3945 (2450)	16
6	19.6 (699)	Ganges (Brahmaputra)	Bay of Bengal (India)	2510 (1560)	23
7	19.4 (692)	Yenisey (Angara, Selenga or Selenge, Ider)	Gulf of Kara Sea (Siberia)	5870 (3650)	5
8	18.2 (650)	Mississippi (Missouri, Ohio, Tennessee, Jefferson, Beaverhead, Red Rock)	Gulf of Mexico/Louisiana	6020 (3740)	4
9	16.0 (568)	Lena	Laptev Sea/Siberia	4400 (2730)	11
17	9.7 (348)	St. Lawrence	Gulf of St. Lawrence/ Canada and United States	3060 (1900)	21
36	2.83 (100)	Nile (Kagera, Ruvuvu, Luvironza)	Mediterranean Sea/Egypt	6690 (4160)	2*

*Measurement in 2007 places the Amazon first in length; see Figure 14.1.

(a) Radar images showing elevation are combined with digitally mapped stream channels in this view of the Amazon River basin. Elevations range from sea level in green to above 4500 m (14,764 ft) in white. In 2007, Brazilian researchers reported finding a new source for the Amazon (white box), near Mount Mismi in southern Peru, and a new length measurement of 6800 km (4225 mi), making it longer than the Nile.

(b) The mouth of the Amazon is 160 km (100 mi) wide and discharges a fifth of all the freshwater that enters the world's oceans. Large islands of sediment are deposited where the flow enters the Atlantic.

▲**Figure 14.1 Amazon River drainage basin and mouth.** [(a) NASA SRTM image by Jesse Allen, University of Maryland, Global Land Cover Facility; stream data World Wildlife Fund, HydroSHEDS project (see http://hydrosheds.cr.usgs.gov/). (b) Terra image, NASA/GSFC/JPL.]

of sediment through the drainage basin, which is as large as the Australian continent.

Drainage Divides

In any drainage basin, water initially moves downslope as *overland flow*, which takes two forms: It can move as **sheetflow**, a thin film spread over the ground surface; and it can concentrate in *rills*, small-scale grooves in the landscape made by the downslope movement of water. Rills may develop into deeper *gullies* and then into stream channels leading to the valley floor.

The high ground that separates one valley from another and directs sheetflow is called an *interfluve* (Figure 14.2). Ridges act as *drainage divides* that define the *catchment*, or water-receiving, area of every drainage basin; such ridges are the dividing lines that control into which basin the surface runoff drains.

A special class of drainage divides, **continental divides**, separate drainage basins that empty into different

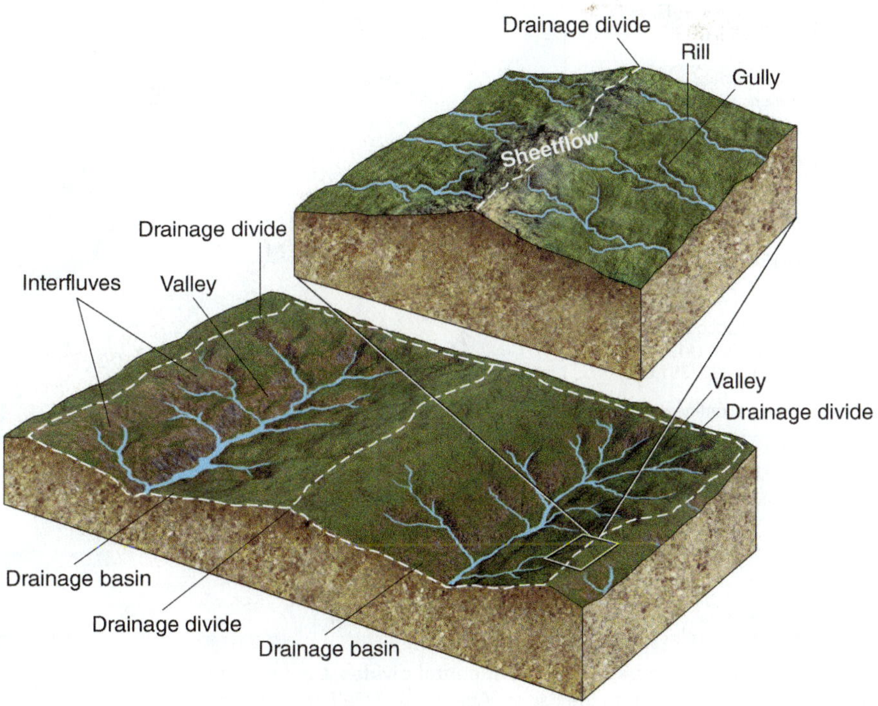

▲**Figure 14.2 Drainage basins.** A drainage divide separates drainage basins.

bodies of water surrounding a continent; for North America, these bodies are the Pacific, the Gulf of Mexico, the Atlantic, Hudson Bay, or the Arctic Ocean. The principal drainage divides and drainage basins in the United States and Canada are mapped in Figure 14.3. These divides form water-resource regions and provide a spatial framework for water-management planning.

In North America, the great Mississippi–Missouri–Ohio River system drains some 3.1 million km^2 (1.2 million mi^2), or 41% of the continental United States (Figure 14.3). Within this basin, rainfall in northern Pennsylvania feeds hundreds of small streams that flow into the Allegheny River. At the same time, rainfall in western Pennsylvania feeds hundreds of streams that flow into the Monongahela River. The two rivers then join at Pittsburgh to form the Ohio River. The Ohio flows southwestward and at Cairo, Illinois, connects with the Mississippi River, which eventually flows on past New Orleans and disperses into the Gulf of Mexico. Each contributing tributary, large or small, adds its discharge, pollution, and sediment load to the larger river. In our example, sediment weathered and eroded in Pennsylvania

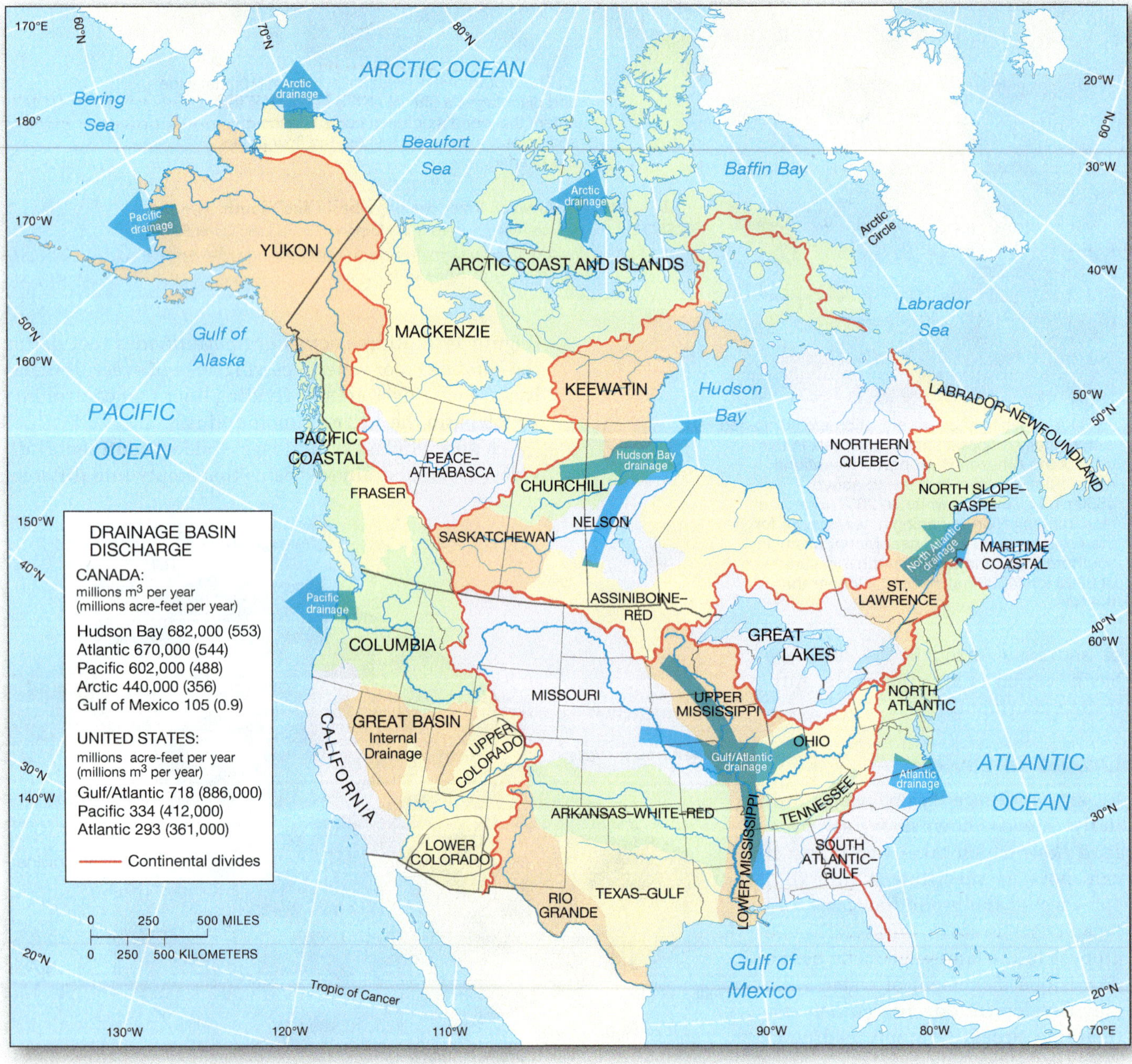

▲Figure 14.3 Drainage basins and continental divides. Continental divides (red lines) separate the major drainage basins that empty through the United States into the Pacific, Atlantic, and Gulf of Mexico and, to the north, through Canada into Hudson Bay and the Arctic Ocean. Subdividing these major drainage basins are major river basins. [After U.S. Geological Survey; *The National Atlas of Canada*, 1985, "Energy, Mines, and Resources Canada"; and Environment Canada, *Currents of Change—Inquiry on Federal Water Policy—Final Report 1986*.]

is transported thousands of kilometers and accumulates on the floor of the Gulf of Mexico, where it forms the Mississippi River delta.

CRITICALthinking 14.1
Locate Your Drainage Basin

Determine the name of the drainage basin within which your campus is located. Where are its headwaters? Where is the river's mouth? If you are in the United States or Canada, use Figure 14.3 to locate the larger drainage basins and divides for your region, and then take a look at this region on Google Earth™. Does any regulatory organization oversee planning and coordination for the drainage basin you identified? Can you find topographic maps online that cover this region? ●

Drainage Basins as Open Systems

Drainage basins are open systems. Inputs include precipitation and the minerals and rocks of the regional geology. Energy and materials are redistributed as the stream constantly adjusts to its landscape. System outputs of water and sediment disperse through the mouth of the stream or river, into a lake, another stream or river, or the ocean, as shown in Figure 14.1.

Change that occurs in any portion of a drainage basin can affect the entire system. If a stream is brought to a geomorphic threshold where it can no longer maintain its present form, the river system may become destabilized, initiating a transition period to a more stable condition. A river system constantly struggles toward equilibrium among the interacting variables of discharge, channel steepness, channel shape, and sediment load, all of which are discussed in the chapter ahead.

International Drainage Basins

The Danube River in Europe, which flows 2850 km (1770 mi) from western Germany's Black Forest to the Black Sea, exemplifies the political complexity of an international drainage basin. The river crosses or forms part of a border of nine countries (Figure 14.4). A total area of 817,000 km^2 (315,000 mi^2) falls within the drainage basin, including some 300 tributaries.

The Danube serves many economic functions: commercial transport, municipal water source, agricultural irrigation, fishing, and hydroelectric power production. An international struggle is under way to save the river from its burden of industrial and mining wastes, sewage, chemical discharge, agricultural runoff, and drainage from ships. The many shipping canals actually spread pollution and worsen biological conditions in the river. All of this pollution passes through Romania and the deltaic ecosystems in the Black Sea. The river is widely regarded as one of the most polluted on Earth.

Political changes in Europe in 1989 allowed the first scientific analysis of the entire Danube River system. The United Nations Environment Programme (UNEP) and the European Union, along with other organizations, are now dedicated to improving water quality and restoring floodplain and delta ecosystems; see http://www.icpdr.org/.

Internal Drainage

As mentioned earlier, the ultimate outlet for most drainage basins is the ocean. In some regions, however, stream drainage does not reach the ocean. Instead, the water leaves the drainage basin by means of evaporation or subsurface gravitational flow. Such basins are described as having **internal drainage**.

Regions of internal drainage occur in Asia, Africa, Australia, Mexico, and the western United States, such as in the Great Basin Physiographic Province in Nevada and Utah, shown in Figure 12.14 and Figure 14.3. An example within this region is the Humboldt River, which flows westward across Nevada and eventually disappears into the Humboldt "sink" as a result of evaporation and seepage losses to groundwater. The area surrounding Utah's Great Salt Lake, outlet for many streams draining the Wasatch Mountains, also exemplifies internal drainage, since its only outlet is evaporation. Internal drainage is also a characteristic of the Dead Sea region in the Middle East and the region around the Aral Sea and Caspian Sea in Asia.

▲**Figure 14.4 An international drainage basin—the Danube River.** The Danube crosses or forms part of a border of nine countries as it flows across Europe to the Black Sea. The river spews polluted discharge into the Black Sea through its arcuate-form delta. [Inset: *Terra* image, June 15, 2002, NASA/GSFC.]

Drainage Patterns

A primary feature of any drainage basin is its **drainage density**, determined by dividing the total length of all stream channels in the basin by the area of the basin. The number and length of channels in a given area reflect the landscape's regional geology and topography.

The **drainage pattern** is the arrangement of channels in an area. Patterns are quite distinctive and are determined by a combination of regional steepness and relief; variations in rock resistance, climate, and hydrology; and structural controls imposed by the underlying rocks. Consequently, the drainage pattern of any land area on Earth is a remarkable visual summary of every geologic and climatic characteristic of that region.

The seven most common types of drainage patterns are shown in Figure 14.5. A most familiar pattern is *dendritic drainage* (Figure 14.5a). This treelike pattern (from the Greek word *dendron*, meaning "tree") is similar to that of many natural systems, such as capillaries in the human circulatory system, or the veins in leaves, or the roots of trees. Energy expenditure in the moving of water and sediment through this drainage system is efficient because the total length of the branches is minimized.

The *trellis drainage* pattern (Figure 14.5b) is characteristic of dipping or folded topography. Such drainage is seen in the nearly parallel mountain folds of the Ridge and Valley Province in the eastern United States (shown in Figure 12.17). Here drainage patterns are influenced by folded rock structures that vary in resistance to erosion. Parallel structures direct the principal streams, while smaller dendritic tributary streams are at work on nearby slopes, joining the main streams at right angles, as in a plant trellis.

The remaining drainage patterns in Figure 14.5c–f are responses to other specific structural conditions:

- A *radial* drainage pattern (c) results when streams flow off a central peak or dome, such as occurs on a volcanic mountain.
- *Parallel* drainage (d) is associated with steep slopes.
- A *rectangular* pattern (e) is formed by a faulted and jointed landscape, which directs stream courses in patterns of right-angle turns.
- *Annular* patterns (f) occur on structural domes, with concentric patterns of rock strata guiding stream courses (discussed in Chapter 12).
- A *deranged* pattern (g) with no clear geometry and no true stream valley occurs in areas such as the glaciated shield regions of Canada, northern Europe, and some parts of Michigan and other states.

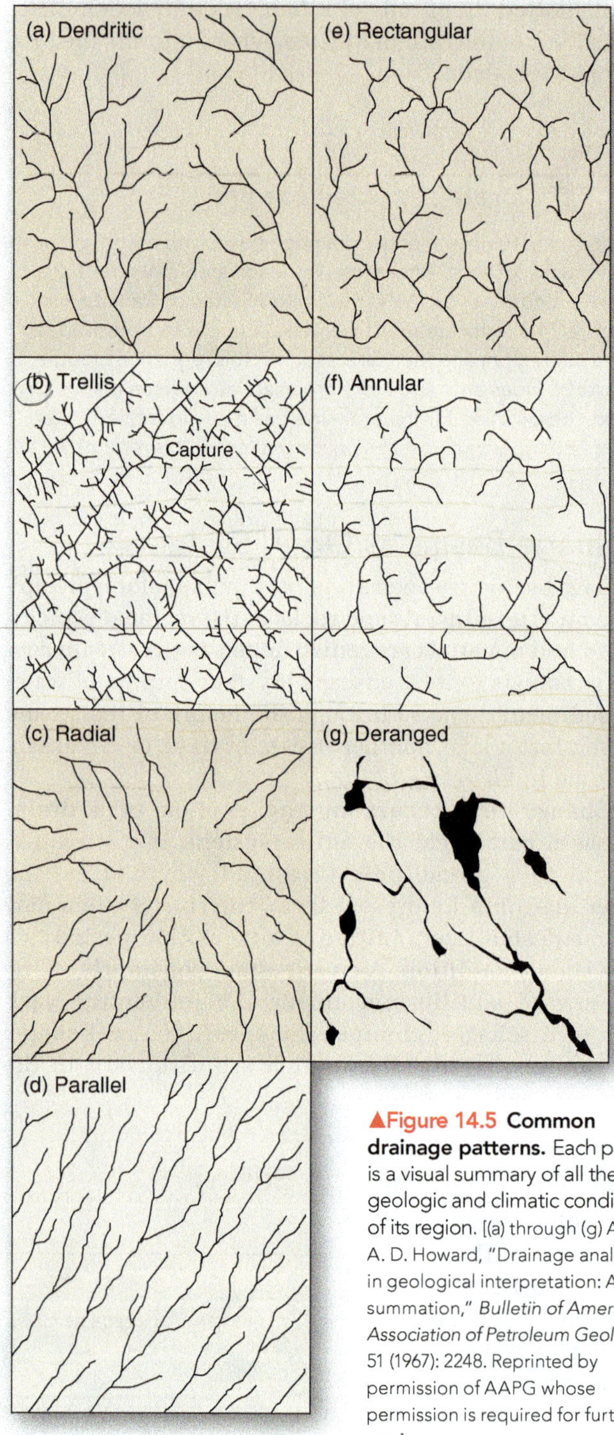

▲**Figure 14.5 Common drainage patterns.** Each pattern is a visual summary of all the geologic and climatic conditions of its region. [(a) through (g) After A. D. Howard, "Drainage analysis in geological interpretation: A summation," *Bulletin of American Association of Petroleum Geologists* 51 (1967): 2248. Reprinted by permission of AAPG whose permission is required for further use.]

Occasionally, drainage patterns occur that seem discordant with the landscape through which they flow. For example, a drainage system may initially develop over horizontal strata that have been deposited on top of uplifted, folded structures. As the streams erode into the older folded strata, they keep their original course, downcutting into the rock in a pattern contrary to its structure. Such a stream is a *superposed stream*, in which a preexisting channel pattern has been imposed upon older underlying rock structures. A few examples are Wills Creek, presently cutting a water gap through Haystack Mountain at Cumberland, Maryland; the Columbia River flowing through the Cascade Mountains of Washington; and the River Arun, which cuts across the Himalayas.

CRITICALthinking 14.2
Identifying Drainage Patterns

Examine the photograph in Figure CT 14.2.1, where you see two distinct drainage patterns. Of the seven types illustrated in Figure 14.5, which two patterns are most like those in the aerial photo? Looking back to Figure 14.1a, which drainage pattern is prevalent in the area around Mount Mismi in Brazil? Explain your answer. The next time you fly in an airplane, look out the window to observe the various drainage patterns across the landscape. ●

▲**Figure CT 14.2.1 Two drainage patterns dominate this scene from central Montana, in response to rock structure and local relief.** [Bobbé Christopherson.]

Basic Fluvial Concepts

Streams, a mixture of water and solids, provide resources and shape landforms. They create fluvial landscapes through the ongoing erosion, transport, and deposition of materials in a downstream direction. The energy of a stream to accomplish this geomorphic work depends on a number of factors, including gradient, base level, and volume of flow (discharge), all discussed in this section.

Gradient

Within its drainage basin, every stream has a degree of inclination or gradient, which is also known as the channel slope. The **gradient** of a stream is defined as the drop in elevation per unit distance, usually measured in meters per kilometer, or feet per mile. Characteristically, a river has a steeper slope nearer the headwaters and a more gradual slope downstream. A stream's gradient affects its energy and ability to move material; in particular, it affects the velocity of the flow (discussed just ahead).

Base Level

The level below which a stream cannot erode its valley is **base level**. In general, the *ultimate base level* is sea level, the average level between high and low tides. Base level can be visualized as a surface extending inland from sea level, inclined gently upward under the continents. In theory, this is the lowest practical level for all denudation processes (Figure 14.6a).

American geologist and explorer John Wesley Powell, mentioned in Focus Study 9.1, put forward the idea of base level in 1875. Powell recognized that not every landscape has degraded all the way to sea level; clearly, other intermediate base levels are in operation. A *local base level,* or temporary one, may determine the lower limit of local or regional stream erosion. A river or lake is a natural local base level; the reservoir behind a dam is a human-caused local base level (Figure 14.6b). In arid landscapes with internal drainage, valleys, plains, or other low points act as local base level.

Stream Discharge

A mass of water situated above base level in a stream has potential energy. As the water flows downslope, or downstream, under the influence of gravity, this energy

▼**Figure 14.6 Ultimate and local base levels.** [(b) Bobbé Christopherson.]

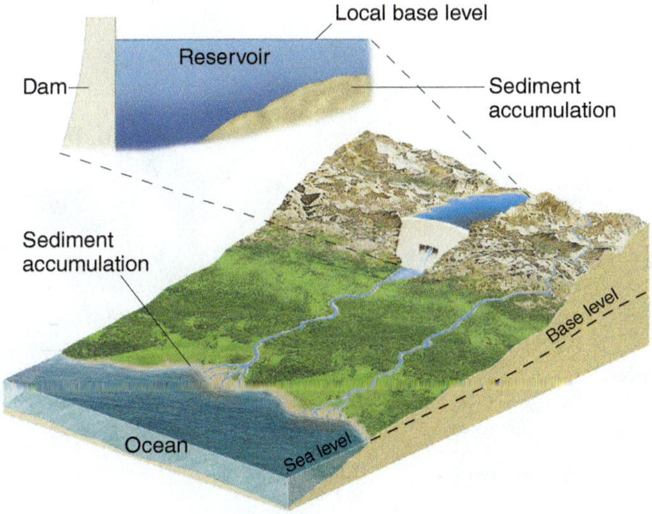

(a) The ultimate base level is sea level. Note how base level curves gently upward from the sea as it is traced inland; this is the theoretical limit for stream erosion. The reservoir behind a dam is a local base level.

(b) Lake Powell behind Glen Canyon Dam is a local base level on the Colorado River.

becomes kinetic energy. The rate of this conversion from potential to kinetic energy determines the ability of the stream to do geomorphic work and depends in part on the volume of water involved.

A stream's volume of flow per unit of time is its **discharge** and is calculated by multiplying three variables measured at a given cross section of the channel. It is summarized in the simple expression

$$Q = wdv$$

where Q = discharge, w = channel width, d = channel depth, and v = stream velocity.

Discharge is expressed in either cubic meters per second (m^3/s) or cubic feet per second (cfs, ft^3/s). According to this equation, as Q increases, one or more of the other variables—channel width, channel depth, stream velocity—must also increase. How these variables interact depends on the climate and geology of the fluvial system.

Changes in Discharge with Distance Downstream In most river basins in humid regions, discharge increases in a downstream direction. The Mississippi River is typical, beginning as many small streams that merge successively with tributaries to form a large-volume river ending in the Gulf of Mexico. However, if a stream originates in a humid region and subsequently flows through an arid region, this relationship may change. High potential evapotranspiration rates in arid regions can cause discharge to decrease with distance downstream, a process that is often exacerbated by water removal for irrigation (Figure 14.7).

The Nile River, one of Earth's longest rivers, drains much of northeastern Africa. But as it flows through the deserts of Sudan and Egypt, it loses water, instead of gaining it, because of evaporation and withdrawal for agriculture. By the time it empties into the Mediterranean Sea, the Nile's flow has dwindled so much that it ranks only 36th in discharge.

In the United States, discharge decreases on the Colorado River with distance from its source; in fact, the river no longer produces enough natural discharge to reach its mouth in the Gulf of California—only some agricultural runoff remains at its delta. The river is depleted by high evapotranspiration and removal of water for agriculture and municipal uses; shifting climatic patterns are adding to the river's water losses (please review Focus Study 9.1).

As discharge increases in a downstream direction, velocity usually increases. Stream velocity is affected by friction between the flow and the roughness of the channel bed and banks. Friction is highest in shallow mountain streams with boulders and other obstacles that add roughness and slow the flow. In streams where contact with the bed and banks is high or where the channel is rough, as in a section of rapids, *turbulent flow* occurs and most of the stream's energy is expended in turbulent eddies. In wide, lowland rivers, where friction is reduced by less contact of the flow with the bed and banks, the apparent smoothness and quietness of the flow mask the increased velocity (Figure 14.8). The energy of these rivers is enough to move large amounts of sediment, discussed in the next section.

Changes in Discharge over Time Discharge varies throughout the year for most streams, depending on precipitation and temperature. Rivers and streams in arid and semiarid regions may have perennial, ephemeral,

▲**Figure 14.7 Declining discharge with distance downstream.** The Virgin River in southwest Utah, a tributary of the Colorado River, is a perennial stream in which discharge decreases in the lower reaches as water is removed for irrigation, is transpired by riparian vegetation (water-loving plants), and is lost to evapotranspiration in this semiarid climate. [Bobbé Christopherson.]

▲**Figure 14.8 Increasing stream velocity and discharge downstream.** The high-discharge, high-velocity, smooth-water flow of the Ocmulgee River near Jacksonville, Georgia. [Bobbé Christopherson.]

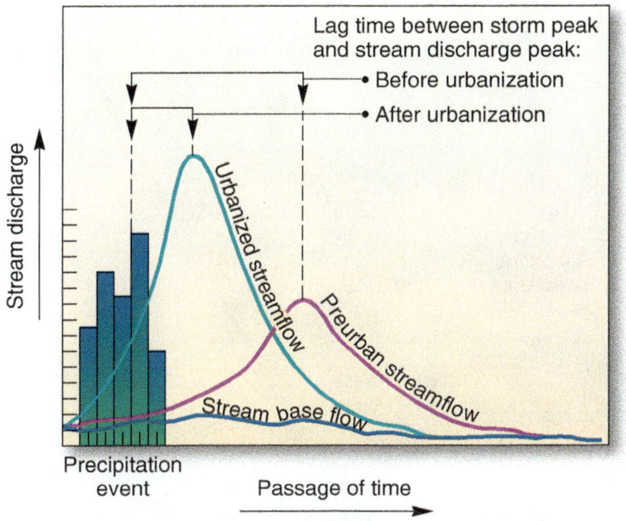

(a) Normal base flow is indicated with a dark blue line. The purple line indicates post-storm discharge prior to urbanization. Following urbanization, stream discharge dramatically increases, as shown by the light blue line.

(b) Increasing urbanization has worsened flooding in many parts of Asia, including Bangkok, Thailand, pictured here in 2011.

▲Figure 14.9 **Effect of urbanization on a typical stream hydrograph.** [(b) Apichart Weerawong/AP.]

or intermittent discharge. *Perennial streams* flow all year, fed by snowmelt, rainfall, groundwater, or some combination of those sources. *Ephemeral streams* flow only after precipitation events and are not connected to groundwater systems. Years may pass between flow events in these usually dry stream channels. *Intermittent streams* flow for several weeks or months each year and may have some groundwater inputs.

Discharge changes over time at any given channel cross section. A graph of stream discharge over time for a specific location is a **hydrograph**. The time scale of a hydrograph can vary. For example, *annual hydrographs* show discharge over the course of an entire year, usually with the highest discharge occurring during the spring snowmelt season. *Storm hydrographs* may cover only a period of days, reflecting changes in discharge caused by specific precipitation events that lead to local flooding. The hydrograph in Figure 14.9a shows the relation between precipitation input (the bar graphs) and stream discharge (the curves). During dry periods, the low discharge is described as *base flow* (dark blue line) and is largely maintained by input from local groundwater (review discussion in Chapter 9).

When rainfall occurs in some portion of the watershed, the runoff is concentrated in streams and tributaries in that area. The amount, location, and duration of the rainfall episode determine the *peak flow*, the highest discharge that occurs during a precipitation event. The nature of the surface in a watershed, whether permeable or impermeable, affects peak flow and the timing of changes recorded in the hydrograph. In deserts, where surfaces have thin, impermeable soils and little vegetation, runoff can be high during rainstorms. A rare or large precipitation event in a desert can fill a stream channel with a torrent known as a **flash flood**. These channels may fill in a few minutes and surge briefly during and after a storm.

Human activities have enormous impact on patterns of discharge in a drainage basin. A hydrograph for a specific portion of a stream changes after a disturbance such as a forest fire or urbanization of the watershed, with peak flows occurring sooner during the precipitation event. The effects of urbanization are quite dramatic, both increasing and hastening peak flow, as you can see by comparing discharge prior to an area's urbanization (purple curve) and discharge after urbanization has occurred (light blue) in Figure 14.9a. In fact, urban areas produce runoff patterns quite similar to those of deserts, since the sealed surfaces of the city drastically reduce infiltration and soil-moisture recharge. These issues will intensify as urbanization continues (Figure 14.9b).

Measuring Discharge Measurements of width, depth, and velocity are needed at a stream cross section in order to calculate discharge. Field measurements of these variables may be difficult to obtain, depending on a stream's size and flow. The common practice is to measure velocity for different subsections of the stream cross section, using a movable current meter (Figure 14.10a). Width and depth for each subsection are then combined with velocity to compute subsection discharge, and then all the subsection discharges for the cross section are totaled (see Figure 14.10). Since channel beds are often composed of soft sediments that may change over short time periods, stream depth is measured as the height of the stream surface above a constant reference elevation (a datum) and is called the *stage*. Scientists may use a *staff gage* (a pole marked with water levels) or a *stilling well* on the stream bank with a gage mounted in it to measure stage (Figure 14.10c).

Approximately 11,000 stream-gaging stations are in use in the United States (an average of more than 200 per state). Of these, 7500 are operated by the U.S. Geological

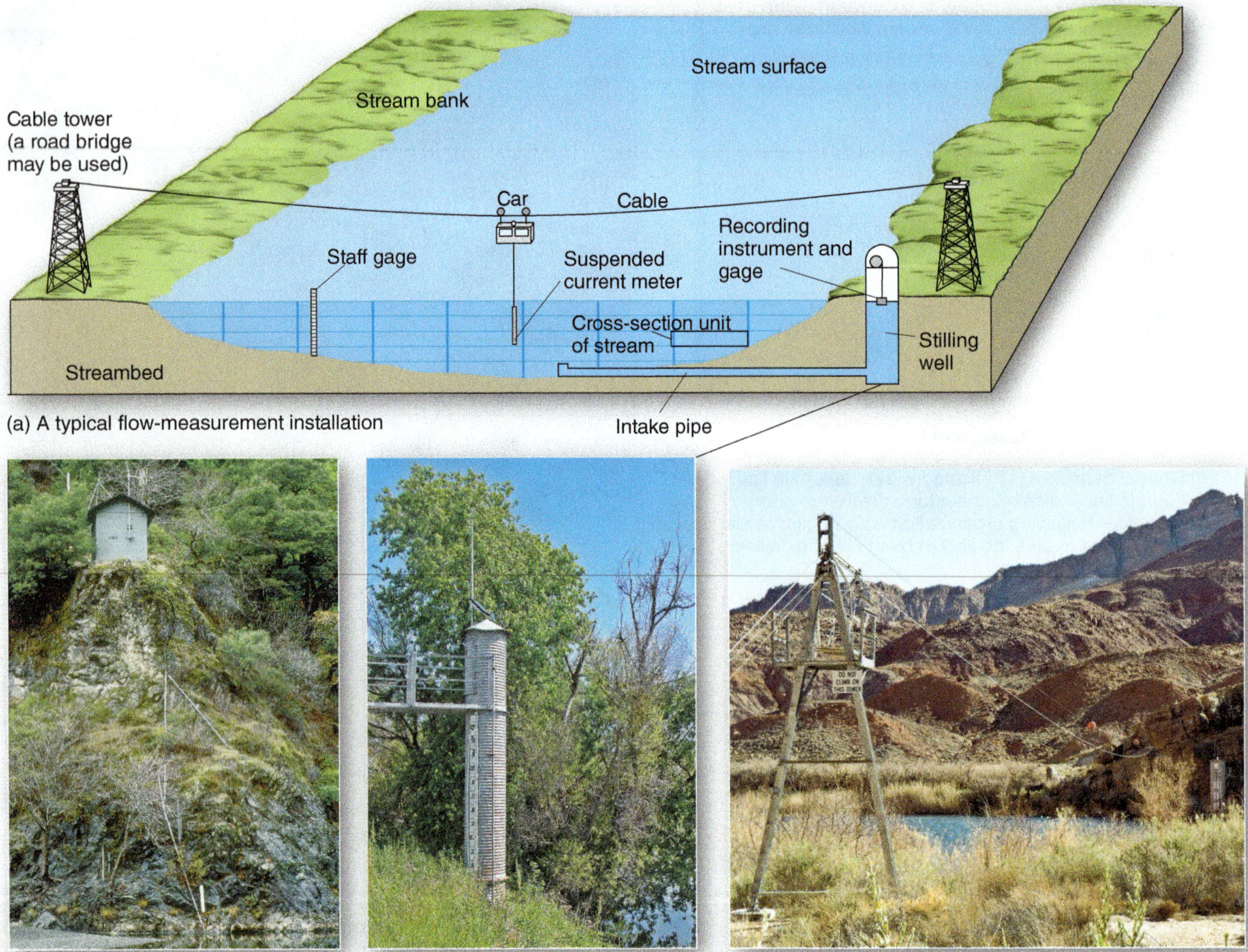

(a) A typical flow-measurement installation

(b) An automated hydrographic station

(c) A stilling well

(d) Cable towers and cable from which current meters are lowered, Lee's Ferry, Arizona

▲Figure 14.10 **Stream-discharge measurement.** [(b) California Department of Water Resources. (c) and (d) by Bobbé Christopherson.]

Survey and have continuous recorders for stage and discharge (see http://pubs.usgs.gov/circ/circ1123/). Many of these stations automatically send telemetry data to satellites, from which information is retransmitted to regional centers (Figure 14.10b and c). The gaging station on the Colorado River at Lees Ferry, just south of Glen Canyon Dam, was established in 1921 and was used to determine flows for the Colorado River Compact, discussed in Focus Study 9.1 (Figure 14.10d).

Fluvial Processes and Landforms

The ongoing interaction between erosion, transportation, and deposition in a river system produces fluvial landscapes. **Erosion** in fluvial systems is the process by which water dislodges, dissolves, or removes weathered surface material. This material is then transported to new locations, where it is laid down in the process of **deposition**. Erosion, transport, and deposition are affected by discharge and channel gradient. Running water is an important erosional force; in fact, in desert landscapes it is the most significant agent of erosion even though precipitation events are infrequent. We discuss processes of erosion and deposition, and their characteristic landforms, in this section.

Stream Channel Processes

The geomorphic work performed by a stream includes erosion and deposition, which depends on the volume of water and the total amount of sediment in the flow. **Hydraulic action** is a type of erosive work performed by flowing water alone, a squeeze-and-release action that loosens and lifts rocks. Hydraulic action is at a maximum in upstream tributaries of a drainage basin, where sediment load is small and flow is turbulent (Figure 14.11). The downstream portions of a river, however, move much

▲**Figure 14.11 A turbulent stream.** The high-gradient, turbulent Maligne River flows through a bedrock canyon in the Canadian Rockies. [Ashley Cooper/Corbis.]

▲**Figure 14.12 Headward erosion.** Headward erosion in several tributaries of an arid-region river in Baja California Sur, Mexico. [RGB Ventures LLC dba SuperStock/Alamy.]

larger volumes of water past a given point and carry larger amounts of sediment. As this debris moves along, it mechanically erodes the streambed further through the process of **abrasion**, with rock and sediment grinding and carving the streambed like liquid sandpaper.

Fluvial erosion by hydraulic action and abrasion causes streams to erode downward (deepen), erode laterally (widen), and erode in an upstream direction (lengthen). The process whereby streams deepen their channel is known as *channel incision* (an example on the San Juan River is discussed ahead). The process of lateral erosion is discussed in the next section with meandering river channels.

The process whereby streams lengthen their channels upstream is called *headward erosion*. This type of erosion occurs when the flow entering a main channel has enough power to downcut, such as occurs at the break in slope where a gully enters a deep valley (Figure 14.12). Headward erosion can also occur from groundwater sapping, in which groundwater seeps out of the ground at the head of the channel and weakens the channel's upstream endpoint. Headward erosion can eventually cause an eroding part of one stream channel to break through a drainage divide and *capture* the headwaters of another stream in an adjacent valley—an event known as *stream piracy*.

Sediment Load When stream energy is high and a supply of sediment is present, streamflow propels sand, pebbles, gravel, and boulders downstream in the process known as **sediment transport**. The material carried by a stream is its *sediment load*, and the sediment supply relates to topographic relief, the nature of rock and soil through which the stream flows, climate, vegetation, and human activity in a drainage basin. Discharge is also closely linked to sediment transport—increased discharge moves a greater amount of sediment, often causing streams to change from clear to murky brown after a heavy or prolonged rainfall. Sediment is moved as dissolved load, suspended load, or bed load by four primary processes: solution, suspension, saltation, and traction (Figure 14.13).

The **dissolved load** of a stream is the material that travels in solution, especially the dissolved chemical compounds derived from minerals such as limestone or dolomite or from soluble salts. The main process contributing material in solution is chemical weathering. Along the San Juan and Little Colorado Rivers, which flow into the Colorado River near the Utah–Arizona border, the salt content of the dissolved load is so high that human use of the water is limited.

The **suspended load** consists of fine-grained clastic particles (bits and pieces of rock). They are held aloft in the stream until the stream velocity slows nearly to zero, at which point even the finest particles are deposited.

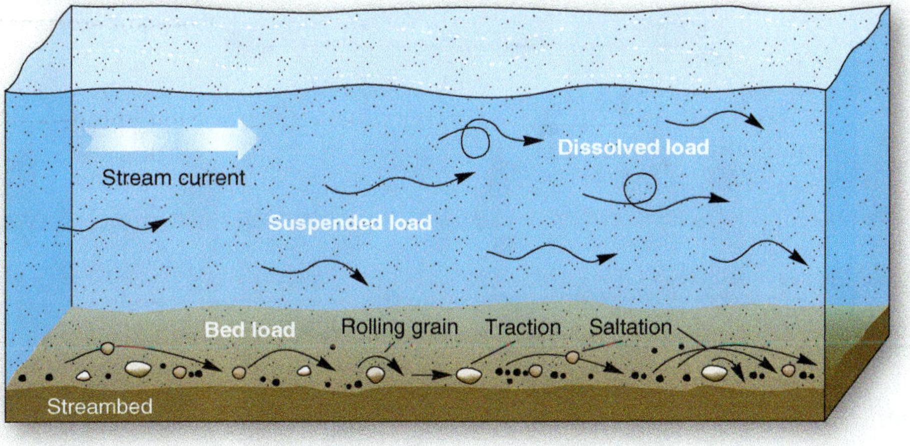

▲**Figure 14.13 Fluvial transport.** Fluvial transportation of eroded materials as dissolved load, suspended load, and bed load.

Turbulence in the water, with random upward motion, is an important mechanical factor in holding a load of sediment in suspension.

Bed load refers to coarser materials that are moved by **traction**, which is the rolling or dragging of materials along the streambed, or by **saltation**, a term referring to the way particles may bounce along in short hops and jumps (from the Latin *saltim,* which means "by leaps or jumps"). Particles transported by saltation are too large to remain in suspension but are not confined to the sliding and rolling motion of traction (see Figure 14.13). Stream velocity affects these processes, particularly the stream's ability to retain particles in suspension. With increased kinetic energy in a stream, parts of the bed load are rafted upward and become suspended load.

During a flood (a high flow that overtops the channel banks), a river may carry an enormous sediment load, as larger material is picked up and carried by the enhanced flow. The *competence* of a stream is its ability to move particles of a specific size and is a function of stream velocity and the energy available to move materials. The *capacity* of a stream is the total possible sediment load that it can transport and is a function of discharge; thus, a large river has higher capacity than a small stream. As flood flows build, stream energy increases and the competence of the stream becomes high enough that sediment transport occurs. As a result, the channel erodes, a process known as **degradation**. With the return of flows to normal, stream energy is reduced, and the sediment transport slows or stops. If the load exceeds a stream's capacity, sediment accumulates in the bed, and the stream channel builds up through deposition; this is the process of **aggradation**.

Sediment Transport During a Flood We saw in the previous section that discharge can change quickly in response to precipitation events in a watershed. Greater discharge increases flow velocity and therefore the competence of the river to transport sediment as the flood progresses. As a result, the river's ability to scour materials from its bed is enhanced.

As an example, Figure 14.14 shows changes in the San Juan River channel in Utah that occurred during a flood. The channel was deepest on October 14, when floodwaters were highest (blue line plotted in Figure 14.14a). During this time, the channel bed eroded. The flood and scouring process moved a depth of about 3 m (10 ft) of sediment from the depicted cross section. By October 26, with the discharge returning to normal, the energy of the river was reduced, and the bed again filled as sediment redeposited. This type of channel adjustment is ongoing, as the system continuously works toward equilibrium, maintaining a balance between discharge, sediment load, and channel form.

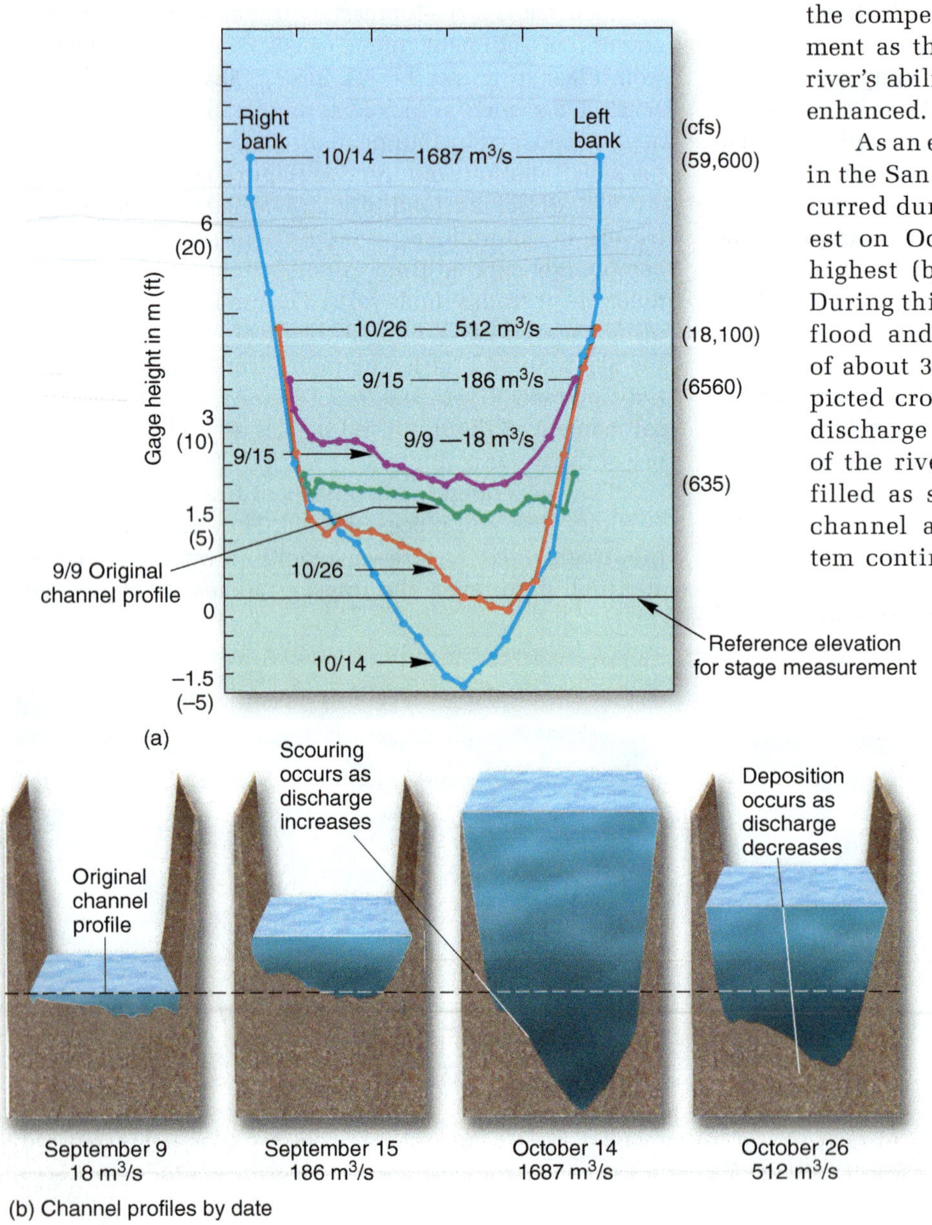

◄Figure 14.14 How a flood affects a stream channel. (a) Channel cross sections showing the effects of a flood on the San Juan River near Bluff, Utah. (b) Details of channel profiles during four stages of the flood. [Adapted from "The Hydraulic Geometry of Stream Channels and Some Physiographic Implications," by L. Leopold and T. Maddock, USGS Professional Paper 252, p. 32, 1941.]

Effects of Dams on Sediment Transport As discussed in the chapter-opening *Geosystems Now*, dams disrupt natural river discharge and sediment regimes, usually with detrimental effects on river systems. For example, Glen Canyon Dam on the Colorado River near the Utah–Arizona border controls discharge and blocks sediment from flowing into the Grand Canyon downstream (see Figure 14.6 and the map in Focus Study 9.1). Consequently, over the years, the river's sediment supply was cut off, starving the river's beaches for sand, disrupting fisheries, and depleting backwater channels of nutrients.

In 1996, 2004, 2008, and 2012, the Grand Canyon was artificially flooded with dam-controlled releases in an unprecedented series of scouring–redistribution–deposition experiments for sediment movement. The first test lasted for 7 days. The later tests were of shorter duration and were timed to coincide with floods in tributaries that supplied fresh sediment to the system.

The results were mixed and the benefits turned out to be limited—the flood releases disrupted ecosystems and eroded some sediment deposits even while building up others. With such a limited sediment supply, not enough sediment is present in the system to build beaches and improve habitat even when high discharge occurs (more information is at http://www.gcmrc.gov/).

Recent dam removals have allowed scientists to study post-dam sediment redistribution. Focus Study 14 discusses dam deconstruction and other stream restoration practices. See Focus Study 14.1, Figure 14.1.2 for a photo of the sediment outflow into the Strait of Juan de Fuca in Washington following removal of two dams on the Elwha River.

▲**Figure 14.15 A braided stream.** The braided Waiho River channel in the glaciated landscape of the west coast of New Zealand's South Island. [David Wall/Alamy.]

Channel Patterns

A number of factors, including the sediment load, affect the channel pattern. Multiple-thread channels, either braided or anabranching (defined below), tend to occur in areas with abundant sediment or in the lowest reaches of large river systems. Single-thread channels are either straight or meandering. Straight channels tend to occur in headwater areas where gradient is high. In lower-gradient areas with finer sediments, meandering is more common; this is the classic river pattern in which a single channel curves from side to side in a valley or canyon.

Multiple-Thread Channels With excess sediment, a stream might become a maze of interconnected channels that form a **braided stream** pattern (Figure 14.15). Braiding often occurs when reduced discharge lowers a stream's transporting ability, such as after flooding, or when a landslide occurs upstream, or when sediment load increases in channels that have weak banks of sand or gravel. Braided rivers commonly occur in glacial environments, where coarse sediment is abundant and slopes are steep, as in New Zealand, Alaska, Nepal, and Tibet. This pattern also occurs in wide, shallow channels with variable discharge, such as in the U.S. Southwest.

In large river systems, an *anabranching channel pattern* sometimes occurs, in which multiple large channels are present across a vast floodplain (Figure 14.16). The

(text continued on page 436)

▲**Figure 14.16 An anabranching channel pattern.** The anabranching channel of the Paraná River in South America. [NASA.]

Focus Study 14.1 Environmental Restoration

Stream Restoration: Merging Science and Practice

As mentioned in Chapter 1, "basic" science is designed to advance knowledge and build scientific theories. "Applied" science solves real-world problems, and by doing so often advances new technologies and develops natural resource management strategies. Beginning in the late 1980s, the restoration of rivers and streams has become a focus for both basic and applied geomorphology.

Stream restoration, also called *river restoration*, is the process that reestablishes the health of a fluvial ecosystem, including channel processes and form, riparian vegetation, and fisheries. Every stream restoration project has a particular focus, which varies with the problems and impacts on that particular stream. Common restoration goals are to reinstate instream flows, restore fish passage, prevent bank erosion, and reestablish vegetation along the channel or on the floodplain. The scale of a stream restoration varies from a few kilometers of stream to hundreds of miles of river to an entire watershed.

Dam Removals

As many small dams overseen by the Federal Energy Regulatory Commission (FERC) have come up for relicensing over the past 25 years, scientists and environmental groups have worked with city, state, and federal agencies to identify dams that should be considered for removal. Numerous dams have been deconstructed because they were unsafe or outdated or their original purpose was no longer valid. In 1999, Edwards Dam was deconstructed from the Kennebec River in Augusta, Maine, marking the first dam removal in the United States for ecological reasons (in this case, primarily to restore passage between the river and sea for migratory fish).

In 2013, the deconstruction of the Elwha and Glines Canyon Dams (both over 80 years old) in northwest Washington became the largest dam removal in U.S. history, restoring fish passage and associated river ecosystems on the 72-km-long (45-mi-long) Elwha River (Figure 14.1.1 and please look back at Figure 1.1 in Chapter 1). A free-flowing Elwha River will enable the return of five species of Pacific salmon to the watershed, with fish numbers expected to increase from 3000 to 390,000 over the next 30 years. Such species are *anadromous* (from the Greek *anadromos*, "running up"), meaning they migrate upstream from the sea into freshwater rivers to spawn.

A year after the dam removal began, scientists reported that native steelhead, as well as chinook, coho, and pink salmon, were already making their way into previously inaccessible stream reaches. Willow and cottonwood saplings were beginning to establish in the newly exposed, silt-laden riverbed. In that first year, an estimated half million tons of sediment, previously trapped behind the dams, began moving downstream and exiting at the river's mouth (Figure 14.1.2). For more information, see http://www.nps.gov/olym/naturescience/elwha-restoration-docs.htm.

A Cooperative Process

Stream restoration involves the cooperation of numerous landowners and regulating agencies within a watershed. For instance, bank erosion at a given location may be linked to processes occurring upstream and in turn can affect downstream processes. Thus, restoring a stream to its natural state involves the entire system rather than just a single, isolated segment. Restoration also involves balancing different water use needs within a drainage basin.

In Maine's largest watershed, the Penobscot River restoration project is attempting to restore fisheries while at the same time maintaining hydropower production. The Penobscot restoration project involves a large coalition of interest groups, including six environmental groups, the Penobscot Indian Nation, the State of Maine, the U.S. Department of the Interior, the National Oceanic and Atmospheric Administration, and hydropower companies. Restoration includes removing two dams, building a fish lift at another dam, and building a fish bypass around a fourth dam, thereby restoring fish access along hundreds of miles of streams. The estimated cost for the entire project is $62 million; the first phase was completed in 2010 with the removal of the Great Works Dam.

(b) A barge-mounted hydraulic hammer chips away at the top of the Glines Canyon Dam in 2012.

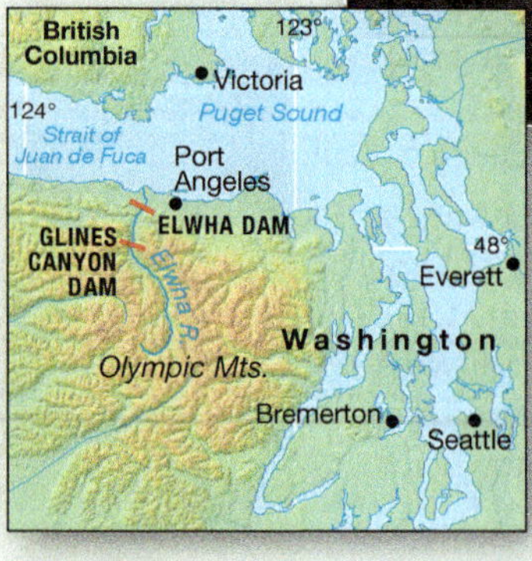

(a) Sites of former dams on the Elwha River.

▲**Figure 14.1.1 Glines Canyon Dam removal, Elwha River.** [(a) NPS/USGS. (b) NPS.]

▲Figure 14.1.2 **Sediment flowing from the mouth of the Elwha River after dam removal.** [Tom Roorda.]

Smaller-scale stream restoration projects may be part of large-scale restoration efforts involving estuaries, bays, and harbors. Near New Bedford Harbor, Massachusetts, state, federal, and local agencies implemented several small stream projects as part of a larger effort to clean up pollution and restore migratory fish passage. In 2007, restoration specialists partially breached the Sawmill Dam on the Acushnet River and constructed a passage structure, or "fishway," consisting of a stone step-pool system designed to mimic natural conditions (Figure 14.1.3). Data collected from 2007 to 2011 show that river herring (alewives and blueback herring) increased over 1000% as fish passage around Sawmill and another nearby dam allowed access to prime spawning grounds.

Stream Restoration Science

Stream restoration practices have become a lucrative business for hundreds of companies throughout the United States, with Americans spending an estimated $1 billion annually on stream projects. However, the science of stream restoration is still in its infancy. At the University of California at Berkeley, scientists are experimenting with a scale model of a meandering gravel-bed river in a laboratory to understand interactions between gravel, sand, vegetation, and meander erosion and deposition processes—a project whose results will find useful application in restoration practices.

At the multidisciplinary National Center for Earth Surface Dynamics at the University of Minnesota, one research goal is to develop a set of free, downloadable scientifically based tools that promote quantitative restoration methods (see http://www.nced.umn.edu/content/streams-science-restoration).

Given the complexities of streams and their ecosystems, some intuitive and artistic license is often used when designing a stream, and not all designs withstand the test of time. Furthermore, the intensive monitoring needed to evaluate success or failure after a project's completion is often neglected due to lack of funds. One example of the long-term ecological studies that are necessary to advance and improve stream restoration science is the Baltimore Ecosystem Study by the Long Term Ecological Research Network, funded by the National Science Foundation (http://www.lternet.edu/sites/bes).

Ongoing dam removals and continued societal and scientific emphasis on ecosystem health have made stream restoration science a growing field of applied fluvial geomorphology, and one to which geographers and other Earth systems scientists are poised to contribute.

▲Figure 14.1.3 **Step-pool fishway construction on the Acushnet River, Massachusetts.** [Steve Block, NOAA.]

Paraná, and other rivers in South America, exhibit this pattern, as do rivers in Canada, Alaska, and parts of Asia.

Single-Thread Channels Although perfectly straight channels are rare or nonexistent in nature, many streams in steep mountain regions or in bedrock-controlled channels have a relatively straight channel pattern (Figure 14.17). Often these are high-gradient streams and have such low sinuosity that they cannot be classified as typical, meandering streams. (*Sinuosity* is the ratio between the distance between two points measured along a stream as it curves and the shortest, straight-line distance between the same two points).

Where channel slope is gradual, streams develop a more sinuous (snakelike) form, weaving back and forth across the landscape in a **meandering stream** pattern and acquiring distinctive flow and channel characteristics. The tendency to meander is evidence of a river system's propensity (like any natural system) to find the path of least effort toward a balance between self-equilibrating order and chaotic disorder.

▲Figure 14.17 **A straight channel pattern.** Relatively straight channel pattern controlled by bedrock on the Ottauquechee River in Quechee Gorge, Vermont. [Fraser Hall/ Robert Harding World Imagery/Corbis.]

Geosystems In Action, Figure GIA 14, illustrates some of the processes associated with meandering streams. A cross-sectional view of a meandering stream channel shows the flow characteristics that produce the channel deposits typical of these streams. In a straight channel or section of channel, the greatest flow velocities are near the surface at the center, corresponding to the deepest part of the stream (Figure GIA 14.1a). Velocities decrease closer to the sides and bottom of the channel because of the frictional drag on the water flow. As the stream flows around a meander curve, the maximum flow velocity shifts from the center of the stream to the outside of the curve. As the stream then straightens, the maximum velocity shifts back to the center, until the next bend, where it shifts to the outside of that meander curve. Thus, the portion of the stream flowing at maximum velocity moves diagonally across the stream from bend to bend.

Because the outer portion of each meandering curve is subject to the fastest water velocity, it undergoes the greatest erosive action, or *scouring*. This action can form a steep **undercut bank**, or *cutbank* (Figure GIA 14.1b). In contrast, the inner portion of a meander experiences the slowest water velocity and thus is a zone of fill (or aggradation) that results in a **point bar**, an accumulation of sediment on the inside of a meander bend. As meanders develop, these scour-and-fill processes gradually work at stream banks, causing them to move laterally across a valley—this is the process of lateral erosion. As a result, the landscape near a meandering river bears marks called *meander scars* that are the residual deposits from previous river channels (seen in the map and image of the Mississippi River's meanders in Figure 14.24 on page 443).

Meandering streams create a remarkable looping pattern on the landscape, as shown in Figure GIA 14.2. Actively meandering streams erode their outside banks as they migrate, often forming a narrow neck of land that eventually erodes through and forms a *cutoff*. A cutoff marks an abrupt change in the stream's lateral movements—the stream becomes straighter.

After the former meander becomes isolated from the rest of the river, the resulting **oxbow lake** may gradually fill with organic debris and silt or may again become part of the river when it floods. The Mississippi River is many miles shorter today than it was in the 1830s because of artificial cutoffs that were dredged across meander necks to improve navigation and safety. How many of these stream features—meanders, oxbow lakes, cutoffs—can you spot in Figure 14.24 just ahead?

Streams often form natural political boundaries, as we saw with the Danube River earlier. Yet problems may arise when boundaries are based on river channels that change course. For example, the Ohio, Missouri, and

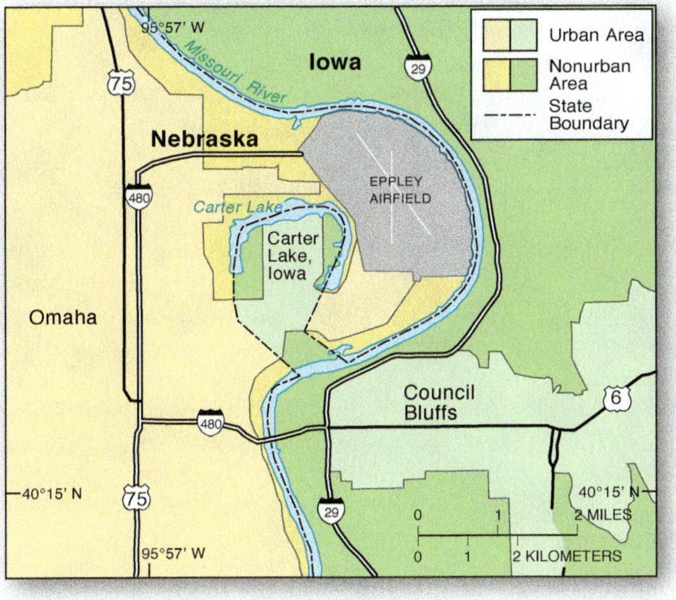

▲Figure 14.18 **A town left stranded by shifting meanders.** Carter lake, Iowa, sits within a curve of a former meander that was cut off by the Missouri River. The city and oxbow lake remain part of Iowa even thought they are now mostly surrounded by Nebraska.

Mississippi Rivers, which form boundaries of several states, can shift their positions quite rapidly during times of flood. Consider the Nebraska–Iowa border, which was originally placed mid-channel in the Missouri River. In 1877, the river cut off the meander loop around a town, leaving the town "captured" by Nebraska (Figure 14.18). The new oxbow lake was called Carter Lake and remains the state boundary. Today, the city of Carter Lake is the only part of Iowa that lies west of the Missouri River. Other states have taken precautions against such events. To avoid boundary disputes along the Rio Grande near El Paso, Texas, and along the Colorado River between Arizona and California, surveys have permanently established political boundaries independent of changing river locations.

Graded Streams

The changes in a river's gradient from its headwater to its mouth are usually represented in a side view called a *longitudinal profile*. The curve of a river's overall gradient is generally concave (Figure 14.19). As mentioned earlier, a river characteristically has a steeper slope nearer the headwaters and a more gradual slope downstream. The causes of this shape are related to the energy available to the stream for transporting the load it receives.

The tendency of natural systems, including streams, to move toward a state of equilibrium causes stream channels, over a period of years, to adjust their channel characteristics so that the flow is able to move the sediment supplied from the drainage basin. A **graded stream** is one in which the channel slope has adjusted, given the discharge and channel conditions, so that stream velocity is just enough to transport the sediment load.

A graded stream has the characteristic longitudinal profile illustrated in Figure 14.19. Any variation, or bump, in the profile, such as the steep drop of a waterfall, will be smoothed out over time as the stream adjusts toward a graded condition. Attainment of a graded condition does not mean that the stream is at its lowest gradient, but rather that it has achieved a state of *dynamic equilibrium* between its gradient and its sediment load. This balance depends on many factors that work together on the landscape and within the river system.

An individual stream can have both graded and ungraded portions, and it can have graded sections without having an overall graded slope. In fact, variations and

(text continued on page 440)

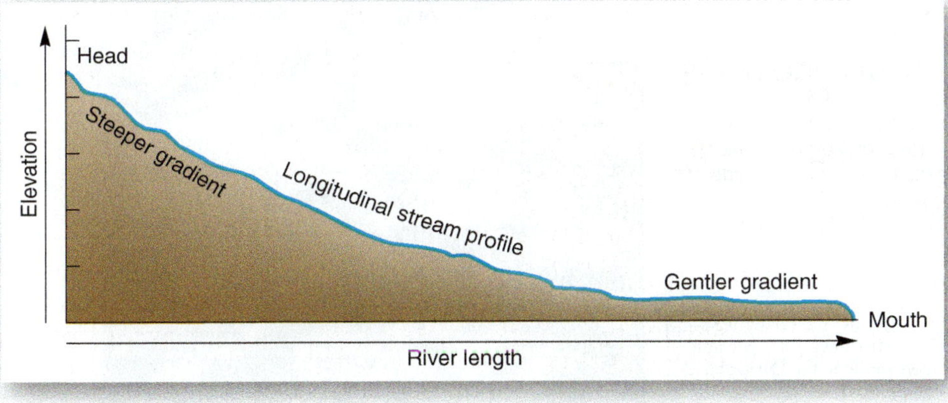

▲Figure 14.19 **A characteristic longitudinal profile.** The characteristic sloping profile of a stream from headwaters to mouth. Upstream segments of the profile have a steeper gradient; downstream, the gradient is gentler.

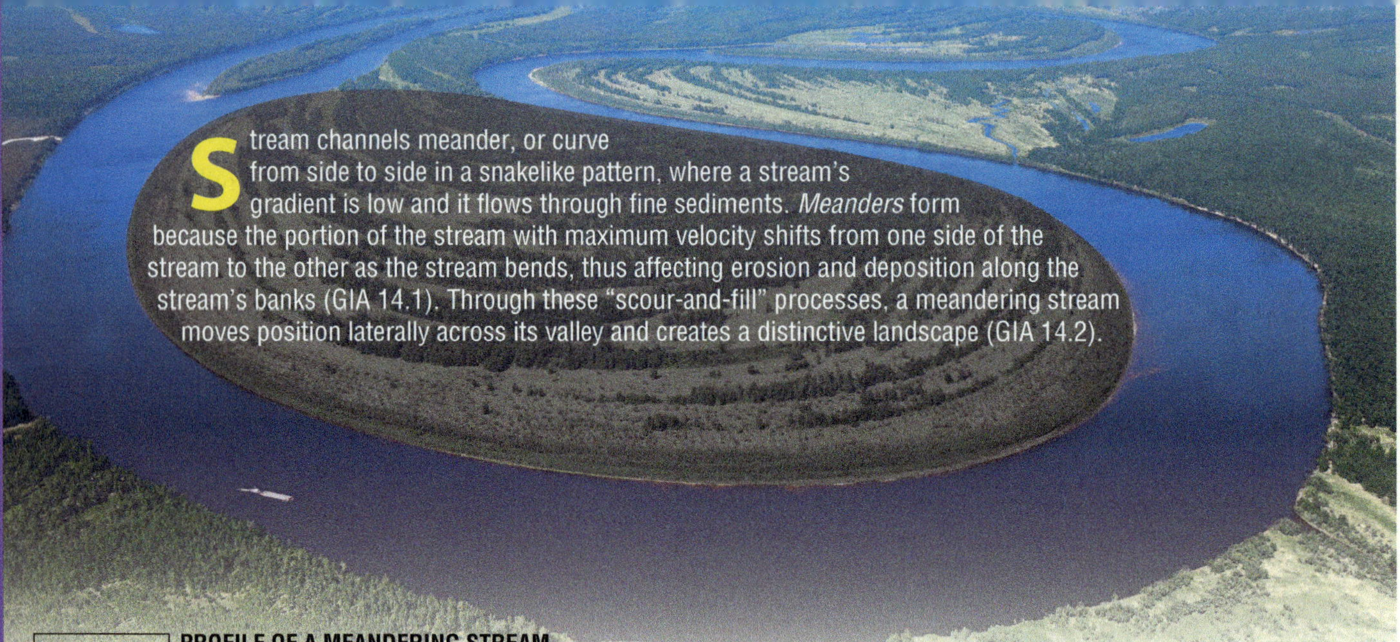

Stream channels meander, or curve from side to side in a snakelike pattern, where a stream's gradient is low and it flows through fine sediments. *Meanders* form because the portion of the stream with maximum velocity shifts from one side of the stream to the other as the stream bends, thus affecting erosion and deposition along the stream's banks (GIA 14.1). Through these "scour-and-fill" processes, a meandering stream moves position laterally across its valley and creates a distinctive landscape (GIA 14.2).

[Vladimir Melnikov/Shutterstock.]

14.1a PROFILE OF A MEANDERING STREAM

The cross sections show how the location of maximum flow velocity shifts from the center along a straight stretch of the stream channel to the outside bend of a meander. The oblique view shows how the stream erodes, or "scours," an *undercut bank* on the outside of a bend, while depositing a *point bar* on the inside of the bend.

Point bar deposition: *On a bend's inner side, stream velocity decreases, leading to deposition of sediment and forming a point bar.*

Undercut bank erosion: *Areas of maximum stream velocity (darker blue) have more power to erode, so they undercut the stream's banks on the outside of a bend.*

14.1b ACTIVE EROSION ALONG A MEANDER

Notice how this stream in Iowa has eroded a steep cutbank on the outside of a bend.

[USDA NRCS.]

Explain: Explain the relationship between stream velocity, erosion, and deposition in the formation of a meander.

MasteringGeography™

Visit the Study Area in MasteringGeography™ to explore meander and oxbow lake formation.

Visualize: Study a geosciences animation of meander and oxbow lake formation.

Assess: Demonstrate understanding of meander and oxbow lake formation (if assigned by instructor).

Animation
Meandering Streams

14.2a STREAM MEANDERING PROCESSES

Over time, stream meanders migrate laterally across a stream valley, eroding the outside of bends and filling the insides of bends. Narrow areas between meanders are *necks*. When discharge increases, the stream may scour through the neck, forming a *cutoff*, as seen in the photograph.

Stream valley landscape:
A neck has recently been eroded, forming a cutoff and straightening the stream channel. The bypassed portion of the stream may become a ***meander scar*** *or an* ***oxbow lake****.*

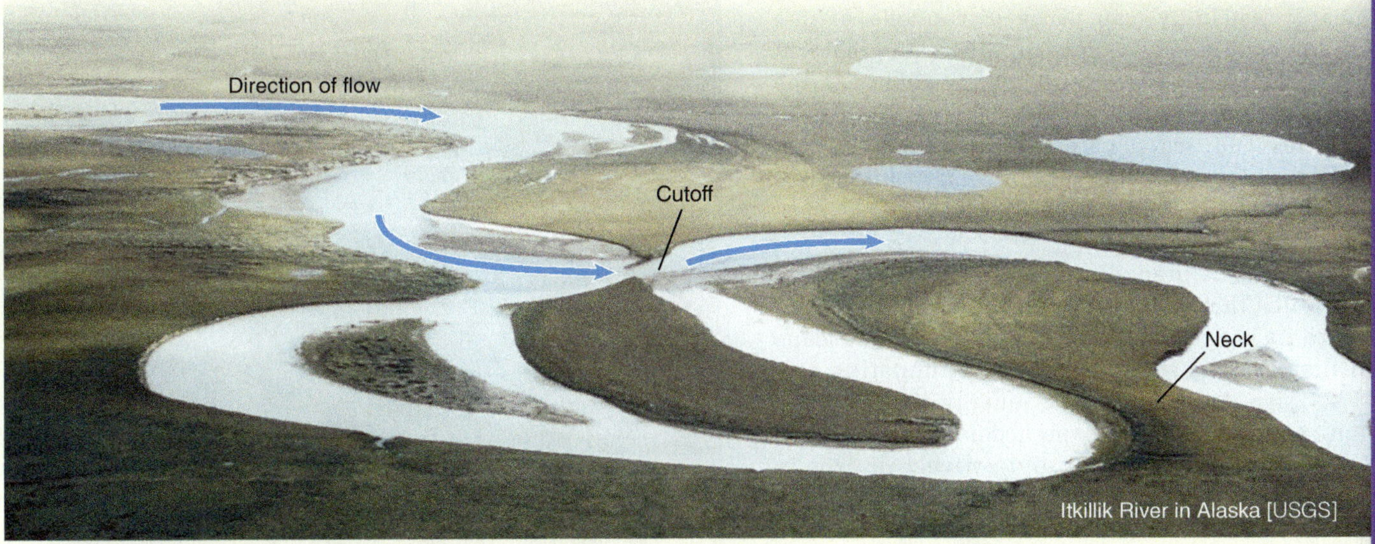

Itkillik River in Alaska [USGS]

14.2b FORMATION OF AN OXBOW LAKE

The diagrams below show the steps often involved in forming an oxbow lake. As stream channels shift, these processes leave characteristic landforms on a floodplain.

Step 1:
A narrow neck is formed where a lengthening meander loops back on itself.

Step 2:
The neck narrows even more due to undercutting of its banks.

Step 3:
The stream erodes through the neck, forming a cutoff.

Step 4:
An oxbow lake forms as sediment fills the area between the new stream channel and its old meander.

MG Animation
Stream Processes, Floodplains, Oxbow Lake Formation

Follow up: In your own words, describe the sequence of steps in the process that forms an oxbow lake.

GEOquiz

1. Explain: Explain the processes that cause a gentle bend along a stream to become a deeply looping meander.

2. Summarize: Summarize the process by which a stream, over time, could produce the landscape in the GIA14.2a photograph.

interruptions are the rule rather than the exception. Disturbances in a drainage basin, such as mass wasting on hillslopes that carries material into stream channels, or overgrazing of riparian vegetation and associated streambank instability, can cause disruptions to this equilibrium condition. The concept of stream gradation is intimately tied to stream gradient; any change in the characteristic longitudinal profile of a river causes the system to respond, seeking a graded condition. The following discussions of tectonic uplift and of nickpoints delve further into this concept.

Tectonic Uplift A graded stream can be affected by tectonic uplift that changes the elevation of the stream relative to its base level. Such lifting of the landscape would increase the stream gradient, stimulating erosional activity. A previously low-energy river flowing through the uplifted landscape becomes *rejuvenated;* that is, the river gains energy and actively returns to downcutting. Degradation of the channel can eventually form *entrenched meanders* that are deeply incised in the landscape (Figure 14.20). Such a stream is called an *antecedent stream* (from the Greek *ante*, meaning before) because it downcuts at the same rate the uplift occurs, thus maintaining its course. Note that superposed streams, mentioned earlier, do not downcut as uplift occurs; instead, they superpose their original course on older rock strata that become exposed by erosion.

▲**Figure 14.20 Aerial view of entrenched meanders.** Aerial view of entrenched meanders of the Escalante River, Utah, on the Colorado Plateau, a high tableland that was uplifted during the Laramide orogeny. [SCPhotos/Alamy.]

Nickpoints When the longitudinal profile of a stream contains an abrupt change in gradient, such as at a waterfall or an area of rapids, the point of interruption is a **nickpoint** (also spelled *knickpoint*). Nickpoints can result when a stream flows across a resistant rock layer or a recent fault line or area of surface deformation. Temporary blockage in a channel, caused by a landslide or a logjam, also could be considered a nickpoint; when the logjam breaks, the stream quickly readjusts its channel to its former grade. Thus, a nickpoint is a relatively temporary and mobile feature on the landscape.

Figure 14.21a shows two nickpoints— an area of rapids (with increased gradient) and a waterfall (even steeper gradient). At a waterfall, the conversion of potential energy in the water at the lip of the falls to concentrated kinetic energy at the base works to eliminate the nickpoint interruption and smooth out the gradient. At the edge of a waterfall, a stream is free-falling, moving at high velocity under the acceleration of gravity, causing abrasion and hydraulic action in the channel below. Over time, the increased erosive action slowly undercuts the waterfall. Eventually, the rock ledge at the lip of the fall collapses, and the height of the waterfall is gradually reduced as debris accumulates at its base (Figure 14.21b). Thus, a nickpoint migrates upstream, sometimes for kilometers, until it becomes a series of rapids and is eventually eliminated.

In the region of Niagara Falls on the Ontario–New York border, glaciers advanced and then receded some 13,000 years ago. In doing so, they exposed resistant rock strata that are underlain by less-resistant shales. The resulting tilted formation is a *cuesta,* which is a ridge with a steep slope on one side (called an escarpment) and beds gently sloping away on the other side (Figure 14.22a). The Niagara escarpment actually stretches across more than 700 km (435 mi); from east of the falls, it extends northward through Ontario, Canada, and the Upper Peninsula of Michigan and then curves south through Wisconsin along the western shore of Lake Michigan and the Door Peninsula. As the less-resistant material continues to weather, the overlying rock strata collapse and Niagara Falls erodes upstream toward Lake Erie (this is the process of headward erosion, described earlier in the chapter). Engineers occasionally use control facilities

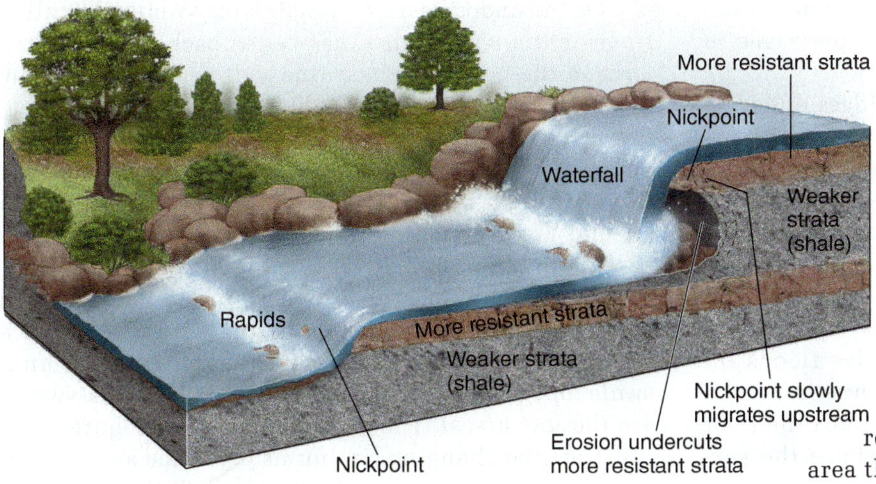

(a) Longitudinal profile of a stream section shows nickpoints produced by resistant rock strata. Stream energy is concentrated at the nickpoint, accelerating erosion, which will eventually eliminate the feature.

(b) Rapids and falls break the flow of the Big Sioux River in Sioux Falls, South Dakota.

▲**Figure 14.21 Nickpoints interrupting a stream profile.** [(b) Bobbé Christopherson.]

upstream to reduce flows over the American Falls at Niagara in order to inspect cliff erosion, which has moved the location of the falls more than 11 km (6.8 mi) upstream from the steep face of the Niagara escarpment (Figures 14.22b and c).

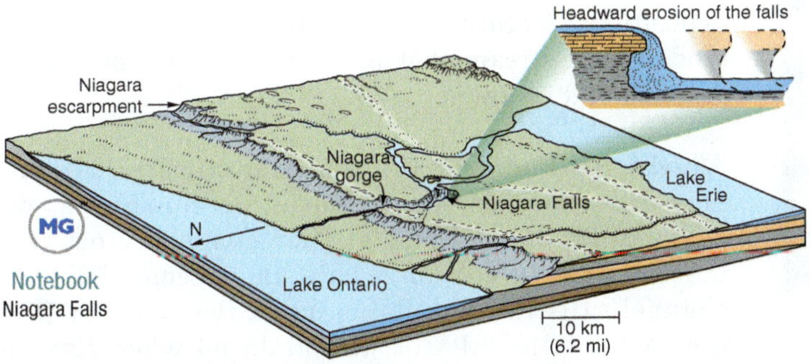

MG Notebook Niagara Falls

(a) Headward retreat of Niagara Falls from the Niagara escarpment has taken about 12,000 years, at a pace of about 1.3 m (4.3 ft) per year.

Depositional Landforms

The general term for the unconsolidated clay, silt, sand, gravel, and mineral fragments deposited by running water is **alluvium**, which may accumulate as sorted or semisorted sediment. The process of *fluvial deposition* occurs when a stream deposits alluvium, thereby creating depositional landforms, such as bars, floodplains, terraces, and deltas.

Floodplains The flat, low-lying area adjacent to a channel and subjected to recurrent flooding is a **floodplain**. It is the area that is inundated when the river overflows its channel during times of high flow. When the water recedes, it leaves behind alluvial deposits that generally mask the underlying rock with their accumulating thickness. The present river channel is embedded in these alluvial deposits. As discussed earlier, stream meanders tend to migrate laterally across a valley; over

▼**Figure 14.22 Retreat of Niagara Falls.** [(a) After W. K. Hamblin, *Earth's Dynamic Systems*, 6th ed., Pearson Prentice Hall, Inc. © 1992, Figure 11.15, p. 246. (b) Courtesy of the New York Power Authority. (c) Bobbé Christopherson.]

(b) Niagara Falls, with the American Falls portion shut off for engineering inspection. Horseshoe Falls in the background is still flowing over its 57-m (188-ft) plunge.

(c) American Falls at full release.

time, they produce characteristic depositional landforms in the floodplain, some of which are portrayed in Figure 14.23.

On either bank of some rivers, low ridges of coarse sediment known as **natural levees** are formed as byproducts of flooding. As discharge increases during a flood, the river overflows its banks, loses stream competence and capacity as it spreads out, and drops a portion of its sediment load. Coarser, sand-sized particles (or larger) are deposited first, forming the principal component of the natural levees; finer silts and clays are deposited farther from the river. Successive floods increase the height of the natural levees (*levée* is French for "raising"). These may grow in height until the river channel becomes elevated, or *perched*, above the surrounding floodplain.

On meandering river floodplains, wetlands called *riparian marshes* (sometimes called backswamps) often form in the poorly drained fine sediments deposited by overbank flows (Figure 14.23b). Another floodplain feature is **yazoo streams**, also known as *yazoo tributaries*, which flow parallel to the main river but are blocked from joining it by the presence of natural levees. (These streams are named after the Yazoo River in the southern part of the Mississippi River floodplain.)

Low-lying ridges of alluvium that accumulate on the inside of meander bends as they migrate across a floodplain often form a landscape referred to as *bar-and-swale topography* (the bars form the higher areas, swales are the low areas). The map and image in Figure 14.24 illustrate the changing landforms over time along a portion of the meandering Mississippi River floodplain.

MG Animation Oxbow Lake Formation

MG Animation Stream Processes, Floodplains

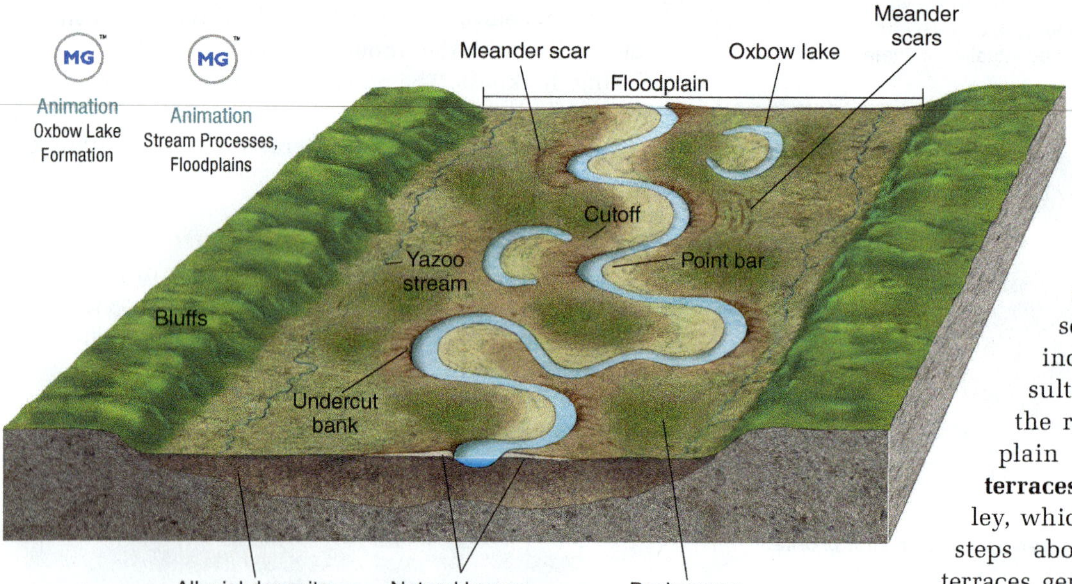

(a) Typical floodplain landscape and related landscape features.

(b) Riparian marshes, or backswamps, are floodplain wetlands that store floodwaters and provide habitat for wildlife. Many such wetlands were filled during the 20th century; restoration is now a priority since wetland water storage feeds streamflow during drought conditions.

▲**Figure 14.23 Floodplain landforms.** [(b) USDA NRCS.]

Stream Terraces

As noted earlier, an uplifting of the landscape or a lowering of base level may rejuvenate stream energy so that a stream again scours downward with increased erosion. The resulting entrenchment of the river into its own floodplain can produce **alluvial terraces** on both sides of the valley, which look like topographic steps above the river. Alluvial terraces generally appear paired at similar elevations on the sides of the valley (Figure 14.25). If more than one set of paired terraces is present, the valley probably has undergone more than one episode of rejuvenation.

If the terraces on the sides of the valley do not match in elevation, then entrenchment actions must have been continuous as the river meandered from side to side, with each meander cutting a terrace slightly lower in elevation. Thus, alluvial terraces represent an original depositional feature, a floodplain, that is subsequently eroded by a stream that has experienced a change in gradient and is downcutting.

Alluvial Fans

In arid and semiarid climates, **alluvial fans** are prominent cone-shaped, or fan-shaped, deposits of fluvial sediments. They commonly occur at the mouth of a canyon where an ephemeral stream channel exits the mountains into a flatter valley (Figure 14.26). Alluvial fans are produced when flowing water (such as a flash flood) abruptly loses velocity as it leaves the constricted channel of a canyon and therefore drops layer upon layer of sediment along the base

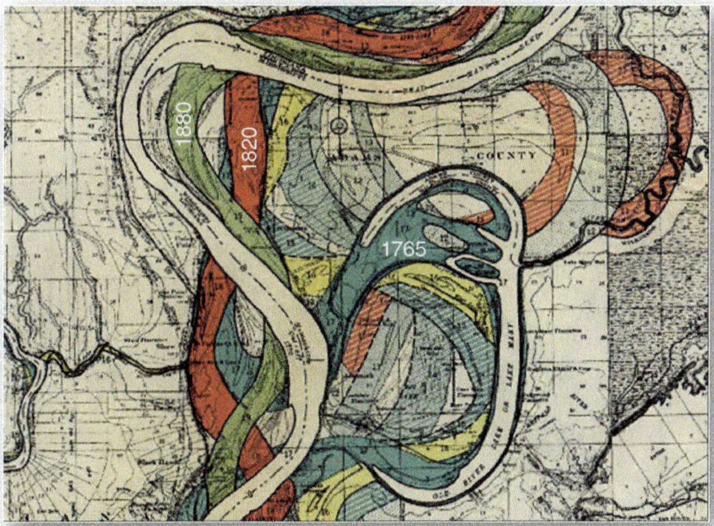

(a) Map of the 1944 channel (white) with former channels for 1765 (blue), 1820 (red), and 1880 (green).

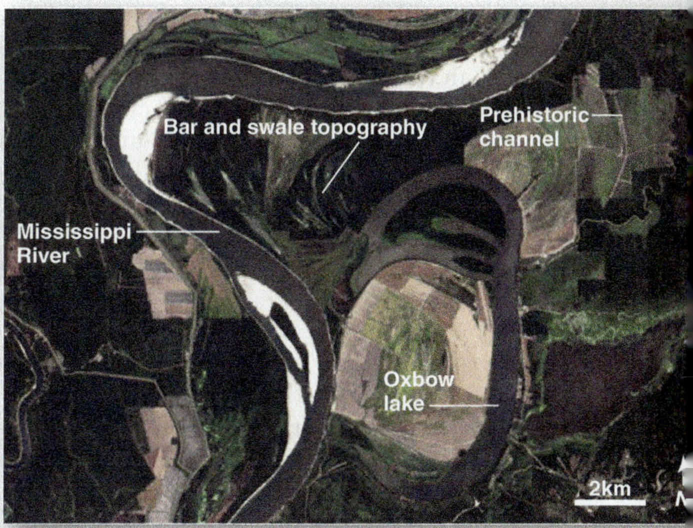

(b) Image of the same portion of the river channel in 1999.

▲**Figure 14.24 Historical shifting of Mississippi River.** The map and image show the portion of the river north of the Old River Control Structure (see Figure 14.29b). [(a) Army Corps of Engineers, *Geological Investigation of the Alluvial Valley of the Lower Mississippi*, 1944. (b) *Landsat* image, NASA.]

of the mountain block. Water then flows over the surface of the fan and produces a braided drainage pattern, sometimes shifting from channel to channel. A continuous apron, or **bajada** (Spanish for "slope"), may form if individual alluvial fans coalesce into one sloping surface (see Figure 12.4b). Alluvial fans also can occur in humid climates along mountain fronts, such as in Japan, Nepal, and Venezuela.

The sediment composing alluvial fans is naturally sorted by size. The coarsest materials (gravels) are deposited near the mouth of the canyon at the apex of the fan, grading slowly to pebbles and finer gravels, then to sands and silts, with the finest clays and dissolved salts carried in suspension and solution all the way to the valley floor. As water evaporates, salt crusts may be left behind on the

MG™
Animation
Stream Terrace Formation

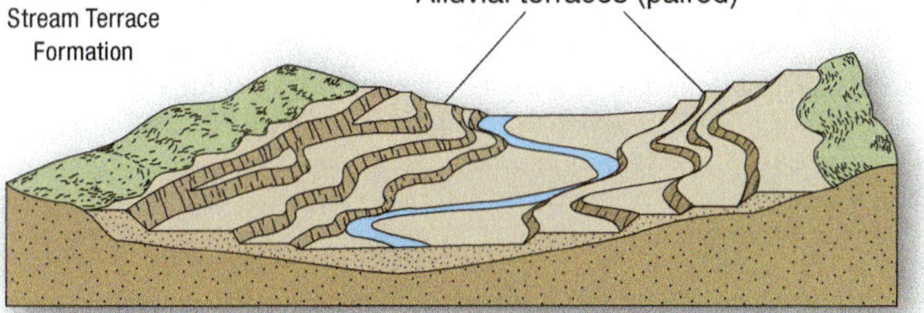

(a) Alluvial terraces are formed as a stream cuts into a valley.

(b) Alluvial terraces along the Rakaia River in New Zealand.

▲**Figure 14.25 Alluvial stream terraces.** [(a) After W. M. Davis, *Geographical Essays* (New York: Dover, 1964 [1909]), p. 541. (b) Bill Bachman/Photo Researchers, Inc.]

▲**Figure 14.26 Badwater alluvial fan, Death Valley, California.** [USGS.]

desert floor in a **playa** (see Figure 12.14c). This intermittently wet and dry lowest area of a closed drainage basin is the site of an *ephemeral lake* when water is present.

Well-developed alluvial fans also can be a major source of groundwater. Some cities—San Bernardino, California, for example—are built on alluvial fans and extract their municipal water supplies from them. In other parts of the world, such water-bearing alluvial fans and water channels are known as *qanat* (Iran), *karex* (Pakistan), or *foggara* (western Sahara).

River Deltas The mouth of a river is where the river reaches a base level. There the river's velocity rapidly decelerates as it enters a larger, standing body of water. The reduced stream energy causes deposition of the sediment load. Coarse sand is deposited closest to the river's mouth. Finer materials, such as silty mud and clays, are carried farther and form the extreme end of the deposit, which may be *subaqueous*, or underwater, even at low tide. The level or nearly level depositional plain that forms at the mouth of a river is a **delta**, named for its characteristic triangular shape, after the Greek letter delta (Δ).

Each flood deposits a new layer of alluvium over portions of the delta, extending the delta outward. As in braided rivers, channels running through the delta divide into smaller courses known as *distributaries*, which appear as a reverse of the dendritic drainage pattern discussed earlier.

The combined delta complex of the Ganges and Brahmaputra Rivers in South Asia is the largest in the world at some 60,000 km^2 (23,165 mi^2). This delta features an extensive lower plain covered by an intricate maze of distributaries formed in an *arcuate* (arc-shaped) pattern (Figure 14.27). Owing to the high sediment load of these rivers, deltaic islands are numerous.

The Nile River delta is another arcuate delta (Figure 14.28 and GeoReport 14.1), as is the Danube River delta in Romania, where the river enters the Black Sea, and the Indus River delta in Pakistan. The Tiber River in Italy has an *estuarine delta,* one that is in the process of filling an **estuary**, the body of water at a river's mouth where freshwater flow encounters seawater.

Numerous rivers throughout the world lack a true delta. In fact, Earth's highest-discharge river, the Amazon, carries sediment far into the deep Atlantic offshore but lacks a delta. Its mouth, 160 km (100 mi) wide, has formed a subaqueous deposit on a sloping continental shelf. As a result, the river ends by braiding into a broad maze of islands and channels (see Figure 14.1).

Other rivers that lack deltas include the Rio de la Plata in Argentina and the Sepik River of Papua New Guinea. Deltaic formations are also absent on rivers that do not produce significant sediment or that discharge into strong erosive currents. The Columbia River of the U.S. Northwest lacks a delta because offshore currents remove sediment as quickly as it is deposited. Examine the delta and its formation in Figure 14.33 at the end of this chapter.

▲**Figure 14.27 The Ganges River delta.** The complex distributary pattern of the "Mouths of the Ganges" in South Asia, also known as the Ganges–Brahmaputra delta or the Sundarbands delta, is the world's largest delta and the largest tract of mangrove forest remaining on Earth. The Sundarbands National Park is a UNESCO World Heritage site, home to Bengal tigers, Ganges and Irrawaddy dolphins, and a rare finless porpoise; go to http://whc.unesco.org/en/list/452 for more information. [*Terra* MODIS, NASA.]

▲**Figure 14.28 The arcuate Nile River delta.** Intensive agricultural activity and small settlements are visible on the delta and along the Nile River floodplain. The two main distributaries are Damietta to the east and Rosetta to the west (see arrows). [*Terra* image, NASA/GSFC/JPL.]

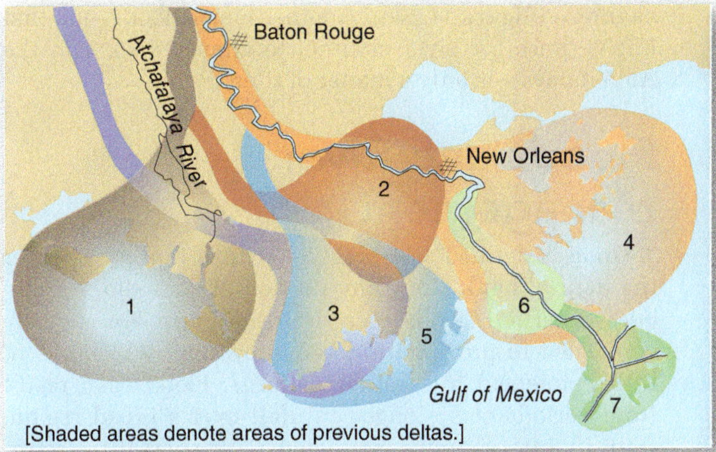

(a) Evolution of the present delta, from 5000 years (1) ago to the present (7)

(b) The bird's-foot delta of the Mississippi River recieves a continuous sediment supply, focused by controlling levees, although subsidence of the delta and rising sea level have diminished the overall surface area.

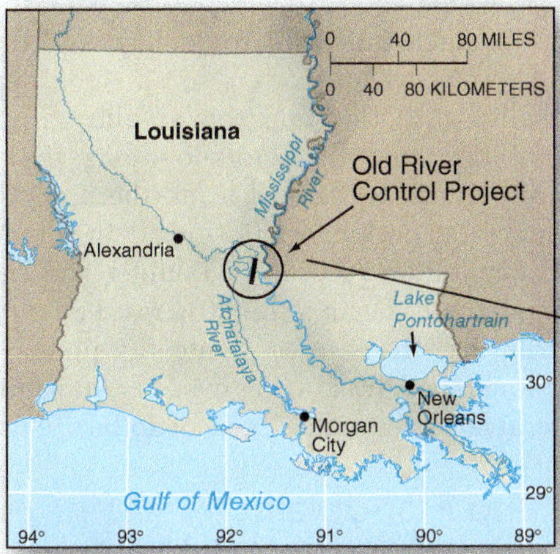

(c) Location of the Old Control Structures and potential capture point where the Atchafalaya River may one day divert the present channel.

(d) Old River Control Auxilliary Structure, one of the dams to keep the Mississippi in its channel.

▲**Figure 14.29 The Mississippi River delta.** [(a) Adapted from C. R. Kolb and J. R. Van Lopik, "Depositional environments of the Mississippi River deltaic plain," in *Deltas in Their Geologic Framework* (Houston, TX: Houston Geological Society, 1966). (b) *Terra* image courtesy of Liam Gumley, Space Science and Engineering Center, University of Wisconsin, and NASA. (d) Bobbé Christopherson.

Mississippi River Delta Over the past 120 million years, the Mississippi River has transported alluvium throughout its vast basin into the Gulf of Mexico. During the past 5000 years, the river has formed a succession of seven distinct deltaic complexes along the Louisiana coast. Each new complex formed after the river changed course, probably during an episode of catastrophic flooding. The first of these deltas was located near the mouth of the Atchafalaya River. The seventh and current delta has been building for at least 500 years and is a classic example of a *bird's-foot delta*—a long channel with many distributaries, and sediments carried beyond the tip of the delta into the Gulf of Mexico (Figure 14.29).

GEOreport 14.1 The disappearing Nile River delta

Over several centuries, more than 9000 km (5500 mi) of canals were built in the Nile River delta in northern Egypt to augment the natural river distributary system carrying water and sediment to the sea. As river discharge enters the network of canals, flow velocity is reduced, stream competence and capacity are decreased, and sediment load is deposited far short of where the delta touches the Mediterranean Sea. In 1964, the completion of the Aswân High Dam blocked sediment movement downstream, decreasing the supply of sediment to the delta. As a result, the delta coastline is actively receding at an alarming 50 to 100 m (165 ft to 330 ft) per year. Seawater is intruding farther inland into both surface water and groundwater. The delta, which provides the fertile soils that produce 60% of the country's food, is also threatened by rising sea level; a 1-m rise, which experts predict is likely during the next 100 years, would inundate about 20% of the delta.

The history of the Mississippi River delta shows a dynamic system with inputs and outputs of sediment and shifting distributaries. The 3.25-million-km^2 (1.25-million-mi^2) Mississippi drainage basin produces 550 million metric tons of sediment annually—enough to extend the Louisiana coast 90 m (295 ft) a year. However, several factors are causing losses to the area of the delta each year.

Compaction and the tremendous weight of the sediment in the Mississippi River create isostatic adjustments in Earth's crust. These adjustments are causing the entire delta region to subside, a natural process that has occurred throughout the evolution of the river basin. In the past, subsidence was balanced by additions of sediment that caused areal growth of the delta. With the onset of human activities such as upstream dam construction and the excavation of canals and waterways through the delta for shipping and oil and gas exploration, the supply of alluvial sediment has decreased. The delta is now subsiding without sediment replenishment. Activities such as the pumping of oil and gas from thousands of onshore and offshore wells are thought to be an additional cause of regional land subsidence.

The present main channel of the Mississippi River persists in large part from the effort and expense directed at maintaining the extensive system of artificial levees. As in the past, a worst-case flood scenario could cause the river to break from its existing channel and seek a new route to the Gulf of Mexico. This process, called *channel avulsion*, occurs as a river suddenly changes its channel—usually to one with a shorter and more direct course—during a flood.

The map in Figure 14.29c and the sediment plume to the west of the main delta in the Figure 14.29b satellite image show that an obvious alternative to the Mississippi's present channel is the Atchafalaya River. The Atchafalaya has a steeper gradient than the Mississippi and would provide a much shorter route to the Gulf of Mexico, less than half the present distance from where the rivers separate. Currently, this distributary carries about 30% of the Mississippi's total discharge. If the Mississippi were to shift to this alternative course, it would bypass New Orleans entirely, removing the flood threat to that urban area. However, this shift would be a financial disaster, as a major U.S. port would consequently fill with silt, and seawater would intrude into freshwater resources.

At present, artificial barriers block the Atchafalaya from reaching the Mississippi at the point shown in Figure 14.29c. The Old River Control Project (1963) maintains three structures and a lock about 320 km (200 mi) from the Mississippi's mouth to keep these rivers in their channels (Figure 14.29d). Another major flood—one that might cause the river channel to change back into the Atchafalaya—is only a matter of time.

Floods and River Management

A **flood** is defined as a high water flow that passes over the natural bank along any portion of a stream. As discussed earlier, floods in a drainage basin are strongly connected to precipitation and snowmelt, which are in turn connected to weather patterns. Floods can result from periods of prolonged rainfall over a broad region, from intense rainfall associated with short-lived thunderstorms, from rapid melting of the snowpack, or from rain-on-snow events that accelerate snowpack melting. Floods vary in magnitude and frequency, and their effects depend on many factors.

In September 2013, flooding from an unusually heavy rainfall event occurred in Colorado along the Rocky Mountain front from the central to northern part of the state. During a 7-day period, over 43 cm (17 in.) of rain fell, shattering precipitation records in Boulder and in other parts of the region. Extensive flooding resulted along the mountain front and further east onto the plains in the Platte River watershed, displacing over 12,000 people and causing several fatalities and billions of dollars in damage.

Humans and Floodplains

Throughout history, civilizations have settled on floodplains and deltas, especially since the agricultural revolution of 10,000 years ago, when the fertility of floodplain soils was discovered. Villages generally were built away from the area of flooding, or on stream terraces, because the floodplain was dedicated exclusively to farming. Over time, as commerce grew and river transportation became more important, development near rivers increased. Also, water is a basic industrial raw material used for cooling and for diluting and removing wastes; thus, industrial sites along rivers became desirable, and remain so. In short, despite our historical knowledge of flood events and their effects, floodplains continue to be important sites of human activity and settlement. These activities place lives and property at risk during floods.

The effects of flooding are especially disastrous in less-developed regions of the world. Bangladesh is perhaps the most persistent example: It is one of the most densely populated countries on Earth, and more than three-fourths of its land area is on a floodplain

GEOreport 14.2 What is a bayou?

Bayou is a general term for a water body in a flat, lowland region, usually a slow-moving stream or secondary waterway. These sometimes stagnant, winding watercourses pass through coastal marshlands and swamps and allow tidal waters access to deltaic lowlands. The story of Bayou Lafourche in the lower Mississippi River system in Louisiana is presented on our *MasteringGeography* website. This area was the main channel for the Mississippi River before 1904.

(a) In 2011, floodwaters flow over part of an intentional breach in the Bird's Point levee in Missouri.

(b) Sheep graze on the slopes of an artificial levee along the Sacramento River in California. Note that the agricultural fields are lower in elevation than the river, caused by subsidence of the Sacramento River delta.

▲**Figure 14.30 Artificial levees.** [(a) Scott Olsen/Getty Images. (b) California Department of Water Resources.]

and delta complex. The country's vast alluvial plain sprawls over an area the size of Alabama (130,000 km^2, or 50,000 mi^2).

In Bangladesh, the severe effects of flooding, both in damage costs and in human fatalities, are a consequence of human economic activities, along with heavy precipitation episodes. Excessive forest harvesting throughout the twentieth century in the upstream portions of the Ganges–Brahmaputra River watersheds increased runoff and sediment load. Over time, the increased sediment load was deposited in the Bay of Bengal, creating new islands. These islands, barely above sea level, became sites for farms and villages. As a result of the 1988 and 1991 floods and storm surges, about 150,000 people in this region perished (revisit Figure 14.27).

Flood Protection

In the United States, floods average about $6 billion in annual losses. The catastrophic floods along the Mississippi River and its tributaries in 1993 and 2011 produced damage that exceeded $30 billion in each occurrence. In 2010, floods occurred in 30 U.S. states and in at least as many countries in the world, as record land and ocean temperatures energized air masses, producing excessive rainfall totals. Flood protection, when in place, generally takes the form of dams (discussed in Chapter 9), spillways, and artificial levee construction along river channels.

Usually, the term *levee* connotes an element of human construction, and these engineered features are common across the United States and throughout the world (Figure 14.30). **Artificial levees** are earthen embankments, often built on top of natural levees. They run parallel to the channel (rather than across it, like a dam) and increase the capacity in the channel by adding to the height of the banks. For efficiency of time and materials, channels are often straightened during levee construction. Levees are intended to hold floods within the channel but not prevent them completely. Eventually, given severe enough conditions, an artificial levee will be overtopped or damaged in a flood. When overtopping (known as levee breaching) or levee failure occurs, extensive flood damage and erosion can result downstream. In the United States, nearly 85% of levees are locally owned, with the remaining 15% maintained by the U.S. Army Corps of Engineers or other state or federal agencies.

The 2011 Mississippi River floods broke records dating to back 1927, exceeding the floods of 1993. Record rainfall throughout the watershed in April coupled with the timing of snowmelt to cause the flood (rated as having a probability of happening only once every 500 years in that region). During the flooding, which extended throughout

▲**Figure 14.31 Opening of the Morganza Spillway, Louisiana.** In May 2011, the U.S. Army Corps of Engineers partially opened the Morganza Spillway to divert water into the Atchafalaya River and alleviate flooding on the Mississippi. [U.S. Army Corps of Engineers.]

the Mississippi River basin across the central United States, some artificial levees were intentionally breached, with the aim of lowering the flood peaks moving downstream toward cities. In Mississippi County, Missouri, near the confluence of the Ohio and Mississippi Rivers, engineers blew a hole in the Bird's Point levee in order to relieve the flood threat in nearby Cairo, Illinois. The levee breach resulted in the flooding of over 100 homes and 518 km^2 (200 mi^2) of farmland (Figure 14.30a).

As the water levels peaked in Mississippi and Louisiana, the Morganza Spillway (just south of the Old River Control Project shown in Figure 14.29) was opened for only the second time in 40 years. The floodplain below this structure is a *bypass channel*, designed to carry discharge from seasonal floods. When not flooded, the bypass channel is forest or farmland, often benefiting from the occasional soil-replenishing inundation. When the river reaches flood stage, large gates are opened, allowing the water to enter the bypass channel. In 2011, the Morganza Spillway gates were opened over a period of days, flooding rural areas in the bypass channel but reducing the downstream flood crest to the cities of Baton Rouge and New Orleans (Figure 14.31).

In many instances, flood protection structures have not protected floodplains as designed—levees, spillways, and even dams themselves have failed. For example, in 2006, the Ka Loko dam on the island of Kauai, Hawai'i, failed suddenly after several weeks of heavy rains; the floodwaters killed seven people. Levee failures are more common, as discussed in Chapter 8 regarding the inundation of New Orleans after Hurricane Katrina.

On Asia's Indus River, which flows through Pakistan into the Arabian Sea, heavy monsoon rains in July 2010 increased the flow of the river and its many tributaries, which led to levee and dam failures that caused extreme flooding in downstream areas. During this event, levee breaches led to channel avulsion. Damage from this event was greater than that from the 2004 Indian Ocean tsunami, as floods inundated entire cities and ruined 3.6 million hectares (8.9 million acres) of productive farmland. More than 2000 people died, and 20 million were left homeless (Figure 14.32). In addition, more than 5.4 million agricultural workers were left unemployed for the 2010–2011 season.

Flood Probability

Maintaining extensive historical records of discharge during precipitation events is critical for predicting the behavior of present streams under similar conditions. The U.S. Geological Survey has detailed records of stream discharge at stream-gaging stations for only about 100 years—in particular, since the 1940s. Flood predictions are based on these relatively short-term data. Scientists hope that adequate funding to support hydrologic monitoring by stream-gaging networks will continue, for the data are essential to flood hazard assessment.

On the basis of these historical data, flood discharges are rated statistically according to the time intervals expected between discharges of similar size. Thus, a "10-year flood" has a recurrence interval of 10 years, a "50-year flood" has a recurrence interval of 50 years, and so on. In other words, a 10-year flood has a discharge that is statistically likely to occur once every 10 years, based on discharge data for that particular stream. This also means that a flood of this size has only a 10% likelihood of occurring in any one year and is likely to occur about 10 times each century. The use of historical data works well where available; however, complications are introduced by urbanization and dam construction, which can change the magnitude and frequency of flood events on a stream or in a watershed.

These statistical estimates are probabilities that events will occur randomly during a specified period; they do not mean that events will occur regularly during that time period. For example, 2 decades might pass without a 50-year flood; or a 50-year level of flooding could occur 3 years in a row. The record-breaking Mississippi River Valley floods in 2011 easily exceeded a 1000-year flood probability.

Recently, scientists are describing floods and precipitation events using the annual exceedance probability (AEP), a term that better represents the statistical likelihood of occurrence. By this measure, a 100-year flood has a 1% annual exceedance probability. Scientists estimate that the AEP of the 7 day rainfall event in September 2013 that caused extensive flooding along the Colorado front range mentioned earlier was 0.1% (a 1000-year event). However, the AEP of the resultant flooding in Boulder was 5%, (only a 50-year event). Had the rain fallen on melting snow, the flood severity would have been much greater.

Floodplain Management

The flood-recurrence interval is useful for floodplain management and hazard assessment. A 10-year flood indicates a moderate threat to a floodplain. A 50-year or 100-year flood is of greater and perhaps catastrophic consequence, but it is also less likely to occur in a given year.

GEOreport 14.3 America's levees

By several estimates, over 100,000 miles of artificial levees exist along rivers and streams in the United States, the vast majority of them privately owned. The U.S. population living in areas protected by levees is estimated to be in the tens of millions; some major urban areas with levee systems are New Orleans, Sacramento, Dallas–Fort Worth, St. Louis, Portland, and Washington, D.C. In fact, over 30 major cities in America are located on floodplains. Currently, no national policy exists concerning the safety of levees. For more information, see http://www.leveesafety.org/docs/NCLS-Recommendation-Report_012009_DRAFT.pdf.

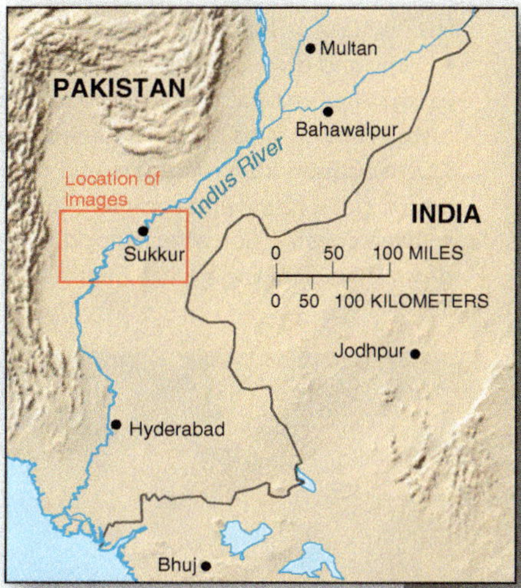

▶**Figure 14.32 Indus River flooding, 2010.** These *Terra* images combine visible light and infrared to enhance contrast between the flooded river and land; water appears blue and clouds blue-green. [NASA/GSFC.]

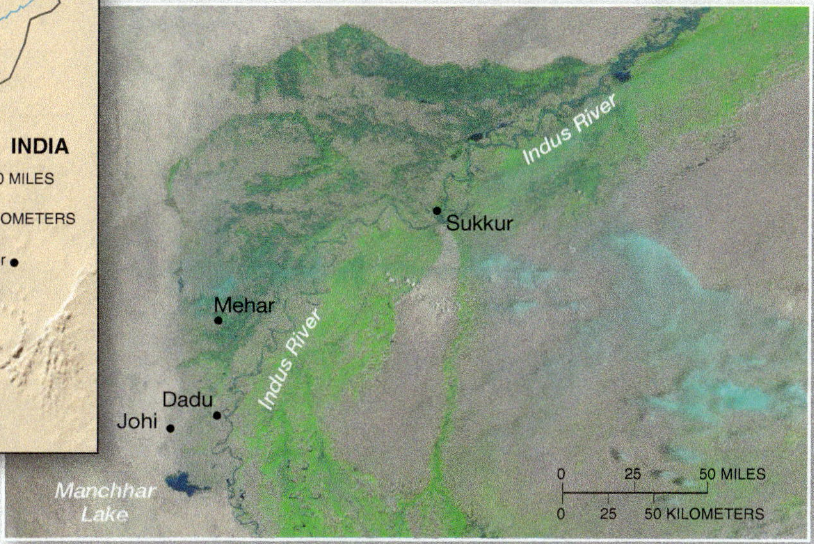

(a) July 19, 2010: The pre-flood braided river channel.

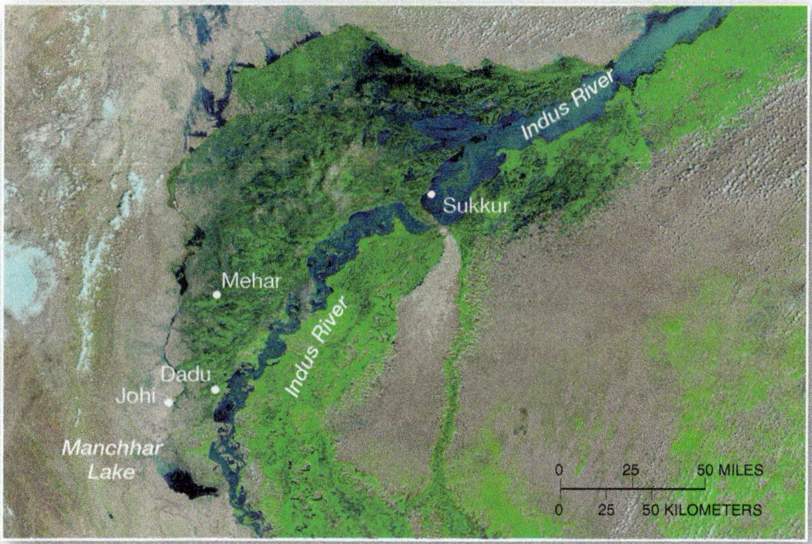

(b) August 11, 2010: Heavy monsoon rains caused increased discharge; flooding was initiated by a levee breach near Sukkur on August 6th, followed by channel avulsion to the west.

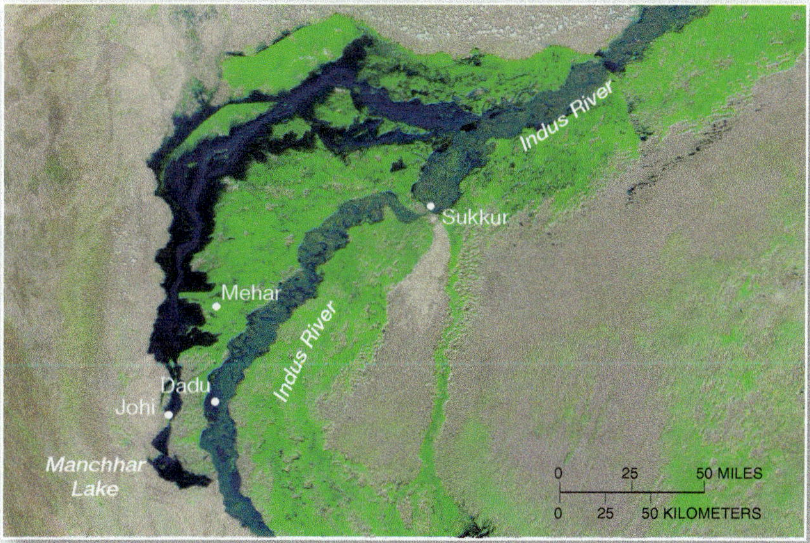

(c) September 7, 2010: The growing channel avulsion forms a separate river flowing into Manchar Lake, threatening the cities of Johi and Dadu.

For a particular river system, or portion of a river system, flood-recurrence intervals can be mapped and used to define floodplains according to flood probability, such as a "50-year floodplain" or a "100-year floodplain." In this way, scientists and engineers can develop the best possible flood-management strategy. Restrictive zoning using these floodplain designations is an effective way of avoiding potential flood damage. Flood hazard mapping shows the different degrees of risk for parts of the floodplain and is used to determine costs for flood insurance (see http://www.fema.gov/floodplain-management).

Restrictive zoning based on flood hazard mapping is not always enforced, and the scenario sometimes goes like this: (1) Minimal zoning precautions are not carefully supervised, (2) a flooding disaster occurs, (3) the public is outraged at being caught off guard, (4) businesses and homeowners are surprisingly resistant to stricter laws and enforcement, and (5) eventually another flood refreshes the memory and promotes more planning meetings and questions. As strange as it seems, there is little indication that human risk perception improves as the risk increases.

For information on floods worldwide, see the Dartmouth Flood Observatory at http://floodobservatory.colorado.edu/. For weather and flood warnings, go to http://www.noaawatch.gov/floods.php.

THE**human**DENOMINATOR 14 Rivers, Floodplains, and Deltas

RIVER SYSTEMS ⇨ HUMANS

- Humans use rivers for recreation and have farmed fertile floodplain soils for centuries.
- Flooding affects human settlements on floodplains and deltas.
- Rivers are transportation corridors, and provide water for municipal and industrial use.

HUMANS ⇨ RIVER SYSTEMS

- Dams and diversions alter river flows and sediment loads, affecting river ecosystems and habitat. River restoration efforts include dam removal to restore ecosystems and threatened species.
- Urbanization, deforestation, and other human activities in watersheds alter runoff, peak flows, and sediment loads in streams.
- Levee construction affects floodplain ecosystems; levee failures cause destructive flooding.

14a

In June 2013, floodwaters following days of heavy rainfall inundated Germany, Austria, Slovakia, Hungary, and the Czech Republic. According to local residents, water levels in Passau, Germany, were higher than any recorded in the past 500 years. [Matthias Schrader/AP Photo.]

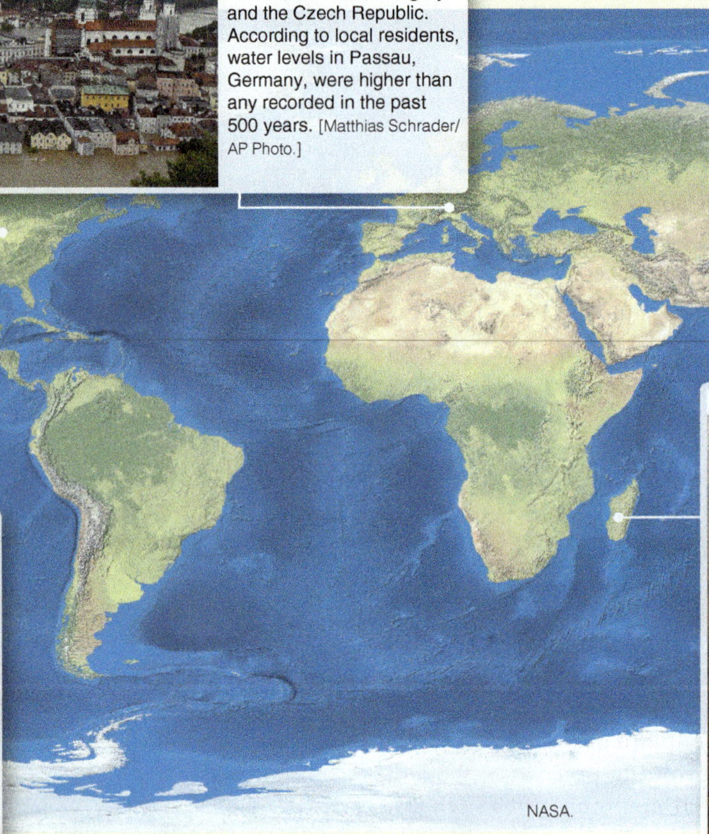

NASA.

14b

Rivers in Madagascar carry an enormous sediment load as a result of deforestation. The trees anchor the soil with their roots; when that stabilizing effect is gone, soil erodes into river channels and is carried into oceans, disrupting coral reefs and other aquatic ecosystems. [Kevin Schafer/Alamy.]

14c

In 2011, Americans spent $42 million on fishing-related activities. Streams in Montana, Missouri, Michigan, Utah, and Wisconsin are of high enough quality that they are designated "blue ribbon fisheries" based on sustainability criteria such as water quality and quantity, accessibility, and the specific species present. [Karl Weatherly/Corbis.]

ISSUES FOR THE 21ST CENTURY

- Increasing population will intensify human settlement on floodplains and deltas worldwide, especially in developing countries, making more people vulnerable to flood impacts.
- Stream restoration will continue, including dam decommissioning and removal, flow restoration, vegetation reestablishment, and restoration of stream geomorphology.
- Global climate change may intensify storm systems, including hurricanes, increasing runoff and flooding in affected regions. Rising sea level will make delta areas more vulnerable to flooding.

GEOSYSTEMS**connection**

While following the flow of water through streams, we examined fluvial processes and landforms and the river-system outputs of discharge and sediment. We saw that a scientific understanding of river dynamics, floodplain landscapes, and related flood hazards is integral to society's ability to perceive hazards in the familiar environments we inhabit. In the next chapter, we examine the erosional activities of waves, tides, currents, and wind as they sculpt Earth's coastlines and desert regions. A significant portion of the human population lives in coastal areas, making the difficulties of hazard perception and the need to plan for the future, given a rising sea level, important aspects of Chapter 16.

KEY LEARNING concepts review

■ *Sketch* a basic drainage basin model, and *identify* different types of drainage patterns by visual examination.

Hydrology is the science of water and its global circulation, distribution, and properties—specifically, water at and below Earth's surface. **Fluvial** processes are stream-related. The basic fluvial system is a **drainage basin**, or *watershed*, which is an open system. *Drainage divides* define the catchment (water-receiving) area of a drainage basin. In any drainage basin, water initially moves downslope in a thin film of **sheetflow**, or *overland flow*. This surface runoff concentrates in *rills*, or small-scale downhill grooves, which may develop into deeper *gullies* and a stream course in a valley. High ground that separates one valley from another and directs sheetflow is an *interfluve*. Extensive mountain and highland regions act as **continental divides** that separate major drainage basins. Some regions, such as the Great Salt Lake Basin, have **internal drainage** that does not reach the ocean, the only outlets being evaporation and subsurface gravitational flow.

Drainage density is determined by the number and length of channels in a given area and is an expression of a landscape's topographic surface appearance. **Drainage pattern** refers to the arrangement of channels in an area as determined by the steepness, variable rock resistance, variable climate, hydrology, relief of the land, and structural controls imposed by the landscape. Seven basic drainage patterns are generally found in nature: dendritic, trellis, radial, parallel, rectangular, annular, and deranged.

hydrology (p. 422)
fluvial (p. 422)
drainage basin (p. 422)
sheetflow (p. 423)
continental divide (p. 423)
internal drainage (p. 425)
drainage density (p. 426)
drainage pattern (p. 426)

1. Define the term *fluvial*. What is a fluvial process?
2. What role is played by rivers in the hydrologic cycle?
3. What are the five largest rivers on Earth in terms of discharge? Relate these to the weather patterns in each area and to regional potential evapotranspiration (POTET) and precipitation (PRECIP)—concepts discussed in Chapter 9.
4. What is the basic organizational unit of a river system? How is it identified on the landscape? Define the several relevant key terms used.
5. In Figure 14.3, follow the Allegheny–Ohio–Mississippi River system to the Gulf of Mexico. Analyze the pattern of tributaries and describe the channel. What role do continental divides play in this drainage?
6. Describe drainage patterns. Define the various patterns that commonly appear in nature. What drainage patterns exist in your hometown? Where you attend school?

■ *Explain* the concepts of stream gradient and base level, and *describe* the relationship between stream velocity, depth, width, and discharge.

The **gradient** of a stream is the slope, or the stream's drop in elevation per unit distance. **Base level** is the lowest-elevation limit of stream erosion in a region. A *local base level* occurs when something interrupts the stream's ability to achieve base level, such as a dam or a landslide that blocks a stream channel.

Discharge, a stream's volume of flow per unit of time, is calculated by multiplying the velocity of the stream by its width and depth for a specific cross section of the channel. Streams may have *perennial*, *ephemeral*, or *intermittent* flow regimes. Discharge usually increases in a downstream direction; however, in rivers in semiarid or arid regions, discharge may decrease with distance downstream as water is lost to evapotranspiration and water diversions.

A graph of stream discharge over time for a specific place is called a **hydrograph**. Precipitation events in urban areas result in higher peak flows during floods. In deserts, a torrent of water that fills a stream channel during or just after a rainstorm is a **flash flood**.

gradient (p. 427)
base level (p. 427)
discharge (p. 428)
hydrograph (p. 429)
flash flood (p. 429)

7. Explain the base level concept. What happens to a stream's base level when a reservoir is constructed?
8. What was the impact of flood discharge on the channel of the San Juan River near Bluff, Utah? Why did these changes take place?
9. Differentiate between a natural stream hydrograph and one from an urbanized area.

■ *Explain* the processes involved in fluvial erosion and sediment transport.

Water dislodges, dissolves, or removes surface material and moves it to new locations in the process of **erosion**. Sediments are laid down by the process of **deposition. Hydraulic action** is the erosive work of water caused by hydraulic squeeze-and-release action to loosen and lift rocks and sediment. As this debris moves along, it mechanically erodes the streambed further through a process of **abrasion**. Streams may deepen their valley by channel incision, they may lengthen in the process of headward erosion, or they may erode a valley laterally in the process of meandering.

When stream energy is high, particles move downstream in the process of **sediment transport**. The sediment load of a stream can be divided into three primary types. The **dissolved load** travels in solution, especially the dissolved chemicals derived from minerals such as limestone or dolomite or from soluble salts. The **suspended load** consists of fine-grained, clastic particles held aloft in the stream, with the finest particles not deposited until the stream velocity slows nearly to zero. **Bed load** refers to coarser materials that are dragged and pushed and rolled along the streambed by **traction** or that bounce and hop along by **saltation**.

Degradation occurs when sediment is eroded and channel incision occurs. If the load in a stream exceeds its capacity, **aggradation** occurs as sediment deposition builds up the stream channel.

erosion (p. 430)
deposition (p. 430)
hydraulic action (p. 430)
abrasion (p. 431)
sediment transport (p. 431)
dissolved load (p. 431)
suspended load (p. 431)
bed load (p. 432)
traction (p. 432)
saltation (p. 432)
degradation (p. 432)
aggradation (p. 432)

10. What is the sequence of events that takes place as a stream dislodges material?
11. How does stream discharge do its erosive work? What are the processes at work in the channel?
12. Differentiate between stream competence and stream capacity.
13. How does a stream transport its sediment load? What processes are at work?

■ *Describe* common stream channel patterns, and *explain* the concept of a graded stream.

With excess sediment, a stream may become a maze of interconnected channels that form a **braided stream** pattern. Where the slope is gradual, stream channels develop a sinuous form called a **meandering stream**. The outer portion of each meandering curve is subject to the fastest water velocity and can be the site of a steep **undercut bank**. The inner portion of a meander experiences the slowest water velocity and forms a **point bar** deposit. When a meander neck is cut off as two undercut banks merge, the meander becomes isolated and forms an **oxbow lake**.

The drop in elevation along a river from headwaters to mouth is usually represented in a side view called a *longitudinal profile*. A **graded stream** condition occurs when the slope is adjusted so that a channel has just enough energy to transport its sediment load; this represents a balance between slope, discharge, channel characteristics, and the load supplied from the drainage basin. Tectonic uplift may cause a stream to develop entrenched meanders as it carves the landscape during uplift. An interruption in a stream's longitudinal profile is called a **nickpoint**. This abrupt change in slope can occur as the stream flows across hard, resistant rock or after tectonic uplift episodes.

braided stream (p. 433)
meandering stream (p. 436)
undercut bank (p. 436)
point bar (p. 436)
oxbow lake (p. 436)
graded stream (p. 437)
nickpoint (p. 440)

14. Describe the flow characteristics of a meandering stream. What is the pattern of flow in the channel? What are the erosional and depositional features and the typical landforms created?
15. Explain these statements: (a) All streams have a gradient, but not all streams are graded. (b) Graded streams may have ungraded segments.
16. Why is Niagara Falls an example of a nickpoint? Without human intervention, what do you think will eventually take place at Niagara Falls?

■ *Describe* the depositional landforms associated with floodplains and alluvial fan environments.

Alluvium is the general term for the clay, silt, sand, gravel, or other unconsolidated rock and mineral fragments deposited by running water. The flat, low-lying area adjacent to a stream channel that is subjected to recurrent flooding is a **floodplain**. On either bank of some streams, **natural levees** develop as by-products of flooding. On a floodplain, riparian marshes are common, and **yazoo streams** may develop, which flow parallel to the river channel but are separated from it by natural levees. Entrenchment of a river into its own floodplain forms **alluvial terraces**.

Along mountain fronts in arid climates, **alluvial fans** develop where ephemeral stream channels exit from canyons into the valley below. A **bajada** may form where individual alluvial fans coalesce along a mountain block. Runoff may flow all the way to the valley floor, where it forms a **playa**, a low, intermittently wet area in a region of internal drainage.

alluvium (p. 441)
floodplain (p. 441)
natural levee (p. 442)
yazoo stream (p. 442)
alluvial terrace (p. 442)
alluvial fan (p. 442)
bajada (p. 443)
playa (p. 444)

17. Describe the formation of a floodplain. How are natural levees, oxbow lakes, backswamps, and yazoo tributaries produced?
18. Describe any floodplains near where you live or where you go to college. Have you seen any of the floodplain features discussed in this chapter? If so, which ones?
19. What processes are involved in the formation of an alluvial fan? What is the arrangement, or sorting, of alluvial material on the fan?

■ *List* and *describe* several types of river deltas, and *explain* flood probability estimates.

A depositional plain formed at the mouth of a river is called a **delta**. Deltas may be arcuate or bird's foot in shape, or estuarine in nature. Some rivers have no deltas. When the mouth of a river enters the sea and is inundated by seawater in a mix with freshwater, it is called an **estuary**. Despite historical devastation by floods, floodplains and deltas are important sites of human activity and settlement. Efforts to reduce flooding include the construction of artificial levees, bypasses, straightened channels, diversions, dams, and reservoirs.

A **flood** occurs when high water overflows the natural bank along any portion of a stream. Human-constructed **artificial levees** are common features along many rivers of the United States, where flood protection is needed for developed floodplains. Both floods and the floodplains they occupy are rated statistically for the expected time interval between floods of given discharges. For example, a 10-year flood has the statistical probability of happening once every 10 years. Flood probabilities are useful for floodplain zoning.

delta (p. 444)
estuary (p. 444)
flood (p. 446)
artificial levee (p. 447)

20. What is a river delta? What are the various deltaic forms? Give some examples.
21. Describe the Ganges River delta. What factors upstream explain its form and pattern? Assess the consequences of settlement on this delta.
22. What is meant by the statement, "The Nile River delta is disappearing"?
23. Specifically, what is a flood? How are such flows measured and tracked, and how are they used in floodplain management?
24. What is channel avulsion, and how does it occur?

MasteringGeography™

Looking for additional review and test prep materials? Visit the Study Area in *MasteringGeography*™ to enhance your geographic literacy, spatial reasoning skills, and understanding of this chapter's content by accessing a variety of resources, including MapMaster interactive maps, geoscience animations, satellite loops, author notebooks, videos, RSS feeds, flashcards, web links, self-study quizzes, and an eText version of *Geosystems*.

◄Figure 14.33 **The Horton River delta, Northwest Territories, Canada.** Prior to about 1800, the Horton River followed a meandering course as it flowed more then 100 km (62 mi) northward along the coast of the Beaufort Sea. As the river scoured the outside of meander bends, it eventually formed a new outlet to Franklin Bay. This *EO–1* (*Earth Observing–1*) satellite image shows the river's fan-shaped delta, forming for the past 200 years along this otherwise straight section of coast. Where the river abandoned its former channel, oxbow lakes have formed. How many can you count in the image? [NASA *EO–1* Team.]

C H A P T E R 1 5

Eolian Processes and Arid Landscapes

Ward Terrace, Coconino County, Arizona, is the bench on the left side of the photo, some 1645 m (5397 ft) in elevation. To the right of the cliff that forms the edge of Ward Terrace, elevations drop to about 1400 m (4593 ft). The center of the aerial photo is approximately 35° 58′N 111° 15′W. The light-colored streaks are wind-blown sand and linear dunes that reactivated about 400 years ago. The scene is on the Navajo Reservation. [Bobbé Christopherson.]

KEY LEARNING CONCEPTS

After reading the chapter, you should be able to:

- ***Characterize*** the unique geomorphic work accomplished by wind and eolian processes.
- ***Describe*** eolian erosion, including deflation, abrasion, and the resultant landforms.
- ***Describe*** eolian transportation, and ***explain*** saltation and surface creep.
- ***Identify*** the major classes of sand dunes, and ***present*** examples within each class.
- ***Define*** loess deposits and their origins, locations, and landforms.
- ***Portray*** desert landscapes, and ***locate*** these regions on a world map.

GEOSYSTEMS NOW

Increasing Desertification and Political Action—A Global Environmental Issue

We are witnessing an unwanted expansion of Earth's deserts in a process known as **desertification**. This is a worldwide phenomenon along the margins of semiarid and arid lands—Earth's dry lands, one of the subjects in this chapter.

Desertification is due in part to poor agricultural practices such as overgrazing and activities that abuse soil structure and fertility, improper soil-moisture management, erosion and salinization, and deforestation. A worsening causative force is global climate change, which is shifting temperature and precipitation patterns and causing poleward movement of Earth's subtropical high-pressure systems, discussed elsewhere in *Geosystems*.

Figure GN 15.1 portrays regions of desertification risk, as defined by the loss of agricultural activity. The severity of this problem is magnified by poverty in many of the affected areas, since most people lack the capital to change agricultural practices and implement conservation strategies. Many of the highest-risk lands are in India and central Asia. In fact, a 2009 mapping project showed that 25% of India is undergoing desertification. Throughout central Asia, overexploitation of water resources has combined with drought to cause the expansion of desert conditions. The Aral Sea, formerly one of the four largest lakes in the world, has steadily shrunk in size since the 1960s as inflowing rivers have been increasingly diverted. As the lake has failed, fine sediment and alkali dust have become available for wind deflation, leading to massive dust storms.

In Africa, the Sahel is the transition region between the Sahara Desert in the subtropics and the wetter equatorial regions. The southward expansion of desert conditions through portions of the Sahel region has left many African peoples on land that no longer experiences the rainfall of just three decades ago. Yet climate change is only part of the story: Other factors contributing to desertification in the Sahel are population increases, poverty, land degradation (deforestation, overgrazing), and the lack of a coherent environmental policy.

The United Nations estimates that degraded lands cover some 800 million hectares (2 billion acres); many millions of additional hectares are added each year. An immediate need is to improve the database of occurrences across the globe for a more accurate accounting of the problem and a better understanding of desertification processes.

One of the successes coming from the 1992 Earth Summit in Rio de Janeiro was the Convention to Combat Desertification, an initiative that began operations in 1994 and is still active. Efforts continue to get these issues to the forefront for action.

In August 2010, the *International Conference: Climate, Sustainability, and Development in the Semi-arid Regions* met for a second time (http://www.unccd.int/). Related issues from more than 100 countries were on the agenda. Launched at this meeting was a global effort for the next decade—the United Nations Decade for Deserts and the Fight Against Desertification (UNDDD).

All this was geared to increase public and government attention to the issue of desertification in preparation for the important 2012 *Rio + 21 Earth Summit* in Rio de Janiero. The 2010 conference called for action on many issues: sustainable development and climate change, political involvement on behalf of dryland peoples, integration of and focus on global environmental initiatives, development of financing, education, and recognition of the urgency of the problem, among other items.

About a billion people are being affected by desertification, with their livelihoods threatened. Further delays in addressing the problem will result in far higher costs when compared to taking action within the next few years. For more on global arid lands, see http://www.undp.org/drylands/.

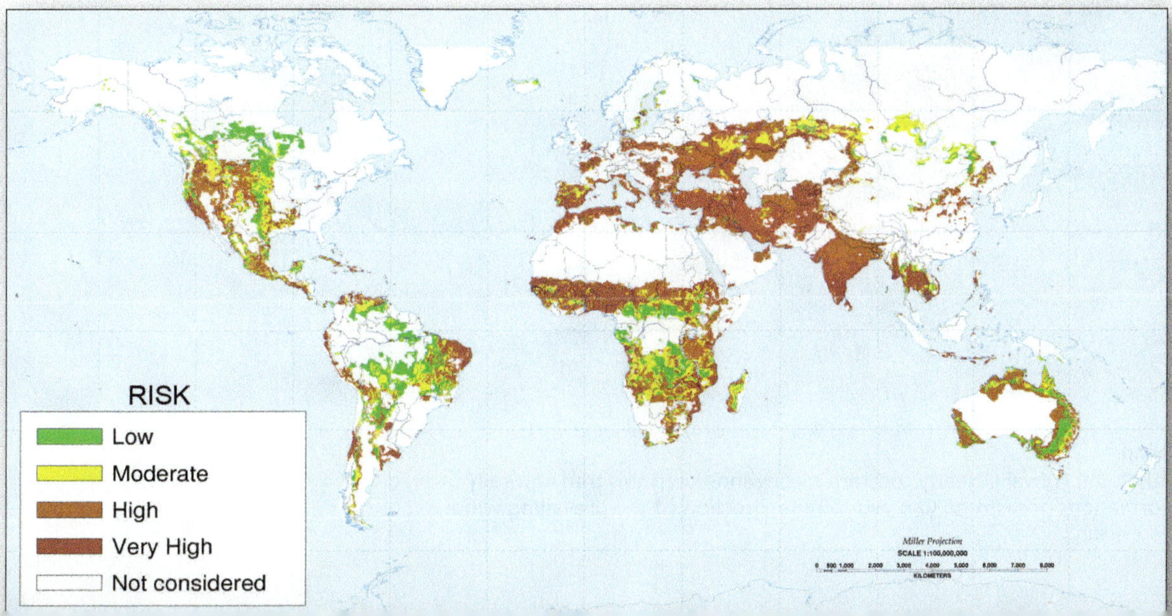

FIGURE GN 15.1 Area at risk of desertification. [Map prepared by U.S. Department of Agriculture–Natural Resources Conservation Service, Soil Survey Division, and includes consideration of population densities in affected regions provided by the National Center for Geographic Information Analysis at University of California, Santa Barbara.]

Wind is an agent of geomorphic change. Like moving water, moving air causes erosion, transportation, and deposition of materials; moving air is a fluid and it behaves similarly, although it has a lower viscosity (it is less dense) than water. Although lacking the lifting ability of water, wind processes can modify and move quantities of materials in deserts, along coastlines, and elsewhere.

Wind contributes to soil formation in distant places, bringing fine particles from regions where glaciers once were active. Fallow fields (those not planted) give up their soil resource to destructive wind erosion. Scientists are only now getting an accurate picture of the amount of windblown dust that fills the atmosphere and crosses the oceans between continents. Chemical fingerprints and satellites trace windblown dust from African soils to South America and Asian landscapes to Europe. Winds even spread living organisms. One study found related mosses, liverworts, and lichens distributed among islands thousands of miles apart in the Southern Ocean. Consistent local wind can prune and shape vegetation and sculpt ice and snow surfaces (Figure 15.1).

Earth's dry lands stand out in stark contrast on the water planet. In desert environments, an overall lack of moisture and stabilizing vegetation allows wind to create extensive sand seas and dunes of infinite variety. The polar regions are deserts as well, so there should be no surprise that weather-monitoring stations in Yuma, Arizona, and in Antarctica report the same amount of annual precipitation. These polar and high-latitude deserts possess unique features related to their cold, dry environment.

Arid landscapes display distinctive landforms and life forms: "Instead of finding chaos and disorder the observer never fails to be amazed at a simplicity of form, an exactitude of repetition and a geometric order."*

*R. A. Bagnold, *The Physics of Blown Sand and Desert Dunes* (London, England: Methuen, 1941, p. xvii).

In this chapter: We examine the work of wind; associated erosion, transport, and depositional processes; and resulting landforms. Windblown fine particles form vast loess deposits, the basis for rich agricultural soils. We include the discussion of arid lands in this chapter. Most deserts are rocky and covered with desert pavement, whereas other dry landscapes are covered in sand dunes. In desert environments, landscapes lacking moisture and stabilizing vegetation are set off in sharp, sun-baked relief. Water is the major erosional agent in the desert, yet water is the limiting resource for human development. We examine the causes and distribution of Earth's desert landscapes. The Colorado River and the western drought are the subject of an important Focus Study.

The Work of Wind

The work of wind—erosion, transportation, and deposition—is **eolian** (also spelled *aeolian*). The word comes from Aeolus, ruler of the winds in Greek mythology. British Army Major Ralph Bagnold, an engineering officer, while stationed in Egypt in 1925 accomplished much eolian research. In the deserts west of the Nile, Bagnold measured, sketched, and developed hypotheses about the wind and desert forms. Imagine the scene in the 1920s: A Model-T Ford chugs across the desert west of Cairo taking Bagnold to his research area. He laid rolls of chicken wire over treacherous stretches of sand to prevent getting stuck. Bagnold's often-cited work *The Physics of Blown Sand and Desert Dunes* was published in 1941.

The ability of wind to move materials is actually small compared with that of other transporting agents such as water and ice because air is so much less dense than those other media. Yet, over time, wind accomplishes enormous work. Bagnold studied the ability of wind

(a) (b)

FIGURE 15.1 The work of wind.
(a) Wind-sculpted tree near South Point, Hawai'i. Nearly constant tradewinds keep this tree naturally pruned.
(b) Blowing wind shapes ice in formations of *sastrugi*. Can you tell the direction of the prevailing winds looking at the ice? [Bobbé Christopherson.]

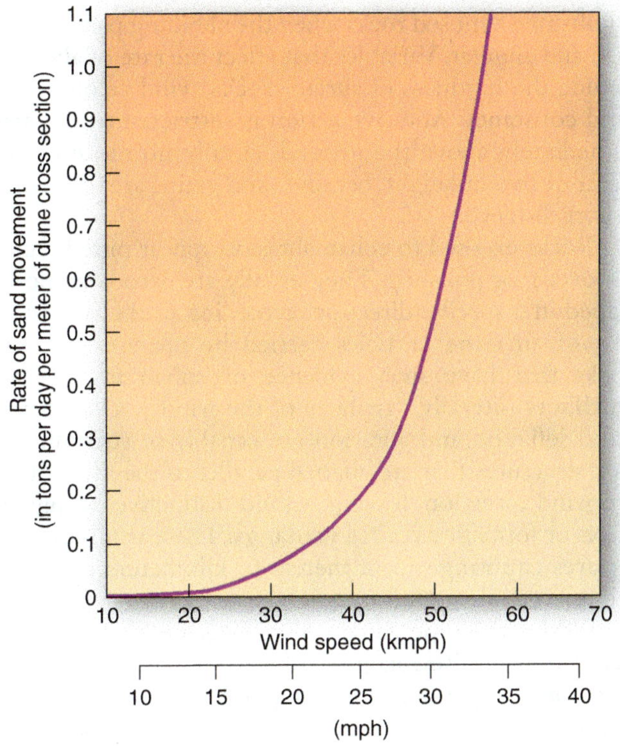

FIGURE 15.2 Sand movement and wind velocity. Sand movement relative to wind velocity, as measured over a meter cross section of ground surface. [After R. A. Bagnold,1941). Adapted by permission.]

to transport sand over the surface of a dune. Figure 15.2 shows that a steady wind of 50 kmph (30 mph) can move approximately one-half ton of sand per day over a 1-m-wide section of dune. The graph also demonstrates how rapidly the amount of transported sand increases with wind speed.

Grain size is important in wind erosion. Intermediate-sized grains move most easily—they bounce along. It is the largest and the smallest sand particles that require the strongest winds to move. The large particles are heavier and thus require stronger winds. The small particles are difficult to move because they exhibit a mutual cohesiveness and because they usually present a smooth (aerodynamic) surface to the wind. Yet the finest dust, once aloft, is carried from continent to continent. In the chapter-opening photo, you can see wind-blown streaks of light-colored sand along the left half of the picture of Ward Terrace.

Eolian Erosion

Two principal wind-erosion processes are **deflation**, the removal and lifting of individual loose particles, and **abrasion**, the grinding of rock surfaces by the "sandblasting" action of particles captured in the air. Deflation and abrasion produce a variety of distinctive landforms and landscapes.

Deflation Deflation literally blows away loose or noncohesive sediment and works with rainwater to form a surface resembling a cobblestone street: a **desert pavement** that protects underlying sediment from further deflation and water erosion. Traditionally, deflation was regarded as a key formative process, eroding fine dust, clay, and sand and leaving behind a concentration of pebbles and gravel as desert pavement (Figure 15.3a).

Another hypothesis that better explains some desert pavement surfaces states that deposition of windblown sediments, not removal, is the formative agent. Windblown particles settle between and below coarse rocks and pebbles that are gradually displaced upward. Rainwater is involved as wetting and drying episodes swell and shrink clay-sized particles. The gravel fragments are gradually lifted to surface positions to form the pavement (Figure 15.3b). Desert pavements are so common that many provincial names are used for them—for example, *gibber plain* in Australia; *gobi* in China; and in Africa, *lag gravels* or *serir*, or *reg* desert if some fine particles remain.

Heavy recreational activity damages fragile desert landscapes, especially in the United States, where more than 15 million off-road vehicles are now in use. Such vehicles crush plants and animals; disrupt desert pavement, leading to greater deflation; and create ruts that easily concentrate sheetwash to form gullies.

Wherever wind encounters loose sediment, deflation may remove enough material to form basins known as **blowout depressions**. These depressions range from small indentations less than a meter wide up to areas hundreds of meters wide and many meters deep. Chemical weathering, although slow in arid regions owing to the lack of water, is important in the formation of a blowout, for it removes the cementing materials that give particles their cohesiveness. Large depressions in the Sahara Desert are at least partially formed by deflation. The enormous Munkhafad el Qaṭṭâra (Qaṭṭâra Depression), which covers 18,000 km^2 (6950 mi^2) just inland from the Mediterranean Sea in the Western Desert of Egypt, is now about 130 m (427 ft) below sea level at its lowest point.

GEO REPORT 15.1 *Bagnold was ahead of his time*

Ralph Bagnold's work in the desert described sand dunes as complex self-organizing systems—with order and regularity arising from chaos and disorder. Today, this is the science of complex system behavior, where system components interact in a self-enhancing manner. For example, consider the ripple patterns in sand: The ripple captures increasing amounts of sand as it grows in size, which reduces the amount of sand grains in the air, forcing the distance span before another ripple forms. We study self-organizing behavior in periglacial regions and how freeze–thaw cycles organize rock sorting and resulting patterns.

FIGURE 15.3 Desert pavement.
(a) Desert pavement may be formed when larger rocks and fragments are left after deflation and sheetwash remove finer dust, silt, and clays. (b) Other desert pavements seem to be formed by wind delivering fine particles that settle and wash downward; through cycles of swelling and shrinking, the gravels gradually migrate upward to form pavement. (c) A typical desert pavement. [Bobbé Christopherson.]

Abrasion You may have seen work crews sandblasting surfaces on buildings, bridges, or streets to clean them or to remove unwanted markings. Sandblasting uses a stream of compressed air filled with sand grains to quickly abrade a surface. Abrasion by windblown particles is nature's slower version of sandblasting, and it is especially effective at polishing exposed rocks when the abrading particles are hard and angular. Variables that affect the rate of abrasion include the hardness of surface rocks, wind velocity, and wind constancy. Abrasive action is restricted to the area immediately above the ground, usually no more than a meter or two in height, because sand grains are lifted only a short distance.

Rocks exposed to eolian abrasion appear pitted, fluted (grooved), or polished. They usually are aerodynamically shaped in a specific direction, according to the consistent flow of airborne particles carried by prevailing winds. Rocks that have such evidence of eolian erosion are **ventifacts** (literally, "artifacts of the wind"). On a larger scale, deflation and abrasion are capable of streamlining rock structures that are aligned parallel to the most effective wind direction, leaving behind distinctive, elongated ridges or formations called **yardangs**. These wind-sculpted features can range from meters to kilometers in length and up to many meters in height.

On Earth, some yardangs are large enough to be detected on satellite imagery. The Ica Valley of southern Peru contains yardangs reaching 100 m (330 ft) in height and several kilometers in length, and yardangs in the Lût Desert of Iran attain 150 m (490 ft) in height. Abrasion is concentrated on the windward end of each yardang, with deflation operating on the leeward portions. Wind-sculpted formations are shown in Figure 15.4.

The Sphinx in Egypt was perhaps partially formed as a yardang, suggesting a head and body. Some scientists think this shape led the ancients to complete the bulk of the sculpture artificially with masonry.

Eolian Transportation

Atmospheric circulation can transport fine material, such as volcanic debris, fire soot and smoke, and dust, worldwide within days. Wind exerts a drag, or frictional pull, on surface particles until they become airborne, just as water in a stream picks up sediment (again, think of air as a fluid). The distance that wind is capable of transporting particles in suspension varies greatly with particle size. Only the finest dust particles travel significant distances, so the finer material suspended in a dust storm is lifted much higher than are the coarser particles of a sandstorm.

People living in areas of frequent dust storms are faced with infiltration of very fine particles into their homes and businesses through even the smallest cracks. Figure 3.8 illustrates such dust storms. People living in desert regions and along sandy beaches, where frequent sandstorms occur, contend with the sandblasting of painted surfaces and etched window glass.

Both human and natural sand erosion and transport from a beach are slowed by conservation measures such as the introduction of stabilizing native plants, the use of fences, and the restriction of pedestrian traffic to walkways (Figure 15.5). As human settlement encroaches on coastal dunes, sand transport becomes problematic.

The term *saltation* was used to describe movement of particles by water. The term also describes the wind

(a) (b)

FIGURE 15.4 A yardang.
(a) A small wind-sculpted rock formation located in Snow Canyon outside St. George, Utah. (b) Consistent winds shape this exposed rock on Bear Island in the North Atlantic. The rock sits on a surface of patterned ground and periglacial features. [Bobbé Christopherson.]

transport of grains, usually larger than 0.2 mm (0.008 in.), along the ground. About 80% of wind transport of particles is accomplished by this skipping and bouncing action (Figure 15.6). Compared with fluvial transport, in which saltation is accomplished by hydraulic lift, eolian saltation is executed by aerodynamic lift, elastic bounce, and impact.

On impact, grains hit other grains and knock them into the air. Saltating particles crash into other particles, knocking them both loose and forward. This type of movement is **surface creep**, which slides and rolls particles too large for saltation and affects about 20% of the material transported by wind. Once in motion, particles continue to be transported by lower wind velocities. In a desert or along a beach, sometimes you can hear a slight hissing sound, almost like steam escaping, produced by the myriad saltating grains of sand as they bounce along and collide with surface particles.

Eolian Depositional Landforms

The smallest features shaped by individual saltating grains are ripples (Figure 15.7). Eolian ripples are similar to fluvial ripples, although the impact of saltating grains is slight in water. Ripples form in crests and troughs, positioned

(a)

(b)

FIGURE 15.5 Preventing sand transport.
(a) Popham Beach State Park, Maine. Further erosion and transport of coastal dunes can be controlled by stabilizing strategies, such as planting native plants and confining pedestrian traffic to walkways. (b) Tufts of beach grass are planted to stabilize this coastal area in Atlantic City, New Jersey. One storm surge can erase such efforts. [Bobbé Christopherson.]

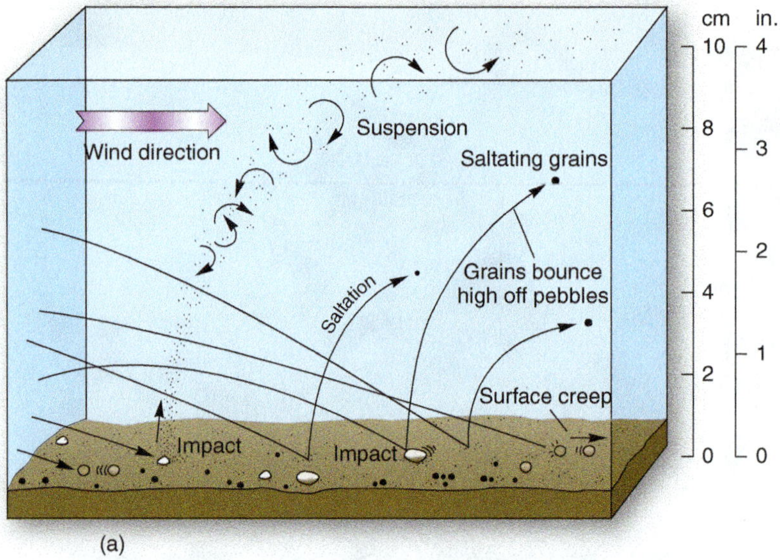

FIGURE 15.6 How the wind moves sand.
(a) Eolian suspension, saltation, and surface creep are transportation mechanisms. Compare with the saltation and traction that occur in another fluid, water. (b) Sand grains saltating along the surface in the Stovepipe Wells dune field, Death Valley, California. [Author.]

MG Animation
How Wind Moves Sand

transversely (at a right angle) to the direction of the wind. The interception of sand grains from the wind stream deprives the downwind area of grains and sets the distance between ripples—this is self-enhancing behavior in a complex system.

Many people know arid landscapes only from the movies, which leave the impression that most deserts are covered by sand. Instead, desert pavements predominate across most subtropical arid landscapes; only about 10% of desert areas are covered with sand. Sand grains generally are deposited as transient ridges or hills called dunes.

FIGURE 15.7 Sand ripples.
Sand ripple patterns later may become lithified into fixed patterns in rock. The area in the photo looks vast, but is only about 1 m (3.3 ft) wide. [Author.]

A **dune** is a wind-sculpted accumulation of sand. An extensive area of dunes, such as that found in North Africa, is characteristic of an **erg desert**, or **sand sea**. The Grand Erg Oriental in the central Sahara exceeds 1200 m (4000 ft) in depth and covers 192,000 km^2 (75,000 mi^2), comparable to the area of Nebraska. This sand sea has been active for more than 1.3 million years. Similar sand seas, such as the Grand Ar Rub'al Khālī Erg, are active in Saudi Arabia.

In eastern Algeria, the Issaouane Erg covers 38,000 km^2 (14,673 mi^2). Winds from varying directions produce star dunes with multiple slipfaces (Figure 15.8a). *Star dunes* are the mountainous giants of the sandy desert. They are pinwheel-shaped, with several radiating arms rising and joining to form a common central peak, where they approach 200 m (650 ft) in height. Some in the Sonoran Desert approach this size. In the photo, you can also see some dunes, which suggest the region has experienced changing wind patterns over time, as they form from a single principal wind flow. Dunes are present on Mars as well as in semiarid areas such as the Great Plains of the United States (Figure 15.8b and c).

Dune Movement and Form Dune fields, whether in arid regions or along coastlines, tend to migrate in the direction of effective, sand-transporting winds. In this regard, stronger seasonal winds or winds from a passing storm may prove more effective than average prevailing winds. When saltating sand grains encounter small patches of sand, their kinetic energy (motion) is dissipated and they accumulate. As height increases above 30 cm (12 in.), a *slipface* and characteristic dune features form.

By studying the dune model in Figure 15.9, you can see that winds characteristically create a gently sloping *windward side* (stoss side), with a more steeply sloped **slipface** on the *leeward side*. A dune usually is asymmetrical

(a) Issaouane Erg, Algeria

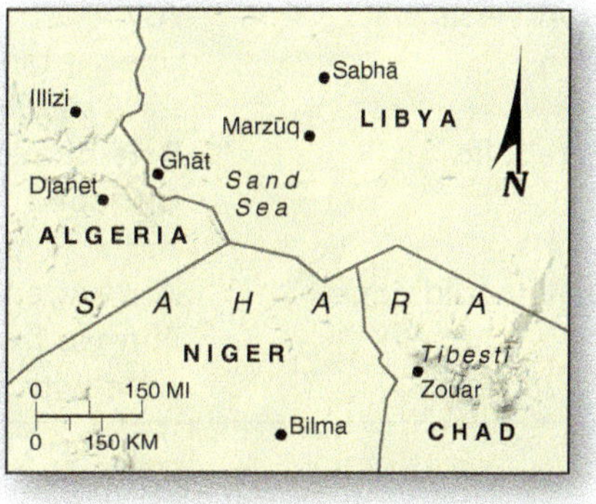

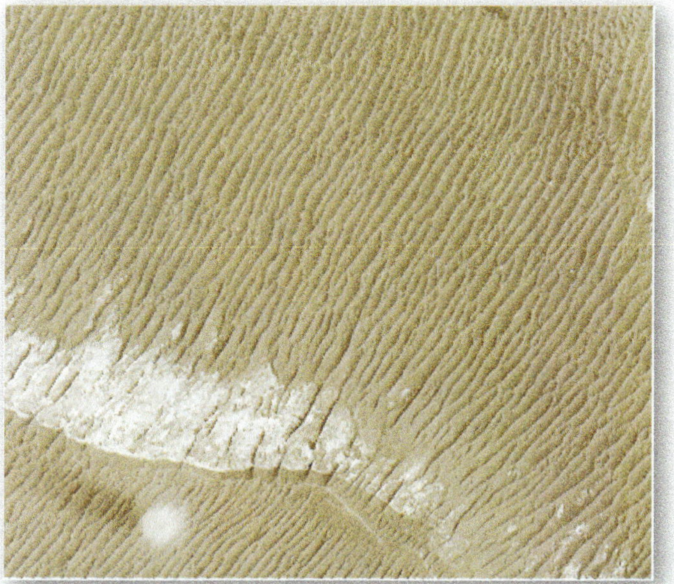

(b) Melas Chasma, Mars

(c) Sand Hills, Nebraska

FIGURE 15.8 Seas of sand.
(a) The Issaouane Erg of eastern Algeria is a blend of principally star dunes, some barchan dunes, and linear dunes in the far upper-left—all disclose the prevailing wind history of the region. (b) Sand dunes are found on the Martian surface in the southern area of Melas Chasma in Valles Marineris. Note the dust devil in the lower left. The area covered is about 2 km wide. (c) Sand Hills, central Nebraska, sit atop the High Plains aquifer, where sand and silt deposits derived from glaciated regions to the north and the Rockies to the west accumulated. These densely packed barchan dunes, inactive for at least 600 years, are fixed (stabilized) by vegetation. [(a) *ISS* astronaut photo, NASA/GSFC. (b) Mars Global Surveyor, Mars Orbiter Camera image courtesy of NASA/JPL/Malin Space Science Systems. (c) Bobbé Christopherson.]

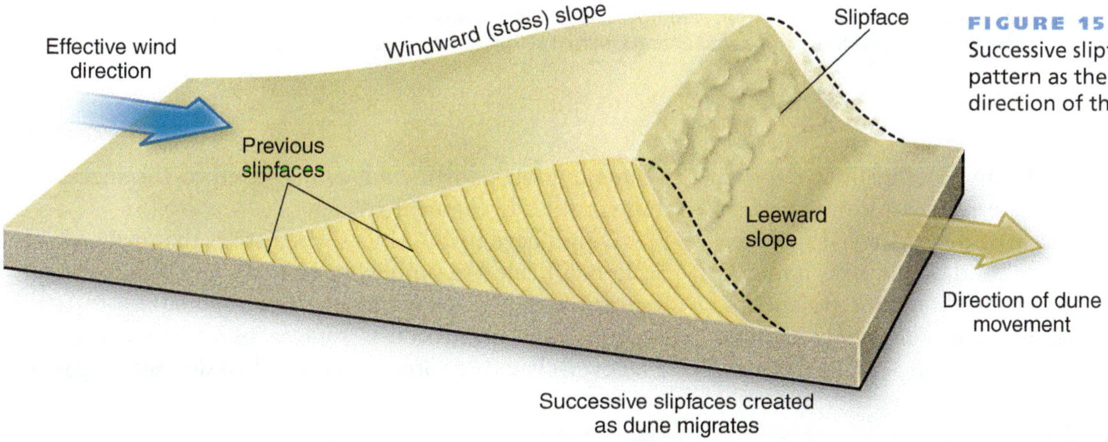

FIGURE 15.9 Dune cross section. Successive slipfaces exhibit a distinctive pattern as the dune migrates in the direction of the effective wind.

MG Animation
Formation of Cross Bedding

Class	Type	Description
Crescentic	Barchan	Crescent-shaped dune with horns pointed downwind. Winds are constant with little directional variability. Limited sand available. Only one slipface. Can be scattered over bare rock or desert pavement or commonly in dune fields.
	Transverse	Asymmetrical ridge, transverse to wind direction (right angle). Only one slipface. Results from relatively ineffective wind and abundant sand supply.
	Parabolic	Role of anchoring vegetation important. Open end faces upwind with U-shaped "blowout" and arms anchored by vegetation. Multiple slipfaces, partially stabilized.
	Barchanoid ridge	A wavy, asymmetrical dune ridge aligned transverse to effective winds. Formed from coalesced barchans; looks like connected crescents in rows with open areas between them.

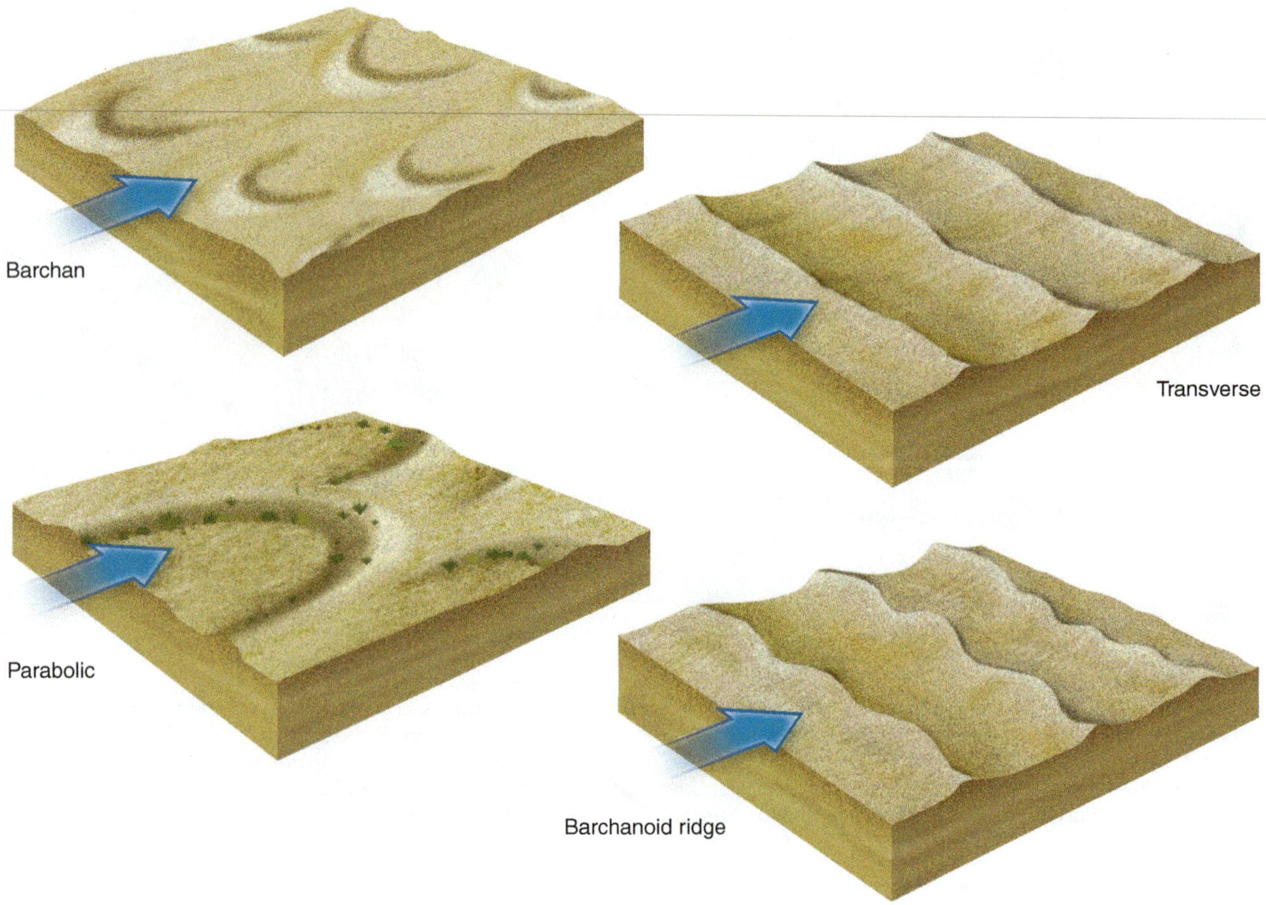

FIGURE 15.10 Major dune forms.
See table for dune descriptions. Arrows show wind direction. [Adapted from E. D. McKee, *A Study of Global Sand Seas,* USGS Professional Paper 1052 (Washington, DC: U.S. Government Printing Office, 1979).]

in one or more directions. The angle of a slipface is the steepest angle at which loose material is stable—its *angle of repose*. Thus, the constant flow of new material makes a slipface, a type of *avalanche slope*. Sand builds up as it moves over the crest of the dune to the brink; then it avalanches, falling and cascading as the slipface continually adjusts, seeking its angle of repose (usually 30° to 34°). In this way, a dune migrates downwind, as suggested by the successive dune profiles in Figure 15.9.

Dunes have many wind-shaped styles that make classification difficult. We simplify dune forms into three classes—*crescentic* (crescent, curved shape), *linear* (straight form), and massive *star dunes*. Each is summarized in Figure 15.10. The ever-changing form of these eolian deposits is part of

Class	Type	Description
Linear	Longitudinal	Long, slightly sinuous, ridge-shaped dune, aligned parallel with the wind direction; two slipfaces. Average 100 m high and 100 km long, and the "draa" at the extreme in size can reach to 400 m high. Results from strong effective winds varying in one direction.
	Seif	After Arabic word for "sword"; a more sinuous crest and shorter than longitudinal dunes. Rounded toward upwind direction and pointed downwind. (Not illustrated.)
Star dune		The giant of dunes. Pyramidal or star-shaped with three or more sinuous, radiating arms extending outward from a central peak. Slipfaces in multiple directions. Results from effective winds shifting in all directions. Tend to form isolated mounds in high, effective winds and connected sinuous arms in low, effective winds.
Other	Dome	Circular or elliptical mound with no slipface. Can be modified into barchanoid forms; vegetation can play role in coastal locales.
	Reversing	Asymmetrical ridge form that is intermediate between star dune and transverse dune. Wind variability can alter shape between forms.

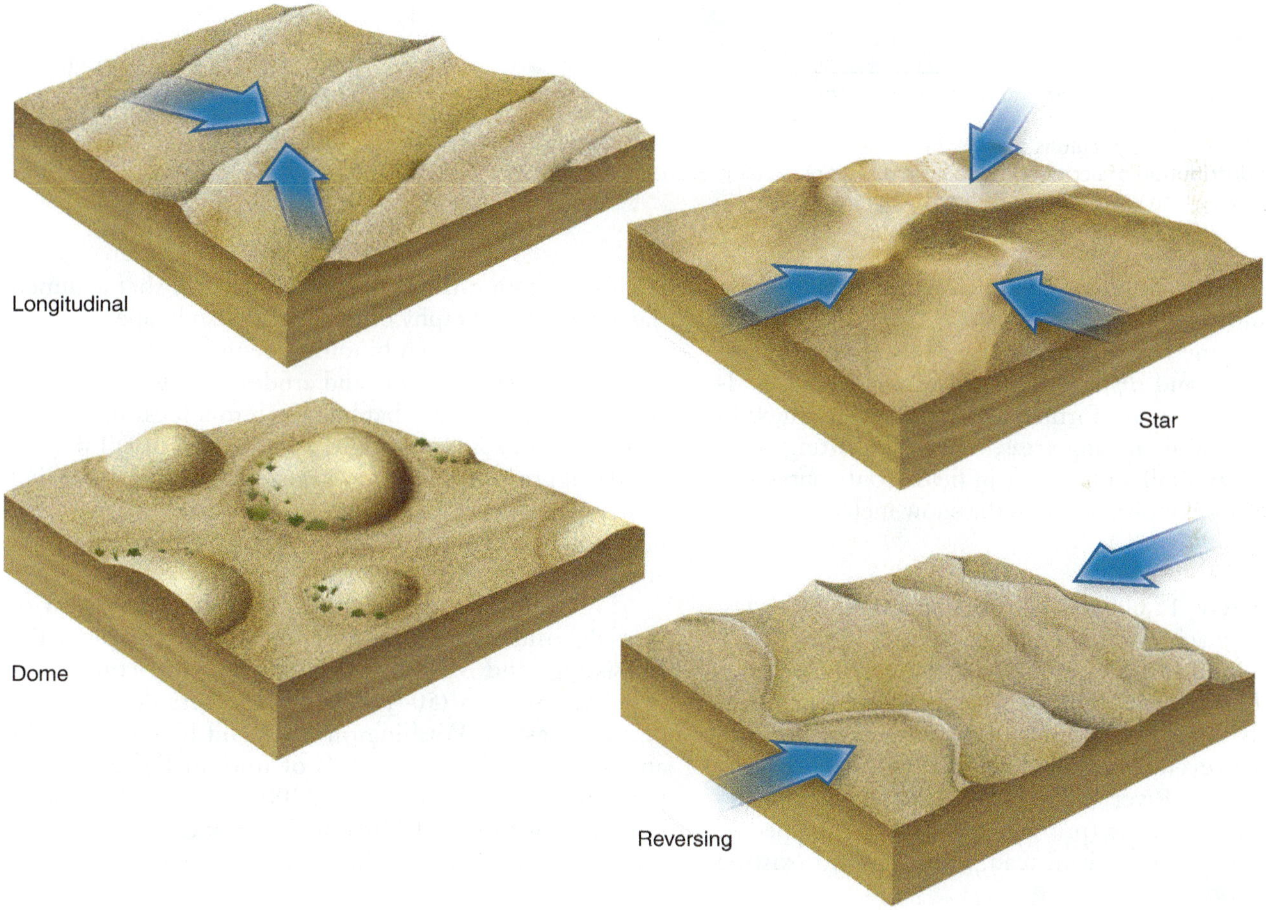

their beauty, eloquently described by one author: "I see hills and hollows of sand like rising and falling waves. Now at midmorning, they appear paper white. At dawn they were fog gray. This evening they will be eggshell brown."*

*J. E. Bowers, *Seasons of the Wind* (Flagstaff, AZ: Northland Press, 1985), p. 1.

The map in Figure 15.11 shows the correlation of active sand regions with deserts (tropical, continental interior, and coastal). Note the limited extent of desert area covered by active sand dunes—only about 10% of all continental land between 30° N and 30° S. Also noted on the map are dune fields in humid climates outside of this range, such as along coastal Oregon, the south shore of

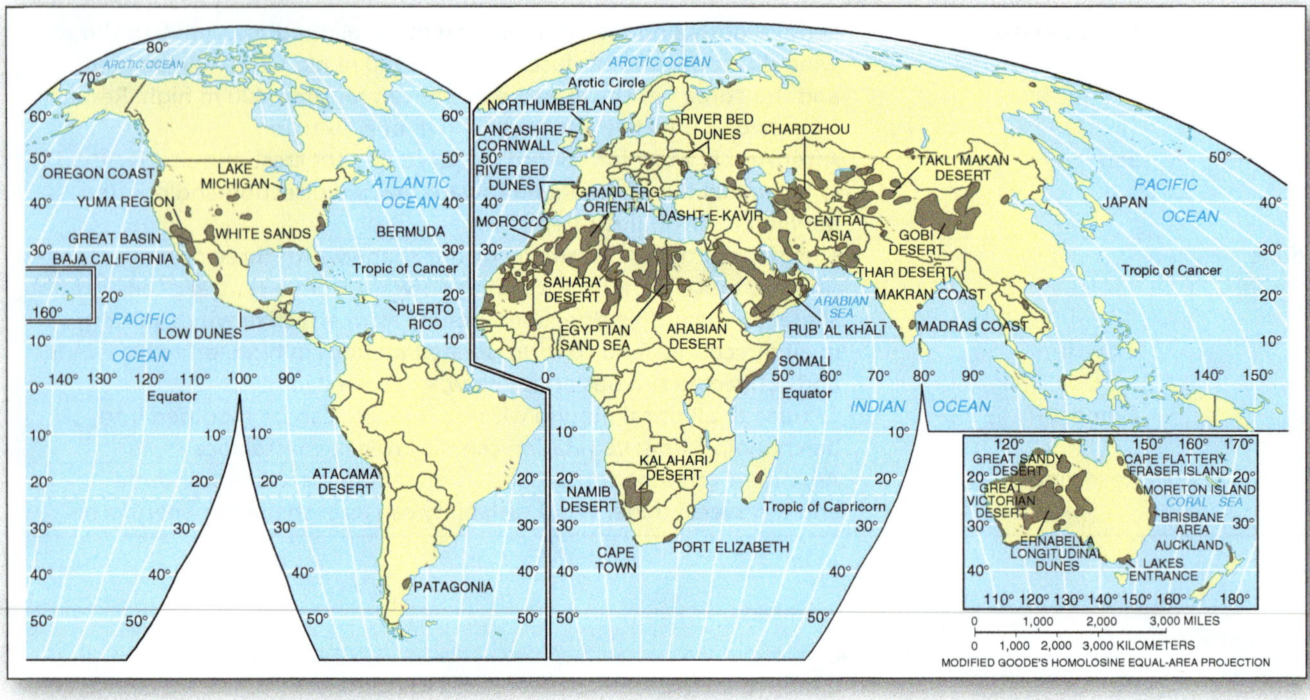

FIGURE 15.11 Sandy regions of the world.
Worldwide distribution of active and stable sand regions. [After R. E. Snead, *Atlas of World Physical Features,* p. 134, © 1972 by John Wiley & Sons. Adapted by permission.]

Lake Michigan, and the Gulf and Atlantic coastlines; in Europe; and elsewhere.

These same dune-forming principles and terms (for example, *dune* and *slipface*) apply to snow-covered landscapes. *Snow dunes* are formed as wind deposits snow in drifts. In semiarid farming areas, capturing drifting snow with fences and tall stubble left in fields contributes significantly to soil moisture when the snow melts.

Loess Deposits

Approximately 15,000 years ago, in several episodes, Pleistocene glaciers retreated in many parts of the world, leaving behind large glacial outwash deposits of fine-grained clays and silts. These materials were blown great distances by the wind and redeposited in unstratified, homogeneous (evenly mixed) deposits. Peasants working along the Rhine River Valley in Germany gave these deposits the name **loess** (pronounced "luss"). No specific landforms were created; instead, loess covered existing landforms with a thick blanket of material that assumed the general topography of the existing landscape.

Because of its own binding strength and its internal coherence, loess weathers and erodes into steep bluffs, or vertical faces. When a bank is cut into a loess deposit, it generally will stand vertically, although it can fail if saturated (Figure 15.12). There are historical accounts of Civil War soldiers in the Vicksburg, Mississippi, area excavating places to live in loess banks; in areas of China, dwellings are carved from the loess.

Figure 15.13a shows the worldwide distribution of loess deposits. Significant accumulations throughout the Mississippi and Missouri river valleys form continuous deposits 15–30 m (50–100 ft) thick. Loess deposits also occur in eastern Washington State and Idaho. The hills and gullies in the Loess Hills of Iowa in Figure 15.13b reach heights of about 61 m (200 ft) above the nearby prairie farmlands and Missouri River and run north and south more than 322 km (200 mi). Only China has deposits

GEO REPORT 15.2 *Conflicts feed the sand seas*

Military activities can threaten desert landscapes. A serious environmental impact of the ongoing wars in Iraq and Afghanistan is the disruption of desert pavement by explosives and the movement of heavy vehicles. As thousands of square kilometers of stable desert pavement break apart, nearby cities and farms experience increased dust and sand invasions.

The Registan Desert of southern Afghanistan, south of the city of Kandahar, is an example of such a sand sea over a flat terrain. A result of drought conditions and war-disrupted desert pavement surfaces, the migrating sand is overtaking areas of sparse agricultural activity and has covered more than 100 villages.

FIGURE 15.12 Example of loess deposit. A loess bluff in western Iowa; note the vertical structural strength. [Bobbé Christopherson.]

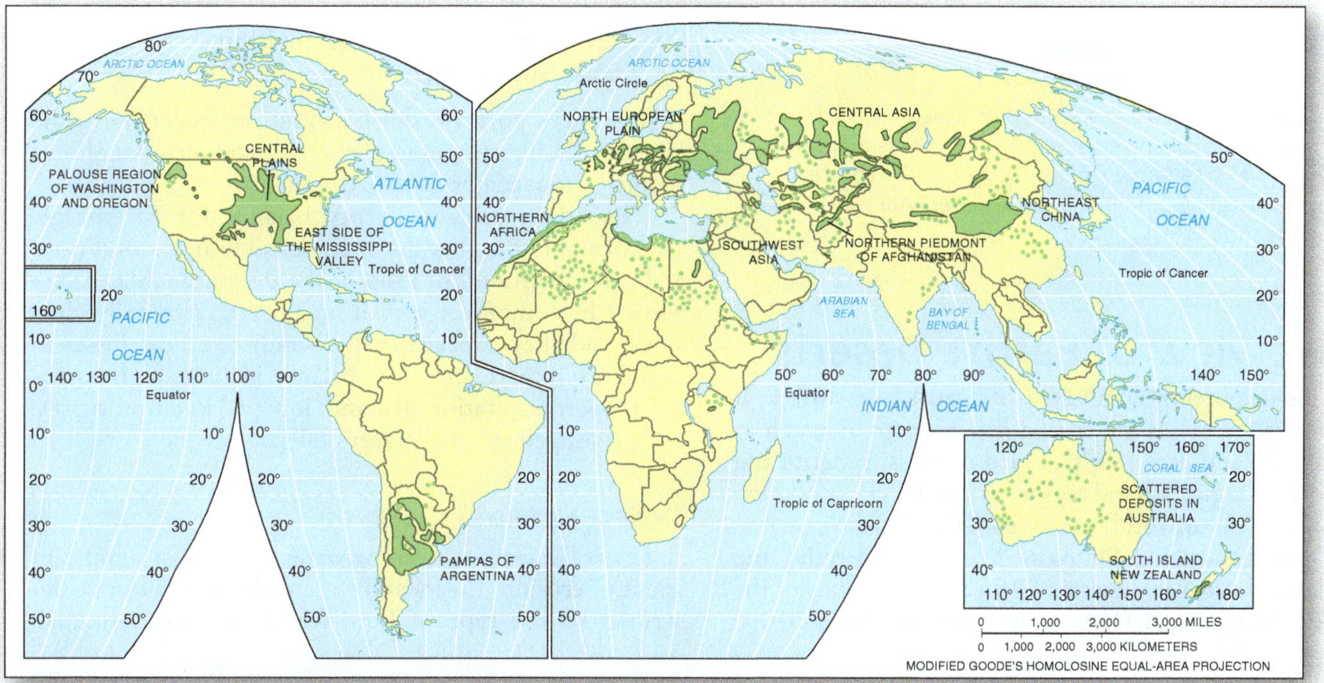

(a)

(b)

FIGURE 15.13 Loess deposits worldwide and in Iowa. (a) Areas of major accumulations of loess. Dots represent small, scattered loess formations. (b) The Loess Hills of western Iowa give evidence of the last ice age, as the finely ground sediments left by the glaciers were windblown to this setting. [(a) After R. E. Snead, *Atlas of World Physical Features,* p. 138, © 1972 by John Wiley & Sons. Adapted by permission. (b) Bobbé Christopherson.]

that exceed these dimensions. This silt explains the fertility of the soils in these regions, for loess deposits are well drained and deep and have excellent moisture retention. Loess deposits also cover much of Ukraine, central Europe, China, the Pampas–Patagonia regions of Argentina, and lowland New Zealand. The soils derived from loess help form some of Earth's "breadbasket" farming regions. Transport of loess soils occurred in the catastrophic 1930s Dust Bowl in the United States.

In Europe and North America, loess is thought to be derived mainly from glacial and periglacial sources. The vast deposits of loess in China, covering more than 300,000 km^2 (116,000 mi^2), are derived from windblown desert sediment rather than glacial sources. Accumulations in the Loess Plateau of China exceed 300 m (1000 ft) in thickness, forming complex weathered badlands and some good agricultural land. These windblown deposits are interwoven with much of Chinese history and society.

CRITICAL THINKING 15.1

The nearest eolian features

What eolian features are nearest to your present location? Are they coastal, lakeshore, or desert dunes, or are they loess deposits? Which causative factors discussed in this chapter explain the location or form of these features? Are you able to visit the site?

Overview of Earth's Deserts

Dry climates occupy about 26% of Earth's land surface. If all semiarid climates are considered, perhaps as much as 35% of all land area tends toward dryness, constituting the largest single climatic region on Earth.

Deserts occur worldwide as topographic plains, such as the Great Sandy and Simpson deserts of Australia, the Arabian and Kalahari deserts of Africa, and portions of the extensive Taklimakan Desert, which covers some 270,000 km^2 (105,000 mi^2) in the central Tarim Basin of China. Deserts also are found in mountainous regions: in interior Asia from Iran to Pakistan and in China and Mongolia. In South America, the rugged Atacama Desert lies between the ocean and the Andes. Desert conditions can also be found on small islands, such as the Cape Verde Islands off the West African coast at around 15° N latitude, an extension of Saharan conditions to the east (Figure 15.14d). Earth's deserts are expanding, as discussed in *Geosystems Now*. Now, let us look at the link between climate and Earth's deserts.

Desert Climates

The spatial distribution of dry lands is related to three climatological settings:

- Areas dominated by subtropical high-pressure cells between 15° and 35°, both N and S latitude;
- Rain shadow areas on the lee side of mountain ranges; and
- Areas at great distance from moisture-bearing air masses, such as central Asia.

Figure 15.14 portrays this distribution according to the climate classification used in this text and includes photographs of four major desert regions.

The daily surface energy balance for El Mirage, California, highlights the high sensible heat conditions and intense ground heating in the desert. Such areas receive a high input of insolation through generally clear skies, and they experience high radiative heat losses at night. A typical desert water balance includes high potential evapotranspiration demand, low precipitation supply, and prolonged summer water deficits.

Desert Fluvial Processes

In desert landscapes, running water is the most significant agent of erosion. Rivers and streams in deserts can be described by the type of flow, which may be perennial,

GEO REPORT 15.3 ***The Dust Bowl***

Wind transport of soils including loess produced a catastrophe in the American Great Plains in the 1930s—the Dust Bowl. More than a century of overgrazing and intensive agriculture left soil susceptible to drought and deflation processes. Reduced precipitation amounts and above-normal temperatures triggered the multiyear period of severe dust storms and loss of farmlands.

The deflation of many centimeters of soil occurred in southern Nebraska, Kansas, Oklahoma, Texas, and eastern Colorado—and even in southern Canada and northern Mexico. The transported dust darkened the skies of midwestern cities, which left the streetlights on throughout the day, and drifted over farmland to depths that covered failing crops. The largest migration in American history occurred as 2 million people moved out of the plains states by 1940.

Tree-ring analyses show that droughts of similar intensity happened about twice a century over the past 400 years. One such intense drought in what is now Nebraska in the 13th century appears to have lasted nearly 40 years. The Dust Bowl weather conditions seem to correlate with sea-surface temperatures that are cool in the Pacific Ocean and warm in the Atlantic Ocean. For more on the Dust Bowl, see http://www.pbs.org/wgbh/americanexperience/films/dustbowl/.

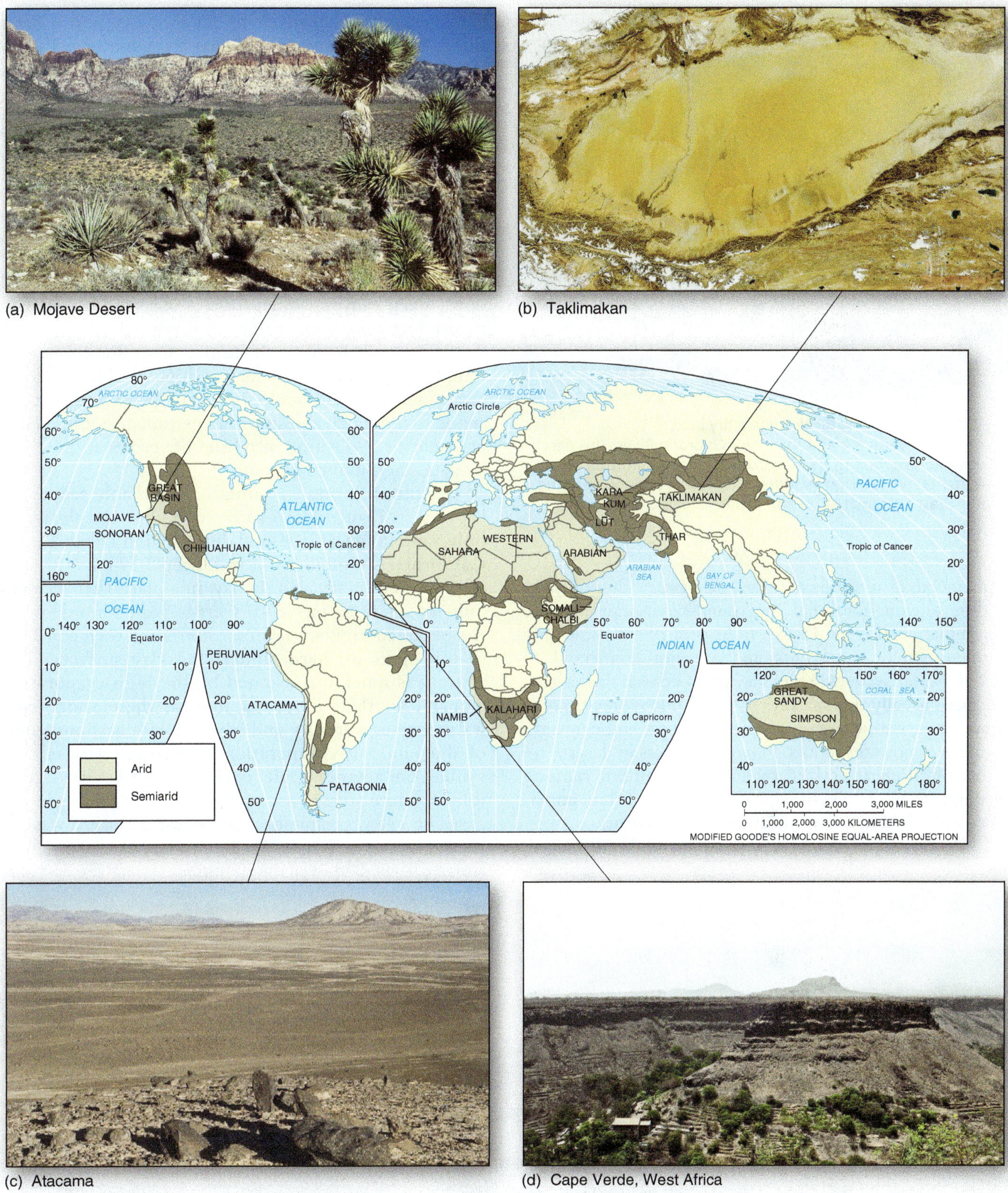

FIGURE 15.14 The world's dry regions.
Worldwide distribution of arid lands (*arid desert* climates) and semiarid lands (*semiarid steppe* climates). (a) Mojave Desert in the American Southwest. (b) Taklimakan Desert in central Asia from orbit. (c) Atacama Desert in subtropical Chile near Baquedano. (d) São Tiago Island desert in Cape Verde; note the light sky filled with dust from Africa. [(a) Author. (b) *Terra* image, NASA/GSFC. (c) Jacques Jangoux/Photo Researchers, Inc. (d) Bobbé Christopherson.]

FIGURE 15.15 Water through a dry land.
The Virgin River is a tributary of the Colorado River. This perennial stream is prone to flash floods during precipitation events. The riparian vegetation includes willow, cottonwood, and the water-drawing tamarisk, or salt cedar. [Bobbé Christopherson.]

ephemeral, or intermittent. *Perennial streams* flow all year, fed by snowmelt, rainfall, or groundwater, often in some combination (Figure 15.15). *Ephemeral streams* flow only after precipitation events and are not connected to groundwater systems. Years may pass between flow events in these usually dry stream channels. *Intermittent streams* flow for several weeks or months each year and may have some groundwater inputs. Fluvial processes in deserts are most often characterized by such intermittent running water, with hard, poorly vegetated surfaces of thin, immature soils yielding high runoff during rainstorms.

Precipitation events in a desert may be rare indeed, even a year or two apart, but when they do occur, an ephemeral stream channel can fill with a torrent, a **flash flood**. These channels may fill in a few minutes and surge briefly during and after a storm. Depending on the region, such a dry channel is known as a **wash**, an *arroyo* (Spanish), or a *wadi* (Arabic). Signs are usually posted where a desert highway crosses a wash to warn drivers not to proceed if rain is in the vicinity, for a flash flood can suddenly arise and sweep away anything in its path. If a long, steady precipitation event occurs, material on slopes may become saturated and flow into canyons. *Debris flows* occur as moving masses of mud, soil, and rock fragments mixed with enough water to keep the material flowing. They can travel at high speed down steep canyons, often leaving unsorted alluvium as their velocity slows and material is deposited.

When washes fill with surging flash flood waters, a unique set of ecological relationships quickly develops. Crashing rocks and boulders break open seeds that respond to the timely moisture and germinate. Other plants and animals also spring into brief life cycles as the water irrigates their limited habitats.

At times of intense rainfall, remarkable scenes fill the desert. Figure 15.16 shows two photographs taken only a month apart in a sand dune field in Death Valley, California. A rainfall event produced 2.57 cm (1.01 in.) of precipitation in a single day, in a place that receives only 4.6 cm (1.83 in.) in an average year. The stream in the photo continued to run for hours and then collected in low spots on hard, underlying clay surfaces. The water was quickly consumed by the high evaporation demand so that, in just a month, these short-lived watercourses were dry and covered with accumulations of alluvium. Such evaporation pans coated with salt are visible on a large scale in Figure 15.8a in Algeria (bluish-white areas).

The 1985 precipitation event just described was exceeded by one on August 15, 2004, when several thunderstorm cells stalled over the mountains along the east side

(a) (b)

FIGURE 15.16 An improbable river in Death Valley.
The Stovepipe Wells dune field of Death Valley, California, shown (a) the day after a 2.57-cm (1.01-in.) rainfall and (b) one month later (identical location). This 1985 rainfall event was exceeded by one in 2004. [Author.]

of Death Valley. Rainfall totals reached 6.35 cm (2.5 in.) in little more than an hour. Cars were swept away from the Furnace Creek Inn parking lot, about 4.8 km (3 mi) of Highways 190 and 178 were washed out, and two fatalities occurred in the Zabriskie Point area as a result of the storm. One annex building had a high-water mark on it at 3 m (9.8 ft). Again, water is sparse in the deserts, yet it is the major erosional force—sometimes dramatically so. For recent information on and maps of Death Valley, see http://www.nps.gov/deva/.

As runoff water evaporates, salt crusts may be left behind on the desert floor in a **playa**. This intermittently wet and dry lowest area of a closed drainage basin is the site of an *ephemeral lake* when water is present. Accompanying our earlier discussion of evaporites.

Permanent lakes and continuously flowing rivers are less common in the desert than in humid regions, although the Nile River and the Colorado River are notable. Both these rivers are *exotic streams*, having their headwaters in a wetter region and the bulk of their course through arid regions. Focus Study 15.1 discusses the Colorado River, the ongoing drought, and the problem of increasing water demands in an arid landscape.

In arid and semiarid climates, a prominent fluvial landform is an **alluvial fan**, which occurs at the mouth of a canyon where an ephemeral stream channel exits into a valley (Figure 15.17). The fan is produced whenever flowing water abruptly loses velocity as it leaves the constricted channel of the canyon, or whenever the stream gradient suddenly decreases, and therefore drops layer upon layer of sediment along the base of the mountain block. Water then flows over the surface of the fan and produces a braided drainage pattern, sometimes shifting from channel to channel. A continuous apron, or **bajada** (Spanish for "slope"), may form if individual alluvial fans coalesce into one sloping surface (see Figure 15.22d). Fan formation of any sort is reduced in humid climates because perennial streams constantly carry away sediment, preventing its deposition.

An aspect of an alluvial fan is the natural sorting of materials by size. Near the mouth of the canyon at the apex of the fan, coarse materials are deposited, grading slowly to pebbles and finer gravels with distance out from the mouth. Then sands and silts are deposited, with the finest clays and dissolved salts carried in suspension and solution all the way to the valley floor. Dissolved minerals accumulate as evaporite deposits left on the valley floor after evaporation of the water from the playa and from salt-laden upward-moving groundwater. Figure 15.18 (p. 450) shows the edge of a playa and active dune field; there is a natural sorting of materials by grain size, with sands dropped in the dunes, silts and clays moving farther toward the playa, and the finest and dissolved materials in the playa.

Well-developed alluvial fans also can be a major source of groundwater. Some cities—San Bernardino, California, for example—are built on alluvial fans and extract their municipal water supplies from them. In other parts of the world, such water-bearing alluvial fans are known as *qanat* (Iran), *karez* (Pakistan), or *foggara* (western Sahara).

(text continued on page 473)

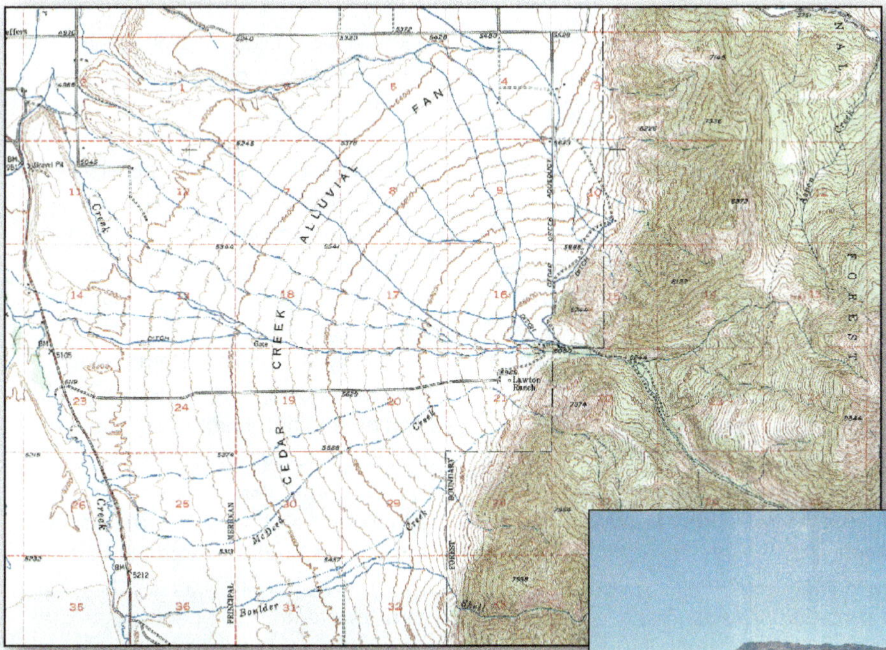

Cedar Creek Fan, Montana

Fan along Black Mountains, Death Valley, California

FIGURE 15.17 An alluvial fan.
The photo shows an alluvial fan in a desert landscape. The topographic map shows the Cedar Creek alluvial fan in southwest Montana, Ennis Quadrangle, 15-minute series, scale 1:62,500, contour interval = 40 ft. [Bobbé Christopherson; U.S. Geological Survey map.]

FOCUS STUDY 15.1

The Colorado River: A System Out of Balance

The headwaters of an exotic stream rise in a region of water surpluses, but the stream flows mostly through arid lands for the rest of its journey to the sea.

In *Geosystems,* the story of the Colorado River began with a *Geosystems Now,* "Water Budgets, Climate Change, and the Southwest." Please revisit this case study.

Orographic precipitation totaling 102 cm (40 in.) per year falls mostly as snow in the Rockies, feeding the Colorado River headwaters in Rocky Mountain National Park. But at Yuma, Arizona, near the river's end, annual precipitation is a scant 8.9 cm (3.5 in.), an extremely small amount when compared with the high annual potential evapotranspiration demand in the Yuma region of 140 cm (55 in.).

From its source region, the Colorado River quickly leaves the relatively wet Rockies and spills out into the dry climates of western Colorado and eastern Utah. At Grand Junction, Colorado, near the Utah border, annual precipitation is only 20 cm (8 in.) (upstream from Figure 15.1.1b, on the map). After carving its way through the Utah canyonlands, the river enters Lake Powell, the reservoir behind Glen Canyon Dam (Figure 15.1.1c), and then flows on through the Grand Canyon. The river then turns southward, marking its final 644 km (400 mi) as the Arizona–California border. Along this stretch sits Hoover Dam, just east of Las Vegas (Figure 15.1.1d); Davis Dam, built to control the releases from Hoover (Figure 15.1.1e); Parker Dam, for the water needs of Los Angeles; three more dams for irrigation water (Palo Verde, Imperial, and Laguna, Figure 15.1.1f); and finally, Morelos Dam at the Mexican border. Mexico owns the end of the river and whatever water is left.

Hydrologically exhausted, the river no longer reaches its mouth in the Gulf of California (Figure 15.1.1g).

Figure 15.1.1h shows the annual water discharge and suspended sediment load for the Colorado River at Yuma, Arizona, from 1905 to 1964. The completion of Hoover Dam in the 1930s dramatically reduced suspended sediment. Subsequent dams upstream from Yuma reduced both discharge and sediment to nearly zero by the mid-1960s.

Overall, the drainage basin encompasses 641,025 km^2 (247,500 mi^2) of mountain, plateau, canyon, basin-and-range, and hot desert landscapes in parts of seven states and two countries. A discussion of the Colorado River is included in this chapter because of its crucial part in the history of the Southwest and its role in the future of this drought-plagued region.

Dividing Up the Colorado's Dammed Water

John Wesley Powell (1834–1902) was the first Euro-American of record to successfully navigate the Colorado River through the Grand Canyon. Lake Powell is named after him. Powell perceived that the challenge of settling the West was too great for individual efforts and believed that solutions to problems such as water availability could be met only through private cooperative efforts. His 1878 study (reprinted in 1962), *Report of the Lands of the Arid Region of the United States,* is a conservation landmark.

Today, Powell probably would be skeptical of the intervention by government agencies in building large-scale reclamation projects. An anecdote in Wallace Stegner's book *Beyond the Hundredth Meridian* tells of an 1893 international irrigation conference held in Los Angeles, where development-minded delegates bragged that the entire West could be conquered and reclaimed from nature and that "rain will certainly follow the plow." Powell spoke against that sentiment: "I tell you, gentlemen, you are piling up a heritage of conflict and litigation over water rights, for there is not sufficient water to supply the land."* He was booed from the hall, but history has shown Powell to be correct.

The Colorado River Compact was signed by six of the seven basin states in 1923. (The seventh, Arizona, signed in 1944, the same year that the United States signed the Mexican Water Treaty.) With this compact, the Colorado River basin was divided into an upper basin and a lower basin, arbitrarily separated for administrative purposes at Lees Ferry near the Utah–Arizona border (noted on the map in Figure 15.1.1). Congress adopted the Boulder Canyon Act in 1928, authorizing Hoover Dam as the first major reclamation project on the river. Also authorized was the All-American Canal into the Imperial Valley, which required an additional dam. Los Angeles then began its project to bring Colorado River water 390 km (240 mi) from still another dam and reservoir on the river to the city.

Shortly after Hoover Dam was finished and downstream enterprises were thus offered flood protection, the other projects were quickly completed. There are now eight major dams on the river and many irrigation works. The last effort to redistribute Colorado River water was the Central Arizona Project, which carries water to the Phoenix area.

*W. Stegner, *Beyond the Hundredth Meridian* (Boston, MA: Houghton Mifflin, 1954), p. 343.

FIGURE 15.1.1 The Colorado River drainage basin. →
The Colorado River basin, showing division of the upper and lower basins near Lees Ferry in northern Arizona. (a) River headwaters near Mount Richthofen in the Colorado Rockies. (b) The river near Moab, Utah. (c) Glen Canyon Dam, a regulatory, administrative facility near Lees Ferry, Arizona. (d) Hoover Dam and Lake Mead with declining water levels. (e) Davis Dam in full release during the 1983 flood. (f) Irrigated fresh-cut flowers in the Imperial Valley, refrigerated trucks in the field, and a cargo jet at a local airport await shipments to distant markets. (g) The Colorado River (far lower left) stops short of its former delta. (h) Dam construction affects river discharge and sediment yields. [(a), (c), (d), and (f) Bobbé Christopherson. (b) and (e) Author. (g) *Terra* image, NASA/GSFC. (h) Data from U.S. Geological Survey, *National Water Summary 1984,* Water Supply Paper 2275 (Washington, DC: Government Printing Office, 1985), p. 55.]

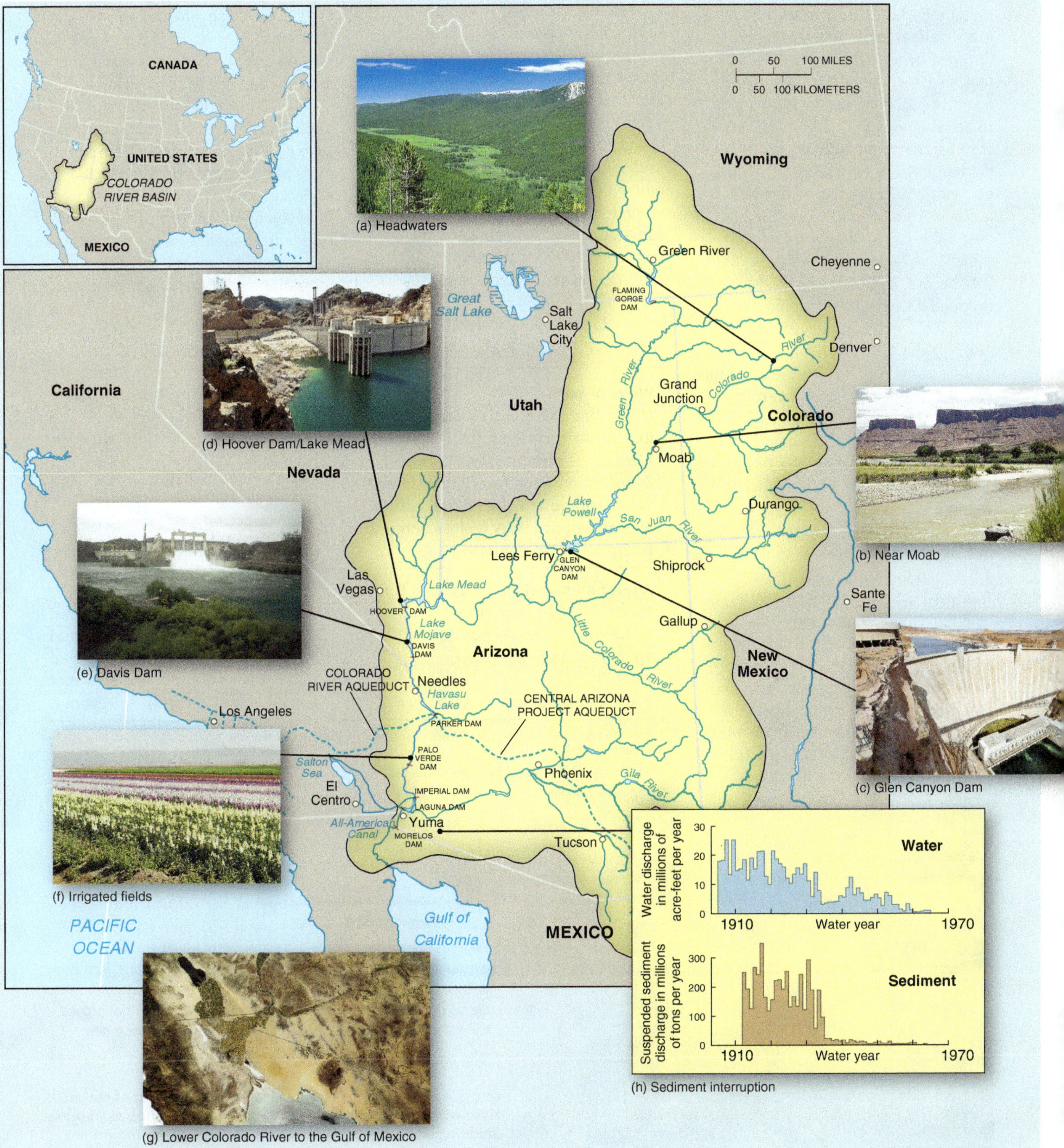

(a) Headwaters
(b) Near Moab
(c) Glen Canyon Dam
(d) Hoover Dam/Lake Mead
(e) Davis Dam
(f) Irrigated fields
(g) Lower Colorado River to the Gulf of Mexico
(h) Sediment interruption

Highly Variable River Flows

The flaw in all this planning and water distribution is that exotic streams have highly variable flows, and the Colorado is no exception. In 1917, the discharge measured at Lees Ferry totaled 24 million acre-feet (maf), whereas in 1934 it dropped by nearly 80%, to only 5.03 maf. In 1977, the discharge dropped again to 5.02 maf, but in 1984 it rose to an all-time high of 24.5 maf. Yet, in contrast, between 2000 and 2007 river discharge fell to exceptionally low levels, with 2002 dropping to a record low of 3.1 maf and 2003 to 6.4 maf. The 2003–2009 water years marked the lowest 7-year average flow in the total record. By 2009, lake levels dropped 43.6 m (143 ft) below the 375-m (1229-ft) capacity reached in 1983.

As a planning basis for the Colorado River Compact, the government used average annual river discharges from 1914 up to the compact signing in 1923, an exceptionally high average of 18.8 maf. That amount was perceived as more than enough for the upper and lower basins, each to receive 7.5 maf, and, later, for Mexico to receive 1.5 maf in the 1944 Mexican Water Treaty.

Should long-range planning be based on such a short time period on a river with highly variable flows? Tree-ring analyses of past climates have disclosed that the only other time Colorado River discharges were at the high 1914–1923 level was between A.D. 1606 and 1625.

Planners have consistently overestimated the dependable flows of the river, a situation that has created shortfalls as demand exceeds supply in the water budget (Table 15.1.1).

Presently, the seven states *ideally* want rights that total as much as 25.0 maf per year. When added to the guarantee for Mexico, this comes up to 26.5 maf of yearly demand (more than in the river by far). And six states share one common opinion: California's right to the water must be limited to a court-ordered 4.4 maf, which it exceeded every year until California's access to surplus Colorado River water was stopped in 2002. At present, no surplus exists during the ongoing western drought.

Colorado River Water Losses

Several natural processes associated with reservoirs and river flow regulation cause water losses and increase overall water demand, as observed in Table 15.1.1. Glen Canyon Dam, located 27.4 km (17 mi) north of Lees Ferry, the division point between the upper and lower basins, serves as an example.

Glen Canyon Dam was completed in 1963 and began impounding water in Lake Powell. Glen Canyon Dam's primary purpose, according to the Bureau of Reclamation, was to regulate flows between the upper and lower basins. Additional benefits are the production of hydroelectric power and a new recreation and tourism industry around Lake Powell, as previously inaccessible

TABLE 15.1.1 Estimated Colorado River Budget Through 2009

Water Demand	Quantity (maf)[a]
Upper basin, 7.5; lower basin, 7.5[b]	15.0
Central Arizona Project (rising to 2.8 maf)	1.0
Mexican allotment (1944 Treaty)	1.5
Evaporation from reservoirs (when full)	1.5
Bank storage at Lake Powell (when full)	0.5
Phreatophytic losses (water-demanding plants)	0.5
Budgeted total demand	**20.0 maf**
Average Annual Flows at Lees Ferry	
1906–1930 (average flow)	17.70 maf
1930–2003 (average flow)	14.10 maf
1990–2006 (average flow)	11.34 maf
2002	3.10 maf
2003–2009 (average flow)	7.34 maf

Source: Bureau of Reclamation and states of Arizona, California, and Nevada. A water year runs from October 1 to September 30.

[a]1 million acre-feet = 325,872 gallons, or 1.24 million liters, or 43,560 ft^3.

[b]In-basin consumptive uses are 75% agricultural.

CRITICAL THINKING 15.2

Water in the desert

Water is the major erosion and transport medium in the desert, although less frequently available than in humid regions. Respond to this concept. How is it possible for this to be true when deserts are so dry? What have you learned from this chapter that proves this statement true? Review some of the examples presented.

FIGURE 15.18 Sand and playa—natural sorting of materials. Playa of dissolved and finer materials (top of photo), sand dunes (middle), and coarser particles on slopes along the mountains (lower area) demonstrate the natural sorting that occurs in desert landscapes. [Bobbé Christopherson.]

desert scenery now can be reached by boat.

Lake Powell inundated 299 km (186 mi) of former river channel, which cuts through porous Navajo sandstone for much of its length. This sandstone absorbs an estimated 0.5 maf of the river's overall annual discharge as bank storage when the reservoir is full—the higher the lake level, the greater the loss into the sandstone. The reservoir also experiences evaporative losses as hot, dry winds flow across an open water body in an arid region. Evaporation accounts for another 0.5 maf of discharge lost annually from the reservoir, equaling about a third of all Colorado River reservoir evaporation losses—again, higher lake levels mean more losses.

In the upper reaches of large reservoirs where water levels fluctuate, as well as along reaches of regulated river flow below dams throughout the basin, sand bars and banks have become near-permanent features. Riparian vegetation has colonized those surfaces, including dense thickets of water-demanding plants called phreatophytes. These species extract and transpire an additional 0.5 maf per year of the river flow. The three water losses discussed here—bank storage, evaporation, and water use by vegetation—are an increasingly large percentage of the overall river discharge as drought conditions persist.

Ongoing Drought in the Colorado River Basin

Geosystems Now, a period of drought began in the Colorado River basin in 2000 and is ongoing. Continuing population growth throughout the region compounds the problem of increasing water demand coupled with decreased river flows. The drought itself appears directly linked to the shifting of subtropical high-pressure systems with climate change. This expansion of subtropical dry zones has given scientists a new approach to understanding drought. In regions such as the American Southwest, application of the term *drought* is changing to consideration of a change in *permanent average climate.* When future droughts occur—for example, during a Pacific Ocean La Niña condition—they will be in addition to this new base state of a drier climate. This is unprecedented in the climate record for this region.

A portion of a quote explains the severity of the drought: "The projected future climate of intensified aridity in the Southwest is caused by ... a poleward expansion of the subtropical dry zones. The drying of subtropical land areas ... imminent or already underway is unlike any climate state we have seen in the instrumental record" (Richard Seager et al., "Model projections of an imminent transition to a more arid climate in southwestern North America," *Science* 316, no. 5828 [May 25, 2007]: 1184).

Government stopgap measures to address water shortages came to head in December 2007 when the Department of the Interior signed a historic Record of Decision, an agreement representing a consensus among the stakeholders. This agreement sets the rules for dealing with water shortages during the drought and determining where water deliveries will be cut; establishes the drought as a basis for planning in times of water shortages; guides the operation rules for Lake Mead and Lake Powell for better coordination during the risk of drought; sets rules for distributing any surpluses, should they ever again arise; and encourages implementation of water conservation initiatives (see http://www.usbr.gov/lc/region/programs/strategies/RecordofDecision.pdf).

Conservation (using less water) and efficiency (using water more effectively) are not yet being stressed as ways to reduce the tremendous demand for water across the Southwest; perhaps this Record of Decision will lead to serious conservation and efficiency strategies. For instance, southern Nevada has launched a campaign to replace lawns with drought-tolerant xeroscaping—desert landscaping—but with 150,000 hotel rooms, we might question planning efforts. Las Vegas receives 90% of its water from the Colorado River and could save 30% of its usage by simply requiring water-efficient fixtures and appliances in hotels and residences.

Limiting or halting further metropolitan construction and population growth in the Colorado River Basin is not part of present strategies to lower demand. The focus remains on water supply. We might wonder what John Wesley Powell would think if he were alive today to witness such errant attempts to control the mighty and variable Colorado River. He foretold such a "heritage of conflict and litigation."

Desert Landscapes

Deserts are not wastelands, for they abound in specially adapted plants and animals. Moreover, the limited vegetation and distant vistas produce starkly beautiful landscapes. The desert's visual enchantment is keyed to sunlight and naked rock formations. The shimmering heat waves and related mirage effects are products of light refraction through layers of air that have developed a temperature gradient near the hot ground. And all deserts are not the same: For example, North American deserts have more vegetation cover than do the generally more barren Asian desert expanses, as we saw in Figure 15.14.

Deserts possess unique landscapes created by the interaction of intermittent precipitation events, weathering processes, and wind with the underlying geology. The buttes, pinnacles, and mesas of the U.S. Southwest are resistant horizontal rock strata that have eroded differentially. Removal of the less-resistant sandstone strata produces unusual desert sculptures—arches, windows, pedestals, and delicately balanced rocks. The upper layers of sandstone along the top of an arch or butte or tower are more resistant to weathering and protect the sandstone rock beneath (Figure 15.19).

The removal of surrounding rock through differential weathering leaves enormous buttes as residuals on the landscape. If you imagine a line intersecting the tops of the Mitten Buttes shown in Figure 15.20, you can gain some idea of the quantity of material that has been removed. These buttes exceed 300 m (1000 ft) in height, similar to the Chrysler Building in New York City or the U.S. Bank Tower (Library Tower) in Los Angeles.

(a) Balanced Rock in Utah

(b) Spider Rock in Arizona

FIGURE 15.19 Differential weathering.
(a) Balanced Rock in Arches National Park, Utah. The overall feature is 39 m (128 ft) tall and composed of Entrada sandstone. The balance-rock portion is 17 m (55 ft) tall and weighs 3255 metric tons (3577 short tons). (b) Spider Rock towers 244 m (800 ft) above the canyon floor of Canyon de Chelly, Arizona—a dramatic monolith; note the lighter color in the capstone rock. [(a) Author. (b) Bobbé Christopherson.]

Desert landscapes are places where stark erosional remnants can stand above the surrounding terrain as knobs or hills. Such a bare, exposed rock is an *inselberg* (island mountain), as exemplified by Uluru (Ayers) Rock in Australia. The formation is 348 m (1145 ft) high and 2.5 km long by 1.6 km wide (1.55 mi by 1.0 mi). Uluru Rock is sacred to Aboriginal peoples and is protected as part of Uluru National Park, established in 1950.

In semiarid regions, weak surface material may weather to a complex, rugged topography, usually of relatively low and varied relief. Such a landscape is a **badland**, probably so named because in the American West it offered little economic value and was difficult to traverse in 19th-century wagons. The Badlands region of the Dakotas and north-central Arizona (the Painted Desert) are so named.

Patterns of cross bedding, or *cross stratification*, are sometimes present in desert sandstone, formed by lithification of sand dunes that existed in ancient deserts. When such a dune was accumulating, sand cascaded down its slipface and distinct bedding planes (layers) were established that remained after the dune lithified (Figure 15.21). Ripple marks, animal tracks, and fossils also are found preserved in these sandstones.

Basin and Range Province

A physiographic province is a large region that is identified by several geologic or topographic traits. We find alternating basins (valleys) and mountain ranges characterizing the **Basin and Range Province** of the western United States (Figure 15.22). This region lies in the rain shadow of the Sierra Nevada, giving it a *midlatitude cold desert* climate. The climate and geology combine to give the province few permanent streams and *internal drainage* that lacks any outlet to the ocean.

The vast Basin and Range Province—almost 800,000 km^2 (300,000 mi^2)—was a major barrier to early settlers in their migration westward. The combination of desert climate and north–south-trending mountain ranges presented harsh challenges. Today, when you follow U.S. Highway 50 across Nevada, you cross five passes of more than 1950 m (6400 ft) and traverse numerous basins. Throughout the drive, you are reminded where you are by prideful signs that plainly state: "The Loneliest Road in America." It is difficult to imagine crossing this topography with wagons and oxen (Figure 15.22e).

John McPhee captured the feel of this desert province in his book *Basin and Range*:

> Supreme over all is silence. Discounting the cry of the occasional bird, the wailing of a pack of coyotes, silence—a great spatial silence—is pure in the Basin and Range. It is a soundless immensity with mountains in it. You stand . . . and look up at a high mountain front, and turn your head and look fifty miles down the valley, and there is utter silence.*

The Basin and Range formed as a result of tectonic processes. As the North American plate lumbered westward, it overrode former oceanic crust and hot spots at such a rapid pace that slabs of subducted material literally were run over. The crust was uplifted and stretched, creating a tensional landscape fractured by many faults. The present landscape consists of nearly parallel sequences: Some are tilted-fault blocks; others have pairs of faults on either side of *horsts* of upward-faulted blocks, which are the "ranges," and *grabens* of downward fault-bounded blocks, which are the "basins." Figure 15.22c shows this pattern of normal faults and tensional stretching.

*J. McPhee, *Basin and Range* (New York, NY: Farrar, Straus, Giroux, 1981), p. 46.

(a)

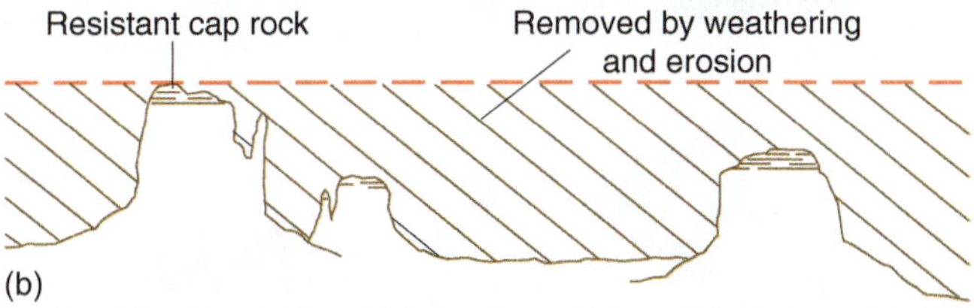

(b)

(c)

(d)

FIGURE 15.20 Monument Valley landscape. (a) Mitten Buttes, Merrick Butte, and rainbow in Monument Valley, Navajo Tribal Park, which have appeared as a backdrop in many movies, along the Utah–Arizona border. (b) A schematic of the tremendous removal of material by weathering, erosion, and transport. (c) Aerial view of Monument Valley. (d) View just east of (c), showing the residual spires of the Yie Bi Chei rock formation that ends with Totem Pole, 137 m (450 ft) tall and only 10 m (33 ft) wide. [(a) and (e) Author. (c) and (d) Bobbé Christopherson.]

(a) Valley of Fire

(b) Red Rocks

FIGURE 15.21 Cross bedding in sedimentary rocks.
The bedding pattern of cross stratification in these sandstone rocks tells us about dune environments before lithification (hardening into rock). [Bobbé Christopherson.]

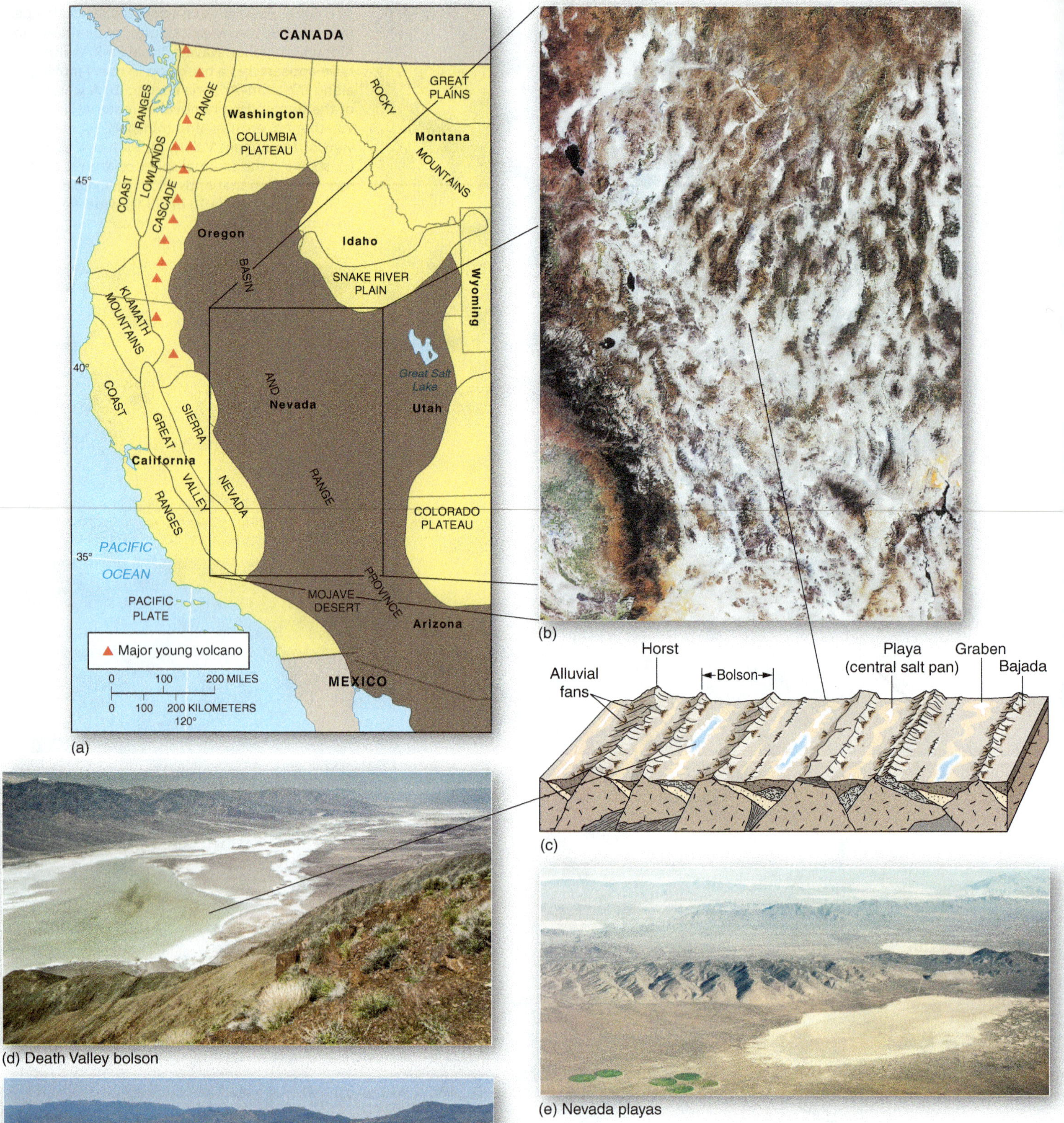

FIGURE 15.22 Basin and Range Province in the western United States.
(a) Map and (b) *Landsat* image of the area. Scientific discoveries extend this province south through northern and central Mexico. (c) Parallel faults produce a series of ranges and basins, including bolsons. (d) Death Valley features a central playa, parallel mountain ranges, alluvial fans, and bajadas along the bases of the ranges. (e) Playas in each basin. (f) "The Loneliest Road in America," Highway U.S. 50, slashes westward through the province. [(b) NASA. (d) Author. (e) Bobbé Christopherson.]

Basin-and-range relief is abrupt, and rock structures are angular and rugged. As the ranges erode, transported materials accumulate to great depths in the basins, gradually producing extensive desert plains. The basins' elevations average roughly 1200–1500 m (4000–5000 ft) above sea level, with mountain crests rising higher by another 900–1500 m (3000–5000 ft). Death Valley, California, is the lowest of these basins, with an elevation of –86 m (–282 ft). However, to the west of the valley, the Panamint Range rises to 3368 m (11,050 ft) at Telescope Peak—almost 3.5 vertical kilometers (2.2 mi) of desert mountain relief.

In Figure 15.22c and d, note the **bolson** label, a slope-and-basin area between the crests of two adjacent ridges in a dry region of internal drainage. Figure 15.22c also identifies a *playa* (central pan), a *bajada* (coalesced alluvial fans), and a mountain front in retreat from weathering and erosional attack. A *pediment* is an area of bedrock that is layered with a thin veneer, or coating, of alluvium. It is an erosional surface, as opposed to the depositional surface of the bajada.

Vast arid and semiarid lands remain an enigma on the water planet. They challenge our need for water and demand technological innovations to access it. Yet these lands hold a fascination, perhaps because they are so lacking in the moisture that infuses our lives.

CRITICAL THINKING 15.3

Relating to the Colorado River crisis

Relative to the Colorado River system, define, compare, and contrast what you think is meant by a *supply strategy* or a *demand strategy*. How does each strategy relate to the present Colorado River water budget (Table 15.1.1, Focus Study 15.1)? What political and economic forces do you think are at play in planning for the river and its dependent region? Who are the stakeholders? With your analysis, briefly speculate as to what you think needs to happen.

GEOSYSTEMS CONNECTION

Wind is an agent of geomorphic change, causing erosion, transportation, and deposition of sediments, sometimes redistributing materials among continents. Sand accumulations form dunes of several types, including crescentic, linear, and star dunes. Wind also transports fine-grained clays and silts to form loess deposits. Wind and scarce water operate in Earth's desert regions, where both sandy and rocky surfaces are found. In the next chapter, we examine coastal processes involved with the agents of waves, tides, and currents as they sculpt Earth's erosional and depositional coastlines. A significant portion of the human population lives in coastal areas, making hazard perception, planning, and the future, given a rising sea level.

KEY LEARNING CONCEPTS REVIEW

Desertification is the ongoing process that leads to an unwanted expansion of the Earth's desert lands; presently, some 1 billion people are affected, and a new international effort is framed under the UNDDD.

desertification (p. 433)

1. What is meant by desertification? Using the maps in Figures GN 15.1 and 15.14 and the text description, locate several of the regions affected by desert expansion.

- ***Characterize*** **the unique geomorphic work accomplished by wind and eolian processes.**

The movement of the atmosphere in response to pressure differences produces wind. Wind is a geomorphic agent of erosion, transportation, and deposition. **Eolian** processes modify and move sand accumulations along coastal beaches and deserts. Wind's ability to move materials is small compared with that of water and ice.

eolian (p. 434)

2. Who was Ralph Bagnold? What was his contribution to eolian studies?
3. Explain the term *eolian* and its application in this chapter. How would you characterize the ability of the wind to move material?

- ***Describe*** **eolian erosion, including deflation, abrasion, and the resultant landforms.**

Two principal wind-erosion processes are **deflation,** the removal and lifting of individual loose particles, and **abrasion,** the

"sandblasting" of rock surfaces with particles captured in the air. Deflation literally blows away loose or noncohesive sediment and works with rainwater to form a surface resembling a cobblestone street: a **desert pavement** that protects underlying sediment from further deflation and water erosion. Wherever wind encounters

loose sediment, deflation may remove enough material to form basins. Called **blowout depressions,** they range from small indentations less than a meter wide up to areas hundreds of meters wide and many meters deep. Rocks that bear evidence of eolian erosion are **ventifacts**. On a larger scale, deflation and abrasion are capable of streamlining rock structures, leaving behind distinctive rock formations or elongated ridges called **yardangs**.

deflation (p. 457)
abrasion (p. 457)
desert pavement (p. 457)
blowout depressions (p. 457)
ventifacts (p. 458)
yardangs (p. 458)

4. Describe the erosional processes associated with moving air.
5. Explain deflation and provide an overview of each hypothesis on the processes that produce desert pavement.
6. How are ventifacts and yardangs formed by the wind?

- ***Describe*** eolian transportation, and ***explain*** saltation and surface creep.

Wind exerts a drag or frictional pull on surface particles until they become airborne. The finer material suspended in a dust storm is lifted much higher than the coarser particles of a sandstorm; only the finest dust particles travel significant distances. Saltating particles crash into other particles, knocking them both loose and forward. The motion of **surface creep** slides and rolls particles too large for saltation.

surface creep (p. 459)

7. Differentiate between a dust storm and a sandstorm.
8. What is the difference between eolian saltation and fluvial saltation?
9. Explain the concept of surface creep.

- ***Identify*** the major classes of sand dunes, and ***present*** examples within each class.

In arid and semiarid climates and along some coastlines where sand is available, dunes accumulate. A **dune** is a wind-sculpted accumulation of sand. An extensive area of dunes, such as that found in North Africa, is characteristic of an **erg desert,** or **sand sea**. When saltating sand grains encounter small patches of sand, their kinetic energy (motion) is dissipated, and they start to accumulate into a dune. As height increases above 30 cm (12 in.), a steeply sloping **slipface** on the lee side and characteristic dune features are formed. Dune forms are broadly classified as *crescentic*, *linear*, and *star*.

dune (p. 460)
erg desert (p. 460)
sand sea (p. 460)
slipface (p. 460)

10. What is the difference between an erg and a reg desert? Which type is a sand sea? Are all deserts covered by sand? Explain.
11. What are the three classes of dune forms? Describe the basic types of dunes within each class. What do you think is the major shaping force for sand dunes?
12. Which form of dune is the mountain giant of the desert? What are the characteristic wind patterns that produce such dunes?

- ***Define*** loess deposits and their origins, locations, and landforms.

Eolian-transported materials contribute to soil formation in distant places. Windblown **loess** deposits occur worldwide and can develop into good agricultural soils. These fine-grained clays and silts are moved by the wind many kilometers and are redeposited as an unstratified, homogeneous blanket of material.

loess (p. 464)

13. How are loess materials generated? What form do they assume when deposited?
14. Name a few examples of significant loess deposits on Earth.

- ***Portray*** desert landscapes, and ***locate*** these regions on a world map.

Dry and semiarid climates occupy about 35% of Earth's land surface. The spatial distribution of these dry lands is related to subtropical high-pressure cells between 15° and 35° N and S, to rain shadows on the lee side of mountain ranges, or to areas at great distance from moisture-bearing air masses, such as central Asia.

Precipitation events are rare, yet running water is still the major erosional agent in deserts. When precipitation events do occur, a dry streambed fills with a torrent, a **flash flood.** Depending on the region, such a dry streambed is known as a **wash,** an *arroyo* (Spanish), or a *wadi* (Arabic). As runoff water evaporates, salt crusts may be left behind on the desert floor. This intermittently wet and dry low area in a region of closed drainage is a **playa,** site of an *ephemeral lake* when water is present.

In arid climates, a prominent landform is the **alluvial fan** at the mouth of a canyon where an ephemeral stream channel exits into a valley. The fan is produced by flowing water that abruptly loses velocity as it leaves the constricted channel of the canyon and deposits a layer of sediment along the mountain block. A continuous apron, or **bajada,** may form if individual alluvial fans coalesce. In a desert area, weak surface material may weather to a complex, rugged topography, usually of relatively low and varied relief. Such a landscape is a **badland.** A *province* is a large region that is characterized by several geologic or physiographic traits. The **Basin and Range Province** of the western United States consists of alternating basins and mountain ranges. A slope-and-basin area between the crests of two adjacent ridges in a dry region of internal drainage is a **bolson.**

flash flood (p. 468)
wash (p. 468)
playa (p. 469)
alluvial fan (p. 469)
bajada (p. 469)
badland (p. 474)
Basin and Range Province (p. 474)
bolson (p. 477)

15. Characterize desert energy and water-balance regimes. What are the significant patterns of occurrence for arid landscapes?
16. How would you describe the water budget of the Colorado River? What was the basis for agreements regarding distribution of the river's water? Why has thinking about the river's discharge been so optimistic? Provide an overview of current conditions of the Colorado River system in light of the ongoing drought and increasing water demands.
17. Describe a desert bolson from crest to crest. Draw a simple sketch with the components of the landscape labelled.
18. Where is the Basin and Range Province? Briefly describe its appearance and character.

MASTERING GEOGRAPHY

Visit www.masteringgeography.com for several figure animations, satellite loops, self-study quizzes, flashcards, visual glossary, case studies, career links, textbook reference maps, RSS Feeds, and an ebook version of *Geosystems*. Included are many geography web links to interesting Internet sources that support this chapter.

16

Oceans, Coastal Systems, and Wind Processes

The Pancake Rocks near Punakaiki on the West Coast of New Zealand's South Island are limestone layers formed about 30 million years ago from the lithified shells of marine organisms. Subsequent uplift along this tectonically active coastline exposed the limestone strata to exogenic processes. The work of wave action, wind, and mildly acidic precipitation in this marine west coast climate have sculpted the layers, forming dramatic vertical "blowholes" through which seawater moves at high tide. This chapter discusses coastal processes and landforms, and the erosive forces of wave action and wind. [Shay Yacobinski/Shutterstock.]

KEY LEARNING concepts

After reading the chapter, you should be able to:

- ***Describe*** the chemical composition and physical structure of the ocean.
- ***Identify*** the components of the coastal environment, ***define*** mean sea level, and ***explain*** the actions of tides.
- ***Describe*** wave motion at sea and near shore, and ***explain*** coastal straightening and coastal landforms.
- ***Describe*** barrier beaches and islands and their hazards as they relate to human settlement.
- ***Describe*** the nature of coral reefs and coastal wetlands, and ***assess*** human impacts on these living systems.
- ***Describe*** eolian transport of dust and sand, and ***discuss*** eolian erosion and the resultant landforms.
- ***Explain*** the formation of sand dunes, and ***describe*** loess deposits and their origins.

Sand Dunes Prevent Coastline Erosion during Hurricane Sandy

During the winter of 2013, in the ongoing aftermath of Hurricane Sandy, many residents along New Jersey's coastline added their discarded Christmas trees to carefully stacked lines of trees intended to act as "seeds" for new sand dune formation along several area beaches. The hope is that the trees will catch windblown sand to begin the dune formation process, in one of many such restoration efforts along the Atlantic coast. In the face of Sandy's winds, houses and neighborhoods with protective dunes in place experienced less damage than those that were more exposed to and closer to the ocean.

Dune Protection versus Ocean Views The effectiveness of dune systems as protection from wave erosion and storm surge during Hurricane Sandy, far from being a subtle statistical phenomenon, was easily observed by local residents. However, the fostering of large and sometimes obtrusive sand dunes near the shoreline is controversial in coastal communities with million-dollar homes. For such dunes to function as barriers to erosion, they must sit between oceanfront property and the sea, thus blocking ocean views and decreasing property values (Figure GN 16.1). For many landowners, establishing dunes for storm protection means financial loss. in the short term, even if long-term protection is the result.

Coastal Dune Geomorphology Coastal sand dunes originate from sediment supplied by the work of ocean waves and by fluvial processes that move sediment onto deltas and estuaries. Once sand is deposited on shore, it is reworked by wind processes into the shape of dunes. Dunes along seacoasts are either *foredunes*, where sand is pushed up the seaward-facing slope, or *backdunes*, which form further away from the beach and are protected from onshore winds; backdunes are more stable and may be hundreds of years old. Most areas of coastal dunes are relatively small in size (especially when compared with desert dune fields that may cover large portions of continents).

Along the Atlantic coast, foredunes are moving inland as sea level rises and storm energy increases with climate change. In developed areas, the foredunes cannot retreat inland without impinging on human development. When storms occur, dune movement is intensified, and either dune erosion or sand deposition, or both, occurs within the developed area of the coast (Figure GN 16.2).

Dune Restoration Efforts The establishment of new foredunes replenishes the sand supply and protects structures and infrastructure, making this a potentially worthwhile investment of money and effort for communities along the New Jersey shoreline. Many experts point out that dunes are *not* a guarantee of storm protection and that Sandy's winds and storm surge were strong enough to erode some large natural dune systems along the Atlantic Seaboard. However, in Bradley Beach, New Jersey, where the storm eroded several miles of restored dunes about 4.6 m (15 ft) in height, the community still escaped excessive damage since the dunes absorbed much of the storm's impact.

Thus, local communities are supporting dune restoration, as evidenced by the Christmas tree initiative. Because vegetation is important for dune stabilization, the planting of grasses is another protective strategy being embraced by New Jersey residents. In this chapter, we discuss coastal systems, wind processes, and dune formation processes.

▲**Figure GN 16.2 Coastal damage from Sandy in Mantoloking, New Jersey.** View looking west before and after Hurricane Sandy. The yellow arrow points to the same feature in each image. [USGS.]

▲**Figure GN 16.1 Constructed dunes.** Restored sand dunes shield homes in Mantoloking, New Jersey, from an incoming nor'easter a few weeks after Hurricane Sandy. [Sharon Karr/FEMA.]

GEOSYSTEMS NOW ONLINE Go to Chapter 16 on the *MasteringGeography* website (www.masteringgeography.com) for resources and activities. For information and links to research on dunes in New Jersey and along the Atlantic coast, see http://marine.rutgers.edu/geomorph/geomorph/_pages/dunes.html. MG

Earth's vast oceanic, atmospheric, and lithospheric systems reach a meeting point along seacoasts. At times, the ocean attacks the coast in a stormy rage of erosive power; at other times, the moist sea breeze, salty mist, and repetitive motion of the water are gentle and calming. The coastlines are areas of dynamic change and beauty.

Commerce and access to sea routes, fishing, and tourism prompt many people to settle near the ocean. In fact, about 40% of Earth's population lives within 100 km (62 mi) of an ocean coast (Figure 16.1). In the United States, about 50% of the people live in areas designated as *coastal* (this includes the Great Lakes). A 2007 study determined that, globally, 634 million people live in low-elevation coastal areas that are less than 30 m (98 ft) above sea level, meaning that 1 in 10 people on Earth live in a zone that is highly vulnerable to tropical storm damage, flooding, and rising sea level. Given this population distribution, an understanding of coastal processes and landforms is important for planning and development.

Pollution is also a major concern in coastal areas. According to the United Nations Environment Program (UNEP), about 6 trillion gallons of sewage is discharged into coastal waters each year, along with about 50,000 tons of toxic organic chemicals and 68,000 tons of toxic metals. Aside from the potentially dangerous biological hazards it poses, coastal and marine pollution affects coastal tourism, which is a large component of the economy in many coastal cities.

Oceans are intricately linked to life on the planet and act as a buffer for changes in other Earth systems, absorbing excess atmospheric carbon dioxide and thermal energy; yet climatic shifts over the past several decades may now be overwhelming oceanic systems. In 2009, the second Conference on the Ocean Observing System for Climate took place in Venice, Italy, summarizing the profound changes that had transpired in the ocean system over the 10 years since the first conference—see http://www.oceanobs09.net/.

Wind is an important geomorphic agent along coastlines as well as in other environments. Although wind's ability to erode, transport, and deposit materials is small compared to that of water and ice, wind processes can move significant quantities of sand and shape landforms. Wind contributes to soil formation (discussed in Chapter 18), fills the atmosphere with dust that crosses the oceans between continents (discussed in Chapter 3), and spreads living organisms. One study found related mosses, liverworts, and lichens distributed among islands thousands of miles apart in the Southern Ocean.

In this chapter: After beginning with a brief look at our global oceans and seas, we discuss the physical and chemical properties of seawater. Next we look at coastal systems, discussing tides, waves, coastal erosion, and depositional landforms such as beaches and barrier islands. A systems framework focusing on inputs (components and driving forces), actions (movements and processes), and outputs (results and consequences) organizes our discussion. We also look at the important organic processes that produce corals, salt marshes, and mangroves. Lastly, we examine wind processes—first, wind erosion and the resulting landforms, and then wind deposition, sand dunes, and sand seas.

▼**Figure 16.1 Bondi Beach, Sydney, Australia.** In Australia, over 85% of the population lives within 50 km (31 mi) of a coast. The country as a whole has over 36,000 km (22,000 mi) of coastline and over 10,000 beaches. Pictured here are surfers and beachgoers at one of Sydney's most popular beaches. [Patrick Ward/Corbis.]

Global Oceans and Seas

The oceans are one of Earth's last great scientific frontiers. Remote sensing from orbiting spacecraft and satellites, aircraft, surface vessels, and submersibles now provides a wealth of data and a new capability for understanding oceanic systems. Earlier chapters have touched on a number of topics related to oceans. We discussed sea-surface temperatures in Chapter 5 (please review Figure 5.9) and ocean currents, both surface and deep, in Chapter 6 (Figures 6.18 and 6.20). The location and surface area of the world's oceans is in Chapter 9 (Figure 9.3).

A *sea* is generally smaller than an ocean and tends to be associated with a landmass. Figure 16.2 shows the world's principal seas (a more detailed map of over 60 seas is on the *MasteringGeography* website). The term *sea* may also refer to a large inland, salty body of water, such as the Black Sea in Europe. The National Ocean Service coordinates many scientific activities related to oceans; information is available at http://www.nos.noaa.gov/.

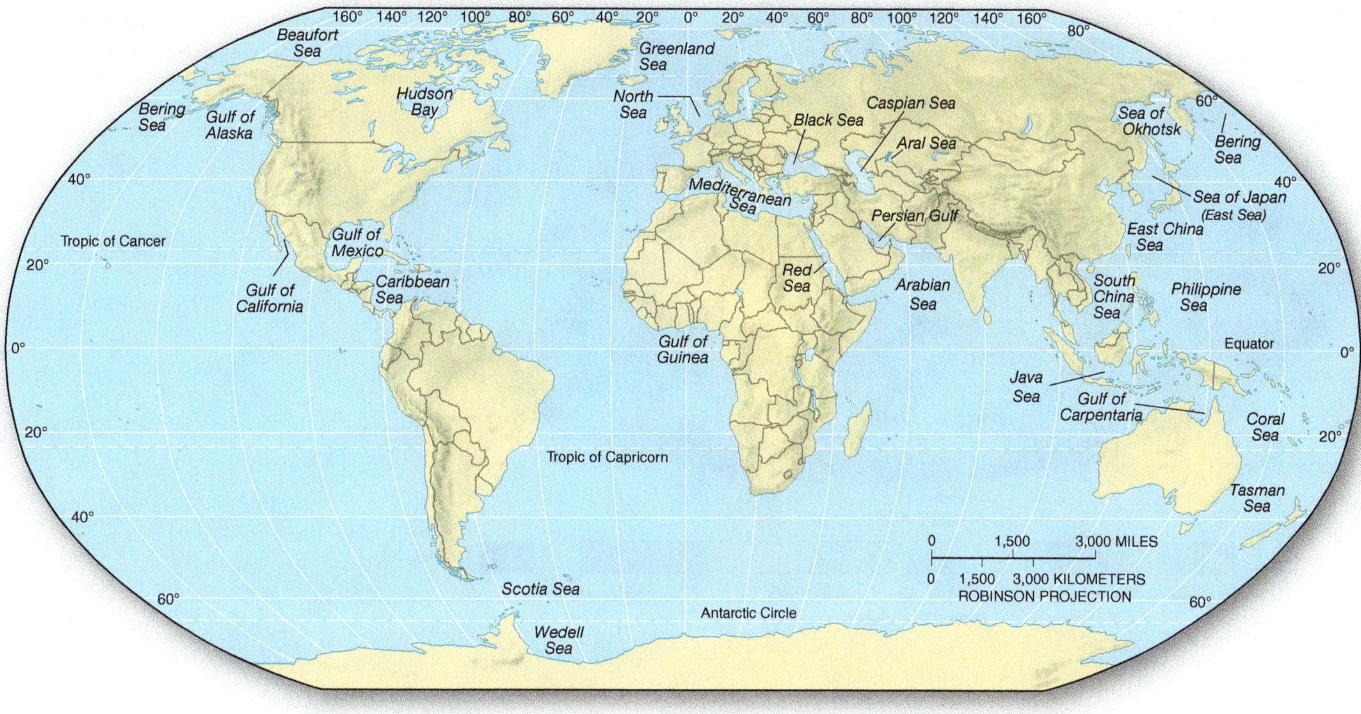

▲Figure 16.2 **Principal seas of the world.**

Properties of Seawater

As mentioned in Chapter 13, water dissolves at least 57 of the 92 elements found in nature and is known as the "universal solvent." In fact, most natural elements and the compounds they form are found in the world's oceans and seas as dissolved solids, or *solutes*. Thus, seawater is a solution, and the concentration of dissolved solids in that solution is known as **salinity**, commonly expressed as dissolved solids per volume. Water, you recall, moves continuously through the hydrologic cycle, driven by energy from the Sun, but the dissolved solids remain in the ocean. The water you drink today may have water molecules in it that not long ago were in the Pacific Ocean, in the Yangtze River, in groundwater in Sweden, or airborne in the clouds over Peru.

Chemical Composition The uniform chemical composition of seawater was first demonstrated in 1874 by scientists sampling seawater as they sailed around the world aboard the British HMS *Challenger*. The ocean continues to be a remarkably homogeneous mixture today—the ratio of individual salts does not change, despite minor fluctuations in overall salinity.

The chemical composition of seawater is affected by the atmosphere, minerals, bottom sediments, and living organisms. For example, the flows of mineral-rich water from hydrothermal (hot water) vents in the ocean floor ("black smokers," as seen in Figure 11.9) alter ocean chemistry in that area. However, the continuous mixing among the interconnected ocean basins keeps the overall chemical composition mostly uniform. Until recently, experts thought the chemistry of seawater to have been fairly constant over the past 500 million years. However, samples of ancient seawater gathered from fluid inclusions in marine formations, such as limestone and evaporite deposits, suggest that slight chemical variations in seawater have occurred over time. The variations are consistent with changes in seafloor spreading rates, volcanic activity, and sea level.

Seven elements account for more than 99% of the dissolved solids in seawater. In solution they take their ionic form (shown here in parentheses): chlorine (as chloride ions, Cl^-), sodium (as Na^+), magnesium (as Mg^{2+}), sulfur (as sulfate ions, SO_4^{2-}), calcium (as Ca^{2+}), potassium (as K^+), and bromine (as bromide ions, Br^-). Seawater also contains dissolved gases (such as carbon dioxide, nitrogen, and oxygen), suspended and dissolved organic matter, and a multitude of trace elements.

Commercially, only sodium chloride (common table salt), magnesium, and bromine are extracted in any significant amount from the ocean. Mining of minerals from the seafloor is technically feasible, although it remains uneconomical.

Average Salinity Scientists express the worldwide average salinity of seawater in several ways:

- 3.5% (parts per hundred)
- 35,000 ppm (parts per million)
- 35,000 mg/L
- 35 g/kg
- 35‰ (parts per thousand); this is the most common notation

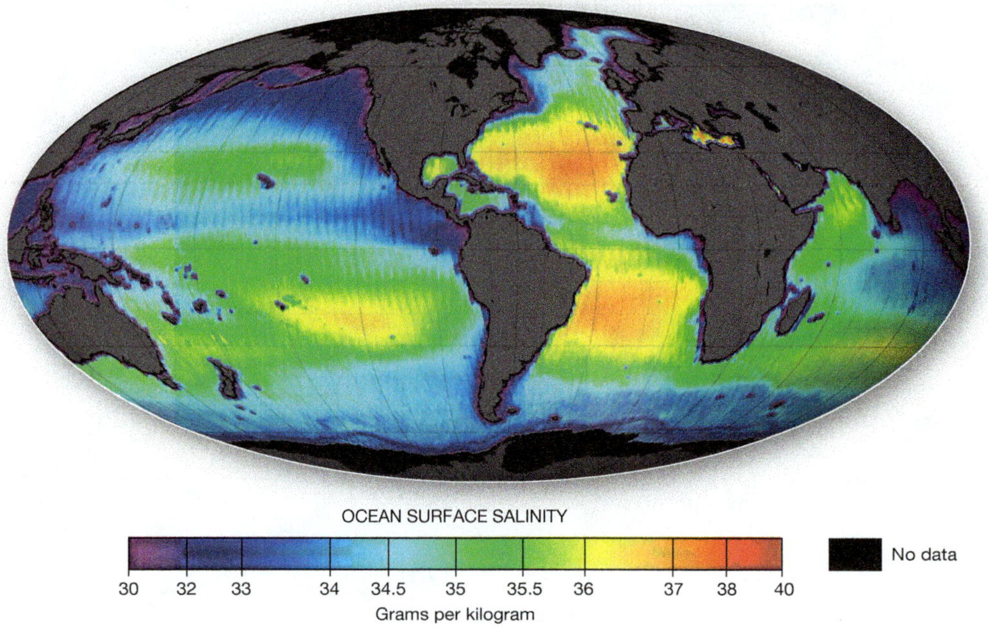

Figure 16.3 Ocean salinity. Composite image of global ocean surface salinity from August 2011 to July 2012 using data from the *Aquarius* satellite, in orbit since 2011. [NASA.]

Salinity worldwide normally varies between 34% and 37%; variations are attributable to atmospheric conditions above the water and to the volume of freshwater inflows. Figure 16.3 shows an image of global variations in salinity (see http://aquarius.nasa.gov/). High annual precipitation over equatorial oceans leads to slightly lower than average salinity values in those regions of about 34.5‰ (note the lower values in the Pacific Ocean along the Intertropical Convergence Zone in Figure 16.3). In subtropical oceans—where evaporation rates are greatest because of the influence of hot, dry subtropical high-pressure cells—salinity is more concentrated. In these regions, salinity is presently increasing as rising temperatures lead to increased evaporation rates.

In general, oceans are lower in salinity near landmasses because of freshwater inputs. The term **brackish** applies to water that is less than 35% salts. Extreme examples include the Baltic Sea (north of Poland and Germany) and the Gulf of Bothnia (between Sweden and Finland), which average 10% or less salinity because of heavy freshwater runoff and low evaporation rates. In general, the high-latitude oceans have been freshening over the past decade with increased melting of glaciers and ice sheets, as mentioned in Chapter 6.

In contrast, the Sargasso Sea, within the North Atlantic subtropical gyre, averages 38‰ (Figure 16.3). The Persian Gulf has a salinity of 40‰ as a result of high evaporation rates in a nearly enclosed basin. The term **brine** is applied to water that exceeds the average of 35‰ salinity. Deep pockets, or "brine lakes," along the floor of the Red Sea and the Mediterranean Sea register up to a salty 225‰.

Physical Structure and Human Impacts

The basic physical structure of the ocean consists of three horizontal layers (Figure 16.4). In the surface layer, warmed by the Sun, mixing is driven by winds. In this *mixing zone*, which represents only 2% of the oceanic mass, variations in water temperature and solutes are blended rapidly. Below the mixing zone is the *thermocline transition zone*, a region more than 1 km deep that lacks the motion of the surface and has a gradient in temperature that decreases with depth. Friction at these depths dampens the effect of surface currents. In addition, colder water temperatures at the thermocline

GEOreport 16.1 The Mediterranean Sea is getting saltier

The Mediterranean Sea is warming at a faster rate than the oceans—notice the high salinity levels in Figure 16.3. Increased salinity and temperatures are occurring in the deep layers, below 600 m (1968 ft) in depth. Saltier conditions change the water density and cause net outflows past the Strait of Gibraltar, thus blocking natural mixing with the Atlantic Ocean. As climate changes, warming is disrupting the natural mixing in large water bodies, including lakes as reported in the *Geosystems Now* in Chapter 9.

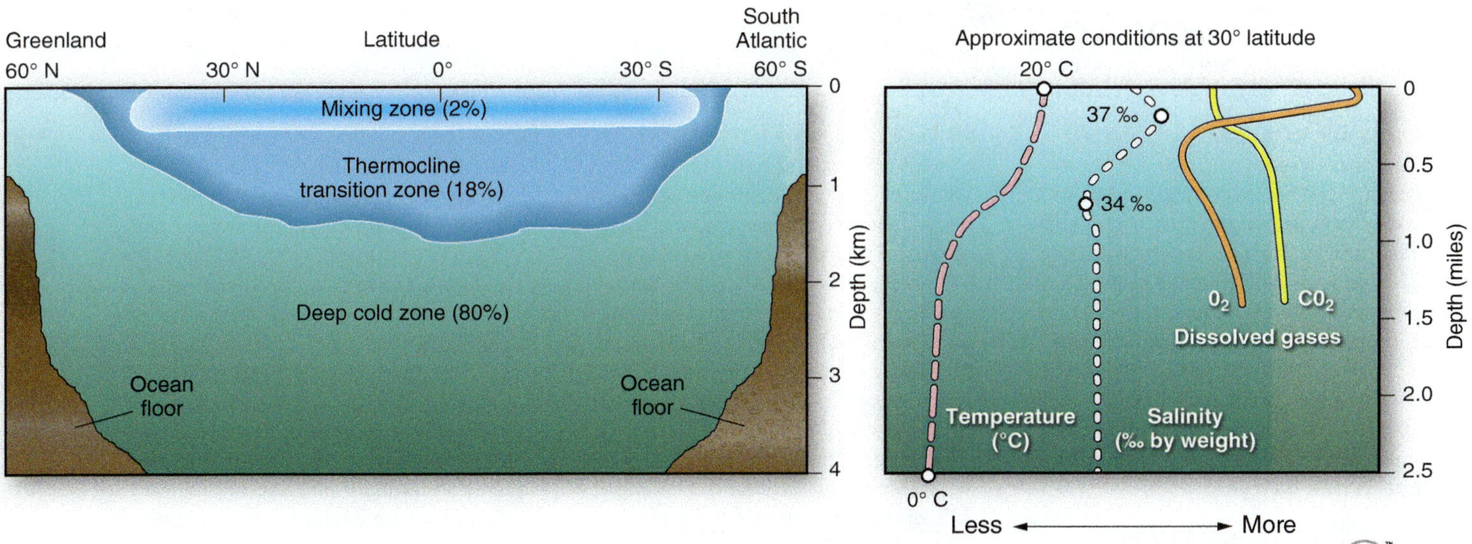

▲**Figure 16.4** **The ocean's physical structure.** Schematic of the average physical structure observed throughout the ocean's vertical profile as sampled along a line from Greenland to the South Atlantic. Temperature, salinity, and dissolved gases are shown plotted by depth.

MG Animation Midlatitude Productivity

transition zone's lower margin tend to inhibit any convective movements.

In the top two layers of the ocean, average temperature, salinity, dissolved carbon dioxide, and dissolved oxygen all vary with increasing depth. In contrast, from a depth of 1–1.5 km (0.6–0.9 mi) to the ocean floor, temperature and salinity values are quite uniform. Temperatures in this *deep cold zone* are near 0°C (32°F), and the coldest water is generally along the ocean bottom. However, seawater in the deep cold zone does not freeze because the freezing point of seawater is lower than that of plain water, owing to the presence of dissolved salts. (At the surface, seawater freezes at about −2°C, or 28.4°F.)

Ocean Acidification The ocean also reflects the changing composition of Earth's atmosphere. As the oceans absorb excess carbon dioxide from the atmosphere, the process of carbonation (discussed in Chapter 13) forms carbonic acid in seawater, resulting in a lowering of the ocean pH—an acidification. A more acid ocean will cause certain marine organisms such as corals and some plankton to have difficulty maintaining external calcium carbonate structures. The ocean's average pH today is 8.2, but pH could decrease by 0.4 to 0.5 units this century. The pH scale is logarithmic, so a decrease of 0.1 equals a 30% increase in acidity (see the pH scale in Chapter 18, Figure 18.7). Oceanic biodiversity and food webs will respond to this change in unknown ways.

Pollution The world's oceans have become repositories for much of the world's waste, whether discarded intentionally into the ocean or accidentally leaked or spilled. Marine and terrestrial oil pollution is a continuing problem in coastal regions, as waste oil seeps and leaks into oceans from improper disposal, and spills into oceans from offshore drilling and transportation problems.

When oil spills into seawater, it first spreads out on the surface, forming an oil slick that may be cohesive or may be broken up by rough seas. The slick may drift over large areas of open ocean, affecting marine habitat, or toward shorelines, impacting coastal wetlands and associated wildlife. The oil may partially evaporate, making the remaining slick denser, may partially dissolve into the water, or may combine with particulate matter and sink to the bottom. Over the long term, some of the oil breaks down through processes driven by sunlight and through decomposition by microorganisms—the rate of this deterioration depends on temperature and the availability of oxygen and nutrients. Along a coastline, oil spreads over beach sediments and drifts into coastal wetlands, contaminating and poisoning aquatic organisms and wildlife, and disrupting human activities such as fishing and recreation. Focus Study 16.1 discusses the system-wide causes and impacts of offshore oil spills.

Coastal System Components

Although many of Earth's surface features, such as mountains and crustal plates, were formed over millions of years, most of Earth's coastlines are relatively young and undergoing continuous change. Land, ocean, atmosphere, Sun, and Moon interact to produce

Focus Study 16.1 Pollution

Coastal Oil Spills: A Systems Perspective

An oil tanker splits open at sea and releases its petroleum cargo. This is carried by ocean currents toward shore, where it coats coastal waters, beaches, and animals. In response, concerned citizens mobilize and try to save as much of the spoiled environment as possible (Figure 16.1.1). But the real problem goes far beyond the immediate consequences of the spill. What are the spatial and systems relationships between an oily bird, coastal restoration, energy demand and consumption, and ongoing global climate change?

Prince William Sound, Alaska

In Prince William Sound off the southern coast of Alaska, in clear weather and calm seas, the *Exxon Valdez,* a single-hulled supertanker operated by Exxon Corporation, struck a reef in 1989. The tanker spilled 42 million liters (11 million gallons) of oil. It took only 12 hours for the *Exxon Valdez* to empty its contents, yet a complete cleanup is impossible, and cleanup costs and private claims have exceeded $15 billion. Scientists are still finding damage and residual spilled oil. Eventually, more than 2400 km (about 1500 mi) of sensitive coastline were ruined, affecting three national parks and eight other protected areas in Alaska.

The death toll of animals was massive: At least 5000 sea otters died, or about 30% of resident otters in the affected areas; about 300,000 birds and uncounted fish, shellfish, plants, and aquatic microorganisms also perished. While some species, such as the Bald Eagle and Common Murre, have recovered, the Pacific herring is still in significant decline, as are the harbor seals. Sublethal effects—namely, mutations—are now appearing in fish. More than two decades later, oil remains in mudflats and marsh soils beneath rocks.

▲Figure 16.1.1 **Oily birds.** Brown Pelicans (*Pelecanus occidentalis*) struggle as they wait for cleaning after the 2010 BP oil spill in the Gulf of Mexico. [Jim Celano, Reuters.]

On average, 27 oil-releasing accidents occur every day, 10,000 a year worldwide, ranging from a few disastrous spills to numerous small ones (Figure 16.1.2a). Fifty spills equal to the *Exxon Valdez* or larger have occurred since 1970. In addition to oceanic oil spills, people improperly dispose of crankcase oil from their automobiles in a volume that annually exceeds all of these tanker spills.

Louisiana Coast, Gulf of Mexico

The largest oil spill in U.S. history occurred in 2010 in the Gulf of Mexico (Figure 16.1.2b). Somewhere between 50,000 and 95,000 barrels of oil a day, for 86 days, exploded from a broken wellhead on the seafloor; this is 2.1 to 4.0 million gallons a day, or the equivalent of a 1989 *Exxon Valdez* oil spill every 4 days. The differences in the spill-rate estimates are a consequence of the concerted efforts of the oil company to cover up the true nature of the spill because of potential financial liabilities.

The *Deepwater Horizon* well, at an ocean depth of 1.6 km (1 mi), was one of the deepest drilling attempts ever, and much of the technology of the operation remains untested or unknown. Scientists are analyzing many aspects of the tragedy to determine the extent of the biological effects on the open water, beaches, wetlands, and wildlife of the Gulf (Figure 16.1.2c). The seafloor in the affected region is essentially dead, covered in oil. Early reports of the effectiveness of microbial digestion in cleaning up the oil in the water column were greatly exaggerated; these processes perhaps eliminated only 10% of the spill mass.

Questions Going Forward

The immediate effect of global oil spills on wildlife is contamination and death. But the issues involved are much bigger than dead birds. Exercising the perspective gained from the systems approach of this text, let us ask some fundamental questions concerning international demand for and trafficking in oil:

- Why was the single-hulled oil tanker in vulnerable Alaskan waters? Why is it necessary to drill in ultra-deep water in the Gulf using untested technology and equipment?
- Why is petroleum imported into the continental United States in such enormous quantities? Is the demand for petroleum products based on real need and maximum efficiency? Do waste and inefficiency occur?

the tides, currents, and waves responsible for the erosional and depositional features along the continental margins.

Inputs to the coastal environment include many elements discussed in previous chapters:

- *Solar energy* input drives the atmosphere and the hydrosphere. Conversion of insolation to kinetic energy produces prevailing winds, weather systems, and climate.
- *Atmospheric winds,* in turn, generate ocean currents and waves, key inputs to the coastal environment.
- *Climatic regimes,* which result from insolation and moisture, strongly influence coastal geomorphic processes.

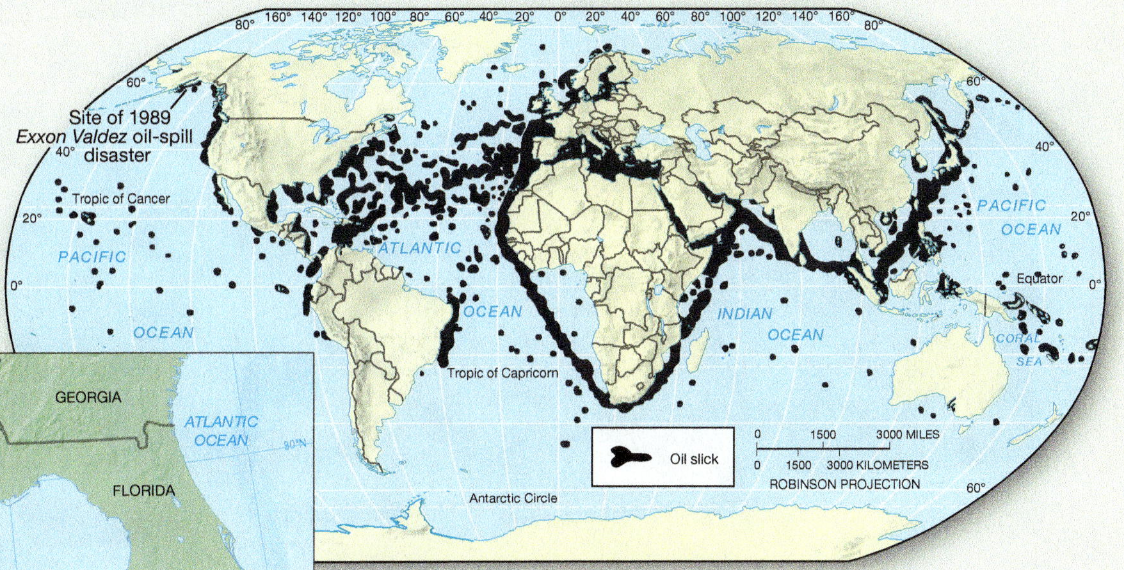

(a) Location of oil slicks worldwide in the 1990s.

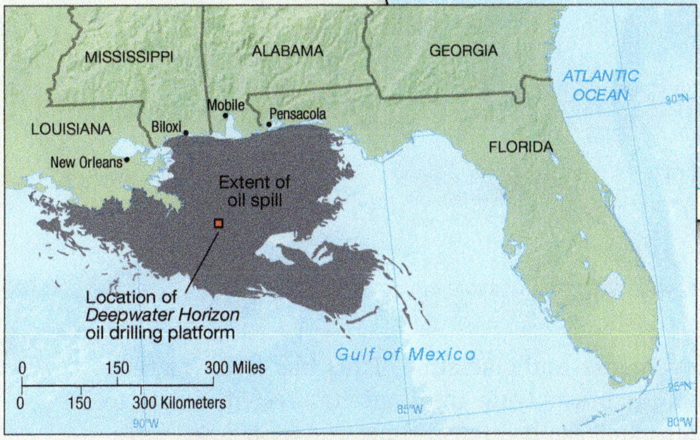

(b) The extent of spreading oil in the Gulf of Mexico on May 24, 2010, just over a month after the Deepwater Horizon spill.

(c) Oil within a Louisiana coastal wetland after the Gulf spill.

Figure 16.1.2 Worldwide oil spills, and the 2010 *Deepwater Horizon* disaster. [(a) Data from Organization for Economic Cooperation and Development. (b and c) NOAA.]

- The U.S. demand for oil on a per capita basis is one of the highest in the world. Efficiencies and mileage ratings in the U.S. transportation sector have improved little for cars, trucks, and SUVs (light trucks) compared to other countries. Why? Has hybrid gas–electric technology made a difference?

The spike in oil and gasoline prices beginning in 2007 coincided with a record decrease in miles driven, as drivers adjusted their driving habits with few difficulties—illustrating the economic principle known as *price elasticity* of demand, as well as the effectiveness of customers' conservation and efficiency strategies. (Strangely, the word *conservation* was not used in the media or in political discourse during these times of record prices and spill accidents.)

Coastal restoration in the Gulf is ongoing, even though the most immediate and dramatic effects have subsided. Oil has the potential to persist in the environment for decades, coating sandy beach sediments and sinking into the muddy bottoms of salt marshes. In May 2013, almost 3 years after the initial Gulf spill, the long-term restoration of Gulf ecosystems and economies is still in the planning stages, with an emphasis on future protection and revitalization. As we examine oil spills and their connections to Earth systems, human societies, and issues of global sustainability, we have much to consider.

- Local characteristics of *coastal rock* and coastal geomorphology are important in determining rates of erosion and sediment production.
- *Human activities* are an increasingly significant input producing coastal change.

All these inputs occur within the ever-present influence of gravity's pull—exerted not only by Earth, but also by the Moon and Sun. Gravity provides the potential energy of position and produces the tides. A dynamic equilibrium among all these components produces coastline features.

The Coastal Environment

The coastal area and shallow offshore environment is the **littoral zone**, from the Latin word *litoris*, for "shore." Figure 16.5 illustrates the littoral zone and includes specific components discussed later in the

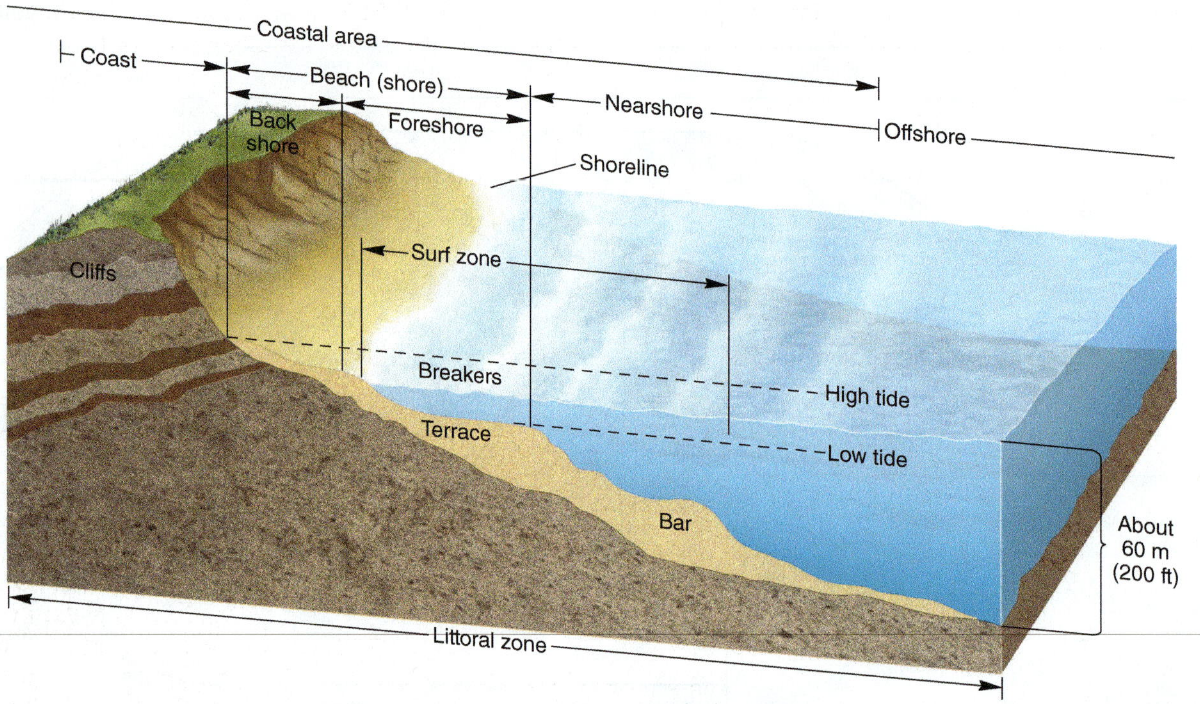

▲**Figure 16.5 The littoral zone.** The littoral zone spans the coast, beach, nearshore, and part of the offshore environment.

chapter. The littoral zone spans land as well as water. Landward, it extends to the highest waterline reached on shore during a storm. Seaward, it extends to where water is too deep for storm waves to move sediments on the seafloor—usually around 60 m, or 200 ft, in depth. The line of actual contact between the sea and the land is the *shoreline*, and it shifts with tides, storms, and sea-level adjustments. The *coast* continues inland from high tide to the first major landform change and may include areas considered to be part of the coast in local usage. The foreshore is often called the *intertidal zone*.

Because the level of the ocean varies, the littoral zone naturally shifts position from time to time. A rise in sea level causes submergence of land, whereas a drop in sea level exposes new coastal areas. In addition, uplift and subsidence of the land itself initiate changes in the littoral zone.

Sea Level

As discussed in Chapter 10B, sea level is an important concept. Average sea level changes daily with the tides and over the long term with changes in climate, tectonic plate movements, and glaciation. Thus, *sea level* is a relative term. At present, no international system exists to determine exact sea level over time. The Global Sea Level Observing System (GLOSS) is an international group actively working on sea-level issues and is part of the larger Permanent Service for Mean Sea Level (see http://www.psmsl.org/).

Elevation on Earth is referenced to mean sea level. **Mean sea level (MSL)** is a value based on average tidal levels recorded hourly at a given site over many years. MSL varies spatially because of ocean currents and waves, tidal variations, air temperature, pressure differences and wind patterns, ocean temperature variations, slight variations in Earth's gravity, and changes in oceanic volume. Over the long term, sea-level fluctuations expose a range of coastal landforms to tidal and wave processes. The present sea-level rise is spatially uneven, as is MSL; for instance, the rate of rise along the coast of Argentina is nearly 10 times the rate along the coast of France.

At present, the overall U.S. MSL is calculated at approximately 40 locations along the coastal margins of the continent. These sites are being upgraded with new equipment in the Next Generation Water Level

GEOreport 16.2 Sea-level variations along the U.S. coastline

At a given instant, sea level varies along the full extent of North American shorelines, as measured by the height of the water relative to a specific point on land. The mean sea level (MSL) of the U.S. Gulf Coast is about 25 cm (10 in.) higher than that of Florida's east coast, which is the lowest in North America. MSL values rise as one moves northward along the eastern coast, being 38 cm (15 in.) higher in Maine than in Florida. Along the U.S. western coast, MSL is higher than Florida's by about 58 cm (23 in.) in San Diego and by about 86 cm (34 in.) in Oregon. Overall, North America's Pacific Coast MSL averages about 66 cm (26 in.) higher than the Atlantic Coast MSL. For the United States, see Sea Levels Online at http://tidesandcurrents.noaa.gov/sltrends/index.shtml.

Measurement System, using next-generation tide gauges, specifically along the U.S. and Canadian Atlantic Coast, Bermuda, and the Hawaiian Islands (see http://co-ops.nos.noaa.gov/levelhow.html). The *NAVSTAR* satellites of the Global Positioning System (GPS) make possible the correlation of data within a network of ground- and ocean-based measurements.

Rising sea level is potentially devastating for many coastal locations. A rise of only 0.3 m (1 ft) would cause shorelines worldwide to move inland an average of 30 m (100 ft), inundating some 20,000 km^2 (7800 mi^2) of land along North American shores, with associated economic losses in the trillions of dollars. A 0.95-m (3.1-ft) sea-level rise could inundate 15% of Egypt's arable land, 17% of Bangladesh, and nearly all of the land area of some island countries.

CRITICALthinking 16.1
Thinking Through a Rising Sea Level

NOAA provides an interactive map for viewing potential sea-level rise and coastal flooding at http://www.csc.noaa.gov/slr/viewer/#. Use the "Sea-Level Rise and Coastal Flooding Impacts Viewer" to observe coastal inundation in parts of the United States under different scenarios. Adjust the height of sea-level rise using the slider on the left side. Briefly analyze what you find. Can the extra cost of developing alternative energy to slow greenhouse gases emissions balance the damage estimates from such coastal inundation? •

Coastal System Actions

The coast is the scene of complex tidal fluctuations, winds, waves, ocean currents, and occasional storms. These forces shape landforms ranging from gentle beaches to steep cliffs, and at the same time sustain delicate ecosystems.

Tides

Tides are complex twice-daily oscillations in sea level, ranging worldwide from barely noticeable to a rise and fall of several meters. They are experienced to varying degrees along every ocean shore around the world. Tidal action is a relentless and energetic agent of geomorphic change. As tides flood (rise) and ebb (fall), the daily migration of the shoreline landward and seaward has significant effects on sediment erosion and transportation.

Tides are important in human activities, including navigation, fishing, and recreation. They are of special concern to ships because the entrance to many ports is limited by shallow water, and thus high tide is required for passage. Conversely, tall-masted ships may need a low tide to clear overhead bridges. Tides also occur in large lakes but are difficult to distinguish from changes caused by wind in those bodies of water because the tidal range is small. Lake Superior, for instance, has a tidal variation of only about 5 cm (2 in.).

Causes of Tides Tides are produced by the gravitational pull of both the Sun and the Moon (Figure 16.6). Chapter 2 discusses Earth's relation to the Sun and Moon and the reasons for the seasons. The Sun's influence is only about half that of the Moon's because of the Sun's greater distance from Earth, although it is a significant force. Figure 16.6 illustrates the relationship between the Moon, the Sun,

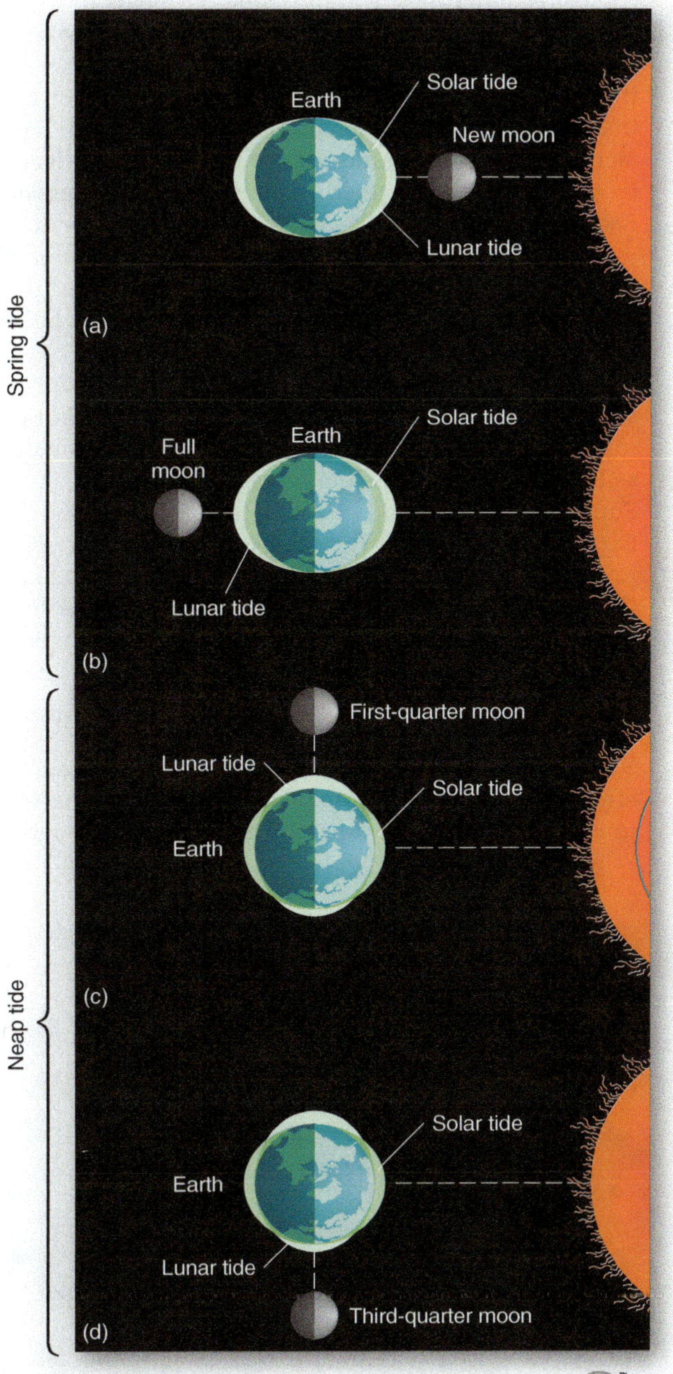

▲**Figure 16.6 The cause of tides.** Gravitational relations of Sun, Moon, and Earth combine to produce spring tides (a, b) and neap tides (c, d). (Tides are greatly exaggerated for illustration.)

MG Animation Monthly Tidal Cycles

and Earth and the generation of variable tidal bulges on opposite sides of the planet.

The gravitational pull of the Moon tugs on Earth's atmosphere, oceans, and lithosphere. The Sun also exerts a gravitational pull, to a lesser extent. Earth's solid and fluid surfaces all experience some stretching as a result of these forces. The stretching raises large *tidal bulges* in the atmosphere (which we can't see), smaller tidal bulges in the ocean, and very slight bulges in Earth's rigid crust. Our concern here is the tidal bulges in the ocean.

Gravity and inertia are essential elements in understanding tides. *Gravity* is the force of attraction between two bodies. *Inertia* is the tendency of objects to stay still if motionless or to keep moving in the same direction if in motion. The gravitational effect on the side of Earth facing the Moon or Sun is greater than that experienced by the far side, where inertial forces are slightly greater. Because of inertia, as the nearside water and Earth are drawn toward the Moon or Sun, the farside water is left behind because of the slightly weaker gravitational pull. This arrangement produces the two opposing tidal bulges on opposite sides of Earth.

Tides appear to move in and out along the shoreline, but they do not actually do so. Instead, Earth's surface rotates into and out of the relatively "fixed" tidal bulges as Earth changes its position in relation to the Moon and Sun. Every 24 hours and 50 minutes, any given point on Earth rotates through two bulges as a direct result of this rotational positioning. Thus, every day, most coastal locations experience two high (rising) tides, known as **flood tides**, and two low (falling) tides, known as **ebb tides**. The difference between consecutive high and low tides is considered the *tidal range*.

Spring and Neap Tides The combined gravitational effect of the Sun and Moon is strongest in the conjunction alignment—when they are on the same side of Earth—and results in the greatest tidal range between high and low tides, known as **spring tides (Figure 16.6a)**. (Spring means to "spring forth"; it has no relation to the season of the year.) Figure 16.6b shows the other alignment that gives rise to spring tides, when the Moon and Sun are at *opposition*—on opposite sides of Earth. In this arrangement, the Moon and Sun cause separate tidal bulges, as each celestial body affects the water nearest to it. In addition, the left-behind water resulting from the pull of the body on the opposite side augments each bulge.

When the Moon and Sun are neither in conjunction nor in opposition, but are more or less in the positions shown in Figure 16.6c and d, their gravitational influences are offset and counteract each other, producing a lesser tidal range known as **neap tide**. (*Neap* means "without the power of advancing.")

Tides also are influenced by other factors, including ocean-basin characteristics (size, depth, and topography), latitude, and shoreline shape. These factors cause a great variety of tidal ranges. For example, some locations may experience almost no difference between high and low tides. The highest tides occur when open water is forced into partially enclosed gulfs or bays. The Bay of Fundy in Nova Scotia records the greatest tidal range on Earth, a difference of 16 m (52.5 ft) (Figure 16.7). For tide predictions in the United States and Caribbean region, see http://tidesandcurrents.noaa.gov/tide_predictions.shtml.

Tidal Power The fact that sea level changes daily with the tides suggests an opportunity: Could these predictable flows be harnessed to generate electricity? The answer is yes, given the right conditions. Bays and estuaries tend to focus tidal energy, concentrating it in a smaller area than in the open ocean. Power generation can be achieved in such locations through the building

(a) Flood tide at Halls Harbor, Nova Scotia, Canada (near the Bay of Fundy).

(b) Ebb tide at Halls Harbor.

▲**Figure 16.7 Tidal range.** [Bobbé Christopherson.]

(a) The La Rance tidal generating station in France harnesses energy of the tides using a tidal barrage, similar to a hydroelectric dam.

(b) Turbine harnesses power from tidal currents at Strangford Lough, Northern Ireland.

▲**Figure 16.8 Tidal power generation.** [(a) Environment Images/UIG/Getty. (b) Robert Harding Picture Library Ltd./Alamy.]

of a dam, called a tidal barrage, that creates a difference in height by holding water at flood tide and releasing it at ebb tide. The first tidal power plant was built on the Rance River estuary on the Brittany coast of France in 1967 using this method of power production. The tides in the La Rance estuary fluctuate up to 13 m (43 ft), providing an electrical-generating capacity of a moderate 240 MW (about 20% of the capacity of Hoover Dam; Figure 16.8a). The first tidal power generation in North America also uses a tidal barrage, at the Annapolis Tidal Generating Station in the Bay of Fundy in Nova Scotia, Canada, built in 1984. Nova Scotia Power Incorporated operates this 20-MW plant.

Tidal power generation can also be achieved through the use of tidal stream generators, underwater turbines that are powered by the movement of flood and ebb tides to produce electricity. This is a more sustainable method with fewer environmental impacts, because a dam is not built within the tidal estuary. The first tidal stream generator was completed in 2007 at Strangford Lough in Northern Ireland (Figure 16.8b). In 2013, the first underwater turbines in the United States began generating power near Eastport, Maine, at the mouth of the Bay of Fundy. The main limitation of tidal power is that only about 30 locations in the world have the tidal energy needed to turn the turbines. However, many scientists suggest that this energy resource has huge potential in some regions.

Waves

Friction between moving air (wind) and the ocean surface generates undulations of water in **waves**, which travel in groups known as *wave trains*. Waves vary widely in scale: On a small scale, a moving boat creates a wake of small waves; at a larger scale, storms generate large wave trains. At the extreme is the wind wake produced by the presence of the Hawaiian Islands, traceable westward across the Pacific Ocean surface for 3000 km (1865 mi). This is a consequence of the islands' disruption of the steady trade winds, which also causes changes in surface temperature.

A stormy area at sea can be a *generating region* for large wave trains, which radiate outward in all directions. The ocean is crisscrossed with intricate patterns of these multidirectional waves. The waves seen along a coast may be the product of a storm center thousands of kilometers away.

Regular patterns of smooth, rounded waves, the mature undulations of the open ocean, are **swells**. As these swells, and the energy they contain, leave the generating region, they can range from small ripples to very large, flat-crested waves. A wave leaving a deep-water generating region tends to extend its wavelength horizontally for many meters (remember from Chapter 2 that wavelength is the distance between corresponding points on any two successive waves). Tremendous energy occasionally accumulates to form unusually large waves. One moonlit night in 1933, the U.S. Navy tanker *Ramapo* reported a wave in the Pacific higher than its mainmast, at about 34 m (112 ft)!

Wave movement in open water suggests to an observer that the water is migrating in the direction of wave travel, but in reality only a slight amount of water is actually advancing. The appearance of movement is produced by the *wave energy* that is moving through the flexible medium of water. The water within a wave in the open ocean is transferring energy from molecule to molecule in simple cyclic undulations known as *waves of transition* (Figure 16.9). Individual water particles move forward only slightly, in a vertical pattern of circles. The diameter of the paths traced by the orbiting water particles decreases with depth.

As a deep-ocean wave approaches the shoreline and enters shallower water (10–20 m, or 30–65 ft), the orbiting water particles are vertically restricted, causing elliptical, flattened orbits of water particles to form near the bottom. This change from circular to elliptical orbits slows the entire wave, although more waves keep arriving. The result is closer-spaced waves, growing in height and steepness, with sharper wave crests. As the crest of each wave rises, a point is reached when its height exceeds its vertical stability, and the wave falls into a characteristic **breaker**, crashing onto the beach (Figure 16.9b).

In a breaker, the orbital motion of transition gives way to elliptical *waves of translation*, in which both energy and water move toward shore. The slope of the shore determines wave type. Plunging breakers indicate a steep bottom profile, whereas spilling breakers indicate a gentle, shallow bottom profile. In some areas, unexpected high waves can arise suddenly, creating unexpected dangers along shorelines.

Another potential danger is the brief, short torrent called a *rip current*, created when the backwash of water produced by breakers flows to the ocean from the beach in a concentrated column, usually at a right angle to the line of breakers (Figure 16.9c). A person caught

▼Figure 16.9 Wave formation and breakers. [(b, c) Bobbé Christopherson.]

MG

Animation
Wave Motion/Wave Refraction

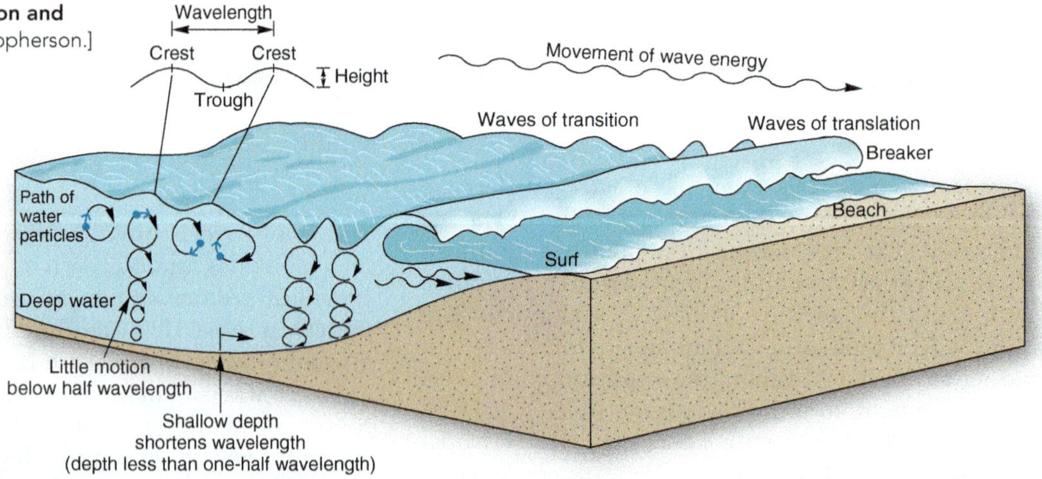

(a) The orbiting tracks of water particles change from circular motions and swells in deep water (waves of transition) to more eliptical orbits near the bottom in shallow water (waves of translation).

(b) Breakers along the coast of Baja California, Mexico.

(c) A dangerous rip current interrupts approaching breakers. Note the churned-up water where the rip current enters the surf.

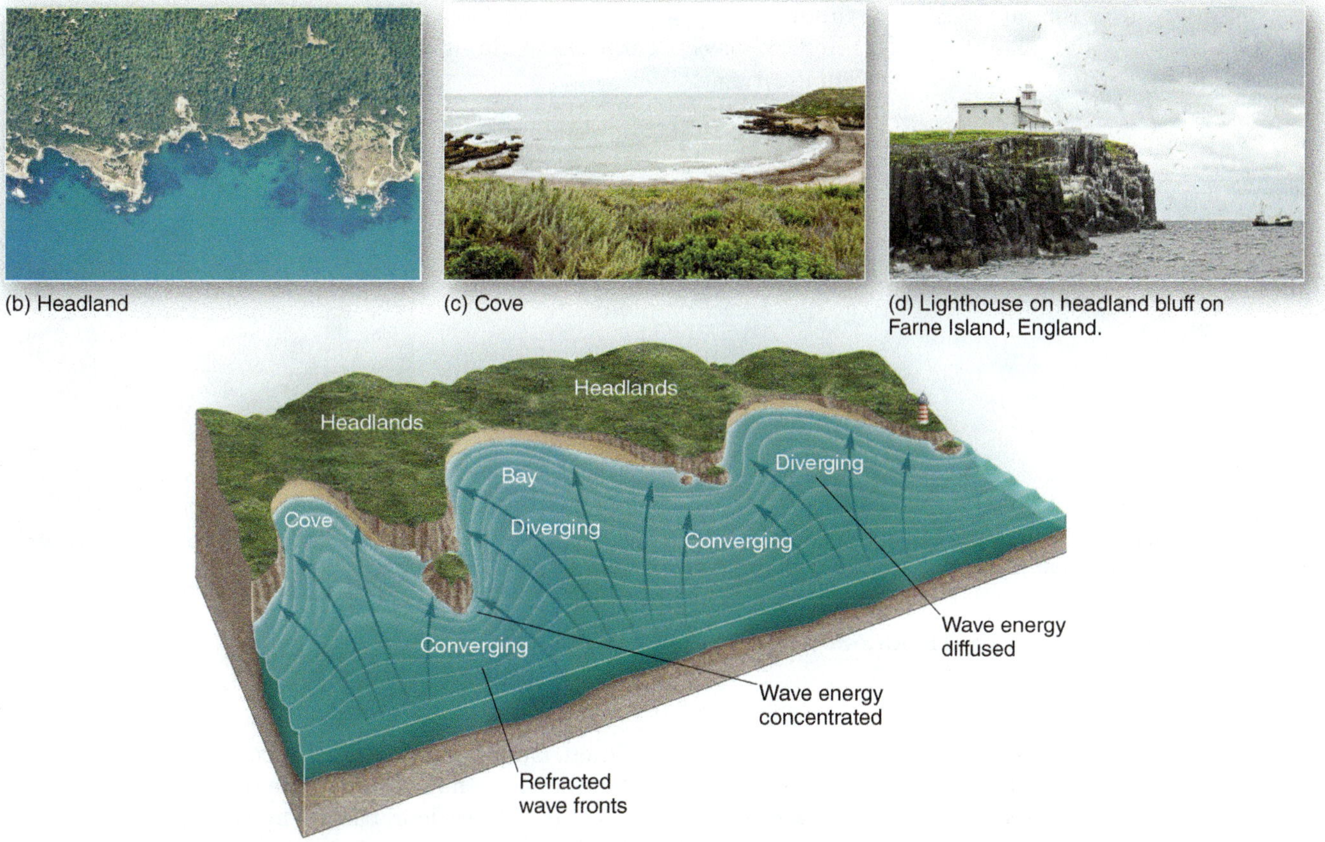

(b) Headland

(c) Cove

(d) Lighthouse on headland bluff on Farne Island, England.

(a) Wave energy is concentrated as it converges on headlands and is diffused as it diverges in coves and bays.

▲**Figure 16.10 Wave refraction and coastal straightening.** [(b, c, d) Bobbé Christopherson.]

in one of these can be swept offshore, but usually only a short distance.

As various wave trains move along in the open sea, they interact by *interference*. When these interfering waves are in alignment, or in phase, so that the crests and troughs from one wave train are in phase with those of another, the height of the waves becomes amplified, sometimes dramatically. The resulting waves, called "killer," "sleeper," "rogue," or "sneaker" waves, can sweep in unannounced and overtake unsuspecting victims. Signs along portions of the California, Oregon, Washington, and British Columbia coastline warn beachgoers to watch for such waves. In November 2012 in northern California, three people were drowned in an incident that began when a family dog was carried away by a sleeper wave. A little over a month later, another person and her dog perished in Shelter Cove, California, victims of a sleeper wave. In 2009, a person was killed as several giant waves hit the shoreline in Maine; in this case the source of the wave energy was Hurricane Bill offshore in the Atlantic.

In contrast, out-of-phase wave trains will dampen wave energy at the shore. When you observe the breakers along a beach, the changing beat of the surf actually is produced by the patterns of wave interference that occurred in far-distant areas of the ocean.

Wave Refraction In general, wave action tends to straighten a coastline. Where waves approach an irregular coast, the submarine topography refracts, or bends, approaching waves around headlands, which are protruding landforms generally composed of resistant rocks (Figure 16.10). The refracted energy becomes focused around the headlands and dissipates in coves, bays, and the submerged coastal valleys between headlands. Thus,

GEOreport 16.3 Surprise waves flood a cruise ship

On March 3, 2010, a large cruise ship in the western Mediterranean Sea, off the coast of Marseilles, France, was struck by three surprise waves about 7.9 m (26 ft) in height. Two passengers were killed and many injured as windows shattered and water flooded parts of the ship's interior. Rescue personnel took the injured to hospitals in Barcelona, Spain. Scientists are studying what causes such abnormal waves, which tend to happen in open ocean; elements include strong winds and wave interference.

Animation
Beach Drift, Coastal Erosion

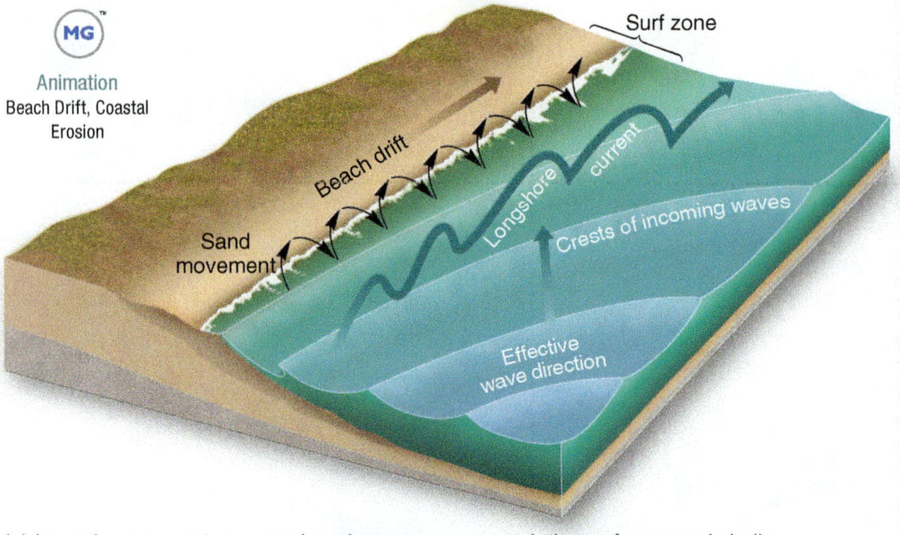

(a) Longshore currents are produced as waves approach the surf zone and shallower water. Longshore drift and beach drift result as substantial volues of material are moved along the shore.

(b) Processes at work along Point Reyes Beach, Point Reyes National Seashore, California (aerial view to the south).

▲**Figure 16.11 Longshore current and beach drift.** [(b) Bobbé Christopherson.]

headlands receive the brunt of wave attack along a coastline. The result of **wave refraction** is a redistribution of wave energy, so that different sections of the coastline vary in erosion potential, with the long-term effect of straightening the coast.

Waves usually approach the coast at a slight angle (Figure 16.11). In consequence, as the shoreline end of the wave enters shallow water and slows down, the portion of the wave in deeper water continues to move at a faster speed. The velocity difference refracts the wave, producing a current that flows parallel to the coast, zigzagging in the prevalent direction of the incoming waves. This **longshore current**, or *littoral current*, depends on wind direction and the resultant wave direction. A longshore current is generated only in the surf zone and works in combination with wave action to transport large amounts of sand, gravel, sediment, and debris along the shore. This process, called **beach drift**, moves particles along a beach with the longshore current by shifting them back and forth between water and land with each *swash* and *backwash* of surf. **Littoral drift** is the term for the combined actions of the longshore current and beach drift.

You have perhaps stood on a beach and heard the sound of littoral drift as the myriad sand grains and seawater mix in the backwash of surf. These dislodged materials are available for transport and eventual deposition in coves and inlets and can represent a significant volume of sediment.

Tsunami A series of waves generated by a large undersea disturbance is known as a **tsunami**, Japanese for "harbor wave" (named for the large size and devastating effects of the waves when their energy is focused in harbors). Often, tsunami are reported incorrectly as "tidal waves," but they have no relation to the tides. Sudden, sharp motions in the seafloor, caused by earthquakes, submarine landslides, eruptions of undersea volcanoes, or meteorite impacts in the ocean, produce tsunami. They are also known as *seismic sea waves* since about 80% of tsunami occur in the tectonically active region associated with the Pacific Ring of Fire. However, tsunami can also be caused by non-seismic events. Often, the first wave of a tsunami is the largest, fostering the misconception that a tsunami is a single wave. However, successive waves may be larger than the first wave, and tsunami danger may last for hours after the first wave's arrival.

Tsunami generally exceed 100 km (60 mi) in wavelength (crest to crest) but are only a meter (3 ft) or so in height. They travel at great speeds in deep-ocean water—velocities of 600–800 kmph (375–500 mph) are not uncommon—but often pass unnoticed on the open sea because their long wavelength makes the rise and fall of water hard to observe.

As a tsunami approaches a coast, the increasingly shallow water forces the wavelength to shorten. As a result, the wave height may increase up to 15 m (50 ft) or more, potentially devastating a coastal area far beyond the tidal zone, and taking many human lives. In 1992, a 12-m (39-ft) tsunami wave killed 270 people in Casares, Nicaragua. A 1998 Papua New Guinea tsunami, launched by a massive undersea landslide of some 4 km^3 (1 mi^3), killed 2000. During the twentieth century, records show 141 damaging tsunami and perhaps 900 smaller ones, with a total death toll of about 70,000. No warning system was in place when these tsunami occurred.

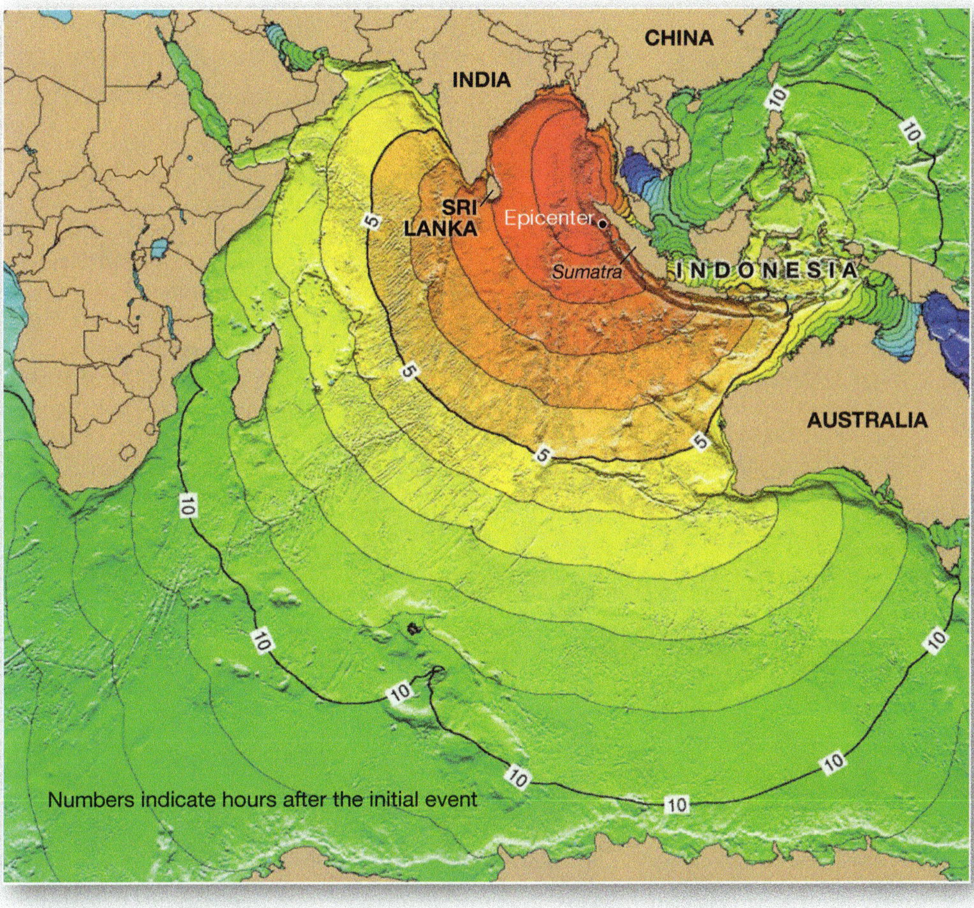

Figure 16.12 Travel times for the 2004 Indian Ocean tsunami. Black circle indicates the earthquake epicenter located 250 km (155 mi) off the west coast of northern Sumatra, Indonesia. Map compiled with integrated data from several sources. [NOAA.]

On December 26, 2004, the M 9.3 Sumatra–Andaman earthquake struck off the west coast of northern Sumatra along the subduction zone formed where the Indo-Australian plate moves beneath the Burma plate along the Sunda Trench. (Please review the Chapter 12 opening map to find this trench along the coast of Indonesia in the eastern Indian Ocean Basin.)

The earthquake caused the island of Sumatra to spring up about 13.7 m (45 ft) from its original elevation, triggering a massive tsunami that traveled across the Indian Ocean (Figure 16.12). Energy from the tsunami waves traveled around the world several times through the global ocean basins before dissipating. With Earth's mid-ocean mountain chain acting as a guide, related large waves arrived at the shores of Nova Scotia, Antarctica, and Peru.

The total human loss from the Indonesian quake and tsunami may never be known but exceeded 150,000 people. In response to this tsunami, the Indian Ocean Tsunami Warning and Mitigation System was created as part of the ongoing United Nations tsunami-warning-system project (see http://itic.ioc-unesco.org/). This event also prompted the addition of 32 ocean stations (pressure sensors on the ocean floor with accompanying surface buoys) as part of the Deep-Ocean Assessment and Reporting of Tsunamis (DART), a global tsunami warning system developed by NOAA in the United States. (For NOAA's tsunami research program, see http://nctr.pmel.noaa.gov/.)

The 2011 Tohoku earthquake in Japan (discussed in Focus Study 12.1) triggered a tsunami that killed over 15,000 people. Despite warning systems in what is one of the most technologically advanced countries in the world, there was little time for evacuation. Focus Study 16.2 describes the tsunami and its effects through the Pacific Ocean basin. Revisit the photos in Focus Study 12.1 to see tsunami-related damage in Japan and the wave as it moved inland.

For Hawai'i and nations surrounding the Pacific, the Pacific Tsunami Warning Center (PTWC) issues tsunami warnings. Alaska and the U.S. West Coast rely on the West Coast/Alaska Tsunami Warning Center. These warning centers, and others throughout the world, use the DART network of 39 stations in the Pacific, Indian, and Atlantic Oceans, discussed in Focus Study 16.2. When a tsunami triggers the DART sensors, regional warning centers issue bulletins to regions that are likely to be affected. The effectiveness of these warnings varies; for those closest to the undersea disturbance, even the most accurate warning cannot help when there are only minutes to reach safety.

Focus Study 16.2 Natural Hazards
The 2011 Japan Tsunami

On March 11, 2011, just minutes after the Tohoku earthquake hit Japan, its epicenter about 129 km (80 mi) offshore of the island of Honshu (discussed in Chapter 12, Focus Study 12.1), tsunami warnings went out across the country. Eight to 10 minutes after the quake, the first tsunami wave hit the northeastern coast of Honshu, the closest shoreline to the earthquake's epicenter.

Tsunami wave heights averaged 10 m (33 ft) in some areas and reached 30 m (98 ft) in narrow harbors. At Ofunato, the tsunami traveled 3 km (1.8 miles) inland; in other areas, waves reached 10 km inland (6.2 miles). Although seawalls and breakwaters designed for typhoon and tsunami waves guard about 40% of Japan's coastline, the deep coastal embayments seaward of the walls worked to magnify the tsunami energy to the point that the walls offered little protection (Figure 16.2.1). In Kamaishi, the $1.5 billion tsunami seawall, anchored to the seafloor and extending 2 km (1.2 mi) in length, was breached by a 6.8-m (22-ft) wave, submerging the city center.

Japan's tsunami early warning system is activated by an earthquake and uses the seismic signals measured during the first minute of the quake as input for computer models designed to estimate the size of the tsunami wave. The Japan Meteorological Agency (JMA) then issues warnings that include the forecasted tsunami wave size for each prefecture.

▲**Figure 16.2.1 Tsunami wave breaks over a seawall, Miyako, Japan.** A tsunami wave breaks over a protective wall onto the streets of Miyako, Iwate Prefecture, in northeastern Japan, triggered by a catastrophic M 9.0 earthquake, March 11, 2011. Buildings, cars, houses, and victims were carried far inland. Miyako is about 120 km (75 mi) north of the quake epicenter and 200 km (125 mi) north of the Fukushima Daiichi nuclear power plant disaster (discussed in Focus Study 12.1). [Mainichi Shimbun/Reuters.]

Coastal System Outputs

Coastlines are active, energetic places, with sediment being continuously delivered and removed. The action of tides, currents, wind, waves, and changing sea level produces a variety of erosional and depositional landforms. We look first at erosional coastlines, such as the U.S. West Coast, where in general more sediment is removed than is deposited. We then look at depositional coastlines, such as the U.S. East and Gulf Coasts, where in general more sediment is deposited, primarily from streams, than eroded. In this era of rising sea level, coastlines are becoming even more dynamic.

Coastal Erosion

The active margin of the Pacific Ocean along North and South America is a typical erosional coastline. *Erosional coastlines* tend to be rugged, of high relief, and tectonically active, as expected from their association with the

Moments after the Tohoku quake, JMA issued a tsunami warning that we now know was an underestimation of the actual wave size. The quake continued for over 2 minutes after the tsunami model calculations began, and during this time the tsunami energy increased. A corrected warning was issued, but not until 20 minutes after the quake and too late for evacuation.

Nine minutes after the initial earthquake, the Pacific Tsunami Warning Center (PTWC) issued tsunami warnings to the Pacific islands and continents around the Pacific basin. The tsunami warning process used by the PTWC begins when ocean-bottom sensors register a change in pressure associated with an ocean disturbance. Data is relayed to a surface buoy and then transmitted via satellite to the PTWC. The network of 32 Pacific Ocean stations, including sensors and buoys, is part of the Deep-Ocean Assessment and Reporting of Tsunamis (DART) that monitors tsunami wave heights. These stations include conventional DART buoys, heavier and more difficult to deploy in high seas, and new, lightweight Easy-to-Deploy (ETD) buoys (Figure 16.2.2).

Over the next hour, tsunami forecasts and warnings continued for the Pacific region, and NOAA issued wave-height predictions. The tsunami moved across the Pacific, its energy guided and deflected by seafloor topography. Four hours after the quake, the tsunami overwashed portions of the Midway Islands, northwest of Hawai'i. Seven hours after the quake, waves ranging from 1 to 2.2 m (3.3 to 7.2 ft) hit Oahu, Maui, and Hawai'i. Finally, nearly 12 hours after the Tohoku quake, waves 2.1 m (7 ft) high reached the northern and central California shorelines, causing several million dollars in damage to harbors, boats, and piers. Waves 30 to 70 cm (1.0 to 2.3 ft) high arrived at New Zealand.

The tsunami warnings generally failed to help the Japanese during this huge-wave event. The people in Japan not only had too little time to flee from vulnerable coastal areas, but also relied for protection on the breakwaters and tsunami walls—structures that in hindsight may have provided a false sense of security. However, the warnings were effective in preparing other Pacific regions for the waves. At the same time, despite warnings in Hawai'i and along the U.S. mainland, some people actually went down to the shorelines with cameras in hand to capture images of the incoming danger.

(a) Conventional DART buoy.

(b) DART ETD buoy.

Figure 16.2.2 NOAA's DART buoys. The sea-surface buoy is anchored above a bottom pressure recorder, the two linked by acoustic telemetry for real-time communication. The buoy transmits readings from the seafloor recorder to land-based surface stations through the *Iridium* satellite system. The National Geophysical Data Center is the data archive for DART. [NOAA.]

leading edge of drifting lithospheric plates (review the plate tectonics discussion in Chapter 11). Figure 16.13 presents features commonly observed along an erosional coast. Some of the landforms within this setting may be formed from depositional processes, despite the erosional nature of the overall landscape.

Sea cliffs are formed by the undercutting action of the sea. As indentations slowly grow at water level, a sea cliff becomes notched and eventually will collapse and retreat (Figure 16.13d). Other erosional forms that evolve along cliff-dominated coastlines include *sea caves* and *sea arches* (Figure 16.13a). As erosion continues, arches may collapse, leaving isolated *sea stacks* in the water (Figure 16.13c).

Wave action can cut a horizontal bench in the tidal zone, extending from the foot of a sea cliff out into the sea. Such a structure is a **wave-cut platform**, or *wave-cut terrace*. In places where the elevation of the land relative

(a) Arch, Ascension Island, Atlantic Ocean.

(b) Wave-cut platform, Monterey County, California.

(c) Stacks, and headland, Gough Island, South Atlantic Ocean.

(d) Notched cliff, Bear Island, Barents Sea.

(e) Collapsing cliffs, California.

▲**Figure 16.13 Erosional coastal landforms.** [Photos by Bobbé Christopherson.]

to sea level has changed over time, multiple platforms or terraces may rise like stair steps back from the coast; some terraces may be more than 370 m (1200 ft) above sea level. A tectonically active region, such as the California coast, has many examples of multiple wave-cut platforms, which at times can be unstable and prone to mass wasting (Figure 16.13b and e).

Coastal Deposition

Depositional coasts generally occur along coastlines where relief is gentle and lots of sediment is available from river systems. Such is the case with the Atlantic and Gulf coastal plains of the United States, which lie along the relatively passive, trailing edge of the North American lithospheric plate. Although the landforms are generally classified as depositional along such coastlines, erosional processes are also at work, especially during storms.

Figure 16.14 illustrates characteristic landforms deposited by waves and currents. **Barrier spits** consist of material deposited in a long ridge extending out from a coast, sometimes partially crossing and blocking the mouth of a bay. A classic barrier spit is Sandy Hook, New Jersey (south of New York City). Such barrier spit formations are also found at Point Reyes (Figure 16.14a) and Morro Bay (Figure 16.14b), in California.

If a spit grows to completely cut off the bay from the ocean, it becomes a **bay barrier**, or *baymouth bar*. Spits and barriers are made up of materials transported by littoral drift. For sediment to accumulate, offshore currents must be weak, since strong currents carry material away

(a) The Limantour barrier spit nearly blocks the entrance to Drakes Estero, along Point Reyes.

(b) A barrier spit forms Morro Bay, with the sound opening to the sea near 178-m (584-ft) Morrow Rock, a volcanic plug.

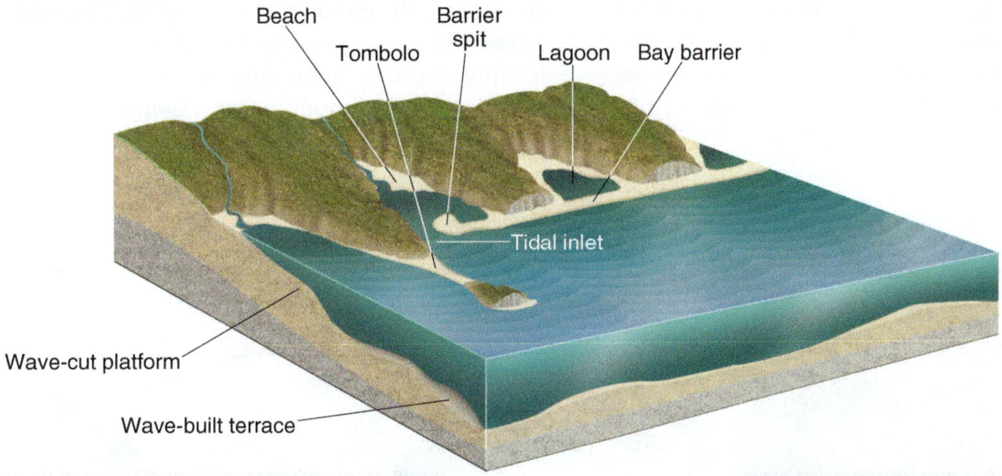

(c) A Tombolo at Point Sur along the central California coast, where sediment deposits connect the shore with an island.

(d) A shell beach deposited by the ocean.

▲**Figure 16.14 Depositional coastal landforms: barrier spits, lagoons, tombolos, and beaches.** [Photos by Bobbé Christopherson.]

before it can be deposited. Bay barriers often surround an inland **lagoon**, a shallow saltwater body that is cut off from the ocean. A **tombolo** occurs when sediment deposits connect the shoreline with an offshore island or sea stack by accumulating on an underwater wave-built terrace (Figure 16.14c).

Beaches Of all the depositional landforms along coastlines, beaches probably are the most familiar. Technically, a **beach** is the relatively narrow strip along a coast where sediment is reworked and deposited by waves and currents. Sediment temporarily resides on the beach while in active transit along the shore. Beaches occur along seacoasts, lakeshores, and rivers. Not all the beaches of the world are composed of sand, for they can be made up of shingles (beach gravel) and shells, among other materials (Figure 16.14d). Gravels reflect the contribution of stream sediments into coastal areas; shells reflect the contribution of materials from oceanic sources. Beaches vary in type and permanence, especially along coastlines dominated by wave action.

On average, the beach zone spans the area from about 5 m (16 ft) above high tide to 10 m (33 ft) below low tide (see Figure 16.5). However, the size and location of the beach zone varies greatly along individual shorelines. Worldwide, quartz (SiO_2) dominates beach sands because it resists weathering and therefore remains after other minerals are removed. In volcanic areas, beaches are derived from wave-processed lava. Hawai'i and Iceland, for example, feature some black-sand beaches.

Many beaches, such as those in southern France and western Italy, lack sand and are composed of pebbles and cobbles—a type of *shingle beach*. Some shores have no beaches at all, but are lined with boulders and cliffs. The coasts of Maine and portions of Canada's Atlantic Provinces are classic examples. These coasts, composed of resistant granite rock, are scenically rugged and have few beaches.

A beach acts to stabilize a shoreline by absorbing wave energy, as is evident from the amount of material that is in almost constant motion (see "Sand movement" in Figure 16.11a). Some beaches are stable; others have seasonal cycles of deposition, erosion, and redeposition. Many beaches accumulate during the summer; are moved offshore by winter storm waves, forming a submerged bar; and are redeposited onshore the following summer. Protected areas along a coastline tend to accumulate sediment, which can lead to large coastal sand dunes. Prevailing winds and storms often drag such coastal dunes inland, sometimes burying trees, highways, and housing developments, as described in *Geosystems Now*.

Beach Protection Changes in coastal sediment transport can disrupt human activities as beaches are lost, harbors are closed, and coastal highways and beach houses are inundated with sediment. Thus, people use various strategies to interrupt littoral drift (Figure 16.15). The goal is either to halt sand accumulation or to force a more desirable type of accumulation, through construction of engineered structures, or "hard" shoreline protection. Common approaches

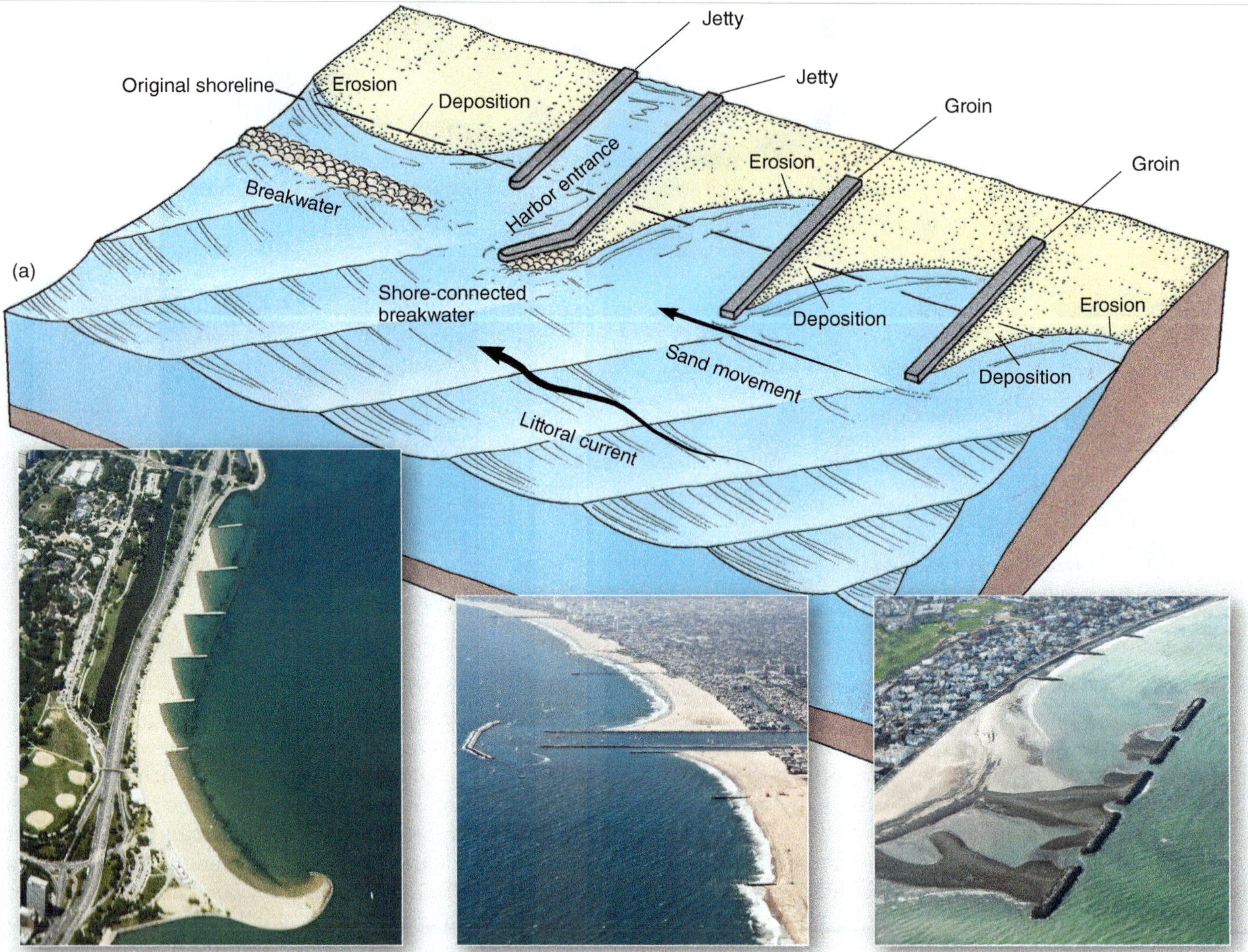

(b) Groins disrupt sediment movement along the coast of Lake Michigan, north of Chicago.

(c) A breakwater and jetties protect the entrance to Marina del Rey, California.

(d) The Five Sisters breakwaters in Winthrop, Massachusetts (near Boston). Coarse gravels and sand accumulated in bars behind the breakwaters since their construction in the 1930s.

▲**Figure 16.15 Interfering with the littoral drift of sand.** Breakwaters, jetties, and groins are constructions that attempt to control littoral drift and beach drift along a coast. [(b), (c), and (d) Bobbé Christopherson.]

MG Animation
Coastal Stabilization Structures

include the building of *groins* to slow drift action along the coast, *jetties* to block material from harbor entrances, and *breakwaters* to create zones of still water near the coastline. However, interrupting the littoral drift disrupts the natural beach replenishment process and may lead to unwanted changes in sediment distribution in areas nearby. Careful planning and impact assessment should be part of any strategy for preserving or altering a beach.

In contrast to "hard" protection, the hauling of sand to replenish a beach is considered "soft" shoreline protection. *Beach nourishment* refers to the artificial replacement of sand along a beach. Theoretically, through such efforts, a beach that normally experiences a net loss of sediment will be fortified with new sand. However, years of human effort and expense to build beaches can be erased by a single storm. In addition, disruption of marine and littoral zone ecosystems may occur if the new sand does not physically and chemically match the existing sand.

In Florida, local, state, and federal agencies spend over $100 million annually on replenishment projects and manage over 200 miles of restored beaches. Until recently, sand was pumped from offshore onto the beach. However, over 30 years of dredging for sand has depleted offshore sand supplies, and now sand transport from far-away source areas is necessary. In Virginia Beach, Virginia, a $9 million beach replenishment project in 2013 rebuilt a strip of sand for the 49th time since 1951. The Army Corps of Engineers, which typically executes such projects on the U.S. East Coast, states that beach replenishment saves money in the long run by preventing damage to coastal development. Others, including scientists and politicians, disagree. At present, the federal government pays for about 65% of all beach replenishment projects, with the remaining cost picked up by state and local communities. As discussed with regard to the impacts of Hurricane Sandy in *Geosystems Now*, the controversy is ongoing.

Barrier Beaches and Islands

Barrier beaches are long, narrow depositional features, generally of sand, that form offshore roughly parallel to the coast. When these features are broader and more extensive, they are called **barrier islands**. The sediment supplied to these beaches often comes from alluvial coastal plains, and tidal variation near these features usually is moderate to low. Barrier beaches and islands are quite common worldwide, lying offshore of nearly 10% of Earth's coastlines. Examples are found off Africa, India's eastern coast, Sri Lanka, Australia, and Alaska's northern slope, as well as offshore in the Baltic and Mediterranean Seas. Earth's most extensive chain of barrier islands is along the U.S. Atlantic and Gulf Coasts, extending some 5000 km (3100 mi) from Long Island to Texas and Mexico.

North Carolina's famed Outer Banks is a 200-mile-long string of barrier islands and peninsulas that separate the Atlantic Ocean from the mainland. The Outer Banks stretch southward from Virginia Beach, Virginia, to Cape Lookout, and are separated from the mainland by Pamlico Sound (*sound* is a general term for a body of water forming an inlet) and two other sounds to the north (Figure 16.16). Mud flats (also called tidal flats) and salt marshes (a type of coastal wetland) are characteristic low-relief environments on the landward side of a barrier formation, where tidal influence is greater than wave action. Typical landforms are foredunes on the seaward side of the formation and backdunes and lagoons on the landward side. Figure 16.17 shows landforms and associated vegetation along with basic human usage and recommendations from a planning perspective, for a typical barrier island along the U.S. East Coast.

▲Figure 16.16 **Barrier island chain along North Carolina coast.** The Outer Banks of North Carolina. The area presently is designated as one of 10 national seashore reserves supervised by the National Park Service. [*Terra* MODIS, NASA/GSFC.]

Barrier Island Processes Various hypotheses explain the formation of barrier islands. They may begin as offshore bars or low ridges of submerged sediment near shore and then gradually migrate toward shore with wave action or rising sea level. Barrier beaches naturally shift position in response to wave action and longshore currents. The name "barrier" is appropriate, for these formations take the brunt of storm energy, migrating over time with erosion and redeposition. For example, during Hurricane Sandy in 2012, Fire Island, a barrier island and popular summer destination off the southern coast of Long Island, New York, moved 19 to 25 m (65 to 85 ft) toward the mainland (Figure 16.18). The processes at work were typical of the effects of storms on barrier

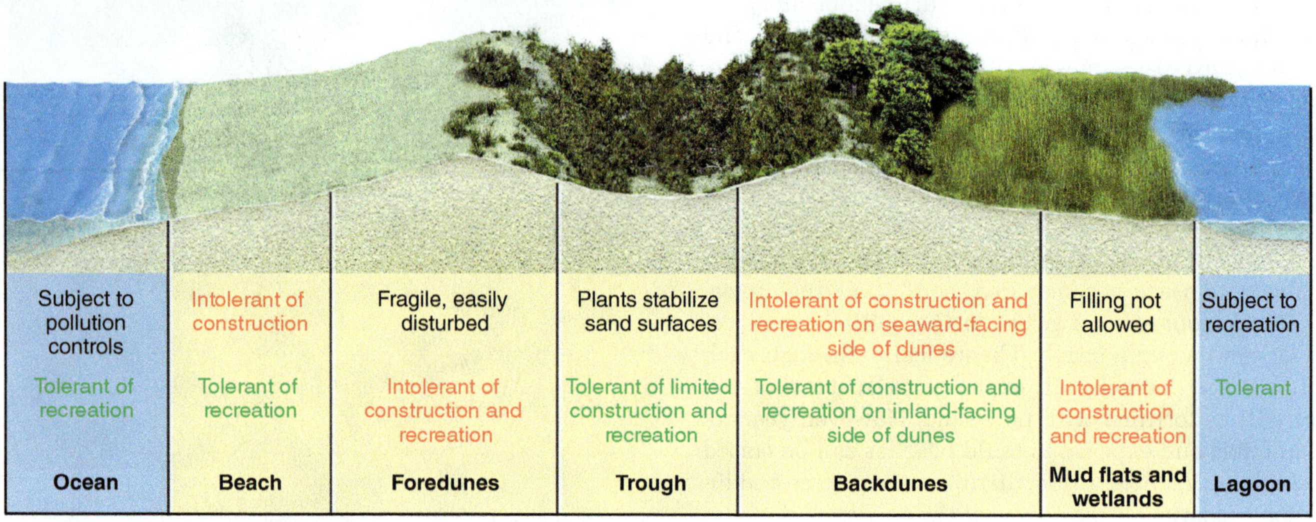

▲Figure 16.17 **Barrier island landforms and ecosystems, with planning guidelines based on the New Jersey coastal environment.** [Planning content after Ian McHarg, *Design with Nature*, Copyright © 1969.]

islands: erosion on the beach and foredune, deposition in the backdune area and in the lagoon, the formation of new inlets, and general shifting of the barrier formation toward shore in response to waves and storm surge.

Development on Barrier Islands Human-built structures on barrier islands are vulnerable to erosion and redeposition of coastal sediments by tropical storms and rising sea level. As proven again in the aftermath of Hurricane Sandy, the effects of hurricanes on barrier islands have inflicted tremendous economic losses, human hardship, and fatalities.

Several hurricanes were notable for their destruction and impacts on barrier island geomorphology. In 1989, Hurricane Hugo hit the South Carolina coast, sweeping away barrier island developments and millions of tons of sand; the hurricane destroyed up to 95% of the single-family homes in one community. In 1998,

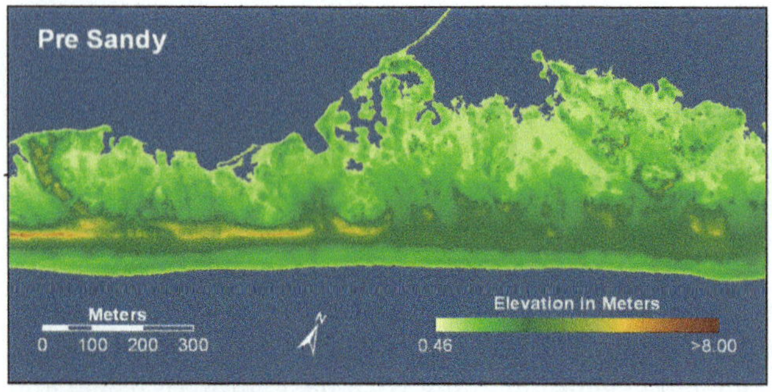

(a) Pre-Sandy LiDAR elevations, Fire Island National Seashore, a narrow portion of the island.

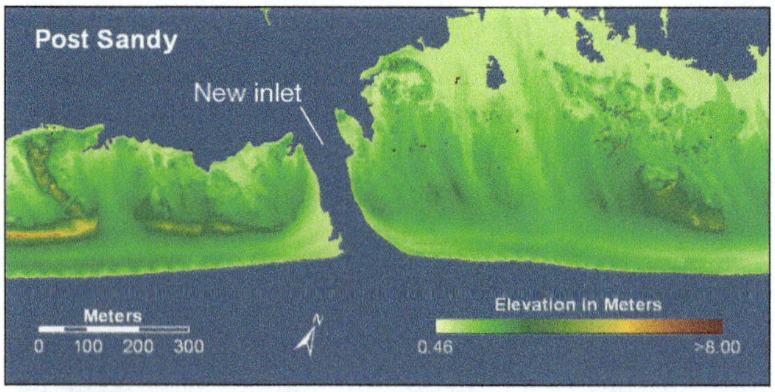

(b) The beach eroded through 4-m-high sand dunes, forming a new inlet.

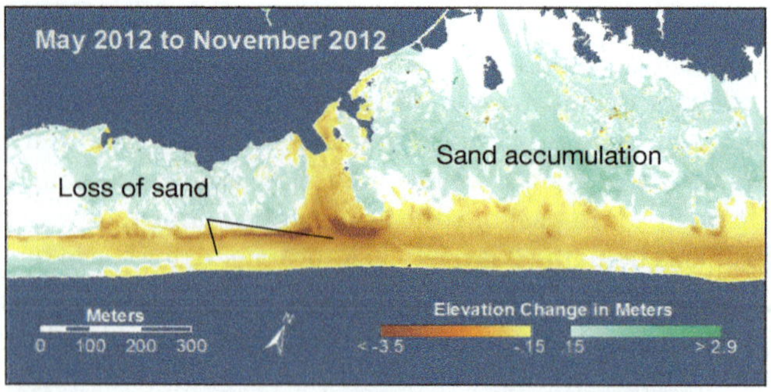

(c) Red-orange colors indicate loss of sand; blue-green colors indicate sand accumulation caused by waves and surge.

MG Satellite Hurricane Isabel 9/6–9/19/04

MG Satellite Hurricane Georges

▲Figure 16.18 **Coastal change at Fire Island, New York, after Hurricane Sandy.** [(a, b, c) USGS.]

Hurricane Georges destroyed large tracts of the Chandeleur Islands offshore from the Louisiana–Mississippi Gulf Coast, and then Hurricane Katrina, and to a lesser extent Dennis, in 2005, swept away much of what remained, leaving only sand bars. Figure 16.35 at the end of the chapter shows the dramatic changes to the Chandeleur Islands over the past 15 years, as some 80% of the land has washed away through storm action. In 2008, Hurricane Ike struck the Texas–Louisiana Gulf Coast, including Galveston Island and the Bolivar Peninsula, a barrier spit near Galveston. In this area, \$30 billion in structures were swept away (80% to 95% of homes) and 195 lives were lost. (See photos of the peninsula and the damage from Hurricane Ike on the *MasteringGeography* website.)

Despite our understanding of beach and barrier island migration and of the effects of storms, along with warnings from scientists and government agencies, coastal development does not always include precautions for limiting erosion. Developers and builders still ignore the scientific evidence that beaches and barrier islands are unstable, temporary features on the landscape. One way to encourage sustainable environmental planning and zoning might be to *allocate responsibility and cost in the event of a disaster.* A system could be set up that would place a hazard tax on land that is based on assessed risk and that restricts the government's responsibility to fund reconstruction or an individual's right to reconstruct on frequently damaged sites. Comprehensive mapping of erosion-hazard areas would help avoid the ever-increasing costs from recurring disasters. (Look back to the planning guidelines on Figure 16.17.)

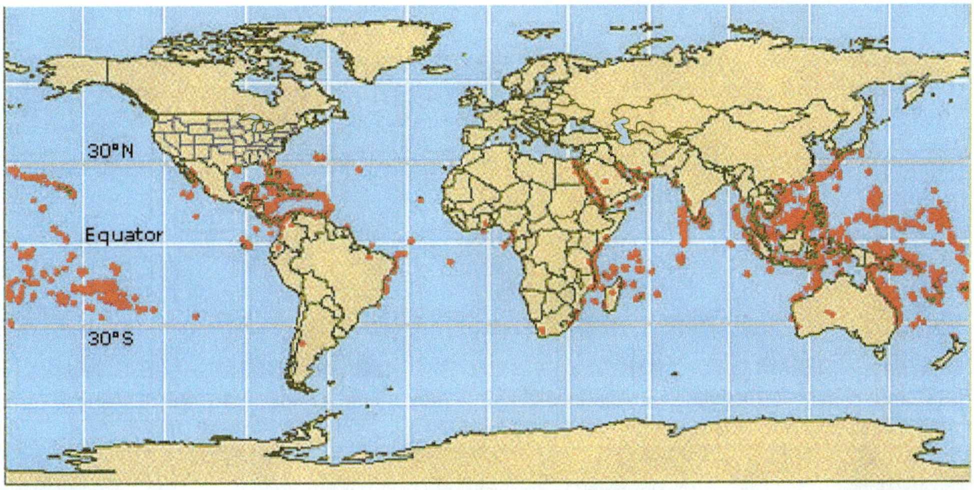

▲**Figure 16.19 Worldwide distribution of living coral formations.** The red dots represent major reef-forming coral colonies. Colonial corals range in distribution from about 30° N to 30° S. [NOS/NOAA, 2008.]

CRITICALthinking 16.2
Allocating Responsibility and Cost for Coastal Hazards

What do you think about the idea of allocating responsibility among individual landowners in the event of a disaster? How would you suggest implementing such a strategy? What special interests might oppose such an approach? What response would you expect from city government or business interests? From banking or the real estate industry? •

Coral Formations

Not all coastlines form by purely physical processes. Some form as the result of biological processes, such as coral growth. A **coral** is a simple marine animal with a small, cylindrical, saclike body called a *polyp*; it is related to other marine invertebrates such as anemones and jellyfish. Corals secrete calcium carbonate ($CaCO_3$) from the lower half of their bodies, forming a hard, calcified external skeleton.

Corals and algae live together in a *symbiotic* relationship, an overlapping arrangement in which each depends on the other for survival. Corals cannot photosynthesize, but they do obtain some of their own nourishment. Algae perform photosynthesis, converting solar energy to chemical energy and providing the coral with about 60% of its nutrition; they also assist the coral with the calcification process. In return, corals provide the algae with certain nutrients. Coral reefs are the most diverse marine ecosystems. Preliminary estimates of species living in coral reefs place the number at a million worldwide, yet, as in most ecosystems in water or on land, biodiversity is declining in these communities.

Figure 16.19 shows the global distribution of living coral formations. Corals mostly thrive in warm tropical oceans, so the difference in ocean temperature between the western coasts and eastern coasts of continents is critical to their distribution. Western coastal waters tend to be cooler, thereby discouraging coral activity, whereas eastern coastal currents are warmer and thus enhance coral growth.

Living colonial corals range in distribution from about 30° N to 30° S. Corals occupy a very specific ecological zone: 10- to 55-m (30- to 180-ft) depth, 27‰ to 40‰ salinity, and 18°C to 29°C (64°F to 85°F) water temperature. Their upper threshold for water temperature is 30°C (86°F); above that limit, the corals begin to bleach and die. Corals require clear, sediment-free water and consequently do not locate near the mouths of sediment-charged freshwater streams. For example, note the lack of these structures along the U.S. Gulf Coast. Corals have low genetic diversity worldwide and long generation times, which together mean that corals are slow to adapt and vulnerable to changing conditions.

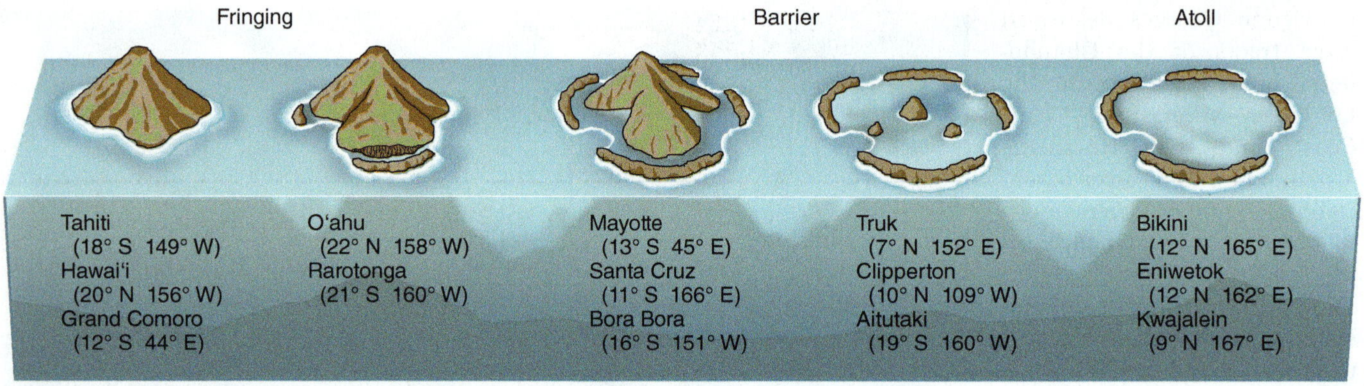

(a) Common coral formations in a sequence of reef growth around a subsiding volcanic island: fringing reefs, barrier reefs, and an atoll.

(b) Tikehau Atoll, French Polynesia.

(c) Tuherahera village and Tikehau airport.

(d) Barrier reefs surround Bora Bora, Tahiti.

▲**Figure 16.20 Coral formations.** [(a) After D. R. Stoddart, *The Geographical Magazine* 63 (1971): 610. (b), (c) *EO-1* image, NASA. (d) wilar/Shutterstock]

However, an interesting exception to these environmental parameters for corals are various species of cool corals that exist in the deep ocean, at temperatures as low as 4°C (39°F) and at depths to 2000 m (6562 ft), well beyond the expected range. Scientists study these remarkable oddities, which do not rely on algae but harvest nutrients from plankton and particulate matter.

Coral Reefs Corals exist as both solitary and colonial formations. The colonial corals produce enormous structures, formed by the accumulation of skeletons into calcium carbonate that is lithified into rock. *Coral reefs* form through many generations, with live corals near the ocean's surface building on the foundation of older coral skeletons, which, in turn, may rest upon a volcanic seamount or some other submarine feature built up from the ocean floor. Thus, a coral reef is a biologically derived sedimentary rock that can assume one of several distinctive shapes.

In 1842, Charles Darwin proposed a hypothesis for reef formation. He suggested that, as reefs develop around a volcanic island and the island itself gradually subsides, equilibrium is maintained between the subsidence of the island and the upward growth of the corals (to keep the living corals at their optimum depth, not too far below the surface). This idea, generally accepted today, is portrayed in Figure 16.20. Note the specific examples of each reef stage: *fringing reefs* (platforms of surrounding coral rock), *barrier reefs* (reefs that enclose lagoons), and *atolls* (circular, ring-shaped reefs).

▲**Figure 16.21 The Bahama Platform.** The Bahama archipelago is made up of two carbonate platforms consisting of shallow-water limestone formations (light blue colors). The slopes of the platforms reach depths of 4000 m (13,100 ft; darker blue colors). [*Terra* image, NASA/GSFC.]

▲**Figure 16.22 Coastal salt marsh.** This wetland along the Long Island coast, New York, is protected from development by a private land trust. Land trusts are non-profit, independent organizations that work with landowners to conserve natural resources and open space. Since the vast majority of coastal areas are privately owned, land trusts have become critical tools for wetland protection. [Brooks Kraft/Corbis.]

Earth's most extensive fringing reef is the Bahama Platform in the western Atlantic (Figure 16.21), covering some 96,000 km^2 (37,000 mi^2). The largest barrier reef, the Great Barrier Reef along the shore of the state of Queensland, Australia, exceeds 2025 km (1260 mi) in length, is 16–145 km (10–90 mi) wide, and includes at least 700 coral-formed islands and keys (coral islets or barrier islands).

Coral Bleaching As mentioned previously, coral reefs may experience a phenomenon known as *bleaching*, in which normally colorful corals turn stark white by expelling their own nutrient-supplying algae. Exactly why the corals eject their symbiotic partner is unknown, for without algae the corals die. Scientists are currently tracking this worldwide phenomenon, which is occurring in the Caribbean Sea and the Indian Ocean as well as off the shores of Australia, Indonesia, Japan, Kenya, Florida, Texas, and Hawai'i. Possible causes include local pollution, disease, sedimentation, changes in ocean salinity, and increasing oceanic acidity.

Since 2000, scientists have acknowledged that the warming of sea-surface temperatures, linked to greenhouse warming of the atmosphere, is a greater threat to corals than local pollution or other environmental problems. Although a natural process, coral bleaching is now occurring at an unprecedented rate as average ocean temperatures climb higher with climate change. The 1998 record El Niño event caused the die-off of perhaps 30% of the world's reefs. In 2010, scientists reported one of the most rapid and severe coral bleaching and mortality events on record near Aceh, Indonesia, on the northern tip of the island of Sumatra. Some species declined 80% in just a few months, in response to increased sea-surface temperatures across the region. Many of these corals previously were resilient in the face of other ecosystem disruptions, including the Sumatra–Andaman tsunami in 2004.

As sea-surface temperatures continue to rise and ocean acidification worsens, coral losses will continue. For more information and Internet links, see the Global Coral Reef Monitoring Network at http://www.gcrmn.org/.

Coastal Wetlands

In some coastal areas, sediments are rich in trapped organic matter and as a result have great *biological productivity*—lush plant growth and spawning grounds for fish, shellfish, and other organisms. A coastal marsh environment of this type can greatly outproduce a wheat field in raw vegetation per acre and provides optimal habitat for varied wildlife. Unfortunately, these wetland ecosystems are quite fragile and are threatened by human development (Figure 16.22).

GEOreport 16.4 Ocean acidification impacts corals

As the oceans absorb more excess carbon dioxide, their acidity increases and potentially attacks coral formations, an interaction that scientists are actively researching. A 2013 study examined Mediterranean red coral (*Corallium rubrum*) colonies under more acidic conditions in a laboratory and discovered reduced growth rates of 59% and abnormal skeleton development when compared with colonies growing under current ocean conditions. The test conditions were at a pH of 7.8 (which would occur with CO_2 levels of 800 ppm, forecasted for the year 2100) as compared to recent conditions of pH 8.1 (380 ppm).

(a) Corals, sponges, anemones, fish and invertebrates live among the roots of the red mangrove (*Rhizophora mangle*) in St. John, U.S. Virgin Islands. No other mangrove system in the Caribbean is known to support such a diversity of corals.

(b) A shrimp farm occupies part of a cleared mangrove forest in Thailand; aquaculture is one of many threats to mangrove ecosystems.

▲**Figure 16.23 Mangroves.** [(a) USGS. (b) think4photop/Shutterstock]

As discussed in Chapter 9, wetlands are permanently or seasonally saturated with water, and as such have hydric soils (with anaerobic, or "oxygen-free" conditions) and support *hydrophytic vegetation* (plants that grow in water or wet soil). Coastal wetlands are of two general types—mangrove swamps (occurring between 30° N and 30° S latitude) and salt marshes (at latitudes of 30°and higher). This distribution is dictated by temperature, specifically, the occurrence of freezing conditions.

In tropical regions, sediment accumulation on coastlines provides sites for mangroves, the name for the trees, shrubs, palms, and ferns that grow in these intertidal areas, as well as for the habitat, which is known as a **mangrove swamp**. These ecosystems have a high diversity of species that are tolerant of saltwater inundation but generally intolerant of freezing temperatures (especially as seedlings). Mangrove roots are typically visible above the waterline, but the root portions that reach below the water surface provide a habitat for a multitude of specialized life forms (Figure 16.23a). The root systems maintain water quality by trapping sediment and taking up excess nutrients and prevent erosion by stabilizing accumulated sediments.

Mangrove ecosystems are threatened by ongoing removal, owing to falsely conceived fears of disease or pestilence; to pollution, especially from agricultural runoff; to overharvesting, especially in developing countries where they supply firewood; to storm surges in areas where protective barrier islands and coral reefs have disappeared; and to climate change, since mangroves require a stable sea level for long-term survival. According to the Food and Agriculture Organization of the United Nations, 20% of the world's mangroves were lost from 1980 to 2005 (Figure 16.23b). A 2011 study using satellite data reported that the remaining extent of global mangroves is 12% less than previously thought. Loss of these ecosystems also affects climate, since mangroves store carbon in greater amounts than other tropical forests.

Salt marshes consist mainly of halophytic (salt-tolerant) plants (mainly grasses) and usually form in estuaries and in the tidal mud flats behind barrier beaches and spits. These marshes occur in the intertidal zone and are often characterized by sinuous, branching water channels produced as tidal waters flood into and ebb from the marsh. Marsh vegetation traps and filters sediment, spreads out floodwaters, and buffers coastlines from storm surges associated with hurricanes. However, in many regions these coastal wetlands are threatened by human activities and the effects of climate change.

Most of coastal Louisiana, which represents about 40% of the coastal marshes in the United States, is profoundly altered and disrupted by dam and levee construction, flow alterations, oil and gas exploration and pumping, pipelines, and dredging for navigation and industrial needs. Wetlands in the Mississippi delta are disappearing at the rate of 65 km^2 (25 mi^2) per year. On the *MasteringGeography* website, you will find a discussion of the impacts of wetland removal, storm surge, and rising sea level on Bayou Lafourche in southern Louisiana.

(a) Wind-sculpted tree near South Point, Hawai`i. Nearly constant tradewinds keep this tree naturally pruned.

(b) Wind-eroded snow, called *sastrugi*, usually forms irregular grooves or ridges that are parallel to the wind direction.

▲**Figure 16.24 The work of wind.** [Bobbé Christopherson.]

Wind Processes

The effects of wind as an agent of geomorphic change are most easily visualized in coastal and desert environments. Moving air is a fluid, like water, and like moving water, causes erosion and deposition, shifting or transporting materials such as dust, sand, and snow. The work of wind is **eolian** (also spelled *aeolian*, for Aeolus, ruler of the winds in Greek mythology).

Since the viscosity and density of air are much lower than those of other transporting agents such as water and ice, the ability of wind to move materials is correspondingly weaker. Yet, over time, wind accomplishes enormous work. Consistent local wind can prune and shape vegetation and sculpt snow surfaces (Figure 16.24).

Eolian Transport of Dust and Sand

Just like water in a stream picking up sediment, wind exerts a drag, or frictional pull, on surface particles until they become airborne. Grain size, or particle size, is important in wind erosion. Intermediate-sized grains move most easily, whereas movement of the largest and the smallest sand particles requires the strongest winds. Stronger wind is needed for the large particles because they are heavier, and for the small particles because they are mutually cohesive and because they usually present a smooth (aerodynamic) surface that minimizes frictional pull. Eolian processes only work on dry surface materials, since wet soils and sediments are too cohesive for movement to occur.

The distance that wind is capable of transporting particles in *suspension* also varies greatly with particle size (for comparison, Figure 14.13 shows the transport of stream sediment in suspension, a similar process). The finer material suspended in a dust storm is lifted much higher than are the coarser particles of a sandstorm (Figure 16.25).

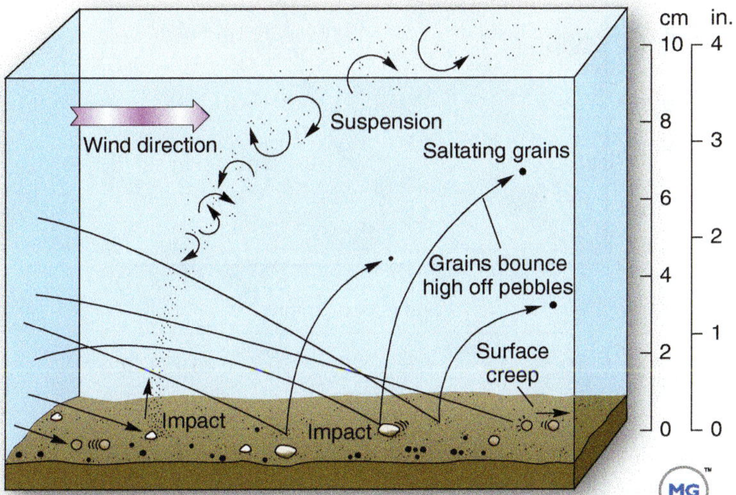

(a) Eolian suspension, saltation, and surface creep are transportation mechanisms.

MG Animation How Wind Moves Sand

(b) Sand grains saltating along the surface in the Stovepipe Wells dune field, Death Valley, California.

▲**Figure 16.25 How the wind moves sand.** [Author.]

▲Figure 16.26 **Dust storm engulfing Phoenix, Arizona.** A massive dust storm known as a haboob passes through Phoenix, Arizona, in July 2011. In dry regions such a wall of dust frequently precedes a thunderstorm, with winds traveling in the opposite direction of the oncoming storm. [Ross D. Franklin/AP Photo.]

Thus, the finest dust particles travel the farthest distances. As discussed in previous chapters, atmospheric circulation can transport fine material, such as volcanic debris, fire soot and smoke, and dust, worldwide within days (please review Chapter 3, Figure 3.7, and Chapter 6, Figure 6.1). In some arid and semiarid regions, winds associated with thunderstorms can cause dramatic dust storms (Figure 16.26) consisting of fine particles that infiltrate even the smallest cracks of homes and businesses.

Eolian processes transport particles larger than about 0.2 mm (0.008 in.) along the ground by *saltation*, the bouncing and skipping action. About 80% of wind transport of particles is accomplished by saltation (Figure 16.25). In fluvial transport, saltation is accomplished by hydraulic lift; in eolian transport, saltation occurs by aerodynamic lift, elastic bounce, and impact with other particles (compare Figure 16.25a with Figure 14.13).

Particles too large for saltation slide and roll along the ground surface, a type of movement called **surface creep**. Saltating particles may collide with sliding and rolling particles, knocking them loose and forward in this process, which affects about 20% of the material transported by wind. In a desert or along a beach, sometimes you can hear a slight hissing sound, almost like steam escaping, produced by the myriad saltating grains of sand as they bounce along and collide with surface particles. Once particles are set in motion, the wind velocity need not be as high to keep them moving.

British Army Major Ralph Bagnold, an engineering officer stationed in Egypt in 1925, pioneered studies of wind transport and authored a classic work in geomorphology, *The Physics of Blown Sand and Desert Dunes*, published in 1941. One result of Bagnold's work was a graph showing the rate at which the amount of transported sand over a dune surface increases with wind speed (Figure 16.27). The graph shows that at lower wind speeds, sand moves only in small amounts; however, beyond a wind speed of about 30 kmph (19 mph), the amount of sand moved increases rapidly. A steady wind of 50 kmph (30 mph) can move approximately one-half ton of sand per day over a 1-m-wide section of dune.

Eolian Erosion

Erosion of the ground surface resulting from the lifting and removal of individual particles by wind is **deflation**. Wherever wind encounters loose sediment, deflation may remove enough material to form depressions in the landscape ranging in size from small indentations less than a meter wide up to areas hundreds of meters wide and many meters deep. The smallest of these are known as *deflation hollows,* or *blowouts.* They commonly occur in dune environments where winds

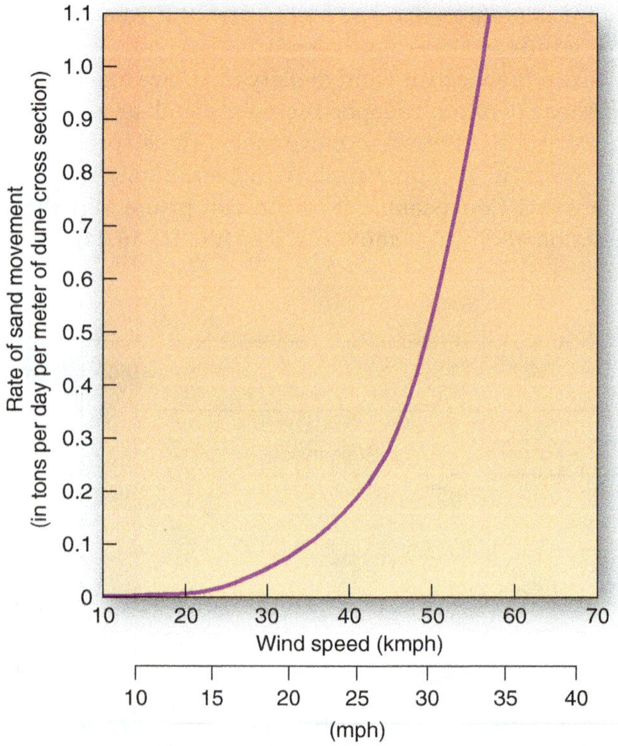

▲Figure 16.27 **Sand movement and wind velocity.** Sand movement relative to wind velocity, as measured over a meter cross section of ground surface. [Created from data in *The Physics of Blown Sand and Desert Dunes*, by R.A. Bagnold, © 1941, 1954 (Methuen and Co., 1954).]

▲**Figure 16.28 A ventifact.** One of the wind-eroded rocks in the Dry Valleys area of Antarctica, a snow-free polar desert with winds reaching speeds of 320 kmph (200 mph). [Scott Darsney/Lonely Planet Images/Getty.]

remove sand from specific areas, often in conjunction with the removal of stabilizing vegetation (possibly by fire, by grazing, or from drought). Large depressions in the Sahara Desert are at least partially formed by deflation but are also affected by large-scale tectonic processes. The enormous Munkhafad el Qattâra (Qattâra Depression), which covers 18,000 km^2 (6950 mi^2) just inland from the Mediterranean Sea in the Western Desert of Egypt, is now about 130 m (427 ft) below sea level at its lowest point.

The grinding and shaping of rock surfaces by the "sandblasting" action of particles captured in the air is **abrasion**. Like the sandblasting of streets and buildings for maintenance, this process is accomplished by a stream of compressed air filled with sand grains that quickly abrade the surface. Since sand grains are not lifted to great heights above the ground surface, abrasive action in nature usually is restricted to a distance of no more than a meter or two above the ground. Variables that affect natural abrasion rates include the hardness of surface rocks, wind velocity, and wind constancy.

Rocks that are pitted, fluted (grooved), or polished from eolian erosion are called **ventifacts** (literally, "artifacts of the wind," shown in Figure 16.28). They usually become aerodynamically shaped in a direction determined by the consistent flow of airborne particles in prevailing winds. On a larger scale, deflation and abrasion together are capable of streamlining multiple rock structures in a landscape in alignments parallel to the most effective wind direction, thus producing distinctive, elongated formations called **yardangs**. Abrasion is concentrated on the windward end of each yardang, with deflation operating on the leeward portions. These wind-sculpted features can range from meters to kilometers in length and up to many meters in height (Figure 16.29).

On Earth, some yardangs are large enough to be detected on satellite imagery. The Ica Valley of southern Peru contains yardangs reaching 100 m (330 ft) in height and several kilometers in length. The Sphinx in Egypt was perhaps partially formed as a yardang, whose natural shape suggested a head and body. Some scientists think this shape led the ancients to complete the bulk of the sculpture artificially with masonry.

▲**Figure 16.29 A field of yardangs.** Abrasion from consistent, unidirectional winds shaped these yardangs in the Qaidam Basin, northwest China. [Xinhua/Photoshot.]

Desert Pavement

The work of wind deflation is important for the formation of **desert pavement**, a hard, stony surface—as opposed to the usual sand—that commonly occurs in arid regions (Figure 16.30a). Scientists have put forth several explanations for the formation of desert pavement. One explanation is that deflation literally blows away loose or noncohesive sediment, eroding fine dust, clay, and sand and leaving behind a compacted concentration of pebbles and gravel (Figure 16.30b).

Another hypothesis that better explains some desert pavement surfaces states that deposition of windblown sediments, not removal, is the formative process. Windblown particles settle between and below coarse rocks and pebbles that are gradually displaced upward. Rainwater plays a part, as wetting and drying episodes swell and shrink clay-sized particles. The gravel fragments are gradually lifted to surface positions to form the pavement (Figure 16.30c).

Desert pavements are so common that many provincial names are used for them—for example, *gibber plain* in Australia; *gobi* in China; and in Africa, *lag gravels* or *serir*, or *reg* desert if some fine particles remain. Most desert pavements are strong enough to support human weight, and some can support motor vehicles; but in general these surfaces are fragile. They are also of critical importance, since they protect underlying sediment from further deflation and water erosion.

(a) A typical desert pavement.

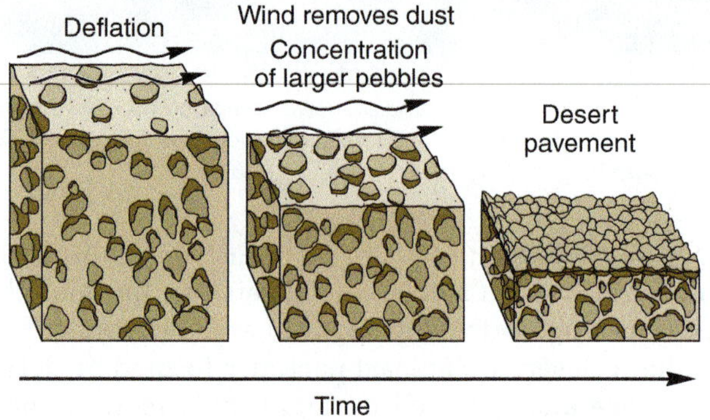

(b) The deflation hypothesis: Wind removes fine particles leaving larger pebbles, gravels, and rocks, which become consolidated into desert pavement.

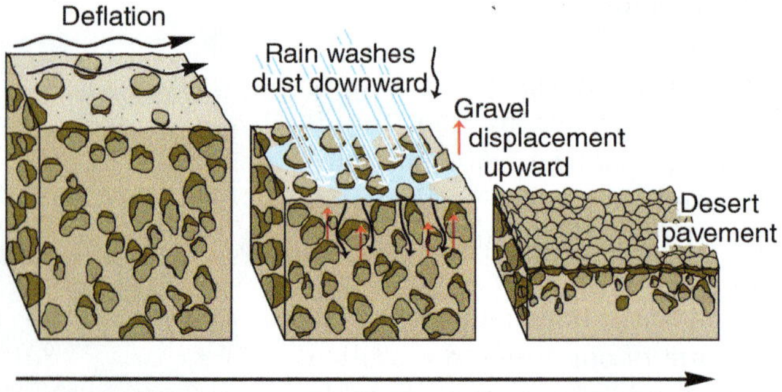

(c) The sediment-accumulation hypothesis: Wind delivers fine particles that settle and wash downward as cycles of swelling and shrinking cause gravels to migrate upward, forming desert pavement.

▲**Figure 16.30 Desert pavement.** [(a) Bobbé Christopherson.]

Eolian Deposition

The smallest features shaped by the movement of windblown sand are ripples, which form in crests and troughs, positioned transversely (at a right angle) to the direction of the wind. Larger deposits of sand grains form **dunes**, defined as wind-sculpted, transient ridges or hills of sand. An extensive area of windblown sand (usually larger than 125 km^2, or 48 mi^2) is an **erg** (after the Arabic word for "dune field"), or a **sand sea**.

The Grand Erg Oriental in the central Sahara, active for more than 1.3 million years, exceeds 1200 m (4000 ft) in depth and covers 192,000 km^2 (75,000 mi^2), comparable to the area of Nebraska. Similar sand seas, such as the Grand Ar Rub'al Khālī Erg, are active in Saudi Arabia. In eastern Algeria, the Issaouane Erg covers 38,000 km^2 (14,673 mi^2) of the Sahara Desert (Figure 16.31a). Extensive dune fields characterize sand seas, which are also present in semiarid regions such as the Great Plains of the United States (Figure 16.31b), as well as on the planet Mars (see photos on the *MasteringGeography* website).

Dune Formation and Movement When saltating sand grains encounter small patches of sand, their kinetic energy (motion) is dissipated and they accumulate. Once the height of such accumulations increases above 30 cm (12 in.), a *slipface* and characteristic dune features form. *Geosystems In Action*, Figure GIA 16, illustrates a dune profile and various dune forms.

A dune usually is asymmetrical in one or more directions. Winds characteristically create a gently sloping *windward side* (stoss side), with a more steeply sloped slipface on the *leeward side* (Figure GIA 16.1).

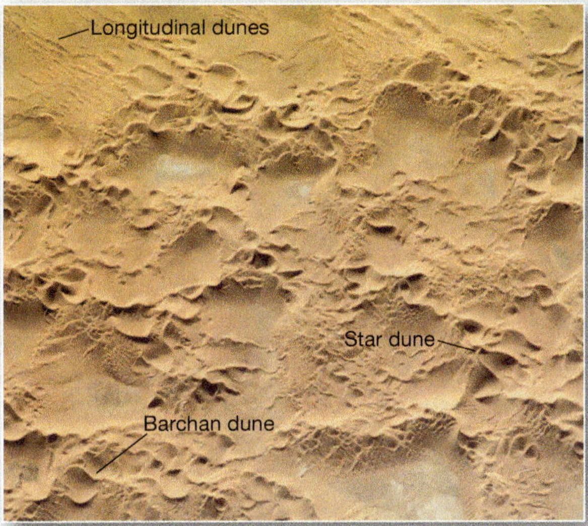

(a) The Issaouane Erg of eastern Algeria consists of star dunes, barchan dunes, and longitudinal dunes, disclosing the prevailing wind history of the region.

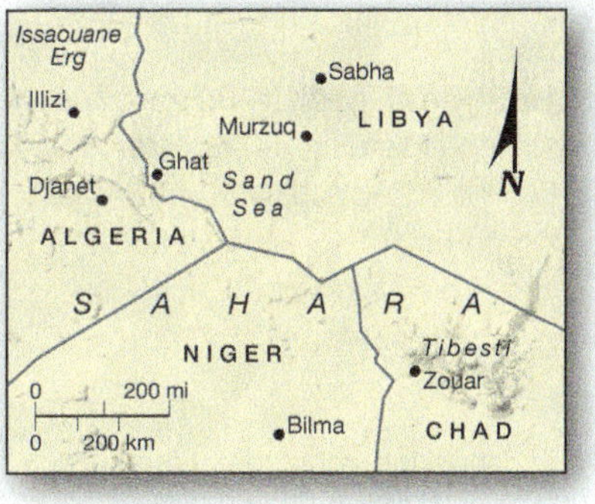

Figure 16.31 Examples of sand seas, or ergs. [(a) ISS astronaut photo, NASA/GSFC. (b) NASA.]

(b) The Sand Hills in central Nebraska are sand and silt deposits derived from glaciated regions to the north and west. These densely packed barchan dunes, inactive for at least 600 years, are now stabilized by vegetation (green). Water is in blue.

The angle of a slipface is the steepest angle at which loose material is stable—its *angle of repose.* Thus, the constant flow of new material makes a slipface a type of avalanche slope: Sand builds up as it moves over the crest of the dune to the brink; then it avalanches, falling and cascading as the slipface continually adjusts, seeking its angle of repose (usually 30° to 34°). In this way, a dune migrates downwind, in the direction in which effective—that is, sand-transporting—winds are blowing, as suggested by the successive dune profiles in GIA 16.1. (Stronger seasonal winds or winds from a passing storm may prove more effective in this regard than average prevailing winds.)

Dunes have many wind-produced shapes that make classification difficult. Scientists generally classify dunes according to three general shapes—*crescentic* (crescent, curved shape), *linear* (straight form), and massive *star dunes.* Figure GIA 16.2 shows eight types of dunes that fall within these classes or are a complex mix of these general shapes. The crescentic class includes *barchan, transverse, parabolic,* and *barchanoid ridge* dunes. Linear dunes include *longitudinal* dunes and *seif* dunes (not pictured). Seif dunes are named after the Arabic word for "sword" and are shorter with a more sinuous crest than longitudinal dunes. Star dunes are the largest in size. *Dome* dunes are rare, and often occur at the margins of sand seas. *Reversing* dunes form where winds frequently reverse direction.

Active sand dunes cover about 10% of Earth's deserts and dune migration can threaten populated areas (see *The Human Denominator,* HD 16d, ahead). Dune fields are also present in humid climates, such as along coastal Oregon, the south shore of Lake Michigan, and the U.S. Gulf and Atlantic coastlines.

GEOreport 16.5 Human activities disturb eolian landforms

In the United States, recreational off-road vehicles (ORVs) and all-terrain vehicles (ATVs) currently number more than 15 million. Such vehicles erode desert dunes, disrupt desert pavement, and create ruts that easily concentrate surface runoff into gullies that deepen with continuing erosion.

Military activities such as explosions and the movement of heavy vehicles also destroy desert pavement, as in Iraq and Afghanistan, where the breakdown of thousands of square kilometers of stable desert pavement has sent dust and sand into nearby cities and onto farmland. In the Registan Desert of southern Afghanistan, south of the city of Kandahar, drought conditions were exacerbated by military activities. The disruption of desert pavement surfaces resulted in sand movement that has overtaken areas of sparse agricultural activity and has covered more than 100 villages.

The dramatic, sculptural shapes of dunes occur in a variety of settings: along shorelines, in sandy parts of deserts, and in semiarid regions. Wherever there is a sufficient supply of loose, dry sand or other fine particles unprotected by plant cover, wind erosion and deposition can build dunes (GIA 16.1). Prevailing winds, along with other factors, create dunes of many sizes and shapes (GIA 16.2).

Sunset in the Sahara Desert [Galyna Andrushko/Shuttershock]

16.1 DUNE PROFILE

Wind erosion and deposition work together to build a dune's characteristic profile. A dune grows as wind-borne particles accumulate on the gentler, windward slope, then cascade down the steep slipface of the leeward slope.

Angle of repose:
The loose particles on the slipface tend to slip and slide downhill until the slope stabilizes at its angle of repose—about 30°–34°—the steepest angle at which the particles are stable.

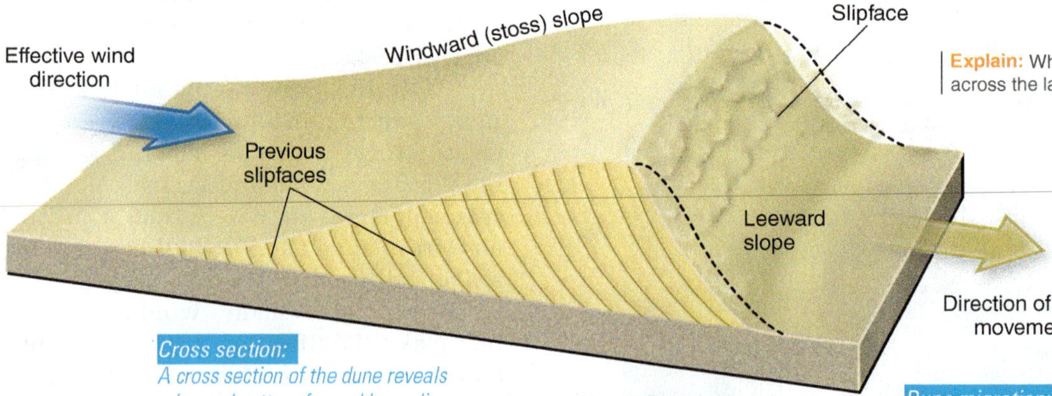

Explain: Why do sand dunes migrate across the land surface?

Cross section:
A cross section of the dune reveals a layered pattern formed by earlier slipfaces as the dune grew.

Dune migration:
Over time, the ongoing process that transfers particles from the windward slope to the leeward slope causes the entire dune to migrate downwind.

16.2 DUNE FORMS

The different types of sand dunes vary in shape and size depending on several factors including:
- directional variability and strength (or "effectiveness") of the wind
- whether the sand supply is limited or abundant
- presence or absence of vegetation

Barchan:
Crescent-shaped dune with horns pointed downwind; found in areas with constant winds and little directional variability, and where limited sand is available.

Transverse:
Long, slightly sinuous dune with an asymmetrical ridge and only one slipface, aligned transverse (or perpendicular) to wind direction; results from relatively ineffective wind and abundant sand supply.

Animation
Dune Formation and Cross-Bedding

MasteringGeography™

Visit the Study Area in MasteringGeography™ to explore dunes.

Visualize: Study a geosciences animation of dunes.

Assess: Demonstrate understanding of dunes (if assigned by instructor).

16.2 DUNE FORMS (continued)

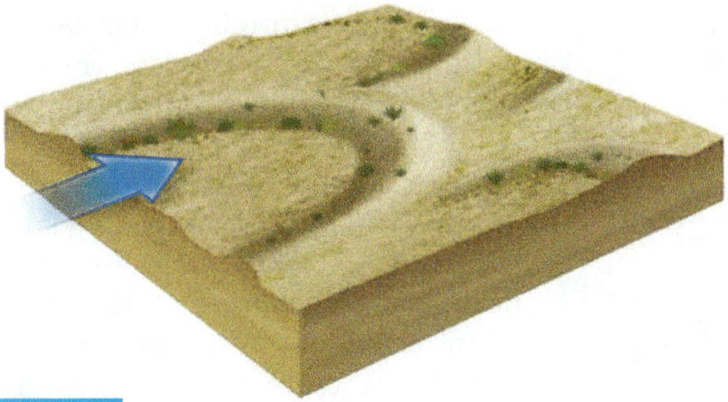

Parabolic:
Crescent-shaped dune with opening end facing upwind; U-shaped "blowout" and arms anchored by vegetation, which stabilizes dune form.

Barchanoid ridge:
Wavy, asymmetrical, dune formed from coalesced barchans with ridges aligned transverse to effective winds; resembles connected crescents in rows with open areas between.

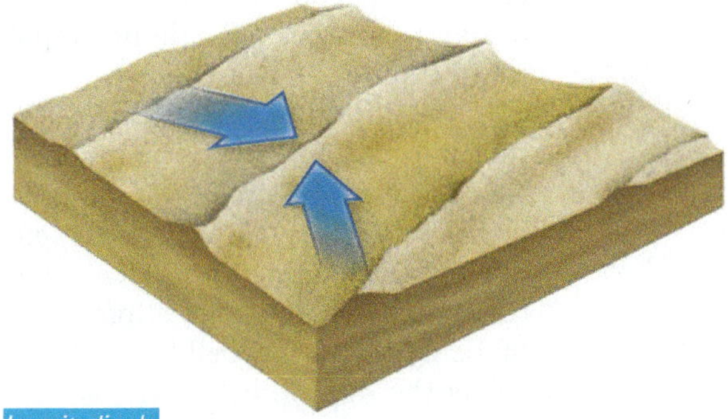

Longitudinal:
Linear, slightly sinuous, ridge-shaped dune, aligned parallel with the wind direction. Averages 100 m (328 ft) high and 100 km (63 mi) long but can reach to 400 m (1312 ft) high.

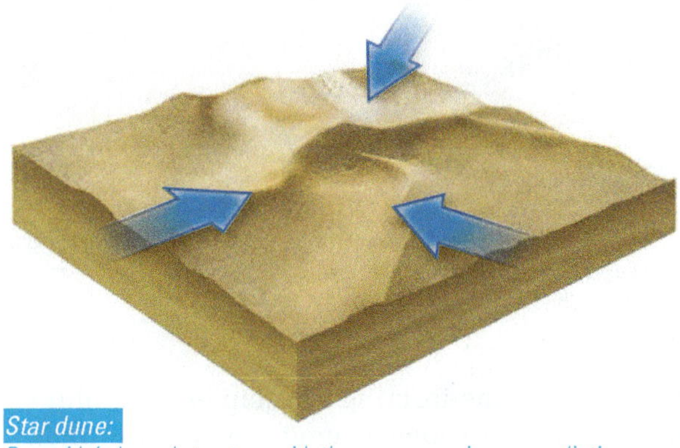

Star dune:
Pyramidal-shaped structure with three or more sinuous, radiating arms extending outward from a central peak; results from effective winds shifting in all directions.

Dome:
Circular or elliptical dune with no slipface; sometimes modified into barchanoid forms, and sometimes stabilized by vegetation.

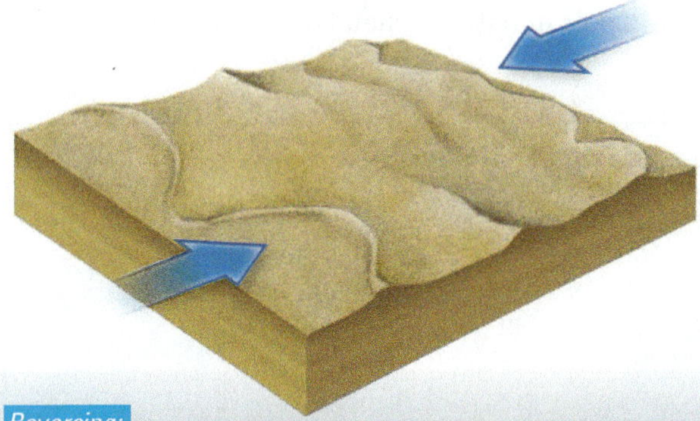

Reversing:
Dune with asymmetrical ridge, intermediate between star dune and transverse dune; wind variability can alter shape between these forms.

| **Compare:** How are barchan and parabolic dunes similar? How are they different? |

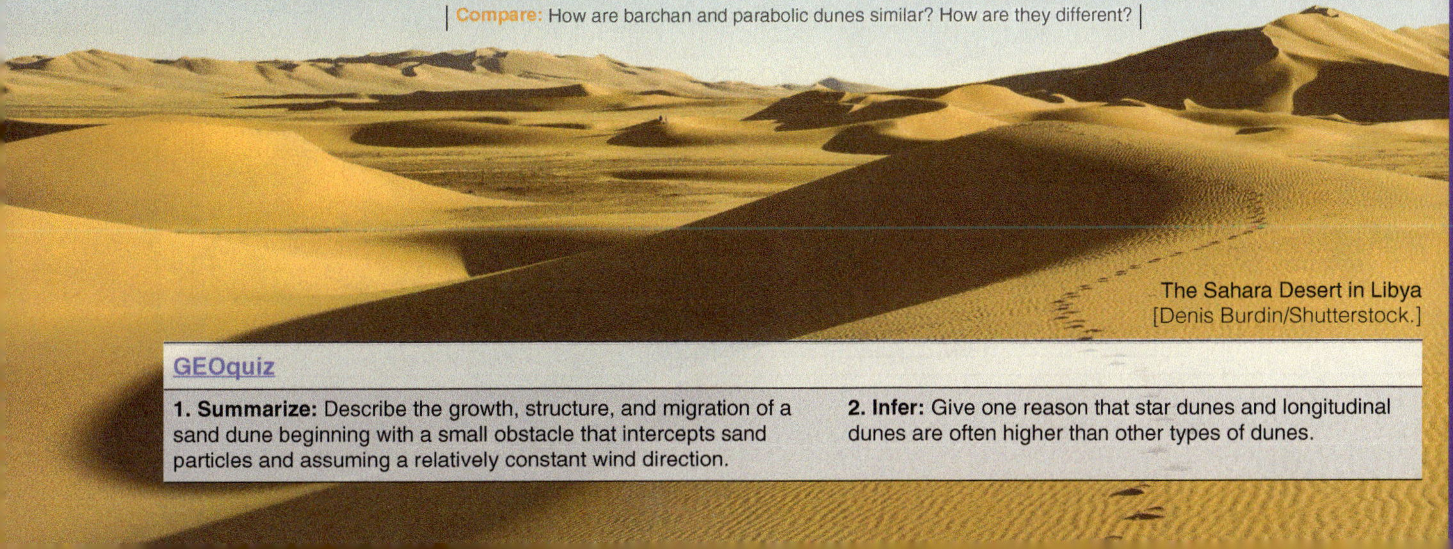

The Sahara Desert in Libya
[Denis Burdin/Shutterstock.]

GEOquiz

1. Summarize: Describe the growth, structure, and migration of a sand dune beginning with a small obstacle that intercepts sand particles and assuming a relatively constant wind direction.

2. Infer: Give one reason that star dunes and longitudinal dunes are often higher than other types of dunes.

▲**Figure 16.32 Cross bedding in sedimentary rocks.** The bedding pattern of cross stratification in these sandstone rocks tells us about dune environments before lithification (hardening into rock). [Bobbé Christopherson.]

In some ergs, winds from varying directions produce star dunes with multiple slipfaces (Figure 16.31a). *Star dunes* are the mountainous giants of the sandy desert. They are pinwheel-shaped, with several radiating arms rising and joining to form a common central peak that can approach 200 m (650 ft) in height. (Some in the Sonoran Desert are close to this size.) In the image, you can also see some crescentic dunes, suggesting that the region has experienced changing wind patterns over time, as, in contrast to star dunes, crescentic dunes form from a single principal wind flow.

Ancient sand dunes can be lithified into sedimentary rock that carries patterns of cross bedding, or *cross stratification*. As the ancient dune was accumulating, the sand that cascaded down its slipface established distinct bedding planes (layers) that remained as the dune lithified. These layers are now visible as cross bedding, so named because they form at an angle to the horizontal layers of the main strata (Figure 16.32). Ripple marks, animal tracks, and fossils also are found preserved in these desert sandstones.

These same dune-forming principles and dune terminology apply to snow-covered landscapes. *Snow dunes* are formed as wind deposits snow in drifts. In semiarid farming areas, drifting snow captured by fences and by tall stubble left in fields contributes significantly to soil moisture when the snow melts.

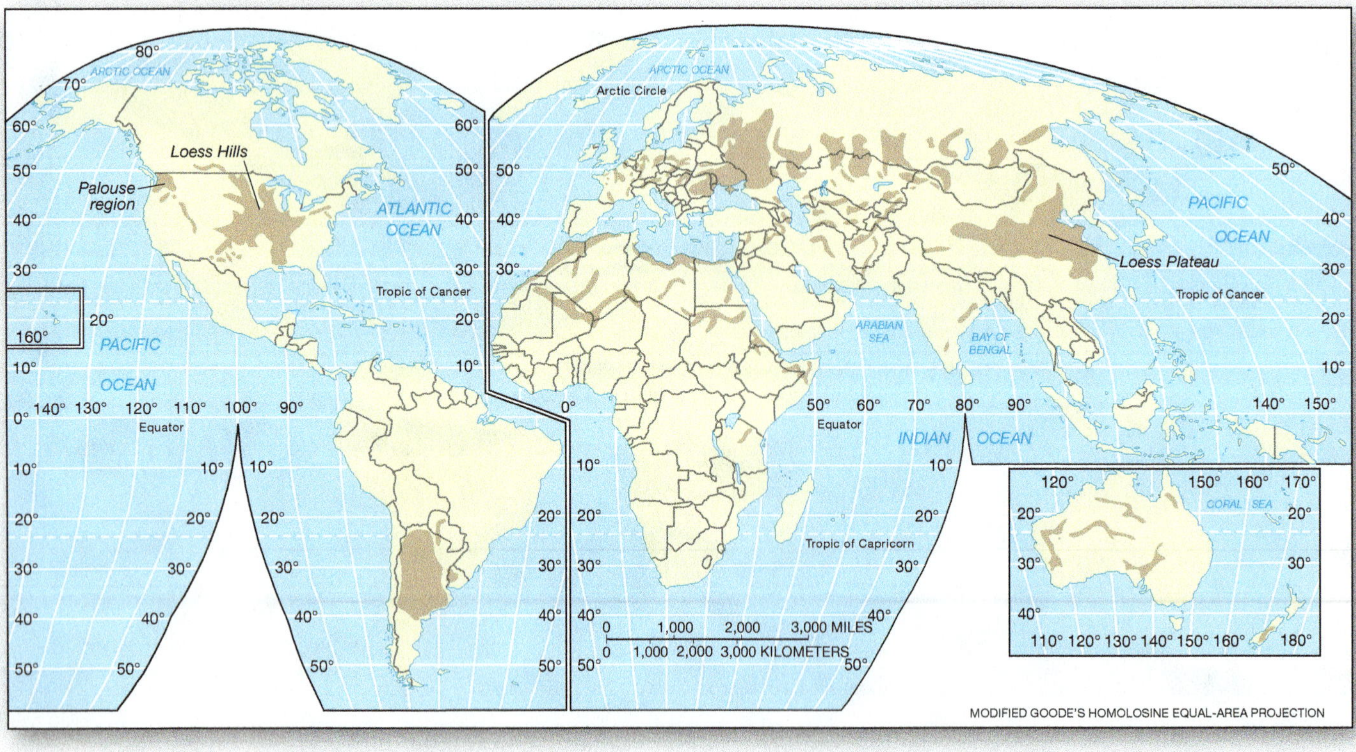

▲**Figure 16.33 Global loess deposits.** Major accumulations of loess. [Adapted from NRCS, FAO, and USGS data.]

Loess Deposits As discussed earlier, wind can transport smaller particles such as dust and silt long distances. In many regions of the world, fine-grained sediment (clays, silts, and fine sand) have accumulated in unstratified, homogeneous (evenly mixed) deposits called **loess** (pronounced "luss"), originally named by peasants working along the Rhine River Valley in Germany. Loess deposits form a thick blanket of material that covers previously existing landforms. Figure 16.33 shows the worldwide distribution of these accumulations.

The loess deposits in Europe and North America are thought to be derived mainly from glacial and periglacial sources, specifically the alluvial deposits from meltwater (called glacial outwash, discussed in Chapter 17). These sediments were left behind by Pleistocene glaciers that retreated in several episodes about 15,000 years ago. In the United States, significant loess accumulations with glacial origins occur throughout the Mississippi and Missouri River Valleys, forming continuous deposits 15–30 m (50–100 ft) thick. The Loess Hills of Iowa reach heights of about 61 m (200 ft) above the nearby prairie farmlands and run north–south for more than 322 km (200 mi; Figure 16.34a). Only China has deposits that exceed these areal dimensions.

The vast deposits of loess in China, covering more than 300,000 km^2 (116,000 mi^2), are derived from windblown desert sediment. Accumulations in the Loess Plateau of China are more than 300 m (1000 ft) thick, forming complex weathered badlands and good agricultural land. These windblown deposits are interwoven through much of Chinese history and social customs; in some areas, dwellings are carved into the strong vertical structure of loess cliffs. Loess deposits also cover much of Ukraine, central Europe, the Pampas region of Argentina, and New Zealand.

Because of its binding strength and internal coherence, loess weathers and erodes into steep bluffs, or vertical faces. When a bank is cut into a loess deposit, it generally will stand vertically, although it can fail if saturated (Figure 16.34b). Loess deposits are well drained and deep and have excellent moisture retention. The soils derived from loess are the basis for some of Earth's "breadbasket" farming regions.

(a) The Loess HIlls of Iowa near the Missouri River are made up of ridges and eroded bluffs.

(b) A loess bluff, or high cliff, exposed by erosion is structurally strong.

▲**Figure 16.34 Loess Hills and bluff, western Iowa.** [(a) Clint Farlinger/Alamy. (b) Bobbé Christopherson.]

CRITICALthinking 16.3
The Nearest Eolian Features

What eolian features are nearest to your present location? Are they coastal, lakeshore, or desert dunes? Which causative factors discussed in this chapter explain the location or form of these features? Are you able to visit the site? •

THE**human**DENOMINATOR 16 Oceans, Coasts, and Dunes

COASTAL SYSTEMS ⇨ HUMANS

- Rising sea level has potential to inundate coastal communities.
- Tsunami cause damage and loss of life along vulnerable coastlines.
- Coastal erosion changes coastal landscapes, affecting developed areas; human development on depositional features such as barrier island chains is at risk from storms, especially hurricanes.

HUMANS ⇨ COASTAL SYSTEMS

- Rising ocean temperatures, pollution, and ocean acidification impact corals and reef ecosystems.
- Human development drains and fills coastal wetlands and mangrove swamps, removing their effect of buffering storms.

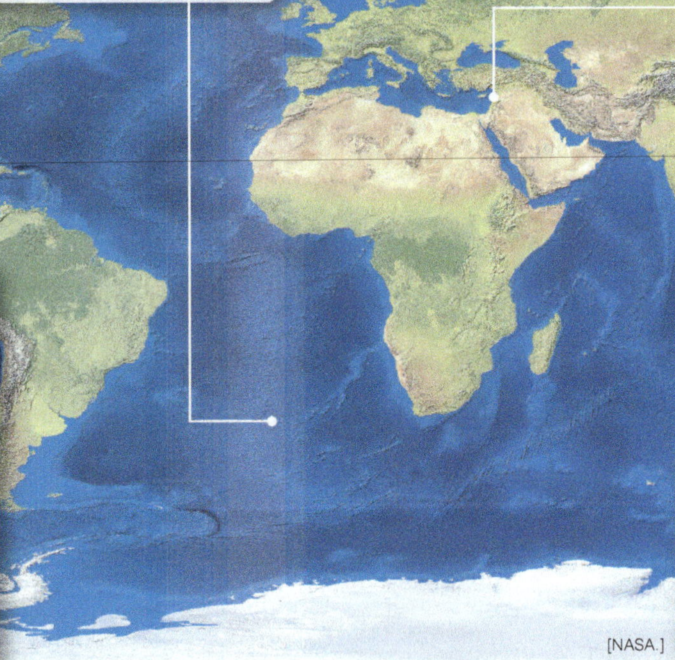

[NASA.]

16a

A tanker ran aground on Nightingale Island in the South Atlantic in 2011, spilling an estimated 800 tons of fuel and coating endangered penguins with oil. Please review the Chapter 6 *Geosystems Now*. [Trevor Glass/AP Photo.]

16b

The Saida Landfill in Lebanon sits on the coast of the Mediterranean. Winds and storms carry garbage out to sea, where it degrades water quality and coastal ecosystems. [Tomasz Grzyb/Demotix/Corbis.]

16c

In Aceh, Indonesia, near the site of the 2004 Indian Ocean tsunami, authorities encourage local people to plant mangroves for protection against future tsunami. [Nani Afrida/epa/Corbis.]

16d

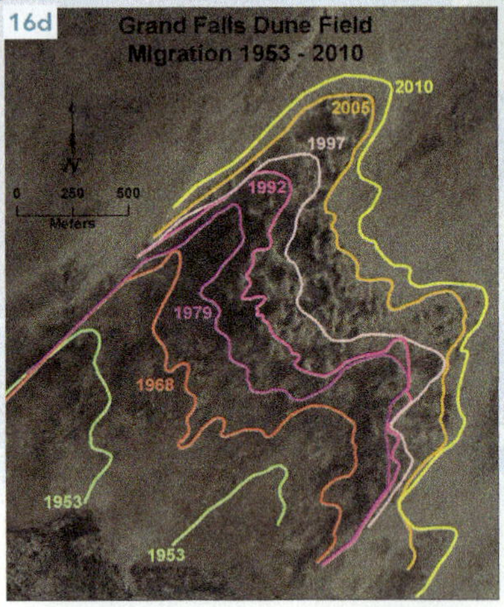

On Navajo Nation lands in the U.S Southwest, dune migration is threatening houses and transportation, and affecting human health. A recent USGS study revealed that the Grand Falls dune field in northeast Arizona increased 70% in areal extent from 1997 to 2007. The increasingly dry climate of this region has caused accelerated dune migration and reactivation of inactive dunes. [USGS.]

ISSUES FOR THE 21ST CENTURY

- Degradation and loss of coastal ecosystems—wetlands, corals, mangroves—will continue with coastal development and climate change.
- Continued building on vulnerable coastal landforms will necessitate expensive recovery efforts, especially if storm systems become more intense with climate change.

GEOSYSTEMS**connection**

In Chapters 13, 14, and 16, we examined aspects of weathering and erosion accomplished by the forces of gravity, streams, waves, and wind. In this chapter, we saw the effects of wind and wave action along coasts, in deserts, and in other environments. Next, in the final chapter of our study of exogenic processes, we look at the geomorphic agents at work in the cryosphere—Earth's snow and ice environments. These cold-region systems are undergoing the highest rate of alteration as Earth's climates change.

KEY LEARNING concepts review

■ *Describe* the chemical composition and physical structure of the ocean.

Because water is the "universal solvent," dissolving at least 57 of the 92 elements found in nature, seawater is a solution, and the concentration of dissolved solids is its **salinity**. **Brackish** water has less than 35‰ (parts per thousand) salinity; **brine** exceeds the average 35‰. The ocean is divided by depth into the narrow mixing zone at the surface, the thermocline transition zone, and the deep cold zone.

salinity (p. 483)
brackish (p. 484)
brine (p. 484)

1. Describe the salinity and composition of seawater, and the distribution of its solutes.
2. Analyze the latitudinal distribution of salinity discussed in the chapter. Why is salinity less along the equator and greater in the subtropics?
3. What are the three general zones in the physical structure of the ocean? Characterize each by temperature, salinity, dissolved oxygen, and dissolved carbon dioxide.

■ *Identify* the components of the coastal environment, *define* mean sea level, and *explain* the actions of tides.

The coastal environment is the **littoral zone** and exists where the tide-driven, wave-driven sea confronts the land. System inputs to the coastal environment include solar energy, wind and weather, climatic variation, coastal geomorphology, and human activities.

Mean sea level (MSL) is based on average tidal levels recorded hourly at a given site over many years. MSL varies over space because of ocean currents and waves, tidal variations, air temperature and pressure differences, ocean temperature variations, slight variations in Earth's gravity, and changes in oceanic volume. MSL worldwide is rising in response to global warming of the atmosphere and oceans.

Tides are complex daily oscillations in sea level, ranging worldwide from barely noticeable to many meters. Tides are produced by the gravitational pull of both the Moon and the Sun. Most coastal locations experience two high (rising) **flood tides** and two low (falling) **ebb tides** every day. The difference between consecutive high and low tides is the tidal range. **Spring tides** exhibit the greatest tidal range, when the Moon and the Sun are in either conjunction or opposition. **Neap tides** produce a lesser tidal range.

littoral zone (p. 487)
mean sea level (MSL) (p. 488)
tide (p. 489)
flood tide (p. 490)
ebb tide (p. 490)
spring tide (p. 490)
neap tide (p. 490)

4. What are the key terms used to describe the coastal environment?
5. Define mean sea level. How is this value determined? Is it constant or variable around the world? Explain.
6. What interacting forces generate the pattern of tides?
7. What characteristic tides are expected during a new Moon or a full Moon? During the first-quarter and third-quarter phases of the Moon? What is meant by a flood tide? An ebb tide?
8. Explain briefly how tidal power is used to produce electricity. Are there any tidal power plants in North America? If so, briefly describe where they are and how they operate.

■ *Describe* wave motion at sea and near shore, and *explain* coastal straightening and coastal landforms.

Friction between moving air (wind) and the ocean surface generates undulations of water that we call **waves**. Wave energy in the open sea travels through water, but the water itself stays in place. Regular patterns of smooth, rounded waves—the mature undulations of the open ocean—are **swells**. Near shore, the restricted depth of water slows the wave, forming *waves of translation*, in which both energy and water actually move forward toward shore. As the crest of each wave rises, the wave falls into a characteristic **breaker**.

Wave refraction redistributes wave energy along a coastline. Headlands are eroded, whereas coves and bays are areas of deposition, with the long-term effect of these differences being a straightening of the coast. As waves approach a shore at an angle, refraction produces a **longshore current** of water moving parallel to the shore. Particles move along the beach as **beach drift**, shifting back and forth between water and land. The combined action of this current produces the **littoral drift** of sand, sediment, gravel, and assorted materials along the shore. A **tsunami** is a seismic sea wave triggered by an undersea landslide or earthquake. It travels at great speeds in the open sea and gains height as it comes ashore, posing a coastal hazard.

An *erosional coast* features wave action that cuts a horizontal bench in the tidal zone, extending from a sea cliff out into the sea. Such a structure is a **wave-cut platform**, or *wave-cut terrace*. In contrast, *depositional coasts* generally are located along land of gentle relief, where depositional sediments are available from many sources. Characteristic landforms deposited by waves and currents are the **barrier spit** (material deposited in a long ridge extending out from a coast); the **bay barrier**, or *baymouth bar* (a spit that cuts the bay off from the ocean and forms an inland **lagoon**); the **tombolo** (where sediment deposits connect the shoreline with an offshore island or sea stack); and the **beach** (land along the shore where sediment is in motion, deposited by waves and currents).

wave (p. 491)
swell (p. 491)
breaker (p. 492)
wave refraction (p. 494)
longshore current (p. 494)
beach drift (p. 494)
littoral drift (p. 494)
tsunami (p. 494)
wave-cut platform (p. 497)
barrier spit (p. 498)
bay barrier (p. 498)
lagoon (p. 499)

tombolo (p. 499)
beach (p. 499)

9. What is a wave? How are waves generated, and how do they travel across the ocean? Does the water travel with the wave? Discuss the process of wave formation and transmission.
10. Describe the refraction process that occurs when waves reach an irregular coastline. Why is the coastline straightened?
11. Describe the process of beach drift and the movement of longshore currents to produce littoral drift.
12. Explain how a seismic sea wave attains such tremendous velocities. Why is it given the name tsunami?
13. What is meant by an erosional coast? What are the expected features of such a coast?
14. What is meant by a depositional coast? What are the expected features of such a coast?
15. How do people attempt to modify littoral drift? What strategies do they use? What are the positive and negative impacts of these actions?
16. Describe a beach—its form, composition, function, and evolution.
17. Is beach replenishment a practical strategy?

■ *Describe* barrier beaches and islands and their hazards as they relate to human settlement.

Barrier chains are long, narrow, depositional features, generally of sand, that form offshore roughly parallel to the coast. Common forms are **barrier beaches** and the broader, more extensive **barrier islands**. Barrier formations are transient coastal features, constantly on the move, and they are a poor, but common, choice for development.

barrier beach (p. 501)
barrier island (p. 501)

18. What types of impacts did Hurricanes Katrina and Sandy have on barrier beaches and islands in the Gulf and along the Atlantic coast?
19. On the basis of the information in the text and any other sources at your disposal, do you think barrier islands and beaches should be used for development? If so, under what conditions? If not, why not?

■ *Describe* the nature of coral reefs and coastal wetlands, and *assess* human impacts on these living systems.

A **coral** is a simple marine invertebrate that forms a hard, calcified, external skeleton. Over generations, corals accumulate in large reef structures. Corals live in a *symbiotic* (mutually helpful) relationship with algae; each is dependent on the other for survival.

Wetlands are lands saturated with water that support specific plants adapted to wet conditions. Coastal wetlands form as **mangrove swamps** equatorward of the 30th parallel in each hemisphere and as **salt marshes** poleward of these parallels.

coral (p. 503)
mangrove swamp (p. 506)
salt marsh (p. 506)

20. How are corals able to construct reefs and islands?
21. Describe a trend in corals that is troubling scientists, and discuss some possible causes.
22. Why are the coastal wetlands poleward of 30° N and S latitude different from those that are equatorward? Describe the differences.

■ *Describe* eolian transport of dust and sand, and *discuss* eolian erosion and the resultant landforms.

Eolian processes modify and move sand accumulations along coastal beaches and deserts. Wind exerts a drag or frictional pull on surface particles until they become airborne. The finer material suspended in a dust storm is lifted much higher than are the coarser particles of a sandstorm; only the finest dust particles travel significant distances. Saltating particles crash into other particles, knocking them both loose and forward. The motion of **surface creep** slides and rolls particles too large for saltation.

Erosion of the ground surface from the lifting and removal of particles by wind is **deflation**. Wherever wind encounters loose sediment, deflation may remove enough material to form depressions called *deflation hollows*, or *blowouts*, ranging in size from small indentations less than a meter wide up to areas hundreds of meters wide and many meters deep. **Abrasion** is the "sandblasting" of rock surfaces with particles captured in the air. Rocks that bear evidence of eolian abrasion are **ventifacts**. On a larger scale, deflation and abrasion are capable of streamlining rock structures, leaving behind distinctive rock formations or elongated ridges called yardangs. **Desert pavement** is the name for the hard, stony surface that forms in some deserts and protects underlying sediment from erosion.

eolian (p. 507)
surface creep (p. 508)
deflation (p. 508)
abrasion (p. 509)
ventifact (p. 509)
yardang (p. 509)
desert pavement (p. 510)

23. Explain the term *eolian*. How would you characterize the ability of the wind to move material?
24. Differentiate between a dust storm and a sandstorm.
25. What is the difference between eolian saltation and fluvial saltation?
26. Explain the concept of surface creep.
27. Who was Ralph Bagnold? What was his contribution to eolian studies?
28. Explain deflation. What role does deflation have in the formation of desert pavement?
29. How are ventifacts and yardangs formed by wind processes?

■ *Explain* the formation of sand dunes, and *describe* loess deposits and their origins.

Dunes are wind-sculpted accumulations of sand that form in arid and semiarid climates and along some coastlines where sand is available. An extensive area of dunes, such as that found in North Africa, is an **erg**, or **sand sea**. When saltating sand grains encounter small patches of sand, kinetic energy is dissipated and the grains start to build into a dune. As the height of the sand pile increases above 30 cm (12 in.), a steeply sloping *slipface* on the lee side and characteristic dune features are formed. Dune forms are broadly classified as *crescentic*, *linear*, and *star*.

Windblown **loess** deposits occur worldwide and can develop into good agricultural soils. These fine-grained clays and silts are moved by the wind many kilometers and are redeposited as an unstratified, homogeneous blanket of material.

dune (p. 510)
erg (p. 510)
sand sea (p. 510)
loess (p. 515)

30. What is an erg? Name an example of a sand sea. Where is the example located?
31. What are the three classes of dune forms? Describe an example within each class. What do you think is the major shaping force for sand dunes?
32. Which form of dune is the mountain giant of the desert? What are the characteristic wind patterns that produce such dunes?
33. Where does the sediment that forms the loess deposits of China come from? What is the origin of loess in Iowa? Name several other significant loess deposits on Earth.

MasteringGeography™

Looking for additional review and test prep materials? Visit the Study Area in *MasteringGeography*™ to enhance your geographic literacy, spatial reasoning skills, and understanding of this chapter's content by accessing a variety of resources, including MapMaster interactive maps, geoscience animations, satellite loops, author notebooks, videos, RSS feeds, flashcards, web links, self-study quizzes, and an eText version of *Geosystems*.

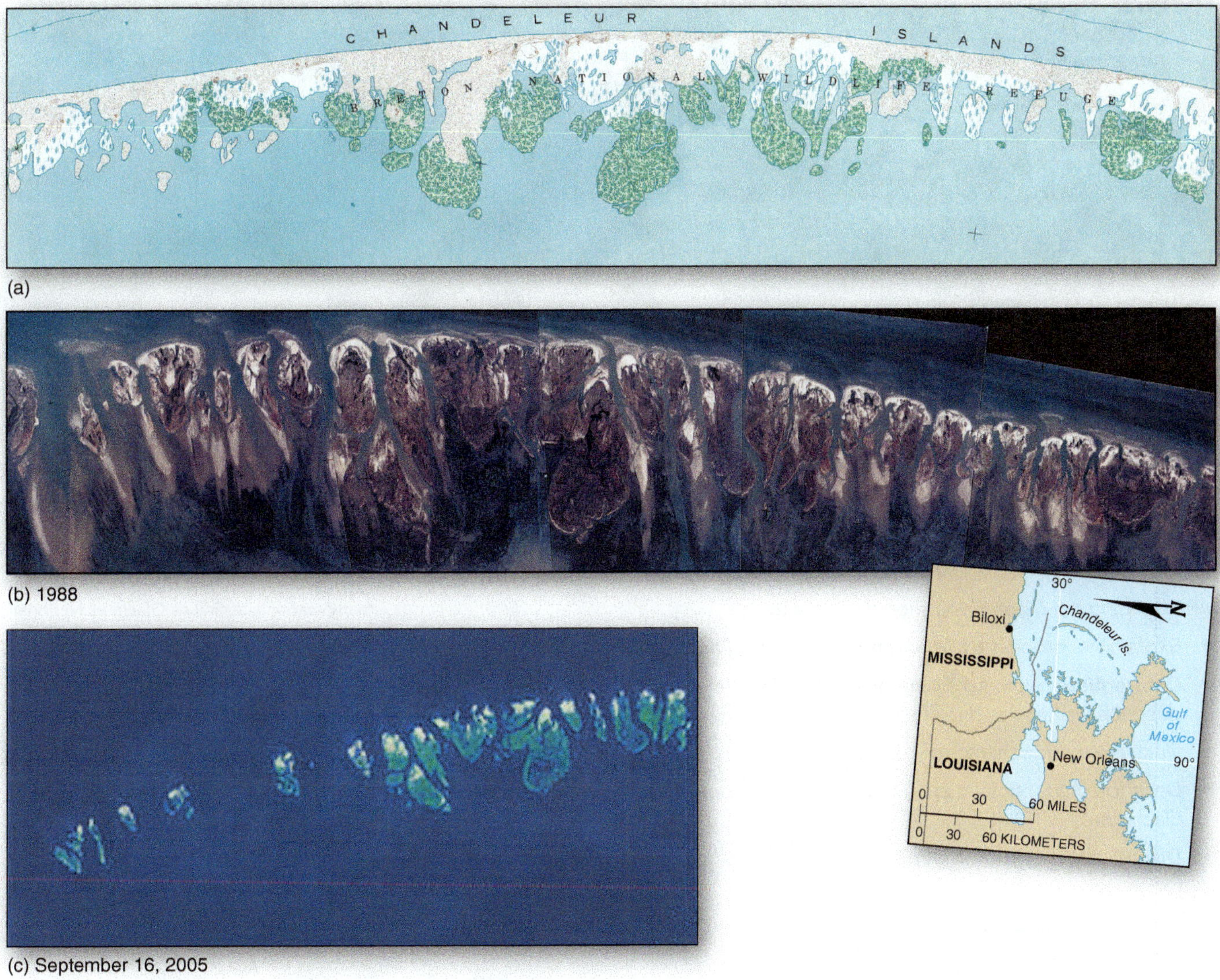

▲**Figure 16.35 Hurricane Katrina takes its toll on a barrier island chain.** Hurricane Katrina in 2005 eroded vast amounts of sand from the Chandeleur Islands just off the Louisiana and Mississippi Gulf coast. Compare the "before" topographic map (a) and aerial photo composite (b) with the 2005 image (c) taken about 3 weeks after Katrina's passage. The storm removed over 80% of the sand, leaving little of the barrier-island chain visible above sea level. The islands do not appear to be rebuilding since Katrina, owing mainly to lack of sediment inputs from the Mississippi River system. [(a) USGS. (b) Aerial Data Service, Earth Imaging. (c) USGS Hurricane Impact Studies.]

17 Glacial and Periglacial Landscapes

Mount Fitz Roy stands at 3375 m (11,073 ft) in elevation in the glacial landscape of the southern Andes in Argentina and Chile. Laguna Torre, in the foreground, is one of many glacial lakes in Los Glaciares National Park, located near Lago Viedma in the South Patagonia Ice Field (see Figure 17.5). Using satellite data from 2000 to 2012, scientists have determined that throughout this ice field, glaciers are thinning, even at the highest elevations, at a rate of about 1.8 m (5.9 ft) per year. Warming temperatures with climate change are causing melting in the Patagonian ice fields that contribute about 0.067 mm/year (0.003 in/year) to rising sea level; these freshwater losses will affect regional water supplies in the coming years. [Pichugan Dmitry/Shutterstock.]

KEY LEARNING concepts

After reading the chapter, you should be able to:

- ***Explain*** the process by which snow becomes glacial ice.
- ***Differentiate*** between alpine glaciers and continental ice sheets, and ***describe*** ice caps and ice fields.
- ***Illustrate*** the mechanics of glacial movement.
- ***Describe*** characteristic erosional and depositional landforms created by glaciation.
- ***Discuss*** the distribution of permafrost, and ***explain*** several periglacial processes.
- ***Describe*** landscapes of the Pleistocene ice-age epoch, and ***list*** changes occurring today in the polar regions.

Tidewater Glaciers and Ice Shelves Give Way to Warming

Warming air and ocean temperatures are causing changes to snow and ice features across the globe, and these changes are perhaps most visible along the coasts of the Arctic region, Greenland, and Antarctica. With ongoing warming, tidewater glaciers—large masses of glacial ice flowing downhill toward the sea—are increasing their flow rate and calving more frequently, breaking off large icebergs into the surrounding ocean. Ice shelves—thick, floating platforms of ice that extend over the sea while still attached to continental ice—are thinning and breaking up, forming large icebergs equivalent in size to small U.S. states. Let us look at some notable examples of these effects.

Petermann Glacier, Greenland In northwest Greenland, calving of the Petermann Glacier has released huge chunks of ice into the sea in recent years. In August 2010, an island of ice measuring 251 km^2 (97 mi^2) broke loose, forming the largest iceberg in the Arctic in a half century (equivalent to 25% of this glacier's entire floating ice shelf). Then in July 2012, another large iceberg broke off the tongue of the glacier, farther upstream than the 2010 event (Figure GN 17.1).

With warmer air melting ice from above and higher sea temperatures melting ice from below, the rate of glacial loss increases. Scientists are studying how these processes apply to the Petermann Glacier. New cracks are opening upstream on the ragged edges of the glacier, demonstrating forward-moving stress. Data show that the Greenland Ice Sheet as a whole, which covers about 80% of the landmass, is now melting at a rate three times faster than in the 1990s. Glaciers such as Petermann along the edge of the ice sheet are melting more rapidly than the ice sheet itself.

▲**Figure GN 17.1 Greenland tidewater glacier breakup, 2012.** An iceberg breaks off the Petermann Glacier. Notice at least five tributary glaciers. [*Aqua* MODIS, NASA.]

Ward Hunt Ice Shelf, Canada Along the northern coast of Canada's Ellesmere Island is the Ward Hunt Ice Shelf, largest in the Arctic. After remaining stable for at least 4500 years, this ice shelf began to break up in 2003 and broke into separate pieces in 2011.

The breakup of an ice shelf does not directly influence sea level because the ice shelf mass has already displaced its own volume in seawater. However, ice shelves hold back flows of grounded ice that, while not yet displacing ocean water, are moving toward the sea. As the ice shelves disappear, their buttressing effect is lost, allowing the glaciers to flow more rapidly. This region of the Canadian Archipelago ranks third behind the Antarctic and Greenland Ice Sheets in terms of ice-mass loss on Earth, and this melting ice is making a significant contribution to global sea-level rise.

Pine Island Glacier, Antarctica To the west of the Antarctic Peninsula, the Pine Island Glacier flows from the West Antarctic Ice Sheet to the Amundsen Sea (noted on Figure 17.7). This glacier is one of Antarctica's largest ice streams, a type of glacier that flows at a faster rate than the surrounding ice mass. The flow rate of the Pine Island Glacier is accelerating, increasing by 30% in the last 10 years.

The Pine Island Glacier's 40-km-wide (25-mi-wide) ice shelf is also calving and thinning at higher rates than previously recorded. This small ice shelf and glacier help buttress downslope movement of the West Antarctic Ice Sheet, which already contributes 0.15 to 0.30 mm/year to sea-level rise.

The warming trends demonstrated in these examples continued through 2013, promoting the rise of sea level across the globe. This chapter examines snow, ice, and frozen ground, and the rapid changes that are occurring in these environments with climate change.

GEOSYSTEMS NOW ONLINE Go to Chapter 17 on the *MasteringGeography* website (www.masteringgeography.com) for more on melting ice shelves and glaciers, or to the National Snow and Ice Data Center at http://nsidc.org/. For the latest ice conditions in Canada, go to http://www.ec.gc.ca/glaces-ice/default.asp?lang=En&n=D32C361E-1.

About three-quarters of Earth's freshwater is frozen. Currently, a volume of more than 32.7 million km^3 (7.8 million mi^3) of water is tied up as ice: in Greenland, Antarctica, and ice caps and mountain glaciers worldwide. The bulk of that snow and ice sits in just two places—Greenland (2.4 million km^3, or 0.6 million mi^3) and Antarctica (30.1 million km^3, or 7.2 million mi^3). The remaining snow and ice (180,000 km^3, or 43,184 mi^3) covers other near-polar regions and various mountains and alpine valleys (Figure 17.1).

Earth's **cryosphere** consists of the portions of the hydrosphere and lithosphere that are perennially frozen, including the freshwater making up snow, ice, glaciers, and frozen ground, and the frozen saltwater in sea ice. These cold regions are generally found at high latitudes and, worldwide, at high elevations on mountains. The extent of the cryosphere changes on a seasonal basis, given that more snow accumulates and more soil and freshwater freezes during the winter.

With rising temperatures causing worldwide glacial and polar ice melts, the cryosphere today is in a state of dramatic change. In 2012, Arctic air temperatures set records of more than 5 C° (9 F°) above normal, and Arctic Ocean sea ice decreased to its smallest areal extent in the past century. The surface ice loss in 2007 was second to this record, with 2008 third.

In this chapter: We focus first on snow and the processes by which permanent snow forms glacial ice. We then look at Earth's extensive ice deposits—their formation, their movement, and the ways in which they produce various erosional and depositional landforms. Glaciers, transient landforms themselves, leave in their wake a variety of landscape features. Glacial processes are intricately tied to changes in global temperature and rising or falling sea level. We also examine the freezing conditions that create permafrost and the periglacial processes such as frost action that shape landscapes. The chapter ends with a look at the changing polar regions.

(a) Alpine glaciers merge from adjoining glacial valleys in the northeast region of Ellesmere Island in the Canadian Arctic.

(b) Cracks indicate movement of the Grrenland Ice Sheet, an accumulation of ice perhaps 100,000 years in the making. The peaks rising above the snow are known as *nunataks*.

▲**Figure 17.1 Rivers and sheets of ice.** [(a) *Terra* MODIS, NASA. (b) Bobbé Christopherson.]

Snow into Ice—The Basis of Glaciers

In previous chapters we discussed some of the important aspects of seasonal and permanent snow cover on Earth. Water stored as snow is released gradually during the summer months, feeding streams and rivers and the resources they provide (discussed in Chapter 9). For example, many western states rely heavily on snowmelt for their municipal water supplies. At the same time, snow can create a hazard in mountain environments (discussed ahead).

Another role of the seasonal snowpack is that it increases Earth's albedo, or reflectivity, affecting the Earth–atmosphere energy balance (discussed in Chapter 4). As temperatures increase with climate change, seasonal snow cover decreases, creating a positive feedback loop in which decreasing snow cover lowers the global albedo, leading to more warming and, in turn, further decreasing the seasonal snow cover.

Properties of Snow

When conditions are cold enough, precipitation falls to the ground as snow. As discussed in Chapter 7, all snowflakes have six sides owing to the molecular structure of water, and yet each snowflake is unique because its growth is dictated by the temperature and humidity conditions in the cloud in which it forms. As snowflakes fall through layers of clouds, their growth follows different patterns, resulting in the intricate shapes that arrive at Earth. Because the temperature at which snowflakes exist is very near their melting point, the flakes can change rapidly once they are on the ground, in a process known as *snow metamorphism*.

When snow falls to Earth it will either accumulate or melt. During the winter in high latitudes or at upper elevations, cold temperatures allow the snow to accumulate seasonally. Each storm is unique, so the snowpack is deposited in distinguishable layers, much like the layered sedimentary rock strata of the Grand Canyon. The properties of each layer and the relationship between them determines the susceptibility of a mountain slope to snow avalanches, which can sometimes be large enough to destroy forests or entire mountain villages (see Focus Study 17.1).

Formation of Glacial Ice

In some regions on Earth, snow is permanent on the landscape, and it is in those regions—both at high latitudes and at high elevations at any latitude—that glaciers form. As mentioned in Chapter 5, a **snowline** is the lowest elevation where snow remains year-round; specifically, it is the lowest line where winter snow accumulation persists throughout the summer. On equatorial mountains, the snowline is around 5000 m (16,400 ft) above sea level; on midlatitude mountains, such as the European Alps, snowlines average 2700 m (8850 ft); and in southern Greenland, snowlines are as low as 600 m (1970 ft).

Glaciers form by the continual accumulation of snow that recrystallizes under its own weight into an ice mass. Ice, in turn, is both a mineral (an inorganic natural compound of specific chemical makeup and crystalline structure) and a rock (a mass of one or more minerals). As mentioned earlier, the accumulation of snow in layered deposits is similar to the layering in sedimentary rock. To give birth to a glacier, snow and ice are transformed under pressure, recrystallizing into a type of metamorphic rock.

Consider that as snow accumulates during the winter, the increasing thickness results in increased weight and pressure on the underlying layers. In summer, rain and snowmelt contribute water, which stimulates further melting, and this meltwater seeps down into the snowfield and refreezes. Air spaces between ice crystals are compressed as the snow packs to a greater density, recrystallizing and consolidating as the pressure continues to increase. Through this process, snow that survives the summer and is still present the following winter begins a slow transformation into glacial ice. In a transition step, the snow becomes **firn**, a granular, partly compacted snow that is intermediate between snow and ice.

Dense **glacial ice** is produced over a period of many years as this process continues. In Antarctica, glacial ice formation may take 1000 years because of the dryness of the climate and minimal snowfall, whereas in wet climates, the time is reduced to several years because of constant heavy snowfall.

Types of Glaciers

A **glacier** is defined as a large mass of ice resting on land or floating in the sea attached to a landmass (the latter is an ice shelf, discussed in *Geosystems Now* and later in the chapter). Glaciers are not stationary; they move under the pressure of their own great weight and the pull of gravity. In fact, they move in streamlike patterns, merging as tributaries into large rivers of ice that slowly flow outward toward the ocean (see Figure 17.1). For an inventory of world glaciers, go to http://glims.colorado.edu/glacierdata/db_summary_stats.php.

Although glaciers are as varied as the landscape itself, they fall within two general groups, based on their form, size, and flow characteristics: alpine glaciers and continental ice sheets (also called continental glaciers). Ice sheets, which occur mainly in Greenland and Antarctica, are extensive areas of glacial ice that cover large landmasses. Today, alpine glaciers and ice sheets cover about 10% of Earth's land area, ranging from the polar regions to midlatitude mountain ranges to some of the high mountains along the equator, such as in the Andes Mountains of South America and on Mount Kilimanjaro in Tanzania, Africa. During colder climate episodes in the past, glacial ice covered as much as 30% of continental land. Throughout these "ice ages," below-freezing temperatures occurred for extended periods at lower latitudes than they do today, allowing snow to accumulate and persist year after year.

Alpine Glaciers

With few exceptions, a glacier in a mountain range is an **alpine glacier**, or *mountain glacier*. The name comes from the Alps of central Europe, where such glaciers abound. Alpine glaciers have several subtypes. *Valley glaciers* are masses of ice confined within a valley that originally was formed by stream action. These glaciers range in length from as little as 100 m (325 ft) to more than 100 km (60 mi). How many valley glaciers do you see joining the main glacier in Figure 17.1a? Figure 17.2 shows a valley glacier in the Tien Shan in central Asia, one of the largest continuous mountain ranges in the world. The two highest peaks in the central part of this range, both shown in the photo, are Xuelian Feng at 6527 m (21,414 ft) and Peak 6231, aptly named at 6231 m (20,443 ft) above sea level.

A glacier that forms within the snow filling a **cirque**, or bowl-shaped recess at the head of a valley, is a *cirque glacier*. Several cirque glaciers may jointly feed a valley

▲Figure 17.2 **Glaciers in the Tien Shan, central Asia, 2011.** Photo location is in the central Tien Shan, north of the Himalayas and just east of the meeting of the borders of China, Kazakhstan, and Kyrgyzstan. [*ISS* Astronaut photo, NASA/GSC.]

Focus Study 17.1 Natural Hazards

Snow Avalanches

A skier moves swiftly down a steep snowy slope. Suddenly the snowpack shatters like a pane of glass and breaks apart in a snow avalanche, capturing the skier and sending thousands of metric tonnes of snow cascading down the mountain side (Figure 17.1.1 and the opening photo of Chapter 1). Snow avalanches move with forces equal to or exceeding those in tornadoes and hurricanes, and can destroy entire towns. For example, an enormous avalanche destroyed the town of Alta, Utah, in 1874, killing more than 60 people. What conditions cause avalanches and how do people trigger them?

▲**Figure 17.1.1 Skier triggering an avalanche in the Wasatch Range, Utah.** [Lee Cohen/Corbis.]

Snow avalanches are defined as the sudden release and movement of massive amounts of snow down a mountain slope. A snow avalanche is a type of mass-wasting event in that the snow moves under the influence of gravity and may carry other materials picked up as it moves downslope. However, these natural hazards are associated with snow rather than soil or rock, and they are understood by examining snow characteristics.

The Role of Terrain, Snowpack, and Weather

Avalanches happen under specific conditions related to terrain, snowpack conditions, and weather. Avalanche terrain consists of mountain slopes that are steeper than 30 degrees and mostly unforested. On these slopes the mountain snowpack accumulates in layers that reflect differences in the atmospheric conditions, temperatures, and wind conditions associated with each storm. Once deposited, these layers change constantly as snow metamorphism processes differentiate them further. The snowpack normally consists of both stronger and weaker layers; when a stronger layer, called a slab, overlies a weaker layer, avalanches are possible.

Weather, especially new snowfall, also has an important role in avalanches. The weight of new snow adds considerable stress to the snowpack, increasing the probability of avalanches. Wind, too, is important in avalanche formation, because it is capable of transporting huge volumes of snow from the windward sides of ridges and gullies onto the lee slopes, where the added weight increases the avalanche danger.

In certain areas—where terrain, snowpack, and weather conditions are right—snow avalanches occur repeatedly. In these mountain regions, *avalanche paths* are visible features of the landscape (Figure 17.1.2). On forested slopes, trees are sometimes completely cleared from these paths; continued avalanche activity prevents new trees from establishing.

Avalanche Triggers

Once conditions are favorable for avalanche release, the only missing ingredient is a trigger. Triggers can be natural, such as the additional load from new snow or the weight of a falling cornice (an overhanging ledge of snow formed by wind, usually at the top of a mountain ridge), or human, such as a skier. A human trigger often becomes an avalanche victim.

glacier (Figure 17.2). A *piedmont glacier* is formed wherever several valley glaciers pour out of their confining valleys and coalesce at the base of a mountain range. A piedmont glacier spreads freely over the lowlands, as demonstrated by the remnants of the Malaspina Glacier, which flows into Yakutat Bay, Alaska.

As a valley glacier flows slowly downhill, it erodes the mountains, canyons, and river valleys beneath its mass, transporting material within or along its base. A portion of the transported debris may also be carried on its icy surface, visible as dark streaks and bands. This surface material is known as *supraglacial debris*, which originates either from rockfalls and other gravity-driven processes that carry material downward from above or from processes that float material upwards from the glacier's bed.

A *tidewater glacier*, or *tidal glacier*, ends in the sea. Such glaciers are characterized by **calving**, a process in which pieces of ice break free to form floating ice masses known as *icebergs* that are usually found wherever glaciers meet an ocean, bay, or fjord (Figure 17.3). Icebergs are inherently unstable, as their center of gravity shifts with melting and further breakup (review the iceberg discussion in Chapter 7).

▲**Figure 17.3 Glacial calving,** Active calving fills the sea with brash ice, bergy bits, and icebergs along the edge of the Austfonna Ice Cap on Nordaustlandet Island in the archipelago of Svalbard, Norway. The glacial front retreated about 2 km (12.4 mi) from 2012 to 2013. [Bobbé Christopherson.]

In the latter part of the 19th century, the typical avalanche casualty in the United States was a miner working in the high mountains of southwestern Colorado or in the Wasatch Range of Utah. Today, a typical avalanche trigger and victim is a backcountry skier, snowboarder, snowmobiler, or climber. Snow avalanches claim about 30 lives each year in the United States.

Snowpack conditions are highly variable across a mountain range and even along a single slope. Unwary people recreating in the winter backcountry tend to trigger avalanches from weaker areas of the slope, where the snow is often thinner. A skier or snowboarder approaching the middle of a slope may cause a cascading fracture that triggers a slab avalanche that is difficult if not impossible to escape. Since snow scientists have not yet developed tools to definitively identify these areas of weaker snowpack, assessing avalanche conditions remains challenging.

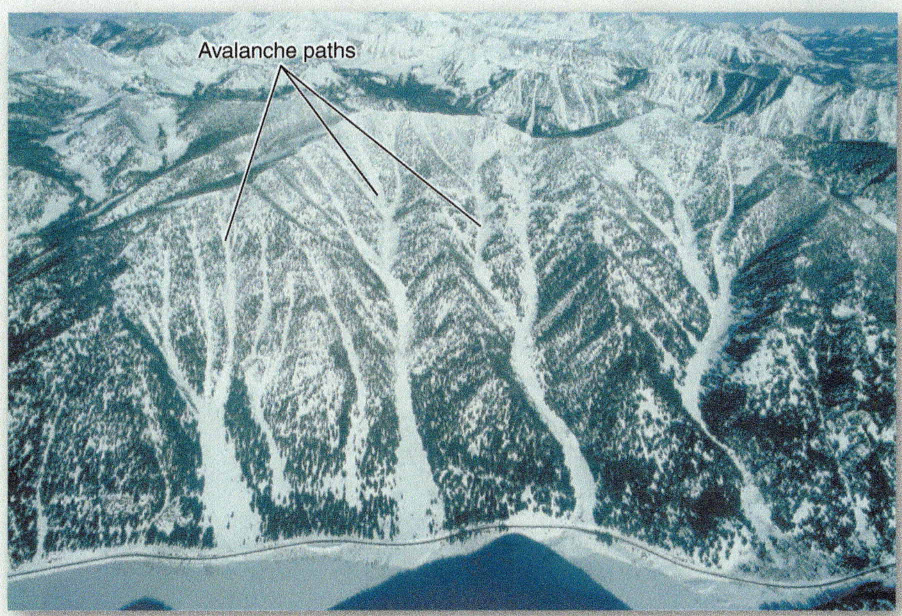

▲**Figure 17.1.2 Avalanche paths in the Madison Range near Hebgen Lake, southwest Montana.** [Karl Birkeland.]

Avalanche Control, Forecasting, and Safety

Avalanche control by ski patrollers at ski areas and by highway workers on mountain passes consists of using hand-thrown explosives and in some cases military artillery to safely trigger avalanches that remove unstable snow at times when people are absent from ski runs and affected roads. After avalanche control work is done, these areas can be safely opened to the public. This work makes it possible for ski resorts to operate with less risk in avalanche terrain and for mountain passes to remain open in winter.

In the United States, a network of avalanche centers provides daily forecasts to mitigate avalanche danger to the public. To remain safe in the winter backcountry, people need to be familiar with the current general conditions, pick terrain appropriate for those conditions, evaluate the specific avalanche danger on the slopes where they are recreating, and expose only one person to the danger at a time so that companions are available for rescue in the event of an avalanche. See http://www.avalanche.org/ for more information.

Continental Ice Sheets

An **ice sheet** is an extensive, continuous mass of ice that may occur on a continental scale. Most of Earth's glacial ice exists in the ice sheets that blanket 81% of Greenland—1,756,000 km^2 (678,000 mi^2) of ice—and 90% of Antarctica—14.2 million km^2 (5.48 million mi^2) of ice. Antarctica alone contains 92% of all the glacial ice on the planet (review the ice volumes in the chapter introduction).

The ice sheets of Antarctica and Greenland have such enormous mass that large portions of each landmass beneath the ice are isostatically depressed (pressed down by weight) below sea level. Each ice sheet reaches thicknesses of more than 3000 m (9800 ft), with average thickness around 2000 m (6500 ft), burying all but the highest peaks of land.

At the edge of ice sheets are *ice shelves*, permanent masses of ice that extend out over the sea. These shelves are often found in protected inlets and bays, cover thousands of square kilometers, and reach thicknesses of 1000 m (3280 ft).

Ice caps and ice fields are two additional types of glaciers with continuous ice cover, on a slightly smaller scale than an ice sheet. An **ice cap** is roughly

GEOreport 17.1 Global glacial ice losses

Worldwide, glacial ice is in retreat, melting at rates exceeding anything previously recorded by scientists. The European Alps have lost more than 50% of their ice mass over the past century, with an acceleration in melt rates since 1980 and nearly 20% of their ice lost during the past 20 years. In Alaska, 98% of surveyed glaciers are in what scientists describe as a "swift retreat." Similar ice losses and reduced snowpacks are reported for the Rockies, Sierra Nevada, Himalayas, and Andes. The USGS is capturing these dramatic changes in a Repeat Photography Project (see http://nrmsc.usgs.gov/repeatphoto/).

▲**Figure 17.4 Ice caps.** The Vatnajökull Ice Cap in southeastern Iceland is the largest of four ice caps on the island (*jökull* means "ice cap" in Danish). Note the ash on the ice cap from the 2004 Grímsvötn eruption [NASA/GSFC.]

circular and, by definition, covers an area of less than 50,000 km^2 (19,300 mi^2), completely burying the underlying landscape.

The volcanic island of Iceland features several ice caps; an example is the Vatnajökull Ice Cap outlined in Figure 17.4. Volcanoes lie beneath these icy surfaces. Iceland's Grímsvötn Volcano erupted in 1996 and 2004, producing large quantities of melted glacial water in a sudden flood called a *jökulhlaup*, an Icelandic term that is now widely used to describe a glacial outburst flood. The most recent eruption in 2011 was the largest in a century but did not produce an outburst flood.

An **ice field** extends in a characteristic elongated pattern over a mountainous region and is not extensive enough to form the dome of an ice cap. The Patagonian ice field of Argentina and Chile is one of Earth's largest. It is only 90 km (56 mi) wide but stretches 360 km (224 mi), from 46° to 51° S latitude (Figure 17.5).

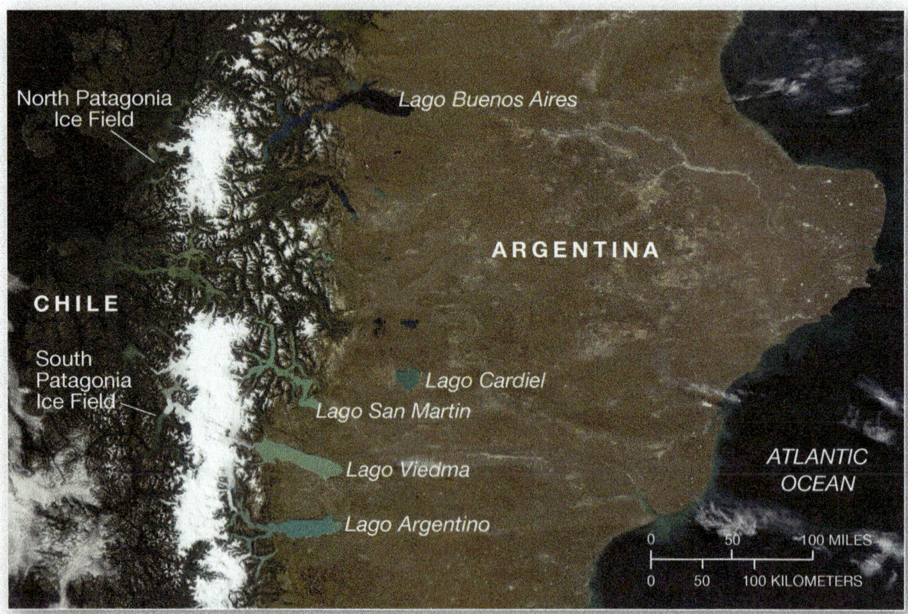

▲**Figure 17.5 The Patagonian ice fields of Argentina and Chile.** [NASA/GSFC.]

Ice sheets and ice caps may be drained by rapidly moving *ice streams*, made up of solid ice that flows at a faster rate than the main ice mass toward lowland areas or the sea. For example, a number of ice streams flow through the periphery of Greenland and Antarctica. An *outlet glacier* is a stream of ice flowing out from an ice sheet or ice cap, usually constrained on each side by the bedrock of a mountain valley.

Glacial Processes

A glacier is a dynamic body, moving relentlessly downslope at rates that vary within its mass, shaping the landscape through which it flows. Like so many of the systems described in this text, glacial processes are linked to the concept of equilibrium. A glacier at equilibrium maintains its size because the incoming snow is approximately equal to the melt rate. In a state of disequilibrium, the glacier either expands (causing its terminus to move downslope) or retreats (causing its terminus to move upslope).

Glacial Mass Balance

A glacier is an open system, with *inputs* of snow and *outputs* of ice, meltwater, and water vapor as illustrated in this chapter's *Geosystems in Action*, Figure GIA 17. Glaciers acquire snow in their accumulation zone, a snowfield at the highest elevation of an ice sheet or ice cap or at the head of a valley glacier, usually in a cirque (Figure GIA 17.1). Snow avalanches from surrounding steep mountain slopes can add to the snowfield depth. The accumulation zone ends at the **firn line**, which marks the elevation above which the winter snow and ice remained intact throughout the summer melting season but below which melting occurs. At the lower end of the glacier, far below the firn line, the glacier undergoes wasting (reduction) through several processes: melting on the surface, internally, and at the base; ice removal by deflation from wind; the calving of ice blocks; and sublimation (recall from Chapter 7 that this is the phase change of solid ice directly into water vapor). Collectively, these processes cause losses to the glacier's mass, known as **ablation**.

These gains (accumulation) and losses (ablation) of glacial ice determine the glacier's *mass balance,* the property that decides whether the glacier will advance (grow larger) or retreat (grow smaller). During cold periods with adequate precipitation, a glacier has a *positive net mass balance,* and advances. In warmer times, a glacier has a *negative net mass balance,* and retreats. Internally, gravity continues to move a glacier forward even though its lower terminus might be in retreat owing to ablation. Within the glacier is a zone where accumulation balances ablation; this is known as the *equilibrium line*, and it generally coincides with the firn line (Figure GIA 17.2).

Illustrating the global trend, the net mass balance of the South Cascade Glacier in Washington State demonstrated significant losses between 1955 and 2010. As a result of this negative mass balance, in some years the terminus retreated tens of meters, and it has retreated every year in the record except 1972. Figure GIA 17.3 presents a photo comparison of the South Cascade Glacier between 1979 and 2010.

A comparison of the trend of this glacier's mass balance with that of others in the world shows that temperature changes apparently are causing widespread reductions in middle- and lower-elevation glacial ice. The present wastage (ice loss) from alpine glaciers worldwide is thought to contribute over 25% to the measured rise in sea level. Figure 17.32 at the end of this chapter shows a retreating glacier and the landscape left behind.

Glacial Movement

Glacial ice is quite different from the small, brittle cubes of ice we find in our freezer. In particular, glacial ice behaves in a plastic (pliable) manner; it distorts and flows in its underlying portions in response to the weight and pressure of overlying snow and the degree of slope below. In contrast, the glacier's upper-surface portions are quite brittle. Rates of flow range from almost no movement to a kilometer or two of movement per year on a steep slope. The rate of accumulation of snow in the glacier's formation area is critical to the speed of forward motion.

Glaciers, then, are not rigid blocks that simply slide downhill. The greatest movement within a valley glacier occurs *internally,* below the rigid surface layer, which fractures as the underlying plastic zone moves forward (Figure 17.6a). At the same time, the base creeps and slides along, varying its speed with temperature and the presence of any lubricating water beneath the ice. This *basal slip* usually is much less rapid than the internal plastic flow of the glacier, so the upper portion of the glacier flows ahead of the lower portion.

Unevenness in the landscape beneath the ice may cause the pressure to vary, melting some of the basal ice by compression at one moment, only to have it refreeze later. This process is *ice regelation,* meaning to refreeze, or re-gel. Such melting/refreezing action incorporates rock debris into the glacier. Consequently, the basal ice layer, which can extend tens of meters above the base of the glacier, has a much higher debris content than the ice above.

A flowing alpine glacier or ice stream can develop vertical cracks known as **crevasses** (Figure 17.6). Crevasses result from friction with valley walls, from tension

(text continued on page 530)

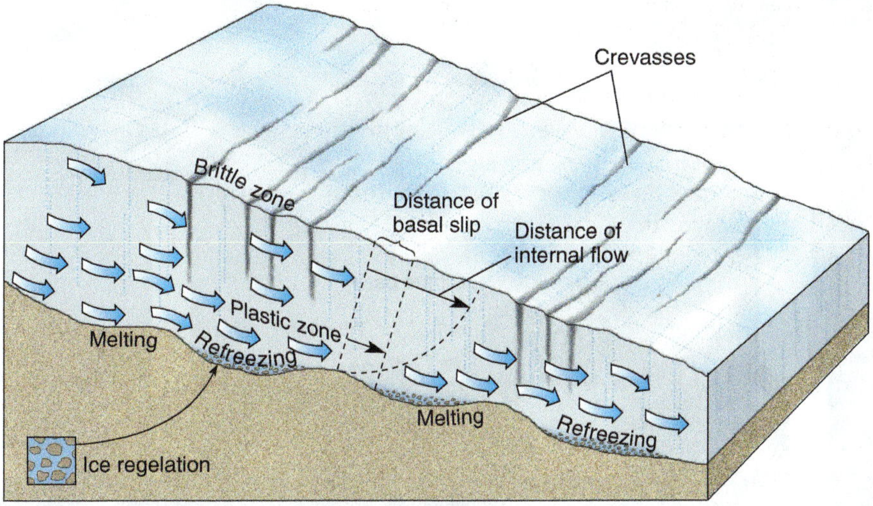

(a) Cross section of a glacier, showing its forward motion, brittle cracking at the surface, and flow along its basal layer.

(b) Surface crevasses are evidence of forward movement on the Fox Glacier, South Island, New Zealand.

▲**Figure 17.6 Glacial movement.** [(b) David Wall/Alamy.]

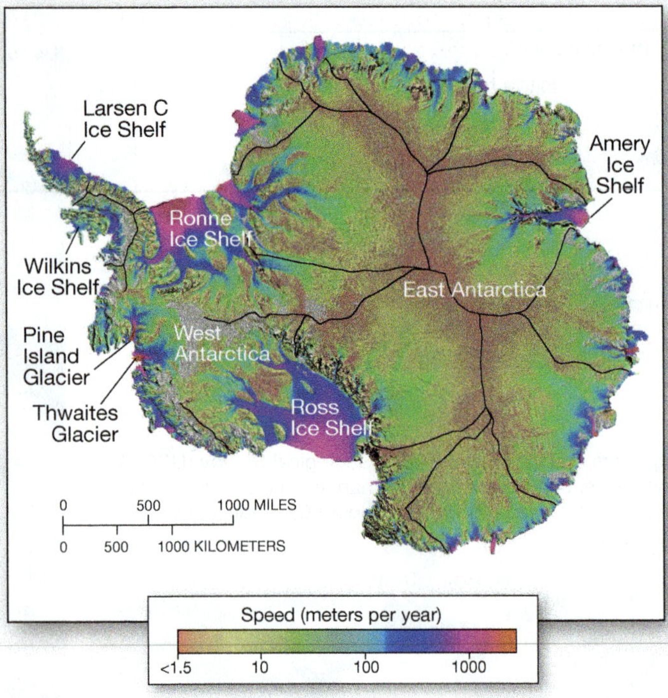

▲**Figure 17.7 First complete map of ice movement speed in Antarctica.** The black lines show ice divides, similar to drainage divides. Colors indicate the speed of ice movement; fastest movement is in red and purple. Note that the fast-flowing ice channels extend far inland, a surprise to scientists. The tributaries shown in blue are moving faster than the ice sheet around them but do not move as quickly as an ice stream. This map was created using data from a group of international satellites from the Canada, Japan, and European Space Agencies. [*RADARSAT–SAR*, NASA/JPL.]

due to stretching as the glacier passes over convex slopes, or from compression as the glacier passes over concave slopes. Traversing a glacier, whether an alpine glacier or an ice sheet, is dangerous because a thin veneer of snow sometimes masks the presence of a crevasse.

Scientists recently completed a map of ice movement on the Antarctic continent based on satellite radar measurements from 1996 to 2009 (Figure 17.7). The map reveals that many of the tributaries around ice shelves extend surprisingly far inland and are moving by basal slip (sliding on the ground) rather than slowly deforming under the weight of the ice. Identifying the areal extent of this type of motion on the ice sheet is important because a loss of ice at the coasts as ice shelves break up may open the tap for massive amounts of ice to flow more quickly from the interior, with implications for sea-level rise.

Glacier Surges Although glaciers flow plastically and predictably most of the time, some will advance rapidly at speeds much faster than normal, in a **glacier surge.** A surge is not quite as abrupt as it sounds; in glacial terms, a surge can be tens of meters per day. The Jakobshavn Glacier on the western Greenland coast, for example, is one of the fastest moving, at between 7 and 12 km (4.3 and 7.5 mi) a year. In 2012 scientists reported that although the overall trend in Greenland was toward glacial surging, in some regions glaciers slowed between 2005 and 2010, pointing to the complexity of glacial behavior.

Scientists are investigating the exact causes of glacier surges. Some surge events result from a buildup of water pressure under the glacier—sometimes enough to actually float the glacier slightly, detaching it from its bed during the surge. Another cause of glacier surges is the presence of a water-saturated layer of sediment, a *soft bed,* beneath the glacier. This is a deformable layer that cannot resist the tremendous sheer stress produced by the moving ice of the glacier. In Antarctica, scientists examining cores taken from several ice streams now accelerating through the West Antarctic Ice Sheet think they have identified the occurrence there of this kind of glacial surge—although water pressure is still important. As any type of surge begins, ice quakes are detectable, and ice faults are visible.

Scientists think that glacial surges in Greenland are related to the meltwater that works its way to the basal layer, lubricating the interface between the glacier and the underlying bedrock. In addition, the warmer surface waters draining beneath the glacier deliver heat that increases basal melt rates. However, a 2013 study reported that the meltwater flows in distinct channels under the ice and is not necessarily lubricating wide areas of the ice sheet.

Glacial Erosion The process by which a glacier erodes the landscape is similar to a large excavation project, with the glacier hauling debris from one site to another for deposition. The passing glacier mechanically picks up rock material and carries it away in a process known as *glacial plucking.* Debris is carried on its surface and is also transported internally, or *englacially,* embedded within the glacier itself.

When a glacier retreats, it can leave behind cobbles or boulders (sometimes house-sized) that are "foreign" in composition and origin to the ground on which they are

GEOreport 17.2 Greenland Ice Sheet melting

The Greenland Ice Sheet is experiencing a greater amount of ice loss and a greater area of surface melting than at any time since systematic satellite monitoring started in the 1970s. Scientists calculated that 98% of the ice sheet's surface melted in July 2012, an occurrence never before seen in the satellite record. A typical summer melt is 50%. Near-surface air temperatures at the highest and coldest station on the ice sheet have increased 0.22 F° per year since 1992, six times faster than the global average temperature rise. In response, the equilibrium line (at which accumulation balances ablation) has been moving up the ice sheet on average 35 m/year (115 ft/year).

▲Figure 17.8 **Glacial sandpapering of rock.** Glacial polish and striations are examples of glacial abrasion and erosion. The polished, marked surface is seen beneath a glacial erratic—a rock left behind by a retreating glacier. [Bobbé Christopherson.]

deposited. These *glacial erratics*, lying in strange locations with no obvious sign of how they got there, were an early clue to the glacial plucking that occurred during times when blankets of ice covered the land (pictured ahead in Figure 17.11).

The rock pieces frozen to the basal layers of the glacier enable the ice mass to scour the landscape like sandpaper as it moves. This process, called **abrasion**, produces a smooth surface on exposed rock, which shines with *glacial polish* when the glacier retreats (Figure 17.8). Larger rocks in the glacier act much like chisels, gouging the underlying surface and producing glacial striations parallel to the flow direction.

Glacial Landforms

Glacial erosion and deposition produce distinctive landforms that differ greatly from those existing before the ice came and went. Alpine glaciers and continental ice sheets each produce characteristic landscapes, although some landforms exist in either type of glacial environment.

Erosional Landforms

A landscape feature produced by both glacial plucking and abrasion is a **roche moutonnée** ("sheep rock" in French), an asymmetrical hill of exposed bedrock. This landform has a characteristic gently sloping upstream side (stoss side) that is polished smooth by glacial action and an abrupt and steep downstream side (lee side) where the glacier plucked rock pieces (Figure 17.9).

Glacial Valleys The effects of alpine glaciation created the dramatic landforms of the Canadian Rockies, the Swiss Alps, and the Himalayan peaks. Geomorphologist William Morris Davis depicted the stages of a valley glacier in drawings published in 1906 and redrawn here in Figures 17.10 and 17.11. Study of these figures reveals the handiwork of ice as sculptor in mountain environments.

Figure 17.10a shows the **V** shape of a typical stream-cut valley as it existed before glaciation. Figure 17.10b shows the same landscape during subsequent glaciation. Glacial erosion actively removes much of the regolith (weathered bedrock) and the soils that covered the stream-valley landscape. When glaciers erode parallel valleys, a thin, sharp ridge forms between them, known as an **arête** ("knife-edge" in French). Arêtes can also form between adjacent cirques as they erode in a headward direction. Two eroding cirques may reduce an arête to a saddlelike depression or pass, forming a **col**. A **horn**, or pyramidal peak, results when several cirque glaciers gouge an individual mountain summit from all sides. Most famous is the Matterhorn in the Swiss Alps, but many others occur worldwide. A **bergschrund**

(a) Lembert Dome in the Tuolumne Meadows area of Yosemite National Park, California.

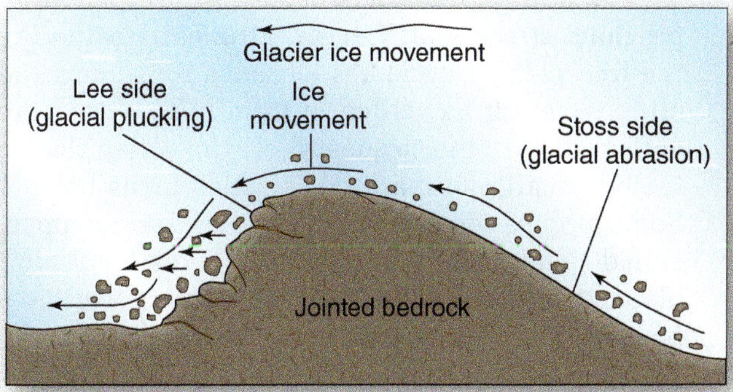

(b) The erosional formation processes at work on Lembert Dome (the white color represents glacial ice).

▲Figure 17.9 **Roche moutonnée.** [Author.]

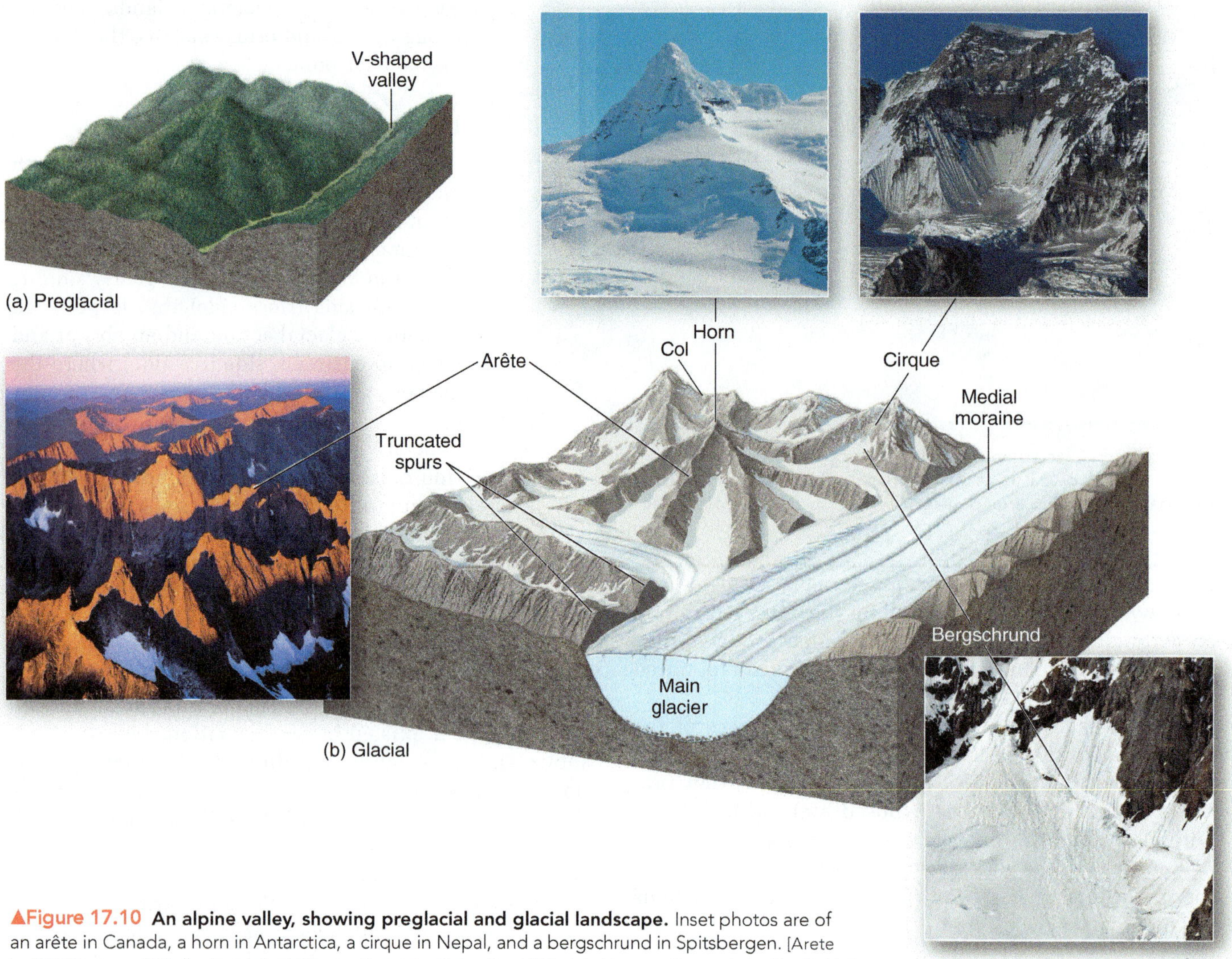

▲**Figure 17.10 An alpine valley, showing preglacial and glacial landscape.** Inset photos are of an arête in Canada, a horn in Antarctica, a cirque in Nepal, and a bergschrund in Spitsbergen. [Arete by RGB Ventures LLC dba Superstock/Alamy. Cirque by Galen Rowell/Corbis. Horn and bergschrund by Bobbé Christopherson.]

is a crevasse, or wide crack, that separates flowing ice from stagnant ice in the upper reaches of a glacier or in a cirque. Bergschrunds are often covered in snow in winter, but become apparent in summer when this snow cover melts.

Figure 17.11 shows the same landscape at a time of warmer climate, when the ice retreated. The glaciated valleys now are **U**-shaped, greatly changed from their previous stream-cut **V** form. Physical weathering from the freeze–thaw cycle has loosened rock along the steep cliffs, where it has fallen to form *talus slopes* along the valley sides. In the cirques where the valley glaciers originated, small mountain lakes called **tarns** have formed. Some cirques contain small, circular, stair-stepped lakes, called **paternoster** ("our father") **lakes** for their resemblance to rosary (religious) beads. Paternoster lakes may have formed from the differing resistance of rock to glacial processes or from damming by glacial deposits.

In some cases, valleys carved by tributary glaciers are left stranded high above the valley floor because the primary glacier eroded the valley floor so deeply. These *hanging valleys* are the sites of spectacular waterfalls. How many of the erosional forms from Figures 17.10 and 17.11 can you identify in Figure 17.12? (Look for arêtes, cols, horns, cirques, cirque glaciers, **U**-shaped valleys, and tarns).

Fjords Where a glacial trough encounters the ocean, the glacier can continue to erode the landscape, even below sea level. As the glacier retreats, the trough floods and forms a deep **fjord** in which the sea extends inland, filling the lower reaches of the steep-sided valley (Figure 17.13). The fjord may be flooded further by rising sea level or by changes in the elevation of the coastal region. All along the glaciated coast of Alaska, retreating alpine glaciers are opening many new fjords that previously were blocked by ice. Coastlines with notable fjords include those of Norway (Figure 17.14), Chile, the South Island of New Zealand, Alaska, and British Columbia.

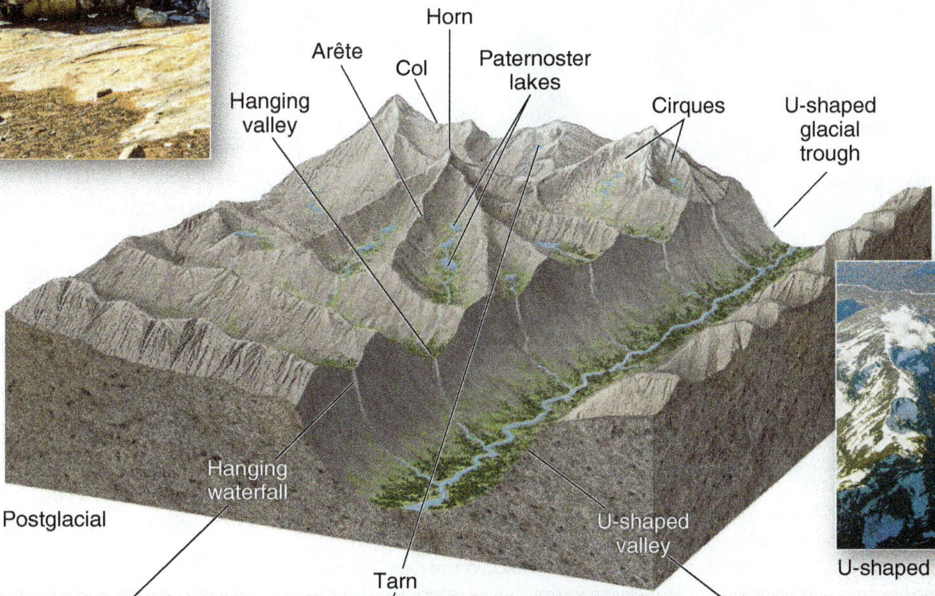

▼**Figure 17.11 The geomorphic handiwork of alpine glaciers.** As the glaciers retreat, the new landscape is unveiled. Inset photos are surface and aerial views from Norway. [Photos by Bobbé Christopherson; waterfall by Author.]

▲**Figure 17.12 Erosional features of alpine glaciation.** How many erosional glacial features can you find in this photo of the Chugach Mountains in Alaska? See Critical Thinking 17.1. [Bruce Molnia, USGS.]

Fjords also occur along the edges of Earth's ice sheets. In Greenland, rising water temperatures in some of the longest fjord systems in the world appear to be accelerating melt rates where the glaciers meet the sea. In Antarctica, recent use of ice-penetrating radar identified numerous fjords beneath the Antarctic Ice Sheet, indicating that the present ice sheet was smaller in areal extent in the past.

CRITICALthinking 17.1
Looking for Glacial Features

After looking at Figure 17.12 to identify glacial features, go back to the photos in the chapter opener and Figures 17.1 and 17.2, and then examine Figure 17.12 again. List all the glacial formations that you can identify in these photos. Are there any erosional landforms you find on all the photos other than the glaciers themselves? ●

▲**Figure 17.13 Fjords on the Pacific Ocean side of the Southern Patagonia Ice Field, Chile.** As ice from the Penguin Glacier and HPS 19 flows into fjords, it calves and forms icebergs. (HPS stands for Hielo Patagónico Sur, or Southern Patagonia Ice Field, in the numbering system for glaciers with no geographic name.) The largest iceberg in the image is about 2 km in width (1.2 mi). [NASA *ISS* Astronaut Photo.]

Depositional Landforms

Glaciers transport materials upon and within the ice, producing unsorted sediment deposits, as well as by the actions of meltwater streams at the glacier's downstream end, producing sorted deposits. The general term for all glacial deposits, both unsorted and sorted, is **glacial drift**.

Moraines

As mentioned earlier, as a glacier flows to a lower elevation, a wide assortment of rock fragments become *entrained* (carried along) on its surface or embedded within its mass or in its base. As the glacier melts, this unsorted and unstratified debris is deposited on the ground as **till**, usually marking the glacier's former margins.

▲**Figure 17.14 Norwegian fjord.** Sediment carried in runoff is visible in this fjord, which fills a U-shaped, glacially carved valley. [Bobbé Christopherson.]

The deposition of glacial sediment also produces a class of landform called a **moraine**, which may take several forms. In areas that have undergone alpine glaciations, **lateral moraines** are lengthy ridges of till along each side of a glacier. If two glaciers with lateral moraines join, a **medial moraine** may form (see Figure 17.1). In areas that were formerly covered by single, large ice sheets, lateral moraines and medial moraines are lacking.

End moraines accumulate at the glacier's *terminus*, or endpoint, and are associated with both alpine and continental-scale glaciation. Eroded debris that is dropped at the glacier's farthest extent is a **terminal moraine** (Figure 17.15). *Recessional moraines* may also be present, having formed at other points where a glacier paused after reaching a new equilibrium between accumulation and ablation.

Till Plains

When the ice sheets retreated from their maximum extent, about 18,000 years ago, during the most recent glaciation in North America and Europe (portrayed in Figure 17.25), they left distinct landscapes that we see today. A **till plain**, also called a *ground moraine*, is a deposition of till that forms behind a terminal moraine as the glacier retreats and is generally spread widely across the ground surface, creating irregular topography but not the characteristic ridges of other moraines. Such plains usually hide the former landscape and are found in parts of the U.S. Midwest. Figure 17.16 illustrates common depositional features associated with the retreat of a continental ice sheet.

Till plains are composed of coarse till, have low and rolling relief, and have a deranged drainage pattern that includes scattered wetlands (Figure 17.16b; see also Figure 15.5g). Common features of till plains are **drumlins**, hills of deposited till that are streamlined in the direction of ice sheet movement, with blunt end upstream and tapered end downstream. The shape of a drumlin sometimes resembles an elongated teaspoon bowl, lying face down. Multiple drumlins, known as *drumlin swarms*, occur across the landscape in portions of New York and Wisconsin, among other areas (Figure 17.17).

Drumlins may attain lengths of 100–5000 m (330 ft–3.1 mi) and heights up to 200 m (650 ft). Although similar in shape to an erosional roche moutonnée, a drumlin is a depositional feature with the blunt, or steep, end upstream and the tapered end downstream. Figure 17.17a shows a portion of a topographic map for the area south of Williamson, New York. Can you identify the numerous drumlins on the map? In what direction do you think the ice sheets moved across this region?

Glacial Outwash Beyond the glacial terminus, meltwater flows downstream and is typically milky in color owing to the sediment load of fine-grained materials, sometimes called "rock flour." This meltwater flow occurs when a glacier is retreating, or during any period of ablation; flow volumes are highest during the warm summer months.

Sediments deposited by glacial meltwater are sorted by size, becoming **stratified drift**. The sorting comes from the combined effect of *glaciofluvial*

▲**Figure 17.15 Terminal moraine.** A terminal moraine of unsorted till forms Isispynten Island, part of the Svalbard archipelago in the Arctic Ocean; the moraine is separated from the present ice cap by more than a kilometer. [Bobbé Christopherson.]

▼**Figure 17.16 Landforms associated with ice sheets.** [(b) Bobbé Christopherson.]

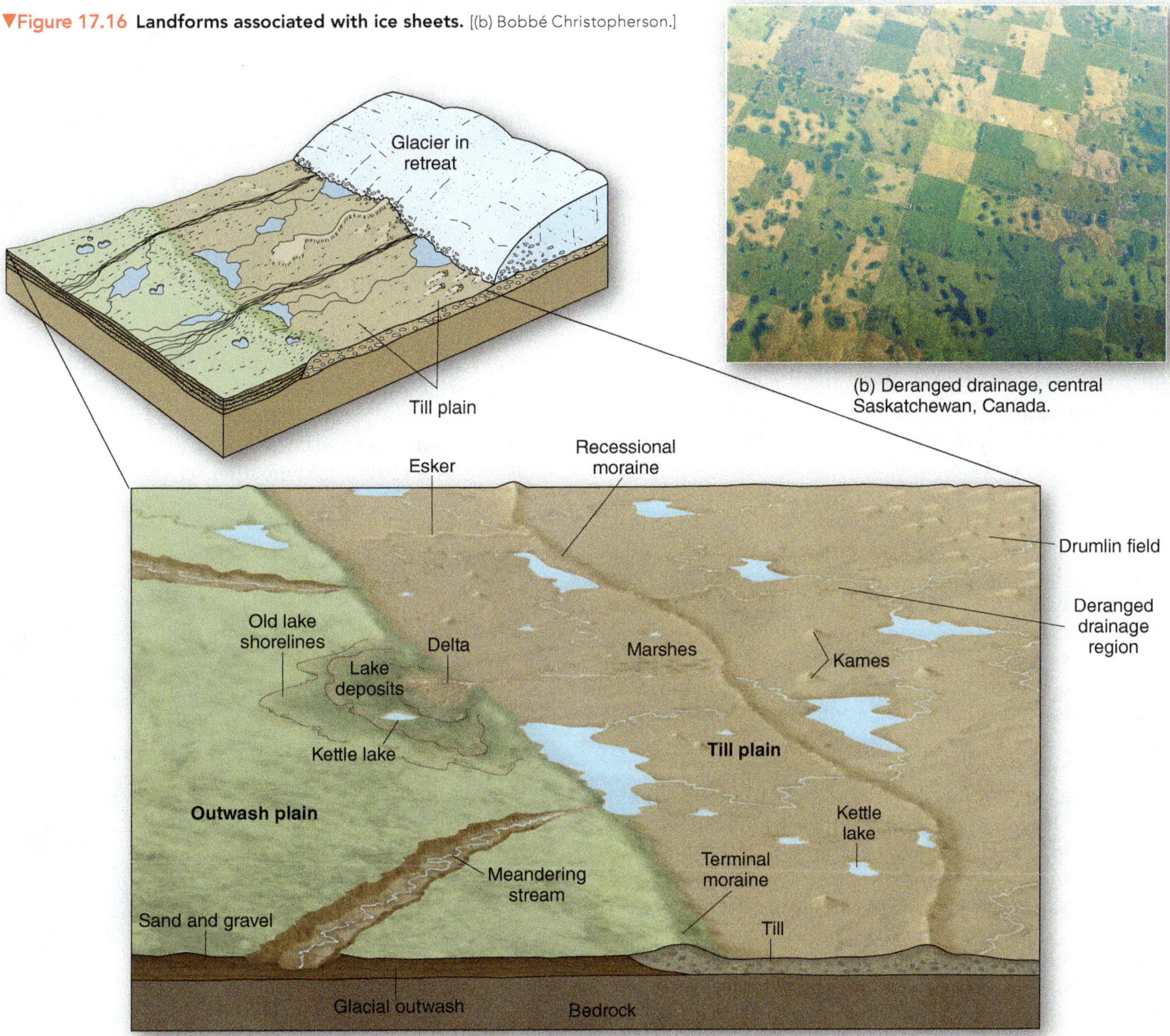

(a) Common depositional landforms produced by continental glaciers.

(b) Deranged drainage, central Saskatchewan, Canada.

(a) Topographic map south of Williamson, near Marion, New York, featuring numerous drumlins (7.5-minute series quadrangle map, originally produced at a 1:24,000 scale; 10-ft contour interval).

Figure 17.17 Drumlins. [(a) USGS map. (b) Bobbé Christopherson.]

(b) Aerial view of drumlin swarm; drumlins with vegetation cover are most visible.

(glacial and fluvial) processes—flowing water sorts sediments according to size, often with the largest particles dropped out of the sediment load closest to the glacial terminus (forming an alluvial fan) and the smaller particles carried farthest downstream. These glaciofluvial sediments are also stratified, with sediments laid down in layers.

The area of sediment deposition beyond the glacial terminus can form an extensive **outwash plain**, also called a *sandur* (a term that originated in Iceland). Outwash plains feature braided stream channels, which typically form when streams carry a large sediment load. When material is moved and deposited by glacial streams in a valley, such as at the terminus of valley glaciers, the outwash forms a *valley train deposit*. Peyto Glacier in Alberta, Canada, produced such a deposit, made up primarily of sand and gravel. This glacier is now actively retreating from Peyto Lake (the glacier is far left in Figure 17.18). The Peyto Glacier has experienced massive ice losses since 1966 and is in retreat due to increased ablation and decreased accumulation related to climate change.

A typical landform that is composed of glacial outwash but located on a till plain is a sinuously curving, narrow ridge of coarse sand and gravel called an **esker**. Eskers form along the channel of a meltwater stream that flows beneath a glacier, in an ice tunnel, or between ice walls. As a glacier retreats, the steep-sided esker is left behind in a pattern roughly parallel to the path of the glacier (Figure 17.16). The ridge may not be continuous and in places may even appear to be branched, following the path set by the subglacial watercourse. Commercially valuable deposits of sand and gravel are quarried from some eskers.

Another landform composed of glaciofluvial deposits is a **kame**, a small hill, knob, or mound of sorted sand and gravel that is deposited by water on the surface of a glacier (for example, after having collected in

Figure 17.18 A valley train deposit. Note the distributary channels of the braided stream, and milky-colored glacial meltwater, below Peyto Glacier in Alberta, Canada. [Author.]

a crevasse) and then is left on the land surface after the glacier retreats. Kames also can be found in deltaic forms, deposits of material that formed deltas at the edges of glacial lakes, and in terraces along valley walls.

Sometimes an isolated block of ice, perhaps more than a kilometer across, remains on a ground moraine, on an outwash plain, or on a valley floor after a glacier has retreated. As much as 20 to 30 years is required for it to melt. In the interim, material continues to accumulate around the melting ice block. When the block finally melts, it leaves behind a steep-sided hole that then frequently fills with water, forming a **kettle**, also known as a *kettle lake*. Thoreau's famous Walden Pond in Massachusetts is such a glacial kettle.

Periglacial Landscapes

In 1909, Polish geologist Walery von Lozinski coined the term **periglacial** to describe processes of frost-action weathering and freeze–thaw rock shattering in the Carpathian Mountains. The term now is used to describe places where geomorphic processes related to freezing water occur. These periglacial regions occupy over 20% of Earth's land surfaces. Periglacial landscapes at high latitudes have a near-permanent ice cover; at high elevation in lower latitudes, these landscapes are seasonally snow-free. These regions are in the *subarctic* and *polar* climate zones, especially in the *tundra* climate either at high latitude or at high elevation in lower-latitude mountains (alpine environments; please review climates types in Chapter 10).

Permafrost and Its Distribution

When soil, sediment, or rock temperatures remain below 0°C (32°F) for at least 2 years, **permafrost** (perennially frozen ground) develops. An area of permafrost that is not covered by glaciers is considered periglacial; the largest extent of such lands is in Russia (Figure 17.19). Approximately 80% of Alaska has permafrost beneath its surface, as do parts of Canada, China, Scandinavia, Greenland, and Antarctica, in addition to alpine mountain regions of the world. Note that the criterion for a permafrost designation is based solely on *temperature* and has nothing to do with how much or how little water is present. Two factors other than temperature also contribute to permafrost conditions and occurrence: the presence of fossil permafrost from previous ice-age conditions and the insulating effect of snow cover or vegetation that inhibits heat loss.

Continuous and Discontinuous Zones

Permafrost regions are divided into two general categories, continuous and discontinuous, which merge along a general transition zone. *Continuous permafrost* is the region of severest cold and is perennial, roughly poleward of the –7°C (19°F) mean annual temperature isotherm (white area in Figure 17.19). Continuous permafrost affects all surfaces except those beneath deep lakes or rivers. The depth of continuous permafrost averages approximately 400 m (1300 ft) and may exceed 1000 m (3300 ft).

Unconnected patches of *discontinuous permafrost* gradually coalesce poleward of the –1°C (30.2°F) mean annual temperature isotherm (light purple area in Figure 17.19), toward the continuous zone. In contrast, equatorward of this isotherm, permafrost becomes scattered or sporadic until it gradually disappears. In the discontinuous zone of the Northern Hemisphere, permafrost is absent on sun-exposed south-facing slopes, in areas of warm soil, and in areas insulated by snow.

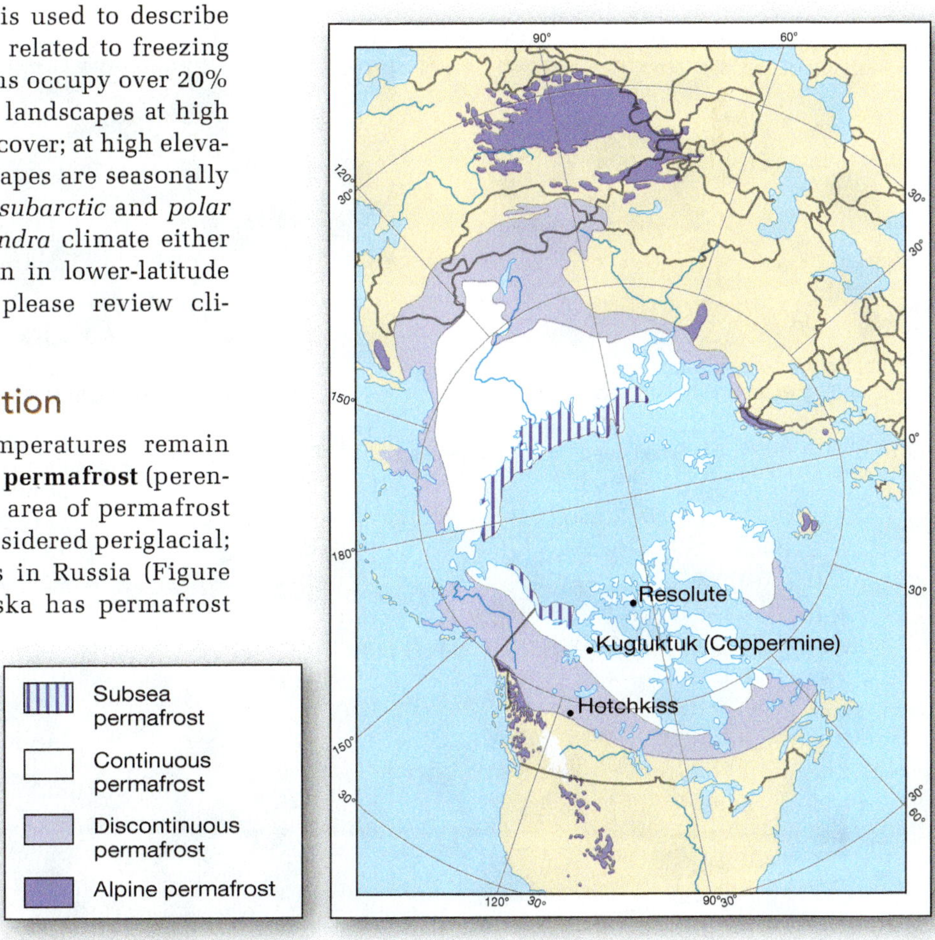

▲**Figure 17.19 Permafrost distribution.** Distribution of permafrost in the Northern Hemisphere. Alpine permafrost is noted except for small occurrences in Hawai'i, Mexico, Europe, and Japan. Subsea permafrost occurs in the ground beneath the Arctic Ocean along the margins of the continents, as shown. Note the towns of Resolute and Kugluktuk (Coppermine) in Nunavut (formerly part of Northwest Territories) and Hotchkiss in Alberta. A cross section of the permafrost beneath these towns is shown in Figure 17.20. [Adapted from T. L. Péwé, "Alpine permafrost in the contiguous United States: A review," *Arctic and Alpine Research* 15, no. 2 (May 1983): 146. © University of Colorado. Used by permission.]

In the Southern Hemisphere, north-facing slopes experience increased warmth.

Discontinuous permafrost zones are the most susceptible to thawing with climate change. Affected peat-rich soils (Gelisols and Histosols, discussed in Chapter 18) contain roughly twice the amount of carbon as is currently in the atmosphere, creating a powerful positive feedback as these soils thaw and oxidation releases more carbon dioxide to the atmosphere. Additionally, the losses of carbon from the thawing soils exceed any increases in carbon uptake that the warmer conditions and higher carbon dioxide levels create, making permafrost thaw an important source of greenhouse gases. (Please review the Chapter 10B *Geosystems Now*.)

Permafrost Behavior Figure 17.20 shows a cross section of a periglacial region in northern Canada, extending from approximately 75° N to 55° N through the three sites located on the map in Figure 17.19. The zone of seasonally frozen ground that exists between the subsurface permafrost layer and the ground surface is called the **active layer** and is subjected to consistent daily and seasonal freeze–thaw cycles. This cyclic thawing of the active layer affects as little as 10 cm (4 in.) of depth in the north of the periglacial region (Ellesmere Island, 78° N), up to 2 m (6.6 ft) in the southern margins (55° N) of the periglacial region, and 15 m (50 ft) in the alpine permafrost of the Colorado Rockies (40° N).

Permafrost actively adjusts to changing climatic conditions: Higher temperatures reduce permafrost thickness and increase the thickness of the active layer; lower temperatures gradually increase permafrost thickness and reduce active-layer thickness. Although somewhat sluggish in response, the active layer is a dynamic, open system driven by energy gains and losses in the subsurface environment.

With the warming temperatures recorded in the Canadian and Siberian Arctic since 1990, more disruption of permafrost surfaces is occurring—leading to highway, railway, and building damage. In Siberia, many lakes have disappeared in the discontinuous permafrost

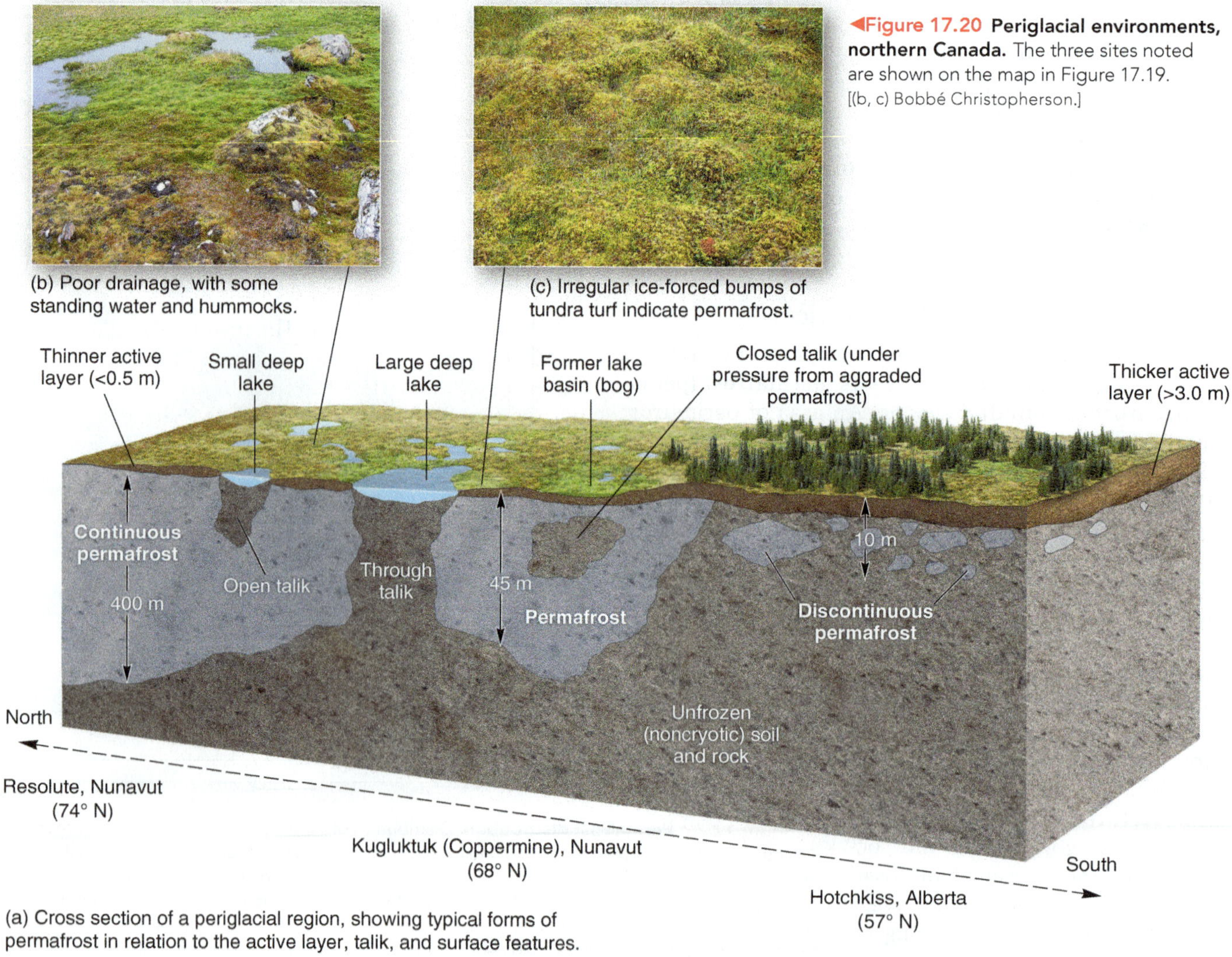

◀**Figure 17.20 Periglacial environments, northern Canada.** The three sites noted are shown on the map in Figure 17.19. [(b, c) Bobbé Christopherson.]

region as thawing of permafrost opens the way for subsurface drainage; yet in the continuous region, new lakes have formed as a result of thawed soils becoming waterlogged. In Canada, hundreds of lakes have disappeared from excessive evaporation into the warming air. These trends are measurable from satellite imagery.

A *talik* is an area of unfrozen ground that may occur above, below, or within a body of discontinuous permafrost or beneath a water body in regions of continuous permafrost. Taliks occur beneath deep lakes and may extend to bedrock and noncryotic soil beneath large, deep lakes (see Figure 17.20). Taliks in areas of discontinuous permafrost form connections between the active layer and groundwater, whereas in continuous permafrost, groundwater is essentially cut off from surface water. In this way, permafrost disrupts aquifers and taliks, leading to water-supply problems.

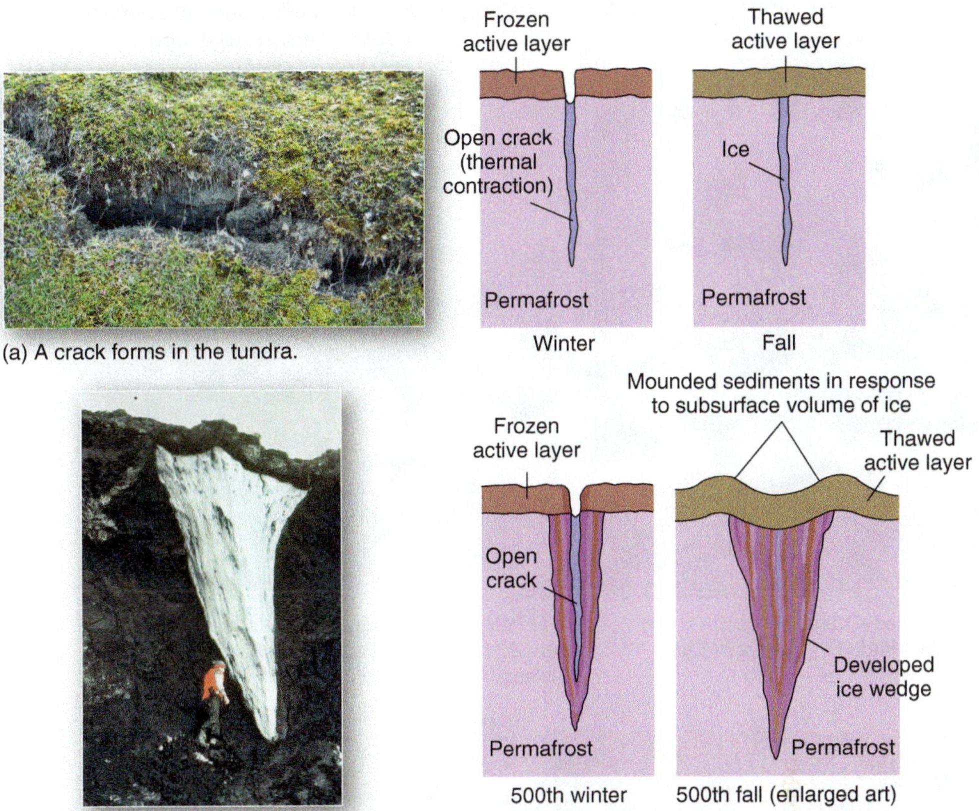

(a) A crack forms in the tundra.

(b) The crack evolves into an ice wedge over a time span of hundreds of years.

(c) An example of an ice wedge and ground ice in northern Canada.

▲**Figure 17.21 Evolution of an ice wedge.** Sequential illustration of ice-wedge formation. [(a) Bobbé Christopherson. (b) Illustration adapted from A. H. Lachenbruch, "Mechanics of thermal contraction and ice-wedge polygons in permafrost," *Geological Society of America Bulletin Special Paper 70* (1962). (c) H. M. French.]

Periglacial Processes

In regions of permafrost, frozen subsurface water is called **ground ice**. The amount of ground ice present varies with moisture content, ranging from only a small percentage in drier regions to almost 100% in regions with saturated soils. The presence of frozen water in the soil initiates geomorphic processes associated with *frost action*.

Frost Action Processes The 9% expansion of water as it freezes produces strong mechanical forces that fracture rock and disrupt soil at or below the surface. If sufficient water freezes, the saturated soil and rocks are subjected to *frost heaving* (vertical movement) and *frost thrusting* (horizontal movement). Boulders and rock slabs may be thrust to the surface. Soil horizons (layers) may be disrupted by frost action and appear to be stirred or churned. Frost action also can produce contractions in soil and rock, opening up cracks in which ice wedges can form.

An *ice wedge* develops when water enters a crack in the permafrost and freezes (Figure 17.21). Thermal contraction in ice-rich soil forms a tapered crack—wider at the top, narrowing toward the bottom. Repeated seasonal freezing and thawing progressively enlarge the wedge, which may widen from a few millimeters to 5–6 m (16–20 ft) and deepen up to 30 m (100 ft). Widening may be small each year, but after many years, the wedge can become significant, as in Figure 17.21c.

In some periglacial regions, the expansion and contraction of frost action result in the movement of soil particles, stones, and small boulders into distinct shapes known as **patterned ground** (Figure 17.22). This freeze–thaw process brings about a self-organization in which stones move toward stone domains (stone-rich areas) and soil particles move toward soil domains (soil-rich areas). The stone-centered polygons in Figure 17.22b indicate higher stone concentrations, and the soil-centered polygons in Figure 17.22c indicate higher soil-particle concentrations with lesser availability of stones. Patterned ground may take centuries to form. Slope angle also affects the arrangement—greater slopes produce striped patterns, whereas lesser slopes result in sorted polygons.

▼Figure 17.22 Patterned-ground phenomena. [(a) Courtesy of Joan Myers. (b), (c) Bobbé Christopherson. (d) Malin Space Science Systems.]

(a) Patterned ground in Beacon Valley of the McMurdo Dry Valleys of East Antarctica.

(b) Polygons and circles (about a meter across) in a stone-dominant area, Nordaustlandet Island, Arctic Ocean.

(c) Polygons and circles in a soil-dominant area, Spitsbergen Island.

(d) On Mars, polygons 100 m (300 ft) across in the northern plains.

Such polygon nets in patterned ground provide vivid evidence of frozen subsurface water on Mars (Figure 17.22d). In this 1999 image, the *Mars Global Surveyor* captured a region of these features on the Martian northern plains. To date, some 600 different sites on Mars demonstrate such patterned ground. In 2008, when the *Phoenix* lander set down in the middle of a Martian arctic plain covered with polygonal forms, the craft dug into the Martian surface and verified the presence of water in the form of ground ice.

Hillslope Processes: Gelifluction and Solifluction Soil drainage is poor in areas of permafrost and ground ice. The active layer of soil and regolith is saturated with soil moisture during the thaw cycle (summer), and the whole layer commences to flow from higher to lower elevation if the landscape is even slightly inclined. This flow of soil is generally called *solifluction*. In the presence of ground ice or permafrost, the more specific term *gelifluction* is applied. In this ice-bound type of soil flow, movement up to 5 cm (2 in.) per year can occur on slopes as gentle as a degree or two.

The cumulative effect of this flow can be an overall flattening of a rolling landscape, combined with identifiable sagging surfaces and scalloped and lobed patterns in the downslope soil movements. Other types of periglacial mass movement include failure in the active layer, producing translational and rotational slides and rapid flows associated with melting ground ice. Periglacial mass-movement processes are related to slope dynamics and processes discussed in Chapter 13.

Humans and Periglacial Landscapes

In areas of permafrost, people face certain problems related to periglacial landforms and phenomena. Because thawed ground above the permafrost zone frequently shifts, highways and rail lines may warp or twist, and utility lines are disrupted. In addition, any building placed directly on frozen ground will "melt" (subside) into the defrosting soil (Figure 17.23).

GEOreport 17.3 Feedback loops from fossil-fuel exploration to permafrost thawing

The lengthening periods of thaw in the active layer of the Alaskan tundra have shortened the annual number of days that oil exploration equipment can operate. Previously, the frozen ground lasted more than 200 days, permitting the heavy rigs and trucks to operate on the solid surface. With the increased thawing and unstable soft surfaces, exploration is now restricted to only 100 days a year. Consider this trend from a systems approach: The exploration is for fossil fuels, which, when burned, increase the carbon dioxide levels in the atmosphere, which enhance greenhouse warming, which increases temperatures and permafrost thawing, which decrease the number of days to explore for fossil fuels—a negative feedback into the system.

▲**Figure 17.23 Permafrost thawing and structure collapse.** Building failure due to improper construction as permafrost thaws south of Fairbanks, Alaska. [Adapted from USGS: photo by Steve McCutcheon; based on U.S. Geological Survey pamphlet "Permafrost" by L. L. Ray.]

In periglacial regions, structures must be suspended slightly above the ground to allow air circulation beneath. The airflow permits the ground to cycle through its normal annual temperature pattern. Utilities such as water and sewer lines must be enclosed aboveground in "utilidors" to protect them from freezing and thawing ground. The trans-Alaska oil pipeline was constructed aboveground on racks for 675 km of its 1285-km length (420 mi of its 800-mi length) to avoid thawing the frozen ground and causing shifting that could rupture the line (Figure 17.24). Where it runs underground, the pipeline uses a cooling system to keep the permafrost around the pipeline stable.

▲**Figure 17.24 Trans-Alaska oil pipeline.** The pipeline is 1.2 m (4 ft) in diameter, and is supported on racks that average 1.5 to 3.0 m (5 to 10 ft) in height above the ground to prevent permafrost thaw. [a96/Zuma Press/Newscom.]

The Pleistocene Epoch

Imagine almost a third of Earth's land surface buried beneath ice sheets and glaciers—most of Canada, the northern Midwest, England, and northern Europe, with many mountain ranges beneath thousands of meters of ice. This occurred at the height of the Pleistocene Epoch of the late Cenozoic Era (see Chapter 11, Figure 11.1). During this last ice age, periglacial regions along the margins of the ice covered about twice the areal extent of periglacial regions today.

The Pleistocene Epoch, thought to have begun about 2.5 million years ago, was one of the more prolonged cold periods in Earth's history. As discussed in Chapter 10B, the term **ice age**, or *glacial age*, is applied to any extended period of cold (not a single brief cold spell), in some cases lasting several million years. An ice age includes one or more *glacials*, characterized by glacial advance, interrupted by brief warm spells known as *interglacials*. The Pleistocene featured not just one glacial advance and retreat but at least 18 expansions of ice over Europe and North America, each obliterating and confusing the evidence from the one before. Apparently, glaciation can take about 100,000 years, whereas deglaciation is rapid, requiring less than about 10,000 years for the accumulation to melt away.

Ice-Age Landscapes

The continental ice sheets that covered portions of Canada, the United States, Europe, and Asia about 18,000 years ago are illustrated on the maps in Figure 17.25. Ice sheets ranged in thickness to more than 2 km (1.2 mi). In North America, the Ohio and Missouri River systems mark the southern terminus of continuous ice at its greatest extent during the Pleistocene Epoch. The ice sheet disappeared by 7000 years ago.

As the glaciers of the last ice age retreated, they exposed a drastically altered landscape: the rocky soils of New England, the polished and scarred surfaces of Canada's Atlantic Provinces, the sharp crests of the Sawtooth Range and Tetons of Idaho and Wyoming, the scenery of the Canadian Rockies and the Sierra Nevada, the Great Lakes of the United States and Canada, the Matterhorn

▼Figure 17.25 Extent of Pleistocene glaciation. Earlier episodes produced ice sheets of slightly greater extent. Note the depth of the continental ice sheets in North America (in meters). [From A. McIntyre, *CLIMAP* (Climate: Long-Range Investigation, Mapping, and Prediction) Project, Lamont-Doherty Earth Observatory. © 1981 by the GSA. Adapted by permission.]

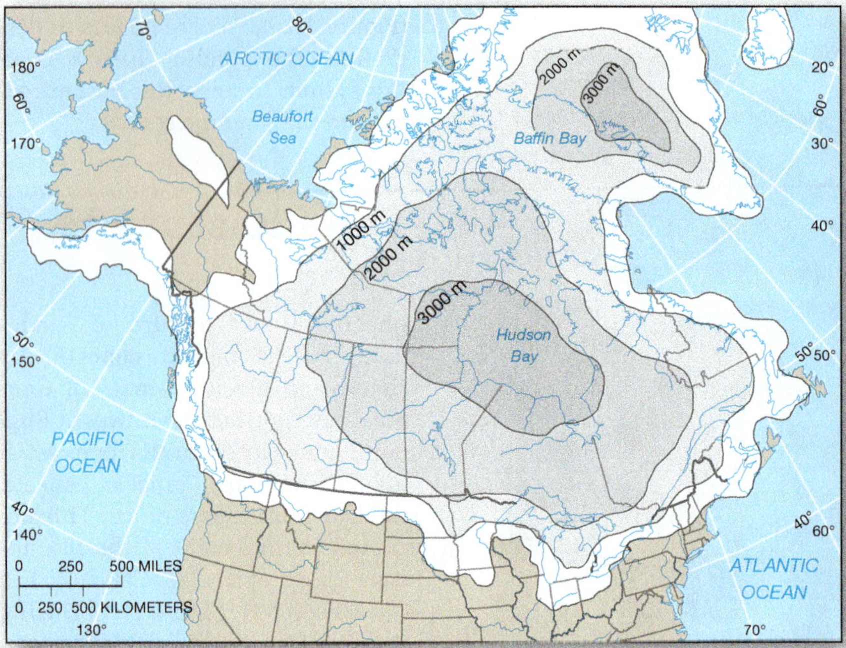

(a) 18,000 years ago.

(c) 9,000 years ago.

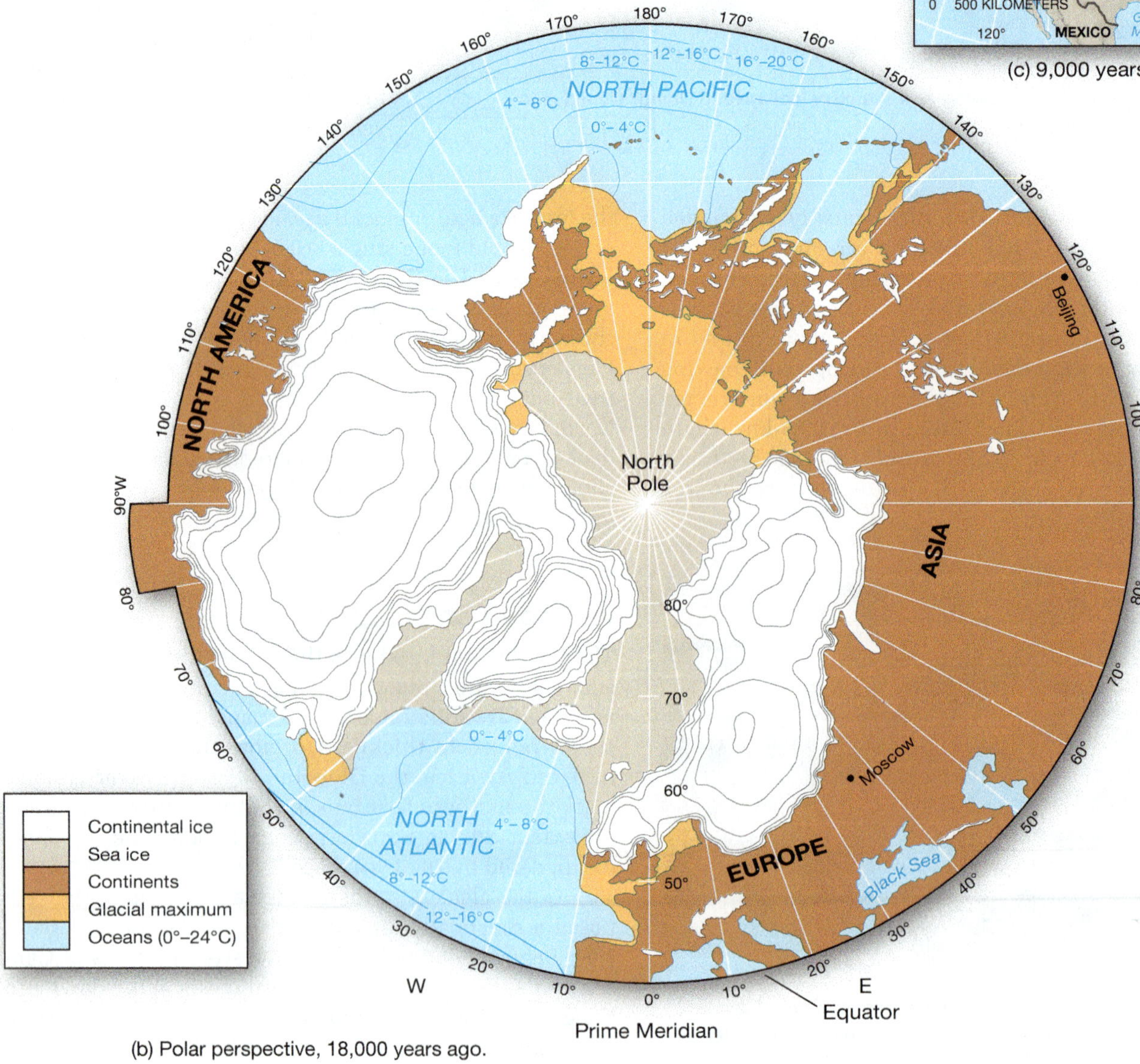

(b) Polar perspective, 18,000 years ago.

of Switzerland, and much more. In the Southern Hemisphere, there is evidence of this ice age in the form of fjords and sculpted mountains in New Zealand and Chile.

Continental glaciation occurred several times over the region we know as the Great Lakes (Figure 17.26). The ice enlarged and deepened stream valleys to form the basins of the future lakes. This complex history produced five lakes that today cover 244,000 km^2 (94,000 mi^2) and hold some 18% of all the lake water on Earth. Figure 17.26 shows the final sequence in the formation of today's Great Lakes, which involved two advancing and two retreating stages—between 13,200 and 10,000 years before the present. During the final retreat, tremendous quantities of glacial meltwater flowed into the isostatically depressed basins—that is, basins that were lowered by the weight of the ice. Drainage at first was to the Mississippi River via the Illinois River, to the St. Lawrence River via the Ottawa River, and to the Hudson River in the east. In recent times, drainage is solely through the St. Lawrence system.

Sea levels 18,000 years ago were approximately 100 m (330 ft) lower than they are today, because so much of Earth's water was frozen in glaciers. Imagine the coastline of New York being 100 km farther east, Alaska and Russia connected by land across the Bering Straits, and England and France joined by a land bridge. In fact, sea ice extended southward into the North Atlantic and Pacific and northward in the Southern Hemisphere about 50% farther than it does today.

Paleolakes

From 12,000 to 30,000 years ago, the American West was dotted with large, ancient lakes—called **paleolakes**, or *pluvial lakes* (Figure 17.27). The term *pluvial* (from the Latin word for "rain") describes any extended period of wet conditions, such as occurred during the Pleistocene Epoch. During pluvial periods in arid regions, lake levels increase in closed basins with internal drainage. The drier periods between pluvials, called *interpluvials*, are often marked by **lacustrine deposits**, the name for lake sediments that form terraces, or benches, along former shorelines. Except for the Great Salt Lake in Utah (a remnant of the former Lake Bonneville; Figure 17.27a) and a few smaller lakes, only dry basins, ancient shorelines, and lake sediments remain today.

Scientists have attempted to correlate pluvial and glacial events, given their coincidence during the Pleistocene. However, few sites actually demonstrate a direct relation. For example, in the western United States, the estimated volume of melted ice from glaciers is only a small portion of the actual water volume that was in the paleolakes. Also, these lakes tend to predate glacial times and are correlated instead with periods of wetter climate or periods thought to have had lower evaporation rates.

Paleolakes existed in North and South America, Africa, Asia, and Australia. Today, the Caspian Sea in Kazakhstan

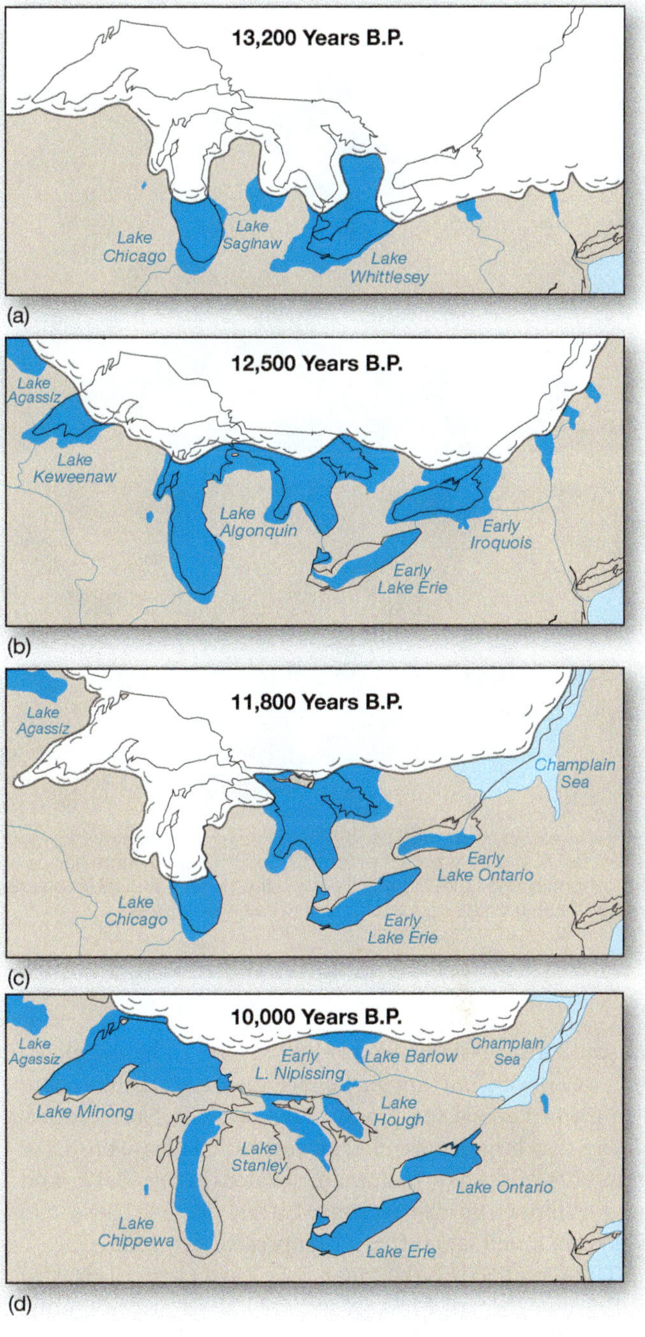

▲**Figure 17.26 Late stages of Great Lakes formation.** Four "snapshots" illustrate the evolution of the Great Lakes during the retreat of the Wisconsinan glaciation. Note the change in stream drainage between (b) and (d). Time is in years before the present (B.P.). [After *The Great Lakes—An Environmental Atlas and Resource Book*, Environment Canada, U.S. EPA, Brock University, and Northwestern University, 1987.]

GEOreport 17.4 Glacial ice might protect underlying mountains

Researchers are studying how glacial ice affected the underlying topography during the last glacial maximum. The temperature at the base of the ice was a determining factor. In the southernmost Patagonian Andes, conditions were so cold that the glacial ice froze to the bedrock. Evidence suggests that this protected the bedrock from erosion and typical glacial excavation, resulting in higher mountain peaks and a wider mountain belt in the southern Andes than in the northern part of the range, where mountain surfaces were ground down and narrowed by glacial action.

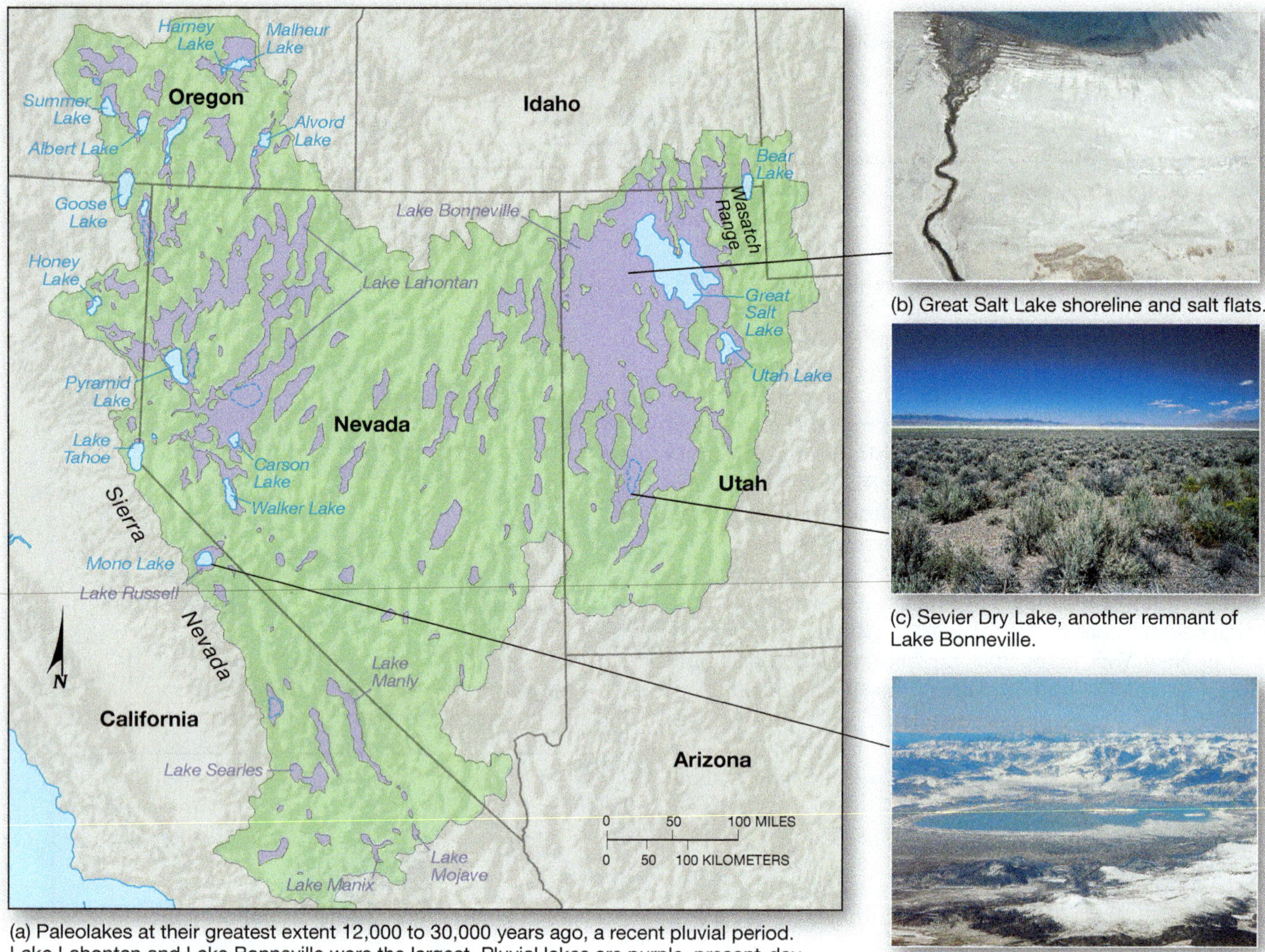

(a) Paleolakes at their greatest extent 12,000 to 30,000 years ago, a recent pluvial period. Lake Lahontan and Lake Bonneville were the largest. Pluvial lakes are purple, present-day lakes are blue, and the region of interior drainage is green.

(b) Great Salt Lake shoreline and salt flats.

(c) Sevier Dry Lake, another remnant of Lake Bonneville.

(d) Mono Lake in California, a remnant of pluvial Lake Russell.

▲**Figure 17.27** **Paleolakes in the western United States.** [(a) After USGS; (b)–(d) Bobbé Christopherson.]

and southern Russia has a level 30 m (100 ft) below global mean sea level, but ancient shorelines are visible about 80 m (265 ft) above the present lake level. In North America, the two largest late Pleistocene paleolakes were Lake Bonneville and Lake Lahontan, located in the Basin and Range Province in the western United States. These two lakes were much larger than their present-day remnants.

The Great Salt Lake, near Salt Lake City, Utah, and the Bonneville Salt Flats in western Utah are remnants of Lake Bonneville; today, the Great Salt Lake is the fourth largest saline lake in the world. At its greatest extent, this paleolake covered more than 50,000 km^2 (19,500 mi^2) and reached depths of 300 m (1000 ft), spilling over into the Snake River drainage to the north. Now, it is a closed-basin terminal lake with no drainage except an artificial outlet to the west, where excess water from the Great Salt Lake can be pumped during rare floods. Lake levels continue to decline in response to climate change to drier conditions.

New evidence suggests that the occurrence of these lakes in North America was related to specific changes in the polar jet stream that steered storm tracks across the region, creating pluvial conditions. The continental ice sheet evidently influenced changes in the position of the jet stream.

Arctic and Antarctic Regions

Climatologists use environmental criteria to designate the Arctic and the Antarctic regions. The 10°C (50°F) isotherm for July defines the **Arctic region** (green line on the map in Figure 17.28a). This line coincides with the visible tree line—the boundary between the northern forests and tundra. The Arctic Ocean is covered by *pack ice* (masses of drifting ice, unattached to shore) of two general types: *floating sea ice* (frozen seawater) and *glacier ice* (frozen freshwater). This pack ice thins in the summer months and sometimes breaks up (Figure 17.28b).

The **Antarctic region** is defined by the Antarctic convergence, a narrow zone that marks the boundary between colder Antarctic water and warmer water at lower latitudes. This boundary extends around the continent, roughly following the 10°C (50°F) isotherm for February, in the Southern Hemisphere summer, and is located near 60° S latitude

(a) Note the 10°C (50°F) isotherm in midsummer, which designates the Arctic region, dominated by pack ice.

(b) Arctic sea ice about 965 km (600 mi) from the North Pole.

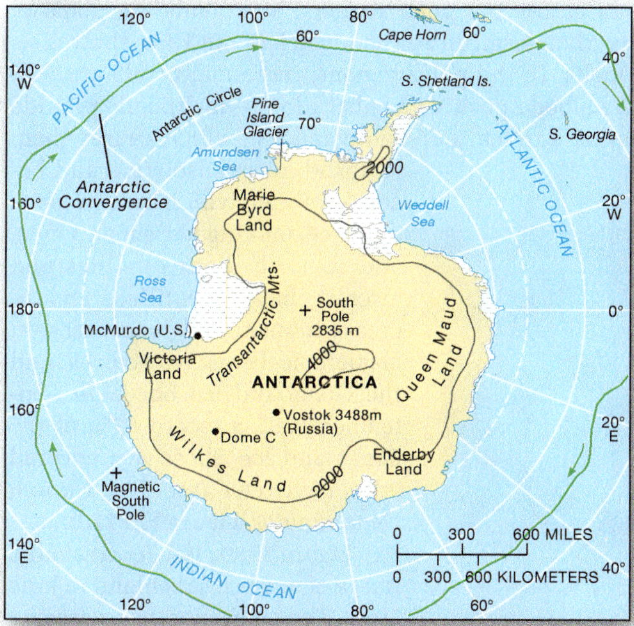

(c) The Antarctic convergence designates the Antarctic region.

▲**Figure 17.28 The Arctic and Antarctic regions.**
[(b) Bobbé Christopherson.]

(green line in Figure 17.28c). The part of the Antarctic region covered just with sea ice represents an area greater than North America, Greenland, and Western Europe combined.

The Antarctic landmass is surrounded by ocean and is much colder overall than the Arctic, which is an ocean surrounded by land. In simplest terms, Antarctica can be thought of as a continent covered by a single enormous glacier, although it contains distinct regions such as the East Antarctic and West Antarctic Ice Sheets, which respond differently to slight climatic variations. These ice sheets are in constant motion.

The fact that Antarctica is so remote from civilization makes it an excellent laboratory for sampling past and present evidence of human and natural variables that are transported by atmospheric and oceanic circulation to this pristine environment. High elevation, winter cold and darkness, and distance from pollution sources make this polar region an ideal location for certain astronomical and atmospheric observations.

CRITICALthinking 17.2
A Sample of Life at the Polar Station

Read some of the posts at http://www.snowbetweenmytoes.blogspot.com/ regarding life at the Amundsen–Scott South Pole Station, Antarctica. More updates, links to blogs, and South Pole news are at http://www.southpolestation.com/. Of the approximately 50 people who winter over at the station (between the last airplane's departure in mid-February to the next scheduled flight arrival in mid-October), many serve as scientists, technicians, and support staff. What are some of the unique aspects of "Life as a Polie"? What do you see as the positives and negatives of living and working there? How would you combat the elements, the isolation, and the dark conditions? •

Recent Polar Region Changes

As mentioned earlier, the smallest extent of Arctic sea ice on record occurred in the year 2012. About half of Arctic sea-ice volume has disappeared since 1970 due to warming throughout the region. As discussed in Chapter 4 *Geosystems Now*, the fabled Northwest Passage across the Arctic from the Atlantic to the Pacific is now ice-free for a portion of the summer as the Arctic ice continues to melt. The Northeast Passage, north of Russia, has been ice-free for the past several years. These changes affect surface albedo, with impacts on global climate (please review the Chapter 4 *Geosystems Now*).

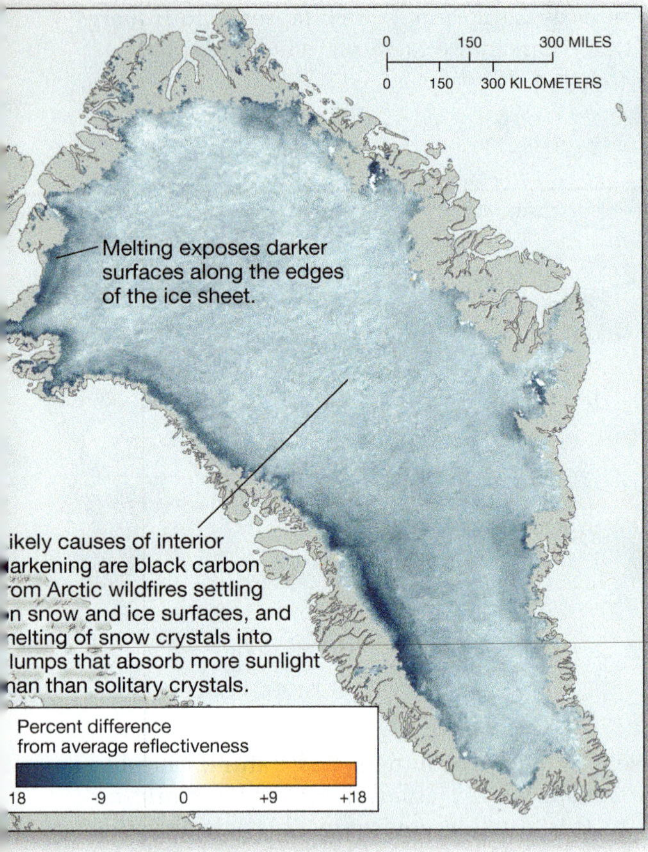

◀**Figure 17.29 Darkening of Greenland ice surfaces.** Data from this 2011 image indicate that some areas of the ice sheet reflect 20% less sunlight than just a decade ago. [*Terra/Aqua* MODIS, NASA.]

Ice Sheet Darkening Satellite measurements show that on the Greenland Ice Sheet, the reflectivity of the snow and ice has decreased over the past decade as the surface has darkened. Along the outer edges, ice melt has exposed darker land, vegetation, and water surfaces. On the interior, black carbon from wildfires in Asia and North America has accumulated on the ice and may be contributing to the overall darkening (Figure 17.29). Another factor may be related to basic processes of snow metamorphism: As temperatures rise, snow crystals clump together in the snow pack, reflecting less light than the smaller, faceted individual crystals. The overall effect is that the ice sheet now absorbs more sunlight, which speeds up melting and causes a positive feedback that accelerates warming.

Meltponds and Supraglacial Lakes Another indicator of changing surface conditions is an increase in meltponds across the polar regions. Meltponds are pools of water that form on sea ice, glaciers, and ice shelves as ice melts. In the polar regions, these meltponds are part of the albedo positive feedback loop—meltponds provide a darker surface than snow and ice, which causes them to absorb more insolation and become warmer, which in turn melts more ice, making more meltponds, and so forth. Both satellites and aircraft have identified an increase in meltpond occurrence on glaciers, icebergs, ice shelves, and the Greenland Ice Sheet. In September 2012, a record 97% of the Greenland Ice Sheet was covered with meltwater (see *The Human Denominator*, Figure HD 17d).

Figure 17.30a is a *Terra* satellite image of western Greenland in June 2003. The snow line has retreated to higher elevations, exposing bare rock. The image shows numerous meltponds and areas of water-saturated ice in the *melt zone*, which advanced 400% from 2001 to 2003. The meltponds look like small blue dots across the ice (also visible in Figure 17.30b and d, taken in southwest Greenland in 2008). Melt streams are also common during periods of warmer temperatures in glaciated environments, especially in summer. The streams can melt through the ice sheet, forming *moulins*, or drainage channels, that work their way to the base of the glacier (Figure 17.30d and e).

) Meltponds are increasing in number on the Greenland e Sheet (shown), as on icebergs and ice shelves roughout the Arctic.

(b) Meltponds.

Close-up of meltpond.

(d) Meltponds and melt streams, southwest Greenland, 2008.

(e) Meltwater flows into a moulin.

▲**Figure 17.30 Meltponds, melt streams, and moulins on the increase in Greenland.** Why are meltponds positive feedback climate indicators? [(a) *Terra* MODIS image, NASA/GSFC. (b), (c) Bobbé Christopherson. (d), (e) Courtesy of JPL/NASA.]

Multiple streams can flow together in summer to form a *supraglacial lake* on top of the ice. When the water pressure builds to a high enough level, the ice below the lake fractures and drains the lake through a near-vertical moulin to the glacier's bed. This drainage may result in a large volume of water arriving suddenly at the base of the glacier. Scientists have witnessed such events and are studying the effects on glacial movement.

Ice Shelves Another recent change in the polar regions is the breakup of ice shelves, as discussed in *Geosystems Now.* Ice shelves surround the margins of Antarctica and constitute about 11% of its surface area (see Figure 17.7 for some of the major ice shelves). Although ice shelves constantly break up to produce icebergs, more large sections have broken free in the past two decades than expected. For example, in March 2000, an iceberg tagged B-15, measuring twice the area of Delaware (300 km by 40 km, or 190 mi by 25 mi), broke off the Ross Ice Shelf (some 3027 km, or 1900 mi, west of the Antarctic Peninsula). In 2013, the Wilkins Ice Shelf underwent further disintegration after major breakup events in 2008 and 2009. Scientists think that the recent breakups left the remaining ice more vulnerable, especially in places where the shelf remnants are in direct contact with open water and the force of ocean waves.

Since 1993, seven ice shelves have disintegrated in Antarctica. More than 8000 km^2 (3090 mi^2) of ice shelf are gone, requiring significant revision of maps, freeing up islands to circumnavigation, and creating thousands of icebergs. The Larsen Ice Shelf, along the east coast of the Antarctic Peninsula, was retreating slowly for years. Larsen A suddenly disintegrated in 1995. Then, in only 35 days in early 2002, Larsen B collapsed into icebergs (Figure 17.31). Larsen B was at least 11,000 years old.

Larsen C, the next segment to the south, is losing mass from both the ocean and the atmosphere sides. Since the water temperature is warmer by 0.65 C° (1.17 F°) than the melting point for ice at a depth of 300 m (984 ft), this ice loss is likely a result of warmer water, as well as of the air temperature increase in the peninsula region during the last 50 years. In response to the increasing warmth, the Antarctic Peninsula is also experiencing previously unseen vegetation growth, reduced sea ice, and disruption of penguin feeding, nesting, and fledging activities. (Among many changes, ticks are a new problem for these animals.)

CRITICALthinking 17.3
The IPY Accomplishment Continues

The International Polar Year (IPY) ran from March 2007 to March 2009, covering two polar summer seasons—it was the fourth IPY conducted since 1882. This global interdisciplinary research, exploration, and discovery effort involved 50,000 scientists in hundreds of collaborative projects. About 65% of the research was in the Arctic region and 35% in the Antarctic region. A systems approach was employed for finding linkages among ecosystems and human activity. The research effort also made use of the traditional knowledge of indigenous peoples across the circumarctic region—the actual experiences of the First Nations people, who are encountering climate change firsthand.

Use your critical-thinking skills in a brief exploration of this IPY. Begin at http://www.ipy.org. You may choose to "Focus On" the atmosphere, ice, people, or other topics in the list at upper left of the home page. •

▲**Figure 17.31 Disintegrating ice shelves along the Antarctic coast.** Disintegration and retreat of the Larsen B ice shelf between January 31 and March 7, 2002. Note the meltponds in the January image. [*Terra* images, NASA.]

THE**human**DENOMINATOR 17 Glaciers and Permafrost

GLACIAL ENVIRONMENTS ⇒ HUMANS

- Glacial ice is a freshwater resource; ice masses affect sea level, which is linked to the security of human population centers along coastlines.
- Snow avalanches are a significant natural hazard in mountain environments.
- Permafrost soils are a carbon sink, estimated to contain half the pool of global carbon.

HUMANS ⇒ GLACIAL ENVIRONMENTS

- Rising temperatures associated with human-caused climate change are accelerating ice sheet losses and glacial melting, and hastening permafrost thaw.
- Particulates in the air from natural and human sources darken snow and ice surfaces, which accelerates melting.

17a

A USGS scientist photographs Grinnell Glacier in Glacier National Park, Montana, as part of a repeat photography project to document the effects of climate change on glacial retreat. [USGS/Lisa McKeon, Northern Rocky Mountain Science Center.]

17c

The northern ice field in the summit crater of Mount Kilimanjaro, Africa's highest mountain, has shrunk in size with dry season melting, and broke into two sections in 2012. Note expedition tents at the base of the ice, and automated meteorological stations on top. Scientists attribute the disappearing glaciers to climate change as well as other factors, such as a drier atmosphere that is depriving the glacier of snowfall to sustain the ice fields. Glacial research is ongoing at the 5803-m (9039-ft) summit. [Photo by Dr. Kimberly Casey.]

17d

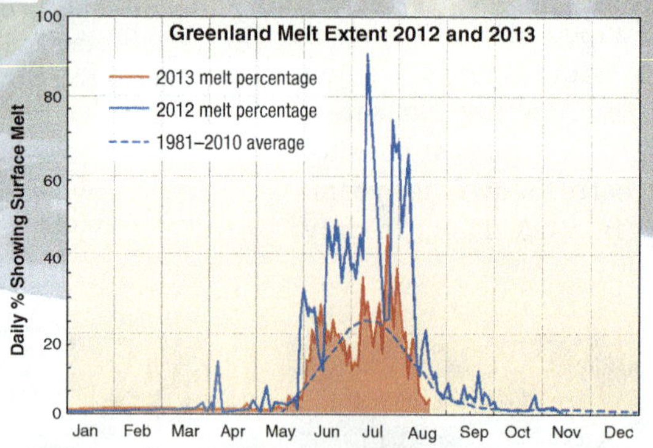

Surface melt on the Greenland Ice Sheet in 2012 reached its greatest extent in the satellite record since 1979. Melt during 2013 was closer to average (dotted line). [Courtesy National Snow and Ice Data Center/Thomas Mote, University of Georgia.]

17b

In the Indian-controlled region of Kashmir, avalanches burie a military camp and killed 16 Indian soldiers in February 2012. Three months later, 100 Pakistani soldiers were killec nearby. This area receives a large volume of snowfall and frequent winds, which combine with steep Himalayan slope to create dangerous avalanche conditions. [Dar Yasin.]

ISSUES FOR THE 21ST CENTURY

- Melting of glaciers and ice sheets will continue to raise sea level, with potentially devastating consequences for coastal communities and low-lying island nations.
- Thawing of permafrost in response to climate change will release vast amounts of carbon into the atmosphere, accelerating global warming.

GEOSYSTEMS**connection**

Our look at the cryosphere and glacial and periglacial processes and landforms ends Part III, "The Earth–Atmosphere Interface." We examined past and present alpine and continental glaciations and the ways in which Earth's landscapes bear the mark of past glacial phases. The polar regions are undergoing rapid change at a pace faster than the lower latitudes. We shift now to Part IV, "Soils, Ecosystems, and Biomes." A synthesis of Parts I through III, these next three chapters focus on the biosphere and the subdiscipline of biogeography.

KEY LEARNING concepts review

Explain the process by which snow becomes glacial ice.

More than 77% of Earth's freshwater is frozen. Ice covers about 11% of Earth's surface, and periglacial features occupy another 20%, in ice-free but cold-dominated landscapes. Earth's **cryosphere** is the portion of the hydrosphere and ground that is perennially frozen, generally at high latitudes and elevations.

A **snowline** is the lowest elevation where snow occurs year-round, and its elevation varies by latitude—higher near the equator, lower poleward. Snow becomes glacial ice through stages of accumulation, increasing thickness, pressure on underlying layers, and recrystallization. Snow progresses through transitional steps from **firn** (compact, granular) to a denser **glacial ice** after many years.

cryosphere (p. 522)
snowline (p. 523)
firn (p. 523)
glacial ice (p. 523)

1. Describe the location of most freshwater on Earth today.
2. Trace the evolution of glacial ice from fresh fallen snow.

Differentiate between alpine glaciers and continental ice sheets, and *describe* ice caps and ice fields.

A **glacier** is a mass of ice sitting on land or floating as an ice shelf in the ocean next to land. Glaciers form in areas of permanent snow. A glacier in a mountain range is an **alpine glacier**. If confined within a valley, it is termed a *valley glacier*. The area of origin is a snowfield, usually in a bowl-shaped erosional landform called a **cirque**. Where alpine glaciers flow down to the sea, the process of **calving** occurs as masses of ice break off the glacier into the sea and form *icebergs*. An **ice sheet** is an extensive continuous mass of ice that may occur on a continental scale. An **ice cap** is a smaller, roughly circular ice mass, less than 50,000 km^2 (19,300 mi^2) in size. An ice mass covering a mountainous region is an **ice field**.

glacier (p. 523)
alpine glacier (p. 523)
cirque (p. 523)
calving (p. 524)
ice sheet (p. 525)
ice cap (p. 525)
ice field (p. 526)

3. What is a glacier? What can we learn about existing climate patterns from conditions in glacial regions and glacial mass balances?
4. Differentiate between an alpine glacier, an ice sheet, an ice cap, and an ice field. Which occurs in mountains? Which covers Antarctica and Greenland?
5. How are icebergs generated? Describe their buoyancy characteristics using this chapter's discussion and the section on ice in Chapter 7. Why do you think an iceberg overturns from time to time as it melts?

Illustrate the mechanics of glacial movement.

A glacier is an open system. The **firn line** is the elevation above which the winter snow and ice remain intact throughout the summer melting season but below which melting occurs. A glacier is fed by snowfall and is wasted by **ablation** (losses from its upper and lower surfaces and along its margins). Accumulation and ablation achieve a mass balance in each glacier.

As a glacier moves downhill, vertical **crevasses** may develop. Sometimes a glacier will move rapidly, an event known as a **glacier surge**. The presence of water along the basal layer appears to be important in glacial movements. As a glacier moves, it plucks rock pieces and debris, incorporating them into the ice, and this debris scours and sandpapers underlying rock through **abrasion**.

firn line (p. 526)
ablation (p. 526)
crevasse (p. 527)
glacier surge (p. 530)
abrasion (p. 531)

6. What is meant by glacial mass balance? What are the basic inputs and outputs contributing to that balance?
7. What is meant by a glacier surge? What do scientists think produces surging episodes?

Describe characteristic erosional and depositional landforms created by glaciation.

A **roche moutonnée** is an erosional landform produced by plucking and abrasion. It is an asymmetrical hill of exposed bedrock, gently sloping on the upstream end and abruptly sloping on the downstream end.

Extensive valley glaciers have profoundly reshaped mountains worldwide, transforming **V**-shaped stream valleys into **U**-shaped glaciated valleys and producing many other distinctive erosional and depositional landforms. As cirque walls erode away, sharp **arêtes** (sawtooth, serrated ridges) form, dividing adjacent cirque basins. Two eroding cirques may reduce an arête to a saddlelike **col**. A **horn** results when several cirque glaciers gouge an individual mountain summit from all sides, forming a pyramidal peak. A **bergschrund** forms when a crevasse or wide crack opens along the boundary between flowing ice and stagnant ice, often near the headwall of a glacier; it is most visible in summer when covering snow is gone. An ice-carved rock basin left as a glacier retreats may fill with water to form a **tarn**; tarns in a string separated by moraines are **paternoster lakes**. Where a glacial trough joins the ocean and the glacier retreats, the sea extends inland to form a **fjord**.

All glacial deposits, whether ice-borne or meltwater-borne, constitute **glacial drift**. Direct deposits from ice consist of unstratified and unsorted **till**. Specific landforms produced by the deposition of till at glacial margins are **moraines.** A **lateral moraine** forms along each side of a glacier; merging glaciers with lateral moraines form a **medial moraine;** and eroded debris dropped at the farthest extent of a glacier's terminus is a **terminal moraine**. Recessional moraines mark temporary endpoints as the glacier advances and retreats over time.

A **till plain** forms behind a terminal moraine, featuring unstratified coarse till, low and rolling relief, and deranged drainage. **Drumlins** are elongated hills of deposited till, streamlined in the direction of continental ice movement (blunt end upstream and tapered end downstream).

Glacial meltwater deposits are sorted and stratified, and called **stratified drift,** forming **outwash plains** featuring braided stream channels that carry a heavy sediment load. An **esker** is a sinuously curving, narrow ridge of coarse sand and gravel that forms along the channel of a meltwater stream beneath a glacier. A **kame** is a small hill, knob, or mound of poorly sorted sand and gravel that is deposited directly on top of glacial ice, and then is deposited on the ground when the glacier melts. An isolated block of ice left by a retreating glacier becomes surrounded by debris; when the block finally melts, it leaves a steep-sided depression called a **kettle** that when filled with water forms a *kettle lake*.

roche moutonnée (p. 531)
arête (p. 531)
col (p. 531)
horn (p. 531)
bergschrund (p. 531)
tarn (p. 532)
paternoster lake (p. 532)
fjord (p. 532)
glacial drift (p. 534)
till (p. 534)
moraine (p. 534)
lateral moraine (p. 534)
medial moraine (p. 534)
terminal moraine (p. 534)
till plain (p. 534)
drumlin (p. 534)
stratified drift (p. 535)
outwash plain (p. 536)
esker (p. 536)
kame (p. 536)
kettle (p. 537)

8. How does a glacier accomplish erosion?
9. Describe the transformation of a V-shaped stream valley into a U-shaped glaciated valley. What features are visible after the glacier retreats?
10. How is an arête formed? A col? A horn? Briefly differentiate among them.
11. Differentiate between two forms of glacial drift—till and glacial outwash.
12. What is a morainal deposit? What specific moraines are created by alpine glaciers?
13. What is a common depositional feature encountered in a till plain?
14. Contrast a roche moutonnée and a drumlin with regard to appearance, orientation, and the way each forms.

■ *Discuss* the distribution of permafrost, and *explain* several periglacial processes.

The term **periglacial** describes cold-climate processes, landforms, and topographic features that exist along the margins of glaciers, past and present. When soil or rock temperatures remain below 0°C (32°F) for at least 2 years, **permafrost** (perennially frozen ground) develops. Note that this defining criterion for permafrost is based solely on temperature and has nothing to do with how much or how little water is present. The **active layer** is the zone of seasonally frozen ground that exists between the subsurface permafrost layer and the ground surface. In regions of permafrost, frozen subsurface water forms **ground ice. Patterned ground** forms in the periglacial environment where freezing and thawing of the ground create polygonal forms of circles, polygons, stripes, nets, and steps.

periglacial (p. 537)
permafrost (p. 537)
active layer (p. 538)
ground ice (p. 539)
patterned ground (p. 539)

15. In terms of climatic types, describe the areas on Earth where periglacial landscapes occur. Include both higher-latitude and higher-altitude climate types.
16. Define two types of permafrost, and differentiate their occurrence on Earth. What are the characteristics of each?
17. Describe the active zone in permafrost regions, and relate the degree of development to specific latitudes.
18. What is a talik? Where might you expect to find taliks, and to what depth do they occur?
19. What is the difference between permafrost and ground ice?
20. Describe the role of frost action in the formation of various landform types in the periglacial region, such as patterned ground.
21. Explain some of the specific problems humans encounter in building on periglacial landscapes.

■ *Describe* landscapes of the Pleistocene ice-age epoch, and *list* changes occurring today in the polar regions.

An **ice age** is any extended period of cold. The late Cenozoic Era featured pronounced ice-age conditions during the Pleistocene. Beyond the ice, **paleolakes** formed because of wetter conditions. **Lacustrine deposits** are lake sediments that form terraces along former shorelines.

The 10°C (50°F) isotherm for July defines the **Arctic region**. This line coincides with the visible tree line—the boundary between the northern forests and tundra. The Antarctic convergence defines the **Antarctic region** in a narrow zone that extends around the continent as a boundary between colder Antarctic water and warmer water at lower latitudes. This boundary follows roughly the 10°C (50°F) isotherm for February, the Southern Hemisphere summer, and is located near 60° S latitude. Changes occurring in the polar regions are causing positive feedback loops related to changes in surface albedo. Warming temperatures are causing the collapse of ice shelves.

ice age (p. 541)
paleolake (p. 543)
lacustrine deposit (p. 543)
Arctic region (p. 544)
Antarctic region (p. 544)

22. Define an ice age. When was the most recent? Explain the terms *glacial* and *interglacial* in your answer.
23. Explain the relationship between the criteria defining the Arctic and Antarctic regions. Is there any coincidence between the Arctic criteria and the distribution of Northern Hemisphere forests?
24. Based on information in this chapter and elsewhere in *Geosystems*, summarize a few of the changing conditions under way in each polar region.

MasteringGeography™

Looking for additional review and test prep materials? Visit the Study Area in *MasteringGeography*™ to enhance your geographic literacy, spatial reasoning skills, and understanding of this chapter's content by accessing a variety of resources, including MapMaster interactive maps, geoscience animations, satellite loops, author notebooks, videos, RSS feeds, flashcards, web links, self-study quizzes, and an eText version of *Geosystems*.

▲Figure 17.32 **Glacial retreat in the Svalbard archipelago, Norway.** This retreating glacier lies along Raudfjorden, a 20-km (12.4-mi) long fjord on the northwest coast of Spitsbergen at 79.5° N latitude. The scoured slopes in the background give you an idea of previous glacial depth about ten years ago. What other evidence do you see in the photo of a decaying glacier? [Bobbé Christopherson.]

IV Soils, Ecosystems, and Biomes

Earth is the home of the Solar System's only known biosphere—a unique, complex, and interactive system of abiotic (nonliving) and biotic (living) components working together to sustain a tremendous diversity of life. Energy enters the biosphere through conversion of solar energy by photosynthesis in the leaves of plants.

Life is organized into a feeding hierarchy from producers to consumers, ending with decomposers. Together, these varied organisms, in concert with the Earth's abiotic components, produce aquatic and terrestrial ecosystems, generally organized into various biomes. Soil is the essential

▼ Forest plantation in the Ashepoo, Combahee, and Edisto Basin, known as the ACE basin, near Charleston along South Carolina's Atlantic coast. [Dave Allen Photography/Shutterstock.]

link connecting the living world to the lithosphere and the rest of Earth's physical systems. Thus, soil is the appropriate bridge between Part III and Part IV of this text.

Today, we face crucial issues, principally the preservation of the diversity of life in the biosphere. Patterns of land and ocean temperatures, precipitation, weather phenomena, and stratospheric ozone, among many other geographic factors that have an impact on life, are changing as global climate systems shift. The resilience of the biosphere as we know it is being tested in a real-time, one-time experiment. These important issues of biogeography are considered in Part IV.

18

The Geography of Soils

KEY LEARNING concepts

After reading the chapter, you should be able to:

- ***Define*** soil and soil science, and ***list*** four components of soil.
- ***Describe*** the principal soil-formation factors, and ***describe*** the horizons of a typical soil profile.
- ***Describe*** the physical properties used to classify soils: color, texture, structure, consistence, porosity, and soil moisture.
- ***Explain*** basic soil chemistry, including cation-exchange capacity, and ***relate*** these concepts to soil fertility.
- ***Discuss*** human impacts on soils, including desertification.
- ***Describe*** the 12 soil orders of the Soil Taxonomy classification system, and ***explain*** their general distribution across Earth.

For over 2000 years, local people have cultivated the Bangaan Rice Terraces rising thousands of meters in elevation on the steep mountain slopes of the island of Luzon, Philippines. The terraces are built using stone and mud walls to create a series of ponded fields called *paddies*. The paddies prevent erosion by holding soil and water obtained from streams originating in the mountains above. Planted, maintained, and harvested by community effort, the terraces exemplify sustainable agriculture in a country that suffers from soil erosion problems. This terraced region was listed as a UNESCO World Heritage in Danger in 2001 as declining traditional practices led to terrace deterioration. Conservation is ongoing. Worldwide, rice paddy fields are a major source of atmospheric methane, a greenhouse gas. [Dave Stamboulis/Alamy.]

Desertification: Declining Soils and Agriculture in Earth's Drylands

In September 2012, on the edge of the Gobi Desert in the Inner Mongolia Autonomous Province of China, a group of volunteers planted the millionth tree in an attempt to fight desertification, the degradation of drylands. In a region devastated by sandstorms and deteriorating land, this forest restoration effort, funded by a private organization since 2007, plants trees (mainly from the genus *Populus*, or poplars), monitors water availability for growth, and educates communities about the importance of trees for preventing erosion, producing oxygen, and storing carbon dioxide. To the west, in the Taklamakan Desert of central Asia, native poplar trees are declining with ongoing drought (Figure GN 18.1).

Desertification is defined by the United Nations (UN) as "the persistent degradation of dryland ecosystems by human activities and climate change." This process along the margins of semiarid and arid lands is caused in part by human abuse of soil structure and fertility—one of the subjects of this chapter (see a map of global desertification risk in Figure 18.10).

▲**Figure GN 18.1 Trees stabilize soils and slow land degradation.** Poplar trees stabilize soils at the edge of the Taklamakan Desert in the Xinjiang Uyghur Autonomous Region, China. These trees help slow desertification, but are declining after years of drought. Restoration efforts are ongoing. [TAO Images Limited/Getty.]

Central Asia Throughout central Asia, overexploitation of water resources has combined with drought to cause desertification. The Aral Sea, formerly one of the four largest lakes in the world, has steadily shrunk in size since the 1960s, when inflowing rivers were diverted for irrigation (see Figure HD 18 at the end of this chapter). Fine sediment and alkali dust on the former lakebed have become available for wind deflation, leading to massive dust storms. This sediment contained fertilizers and other pollutants from agricultural runoff, so its mobilization and spread over the land has caused crop damage and human health problems, including increased cancer rates.

Africa's Sahel In Africa, the Sahel is the transition region between the Sahara Desert in the subtropics and the wetter equatorial regions. The southward expansion of desert conditions through portions of the Sahel region has left many African peoples on land that no longer experiences the rainfall of just three decades ago (Figure GN 18.2). Yet climate change is only part of the story: Other factors contributing to desertification in the Sahel are population increases, land degradation from deforestation and overgrazing of cattle, poverty, and the lack of a coherent environmental policy.

A Growing Problem The UN estimates that degraded lands worldwide cover some 1.9 billion hectares (4.7 billion acres) and affect 1.5 billion people; many millions of additional hectares are added each year. The primary causes of desertification are overgrazing, unsustainable agricultural practices, and forest removal. However, desertification is a complex phenomenon that is related to population issues, poverty, resource management, and government policies.

▲**Figure GN 18.2 Desertification in the Sahel.** Herders carry straw to cattle outside a village in Mali. Ethnic tensions in this country located in the west African Sahel are interfering with agricultural activities, worsening food shortages, and slowing efforts aimed toward sustainable land stewardship. [Nic Bothma/epa/Corbis.]

Initiatives to combat desertification began in 1994 with the Convention to Combat Desertification, an effort that is still active. In August 2010, the *International Conference: Climate, Sustainability, and Development in the Semi-arid Regions* met for a second time (http://www.unccd.int/). Launched at this meeting was a global effort for the next decade—the United Nations Decade for Deserts and the Fight Against Desertification (UNDDD). Desertification threatens livelihoods and food resources in many areas of the world. Further delays in addressing the problem will result in far higher costs when compared to the cost of taking action now.

GEOSYSTEMS NOW ONLINE Go to Chapter 18 on the *MasteringGeography* (www.masteringgeography.com) website for more information on desertification or see http://www.un.org/en/events/desertificationday/background.shtml. More on declining poplar forests and related land degradation in the Taklamakan–Gobi Desert region is at http://whc.unesco.org/en/tentativelists/5532/.

Earth's landscape generally is covered with **soil**, a dynamic natural material composed of water, air, and fine particles—both mineral fragments (sands, silts, clays) and organic matter—in which plants grow. Soil is the basis for functioning ecosystems: It retains and filters water; is habitat for a host of microbial organisms, many of which produce antibiotics that fight human diseases; serves as a source of slow-release nutrients; and stores carbon dioxide and other greenhouse gases. Nearly 80% of terrestrial organic and inorganic carbon is stored in soil, and about a quarter of this amount is stored in wetlands, which are defined by the presence of hydric soils (waterlogged soils, discussed in Chapter 9 and later in this chapter). The soil carbon pool is about three times larger than the atmospheric carbon pool; only the ocean stores more carbon than soil. Remember from Chapter 10B that although about two-thirds of the recent increase in atmospheric carbon dioxide comes from fossil-fuel burning, one-third comes from the loss of soil associated with land-use changes.

Soils develop over long periods of time; in fact, many soils bear the legacy of climates and geological processes over the last 15,000 years or more. Soils do not reproduce, nor can they be re-created—they are a nonrenewable natural resource. This fact means that human use and abuse of soils is happening at rates much faster than soils form or can be replaced.

Soil is a complex substance whose characteristics vary from kilometer to kilometer—and even centimeter to centimeter. Physical geographers are interested in the spatial distributions of soil types and the physical factors that interact to produce them. Knowledge of soils is also critical for agriculture and food production. **Soil science** is the interdisciplinary study of soil as a natural resource on Earth's surface; the field draws on aspects of physics, chemistry, biology, mineralogy, hydrology, taxonomy, climatology, and cartography. *Pedology* deals with the origin, classification, distribution, and description of soils (*ped* is from the Greek *pedon*, meaning "soil" or "earth"). The soil layer is sometimes called the edaphosphere (*edaphos* means "soil" or "ground"). *Edaphology* specifically focuses on the study of soil as a medium for sustaining the growth of higher plants.

In this chapter: We begin with an examination of soil development and the soil horizons of a typical soil profile. We look at the properties that affect soil fertility and determine soil classification, including color, texture, structure, consistence, porosity, moisture, and chemistry. We also discuss human impacts on soils, and desertification. We conclude with a brief examination of the United States Soil Taxonomy, focusing on the 12 principal soil orders and their spatial distribution.

Soil-Formation Factors and Soil Profiles

Soil is composed of about 50% mineral and organic matter; the other 50% is air and water stored in the pore spaces around soil particles. The organic matter, although making up only about 5% of a given soil volume, is critical for soil function and includes living microorganisms and plant roots, dead and partially decomposed plant matter, and fully decomposed plant material that forms a nutrient-rich mixture called humus (discussed ahead).

Soil is an open system with physical inputs of insolation, water, rock and sediment, and microorganisms, and outputs of plant ecosystems that sustain animals and human societies, and improved air and water quality. Soil scientists recognize five primary natural soil-forming factors: parent material, climate, organisms, topography and relief, and time. Human activities, especially those related to agriculture and livestock grazing, also affect soil development and are discussed later in the chapter. Soils are assessed and classified using soil cross sections, usually extending from the ground surface to the bedrock or sediments beneath.

Natural Factors in Soil Development

As discussed in Chapter 13, physical and chemical weathering of rocks in the upper lithosphere provides the raw mineral ingredients for soil formation. Bedrock, rock fragments, and sediments are the parent material, and their composition, texture, and chemical nature help determine the type of soil that forms. Clay minerals are the principal weathered by-products in soil.

Climate also influences soil development; in fact, soil types correlate closely with climate types worldwide. The temperature and moisture regimes of climates determine the chemical reactions, organic activity, and movement of water within soils. The present-day climate is important, but many soils also exhibit the imprint of past climates, sometimes over thousands of years. Most notable is the effect of glaciations. Among other contributions, glaciation produced the loess soil materials that were windblown thousands of kilometers to their present locations (see discussion in Chapter 16).

Biological activity is an essential factor in soil development (see the *Geosystems In Action*, Figure GIA 18, on page 561). Vegetation and the activities of animals and bacteria—all the organisms living in, on, and over the soil, such as algae, fungi, worms, insects—determine the organic content of soil. The chemical characteristics of the vegetation and many other life forms contribute to the acidity or alkalinity of the soil solution (soil pH is discussed in the next section). For example, broadleaf trees tend to increase alkalinity, whereas needleleaf trees tend to produce higher acidity. When humans move into new areas and alter natural vegetation by logging or plowing, the affected soils are likewise altered, often permanently.

Relief and topography also affect soil formation (Figure 18.1). Slopes that are too steep cannot have full soil development, because gravity and erosional processes remove materials. Lands that are level or nearly level tend to develop thicker soils, but may be subject to soil drainage issues such as waterlogging. The orientation of slopes relative to the Sun is also important, because it controls exposure to sunlight. In the Northern

▲**Figure 18.1 Topography and soil development, Blue Mountains, Oregon.** [Kevin Ebi/Alamy.]

Hemisphere, a south-facing slope is warmer overall through the year because it receives direct sunlight. North-facing slopes are colder, causing slower snowmelt and a lower evaporation rate, thus providing more moisture for plants than is available on south-facing slopes, which tend to dry out faster.

All of the identified natural factors in soil development require *time* to operate. Over geologic time, plate tectonics has redistributed landscapes and thus subjected soil-forming processes to diverse conditions. The rate of soil development is closely tied to the nature of the parent material (soils develop more quickly from sediments than from bedrock) and to climate (soils develop at a faster rate in warm, humid climates).

Soil Horizons

As a book cannot be judged by its cover, so soils cannot be evaluated at the ground surface only. Instead, scientists evaluate soils using a **soil profile**, a vertical section of soil that extends from the surface to the deepest extent of plant roots or to the point where regolith or bedrock is encountered. Soil profiles may be exposed by human activities, such as at a construction site or excavation, or along a highway road cut. When soil is not exposed by natural processes or human activity, scientists dig soil pits to expose a soil profile for analysis.

For soil classification, pedologists use a three-dimensional representation of the soil profile, known as a *pedon*. A soil pedon is the smallest unit of soil that represents all the characteristics and variability used for classification (discussed later in the chapter). A soil profile represents one side of a pedon, as shown in Figure 18.2a.

Within a soil profile, soils are generally organized into distinct horizontal layers known as **soil horizons**. These horizons are roughly parallel to the land surface and have characteristics recognizably different from horizons directly above or below. The four "master" horizons in most agricultural soils are known as the O, A, B, and C horizons (Figure 18.2).

The boundary between horizons usually is distinguishable when viewed in profile, owing to differences in one or more physical soil characteristics, such as color, texture, structure, consistence (meaning soil consistency or cohesiveness), porosity, or moisture. These and other soil properties, which all affect soil function, are discussed in the next section.

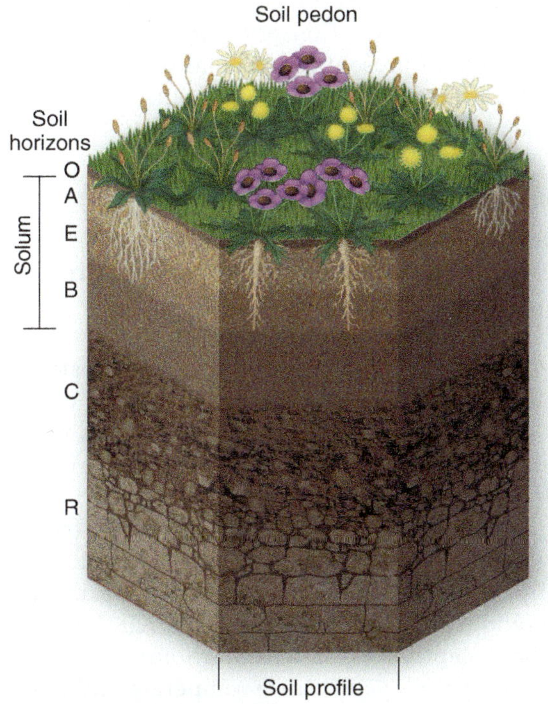

(a) An idealized soil profile within a pedon.

(b) Profile of a well-drained soil with till as parental material (a Mollisol) in southeastern South Dakota. Carbonate nodules are visible in the lower B and upper C horizons.

◄**Figure 18.2 A typical soil profile within a pedon, and example.** [Marbut Collection, Soil Science Society of America, Inc.]

O Horizon At the top of the soil profile is the *O horizon*, named for its organic composition, derived from plant and animal litter that was deposited on the surface and transformed into **humus**, a mixture of decomposed and synthesized organic materials that is usually dark in color. Microorganisms work busily on this organic debris, performing a portion of the *humification* (humus-making) process. The O horizon is 20%–30% or more organic matter, which is important because of its ability to retain water and nutrients and because of the way its behavior complements that of clay minerals.

The A, E, B, and C horizons extend below the O horizon to the R horizon, which is composed of sediment or bedrock. These middle layers are composed of sand, silt, clay, and other weathered by-products.

A Horizon In the *A horizon*, humus and clay particles are particularly important as they provide essential chemical links between soil nutrients and plants. This horizon usually is richer in organic content, and hence darker, than lower horizons. Human disruption through plowing, pasturing, and other uses takes place in the A horizon. This horizon is commonly called *topsoil*.

E Horizon The A horizon grades downward into the *E horizon*, which is made up mainly of coarse sand, silt, and leaching-resistant minerals. From the lighter-colored E horizon, silicate clays and oxides of aluminum and iron are leached (removed by water) and carried to lower horizons with the water as it percolates through the soil. This process of removing fine particles and minerals by water, leaving behind sand and silt, is **eluviation**—thus, the E designation for this horizon. As precipitation increases, so does the rate of eluviation.

B Horizon In contrast to the A and E horizons, *B horizons* accumulate clays, aluminum, and iron. B horizons are dominated by **illuviation**, in which materials leached by water from one layer enter and accumulate in another. Both eluviation and illuviation are types of *translocation*, in which material (such as nutrients, salts, clays) is moved downward in the soil. In contrast to eluviation, which is erosional, illuviation is a depositional process. B horizons may exhibit reddish or yellowish hues because of the presence of illuviated minerals (silicate clays, iron and aluminum, carbonates, gypsum) and organic oxides. Some materials occurring in the B horizon may have formed in place from weathering processes rather than arriving there by translocation, especially in the humid tropics.

Together, the A, E, and B horizons are designated the **solum**, considered the true definable soil of the profile (and labeled in Figure 18.2). The horizons of the solum experience active soil processes.

C Horizon Below the solum is the *C horizon*, made up of weathered bedrock or weathered parent material. This zone is identified as *regolith* (although the term sometimes is used to include the solum as well). The C horizon is not much affected by soil operations in the solum and lies outside the biological influences experienced in the shallower horizons. Plant roots and soil microorganisms are rare in the C horizon. It lacks clay concentrations and generally is made up of carbonates, gypsum, or soluble salts of iron and silica, which can form cementing agents. In dry climates, calcium carbonate commonly forms the cementing material that causes hardening of this layer.

R Horizon At the bottom of the soil profile is the *R* (rock) *horizon*, consisting of either unconsolidated (loose) material or consolidated bedrock. When bedrock physically and chemically weathers into regolith, it may or may not contribute to overlying soil horizons.

Soil scientists using the U.S. classification system also employ lowercase letters to designate subhorizons within each master horizon, indicating particular conditions. For example, the Ap horizon refers to an A horizon that has undergone plowing; the Bh horizon refers to the presence of organics (humic material).

Soil Characteristics

A number of physical and chemical characteristics differentiate soils and affect soil fertility and the resistance of soils to erosion. **Soil fertility** is the ability of soil to sustain plants. Billions of dollars are expended to create fertile soil conditions, yet the future of Earth's most fertile soils is threatened because soil erosion is on the increase worldwide.

Here we discuss the most widely applicable properties for describing and classifying soils; however, other properties exist and may be of value depending on the particular site. Chapter 3 of the U.S. Department of Agriculture, Natural Resources Conservation Service (NRCS) *Soil Survey Manual* presents additional information on soil properties (http://soils.usda.gov/technical/manual/contents/chapter3.html).

Physical Properties

The physical properties that distinguish soils and can be observed in soil profiles are color, texture, structure, consistence, porosity, and moisture.

Soil Color Color is important because it suggests the composition and chemical makeup of a soil. Among the many possible hues are the reds and yellows found in soils of the southeastern United States (high in iron oxides), the blacks of prairie soils in portions of the U.S. grain-growing regions and in Ukraine (richly organic), and the white to pale hues found in soils containing silicates and aluminum oxides. Color may be the most obvious trait in an exposed soil. All the same, color can be deceptive: Soils with high humus content are often dark, yet clays of warm temperate and tropical regions with less than 3% organic content are some of the world's blackest soils.

▲Figure 18.3 A Munsell Soil Color Chart page. A soil sample is viewed through holes in the page to match it with a color on the chart. Hue, value, and chroma are the characteristics of color assessed by this system. [Courtesy of Gretag Macbeth, Munsell Color.]

To standardize their descriptions, soil scientists describe a soil's color at various depths within the soil profile by comparing it with a *Munsell Color Chart*, developed by artist and teacher Albert Munsell in 1913 (Figure 18.3). These charts display 175 colors arranged by *hue* (the dominant spectral color, such as red), *value* (the degree of darkness or lightness), and *chroma* (the purity and saturation of the color, which increase with decreasing grayness). A Munsell notation identifies each color by a name, so soil scientists can make worldwide comparisons of soil color.

Soil Texture Texture refers to the mixture and proportions of different particle sizes and is perhaps a soil's most permanent attribute. Individual mineral particles are *soil separates*. All particles smaller in diameter than 2 mm (0.08 in.), the size of very coarse sand, are considered part of the soil. Larger particles, such as pebbles, gravel, or cobbles, are not part of the soil. (Sands are graded from coarse to medium to fine, down to 0.05 mm; silt is finer, to 0.002 mm; and clay is finer still, at less than 0.002 mm.)

Figure 18.4 is a *soil texture triangle* showing the relation of sand, silt, and clay concentrations in soils. Each corner of the triangle represents a soil consisting solely of the particle size noted (although rarely are true soils composed of a single separate). Every soil on Earth is defined somewhere in this triangle.

Loam is the common designation for the balanced mixture of sand, silt, and clay that is beneficial to plant growth (Figure 18.4). Farmers consider a sandy loam with clay content below 30% (lower left) as the ideal soil because of its water-holding characteristics and ease of cultivation. Soil texture is important in determining water-retention and water-transmission traits.

To see how the soil texture triangle works, consider *Miami silt loam*, a soil type that is common in Indiana. Samples from this soil type are plotted on the soil texture triangle in Figure 18.4 as points 1, 2, and 3. Point 1

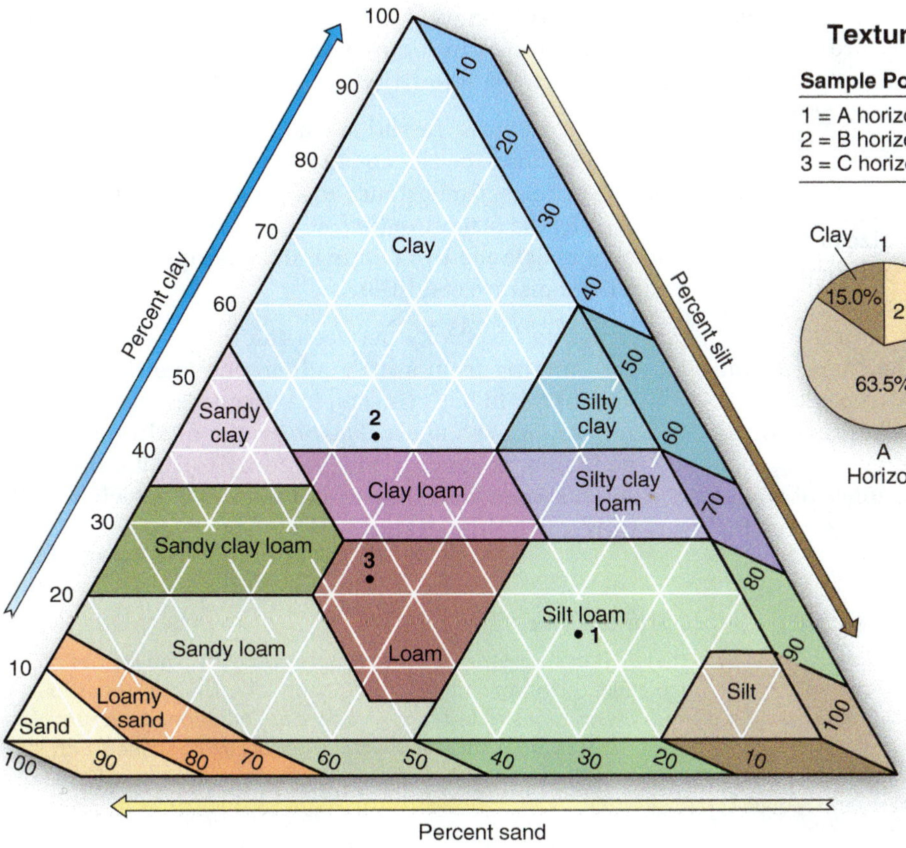

Textual Analysis of Miami Silt Loam

Sample Points	% Sand	% Silt	% Clay
1 = A horizon	21.5	63.5	15.0
2 = B horizon	31.5	25.1	43.4
3 = C horizon	42.4	34.1	23.5

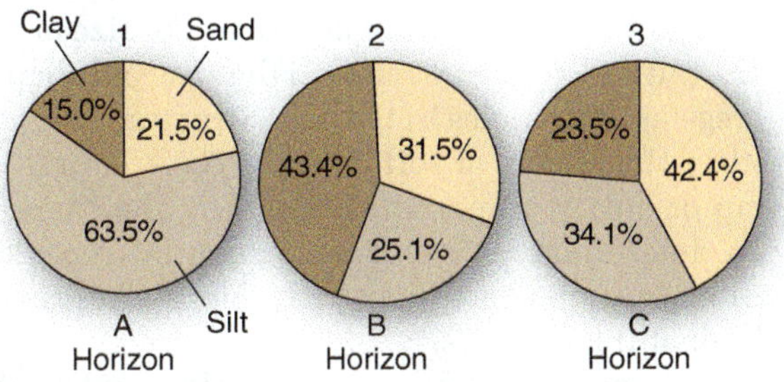

▲Figure 18.4 Soil texture triangle. The ratio of clay, silt, and sand determines soil texture. As an example, points 1 (horizon A), 2 (horizon B), and 3 (horizon C) designate samples taken from three different horizons in the Miami silt loam in Indiana. Note the ratios of sand to silt to clay shown in the three pie diagrams and table. [After USDA–NRCS, *Soil Survey Manual*, Agricultural Handbook No. 18, p. 138 (1993).]

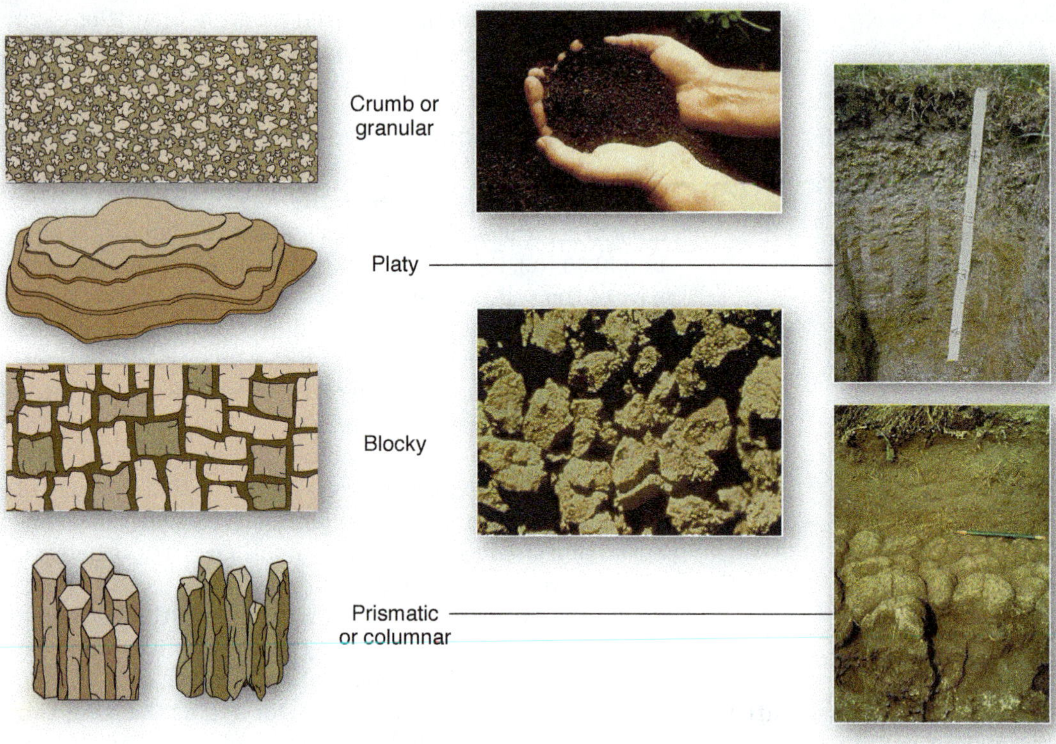

Figure 18.5 Types of soil structure. Structure is important because it controls drainage, rooting of plants, and how well the soil delivers nutrients to plants. The shape of individual peds, shown here, controls a soil's structure. [USDA–NRCS, National Soil Survey Center.]

describes a sample taken near the surface in the A horizon; point 2 describes a sample taken from the B horizon; and point 3 describes a sample from the C horizon. Textural analyses of these samples are summarized in the table and in the three pie diagrams to the right of the triangle. Note that silt dominates the surface, clay the B horizon, and sand the C horizon. The *Soil Survey Manual* presents guidelines for estimating soil texture by feel, a relatively accurate method when used by an experienced person. However, laboratory methods using graduated sieves and separation by mechanical analysis in water allow more precise measurements.

Soil Structure Soil texture describes the size of soil particles, but soil *structure* refers to the size and shape of the aggregates of particles in the soil. Structure can partially modify the effects of soil texture. The smallest natural lump or cluster of particles is a *ped*. The shape of soil peds determines which of the structural types the soil exhibits: crumb or granular, platy, blocky, or prismatic or columnar (Figure 18.5).

Peds separate from each other along zones of weakness, creating voids, or pores, that are important for moisture storage and drainage. Rounded peds have more pore space between them and greater permeability than do other shapes. They are therefore better for plant growth than are blocky, prismatic, or platy peds, despite comparable fertility. Terms used to describe soil structure include *fine*, *medium*, and *coarse*. Adhesion among peds ranges from weak to strong. The work of soil organisms, illustrated in Figure GIA 18, affects soil structure and increases soil fertility.

Soil Consistence In soil science, the term *consistence* is used to describe the consistency of a soil or cohesion of its particles. Consistence is a product of texture (particle size) and structure (ped shape). Consistence reflects a soil's resistance to breaking and manipulation under varying moisture conditions:

- A *wet soil* is sticky between the thumb and forefinger, ranging from a little adherence to either finger, to sticking to both fingers, to stretching when the fingers are moved apart. *Plasticity*, the quality of being moldable, is roughly measured by rolling a piece of soil between your fingers and thumb to see whether it rolls into a thin strand.

(text continued on page 562)

GEOreport 18.1 Soil compaction—causes and effects

Soil compaction is the physical consolidation of the soil that destroys soil structure and reduces porosity. The increasing weight of today's heavy agricultural machinery, in addition to earlier planting and the arrangement of row crops, tends to increase soil compaction and can result in a 50% reduction in crop yields owing to restricted root growth, poor aeration of the root zone, and poor drainage. Scientists now suggest that no-till agricultural practices (in which plowing does not occur) combined with maintaining a continuous cover of actively growing plants is the best way to reduce soil compaction, since roots increase porosity and water availability, preserve organic matter content, and reduce surface erosion.

A diverse collection of organisms inhabit soil environments (GIA 18.1). Organisms play a vital role in soil-forming processes, helping to weather rock both mechanically and chemically, breaking up and mixing soil particles, and enriching soil with organic matter from their remains and wastes.

Moles can cause extensive soil disturbances.
[D. Kucharski K. Kucharska /Shutterstock]
[Background photo: DJTaylor/Shutterstock.]

18.1 SOIL ORGANISMS

Soil organisms range in size from land mammals that burrow into the ground, such as badgers, prairie dogs, and voles; to earthworms that ingest and secrete soil; to microscopic organisms that break down organic matter. The actions of these living organisms help maintain soil fertility.

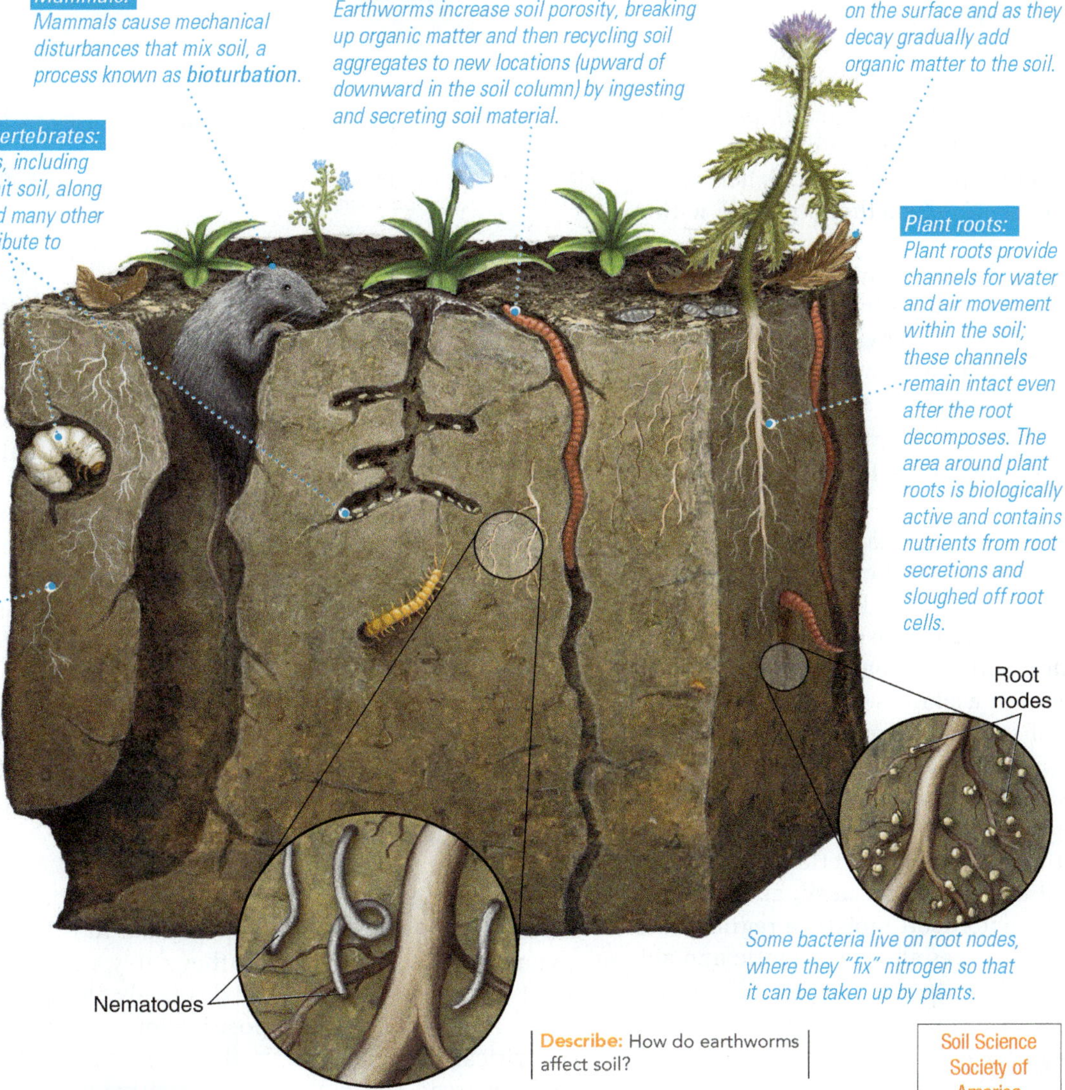

Video
The Soil Moisture Active Passive (SMAP) mission

Describe: How do earthworms affect soil?

Soil Science Society of America

MasteringGeography™

Visit the Study Area in MasteringGeography™ to explore biological activity in soil.
Visualize: Study a video of Soil Moisture Active Passive mission.
Assess: Demonstrate understanding of biological activity in soil. (if assigned by instructor).

GEOquiz

1. **Compare:** How are the effects of mammals and plant roots on soil similar? How are they different?
2. **Explain:** Explain three ways in which organisms improve soil fertility.

- A *moist soil* is filled to about half of field capacity (the usable water capacity of soil), and its consistence grades from loose (noncoherent) to *friable* (easily pulverized) to firm (not crushable between the thumb and forefinger).
- A *dry soil* is typically brittle and rigid, with consistence ranging from loose to soft to hard to extremely hard.

Soil Porosity Porosity refers to the available air spaces within a material; **soil porosity** denotes the part of a volume of soil that is filled with air, gases, or water (as opposed to soil particles or organic matter). We discussed soil porosity, permeability, and moisture storage in Chapter 9.

Pores in the soil horizon control the movement of water—its intake, flow, and drainage—and air ventilation. Important porosity factors are pore *size*, pore *continuity* (whether pores are interconnected), pore *shape* (whether pores are spherical, irregular, or tubular), pore *orientation* (whether pore spaces are vertical, horizontal, or random), and pore *location* (whether pores are within or between soil peds).

Porosity is improved by the presence of plant roots; by animal activity such as the tunneling actions of gophers or worms (see Figure GIA 18); and by human actions such as supplementing the soil with humus or sand, or planting soil-building crops. Much of the soil-preparation work done by farmers before they plant—and by home gardeners as well—is done to improve soil porosity.

Soil Moisture As discussed in Chapter 9 and shown in Figure 9.9, plants operate most efficiently when the soil is at *field capacity*, which is the maximum water availability for plant use after large pore spaces have drained of gravitational water. Soil type determines field capacity. The depth to which a plant sends its roots determines the amount of soil moisture to which the plant has access. If soil moisture is below field capacity, plants must exert increased energy to obtain available water. This moisture-removal inefficiency worsens until the plant reaches its wilting point. Beyond this point, plants are unable to extract the water they need, and they die. More than any other factor, soil moisture regimes and their associated climate types shape the biotic and abiotic properties of the soil.

Chemical Properties

Recall that soil pores may be filled with air, water, or a mixture of the two. Consequently, soil chemistry involves both air and water. The atmosphere within soil pores is mostly nitrogen, oxygen, and carbon dioxide. Nitrogen concentrations are about the same as in the atmosphere, but oxygen is less and carbon dioxide is greater because of ongoing respiration processes in the ground.

Water present in soil pores is called the *soil solution* and is the medium for chemical reactions in soil.

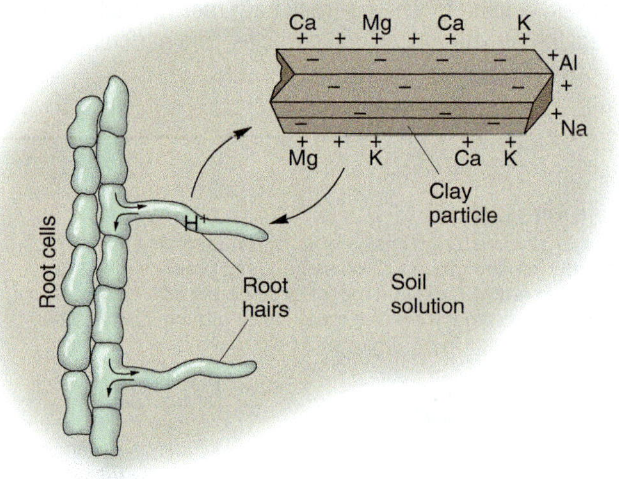

▲Figure 18.6 **Soil colloids and cation-exchange capacity (CEC).** This typical soil colloid retains mineral ions by adsorption to its surface (opposite charges attract). This process holds the ions until they are absorbed by root hairs.

This solution is a critical source of nutrients for plants, providing the foundation of soil fertility. Carbon dioxide combines with the water to produce carbonic acid, and various organic materials combine with the water to produce organic acids. These acids are then active participants in soil processes, as are dissolved alkalis and salts.

A brief review of chemistry basics helps us understand how the soil solution behaves. An *ion* is an atom or group of atoms that carries an electrical charge (examples: Na^+, Cl^-, HCO_3^-). An ion has either a positive charge or a negative charge. For example, when NaCl (sodium chloride) dissolves in solution, it separates into two ions: Na^+, which is a *cation* (positively charged ion), and Cl^-, which is an *anion* (negatively charged ion). Some ions in soil carry single charges, whereas others carry double or even triple charges (e.g., sulfate, SO_4^{2-}; and aluminum, Al^{3+}).

Soil Colloids and Mineral Ions The tiny particles of clay or organic material (humus) suspended in the soil solution are **soil colloids.** Because they carry a negative electrical charge, they attract any positively charged ions in the soil (Figure 18.6). The positive ions, many metallic, are critical to plant growth. If it were not for the negatively charged soil colloids, the positive ions would be leached away in the soil solution and thus would be unavailable to plant roots.

Individual clay colloids are thin and platelike, with parallel surfaces that are negatively charged. They are more chemically active than silt and sand particles, but less active than organic colloids. Metallic cations attach to the surfaces of the colloids by *adsorption* (not *ab*sorption, which means "to enter"). Colloid surfaces can

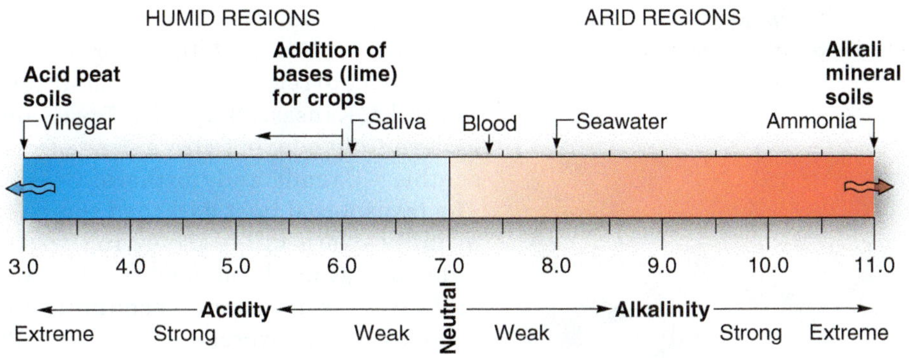

▲**Figure 18.7** **pH scale.** The pH scale measures acidity (lower pH) and alkalinity (higher pH). (The complete pH scale ranges between 0 and 14.)

exchange cations with the soil solution, an ability called **cation-exchange capacity (CEC)**, which is the measure of soil fertility. A high CEC means that the soil colloids can store or exchange a relatively large amount of cations from the soil solution, an indication of good soil fertility (unless a complicating factor exists, such as a soil that is too acid). Soil is fertile when it contains organic substances and clay minerals that *absorb* water and *adsorb* certain elements needed by plants.

Soil Acidity and Alkalinity A soil solution may contain a significant amount of hydrogen ions (H^+), the cations that stimulate acid formation. The result is a soil rich in hydrogen ions, or an *acid soil*. A soil high in base cations (calcium, magnesium, potassium, sodium) is a *basic* or *alkaline soil*. Such acidity or alkalinity is expressed on the pH scale (Figure 18.7).

Pure water has close to a neutral pH of 7.0. Readings below 7.0 represent increasing acidity. Readings above 7.0 indicate increasing alkalinity. Acidity usually is regarded as strong at 5.0 or lower on the pH scale, whereas 10.0 or above is considered strongly alkaline.

The major contributor to soil acidity in this modern era is acid precipitation (rain, snow, fog, or dry deposition), as discussed in Chapter 3. Scientists have measured acid precipitation values below pH 2.0 (the acidity of lemon juice)—incredibly low for natural precipitation. Increased acidity in the soil solution accelerates the chemical weathering of mineral nutrients and increases their depletion rates. Because most crops are sensitive to specific pH levels, acid soils below pH 6.0 require treatment to raise the pH. This soil treatment is accomplished by the addition of bases in the form of minerals that are rich in base cations, usually lime (calcium carbonate, $CaCO_3$).

Human Impacts on Soils

Unlike living species, soils do not reproduce; nor can they be re-created. A few centimeters' thickness of prime farmland soil may require *500 years* to mature. Yet this same thickness is being lost annually through soil erosion that occurs when humans remove vegetation and plow the land, regardless of topography (Figure 18.8). Additional losses occur when flood control structures block sediments and nutrients from replenishing floodplain soils. As a result of human intervention and unsustainable agricultural practices, some 35% of farmlands are losing soil faster than it can form—a loss exceeding 23 billion metric tons (25 billion tons) per year. Soil depletion (such as the loss of fertility that occurs when soils are leached of cations) and soil loss are at record levels from Iowa to China, Peru to Ethiopia, and the Middle East to the Americas. The impact on society is potentially disastrous as population and food demands increase.

Soil Erosion

The NRCS describes *soil erosion* as "the breakdown, detachment, transport, and redistribution of soil particles by forces of wind, water, or gravity." Overcultivation and excessive tilling, overgrazing, and the clearing of forested slopes are some of the main human activities that make soils more prone to erosion. Soil erosion removes topsoil, the layer that is richest in organic matter and nutrients. Millennia ago, farmers in most cultures learned to plant slopes "on the contour"—to sow seeds in rows or mounds that run around a slope at the same elevation, rather than vertically up and down the slope. Planting on the contour prevents water from flowing straight down the slope and thus reduces soil erosion. Contour farming, on land with gradual slopes, and terracing (cutting level platforms into steep

▲**Figure 18.8** **Soil degradation.** An example of soil loss through sheet and gully erosion on a northwest Iowa farm. One millimeter of soil lost from an acre weighs about 5 tons. [USDA–NRCS, National Soil Survey Center.]

▼**Figure 18.9 Roots of wheat plants in tilled and untilled soils.** [NRCS.]

(a) No-till farming leads to greater moisture retention and looser soil, allowing wheat to develop longer root systems that access moisture and nutrients farther below the surface.

(b) Conventional farming leads to soil compaction and shorter plant root systems.

terrain), in mountainous regions, are used today to combat erosion (see the chapter-opening photo).

An expanding practice for slowing soil erosion is *no-till agriculture* (also called no-till farming). In this approach, farmers no longer till, or plow, the soil after a harvest. Instead, they leave crop residue on the field between plantings, thus preventing soil erosion by wind and water. Seeds are then inserted into the ground without disturbing the soil. Planting the new crop on top of the old crop also preserves moisture. In Texas, where severe drought conditions have persisted in recent years, some farmers are successfully using no-till practices. The photos in Figure 18.9 compare roots of wheat planted using no-till practices with those of wheat planted using conventional agriculture. Because the soil is less compacted, roots in the untilled fields can grow longer, reaching moisture and nutrients farther below the surface.

The Dust Bowl Large-scale removal of native vegetation associated with the expansion of farming into the American Great Plains in the late 1800s and early 1900s led to a catastrophic soil erosion event known as the Dust Bowl. Intensive agriculture and overgrazing combined with reduced precipitation and above-normal temperatures to trigger a multiyear period of severe wind erosion and loss of farmlands.

The deflation of soil—in some places as much as 10 cm (4 in.) over several years—occurred mainly in southern Nebraska, Kansas, Oklahoma, Texas, and eastern Colorado, but also extended into southern Canada and northern Mexico. The transported dust darkened the skies of midwestern cities (requiring streetlights to stay on throughout the day) and drifted over farmland, accumulating in depths that covered failing crops. By 1940, 2 million people had moved out of the plains states—the largest migration in American history. The recent prolonged drought through 2012 in the American Southwest and into west and central Texas and Oklahoma is similar climatically to the conditions that led up to the Dust Bowl (see http://www.pbs.org/wgbh/americanexperience/films/dustbowl/ for more information).

Erosion Rates and Costs In 2007, the NRCS reported a 43% decrease in soil erosion on U.S. cropland between 1982 and 2007, including declines in both water and wind erosion. Most of this decline occurred between 1982 and 1992, with erosion rates showing less variation since 1992. The study also revealed that erosion by water in rills and gullies in the United States varies geographically, with the majority (54%) concentrated in the Corn Belt, stretching from Ohio in the east to Iowa in the west, and in the Northern Plains, stretching from Nebraska in the south into North Dakota. Ninety-seven percent of wind erosion occurred in the Great Plains, Rocky Mountain Region, and Lake Region of Minnesota and Wisconsin.

In 2010, the Environmental Working Group reported that according to scientists at Iowa State University, soil erosion rates are far higher than the NRCS estimates. According to these new studies, the effects of individual storm events are more significant for erosion than the long-term, statewide averages calculated by the NRCS. As the intensity of storm events increases with climate change, erosion may accelerate.

Soil erosion can be compensated for in the short run by using more fertilizer, increasing irrigation, and planting higher-yielding crops. However, agricultural fertilizers pollute runoff, with devastating effects on streams, rivers, and river deltas. The "dead zone" created by nitrogen and other agricultural fertilizers in the coastal waters of

GEOreport 18.2 Slipping through our fingers

The U.S. Department of Agriculture estimated that more than 5 million acres of prime farmland are lost each year in the United States through mismanagement or conversion to nonagricultural uses. About half of all cropland in the United States and Canada (two of the few countries that monitor loss of topsoil) is experiencing excessive rates of soil erosion. Worldwide, about one-third of potentially farmable land has been lost to erosion, much of that in the past 40 years. The causes for degraded soils, in order of severity, include overgrazing, vegetation removal, agricultural practices, conversion to nonagricultural activities, overexploitation, and industrial and bioindustrial use.

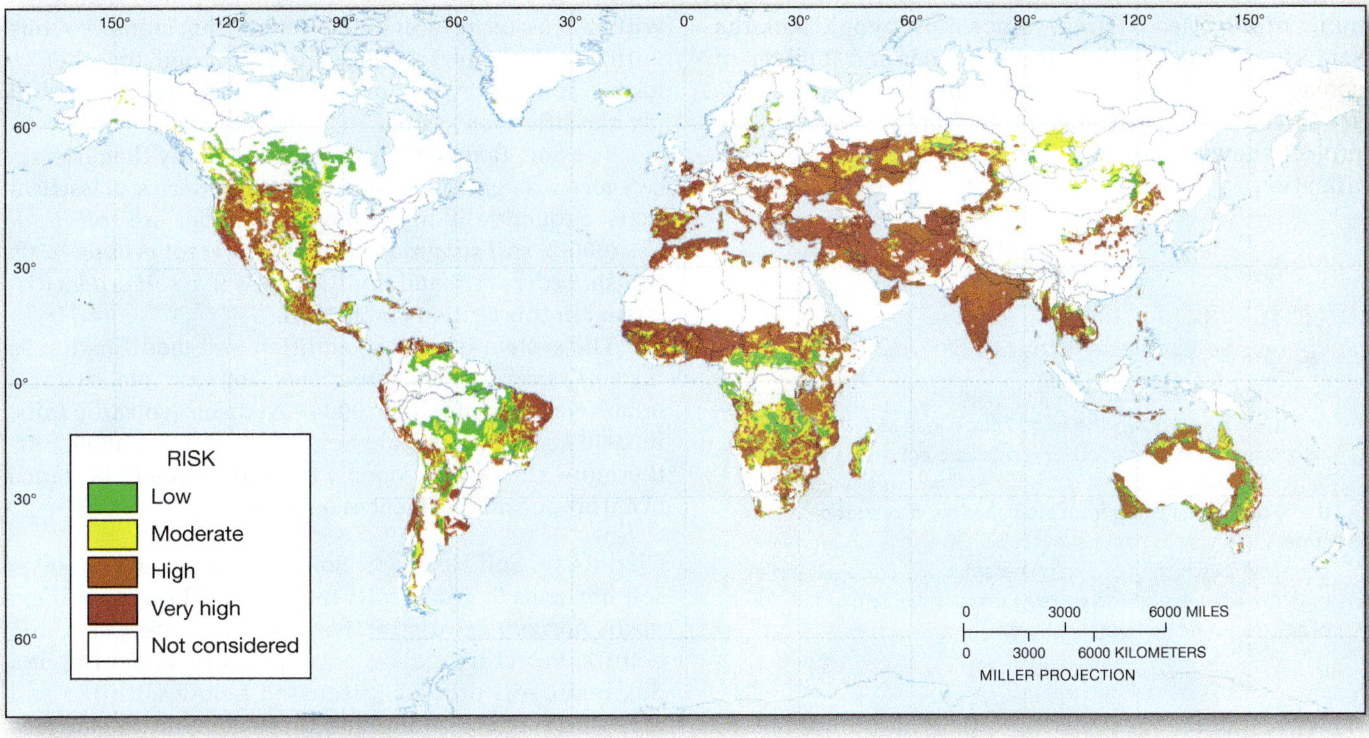

▲**Figure 18.10 Areas at risk of desertification.** [Map prepared by USDA–NRCS, Soil Survey Division; includes consideration of population densities in affected regions provided by the National Center for Geographic Information Analysis at University of California, Santa Barbara.]

the Mississippi River Delta is reaching its largest size ever, comparable to the area of New Jersey (see discussion and illustrations in Chapter 19).

One recent study tabulated the monetary impact of lost soil nutrients and of other variables affected by erosion, showing that the sum of direct damage (to agricultural land) and indirect damage (to streams, society's infrastructure, and human health) was estimated at more than $25 billion a year in the United States and hundreds of billions of dollars worldwide. Of course, this is a controversial assessment in the agricultural industry. The cost to bring erosion under control in the United States is estimated at approximately $8.5 billion, or about 30 cents on every dollar of damage and loss.

Desertification

Land degradation that occurs in dry regions is known as **desertification**, the expansion of deserts. This worldwide phenomenon along the margins of semiarid and arid lands is caused in part by human activities that degrade soils, leading to losses of topsoil and declines in food production.

Desertification results from a combination of factors: poor agricultural practices, such as overgrazing and activities that abuse soil structure and fertility; improper soil-moisture management; salinization (the accumulation of salts on the soil surface, discussed later in the chapter) and nutrient depletion; and deforestation. A worsening causative force is global climate change, which is shifting temperature and precipitation patterns and causing poleward movement of Earth's subtropical high-pressure systems, discussed in previous chapters. Desertification is now affecting over a billion people worldwide. (Please revisit this chapter's *Geosystems Now* feature for more on this topic.)

Figure 18.10 portrays regions of desertification risk, as defined by the loss of agricultural activity. The severity of this problem is magnified by poverty in

GEOreport 18.3 Overgrazing effects on Argentina's grasslands

Overgrazing occurs when the number of animals is greater than the productive capacity of the land. Specifically, overgrazing results in reduced plant leaf area, which weakens plants and leads to deterioration of vegetation and soils. In southern Argentina, sheep ranching is the primary economic activity in the expansive grasslands, once home to only one native grazing mammal: the guanaco, a relative of the llama. Although sheep numbers have declined since the 1950s, overgrazing over the past century has led to widespread desertification. Sustainable grazing practices are becoming a priority as this region seeks to preserve its economic base of wool and meat production, and reverse the desertification trend.

many of the affected areas, since most people lack the capital to change agricultural practices and implement conservation strategies. Many of the highest-risk lands are in India and central Asia. In fact, a 2009 mapping project showed that 25% of India is undergoing desertification.

CRITICALthinking 18.1
Soil Losses—What to Do?

Using the information you just read in "Human Impacts on Soils," break down the issue of soil loss into the forcing causes, impacts produced, and possible actions to slow soil degradation and loss. Next, consider the possible actions in terms of scale, moving from individual to local to regional to state and finally to national in scope. Given the problem and the scale at which you determine solutions are best implemented, what actions do you suggest to reverse the situation of soil degradation here and abroad? •

Soil Classification

Soil classification is complicated by the variety of interactions that create thousands of distinct soils—well over 15,000 soil types in the United States and Canada alone. Not surprisingly, a number of different classification systems are in use worldwide, including those from the United States, Canada, the United Kingdom, Germany, Australia, Russia, and the United Nations Food and Agricultural Organization (FAO). Each system reflects the environment of the country or countries in which it originated. Because of the involvement of interacting variables, classifying soils is similar to classifying climates.

For many locales in the United States, an *agricultural extension service* can provide specific information about and perform a detailed analysis of local soils. Soil surveys and local soil maps are available through the NRCS for most counties in the United States; see http://www.nrcs.usda.gov/wps/portal/nrcs/main/national/soils/).

Soil Taxonomy

The U.S. soil classification system, first developed by the NRCS in 1975, is the **Soil Taxonomy**. Information in this chapter is based on the publication *Keys to Soil Taxonomy*, now in its 11th edition (2010), available for free download from the NRCS at http://soils.usda.gov/technical/classification/tax_keys/.

The basis for the Soil Taxonomy system is field observation of soil properties and morphology (appearance, form, and structure). The smallest unit of soil used in soil surveys is a **pedon**, a hexagonal column measuring 1 to 10 m^2 in top surface area (see Figure 18.2a). A pedon is considered a soil individual, and a soil profile within it is used to evaluate its soil horizons. Pedons with similar characteristics are grouped together to form a *soil series*, the lowest and most precise level of the classification system.

The Soil Taxonomy is a sorted hierarchy that uses six categories, beginning with 15,000 soil series classifications. Sequentially higher-level categories are *soil families* (6000), *soil subgroups* (1200), *soil great groups* (230), *soil suborders* (47), and finally, the twelve *soil orders* discussed in this section (Table 18.1).

The system is open to addition and modification as the soil sampling database grows. For example, two soil orders were added in the 1990s—Andisols (volcanic soils) in 1990 and Gelisols (cold and frozen soils) in 1998. Furthermore, the Soil Taxonomy recognizes the importance of anthropogenic influences on soils.

Diagnostic Soil Horizons Soil scientists use diagnostic soil horizons to group soils into each soil series. A *diagnostic horizon* has distinctive physical properties (color, texture, structure, consistence, porosity, moisture) or a dominant soil process (discussed below with the soil types). A diagnostic horizon that occurs at the surface or just below it is an **epipedon**. It may extend downward through the A horizon and may even include all or part of an illuviated B horizon. It is visibly darkened by organic matter and sometimes is leached of minerals. Excluded from consideration as part of the epipedon are alluvial deposits, eolian deposits, and cultivated areas, because these are relatively short-lived surfaces that soil-forming processes would eventually erase. A diagnostic horizon that forms below the soil surface at varying depths is called a **diagnostic subsurface horizon**. It may include part of the A or B horizon or both.

Pedogenic Regimes Prior to the U.S. Soil Taxonomy system, scientists used **pedogenic regimes** to describe soils. These regimes attach specific soil-forming processes to climatic regions. Although such climate-based regimes are convenient for relating climate and soil processes, the use of climatic variables as the sole basis of soil classification leads to uncertainty and inconsistency. In this chapter, we discuss each pedogenic regime with the soil order in which it most commonly occurs, even though each pedogenic process may be active in several soil orders and in different climates.

The five pedogenic regimes are

- **Laterization**—a leaching process in humid and warm climates (discussed with Oxisols and shown in Figure 18.13)
- **Salinization**—a process that concentrates salts in soils in climates with high potential evapotranspiration (POTET) rates (discussed with Aridisols)
- **Calcification**—a process that produces an illuviated accumulation of calcium carbonates in continental climates (discussed with Mollisols and Aridisols and shown in Figure 18.17)

TABLE 18.1 Soil Orders

Order	General Location and Climate	Description
Oxisols	Tropical soils; hot, humid areas	Maximum weathering of Fe and Al and eluviation; continuous plinthite layer.
Aridisols	Desert soils; hot, dry areas	Limited alteration of parent material; low climate activity; light color; low humus content; subsurface illuviation of carbonates.
Mollisols	Grassland soils; subhumid, semiarid lands	Noticeably dark with organic material; humus-rich; high, friable surface with well-structured horizons.
Alfisols	Moderately weathered forest soils; humid temperate forests	B horizon high in clays; illuviated clay accumulation; no pronounced color change with depth.
Ultisols	Highly weathered forest soils; subtropical forests	Similar to Alfisols; B horizon high in clays; strong weathering in subsurface horizons; redder than Alfisols.
Spodosols	Northern conifer forest soils; cool, humid forests	Illuvial B horizon of Fe/Al clays; humus accumulation; without structure; partially cemented; highly leached; strongly acid.
Entisols	Recent soils; profile undeveloped, all climates	Limited development; inherited properties from parent material; pale color; low humus; few specific properties; hard and massive when dry.
Inceptisols	Weakly developed soils; humid regions	Intermediate development; embryonic soils, but few diagnostic features; further weathering possible in altered or changed subsurface horizons.
Gelisols	Permafrost-affected soils; high latitudes in Northern Hemisphere, southern limits near tree line, high elevations	Permafrost within 100 cm of the soil surface; evidence of cryoturbation (frost churning) and/or an active layer; patterned ground.
Andisols	Soils formed from volcanic activity; areas affected by frequent volcanic activity, especially the Pacific Rim	Volcanic parent materials, particularly ash and volcanic glass; weathering and mineral transformation important; high CEC and organic content; generally fertile.
Vertisols	Expandable clay soils; subtropics, tropics; sufficient dry period	Forms large cracks on drying; self-mixing action; contains >30% swelling clays; light color; low humus content.
Histosols	Organic soils; wetlands	Peat or bog; >20% organic matter; much with clay >40 cm thick; surface organic layers; no diagnostic horizons.

- **Podzolization**—a process of soil acidification associated with forest soils in cool climates (discussed with Spodosols and shown in Figure 18.21)
- **Gleization**—a process that results in an accumulation of humus and a thick, water-saturated gray layer of clay beneath, usually in cold, wet climates and poor drainage conditions

The 12 Soil Orders of the Soil Taxonomy

The Soil Taxonomy includes 12 general soil orders, listed in Table 18.1. Their worldwide distribution is shown in Figure 18.11 and in individual maps provided with each description in the following discussion. Please consult the table and maps as you read the descriptions; further information on each soil order is at http://soils.ag.uidaho.edu/soilorders/. Because the Soil Taxonomy evaluates each soil order on the basis of its own characteristics, there is no priority to the classification. However, you will find a progression arranged loosely by latitude; we begin, as in Chapters 10 (on climates) and 20 (on terrestrial biomes), along the equator.

CRITICALthinking 18.2
Soil Observations

Select a small soil sample from your campus or an area near where you live. Based on the sections in this chapter on soil characteristics, properties, and formation, describe the sample as completely as you can. Using the general soil map and any other sources available (e.g., local agriculture extension agent, Internet, related department on campus), are you able to roughly place this sample in one of the soil orders? •

Oxisols The intense moisture, high temperature, and uniform daylength of equatorial latitudes profoundly affect soils. In these generally old landscapes, exposed to tropical conditions for millennia or hundreds of millennia, soils are well developed, and their minerals are altered (except in certain newer volcanic soils in Indonesia—the Andisols). Thus, Oxisols are among the most mature soils on Earth. Distinct horizons usually are

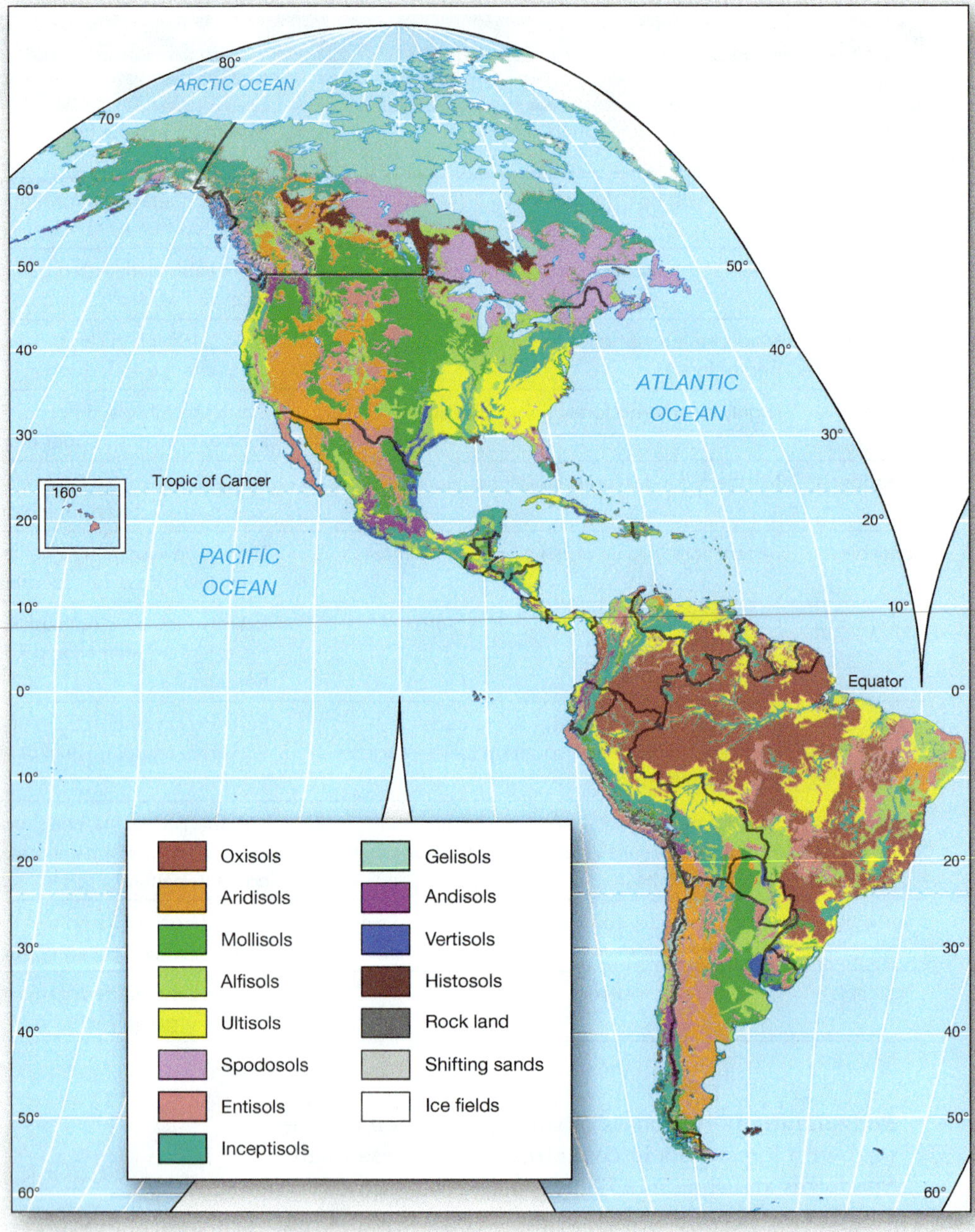

Figure 18.11 Soil Taxonomy. Worldwide distribution of the Soil Taxonomy's 12 soil orders. [Adapted from maps prepared by World Soil Resources Staff, Natural Resources Conservation Service, 1999, 2006.]

lacking where these soils are well drained (Figure 18.12a). Related vegetation is the luxuriant and diverse tropical and equatorial rain forest.

Oxisols (tropical soils) are so named because they have a distinctive horizon of iron and aluminum oxides. The concentration of oxides results from heavy precipitation, which leaches soluble minerals and soil constituents from the A horizon. Typical Oxisols are reddish (from the iron oxide) or yellowish (from the aluminum oxides), with a weathered claylike texture, sometimes in a granular soil structure that is easily broken apart. The high degree of eluviation removes basic cations and colloidal material to lower illuviated horizons. Thus, Oxisols are low in CEC and fertility, except in regions augmented by alluvial or volcanic materials. In short, Oxisols have a diagnostic subsurface horizon that is highly weathered, contains iron and aluminum oxides, is at least 30 cm (12 in.) thick, and lies within 2 m (6.5 ft) of the surface (see Figure 18.12).

The world's lush rain forests are found in regions of Oxisols, even though these soils are poor in inorganic nutrients. This forest system relies on the recycling of nutrients from soil organic matter to sustain fertility; however, this nutrient-recycling ability is quickly lost when the ecosystem is disturbed.

Figure 18.13 illustrates *laterization*, the leaching process that operates in well-drained soils in warm,

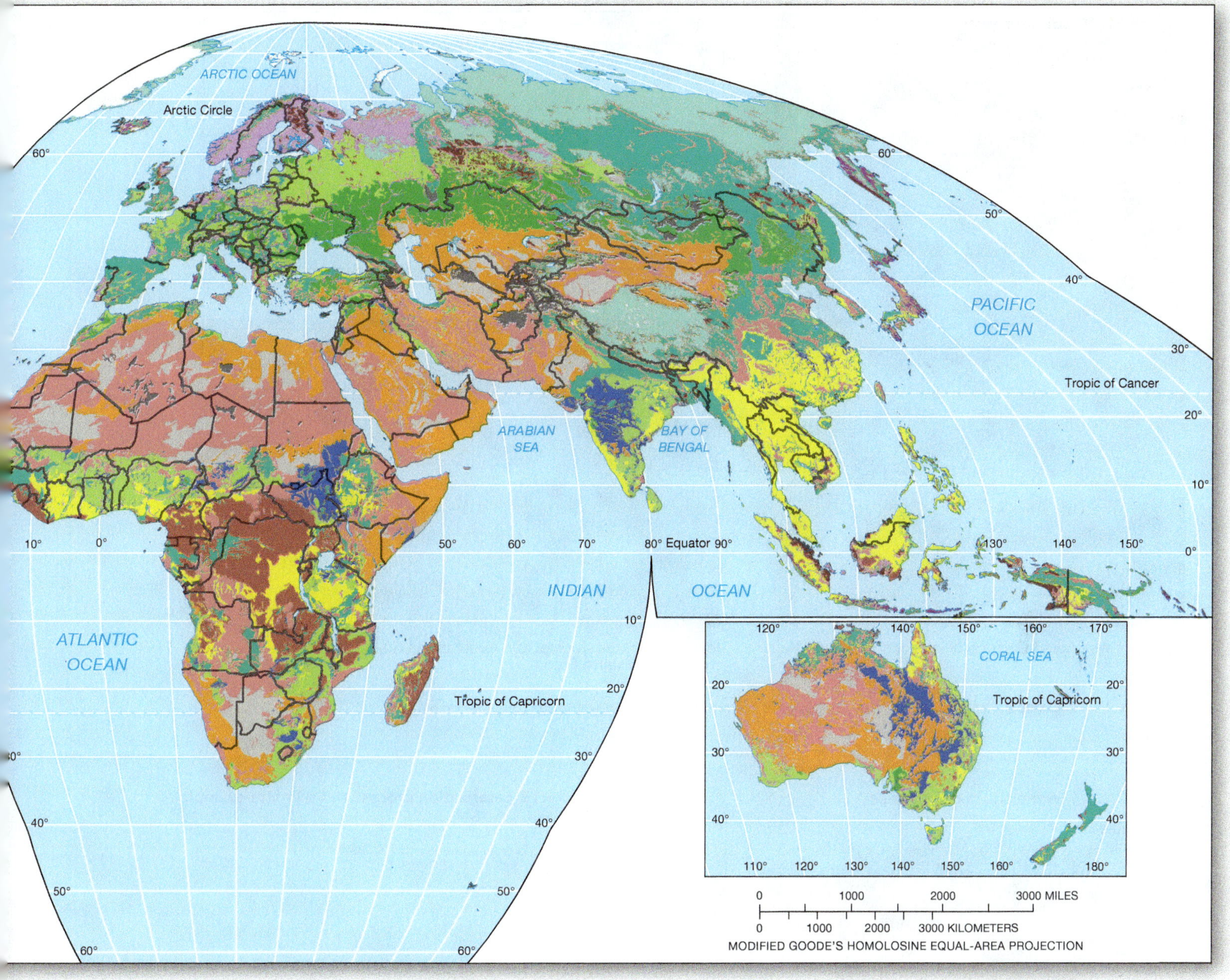

humid tropical and subtropical climates. If the soils are subjected to repeated wetting and drying, a *hardpan* develops (a hardened soil layer)—in this case, an iron-rich and humus-poor clay with quartz and other minerals—in the lower A or in the B horizon. This process occurs in Oxisols and Vertisols (discussed ahead), forming *plinthite* (from the Greek *plinthos*, meaning "brick"), also known as a *laterite*, which can be quarried in blocks and used as a building material (Figure 18.13b).

The traditional agricultural practice of shifting cultivation, also known as *slash-and-burn agriculture* or *swidden agriculture*, is common in parts of Asia, Africa, and South America. This style of crop rotation begins with the cutting of small tracts of tropical forests into *slash* (cut vegetation) that is then dried and burned. The resultant ash provides a soil environment that is rich in nutrients for crops, usually maize, beans, and squash. However, after 3 to 5 years, soil fertility declines through leaching by intense rainfall, causing farmers to shift cultivation to another tract, where the process is repeated. After a period of recovery, a previously used tract can again be cut and burned as the cycle repeats.

Although this practice has occurred for thousands of years, the orderly land rotation inherent to its success is now disrupted by the influx of foreign plantation interests, development by local governments, vastly increased population pressures, and conversion of vast forest tracts

(a) Highly weathered Oxisol profile in central Puerto Rico.

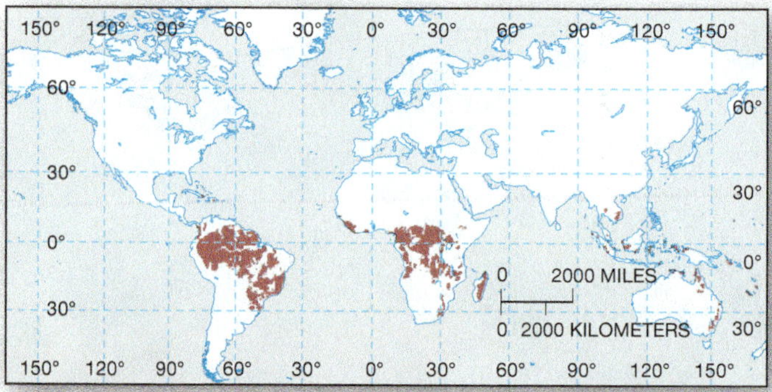

(b) Worldwide Oxisol distribution.

(c) Oxisols in the tropical rain forest of the Sierra de Luquillo mountains, Puerto Rico.

▲**Figure 18.12 Oxisols.** [(a) Marbut Collection, Soil Science Society of America, Inc. (c) [Bobbé Christopherson.]

▼**Figure 18.13 Laterization process and plinthite example.** [Courtesy of Henry D. Foth.]

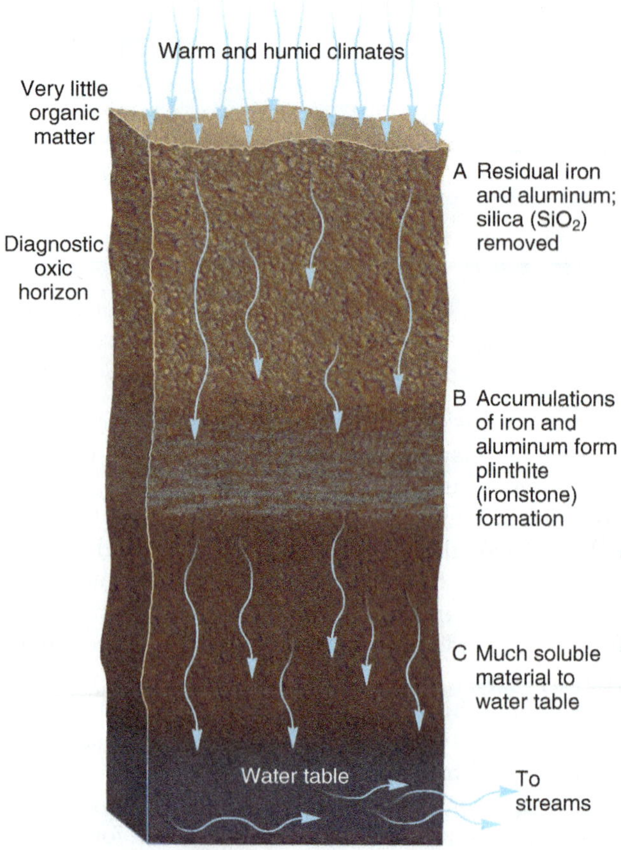

(a) Laterization is characteristic of moist tropical and subtropical climate regimes.

(b) Laterization processes form plinthite, here quarried in India for building material. Inset photo shows close-up of plinthite bricks.

(a) Aridisol profile in central Arizona.

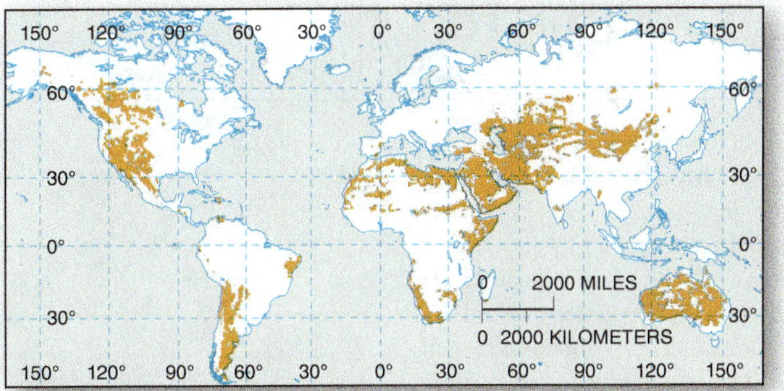

(b) Worldwide Aridisol distribution.

(c) Irrigated cropland and surrounding desert, Imperial Valley, southern California.

(d) Salinization in an irrigated vineyard, San Juan, Argentina.

▲**Figure 18.14 Aridisols.** [(a) Marbut Collection, Soil Science Society of America, Inc. (c) Bobbé Christopherson. (d) Eduardo Pucheta/Alamy.]

to pasturage. Permanent tracts of cleared land, taken out of the former rotation mode, have been severely eroded. When Oxisols are disturbed, soil loss can exceed 1000 tons per square kilometer per year, not to mention the greatly increased extinction rates of plant and animal species that accompany soil depletion and rain forest destruction. The regions dominated by the Oxisols and rain forests are rightfully the focus of much worldwide environmental attention.

Aridisols The largest single soil order occurs in the world's dry regions. **Aridisols** (desert soils) occupy approximately 19% of Earth's land surface (see Figure 18.11). A pale, light soil color near the surface is diagnostic (Figure 18.14a).

Not surprisingly, the water balance in Aridisol regions is characterized by periods of soil-moisture deficit and generally inadequate soil moisture for plant growth. High potential evapotranspiration and low precipitation produce very shallow soil horizons. Usually, these soils have adequate moisture for only 3 months a year. Lacking water and therefore lacking vegetation, Aridisols also lack organic matter of any consequence. Low precipitation means infrequent leaching, yet Aridisols are leached easily when exposed to excessive water, for they lack a significant colloidal structure.

Aridisols can be made productive for agriculture using irrigation. Two related problems common in irrigated lands are salinization and water logging. *Salinization* occurs as salts dissolved in soil water migrate to surface horizons and are deposited as the water evaporates. These deposits appear as subsurface salty horizons, which will damage or kill plants if the horizons occur near the root zone. Salinization is common in Aridisols and results from the excessive potential evapotranspiration rates in arid and semiarid regions. Waterlogging (saturation of the soil that interferes with plant growth) occurs with the introduction of irrigation water for farming, especially in soils that are poorly drained.

Focus Study 18.1 Pollution
Selenium Concentration in Soils: The Death of Kesterson

About 95% of the irrigated acreage in the United States lies west of the 98th meridian, a region that is increasingly troubled with salinization and waterlogging problems. In these semiarid and arid lands, river discharge is inadequate to dilute and remove wastewater from the fields, which are often purposely overwatered to keep salts away from the effective rooting depth of the crops. One solution is to place field drains beneath the soil to collect gravitational water (Figure 18.1.1). But the water must then go somewhere, and for the San Joaquin Valley of central California, this problem triggered a 25-year controversy surrounding toxic levels of selenium in the wetland ecosystem of Kesterson Reservoir.

Selenium Toxicity

Selenium is a trace element that occurs naturally in bedrock, particularly Cretaceous shales found throughout the U.S. West. Toxic effects of selenium were reported during the 1980s in some domestic animals grazing on grasses grown in selenium-rich soils in the Great Plains. In California, the coastal mountain ranges are a significant source region for selenium. As parent materials weather, selenium-rich alluvium washes into valleys, forming Aridisols that become productive with the addition of irrigation water. Selenium then becomes concentrated by evaporation in farmlands and from there may be transported by irrigation drainage into wetlands, where it bioaccumulates to toxic levels. California's western San Joaquin Valley is one of at least nine sites in the western United States experiencing contamination from increasing selenium concentrations.

▲**Figure 18.1.1 Drainage canal.** A drainage canal collects contaminated water from field drains and directs it into the Salton Sea. Such soil-moisture tile drains and collection channels are also in use in the San Joaquin Valley. [Author.]

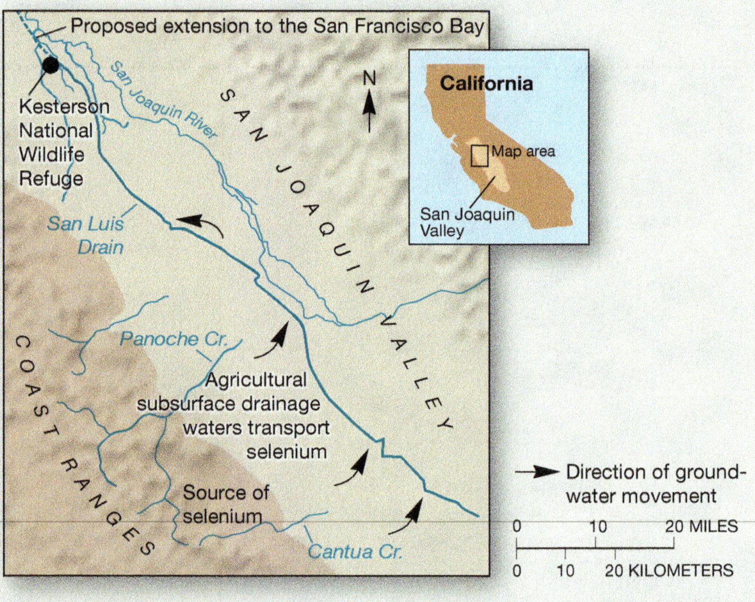

▲**Figure 18.1.2 Kesterson locator map.** The source of selenium is in the Coast Ranges; surface runoff delivers this trace element over thousands of years to the soils of the region. The agricultural drains complete the delivery to the wildlife refuge. [USGS.]

Contamination at Kesterson

Central California's potential drain outlets for agricultural wastewater are limited. Yet, by the late 1970s, about 80 miles of a drain were built in the western San Joaquin Valley, without an outlet or even a formal plan to complete the drain. Large-scale irrigation of corporate-owned farms continued, supplying the field drains with salty, selenium-laden runoff that flowed through the unfinished drain, which stopped abruptly at the boundary of the Kesterson National Wildlife Refuge (Figure 18.1.2).

The selenium-tainted drainage took only 3 years to contaminate the wildlife refuge, which was officially declared a toxic waste site. Selenium was first taken in by aquatic life forms (e.g., marsh plants, plankton, and insects) and then made its way up the food chain and into the diets of higher life forms in the refuge. According to U.S. Fish and Wildlife Service scientists, the toxicity causes genetic abnormalities and death in wildlife, including all varieties of birds that nested at Kesterson; approximately 90% of the exposed birds perished or suffered abnormalities. Because this refuge was a major migration flyway and stopover point for birds from throughout the Western Hemisphere, its contamination also violated several multinational wildlife protection treaties.

Restoration Efforts

The field drains were sealed and removed in 1986, following a court order that forced the federal government to uphold existing laws. Irrigation water then immediately began backing up in the corporate farmlands, producing both waterlogged and selenium-contaminated soils.

The Kesterson Unit became part of the San Luis National Wildlife Refuge—and selenium control and restoration became a priority. In 1996, the U.S. Bureau of Reclamation implemented the Grasslands Bypass Project to protect the water quality of central California wetlands and wildlife refuges. This project redirected agricultural drain water through the old San Luis Drain to Mud Slough, a natural waterway through the San Luis National Wildlife Refuge, and then on to the San Joaquin River, the San Joaquin–Sacramento delta, and eventually San Francisco Bay. Initially, selenium levels increased in those channels, although fluctuations downward in concentrations also were noted. Biological monitoring continues today.

The Kesterson tragedy provided important lessons regarding irrigation practices and the movement of selenium from sedimentary rock formations into aquatic ecosystems. The pathway of selenium from geologic source through agricultural drainage and ultimately into waterfowl is today known as the "Kesterson Effect."

Irrigated agriculture has increased greatly since 1800, when only 8 million hectares (about 20 million acres) were irrigated worldwide. Today, approximately 255 million hectares (about 630 million acres) are irrigated, many of them Aridisols, and this figure is on the increase in some parts of the world (Figure 18.14c). However, in many areas, crop production has decreased and even ended because of salt buildup in the soils. Examples include areas along the Tigris and Euphrates Rivers in Iraq, the Indus River Valley in Pakistan, sections of South America and Africa, and the western United States.

In California, the former Kesterson National Wildlife Refuge, located on Aridisols, was reduced to a toxic waste dump in the early 1980s by contaminated agricultural drainage. Focus Study 18.1 elaborates on the Kesterson tragedy.

Mollisols Some of Earth's most significant agricultural soils are **Mollisols** (grassland soils). This group includes seven recognized suborders that vary in fertility. The dominant diagnostic horizon is a dark, organic surface layer some 25 cm (10 in.) thick (Figure 18.15). As the Latin origin of the name implies (words using the same root, *mollis*, meaning "soft," are *mollify* and *emollient*), Mollisols are soft, even when dry. They have granular or crumbly peds, loosely arranged when dry. These humus-rich soils are high in basic cations (calcium, magnesium, and potassium) and have a high CEC and therefore high fertility. In soil moisture, these soils are intermediate between humid and arid.

Soils of the steppes and prairies of the world—the North American Great Plains, the Palouse of Washington State, the Pampas of Argentina, and the region stretching from Manchuria in China through to Europe—belong to this group. Agriculture in these areas ranges from large-scale commercial grain farming to grazing along the drier portions. With fertilization or soil-building practices, high crop yields are common. The "fertile triangle" of Ukraine, Russia, and western portions of the former Soviet Union is of this soil type.

In North America, the Great Plains straddle the 98th meridian, which is coincident with the 51-cm (20-in.) isohyet of annual precipitation—wetter to the east and drier to the west. The Mollisols here mark the historic division between the short- and tall-grass prairies (Figure 18.16).

Calcification is a soil process characteristic of some Mollisols and adjoining marginal areas of Aridisols. Calcification is the accumulation of calcium carbonate or magnesium carbonate in the B and C horizons. Calcification by calcium carbonate ($CaCO_3$)

(a) Mollisol profile in eastern Idaho, from loess that is high in calcium carbonate, related to the soils of the Palouse agricultural region.

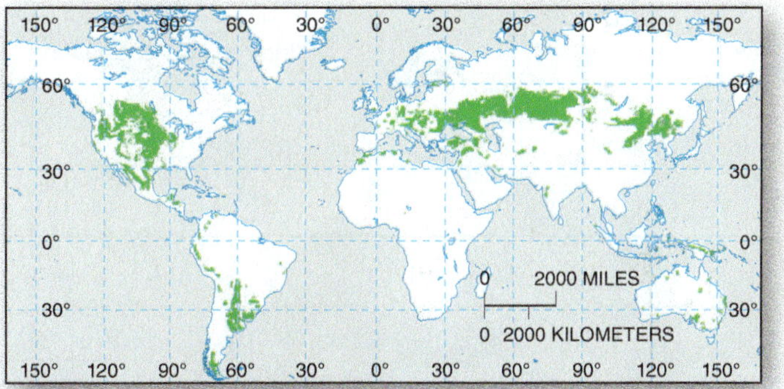

(b) Worldwide Mollisol distribution.

(c) Wheat flourishes in the fertile Palouse of eastern Washington.

▲**Figure 18.15 Mollisols.** [(a) Marbut Collection, Soil Science Society of America, Inc. (c) Bobbé Christopherson.]

forms a diagnostic subsurface horizon that is thickest along the boundary between dry and humid climates (Figure 18.17).

Alfisols Spatially, **Alfisols** (moderately weathered forest soils) are the most widespread of the soil orders, extending in five suborders from near the equator to high latitudes. Representative Alfisol areas include Boromo and Burkina Faso (interior western Africa); Fort Nelson, British Columbia; the states near the Great Lakes; and the valleys of central California. Most Alfisols are grayish brown to reddish and are considered moist versions of the Mollisol soil group. Moderate eluviation is present as well as a subsurface horizon of illuviated clays and clay formation because of a pattern of increased precipitation (Figure 18.18).

Alfisols have moderate to high reserves of basic cations and are fertile. However, productivity depends

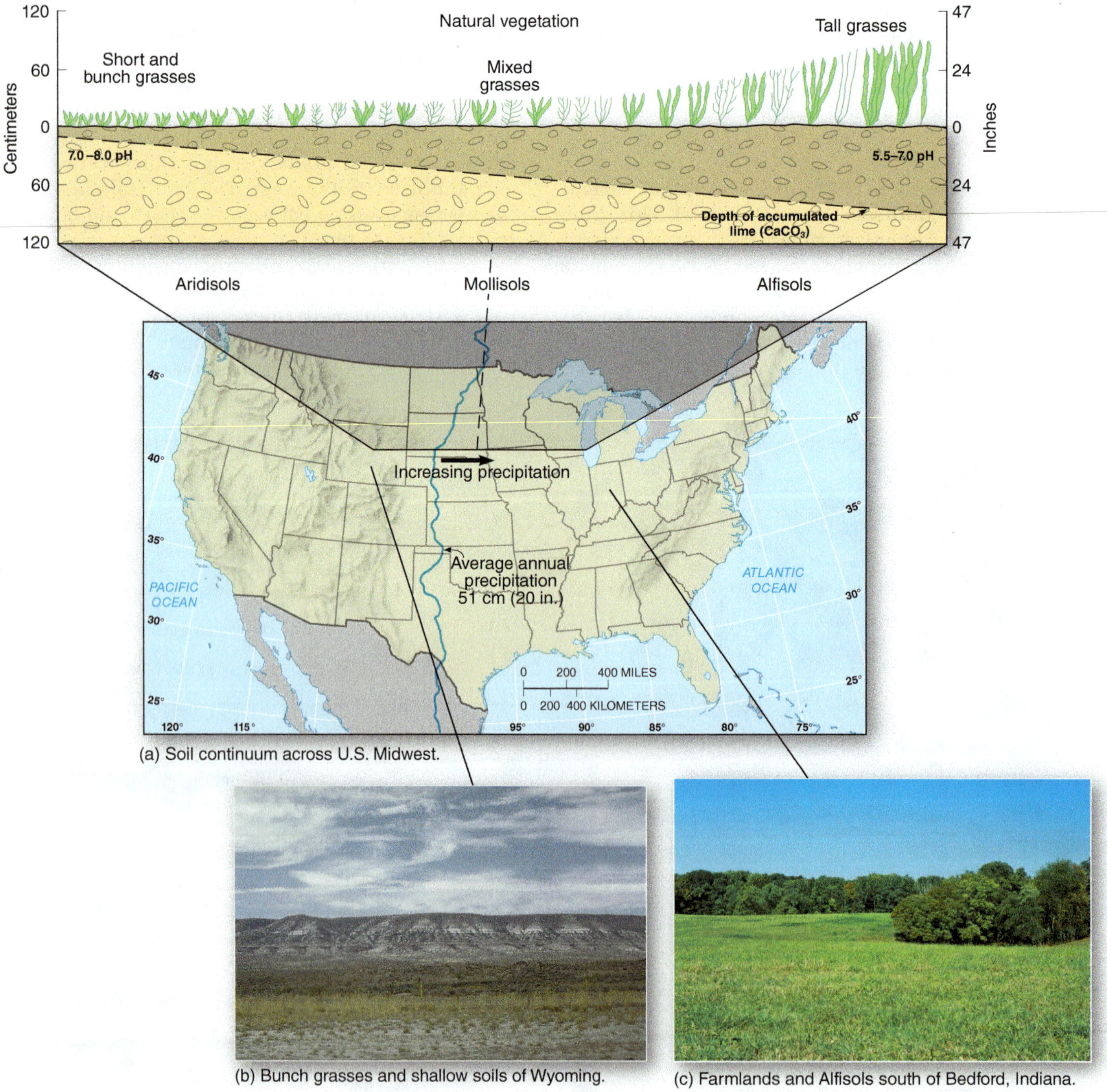

(a) Soil continuum across U.S. Midwest.
(b) Bunch grasses and shallow soils of Wyoming.
(c) Farmlands and Alfisols south of Bedford, Indiana.

▲**Figure 18.16 Soils of the Midwest.** Aridisols (to the west), Mollisols (central), and Alfisols (to the east) are part of a soil continuum in the north-central United States and southern Canadian prairies. Note the changes that occur in soil pH and the depth of accumulated lime.
[(a) Illustration adapted from N. C. Brady, *The Nature and Properties of Soils*, 10th ed., © 1990 by Macmillan Publishing Company, adapted by permission. (b) Author. (c) Bobbé Christopherson.]

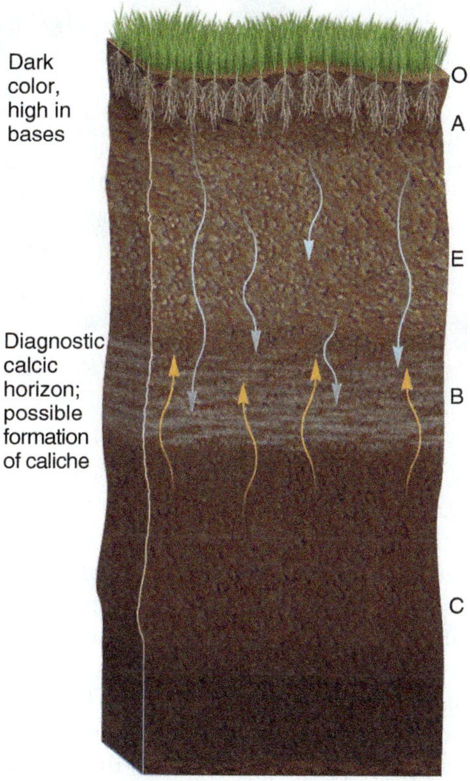

▲**Figure 18.17 Calcification process in soil.** The calcification process in desert and grassland soils occurs in climatic regimes that have potential evapotranspiration equal to or greater than precipitation.

on moisture and temperature. Alfisols usually are supplemented by a moderate application of lime and fertilizer in areas of active agriculture. Some of the best U.S. farmland occurs in the humid continental hot summer climates surrounding the Great Lakes. These Alfisols produce grains, hay, and dairy products. The moist winter, dry summer pattern of the Mediterranean climate also produces Alfisols. These naturally productive soils are farmed intensively for subtropical fruits, nuts, and special crops that grow in only a few locales worldwide—for example, California olives, grapes, citrus, artichokes, almonds, and figs (Figure 18.18c).

Ultisols Farther south in the United States are the **Ultisols** (highly weathered forest soils). An Alfisol might degenerate into an Ultisol, given time and exposure to increased weathering under moist conditions. These soils tend to be reddish because of residual iron and aluminum oxides in the A horizon (Figure 18.19).

(a) Alfisol profile in northern Idaho loess.

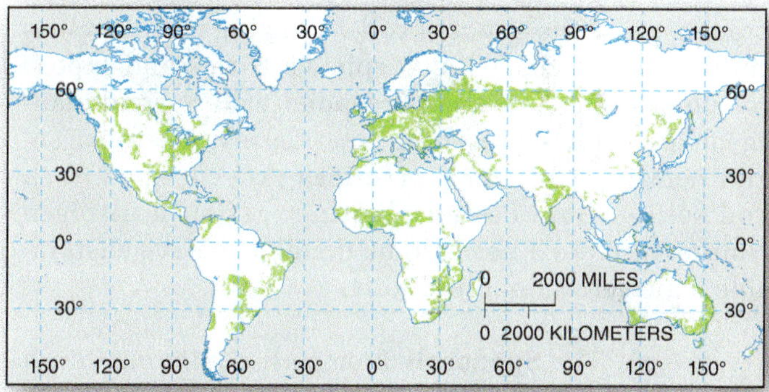

(b) Worldwide Alfisol distribution.

(c) An olive orchard in northern California, where virtually all U.S. olive production occurs.

▲**Figure 18.18 Alfisols.** [(a) Marbut Collection, Soil Science Society of America, Inc. (c) Bobbé Christopherson.]

GEOreport 18.4 Loss of marginal lands puts pressure on prime lands

Since 1985, more than 0.6 million hectares (1.5 million acres) of irrigated Aridisols and Alfisols have been removed from production in California, owing to water shortages and soil-quality problems, marking the end of several decades of irrigated farming in climatically marginal lands. Severe cutbacks in irrigated acreage no doubt will continue, underscoring the need to preserve prime farmlands in California and in wetter regions elsewhere in the country.

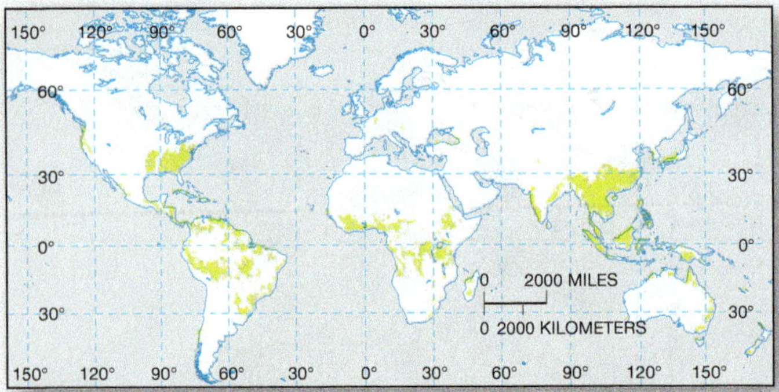

(a) Worldwide Ultisol distribution.

(b) Ultisols planted in rows of peanuts in west-central Georgia have the characteristic reddish color.

▲**Figure 18.19 Ultisols.** [(b) Bobbé Christopherson.]

The relatively high precipitation in Ultisol regions causes greater mineral alteration and more eluvial leaching than in other soils. Therefore, the level of basic cations is lower, and the soil fertility is lower. Fertility is further reduced by certain agricultural practices and the effect of soil-damaging crops such as cotton and tobacco, which deplete nitrogen and expose soil to erosion. These soils respond well to good management—for example, crop rotation restores nitrogen, and certain cultivation practices prevent sheetwash and soil erosion. Peanut plantings assist in nitrogen restoration. Much needs to be done to achieve sustainable management of these soils.

Spodosols The **Spodosols** (northern coniferous forest soils) occur generally to the north and east of the Alfisols, mainly in forested areas in the humid continental mild-summer climates of northern North America and Eurasia, Denmark, the Netherlands, and southern England. Because no comparable climates exist in the Southern Hemisphere, this soil type is not identified there. Spodosols form from sandy parent materials, shaded under evergreen forests of spruce, fir, and pine. Spodosols with more moderate properties form under mixed or deciduous forests (Figure 18.20).

Spodosols lack humus and clay in the A horizon. Characteristics of the podzolization process are present: The A horizon is sandy and bleached in color and is eluviated, leached of clays and iron; the B horizon is composed of illuviated organic matter and iron and aluminum oxides (Figure 18.21). The surface horizon receives organic litter from base-poor, acid-rich evergreen trees, which contribute to acid accumulations in the soil. The solution in acidic soils effectively leaches clays, iron, and aluminum, which are passed to the upper diagnostic horizon. An ashen-gray color is common in these subarctic forest soils and is characteristic of podzolization (Figure 18.20a).

When agriculture is attempted, the low basic cation content of Spodosols requires the addition of nitrogen, phosphate, and potash (potassium carbonate)—and perhaps crop rotation as well. A soil *amendment* such as limestone can significantly increase crop production by raising the pH of these acidic soils. For example, the yields of several crops (corn, oats, wheat, and hay) grown in specific Spodosols in New York State were increased up to a third with the application of 1.8 metric tons (2 tons) of limestone per 0.4 hectare (1.0 acre) during each 6-year rotation.

Entisols The **Entisols** (recent, undeveloped soils) lack vertical development of their horizons. The presence of Entisols is not climate-dependent, for they occur in many climates worldwide. Entisols are true soils, but they have not had sufficient time to generate the usual horizons.

Entisols generally are poor agricultural soils, although those formed from river silt deposits are quite fertile. The same conditions that inhibit complete development also prevent adequate fertility—too much or too little water, poor structure, and insufficient accumulation of weathered nutrients. These soils are characteristically found on active slopes, alluvium-filled floodplains, poorly drained tundra, tidal mudflats, dune sands and erg (sandy) deserts, and plains of glacial outwash. Figure 18.22 shows an Entisol in a desert climate where shales formed the parent material.

Inceptisols Young soils are **Inceptisols** (weakly developed soils), although they are more developed than the Entisols. Since they have not reached a mature condition, they are inherently infertile. This soil order includes a wide variety of different soils, all having in common a lack of maturity, with evidence of weathering just beginning.

Inceptisols are associated with moist soil regimes and are regarded as eluvial because they demonstrate a loss of soil constituents throughout their profile; however, they do retain some weatherable minerals. This soil group has no distinct illuvial horizons. Most of the glacially derived till and outwash materials from New York down through the Appalachians are Inceptisols, as is the alluvium on the Mekong and Ganges floodplains.

Gelisols The **Gelisols** (cold and frozen soils) contain permafrost within 2 m (6.5 ft) of the surface and are found at high latitudes (Canada, Alaska, Russia, Arctic Ocean islands, and the Antarctic Peninsula) and high elevations (mountains). Temperatures in these regions are at or below 0°C (32°F), making soil development a slow process and disturbances of the soil long-lasting. Cold temperatures slow the decomposition of materials in the soil, so Gelisols can store large amounts of organic matter (Figure 18.23). Only Histosols

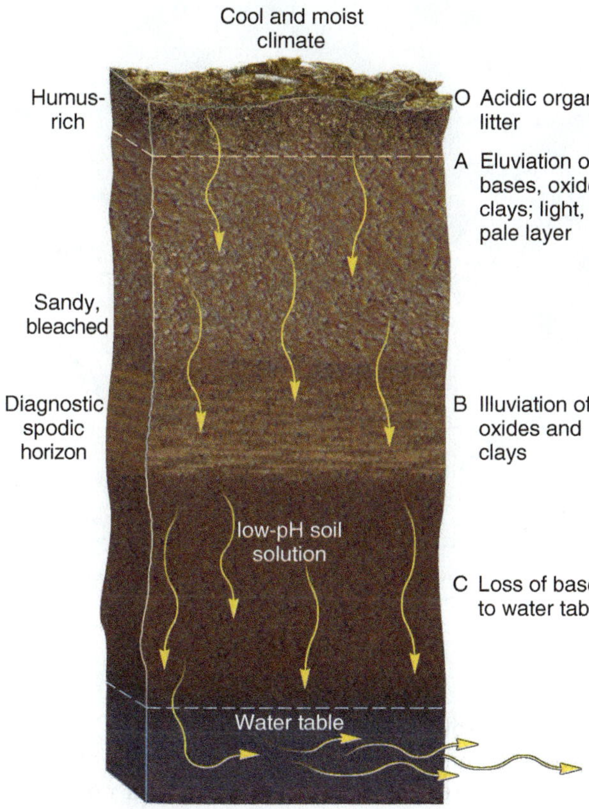

▲**Figure 18.21 Podzolization process in soil.** The podzolization process is typical in cool and moist climatic regimes and Spodosols.

▼**Figure 18.20 Spodosols.** [(a) Marbut Collection, Soil Science Society of America, Inc. (c), (d) Bobbé Christopherson.]

(a) Spodosol profile from northern New York.

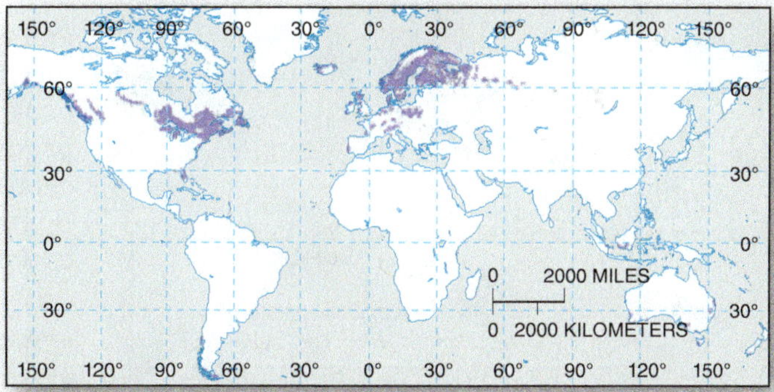

(b) Worldwide Spodosol distribution.

(c) Characteristic temperate forest and Spodosols in the cool, moist climate of central Vancouver Island.

(d) Freshly plowed Spodosols (Podzolic soils, in the Canadian System) near Lakeville, Nova Scotia. The soils are formed beneath coniferous forest before the land is cleared for agriculture.

▲**Figure 18.22 Entisols.** A characteristic Entisol: young, undeveloped soil forming from a shale parent material in the Anza–Borrego Desert, California. [Bobbé Christopherson.]

(discussed ahead) have as high a content of organic matter as Gelisols.

Gelisols contain about half of the pool of global carbon. The latest estimate of the carbon contained in these periglacial soils is 1.7 trillion tons. When permafrost thaws, substantial amounts of greenhouse gases are released into the atmosphere. The process begins as warming occurs in the higher latitudes. Gelisols can quickly become wet and soggy with only a slight shift in their thermal balance. As this thawing progresses, the poorly decomposed organic content in the soil begins to decay, and its decomposition releases enormous quantities of carbon dioxide into the atmosphere through increased respiration. Another greenhouse gas, methane, is released as well (discussed in Chapter 10B).

Gelisols are subject to *cryoturbation* (frost churning and mixing) in the freeze–thaw cycle in the active layer (see Chapter 17). This process disrupts soil horizons, pulling organic material to lower layers and drawing rocky C-horizon material to the surface, a process that often forms patterned ground.

Andisols Areas of volcanic activity feature **Andisols** (soils with volcanic parent materials). Andisols are derived from volcanic ash and glass, and frequently bury previous soil horizons with materials from repeated eruptions. Volcanic soils are unique in their mineral content because they are recharged by eruptions.

Weathering and mineral transformations are important in this soil order. For example, volcanic glass weathers readily into a clay colloid and oxides of aluminum and iron. Andisols have a high CEC and high water-holding ability and develop moderate fertility, although phosphorus availability is an occasional problem. In Hawai'i, fields of Andisols produce coffee, pineapples, macadamia nuts, and small amounts of sugar cane as important cash crops (Figure 18.24a). Andisol distribution is small in areal extent; however, such soils are locally important in the volcanic ring of fire surrounding the Pacific Rim (Figure 18.24b).

Vertisols Heavy clay soils are the **Vertisols** (expandable clay soils). They contain more than 30% swelling clays (clays that swell significantly when they absorb water), such as *montmorillonite*. They are located in regions experiencing highly variable soil-moisture balances through

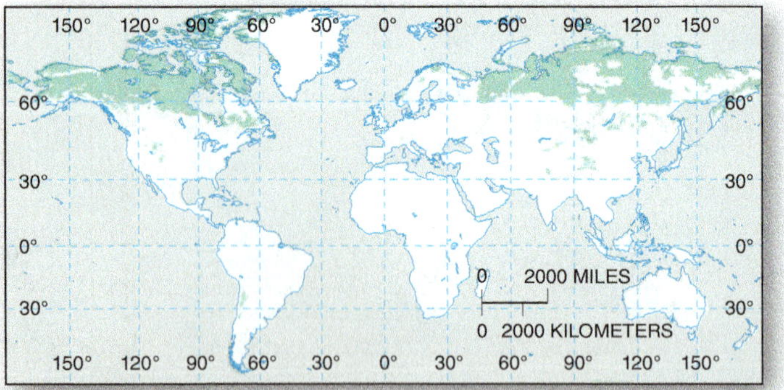

(a) Worldwide Gelisol distribution.

(b) The tundra is green in the brief summer season on Spitsbergen Island, as the active layer thaws.

(c) Fibrous organic content exposed on the underside of a soil clod, Spitsbergen.

▲**Figure 18.23 Gelisols.** [(b), (c) Bobbé Christopherson.]

the seasons. These soils occur in areas of subhumid to semiarid moisture and moderate to high temperature. Vertisols frequently form under savanna and grassland vegetation in tropical and subtropical climates and are sometimes associated with a distinct dry season following a wet season. Although widespread in their distribution, individual Vertisol units are limited in extent.

Vertisol clays are black when wet due to specific mineral content rather than high amounts of organic matter. When dry, they range from brown to dark gray. These deep clays swell when moistened and shrink when dried. In the drying process, they may form vertical cracks as wide as 2–3 cm (0.8–1.2 in.) and up to 40 cm (16 in.) deep. Loose material falls into these cracks, only to disappear when the soil again expands and the cracks close. After many such cycles, soil contents tend to invert, or mix vertically, bringing lower horizons to the surface (Figure 18.25).

Despite the fact that clay soils become plastic and heavy when wet, and leave little soil moisture available for plants, Vertisols are high in bases and nutrients and thus are some of the better farming soils where they occur. Vertisols often are planted with grain sorghums, corn, and cotton (Figure 18.25c).

Histosols Accumulations of thick organic matter can form **Histosols** (organic soils). In the midlatitudes, when conditions are right, beds of former lakes may turn into Histosols, with water gradually replaced by organic material to form a bog (Figure 18.26). (Lake succession and bog or marsh formation are discussed in Chapter 19.) Histosols also form in small, poorly drained depressions, where conditions can be ideal for significant deposits of sphagnum peat to form.

Peat beds, often more than 2 m thick, can be cut by hand with a spade into blocks, which are then dried, baled, and sold as a soil amendment (Figure 18.26a). Once dried, the peat blocks burn hot and smoky. Peat is the first stage in the natural formation of lignite, an intermediate step toward coal. The Histosols that formed in lush swamp environments in the Carboniferous Period (359 to 299 million years ago) eventually underwent coalification to become coal deposits.

Most Histosols are **hydric soils**, defined as soils that are saturated or flooded for long enough periods of time to develop anaerobic (oxygen-free) conditions during the growing season. The presence of hydric soils is the basis for the legal delineation of wetlands, which are protected from dredging, filling, and the discharge of pollutants under the U.S. Clean Water Act. Bogs are a type of wetland, and many Histosols develop in wetland environments. Entisols are another soil order that frequently develop as hydric soils in large river valleys and along coastlines.

▼**Figure 18.24 Andisols in agricultural production.** [(a) Bobbé Christopherson. (b) Alfredo Maiquez/Getty Images.]

(a) Groves of macadamia nuts planted in fertile Andisols on the Big Island of Hawai`i.

(b) An onion plantation in the Chiriqui province on the west coast of Panama in Central America.

(a) Vertisol profile in the Lajas Valley of Puerto Rico.

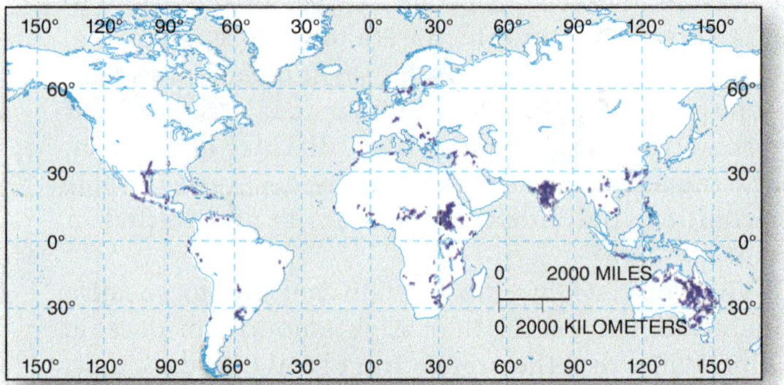

(b) Worldwide Vertisol distribution.

(c) Commercial sorghum crop planted in Vertisols on the Texas coastal plain, northeast of Palacios. Note the characteristic dark soil color.

▲**Figure 18.25 Vertisols.** [(a) Marbut Collection, Soil Science Society of America, Inc. (c) Bobbé Christopherson.]

(a) A Histosol profile on Mainland Island in the Orkneys, north of Scotland. The inset photo shows drying blocks of peat, used as fuel. Note the fibrous texture of the sphagnum moss growing on the surface and the darkening layers with depth in the soil profile as the peat is compressed and chemically altered.

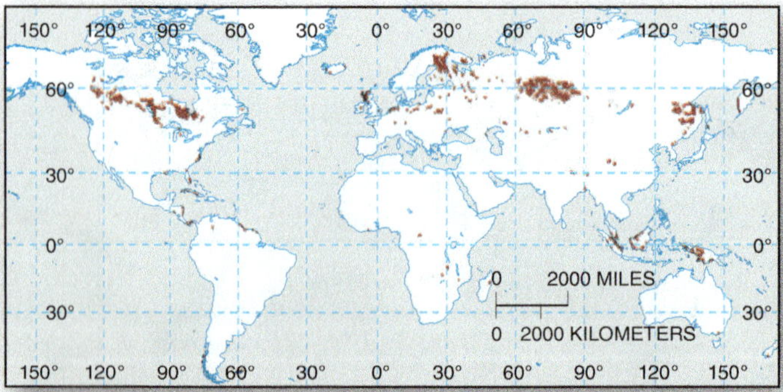

(b) Worldwide Histosol distribution.

(c) A bog in coastal Maine, near Popham Beach State Park.

▲**Figure 18.26 Histosols.** [(a), (c) Bobbé Christopherson.]

THE human DENOMINATOR 18 Soils and Land Use

SOILS ⇨ HUMANS

- Soils are the foundation of basic ecosystem function and are a critical resource for agriculture.
- Soils store carbon dioxide and other greenhouse gases in soil organic matter.

HUMANS ⇨ SOILS

- Humans have modified soils through agricultural activities. Recently, fertilizer use, nutrient depletion, and salinization have increased soil degradation.
- Poor land-use practices are combining with changing climate to cause desertification, soil erosion, and the loss of prime farmland.

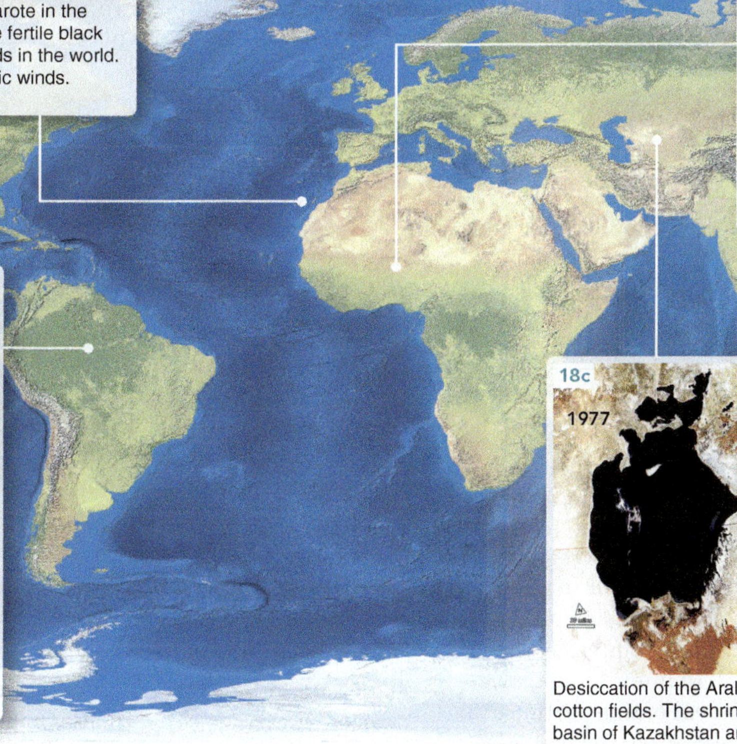

18a Grapevines grow in the Andisols of Lanzarote in the Canary Islands, producing wines from the fertile black ash of some of the most isolated vineyards in the world. Stone walls protect plants from the Atlantic winds. [Raul Mateos Fotografia/Getty Images.]

18b Nigerian women dig a trench to collect rainwater in the Sahel region of Africa. Although above-average rainfall in 2012 led to a successful harvest, the effects of desertification are ongoing throughout the region. [AFP/Getty Images.]

18c Desiccation of the Aral Sea began when rivers were diverted to irrigate cotton fields. The shrinking lake has accelerated desertification in the Aral basin of Kazakhstan and Uzbekistan, and affected the local climate, which is now hotter in summer without the moderating influence of the former large water body. [USGS EROS Data Center.]

18d Soybean fields are readied for planting in Mato Grosso state, Brazil, where vast tracts of rain forest are being cleared for industrial agriculture. Oxisols, the soils of the tropics, have low fertility, requiring the use of fertilizers, which along with pesticides are affecting water quality throughout the region. [Latin Content/Getty Images.]

ISSUES FOR THE 21ST CENTURY

- Continued soil erosion and degradation will cause lowered agricultural productivity worldwide and possible food shortages.
- Thawing of frozen soils (Gelisols) in the northern latitudes emits carbon dioxide and methane into the atmosphere, creating a positive feedback loop that leads to further warming.
- Increased use of sustainable soil management and land use practices will improve soil quality and agriculture.

GEOSYSTEMS connection

Soil science forms the bridge between Parts I, II, and III, which cover the abiotic systems, and Part IV, which explores the biotic systems of Earth. Soil formation is affected by temperature, moisture, parent material, topography, and living organisms, including humans. Thus, this chapter combines aspects of the energy–atmosphere and the water, weather, and climate systems, as well as of the weathering processes that are the source for soil particles and minerals, into a study of the soils that cover Earth's surface. Next, we move to the essential components of Earth's ecosystems, and from there to the biotic operations that fuel living systems and the communities that organize life.

KEY LEARNING concepts review

■ *Define* soil and soil science, and *list* four components of soil.

Soil is a dynamic natural mixture of fine materials, including both mineral and organic matter. **Soil science** is the interdisciplinary study of soils involving physics, chemistry, biology, mineralogy, hydrology, taxonomy, climatology, and cartography. *Pedology* deals with the origin, classification, distribution, and description of soil. *Edaphology* specifically focuses on the study of soil as a medium for sustaining the growth of plants. Soil is composed of about 50% mineral and organic matter, and 50% water and air contained in pore spaces between the soil particles.

soil (p. 556)
soil science (p. 556)

1. Soils provide the foundation for animal and plant life and therefore are critical to Earth's ecosystems. Why is this true?
2. What are the differences between soil science, pedology, and edaphology?

■ *Describe* the principal soil-formation factors, and *describe* the horizons of a typical soil profile.

Environmental factors that affect soil formation include parent materials, climate, vegetation, topography, and time. To evaluate soils, scientists use a **soil profile**, a vertical section of soil that extends from the surface to the deepest extent of plant roots or to the point where regolith or bedrock is encountered. Each discernible layer in a soil profile is a **soil horizon**. The horizons are designated O (contains **humus**, a complex mixture of decomposed and synthesized organic materials), A (rich in humus and clay, darker), E (zone of **eluviation**, the removal of fine particles and minerals by water), B (zone of **illuviation**, the deposition of clays and minerals translocated from elsewhere), C (*regolith*, weathered bedrock), and R (bedrock). Soil horizons A, E, and B experience the most active soil processes and together are designated the **solum**.

soil profile (p. 557)
soil horizon (p. 557)
humus (p. 558)
eluviation (p. 558)
illuviation (p. 558)
solum (p. 558)

3. Briefly describe the contributions of the following factors and their effects on soil formation: parent material, climate, vegetation, landforms, time, and humans.
4. Characterize the principal aspects of each soil horizon. Where does the main accumulation of organic material occur? Where does humus form? Which horizons constitute the solum?
5. Explain the difference between the processes of eluviation and illuviation.

■ *Describe* the physical properties used to classify soils: color, texture, structure, consistence, porosity, and soil moisture.

We use several physical properties to assess **soil fertility** (the ability of soil to sustain plants) and classify soils. Color suggests composition and chemical makeup. Soil texture refers to the size of individual mineral particles and the ratios of different sizes. For example, **loam** is a balanced mixture of sand, silt, and clay. Soil structure refers to the shape and size of the soil *ped*, which is the smallest natural cluster of particles in a given soil. The cohesion of soil particles to each other is called soil consistence. **Soil porosity** refers to the size, alignment, shape, and location of spaces in the soil that are filled with air, gases, or water. Soil moisture refers to water in the soil pores and its availability to plants.

soil fertility (p. 558)
loam (p. 559)
soil porosity (p. 562)

6. How can soil color be an indication of soil qualities? Give a couple of examples.
7. Define a soil separate. What are the various sizes of particles in soil? What is loam? Why is loam regarded so highly by agriculturists?
8. What is a quick, hands-on method for determining soil consistence?
9. Summarize the role of soil moisture in mature soils.

■ *Explain* basic soil chemistry, including cation-exchange capacity, and *relate* these concepts to soil fertility.

Particles of clay and organic material form negatively charged **soil colloids** that attract and retain positively charged mineral ions in the soil. The capacity to exchange ions between colloids and roots is called the **cation-exchange capacity (CEC)**.

soil colloid (p. 562)
cation-exchange capacity (CEC) (p. 563)

10. What are soil colloids? How are they related to cations and anions in the soil? Explain cation-exchange capacity.
11. What is meant by the concept of soil fertility, and how does soil chemistry affect fertility?

■ *Discuss* human impacts on soils, including desertification.

Essential soils for agriculture and their fertility are threatened by human activities and mismanagement. *Soil erosion* is the breakdown and redistribution of soils by wind, water, and gravity. To slow soil erosion, farmers in the United States and elsewhere use *no-till agriculture*, a practice in which the land is not plowed after a harvest. **Desertification** is the ongoing degradation of drylands caused by human activities and climate change; presently, this process affects some 1.5 billion people.

desertification (p. 565)

12. What is meant by desertification? What world regions are affected by this phenomenon?

13. Explain some of the details that support the concern over loss of our most fertile soils. What cost estimates have been placed on soil erosion?

■ *Describe* the 12 soil orders of the Soil Taxonomy classification system, and *explain* their general distribution across Earth.

The U.S. **Soil Taxonomy** classification system is built around an analysis of various diagnostic horizons and 12 soil orders, as actually seen in the field. The basic sampling unit used in soil surveys is the **pedon**. The system divides soils into six hierarchical categories. From smallest to largest, they are series, families, subgroups, great groups, suborders, and orders. The Soil Taxonomy system uses two diagnostic horizons to identify soils: the **epipedon**, or the surface soil, and the **diagnostic subsurface horizon**, or the soil layer below the surface (at various depths) having properties specific to the type of soil.

Specific soil-forming processes keyed to climatic regions (not a basis for classification) are called **pedogenic regimes**. These include **laterization** (leaching in warm and humid climates), **salinization** (collection of salt residues in surface horizons in hot, dry climates), **calcification** (accumulation of carbonates in the B and C horizons in drier continental climates), **podzolization** (soil acidification in forest soils in cool climates), and **gleization** (humus and clay accumulation in cold, wet climates with poor drainage).

The 12 soil orders are **Oxisols** (tropical soils), **Aridisols** (desert soils), **Mollisols** (grassland soils), **Alfisols** (moderately weathered, temperate forest soils), **Ultisols** (highly weathered, subtropical forest soils), **Spodosols** (northern conifer forest soils), **Entisols** (recent, undeveloped soils), **Inceptisols** (weakly developed, humid-region soils), **Gelisols** (cold soils underlain by permafrost), **Andisols** (soils formed from volcanic materials), **Vertisols** (expandable clay soils), and **Histosols** (organic soils). Soils saturated for long enough periods to develop anaerobic, or "oxygen-free," conditions are **hydric soils**.

Soil Taxonomy (p. 566)
pedon (p. 566)
epipedon (p. 566)
diagnostic subsurface horizon (p. 566)
pedogenic regime (p. 566)
laterization (p. 566)
salinization (p. 566)
calcification (p. 566)
podzolization (p. 567)
gleization (p. 567)
Oxisols (p. 568)
Aridisols (p. 571)
Mollisols (p. 573)
Alfisols (p. 574)
Ultisols (p. 575)
Spodosols (p. 576)
Entisols (p. 576)
Inceptisols (p. 576)
Gelisols (p. 577)
Andisols (p. 578)
Vertisols (p. 578)
Histosols (p. 579)
hydric soil (p. 579)

14. What is the basis of the Soil Taxonomy system? What is the number of classifications in each of the following: soil orders, suborders, great groups, subgroups, families, series?
15. Define an epipedon and a diagnostic subsurface horizon. Give a simple example of each.
16. Which two orders were most recently added to the classification system, and what soils are in these new orders?
17. Locate each soil order on the world map and on the U.S. map as you give a general description of it.
18. How was slash-and-burn shifting cultivation, as practiced in the past, a form of crop and soil rotation and conservation of soil properties?
19. Describe the salinization process in arid and semiarid soils. What associated soil horizons develop?
20. Which of the soil orders are associated with Earth's most productive agricultural areas?
21. Why is the 51-cm (20-in.) isohyet of annual precipitation in the Midwest significant to plants? How do the pH and lime content of soils change on either side of this isohyet?
22. Describe the podzolization process associated with northern coniferous forest soils. What characteristics are associated with the surface horizons? What strategies might enhance these soils?
23. Describe the location, nature, and formation processes of Gelisols. What is the linkage between these soils and climate change?
24. What type of soil is used to define wetlands?
25. Why has a selenium contamination problem arisen in some western U.S. soils? Explain the impact of agricultural practices on selenium concentrations.

19 Ecosystem Essentials

KEY LEARNING concepts

After reading the chapter, you should be able to:

- ***Define*** ecology, biogeography, and ecosystem.
- ***Explain*** photosynthesis and respiration, and ***describe*** the world pattern of net primary productivity.
- ***Discuss*** the oxygen, carbon, and nitrogen cycles, and ***explain*** trophic relationships.
- ***Describe*** communities and ecological niches, and ***list*** several limiting factors on species distributions.
- ***Outline*** the stages of ecological succession in both terrestrial and aquatic ecosystems.
- ***Explain*** how biological evolution led to the biodiversity of life on Earth.

The Atlantic Forest of Brazil consists of highly vulnerable tropical ecosystems that rival the Amazon rain forest in biodiversity. Extending along the coastal region through the most populous part of Brazil and into portions of Paraguay and Argentina, the Atlantic Forest is disappearing. Current estimates place this forest at 7% of its size when Europeans arrived in the 1500s. The remaining area is fragmented, with many species critically endangered by the effects of logging, agriculture, industry, and development. Protected areas, including Serra dos Órgãos National Park pictured here, are helping to preserve species diversity. This chapter discusses the essential processes of functioning ecosystems, including impacts on biodiversity from human activities and human-caused climate change. [Kevin Schafer/Corbis.]

Species' Distributions Shift with Climate Change

Every plant or animal species in nature has a range of tolerance for variations in the physical characteristics of its environment, and this range of tolerance affects its geographical distribution. In particular, temperature and moisture requirements are often critical for determining a species range. Thus, changing climate can have a huge effect on the fate of a biological community. If conditions become too difficult (for example, temperatures too high or too low; rainfall too much or too little), species must adapt, seek out new habitat elsewhere, or become extinct.

Range Changes by Elevation A significant body of research from tropical and temperate mountains now shows that species are shifting their distributions to higher elevations. A long-term study in Yosemite National Park, California, found that small-mammal communities shifted their distributions in response to rising temperatures, notably the 3 C° (5.5 F°) increase in minimum temperature during the last century. Half the species monitored showed substantial movement toward cooler temperatures at higher elevations. Formerly-low-elevation species expanded their ranges higher, and high-elevation species contracted their ranges away from lower areas. These species also risk mountaintop extinction as warming pushes climatically suitable conditions beyond the reaches of mountain peaks.

This study is consistent with others suggesting that high-elevation animal species are losing available habitat as temperatures increase, a situation that ultimately threatens these species' survival. In the southern Sierra Nevada, such a shift was under way with the endangered Inyo chipmunk (*Tamias umbrinus inyoensis*), a subspecies of the Uinta chipmunk. This animal has not been seen for more than 4 years in the Sierra, raising fears of extinction. In these western mountains, species of voles, bats, and pikas, as well as other squirrels, are at risk of extinction as their ranges collapse.

A 2013 study reported that almost 90% of native plants in the Indian Himalayas shifted their ranges upward in elevation over the last century. One notable finding was that low-elevation species shifted their ranges upward to a greater extent than high-elevation species, suggesting that high-elevation species are already reaching the limits of available mountain habitat.

Range Changes by Latitude As temperatures increase, many plants and animals are shifting their distributions into higher-latitude areas with more favorable temperature conditions. Numerous studies spanning the past 40 years show that maximum range shifts vary from 200 to 1000 km (125 to 620 mi); a more recent study documented range shifts averaging 6.1 km (3.8 mi) per year.

In general, plant communities are slower than animal species to respond to environmental change; however, research suggests that global climate change is already impacting forest distributions. A recent study of 130 species of North American trees suggests that ranges will shift northward between 330 and 700 km (205 and 435 mi) depending on the success of dispersal into new habitats. Thus, deciduous forests that are now common in the United States would be found in Canada by century's end, and their present range would be occupied by grasslands in some areas and by a different mix of tree species in other regions.

To illustrate the magnitude of these shifting distributions, Figure GN 19.1 shows the predicted "migration" of what is now Illinois during this century based on forecasted changes in temperature and precipitation—in summer conditions, in particular. By mid-century, Illinois' summers will be similar to those of Arkansas and northern Louisiana; by 2090, Illinois' climate will be similar to that of Texas and Louisiana. The plants and animals that live in Illinois today must adapt quickly to survive such thermal and moisture-balance changes; these communities will shift northward to stay within their tolerance limits. We can expect present Illinois environmental conditions to be in central Manitoba and Ontario, Canada, by century's end.

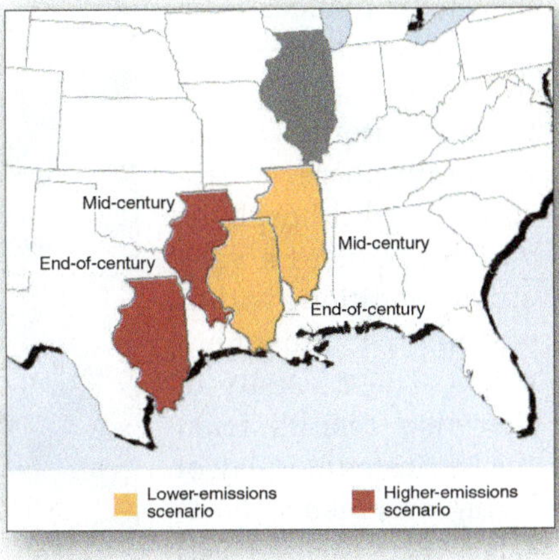

▲**Figure GN 19.1 Climatic shifts forecasted for the Midwest.** The relative future climatic conditions of Illinois forecasted by 2090 for two scenarios. [U.S. Global Change Research Program, 2009.]

The Intergovernmental Panel on Climate Change's 2007 *Fourth Assessment Report* estimated that a warming of 1.9 to 3.0 C° (3.5 to 5.5 F°) would leave more than 30% of species outside of their historic range, placing them at risk of extinction if they are unable to adapt or shift their ranges northward or upslope. A physical geography text written 20 to 70 years from now may have different vegetation maps from the ones depicting distributions and vegetation patterns today. In this chapter, we discuss ecosystems and species distributions; world vegetation patterns are the focus of Chapter 20.

GEOSYSTEMS NOW ONLINE Go to Chapter 19 on the *Mastering Geography* website for more information on changing species distributions in California or see the U.C. Berkeley News Center at http://newscenter.berkeley.edu/2012/08/15/climate-change-range-shift/. For a study on the effects of climate change on genetic variation (discussed in this chapter), go to http://evolution.berkeley.edu/evolibrary/news/120301_chipmunks. MG

The diversity of organisms on the living Earth is one if our planet's most impressive features. This diversity is a response to the interaction of the atmosphere, hydrosphere, and lithosphere, which produces a variety of conditions within which the biosphere exists. The diversity of life also results from the intricate interplay of living organisms themselves. Each species uses strategies that maintain biodiversity and species coexistence.

(a) The Barataria Perserve Wetland is a swamp in the Mississippi delta with bald cypress trees.

(b) A Little Blue Heron in coastal wetland habitat near Sanibel Island, Florida Gulf Coast.

▲**Figure 19.1 Gulf Coast wetland ecosystem.** [Bobbé Christopherson.]

The biosphere, the sphere of life and organic activity, extends from the ocean floor to an altitude of about 8 km (5 mi) into the atmosphere. The biosphere includes myriad ecosystems, from simple to complex, each operating within general spatial boundaries. An **ecosystem** is a self-sustaining association of living plants and animals and their nonliving physical environment. Earth's biosphere itself is a collection of ecosystems within the natural boundary of the atmosphere and Earth's crust. Natural ecosystems are open systems with regard to both solar energy and matter, with almost all ecosystem boundaries functioning as transition zones rather than as sharp demarcations. Ecosystems vary in size from small-scale, such as the ecosystem of a city park or pond, to mid-scale, as a mountaintop or beach, to large-scale, as a forest or desert. Internally, every ecosystem is a complex of many interconnected variables, all functioning independently yet in concert, with complicated flows of energy and matter.

Ecology is the study of the relationships between organisms and their environment and among the various ecosystems in the biosphere. The word *ecology*, developed by German naturalist Ernst Haeckel in 1869, is derived from the Greek *oikos* ("household" or "place to live") and *logos* ("study of"). **Biogeography** is the study of the distribution of plants and animals, the diverse spatial patterns they create, and the physical and biological processes, past and present, that produce Earth's species richness.

Earth's most influential biotic agents are humans. This is not arrogance; it is fact—humans powerfully affect every ecosystem on Earth. From the time humans first developed agriculture, tended livestock, and used fire, the influence of humans over Earth's physical systems has been increasing. For example, the Gulf Coast wetlands and associated coastal habitat, shown in Figure 19.1, bore the brunt of the 2010 BP oil spill; restoration of coastal marshes is ongoing. The degree to which modern society understands Earth's biogeography and conserves Earth's living legacy will determine the extent of our success as a species and the long-term survival of a habitable Earth.

In this chapter: We explore the methods by which plants use photosynthesis and respiration to translate solar energy into usable forms to energize life. We then examine relevant nonliving systems and important biogeochemical cycles, and look at the organization of living ecosystems into complex food chains and webs. We also examine communities and species interactions. Next we consider how the biodiversity of living organisms results from biological evolution over the past 3.6+ billion years. We conclude with a discussion of ecosystem stability and resilience, and how living landscapes change over space and time through the process of succession, now influenced by the effects of global climate change.

Energy Flows and Nutrient Cycles

By definition, an ecosystem includes both biotic and abiotic components. Chief among the abiotic components is the direct input of solar energy, on which nearly all ecosystems depend; the few limited ecosystems that exist in dark caves, in wells, or on the ocean floor depend on chemical reactions (chemosynthesis) for energy.

Ecosystems are divided into subsystems. The biotic tasks are performed by primary producers (plants, cyanobacteria, and some other unicellular organisms), consumers (animals), and detritus feeders and decomposers (worms, mites, bacteria, and fungi). The abiotic processes include

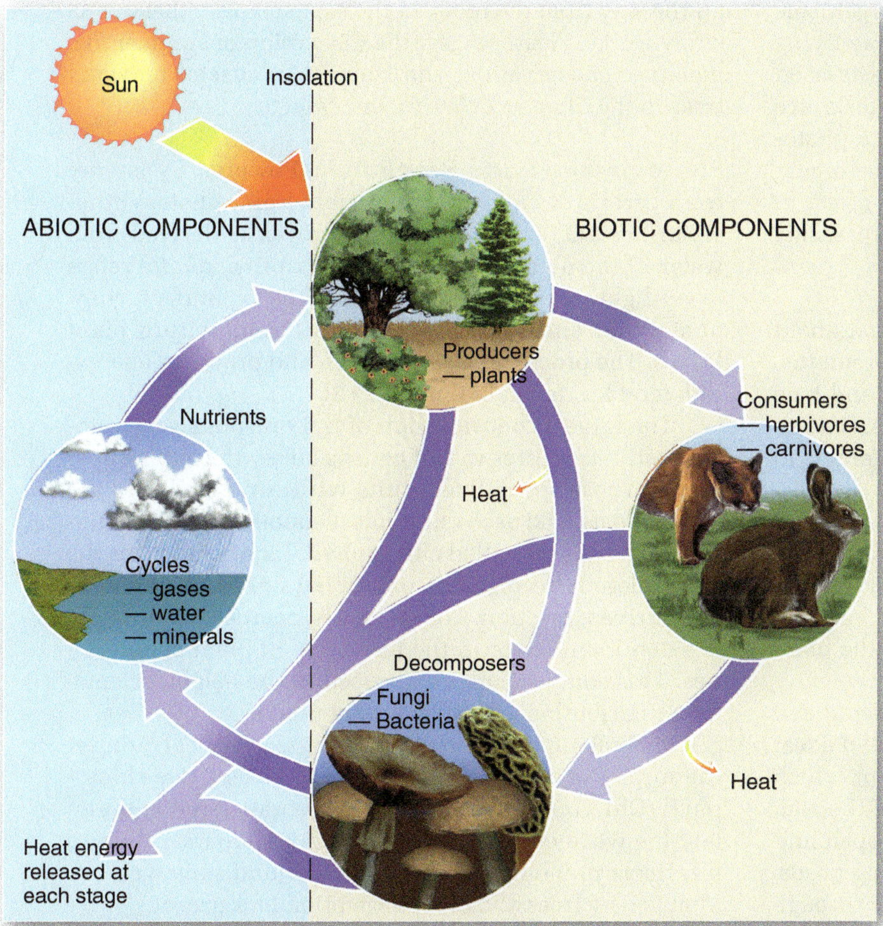

(a) Solar energy input drives biotic and abiotic ecosystem processes. Heat energy and biomass are the outputs from the biosphere.

(b) Biotic and abiotic ingredients operate together to form this rainforest floor ecosystem in Puerto Rico.

(c) Five or six species of lichen live in extreme Arctic climate conditions on Bear Island in the Barents Sea. Each little indentation in the rock provides some advantage to the lichen.

(d) Brain coral in the Caribbean Sea, at a 3-m (9.8-ft) depth, lives in a symbiotic relationship with algae.

▲**Figure 19.2 Biotic and abiotic components of ecosystems.** [Bobbé Christopherson.]

gaseous, hydrologic, and mineral cycles. Figure 19.2 illustrates these essential elements of an ecosystem and how they operate together.

Converting Energy to Biomass

The energy that powers the biosphere comes primarily from the Sun. Solar energy enters the ecosystem energy flow by way of photosynthesis; heat energy is dissipated from the system, as an output, at many points. Of the total energy intercepted at Earth's surface and available for work, only about 1.0% is actually fixed by photosynthesis as carbohydrates in plants, which then become the source of energy or the construction materials for the rest of the ecosystem. "Fixed" means chemically bound into plant tissues.

Plants (in terrestrial ecosystems) and algae (in aquatic ecosystems) are the critical biotic link between solar energy and the biosphere. Organisms that are capable of using the Sun's energy directly to produce their own food (using carbon dioxide (CO_2) as their sole source of carbon) are *autotrophs* (self-feeders), or **producers**. These include plants, algae, and cyanobacteria (a type of blue-green algae). Autotrophs accomplish this transformation of light energy into chemical energy by the process of photosynthesis, as previously mentioned. Ultimately, the fate of all members of the biosphere, including humans, rests on the success of these organisms and their ability to turn sunlight into food.

The oxygen gas in Earth's atmosphere was produced as a by-product of photosynthesis. The first photosynthesizing bacteria to produce oxygen appeared in oceans on Earth about 2.7 billion years ago. These *cyanobacteria*—microscopic, usually unicellular, blue-green algae that can form large colonies—were fundamental to the creation of Earth's modern atmosphere. These organisms were also critical in the origin of plants, since free-living cyanobacteria eventually became the chloroplasts used in plant photosynthesis. Although these bacteria are called blue-green algae (mainly because they are photosynthetic and aquatic), they are not related to other organisms we know as algae. True *algae* are a large group of single-celled or multi-celled photosynthetic organisms that range in size from microscopic diatoms (a type of phytoplankton) to giant sea kelp.

Land plants (and animals) became common about 430 million years ago, according to fossilized remains. **Vascular plants** developed conductive tissues and true roots for internal transport of fluid and nutrients. (*Vascular* is from a Latin word for "vessel-bearing," referring to the conducting cells.)

In plants, leaves are solar-powered chemical factories wherein photochemical reactions take place. Veins in the leaf bring in water and nutrient supplies and carry off the sugars (food) produced by photosynthesis. The veins in each leaf connect to the stems and branches of the plant and to the main circulation system.

Flows of CO_2, water, light, and oxygen enter and exit the surface of each leaf. Gases move into and out of a leaf through small pores, the **stomata** (singular: *stoma*), which usually are most numerous on the lower side of the leaf. Each stoma is surrounded by guard cells that open and close the pore, depending on the plant's changing needs. Water that moves through a plant exits the leaves through the stomata in the process of transpiration, thereby assisting the plant's temperature regulation. As water evaporates from the leaves, a pressure gradient is created that allows atmospheric pressure to push water up through the plant all the way from the roots, in the same manner that a soda straw works. Please review the discussion of summer fog, moisture conservation, and stomata in coastal redwood trees in the Chapter 7 *Geosystems Now.*

Photosynthesis and Respiration Powered by energy from certain wavelengths of visible light, **photosynthesis** unites CO_2 and hydrogen (hydrogen is derived from water in the plant). The term is descriptive: *photo-* refers to sunlight and *-synthesis* describes the "manufacturing" of starches and sugars through reactions within plant leaves. The process releases oxygen and produces energy-rich food for the plant (Figure 19.3).

The largest concentration of light-responsive, photosynthetic structures in leaf cells is below the leaf's upper layers. These specialized units within the cells are the *chloroplasts*, and each chloroplast contains a green, light-sensitive pigment called **chlorophyll.** Light stimulates the molecules of this pigment, producing a photochemical, or light-driven, reaction. Consequently, competition for light is a dominant factor in the formation of plant communities. This competition is expressed in the height, orientation, distribution, and structure of plants.

Only about one-quarter of the light energy arriving at the surface of a leaf is useful to the light-sensitive chlorophyll. Chlorophyll absorbs only the orange-red and violet-blue wavelengths for photochemical operations, and it reflects predominantly green hues (and some yellow). That is why trees and other vegetation look green.

Photosynthesis essentially follows this equation:

$$\underset{\text{(carbon dioxide)}}{6CO_2} + \underset{\text{(water)}}{6H_2O} + \underset{\text{(solar energy)}}{\text{Light}} \rightarrow \underset{\text{(glucose, carbohydrate)}}{C_6H_{12}O_6} + \underset{\text{(oxygen)}}{6O_2}$$

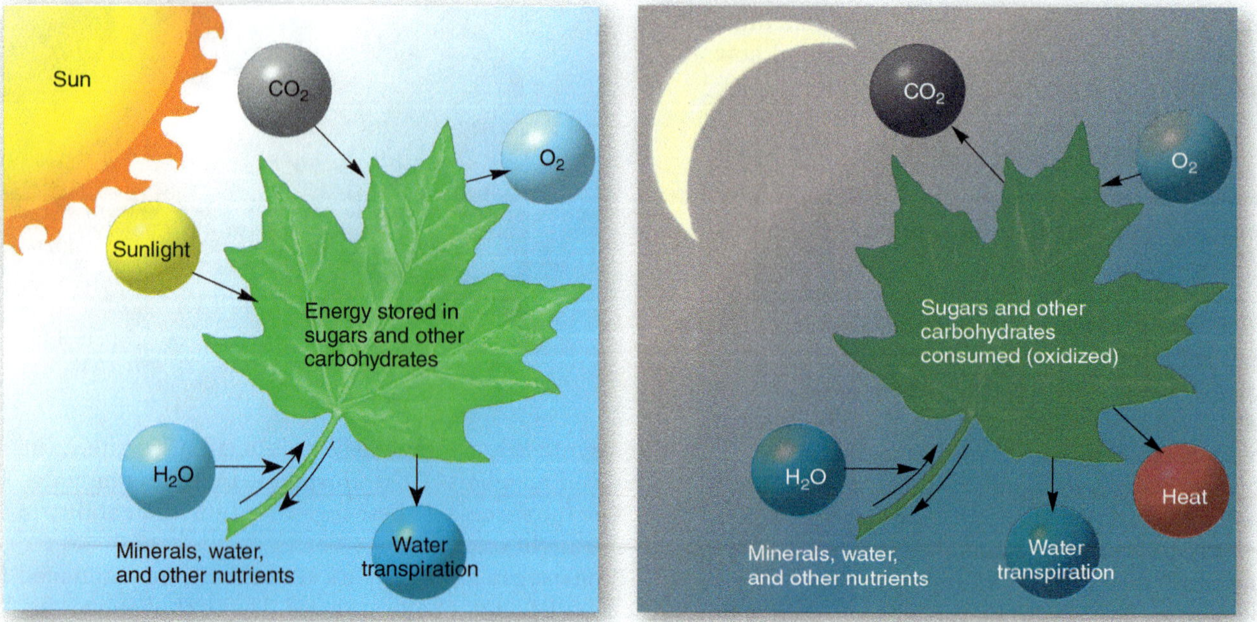

(a) Plant photosynthesis (b) Plant respiration

▲**Figure 19.3 Photosynthesis and respiration.** In the process of photosynthesis, plants consume light, carbon dioxide (CO_2), nutrients, and water (H_2O) and produce outputs of oxygen (O_2) and carbohydrates (sugars) as stored chemical energy. Plant respiration, illustrated here at night, approximately reverses this process. The balance between photosynthesis and respiration determines net photosynthesis and plant growth.

From the equation, you can see that photosynthesis removes carbon (in the form of CO_2) from Earth's atmosphere. The quantity is enormous: approximately 91 billion metric tons (100 billion tons) of CO_2 per year. Carbohydrates, the organic result of the photosynthetic process, are combinations of carbon, hydrogen, and oxygen. The simple sugar *glucose* ($C_6H_{12}O_6$) is an example. Plants use glucose to build starches, which are more-complex carbohydrates and the principal food stored in plants.

Plants store energy (in the bonds within carbohydrates) for later use. They consume this energy as needed through respiration, which converts the carbohydrates to energy for their other operations. Thus, **respiration** is essentially a reverse of the photosynthetic process:

$C_6H_{12}O_6$	+	$6O_2$	→	$6CO_2$	+	$6H_2O$	+	energy
(glucose, carbohydrate)		(oxygen)		(carbon dioxide)		(water)		(heat energy)

In respiration, plants oxidize carbohydrates (break them down through reaction with oxygen), releasing CO_2, water, and energy as heat. The difference between photosynthetic production of carbohydrates and respiration loss of carbohydrates is *net photosynthesis*. The overall growth of a plant depends on the amount of net photosynthesis, a surplus of carbohydrates beyond those lost through plant respiration.

The *compensation point* is the break-even point between the production and consumption of organic material. Each leaf must operate on the production side of the compensation point, or else the plant eliminates it—you may have observed a plant shedding leaves if it receives inadequate water or light.

The effects of today's rapidly increasing atmospheric CO_2 concentrations on plant photosynthesis and growth are under study at a number of Free-Air CO_2 Enrichment (FACE) facilities in the United States. This research shows that fumigating natural and agricultural ecosystems with elevated concentrations of CO_2 stimulates the photosynthetic processes, increasing plant growth, but also appears to have some complex and varied side effects. At the same time, the presence of ground-level ozone (O_3), a toxic gas (discussed in Chapter 3) that is increasing in the lower atmosphere, decreases plant growth, offsetting the effects of increased CO_2 when the two occur together in the atmosphere. (For more information, see http://aspenface.mtu.edu/ or http://climatechangescience.ornl.gov/content/free-air-co2-enrichment-face-experiment.)

Net Primary Productivity The net photosynthesis for an entire ecosystem is its **net primary productivity**. This is the amount of stored chemical energy that the ecosystem generates. The total organic matter (living and recently living, both animal and plant) in an ecosystem, with its associated chemical energy, is the ecosystem's **biomass** and is often measured as the net dry weight of all organic material. Net primary productivity is an important aspect of any type of ecosystem because it determines the biomass available for consumption by *heterotrophs*, or **consumers**—organisms that feed on others. The distribution of productivity over Earth's surface is an important aspect of biogeography.

Net primary productivity is measured as fixed carbon per square meter per year. The map in Figure 19.4 shows that on land, net primary production tends to be highest between the Tropics of Cancer and Capricorn at sea level and decreases toward higher latitudes and elevations. Productivity levels are tied to both sunlight and precipitation, as evidenced by the correlations of abundant precipitation

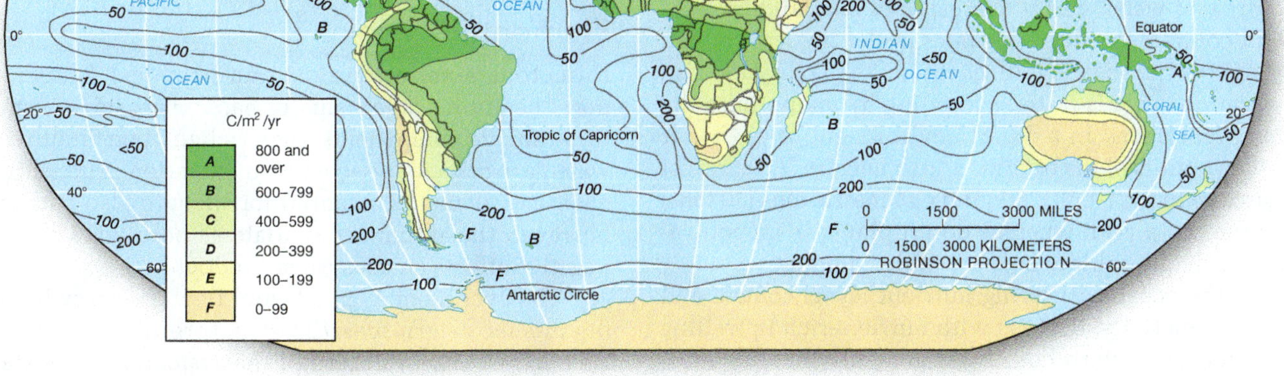

▲**Figure 19.4 Net primary productivity.** Worldwide net primary productivity in grams of carbon per square meter per year (approximate values). [Adapted from D. E. Reichle, *Analysis of Temperate Forest Ecosystems* (Heidelberg, Germany: Springer, 1970).]

TABLE 19.1 Net Primary Productivity and Plant Biomass on Earth

		Net Productivity Area ($g/m^2/yr$)[b]	Primary per Unit	
Ecosystem	**Area (10^6 km^2)[a]**	**Normal Range**	**Mean**	**World Net Biomass (10^9 metric tons/yr)[c]**
Tropical rain forest	17.0	1000–3500	2200	37.4
Tropical seasonal forest	7.5	1000–2500	1600	12.0
Temperate evergreen forest	5.0	600–2500	1300	6.5
Temperate deciduous forest	7.0	600–2500	1200	8.4
Boreal forest	12.0	400–2000	800	9.6
Woodland and shrubland	8.5	250–1200	700	6.0
Savanna	15.0	200–2000	900	13.5
Temperate grassland	9.0	200–1500	600	5.4
Tundra and alpine region	8.0	10–400	140	1.1
Desert and semidesert scrub	18.0	10–250	90	1.6
Extreme desert, rock, sand, ice	24.0	0–10	3	0.07
Cultivated land	*14.0*	*100–3500*	*650*	*9.1*
Swamp and marsh	2.0	800–3500	2000	4.0
Lake and stream	2.0	100–1500	250	0.5
Total continental	**149.0**	—	**773**	**115.17**
Open ocean	332.0	2–400	125	41.5
Upwelling zones	0.4	400–1000	500	0.2
Continental shelf	26.6	200–600	360	9.6
Algal beds and reefs	0.6	500–4000	2500	1.6
Estuaries	1.4	200–3500	1500	2.1
Total marine	**361.0**	—	**152**	**55.0**
Grand total	**510.0**	—	**333**	**170.17**

Source: From Whittaker, Robert C., *Communities and Ecosystems* 2nd Ed, © 1975, p. 224. Reprinted and electronically reproduced by permission of Pearson Education, Inc., Upper Saddle River, New Jersey.

[a]1 km^2 = 0.38 mi^2.

[b]1 g/m^2 = 8.92 lb/acre.

[c]1 metric ton (10^6 g) = 1.1023 tons.

with high productivity adjacent to the equator and reduced precipitation with low productivity in the subtropical deserts. Even though deserts receive high amounts of solar radiation, water availability and other controlling factors, such as soil conditions, limit productivity.

In temperate and high latitudes, the rate at which carbon is fixed by vegetation varies seasonally. It increases in spring and summer as plants flourish with increasing solar input and, in some areas, with more available (nonfrozen) water, and it decreases in late fall and winter. In contrast, productivity rates in the tropics are high throughout the year, and turnover in the photosynthesis–respiration cycle is faster, exceeding by many times the rates experienced in a desert environment or in the far northern limits of the tundra. A lush hectare (2.5 acres) of sugarcane in the tropics might fix 45 metric tons (50 tons) of carbon in a year, whereas desert plants in an equivalent area might achieve only 1% of that amount.

In the oceans, differing nutrient levels control and limit productivity. Regions with nutrient-rich upwelling currents off western coastlines generally are the most productive. Figure 19.4 shows that the tropical oceans and areas of subtropical high pressure are quite low in productivity.

Table 19.1 lists the net primary productivity of various ecosystems and provides an estimate of net total biomass worldwide—170 billion metric tons (187 billion tons) of dry organic matter per year. Compare the productivity of the various ecosystems in the table, noting especially how the productivity of cultivated land compares with that of natural communities.

Elemental Cycles

Numerous abiotic physical and chemical factors support the living organisms of each ecosystem. Some of these abiotic components, such as light, temperature, and water, are critical for ecosystem operation. Nutrients (the chemical elements essential for life) are also necessary; however, those that accumulate as pollutants can have negative effects on ecosystem function.

The cycling of nutrients and flow of energy between organisms determines the structure of an ecosystem. As energy cascades through the system, it is constantly replenished by the Sun. But nutrients and minerals

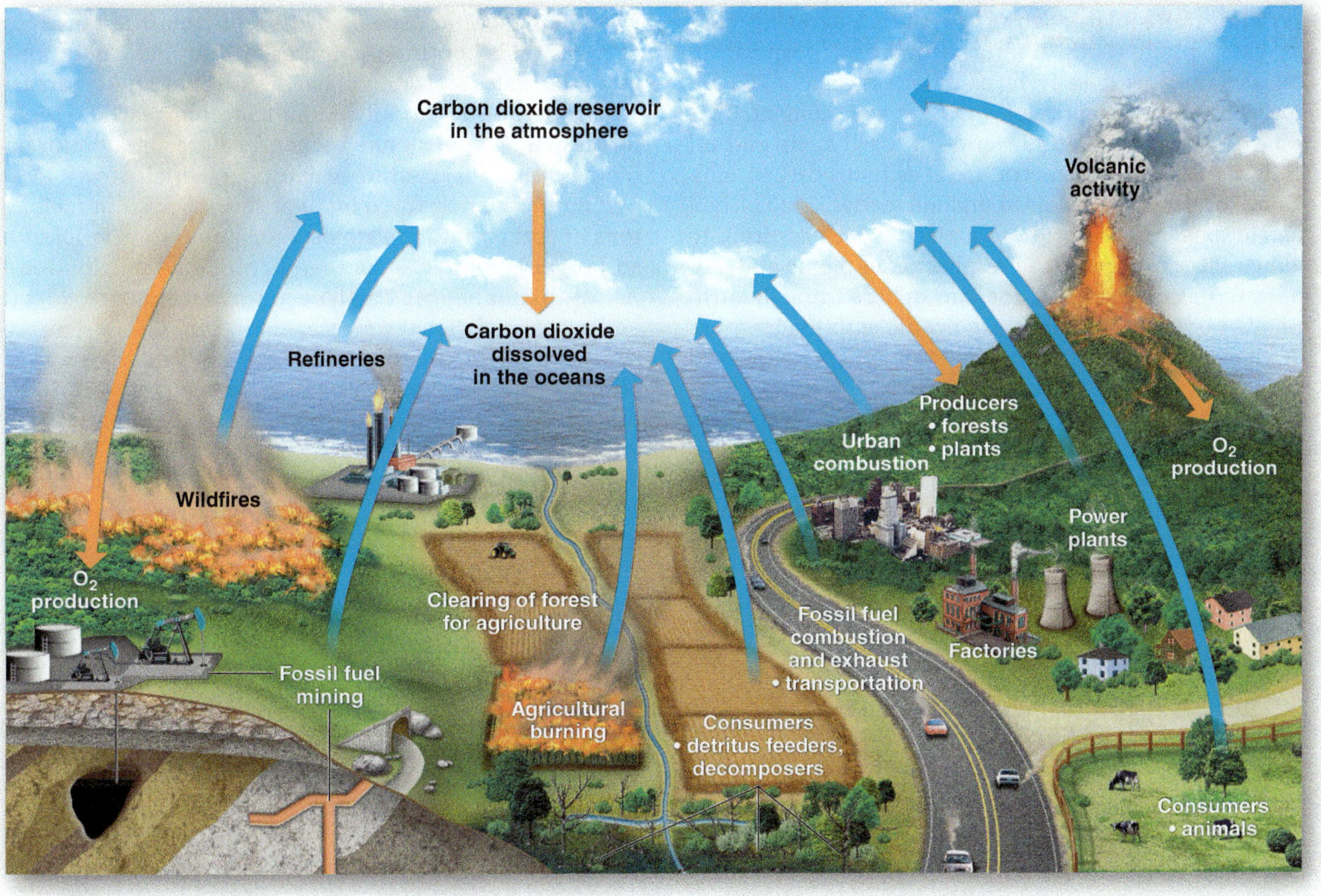

▲**Figure 19.5 The carbon and oxygen cycles.** Carbon is fixed (orange arrows) through photosynthesis, with oxygen as a by-product. Respiration by living organisms, the burning of forests and grasslands, and the combustion of fossil fuels release carbon to the atmosphere (blue arrows). These cycles are greatly influenced by human activities.

cannot be replenished from an external source, so they constantly cycle within each ecosystem and through the biosphere.

The most abundant natural elements in living matter are hydrogen (H), oxygen (O), and carbon (C). Together, these elements make up more than 99% of Earth's biomass; in fact, all life (being composed of organic molecules) contains hydrogen and carbon. In addition, nitrogen (N), calcium (Ca), potassium (K), magnesium (Mg), sulfur (S), and phosphorus (P) are significant nutrients, elements necessary for the growth of a living organism.

These key elements flow through the natural world in various chemical cycles. Oxygen, carbon, and nitrogen each have *gaseous cycles*, parts of which take place in the atmosphere. Other major elements, including phosphorus, calcium, potassium, and sulfur, have *sedimentary cycles*, which principally involve mineral and solid phases. Some elements cycle through both gaseous and sedimentary stages. The recycling of gases and nutrient sedimentary materials forms Earth's **biogeochemical cycles**, so called because they involve chemical reactions necessary for growth and development of living systems. The chemical elements themselves recycle over and over again in life processes.

Oxygen and Carbon Cycles

We consider the oxygen and carbon cycles together because they are so closely intertwined through photosynthesis and respiration (Figure 19.5). The atmosphere is the principal reservoir of available oxygen. Larger reserves of oxygen exist in Earth's crust, but they are unavailable, being chemically bound with other elements, especially the silicate (SiO_2) and carbonate (CO_3) mineral families. Unoxidized reserves of fossil fuels and sediments also contain oxygen.

The oceans are enormous pools of carbon—about 42,900 billion metric tons, or 47,190 billion tons (metric tons times 1.10 equals short tons). However, all of this carbon is bound chemically in CO_2, calcium carbonate, and other compounds. The ocean initially absorbs CO_2 by means of the photosynthesis carried on by phytoplankton; it becomes part of the living organisms and through them is fixed in certain carbonate minerals, such as limestone ($CaCO_3$). The ocean water can also absorb CO_2 directly from the atmosphere. In Chapter 16, we discussed ocean acidification, the lowering of ocean pH caused by excessive absorption of atmospheric CO_2 into the water. This condition of acidity makes it harder for plankton, corals, and other organisms to maintain calcium carbonate skeletons.

The atmosphere, which is the integrating link between photosynthesis (fixation) and respiration (release) in the carbon cycle, contains only about 700 billion metric tons of carbon (as CO_2) at any moment. This is far less carbon than is stored in fossil fuels and oil shales (13,200 billion metric tons, as hydrocarbon molecules) or in living and dead organic matter (2500 billion metric tons, as carbohydrate molecules). In addition to being released through the respiration of plants and animals, CO_2 is released into the atmosphere through burning of grasslands and forests, volcanic activity, land-use changes, and fossil-fuel combustion by industry and transportation.

The carbon released into the atmosphere by human activity constitutes a vast, real-time geochemical experiment, using the Earth's actual and only atmosphere as a laboratory. Annually, we are adding carbon to the atmospheric pool in an amount 400% greater than we were in 1950. Global emissions of carbon from the burning of fossil fuels continue to increase. The removal of forests eliminates a portion of Earth's carbon "sinks" (areas where carbon is stored). Land-use changes in which forest (which stores more carbon in a greater amount of biomass) is converted to agriculture (which stores less carbon) is causing the release of several million metric tons of carbon into the atmosphere annually; the loss of soil organic matter is another significant contribution. All of this enhances Earth's natural greenhouse effect and associated global warming.

Nitrogen Cycle Nitrogen, which accounts for 78.084% of each breath we take, is the major constituent of the atmosphere. Nitrogen also is important in the makeup of organic molecules, especially proteins, and therefore is essential to living processes. A simplified view of the nitrogen cycle is portrayed in Figure 19.6.

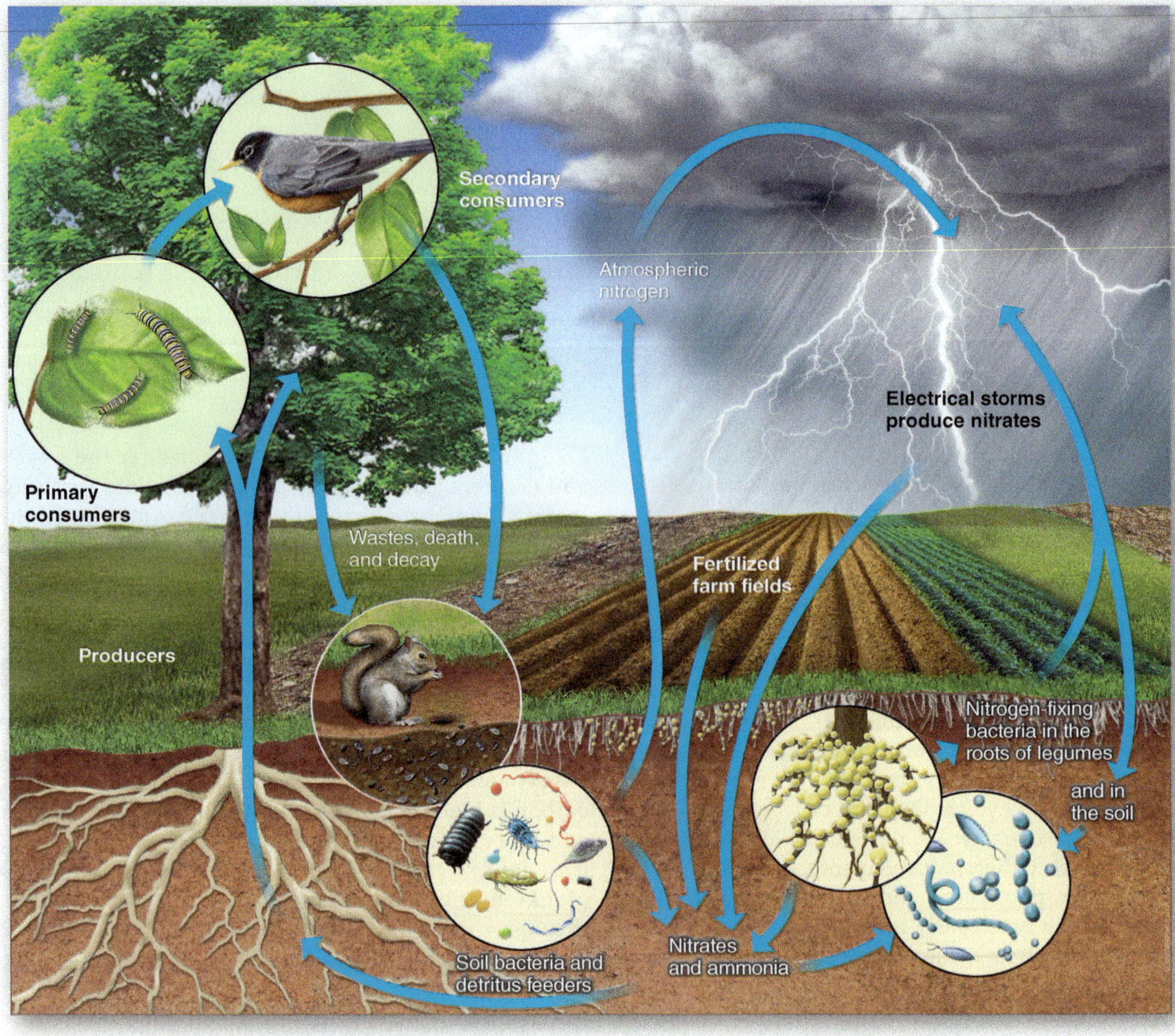

▲Figure 19.6 **The nitrogen cycle.** The atmosphere is the reservoir of gaseous nitrogen. Atmospheric nitrogen gas is chemically fixed by bacteria in producing ammonia. Lightning and forest fires produce nitrates, and fossil-fuel combustion forms nitrogen compounds that are washed from the atmosphere by precipitation. Plants absorb nitrogen compounds and incorporate the nitrogen into organic material.

Nitrogen-fixing bacteria, which live principally in the soil and are associated with the roots of certain plants, are critical for bringing atmospheric nitrogen into living organisms. Colonies of these bacteria reside in nodules on the roots of legumes (plants such as clover, alfalfa, soybeans, peas, beans, and peanuts) and chemically combine the nitrogen from the air into nitrates (NO_3) and ammonia (NH_3). Plants use the nitrogen from these molecules to produce their own organic matter. Anyone or anything feeding on the plants thus ingests the nitrogen. Finally, the nitrogen in the organic wastes of the consuming organisms is freed by denitrifying bacteria, which recycle it to the atmosphere.

To improve agricultural yields, many farmers enhance the available nitrogen in the soil by means of synthetic inorganic fertilizers, as opposed to soil-building organic fertilizers (manure and compost). Inorganic fertilizers are chemically produced through artificial nitrogen fixation at factories. Humans presently fix more nitrogen as synthetic fertilizer per year than is found in all terrestrial sources combined—and the present production of synthetic fertilizers now is doubling every 8 years; some 1.82 million metric tons (2 million tons) are produced per week, worldwide. The crossover point at which anthropogenic sources of fixed nitrogen exceeded the normal range of naturally fixed nitrogen occurred in 1970.

This surplus of usable nitrogen accumulates in Earth's ecosystems. Some is present as excess nutrients, washed from soil into waterways and eventually to the ocean. This excess nitrogen load begins a water pollution process that feeds an excessive growth of algae and phytoplankton, increases biochemical oxygen demand, diminishes dissolved oxygen reserves, and eventually disrupts the aquatic ecosystem. In addition, excess nitrogen compounds in air pollution are a component in acid deposition, further altering the nitrogen cycle in soils and waterways.

Dead Zones The Mississippi River receives runoff from 41% of the area of the continental United States. It carries agricultural fertilizers, farm sewage, and other nitrogen-rich wastes to the Gulf, causing huge spring blooms of phytoplankton: an explosion of primary productivity. By summer, the biological oxygen demand of bacteria feeding on the decay of the spring bloom exceeds the dissolved oxygen content of the water; hypoxia (oxygen depletion) develops, killing any fish that venture into the area. These low-oxygen conditions create **dead zones** that limit marine life. *Geosystems In Action*, Figure GIA 19, illustrates dead zones in the Gulf of Mexico and elsewhere. From 2002 on, the Gulf Coast dead zone has expanded to an area of more than 22,000 km^2 (8500 mi^2) each year. The agricultural, feedlot, and fertilizer industries dispute the connection between their nutrient input and this extensive dead zone. The human-caused creation of dead zones in water bodies is cultural eutrophication, discussed later in the chapter.

Similar coastal dead zones occur as the result of nutrient outflows from more than 400 river systems worldwide, affecting almost 250,000 km^2 (94,600 mi^2) of offshore oceans and seas (Figure GIA 19.2). In Sweden and Denmark, however, a concerted effort to reduce nutrient flows into rivers reversed hypoxic conditions in the Kattegat strait (between the Baltic and North Seas). Also, fertilizer use has decreased more than 50% in the former Soviet Republics since the fall of state agriculture in 1990. The Black Sea no longer undergoes year-round hypoxia at river deltas, as the dead zones in those areas now disappear for several months each year.

Dead zones are occurring in lakes as well, such as those that appeared in Lake Erie, one of the Great Lakes, in the 1960s. In 2011, the dead zone in this lake reached its largest extent in recorded history, caused by fertilizers (mainly phosphorus) flowing into the lake combined with slower natural mixing attributed to climate change (see Figure GIA 19.6 and the discussion in Focus Study 19.2).

As with most environmental pollution, the cost of mitigation is cheaper than the cost of continued damage to marine ecosystems. Experts estimate that a 20%–30% cut in nitrogen inflow upstream would increase dissolved oxygen levels by more than 50% in the dead zone region of the Gulf. One government study estimated the level of application of nitrogen fertilizer to be 20% more than soils and plants needed in Iowa, Illinois, and Indiana. The initial step to resolving this issue might be to mandate applying only the levels of fertilizer needed—thus also reaping a savings in overhead costs of agriculture—and, as a second step, to begin dealing with animal wastes from feedlots.

Energy Pathways

The feeding relationships among organisms make up the energy pathways in an ecosystem. These *trophic relationships*, or feeding levels, consist of food chains and food

(text continued on page 596)

GEOreport 19.1 Carbon cycle response to the Mount Pinatubo eruption

One month after the 1991 Mount Pinatubo eruption, the second largest volcanic eruption of the 20th century, temperatures decreased in the Northern Hemisphere and global carbon dioxide levels declined sharply. Scientists initially thought that the CO_2 decrease was caused by a decline in plant respiration linked to cooling temperatures. However, research now suggests that the globally spread atmospheric aerosols from the eruption caused an increase in diffuse light, allowing sunlight to reach more plant leaves (as opposed to direct sunlight that creates shadows). This change increased plant photosynthesis and removed more CO_2 from the air. In one deciduous forest, photosynthesis increased by 23% in 1992 and 8% in 1993 under cloudless conditions. Thus, the eruption's aerosols affected the global carbon cycle, lowering atmospheric carbon levels and enhancing the terrestrial carbon sink. (Please review other effects of the eruption in Chapter 1, Figure 1.11.)

Coastal ocean waters are often highly productive ecosystems teeming with marine life. Yet they can become *dead zones* where organisms die because of low-oxygen conditions (GIA 19.1). Dead zones result from a process that begins with agricultural runoff of fertilizers and farm-animal wastes (GIA 19.2, 19.3, and 19.4). The size of dead zones can vary from year to year (GIA 19.5). Freshwater lakes are also subject to "blooms" of the algae that form dead zones (GIA 19.6).

19.1

FORMATION OF A DEAD ZONE

In the water's surface layer, agricultural runoff delivers nitrogen and phosphorus, nutrients that greatly boost the growth of algae, producing an algal bloom. When the algae die, they sink into the bottom layer. Bacteria feed on the dead algae and deplete the water of oxygen, forming a dead zone. Marine organisms that cannot leave the dead zone will die.

Explain: How does the water in a dead zone become depleted of oxygen?

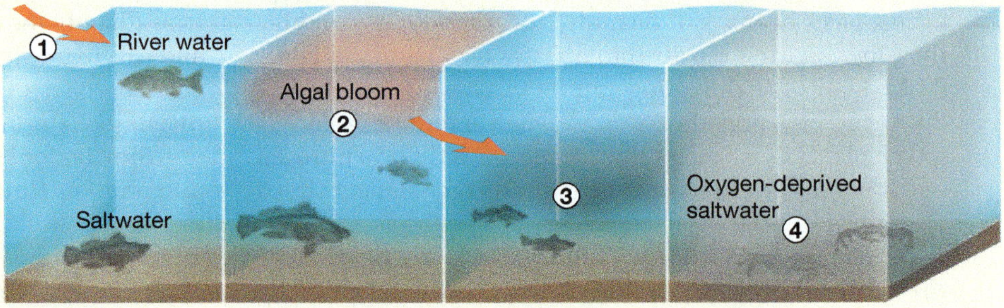

① *Agricultural runoff enters rivers, which then moves downstream to the ocean or to a lake.*

② *The nutrients, mainly nitrogen and phosphorus, cause algal blooms.*

③ *The algae die, sink into the bottom layer, and are decomposed by bacteria, using up the oxygen in the water.*

④ *A dead zone (defined as water with less than 2 mg/l dissolved oxygen) is formed, killing organisms that cannot flee.*

19.2

DEAD ZONES: A GLOBAL PROBLEM

Generally, dead zones correlate with large human populations, mouths of rivers that deliver nitrogen-rich pollutants, and semi-enclosed bodies of water such as seas, bays, and estuaries. [NASA.]

Reading the map: The amount of organic matter in the ocean surface layer is shown in shades of blue. Dead zones, which form where organic matter is high, are shown as red circles.

Infer: What do you think explains the large number of dead zones along the east coast of the United States?

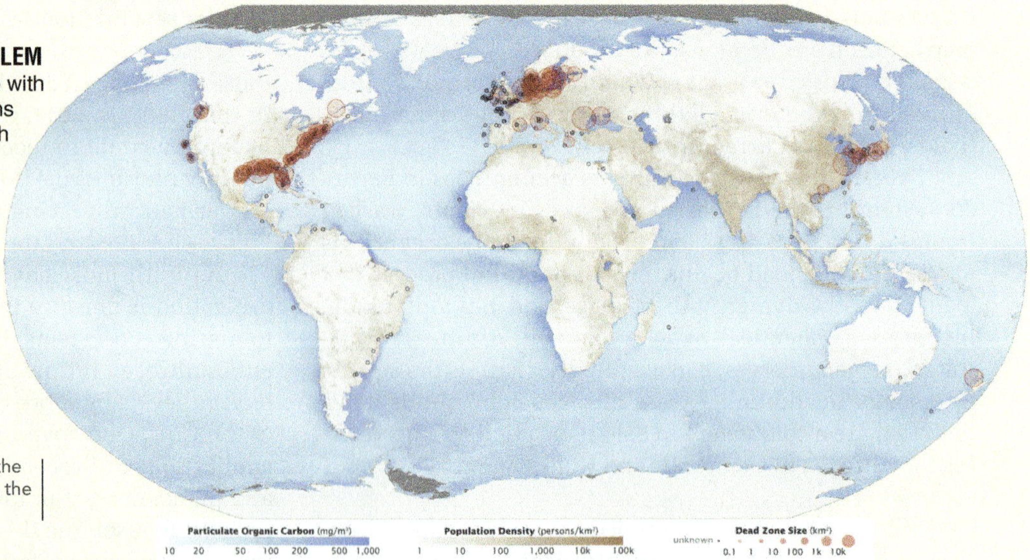

19.3

VIEWS OF DEAD ZONES

Satellite images reveal the effects of nutrient-rich water reaching the ocean (GIA 19.3a and 19.3b). Dead zone formation is seasonal: algal blooms occur in the spring, and dead zones can last through the summer. [NASA.]

19.3a

Gulf of Mexico dead zone: Channels in the Mississippi River's delta bring nutrients that fuel the green algal blooms seen in this image and help to form the Gulf's large, and expanding, dead zone.

19.3b

Baltic Sea dead zone: Swirling, green plumes of algae are visible in the Baltic Sea between Sweden and Latvia. Countries around the sea have reduced nitrogen fertilizer runoff and the duration of the dead zone.

Video
The Ocean's Green Machines

MasteringGeography™

Visit the Study Area in MasteringGeography™ to explore dead zones.

Visualize: Study the video of phytoplankton.

Assess: Demonstrate understanding of dead zones (if assigned by instructor).

19.4 NITROGEN SOURCE FOR THE GULF OF MEXICO DEAD ZONE

Agricultural runoff from the Mississippi River watershed provides the nitrogen for the Gulf of Mexico dead zone (GIA 19.4a). In 2013, the dead zone extended along much of the Louisiana coast (GIA 19.4b).

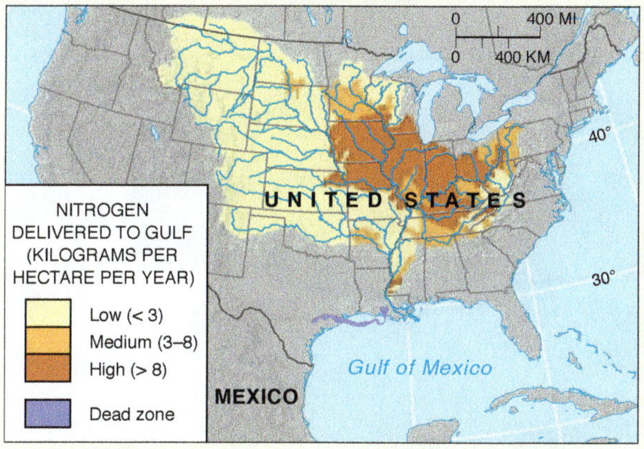

[Adapted from R.B. Alexander, R.A. Smith, and G.E. Schwatz, 2000, "Effect of stream channel size on the delivery of nitrogen to the Gulf of Mexico," Nature 403: 761.]

19.4a *The map shows total nitrogen in the upstream portion of the Mississippi River watershed.*

Bottom-water dissolved oxygen across the Louisiana shelf from July 22-28, 2013

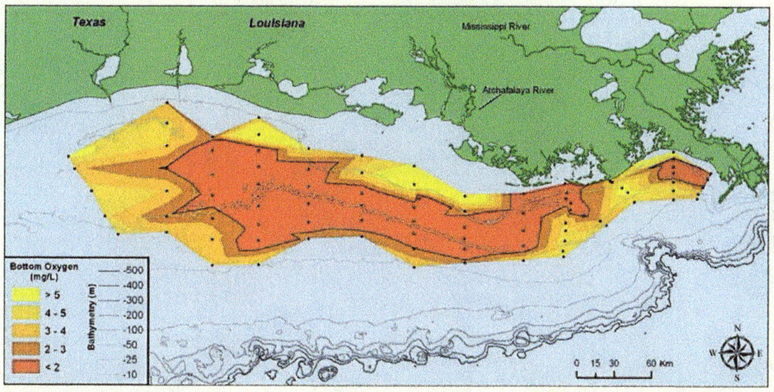

[Data source: N.N. Rabafais, Louisiana Universities Marine Consortium, R.E. Turner, Louisiana State University. Funded by: NOAA, Center for Sponsored Coastal Ocean Research.]

19.4b *Nutrients from the Mississippi River enrich the offshore waters of the Gulf, causing huge algal blooms in early spring that later form the dead zone (red areas).*

19.5 ANNUAL VARIATIONS IN THE GULF OF MEXICO DEAD ZONE

The graph shows annual variations in the size of the dead zone, (or "hypoxic area"—hypoxic means "low oxygen"), reflecting the input of nitrogen from the Mississippi River watershed. Drought reduces the size of the dead zone (by decreasing runoff), while flooding increases it. [NOAA.]

Calculate: How many times larger is the largest dead zone on the graph than the smallest?

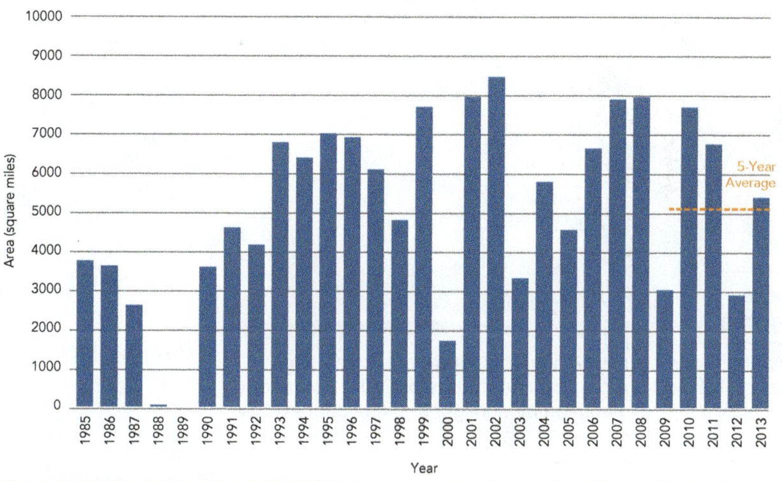

19.6 LAKE ERIE ALGAL BLOOM, 2011

The bright green areas in the lake are an algal bloom (GIA 19.6a and 19.6b). The 2011 bloom was 2.5 times larger than any previously observed on Lake Erie. [*Aqua* MODIS, NASA.]

Explain: What must happen to transform the phytoplankton bloom in the satellite image into a dead zone?

19.6a

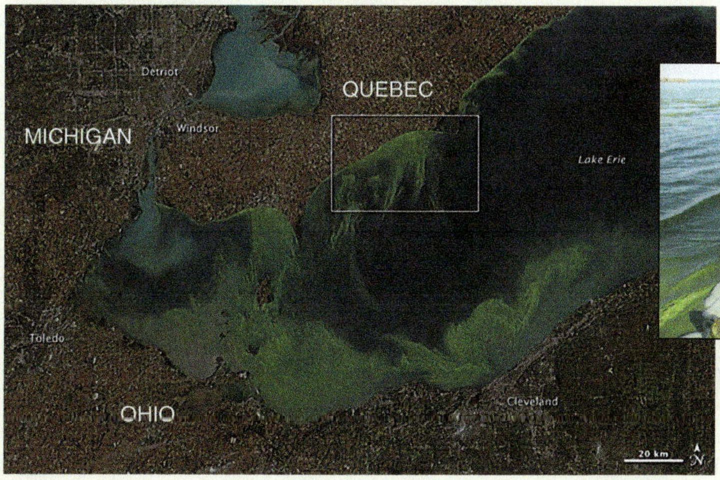

19.6b Algae deposits along the shore of Lake Erie during the 2011 bloom. [Brenda Culler/ODNR Coastal Management.]

GEOquiz

1. Interpret Maps: Look at the map in GIA 19.2. Why do you think the dead zones in the Southern Hemisphere are so much smaller and fewer in number than those in the Northern Hemisphere?

2. Solve Problems: Suppose you are a member of a commission charged with developing a plan to reduce the size of the Gulf of Mexico dead zone. Describe strategies you would suggest and the changes needed to achieve this goal.

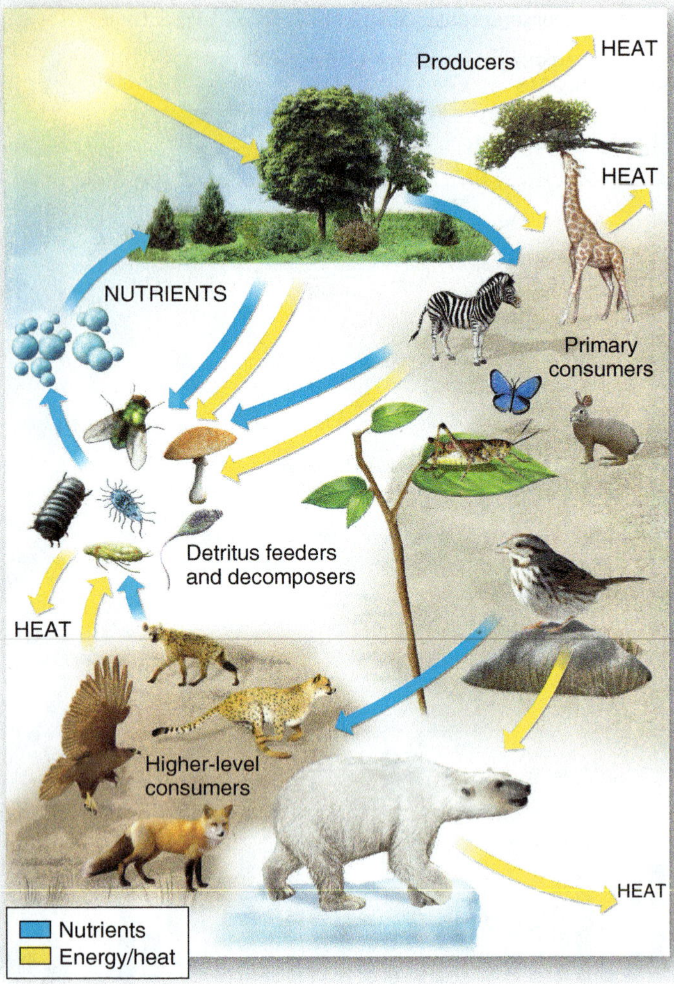

(a) The flow of energy, cycling of nutrients, and trophic (feeding) relationships portrayed for a generalized ecosystem. The operation is fueled by radiant energy supplied by sunlight and first captured by the plants.

(b) A bearded seal on a bergy bit in the Arctic Ocean.

(c) A solitary male polar bear drags its prey, a seal, across pack ice.

(d) A mother polar bear and cubs eat a seal on an iceberg. Glaucous Gulls and Ivory Gulls consume part of the leftovers.

▲**Figure 19.7 Energy, nutrient, and food pathways in the environment.** [Photos by Bobbé Christopherson.]

webs that range from simple to complex. As discussed earlier, autotrophs are the producers. Organisms that depend on producers as their carbon source are heterotrophs, or consumers, and are generally animals.

Trophic Relationships The producers in an ecosystem capture sunlight and convert it to chemical energy, incorporating carbon, forming new plant tissue and biomass, and freeing oxygen. From the producers, which manufacture their own food, energy flows through the system along an idealized unidirectional pathway called a **food chain**. Solar energy enters each food chain through the producers, either plants or phytoplankton, and subsequently flows to higher and higher levels of consumers. Organisms that share the same feeding level in a food chain are said to be at the same *trophic level*. Food chains usually have between three and six levels, beginning with primary producers and ending with *detritivores*, which break down organic matter and are the final link in the theoretical chain (Figure 19.7).

The actual trophic relationships between species in an ecosystem are usually more complex than the simple food chain model might suggest. The more common arrangement of feeding relationships is a **food web**, a complex network of interconnected food chains with multidirectional branches. In a food web, consumers often participate in several different food chains.

Nutrient cycling is continuous within a food web, aided by **detritivores** (also known as detritus feeders), the organisms that feed on *detritus*—dead organic debris (dead bodies, fallen leaves, and waste products) produced

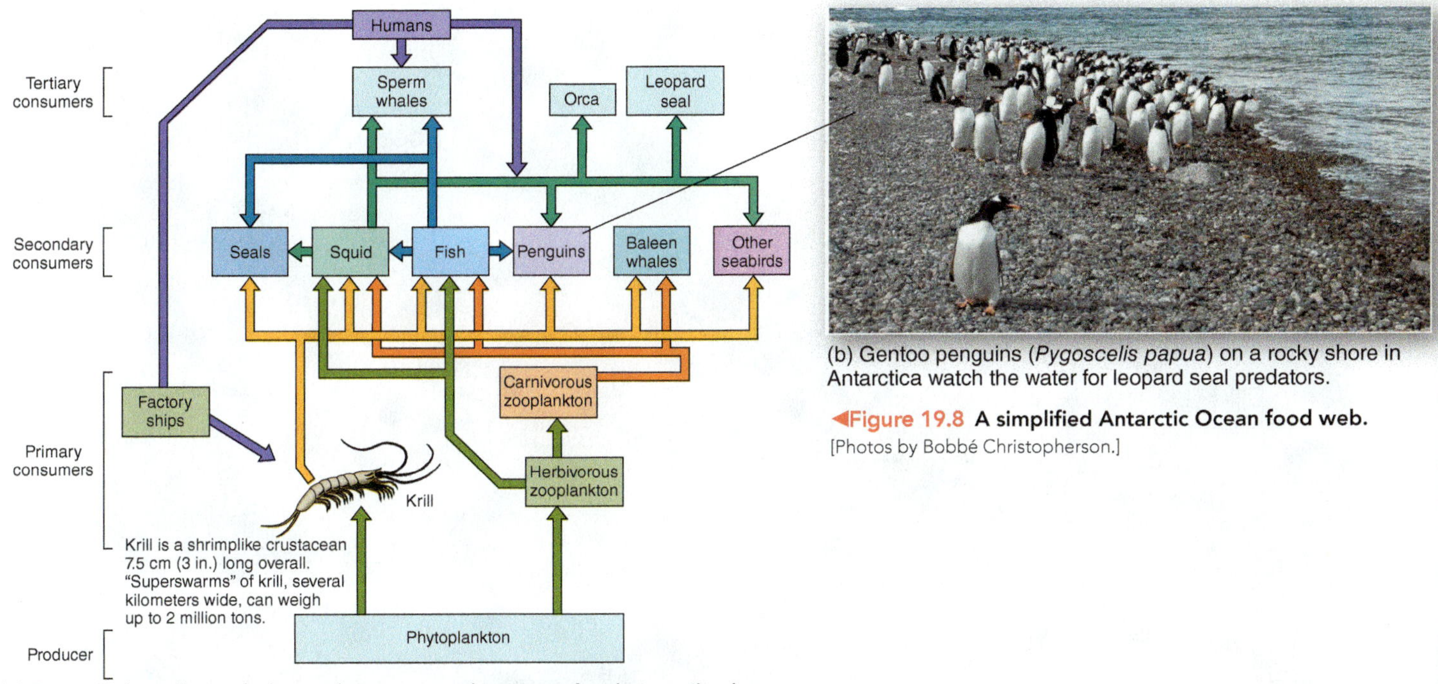

(a) Phytoplankton (bottom), the producers, use solar energy for photosynthesis. Krill and other herbivorous zooplankton eat the phytoplankton. Krill, in turn, are consumed by organisms at the next trophic level.

(b) Gentoo penguins (*Pygoscelis papua*) on a rocky shore in Antarctica watch the water for leopard seal predators.

◀Figure 19.8 A simplified Antarctic Ocean food web. [Photos by Bobbé Christopherson.]

by living organisms. Detritivores include worms, mites, termites, centipedes, snails, slugs, and crabs in terrestrial environments and bottom feeders in marine environments. These organisms renew the entire system by breaking down these organic materials and releasing simple inorganic compounds and nutrients. **Decomposers** are primarily bacteria and fungi that digest organic debris outside their bodies and absorb and release nutrients in the process. The metabolic work of microbial decomposers produces the "rotting" action that breaks down detritus. Detritus feeders and decomposers, although operating differently, have a similar function in an ecosystem.

In a food web, the organisms that feed on producers are *primary consumers*. Because producers are always plants, the primary consumer is a **herbivore**, or plant eater. A *secondary consumer* mainly eats primary consumers (herbivores) and is therefore a **carnivore**. A *tertiary consumer* eats primary and secondary consumers and is referred to as the "top carnivore" in the food chain; examples are the polar bear in the Arctic and the leopard seal and orca in Antarctica (discussed just ahead). Orca, an oceanic dolphin, feeds on fish, seals, penguins, and other whales and occurs in both Arctic and Antarctic waters. A consumer that feeds on both producers (plants) and consumers (meat) is an **omnivore**—a category occupied by humans, among other animals.

Several examples illustrate food webs. In the Arctic waters, bearded seals (*Erignathus barbatus*), with body fat often exceeding 30%, are consumers that feed on fish and clams (Figure 19.7b). In turn, the seal is preyed on by the polar bear (*Ursus maritimus*), another marine mammal and the dominant Arctic predator. Polar bears consume most of the seal except for the bones and intestines, which are quickly eaten by scavenging birds (Figure 19.7c and d). Figure 19.24 at the end of the chapter shows several large Arctic mammals, and discusses the effect of sea-ice losses on Arctic food webs.

In the Antarctic region, the food web begins with phytoplankton, the microscopic algae that harvest solar energy in photosynthesis (Figure 19.8). Herbivorous zooplankton, such as the shrimplike crustacean called krill (*Euphausia*), eat phytoplankton and are thus the primary consumers. Secondary consumers such as whales, fish, seabirds, seals, and squid eat krill, forming the next trophic level. Many Antarctic-dwelling seabirds depend on krill and on fish that eat krill. All of these organisms participate in other food chains, some as consumers and some being consumed.

Figure 19.9 shows an example of a temperate forest food web in eastern North America. Like the previous diagrams, the figure is simplified compared to the actual complexity of nature.

Energy Pyramids The overall amount of energy moving through trophic levels decreases from lower to higher levels, a pattern that can be illustrated in an *energy pyramid* in which horizontal bars represent each trophic level. (Ecological pyramids also include *biomass pyramids*, discussed ahead.)

At the bottom of the pyramid are the producers, which have the most energy and usually (but not always) the highest biomass and numbers of organisms. The next level, primary consumers, represents less energy because energy is used up (by metabolism and given off as heat) as one organism eats another. Energy decreases again at the

▲**Figure 19.9 Temperate forest food web.** Based on the text discussion, can you find the primary producers and then locate the primary, secondary, and tertiary consumers in this web? What role do the earthworms and bacteria play?

next level (secondary consumers), with each trophic level having (usually) less biomass and fewer organisms than the one beneath (Figure 19.10a). Although the pattern generally holds for numbers of organisms and biomass, exceptions exist and pyramids can be inverted, such as when the number of large trees (producers) is less than the number of small insects (primary consumers) or when the biomass of phytoplankton (which have short life spans) is less than that of the zooplankton that eat them. Only energy consistently decreases between lower and higher trophic levels, maintaining the true pyramid shape.

Food Web Efficiency In terms of energy, only about 10% of the kilocalories (food calories, not heat calories) in plant matter are passed from primary producers to primary consumers. In turn, only about 10% of the energy for primary consumers is passed to secondary consumers, and so on. Thus, the most efficient consumption of resources happens at the bottom of the food chain where plant biomass is higher and the energy input toward food production is lowest.

This concept applies to human eating habits and, on a broader scale, to world food resources. If humans take the role of herbivores, or primary consumers, they eat food with the highest energy available in the food chain. If humans take the role of carnivores, or secondary consumers, they eat food in which the available energy has been cut by 90% (the grain is fed first to cattle, then the cattle are consumed by humans). In terms of biomass, 810 kg of grain is reduced to 82 kg of meat. In terms of the numbers of organisms, if 1000 people can be fed as primary consumers, only 100 people can be fed as secondary consumers. By the latter analysis, far more people can be fed from the same land area producing grain than producing meat (Figure 19.10b).

Today, approximately half of the cultivated acreage in the United States and Canada is planted for animal consumption—beef and dairy cattle, hogs, chickens, and turkeys. Much of U.S. grain production goes to livestock

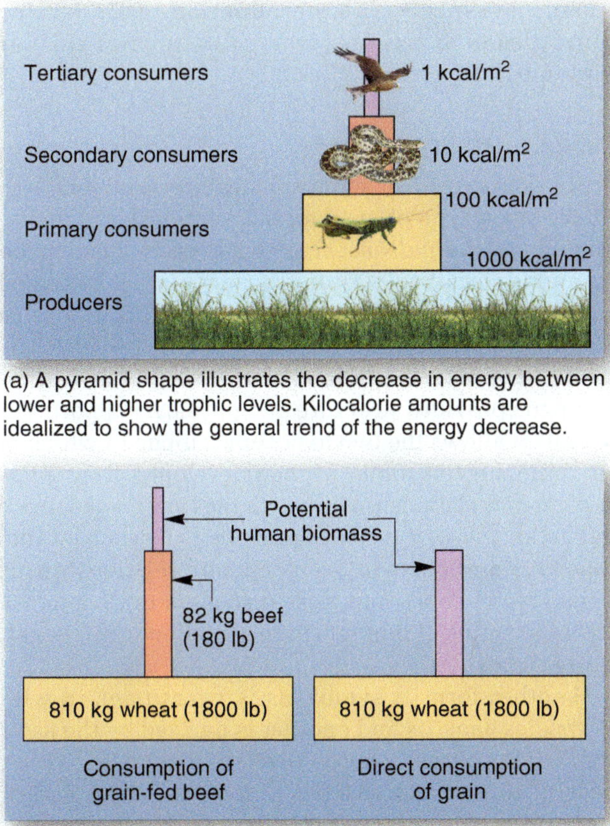

(a) A pyramid shape illustrates the decrease in energy between lower and higher trophic levels. Kilocalorie amounts are idealized to show the general trend of the energy decrease.

(b) Biomass pyramids illustrate the difference in efficiency between direct and indirect consumption of grain.

▲Figure 19.10 **Energy pyramids and biomass pyramids.**

feed rather than to human consumption. In some areas of the world, forests are being cleared and converted to pasture for beef production—in most cases for export to developed countries. Thus, dietary patterns in North America and Europe are perpetuating inefficiency, since consumption of animal products requires much more energy for each calorie produced than consumption of plant products.

Biological Amplification When chemical pesticides are applied to an ecosystem of producers and consumers, the food web concentrates some of these chemicals. Many chemicals are degraded or diluted in air and water and thus are rendered relatively harmless. Other chemicals, however, are long-lived, stable, and soluble in the fatty tissues of consumers. They become increasingly concentrated at each higher trophic level. This is called *biological amplification*, or *biomagnification*. In the 1970s, scientists suspected that the pesticide DDT was biomagnifying, especially in birds, building up in their fat tissues and causing a thinning of eggshells that caused hatchling mortality. The subsequent ban on DDT for agricultural use is now credited by many experts as saving the Brown Pelican and Peregrine Falcon from extinction.

Thus, pollution in a food web can efficiently poison the organism at the top. The polar bears of the Barents Sea near northern Europe have some of the highest levels of *persistent organic pollutants* (POPs) in any animal in the world, despite their remoteness from civilization. Many species are threatened in this manner (see the orcas in *The Human Denominator*, Figure HD 19), and, of course, humans are at the top of many food chains and therefore at risk of ingesting chemicals concentrated in this way.

Communities and Species Distributions

The levels of organization within ecology and biogeography range from the biosphere, at the top, encompassing all life on Earth, down to single living organisms at the bottom. The biosphere can be broadly grouped into ecosystems (including biomes, discussed in Chapter 20), each of which can then be grouped into **communities**, made up of interacting populations of living plants and animals in a particular place. A community may be identified in several ways—by its physical appearance, by the species present and the abundance of each, or by the complex patterns of their interdependence, such as the trophic (feeding) structure.

For example, in a forest ecosystem, a specific community may exist on the forest floor, while another community may function in the canopy of leaves high above. Similarly, within a lake ecosystem, the plants and animals that flourish in the bottom sediments form one community, whereas those near the surface form another.

Whether viewed in terms of its ecosystem or in terms of its community within an ecosystem, each species has a **habitat**, defined as the environment in which an organism resides or is biologically adapted to live. A habitat includes both biotic and abiotic elements of the environment, and habitat size and character vary with each species' needs. For example, Black-legged Kittiwakes (*Rissa tridactyla*) are a type of gull that prefers small nesting habitats on the sheer cliff faces of offshore islands or sea stacks where their young are safe from predators (Figure 19.11). When breeding season is over, these gulls return to the open ocean for the rest of the year.

The Niche Concept

An **ecological niche** (from the French word *nicher*, meaning "to nest") is the function, or occupation, of a life form within a given community. A niche is determined by the physical, chemical, and biological needs of the organism. This is not the same concept as habitat. Niche and habitat are different in that habitat is an environment that can be shared by many species, whereas niche is the specific, unique role that a species performs within that habitat.

For example, the White-breasted Nuthatch (*Sitta carolinensis*) is a small bird that occurs throughout the United States and in parts of Canada and Mexico in forest habitats, especially in *deciduous forests* (those that

▲Figure 19.11 **Black-legged Kittiwake in nesting habitat, northern coast of Iceland.** [Bobbé Christopherson.]

drop their leaves in winter). Like other nuthatches, this species occupies a particular ecological niche by foraging for insects up and down tree trunks, probing into the bark with their sharp bills and often turning upside down and sideways as they move (Figure 19.12). This behavior enables them to find and extract insects that are overlooked by other birds. They also jam nuts and acorns into the bark, and then bang on them with their bill to extract the seeds. Although nuthatches and woodpeckers occupy a similar habitat, the nuthatch's distinctive foraging behavior causes it to occupy a specific niche that is different from a woodpecker's.

The *competitive exclusion principle* states that no two species can occupy the same niche (using the same food or space) because one species will always outcompete the other. Thus, closely related species are spatially separated either by distance or by species-specific strategies. In other words, each species operates to reduce competition and to maximize its own reproduction rate—because, literally, species survival depends on successful reproduction. This strategy, in turn, leads to greater diversity as species shift and adapt to fill different niches.

▲Figure 19.12 **White-breasted Nuthatch in its ecological niche.** [Cally/Alamy.]

Species Interactions

Within communities, some species are *symbiotic*—that is, have some type of overlapping relationship. One type of symbiosis, *mutualism*, occurs when each organism benefits and is sustained over an extended period by the relationship. For example, lichen (pronounced "liken") is made up of algae and fungi living together (Figure 19.2c). The alga is the producer and food source for the fungus, and the fungus provides structure and physical support. Their mutualism allows the two to occupy a niche in which neither could survive alone. Lichen developed from an earlier parasitic relationship in which the fungi broke into the algal cells. Today, the two organisms have evolved into a supportive and harmonious symbiotic relationship. The partnership of corals and algae discussed in Chapter 16 is another example of mutualism in a symbiotic relationship (Figure 19.2d).

Another form of symbiosis is *parasitism*, in which one species benefits and another is harmed by the association. Often this association involves a parasite living off a host organism, such as a flea living on a dog. A parasitic relationship may eventually kill the host—an example is parasitic mistletoe (*Phoradendron*), which lives on and can kill various kinds of trees (Figure 19.13).

A third form of symbiosis is *commensalism*, in which one species benefits and the other experiences

▲Figure 19.13 **Dwarf mistletoe growing on a Douglas fir in the Rocky Mountains, Colorado.** Dwarf mistletoes (*Arceuthobium*) are parasitic plants found on conifers throughout the U.S. West. The host trees suffer reductions in growth and seed production, and are prone to infectious diseases. At least five species occur in the Rocky Mountain region. [USFS.]

▲**Figure 19.14 Epiphytes using a tree trunk for support, Washington.** Epiphytic club mosses are common in the temperate rain forest of Olympic National Park. [Don Johnston/Alamy.]

neither harm nor benefit. An example is the remora (a sucker fish) that lives attached to sharks and consumes the waste produced as the shark eats its prey. Epiphytic plants, such as orchids, are another example; these "air plants" grow on the branches and trunks of trees, using them for physical support (Figure 19.14).

A final symbiotic relationship is *amensalism*, in which one species harms another but is not affected itself. This typically occurs either as competition, when one organism deprives another of food or habitat, or when a plant produces chemical toxins that damage or kill other plants. For example, black walnut trees excrete a chemical toxin through their root systems into the soil that inhibits the growth of other plants beneath them.

CRITICAL**thinking** 19.1 Mutualism? Parasitism? Where Do We Fit in?

Some scientists are asking whether our human society and the physical systems of Earth constitute a global-scale symbiotic relationship of mutualism, which is sustainable, or of parasitism, which is unsustainable. After reviewing the definitions of these terms, what is your response to that statement? How well do our human economic systems coexist with the need to sustain the planet's life-supporting natural systems? Do you characterize this as mutualism, parasitism, or something else? ●

Abiotic Influences

A number of abiotic environmental factors influence species distributions, interactions, and growth. For example, the distribution of some plants and animals depends on *photoperiod*, the duration of light and dark over a 24-hour period. Many plants require longer days for flowering and seed germination, such as ragweed (*Ambrosia*). Other plants require longer nights to stimulate seed production, such as poinsettia (*Euphorbia pulcherrima*), which needs at least 2 months of 14-hour nights to start flowering. These species cannot survive in equatorial regions with little daylength variation; they are instead restricted to latitudes with appropriate photoperiods, although other factors may also affect their distribution.

In terms of entire ecosystems, air and soil temperatures are important since they determine the rates at which chemical reactions proceed. Precipitation and water availability are also critical, as is water quality—its mineral content, salinity, and levels of pollution and toxicity. All of these factors work together to determine the distributions of species and communities in a given location.

Pioneering work in the study of species distribution was done by geographer and explorer Alexander von Humboldt (1769–1859), the first scientist to write about the distinct zonation of plant communities with changing elevation. After several years of study in the Andes Mountains of Peru, von Humboldt hypothesized that plants and animals occur in related groupings wherever similar climatic conditions occur. His ideas were the basis for the *life zone concept*, which describes this zonation of flora and fauna along an altitudinal transect (Figure 19.15). Each **life zone** possesses its own temperature and precipitation regime and, therefore, its own biotic communities.

The life zone concept became prominent in the 1890s with the work of ecologist C. Hart Merriam, who mapped 12 life zones with distinct plant associations in the San Francisco Peaks in northern Arizona. Merriam also expanded the concept to include the changing zonation from the equator toward higher latitudes. Chapter 20 further discusses plant associations in relationship to climatic conditions.

As discussed in this chapter's *Geosystems Now*, recent scientific studies show that climate change is causing plants and animals to move their ranges to higher elevations with more suitable climates as established life zones shift. Evidence exists that some species have run

GEO**report** 19.2 Sea turtles navigate using Earth's magnetic field

The fact that birds and bees can detect Earth's magnetic field and use it for finding direction is well established. Small amounts of magnetically sensitive particles in the skull of the bird and the abdomen of the bee provide compass directions. Recently, scientists found that sea turtles detect magnetic fields of different strengths and inclinations (angles). This means that the turtles have a built-in navigation system that helps them find certain locations on Earth. Loggerhead turtles hatch in Florida, crawl into the water, and spend the next 70 years traveling thousands of miles between North America and Africa around the subtropical high-pressure gyre in the Atlantic Ocean. The females return to where they were hatched to lay their eggs. In turn, the hatchlings are imprinted with magnetic data unique to the location of their birth and then develop a more global sense of position as they live a life swimming across the ocean.

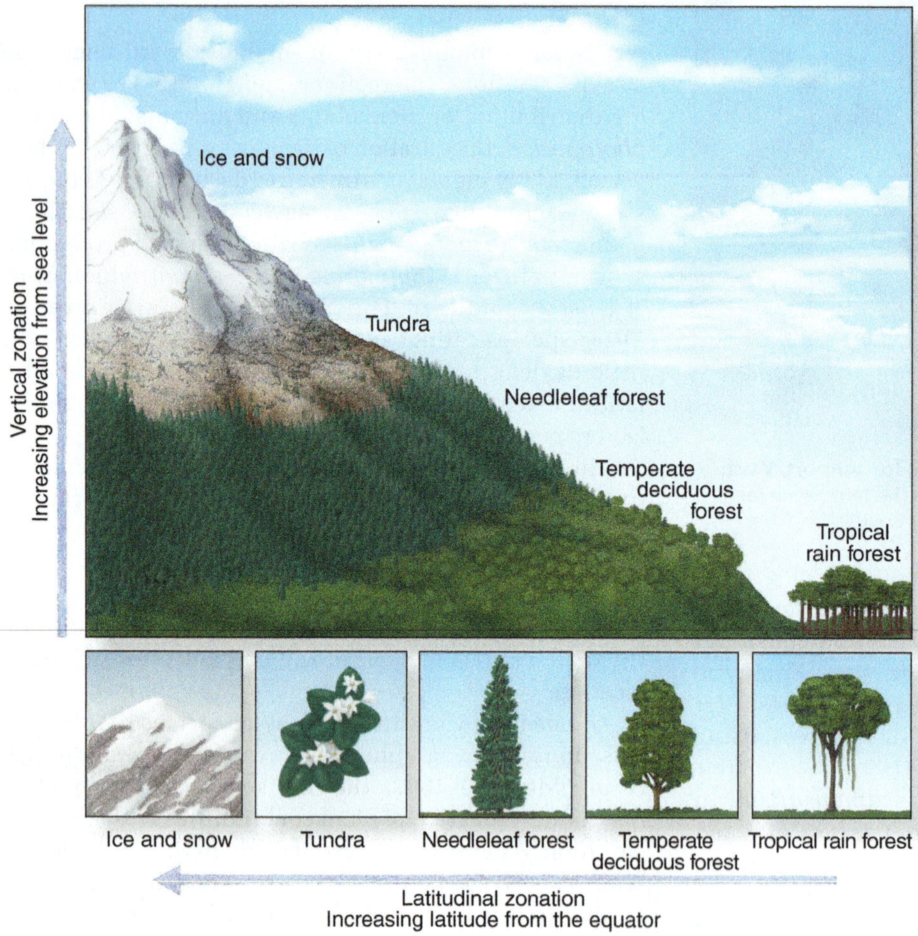

(a) Progression of plant community life zones with changing elevation or latitude.

(b) Treeline for a needleleaf forest in the Canadian Rockies marks the point above which trees cannot grow.

▲**Figure 19.15 Vertical and latitudinal zonation of plant communities.** [Author.]

out of space on mountains, as environmental conditions are pushing them to elevations beyond their mountains' reach, essentially taking them "out of bounds," forcing them to either move elsewhere or into extinction.

Limiting Factors

The term **limiting factor** refers to physical, chemical, or biological characteristics of the environment that determine species distributions and population size. For example, in some ecosystems, precipitation is a limiting factor on plant growth, through either its lack or its excess. Temperature, light levels, and soil nutrients all affect vegetation patterns and abundance:

- Low temperatures limit plant growth at high elevations.
- Lack of water limits growth in a desert; excess water limits growth in a bog.
- Changes in salinity levels affect aquatic ecosystems.
- Low phosphorus content of soils limits plant growth.
- The general lack of active chlorophyll above 6100 m (20,000 ft) limits primary productivity.

For animal populations, limiting factors may be the number of predators, availability of suitable food and habitat, availability of breeding sites, and prevalence of disease. The Snail Kite (*Rostrhamus sociabilis*), a tropical raptor with a small habitat in the Florida Everglades, is a specialist that feeds on only one specific type of snail. In contrast, the Mallard Duck (*Anas platyrhynchos*) is a generalist, feeds from a variety of widely diverse sources, is easily domesticated, and is found throughout most of North America (Figure 19.16a).

For some species, one critical limiting factor determines survival and growth; for other species, a combination of factors is at play, with no one single factor being dominant. When taken together, limiting factors determine the environmental resistance, which eventually stabilizes populations in an ecosystem.

Each organism possesses a *range of tolerance* for physical and chemical environmental characteristics. Within that range, species abundance is high; at the edges of the range, the species is found infrequently; and beyond the range limits, the species is absent. For example, the coast redwood (*Sequoia sempervirens*) is abundant within a narrow range along the California and Oregon coast where foggy conditions provide condensation to meet the tree's water needs (discussed in the Chapter 7 *Geosystems Now*). Redwoods at the limit of their range—for example, at higher elevations above the fog layer—are shorter, smaller, and less abundant. The red maple (*Acer rubrum*) has a wide tolerance range and is distributed over a large area with varying moisture and temperature conditions (Figure 19.16b).

Disturbance and Succession

Over time, communities undergo natural disturbance events such as windstorms, severe flooding, a volcanic eruption, or an insect infestation. Human activities, such as the logging of a forest or the overgrazing of a rangeland,

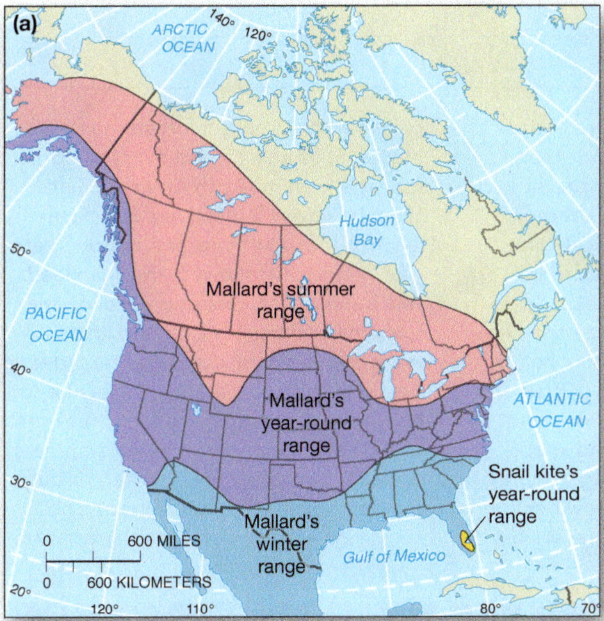

(a) The Snail Kite's small range in North America depends on a single food source; the Mallard Duck is a generalist and feeds widely.

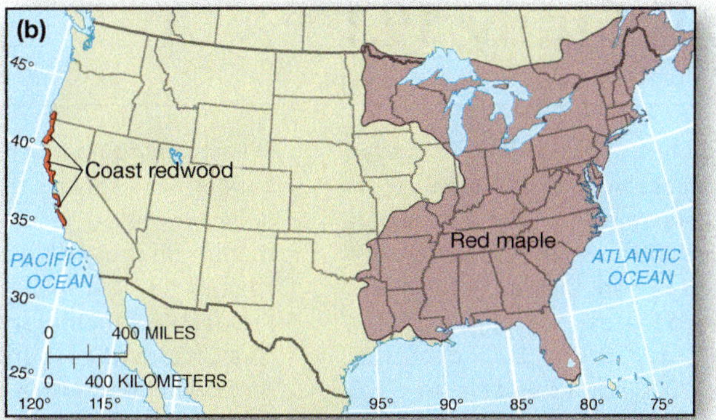

(b) The coast redwood is limited by the presence of fog as a moisture source; red maple tolerates a variety of environmental conditions.

▲Figure 19.16 **Limiting factors affecting species' distributions.**

MG Notebook
Limiting Factors Affect Plant and Animal Species

also create disturbance (Figure 19.17). Such events damage or remove existing organisms, making way for new communities.

Wildfires are a natural component of many ecosystems and a common cause of ecosystem disturbance. The science of **fire ecology** examines the role of fire in ecosystems, including the adaptations of individual plants to the effects of fire and the human management of fire-adapted ecosystems. Focus Study 19.1 takes a look at this important subject.

When a community is disturbed enough that most, or all, of its species are eliminated, a process known as **ecological succession** occurs, in which the cleared area undergoes a series of changes in species composition as newer communities of plants and animals replace older ones. Each successive community modifies the physical environment in a manner that favors a different community. During the transitions between communities, species having an adaptive advantage, such as the ability to produce lots of seeds or disperse them over great distances, will outcompete other species for space, light, water, and nutrients. Successional processes occur in both terrestrial and aquatic ecosystems.

CRITICALthinking 19.2
Observe Ecosystem Disturbances

Over the next several days, observe the landscape as you travel between home and campus, a job, or other localities. What types of ecosystem disturbances do you see? Imagine that several acres in the area you are viewing escape any disruptions for a century or more. Describe what some of these ecosystems and communities might then be like. ●

(a) Damage from a debris flow in North Carolina, triggered by the rainfall associated with Hurricane Ivan in 2004.

(b) Thunderstorm winds damaged forests in eastern Minnesota and northwest Wisconsin in July 2011; wind speeds reached over 160 kmph (100 mph).

▲Figure 19.17 **Ecosystem disturbance, making way for new communities.** [NOAA.]

Focus Study 19.1 Natural Hazards
Wildfire and Fire Ecology

Fire is one of Earth's significant ecosystem processes, and in some areas of the world has become an economic burden. In 2006 and 2007, and again in 2012, over 9 million acres were burned in wildfires in the United States. In Australia, the "Black Saturday" fires in 2009 destroyed more than 2000 homes and claimed 173 lives, the most destructive in the country's history. The cost of fire suppression in the United States reached almost $2 billion in 2012. (See the National Interagency Fire Center at http://www.nifc.gov/fireInfo/fireInfo_main.html.)

Lightning-caused wildfire is a natural disturbance, with a dynamic role in community succession. Many ecosystems have properties that influence the intensity and size of wildfires, which in turn create a mosaic of habitats, ranging from totally burned to partially burned to unburned areas. This patchwork of habitats ultimately benefits biodiversity. Fire affects soils, making them more nutrient-rich in some cases, yet more susceptible to erosion in cases of intensely hot fires. Fire also affects plants and animals; in fact, some species are adapted to, even dependent on, frequent fire occurrence.

Fire-Adapted Ecosystems

A number of Earth's grasslands, forests, and scrublands have evolved through interaction with fire and are known as *fire-adapted ecosystems*. Plant species in such environments may have dense bark, which protects them from heat, or lack lower branches, which protects them from ground fires. Fire-adapted species typically resprout quickly after fire destroys their branches or trunks (see Chapter 20, Figure 20.15b).

Several North American tree species depend on fire for reproduction. For example, lodgepole pine and jack pine rely on fire to crack and open the resin that otherwise seals their cones and keeps seeds from being released (Figure 19.1.1). Seedlings of giant sequoia grow best on open, burned sites, without grasses or other vegetation competing for resources. Fire-disturbed areas quickly recover with stimulated seed production, protein-rich woody growth, and young plants that provide abundant food for animals (Figure 19.1.2).

Fire Management

Modern society's demand for fire prevention to protect property began with European forestry practices of the 1800s and carried over into forest management in North America. Since then, however, forestry experts have learned that when fire-prevention strategies are rigidly followed, they can lead to a buildup of forest undergrowth that fuels major fires. For example, in Yellowstone National Park, after decades of fire suppression, forest managers began a new policy of letting natural fires burn. In 1988, after one of the driest summers on record, massive fires burned through the park, destroying buildings and about 1.2 million acres of forest and grassland.

▲**Figure 19.1.1 Recovery after the 1988 Yellowstone National Park wildfires.** Decades of fire suppression in Yellowstone created a buildup of undergrowth that fueled the massive, uncontrollable 1988 fires. Ten years later, young lodgepole pine, a fire-adapted species, grow among the burned stands. [Jim Peaco, NPS.]

Terrestrial Succession An area of bare rock or a disturbed site with no vestige of a former community can be a site for **primary succession**, the beginning stage of an ecosystem. Primary succession can occur on new surfaces created by mass movement of land, glacial retreat, volcanic eruptions, surface mining, clear-cut logging, or the movement of sand dunes. In terrestrial ecosystems, primary succession begins with the arrival of organisms that are well adapted for colonizing new substrates, forming a **pioneer community**. For example, a pioneer community of lichens, mosses, and ferns may establish on bare rock (Figure 19.18). These early inhabitants prepare the way for further succession: lichens secrete acids that break down rock, which begins the process of soil formation, which enhances habitat for other organisms. As new organisms colonize

▲**Figure 19.18 Primary succession.** Plants establishing on recently cooled lava flows from the Kīlauea volcano in Hawai'i illustrate primary succession. [Bobbé Christopherson.]

▲**Figure 19.1.2 Native grasses after a prescribed fire, South Africa.** Impala graze on the new grasses a year after a prescribed fire in Kruger National Park in 2010. [Navashni Govender, SANParks/NASA.]

▲**Figure 19.1.3 Prescribed burn for grassland restoration, Colorado.** Prescribed burn near Fort Lewis, Colorado, part of nearly 12,000 acres of prairie burned in 2009 to eliminate invasive species such as scotch broom. [Ingrid Barrantine/U.S. Army.]

Today, following the principles of fire ecology, fire specialists worldwide use controlled ground fires, deliberately set to prevent undergrowth accumulation and maintain ecosystem health. These controlled "cool fires" remove fuel and prevent catastrophic and destructive "hot fires" that burn through the forest crown. In grasslands, frequent fires prevent the growth of woody species that would compete for light, moisture, and other resources. Scientists and forest managers use prescribed burns in grasslands to control invasive species and restore natural habitat (Figure 19.1.3).

Wildfire and Climate Change

With ongoing climate change, the wildfire threat is worsening. Scientists have linked record wildfires in the American West since 2000 to increased spring and summer temperatures and an earlier spring snowmelt. Perhaps more important is the fact that these climatic changes are occurring after 150 years or more of fire suppression across the United States. Lightning-caused fires related to more intense weather systems are also on the increase in some regions. In June 2013, the West Fork Complex fire, ignited by lightning, burned explosively through the rugged terrain of southwest Colorado, fueled by forests desiccated from drought and from extensive die-off caused by spruce beetle.

The destruction caused by wildfires is increasing as urban development encroaches on forests, putting homes at risk and threatening public safety. In 2013, the most destructive fire in Colorado's history burned over 500 homes north of Colorado Springs. Wildfires have devastated communities in southern California, especially those developed in the "wilds" of chaparral country. In 2007, fire destroyed more than 2000 homes in that region and burned over 500,000 acres in two dozen wildfires. This is part of the overall pattern continuing today.

soil surfaces, they bring nutrients that further change the habitat, eventually leading to the growth of grasses, shrubs, and trees.

More commonly encountered in nature is **secondary succession**, which occurs when some aspect of a previously functioning community is still present; for example, a disturbed area where the underlying soil remains intact. As secondary succession begins, new plants and animals having niches that differ from those of the previous community colonize the area; species assemblages may shift as soil develops, habitats change, and the community matures.

Most of the areas affected by the Mount St. Helens eruption and blast in 1980, which burned or blew down about 38,450 hectares (95,000 acres) of trees, underwent secondary succession (Figure 19.19). Some soils, young trees, and plants were protected under ash and snow, so community development began almost immediately after the event. The areas completely destroyed near the Mount St. Helens volcano and those buried beneath the massive landslide north of the mountain became candidates for primary succession.

Traditionally, communities of plants and animals were thought to pass through several successional stages, eventually reaching a mature state with a predictable *climax community*—a stable, self-sustaining assemblage of species that would remain until the next major disturbance. However, contemporary biogeography and ecology assume that disturbances constantly disrupt the sequence and that a community may never reach some climax stage. Mature communities are in a state of constant adaptation—a

(a) Mount St. Helens, 2008.

(b) Pre-eruption landscape north of volcano, 1979.

(c) Post-eruption landscape in same region, 1980.

(d) Repeat photography of post-eruption recovery at Meta Lake, 1983 and 1999.

e) Repeat photography showing secondary succession, 1983 and 1999.

▲**Figure 19.19 The pace of change in the region of Mount St. Helens.** [(a) *ISS* astronaut photo, NASA/GSFC; (a) all 1983 photos by author; (b) all 1999 photos by Bobbé Christopherson.]

dynamic equilibrium—sometimes with a lag time in their adjustment to environmental changes. Scientists now know that successional processes are driven by a dynamic set of interactions with sometimes unpredictable outcomes.

Disturbance often occurs in discrete spatial units across the landscape, creating habitats, or *patches*, at different successional stages. The concept of *patch dynamics* refers to the interactions between and within this mosaic of habitats, which add to complexity across the landscape. The overall biodiversity of an ecosystem is, in part, the result of such patch dynamics.

Aquatic Succession Aquatic ecosystems occur in lakes, estuaries, wetlands, and along shorelines, and the communities in these systems also undergo succession. For example, lakes and ponds exhibit successional stages as they fill with sediment and nutrients and as aquatic plants take root and grow. The plant growth captures more sediment and adds organic debris to the system (Figure 19.20). This gradual enrichment in water bodies is known as **eutrophication** (from the Greek *eutrophos*, meaning "well nourished").

In moist climates, a lake will develop a floating mat of vegetation that grows outward from the shore to form a bog. Cattails and other marsh plants become established, and partially decomposed organic material accumulates in the basin, with additional vegetation bordering the remaining lake surface. Vegetation and soil and a meadow may fill in as water is displaced; willow trees follow, and perhaps cottonwood trees; eventually, the lake may evolve into a forest community. Thus, when viewed across geologic time, a lake or pond is really a temporary feature on the landscape.

The stages in lake succession are named for their nutrient levels: *oligotrophic* (low nutrients), *mesotrophic* (medium nutrients), and *eutrophic* (high nutrients). Greater primary productivity and resultant decreases in water transparency mark each stage, so that photosynthesis becomes concentrated near the surface. Energy flow shifts from production to respiration in the eutrophic stage, with oxygen demand exceeding oxygen availability.

Nutrient levels also vary spatially within a lake: Oligotrophic conditions occur in deep water, whereas eutrophic conditions occur along the shore, in shallow

(a) A lake gradually fills with organic and inorganic sediments, shrinking the area of open water. A bog forms, then a marsh, and finally a meadow, the last of the successional stages.

(b) Spring Mill Lake, Indiana

(c) Organic content increases as succession progresses in a mountain lake.

(d) Peat bog with acidic soils, Richmond Nature Park, near Vancouver, British Columbia.

▲**Figure 19.20 Idealized lake–bog–meadow succession in temperate conditions, with read-world examples.** [Bobbé Christopherson.]

bays, or where sewage, fertilizer, or other nutrient inputs occur. Even large bodies of water may have eutrophic areas along the shore. As humans dump sewage, agricultural runoff, and pollution into waterways, the nutrient load is enhanced beyond the cleansing ability of natural biological processes. This human-caused eutrophication, known as *cultural eutrophication*, hastens succession in aquatic systems.

Biodiversity, Evolution, and Ecosystem Stability

Far from being static, Earth's ecosystems have been dynamic—vigorous, energetic, and ever-changing—from the beginning of life on the planet. Over time, communities of plants and animals have adapted and evolved to produce great diversity and, in turn, have shaped their

GEOreport 19.3 Another take on lake–bog succession

In areas that have been recently deglaciated, scientists have uncovered a pattern that seems the opposite of lake–bog succession, underscoring the poor understanding of successional processes in aquatic ecosystems. Evidence from lakes in the Glacier Bay, Alaska, area suggests that these water bodies became more dilute, acidic, and unproductive—in other words, less eutrophic—over the past 10,000 years. Successional changes in the surrounding vegetation and soils seem to be linked to these changes in the aquatic ecosystem. The studies suggest that in various cool, moist, temperate rainforest climates with landscapes created by glacial retreat, different processes are at work in lakes than the ones described as typical in this chapter.

environments. Each ecosystem is constantly adjusting to changing conditions and disturbances. Ironically, the concept of change is key to understanding ecosystem stability.

The dynamics of change in natural ecosystems can range from gradual transitions between equilibrium states to abrupt changes caused by extreme catastrophes such as an asteroid impact or severe volcanism. For most of the last century, scientists thought that an undisturbed ecosystem—whether a forest, a grassland, or a lake—would progress to a stage of equilibrium, a stable point, with maximum chemical storage and biomass. Modern research has determined, however, that ecosystems do not progress to some static conclusion. In other words, there is no "balance of nature"; instead, the "balance" is best thought of as a constant interplay of physical, chemical, and living factors in a dynamic equilibrium.

Biological Evolution Delivers Biodiversity

A critical aspect of ecosystem stability and vitality is **biodiversity**, or variation of life (a combination of *bio*logical and *diversity*). The concept of biodiversity encompasses species diversity, the number and variety of different species; genetic diversity, the number of genetic variations within these species; and ecosystem diversity, the number and variety of ecosystems, habitats, and communities on a landscape scale.

The origin of this diversity is embodied in the *theory of evolution.* (The definition of a scientific theory is given in Chapter 1.) **Evolution** states that single-cell organisms adapted, modified, and passed along inherited changes to multicellular organisms. The genetic makeup of successive generations was shaped by environmental factors, physiological functions, and behaviors that led to a greater rate of survival and reproduction. Thus, in this continuing process, traits that help a species survive and reproduce are passed along more frequently than are those that do not. A *species* is a population that can reproduce sexually and produce viable offspring. By definition, this means reproductive isolation from other species.

Inherited traits are guided by an array of genes, part of an organism's primary genetic material, *DNA* (deoxyribonucleic acid), which resides in the nest of chromosomes in every cell nucleus. These traits—and especially the ones that are most successful at exploiting niches different from those of other species or that help the species adapt to environmental changes—are passed to successive generations. Such differential reproduction and adaptation pass along successful genes or groups of genes in a process of genetic favoritism known as *natural selection.* The process of evolution continues generation after generation and tracks the passage of inherited characteristics that were successful—the failures passed into extinction. Thus, today's humans are the result of billions of years of affirmative natural selection.

New genes in the *gene pool*, the collection of all genes possessed by individuals in a given population, result from mutations. *Mutation* is a process that takes place when a random occurrence, perhaps an error as the DNA reproduces, produces altered genetic material and inserts new traits into the inherited stream. Geography also comes into play, since spatial variation in physical environments affects natural selection. For example, a species may disperse through migration, such as across ice bridges or land connections at times of low sea level, to a different environment where new traits are favored. A species may also be separated from other species by a natural *vicariance* (fragmentation of the environment) event. An example is continental drift, which established natural barriers to species movement and resulted in the evolution of new species. The physical and chemical evolution of Earth's systems is therefore closely linked to the biological evolution of life.

Biodiversity Fosters Ecosystem Stability

In the 1990s, field experiments in the prairie ecosystems of Minnesota began to confirm an important scientific assumption: Greater biological diversity in an ecosystem leads to greater long-term stability and productivity. For instance, during a drought, some species of plants will be damaged from water stress. In a diverse ecosystem, however, other species with deeper roots and better water-obtaining ability will thrive. Ongoing experiments also suggest that a more diverse plant community retains and uses soil nutrients more efficiently than one with less diversity. (More about this research at the Cedar Creek Ecosystem Science Reserve is at http://www.cedarcreek.umn.edu/about/.)

Stability and Resilience In the context of ecosystems, "stable" does not mean unchanging; stable ecosystems are constantly changing. A stable ecosystem is one that does not deviate greatly from its original state, despite changing environmental conditions (the environmental resistance, mentioned earlier). *Resilience* is the ecosystem's ability to recover from disturbance quickly and return to its original state. However, some disturbances are too extreme for even a highly resilient ecosystem. For example, several of the dramatic asteroid-impact episodes that triggered partial extinctions over the past 440 million years overcame the resilience of plant and animal communities, destabilizing the ecosystem. When an ecosystem crosses such a threshold, it moves toward a new stable state.

In an ecosystem, a population of organisms can be stable yet not resilient. A tropical rain forest is a diverse, stable community that can withstand most natural disturbance. (An ecosystem with *inertial stability* has the ability to resist some low-level disturbance.) Yet this ecosystem has low resilience in terms of severe events; a cleared tract of rain forest will recover at a slower rate than many other communities because most of the nutrients are stored in the vegetation rather than in the soil. Furthermore, changes in microclimates may make regrowth of the same species difficult. In contrast, a midlatitude grassland, although less diverse than a rain forest, has high resilience because the system can cope with

◀Figure 19.21 Disruption of a forest community. An example of clear-cut timber harvesting that devastated a stable forest community and produced drastic changes in microclimatic conditions. About 10% of the Northwest's old-growth forests remain, as identified through satellite images and GIS analysis. [Bobbé Christopherson.]

a range of disturbances and recover quickly. For example, after a fire, rapid regrowth occurs from extensive root systems of the grassland species.

In the context of ecosystem stability and resilience, consider the elimination of a full section of forest ecosystem on private land in southern Oregon, south-southeast of Crater Lake, shown in Figure 19.21. The practice of clear-cutting (the complete removal of timber) can cause a disturbance that overcomes a forest community's resilience and prevents it from returning to its natural stable state. Adjacent lands administered by the U.S Forest Service use more sustainable harvesting regimes, including partial tree removal and forest thinning, that are not as likely to destabilize forest communities.

Agricultural Ecosystems When humans purposely eliminate biodiversity from an area, as they do in most agricultural practices, the area becomes more vulnerable to disturbance. An artificially produced monoculture community, such as a field of wheat or a tree plantation, is vulnerable to insect infestations or plant diseases (Figure 19.22). In some regions, simply planting multiple crops brings more stability to the ecosystem—this is an important principle of sustainable agriculture.

A modern agricultural ecosystem requires enormous amounts of energy, chemical pesticides and herbicides, artificial fertilizer, and irrigation water. The practice of harvesting and removing biomass from the land interrupts the cycling of materials into the soil and depletes soil nutrients over time, a loss that must be artificially replenished.

▲Figure 19.22 Tree plantation, southern Georgia. [Bobbé Christopherson.]

Biodiversity on the Decline

Human activities have great impact on global biodiversity; the present loss of species is irreversible and is accelerating. Extinction is final no matter how complex the organism or how long its existence. We are now facing a loss of genetic diversity that may be unparalleled in Earth's history, even compared with the major extinctions that punctuate the geologic record.

Since life arose on the planet, six major extinctions have occurred. The fifth one was 65 million years ago, whereas the sixth is happening over the present decades (see Figure 11.1). Of all these extinction episodes, this is the only one of biotic origin, caused for the most part by human activity.

Presently, about 270,000 species of plants are known to exist, with many more species yet to be identified. They represent a great untapped resource base. Only about 20 species of plants provide 90% of food for humans; just three—wheat, maize (corn), and rice—make up half of that supply. Plants are also a major source of new medicines and chemical compounds that benefit humanity.

Table 19.2 summarizes the numbers of known and estimated species on Earth. Scientists have classified only 1.75 million species of plants and animals out of an estimated 13.6 million overall; this figure represents an increase in what scientists once thought to be the diversity of life on Earth. The wide range of estimates places the expected species count between a low of 3.6 million and a high of 111.7 million. Estimates of annual species loss range between 1000 and 30,000 species, although this range might be conservative. The possibility exists that over half of Earth's present species could be extinct within the next 100 years. *The Human Denominator,* Figure HD 19 on page 614, offers examples of some human causes of biodiversity decline.

TABLE 19.2 Known and Estimated Species on Earth

Categories of Living Organisms	Number of Known Species	Estimated Number of Species			Accuracy
		High (x1000)	Low (x1000)	Working Estimate (x1000)	
Viruses	4000	1000	50	400	Very poor
Bacteria	4000	3000	50	1000	Very poor
Fungi	72,000	27,000	200	1500	Moderate
Protozoa	40,000	200	60	200	Very poor
Algae	40,000	1000	150	400	Very poor
Plants	270,000	500	300	320	Good
Nematodes	25,000	1000	100	400	Poor
Arthropods:					
Crustaceans	40,000	200	75	150	Moderate
Arachnids	75,000	1000	300	750	Moderate
Insects	950,000	100,000	2000	8000	Moderate
Mollusks	70,000	200	100	200	Moderate
Chordates	45,000	55	50	50	Good
Others	115,000	800	200	250	Moderate
Total	**1,750,000**	**111,655**	**3635**	**13,620**	**Very poor**

Source: United Nations Environment Programme, *Global Biodiversity Assessment* (Cambridge, England: Cambridge University Press, 1995), Table 3.1–3.2, p. 118, used by permission.

Five categories of human impact represent the greatest threat to biodiversity:

- Habitat loss, degradation, and fragmentation as natural areas are converted for agriculture and urban development
- Pollution of air, water, and soils
- Resource exploitation and harvesting of plants and animals at unsustainable levels
- Human-induced climate change, discussed in this chapter's *Geosystems Now* and in Chapter 10B
- Introduction of non-native plants and animals, discussed in Chapter 20

The World Conservation Monitoring Centre and its International Union for Conservation of Nature (IUCN) maintain a global "Red List" of endangered species at http://www.iucnredlist.org/. See also the endangered species home page of the Fish and Wildlife Service at http://www.fws.gov/endangered/. Let us now look at some specific examples of species in decline.

Threatened Species—Examples Research indicates that amphibians are vulnerable to changes in both terrestrial and aquatic ecosystems, such as habitat destruction, pollution, invasive species, and changing climate—this puts them at higher risk than mammals, fish, and birds. Although amphibian declines can also be attributed to natural causes such as competition, predation, and disease, the bottom line is that these species are not evolving fast enough to keep up with the rate of change.

In the Arctic region, the polar bear (*Ursus maritimus*) faces melting sea-ice habitat associated with climate change. The IUCN in 2006 listed the species as "vulnerable" to extinction, and it was designated as "threatened" under the U.S. Endangered Species Act (ESA) in 2008. Research released by the U.S. Geological Survey in September 2007 predicted that with the loss of Alaskan, Canadian, and Russian sea-ice habitat, some two-thirds of the world's 23,000 polar bears will die off by 2050 or earlier. The remaining 7500 bears will be struggling.

In Africa, black rhinos (*Diceros bicornis*) and white rhinos (*Ceratotherium simum*) exemplify species in jeopardy from declining habitat and overharvesting. Rhinos once grazed over much of the savannas and woodlands. Today, they survive only in protected districts in heavily guarded sanctuaries (Figure 19.23). Consider these statistics:

- Black rhinos: The population of 70,000 in 1960 dropped to 2599 in 1998—a decline of 96%. South Africa guards about 50% of the herd. Slow recovery is under way; numbers rose to 4880 in 2010. The western black rhino, a subspecies, has not been seen since 2006 and is considered extinct.

GEOreport 19.4 Will species adapt to climate change?

A 2013 study reveals that in order for vertebrate species to adapt to projected changes in climate by the year 2100, they will need to evolve their niche requirements 10,000 times faster than rates in the past. Using genetic data for over 500 species of terrestrial vertebrates, including frogs, snakes, birds, and mammals, spread out over 17 evolutionary trees, the scientists examined how long each species took to shift its climatic niche under past environmental conditions. They found that over about a million years, species were able to adapt to a temperature difference of 1 C°. These results suggest that adaptation may not be an option for species survival in today's rapidly warming climate.

▲**Figure 19.23 The rhinoceros in Africa.** White rhinoceros (*Ceratotherium simum*) with young, from the southern population, Lake Nakuru National Park, Kenya. [Ingo Arndt/naturepl.com.]

- White rhinos: The 11 northern white rhinos surviving in 1984 increased to over 25 by 1998 but then dropped as political unrest in the Congo slowed protection efforts. In 2013, only 4 remain in the Ol Pejeta Conservancy in Kenya. The southern white rhinos are on the increase; their population topped 20,000 in 2010.

Rhinoceros horn sells for $29,000 per kilogram as an aphrodisiac (but in reality has no medicinal effect). These large land mammals are nearing extinction and will survive only as a dwindling zoo population. The limited genetic pool that remains complicates further reproduction.

Relating Systems Analysis to Species Extinction Scientists are using systems analysis (as discussed in Chapter 1) to understand extinctions of harlequin frog (*Atelopus varius*) and golden toad (*Bufo periglenes*) species in the Monteverde Cloud Forest Reserve, Costa Rica, and elsewhere in Central and South America. Specifically, they are analyzing how climate change is altering the frogs' exposure to diseases and pathogens. One infectious agent, the *chytrid fungus* (*Batrachochytrium dendrobatidis*), has spread into mid-elevation regions (1000 m to 2400 m; 3280 ft to 7874 ft) of the mountain cloud forest, where optimum temperature conditions for the fungus now exist. As a result, this non-native fungus today coexists in habitats of the harlequin frog.

To understand the amphibian die-offs, scientists analyzed ocean and air temperature records and used a systems perspective to make connections. They propose that human-forced climate change has increased the temperature of the ocean and atmosphere, which causes higher evaporation rates, which in turn affects condensation (discussed in Chapter 7). As the warm, moist air moves onshore and reaches the mountains, the air lifts and cools, and condensation occurs at higher elevations than before. Extra moisture means more cloud cover, and these clouds affect the daily temperature range: Clouds at night, acting like insulation, raise nighttime minimum temperatures, whereas clouds during the day act as reflectors, lowering daytime maximum temperatures (discussed in Chapter 4).

Optimal conditions for the non-native *chytrid* fungus occur when temperatures are between 17°C and 25°C (63°F and 77°F). The new cloud cover keeps the daytime maximum below 25°C (77°F) at the forest floor. In these more favorable conditions, the disease pathogen flourishes. Harlequin frogs have moist, porous skin that the fungus penetrates, killing the frog. Some researchers suggest that the spread of this fungus may be related to other factors in addition to climate change, including human population growth. More research is clearly needed to find a definitive cause. Between 1986 and 2006, approximately 67% of the 110 known species of harlequin frogs went extinct.

Species and Ecosystem Restoration Since the 1990s, species restoration efforts in North America have focused on returning predators such as wolves and condors to parts of the American West and, recently, jaguars to the Southwest. Other efforts have resulted in rising populations of black-footed ferrets and Whooping Cranes in the prairie regions and shortnose sturgeon along the Atlantic seaboard. These projects have reintroduced captive-bred animals or relocated wild animals into their former habitats, while at the same time limiting practices such as hunting that once caused species decline. The preservation of large habitats has also played a critical role in restoring these, and other, endangered species worldwide.

Recent efforts at ecosystem restoration have had some success in restoring or preserving biodiversity, although questions remain about the effects of such work on overall ecosystem functioning. Numerous river restoration projects, such as the dam removals discussed in Chapter 14, are successfully restoring natural conditions for fisheries and riparian wetlands in the short term. In the Florida Everglades, a $9.5 billion restoration project began in 2000 and is ongoing. The goal is to return freshwater flow into the south Florida swamplands to revive the dying ecosystem. The Everglades restoration is the largest and most ambitious watershed restoration project in history (see http://www.evergladesplan.org/index.aspx).

Restoration of the Great Lakes ecosystems began in the 1970s and is ongoing. Focus Study 19.2 discusses the various environmental issues that have plagued this region of intense human use and development, and the management strategies in use to restore natural conditions. However, the question remains in these and many other ecosystems as to what is "natural." The goal of returning ecosystems to the conditions that prevailed before human intervention is now being expanded to include the possibility of creating "novel ecosystems," human-built or human-modified ecosystems that may have species and habitats that have never occurred together. These novel systems have no natural analogs on which to base scientific hypotheses or restoration strategies—and yet, to sustain biodiversity and ecosystem function in our changing world it may be necessary to consider and manage such ecosystems.

Focus Study 19.2 Environmental Restoration

Ecosystems of the Great Lakes

The Great Lakes—Superior, Michigan, Huron, Erie, and Ontario—contain 18% of the total volume of all freshwater lakes in the world. With their connecting rivers, they form an international waterway that spans more than 2771 km (1722 mi) from west to east (Figure 19.2.1). The 18,000 km (11,000 mi) of coastline in the lakes includes diverse environments featuring dunes, beaches of sand and gravel, bedrock shorelines, and wetlands.

About 10% of the U.S. population and 25% of Canada's population live in the Great Lakes drainage basin. Land use around the lakes includes agriculture, industrial activity, maritime commerce, and tourism. The eight Great Lakes states generate $18 billion per year in revenue, mainly from agriculture, pulp and paper, fisheries, transportation, and tourism.

Human impacts on the aquatic and terrestrial lake ecosystems have resulted in severe ecological degradation—virtually none of the original land cover in the region is undisturbed. Restoration and recovery in the Great Lakes Basin began in the 1970s and is ongoing today, emphasizing an ecosystem management approach aimed at balancing ecosystem functioning with human use. In 2010, the U.S. government approved $475 million in funding for the Great Lakes Restoration Initiative, the largest investment of federal dollars in the Great Lakes in 20 years. This initiative targets some of the greatest threats to the lakes—namely, climate change, pollution of water and sediment, and invasive species (discussed in Chapter 20). Priorities include restoring wetlands and other critical habitats (see http://greatlakesrestoration.us/).

Lake Levels and Stratification

Figure 19.2.2 profiles the dimensions, relative elevations, and connections between the lakes. Compared to an average water-retention time of 191 years in Lake Superior (the largest of the lakes), Lake Erie has the shortest retention time of 2.6 years. Water flowing into the St. Lawrence River drains eventually into the Gulf of St. Lawrence along the Canadian coast.

Lake levels fluctuate annually with precipitation and evaporation, and are higher in the summer owing to inputs from snowmelt and increased summer maximum precipitation. Recently, water levels have been lower than the average level (based on records extending to 1860), and climate change is forecasted to lower lake levels further as evaporation increases from lake surfaces. Climate models forecast reductions from 0.3 m (1 ft) to as much as 1.5 m (5 ft), reducing outflow into the St. Lawrence by 20% to 40%. Increased human consumption and water demand across the region will add to the water withdrawals.

Climate change is also affecting lake stratification. In the summer, the Great Lakes stratify, as most lakes do: The surface waters warm and become less dense, and the cooler waters form the lower layer in the lake. Productivity accelerates in the top layers owing to increased nutrients and light penetration, and little mixing occurs between the layers. With the onset of cooler temperatures in the fall, the normal pattern is for surface water to cool and sink, displacing deeper water and creating a turnover of the lake mass.

In July 2007, Lake Superior surface water reached 23.9°C (75°F). During August 2010, Lake Superior was 8 C° (14 F°) above normal water temperature and Lake Michigan 4 C° (7 F°) above normal—both new records. As discussed in Chapter 9, warming temperatures prolong the period of lake stratification, which interrupts normal lake turnover,

▲Figure 19.2.1 **Map of Great Lakes basin, with primary land use.** Map shows the drainage basin for the Great Lakes system, along with primary nonagricultural and agricultural land uses. [(a) Map based on Environment Canada, U.S. EPA and Brock University cartography.]

isolating the cooler bottom waters for a longer time each year.

Wetland Ecosystems

Wetlands occupy a critical land–water interface, storing and cycling organic material and nutrients between terrestrial and aquatic ecosystems, filtering pollutants, anchoring soils and sediment against wave erosion, and providing seasonal reproductive habitat for migratory waterfowl, amphibians, and fish. The health of wetlands in the Great Lakes is tied to lake levels, which determine wetland distribution, vegetation composition, and ecological function. Freshwater wetland ecosystems—primarily marshes, swamps, bogs, and fens—exist throughout the Great Lakes along lake shorelines and connecting river channels and floodplains. Lower lake levels can effectively drain wetlands, causing habitat loss for many species.

Over two-thirds of Great Lakes wetlands have been filled or drained for a variety of uses (agriculture, development, recreation) during this century, and those remaining are threatened by development, drainage, or pollution. The loss of these ecosystems affects water quality, since they dilute wastes from cities and industry, dissipate thermal pollution from power plants, and at the same time provide municipal drinking water and irrigation water.

Aquatic Ecosystems and Water Quality

Like any aquatic ecosystem, the Great Lakes support a food web of producers and consumers, and this web has suffered from human impacts over the years. Native fish populations have declined from overfishing, introductions of non-native species, pollution from excess nutrients, toxic contamination, and disruption of spawning habitats. Commercial fishing peaked in the 1880s, but it wasn't until water pollution peaked in the 1960s and early 1970s that fisheries declined to their lowest levels.

Toxic chemicals such as polychlorinated biphenyls (PCBs) and DDT—both banned in the United States in the 1970s—accumulate and persist in lake sediments and organisms and are subject to biological amplification as they move up the food chain. In the Great Lakes, scientists have reported PCBs in phytoplankton, rainbow smelt, and lake trout, as well as higher in the food chain, as in eggshells of herring gulls.

In 1972, Canada and the United States signed the Great Lakes Water Quality Agreement, which implemented strong pollution-control programs, run by governments working in concert with citizens, industries, and private organizations. These efforts resulted in declining levels of PCBs and DDT in the lakes up until 1990, when rates of decline leveled off. Today, the fishery has mostly been restored. However, health advisories are occasionally issued concerning consumption of fish of certain species, sizes, and locales.

The Lake Erie Dead Zone

About a third of the overall population in the Great Lakes Basin lives in the Lake Erie drainage basin, making it the major recipient of sewage effluent from treatment plants. As the shallowest and warmest of the Great Lakes, Lake Erie was the first to show severe effects of eutrophication.

Over the years, dissolved oxygen levels in the lake dropped, resulting in the formation of dead zones. Algae flourished in these eutrophic conditions and coated the beaches, turning the lake a greenish brown. In the 1950s and 1960s, the oily contaminated surface of the Cuyahoga River, flowing into Lake Erie through Cleveland, Ohio, caught fire several times. The 1969 fire finally spurred the public to demand action.

In the 1970s, environmental restrictions reduced inputs of phosphorus into Lake Erie, eventually by about 90%. The dead zone steadily declined. However, in the 1990s, scientists again identified a significant Lake Erie dead zone, and in 2011, this zone expanded to its largest area ever recorded, attributed to severe spring storms that carried fertilizer runoff, primarily phosphorus, into the lake. (see Figure GIA 19.6) The warmer water and generally lower wind speeds slowed the natural mixing; scientists expect that the agricultural and meteorological conditions that produced this bloom will continue in the future.

More information on Great Lakes ecosystems is at http://www.ec.gc.ca/greatlakes/ and http://www.glerl.noaa.gov/.

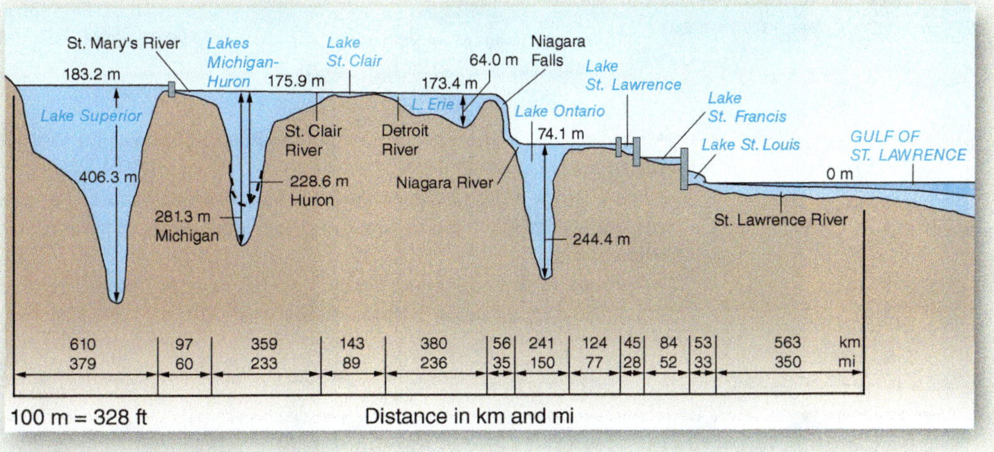

▲**Figure 19.2.2 Great Lakes elevation profile and lake levels.** Average lake surface levels and depths, 2013. The International Great Lakes Datum of 1985 is the standard baseline (and was last adjusted in 1992). [Data courtesy of the Canadian Hydrographic Service, Central Region, and the International Coordinating Committee on the Great Lakes Basin Hydraulic and Hydrographic Data.]

MG Notebook
Ecosystems of the Great Lakes

THE**human**DENOMINATOR 19 Ecosystems and Biodiversity

ECOSYSTEM PROCESSES ⇨ HUMANS

- All life, including humans, depends on healthy, functioning ecosystems, which provide the food and all other natural resources that humans use.

HUMANS ⇨ ECOSYSTEM PROCESSES

Human activities cause declining biodiversity. For example,

- Habitat loss occurs as natural lands are converted for agriculture and urban development.
- Pesticides and other pollutants poison organisms and food webs.
- Overharvesting of plants and animals leads to extinction.
- Climate change affects plant and animal distributions and overall ecosystem function.
- Fertilizer use and industrial activities alter biogeochemical cycles, as when dead zones disrupt the nitrogen cycle.

19a Beaver are large, semi-aquatic rodents known for building dams that modify landscapes and create wetland habitat for many plants and other animals. Hunted to near-extinction by the early 1900s, the Eurasian beaver (*Castor fiber*) was successfully reintroduced throughout most of its former range. The North American beaver (*Castor canadensis*) also declined, although populations have today recovered in most regions. [Danita Delimont/Getty Images.]

19b Scientists outfitted the first satellite tracking tags for baby loggerhead sea turtles in 2012. Past efforts at developing such devices were limited by the animal's small size and rapid growth. The tags will allow experts to follow migration routes throughout all life stages, providing critical information for sea turtle conservation. (See Geo Report 19.2.) [Jim Abernathy, NOAA.]

19d Orcas (*Orcinus orca*), also known as killer whales, are threatened by high levels of polychlorinated biphenyls (PCBs), as well as other contaminants, in Puget Sound, Washington. Persistent organic pollutants are stored in the whale's tissues and may be a primary cause for declining resident populations of these marine mammals. [Danita Delimont/Getty Images.]

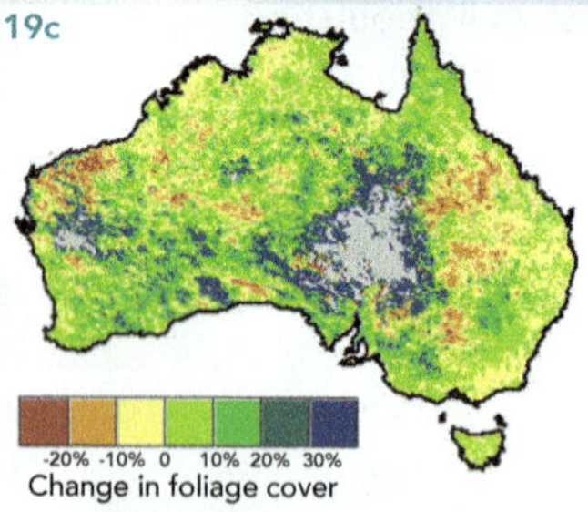

19c Scientists are attributing increased foliage cover since 1982 throughout parts of Australia to the "CO_2 fertilization effect"—the increase in photosynthesis caused by rising atmospheric CO_2 levels. In Australia's warm, dry climates, leaf cover is more responsive to increased CO_2 than in other regions; as leaves draw in extra CO_2, they lose less water so that the plant puts out more leaves, producing a "greening" that shows in satellite images. Other warm, arid regions of the world show the same trend. (See the world map of foliar cover change at http://www.csiro.au/Portals/Media/Deserts-greening-from-rising-CO2.aspx.) [© Copyright CSIRO, 2013. Used by permission.]

ISSUES FOR THE 21ST CENTURY

- Species and ecosystem conservation and restoration will be essential for saving species from extinction.
- Fire ecology will become increasingly important as climate change leads to prolonged drought in some areas and as human populations spread further into wildlands.
- Addressing and mitigating climate change may become essential to preserving a future for all species, including humans.

GEOSYSTEMS**connection**

Earth's biosphere is a remarkable functioning entity of abiotic and biotic components, all interacting and interrelated through some 13.6 million species. Plants harvest sunlight through photosynthesis and thus begin vast food webs of energy and nutrition. Life on Earth evolved into this biologically diverse structure of organisms, communities, and ecosystems that gain strength and resilience through their biodiversity. We next move to a discussion of biomes and to a synthesis of the subjects from Chapters 2 through 19, as we bring all the book's topics together to form a portrait of our planet.

KEY LEARNING concepts review

■ *Define* ecology, biogeography, and ecosystem.

Earth's biosphere is made up of **ecosystems**, self-sustaining associations of living plants and animals and their nonliving physical environment. **Ecology** is the study of the relationships between organisms and their environment and among the various ecosystems in the biosphere. **Biogeography** is the study of the distribution of plants and animals and the diverse spatial patterns they create.

ecosystem (p. 586)
ecology (p. 586)
biogeography (p. 586)

1. What is the relationship between the biosphere and an ecosystem? Define ecosystem, and give some examples.
2. What does biogeography include? Describe its relationship to ecology.
3. Briefly summarize what ecosystem operations imply about the complexity of life.

■ *Explain* photosynthesis and respiration, and *describe* the world pattern of net primary productivity.

Producers, which fix the carbon they need from CO_2, are the plants, algae, and cyanobacteria (a type of blue-green algae). As plants evolved, the **vascular plants** developed conductive tissues. **Stomata** on the underside of leaves are the portals through which the plant participates with the atmosphere and hydrosphere. Plants (primary producers) perform **photosynthesis** as sunlight stimulates a light-sensitive pigment called **chlorophyll**. This process produces food sugars and oxygen to drive biological processes. **Respiration** is essentially the reverse of photosynthesis and is the way the plant derives energy by oxidizing carbohydrates. **Net primary productivity** is the net photosynthesis (photosynthesis minus respiration) of an entire community. **Biomass** is the total organic matter derived from all living and recently living organisms and is measured as the net dry weight of organic material. Net primary productivity produces the energy needed for **consumers**—generally animals (including zooplankton in aquatic ecosystems)—that depend on producers as their carbon source.

producer (p. 587)
vascular plant (p. 588)
stomata (p. 588)
photosynthesis (p. 588)
chlorophyll (p. 588)
respiration (p. 569)
net primary productivity (p. 569)
biomass (p. 569)
consumer (p. 569)

4. Define a vascular plant. How many plant species are there on Earth?
5. How do plants function to link the Sun's energy to living organisms? What is formed within the light-responsive cells of plants?
6. Compare photosynthesis and respiration with regard to the concept of net photosynthesis, which is the result of deducting respiration from photosynthesis. What is the importance of knowing the net primary productivity of an ecosystem and how much biomass an ecosystem has accumulated?
7. Briefly describe the global pattern of net primary productivity.

■ *Discuss* the oxygen, carbon, and nitrogen cycles, and *explain* trophic relationships.

Life is sustained by **biogeochemical cycles**, through which circulate the gases and nutrients necessary for growth and development of living organisms. Excessive nutrient inputs into oceans or lakes can create **dead zones** in the water, areas with low-oxygen conditions that limit underwater life.

Energy in an ecosystem flows through *trophic levels*, or feeding levels, which are the links that make up a **food chain**, the linear energy flow from producers through various consumers. Producers, at the lowest trophic level, build sugars (using sunlight, carbon dioxide, and water) to use for energy and tissue components. Within ecosystems, the feeding relationships are arranged in a complex network of interconnected food chains called a **food web**.

Herbivores (plant eaters) are primary consumers. **Carnivores** (meat eaters) are secondary consumers. A consumer that eats both producers and other consumers is an **omnivore**—a role occupied by humans. **Detritivores** are detritus feeders (including worms, mites, termites, and centipedes) that ingest dead organic material and waste products, and release simple inorganic compounds and nutrients. **Decomposers** are the bacteria and fungi that process organic debris outside their bodies and absorb nutrients in the process, producing the rotting action that breaks down detritus. Energy and biomass pyramids illustrate the flow of energy between trophic levels; energy always decreases with movement from lower to higher feeding levels in an ecosystem.

biogeochemical cycle (p. 591)
dead zone (p. 593)
food chain (p. 596)
food web (p. 596)
detritivore (p. 596)
decomposer (p. 597)
herbivore (p. 597)
carnivore (p. 597)
omnivore (p. 597)

8. What are biogeochemical cycles? Describe several of the essential cycles.
9. What roles are played in an ecosystem by producers and consumers?
10. Describe the usual trophic relationships between producers, consumers, and detritivores in an ecosystem. What is the place of humans in a trophic system?
11. What is an energy pyramid? Describe how it relates to the nature of trophic levels.

■ *Describe* communities and ecological niches, and *list* several limiting factors on species distributions.

A **community** is formed by the interactions among populations of living animals and plants. Within a community, a **habitat** is the specific physical location of an organism—its address. An **ecological niche** is the function or operation of a life form within a given community—its profession.

Light, temperature, water, and nutrients constitute the life-supporting abiotic components of ecosystems. The zonation of plants and animal communities with altitude is the **life zone** concept, based on visible differences between ecosystems at different elevations. Each species has a *range of tolerance* that determines distribution. Species populations are stabilized by **limiting factors**, which may be physical, chemical, or biological characteristics of the environment.

community (p. 599)
habitat (p. 599)
ecological niche (p. 599)
life zone (p. 601)
limiting factor (p. 602)

12. Define a community within an ecosystem.
13. What do the concepts of habitat and niche involve? Relate them to some specific plant and animal communities.
14. Describe mutualistic and parasitic relationships in nature. Draw an analogy between these relationships and the relationship of humans to our planet. Explain.
15. Discuss several abiotic influences on the function and distribution of species and communities.
16. Describe what Alexander von Humboldt found that led him to propose the life zone concept. What are life zones? Explain the interaction between elevation, latitude, and the types of communities that develop.
17. What is a limiting factor? How does it function to control populations of plant and animal species?

■ *Outline* the stages of ecological succession in both terrestrial and aquatic ecosystems.

Natural and anthropogenic disturbance are common in most ecosystems. Wildfire can have far-ranging effects on communities; the science of **fire ecology** examines the role of fire in ecosystem maintenance. **Ecological succession** describes the process whereby communities of plants and animals change over time, often after an initial disturbance. An area of bare rock and soil with no trace of a former community can be a site for **primary succession**. The species that first establish in a disturbed area make up the **pioneer community** that then alters the habitat such that different species arrive. **Secondary succession** begins in an area that has a vestige of a previously functioning community in place. Rather than progressing smoothly to a definable stable endpoint, ecosystems tend to operate in a dynamic condition, with intermittent disturbance that forms a mosaic of habitats at different successional stages. Aquatic ecosystems also undergo succession; **eutrophication** is the gradual enrichment of water bodies that occurs with nutrient inputs, either natural or human-caused.

fire ecology (p. 603)
ecological succession (p. 603)
primary succession (p. 604)
pioneer community (p. 604)
secondary succession (p. 605)
eutrophication (p. 606)

18. How does ecological succession proceed? Describe the character of a pioneer community. What is the difference between primary and secondary succession?
19. How are wildfires important for ecological succession? How have species and ecosystems adapted for frequent wildfire?
20. Assess the impact of climate change on natural communities and ecosystems. Possible examples are changes in species' distributions or the changes and effects of wildfire.
21. Summarize the process of succession in a body of water. What is meant by eutrophication?
22. Is eutrophication occurring in the Great Lakes ecosystems? How does eutrophication relate to the formation of dead zones in freshwater ecosystems?

■ *Explain* how biological evolution led to the biodiversity of life on Earth.

Biodiversity (a combination of *bio*logical and *diversity*) includes the number and variety of different species, the genetic diversity within species, and ecosystem and habitat diversity. The greater the biodiversity is within an ecosystem, the more stable and resilient the system is and the more productive it will be. Modern agriculture often creates a nondiverse monoculture that is particularly vulnerable to failure.

Evolution states that single-cell organisms adapted, modified, and passed along inherited changes to multicellular organisms. The genetic makeup of successive generations was shaped by environmental factors, physiological functions, and behaviors that created a greater rate of survival and reproduction; these successful traits were passed along through natural selection.

biodiversity (p. 608)
evolution (p. 608)

23. Give some of the reasons why biodiversity infers more stable, efficient, and sustainable ecosystems.
24. Referring back to Chapter 1, define the scientific method and a theory, and describe the progressive stages that lead to the development of a theory.
25. What is meant by ecosystem stability?
26. What do prairie ecosystems teach us about communities and biodiversity?

MasteringGeography™

Looking for additional review and test prep materials? Visit the Study Area in *MasteringGeography*™ to enhance your geographic literacy, spatial reasoning skills, and understanding of this chapter's content by accessing a variety of resources, including MapMaster interactive maps, geoscience animations, satellite loops, author notebooks, videos, RSS feeds, flashcards, web links, self-study quizzes, and an eText version of *Geosystems*.

▲Figure 19.24 **Declining food resources due to a loss of sea ice in an Arctic food web.** A male polar bear (*Ursus maritimus*) walks between the ocean and a walrus (*Odobenus rosmarus*) haul along the ice-free shore of Phippsøya (Phipps Island), Arctic Ocean, in 2013. Sea ice, including the multiyear ice preferred by polar bears in their hunt for seals, dominated this same scene in the early 2000s. Without sea ice for their hunting platform, polar bears must forage for food on land, searching for bird eggs, injured walruses or seals, and even chewing kelp. Meanwhile, walruses feed on clams in the shallows near land, brushing aside sand and sediment with their whiskers, pulling the clam meat into their mouths with a powerful sucking action. These walrus males weigh more than 1000 kg (2200 lbs) each. The hungry bear keeps its distance, instinctively avoiding exposure to the dangerous walrus tusks. [Bobbé Christopherson.]

20 Terrestrial Biomes

The vegetation of almost half of Africa falls within the tropical savanna biome, consisting of grasses with scattered trees and shrubs. In northern Namibia, the southern giraffe (*Giraffa camelopardalis giraffa*) is one of many herbivores ranging throughout 233,000 km^2 (90,000 mi^2) of tropical savanna. An estimated 12,000 individuals of this subspecies remain in the wild, but numbers are declining as the animals are hunted for their hides and meat. The tropical savannas of Africa support the largest diversity of hoofed animals of any biome on Earth. [Jim Zuckerman/Corbis.]

KEY LEARNING concepts

After reading the chapter, you should be able to:

- ***Locate*** the world's biogeographic realms, and ***discuss*** the basis for their delineation.
- ***Explain*** the basis for grouping plant communities into biomes, and ***list*** the major aquatic and terrestrial biomes on Earth.
- ***Explain*** the potential impact of non-native species on biotic communities, using several examples.
- ***Summarize*** the characteristics of Earth's 10 major terrestrial biomes, and ***locate*** them on a world map.
- ***Discuss*** strategies for ecosystem management and biodiversity conservation.

Invasive Species Arrive at Tristan da Cunha

Geosystems Now in Chapter 6 describes an accident that left an oil-drilling platform adrift in the currents of the South Atlantic Ocean. Eventually the rig ran aground in Trypot Bay on the remote island of Tristan da Cunha, carrying new organisms into an existing marine ecosystem (Figure GN 20.1).

Aquatic Species on the Rig Because the rig had not been cleaned before towing, the platform was carefully checked for unwelcome organisms after making landfall in Tristan. Although the living quarters inside the rig showed no rodents or other terrestrial animals, the underwater portion of the rig carried a virtually intact subtropical reef community of 62 species, all non-native to the island.

A non-native species, also called an exotic or alien species, is one that originates outside of a region of interest. If a non-native species becomes invasive in a new environment, it outcompetes native species for resources and can introduce new predators, pathogens, or parasites into an ecosystem. Invasive species can devastate biodiversity, especially in isolated island ecosystems. The arrival of a marine community on the drilling platform thus provided an opportunity for scientists to assess a potential biological invasion.

The Scientific Assessment Scientists made numerous dives to survey the newly arrived organisms to Tristan da Cunha. They developed a system of four assessed risk levels for a potential invasion: (1) no threat from species that perished in transit; (2) low threat from species that persist on the rig alone; (3) medium threat from species that could spread from the rig; and (4) high threat from breeding species with strong invasive potential.

At level 1, scientists found corals that died in transit, providing skeleton microhabitats for various non-native worms, small crabs, amphipods, and other species that posed a level-2 threat. Also present were living and dead barnacles; the living acorn barnacles, the largest on the rig, were assessed at level 3. Among the shells left by the dead barnacles (assessed at level 1) lived small sponges, brown mussels, and urchins, assessed at level 2 (Figure GN 20.2).

A population of free-swimming finfish species was found around the rig, apparently having swum along with the rig as it drifted—the first report of such finfish becoming established after this type of unintentional movement. Two of these species, the silver porgy (with more than 60 counted around the rig) and the variable blenny (having a reproductive doubling time of 15 months), pose the greatest threat to native communities—a level-4 risk. Scientists also found porcelain crabs and a balanidae (a type of barnacle), both assessed as presenting a level-4 invasion risk.

Effects and Lessons Learned Tristan's population of about 300 people depends on rock lobster (a crawfish) for export, the island's main source of income (Figure GN 20.1b). In 2007, to prevent invasive species from threatening the rock lobster industry, the drilling platform was salvaged and then towed to distant deep water and sunk, at a total cost of $20 million. Nine months after the rig's removal, scientists failed to find any exotic species at the grounding site. However, the risk of invasion will continue for an unknown period. The scientific team concluded that the towing of rigs that have not been properly cleaned presents "unexcelled opportunities for invasion to a wide diversity of marine species."

The arrival of the oil platform and its associated species into Tristan's waters demonstrates how globalization affects even the most distant archipelago. Aquatic and terrestrial biomes and the effects of invasive species are the subjects of this chapter.

▲**Figure GN 20.2 Coral skeletons and dead barnacle shells on the submerged portion of the oil-drilling rig.** [Sue Scott. All rights reserved.]

(a) Natural kelp forest.

(b) Native Tristan rock lobster.

▲**Figure GN 20.1 Native marine environment offshore from Tristan da Cunha.** [Sue Scott. All rights reserved.]

GEOSYSTEMS NOW ONLINE Go to Chapter 20 on the *MasteringGeography* website (www.masteringgeography.com) for more on non-native and invasive species, and see the Tristan website at http://www.tristandc.com/. Also, see R. M. Wanless, S. Scott, et al., 2009, Semi-submersible rigs: A vector transporting entire marine communities around the world, *Biological Invasions* 10 (8): 2573-2583. MG

The patterns of species distributions on Earth are important subjects of biogeography. Earth's biodiversity is spread unevenly across the planet and is related to geology, climate, and the evolutionary history of particular species and species assemblages. The branch of biogeography that is concerned with the past and present distributions of animals is called *zoogeography*; the corresponding branch for plants is *phytogeography*.

Plant and animal communities are commonly grouped into *biomes*, also known as *ecoregions*, representing the major ecosystems of Earth. A biome is a large, stable community of plants and animals whose boundaries are closely linked to climate. Ideally, biomes are defined by mature, natural vegetation; however, most of Earth's biomes have been affected by human activities, and many are now experiencing accelerated rates of change that could produce dramatic alterations in the biosphere within our lifetime.

In this chapter: We begin with a discussion of Earth's biogeographic realms, the broadest groupings of species. We then examine biomes, explaining the basis for their classification and considering invasive species and their impacts on plant and animal communities within biomes. The bulk of this chapter explores Earth's 10 major terrestrial biomes, including their location, community structure, and sensitivity to human impacts. Table 20.1 summarizes the connections between vegetation, climate, soils, and water-budget characteristics for each biome.

Biogeographic Divisions

Earth's biosphere can be divided geographically based on assemblages of similar plant and animal communities. One class of geographic division—the biogeographic region, or realm—is determined by species distributions and their evolutionary history. Another class—the biome—is based on plant communities; it is determined mainly by vegetation life form and community characteristics as they relate to climate and soils.

Biogeographic Realms

The recognition that distinct regions of broadly similar flora (plants) and fauna (animals) exist was the earliest beginning of *biogeography* as a discipline. (*Flora* and *fauna* are general terms for the typical collections of animals and plants throughout a region or ecosystem.) A **biogeographic realm** (sometimes called an *ecozone*) is a geographic region where a group of associated plant and animal species evolved. Alfred Wallace (1823–1913), the first scholar of zoogeography, developed a map delineating six zoogeographical regions in 1860, building on earlier work by others regarding bird distributions (Figure 20.1a). Wallace's realms correspond generally to the continental plates, although Wallace knew nothing of the theory of plate tectonics at the time. Biogeographic realms were also developed based on plant associations

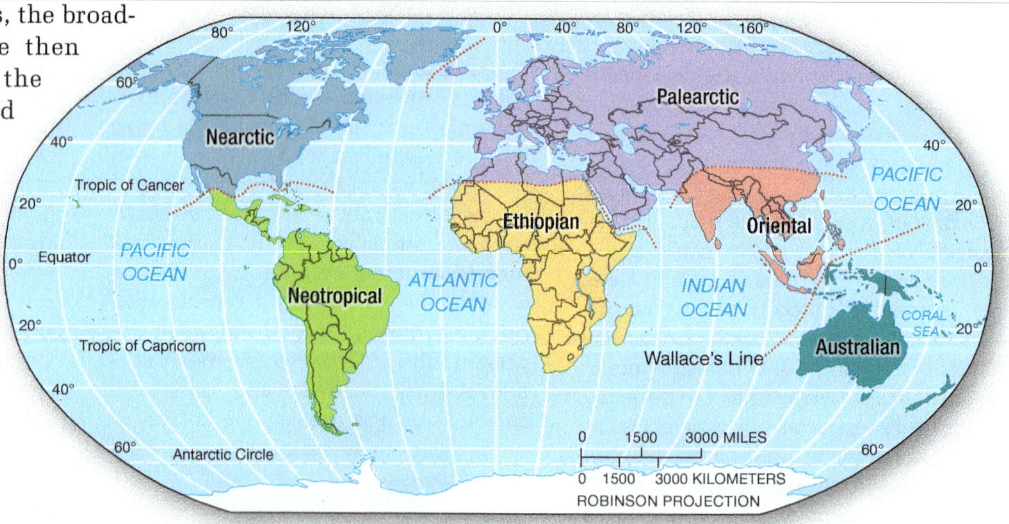

(a) The six animal realms as defined by biographer Alfred Wallace in 1860.

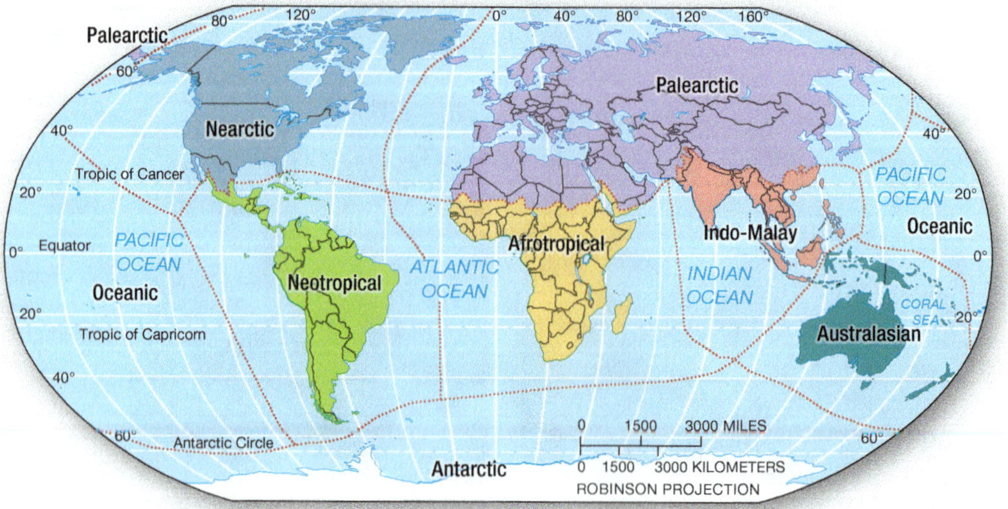

(b) The eight realms in use today, based on plant and animal associations and evolution.

▲**Figure 20.1 Biogeographic realms.** [(b) After Olsen, et al., "Terrestrial Ecoregions of the World: A New Map of Life on Earth," *Bioscience* 51:933-938 (2004); modified by UNEP/WCMC, 2011.]

▲Figure 20.2 **The unique Australian biogeographic realm.** Australia's flora and fauna evolved in isolation; here a western gray kangaroo stands near a dry river with eucalyptus trees. [A. Held/www.agefotostock.com.]

and modified over time, resulting in the similar though slightly more detailed divisions in use today, shown in Figure 20.1b.

Species interactions occurred as continents collided and remained attached; species became separated when continents drifted apart. Consequently, the organisms within each realm are a product of plate tectonics and evolutionary processes. For example, the Australian realm is unique for its approximately 450 species of *Eucalyptus* among its plants and for its 125 species of marsupials—animals, such as kangaroos, that carry their young in pouches, where gestation is completed (Figure 20.2). The presence of monotremes, egg-laying mammals such as the platypus, adds further distinctiveness to this realm.

The boundaries of biogeographic regions are usually determined by climatic and topographic barriers, such as deserts, rivers, mountain ranges, and oceans. Australia's unique native flora and fauna are the result of its early isolation from the other continents. During critical evolutionary times, Australia drifted away from Pangaea (see Chapter 11, Figure 11.13) and never again was reconnected by a land bridge, even when sea level lowered during repeated glacial ages. New Zealand, although relatively close in location, is isolated from Australia, explaining why it has no native marsupials. However, other factors resulted in the grouping of New Zealand within the Australian realm in the most recent classification of biogeographic realms.

Wallace noted the stark contrast in animal species between several of the islands of present-day Indonesia—Borneo and Sulawesi, in particular—and those of Australia. This led him to delineate a dividing line between the Oriental and Australian realms over which he believed species did not cross. This deep water barrier existed even during the lower sea levels of the last glacial maximum, when land connections existed between the two realms. His boundary is today known as "Wallace's line." Modern biogeographers have modified this line so that it now encircles the island region between Java and Papua New Guinea, an area that never had a land connection to the mainland and is now sometimes referred to as *Wallacea.*

Biomes

A **biome** is defined as a large, stable, terrestrial or aquatic ecosystem classified according to the predominant vegetation type and the adaptations of particular organisms to that environment. Although scientists identify and describe aquatic biomes (the largest of which are separated into freshwater and marine), biogeographers have applied the biome concept much more extensively to *terrestrial ecosystems*, associations of land-based plants and animals and their abiotic environment.

Vegetation Types

Scientists determine biomes based on easily identifiable vegetation characteristics (*vegetation* includes the entire flora in a region). Earth's vegetation types can be grouped according to the *growth form* (sometimes called *life form*) of the dominant plants. Examples of growth forms include:

- *Winter-deciduous trees*—large, woody, perennial plants that lose leaves during the cold season in response to temperature
- *Drought-deciduous shrubs*—smaller woody plants with stems branching at the ground that lose leaves during the dry season
- *Annual herbs*—small seed-producing plants without woody stems that live for one growing season
- *Bryophytes*—nonflowering, spore-producing plants, such as mosses (see Geo Report 20.2)
- *Lianas*—woody vines
- *Epiphytes*—plants growing above the ground on other plants and using them for support (see Figure 19.14)

These are just a few examples among many growth forms that are based on size, woodiness, life span, leaf traits, and general plant morphology. Vegetation can

GEOreport 20.1 A new look at Wallace's zoogeographic regions

In 2013, a group of scientists published a new map of zoogeographic realms based on the present distributions of amphibians, birds, and nonmarine mammals, as well as their phylogeny (the evolutionary relationships between organisms based on their ancestors and descendants). The research group identified 11 zoogeographic realms, which are roughly similar to but more detailed than Wallace's original six realms. This updated map provides a baseline for a variety of biogeographical and conservation-oriented studies; see the map, published in the journal *Science*, at http://macroecology.ku.dk/resources/wallace/credit_journal_science_aaas.jpg/.

also be characterized by the structure of the canopy, especially in forested regions. Together, the dominant growth form and the canopy structure characterize the vegetation type; the dominant vegetation type then characterizes the biome. (The dominant vegetation type extending across a region is also sometimes called the *formation class.*)

Biogeographers often designate six major groups of terrestrial vegetation: forest, savanna, shrubland, grassland, desert, and tundra. However, most biome classifications are more specific, with the total number of biomes usually ranging from 10 to 16, depending on the particular classification system being used. The specific vegetation types of each biome foster related animal associations that also help define its geographic area.

For example, forests can be subdivided into several biomes—rain forests, seasonal forests, broadleaf mixed forests, and needleleaf forests—based on moisture regime, canopy structure, and leaf type. *Rain forests* occur in areas with high rainfall; tropical rain forests are composed of mainly evergreen *broadleaf trees* (having broad leaves, as opposed to needles), and temperate rain forests are composed of both broadleaf and *needleleaf trees* (having needles as leaves). *Seasonal forests*, also known as dry forests, are characterized by distinct wet and dry seasons during the year, with trees that are mainly *deciduous* (shedding their leaves for some season of the year) during the dry season. *Broadleaf mixed forests* occur in temperate regions and include broadleaf deciduous trees, as well as needleleaf trees. *Needleleaf forests* are the coniferous forests of Earth's high-latitude and high-elevation mountain regions. *Coniferous forests* are cone-bearing trees with needles or scaled evergreen leaves, such as pines, spruces, firs, and larches.

In their form and distribution, plants reflect Earth's physical systems (the abiotic factors discussed in Chapter 19), including its energy patterns; atmospheric composition; temperature and winds; precipitation quantity, quality, and seasonal timing; soils and nutrients; chemical pathways; and geomorphic processes. Biomes usually correspond directly to moisture and temperature regimes (Figure 20.3). In addition, plant communities also reflect the growing influence of humans.

Biomes are defined by species that are native to a region, meaning that their occurrence is a result of natural processes. Today, few natural communities of plants and animals remain; most biomes have been greatly altered by human intervention. Thus, the "natural vegetation" identified on many biome maps reflects idealized potential mature vegetation given the environmental characteristics in a region. For example, in Norway, the former needleleaf forest is today a mix of second-growth forests, farmlands, and altered landscapes (Figure 20.4). However, the boreal forest biome designation for this region remains, based on idealized conditions before human impacts (discussed later in the chapter).

Ecotones Boundaries between natural systems, whether they separate biomes, ecosystems, or small habitats, are often zones of gradual transition in species composition, rather than rigidly defined frontiers marked by abrupt change. A boundary zone between different but adjoining ecosystems at any scale is an **ecotone**. These are often "zones of shared traits" between different communities.

Because different ecotones are often defined by different physical factors, they are likely to vary in width. Ecosystems separated by different climatic conditions usually have gradual ecotones, whereas those separated by differences in soils or topography may have abrupt boundaries. For example, the climatic boundary between grasslands and forests can occupy many kilometers of land, while a boundary in the form of a landslide, a river, a lakeshore, or a mountain ridge may occupy only a few meters. As human impacts cause ecosystem fragmentation, ecotones between habitats and ecosystems are becoming more numerous across the landscape (Figure 20.4).

The range of environmental conditions frequently found in ecotones can make them areas of high biodiversity; often they have larger population densities than communities on either side. Scientists have defined certain plant and animal species as having a range of tolerance for varying habitats; these "edge" species are often able to occupy territory within and on either side of the ecotone.

Invasive Species

The native species of natural biomes have come to inhabit those areas as a consequence of the evolutionary and physical factors discussed previously in this chapter and in Chapter 19. However, communities, ecosystems, and biomes can also be inhabited by species that are introduced from elsewhere by humans, either intentionally or accidentally, as described for Tristan da Cunha in

GEOreport 20.2 Plant communities survive under glacial ice

Glacial retreat has exposed communities of bryophytes that lived 400 years ago, during the warmer interglacial period known as the Little Ice Age. Recently, scientists collected and dated samples of these communities in the Canadian Arctic. They also successfully cultured the plants in a laboratory, using a single cell of the exhumed material to regenerate the entire original organism. Thus, bryophytes can survive long periods of burial under thick glacial ice, and under the right conditions, potentially recolonize a landscape after glaciation.

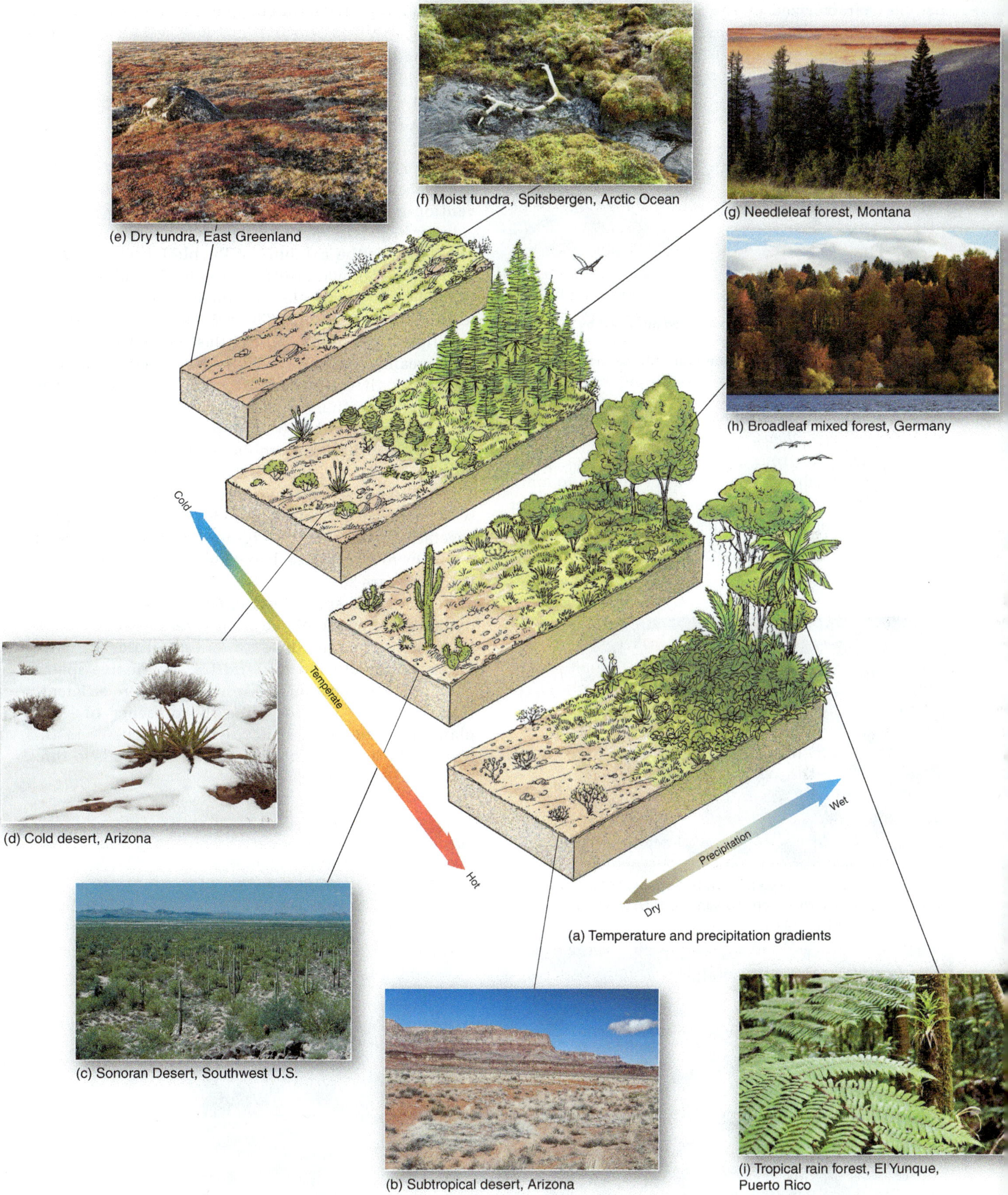

▲**Figure 20.3 Vegetation patterns in relationship to temperature and precipitation.** [(b, d, e, f, i) Bobbé Christopherson. (c) Author. (g) SNEHIT/Shutterstock. (h) blickwinkel/Alamy.]

▲Figure 20.4 **Needleleaf forest landscape modified by human activity, Norway.** Edge species often occupy the varied habitats where natural habitat borders disturbed land. [Bobbé Christopherson.]

Geosystems Now. Such non-native species are also known as *exotic species*, or *aliens*.

After arriving from a different ecosystem, probably 90% of introduced non-native species fail to move into established niches in their new community or habitat. However, some species are able to do so, taking over niches already occupied by native species and thus becoming **invasive species**. The 10% that become invasive can alter community dynamics and lead to declines in native species. Prominent examples are Africanized "killer bees" in North and South America; brown tree snakes in Guam zebra and quagga mussels in the Great Lakes (Figure 20.5a); Russian olive and tamarisk trees along streams in the U.S. Southwest (Figure 20.5b); and kudzu in the U.S. Southeast (Figure 20.5c). For information on invasive species prevention and management, see http://invasions.bio.utk.edu/ or http://www.invasivespecies.gov/.

Consider the example of Purple loosestrife (*Lythrum salicaria*), which was introduced from Europe in the 1800s as a desirable ornamental plant with some medicinal applications. The plant's seeds also arrived in ships that used soil for ballast. This hardy *perennial*, meaning a plant that lives for more than 2 years, escaped cultivation and invaded wetlands across the eastern portions of the United States and Canada, through the upper Midwest, and as far west as Vancouver Island, British Columbia, replacing native plants on which wildlife depend. The plant's invasive characteristics are its ability to produce vast quantities of seed during an extended flowering season and to spread vegetatively through underground stems, as well as its tendency to form dense, homogenous stands once established (Figure 20.6).

On a large spatial scale, invasions can alter the dynamics of entire biomes. For example, in the Mediterranean shrubland of southern California, the native vegetation is adapted for wildfire. Non-native species often are able to colonize burned areas more efficiently than do native species; thus the presence of exotic plants can change the successional processes in this biome. The establishment of non-natives leads to thick

(a) Zebra mussels cover most hard surfaces in the Great Lakes; they rapidly colonize on any surface, even sand, in freshwater environments.

(b) Invasive Russian olive (green-gray color) and tamarisk (dark green color in the shade) along Chinle Wash, New Mexico, in riparian habitat formerly occupied by native cottonwoods.

(c) Kudzu, originally imported for cattle feed, spread from Texas to Pennsylvania; here it overruns pasture and forest in western Georgia.

◄Figure 20.5 **Exotic species.** [(a) Purestock/Alamy (b) Lindsay Reynolds, Colorado State University/USGS. (c) Bobbé Christopherson.]

▲**Figure 20.6 Purple loosestrife in southern Ontario, Canada.** [Gaertner/Alamy.]

undergrowth, providing more fuel for fires that are increasing in frequency. Native vegetation is adapted for fires that occur at intervals of 30 to 150 years; the increase in fire frequency with climate change puts these species at a disadvantage. More frequent fires in the region combine with the increasing numbers of non-native species to cause the conversion of southern California's shrubland to grassland.

In the United States and Canada, humans are perpetuating a new type of terrestrial plant community in developed areas. This new community, a mix of native and non-native species used for landscaping, is somewhere between a grassland and a forest. Investments of water, energy, and capital are required to sustain the new species. Additionally, on rangelands and in agricultural areas, humans alter natural biomes by grazing non-native animals and planting crops from other regions (we discuss anthropogenic biomes further at the end of the chapter). Whether the land would return to its natural vegetation if human influences were removed is unknown.

Earth's Terrestrial Biomes

Given that extensive transition zones separate many of Earth's biomes, the classification of biomes according to distinct vegetation associations is difficult and somewhat arbitrary. The result is a number of classification systems—similar in concept but different in detail—used in biology, ecology, and geography textbooks. In *Geosystems*, we describe 10 biomes that are common to most classification systems: tropical rain forest, tropical seasonal forest and scrub, tropical savanna, midlatitude broadleaf forest, boreal and montane forest, temperate rain forest, Mediterranean shrubland, midlatitude grassland, desert, and arctic and alpine tundra.

The global distribution of these biomes is portrayed in Figure 20.7 and summarized in Table 20.1, which also includes pertinent information regarding climate, soils, and water availability. The following pages provide descriptions of each biome, synthesizing all we have learned in previous chapters about the interactions of atmosphere, hydrosphere, lithosphere, and biosphere. Because plant distributions respond to environmental conditions and reflect variations in climate and soil, the world climate map in Chapter 10, Figure 10.2, is a helpful reference for this discussion.

CRITICALthinking 20.1
Reality Check

Using the map in Figure 20.7, the information in Table 20.1, and the discussion in this chapter, describe the biome in which you are located. What changes in the natural vegetation do you see, as we all live in altered environments brought on by human activities. Consider the climate classification information in Ch 10, as you evaluate your biome. What generalizations can you make? ●

Tropical Rain Forest

The lush biome covering Earth's equatorial regions is the **tropical rain forest**. In the tropical climates of these forests, with consistent year-round daylength (12 hours), high insolation, average annual temperatures around 25°C (77°F), and plentiful moisture, plant and animal populations have responded with the most diverse expressions of life on the planet. Rainforest species evolved during the long-term residence of the continental plates near equatorial latitudes. Although this biome is stable in its natural state, undisturbed tracts of rain forest are becoming increasingly rare; deforestation is perhaps the most pervasive human impact.

The largest tract of tropical rain forest occurs in the Amazon region, where it is called the *selva*. Tropical rain forests also cover the equatorial regions of Africa, parts of Indonesia, the margins of Madagascar and Southeast Asia, the Pacific coast of Ecuador and Colombia, and the east coast of Central America, with small discontinuous patches elsewhere. The cloud forests of western Venezuela are high-elevation tropical rain forests, perpetuated by high humidity and cloud cover. Rain forests occupy about 7% of the world's total land area but represent approximately 50% of Earth's species and about half of its remaining forests.

Rain Forest Flora and Fauna The rainforest canopy forms three levels (Figure GIA 20.1). The upper level, called the *overstory*, is not continuous, but features emergent tall trees whose high crowns rise above the middle canopy, which is continuous. Biomass in a rain forest is concentrated in the dense mass of overhead leaves in these two areas of the canopy. The lower level is the *understory*, where broad leaves block much of the light so that the forest floor receives only about 1% of the sunlight arriving at the canopy. The lower level of vegetation is composed of seedlings, ferns, and bamboo, leaving the litter-strewn ground surface in deep shade and fairly open. The constant moisture, odors of mold and rotting vegetation, strings of thin roots and vines dropping down from above, windless air, and echoing sounds of life in the trees together create a unique environment.

Infertile soils, principally Oxisols, support these biologically rich forests. Rainforest trees have adapted to the poor soils with root systems able to capture nutrients from litter decay at the soil surface.

Rain forests feature ecological niches that are distributed vertically rather than horizontally because of the competition for light. The canopy is filled with a rich variety of plants and animals. Lianas (woody vines that are rooted in the soil) stretch from tree to tree, entwining them with cords that can reach 20 cm (8 in.) in diameter. Epiphytes flourish there, too; these plants, such as orchids, bromeliads, and ferns, live entirely aboveground, supported physically, but not nutritionally, by the structures of other plants. On the forest floor, the smooth, slender trunks of rainforest trees are covered with thin bark and buttressed by large, wall-like flanks that grow out from the trees to brace the trunks. These buttresses form angular hollows that are ready habitat for various animals. Branches are usually absent on at least the lower two-thirds of the tree trunks.

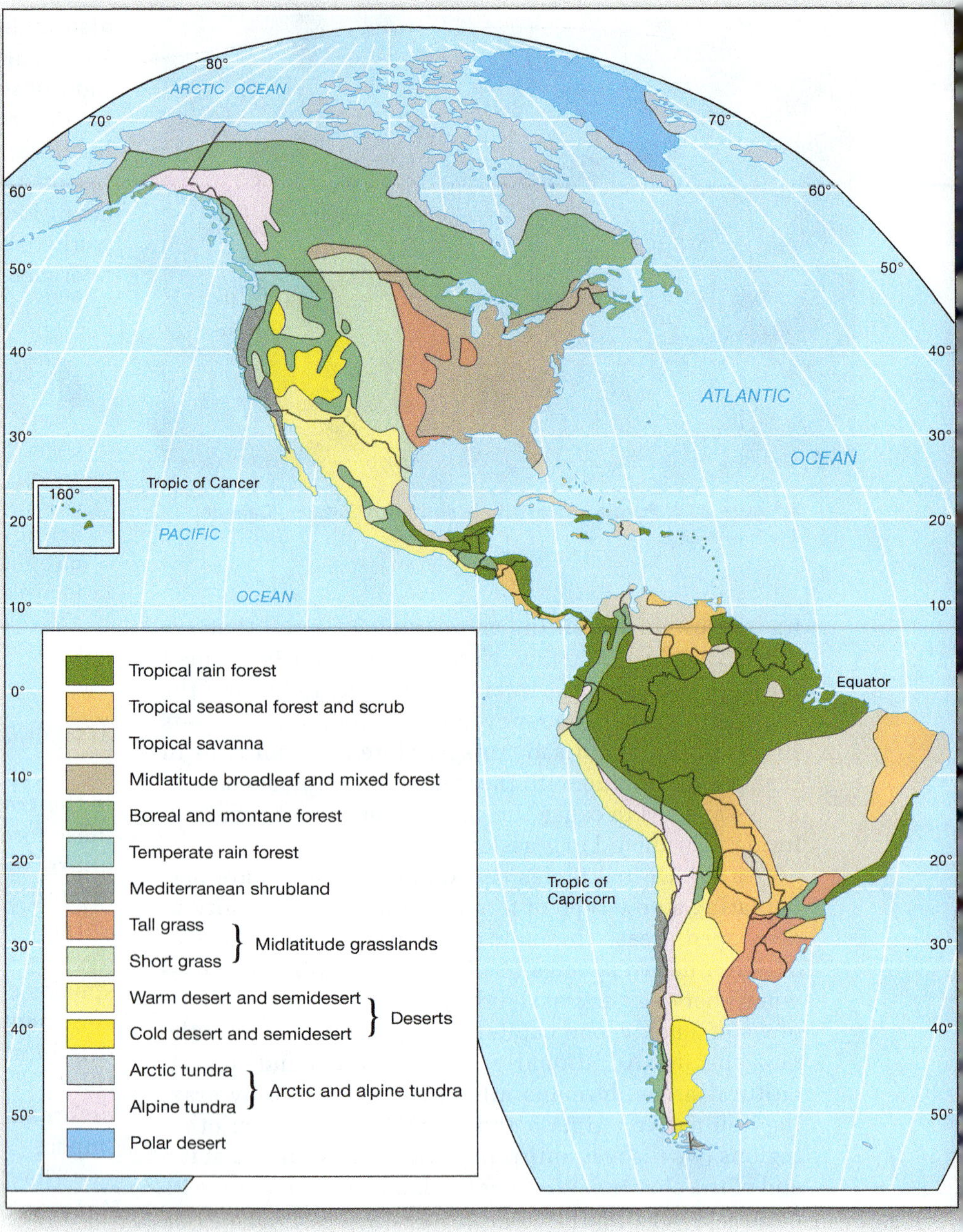

▲**Figure 20.7 The 10 major global terrestrial biomes.** Biomes are described in Table 20.1.

The animal and insect life of the rain forest is diverse, ranging from animals living exclusively in the upper stories of the trees to decomposers (bacteria) working the ground surface. *Arboreal* (from the Latin word meaning "tree") species, those dwelling in the trees, include sloths, monkeys, lemurs, parrots, and snakes. Throughout the canopy are multicolored birds, tree frogs, lizards, bats, and a rich insect community that includes more than 500 species of butterflies. On the forest floor, animals include pigs (the bushpig and giant forest hog in Africa, wild boar and bearded pig in Asia, and peccary in South America), small antelope, and mammalian predators (the tiger in Asia, jaguar in South America, and leopard in Africa and Asia).

Deforestation of the Tropics More than half of Earth's old-growth tropical rain forest is gone, cleared for pasture, timber, fuel wood, and farming. Tropical deforestation is threatening native rainforest species, and in the bigger picture of global systems and societies, is jeopardizing an important recycling system for atmospheric carbon dioxide, as well as potential sources of valuable pharmaceuticals and new foods—so much is still unknown and undiscovered.

Worldwide, an area nearly the size of Wisconsin is cleared each year (169,000 km^2, or 65,000 mi^2), and about a third more is disrupted by selective cutting of canopy trees that occurs along the edges of deforested areas. The economically valuable varieties of trees include mahogany, ebony, and rosewood. Selective cutting required for species-specific logging is difficult because individual species are widely scattered; a species may occur only once or twice per square kilometer. As a result, most logging in rain forests is for pulpwood production, which takes all species.

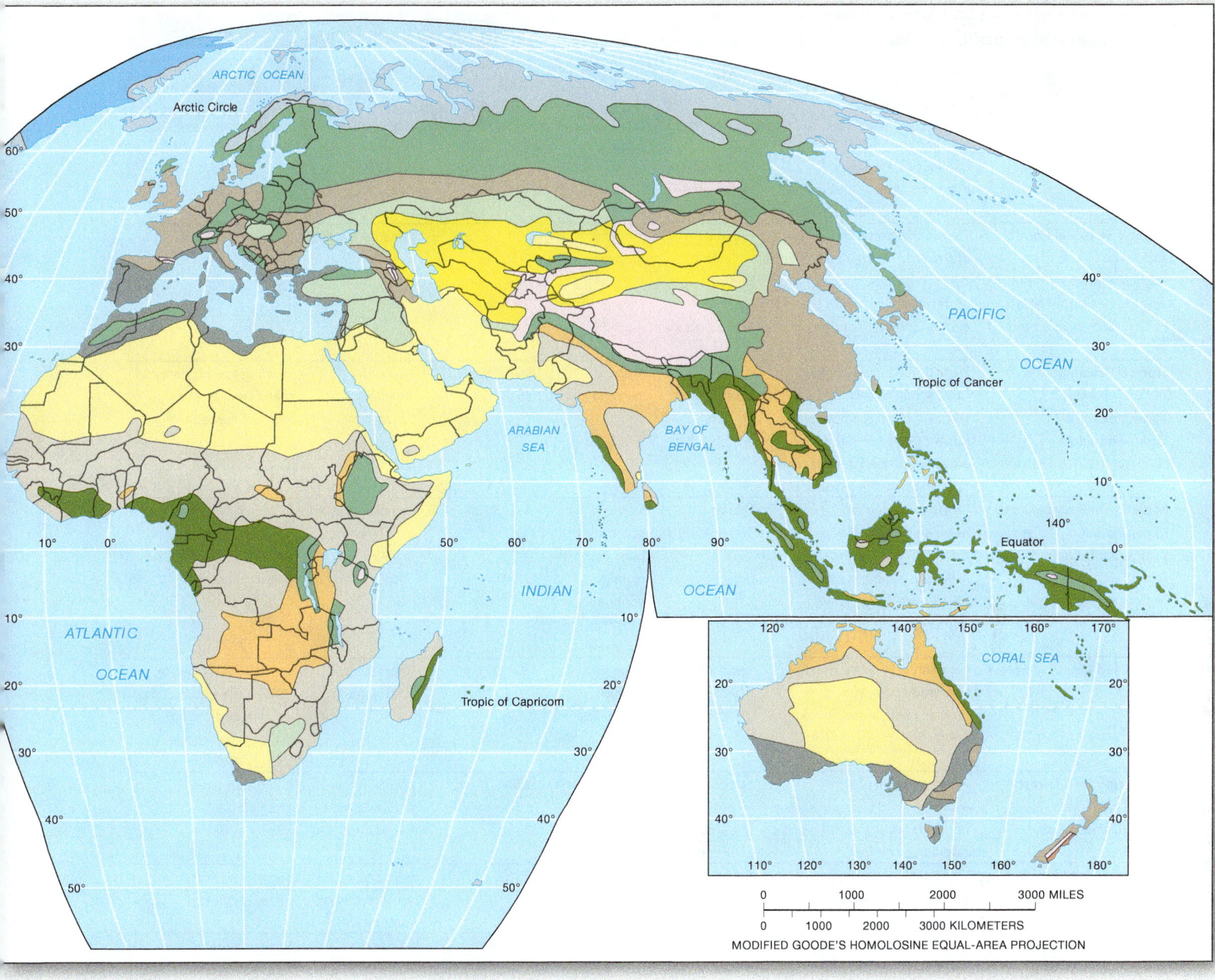

Fires are used to clear forested lands for agriculture, which is intended to feed the domestic population as well as to produce accelerating cash exports of beef, soybeans, rubber, coffee, palm oil, and other commodities. When orbiting astronauts look down on the rain forests at night, they see thousands of human-set fires. During the day, the lower atmosphere in these regions is choked with the smoke. Forest clearing and burning release millions of metric tons of carbon into the atmosphere each year.

Because of the poor soil fertility, intensive farming quickly exhausts the productivity of cleared lands, which are then generally abandoned in favor of newly burned lands (unless fertility is maintained artificially by chemical fertilizers). The dominant trees require from 100 to 250 years to reestablish after a major disturbance. Once cleared and abandoned, the former forest becomes a mass of low bushes intertwined with vines and ferns, slowing the return of trees.

Satellite images show rainforest destruction over the past several decades. For example, new roads and forest removal have significantly changed the landscape of the state of Rondônia in western Brazil (Figure GIA 20.2; note the encroachment along new roads branching from highway BR364). The edges of every road and cleared area represent a significant portion of the habitat disturbance—more impact occurs along these edges than in the tracts of clear-cut logging and burning. The hotter, drier, and windier edge conditions penetrate the forest up to 100 m (328 ft) and affect species and community dynamics.

In Brazil, deforestation has decreased in recent years as the government has stepped up enforcement

TABLE 20.1 Terrestrial Biomes and Their Characteristics

Biome and Ecosystems	Vegetation Characteristics	Soil Orders	Climate Type	Annual Precipitation Range	Temperature Patterns	Water Balance
Tropical Rain Forest Evergreen broadleaf forest Selva	Leaf canopy thick and continuous; broadleaf evergreen trees, vines (lianas), epiphytes, tree ferns, palms	Oxisols Ultisols (on well-drained uplands)	*Tropical*	180–400 cm (>6 cm/mo)	Always warm (21°–30°C; avg. 25°C)	Surpluses all year
Tropical Seasonal Forest and Scrub Tropical monsoon forest Tropical deciduous forest Scrub woodland and thorn forest	Transitional between rain forest and grassland; broadleaf, some deciduous trees; open parkland to dense undergrowth; acacias and other thorn trees in open growth	Oxisols Ultisols Vertisols (in India) Some Alfisols	*Tropical monsoon, tropical savanna*	130–200 cm (>40 rainy days during 4 driest months)	Variable, always warm (>18°C)	Seasonal surpluses and deficits
Tropical Savanna Tropical grassland Thorn tree scrub Thorn woodland	Transitional between seasonal forests, rain forests, and semiarid tropical steppes and desert; trees with flattened crowns, clumped grasses, and bush thickets; fire association	Alfisols Ultisols Oxisols	*Tropical savanna*	9–150 cm, seasonal	No cold-weather limitations	Tending toward deficits, therefore fire- and drought-susceptible
Midlatitude Broadleaf and Mixed Forest Temperate broadleaf Midlatitude deciduous Temperate needleleaf	Mixed broadleaf and needleleaf trees; deciduous broadleaf, losing leaves in winter; southern and eastern evergreen pines demonstrating fire association	Ultisols Some Alfisols	*Humid subtropical* (warm summer) *Humid* continental (warm summer)	75–150 cm	Temperate with cold season	Seasonal pattern with summer-maximum PRECIP and POTET; no irrigation needed
Boreal and Montane Forest Taiga	Needleleaf conifers, mostly evergreen pine, spruce, fir; Russian larch, a deciduous needleleaf	Spodosols Histosols Inceptisols Alfisols	*Subarctic* *Humid continental* (cool summer) *Highland*	30–100 cm	Short summer, cold winter	Low POTET, moderate PRECIP; moist soils, some waterlogged and frozen in winter; no deficits
Temperate Rain Forest West Coast forest U.S. coast redwoods	Narrow margin of lush evergreen and deciduous trees on windward slopes; redwoods, tallest trees on Earth	Spodosols Inceptisols (mountainous environs)	*Marine west coast*	150–500 cm	Mild summer and mild winter for latitude	Large surpluses and runoff
Mediterranean Shrubland Sclerophyllous shrub Australian eucalyptus forest	Short shrubs, drought-adapted, tending to grassy woodlands and chaparral	Alfisols Mollisols	*Mediterranean* (dry summer)	25–65 cm	Hot, dry summers, cool winters	Summer deficits, winter surpluses
Midlatitude Grassland Temperate grassland Sclerophyllous shrub	Tallgrass prairies and shortgrass steppes, highly modified by human activity; major areas of commercial grain farming; plains, pampas, and veld; fire association	Mollisols Aridisols	*Humid subtropical* *Humid continental* (hot summer)	25–75 cm	Temperate continental regimes	Soil-moisture utilization and recharge balanced; irrigation and dry farming in drier areas
Warm Desert and Semidesert Subtropical desert and scrubland	Bare ground graduating into xerophytic plants including succulents, cacti, and dry shrubs	Aridisols Entisols (sand dunes)	*Arid desert*	< 2 cm	Average annual temperature, around 18°C, highest temperatures on Earth	Chronic deficits, irregular precipitation events, PRECIP < 1/2 POTET
Cold Desert and Semidesert Midlatitude desert, scrubland, and steppe	Temperate desert vegetation including short grass and dry shrubs	Aridisols Entisols	*Semiarid steppe*	2–25 cm	Average annual temperature around 18°C	PRECIP >1/2 POTET
Arctic and Alpine Tundra	Treeless; dwarf shrubs, stunted sedges, mosses, lichens, and short grasses; alpine, grass meadows	Gelisols Histosols Entisols (permafrost)	*Tundra* *Subarctic* (very cold)	15–180 cm	Warmest months >10°C, only 2 or 3 months above freezing	Not applicable most of the year, poor drainage in summer
Polar Desert	Mosses, lichens	Permafrost	*Ice sheet, ice cap*	<25 cm	Warmest months <10°C	Not applicable

to preserve the forest; 2009–2010 combined losses were down to approximately 18,000 km^2 (6950 mi^2), or about 80% of the area of New Jersey (Figure GIA 20.2c). By continent, rainforest losses are estimated at more than 50% in Africa, more than 40% in Asia, and 40% in Central and South America.

Deforestation is a highly charged issue, especially for Brazil's growing cattle industry, which uses deforested lands for pasture. Cattle herds reached over 60 million head by 2010, generating $3 billion in revenue. Among many available websites, see the Tropical Rainforest Coalition at http://www.rainforest.org/ or Rainforest Action Network at http://www.ran.org/.

CRITICALthinking 20.2
Tropical Forests: A Global or Local Resource?

Given the information presented in this chapter about deforestation in the tropics, and in the previous chapter about declining biodiversity, assess the present controversy over rainforest resources. What are the main issues? How do developing countries, which possess most of the world's rain forests, view rainforest destruction? How do developed countries, with their transnational corporations, view rainforest destruction? How do concerns for planetary natural resources balance against the needs of local peoples and sovereign state rights? How is climate change related to these issues? What kind of action plan would you develop to accommodate all parties? •

Tropical Seasonal Forest and Scrub

The area of seasonal changes in precipitation at the margins of the world's rain forests is the **tropical seasonal forest and scrub** biome, which occupies regions of lower and more erratic rainfall than occurs in the equatorial regions. The biome encompasses the tropical monsoon and tropical savanna climates, with fewer than 40 rainy days during their four consecutive driest months and heavy monsoon downpours during their summers (see Chapter 6, Figures 6.15 and GIA 6). The shifting intertropical convergence zone (ITCZ) affects precipitation regimes, bringing moisture with the high Sun of summer and then a season of dryness with the low Sun of winter. This shift produces a seasonal pattern of moisture deficits, which affects vegetation leaf loss and flowering. The term *semideciduous* applies to many broadleaf trees that lose some of their leaves during the dry season.

Thus, the tropical seasonal forest and scrub is a varied biome that occupies a transitional area from wetter to drier tropical climates. Natural vegetation ranges from monsoon forests to open woodlands to thorn forests to semiarid shrublands. The monsoonal forests have an average height of 15 m (50 ft) with no continuous canopy of leaves, transitioning into drier areas with open grassy areas or into areas choked by dense undergrowth. In the more-open tracts, a common tree is the acacia, with its flat top and usually thorny stems. *Scrub vegetation* consists of low shrubs and grasses with some adaptations to semiarid conditions.

Local names are given to these communities: the *Caatinga* of the Bahia State of northeastern Brazil; the *Chaco* (or *Gran Chaco*) of southeastern Brazil, Paraguay, and northern Argentina (Figure 20.8a, page 631); the *brigalow* scrub of Australia; and the *dornveld* of southern Africa. In Africa, this biome extends west to east from Angola through Zambia to Tanzania and Kenya. Tropical seasonal forests are also present in Southeast Asia and portions of India, from interior Myanmar through northeastern Thailand; and in parts of Indonesia.

The trees throughout most of this biome make poor lumber, but some, especially teak, may be valuable for fine cabinetry and furniture. In addition, some of the plants with dry-season adaptations produce usable waxes and gums, such as the carnauba wax produced by the Brazilian palm tree. Animal life includes the koalas and cockatoos of Australia and the elephants, large cats, anteaters, rodents, and ground-dwelling birds in other occurrences of this biome. Worldwide, humans use these areas for ranching (Figure 20.8b); in Africa, this biome includes numerous wildlife parks and preserves.

Tropical Savanna

The **tropical savanna** biome consists of large expanses of grassland, interrupted by scattered trees and shrubs. Treeless tracts of grassland can also occur in this biome, with grasses in discontinuous clumps separated by bare ground. Tropical savannas receive precipitation during approximately 6 months of the year, when they are influenced by the shifting ITCZ. The rest of the year they are under the drier influence of shifting subtropical high-pressure cells. This is a transitional biome between the tropical seasonal forests and the semiarid tropical steppes and deserts.

Shrubs and trees of the savanna biome are adapted to drought, grazing by large herbivores, and fire. Most species are *xerophytic*, or drought resistant, with

(text continued on page 631)

GEOreport 20.3 Tropical rain forests as nature's medicine cabinet

Nature's biodiversity is like a full medicine cabinet. Since 1959, 25% of all prescription drugs were originally derived from higher plants. Scientists have identified 3000 plants, many of them in tropical rain forests, as having anticancer properties. The rosy periwinkle (*Catharanthus roseus*) of Madagascar contains two alkaloids that combat two forms of cancer. Yet less than 3% of flowering plants have been examined for alkaloid content. It defies common sense to throw away the medicine cabinet before we even open the door to see what is inside.

Earth's rain forests are the natural vegetation of tropical regions with abundant precipitation. Rain forests are also a vast reservoir of biodiversity. The rain forest's layered structure reflects intense competition for sunlight and space among numerous species of trees and other plants (GIA 20.1). For several decades, humans have cleared rain forests for agriculture, cattle ranching, timber export, and palm oil production. (GIA 20.2).

Rain forest in the mountains of Costa Rica [Ivalin/Shutterstock.]

20.1 VERTICAL STRUCTURE OF A RAIN FOREST

Trees create the distinctive "architecture" of the rain forest, made up of overstory, canopy, and understory, shown below. Long vines called lianas, rooted in the soil, connect these layers, while dead leaves form litter on the deeply shaded forest floor.

20.1a

Describe: What are the characteristics of the three main layers of the rain forest?

20.1b *Lianas, here shown in the middle canopy, drape over branches and trunks to the forest floor.* [Bobbé Christopherson]

20.1c *Leaf litter covers the rainforest floor, seen here with a typical buttressed tree trunk, and lianas in the background.* [Bobbé Christopherson]

Rainforest soil:
Rainforest soil is poor in nutrients. Most soil nutrients have been taken up to help form the biomass of trees and other rainforest organisms. These nutrients are recycled rapidly as plant and animal remains decay on the forest floor and are reabsorbed by tree roots.

Video
Plant Productivity in a Warming World

MasteringGeography™

Visit the Study Area in MasteringGeography™ to explore rain forests.

Visualize: Study a video of plant productivity in a warming world.

Assess: Demonstrate understanding of rain forests (if assigned by instructor).

Video
Amazon Deforestation

20.2 AMAZON RAIN FOREST DESTRUCTION

In Brazil, the vast Amazon rain forest is cleared to make way for agriculture and ranching, and for selective timber export, some illegal, of species such as mahogany. Roads that penetrate undeveloped areas facilitate this destruction, increasing habitat fragmentation and loss of biodiversity. GIA 20.2a and 20.2b show how one area changed between 2000 and 2009. During roughly the same time span, Brazil lost rain forest equal in area to all of the New England states plus New Jersey (GIA 20.2c)

20.2a *True-color satellite image of Rondônia in western Brazil in 2000 shows deforestation along highway BR364, the main artery of the region.*

[Eco Images/Getty.]

Analyze: Refer to the satellite images in GIA 20.2. How did the area between the Madeira River and highway BR364 change between 2000 and 2009? How might these changes have affected wildlife populations? Explain.

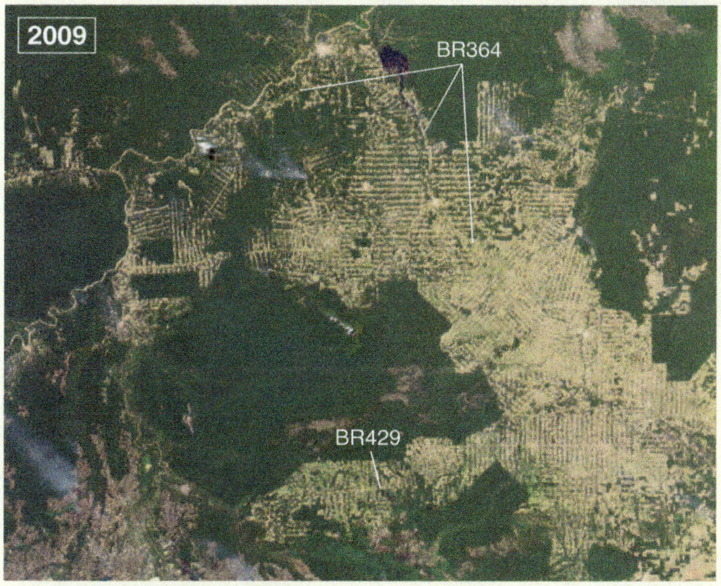

20.2b *The same region in 2009. The branching pattern of feeder roads causes habitat fragmentation.* (a, b) [Terra MODIS, NASA.]

20.2c *Extent of deforestation in Brazil (yellow) shown relative to the equivalent U.S. area.*

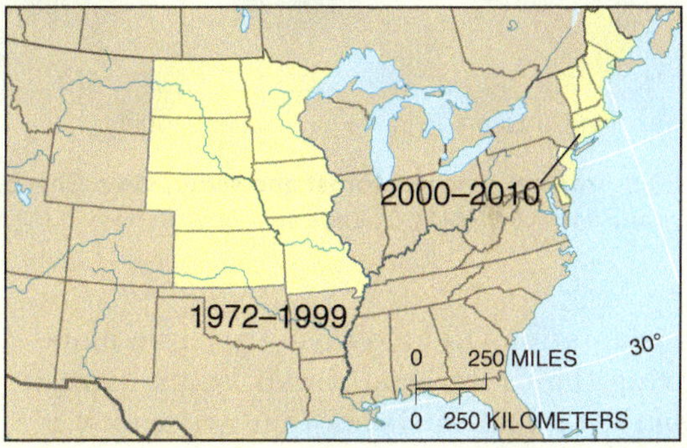

Recently burned rain forest in the Amazon. [Brasil2/Getty Images]

GEOquiz

1. Infer: How does the physical structure of the rain forest help to account for the great variety of organisms found there?

2. Predict: Identify an area of rain forest on the 2009 satellite image that is likely to be deforested in the near future and explain how and why the area will change.

(a) Trumpet trees (*Tabebuia caraiba*) are dry-season deciduous trees that are common in the seasonal forest and scrub of Paraguay.

(b) Cattle ranching is common in the Gran Chaco region.

▲**Figure 20.8 Tropical seasonal forest and scrub, Gran Chaco, Paraguay.** [(a) Imagebroker/Alamy. (b) Universal Images/DeAgostini/Alamy.]

various adaptations to help them conserve moisture during the dry season: small, thick leaves; rough bark; or leaf surfaces that are waxy or hairy. Many trees of the savanna woodlands have small leaves to retain moisture and are characteristically flat-topped or umbrella shaped in order to capture the maximum amount of sunlight on their small leaf surfaces (Figure 20.9).

Savanna vegetation is maintained by fire, both a natural and a human-caused disturbance in this biome. During the wet season, grasses flourish, and as rainfall diminishes, this thick growth provides fuel for fires, which are often intentionally set to maintain the open grasslands and suppress the growth of trees. Hot-burning dry-season fires kill trees and seedlings and deposit a layer of nutrient-rich ash over the landscape. These conditions foster the regrowth of grasses, which again grow vigorously as the wet season returns, sprouting from extensive underground root systems that are an adaptation for surviving fire disturbance. In northern Australia, the aboriginal people are credited with creating and maintaining many of the region's tropical savannas; as the traditional practice of setting annual fires declines, many savannas are reverting to forest.

Tropical savanna soils are much richer in humus than the soils of the wetter tropics and better drained, thereby providing a strong base for agriculture and grazing. Sorghums, wheat, and groundnuts (peanuts) are common crops of this biome.

Africa has the largest area of tropical savanna on Earth, including the famous Serengeti Plains of Tanzania and Kenya, and the Sahel region, south of the Sahara. Portions of Australia, India, and South America also are part of the savanna biome. Local names for tropical savannas include the *Llanos* in Venezuela, stretching along the coast and inland east of Lake Maracaibo and the Andes; the *Campo Cerrado* of Brazil and Guiana; and the *Pantanal* of southwestern Brazil.

Particularly in Africa, savannas are the home of large land mammals—zebra, giraffe, buffalo, gazelle, wildebeest, antelope, rhinoceros, and elephant. These animals graze on savanna grasses, while others (lion, cheetah) feed upon the grazers themselves. Birds include the ostrich, Martial Eagle (largest of all eagles), and Secretary Bird. Many species of venomous snakes, as well as the crocodile, are present in this biome.

Midlatitude Broadleaf and Mixed Forest

Moist continental climates support a mixed forest in areas of warm to hot summers and cool to cold winters. This **midlatitude broadleaf and mixed forest** biome includes several distinct communities in North

▲**Figure 20.9 Savanna landscape of the Serengeti Plains, east Africa.** Wildebeest, zebras, and acacia trees. [Stephen F. Cunha.]

America, Europe, and Asia. In the United States, relatively lush evergreen broadleaf forests occur along the Gulf of Mexico. To the north are the mixed deciduous broadleaf and needleleaf trees associated with sandy soils and frequent fires—pines (longleaf, shortleaf, pitch, loblolly) predominate in the southeastern and Atlantic coastal plains. In areas of this region protected from fire, broadleaf trees are dominant. Into New England and westward in a narrow belt to the Great Lakes, white and red pines and eastern hemlock are the principal conifers, mixed with broadleaf deciduous oak, beech, hickory, maple, elm, chestnut, and many others (Figure 20.10).

These mixed stands contain valuable timber, and logging has altered their distribution. Native stands of white pine in Michigan and Minnesota were removed before 1910, although reforestation sustains their presence today. In northern China, these forests have almost disappeared as a result of centuries of harvest. The forest species that once flourished in China are similar to species in eastern North America: oak, ash, walnut, elm, maple, and birch. This biome is quite consistent in appearance from continent to continent and at one time represented the principal vegetation of the humid subtropical (hot summer) regions of North America, Europe, and Asia.

A wide assortment of mammals, birds, reptiles, and amphibians is distributed throughout this biome. Representative animals (some migratory) include red fox, white-tailed deer, southern flying squirrel, opossum, bear, and a great variety of birds, including Tanager and Cardinal. To the west of this biome in North America are the rich soils and midlatitude climates that favor grasslands, and to the north is the gradual transition to the poorer soils and colder climates that favor the coniferous trees of the northern boreal forests.

▲Figure 20.10 **Mixed broadleaf forest, southeastern United states.** Great Smokey Mountains National Park in the southern Appalachian Mountains contains one of the largest remaining tracts of old-growth forest in North America, including nearly 100 native tree species, 66 mammal species, and over 200 species of birds. [BioLife Pics/Shutterstock.]

Boreal and Montane Forest

Stretching from the east coast of Canada and the Atlantic Provinces westward to the Canadian Rockies and portions of Alaska, and from Siberia across the entire extent of Russia to the European Plain is the **boreal forest** biome, also known as the northern **needleleaf forest** (Figure 20.11). The northern, less densely forested part of this biome, transitional to the arctic tundra biome, is called the **taiga**. This biome is characteristic of microthermal climates (having a cold winter season and also some summer warmth); the Southern Hemisphere has no such biome except in a few mountainous locales. The needleleaf forests at high elevations on mountains worldwide are called **montane forests**.

Boreal forests of pine, spruce, fir, and larch occupy most of the subarctic climates on Earth that are dominated by trees. Although these forests have similar vegetation life forms, individual species vary between North America and Eurasia. The larch (*Larix*) is one of only a few needleleaf trees that drop needles in the winter months, perhaps as a defense against the extreme cold of its native Siberia (see the Verkhoyansk climograph and photograph in Figure 10.16). Larches also occur in North America.

▲Figure 20.11 **Boreal forest of Canada (*boreal* means "northern").** [Author.]

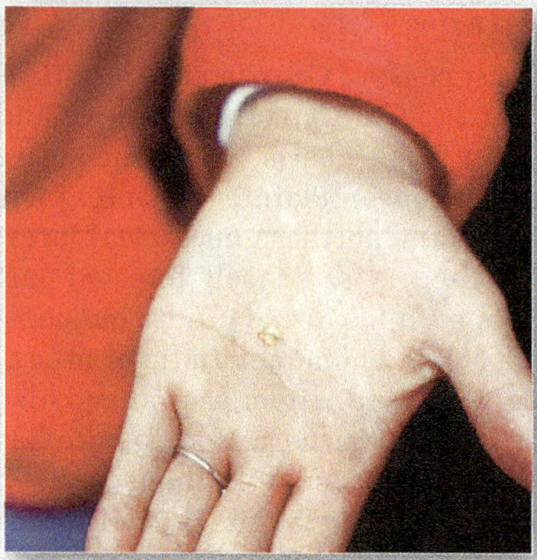

(a) A sequoia seed. About 300 seeds are in each sequoia cone.

(b) Seedling at approximately 50 years of age.

(c) The General Sherman tree, probably wider than a standard classroom. The first branch is 45 m (150 ft) off the ground and 2 m (6.5 ft) in diameter.

▲**Figure 20.12 Sequoia life stages, in Sequoia National Park, California.** [Author.]

This biome also occurs at high elevations at lower latitudes, such as in the Sierra Nevada, Rocky Mountains, Alps, and Himalayas. Douglas fir and white fir grow in the western mountains of the United States and Canada. Economically, these forests are important for lumber in the southern margins of the biome and for pulpwood throughout the middle and northern portions. Present logging practices and the sustainability of these yields are issues of increasing controversy.

In the montane forests of California's Sierra Nevada, giant sequoia (*Sequoia gigantean*) naturally occur in 70 isolated groves. The small seeds of these trees eventually become Earth's largest living organisms (in terms of biomass), some exceeding 8 m (28 ft) in diameter and 83 m (270 ft) in height (Figure 20.12). The largest of these is the General Sherman tree in Sequoia National Park, estimated to be 3500 years old. The bark is fibrous and a half-meter thick and lacks resins, so it effectively resists fire. Imagine the lightning strikes and fires that must have passed by the Sherman tree in 35 centuries! Standing among these giant trees is an overwhelming experience and creates a sense of the majesty of the biosphere.

Soils of the boreal forest are typically Spodosols (subject to podzolization), characteristically acidic and leached of humus and clays. In certain regions, permafrost (discussed in Chapter 17) is coupled with rocky and poorly developed soils so that only trees with shallow root systems are present. The poor drainage conditions associated with summer thaw of the active layer results in the characteristic presence of muskeg (moss-covered) bogs. Global warming is causing increased thawing of permafrost in forested regions; affected forests are dying in response to the resultant waterlogged soils.

Representative fauna in this biome include wolf, elk, moose (the largest member of the deer family), bear, lynx, beaver, wolverine, marten, small rodents, and migratory birds during the brief summer season (Figure 20.13). Birds include hawks and eagles, several species of grouse, Pine Grosbeak, Clark's Nutcracker, and several species of owls. About 50 species of insects particularly adapted to the presence of coniferous trees inhabit the biome.

Temperate Rain Forest

The lush forests in wet, humid regions make up the **temperate rain forest** biome. These forests of broadleaf and needleleaf trees, epiphytes, huge ferns, and thick undergrowth correspond generally to the Marine West Coast climates (occurring along middle to high-latitude west coasts), with precipitation approaching 400 cm (160 in.)

▲**Figure 20.13 A male elk (*Cervus canadensis*) in the boreal forest.** [Steven K. Huhtala.]

(a) Old-growth Douglas fir, redwoods, cedars, and a mix of deciduous trees, ferns, and mosses in the Gifford Pinchot National Forest, Washington. Only a small percentage of these old-growth forests remain in the Pacific Northwest.

(b) The Northern Spotted Owl, here with fledgling, is an "indicator species" representing the health of the temperate rainforest ecosystem.

▲Figure 20.14 **Temperate rain forest.** [(a) Bobbé Christopherson. (b) Woodfall Wild Images/Photoshot.]

per year, moderate air temperatures, summer fog, and an overall maritime influence. In North America, this biome occurs only along narrow margins of the Pacific Northwest. Similar temperate rain forests exist in southern China, small portions of southern Japan, New Zealand, and a few areas of southern Chile.

The biome is home to bear, badger, deer, wild pig, wolf, bobcat, fox, and numerous bird species, including the Northern Spotted Owl (Figure 20.14). In the 1990s, this owl became a symbol for the conflict between species-preservation efforts and the use of resources to fuel local economies. In 1990, the U.S. Fish and Wildlife Service listed the owl as a "threatened" species under the U.S. Endangered Species Act, citing the loss of old-growth forest habitat as the primary cause for its decline. The next year, logging practices in areas with spotted owl habitat were halted by court order. The ensuing controversy pitted environmentalists against loggers and other forest users, with the end result being large-scale changes in forest management throughout the Pacific Northwest.

Later research by the U.S. Forest Service and independent scientists noted the failing health of temperate rain forests and suggested that timber-management plans balance resource use with ecosystem preservation. Sustainable forestry practices emphasize the continuing health and productivity of forests into the future and are increasingly based on a multi-use ethic that serves local, national, and global interests.

The tallest trees in the world occur in this biome—the coast redwoods (*Sequoia sempervirens*) of the California and Oregon coasts (discussed in the Chapter 7 *Geosystems Now*). These trees can exceed 1500 years in age and typically range in height from 60 to 90 m (200 to 300 ft), with some exceeding 100 m (330 ft). Virgin stands of other representative trees, such as Douglas fir, spruce, cedar, and hemlock, have been reduced by timber harvests to a few remaining valleys in Oregon and Washington, less than 10% of the original forest that existed when Europeans first arrived. Most forests in this biome are secondary-growth forests, which have regrown from a major disturbance, usually human-caused. In similar forests in Chile, large-scale timber harvests and new mills began operations in 2000. Currently, U.S. corporations are shifting their logging operations to these forests, many of which are located in the Chilean Lake District and northern Patagonia.

Mediterranean Shrubland

The **Mediterranean shrubland** biome, also referred to as a temperate shrubland, occupies temperate regions that have dry summers, generally corresponding to the Mediterranean climates. The dominant shrub formations that occupy these regions are low growing and able to withstand hot-summer drought. The vegetation is *sclerophyllous* (from *sclero*, for "hard," and *phyllos*, for "leaf"). Most shrubs average a meter or two in height, with deep, well-developed roots, leathery leaves, and uneven low branches.

Typically, the vegetation varies between woody shrubs covering more than 50% of the ground and grassy woodlands covering 25%–60% of the ground. In California, the Spanish word *chaparro* for "scrubby evergreen" gives us the name **chaparral** for this vegetation type (Figure 20.15a). This scrubland includes species such as manzanita, toyon, red bud, ceanothus, mountain mahogany, blue and live oaks, and the dreaded poison oak.

(a) Chaparral vegetation, southern California.

(b) Fire-adapted chaparral sends out sprouts from roots, a few months after a wildfire in the San Jacinto Mountains, southern California.

▲**Figure 20.15 Mediterranean chaparral and fire adaptations.** [Bobbé Christopherson.]

This biome is located poleward of the shifting subtropical high-pressure cells in both hemispheres. The stable high pressure produces the characteristic dry-summer climate and establishes conditions conducive to fire. The vegetation is adapted for rapid recovery after fire—many species are able to resprout from roots or burls after a burn, or have seeds that require fire for germination (Figure 20.15b).

A counterpart to the California chaparral in North America is the *maquis* of the Mediterranean region of Europe, which includes live and cork oak trees (the source of cork) as well as pine and olive trees. In Chile, this biome is known as the *mattoral,* and in southwestern Australia, it is *mallee scrub.* In Australia, the bulk of the eucalyptus species are sclerophyllous in form and structure in whichever area they occur.

As described in Chapter 10, commercial agriculture of the *Mediterranean* climates includes subtropical fruits, vegetables, and nuts, with many food types (e.g., artichokes, olives, almonds) produced only in these climates. Animals include several types of deer, coyote, wolf, bobcat, a variety of rodents, other small animals, and various birds. In Australia, this biome is home to Malleefowl (*Leipoa ocellata*), a ground-dwelling bird, and numerous marsupials.

Midlatitude Grassland

Of all the natural biomes, the **midlatitude grassland** is the most modified by human activity. This biome includes the world's "breadbaskets"—regions that produce bountiful grain (wheat and corn), soybeans, and livestock (hogs and cattle). In these regions, the only naturally occurring trees were deciduous broadleaf trees along streams and other limited sites. These regions are called grasslands because of the predominance of grasslike plants before human intervention (Figure 20.16).

In North America, tallgrass prairies once rose to heights of 2 m (6.5 ft) and extended westward to about the 98th meridian, with shortgrass prairies in the drier lands farther west. The 98th meridian is roughly the location of the 51-cm (20-in.) isohyet, with wetter conditions to the east and drier conditions to the west (see Figure 18.16).

The deep, tough sod of these grasslands posed problems for the first European settlers, as did the climate. The self-scouring steel plow, introduced in 1837 by John Deere, allowed the interlaced grass sod to be broken apart, freeing the soils for agriculture. Other inventions were critical to opening this region and solving its unique spatial problems: barbed wire (the fencing material for a treeless prairie); well-drilling techniques developed by Pennsylvania

▲**Figure 20.16 Protected grasslands in the United States.** Buffalo Gap National Grassland in western South Dakota is one of 20 national grasslands administered by the U.S. Forest Service (see http://www.fs.fed.us/grasslands/). [Jason Patrick Ross/Shutterstock.]

oil drillers, but used for water wells; windmills for pumping; and railroads to transport materials.

Few patches of the original prairies (tall grassland) or steppes (short grassland) remain within this biome. In the prairies alone, the natural vegetation was reduced from 100 million hectares (250 million acres) down to a few areas of several hundred hectares each. The map in Figure 20.7 shows the natural location of this former prairie and steppe grassland.

Characteristic midlatitude grasslands outside North America are the *Pampas* of Argentina and Uruguay and the grassland of Ukraine. In most regions where these grasslands were the natural vegetation, human development of them was critical to territorial expansion.

This biome is the home of large grazing animals, including deer, pronghorn, and bison (the almost complete annihilation of the latter is part of American history, see Figure 20.17). Gophers, prairie dogs, ground squirrels, Turkey Vultures, grouse, and Prairie Chickens are common, as well as grasshoppers and other insects. Predators include the coyote, nearly extinct black-footed ferret, badger, and birds of prey—hawks, eagles, and owls.

Deserts

Earth's desert biomes cover more than one-third of its land area (Figure 20.7). We subdivide the desert biomes into **warm desert and semidesert**, caused by the dry air and low precipitation of subtropical high-pressure cells, and **cold desert and semidesert**, which tend toward higher latitudes, where subtropical high pressure affects climate for less than 6 months of the year. A third subdivision, Earth's **polar deserts**, occur in high-latitude regions, including most of Antarctica and Greenland, with very cold, dry climates. Vegetation, sparse in these predominantly ice- and rock-covered regions, is mainly lichens and mosses.

▲**Figure 20.17 Bison grazing in the Big Basin Prairie Preserve, western Kansas.** Inset shows a male and female shedding winter coats. [Bobbé Christopherson.]

Desert vegetation includes numerous xerophytes, plants that are adapted to dry conditions with mechanisms that prevent water loss, such as cacti and other succulents (plants that store water in their thick and fleshy tissues). Xerophytic plants have a range of adaptations, including long taproots to access groundwater (mesquite trees); shallow, spreading root systems to maximize water uptake (palo verde trees); small leaves to minimize surface area for water loss (acacia); waxy leaf coatings to retard water loss (creosote bush); leaf drop during dry periods (ocotillo); and as previously mentioned, the succulent tissue to store water. A number of plants have also developed spines, thorns, or bad-tasting tissue to discourage herbivory. Finally, some plants that grow along desert washes produce seeds that require *scarification*—abrasion or weathering of the surface—for the seed to open and permit germination. This can occur from the tumbling, churning action of a flash flood flowing down a desert wash; such an event also produces the moisture for seed germination.

Some desert plants are *ephemeral*, or short-lived, an adaptation that takes advantage of a short wet season or even a single rainfall event in desert environments. The seeds of desert ephemerals lie dormant on the ground until a rainfall event stimulates the seed germination. Seedlings grow rapidly, mature, flower, and produce large numbers of new seeds, which are then dispersed long distances by wind or water. Seeds then go dormant until the next rainfall event.

The vegetation of the lower Sonoran Desert of southern Arizona is an example of the warm desert biome (Figure 20.18). This landscape features the unique saguaro cactus (*Carnegiea gigantea*), which grows to many meters in height and up to 200 years in age if undisturbed. First blooms do not appear until it is 50 to 75 years old. In cold deserts, where precipitation is greater and temperatures are colder, characteristic vegetation includes grasses, xerophytic shrubs, such as creosote bush, and woody shrubs, such as sagebrush (*Artemisia tridentata*). Succulents that hold large amounts of water, such as the saguaro cactus, cannot survive in cold deserts that experience consecutive days or nights with freezing winter temperatures.

GEOreport 20.4 Biodiversity and food sources

Wheat, maize (corn), and rice—just three grains—fulfill about 50% of human planetary food demands. About 7000 plant species have been gathered for food throughout human history, but more than 30,000 plant species have edible parts. Natural, as yet undiscovered potential food resources are waiting to be found and developed. Biodiversity, if preserved in each of the biomes, provides us a potential cushion for all future food needs, but only if species are identified, inventoried, and protected.

▲Figure 20.18 **Sonoran Desert scene.** A saguaro cactus and other characteristic vegetation in the Lower Sonoran Desert west of Tucson, Arizona, at an elevation of about 900 m (2950 ft). [Author.]

The faunas of both warm and cold deserts are limited by the extreme conditions and include only a few resident large animals. Camels, occurring in the deserts of Africa and the Middle East, are so well adapted to warm desert conditions that they can lose up to 30% of their body weight in water without harm (for humans, a 10%–12% loss is dangerous). Desert bighorn sheep are another large animal, occurring in scattered populations in inaccessible mountains and canyons, such as the inner Grand Canyon but not at the rim. Desert bighorn declined precipitously from about 1850 to 1900 due to competition with livestock for food and water, as well as exposure to parasites and disease. In an effort to reestablish bighorn populations, several states are transplanting the animals to their former ranges.

Other representative desert animals are the ring-tailed cat, kangaroo rat, lizards, scorpions, and snakes. Most of these animals become active only at night, when temperatures are lower. In addition, various birds have adapted to desert conditions and available food sources—for example, roadrunners, thrashers, ravens, wrens, hawks, grouse, and nighthawks.

Arctic and Alpine Tundra

The **arctic tundra** biome is located in the extreme northern area of North America and Russia, bordering the Arctic Ocean and generally north of the 10°C (50°F) isotherm for the warmest month. Daylength varies greatly throughout the year, seasonally changing from almost continuous day to continuous night. The region, except for a few portions of Alaska and Siberia, was covered by ice during all of the Pleistocene glaciations.

(a) Tundra mosses with a glacially eroded roche moutonnée in the background (shape denotes glacial movement from left to right).

(b) Grasses, mosses, and dwarf willow flourish in the cold high-latitude climates.

▲Figure 20.19 **Arctic tundra.** [Bobbé Christopherson.]

With recent climate change, these regions have been warming at more than twice the rate of the rest of the planet over the past few decades.

This biome corresponds to the tundra climates; winters are cold and long; summers are cool and brief. A growing season of sorts lasts only 60–80 days, and even then frosts can occur at any time. Soils are poorly developed periglacial surfaces that are underlain by permafrost. In the summer months, the surface horizons thaw, thus producing a mucky surface of poor drainage (Figure 20.19a). Roots can penetrate only to the depth of thawed ground, usually about a meter (3 ft).

Arctic tundra vegetation consists of low, ground-hugging herbaceous plants such as sedges, mosses, arctic meadow grass, and snow lichen and some woody species such as dwarf willow (Figure 20.19b). Owing to the short growing season, some perennials form flower buds one summer and open them for pollination the next. Animals of the tundra biome include musk ox, caribou, reindeer, rabbit, Ptarmigan, lemming, and other small rodents, which are important food for the larger carnivores—the wolf, fox, weasel, Snowy Owl, polar bear, and, of course, mosquito. The tundra is an important breeding ground for geese, swans, and other waterfowl.

▲**Figure 20.20 Alpine tundra conditions.** An alpine tundra and grazing mountain goats near Mount Evans, Colorado, at an elevation of 3660 m (12,000 ft). [Bobbé Christopherson.]

Alpine tundra is similar to arctic tundra, but it can occur at lower latitudes because it is associated with high elevations. This biome usually occurs above the treeline (the elevation above which trees cannot grow), which shifts to higher elevations closer to the equator. Alpine tundra communities occur in the Andes near the equator, the White Mountains and Sierra of California, the American and Canadian Rockies, the Alps, and Mount Kilimanjaro of equatorial Africa as well as in mountains from the Middle East to Asia.

Alpine tundra features grasses, herbaceous annuals (small plants), and low-growing shrubs, such as willows and heaths. Because alpine locations are frequently windy sites, many plants there have forms that are shaped by the wind. Alpine tundra can experience permafrost conditions. Characteristic fauna include mountain goats, Rocky Mountain bighorn sheep, elk, and voles (Figure 20.20).

Vegetation of the tundra biome is slow-growing, has low productivity, and is easily disturbed. Hydroelectric projects, mineral exploitation, and even tire tracks leave marks on the landscape that persist for hundreds of years. As growth in population and in energy demand continue, the region will face even greater challenges from petroleum resource development, including oil spills and the associated contamination, and landscape disruption.

CRITICALthinking 20.3
A Shifting-Climate Hypothetical

Using Figure 20.7 (biomes), Figure 10.2 (climates), Figures 9.7 and 10.1 (precipitation), and Figure 8.1 (air masses), and noting the printed graphic scales on these maps, consider the following hypothetical situation. Assume a northward climatic shift in the United States and Canada of 500 km (310 mi); in other words, imagine moving North America 500 km south to simulate climatic categories shifting north. Describe your analysis of conditions through the Midwest from Texas to the prairies of Canada. Describe your analysis of conditions from New York through New England and into the Maritime Provinces. How will biomes change? What economic relocations do you envision? Expand your thinking to another region of the world: If the subtropical high-pressure cell over Australia expanded and intensified, consider the new pattern of climate categories and ecosystems. •

GEOreport 20.5 The Porcupine caribou herd

The Porcupine caribou herd is a population of tundra caribou that inhabit Alaska, Yukon, and Northwest Territories. Most of the herd migrates to the coastal plain on Alaska's North Slope to calve in early June. The calving area is fairly small, and 80%–85% of the herd uses it year after year. Oil exploration is not permitted in these areas because it degrades habitat and affects animal migrations. However, intense political pressure to relax any protection is ever present, and an aggressive campaign to overturn bans on exploration and drilling is active. The possibility of drilling for shale gas and methane hydrates is a new threat to the region.

Conservation, Management, and Human Biomes

In the early 2000s, biogeographers defined a new, emerging scientific field called *conservation biogeography*. This subdiscipline applies biogeographic principles, theories, and analyses to solve problems in biodiversity conservation. Among the hot research topics in this field are the distribution and effects of invasive species, the impacts of rapid climate change on biodiversity, and the implementation of conservation planning and establishment of protected areas. Focus Study 20.1 discusses specific efforts to preserve biodiversity, all of which are based loosely on the concept of natural biomes and principles of island biogeography.

Island Biogeography for Species Preservation

When the first European settlers landed on the Hawaiian Islands in the late 1700s, they counted 43 species of birds. Today, 15 of those species are extinct, and 19 more are threatened or endangered, with only one-fifth of the original species populations relatively healthy. In most of Hawai'i, native species no longer exist below elevations of 1220 m (4000 ft) because of an introduced avian virus. This is one example among thousands of species extinctions and declining biodiversity across the globe. Island ecosystems are particularly vulnerable because their unique ecosystems evolved in isolation from mainland species.

Early explorers in the Atlantic Ocean saw remote islands as places to cut trees for masts and repairs and to obtain water and food supplies. Frequently, they released goats and rabbits on islands to proliferate and be a ready food source for future visits. These animals decimated native species populations, as did the introduced rats that attacked nesting birds. Today, restoration efforts on many islands include eradication of non-native species and the breeding of endangered native plant species in greenhouses and nurseries for reintroduction into natural island ecosystems (Figure 20.21). On some islands in the Atlantic, scientists found surviving native species, once thought extinct, growing on remote cliffs where they were retrieved for breeding stock.

The principles developed in the study of isolated species' evolution and decline with the introduction of non-native species on islands have become useful concepts informing global conservation efforts. One strategy for conservation of species is to focus on habitat preservation, such as setting aside parks and wildlife refuges. Yet these protected areas are often isolated, surrounded by human development and disconnected from other natural habitat. Such habitat fragmentation is problematic for species requiring a large range for survival. In the 1980s, researchers discovered that a number of U.S. national parks had become isolated "islands" of biodiversity and some species within them were declining or disappearing completely.

(a) On St. Helena, endangered native species grow in greenhouses and will be eventually planted in the wild.

(b) Endemic species across the island were severely depleted during the age of exploration; shown here is Jamestown, the capital city on the northwest shore.

▲**Figure 20.21 Ecological restoration on Saint Helena island, located at 16° S latitude in the Atlantic Ocean.** [(a) Author. (b) Bobbé Christopherson.]

Focus Study 20.1 Environmental Restoration

Global Conservation Strategies

One goal of conservation biogeography, restoration ecology, and other related scientific fields is the preservation of biodiversity through habitat conservation. Habitat fragmentation is a major cause of species decline and extinction, and protecting and restoring habitat has become a conservation focus. Climate change is an important variable to consider in setting aside protected areas, since temperature and precipitation regimes in parks and reserves could eventually end up outside the natural range of the species in question.

Several concepts are used as the basis for conservation strategies for maintaining biodiversity. First implemented by Conservation International in 1989, the idea of biodiversity "hotspots" has provided a focus for conservation efforts and has received scientific attention and financial support. To qualify as a hotspot, a community or ecosystem must contain at least 1500 endemic (native) plant species, and it must have lost at least 70% of its original habitat (more information is at http://www.biodiversityhotspots.org/).

The World Wildlife Fund (WWF) uses the concept of *ecoregions* as the basis for its conservation strategies, with the goal of implementing conservation on the scale of large natural habitats similar to biomes. In 2000, a team of WWF-sponsored scientists identified 238 ecoregions, known as the "Global 200," as the focus of conservation efforts. The protection of these representative habitats could save a broad diversity of Earth's species (Figure 20.1.1).

The Man and the Biosphere Program, launched in 1971 by the United Nations Educational, Scientific, and Cultural Organization, establishes biosphere reserves with the goal of preserving biodiversity as well as promoting economic and social development and maintaining the cultural values of local communities. This integrated approach to species conservation attempts to mesh biological assets with human activity. Its world network of reserves ranges from relatively undisturbed ecosystems, such as Glacier Bay in Alaska, to the mixed towns and green space of southern Germany.

The intent of the biosphere reserves in the UN program is to promote sustainable development by establishing a core in which natural features are protected from outside disturbances, surrounded by zones of local development, natural and cultural resource management, and scientific experimentation. Some reserves remain in the planning stage, although they are officially designated. In the United States, several biosphere reserves include national parks, such as Everglades National Park in Florida and Olympic National Park in Washington, both established in the 1970s.

The ultimate goal, about half achieved, is to create at least one reserve in each of the 194 distinctive biogeographic communities presently identified. Scientists predict that by about 2025, designating new, undisturbed reserves may no longer be possible because pristine areas will be gone. More than 621 biosphere reserves now exist in 117 countries (see http://www.unesco.org/new/en/natural-sciences/environment/ecological-sciences/man-and-biosphere-programme/).

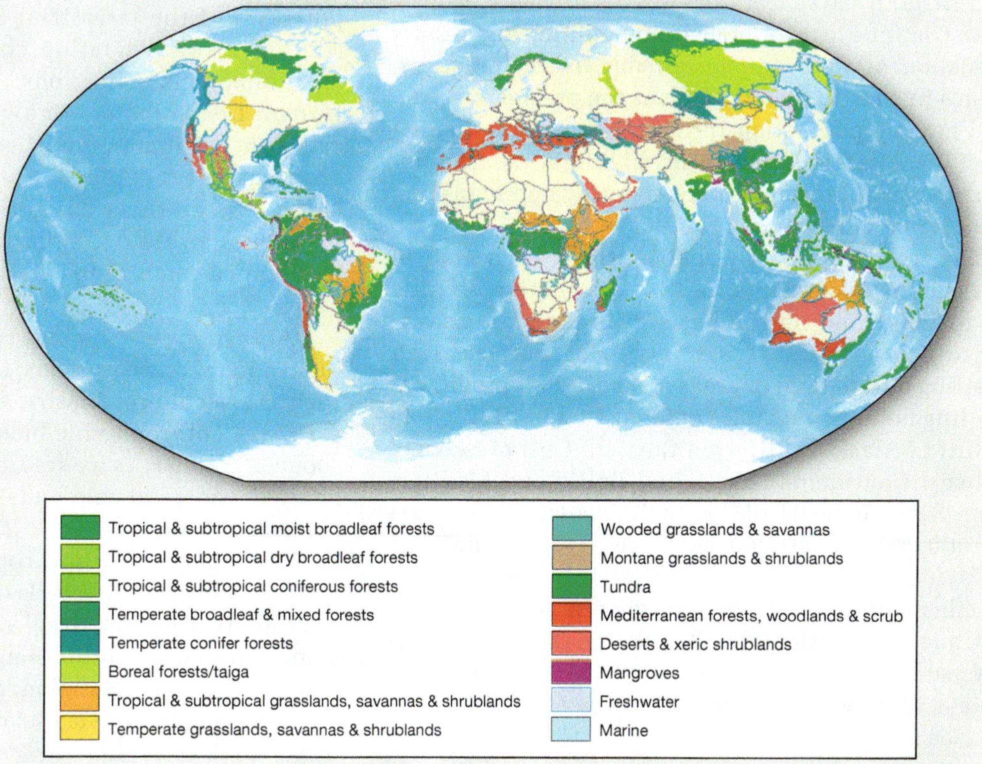

Figure 20.1.1 Biomes containing the WWF "Global 200" ecoregions important for biodiversity preservation. The 14 biomes shown, plus the freshwater and marine biomes, are the broad divisions containing a total of 238 specific ecoregions that, if protected, could preserve a large percentage of Earth's biodiversity. [World Wildlife Fund; http://www.worldwildlife.org/science/ecoregions/global200.html.]

A key conceptual model for understanding the effects of habitat fragmentation was Robert MacArthur and E. O. Wilson's theory of **island biogeography**, published as a book by the same name in 1967. The theory, based on scientific work on small, isolated mangrove islands in the Florida Keys, links the number of species on an island to the island's size and distance from the mainland.

The theory summarized three patterns of species distributions on islands: (1) The number of species increases with island area, (2) the number of species decreases with island isolation (distance from the mainland), and (3) the number of species on an island represents an equilibrium between the rates of immigration and extinction. Larger islands have a wider variety of habitat and niches, and thus lower extinction rates. This theory provided the foundation for understanding "islands" of fragmented habitat, inspired thousands of studies in biogeography and ecology, and increased awareness of the importance of landscape-scale thinking for species preservation. Although present research goes beyond the original theory, the basic conceptual ideas inform conservation science, especially with regard to proper formation of parks and reserves.

Aquatic Ecosystem Management

Human activities have affected aquatic ecosystems, whether freshwater or marine, in roughly similar ways to terrestrial ecosystems. Coastal ocean waters, in particular, continue to deteriorate from pollution and habitat degradation, as well as unsustainable fishing practices. Declines in aquatic species, such as the precipitous drop of the herring population in the Georges Bank fishing area of the Atlantic in the 1970s, highlight the need for an ecosystem approach to understanding and managing these international waters.

This need was partly met by the designation of **large marine ecosystems** (LMEs), distinctive oceanic regions identified on the basis of organisms, ocean-floor topography, currents, areas of nutrient-rich upwelling circulation, or areas of significant predation, including human. Examples of identified LMEs include the Gulf of Alaska, California Current, Gulf of Mexico, Northeast Continental Shelf, and Baltic and Mediterranean Seas. Some 64 LMEs, each encompassing more than 200,000 km^2 (77,200 mi^2), are presently defined worldwide (see the list at http://www.lme.noaa.gov/). A number of these LMEs include government-protected areas, such as the Monterey Bay National Marine Sanctuary within the California Current LME and the Florida Keys Marine Sanctuary within the Gulf of Mexico LME (see http://sanctuaries.NOAA.gov/).

Anthropogenic Biomes

In 1874, Earth was viewed as offering few limits to human enterprise. But visionary American diplomat and conservationist George Perkins Marsh perceived the need to manage and to conserve the environment. He warned that humans "have now felled forest enough everywhere, in many districts far too much."

Marsh's warning was ignored, and the clearing of the forests is perhaps the most fundamental change wrought by humans on Earth. The development and expansion of civilization was fueled by and built on the consumption of forests and other natural resources. The natural "work" supplied by Earth's physical, chemical, and biological systems has an estimated annual worth of some $35 trillion to the global economy.

Humans now have become the most powerful biotic agent on Earth, influencing all ecosystems on a planetary scale. Even in many of the most pristine ecosystems on Earth, evidence of early human settlement exists. Scientists are measuring ecosystem properties and building elaborate computer models to simulate the evolving human–environment experiment on our planet—in particular, the shifting patterns of environmental factors (temperatures and changing frost periods; precipitation timing and amounts; air, water, and soil chemistry; and nutrient redistribution) wrought by human activities.

In 2008, two geographers presented the concept of "anthropogenic biomes," based on today's human-altered ecosystems, as an updated and more accurate portrayal of the terrestrial biosphere than the "pristine" natural vegetation communities described in most biome classifications. Their map, shown in *The Human Denominator*, Figure HD 20, shows five broad categories of human-modified landscapes: settlements, croplands, rangelands, forested lands, and wildlands. Within these categories, the scientists defined 21 biomes, which summarize the current mosaic of landscapes combining a variety of land uses and land cover.

Anthropogenic biomes result from ongoing human interaction with ecosystems, linked to land-use practices such as agriculture, forestry, and urbanization. The most extensive anthropogenic biome is rangelands, covering about 32% of Earth's ice-free land; followed by croplands, forested lands, and wildlands, each at about 20%; and settlements at about 7%.

The concept of anthropogenic biomes does not replace terrestrial biome classifications but instead presents another perspective. Understanding of the natural biomes presented in this chapter is essential for ecosystem and species restoration, as well as for the advancement of basic and applied science.

THE**human**DENOMINATOR 20 Anthropogenic Environments

BIOMES ⇨ HUMANS

- Natural plant and animal communities are linked to human cultures, providing resources for food and shelter.
- Earth's remaining undisturbed ecosystems are becoming a focus for tourism, recreation, and scientific attention.

HUMANS ⇨ BIOMES

- Invasive species, many introduced by humans, disrupt native ecosystems.
- Tropical deforestation is ongoing, with more than half Earth's original rain forest already cleared.

20a **Residential irrigated cropland**. Prince Edward Island, Canada. [All Canada Photos/Alamy.]

20b **Urban**. London, England. [Justin Kaze zsixz /Alamy.]

20c **Irrigated villages**. Satpara, Pakistan. [Dave Stamboulis/Alamy.]

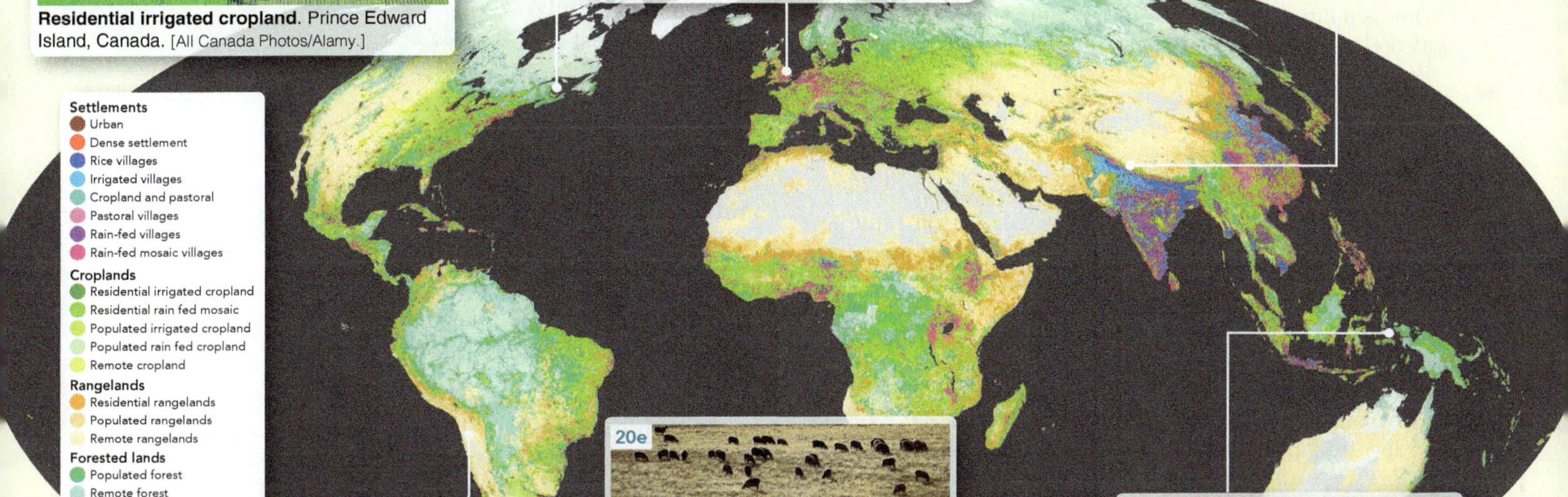

Anthropogenic Biomes. [Courtesy Erle Ellis, University of Maryland, Baltimore County, and Navin Ramankutty, McGill University/NASA.]

20e **Remote rangelands**. Northern Chile. [Independent Picture Service/Alamy.]

20d **Populated forest**. Raja Ampat Islands, Indonesia. [Images & Stories/Alamy.]

ISSUES FOR THE 21ST CENTURY

- Species and ecosystem management for conserving biodiversity must become a priority to avoid species extinctions.
- Shifting of species distributions in response to environmental factors will continue with ongoing climate change.
- Population control and global education (including women and minorities in all countries) is key for sustaining natural and anthropogenic biomes.

GEOSYSTEMS**connection**

All of Earth's physical systems combine to produce the biomes into which all regions on the Earth can be classified. We see the patterns of these interactions throughout the biosphere and in the diversity of life. Real issues abound regarding preservation of this diversity and the influence of climate change on communities, ecosystems, and biomes. Physical geography and Earth systems science are especially well suited to study these changes.

As our exploration of *Geosystems* concludes, your journey now leads onward into the 21st century. Study well and travel safely—with the wind at your back.

KEY LEARNING concepts review

■ ***Locate*** **the world's biogeographic realms, and *discuss* the basis for their delineation.**

The interplay of evolutionary and abiotic factors within Earth's ecosystems determines biodiversity and the distribution of plant and animal communities. A **biogeographic realm** is a geographic region in which a group of plant or animal species evolved. This recognition laid the groundwork for understanding communities of flora and fauna known as biomes.

biogeographic realm (p. 620)

1. What is a biogeographic realm? What are the zoological realms? What is Wallace's line?

■ ***Explain*** **the basis for grouping plant communities into biomes, and *list* the major aquatic and terrestrial biomes on Earth.**

A **biome** is a large, stable, terrestrial or aquatic ecosystem classified according to the predominant vegetation type and the adaptations of particular organisms to that environment. Biomes carry the name of the dominant vegetation because it is the most easily identified feature. The six main terrestrial vegetation classifications are forest, savanna, grassland, shrubland, desert, and tundra. Within these general groups, biome designations are based on more specific growth forms; for example, forests are subdivided into rain forests, seasonal forests, broadleaf mixed forests, and needleleaf forests. Ideally, a biome represents a mature community of natural vegetation. A boundary transition zone between adjoining ecosystems is an **ecotone**.

biome (p. 621)
ecotone (p. 622)

2. Define biome. What is the basis for the designation?
3. Give some examples of vegetation growth forms.
4. Describe a transition zone between two ecosystems. How wide is an ecotone? Explain.

■ ***Explain*** **the potential impact of non-native species on biotic communities, using several examples.**

Communities, ecosystems, and biomes can be affected by species that are introduced from elsewhere by humans, either accidentally or intentionally. These non-native species are also called *exotic species* or *aliens*. After arriving in the new ecosystem, some species may disrupt native ecosystems and become **invasive species**.

invasive species (p. 624)

5. Give several examples of invasive species described in the text, and describe their impact on natural systems.
6. What happened in the waters of Tristan da Cunha? Why was this subject first introduced in Chapter 6 of the text and then concluded in this chapter? What are the linkages between the chapters? What economic damage might evolve in Tristan's marine ecosystems?

■ ***Summarize*** **the characteristics of Earth's 10 major terrestrial biomes, and *locate* them on a world map.**

For an overview of Earth's 10 major terrestrial biomes and their vegetation characteristics, soil orders, climate-type designations, annual precipitation ranges, temperature patterns, and water-balance characteristics, review Table 20.1. The tropical rainforest biome is undergoing rapid deforestation. Because the rain forest is Earth's most diverse biome and is important to the climate system, this loss is creating great concern among citizens, scientists, and nations.

tropical rain forest (p. 625)
tropical seasonal forest and scrub (p. 629)
tropical savanna (p. 629)
midlatitude broadleaf and mixed forest (p. 632)
boreal forest (p. 633)
needleleaf forest (p. 633)
taiga (p. 633)
montane forest (p. 633)
temperate rain forest (p. 634)
Mediterranean shrubland (p. 635)
chaparral (p. 635)
midlatitude grassland (p. 636)
warm desert and semidesert (p. 637)
cold desert and semidesert (p. 637)
polar desert (p. 637)
arctic tundra (p. 638)
alpine tundra (p. 639)

7. Using the integrative chart in Table 20.1 and the world map in Figure 20.7, select any two biomes and study the correlation of vegetation characteristics, soil, moisture, and climate with their spatial distribution. Then contrast the two using each characteristic.
8. Describe the tropical rain forests. Why is the rainforest floor somewhat clear of plant growth? Why is logging of individual tree species so difficult there?
9. What issues surround the deforestation of the rain forest? What is the impact of these losses on the rest of the biosphere? What new threat to the rain forest has emerged?
10. What do *Caatinga, Chaco, brigalow,* and *dornveld* refer to? Explain.
11. Describe the role of fire in the tropical savanna biome and in the midlatitude broadleaf and mixed forest biome.
12. Why does the boreal forest biome not exist in the Southern Hemisphere, except in mountainous regions? Where is this biome located in the Northern Hemisphere, and what is its relationship to climate type?
13. In which biome do we find Earth's tallest trees? Which biome is dominated by small, stunted plants, lichens, and mosses?
14. What type of vegetation predominates in the Mediterranean (dry summer) climate? Describe the adaptations necessary for these plants to survive.
15. What is the significance of the 98th meridian in terms of North American grassland? What types of inventions enabled agriculture in this grassland?
16. Describe some of the unique adaptations of xerophytes.

17. What types of plants and animals are found in the tundra biome?
18. As an example of shifting-climate impacts, we tracked temperature and precipitation conditions for Illinois in *Geosystems Now* in Chapter 19. What impacts do you think climate change will have on biomes in the United States and in other countries?

■ *Discuss* strategies for ecosystem management and biodiversity conservation.

Efforts are under way worldwide to set aside and protect remaining representative sites within most of Earth's principal biomes. Principles of **island biogeography** used in the study of isolated ecosystems are important in setting up biosphere reserves. Island communities are special places for study because of their spatial isolation and the relatively small number of species present. Protection of aquatic ecosystems includes the designation of **large marine ecosystems (LMEs)**. In reality, few undisturbed biomes exist in the world, for most have been modified by human activity. The new concept of **anthropogenic biomes** considers the impacts of human settlement, agriculture, and forest practices on vegetation patterns.

island biogeography (p. 642)
large marine ecosystem (LME) (p. 642)
anthropogenic biome (p. 642)

19. Describe the theory of island biogeography. How has this theory been important for preserving biodiversity? What are the goals of a biosphere reserve?
20. What are threats to coastal aquatic ecosystems? Name an example of a protected large marine ecosystem in the United States.
21. Describe the concept of anthropogenic biomes. According to the categories presented on the map in Human Denominator 20, how would you classify the area in which you live?

MasteringGeography™

Looking for additional review and test prep materials? Visit the Study Area in *MasteringGeography*™ to enhance your geographic literacy, spatial reasoning skills, and understanding of this chapter's content by accessing a variety of resources, including MapMaster interactive maps, geoscience animations, satellite loops, author notebooks, videos, RSS feeds, flashcards, web links, self-study quizzes, and an eText version of *Geosystems*.

▲Figure 20.22 **Human modification alters natural biomes.** In rural Poland near the village of Lanckorona, southwest of Krakov, more than a millennium of human intervention altered the original natural biome. Krakov was settled in the seventh century and was a major European city by A.D. 1000. Originally this countryside was in the ecotone between midlatitude broadleaf mixed forest and boreal forest. Today, the landscape shows active farming on fragmented tracts of land, fallow fields, and some protected areas; to the right in the photo is a national forest preserve that is many generations removed from the original landscape. As distant as this rural setting seems from North America, we are all linked through Earth's systems, interconnected for sustained life support—thus, our journey through *Geosystems*. [Bobbé Christopherson.]

A Maps in This Text and Topographic Maps

Maps Used in This Text

Geosystems uses several map projections: Goode's homolosine, Robinson, and Miller cylindrical, among others. Each was chosen to best present specific types of data. **Goode's homolosine projection** is an interrupted world map designed in 1923 by Dr. J. Paul Goode of the University of Chicago. Rand McNally *Goode's Atlas* first used it in 1925. Goode's homolosine equal-area projection (Figure A.1) is a combination of two oval projections (*homolo*graphic and *sin*usoidal projections).

Two equal-area projections are cut and pasted together to improve the rendering of landmass shapes. A *sinusoidal projection* is used between 40° N and 40° S latitudes. Its central meridian is a straight line; all other meridians are drawn as sinusoidal curves (based on sine-wave curves) and parallels are evenly spaced. A *Mollweide projection*, also called a *homolographic projection*, is used from 40° N to the North Pole and from 40° S to the South Pole. Its central meridian is a straight line; all other meridians are drawn as elliptical arcs and parallels are unequally spaced—farther apart at the equator, closer together poleward. This technique of combining two projections preserves areal size relationships, making the projection excellent for mapping spatial distributions when interruptions of oceans or continents do not pose a problem.

We use Goode's homolosine projection throughout this book. Examples include the world climate map and smaller climate type maps in Chapter 10, topographic regions and continental shields maps (Figures 12.3 and 12.4), world karst map (Figure 13.15), world loess deposits (Figure 16.33), and the terrestrial biomes map in 20.7.

Another projection we use is the **Robinson projection**, designed by Arthur Robinson in 1963 (Figure A.2). This projection is neither equal area nor true shape, but is a compromise between the two. The North and South poles appear as lines slightly more than half the length of the equator; thus higher latitudes are exaggerated less than on other oval and cylindrical projections. Some of the Robinson maps employed include the latitudinal geographic zones map in Chapter 1 (Figure 1.15), the daily net radiation map (Figure 2.10), the world temperature range map in Chapter 5 (Figure 5.16), the maps of lithospheric plates and volcanoes and earthquakes in Chapter 11 (Figures 11.18 and 11.20), and the global oil spills map in Chapter 16 (Figure 16.1.2).

Another compromise map, the **Miller cylindrical projection**, is used in this text (Figure A.3). Examples of this projection include the world time zone map (Figure 1.19), global temperature maps in Chapter 5 (Figures 5.12 and 5.13), and the two global pressure maps in Figure 6.10. This projection is neither true shape nor true area, but is a compromise that avoids the severe scale distortion of the Mercator. The Miller projection frequently appears in world atlases. The American Geographical Society presented Osborn Miller's map projection in 1942.

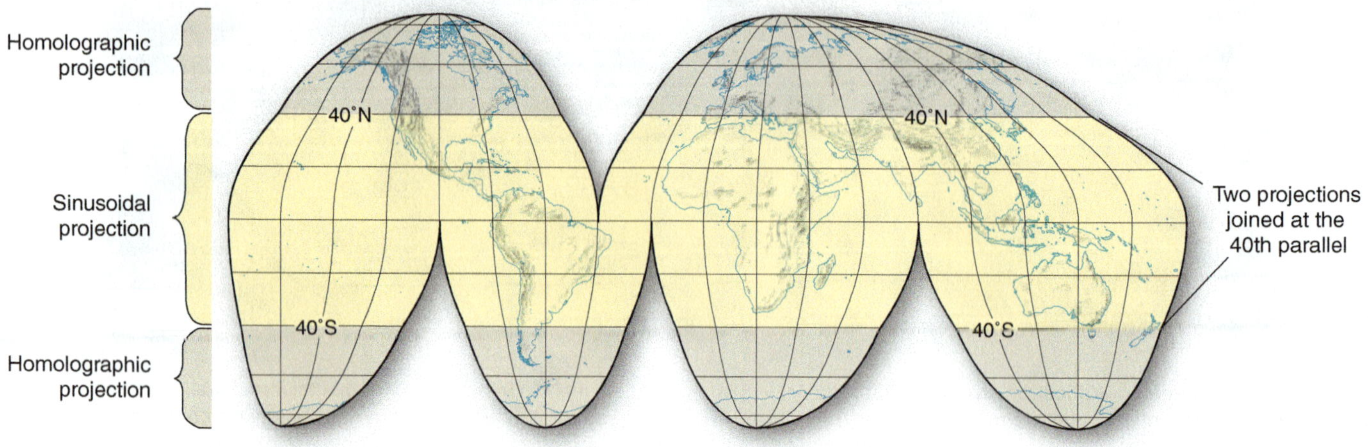

Figure A.1 Goode's homolosine projection.
An equal-area map. [Copyright by the University of Chicago. Used by permission of the University of Chicago Press.]

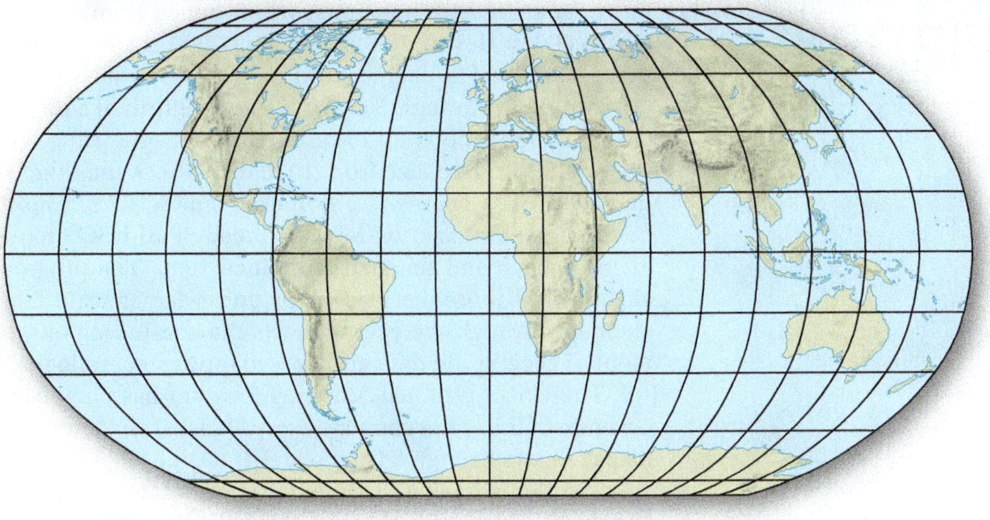

Figure A.2 Robinson projection. A compromise between equal area and true shape. [Developed by Arthur H. Robinson, 1963.]

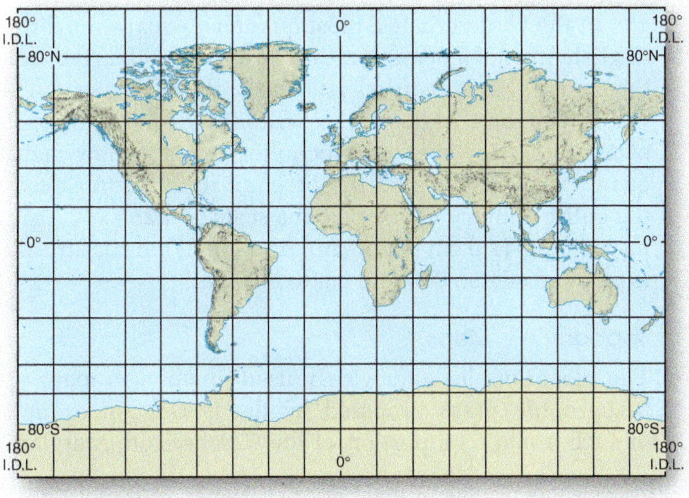

Figure A.3 Miller cylindrical projection. A compromise map projection between equal area and true shape. [Developed by Osborn M. Miller, American Geographical Society, 1942.]

Mapping, Quadrangles, and Topographic Maps

The westward expansion across the vast North American continent demanded a land survey for the creation of accurate maps. Maps were needed to subdivide the land and to guide travel, exploration, settlement, and transportation. In 1785, the Public Lands Survey System began surveying and mapping government land in the United States. In 1836, the Clerk of Surveys in the Land Office of the Department of the Interior directed public-land surveys. The Bureau of Land Management replaced this Land Office in 1946. The actual preparation and recording of survey information fell to the U.S. Geological Survey (USGS), also a branch of the Department of the Interior (see http://www.usgs.gov/pubprod/maps.html).

In Canada, National Resources Canada conducts the national mapping program. Canadian mapping includes base maps, thematic maps, aeronautical charts, federal topographic maps, and the National Atlas of Canada, now in its fifth edition (see http://atlas.nrcan.gc.ca/).

Quadrangle Maps

The USGS depicts survey information on quadrangle maps, so called because they are rectangular maps with four corner angles. The angles are junctures of parallels of latitude and meridians of longitude rather than political boundaries. These quadrangle maps utilize the Albers equal-area projection, from the conic class of map projections.

The accuracy of conformality (shape) and scale of this base map is improved by the use of not one but two standard parallels. (Remember from Chapter 1 that standard lines are where the projection cone touches the globe's surface, producing greatest accuracy.) For the conterminous United States (the "lower 48"), these parallels are 29.5° N and 45.5° N latitude (noted on the Albers projection shown in Figure 1.23). The standard parallels shift for conic projections of Alaska (55° N and 65° N) and for Hawai'i (8° N and 18° N).

Because a single map of the United States at 1:24,000 scale would be more than 200 m wide (more than 600 ft), some system had to be devised for dividing the map into a manageable size. Thus, a quadrangle system using latitude and longitude coordinates developed. Note that these maps are not perfect rectangles, because meridians converge toward the poles. The width of quadrangles narrows noticeably as you move north (poleward).

Quadrangle maps are published in different series, covering different amounts of Earth's surface at different scales. You see in Figure A.4 that each series is referred

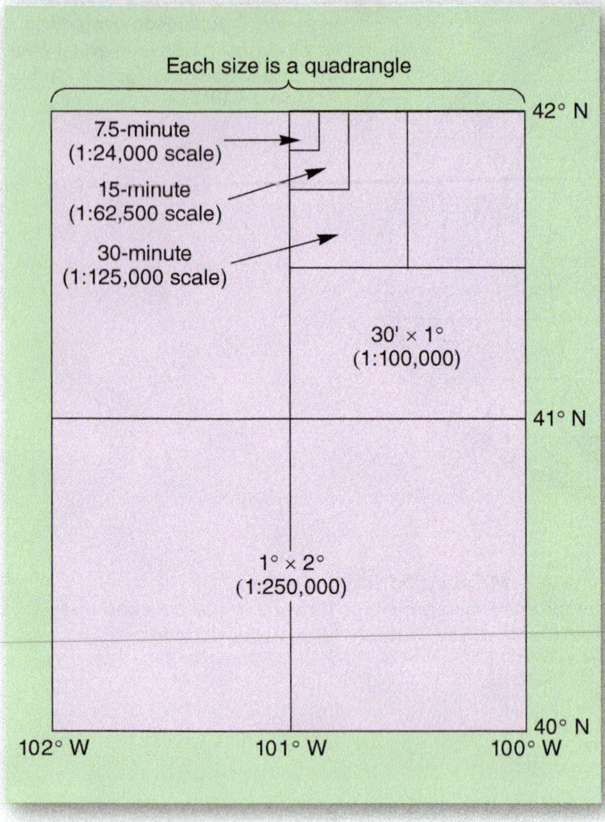

Figure A.4 **Quadrangle system of maps used by the USGS.**

to by its angular dimensions, which range from 1° × 2° (1:250,000 scale) to 7.5′ × 7.5′ (1:24,000 scale). A map that is one-half a degree (30′) on each side is called a "30-minute quadrangle," and a map one-fourth of a degree (15′) on each side is a "15-minute quadrangle" (this was the USGS standard size from 1910 to 1950). A map that is one-eighth of a degree (7.5′) on each side is a 7.5-minute quadrangle, the most widely produced of all USGS topographic maps and the standard since 1950. The progression toward more-detailed maps and a larger-scale map standard through the years reflects the continuing refinement of geographic data and new mapping technologies.

The USGS National Mapping Program has completed coverage of the entire country (except Alaska) on 7.5-minute maps (1 in. to 2000 ft, a large scale). It takes 53,838 separate 7.5-minute quadrangles to cover the lower 48 states, Hawai'i, and the U.S. territories. A series of smaller-scale, more-general 15-minute topographic maps offer Alaskan coverage.

In the United States, most quadrangle maps remain in English units of feet and miles. The eventual changeover to the metric system requires revision of the units used on all maps, with the 1:24,000 scale eventually changing to a scale of 1:25,000. However, after completing only a few metric quads, the USGS halted the program in 1991. In Canada, the entire country is mapped at a scale of 1:250,000, using metric units (1.0 cm to 2.5 km). About half the country also is mapped at 1:50,000 (1.0 cm to 0.50 km).

Topographic Maps

The most popular and widely used quadrangle maps are **topographic maps** prepared by the USGS. An example of such a map is a portion of the Cumberland, Maryland,

Figure A.5 **An example of a topographic map from the Appalachians.**
Cumberland, MD, PA, WV 7.5-minute quadrangle topographic map prepared by the USGS. Note the water gap through Haystack Mountain. In the *Applied Physical Geography*, 8/e, lab manual this topographic map is accessible through Google Earth™ mapping services, where you experience the map as a 3-D landscape and maneuver with your computer to see the topography at any angle or detail you choose.

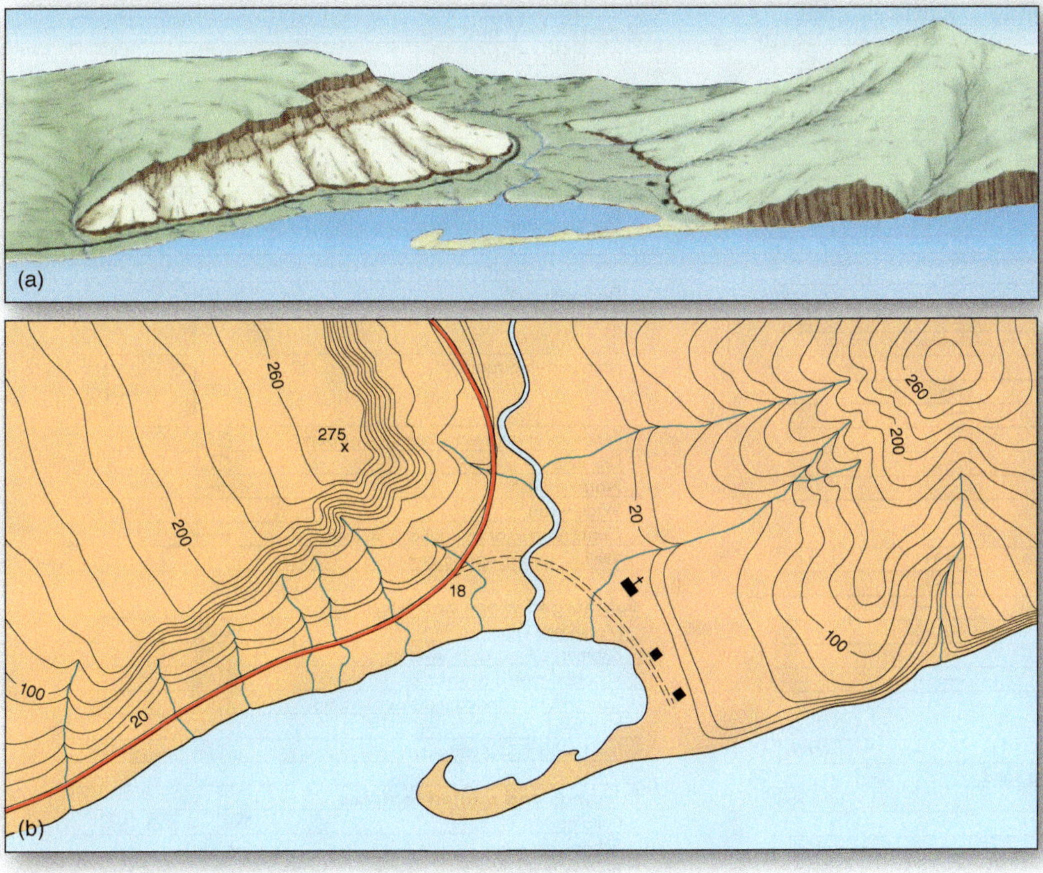

Figure A.6 Topographic map of a hypothetical landscape.
(a) Perspective view of a hypothetical landscape. (b) Depiction of that landscape on a topographic map. The contour interval on the map is 20 feet (6.1 m). [After the U.S. Geological Survey.]

quad shown in Figure A.5. You will find several topographic maps throughout *Geosystems* because they portray landscapes so effectively. As examples, see Figure 13.18, karst landscapes and sinkholes near Orleans, Indiana; and Figure 17.17, drumlins in New York.

A **planimetric map** shows the horizontal position (latitude/longitude) of boundaries, land-use aspects, bodies of water, and economic and cultural features. A highway map is a common example of a planimetric map.

A topographic map adds a vertical component to show topography (configuration of the land surface), including slope and relief (the vertical difference in local landscape elevation). These fine details are shown through the use of elevation contour lines (Figure A.6). A *contour line* connects all points at the same elevation. Elevations are shown above or below a vertical datum, or reference level, which usually is mean sea level. The contour interval is the vertical distance in elevation between two adjacent contour lines (20 ft, or 6.1 m in Figure A.6b).

The topographic map in Figure 6b shows a hypothetical landscape, demonstrating how contour lines and intervals depict slope and relief, which are the three- and dimensional aspects of terrain. The pattern of lines and the spacing between them indicate slope. The steeper a slope or cliff, the closer together the contour lines appear—in the figure, note the narrowly spaced contours that represent the cliffs to the left of the highway. A wider spacing of these contour lines portrays a more gradual slope, as you can see from the widely spaced lines on the beach and to the right of the river valley.

In Figure A.7 are the standard symbols commonly used on these topographic maps. These symbols and the colors used are standard on all USGS topographic maps: black for human constructions, blue for water features, brown for relief features and contours, pink for urbanized areas, and green for woodlands, orchards, brush, and the like.

The margins of a topographic map contain a wealth of information about its concept and content. In the margins of topographic maps, you find the quadrangle name, names of adjoining quads, quad series and type, position in the latitude-longitude and other coordinate systems, title, legend, magnetic declination (alignment of magnetic north) and compass information, datum plane, symbols used for roads and trails, the dates and history of the survey of that particular quad, and more.

Topographic maps may be purchased directly from the USGS or Centre for Topographic Information, NRC (http://maps.nrcan.gc.ca/). Many state geological survey offices, national and state park headquarters, outfitters, sports shops, and bookstores also sell topographic maps to assist people in planning their outdoor activities.

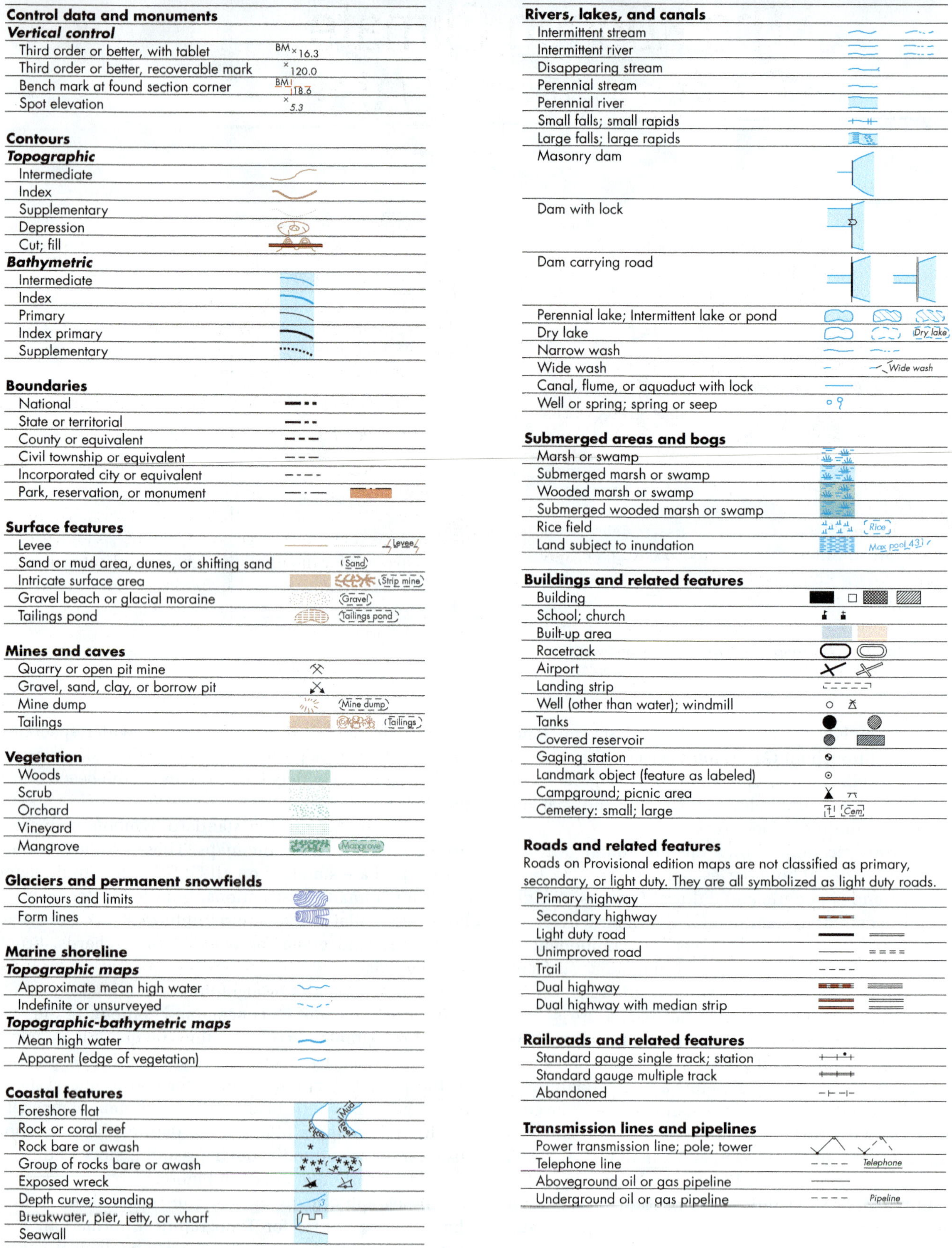

Figure A.7 **Standardized topographic map symbols used on USGS maps.**
English units still prevail, although a few USGS maps are in metric. [From USGS, Topographic Maps, 1969.]

B The Köppen Climate Classification System

The Köppen climate classification system was designed by Wladimir Köppen (1846–1940), a German climatologist and botanist, and is widely used for its ease of comprehension. The basis of any empirical classification system is the choice of criteria used to draw lines on a map to designate different climates. The Köppen–Geiger climate classification uses *average monthly temperatures, average monthly precipitation*, and *total annual precipitation* to devise its spatial categories and boundaries. But we must remember that boundaries are transition zones of gradual change. The trends and overall patterns of boundary lines are more important than their precise placement, especially with the small scales generally used on world maps.

Take a few minutes and examine the Köppen system, the criteria, and the considerations for each of the principal climate categories. The modified Köppen–Geiger system has its drawbacks, however. It does not consider winds, temperature extremes, precipitation intensity, quantity of sunshine, cloud cover, or net radiation. Yet the system is important because its correlations with the actual world are reasonable and the input data are standardized and readily available.

Köppen's Climatic Designations

The genetic, or causative, factors that form the basis of Köppen's climate categories are presented on a world climate map in Figure 10.2, pages 488–89. Here we present the actual criteria that Köppen developed. The Köppen system uses capital letters (A, B, C, D, E, H) to designate climatic categories from the equator to the poles. The guidelines for each of these categories are in the margin of Figure B.1.

Five of the climate classifications are based on thermal criteria:

A. Tropical (equatorial regions)

C. Mesothermal (Mediterranean, humid subtropical, marine west coast regions)

D. Microthermal (humid continental, subarctic regions)

E. Polar (polar regions)

H. Highland (compared to lowlands at the same latitude, highlands have lower temperatures—recall the normal lapse rate—and more efficient precipitation due to lower moisture demand)

Only one climate classification is based on moisture as well:

B. Dry (deserts and semiarid steppes)

Within each climate classification additional lowercase letters are used to signify temperature and moisture conditions. For example, in a tropical rain forest *Af* climate, the *A* tells us that the average coolest month is above 18°C (64.4°F, average for the month), and the *f* indicates that the weather is constantly wet, with the driest month receiving at least 6 cm (2.4 in.) of precipitation. (The designation *f* is from the German *feucht*, for "moist.") As you can see on the climate map, the tropical rain forest *Af* climate dominates along the equator and equatorial rain forest.

In a *Dfa* climate, the *D* means that the average warmest month is above 10°C (50°F), with at least one month falling below 0°C (32°F); the *f* says that at least 3 cm (1.2 in.) of precipitation falls during every month; and the *a* indicates a warmest summer month averaging above 22°C (71.6°F). Thus, a *Dfa* climate is a humid-continental, hot-summer climate in the microthermal category.

Köppen Guidelines

The Köppen guidelines and map portrayal are in Figure B.1. First, check the guidelines for a climate type, then the color legend for the subdivisions of the type, then check out the distribution of that climate on the map. You may want to compare this with the climate map in Figure 10.2 that presents causal elements that produce these climates.

Five of the climate classifications are based on thermal criteria:

A. Tropical (equatorial regions)

C. Mesothermal (Mediterranean, humid subtropical, marine west coast regions)

D. Microthermal (humid continental, subarctic regions)

E. Polar (polar regions)

H. Highland (compared to lowlands at the same latitude, highlands have lower temperatures—recall the normal lapse rate—and more efficient precipitation due to lower moisture demand)

Only one climate classification is based on moisture as well:

B. Dry (deserts and semiarid steppes)

Figure B.1 World climates and their guidelines according to the Köppen classification system.

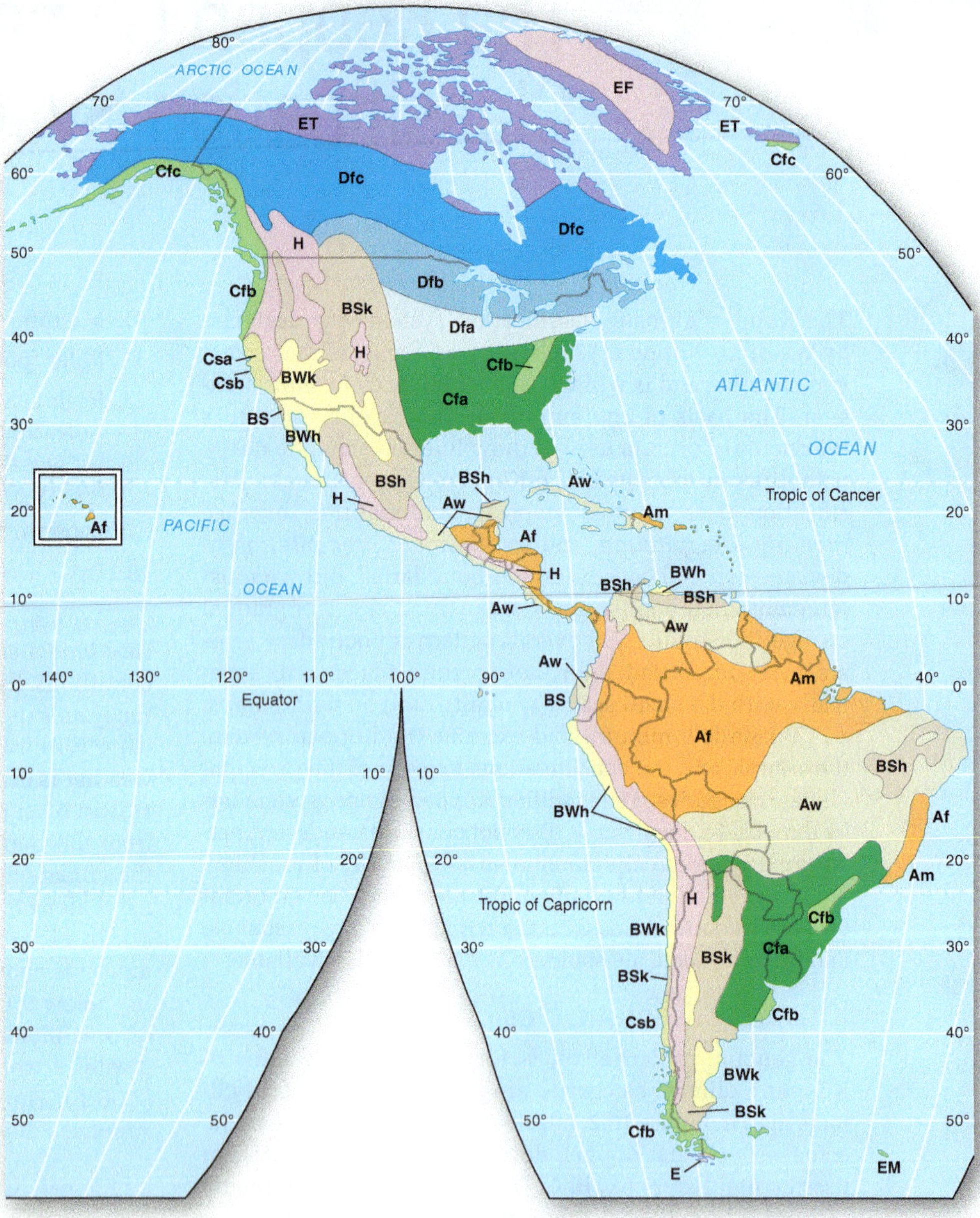

Köppen Guidelines

Tropical Climates — A

Consistently warm with all months averaging above 18°C (64.4°F); annual water supply exceeds water demand.

Af — Tropical rain forest:
- f = All months receive precipitation in excess of 6 cm (2.4 in.).

Am — Tropical monsoon:
- m = A marked short dry season with 1 or more months receiving less than 6 cm (2.4 in.) precipitation; an otherwise excessively wet rainy season. ITCZ 6–12 months dominant.

Aw — Tropical savanna:
- w = Summer wet season, winter dry season; ITCZ dominant 6 months or less, winter water-balance deficits.

Mesothermal Climates — C

Warmest month above 10°C (50°F); coldest month above 0°C (32°F) but below 18°C (64.4°F); seasonal climates.

Cfa, Cwa — Humid subtropical:
- a = Hot summer; warmest month above 22°C (71.6°F).
- f = Year-round precipitation.
- w = Winter drought, summer wettest month 10 times more precipitation than driest winter month.

Cfb, Cfc — Marine west coast, mild-to-cool summer:
- f = Receives year-round precipitation.
- b = Warmest month below 22°C (71.6°F) with 4 months above 10°C.
- c = 1–3 months above 10°C.

Csa, Csb — Mediterranean summer dry:
- s = Pronounced summer drought with 70% of precipitation in winter.
- a = Hot summer with warmest month above 22°C (71.6°F).
- b = Mild summer; warmest month below 22°C.

Microthermal Climates — D

Warmest month above 10°C (50°F); coldest month below 0°C (32°F); cool temperature-to-cold conditions; snow climates. In Southern Hemisphere, occurs only in highland climates.

Dfa, Dwa — Humid continental:
- a = Hot summer; warmest month above 22°C (71.6°F).
- f = Year-round precipitation.
- w = Winter drought.

Dfb, Dwb — Humid continental:
- b = Mild summer; warmest month below 22°C (71.6°F).
- f = Year-round precipitation.
- w = Winter drought.

Dfc, Dwc, Dwd — Subarctic:
Cool summers, cold winters.
- f = Year-round precipitation.
- w = Winter drought.
- c = 1–4 months above 10°C.
- b = Coldest month below −38°C (−36.4°F), in Siberia only.

Dry Arid and Semiarid Climates — B

Potential evapotranspiration* (natural moisture demand) exceeds precipitation (natural moisture supply) in all B climates. Subdivisions based on precipitation timing and amount and mean annual temperature.

Earth's arid climates.

BWh — Hot low-latitude desert

BWk — Cold midlatitude desert
- BW = Precipitation less than 1/2 natural moisture demand.
- h = Mean annual temperature >18°C (64.4°F).
- k = Mean annual temperature <18°C.

Earth's semiarid climates.
BSh — Hot low-latitude steppe
BSk — Cold midlatitude steppe
- BS = Precipitation more than 1/2 natural moisture demand but not equal to it.
- h = Mean annual temperature >18°C.
- k = Mean annual temperature <18°C.

Polar Climates — E

Warmest month below 10°C (50°F); always cold; ice climates.

ET — Tundra:
Warmest month 0–10°C (32–50°F); precipitation exceeds small potential evapotranspiration demand*; snow cover 8–10 months.

EF — Ice cap:
Warmest month below 0°C (32°F); precipitation exceeds a very small potential evapotranspiration demand; the polar regions.

EM — Polar marine:
All months above −7°C (20°F), warmest month above 0°C; annual temperature range <17 C° (30 F°).

*Potential evapotranspiration = the amount of water that would evaporate or transpire if it were available—the natural moisture demand in an environment; see Chapter 9.

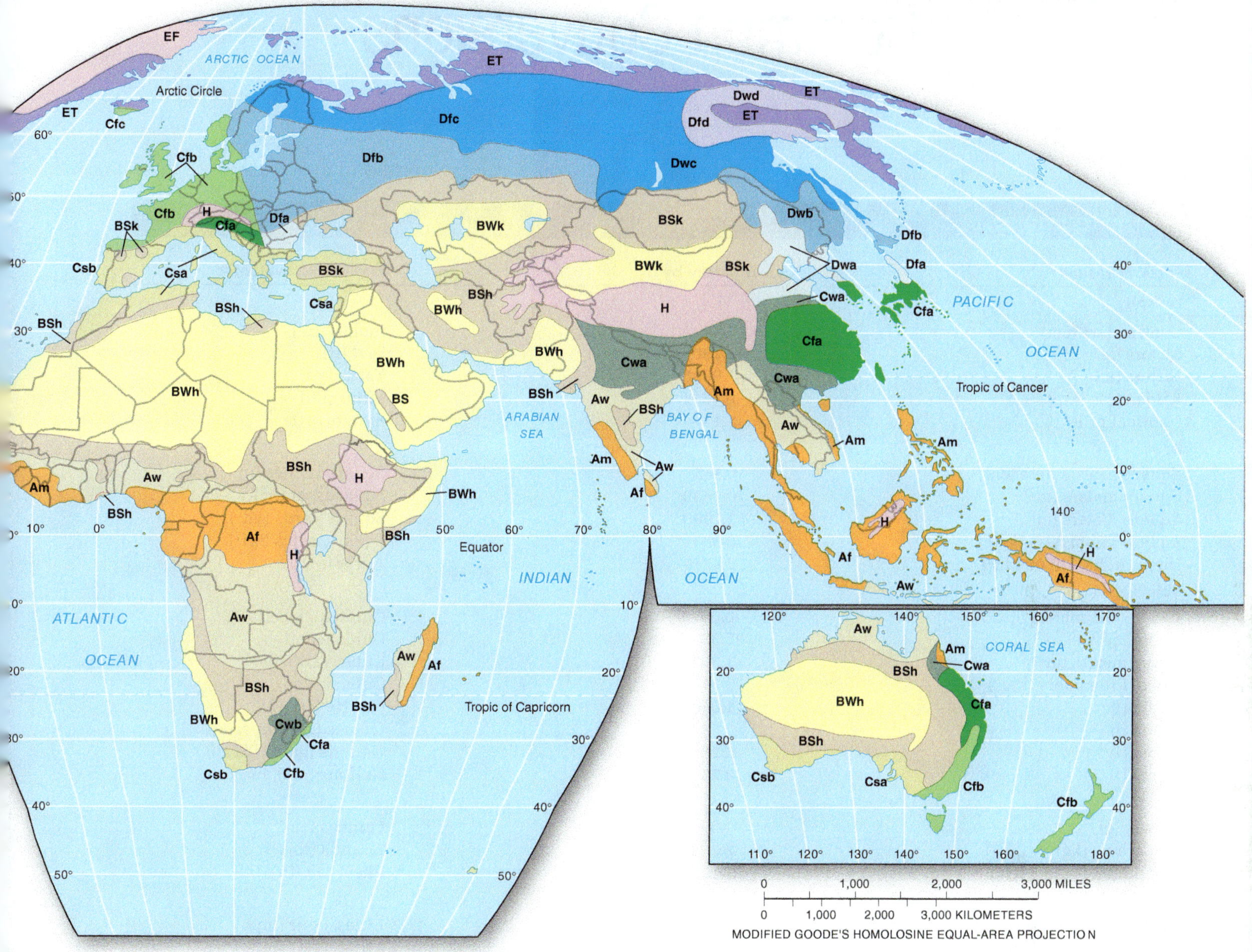

A TROPICAL CLIMATES

- **Af** Tropical rain forest climate
- **Am** Tropical monsoon climate
- **Aw** Tropical savanna climate

C MESOTHERMAL CLIMATES

- **Cfa** Humid subtropical, without dry season, hot summers
- **Cwa Cwb** Humid subtropical, winter-dry
- **Cfb Cfc** Marine west coast, without dry season, warm to cool summers
- **Csa Csb** Mediterranean summer-dry

D MICROTHERMAL CLIMATES

- **Dfa Dwa** Humid continental, hot summers
- **Dfb Dwb** Humid continental, warm summers
- **Dfc Dwc** Subarctic, cool summers
- **Dfd Dwd** Subarctic, very cold winter

w Winter dry
f Without a dry season

B DRY ARID AND SEMIARID CLIMATES

- **BW** Desert climate
- **BS** Steppe climate

h Low-latitude hot
k Midlatitude cold

E POLAR CLIMATES
H HIGHLAND

- **ET** Tundra climate
- **EF** Ice cap and sheets
- **H** Denotes cold climate due to elevation

C Common Conversions

Metric to English

Metric Measure	Multiply by	English Equivalent
Length		
Centimeters (cm)	0.3937	Inches (in.)
Meters (m)	3.2808	Feet (ft)
Meters (m)	1.0936	Yards (yd)
Kilometers (km)	0.6214	Miles (mi)
Nautical mile	1.15	Statute mile
Area		
Square centimeters (cm^2)	0.155	Square inches ($in.^2$)
Square meters (m^2)	10.7639	Square feet (ft^2)
Square meters (m^2)	1.1960	Square yards (yd^2)
Square kilometers (km^2)	0.3831	Square miles (mi^2)
Hectare (ha) (10,000 m^2)	2.4710	Acres (a)
Volume		
Cubic centimeters (cm^3)	0.06	Cubic inches ($in.^3$)
Cubic meters (m^3)	35.30	Cubic feet (ft^3)
Cubic meters (m^3)	1.3079	Cubic yards (yd^3)
Cubic kilometers (km^3)	0.24	Cubic miles (mi^3)
Liters (l)	1.0567	Quarts (qt), U.S.
Liters (l)	0.88	Quarts (qt), Imperial
Liters (l)	0.26	Gallons (gal), U.S.
Liters (l)	0.22	Gallons (gal), Imperial
Mass		
Grams (g)	0.03527	Ounces (oz)
Kilograms (kg)	2.2046	Pounds (lb)
Metric ton (tonne) (t)	1.10	Short ton (tn), U.S.
Velocity		
Meters/second (mps)	2.24	Miles/hour (mph)
Kilometers/hour (kmph)	0.62	Miles/hour (mph)
Knots (kn) (nautical mph)	1.15	Miles/hour (mph)
Temperature		
Degrees Celsius (°C)	1.80 (then add 32)	Degrees Fahrenheit (°F)
Celsius degree (C°)	1.80	Fahrenheit degree (F°)
Additional water measurements:		
Gallon (Imperial)	1.201	Gallon (U.S.)
Gallons (gal)	0.000003	Acre-feet
1 cubic foot per second per day = 86,400 cubic feet = 1.98 acre-feet		

Additional Energy and Power Measurements

1 watt (W) = 1 joule/s

1 joule = 0.239 calorie

1 calorie = 4.186 joules

1 W/m^2 = 0.001433 cal/min

697.8 W/m^2 = 1 cal/cm^2min^{-1}

1 W/m^2 = 2.064 cal/cm^2day^{-1}

1 W/m^2 = 61.91 cal/cm^2month^{-1}

1 W/m^2 = 753.4 cal/cm^2year^{-1}

100 W/m^2 = 75 $kcal/cm^2year^{-1}$

Solar constant:
1372 W/m^2
2 cal/cm^2min^{-1}

English to Metric

English Measure	Multiply by	Metric Equivalent
Length		
Inches (in.)	2.54	Centimeters (cm)
Feet (ft)	0.3048	Meters (m)
Yards (yd)	0.9144	Meters (m)
Miles (mi)	1.6094	Kilometers (km)
Statute mile	0.8684	Nautical mile
Area		
Square inches (in.^{2})	6.45	Square centimeters (cm^2)
Square feet (ft^2)	0.0929	Square meters (m^2)
Square yards (yd^2)	0.8361	Square meters (m^2)
Square miles (mi^2)	2.5900	Square kilometers (km^2)
Acres (a)	0.4047	Hectare (ha)
Volume		
Cubic inches (in.^{3})	16.39	Cubic centimeters (cm^3)
Cubic feet (ft^3)	0.028	Cubic meters (m^3)
Cubic yards (yd^3)	0.765	Cubic meters (m^3)
Cubic miles (mi^3)	4.17	Cubic kilometers (km^3)
Quarts (qt), U.S.	0.9463	Liters (l)
Quarts (qt), Imperial	1.14	Liters (l)
Gallons (gal), U.S.	3.8	Liters (l)
Gallons (gal), Imperial	4.55	Liters (l)
Mass		
Ounces (oz)	28.3495	Grams (g)
Pounds (lb)	0.4536	Kilograms (kg)
Short ton (tn), U.S.	0.91	Metric ton (tonne) (t)
Velocity		
Miles/hour (mph)	0.448	Meters/second (mps)
Miles/hour (mph)	1.6094	Kilometers/hour (kmph)
Miles/hour (mph)	0.8684	Knots (kn) (nautical mph)
Temperature		
Degrees Fahrenheit (°F)	0.556 (after subtracting 32)	Degrees Celsius (°C)
Fahrenheit degree (F°)	0.556	Celsius degree (C°)
Additional water measurements:		
Gallon (U.S.)	0.833	Gallons (Imperial)
Acre-feet	325,872	Gallons (gal)

Additional Notation

Multiples	Prefixes	
1,000,000,000 = 10^9	giga	G
1,000,000 = 10^6	mega	M
1,000 = 10^3	kilo	k
100 = 10^2	hecto	h
10 = 10^1	deka	da
1 = 10^0		
0.1 = 10^{-1}	deci	d
0.01 = 10^{-2}	centi	c
0.001 = 10^{-3}	milli	m
0.000001 = 10^{-6}	micro	μ

glossary

The chapter in which each term appears **boldfaced** in parentheses is followed by a specific definition relevant to the term's usage in the chapter.

Aa (12) Rough, jagged, and clinkery basaltic lava with sharp edges. This texture is caused by the loss of trapped gases, a slow flow, and the development of a thick skin that cracks into the jagged surface.

Abiotic (1) Nonliving; Earth's nonliving systems of energy and materials.

Ablation (17) Loss of glacial ice through melting, sublimation, wind removal by deflation, or the calving of blocks of ice. (*See* Deflation.)

Abrasion (16, 17) Mechanical wearing and erosion of bedrock accomplished by the rolling and grinding of particles and rocks carried in a stream, removed by wind in a "sandblasting" action, or imbedded in glacial ice.

Absorption (4) Assimilation and conversion of radiation from one form to another in a medium. In the process, the temperature of the absorbing surface is raised, thereby affecting the rate and wavelength of radiation from that surface.

Active layer (17) A zone of seasonally frozen ground that exists between the subsurface permafrost layer and the ground surface. The active layer is subject to consistent daily and seasonal freeze–thaw cycles. (*See* Permafrost, Periglacial.)

Actual evapotranspiration (9) ACTET; the actual amount of evaporation and transpiration that occurs; derived in the water balance by subtracting the deficit (DEFIC) from potential evapotranspiration (POTET).

Adiabatic (7) Pertaining to the cooling of an ascending parcel of air through expansion or the warming of a descending parcel of air through compression, without any exchange of heat between the parcel and the surrounding environment.

Advection (4) Horizontal movement of air or water from one place to another. (*Compare* Convection.)

Advection fog (7) Active condensation formed when warm, moist air moves laterally over cooler water or land surfaces, causing the lower layers of the air to be chilled to the dew-point temperature.

Aerosols (3) Small particles of dust, soot, and pollution suspended in the air.

Aggradation (14) The general building of land surface because of deposition of material; opposite of degradation. When the sediment load of a stream exceeds the stream's capacity to carry it, the stream channel becomes filled through this process.

Air (3) A simple mixture of gases (N, O, Ar, CO_2, and trace gases) that is naturally odorless, colorless, tasteless, and formless, blended so thoroughly that it behaves as if it were a single gas.

Air mass (8) A distinctive, homogeneous body of air that has taken on the moisture and temperature characteristics of its source region.

Air pressure (3, 6) Pressure produced by the motion, size, and number of gas molecules in the air and exerted on surfaces in contact with the air; an average force at sea level of 1 kg/cm^3 (14.7 lb/in^2). Normal sea-level pressure, as measured by the height of a column of mercury (Hg), is expressed as 1013.2 millibars, 760 mm of Hg, or 29.92 inches of Hg. Air pressure can be measured with mercury or aneroid barometers (*see listings for both*).

Albedo (4) The reflective quality of a surface, expressed as the percentage of reflected insolation to incoming insolation; a function of surface color, angle of incidence, and surface texture.

Aleutian low (6) *See* Subpolar low-pressure cell.

Alfisols (18) A soil order in the Soil Taxonomy. Moderately weathered forest soils that are moist versions of Mollisols, with productivity dependent on specific patterns of moisture and temperature; rich in organics. Most wide-ranging of the soil orders.

Alluvial fan (14) Fan-shaped fluvial landform at the mouth of a canyon; generally occurs in arid landscapes where streams are intermittent. (*See* bajada.)

Alluvial terraces (14) Level areas that appear as topographic steps above a stream, created by the stream as it scours with renewed downcutting into its floodplain; composed of unconsolidated alluvium. (*See* Alluvium.)

Alluvium (14) General descriptive term for clay, silt, sand, gravel, or other unconsolidated rock and mineral fragments transported by running water and deposited as sorted or semisorted sediment on a floodplain, delta, or streambed.

Alpine glacier (17) A glacier confined in a mountain valley or walled basin, consisting of three subtypes: valley glacier (within a valley), piedmont glacier (coalesced at the base of a mountain, spreading freely over nearby lowlands), and outlet glacier (flowing outward from a continental glacier; *compare to* Continental glacier).

Alpine tundra (20) Tundra conditions at high elevation. (*See* Arctic tundra.)

Altitude (2) The angular distance between the horizon (a horizontal plane) and the Sun (or any point in the sky).

Altocumulus (7) Middle-level, puffy clouds that occur in several forms: patchy rows, wave patterns, a "mackerel sky," or lens-shaped "lenticular" clouds.

Andisols (18) A soil order in the Soil Taxonomy; derived from volcanic parent materials in areas of volcanic activity. A new order, created in 1990, of soils previously considered under Inceptisols and Entisols.

Anemometer (6) A device that measures wind velocity.

Aneroid barometer (6) A device that measures air pressure using a partially evacuated, sealed cell. (*See* Air pressure.)

Angle of repose (13) The steepness of a slope that results when loose particles come to rest; an angle of balance between driving and resisting forces, ranging between 33° and 37° from a horizontal plane.

Antarctic Circle (2) This latitude (66.5° S) denotes the northernmost parallel (in the Southern Hemisphere) that experiences a 24-hour period of darkness in winter or daylight in summer.

Antarctic high (6) A consistent high-pressure region centered over Antarctica; source region for an intense polar air mass that is dry and associated with the lowest temperatures on Earth.

Antarctic region (17) The Antarctic convergence defines the Antarctic region in a narrow zone that extends around the continent as a boundary between colder Antarctic water and warmer water at lower latitudes.

Anthropogenic atmosphere (3) Earth's future atmosphere, so named because humans appear to be the principal causative agent.

Anthropogenic biome (20) A recent conceptual term for large-scale, stable ecosystems that result from ongoing human interaction with natural environments. Human modifications are often linked to land-use practices such as agriculture, forestry, and urbanization.

Anticline (12) Upfolded rock strata in which layers slope downward from the axis of the fold, or central ridge. (*Compare* Syncline.)

Anticyclone (6) A dynamically or thermally caused area of high atmospheric pressure with descending and diverging airflows that rotate clockwise in the Northern Hemisphere and counterclockwise in the Southern Hemisphere. (*Compare* Cyclone.)

Aphelion (2) The point of Earth's greatest distance from the Sun in its elliptical orbit; reached on July 4 at a distance of 152,083,000 km (94.5 million mi); variable over a 100,000-year cycle. (*Compare* Perihelion.)

Aquiclude (9) An impermeable rock layer or body of unconsolidated materials that blocks groundwater flow and forms the boundary of a confined aquifer. (*Compare* Aquifer.)

Aquitard (9) A body of rock or unconsolidated sediments with low permeability that does not conduct groundwater in useable amounts.

Aquifer (9) A body of rock that conducts groundwater in usable amounts; a permeable rock layer. (*Compare* Aquiclude.)

Aquifer recharge area (9) The surface area where water enters an aquifer to recharge the water-bearing strata in a groundwater system.

Arctic Circle (2) This latitude (66.5° N) denotes the southernmost parallel (in the Northern Hemisphere) that experiences a 24-hour period of darkness in winter or daylight in summer.

Arctic region (17) The 10°C (50°F) isotherm for July defines the Arctic region; coincides with the visible tree line—the boundary between the northern forests and tundra.

Arctic tundra (20) A biome in the northernmost portions of North America and northern Europe and Russia, featuring low, ground-level herbaceous plants as well as some woody plants. (*See* Alpine tundra.)

Arête (17) A sharp ridge that divides two cirque basins. Derived from "knife edge" in French, these form sawtooth and serrated ridges in glaciated mountains.

Aridisols (18) A soil order in the Soil Taxonomy; largest soil order. Typical of dry climates; low in organic matter and dominated by calcification and salinization.

Artesian water (9) Pressurized groundwater that rises in a well or a rock structure above the local water table; may flow out onto the ground without pumping. (*See* Potentiometric surface.)

Artificial levee (14) Human-built earthen embankments along river channels, often constructed on top of natural levees.

Asthenosphere (11) Region of the upper mantle just below the lithosphere; the least rigid portion of Earth's interior and known as the plastic layer, flowing very slowly under extreme heat and pressure.

Atmosphere (1) The thin veil of gases surrounding Earth, which forms a protective boundary between outer space and the biosphere; generally considered to extend out about 480 km (300 mi) from Earth's surface.

Atmosphere-Ocean General Circulation Model (AOGCM) (10B) A sophisticated general circulation model that couples atmosphere and ocean submodels to simulate the effects of linkages between specific climate components over different time frames and at various scales.

Aurora (2) A spectacular glowing light display in the ionosphere, stimulated by the interaction of the solar wind with principally oxygen and nitrogen gases and few other atoms at high latitudes; called *aurora borealis* in the Northern Hemisphere and *aurora australis* in the Southern Hemisphere.

Autumnal (September) equinox (2) The time around September 22–23 when the Sun's declination crosses the equatorial parallel (0° latitude) and all places on Earth experience days and nights of equal length. The Sun rises at the South Pole and sets at the North Pole. (*Compare* Vernal [March] equinox.)

Axial parallelism (2) Earth's axis remains aligned the same throughout the year (it "remains parallel to itself"); thus, the axis extended from the North Pole points into space always near Polaris, the North Star.

Axial tilt (2) Earth's axis tilts 23.5° from a perpendicular to the plane of the ecliptic (plane of Earth's orbit around the Sun).

Axis (2) An imaginary line, extending through Earth from the geographic North Pole to the geographic South Pole, around which Earth rotates.

Azores high (6) A subtropical high-pressure cell that forms in the Northern Hemisphere in the eastern Atlantic (*see* Bermuda high); associated with warm, clear water and large quantities of sargassum, or gulf weed, characteristic of the Sargasso Sea.

Bajada (14) A continuous apron of coalesced alluvial fans, formed along the base of mountains in arid climates; presents a gently rolling surface from fan to fan. (*See* Alluvial fan.)

Barrier beach (16) Narrow, long, depositional feature, generally composed of sand, that forms offshore roughly parallel to the coast; may appear as barrier islands and long chains of barrier beaches. (*See* Barrier island.)

Barrier island (16) Generally, a broadened barrier beach offshore. (*See* Barrier beach.)

Barrier spit (16) A depositional landform that develops when transported sand or gravel in a barrier beach or island is deposited in long ridges that are attached at one end to the mainland and partially cross the mouth of a bay.

Basalt (11) A common extrusive igneous rock, fine-grained, comprising the bulk of the ocean-floor crust, lava flows, and volcanic forms; gabbro is its intrusive form.

Base flow (9) The portion of streamflow that consists of groundwater.

Base level (14) A hypothetical level below which a stream cannot erode its valley—and thus the lowest operative level for denudation processes; in an absolute sense, it is represented by sea level, extending under the landscape.

Basin and Range Province (12) A region of dry climates, few permanent streams, and interior drainage patterns in the western United States; a faulted landscape composed of a sequence of horsts and grabens.

Batholith (11) The largest plutonic form exposed at the surface; an irregular intrusive mass; it invades crustal rocks, cooling slowly so that large crystals develop. (*See* Pluton.)

Bay barrier (16) An extensive barrier spit of sand or gravel that encloses a bay, cutting it off completely from the ocean and forming a lagoon; produced by littoral drift and wave action; sometimes referred to as a baymouth bar. (*See* Barrier spit, Lagoon.)

Beach (16) The portion of the coastline where an accumulation of sediment is in motion.

Beach drift (16) Material, such as sand, gravel, and shells, that is moved by the longshore current in the effective direction of the waves.

Bed load (14) Coarse materials that are dragged along the bed of a stream by traction or by the rolling and bouncing motion of saltation; involves particles too large to remain in suspension. (*See* Traction, Saltation.)

Bedrock (13) The rock of Earth's crust that is below the soil and is basically unweathered; such solid crust sometimes is exposed as an outcrop.

Bergschrund (17) This forms when a crevasse or wide crack opens along the headwall of a glacier; most visible in summer when covering snow is gone.

Bermuda high (6) A subtropical high-pressure cell that forms in the western North Atlantic. (*See* Azores high.)

Biodiversity (19) A principle of ecology and biogeography: The more diverse the species population in an ecosystem (in number of species, quantity of members in each species, and genetic content), the more risk is spread over the entire community, which results in greater overall stability, greater productivity, and increased use of nutrients, as compared to a monoculture of little or no diversity.

Biogeochemical cycle (19) One of several circuits of flowing elements and materials (carbon, oxygen, nitrogen, phosphorus, water) that combine Earth's biotic (living) and abiotic (nonliving) systems; the cycling of materials is continuous and renewed through the biosphere and the life processes.

Biogeographic realm (20) One of eight regions of the biosphere, each representative of evolutionary core areas of related flora (plants) and fauna (animals); a broad geographical classification scheme.

Biogeography (19) The study of the distribution of plants and animals and related ecosystems; the geographical relationships with their environments over time.

Biomass (19) The total mass of living organisms on Earth or per unit area of a landscape; also the weight of the living organisms in an ecosystem.

Biome (20) A large-scale, stable, terrestrial or aquatic ecosystem classified according to the predominant vegetation type and the adaptations of particular organisms to that environment.

Biosphere (1) That area where the atmosphere, lithosphere, and hydrosphere function together to form the context within which life exists; an intricate web that connects all organisms with their physical environment.

Biotic (1) Living; referring to Earth's living system of organisms.

Blowout depression (14) Eolian (wind) erosion in which deflation forms a basin in areas of loose sediment. Diameter may range up to hundreds of meters. (*See* Deflation.)

Bolson (12) The slope and basin area between the crests of two adjacent ridges in a dry region.

Boreal forest (20) *See* Needleleaf forest.

Brackish (16) Descriptive of seawater with a salinity of less than 35‰; for example, the Baltic Sea. (*Compare* Brine.)

Braided stream (14) A stream that becomes a maze of interconnected channels laced with excess sediment. Braiding often occurs with a reduction of discharge that reduces a stream's transporting ability or with an increase in sediment load.

Breaker (16) The point where a wave's height exceeds its vertical stability and the wave breaks as it approaches the shore.

Brine (16) Seawater with a salinity of more than 35‰; for example, the Persian Gulf. (*Compare* Brackish.)

Calcification (18) The illuviated (deposited) accumulation of calcium carbonate or magnesium carbonate in the B and C soil horizons.

Caldera (12) An interior sunken portion of a composite volcano's crater; usually steep-sided and circular, sometimes containing a lake; also can be found in conjunction with shield volcanoes.

Calving (17) The process in which pieces of ice break free from the terminus of a tidewater glacier or ice sheet to form floating ice masses (icebergs) where glaciers meet an ocean, bay, or fjord.

Capillary water (9) Soil moisture, most of which is accessible to plant roots; held in the soil by the water's surface tension and cohesive forces between water and soil. (*See also* Available water, Field capacity, Hygroscopic water, Wilting point.)

Carbonation (13) A chemical weathering process in which weak carbonic acid (water and carbon dioxide) reacts with many minerals that contain calcium, magnesium, potassium, and sodium (especially limestone), transforming them into carbonates.

Carbon monoxide (CO) (3) An odorless, colorless, tasteless combination of carbon and oxygen produced by the incomplete combustion of fossil fuels or other carbon-containing substances; toxicity to humans is due to its affinity for hemoglobin, displacing oxygen in the bloodstream.

Carbon sink (10B) An area in Earth's atmosphere, hydrosphere, lithosphere, or biosphere where carbon is stored; also called a *carbon reservoir*.

Carnivore (19) A secondary consumer that principally eats meat for sustenance. The top carnivore in a food chain is considered a tertiary consumer. (*Compare* Herbivore.)

Cartography (1) The making of maps and charts; a specialized science and art that blends aspects of geography, engineering, mathematics, graphics, computer science, and artistic specialties.

Cation-exchange capacity (CEC) (18) The ability of soil colloids to exchange cations between their surfaces and the soil solution; a measured potential that indicates soil fertility. (*See* Soil colloid, Soil fertility.)

Chaparral (20) Dominant shrub formations of *Mediterranean* (dry summer) climates; characterized by sclerophyllous scrub and short, stunted, tough forests; derived from the Spanish *chaparro*; specific to California. (*See* Mediterranean shrubland.)

Chemical weathering (13) Decomposition and decay of the constituent minerals in rock through chemical alteration of those minerals. Water is essential, with rates keyed to temperature and precipitation values. Chemical reactions are active at microsites even in dry climates. Processes include hydrolysis, oxidation, carbonation, and solution.

Chinook wind (8) North American term for a warm, dry, downslope airflow; characteristic of the rain-shadow region on the leeward side of mountains; known as *föhn* or *foehn* winds in Europe. (*See* Rain shadow.)

Chlorofluorocarbons (CFC) (3) A manufactured molecule (polymer) made of chlorine, fluorine, and carbon; inert and possessing remarkable heat properties; also known as one of the halogens. After slow transport to the stratospheric ozone layer, CFCs react with ultraviolet radiation, freeing chlorine atoms that act as a catalyst to produce reactions that destroy ozone; manufacture banned by international treaties.

Chlorophyll (19) A light-sensitive pigment that resides within chloroplasts (organelles) in leaf cells of plants; the basis of photosynthesis.

Cinder cone (12) A volcanic landform of pyroclastics and scoria, usually small and cone-shaped and generally not more than 450 m (1500 ft) in height, with a truncated top.

Circle of illumination (2) The division between light and dark on Earth; a day–night great circle.

Circum-Pacific belt (12) A tectonically and volcanically active region encircling the Pacific Ocean; also known as the "ring of fire."

Cirque (17) A scooped-out, amphitheater-shaped basin at the head of an alpine glacier valley; an erosional landform.

Cirrus (7) Wispy, filamentous ice-crystal clouds that occur above 6000 m (20,000 ft); appear in a variety of forms, from feathery hairlike fibers to veils of fused sheets.

Classification (10) The process of ordering or grouping data or phenomena in related classes; results in a regular distribution of information; a taxonomy.

Climate (10) The consistent, long-term behavior of weather over time, including its variability; in contrast to weather, which is the condition of the atmosphere at any given place and time.

Climate change science (10B) The interdisciplinary study of the causes and consequences of changing climate for all Earth systems and the sustainability of human societies.

Climate feedback (10B) A process that either amplifies or reduces a climatic trend toward either warming or cooling.

Climatic region (10) An area of homogenous climate that features characteristic regional weather and air mass patterns.

Climatology (10) The scientific study of climate and climatic patterns and the consistent behavior of weather, including its variability and extremes, over time in one place or region; includes the effects of climate change on human society and culture.

Climograph (10) A graph that plots daily, monthly, or annual temperature and precipitation values for a selected station; may also include additional weather information.

Closed system (1) A system that is shut off from the surrounding environment, so that it is entirely self-contained in terms of energy and materials; Earth is a closed material system. (*Compare* Open system.)

Cloud (7) An aggregate of tiny moisture droplets and ice crystals; classified by altitude of occurrence and shape.

Cloud-albedo forcing (4) An increase in albedo (the reflectivity of a surface) caused by clouds due to their reflection of incoming insolation.

Cloud-condensation nuclei (7) Microscopic particles necessary as matter on which water vapor condenses to form moisture droplets; can be sea salts, dust, soot, or ash.

Cloud-greenhouse forcing (4) An increase in greenhouse warming caused by clouds because they can act like insulation, trapping longwave (infrared) radiation.

Col (17) Formed by two headward-eroding cirques that reduce an arête (ridge crest) to form a high pass or saddlelike narrow depression.

Cold desert and semidesert (20) A type of desert biome found at higher latitudes than warm deserts. Interior location and rain shadows produce these cold deserts in North America.

Cold front (8) The leading edge of an advancing cold air mass; identified on a weather map as a line marked with triangular spikes pointing in the direction of frontal movement. (*Compare* Warm front.)

Community (19) A convenient biotic subdivision within an ecosystem; formed by interacting populations of animals and plants in an area.

Composite volcano (12) A volcano formed by a sequence of explosive volcanic eruptions; steep-sided, conical in shape; sometimes referred to as a stratovolcano, although composite is the preferred term. (*Compare* Shield volcano.)

Conduction (4) The slow molecule-to-molecule transfer of heat through a medium, from warmer to cooler portions.

Cone of depression (9) The depressed shape of the water table around a well after active pumping. The water table adjacent to the well is drawn down by the water removal.

Confined aquifer (9) An aquifer that is bounded above and below by impermeable layers of rock or sediment. (*See* Artesian water, Unconfined aquifer.)

Constant isobaric surface (6) An elevated surface in the atmosphere on which all points have the same pressure, usually 500 mb. Along this constant-pressure surface, isobars mark the paths of upper-air winds.

Consumer (19) Organism in an ecosystem that depends on producers (organisms that use carbon dioxide as their sole source of carbon) for its source of nutrients; also called a *heterotroph*. (*Compare* Producer.)

Consumptive use (9) A use that removes water from a water budget at one point and makes it unavailable farther downstream. (*Compare* Withdrawal.)

Continental divide (14) A ridge or elevated area that separates drainage on a continental scale; specifically, that ridge in North America that separates drainage to the Pacific on the west side from drainage to the Atlantic and Gulf on the east side and to Hudson Bay and the Arctic Ocean in the north.

Continental drift (11) A proposal by Alfred Wegener in 1912 stating that Earth's landmasses have migrated over the past 225 million years from a supercontinent he called Pangaea to the present configuration; the widely accepted plate tectonics theory today. (*See* Plate tectonics.)

Continental effect (5) A quality of regions that lack the temperature-moderating effects of the ocean and that exhibit a greater range of minimum and maximum temperatures, both daily and annually, than do marine stations. (*See* Marine effect, Land–water heating difference.)

Continental landmasses (12) The broadest category of landform, including those masses of crust that reside above or near sea level and the adjoining undersea continental shelves along the coastline; sometimes synonymous with *continental platforms*.

Continental shield (12) Generally, old, low-elevation heartland regions of continental crust; various cratons (granitic cores) and ancient mountains are exposed at the surface.

Contour lines (Appendix A) Isolines on a topographic map that connect all points at the same elevation relative to a reference elevation called the vertical datum.

Convection (4) Transfer of heat from one place to another through the physical movement of air; involves a strong vertical motion. (*Compare* Advection.)

Convectional lifting (8) Air passing over warm surfaces gains buoyancy and lifts, initiating adiabatic processes.

Convergent lifting (8) Air flowing from different directions forces lifting and displacement of air upward, initiating adiabatic processes.

Coordinated Universal Time (UTC) (1) The official reference time in all countries, formerly known as Greenwich Mean Time; now measured by primary standard atomic clocks, the time calculations are collected in Paris at the International Bureau of Weights and Measures (BIPM); the legal reference for time in all countries and broadcast worldwide.

Coral (16) A simple, cylindrical marine animal with a saclike body that secretes calcium carbonate to form a hard external skeleton and, cumulatively, landforms called reefs; lives symbiotically with nutrient-producing algae; presently in a worldwide state of decline due to bleaching (loss of algae).

Core (11) The deepest inner portion of Earth, representing one-third of its entire mass; differentiated into two zones—a solid-iron inner core surrounded by a dense, molten, fluid metallic-iron outer core.

Coriolis force (6) The apparent deflection of moving objects (wind, ocean currents, missiles) from travelling in a straight path, in proportion to the speed of Earth's rotation at different latitudes. Deflection is to the right in the Northern Hemisphere and to the left in the Southern Hemisphere; maximum at the poles and zero along the equator.

Crater (12) A circular surface depression formed by volcanism; built by accumulation, collapse, or explosion; usually located at a volcanic vent or pipe; can be at the summit or on the flank of a volcano.

Crevasse (17) A vertical crack that develops in a glacier as a result of friction between valley walls, or tension forces of extension on convex slopes, or compression forces on concave slopes.

Crust (11) Earth's outer shell of crystalline surface rock, ranging from 5 to 60 km (3 to 38 mi) in thickness from oceanic crust to mountain ranges. Average density of continental crust is $2.7 g/cm^3$, whereas oceanic crust is $3.0 g/cm^3$.

Cryosphere (1, 17) The frozen portion of Earth's waters, including ice sheets, ice caps and fields, glaciers, ice shelves, sea ice, and subsurface ground ice and frozen ground (permafrost).

Cumulonimbus (7) A towering, precipitation-producing cumulus cloud that is vertically developed across altitudes associated with other clouds; frequently associated with lightning and thunder and thus sometimes called a *thunderhead*.

Cumulus (7) Bright and puffy cumuliform clouds up to 2000 m (6500 ft) in altitude.

Cyclogenesis (8) An atmospheric process that describes the birth of a midlatitude wave cyclone, usually along the polar front. Also refers to strengthening and development of a midlatitude cyclone along the eastern slope of the Rockies, other north–south mountain barriers, and the North American and Asian east coasts. (*See* Midlatitude cyclone, Polar front.)

Cyclone (6) A dynamically or thermally caused area of low atmospheric pressure with ascending and converging airflows that rotate counterclockwise in the Northern Hemisphere and clockwise in the Southern Hemisphere. (*Compare* Anticyclone; *see* Midlatitude cyclone, Tropical cyclone.)

Daylength (2) Duration of exposure to insolation, varying during the year depending on latitude; an important aspect of seasonality.

Daylight saving time (1) Time is set ahead 1 hour in the spring and set back 1 hour in the fall in the Northern Hemisphere. In the United States and Canada, time is set ahead on the second Sunday in March and set back on the first Sunday in November—except in Hawai'i, Arizona, and Saskatchewan, which exempt themselves.

Dead zone (19) Low-oxygen conditions and limited marine life caused by excessive nutrient inputs in coastal oceans and lakes.

Debris avalanche (13) A mass of falling and tumbling rock, debris, and soil; can be dangerous because of the tremendous velocities achieved by the onrushing materials.

December solstice (2) *See* Winter (December) solstice.

Declination (2) The latitude that receives direct overhead (perpendicular) insolation on a particular day; the subsolar point migrates annually through 47° of latitude between the Tropics of Cancer (23.5° N) and Capricorn (23.5° S).

Decomposers (19) Bacteria and fungi that digest organic debris outside their bodies and absorb and release nutrients in an ecosystem. (*See* Detritivores.)

Deficit (9) DEFIC; in a water balance, the amount of unmet (unsatisfied) potential evapotranspiration (POTET, or PE); a natural water shortage. (*See* Potential evapotranspiration.)

Deflation (16) A process of wind erosion that removes and lifts individual particles, literally blowing away unconsolidated, dry, or noncohesive sediments. (*See* Blowout depression.)

Degradation (14) The process occurring when sediment is eroded along a stream, causing channel incision.

Delta (14) A depositional plain formed where a river enters a lake or an ocean; named after the triangular shape of the Greek letter delta, **Δ**.

Dendroclimatology (10B) The study of past climates using tree rings. The dating of tree rings by analysis and comparison of ring widths and coloration is *dendrochronology*.

Denudation (13) A general term that refers to all processes that cause degradation of the landscape: weathering, mass movement, erosion, and transport.

Deposition (14) The process whereby weathered, wasted, and transported sediments are laid down by air, water, and ice.

Derechos (8) Strong linear winds in excess of 26 m/sec (58 mph), associated with thunderstorms and bands of showers crossing a region.

Desalination (9) In a water resources context, the removal of organics, debris, and salinity from seawater through distillation or reverse osmosis to produce potable water.

Desertification (18) The expansion of deserts worldwide, related principally to poor agricultural practices (overgrazing and inappropriate agricultural practices), improper soil-moisture management, erosion and salinization, deforestation, and the ongoing climatic change; an unwanted semipermanent invasion into neighboring biomes.

Desert pavement (16) On arid landscapes, a surface formed when wind deflation and sheetflow remove smaller particles, leaving residual pebbles and gravels to concentrate at the surface; an alternative sediment-accumulation hypothesis explains some desert pavements; resembles a cobblestone street. (*See* Deflation, Sheetflow.)

Detritivores (19) Detritus feeders and decomposers that consume, digest, and destroy organic wastes and debris. *Detritus feeders*—worms, mites, termites, centipedes, snails, crabs, and even vultures, among others—consume detritus and excrete nutrients and simple inorganic compounds that fuel an ecosystem. (*Compare* Decomposers.)

Dew-point temperature (7) The temperature at which a given mass of air becomes saturated, holding all the water it can hold. Any further cooling or addition of water vapor results in active condensation.

Diagnostic subsurface horizon (18) A soil horizon that originates below the epipedon at varying depths; may be part of the A and B horizons; important in soil description as part of the Soil Taxonomy.

Differential weathering (13) The effect of different resistances in rock, coupled with variations in the intensity of physical and chemical weathering.

Diffuse radiation (4) The downward component of scattered incoming insolation from clouds and the atmosphere.

Discharge (14) The measured volume of flow in a river that passes by a given cross section of a stream in a given unit of time; expressed in cubic meters per second or cubic feet per second.

Dissolved load (14) Materials carried in chemical solution in a stream, derived from minerals such as limestone and dolomite or from soluble salts.

Downwelling current (6) An area of the sea where a convergence or accumulation of water thrusts excess water downward; occurs, for example, at the western end of the equatorial current or along the margins of Antarctica. (*Compare* Upwelling current.)

Drainage basin (14) The basic spatial geomorphic unit of a river system; distinguished from a neighboring basin by ridges and highlands that form divides, marking the limits of the catchment area of the drainage basin.

Drainage density (14) A measure of the overall operational efficiency of a drainage basin, determined by the ratio of combined channel lengths to the unit area.

Drainage pattern (14) A distinctive geometric arrangement of streams in a region, determined by slope, differing rock resistance to weathering and erosion, climatic and hydrologic variability, and structural controls of the landscape.

Drawdown (9) *See* Cone of depression.

Drought (9) Does not have a simple water-budget definition; rather, it can occur in at least four forms: *meteorological drought, agricultural drought, hydrologic drought,* and/or *socioeconomic drought.*

Drumlin (17) A depositional landform related to glaciation that is composed of till (unstratified, unsorted) and is streamlined in the direction of continental ice movement—blunt end upstream and tapered end downstream with a rounded summit.

Dry adiabatic rate (DAR) (7) The rate at which an unsaturated parcel of air cools (if ascending) or heats (if descending); a rate of 10 C° per 1000 m (5.5 F° per 1000 ft). (*See* Adiabatic; *compare* Moist adiabatic rate.)

Dune (16) A depositional feature of sand grains deposited in transient mounds, ridges, and hills; extensive areas of sand dunes are called sand seas.

Dust dome (4) A dome of airborne pollution associated with every major city; may be blown by winds into elongated plumes downwind from the city.

Dynamic equilibrium (1) Increasing or decreasing operations in a system demonstrate a trend over time, a change in average conditions.

Dynamic equilibrium model (13) The balancing act between tectonic uplift and erosion, between the resistance of crust materials and the work of denudation processes. Landscapes evidence ongoing adaptation to rock structure, climate, local relief, and elevation.

Earthquake (12) A sharp release of energy that sends waves travelling through Earth's crust at the moment of rupture along a fault or in association with volcanic activity. The moment magnitude scale (formerly the Richter scale) estimates earthquake magnitude; intensity is described by the Mercalli scale.

Earth systems science (1) An emerging science of Earth as a complete, systematic entity. An interacting set of physical, chemical, and biological systems that produce the processes of a whole Earth system. A study of planetary change resulting from system operations; includes a desire for a more quantitative understanding among components rather than a qualitative description.

Ebb tide (16) Falling or lowering tide during the daily tidal cycle. (*Compare* Flood tide.)

Ecological niche (19) The function, or operation, of a life form within a given ecological community.

Ecological succession (19) The process whereby different and usually more complex assemblages of plants and animals replace older and usually simpler communities; communities are in a constant state of change as each species adapts to changing conditions. Ecosystems do not exhibit a stable point or successional climax condition as previously thought. (*See* Primary succession, Secondary succession.)

Ecology (19) The science that studies the relations between organisms and their environment and among various ecosystems.

Ecosphere (1) Another name for the biosphere.

Ecosystem (19) A self-regulating association of living plants and animals and their nonliving physical and chemical environments.

Ecotone (20) A boundary transition zone between adjoining ecosystems that may vary in width and represent areas of tension as similar species of plants and animals compete for the resources. (*See* Ecosystem.)

Effusive eruption (12) A volcanic eruption characterized by low-viscosity basaltic magma and low-gas content, which readily escapes. Lava pours forth onto the surface with relatively small explosions and few pyroclastics; tends to form shield volcanoes. (*See* Shield volcano, Lava, Pyroclastics; *compare* Explosive eruption.)

Elastic-rebound theory (12) A concept describing the faulting process in Earth's crust, in which the two sides of a fault appear locked despite the motion of adjoining pieces of crust, but with accumulating strain, they rupture suddenly, snapping to new positions relative to each other, generating an earthquake.

Electromagnetic spectrum (2) All the radiant energy produced by the Sun placed in an ordered range, divided according to wavelengths.

El Niño–Southern Oscillation (ENSO) (6) Sea-surface temperatures increase, sometimes more than 8 C° (14 F°) above normal in the central and eastern Pacific, replacing the normally cold, nutrient-rich water along Peru's coastline. Pressure patterns and surface ocean temperatures shift from their usual locations across the Pacific, forming the Southern Oscillation.

Eluviation (18) The removal of finer particles and minerals from the upper horizons of soil; an erosional process within a soil body. (*Compare* Illuviation.)

Empirical classification (10) A climate classification based on weather statistics or other data; used to determine general climate categories. (*Compare* Genetic classification.)

Endogenic system (11) The system internal to Earth, driven by radioactive heat derived from sources within the planet. In response, the surface fractures, mountain building occurs, and earthquakes and volcanoes are activated. (*Compare* Exogenic system.)

Entisols (18) A soil order in the Soil Taxonomy. Specifically lacks vertical development of horizons; usually young or undeveloped. Found in active slopes, alluvial-filled floodplains, and poorly drained tundra.

Environmental lapse rate (3) The actual rate of temperature decrease with increasing altitude in the lower atmosphere at any particular time under local weather conditions; may deviate above or below the normal lapse rate of 6.4 C° per km, or 1000 m (3.5 F° per 1000 ft). (*Compare* Normal lapse rate.)

Eolian (16) Caused by wind; refers to the erosion, transportation, and deposition of materials; spelled *aeolian* in some countries.

Epipedon (18) The diagnostic soil horizon that forms at the surface; not to be confused with the A horizon; may include all or part of the illuviated B horizon.

Equal area (1, Appendix A) A trait of a map projection; indicates the equivalence of all areas on the surface of the map, although shape is distorted. (*See* Map projection.)

Equatorial low (6) A thermally caused low-pressure area that almost girdles Earth, with air converging and ascending all along its extent; also called the *intertropical convergence zone (ITCZ).*

Erg (16) An extensive area of sand and dunes; from the Arabic word for "dune field." (*Compare* Sand sea.)

Erosion (14) Denudation by wind, water, or ice, which dislodges, dissolves, or removes surface material.

Esker (17) A sinuously curving, narrow deposit of coarse gravel that forms along a meltwater stream channel, developing in a tunnel beneath a glacier.

Estuary (14) The point at which the mouth of a river enters the sea, where freshwater and seawater are mixed; a place where tides ebb and flow.

Eustasy (9) Refers to worldwide changes in sea level that are related not to movements of land, but rather to changes in the volume of water in the oceans.

Eutrophication (19) The gradual enrichment of water bodies that occurs with nutrient inputs, either natural or human-caused.

Evaporation (9) The movement of free water molecules away from a wet surface into air that is less than saturated; the phase change of water to water vapor.

Evaporation fog (7) A fog formed when cold air flows over the warm surface of a lake, ocean, or other body of water; forms as the water molecules evaporate from the water surface into the cold, overlying air; also known as steam fog or sea smoke.

Evaporation pan (9) A weather instrument consisting of a standardized pan from which evaporation occurs, with water automatically replaced and measured; an evaporimeter.

Evapotranspiration (9) The merging of evaporation and transpiration water loss into one term. (*See* Potential evapotranspiration, Actual evapotranspiration.)

Evolution (19) A theory that single-cell organisms adapted, modified, and passed along inherited changes to multicellular organisms. The genetic makeup of successive generations is shaped by environmental factors, physiological functions, and behaviors that created a greater rate of survival and reproduction and were passed along through natural selection.

Exfoliation (13) The physical weathering process that occurs as mechanical forces enlarge joints in rock into layers of curved slabs or plates, which peel or slip off in sheets; also called *sheeting.*

Exfoliation dome (13) A dome-shaped feature of weathering, produced by the response of granite to the overburden removal process, which relieves pressure from the rock. Layers of rock slough (pronounced "sluff") off in slabs or shells in a sheeting process.

Exogenic system (11) Earth's external surface system, powered by insolation, which energizes air, water, and ice and sets them in motion, under the influence of gravity. Includes all processes of landmass denudation. (*Compare* Endogenic system.)

Exosphere (3) An extremely rarefied outer atmospheric halo beyond the thermopause at an altitude of 480 km (300 mi); probably composed of hydrogen and helium atoms, with some oxygen atoms and nitrogen molecules present near the thermopause.

Exotic stream (9) A river that rises in a humid region and flows through an arid region, with discharge decreasing toward the mouth; for example, the Nile River and the Colorado River.

Explosive eruption (12) A violent and unpredictable volcanic eruption, the result of magma that is thicker (more viscous), stickier, and higher in gas and silica content than that of an effusive eruption; tends to form blockages within a volcano; produces composite volcanic landforms. (*See* Composite volcano; *compare* Effusive eruption.)

Extrusive igneous rock (11) A rock that solidifies and crystallizes from a molten state as it extrudes onto the surface, such as basalt.

Faulting (12) The process whereby displacement and fracturing occur between two portions of Earth's crust; usually associated with earthquake activity.

Feedback loop (1) Created when a portion of system output is returned as an information input, causing changes that guide further system operation. (*See* Negative feedback, Positive feedback.)

Field capacity (9) Water held in the soil by hydrogen bonding against the pull of gravity, remaining after water drains from the larger pore spaces; the available water for plants. (*See* Available water, Capillary water.)

Fire ecology (19) The study of fire as a natural agent and dynamic factor in community succession.

Firn (17) Snow of a granular texture that is transitional in the slow transformation from snow to glacial ice; snow that has persisted through a summer season in the zone of accumulation.

Firn line (17) The snow line that is visible on the surface of a glacier, where winter snows survive the summer ablation season; analogous to a snow line on land. (*See* Ablation.)

Fjord (17) A drowned glaciated valley, or glacial trough, along a seacoast.

Flash flood (14) A sudden and short-lived torrent of water that exceeds the capacity of a stream channel; associated with desert and semiarid washes.

Flood (14) A high water level that overflows the natural riverbank along any portion of a stream.

Flood basalt (12) An accumulation of horizontal flows formed when lava spreads out from elongated fissures onto the surface in extensive sheets; associated with effusive eruptions. (*See* Basalt.)

Floodplain (14) A flat, low-lying area along a stream channel, created by and subject to recurrent flooding; alluvial deposits generally mask underlying rock.

Flood tide (16) Rising tide during the daily tidal cycle. (*Compare* Ebb tide.)

Fluvial (14) Stream-related processes; from the Latin *fluvius* for "river" or "running water."

Fog (7) A cloud, generally stratiform, in contact with the ground, with visibility usually reduced to less than 1 km (3300 ft).

Folding (12) The bending and deformation of beds of rock strata subjected to compressional forces.

Food chain (19) The circuit along which energy flows from producers (plants), which manufacture their own food, to consumers (animals); a one-directional flow of chemical energy, ending with decomposers.

Food web (19) A complex network of interconnected food chains. (*See* Food chain.)

Formation class (20) The dominant vegetation type extending across a region.

Friction force (6) The effect of drag by the wind as it moves across a surface; may be operative through 500 m (1600 ft) of altitude. Surface friction slows the wind and therefore reduces the effectiveness of the Coriolis force.

Frost action (13) A powerful mechanical force produced as water expands up to 9% of its volume as it freezes. Water freezing in a cavity in a rock can break the rock if it exceeds the rock's tensional strength.

Funnel cloud (8) The visible swirl extending from the bottom side of a cloud, which may or may not develop into a tornado. A tornado is a funnel cloud that has extended all the way to the ground. (*See* Tornado.)

Fusion (2) The process of forcibly joining positively charged hydrogen and helium nuclei under extreme temperature and pressure; occurs naturally in thermonuclear reactions within stars, such as our Sun.

Gelisols (18) A new soil order in the Soil Taxonomy, added in 1998, describing cold and frozen soils at high latitudes or high elevations; characteristic tundra vegetation.

General circulation model (GCM) (10B) Complex, computer-based climate model that produces generalizations of reality and forecasts of future weather and climate conditions. Complex GCMs (three-dimensional models) are in use in the United States and in other countries.

Genetic classification (10) A climate classification that uses causative factors to determine climatic regions; for example, an analysis of the effect of interacting air masses. (*Compare* Empirical classification.)

Geodesy (1) The science that determines Earth's shape and size through surveys, mathematical means, and remote sensing. (*See* Geoid.)

Geographic information system (GIS) (1) A computer-based data processing tool or methodology used for gathering, manipulating, and analyzing geographic information to produce a holistic, interactive analysis.

Geography (1) The science that studies the interdependence and interaction among geographic areas, natural systems, processes, society, and cultural activities over space—a spatial science. The five themes of geographic education are location, place, movement, regions, and human–Earth relationships.

Geoid (1) A word that describes Earth's shape; literally, "the shape of Earth is Earth-shaped." A theoretical surface at sea level that extends through the continents; deviates from a perfect sphere.

Geologic cycle (11) A general term characterizing the vast cycling that proceeds in the lithosphere. It encompasses the hydrologic cycle, tectonic cycle, and rock cycle.

Geologic time scale (11) A depiction of eras, periods, and epochs that span Earth's history; shows both the sequence of rock strata and their absolute dates, as determined by methods such as radioactive isotopic dating.

Geomagnetic reversal (11) A polarity change in Earth's magnetic field. With uneven regularity, the magnetic field fades to zero and then returns to full strength, but with the magnetic poles reversed. Reversals have been recorded nine times during the past 4 million years.

Geomorphic threshold (13) The threshold up to which landforms change before lurching to a new set of relationships, with rapid realignments of landscape materials and slopes.

Geomorphology (11) The science that analyzes and describes the origin, evolution, form, classification, and spatial distribution of landforms.

Geostrophic wind (6) A wind moving between areas of different pressure along a path that is parallel to the isobars. It is a product of the pressure gradient force and the Coriolis force. (*See* Isobar, Pressure gradient force, Coriolis force.)

Geothermal energy (11) The energy in steam and hot water heated by subsurface magma near groundwater. Geothermal energy literally refers to heat from Earth's interior, whereas *geothermal power* relates to specific applied strategies of geothermal electric or geothermal direct applications. This energy is used in Iceland, New Zealand, Italy, and northern California, among other locations.

Glacial drift (17) The general term for all glacial deposits, both unsorted (till) and sorted (stratified drift).

Glacial ice (17) A hardened form of ice, very dense in comparison to normal snow or firn.

Glacier (17) A large mass of perennial ice resting on land or floating shelflike in the sea adjacent to the land; formed from the accumulation and recrystallization of snow, which then flows slowly under the pressure of its own weight and the pull of gravity.

Glacier surge (17) The rapid, lurching, unexpected forward movement of a glacier.

Glacio-eustatic (7) Changes in sea level in response to changes in the amount of water stored on Earth as ice; the more water that is bound up in glaciers and ice sheets, the lower the sea level. (*Compare* Eustasy.)

Gleization (18) A process of humus and clay accumulation in cold, wet climates with poor drainage.

Global carbon budget (10B) The exchange of carbon between sources and sinks in Earth's atmosphere, hydrosphere, lithosphere, and biosphere.

Global dimming (4) The decline in sunlight reaching Earth's surface due to pollution, aerosols, and clouds.

Global Positioning System (GPS) (1) Latitude, longitude, and elevation are accurately calibrated using a handheld instrument that receives radio signals from satellites.

Goode's homolosine projection (Appendix A) An equal-area projection formed by splicing together a sinusoidal and a homolographic projection.

Graben (12) Pairs or groups of faults that produce downward-faulted blocks; characteristic of the basins of the interior western United States. (*Compare* Horst; *see* Basin and Range Province.)

Graded stream (14) An idealized condition in which a stream's load and the landscape mutually adjust. This forms a dynamic equilibrium among erosion, transported load, deposition, and the stream's capacity.

Gradient (14) The drop in elevation from a stream's headwaters to its mouth, ideally forming a concave slope.

Granite (11) A coarse-grained (slow-cooling) intrusive igneous rock of 25% quartz and more than 50% potassium and sodium feldspars; characteristic of the continental crust.

Gravitational water (9) That portion of surplus water that percolates downward from the capillary zone, pulled by gravity to the groundwater zone.

Gravity (2) The mutual force exerted by the masses of objects that are attracted one to another and produced in an amount proportional to each object's mass.

Great circle (1) Any circle drawn on a globe with its center coinciding with the center of the globe. An infinite number of great circles can be drawn, but only one parallel of latitude—the equator—is a great circle. (*Compare* Small circle.)

Greenhouse effect (4) The process whereby radiatively active gases (carbon dioxide, water vapor, methane, and CFCs) absorb and emit the energy at longer wavelengths, which are retained longer, delaying the loss of infrared to space. Thus, the lower troposphere is warmed through the radiation and re-radiation of infrared wavelengths. The approximate similarity between this process and that of a greenhouse explains the name.

Greenhouse gases (4) Gases in the lower atmosphere that delay the passage of longwave radiation to space by absorbing and reradiating specific wavelengths. Earth's primary greenhouse gases are carbon dioxide, water vapor, methane, nitrous oxide, and flourinated gases, such as chlorofluorocarbons (CFCs).

Greenwich Mean Time (GMT) (1) Former world standard time, now reported as Coordinated Universal Time (UTC). (*See* Coordinated Universal Time.)

Ground ice (17) The subsurface water that is frozen in regions of permafrost. The moisture content of areas with ground ice may vary from nearly absent in regions of drier permafrost to almost 100% in saturated soils.

Groundwater (9) Water beneath the surface that is beyond the soil-root zone; a major source of potable water.

Groundwater mining (9) Pumping an aquifer beyond its capacity to flow and recharge; an overuse of the groundwater resource.

Gulf Stream (5) A strong, northward-moving, warm current off the east coast of North America, which carries its water far into the North Atlantic.

Habitat (19) A physical location to which an organism is biologically suited. Most species have specific habitat parameters and limits. (*Compare* Niche.)

Hail (8) A type of precipitation formed when a raindrop is repeatedly circulated above and below the freezing level in a cloud, with each cycle freezing more moisture onto the hailstone until it becomes too heavy to stay aloft.

Hair hygrometer (7) An instrument for measuring relative humidity; based on the principle that human hair will change as much as 4% in length between 0% and 100% relative humidity.

Heat (4) The flow of kinetic energy from one body to another because of a temperature difference between them.

Heat wave (5) A prolonged period of abnormally high temperatures, usually, but not always, in association with humid weather.

Herbivore (19) The primary consumer in a food web, which eats plant material formed by a producer (plant) that has photosynthesized organic molecules. (*Compare* Carnivore.)

Heterosphere (3) A zone of the atmosphere above the mesopause, from 80 km (50 mi) to 480 km (300 mi) in altitude; composed of rarefied layers of oxygen atoms and nitrogen molecules; includes the ionosphere.

Histosols (18) A soil order in the Soil Taxonomy. Formed from thick accumulations of organic matter, such as beds of former lakes, bogs, and layers of peat.

Homosphere (3) A zone of the atmosphere from Earth's surface up to 80 km (50 mi), composed of an even mixture of gases, including nitrogen, oxygen, argon, carbon dioxide, and trace gases.

Horn (17) A pyramidal, sharp-pointed peak that results when several cirque glaciers gouge an individual mountain summit from all sides.

Horst (12) Upward-faulted blocks produced by pairs or groups of faults; characteristic of the mountain ranges of the interior of the western United States. (*See* Graben, Basin and Range Province.)

Hot spot (11) An individual point of upwelling material originating in the asthenosphere, or deeper in the mantle; tends to remain fixed relative to migrating plates; some 100 are identified worldwide, exemplified by Yellowstone National Park, Hawai'i, and Iceland.

Human–Earth relationships (1) One of the oldest themes of geography (the human–land tradition); includes the spatial analysis of settlement patterns, resource utilization and exploitation, hazard perception and planning, and the impact of environmental modification and artificial landscape creation.

Humidity (7) Water vapor content of the air. The capacity of the air for water vapor is mostly a function of the temperature of the air and the water vapor.

Humus (18) A mixture of organic debris in the soil worked by consumers and decomposers in the humification process; characteristically formed from plant and animal litter deposited at the surface.

Hurricane (8) A tropical cyclone that is fully organized and intensified in inward-spiraling rainbands; ranges from 160 to 960 km (100 to 600 mi) in diameter, with wind speeds in excess of 119 kmph (65 knots, or 74 mph); a name used specifically in the Atlantic and eastern Pacific. (*Compare* Typhoon.)

Hydration (13) A chemical weathering process involving water that is added to a mineral, which initiates swelling and stress within the rock, mechanically forcing grains apart as the constituents expand. (*Compare* Hydrolysis.)

Hydraulic action (14) The erosive work accomplished by the turbulence of water; causes a squeezing and releasing action in joints in bedrock; capable of prying and lifting rocks.

Hydric soil (18) A soil that is saturated for long enough periods to develop anaerobic, or "oxygen-free," conditions. Hydric soils are characteristic of wetlands.

Hydrograph (14) A graph of stream discharge (in m^3/s or cfs, ft^3/s) over a period of time (minutes, hours, days, years) at a specific place on a stream. The relationship between stream discharge and precipitation input is illustrated on the graph.

Hydrologic cycle (9) A simplified model of the flow of water, ice, and water vapor from place to place. Water flows through the atmosphere and across the land, where it is stored as ice and as groundwater. Solar energy empowers the cycle.

Hydrology (14) The science of water, including its global circulation, distribution, and properties—specifically water at and below Earth's surface.

Hydrolysis (13) A chemical weathering process in which minerals chemically combine with water; a decomposition process that causes silicate minerals in rocks to break down and become altered. (*Compare* Hydration.)

Hydropower (9) Electricity generated using the energy of moving water, usually flowing downhill through the turbines at a dam; also called *hydroelectric power*.

Hydrosphere (1) An abiotic open system that includes all of Earth's water.

Hygroscopic water (9) That portion of soil moisture that is so tightly bound to each soil particle that it is unavailable to plant roots; the water, along with some bound capillary water, that is left in the soil after the wilting point is reached. (*See* Wilting point.)

Ice age (17) A cold episode, with accompanying alpine and continental ice accumulations, that has repeated roughly every 200 to 300 million years since the late Precambrian Era (1.25 billion years ago); includes the most recent episode during the Pleistocene Ice Age, which began 1.65 million years ago.

Ice cap (17) A large, dome-shaped glacier, less extensive than an ice sheet although it buries mountain peaks and the local landscape; generally, less than 50,000 km^2 (19,300 mi^2).

Ice field (17) The least extensive form of a glacier, with mountain ridges and peaks visible above the ice; less than an ice cap or ice sheet.

Icelandic low (6) *See* Subpolar low-pressure cell.

Ice sheet (17) A continuous mass of unconfined ice, covering at least 50,000 km^2 (19,500 mi^2). The bulk of glacial ice on Earth covers Antarctica and Greenland in two ice sheets (*Compare* Alpine glacier.)

Igneous rock (11) One of the basic rock types; it has solidified and crystallized from a hot molten state (either magma or lava). (*Compare* Metamorphic rock, Sedimentary rock.)

Illuviation (18) The downward movement and deposition of finer particles and minerals from the upper horizon of the soil; a depositional process. Deposition usually is in the B horizon, where accumulations of clays, aluminum, carbonates, iron, and some humus occur. (*Compare* Eluviation; *see* Calcification.)

Inceptisols (18) A soil order in the Soil Taxonomy. Weakly developed soils that are inherently infertile; usually, young soils that are weakly developed, although they are more developed than Entisols.

Industrial smog (3) Air pollution associated with coal-burning industries; it may contain sulfur oxides, particulates, carbon dioxide, and exotics.

Infiltration (9) Water access to subsurface regions of soil moisture storage through penetration of the soil surface.

Insolation (2) Solar radiation that is incoming to Earth systems.

Interception (9) A delay in the fall of precipitation toward Earth's surface caused by vegetation or other ground cover.

Internal drainage (14) In regions where rivers do not flow into the ocean, the outflow is through evaporation or subsurface gravitational flow. Portions of Africa, Asia, Australia, and the western United States have such drainage.

International Date Line (IDL) (1) The 180° meridian, an important corollary to the prime meridian on the opposite side of the planet; established by an 1884 treaty to mark the place where each day officially begins.

Intertropical convergence zone (ITCZ) (6) *See* Equatorial low-pressure trough.

Intrusive igneous rock (11) A rock that solidifies and crystallizes from a molten state as it intrudes into crustal rocks, cooling and hardening below the surface, such as granite.

Invasive species (20) Species that are brought, or introduced, from elsewhere by humans, either accidentally or intentionally. These non-native species are also known as *exotic species* or *alien species*.

Ionosphere (3) A layer in the atmosphere above 80 km (50 mi) where gamma, X-ray, and some ultraviolet radiation is absorbed and converted into longer wavelengths and where the solar wind stimulates the auroras.

Island biogeography (20) Island communities are special places for study because of their spatial isolation and the relatively small number of species present. Islands resemble natural experiments because the impact of individual factors, such as civilization, can be more easily assessed on islands than over larger continental areas.

Isobar (6) An isoline connecting all points of equal atmospheric pressure.

Isostasy (11) A state of equilibrium in Earth's crust formed by the interplay between portions of the less-dense lithosphere and the more-dense asthenosphere and the principle of buoyancy. The crust depresses under weight and recovers with its removal—for example, with the melting of glacial ice. The uplift is known as isostatic rebound.

Isotherm (5) An isoline connecting all points of equal temperature.

Isotope analysis (10B) A technique for long-term climatic reconstruction that uses the atomic structure of chemical elements, specifically the relative amounts of their isotopes, to identify the chemical composition of past oceans and ice masses.

Jet contrails (4) Condensation trails produced by aircraft exhaust, particulates, and water vapor can form high cirrus clouds, sometimes called *false cirrus clouds*.

Jet stream (6) The most prominent movement in upper-level westerly wind flows; irregular, concentrated, sinuous bands of geostrophic wind, travelling at 300 kmph (190 mph).

Joint (13) A fracture or separation in rock that occurs without displacement of the sides; increases the surface area of rock exposed to weathering processes.

June solstice (2) *See* Summer (June) solstice.

Kame (17) A depositional feature of glaciation; a small hill of poorly sorted sand and gravel that accumulates in crevasses or in ice-caused indentations in the surface.

Karst topography (13) Distinctive topography formed in a region of chemically weathered limestone with poorly developed surface drainage and solution features that appear pitted and bumpy; originally named after the Krš Plateau in Slovenia.

Katabatic winds (6) Air drainage from elevated regions, flowing as gravity winds. Layers of air at the surface cool, become denser, and flow downslope; known worldwide by many local names.

Kettle (17) Forms when an isolated block of ice persists in a ground moraine, an outwash plain, or a valley floor after a glacier retreats; as the block finally melts, it leaves behind a steep-sided hole that frequently fills with water.

Kinetic energy (3) The energy of motion in a body; derived from the vibration of the body's own movement and stated as temperature.

Köppen–Geiger climate classification (Appendix B) An empirical classification system that uses average monthly temperatures, average monthly precipitation, and total annual precipitation to establish regional climate designations.

Lacustrine deposit (17) Lake sediments that form terraces, or benches, along former lake shorelines and often mark lake-level fluctuations over time.

Lagoon (16) An area of coastal seawater that is virtually cut off from the ocean by a bay barrier or barrier beach; also, the water surrounded and enclosed by an atoll.

Landfall (8) The location along a coast where a storm moves onshore.

Land–sea breeze (6) Wind along coastlines and adjoining interior areas created by different heating characteristics of land and water surfaces—onshore (landward) breeze in the afternoon and offshore (seaward) breeze at night.

Landslide (13) A sudden rapid downslope movement of a cohesive mass of regolith and/or bedrock in a variety of mass-movement forms under the influence of gravity; a form of mass movement.

Land–water heating difference (5) Differences in the degree and way that land and water heat, as a result of contrasts in transmission, evaporation, mixing, and specific heat capacities. Land surfaces heat and cool faster than water and have continentality, whereas water provides a marine influence.

Large marine ecosystem (LME) (20) Distinctive oceanic regions identified for conservation purposes on the basis of organisms, ocean-floor topography, currents, areas of nutrient-rich upwelling circulation, or areas of significant predation, including human. The LME system is managed by the U.S. National Oceanic and Atmospheric Administration (NOAA).

Latent heat (7) Heat energy is stored in one of three states—ice, water, or water vapor. The energy is absorbed or released in each phase change from one state to another. Heat energy is absorbed as the latent heat of melting, vaporization, or evaporation. Heat energy is released as the latent heat of condensation and freezing (or fusion).

Latent heat of condensation (7) The heat energy released to the environment in a phase change from water vapor to liquid; under normal sea-level pressure, 540 calories are released from each gram of water vapor that changes phase to water at boiling, and 585 calories are released from each gram of water vapor that condenses at 20°C (68°F).

Latent heat of sublimation (7) The heat energy absorbed or released in the phase change from ice to water vapor or water vapor to ice—no liquid phase. The change from water vapor to ice is also called deposition.

Latent heat of vaporization (7) The heat energy absorbed from the environment in a phase change from liquid to water vapor at the boiling point; under normal sea-level pressure, 540 calories must be added to each gram of boiling water to achieve a phase change to water vapor.

Lateral moraine (17) Debris transported by a glacier that accumulates along the sides of the glacier and is deposited along these margins.

Laterization (18) A pedogenic process operating in well-drained soils that occurs in warm and humid regions; typical of Oxisols. Plentiful precipitation leaches soluble minerals and soil constituents. Resulting soils usually are reddish or yellowish.

Latitude (1) The angular distance measured north or south of the equator from a point at the center of Earth. A line connecting all points of the same latitudinal angle is a parallel. (*Compare* Longitude.)

Lava (11) Magma that issues from volcanic activity onto the surface; the extrusive rock that results when magma solidifies. (*See* Magma.)

Life zone (19) A zonation by altitude of plants and animals that form distinctive communities. Each life zone possesses its own temperature and precipitation relations.

Lightning (8) Flashes of light caused by tens of millions of volts of electrical charge heating the air to temperatures of 15,000°C to 30,000°C.

Limestone (11) The most common chemical sedimentary rock (nonclastic); it is lithified calcium carbonate; very susceptible to chemical weathering by acids in the environment, including carbonic acid in rainfall.

Limiting factor (19) The physical or chemical factor that most inhibits biotic processes, through either lack or excess.

Lithification (11) The compaction, cementation, and hardening of sediments into sedimentary rock.

Lithosphere (1, 11) Earth's crust and that portion of the uppermost mantle directly below the crust, extending down about 70 km (45 mi). Some sources use this term to refer to the entire Earth.

Littoral drift (16) Transport of sand, gravel, sediment, and debris along the shore; a more comprehensive term that considers *beach drift* and *longshore drift* combined.

Littoral zone (16) A specific coastal environment; that region between the high-water line during a storm and a depth at which storm waves are unable to move sea-floor sediments.

Loam (18) A soil that is a mixture of sand, silt, and clay in almost equal proportions, with no one texture dominant; an ideal agricultural soil.

Location (1) A basic theme of geography dealing with the absolute and relative positions of people, places, and things on Earth's surface.

Loess (16) Large quantities of fine-grained clays and silts left as glacial outwash deposits; subsequently blown by the wind great distances and redeposited as a generally unstratified, homogeneous blanket of material covering existing landscapes; in China, loess originated from desert lands.

Longitude (1) The angular distance measured east or west of a prime meridian from a point at the center of Earth. A line connecting all points of the same longitude is a meridian. (*Compare* Latitude.)

Longshore current (16) A current that forms parallel to a beach as waves arrive at an angle to the shore; generated in the surf zone by wave action, transporting large amounts of sand and sediment. (*See* Beach drift.)

Lysimeter (9) A weather instrument for measuring potential and actual evapotranspiration; isolates a portion of a field so that the moisture moving through the plot is measured.

Magma (11) Molten rock from beneath Earth's surface; fluid, gaseous, under tremendous pressure, and either intruded into existing crustal rock or extruded onto the surface as lava. (*See* Lava.)

Magnetosphere (2) Earth's magnetic force field, which is generated by dynamo-like motions within the planet's outer core; deflects the solar wind flow toward the upper atmosphere above each pole.

Mangrove swamp (16) A wetland ecosystem between 30° N and 30° S; tends to form a distinctive community of mangrove plants. (*Compare* Salt marsh.)

Mantle (11) An area within the planet representing about 80% of Earth's total volume, with densities increasing with depth and averaging $4.5 g/cm^3$; occurs between the core and the crust; is rich in iron and magnesium oxides and silicates.

Map (1, Appendix A) A generalized view of an area, usually some portion of Earth's surface, as seen from above at a greatly reduced size. (*See* Scale, Map projection.)

Map projection (1, Appendix A) The reduction of a spherical globe onto a flat surface in some orderly and systematic realignment of the latitude and longitude grid.

March equinox (2) *See* Vernal (March) equinox.

Marine effect (5) A quality of regions that are dominated by the moderating effect of the ocean and that exhibit a smaller range of minimum and maximum temperatures, both daily and annually, than do continental stations. (*See* Continental effect, Land–water heating difference.)

Mass movement (13) All unit movements of materials propelled by gravity; can range from dry to wet, slow to fast, small to large, and free-falling to gradual or intermittent.

Mass wasting (13) Gravitational movement of nonunified material downslope; a specific form of mass movement.

Maunder Minimum (10B) A solar minimum (a period with little sunspot activity and reduced solar irradiance) that lasted from about 1645 to 1715, corresponding with one of the coldest periods of the Little Ice Age. This relationship suggests a causal effect between decreased sunspot numbers and cooling temperatures in the North Atlantic region. However, research has repeatedly refuted this hypothesis (for example, recent temperature warming corresponds with a prolonged solar minimum).

Meandering stream (14) The sinuous, curving pattern common to graded streams, with the energetic outer portion of each curve subjected to the greatest erosive action and the lower-energy inner portion receiving sediment deposits. (*See* Graded stream.)

Mean sea level (MSL) (16) The average of tidal levels recorded hourly at a given site over a long period, which must be at least a full lunar tidal cycle.

Medial moraine (17) Debris transported by a glacier that accumulates down the middle of the glacier, resulting from two glaciers merging their lateral moraines; forms a depositional feature following glacial retreat.

Mediterranean shrubland (20) A major biome dominated by the *Mediterranean* (dry summer) climate and characterized by sclerophyllous scrub and short, stunted, tough forests. (*See* Chaparral.)

Mercator projection (1, Appendix A) A true-shape projection, with meridians appearing as equally spaced straight lines and parallels appearing as straight lines that are spaced closer together near the equator. The poles are infinitely stretched, with the 84th north parallel and 84th south parallel fixed at the same length as that of the equator. It presents false notions of the size (area) of midlatitude and poleward landmasses, but presents true compass direction. (*See* Rhumb line.)

Mercury barometer (6) A device that measures air pressure using a column of mercury in a tube; one end of the tube is sealed, and the other end is inserted in an open vessel of mercury. (*See* Air pressure.)

Meridian (1) A line designating an angle of longitude. (*See* Longitude.)

Mesocyclone (8) A large, rotating atmospheric circulation, initiated within a parent cumulonimbus cloud at midtroposphere elevation; generally produces heavy rain, large hail, blustery winds, and lightning; may lead to tornado activity.

Mesosphere (3) The upper region of the homosphere from 50 to 80 km (30 to 50 mi) above the ground; designated by temperature criteria; atmosphere extremely rarified.

Metamorphic rock (11) One of three basic rock types, it is existing igneous and sedimentary rock that has undergone profound physical and chemical changes under increased pressure and temperature. Constituent mineral structures may exhibit foliated or nonfoliated textures. (*Compare* Igneous rock, Sedimentary rock.)

Meteorology (8) The scientific study of the atmosphere, including its physical characteristics and motions; related chemical, physical, and geological processes; the complex linkages of atmospheric systems; and weather forecasting.

Microclimatology (4) The study of local climates at or near Earth's surface or up to that height above the Earth's surface where the effects of the surface are no longer determinative.

Midlatitude broadleaf and mixed forest (20) A biome in moist *continental* climates in areas of warm-to-hot summers and cool-to-cold winters; relatively lush stands of broadleaf forests trend northward into needleleaf evergreen stands.

Midlatitude cyclone (8) An organized area of low pressure, with converging and ascending airflow producing an interaction of air masses; migrates along storm tracks. Such lows or depressions form the dominant weather pattern in the middle and higher latitudes of both hemispheres.

Midlatitude grassland (20) The major biome most modified by human activity; so named because of the predominance of grasslike plants, although deciduous broadleafs appear along streams and other limited sites; location of the world's breadbaskets of grain and livestock production.

Mid-ocean ridge (11) A submarine mountain range that extends more than 65,000 km (40,000 mi) worldwide and averages more than 1000 km (600 mi) in width; centered along sea-floor spreading centers. (*See* Sea-floor spreading.)

Milankovitch cycles (10B) The consistent orbital cycles—based on the irregularities in Earth's orbit around the Sun, its rotation on its axis, and its axial tilt—that relate to climatic patterns and may be an important cause of glacials and interglacials. Milutin Milankovitch (1879–1958), a Serbian astronomer, was the first to correlate these cycles to changes in insolation that affected temperatures on Earth.

Milky Way Galaxy (2) A flattened, disk-shaped mass in space estimated to contain up to 400 billion stars; a barred-spiral galaxy; includes our Solar System.

Miller cylindrical projection (Appendix A) A compromise map projection that avoids the severe distortion of the Mercator projection. (*See* Map projection.)

Mineral (11) An element or combination of elements that forms an inorganic natural compound; described by a specific formula and crystal structure.

Mirage (4) A refraction effect when an image appears near the horizon where light waves are refracted by layers of air at different temperatures (and consequently of different densities).

Model (1) A simplified version of a system, representing an idealized part of the real world.

Mohorovičić discontinuity, or Moho (11) The boundary between the crust and the rest of the lithospheric upper mantle; named for the Yugoslavian seismologist Mohorovičić; a zone of sharp material and density contrasts.

Moist adiabatic rate (MAR) (7) The rate at which a saturated parcel of air cools in ascent; a rate of 6 C° per 1000 m (3.3 F° per 1000 ft). This rate may vary, with moisture content and temperature, from 4 C° to 10 C° per 1000 m (2 F° to 6 F° per 1000 ft). (*See* Adiabatic; *compare* Dry adiabatic rate.)

Moisture droplet (7) A tiny water particle that constitutes the initial composition of clouds. Each droplet measures approximately 0.002 cm (0.0008 in.) in diameter and is invisible to the unaided eye.

Mollisols (18) A soil order in the Soil Taxonomy. These have a mollic epipedon and a humus-rich organic content high in alkalinity. Some of the world's most significant agricultural soils are Mollisols.

Moment magnitude (M) scale (12) An earthquake magnitude scale. Considers the amount of fault slippage, the size of the area that ruptured, and the nature of the materials that faulted in estimating the magnitude of an earthquake—an assessment of the seismic moment. Replaces the Richter scale (amplitude magnitude); especially valuable in assessing larger-magnitude events.

Monsoon (6) An annual cycle of dryness and wetness, with seasonally shifting winds produced by changing atmospheric pressure systems; affects India, Southeast Asia, Indonesia, northern Australia, and portions of Africa. From the Arabic word *mausim*, meaning "season."

Montane forest (20) Needleleaf forest associated with mountain elevations. (*See* Needleleaf forest.)

Moraine (17) Marginal glacial deposits (lateral, medial, terminal, ground) of unsorted and unstratified material.

Mountain–valley breeze (6) A light wind produced as cooler mountain air flows downslope at night and as warmer valley air flows upslope during the day.

Movement (1) A major theme in geography involving migration, communication, and the interaction of people and processes across space.

Mudflow (13) Fluid downslope flows of material containing more water than earthflows.

Natural levee (14) A long, low ridge that forms on both sides of a stream in a developed floodplain; a depositional product (coarse gravels and sand) of river flooding.

Neap tide (16) Unusually low tidal range produced during the first and third quarters of the Moon, with an offsetting pull from the Sun. (*Compare* Spring tide.)

Needleleaf forest (20) Consists of pine, spruce, fir, and larch and stretches from the east coast of Canada westward to Alaska and continuing from Siberia westward across the entire extent of Russia to the European Plain; called the *taiga* (a Russian word) or the *boreal forest*; principally in the microthermal climates. Includes montane forests that may be at lower latitudes at higher elevations.

Negative feedback (1) Feedback that tends to slow or dampen responses in a system; promotes self-regulation in a system; far more common than positive feedback in living systems. (*See* Feedback loop; *compare* Positive feedback.)

Net primary productivity (19) The net photosynthesis (photosynthesis minus respiration) for a given community; considers all growth and all reduction factors that affect the amount of useful chemical energy (biomass) fixed in an ecosystem.

Net radiation (NET R) (4) The net all-wave radiation available at Earth's surface; the final outcome of the radiation balance process between incoming shortwave insolation and outgoing longwave energy.

Nickpoint (knickpoint) (14) The point at which the longitudinal profile of a stream is abruptly broken by a change in gradient; for example, a waterfall, rapids, or cascade.

Nimbostratus (7) Rain-producing, dark, grayish stratiform clouds characterized by gentle drizzle.

Nitrogen dioxide (3) A noxious (harmful) reddish-brown gas produced in combustion engines; can be damaging to human respiratory tracts and to plants; participates in photochemical reactions and acid deposition.

Noctilucent cloud (3) A rare, shining band of ice crystals that may glow at high latitudes long after sunset; formed within the mesosphere, where cosmic and meteoric dust act as nuclei for the formation of ice crystals.

Normal fault (12) A type of geologic fault in rocks. Tension produces strain that breaks a rock, with one side moving vertically relative to the other side along an inclined fault plane. (*Compare* Reverse fault.)

Normal lapse rate (3) The average rate of temperature decrease with increasing altitude in the lower atmosphere; an average value of 6.4 C° per km, or 1000 m (3.5 F° per 1000 ft). (*Compare* Environmental lapse rate.)

Occluded front (8) In a cyclonic circulation, the overrunning of a surface warm front by a cold front and the subsequent lifting of the warm air wedge off the ground; initial precipitation is moderate to heavy.

Ocean basin (12) The physical container (a depression in the lithosphere) holding an ocean.

Omnivore (19) A consumer that feeds on both producers (plants) and consumers (meat)—a role occupied by humans, among other animals. (*Compare* Consumer, Producer.)

Open system (1) A system with inputs and outputs crossing back and forth between the system and the surrounding environment. Earth is an open system in terms of energy. (*Compare* Closed system.)

Orogenesis (12) The process of mountain building that occurs when large-scale compression leads to deformation and uplift of the crust; literally, the birth of mountains.

Orographic lifting (8) The uplift of a migrating air mass as it is forced to move upward over a mountain range—a topographic barrier. The lifted air cools adiabatically as it moves upslope; clouds may form and produce increased precipitation.

Outgassing (9) The release of trapped gases from rocks, forced out through cracks, fissures, and volcanoes from within Earth; the terrestrial source of Earth's water.

Outwash plain (17) Area of glacial stream deposits of stratified drift with meltwater-fed, braided, and overloaded streams; occurs beyond a glacier's morainal deposits.

Overland flow (9) Surplus water that flows across the land surface toward stream channels. Together with precipitation and subsurface flows, it constitutes the total runoff from an area.

Oxbow lake (14) A lake that was formerly part of the channel of a meandering stream; isolated when a stream eroded its outer bank, forming a cutoff through the neck of the looping meander (*see* Meandering stream). In Australia, known as a billabong (the Aboriginal word for "dead river").

Oxidation (13) A chemical weathering process in which oxygen dissolved in water oxidizes (combines with) certain metallic elements to form oxides; most familiar is the "rusting" of iron in a rock or soil (Ultisols, Oxisols), which produces a reddish-brown stain of iron oxide.

Oxisols (18) A soil order in the Soil Taxonomy. Tropical soils that are old, deeply developed, and lacking in horizons wherever well drained; heavily weathered, low in cation-exchange capacity, and low in fertility.

Ozone layer (3) *See* Ozonosphere.

Ozonosphere (3) A layer of ozone occupying the full extent of the stratosphere (20 to 50 km, or 12 to 30 mi, above the surface); the region of the atmosphere where ultraviolet wavelengths of insolation are extensively absorbed and converted into heat.

Pacific high (6) A high-pressure cell that dominates the Pacific in July, retreating southward in the Northern Hemisphere in January; also known as the *Hawaiian high*.

Pahoehoe (12) Basaltic lava that is more fluid than aa. Pahoehoe forms a thin crust that forms folds and appears "ropy," like coiled, twisted rope.

Paleoclimatology (10B) The science that studies the climates, and the causes of variations in climate, of past ages, throughout historic and geologic time.

Paleolake (17) An ancient lake, such as Lake Bonneville or Lake Lahonton, associated with former wet periods when the lake basins were filled to higher levels than today.

PAN (3) *See* Peroxyacetyl nitrate.

Pangaea (11) The supercontinent formed by the collision of all continental masses approximately 225 million years ago; named in the continental drift theory by Wegener in 1912. (*See* Plate tectonics.)

Parallel (1) A line, parallel to the equator, that designates an angle of latitude. (*See* Latitude.)

Parent material (13) The unconsolidated material, from both organic and mineral sources, that is the basis of soil development.

Particulate matter (PM) (3) Dust, dirt, soot, salt, sulfate aerosols, fugitive natural particles, or other material particles suspended in air.

Paternoster lake (17) One of a series of small, circular, stair-stepped lakes formed in individual rock basins aligned down the course of a glaciated valley; named because they look like a string of rosary (religious) beads.

Patterned ground (17) Areas in the periglacial environment where freezing and thawing of the ground create polygonal forms of arranged rocks at the surface; can be circles, polygons, stripes, nets, and steps.

Pedogenic regime (18) A specific soil-forming process keyed to a specific climatic regime: laterization, calcification, salinization, and podzolization, among others; not the basis for soil classification in the Soil Taxonomy.

Pedon (18) A soil profile extending from the surface to the lowest extent of plant roots or to the depth where regolith or bedrock is encountered; imagined as a hexagonal column; the basic soil sampling unit.

Percolation (9) The process by which water permeates the soil or porous rock into the subsurface environment.

Periglacial (17) Cold-climate processes, landforms, and topographic features along the margins of glaciers, past and present; periglacial characteristics exist on more than 20% of Earth's land surface; includes permafrost, frost action, and ground ice.

Perihelion (2) The point of Earth's closest approach to the Sun in its elliptical orbit, reached on January 3 at a distance of 147,255,000 km (91,500,000 mi); variable over a 100,000-year cycle. (*Compare* Aphelion.)

Permafrost (17) Forms when soil or rock temperatures remain below 0°C (32°F) for at least 2 years in areas considered periglacial; criterion is based on temperature and not on whether water is present. (*See* Periglacial.)

Permeability (9) The ability of water to flow through soil or rock; a function of the texture and structure of the medium.

Peroxyacetyl nitrate (PAN) (3) A pollutant formed from photochemical reactions involving nitric oxide (NO) and volatile organic compounds (VOCs). PAN produces no known human health effect, but is particularly damaging to plants.

Phase change (7) The change in phase, or state, among ice, water, and water vapor; involves the absorption or release of latent heat. (*See* Latent heat.)

Photochemical smog (3) Air pollution produced by the interaction of ultraviolet light, nitrogen dioxide, and hydrocarbons; produces ozone and PAN through a series of complex photochemical reactions. Automobiles are the major source of the contributive gases.

Photogrammetry (1) The science of obtaining accurate measurements from aerial photos and remote sensing; used to create and to improve surface maps.

Photosynthesis (19) The process by which plants produce their own food from carbon dioxide and water, powered by solar energy. The joining of carbon dioxide and hydrogen in plants, under the influence of certain wavelengths of visible light; releases oxygen and produces energy-rich organic material, sugars, and starches. (*Compare* Respiration.)

Physical geography (1) The science concerned with the spatial aspects and interactions of the physical elements and process systems that make up the environment: energy, air, water, weather, climate, landforms, soils, animals, plants, microorganisms, and Earth.

Physical weathering (13) The breaking up and disintegrating of rock without any chemical alteration; sometimes referred to as *mechanical* or *fragmentation weathering*.

Pioneer community (19) The initial plant community in an area; usually is found on new surfaces or those that have been stripped of life, as in beginning primary succession, and includes lichens, mosses, and ferns growing on bare rock.

Place (1) A major theme in geography, focused on the tangible and intangible characteristics that make each location unique; no two places on Earth are alike.

Plane of the ecliptic (2) A plane (flat surface) intersecting all the points of Earth's orbit.

Planetesimal hypothesis (2) Proposes a process by which early protoplanets formed from the condensing masses of a nebular cloud of dust, gas, and icy comets; a formation process now being observed in other parts of the galaxy.

Planimetric map (Appendix A) A basic map showing the horizontal position of boundaries; land-use activities; and political, economic, and social outlines.

Plateau basalt (12) An accumulation of horizontal flows formed when lava spreads out from elongated fissures onto the surface in extensive sheets; associated with effusive eruptions; also known as *flood basalts*. (*See* Basalt.)

Plate tectonics (11) The conceptual model and theory that encompass continental drift, sea-floor spreading, and related aspects of crustal movement; accepted as the foundation of crustal tectonic processes. (*See* Continental drift.)

Playa (14) An area of salt crust left behind by evaporation on a desert floor, usually in the middle of a desert or semiarid bolson or valley; intermittently wet and dry.

Pluton (11) A mass of intrusive igneous rock that has cooled slowly in the crust; forms in any size or shape. The largest partially exposed pluton is a batholith. (*See* Batholith.)

Podzolization (18) A pedogenic process in cool, moist climates; forms a highly leached soil with strong surface acidity because of humus from acid-rich trees.

Point bar (14) In a stream, the inner portion of a meander, where sediment fill is redeposited. (*Compare* Undercut bank.)

Polar easterlies (6) Variable, weak, cold, and dry winds moving away from the polar region; an anticyclonic circulation.

Polar desert (20) A type of desert biome found at higher latitudes than cold deserts, occurring mainly in the very cold, dry climates of Greenland and Antartica.

Polar front (6) A significant zone of contrast between cold and warm air masses; roughly situated between 50° and 60° N and S latitude.

Polar high (6) Weak, anticyclonic, thermally produced pressure systems positioned roughly over each pole; that over the South Pole is the region of the lowest temperatures on Earth. (*See* Antarctic high.)

Pollutants (3) Natural or human-caused gases, particles, and other substances in the troposphere that accumulate in amounts harmful to humans or to the environment.

Polypedon (18) The identifiable soil in an area, with distinctive characteristics differentiating it from surrounding polypedons that form the basic mapping unit; composed of many pedons. (*See* Pedon.)

Positive feedback (1) Feedback that amplifies or encourages responses in a system. (*Compare* Negative feedback; *see* Feedback loop.)

Potential evapotranspiration (9) POTET, or PE; the amount of moisture that would evaporate and transpire if adequate moisture were available; it is the amount lost under optimum moisture conditions, the moisture demand. (*Compare* Actual evapotranspiration.)

Potentiometric surface (9) A pressure level in a confined aquifer, defined by the level to which water rises in wells; caused by the fact that the water in a confined aquifer is under the pressure of its own weight; also known as a *piezometric surface*. This surface can extend above the surface of the land, causing water to rise above the water table in wells in confined aquifers. (*See* Artesian water.)

Precipitation (9) Rain, snow, sleet, and hail—the moisture supply; called PRECIP, or P, in the water balance.

Pressure gradient force (6) Causes air to move from an area of higher barometric pressure to an area of lower barometric pressure due to the pressure difference.

Primary succession (19) Succession that occurs among plant species in an area of new surfaces created by mass movement of land, cooled lava flows and volcanic eruption landscapes, or surface mining and clear-cut logging scars; exposed by retreating glaciers, or made up of sand dunes, with no trace of a former community.

Prime meridian (1) An arbitrary meridian designated as 0° longitude, the point from which longitudes are measured east or west; established at Greenwich, England, by international agreement in an 1884 treaty.

Process (1) A set of actions and changes that occur in some special order; analysis of processes is central to modern geographic synthesis.

Producer (19) Organism (plant) in an ecosystem that uses carbon dioxide as its sole source of carbon, which it chemically fixes through photosynthesis to provide its own nourishment; also called an *autotroph*. (*Compare* Consumer.)

Proxy method (10B) Information about past environments that represent changes in climate, such as isotope analysis or tree ring dating; also called a *climate proxy*.

Pyroclastic (12) An explosively ejected rock fragment launched by a volcanic eruption; sometimes described by the more general term *tephra*.

Radiation fog (7) Formed by radiative cooling of a land surface, especially on clear nights in areas of moist ground; occurs when the air layer directly above the surface is chilled to the dew-point temperature, thereby producing saturated conditions.

Radiative forcing (10B) The amount by which some perturbation causes Earth's energy balance to deviate from zero; a positive forcing indicates a warming condition, a negative forcing indicates cooling; also called *climate forcing*.

Radioactive isotope (10B) An unstable isotope that decays, or breaks down, into a different element, emitting radiation in the process. The unstable isotope carbon-14 has a constant rate of decay known as a *half-life* that can be used to date plant material in a technique called *radiocarbon dating*.

Rain gauge (9) A weather instrument; a standardized device that captures and measures rainfall.

Rain shadow (8) The area on the leeward slope of a mountain range where precipitation receipt is greatly reduced compared to the windward slope on the other side. (*See* Orographic lifting.)

Reflection (4) The portion of arriving insolation that is returned directly to space without being absorbed and converted into heat and without performing any work. (*See* Albedo.)

Refraction (4) The bending effect on electromagnetic waves that occurs when insolation enters the atmosphere or another medium; the same process disperses the component colors of the light passing through a crystal or prism.

Region (1) A geographic theme that focuses on areas that display unity and internal homogeneity of traits; includes the study of how a region forms, evolves, and interrelates with other regions.

Regolith (13) Partially weathered rock overlying bedrock, whether residual or transported.

Relative humidity (7) The ratio of water vapor actually in the air (content) to the maximum water vapor possible in air (capacity) at that temperature; expressed as a percentage. (*Compare* Vapor pressure, Specific humidity.)

Relief (12) Elevation differences in a local landscape; an expression of local height differences of landforms.

Remote sensing (1) Information acquired from a distance, without physical contact with the subject—for example, photography, orbital imagery, and radar.

Respiration (19) The process by which plants oxidize carbohydrates to derive energy for their operations; essentially, the reverse of the photosynthetic process; releases carbon dioxide, water, and heat energy into the environment. (*Compare* Photosynthesis.)

Reverse fault (12) Compressional forces produce strain that breaks a rock so that one side moves upward relative to the other side; also called a *thrust fault*. (*Compare* Normal fault.)

Revolution (2) The annual orbital movement of Earth about the Sun; determines the length of the year and the seasons.

Rhumb line (1) A line of constant compass direction, or constant bearing, that crosses successive meridians at the same angle; appears as a straight line only on the Mercator projection.

Richter scale (12) An open-ended, logarithmic scale that estimates earthquake amplitude magnitude; designed by Charles Richter in 1935; now replaced by the moment magnitude scale. (*See* Moment magnitude scale.)

Ring of fire (12) *See* Circum-Pacific belt.

Robinson projection (Appendix A) A compromise (neither equal area nor true shape) oval projection developed in 1963 by Arthur Robinson.

Roche moutonnée (17) A glacial erosion feature; an asymmetrical hill of exposed bedrock; displays a gently sloping upstream side that has been smoothed and polished by a glacier and an abrupt, steep downstream side.

Rock (11) An assemblage of minerals bound together, or sometimes a mass of a single mineral.

Rock cycle (11) A model representing the interrelationships among the three rock-forming processes: igneous, sedimentary, and metamorphic; shows how each can be transformed into another rock type.

Rockfall (13) Free-falling movement of debris from a cliff or steep slope, generally falling straight or bounding downslope.

Rossby wave (6) An undulating horizontal motion in the upper-air westerly circulation at middle and high latitudes.

Rotation (2) The turning of Earth on its axis, averaging about 24 hours in duration; determines day–night relation; counterclockwise when viewed from above the North Pole and from west to east, or eastward, when viewed from above the equator.

Salinity (16) The concentration of natural elements and compounds dissolved in solution, as solutes; measured by weight in parts per thousand (‰) in seawater.

Salinization (18) A pedogenic process that results from high potential evapotranspiration rates in deserts and semiarid regions. Soil water is drawn to surface horizons, and dissolved salts are deposited as the water evaporates.

Saltation (14) The transport of sand grains (usually larger than 0.2 mm, or 0.008 in.) by stream or wind, bouncing the grains along the ground in asymmetrical paths.

Salt marsh (16) A wetland ecosystem characteristic of latitudes poleward of the 30th parallel. (*Compare* Mangrove swamp.)

Sand sea (16) An extensive area of sand and dunes; characteristic of Earth's erg deserts. (*Compare* Erg desert.)

Saturation (7) State of air that is holding all the water vapor that it can hold at a given temperature, known as the dew-point temperature.

Scale (1) The ratio of the distance on a map to that in the real world; expressed as a representative fraction, graphic scale, or written scale.

Scarification (13) Human-induced mass movement of Earth materials, such as large-scale open-pit mining and strip mining.

Scattering (4) Deflection and redirection of insolation by atmospheric gases, dust, ice, and water vapor; the shorter the wavelength, the greater the scattering; thus, skies in the lower atmosphere are blue.

Scientific method (1) An approach that uses applied common sense in an organized and objective manner; based on observation, generalization, formulation, and testing of a hypothesis, ultimately leading to the development of a theory.

Seafloor spreading (11) As proposed by Hess and Dietz, the mechanism driving the movement of the continents; associated with upwelling flows of magma along the worldwide system of mid-ocean ridges. (*See* Mid-ocean ridge.)

Secondary succession (19) Succession that occurs among plant species in an area where vestiges of a previously functioning community are present; an area where the natural community has been destroyed or disturbed, but where the underlying soil remains intact.

Sediment (11) Fine-grained mineral matter that is transported and deposited by air, water, or ice.

Sediment transport (14) The movement of rocks and sediment downstream when energy is high in a river or stream.

Sedimentary rock (11) One of the three basic rock types; formed from the compaction, cementation, and hardening of sediments derived from other rocks. (*Compare* Igneous rock, Metamorphic rock.)

Seismic wave (11) The shock wave sent through the planet by an earthquake or underground nuclear test. Transmission varies according to temperature and the density of various layers within the planet; provides indirect diagnostic evidence of Earth's internal structure.

Seismograph (11) A device that measures seismic waves of energy transmitted throughout Earth's interior or along the crust (also called a seismometer).

Seismometer (12) An instrument used to detect and record the ground motion during an earthquake caused by seismic waves traveling through Earth's interior to the surface; the instrument records the waves on a graphic plot called a *seismogram*.

Sensible heat (4) Heat that can be measured with a thermometer; a measure of the concentration of kinetic energy from molecular motion.

September equinox (2) *See* Autumnal (September) equinox.

Sheetflow (14) Surface water that moves downslope in a thin film as overland flow; not concentrated in channels larger than rills.

Sheeting (13) A form of weathering associated with fracturing or fragmentation of rock by pressure release; often related to exfoliation processes. (*See* Exfoliation dome.)

Shield volcano (12) A symmetrical mountain landform built from effusive eruptions (low-viscosity magma); gently sloped and gradually rising from the surrounding landscape to a summit crater; typical of the Hawaiian Islands. (*Compare* Effusive eruption, Composite volcano.)

Sinkhole (13) Nearly circular depression created by the weathering of karst landscapes with subterranean drainage; also known as a *doline* in traditional studies; may collapse through the roof of an underground space. (*See* Karst topography.)

Sleet (8) Freezing rain, ice glaze, or ice pellets.

Sling psychrometer (7) A weather instrument that measures relative humidity using two thermometers—a dry bulb and a wet bulb—mounted side by side.

Slipface (16) On a sand dune, formed as dune height increases above 30 cm (12 in.) on the leeward side at an angle at which loose material is stable—its angle of repose (30° to 34°).

Slope (13) A curved, inclined surface that bounds a landform.

Small circle (1) A circle on a globe's surface that does not share Earth's center—for example, all parallels of latitude other than the equator. (*Compare* Great circle.)

Snowline (17) A temporary line marking the elevation where winter snowfall persists throughout the summer; seasonally, the lowest elevation covered by snow during the summer.

Soil (18) A dynamic natural body made up of fine materials covering Earth's surface in which plants grow, composed of both mineral and organic matter.

Soil colloid (18) A tiny clay and organic particle in soil; provides a chemically active site for mineral ion adsorption. (*See* Cation-exchange capacity.)

Soil creep (13) A persistent mass movement of surface soil where individual soil particles are lifted and disturbed by the expansion of soil moisture as it freezes or by grazing livestock or digging animals.

Soil fertility (18) The ability of soil to support plant productivity when it contains organic substances and clay minerals that absorb water and certain elemental ions needed by plants through adsorption. (*See* Cation-exchange capacity.)

Soil horizons (18) The various layers exposed in a pedon; roughly parallel to the surface and identified as O, A, E, B, C, and R (bedrock).

Soil-moisture recharge (9) Water entering available soil storage spaces.

Soil-moisture storage (9) STRGE; the retention of moisture within soil; it is a savings account that can accept deposits (soil-moisture recharge) or allow withdrawals (soil-moisture utilization) as conditions change.

Soil porosity (18) The total volume of space within a soil that is filled with air, gases, or water (as opposed to soil particles or organic matter).

Soil-moisture utilization (9) The extraction of soil moisture by plants for their needs; efficiency of withdrawal decreases as the soil-moisture storage is reduced.

Soil profile (18) A vertical section of soil extending from the surface to the deepest extent of plant roots or to regolith or bedrock.

Soil-moisture zone (9) The area of water stored in soil between the ground surface and the water table. Water in this zone may be available or unavailable to plant roots, depending on soil texture characteristics.

Soil science (18) Interdisciplinary science of soils. Pedology concerns the origin, classification, distribution, and description of soil. Edaphology focuses on soil as a medium for sustaining higher plants.

Soil Taxonomy (18) A soil classification system based on observable soil properties actually seen in the field; published in 1975 by the U.S. Soil Conservation Service and revised in 1990 and 1998 by the Natural Resources Conservation Service to include 12 soil orders.

Solar constant (2) The amount of insolation intercepted by Earth on a surface perpendicular to the Sun's rays when Earth is at its average distance from the Sun; a value of 1372 W/m^2 (1.968 $calories/cm^2$) per minute; averaged over the entire globe at the thermopause.

Solar wind (2) Clouds of ionized (charged) gases emitted by the Sun and travelling in all directions from the Sun's surface. Effects on Earth include auroras, disturbance of radio signals, and possible influences on weather.

Solum (18) A true soil profile in the pedon; ideally, a combination of O, A, E, and B horizons. (*See* Pedon.)

Spatial (1) The nature or character of physical space, as in an area; occupying or operating within a space. Geography is a spatial science; spatial analysis its essential approach.

Spatial analysis (1) The examination of spatial interactions, patterns, and variations over area and/or space; a key integrative approach of geography.

Specific heat (5) The increase of temperature in a material when energy is absorbed; water has a higher specific heat (can store more heat) than a comparable volume of soil or rock.

Specific humidity (7) The mass of water vapor (in grams) per unit mass of air (in kilograms) at any specified temperature. The maximum mass of water vapor that a kilogram of air can hold at any specified temperature is termed its maximum specific humidity. (*Compare* Vapor pressure, Relative humidity.)

Speed of light (2) Specifically, 299,792 kilometers (186,282 miles) per second, or more than 9.4 trillion kilometers (5.9 trillion miles) per year—a distance known as a light-year; at light speed, Earth is 8 minutes and 20 seconds from the Sun.

Speleothem (10B) A calcium carbonate mineral deposit in a cave or cavern, such as a stalactite or stalagmite, that forms as water drips or seeps from rock and subsequently evaporates, leaving behind a residue of calcium carbonate that builds up over time.

Spheroidal weathering (13) A chemical weathering process in which the sharp edges and corners of boulders and rocks are weathered in thin plates that create a rounded, spheroidal form.

Spodosols (18) A soil order in the Soil Taxonomy. Occurs in northern coniferous forests; best developed in cold, moist, forested climates; lacks humus and clay in the A horizon, with high acidity associated with podzolization processes.

Spring tide (16) The highest tidal range, which occurs when the Moon and the Sun are in conjunction (at new Moon) or in opposition (at full Moon) stages. (*Compare* Neap tide.)

Squall line (8) A zone slightly ahead of a fast-advancing cold front where wind patterns are rapidly changing and blustery and precipitation is strong.

Stability (7) The condition of a parcel of air with regard to whether it remains where it is or changes its initial position. The parcel is stable if it resists displacement upward and unstable if it continues to rise.

Stationary front (8) A frontal area of contact between contrasting air masses that shows little horizontal movement; winds in opposite directions on either side of the front flow parallel along the front.

Steady-state equilibrium (1) The condition that occurs in a system when the rates of input and output are equal and the amounts of energy and stored matter are nearly constant around a stable average.

Steppe (10) A regional term referring to the vast semiarid grassland biome of Eastern Europe and Asia; the equivalent biome in North America is shortgrass prairie, and in Africa, it is the savanna. Steppe in a climatic context is considered too dry to support forest, but too moist to be a desert.

Stomata (19) Small openings on the undersides of leaves through which water and gasses pass.

Storm surge (8) A large quantity of seawater pushed inland by the strong winds associated with a tropical cyclone.

Storm tracks (8) Seasonally shifting paths followed by migrating low-pressure systems.

Stratified drift (17) Sediments deposited by glacial meltwater that appear sorted; a specific form of glacial drift. (*Compare* Till.)

Stratigraphy (11) A science that analyzes the sequence, spacing, geophysical and geochemical properties, and spatial distribution of rock strata.

Stratocumulus (7) A lumpy, grayish, low-level cloud, patchy with sky visible, sometimes present at the end of the day.

Stratosphere (3) That portion of the homosphere that ranges from 20 to 50 km (12.5 to 30 mi) above Earth's surface, with temperatures ranging from −57°C (−70°F) at the tropopause to 0°C (32°F) at the stratopause. The functional ozonosphere is within the stratosphere.

Stratus (7) A stratiform (flat, horizontal) cloud generally below 2000 m (6500 ft).

Strike-slip fault (12) Horizontal movement along a fault line—that is, movement in the same direction as the fault; also known as a *transcurrent* fault. Such movement is described as right lateral or left lateral, depending on the relative motion observed across the fault. (*See* Transform fault.)

Subduction zone (11) An area where two plates of crust collide and the denser oceanic crust dives beneath the less dense continental plate, forming deep oceanic trenches and seismically active regions.

Sublimation (7) A process in which ice evaporates directly to water vapor or water vapor freezes to ice (deposition).

Subpolar low (6) A region of low pressure centered approximately at 60° latitude in the North Atlantic near Iceland and in the North Pacific near the Aleutians as well as in the Southern Hemisphere. Airflow is cyclonic; it weakens in summer and strengthens in winter. (*See* Cyclone.)

Subsolar point (2) The only point receiving perpendicular insolation at a given moment—that is, the Sun is directly overhead. (*See* Declination.)

Subtropical high (6) One of several dynamic high-pressure areas covering roughly the region from 20° to 35° N and S latitudes; responsible for the hot, dry areas of Earth's arid and semiarid deserts. (*See* Anticyclone.)

Sulfate aerosols (3) Sulfur compounds in the atmosphere, principally sulfuric acid; principal sources relate to fossil fuel combustion; scatter and reflect insolation.

Sulfur dioxide (SO_2) (3) A colorless gas detected by its pungent odor; produced by the combustion of fossil fuels, especially coal, that contain sulfur as an impurity; can react in the atmosphere to form sulfuric acid, a component of acid deposition.

Summer (June) solstice (2) The time when the Sun's declination is at the Tropic of Cancer, at 23.5° N latitude, June 20–21 each year. The night is 24 hours long south of the Antarctic Circle. The day is 24 hours long north of the Arctic Circle. (*Compare* Winter [December] solstice.)

Sunrise (2) That moment when the disk of the Sun first appears above the horizon.

Sunset (2) That moment when the disk of the Sun totally disappears below the horizon.

Sunspots (2) Magnetic disturbances on the surface of the Sun, occurring in an average 11-year cycle; related flares, prominences, and outbreaks produce surges in solar wind.

Surface creep (16) A form of eolian transport that involves particles too large for saltation; a process whereby individual grains are impacted by moving grains and slide and roll.

Surface runoff (9) Surplus water that flows across the ground surface toward stream channels when soils are saturated or when the ground is impermeable; also called *overland flow.*

Surplus (9) SURPL; the amount of moisture that exceeds potential evapotranspiration; moisture oversupply when soil-moisture storage is at field capacity; extra or surplus water.

Suspended load (14) Fine particles held in suspension in a stream. The finest particles are not deposited until the stream velocity nears zero.

Sustainability science (1) An emerging, integrated scientific discipline based on the concepts of sustainable development related to functioning Earth systems.

Swell (16) Regular patterns of smooth, rounded waves in open water; can range from small ripples to very large waves.

Syncline (12) A trough in folded strata, with beds that slope toward the axis of the downfold. (*Compare* Anticline.)

System (1) Any ordered, interrelated set of materials or items existing separate from the environment or within a boundary; energy transformations and energy and matter storage and retrieval occur within a system.

Taiga (20) *See* Needleleaf forest.

Talus slope (13) Formed by angular rock fragments that cascade down a slope along the base of a mountain; poorly sorted, cone-shaped deposits.

Tarn (17) A small mountain lake, especially one that collects in a cirque basin behind risers of rock material or in an ice-gouged depression.

Temperate rain forest (20) A major biome of lush forests at middle and high latitudes; occurs along narrow margins of the Pacific Northwest in North America, among other locations; includes the tallest trees in the world.

Temperature (5) A measure of sensible heat energy present in the atmosphere and other media; indicates the average kinetic energy of individual molecules within a substance.

Temperature inversion (3) A reversal of the normal decrease of temperature with increasing altitude; can occur anywhere from ground level up to several thousand meters; functions to block atmospheric convection and thereby trap pollutants.

Terminal moraine (17) Eroded debris that is dropped at a glacier's farthest extent.

Terrane (12) A migrating piece of Earth's crust, dragged about by processes of mantle convection and plate tectonics. Displaced terranes are distinct in their history, composition, and structure from the continents that accept them.

Thermal equator (5) The isoline on an isothermal map that connects all points of highest mean temperature.

Thermohaline circulation (6) Deep-ocean currents produced by differences in temperature and salinity with depth; Earth's deep currents.

Thermopause (2, 3) A zone approximately 480 km (300 mi) in altitude that serves conceptually as the top of the atmosphere; an altitude used for the determination of the solar constant.

Thermosphere (3) A region of the heterosphere extending from 80 to 480 km (50 to 300 mi) in altitude; contains the functional ionosphere layer.

Threshold (1) A moment in which a system can no longer maintain its character, so it lurches to a new operational level, which may not be compatible with previous conditions.

Thrust fault (12) A reverse fault where the fault plane forms a low angle relative to the horizontal; an overlying block moves over an underlying block.

Thunder (8) The violent expansion of suddenly heated air, created by lightning discharges, which sends out shock waves as an audible sonic bang.

Tide (16) A pattern of twice-daily oscillations in sea level produced by astronomical relations among the Sun, the Moon, and Earth; experienced in varying degrees around the world. (*See* Neap tide, Spring tide.)

Till (17) Direct ice deposits that appear unstratified and unsorted; a specific form of glacial drift. (*Compare* Stratified drift.)

Till plain (17) A large, relatively flat plain composed of unsorted glacial deposits behind a terminal or end moraine. Low-rolling relief and unclear drainage patterns are characteristic.

Tombolo (16) A landform created when coastal sand deposits connect the shoreline with an offshore island outcrop or sea stack.

Topographic map (Appendix A) A map that portrays physical relief through the use of elevation contour lines that connect all points at the same elevation above or below a vertical datum, such as mean sea level.

Topography (12) The undulations and configurations, including its relief, that give Earth's surface its texture, portrayed on topographic maps.

Tornado (8) An intense, destructive cyclonic rotation, developed in response to extremely low pressure; generally associated with mesocyclone formation.

Traction (14) A type of sediment transport that drags coarser materials along the bed of a stream. (*See* Bed load.)

Trade winds (6) Winds from the northeast and southeast that converge in the equatorial low-pressure trough, forming the intertropical convergence zone.

Transform fault (11) A type of geologic fault in rocks. An elongated zone along which faulting occurs between mid-ocean ridges; produces a relative horizontal motion with no new crust formed or consumed; strike-slip motion is either left or right lateral. (*See* strike-slip fault.)

Transmission (4) The passage of shortwave and longwave energy through space, the atmosphere, or water.

Transparency (5) The quality of a medium (air, water) that allows light to easily pass through it.

Transpiration (9) The movement of water vapor out through the pores in leaves; the water is drawn by the plant roots from soil-moisture storage.

Tropical cyclone (8) A cyclonic circulation originating in the tropics, with winds between 30 and 64 knots (39 and 73 mph); characterized by closed isobars, circular organization, and heavy rains. (*See* Hurricane, Typhoon.)

Tropical rain forest (20) A lush biome of tall broadleaf evergreen trees and diverse plants and animals, roughly between 23.5° N and 23.5° S. The dense canopy of leaves is usually arranged in three levels.

Tropical savanna (20) A major biome containing large expanses of grassland interrupted by trees and shrubs; a transitional area between the humid rain forests and tropical seasonal forests and the drier, semiarid tropical steppes and deserts.

Tropical seasonal forest and scrub (20) A variable biome on the margins of the rain forests, occupying regions of lesser and more erratic rainfall; the site of transitional communities between the rain forests and tropical grasslands.

Tropic of Cancer (2) The parallel that marks the farthest north the subsolar point migrates during the year; 23.5° N latitude. (*See* Tropic of Capricorn, Summer [June] solstice.)

Tropic of Capricorn (2) The parallel that marks the farthest south the subsolar point migrates during the year; 23.5° S latitude. (*See* Tropic of Cancer, winter [December] solstice.)

Tropopause (3) The top zone of the troposphere defined by temperature; wherever −57°C (−70°F) occurs.

Troposphere (3) The home of the biosphere; the lowest layer of the homo-sphere, containing approximately 90% of the total mass of the atmosphere; extends up to the tropopause; occurring at an altitude of 18 km (11 mi) at the equator, at 13 km (8 mi) in the middle latitudes, and at lower altitudes near the poles.

True shape (1) A map property showing the correct configuration of coastlines; a useful trait of conformality for navigational and aeronautical maps, although areal relationships are distorted. (*See* Map projection; *compare* Equal area.)

Tsunami (16) A seismic sea wave, travelling at high speeds across the ocean, formed by sudden motion in the seafloor, such as a sea-floor earthquake, submarine landslide, or eruption of an undersea volcano.

Typhoon (8) A tropical cyclone with wind speeds in excess of 119 kmph (65 knots, or 74 mph) that occurs in the western Pacific; same as a hurricane except for location. (*Compare* Hurricane.)

Ultisols (18) A soil order in the Soil Taxonomy. Features highly weathered forest soils, principally in the humid subtropical climatic classification. Increased weathering and exposure can degenerate an Alfisol into the reddish color and texture of these Ultisols. Fertility is quickly exhausted when Ultisols are cultivated.

Unconfined aquifer (9) An aquifer that is not bounded by impermeable strata. It is simply the zone of saturation in water-bearing rock strata with no impermeable overburden, and recharge is generally accomplished by water percolating down from above. (*Compare* to Confined aquifer.)

Undercut bank (14) In streams, a steep bank formed along the outer portion of a meandering stream; produced by lateral erosive action of a stream; sometimes called a *cutbank*. (*Compare* Point bar.)

Uniformitarianism (11) An assumption that physical processes active in the environment today are operating at the same pace and intensity that has characterized them throughout geologic time; proposed by Hutton and Lyell.

Upslope fog (7) Forms when moist air is forced to higher elevations along a hill or mountain and is thus cooled. (*Compare* Valley fog.)

Upwelling current (6) An area of the sea where cool, deep waters, which are generally nutrient-rich, rise to replace vacating water, as occurs along the west coasts of North and South America. (*Compare* Downwelling current.)

Urban heat island (4) An urban microclimate that is warmer on average than areas in the surrounding countryside because of the interaction of solar radiation and various surface characteristics.

Valley fog (7) The settling of cooler, more dense air in low-lying areas; produces saturated conditions and fog. (*Compare* Upslope fog.)

Vapor pressure (7) That portion of total air pressure that results from water vapor molecules, expressed in millibars (mb). At a given dew-point temperature, the maximum capacity of the air is termed its saturation vapor pressure.

Vascular plant (19) A plant having internal fluid and material flows through its tissues; almost 270,000 species exist on Earth.

Ventifact (16) A piece of rock etched and smoothed by eolian erosion—that is, abrasion by windblown particles.

Vernal (March) equinox (2) The time around March 20–21 when the Sun's declination crosses the equatorial parallel (0° latitude) and all places on Earth experience days and nights of equal length. The Sun rises at the North Pole and sets at the South Pole. (*Compare* Autumnal [September] equinox.)

Vertisols (18) A soil order in the Soil Taxonomy. Features expandable clay soils; composed of more than 30% swelling clays. Occurs in regions that experience highly variable soil moisture balances through the seasons.

Volatile organic compounds (VOCs) (3) Compounds, including hydrocarbons, produced by the combustion of gasoline, from surface coatings, and from combustion to produce electricity; participate in the production of PAN through reactions with nitric oxides.

Volcano (12) A mountainous landform at the end of a magma conduit, which rises from below the crust and vents to the surface. Magma rises and collects in a magma chamber deep below, erupting effusively or explosively and forming composite, shield, or cinder-cone volcanoes.

Warm desert and semidesert (20) A desert biome caused by the presence of subtropical high-pressure cells; characterized by dry air and low precipitation.

Warm front (8) The leading edge of an advancing warm air mass, which is unable to push cooler, passive air out of the way; tends to push the cooler, underlying air into a wedge shape; identified on a weather map as a line marked with semicircles pointing in the direction of frontal movement. (*Compare* Cold front.)

Waterspout (8) An elongated, funnel-shaped circulation formed when a tornado exists over water.

Water budget (9) A water accounting system for an area of Earth's surface using inputs of precipitation and outputs of evapotranspiration (evaporation from ground surfaces and transpiration from plants) and surface runoff. Precipitation "income" balances evaporation, transpiration, and runoff "expenditures"; soil moisture storage acts as "savings" in the budget.

Water table (9) The upper surface of groundwater; that contact point between the zone of saturation and the zone of aeration in an unconfined aquifer. (*See* Zone of aeration, Zone of saturation.)

Water withdrawal (9) Sometimes called *offstream use*, the removal of water from the natural supply, after which it is used for various purposes and then is returned to the water supply.

Wave (16) An undulation of ocean water produced by the conversion of solar energy to wind energy and then to wave energy; energy produced in a generating region or a stormy area of the sea.

Wave-cut platform (16) A flat or gently sloping, tablelike bedrock surface that develops in the tidal zone where wave action cuts a bench that extends from the cliff base out into the sea.

Wave cyclone (8) *See* Midlatitude cyclone.

Wavelength (2) A measurement of a wave; the distance between the crests of successive waves. The number of waves passing a fixed point in 1 second is called the frequency of the wavelength.

Wave refraction (16) A bending process that concentrates wave energy on headlands and disperses it in coves and bays; the long-term result is coastal straightening.

Weather (8) The short-term condition of the atmosphere, as compared to climate, which reflects long-term atmospheric conditions and extremes. Temperature, air pressure, relative humidity, wind speed and direction, daylength, and Sun angle are important measurable elements that contribute to the weather.

Weathering (13) The processes by which surface and subsurface rocks disintegrate, or dissolve, or are broken down. Rocks at or near Earth's surface are exposed to physical and chemical weathering processes.

Westerlies (6) The predominant surface and aloft wind-flow pattern from the subtropics to high latitudes in both hemispheres.

Western intensification (6) The piling up of ocean water along the western margin of each ocean basin, to a height of about 15 cm (6 in.); produced by the trade winds that drive the oceans westward in a concentrated channel.

Wetland (9) An area that is permanently or seasonally saturated with water and characterized by vegetation adapted to hydric soils; highly productive ecosystem with an ability to trap organic matter, nutrients, and sediment.

Wilting point (9) That point in the soil-moisture balance when only hygroscopic water and some bound capillary water remain. Plants wilt and eventually die after prolonged stress from a lack of available water.

Wind (6) The horizontal movement of air relative to Earth's surface; produced essentially by air pressure differences from place to place; turbulence, wind updrafts and downdrafts, adds a vertical component; its direction is influenced by the Coriolis force and surface friction.

Wind vane (6) A weather instrument used to determine wind direction; winds are named for the direction from which they originate.

Winter (December) solstice (2) The time when the Sun's declination is at the Tropic of Capricorn, at 23.5° S latitude, December 21–22 each year. The day is 24 hours long south of the Antarctic Circle. The night is 24 hours long north of the Arctic Circle. (*Compare* Summer [June] solstice.)

Yardang (16) A streamlined rock structure formed by deflation and abrasion; appears elongated and aligned with the most effective wind direction.

Yazoo stream (14) A small tributary stream draining alongside a floodplain; blocked from joining the main river by its natural levees and elevated stream channel. (*See* Backswamp.)

Zone of aeration (9) A zone above the water table that has air in its pore spaces and may or may not have water.

Zone of saturation (9) A groundwater zone below the water table in which all pore spaces are filled with water.

Index

World – Physical

Great Basin	Land features
Caribbean Sea	Water bodies
Aleutian Trench	Underwater features